ENERGY REVOLUTION AND CHEMICAL RESEARCH

The primary goal of the book is to promote research and developmental activities in energy, power technology and chemical technology. Besides, it aims to promote scientific information interchange between scholars from top universities, business associations, research centers and high-tech enterprises working all around the world.

The conference conducted in-depth exchanges and discussions on relevant topics such as energy engineering and chemical engineering, aiming to provide an academic and technical communication platform for scholars and engineers engaged in scientific research and engineering practice in the field of energy materials, energy equipment and electrochemistry. By sharing the research status of scientific research achievements and cutting-edge technologies, it helps scholars and engineers all over the world comprehend the academic development trends and broaden research ideas. So as to strengthen international academic research, academic topics exchange and discussion, and promote the industrialization cooperation of academic achievements.

PROCEEDINGS OF THE 8TH INTERNATIONAL CONFERENCE ON ENERGY SCIENCE AND CHEMICAL ENGINEERING (ICESCE 2022), ZHANGJIAJIE, CHINA, 22–24 APRIL 2022

Energy Revolution and Chemical Research

Edited by

Kok-Keong Chong

Lee Kong Chian Faculty of Engineering and Science, Universiti Tunku Abdul Rahman, Malaysia

Zhongliang Liu

College of Energy and Power Engineering, Beijing University of Technology, China

CRC Press
Taylor & Francis Group
Boca Raton London New York Leiden

CRC Press is an imprint of the
Taylor & Francis Group, an **informa** business
A BALKEMA BOOK

First published 2023
by CRC Press/Balkema
4 Park Square, Milton Park / Abingdon, Oxon OX14 4RN / UK
e-mail: enquiries@taylorandfrancis.com
www.routledge.com – www.taylorandfrancis.com

CRC Press/Balkema is an imprint of the Taylor & Francis Group, an informa business

Library of Congress Cataloging-in-Publication Data
A catalog record has been requested for this book

ISBN: 978-1-032-36554-1 (hbk)
ISBN: 978-1-032-36558-9 (pbk)
ISBN: 978-1-003-33265-7 (ebk)

DOI: 10.1201/9781003332657

Typeset in Times New Roman
by MPS Limited, Chennai, India

Energy Revolution and Chemical Research – Kok-Keong Chong and Zhongliang Liu (Eds)
© 2023 The Editors and Contributors, ISBN 978-1-032-36554-1

Table of contents

Energy Revolution and Chemical Research – Kok-Keong Chong and Zhongliang Liu (Eds)
© 2023 The Editors and Contributors, ISBN 978-1-032-36554-1

Preface

It is over two years since COVID-19 broke out. The whole world is still struggling with the virulent pandemic COVID-19. It is still difficult to take international travel. Many conferences are held virtually, for the sake of protecting all the participants and conference staff from getting infected by the virus.

It is uncertain when the COVID-19 will end, so it remains unclear how long the meeting needed to be postponed, while many scholars and researchers wanted to attend this long-waited conference and have academic exchanges with their peers. Therefore, in order to actively respond to the call of the government, and meet author's request, the 8th International Conference on Energy Science and Chemical Engineering (ICESCE 2022), which was planned to be held in Zhangjiajie, China from April 22–24,2022 was changed to be held online through Zoom software. This approach not only avoids people gathering, but also meets their communication needs.

ICESCE 2022 is to bring together innovative academics and industrial experts in the field of energy science and chemical engineering to a common forum. The primary goal of the conference is to promote research and developmental activities in energy science and chemical engineering and another goal is to promote scientific information interchange between researchers, developers, engineers, students, and practitioners working all around the world. The conference will be held every year to make it an ideal platform for people to share views and experiences in energy science and chemical engineering and related areas.

The conference brings together about 170 well-known scholars in the field of energy science and chemical engineering from China and abroad. The reports were divided into keynote speeches, oral presentations, and poster presentations to share their latest research results and experiences in related research fields. In the first part, all keynote speakers were allocated 30 minutes to present their talks via Zoom. After the keynote talks, all participants joined in a WeChat communication group to discuss more about the talks and presentations.

We were greatly honored to have invited six distinguished experts as our keynote speakers. Prof. Zhou Guojiang, Heilongjiang University of Science and Technology, China. He presented an insightful speech: *New materials of coal chemical industry and energy*. And then we had Prof. Liguo Wei, Heilongjiang University of Science and Technology, China. He delivered a speech: *Strategy and Thinking on Performance Improvement of DSSCs*. Prof. Wu Peng, Heilongjiang University of Science and Technology, China. She mainly engaged in the teaching and scientific research of the performance and application of new carbon materials and clean utilization of low-rank coal. Prof. Jordi Arbiol, ICREA and Catalan Institute of Nanoscience and Nanotechnology (ICN2). He performed a thought-provoking speech: *A Close Look to Single Atom Metal Catalysts*. A. Prof. Xiaoming Zhang, Hebei University of Technology. Prof. Zhang's main research interests include topological electronic materials, electronic crystals, new electronic functional materials and other new materials and physical properties. Our finale keynote speaker, Prof. Kun Liang, Ningbo Institute of Materials Technology and Engineering, CAS. Prof. Liang's research focuses on developing new two-dimensional nanomaterials with precisely controlled chemical composition and morphology and exploration of novel applications in energy generation and storage and flexible electronics. Their insightful speeches had triggered heated discussion in the second session of the conference. The WeChat discussion lasted for about 30 minutes. Every participant praised this conference for disseminating useful and insightful knowledge.

We are glad to share with you that we received lots of submissions from the conference and we selected a number of high-quality papers and compiled them into the proceedings after rigorously review. These papers feature the following topics but are not limited to: Energy Science and Engineering, Chemical Engineering and Technology and other relevant directions. All the papers have been through rigorous review and process to meet the requirements of international publication standard.

Lastly, we would like to warmly thank all the authors who, with their presentations and papers, generously contributed to the lively exchange of scientific information that is so vital to the endurance of scientific conferences of this kind.

The Committee of ICESCE 2022

Energy Revolution and Chemical Research – Kok-Keong Chong and Zhongliang Liu (Eds)
© 2023 The Editors and Contributors, ISBN 978-1-032-36554-1

Committee members

Conference General Chair
Prof. Zhou Guojiang, *Heilongjiang University of Science and Technology, China*
Prof. Chong Kok Keong, *Fellow of Academy of Science Malaysia, Fellow of ASEAN Academy of Engineering & Technology (AAET), Malaysia*

Technical Program Committee Chairs
Prof. Zhongliang Liu, *College of Energy and Power Engineering, Beijing University of Technology, China*

Local Organizing Chairs
Prof. Zhongliang Liu, *College of Energy and Power Engineering, Beijing University of Technology, China*

Publication Chairs
Prof. Hongwei Li, *School of Electrical Engineering and Information, Southwest Petroleum University, China*

Committee members
Prof. Wenxue Chen, *Petroleum exploration & development research institute, Sinopec, China*
Prof. Gang Chen, *Xi'an Shiyou University, China*
Prof. Gang Chen, *Xi'an Shiyou University, China*
Prof. Jianhua Zhang, *North China Electric Power University, China*
Prof. Guohe Huang, *Beijing Normal University, China*
Engineer Junwen Chen, *China Petroleum Engineering & Construction Corporation, Southwest Branch*
Prof. Chunfeng Shi, *Research Institute of Petroleum Processing, Sinopec, China*
Prof. Baiping Xu, *Guangdong Industry Polytechnic, China*
Prof. Hui Gao, *Xian Shiyou University, China*
Prof. Liguang Wu, *Zhejiang Gongshang University, China*
Prof. Shizhao Yang, *Air Force Logistics College, China*
Prof. Huiwei Liao, *Southwest University of Science and Technology, China*
Prof. Zhijia Yu, *Dalian University of Technology, China*
A. Prof. Xianghua Yang, *Guangdong University of Technology, China*
Prof. Huaili Zheng, *Chongqing University, China*
Prof. Huaili Zheng, *Chongqing University, China*
Dr. Xiaolong Li, *China University of Petroleum (East China), China*
Dr. Xinrui Feng, *Hainan University, China*
Dr. Qinghua Wei, *Northwestern Polytechnical University, China*
Dr. Lingjun Kong, *Guangzhou University, China*
Dr. Yanjun Li, *Florida Atlantic University, USA*
Dr. Zhenjun Ma, *University of Wollongong, Australia*
Dr. Bin Li, *Qingdao Institute of Bioenergy and Bioprocess Technology, Chinese Academy of Sciences, China*
Dr. Liying Guo, *Shenyang University of Technology, China*

Dr. Kangxu Ren, *PetroChina Research Institute of Petroleum Exploration & Development, China*
Dr. Bo Cai, *Petroleum exploration & development research institute, Petrochina, China*

Technical Program Committee
Prof. Christo Boyanov Boyadjiev, *Institute of Chemical Enginerring, Bulgarian*
Prof. Kesen Ma, *University of Waterloo, Canada*
Prof. Sanette, *Centre of Excellence in Carbon Based Fuels, South Africa*
Prof. T. Satyanarayana, *University of Delhi, India*
Prof. Tianshou Ma, *Southwest Petroleum University, China*
Prof. Fuchuan Huang, *Guangxi University, China*
Prof. Xiuwen Cheng, *Lanzhou University, China*
Prof. Yongbin Yang, *Central South University, China*
Prof. Wei Lu, *Guangxi University, China*
Prof. Masakazu Anpo, *Fuzhou University, China*
Prof. Jianjun Dai, *Beijing University of Chemical Technology, China*
Prof. Xiangzhou Yuan, *Korea University, Republic of Korea*
A. Prof. Chunni Tang, *Shaanxi Institute of Technology, China*
Dr. Hwai, *Chyuan Ong University of Malaya, Malaysia*
Dr. Rodrigo Soto, *University of Limerick, Ireland*
Dr. N. Gokarneshan, *Textile Technology, India*
Dr. Papurello Davide, *Denerg-Polito, Italy*
Dr. Jingshou Liu, *China University of Geosciences, China*
Raheleh Mohammadpour, *Sharif University of Technology, Iran*

Energy development and utilization and green environmental protection and energy saving

Energy Revolution and Chemical Research – Kok-Keong Chong and Zhongliang Liu (Eds)

Research on energy-saving technologies and countermeasures for public buildings

Dexia Kong* & Jun Liang
Shangdong Huayu University of Technology, Dezhou Key Laboratory of High-Efficiency Heat Pump Air Conditioning Equipment and System Energy Saving Technology, Dezhou, Shandong, China

ABSTRACT: This paper analyzes the existing problems of energy consumption in public buildings from the perspective of energy consumption of public buildings and the application of energy-saving technologies in public buildings and gives appropriate energy-saving measures and suggestions to reduce building energy consumption, accelerate the transformation and upgrading of the construction industry, and promote the construction industry.

1 INTRODUCTION

With the continuous development of productive forces, the energy problem has become increasingly prominent, which has become one of the important factors affecting the development of the times. In my country, building energy consumption accounts for about 30% of the country's total energy consumption. With the construction of urbanization and the improvement of people's requirements for living and living environment, the total building energy consumption continues to increase, accounting for an increasing proportion of social terminal energy. The application of building energy-saving technology and energy-saving measures are of great significance to reduce energy consumption in public buildings, and can also alleviate energy problems and protect the environment.

2 ENERGY CONSUMPTION ANALYSIS

The energy consumption in public buildings mainly comes from the use of energy-consuming equipment in the building and the energy needed by people to improve the comfort of the indoor environment of the building. Table 1 (GB 50189-2005, Design standard for energy conservation of public buildings [S]) shows the proportion of annual energy consumption of public buildings, of which the energy consumption of heating and air-conditioning systems is the highest, followed by lighting energy consumption, and other energy-consuming equipment energy consumption is lower.

2.1 *Heating and air conditioning system*

Among the annual energy consumption of public buildings, the proportion of heating and the air-conditioning energy consumption is the largest, and the energy consumption caused by the heat transfer of the envelope structure accounts for about 20% to 50% of the annual energy consumption. According to the formula of the basic heat consumption of the building envelope (2-1) (Lu et al. 2015) and the instantaneous cooling load of the building envelope (2-2)(Lu et al. 2015), the thermal

*Corresponding Author: 369709294@qq.com

DOI 10.1201/9781003332657-1

Table 1. The proportion of annual energy consumption of public buildings.

The type of energy consumption	The proportion of annual energy consumption
Heating and air conditioning	40%~50%
lighting	30%~40%
Other energy-using equipment	10%~20%

performance of the building envelope directly affects the energy efficiency of the HVAC system and the energy consumption in the building The good thermal insulation performance of the envelope structure can reduce the cooling load in summer and the heating load in winter, and reduce the energy consumption of the building.

$$Q_1 = AK(t_R - t_o)\alpha \tag{2-1}$$

$$Q_2 = AK(t_c - t_R) \tag{2-2}$$

Q_1—The basic heat consumption of the building envelope, W;
Q_2—hourly cooling load, W;
A—heat transfer surface area, m^2;
K—Heat transfer coefficient, $W/(m^2 \cdot °C)$;
t_R—winter indoor temperature, °C;
t_o—winter outdoor temperature, °C;
t_c—Hourly value of outdoor temperature, °C;
α—Temperature difference correction factor.

The integration of the HVAC system and the overall structure of the building also affects the energy consumption of public buildings. Scientifically and rationally optimizing the HVAC system and optimizing the HVAC system and public buildings in various aspects can reduce energy consumption as a whole.

2.2 *Electrical energy*

Among the annual energy consumption of public buildings, the energy consumption of architectural lighting is relatively high. The electric energy in public buildings is mainly the consumption of lighting and energy-consuming equipment and the loss of the distribution power system. Regarding power loss, the total loss of the transformer accounts for about 8% of the total power generation (Liu & Yang 2008). Energy consumption should be reduced in two aspects: saving electric energy and reducing power distribution system losses.

3 ENERGY-SAVING APPLICATIONS IN PUBLIC BUILDINGS

The energy consumption of public buildings is huge. To alleviate the energy shortage and protect the environment, clean energy has gradually entered people's field of vision, has received attention, and is widely used in public buildings.

3.1 *Application of solar technology*

Solar energy is a pure natural clean energy. With the continuous progress of solar energy technology, the technology of using solar energy for power supply, heating, cooling, and providing hot water is gradually applied to public buildings (Cheng 2014).

3.1.1 *Solar photovoltaic power generation*

Solar photovoltaic power generation is widely used in public buildings. An important market for photovoltaic power generation technology in the future is photovoltaic building integration. Photovoltaic power generation converts solar energy into electrical energy through the photovoltaic effect. Photovoltaic technology is applied in public buildings to meet the electricity demand of buildings, and at the same time, sunlight passes through photovoltaic modules to meet the lighting needs of buildings. The application of building photovoltaic integrated technology to public buildings is both energy-saving and environmentally friendly.

3.1.2 *Solar cooling system*

Solar cooling is combined with the cooling system of public buildings to meet the cooling load demand in the building together. Solar cooling is mainly driven by the heat energy generated by the collector plate to drive the refrigeration machine to cool. There are three forms of adsorption refrigerator, absorption refrigerator, and dehumidification cooling, among which the absorption refrigeration technology is the most mature.

3.1.3 *Solar hot water system and heating*

Solar water heating systems are relatively mature and widely used in public buildings, mainly using collectors to heat domestic water through photothermal conversion to meet people's normal water needs (Xuan 2019). The heat collection efficiency is an important factor affecting the system. During the installation of the system, the angles of the collector and the solar radiation are adjusted according to the local climatic conditions to enhance the heat collection efficiency.

The difference between a solar heating system and a hot water system is that the heating system is added with a heating device. But the hot water system and the heating system are also very different, mainly because the heat load required by the heating system is larger than that of the hot water system.

With the advancement of science and technology, various clean energy sources have been developed and utilized in public buildings. Due to the instability of solar energy resources, solar energy combined with air source heat pumps, ground-source heat pumps, biomass energy, electricity, and other clean energy heating methods is applied in real life.

3.2 *Clean energy applications*

In the context of energy conservation and emission reduction, more and more clean energy has been paid attention to. Public buildings use a variety of clean energy to improve the comfort of the indoor environment, such as wind energy, geothermal energy, air energy, etc. New technologies such as heat pumps and air source heat pumps are also constantly being incorporated into public buildings (Xu 2020).

4 PROBLEM ANALYSIS

4.1 *Building energy consumption is large and energy efficiency is low*

China is in the stage of urbanization and industrialization, and the energy consumption of buildings is large. The thermal performance of the public building envelope is poor, and there is no good thermal insulation performance. There are problems in the design of the energy system and the structure of the public building, resulting in low energy utilization efficiency of the energy system and substandard energy efficiency.

4.2 *Less energy-saving technologies are used, and energy consumption continues to increase*

The total scale of public buildings has increased, and people have high requirements for the comfort of the indoor environment of buildings. However, the application of energy-saving technologies in public buildings is less, and energy consumption continues to increase.

4.3 *Lack of intelligent energy system monitoring*

No intelligent energy monitoring system has been established in public buildings, and there is no classification, measurement, summary, and analysis of the consumption of water, electricity, and heating in public buildings.

4.4 *Energy system design issues*

Due to some reasons in the design of public buildings, the equipment room is far away from the load center, for example, the energy equipment center is far away from the cooling and heating load center, and the substation is far away from the electric well and the load center, which increases the energy consumption of transmission.

4.5 *The application of energy-saving technology is not widely used, and the matching between equipment and system is poor*

Building energy-saving technology is constantly updated with the development of science and technology, the matching of new energy-saving products and energy systems is relatively poor, and the energy efficiency of energy-saving equipment cannot be exerted. The energy-saving effect of the energy system is also not significant.

4.6 *Weak awareness of energy saving*

People's awareness of energy conservation is weak, the meaning of green buildings is not clear, and energy conservation awareness has not penetrated people's daily life.

4.7 *Energy-saving regulations and systems are not perfect*

In recent years, the total amount of green buildings has been increasing, but the establishment of regulations and systems for energy-efficient buildings is not perfect, lacking technical details and operational standards. Compared with traditional buildings, green buildings have a larger investment. To reduce cost control, some public buildings do not meet energy-saving standards.

5 ENERGY-SAVING MEASURES AND SUGGESTIONS

5.1 *Building energy efficiency*

When designing public buildings, consider the climatic environment around the building, make full use of passive energy-saving methods such as natural lighting or natural ventilation, or change the thermal environment around the building through greening to achieve the purpose of energy conservation.

5.2 *Use new energy-saving and environmentally friendly materials*

The envelope structure of public buildings affects the cooling load in summer and the heat load in winter, and new energy-saving and environmentally friendly envelope structures, such as molded polystyrene foam boards, are used to enhance the thermal insulation performance of the envelope structure. Public building energy consumption can be reduced by reducing the summer cooling load and winter heating load (Yan 2021).

5.3 *Optimizing HVAC systems*

The energy equipment room is too far from the cooling and heating load center, which shortens the energy supply and transportation distance and reduces the energy consumption of transportation. The HVAC system should reasonably select the cold and heat sources according to the local climatic conditions and resources, determine the appropriate heating, ventilation, and air systems, and at the same time recover the waste heat such as exhaust air and equipment heat in the system to utilize the waste heat and improve the energy utilization efficiency.

5.4 *Optimize the power distribution system*

Substations and power wells should be close to the load center to reduce the power supply radius, reduce line losses, and reduce copper usage. Reasonable selection of cable cross-section and laying path to reduce line loss. Electrical devices with high efficiency, low energy consumption, advanced performance, durability, and reliability, and those made of green environmental protection materials should be selected.

5.5 *Intelligent system control*

The intelligent energy management system is adopted to automatically monitor the high-voltage and low-voltage power distribution systems and set the standard value of electricity consumption. It is needed to measure power loads of different natures such as lighting, air conditioning, power, etc., in low-voltage distribution cabinets and terminal distribution cabinets in substations. Different types of electricity classification metering and summarization are carried out through smart metering instruments. The overvalued alarm can realize the automatic operation of the power supply and distribution system. The automatic metering system of water, electricity, and the heating energy consumption is adopted to improve energy consumption management and enhance users' awareness of energy conservation (Gao 2021).

5.6 *Actively publicize green buildings and strengthen energy conservation awareness*

It should be committed to popularizing the meaning of green buildings and actively publicizing green and energy-saving buildings. Green and environmentally friendly energy-saving materials and equipment are given priority when developing and designing. Efforts should be made to actively publicize the significance of energy conservation and emission reduction, and enhance people's energy conservation awareness.

5.7 *Improve energy conservation regulations and standards*

Green energy-saving buildings should improve the corresponding laws, regulations, and standards, boost the building energy-saving technology system and green energy-saving incentive policies, enhance the corresponding regulatory mechanism, pay attention to the application of renewable energy in the field of construction, and actively promote green energy-saving technologies (Liang 2021).

6 SIGNIFICANCE

6.1 *Alleviate energy shortage and promote the development of clean energy*

Building energy consumption is large, and the development of building energy-saving technologies can reduce energy consumption. The use of clean energy to provide energy for public buildings can not only alleviate the problem of dependence on traditional energy and energy shortage but also promote the development of clean energy (Yan 2020).

6.2 *Improve the ecological environment and reduce pollution*

Using clean energy not only provides energy for buildings and not only reduces the consumption of traditional energy but also reduces the emission of pollutants. Using clean energy to reduce the emission of dangerous substances can improve the ecological environment and human settlements.

7 CONCLUSION

Building energy efficiency is the systematic engineering of building and energy system optimization. The energy conservation of public buildings must be combined with local climatic conditions, while ensuring the quality of the indoor environment and meeting people's requirements for indoor comfort, to improve the energy efficiency of lighting equipment and heating and air conditioning equipment. For example, it can improve the thermal performance of the envelope structure, optimize the energy system or use energy-saving technologies and other methods to reduce the energy demand of public buildings and improve the utilization efficiency of the energy system.

ACKNOWLEDGMENTS

This work was supported by Dezhou Key Laboratory of High-efficiency Heat Pump Air Conditioning Equipment and System Energy Saving Technology.

This paper is one of the phased achievements of Dezhou Key Laboratory of high-efficiency heat pump air conditioning equipment and system energy-saving technology (Project No. 26)

REFERENCES

Cheng Yun. Chinese Ningbo University, 2014.

Gaoming. Study on the difficulties and countermeasures of building energy conservation and emission reduction [J]. *Resource-saving and environmental protection*, 2021, (04): 3–4.

GB 50189-2005, *Design standard for energy conservation of public buildings*[S].

Liang Kong Ming. Research on the Difficulties and Countermeasures of my country's Building Energy Conservation and Emission Reduction [J]. *Intelligent Building and Smart City*, 2021, (12): 130–131.

Liu Chengjun, Yang Rengang. Calculation and Analysis of Transformer Harmonic Loss[J]. *Power System Protection and Control*, 2008, (13):33–36+42

Lu Yajun, Ma Zuiliang, Zou Pinghua. HVAC, 3rd edition, 2015

Shangyu Xuan. *Research on the energy-saving renovation technology of public buildings in summer hot winter-cooled areas* [D]. Hunan University, 2019.

Xu Xin. *Study on the application of the composite heating system in villages and towns in the severe cold area* [D]. Shihezi University, 2020.

Yan Baihui. Application of Building Energy Saving and Green Building Technology [J]. *China High-tech*, 2021, (14): 69–70.

Yan Wei. Significance and technical measures for civil construction projects [J]. *Volkswagen Standardization*, 2020, (03): 27–28.

Energy Revolution and Chemical Research – Kok-Keong Chong and Zhongliang Liu (Eds)
 ISBN: 978-1-032-36554-1

Energy consumption and carbon emission study of heat pumps with different working fluids for residential buildings heating

Yuan Zhao
Powerchina HuaDong Engineering Corporation Limited, Hangzhou, China
Key Laboratory of Efficient Utilization of Low and Medium Grade Energy (Tianjin University), MOE, Tianjin, China

Ke Sun & Chenghao Gao
Powerchina HuaDong Engineering Corporation Limited, Hangzhou, China

Dabiao Wang
National Engineering Research Center of Chemical Fertilizer Catalyst (NERC-CFC), School of Chemical Engineering, Fuzhou University, Fujian, China

Chen Liu, Ruirui Zhao & Baomin Dai*
Tianjin Key Laboratory of Refrigeration Technology, Tianjin University of Commerce, Tianjin, China

ABSTRACT: The heat pump is the best way to convert electricity and heat under the background of zero-carbon energy, and the application of the heat pump represents the advanced development direction of heating in the field of dual-carbon architecture in the future. In this paper, heat pump systems with eight kinds of different working fluids are proposed and applied to heating residential buildings. The overall energy consumption and carbon emission during the life cycle of a heat pump with different working fluids are analyzed and compared with traditional coal-fired boilers. The results demonstrate that the COP of the heat pump using R32 as the working fluid is 1.74~3.52, which is 0.13~25.39% higher than that of other working fluids. For energy and the environment, the primary energy consumption and carbon emissions generated by the heat pump using R32 can be reduced by 8.36% and 8.38% compared with that of the coal-fired boiler, respectively. In addition, the CO_2 heat pump is more suitable for areas with low ambient temperature for heating with promising environmental conservation.

1 INTRODUCTION

The burning of fossil fuels has adversely affected the environment, releasing 2.2 trillion carbon dioxide into the atmosphere over the past 200 years. On September 22, 2020, China proposed at the 75th Session of the United Nations General Assembly that it would strive to hit the carbon peak by 2030 and achieve carbon neutrality by 2060 (Jian & S 2021). Heat pump technology will become an effective technical route to achieve carbon peak and carbon neutrality. In 2016, parties to the Montreal Protocol agreed on the Kigali Amendment, which aims to limit emissions of hydrofluorocarbons (HFCS) (Ministry of Ecology and Environment 2021). Therefore, the use of environmentally friendly working fluid with low GWP is the trend for the development of heat pump systems in the future.

Owing to the advantage of the heat pump, lots of studies are reported about the heat pump for heating or making hot water by using different refrigerants. Li et al. (Li 2015) compared and

*Corresponding Author: dbm@tjcu.edu.cn

DOI 10.1201/9781003332657-2

analyzed the performance of R22, R134a, and CO_2 refrigerants in solar-air hybrid heat source heat pump water heater and the results showed that the COP of CO_2 was poor, and the difference between R22 and R134a was not significant. Wang et al. (Wang 2017) introduced a new frost-free air source heat pump system using R134a and R407C as R22 substitutes, and the results showed that the average COP of R134a was higher than that of R22 and R407C. Nie et al. (Nie 2017) used different pure HFC refrigerants (R32, R134a, R143a, and R152a) to further optimize the air source heat pump for space heating in northern China, and the results showed that R152a had higher COP values. Shen et al. (Shen 2017) used R1234yf and R1234ze(E) to simulate a residential heat pump water heater and proved that R1234yf and R1234ze(E) have equal potential to replace R134a. Zhou (Zhou 2015) found that R32, R1234yf, and their mixtures are promising low GWP substitutes compared with R410A.

Nowadays, building heating occupies a considerable proportion of the total energy consumption of buildings. There are few studies on heat pumps with different refrigerants in specific application cases and no comparison with traditional methods. Therefore, heat pump systems with eight kinds of different working fluids are compared in this paper and applied in residential building heating. After that, the energy consumption and carbon emission performances throughout the lifetime are analyzed and compared with the traditional coal-fired boiler, which provides the theoretical reference for the application of heat pumps in residential buildings.

2 DESCRIPTION OF SYSTEM

2.1 *Basic information*

The heating area layout plan is shown in Figure 1. The residential buildings are located in Beijing, consisting of 30 buildings, each with an area of 85 m^2. The total heating area is 35700 m^2, and the annual total heating load is 1539740 kW. The heat pump is used for heating residential buildings in winter in this study.

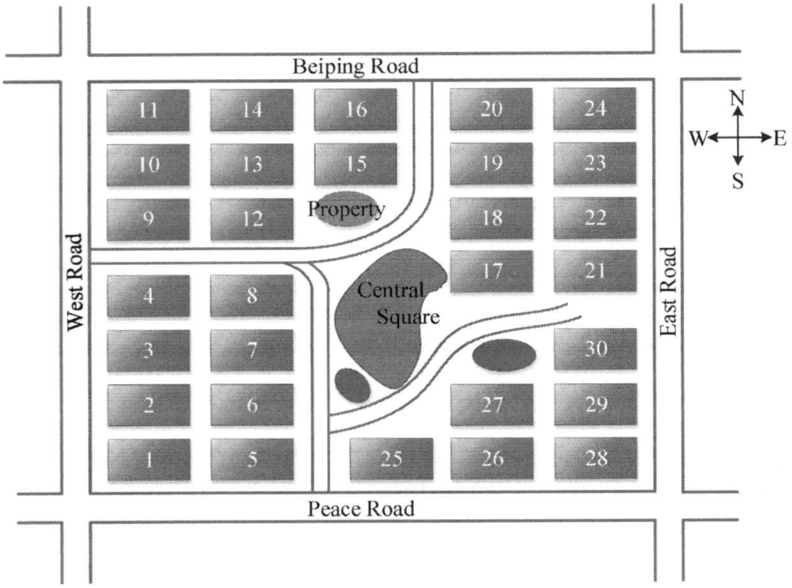

Figure 1. Heating area layout plan.

2.2 *System description*

The heat pump is used for heating residential buildings in winter in this paper. The system diagram and *T-S* diagram of the heat pump system are described in Figure 2. The refrigerant enters the compressor as dry saturated steam from the evaporator outlet and is compressed to superheated steam (1-2). Then it flows into the condenser or gas cooler, and the refrigerant transfers heat with

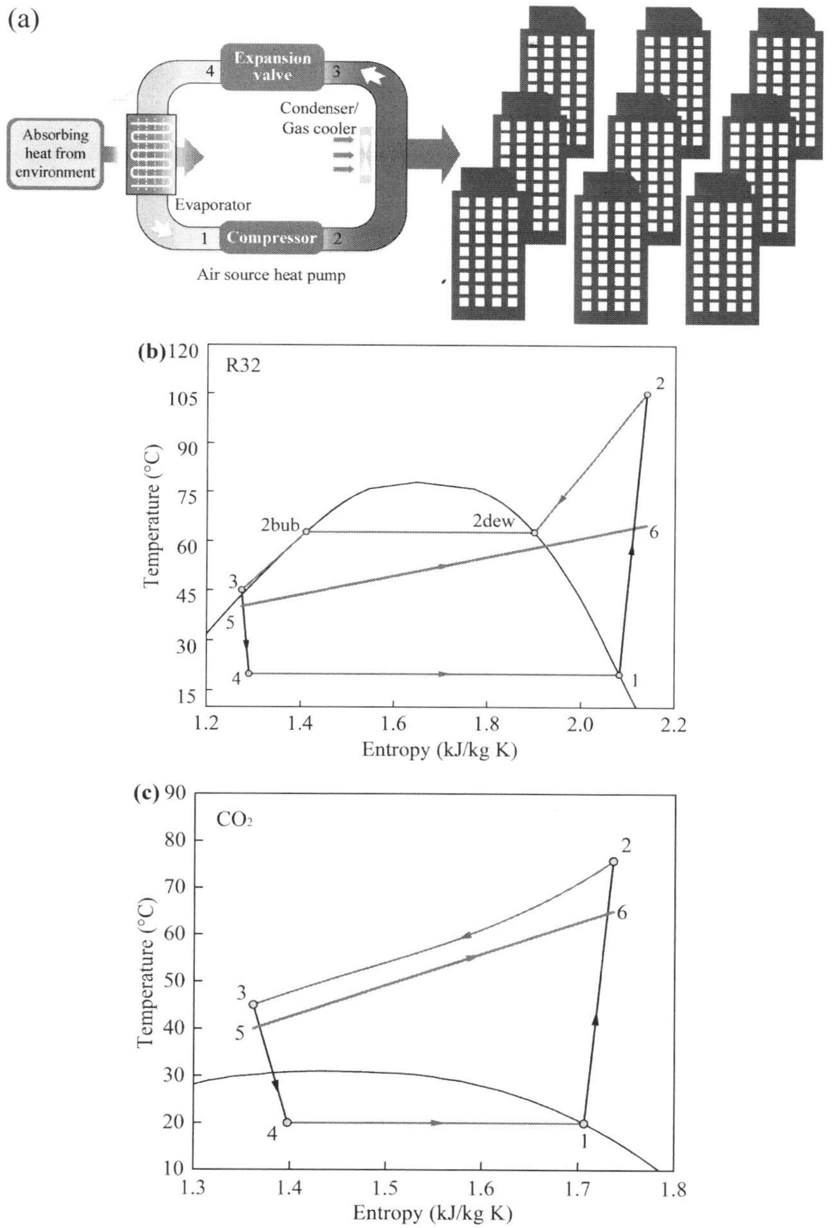

Figure 2. Heat pump system. (a) Schematic. (b) *T-s* diagram of the regular working fluid. (c) *T-s* diagram of CO_2.

the fluid in the gas cooler to cool the refrigerant (2-3) and heat the fluid. The refrigerant comes out of the gas cooler directly into the throttle valve and becomes a two-phase fluid (3-4). Finally, it flows into the evaporator (4-1) and absorbs heat from the environment.

3 MATHEMATICAL MODELING

The residential buildings in Beijing are fabricated by using DeST software (Yan, Xia, Tang, Song, Zhang, Jiang). Eight kinds of the working fluid are selected as the refrigerant of the heat pump, which are R134a, R22, R410A, R32, R1234yf, R1234ze(E), R1234ze(Z), and CO_2. The heating water supply and return temperature is 65°C and 40°C. The fluid properties are determined by REFPROP 10.0 (Lemmon E 2018). The pinch point temperature difference between the evaporator, gas cooler, and condenser is set at 5°C (Bai 2015). The refrigerant from the outlet of the evaporator is in a saturated state. The plant lifetime (n) is set as 15 years (Fazelpour F 2014).

The global efficiency of the compressor for different refrigerants are shown as follows (Gullo P 2016); (Botticella F 2018):

$$\eta_{g,CO2} = -0.0021\left(\frac{p_2}{p_1}\right)^2 - 0.0155\left(\frac{p_2}{p_1}\right) + 0.7325 \tag{1}$$

$$\eta_{g,\,Conv} = \begin{cases} 0.0087\left(\frac{p_2}{p_1}\right)^3 - 0.107\left(\frac{p_2}{p_1}\right)^2 \\ +0.403\left(\frac{p_2}{p_1}\right) + 0.1805\left(\frac{p_2}{p_1}\right) \le 4 \\ 0.00049\left(\frac{p_2}{p_1}\right)^3 - 0.0109\left(\frac{p_2}{p_1}\right)^2 \\ +0.0284\left(\frac{p_2}{p_1}\right) + 0.6643\left(\frac{p_2}{p_1}\right) > 4 \end{cases} \tag{2}$$

The COP of the system is calculated as the heat capacity to compressor power consumption:

$$COP = \frac{Q_h}{W_{Comp}} \tag{3}$$

The consumptions of coal and electricity by the coal-fired boiler (CFB) and heat pump (HP) are calculated as follows, respectively:

$$M_{coal} = \frac{AHD}{\eta_{CFB}\eta_{coal,trans}LHV_{coal}} \tag{4}$$

$$E_{HP} = \frac{AHD}{\eta_{grid,trans}COP} \tag{5}$$

where, AHD is the annual heating demand; η_{CFB} is the heating efficiency of CFB, which is set as 75% (Zhang 2017); $\eta_{coal,\,trans}$, and η_{grid} are the transmission efficiency of coal and electricity set as 80% and 92%, respectively (Liu 2019); LHV_{coal} is the lower heating value of coal, which is set as 29307 kJ/kg (Hu 2008).

Primary energy consumption is selected to evaluate the life cycle energy consumption of different systems, and both electric energy and coal are converted into equivalent standard coal for comparison as follows:

$$PEC_{HP} = p_{el} \cdot E_{HP} \tag{6}$$

where p_{el} is coal convert factors for electricity, which is set as 0.4 kgce/kW·h (Zhang 2017).

The emissions of CO_2 are evaluated to assess the environmental effects of the heat pump systems and coal-fired boilers by:

$$m_{CO2} = E \cdot \mu_{CO2} \tag{7}$$

$$m_{CO2} = M \cdot \mu_{CO2} \tag{8}$$

where μ_{CO2} is the emission conversion factor, which is set as 0.997 kg/kW·h and 2.493 kg/kg for electric or coal, respectively (Dai 2020).

4 RESULTS AND DISCUSSION

4.1 *Basic data*

The heating load of the community is shown in Figure 3. The simulation time is set to one year, and the simulation step is 1 h. The lifetime is 15 years. The heating load demand mainly concentrated from November to March of the following year, with the highest heat load date reaching 3929.4 kW on January 18. Figure 4. displays the variation of ambient temperature hourly throughout the year. It can be noted that the ambient temperature varies greatly with the seasons and it is a typical temperate monsoon climate. Moreover, the ambient temperature is in the range of 14.2~37.2°C throughout the year, with the lowest temperature on January 19, and the highest temperature on June 21.

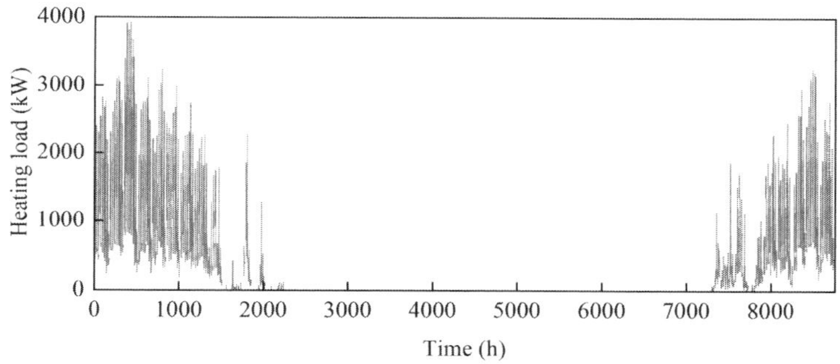

Figure 3. Hourly heating load profiles for the community.

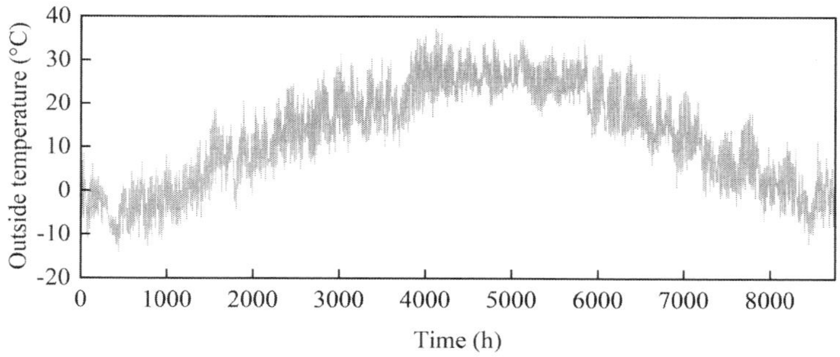

Figure 4. Hourly ambient temperature profiles.

4.2 *COP*

The COP of the heat pump by using different working fluids is demonstrated in Figure 5. It can be observed that COP rises with the increase in ambient temperature. Furthermore, the COP of

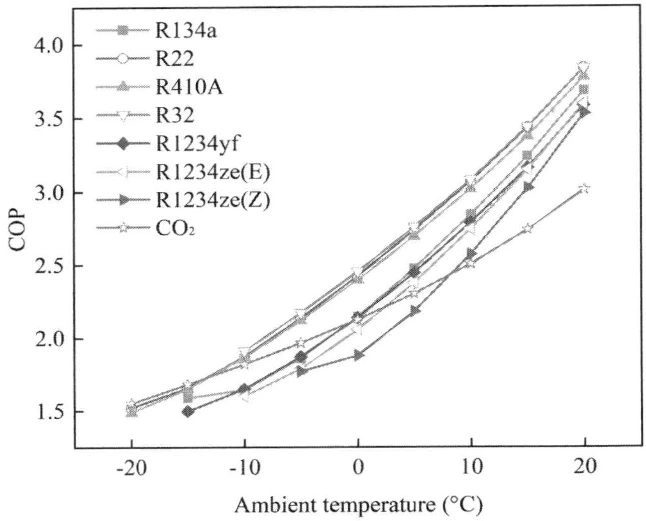

Figure 5. Variation of COP with ambient temperature for the different working fluids.

the heat pump with R32 as working fluid is the highest, which is 1.64~27.38% higher than that of other working fluids at the ambient temperature of −10~20°C. Although COP of CO_2 heat pump is lower at a higher temperature, COP of CO_2 is 2.04~4.15% higher than that of other conventional working fluid heat pumps when the temperature is lower than −12°C, while R134a, R32, R1234yf, R1234ze(E), and R1234ze(Z) are not applicable to such low temperature. Therefore, in areas where the ambient temperature is lower than −12°C in winter, CO_2 can be selected as the working fluid of the heat pump, which can make the system perform better.

Heating is demanded in the residential buildings when there is a heating load demand. The annual hourly COP of eight different working fluid heat pumps is compared in Figure 6. The COP of the heat pump using R32 as the working fluid is 1.74~3.52, which is 0.13~25.39% higher than that of other working fluids. It can be explained by Figure 4. that the ambient temperature in Beijing when there is heat load demand is between −15~20°C, and the heat pump performance can be optimized by using R32 as the working fluid within this temperature range from Figure 5. Moreover, the COP

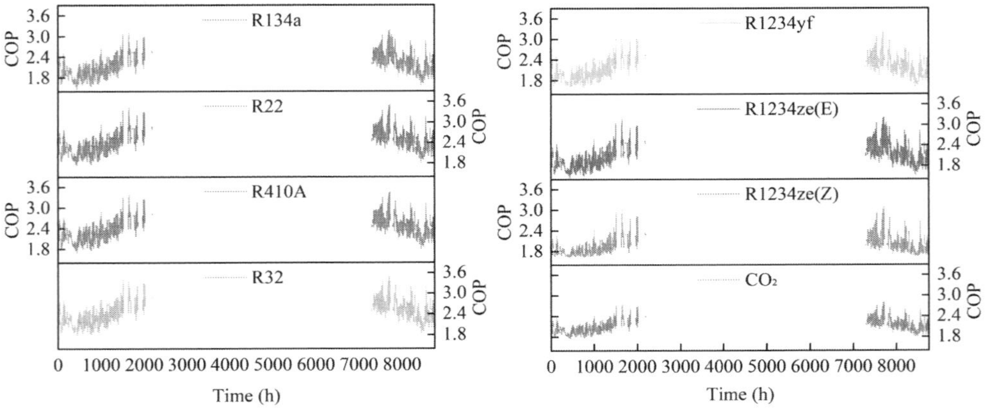

Figure 6. Hourly variation of COP for the different working fluids.

difference among R22, R410A, and R32 heat pumps is small. Although the COP of the CO_2 heat pump is the lowest, only 1.75~2.8, it fluctuates with the temperature significantly less than that of other conventional working fluid heat pumps, indicating that it is less affected by ambient temperature than other working fluids and is more stable.

4.3 *Primary energy consumption and carbon emissions*

The primary energy consumption during the whole lifetime is selected as a comparison baseline considering the fuels are different for different heating solutions, and the energy consumption of the heat pump with different working fluids and the coal-fired boiler is shown in Figure 7. The heat pump with R32 has the lowest PEC, which is 4333.27 tce, and is reduced by 1.39~19.69% in comparison with the heat pumps by using R134a, R22, R410A, R1234yf, R1234ze(E), R1234ze(Z), and CO_2 due to the highest energy efficiency. Moreover, the coal-fired boiler is widely used as a traditional heating method. For the use of the R32 heat pump, the primary energy consumption can be cut down by 8.36% compared with CFB.

The life cycle carbon emissions of heat pumps and coal-fired boilers are also shown in Figure 7. The heat pump with R32 shows the least carbon emissions, which is 10800.67 t and is 1.39~19.69% lower than the heat pump systems by other working fluid and 8.38% lower than the coal-fired boiler.

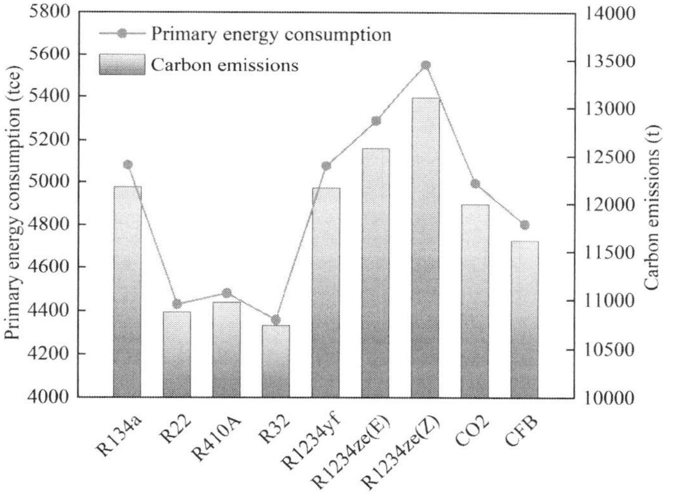

Figure 7. Primary energy consumption and carbon emissions by adopting heat pumps with different working fluids and CFB as heating solutions.

To sum up, in this study scenario, the R32 heat pump system is obviously superior to the heat pump with other working fluids and the coal-fired boiler in terms of primary energy consumption and carbon emissions. It is worth noting that R32, R22, and R410A performed well in the above study. The ODP of R22 is greater than 0, and the GWP of R410A is as high as 1730. Despite the GWP of R32 being smaller than that of R410A, these three kinds of working fluids have different degrees of pollution to the environment. As can be seen from Figure 7, CO_2, as an environmentally friendly natural refrigerant, also has certain advantages in primary energy consumption and carbon emissions. Therefore, CO_2 can also be considered an excellent refrigerant for heat pumps.

5 CONCLUSION

In this study, for residential building heating, a heat pump system model with eight kinds of working fluids, including R134a, R22, R410A, R32, R1234yf, R1234ze(E), R1234ze(Z), and

CO_2 is established. The optimization analysis and energy efficiency comparison are carried out. In addition, the primary energy consumption and carbon emission are compared with traditional coal-fired boilers for the whole life cycle. It can be concluded that COP rises with the increase of ambient temperature and the COP of the heat pump using R32 as the working fluid is 1.74~3.52, which is 0.13~25.39% higher than that of other working fluids. For the use of the R32 heat pump, the life cycle primary energy consumption and carbon emissions can be cut down by 8.36% and 8.38% compared with that of CFB, respectively. In addition, using the CO_2 heat pump for heating is more suitable for areas with low ambient temperature, and has a strong advantage in environmental protection.

ACKNOWLEDGMENT

The authors gratefully acknowledge the financial support from the National Key Research and Development Program of China (No. 2021YFE0116100) and the China Postdoctoral Science Foundation (2020M681799).

REFERENCES

Bai T, Yan G, Yu J. Thermodynamic analyses on an ejector enhanced CO2 transcritical heat pump cycle with vapor-injection. *International Journal of Refrigeration*. 2015; 58:22–34.

Botticella F, de Rossi F, Mauro AW, Vanoli GP, Viscito L. Multi-criteria (thermodynamic, economic, and environmental) analysis of possible design options for residential heating split systems working with low GWP refrigerants. *International Journal of Refrigeration*. 2018; 87: 131–53.

Dai B, Zhao X, Liu S, Yang Q, Zhong D, Cao Y, et al. Heating and cooling of the residential annual application using DMS transcritical CO2 reversible system and traditional solutions: An environment and economic feasibility analysis. *Energy Conversion and Management*. 2020; 210: 112714.

Fazelpour F, Morosuk T. Exergoeconomic analysis of carbon dioxide transcritical refrigeration machines. *International Journal of Refrigeration*. 2014; 38:128–39.

Gullo P, Elmegaard B, Cortella G. Energy and environmental performance assessment of R744 booster supermarket refrigeration systems operating in warm climates. *International Journal of Refrigeration*. 2016; 64:61–79.

Hu X, Li A, Chen H, Xin D, Zhang G, Zheng B. *General principles for the calculation of comprehensive energy consumption*. National Standards of China Beijing: China Standards Press GB/T 2589–20082008.

Jian, S., Shicai, M., Cheng, H., Zhihua, G., Shaoxiang, Z. (2021) Analysis on application status and the prospect of heat pump technology under carbon neutral target. *J. HT. Commun.*, 22–30.

Jinzhe Nie Z L, Xiangrui Kong, Deying Li. Analysis and Comparison Study on Different HFC Refrigerants for Space Heating Air Source Heat Pump in Rural Residential Buildings of North China. *Science Direct*, 2017.

Lemmon E, Bell IH, Huber M, McLinden M. *NIST Standard Reference Database 23: Reference Fluid Thermodynamic and Transport Properties-REFPROP*, Version 10.0, National Institute of Standards and Technology. Standard Reference Data Program, Gaithersburg. 2018.

Liu S, Li Z, Dai B, Zhong Z, Li H, Song M, et al. Energetic, economic and environmental analysis of air source transcritical CO2 heat pump system for residential heating in China. *Applied Thermal Engineering*. 2019; 148: 1425–39.

Ministry of Ecology and Environment. *Kigali Amendment to the Montreal Protocol on Substances That Deplete the Ozone layer*. http://www.mee.gov.cn/ywdt/hjywnews/202106/t20210621_841062.shtml,2021-6-17 (Ministry of Ecology and Environment. Our Country formally accepts the Kigali Amendment to the Montreal Protocol on Substances That Deplete the Ozone Layer[R]. http://www.mee.gov.cn/ywdt/hjywnews/202106/t20210621_841062.shtml,2021-6-17).

Shen B. R1234yf and R1234ze(E) as low-GWP refrigerants for residential heat pump water heaters. *International journal of refrigeration*, 2017.

Yan D, Xia J, Tang W, Song F, Zhang X, Jiang Y. *DeST—An integrated building simulation toolkit Part I: Fundamentals*. Conference DeST—An integrated building simulation toolkit Part I: Fundamentals, vol. 1. Springer, p. 95–110.

Zhang S L X. Comparison analysis of different refrigerants in solar-air hybrid heat source heat pump water heater. *International journal of refrigeration*, 2015.
Zhihua Wang F W, Zhenjun Ma, Mengjie Song Numerical study on the operating performances of a novel frost-free air-source heat pump unit using three different types of refrigerant. *Applied Thermal Engineering*, 2017.
Zhou Zicheng. The Performance of Low GWP Refrigerants in a Residential Heat Pump System [J]. *Refrigeration*, 2015, 34(02):72–79.
Zhang Q, Zhang L, Nie J, Li Y. Techno-economic analysis of air source heat pump applied for space heating in northern China. *Applied Energy*. 2017; 207: 533–42.

Energy Revolution and Chemical Research – Kok-Keong Chong and Zhongliang Liu (Eds)
© 2023 The Authors, ISBN: 978-1-032-36554-1

Research on the equilibrium of carbon trading income under the "source-load" scenario incorporated into the aggregator

Ren Qiang, Zhiruo Meng*, Qinglin Yuan & Linlin Yuan
North China Electric Power University, Beijing Engineering Research Center of Energy Electric Power Information Security, Beijing, China

Xiying Gao
State Grid Liaoning Electric Power Co., Ltd., Marketing Service Center, Shenyang, China

Xinyi Lu
North China Electric Power University, State Grid Liaoning Electric Power Co., Ltd., Marketing Service Center, Shenyang, China

Gangjun Gong
North China Electric Power University, Beijing Engineering Research Center of Energy Electric Power Information Security, Beijing, China

ABSTRACT: The carbon emission rights trading market is to effectively promote the emission reduction of carbon dioxide and other greenhouse gases through the financial market mechanism, so as to achieve the coordinated and sustainable development of the economy and the ecological environment. Carbon trading has traditionally been identified as the transaction of emission allowances between high-energy-consuming and high-emitting enterprises at the source end, and carbon trading with load-end users has become a subject that can be studied, this topic makes it the key to maximizing the benefits of both the source and the load and minimize carbon emissions. Therefore, establishing a quota, cost, and benefit mathematical model based on emission control enterprise clusters, clean energy enterprise clusters and aggregators is the key basis for evaluating project feasibility.

1 INTRODUCTION

President Xi proposed that China will achieve the carbon peak target by 2030 and the carbon neutrality target by 2060, which has been highly praised and widely responded to by the international community (Zhang 2021). At present, carbon emissions from the combustion of fossil energy in my country account for nearly 90% of the total carbon emissions of the whole society. The key to solving the problem of carbon emissions is to reduce carbon emissions from energy sources, and the fundamental solution is to transform the way of energy development (Wang 2021). Carbon emission trading can achieve the established carbon emission reduction goals at a lower cost, make full use of the carbon trading mechanism, and establish a complete carbon market, which is an effective policy tool for China to fulfill its climate change commitments (Li 2021). The effective operation of the carbon market requires a sound mechanism design as a guarantee, and the allocation of carbon allowances is a key element of the carbon market mechanism design (Qiu 2021). Based on the above description, under the background of dual carbon goals, carbon trading has become one of the feasible carbon emission reduction schemes (Zhang 2021). Therefore, carbon emission rights trading enables both parties to obtain benefits from it. Under the condition that both the

*Corresponding Author: 1946598965@qq.com

 DOI 10.1201/9781003332657-3

total quota and the total carbon emission target are determined, carbon emission rights trading can achieve the goal of minimizing the cost of social emission reduction.

The above introduction on carbon trading is aimed at national policies and is a traditional trading method. This paper hopes that the load-side users will be absorbed into the carbon trading sequence so that the main body of energy consumption can participate in the ranks of carbon emission reduction based on carbon trading. Therefore, the concept of the aggregator is introduced, and the aggregator is the carbon emission reduction and conversion amount collection agency of the load-side user. The participation of aggregators enables load-side users to be indirectly included in the carbon trading market, further building a transparent and expanded carbon trading market.

2 CARBON ALLOWANCE TRADING UNDER THE "SOURCE-LOAD" SCENARIO THAT INCORPORATES AGGREGATORS

In the source-end carbon trading market, the government has allocated a certain amount of carbon allowances to a single enterprise. Among them, there will be some companies with excess carbon allowances, and some companies still have a carbon allowance surplus, which constitutes a carbon allowance transaction between the excess and the surplus.

The aggregator is introduced at the load side, and the carbon emission reduction per unit time of users on the load side is collected, and the aggregator is converted into carbon emission rights and included in the market transaction. Another important role of the aggregator is to propose an incentive mechanism to attract many source-end enterprises and load-side users to join the mechanism so that the aggregator will play a leading role in the transaction.

The aggregator collects the carbon emission reduction per unit time of the user at the load end, proposes a conversion mechanism between the user's carbon emission reduction and carbon emission quota, and participates in the carbon emission transaction with other companies. Therefore, the majority of load-side users do not need to directly join the industrial carbon trading platform, but directly reach a mutual trading agreement with aggregators. Aggregators dominate the industrial carbon trading platform. For load-side users, the conversion of carbon emission reduction per unit time itself is the incentive mechanism proposed by the aggregator. Therefore, aggregators have the pricing power to convert carbon emission reductions for load-side users, and in order to limit the power boundaries of the aggregators themselves, the bidding mode proposed by the aggregator in the transaction process is recognized by all trading nodes, so that all trading nodes have the right to know and speak. Finally, a transaction model of co-governance and sharing of the aggregator-enterprise cluster alliance will be constructed to achieve fair transactions and achieve a win-win situation.

3 CALCULATION MODEL OF CARBON QUOTA FOR ENTERPRISE CLUSTER NODES AND AGGREGATOR NODES

3.1 *Carbon quota calculation model for coal power companies*

At this stage, there is a fixed calculation model for the carbon quota of coal power enterprises. Figure 1 shows the overall model of carbon allowance calculation for coal power.

Estimated total coal power generation at the source,

$$P_{all} = \alpha P_{yall} \tag{1}$$

Among them, P_{all} is the total installed power generation of coal power; P_{yall} is the total power generation of the source; α is the proportion of coal power at the source.

Estimated total coal consumption of source coal power,

$$T_{all} = P_{all} P_{per} \tag{2}$$

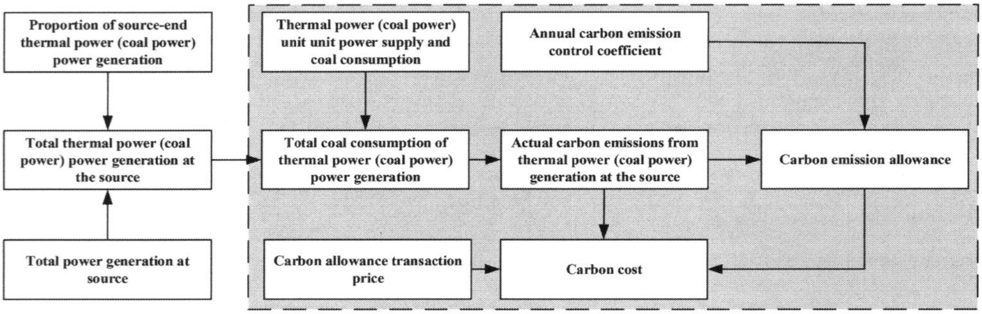

Figure 1. The overall model of carbon allowance calculation for coal power.

Among them, T_{all} is the actual total coal consumption in the furnace; P_{per} is the coal consumption per unit of power supply of the coal power unit.

Actual carbon emissions from coal power at the source,

$$D_{all} = KT_{all} \tag{3}$$

According to the 13th Five-Year Power Plan, when calculating carbon emissions, the emission coefficient of standard coal carbon dioxide is selected as, K = 2.8.

Source carbon allowance,

$$Q_{all} = \delta D_{all} \tag{4}$$

Among them, Q_{all} is the total amount of carbon allowances at the source, and δ is the annual carbon emission control coefficient.

For any coal-fired power company at the source, there are,

$$\begin{cases} P_{all} = \sum\limits_{i=1}^{n} P_i \\ T_{all} = \sum\limits_{i=1}^{n} T_i \\ D_{all} = \sum\limits_{i=1}^{n} D_i \\ Q_{all} = \sum\limits_{i=1}^{n} Q_i, Q_i = \delta D_i \end{cases} \tag{5}$$

In the formula, P_i is the estimated power generation of a coal-fired power company; T_i is the estimated actual coal consumption of the coal-fired power company; D_i is the estimated actual carbon emissions of the coal-fired power company; Q_i is the carbon allowance of the coal power company; record the source end There are n coal-fired power companies, $n = \{i_1, i_2, i_3, \ldots, i_n\}$.

3.2 *Calculation model of node carbon quota for clean energy enterprises*

The carbon allowances of clean energy companies are converted based on their estimated power generation in the current year and converted into carbon emission reductions. which can be recorded as the conversion coefficient between the power generation of a clean energy company and its carbon emission reduction is ω, the conversion coefficient between the carbon emission reduction and the carbon allowance is ψ, and the carbon emission reduction of the clean energy company can be obtained as,

$$L_I = \omega P_I \tag{6}$$

In the formula, P_I is the power generation of the clean energy company.

And, the carbon allowance of the clean energy company is,

$$Q_I = \psi L_I \tag{7}$$

Therefore, for all clean energy companies that can participate in the transaction at the source, the total carbon emission reduction, and the total carbon allowance can be expressed by the following formula,

$$\begin{cases} L_{ALL} = \sum_{I=1}^{N} L_I = \omega \sum_{I=1}^{N} P_I \\ Q_{ALL} = \sum_{I=1}^{N} Q_I = \psi \sum_{I=1}^{N} L_I \end{cases} \tag{8}$$

Among them, there are N total clean energy companies at the source end, and $N = \{I_1, I_2, I_3, \ldots, I_N\}$.

3.3 *Calculation model of carbon quota for aggregator nodes*

The electricity saved by the load-side user in a period of time is used as the carbon quota conversion index. The user's (n + 1)th year's power consumption can be set as $EC_{U_i}^{n+1}$, which can be obtained for a certain period of time. The difference in user power consumption is,

$$\Delta EC_{U_i} = EC_{U_i}^{n+1} - EC_{U_i}^{n} \tag{9}$$

$$s.t. \begin{cases} \Delta EC_{U_i} > 0, \textit{the user node participates in the} \\ \qquad \textit{transaction with the aggregator} \\ \Delta EC_{U_i} \leq 0, \textit{the user node does not participate} \\ \qquad \textit{in the same aggregator transaction} \end{cases}$$

The conversion coefficient between the difference between the user's electricity consumption and the user's carbon emission reduction can be recorded as β, and the conversion coefficient between the carbon emission reduction and the carbon allowance can be recorded as γ, then we can get:

Carbon emission reduction of a user on the load side,

$$L_{U_i} = \beta \Delta EC_{U_i} \tag{10}$$

A user on the load side converts carbon allowances,

$$Q_{U_i} = \gamma L_{U_i} \tag{11}$$

The aggregator can gather M users, denoted as $M = \{U_1, U_2, U_3, \ldots, U_M\}$, then the aggregate carbon quota of the aggregator can be expressed as,

$$Q_{AG} = \sum_{i=1}^{M} Q_{U_i} \tag{12}$$

4 ENTERPRISE CLUSTER, AGGREGATOR NODE COST AND PROFIT CALCULATION MODEL

4.1 *The expression of the node cost function, carbon allowance income function, and profit function of coal power enterprises*

At this stage, the introduction of carbon quota trading can enable coal-fired power companies with shortfalls to maintain the original power generation by purchasing carbon quotas, avoiding unnecessary large-scale boiler shutdowns and unit shutdowns. The additional carbon allowances purchased by the company with other companies or aggregators will be included in the company's carbon emission cost for the current year. The carbon emission cost is calculated as follows,

$$\begin{cases} FC_i = (D_i - D_s)V \\ D_i - D_s > 0 \end{cases} \tag{13}$$

Where FC_i is the carbon emission cost of a coal-fired power company; D_i is the actual carbon emission of the company; D_s is the corresponding carbon emission after the company gets the carbon allowance; V is the final transaction price of the unit carbon allowance. It can be obtained that the total production cost of the coal-fired power plant is,

$$FC_{i_{all}} = FC_o + FC_i \tag{14}$$

Where FC_o is the cost of power generation energy consumption of a coal power company, including other costs such as coal consumption, employee wages, and equipment maintenance.

Some coal-fired power companies have not completed their carbon quota plan and therefore have a carbon quota surplus. Under ideal circumstances, the carbon quota revenue function of these companies is expressed as,

$$\begin{cases} IC_i = (D_s - D_i)V \\ D_s - D_i > 0 \end{cases} \tag{15}$$

Under normal circumstances, a company with surplus carbon allowances will not necessarily sell all of its carbon allowances. Its income still depends on the demand of the company that buys carbon allowances from him. Therefore, in actual situations, its carbon allowance income expression should be,

$$IC_i' = \sum_{i=1}^{n} Q_{A_i \to A_{i-1}} V \tag{16}$$

4.2 *Aggregator node cost function expression and carbon allowance revenue function expression*

ρ can be recorded as the quota-revenue transaction coefficient between the load-side user and the aggregator, and we can get,

$$\rho = \frac{EARN_{U_i}}{Q_{U_i}}, \rho \geq 0 \tag{17}$$

In the formula, $EARN_{U_i}$ is the quota sales revenue of any user on the load side. Then the total cost of the aggregator to purchase load-side users is,

$$FC_U = \sum_{i=1}^{M} EARN_{U_i} \tag{18}$$

The expression of the cost function of the aggregator node is as follows,

$$FC_{AG} = FC_{AG_{fix}} + FC_U + FC_{ENT} \tag{19}$$

Among them, FC_{AG} represents the total cost of the aggregator node; $FC_{AG_{fix}}$ represents the fixed cost of the aggregator node, including labor costs, equipment purchase, maintenance, etc., all necessary expenses to maintain the normal operation of the aggregator node; FC_U means the cost of carbon emission reduction conversions purchased by the aggregator to the load-side users; FC_{ENT} means the cost of carbon allowances purchased by the aggregator from other companies. These include the cost of carbon allowances purchased from coal-fired power companies with surpluses and the cost of carbon allowances purchased from clean energy companies.

When the aggregator integrates the carbon allowances obtained by all parties and sells them to the coal-fired power companies that are in short supply, the aggregator generates revenue. Ideally, the aggregator's revenue can be expressed as,

$$IC_{AG} = \sum_{i=1}^{n} FC_i \tag{20}$$

In the formula, FC_i is the carbon emission cost of coal-fired power companies mentioned above. Therefore, the carbon emission cost of each coal-fired power company is the revenue of aggregators.

However, coal-fired power companies do not necessarily buy all the carbon allowances they need from aggregators. In order to minimize the cost of carbon emissions, under normal circumstances. Under normal circumstances, these companies will purchase the carbon allowances they need in batches from merchants, so the general expression of the aggregator's income should be,

$$IC'_{AG} = \sum_{i=1}^{n} Q_{A\to AG} V \tag{21}$$

In the formula, $Q_{A\to AG}$ represents the quantity of carbon allowances to be purchased by the coal-fired power company and the aggregator to reach a transaction agreement, and V is the final transaction price of this part of the carbon allowance transaction in the node transaction agreement between the two parties (In this article, all the transaction price corresponding to the carbon allowance transaction volume that has reached the transaction agreement is denoted by V.)

4.3 *Expression of clean energy enterprise cost and carbon allowance income function*

The cost consumption of clean energy enterprises can be expressed as FC_1, including costs such as equipment maintenance and personnel expenses.

Under normal circumstances, clean energy companies obtain carbon allowances based on their carbon emission reductions, and sell this part of the carbon allowances as additional income for the company. In an ideal situation, that is, if all the carbon allowances it owns are sold, the carbon allowance income will be,

$$IC_I = Q_I V \tag{22}$$

It can be seen from the above that Q_I is the quantity of carbon allowances owned by any clean energy company.

However, just as a coal-fired power company wants to purchase quotas from aggregators, a coal-fired power company may not purchase all the carbon quotas of the clean energy company, and the clean energy company may not be able to sell all of its carbon quotas. Therefore, in reality, its carbon allowance income expression should be,

$$IC'_I = \sum_{i=1}^{n} Q_{A\to B} V \tag{23}$$

Among them, $Q_{A\to B}$ represents the quantity of carbon allowances purchased by a coal-fired power company that has reached a deal with the clean energy company.

4.4 *Expression of node profit function of coal power enterprises, aggregators, and clean energy enterprises*

$$\begin{cases} P_i = p_i + IC'_i - FC_{i_a} \\ P_{AG} = IC'_{AG} - FC_{AG} \\ P_I = p_I + IC'_I - FC_I \end{cases} \tag{24}$$

In the formula, P_i, P_{AG}, and P_I are the profits of coal power companies, aggregators, and clean energy companies, respectively, and p_i and p_I represent the profits from coal power companies and the profit from the sale of electricity by clean energy companies.

4.5 *Constraints*

1) User benefit constraints on the load side,

$$\begin{cases} EARN_U = \sum_{i=1}^{n} EARN_{U_i} \\ EARN_U > 0 \end{cases} \tag{25}$$

In order to achieve the effect of motivating users, all load-side users who participate in transactions with aggregators should have corresponding benefits.

2) Constraints on the rationing of total source carbon allowances,

$$\begin{cases} Q_{source} = (0.0017Q_{counteract} - 0.0056Q_{summit} \\ \quad - 0.0444Q_{first})t^2 + (0.1611Q_{summit} \\ \quad - 0.0155Q_{counteract} - 0.1456Q_{first})t + Q_{first} \\ Q_{source} = Q_{all} + Q_{ALL} = \sum_{i=1}^{n} Q_i + \sum_{I=1}^{N} Q_I \end{cases} \tag{26}$$

According to the "30·60" national carbon emission reduction target, carbon emissions will reach a peak in 2030, and then gradually decline until they reach relatively zero emissions in 2060, that is, offsetting carbon emissions from the source through other emission reduction methods. It is possible to set the total source-end carbon allowances of this year as Q_{first}, the total source-end carbon allowances in the "peak" year (2030) as Q_{summit}, and the total source-end carbon allowances in the "neutralization" year (2060). The amount is $Q_{counteract}$. In the formula, Q_{all} and Q_{ALL} are the total carbon allowances of coal-fired power companies and clean energy companies respectively.

5 CONCLUSIONS

The carbon trading market aims to effectively promote the emission reduction of carbon dioxide and other greenhouse gases through the financial market mechanism, so as to achieve the coordinated and sustainable development of the economy and the ecological environment. At present, preliminary progress has been made in the pilot carbon emissions trading, but of course, there are some deficiencies, which is the only way for the development process. Therefore, this paper proposes a new carbon trading model that incorporates aggregators based on traditional source-side carbon trading. Aggregators enable the majority of load-side users to indirectly participate in the carbon trading market. In this paper, the energy and power industry is selected as the research focus. Through the calculation of carbon allowances and costs and benefits of power generation enterprise clusters and aggregators, the interaction of carbon trading in the "source-load" scenario is favorably explained. It is true that the carbon trading market still has extensive development space and amazing development potential. With the unremitting efforts of researchers or government agencies, the construction of the carbon trading market will be more perfect.

ACKNOWLEDGMENT

We would like to express our sincere gratitude to everyone who helped in this research work. This project was supported by State Grid Liaoning Province Electric Power Co. Ltd., Shenyang, China, and State Grid Liaoning Electric Power Co., LTD. Marketing Service Center, Shenyang, China under the company's project "Research on the business model of precision marketing service based on the analysis of users' electricity behavior."

REFERENCES

Li X.Y, Wang D.P, Research on Supply Chain Coordination Considering Competition and Information Asymmetry under Carbon Trading Mechanism. *J. Operations Research and Management*, 2021 pp. 1–9.

Qiu B, Song S.X, Wang K., Yang Z, Optimal operation of regionally integrated energy system considering demand response and tiered carbon trading mechanism. *J. Journal of Electric Power System and Automation*, 2021, pp. 1–16.

Wang X.J, Liu P, Li R.C, Feng J, Jiang D.F, Research progress and prospects of advanced power generation technology under the "dual carbon" goal. *J. Thermal power generation*, 2021, pp. 1–8.

Zhang Y.G, Bai Y.J, The path to achieving the "dual carbon" goal of regional differentiation. *J. Reform*, 2021, pp. 1–18.

Zhang N, Pang J, Feng X.Z, Research on the economic impact of introducing a quota auction mechanism and implementing carbon tax supporting measures in the national carbon market. *J. China Environmental Science*, 2021, pp. 1–13.

Energy Revolution and Chemical Research – Kok-Keong Chong and Zhongliang Liu (Eds)
© 2023 The Authors, ISBN: 978-1-032-36554-1

Experimental research on the effects of different aquatic plants and their combinations on the removal of nitrogen and phosphorus in polluted water in the Poyang Lake area

Jia Wang*
Jiangxi Academy of Water Science and Engineering, Nanchang, China
College of Forestry, Jiangxi Agricultural University, Nanchang, China

Minghao Mo*
Jiangxi Academy of Water Science and Engineering, Nanchang, China

Xiulong Chen
College of Land Resource and Environment, Jiangxi Agricultural University, Nanchang, China

Qing Li
School of Geography and Environment, Jiangxi Normal University, Nanchang, China

Lingyun Wang
Jiangxi Academy of Water Science and Engineering, Nanchang, China

ABSTRACT: Based on the actual situation of rural areas in De'an County of Poyang Lake area, the indoor artificial simulation of the rural channel was carried out, and different aquatic plants (Iris tectorum, Zizania aquatica, Nymphaea tetragona, Myriophyllum verticillatum L.) and their combinations were set to explore the effects of aquatic plants on the removal of nitrogen and phosphorus them in water. The results showed that in terms of the treatments of Iris tectorum, Zizania aquatica, Nymphaea tetragona, Myriophyllum verticillatum L, iris + Nymphaea, Zizania + Nymphaea, whether they are single plant or plant combination, in which the removal rate of total nitrogen was more than 75%, the removal rate of total phosphorus was more than 57%, and the effect was good. Among them, the treatment effect of Iris tectorum + Nymphaea was the best, TN removal rate was 89.4%, TP removal rate was 68.87%, and microbial content is the highest. The research results can not only effectively prevent and control rural non-point source pollution and improve the rural water ecological environment, but also have important theoretical value and practical significance for promoting the construction a of clean small watershed in Jiangxi Province and the construction of beautiful countryside in Jiangxi Province.

1 INTRODUCTION

At present, with the improvement of rural living standards, rural domestic wastewater is increasing year by year. The use of pesticides and fertilizers, soil erosion, and unreasonable farming methods make a large number of pollutants enter the water body without ecological treatment, which makes the water body in rural areas of China seriously polluted. As an important part of the rural water pollution control technology system, the ecological restoration technology the of rural water environment has important practical significance (Chen 2017; Pei 2014; Wang 2014). Phytoremediation technology, as one of the means of sewage treatment, has the advantage of being cheap and efficient in treating polluted water. Aquatic plants will absorb a large number of nutrients such

*Corresponding Authors: mominghao@126.com and 861663497@qq.com

 DOI 10.1201/9781003332657-4

as nitrogen and phosphorus during their growth (Hu 2011), and their developed roots are a good environment for microbial survival (Fu 2014), thereby promoting the degradation of pollutants. Many studies have shown that aquatic plants have the effect of removing nitrogen and phosphorus in sewage, and can effectively improve the water quality of eutrophic water bodies.

There are many experiments on nitrogen and phosphorus removal by the single aquatic plant at home and abroad, in order to screen the plant species with the highest purification rate of nitrogen and phosphorus in the water. Among them, the plants to be screened are mainly concentrated in ecological aquatic plants, and there are few experimental studies on different combinations of aquatic plants such as economic, landscape, and ecological types. Because of its good purification effect, special economic benefits, low energy consumption, and easiness to rebuild and restore a good aquatic ecosystem, aquatic plant restoration technology is increasingly being concerned (Lei 2005). According to the *Statistical Yearbook of Jiangxi Province*, the emission of agricultural chemical fertilizer alone in rural areas of Jiangxi Province reached 27105 t/a (nitrogen) and 1369 t/a (phosphorus), accounting for 44.55% and 21.31% of a total load of nitrogen and phosphorus into Poyang Lake, which is one of the important sources of nitrogen and phosphorus pollution in Poyang Lake area (Tang 2014). In view of this, based on the actual situation of rural areas in the Poyang Lake area, the indoor experiments of nitrogen and phosphorus removal by different aquatic plants and combinations were carried out in order to provide a reference for rural aquatic ecological restoration.

2 MATERIALS AND METHODS

2.1 *Selection of test plants*

The selection of aquatic plants in this experiment is based on the principle of economic value, landscape viewing, and strong ability to absorb pollution, which are common native plants, emergent, floating, and submerged plants are selected. Wild aquatic plants were used as experimental materials, including emergent plants Iris tectorum, Zizania aquatica, floating plants Nymphaea tetragonal, and submerged plants Myriophyllum verticillatum L. The nitrogen and phosphorus removal experiments were conducted by aeration treatment which was Iris tectorum (Y), Zizania aquatica (J), Nymphaea tetragona (S), Myriophyllum verticillatum L (H), Iris + Nymphaea (Y + S), Zizania latifolia + Nymphaea (J + S), blank aeration (CK1) and blank control (CK2).

2.2 *Treatment of test water*

250 ml plant nutrient solution (nitrogen 32.0 g/L, phosphorus anhydride 12.0 g/L, potassium oxide 20.0 g/L, magnesium oxide 1.3 g/L, sulfur 1.7 g/L, total trace element content >0.2 g/L, etc.) was added to each treatment to ensure the growth of aquatic plants. On this basis, the mixture of potassium nitrate (KNO_3) and potassium dihydrogen phosphate (KH_2PO_4) was added to the test water to simulate the water concentration of total nitrogen (TN) and total phosphorus (TP) in the rural channel of Baota Township, De'an County, Poyang Lake area. The water quality indexes of test water are shown in Table 1. Referring to Environmental Quality Standards for Surface Water (GB3838-2002) and evaluated by a single factor, the water quality of the test water is worse than Grade V.

Table 1. The quality concentration of test water.

TN (mg/L)	TN (mg/L)	NH_3-N (mg/L)	DO (mg/L)	pH
6.340±0.198	0.638±0.013	1.222±0.186	5.150±0.173	7.663±0.124

2.3 *Test method*

The experiment was carried out in the greenhouse of Jiangxi Provincial Eco-Science Park of Soil and Water Conservation (29°16′–29°17′N, 115°42′–115°43′E) in De'an County, Jiangxi Province.

The plastic box with a length of 80 cm × width of 60 cm × height of 60 cm was selected for the experiment of removing nitrogen and phosphorus in the water. A total of six plant treatments (4 plants per box for emergent plants and floating plants, and the coverage of submerged plants on the water surface is 80%) and two control treatments (pure test water, no aeration, and pure test water + aeration, no plants) were set up in the experiment. Both emergent plants and floating plants were planted in small flowerpots filled with small gravel after cleaning their roots. The plant heights of emergent plants are treated to be the same, the height of Zizania aquatica was 45 cm and Iris tectorum was 35 cm. Set up 3 replicates for each treatment, a total of 24 test water tanks, as far as possible to ensure that the lighting, ventilation, and other conditions were consistent. During the test, tap water was added to supplement the water consumed by evaporation, plant transpiration, and sampling, and the water level in the tanks was maintained at 35 cm. Except for three replicates of CK2, all the experiments were treated with aeration, and the aeration mode was continuous aeration.

During the experiment, the second sampling was carried out one day after the first sampling, and the water sample was taken once every five days thereafter. The water was stirred many times before sampling so that the water in the tank was mixed evenly, and a 100 mL water sample was taken each time. To reduce the test error, the sampling time was around 17:00. The trial period was 32 days from 31 May to 1 July 2018. According to Water and Wastewater Monitoring and Analysis Method, total nitrogen was determined by potassium persulfate oxidation-ultraviolet spectrophotometry, and total phosphorus was determined by potassium persulfate digestion-molybdenum antimony spectrophotometry. After the end of the test cycle, water samples, filler samples, and root samples were taken to determine the number of microorganisms. The bacteria were cultured in beef extract peptone agar medium, the actinomycetes were cultured in modified Gaoshi No. 1 medium, and the fungi were cultured in rose Bengal medium (Li 2018); (Tian 2011).

2.4 *Data processing*

The data were analyzed by EXCEL and SPSS. All data were expressed by mean ± standard deviation, and the difference was analyzed by variance analysis LSD test (Levene's-test was used to test the homogeneity of variance between different groups), and the significance level was set to $P < 0.05$.

The calculation formula for the removal rate of pollutants in the water body is: removal rate (%) $= (C_0 - C_i)/C_0 \times 100\%$, where C_0 is the initial concentration and Ci is the concentration on the i-th day (Cai 2011).

3 TEST RESULTS AND DISCUSSIONS

3.1 *Total N content and N removal rate in water*

Plants need to absorb inorganic nitrogen as their own nutrients for the synthesis of organic nitrogen such as plant proteins. At the same time, an aerobic-anoxic-anaerobic microenvironment can be formed near the roots of plants, which is conducive to the coexistence of nitrifying bacteria and denitrifying bacteria, thereby enhancing the nitrification and denitrification of microorganisms, and improving the purification efficiency of nitrogen in sewage. The total nitrogen concentration and removal rate in different treated water bodies are shown in Figure 1 and Table 2.

From the perspective of water quality, the initial concentration of TN in water was 6.340 ± 0.198 mg/L. After 32 days, the concentrations of TN in Y, J, S, H, Y + S, J + S, CK1 and CK2 were as follows: 0.879 ± 0.577 mg/L, 0.922 ± 0.205 mg/L, 0.553 ± 0.273 mg/L, 1.470±0.053 mg/L, 0.682 ± 0.090 mg/L, 0.744 ± 0.026 mg/L, 1.920 ± 0.319 mg/L, 3.682 ± 0.551 mg/L. The initial water quality of the test water is worse than Grade V. Although the prepared sewage CK2 can be degraded without any treatment, the TN concentration is still 3.682 mg/L after 32 days, which is still worse than Grade V. The TN concentration of CK1 treated with aeration was lower, which was 1.920 mg/L, and it was generally Grade V. After the water treated by Y, J, S, Y + S, and J + S for 32 days, the water quality was in Grade III, and the water treated by H and CK1 was generally in Grade IV, indicating that aquatic plants and their fillers had good effects on TN removal.

From the perspective of TN concentrations, after 32 days of the experiment, the order of TN concentration in water is CK2 > CK1 > H > J > Y > J + S > Y + S > S. As can be seen from Figure 1, TN concentrations in all treated waters decreased overtime during the month. The decreasing trend of TN concentration in CK2 and CK1 treatments was not obvious, and the decreasing trend of TN concentration in other treatments with aquatic plants was obvious. Compared with the blank control CK2 without any treatment, the final TN concentrations of CK1, J, S, Y, J + S, Y + S and H were 52.15%, 25.05%, 15.02%, 23.86%, 20.22%, 18.52% and 39.91% of CK2, respectively. Compared with the aerated blank control CK1, the final TN concentrations of J, S, Y, J + S, Y + S, and H were 48.03%, 28.81%, 45.76%, 38.77%, 35.51%, and 76.54% of CK1, respectively. It can be seen that both aquatic plants and aeration treatment play an important role in reducing TN concentration.

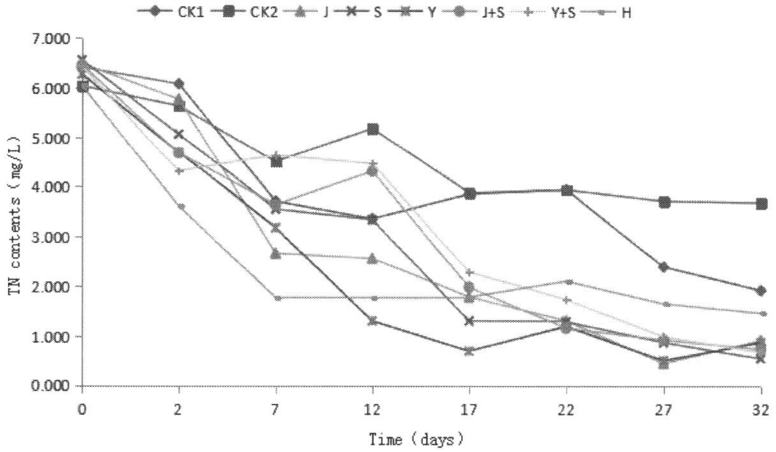

Figure 1. Removal effect of TN in test water under different treatments.

Table 2. Removal rate of TN in sewage by different treatments over time (%).

treatment	2d	7d	12d	17d	22d	27d	32d
CK1	5.09±2.18 d	42.22±5.92 c	47.66±2.35 c	39.99±1.71 d	38.52±5.50 b	62.69±11.08 c	70.06±4.23 d
CK2	6.57±3.01 d	25.31±0.53 d	14.01±4.54 e	35.89±4.37 d	34.84±2.58 b	38.57±8.94 d	39.07±9.80 c
J	10.86±3.97 d	58.88±1.43 b	60.52±3.27 b	72.37±5.79 bc	79.70±12.21 a	92.97±2.47 a	85.78±2.44 ab
S	22.57±5.52 c	45.87±7.09 bc	49.06±4.07 c	80.03±5.97 ab	80.25±8.76 a	86.56±5.27ab	91.57±4.20 a
Y	25.06±2.57 bc	49.36±5.37 bc	79.16±7.41 a	88.73±5.46 a	80.83±8.42 a	91.80±1.05 a	86.00±8.08 ab
J+S	27.39±5.37 bc	43.76±9.42 c	33.2±5.79 d	69.30±7.06 bc	81.97±3.95 a	85.60±5.75 ab	88.47±0.54 a
Y+S	32.90±5.09 b	27.77±8.87 d	30.58±8.20 d	64.46±6.81 c	72.91±3.94 a	84.52±6.13 ab	89.40±1.24 a
H	40.34±5.18 a	70.64±5.69 a	70.68±1.55 a	70.47±0.99 bc	65.21±8.75 a	72.61±4.96 bc	75.71±0.48 bc

Note: There were significant differences between treatments without the same letters. $P < 0.05$, LSD. Test

From the perspective of TN removal rate, the initial concentration of each treatment was taken as the reference, the order of TN removal rate from high to low was S (91.57%) > Y + S (89.40%) > J + S (88.47%) > Y (86.00%) > J (85.78%) > H (75.71%) > CK1 (70.06%) > CK2 (39.07%), and the TN removal rate by S was the highest. It can be inferred that for the sewage with low nitrogen concentration, the treatment effect of S is better. However, H may have a good effect on sewage treatment with a large pollution degree, and it was not the best for sewage treatment with low nitrogen concentration. In addition, the emergent plants and floating plants used in this experiment have flowerpots of broken stones as the substrate, and the roots grew in the substrate, while the submerged plant H did not have the substrate, and the removal effect of nitrogen was not as good as that of S, Y and J may also be related to the absence of substrate. However, the removal rate of TN by H can still reach 75.71% in a month, and the effect is also good. The significant difference analysis of SPSS showed that there were significant differences in TN removal rate among the three groups of blank, blank aeration, and aquatic plants. In the aquatic plant group, except for the significant difference between Y and Y + S, there was no significant difference in TN removal rate among other plants. It showed that aquatic plant treatment has an obvious effect on TN removal in sewage, and the effect of plant combination Y + S was better than that of single plant Y. Similar to the change of TN concentration with time, the removal rate of TN was higher before 17 days, and then the increase of removal rate was relatively small, indicating that aquatic plants and aeration treatment had a better effect in the early stage. In addition, the TN removal rate of H reached 70% in the first 7 days, and then the removal rate changed little, indicating that H had a rapid removal of nitrogen in sewage.

3.2 *Total P content and P removal rate in water*

The removal of phosphorus in water, on the one hand, was in the form of phosphate deposition and consolidation on the substrate, on the other hand, the available phosphorus was absorbed by plants, so the phosphorus content in water decreased after the experiment, but when it decreased to a certain extent, a small amount of phosphorus may be released from the substrate. The concentration and removal rate of TP in different treated water were shown in Figure 2 and Table 3.

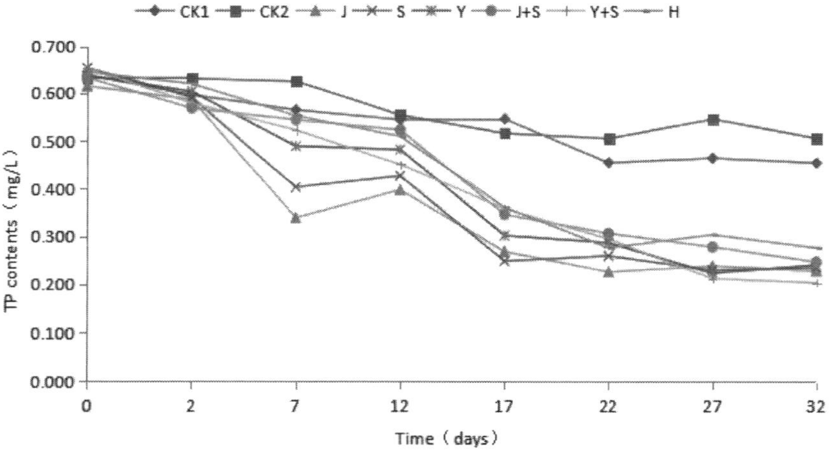

Figure 2. Removal effect of TP in test water under different treatments.

Table 3. Removal rate of TP in sewage by different treatments over time (%).

treatment	2d	7d	12d	17d	22d	27d	32d
CK1	6.54±5.68 a	11.46±7.57 bc	14.62±7.79 c	14.33±6.30 e	28.63±8.98 b	27.10±8.59 c	28.71±7.62 c
CK2	0.23±0.36 a	1.16±1.83 c	12.28±6.91 c	18.23±2.46 e	19.93±1.46 c	13.55±4.91 d	19.87±5.05 d
J	4.61±1.11 a	44.72±6.55 a	35.27±3.20 a	56.15±1.51 ab	63.07±1.59 a	61.08±4.44 ab	62.78±0.99 ab
S	9.63±1.08 a	38.37±1.25 a	34.72±4.82 a	61.85±1.90 a	60.18±9.53 a	64.74±2.02 ab	63.92±4.21 ab
Y	5.41±2.50 a	23.16±7.06 b	24.25±4.88 abc	52.48±4.80 bc	54.86±6.50 a	64.81±7.79 ab	62.09±2.91 ab
J+S	10.08±7.61 a	13.86±6.69 bc	17.17±4.89 c	45.15±3.93 cd	51.37±7.17 a	55.85±9.25 ab	60.75±4.27 ab
Y+S	11.40±5.09 a	19.96±6.39 b	30.93±2.73 ab	45.15±4.07 cd	54.73±3.17 a	67.46±6.73 a	68.87±4.58 a
H	4.07±4.70 a	14.35±7.05 bc	20.95±5.58 bc	43.95±6.91 d	57.04±1.91 a	52.81±5.49 b	57.08±2.38 b

Note: There are significant differences between treatments without the same letters. $P < 0.05$, LSD. Test

From the perspective of the water quality category, the initial concentration of TP in the test water was 0.638 ± 0.013 mg/L, and 32 days later, the water concentrations of Y, J, S, H, Y + S, J + S, CK1 and CK2 were 0.242 ± 0.031 mg/L, 0.229 ± 0.011 mg/L, 0.236 ± 0.031 mg/L, 0.277 ± 0.008 mg/L, 0.203 ± 0.048 mg/L, 0.248 ± 0.041 mg/L, 0.455 ± 0.062 mg/L and 0.506 ± 0.010 mg/L, respectively. It can be seen that the initial water quality of the test water was worse than Grade V. Although the prepared sewage CK2 could be degraded without any treatment, the TP concentration was still 0.506 mg/L after 32 days, which was still worse than Grade V. The TP concentration of CK1 treated with aeration was lower, which was 0.455 mg/L, but it was still worse than Grade V. The water quality treated with Y, J, S, H, and J + S was generally in Grade IV water level after 32 days. The water quality treated with Y + S was in Grade III or IV water level after 32 days, indicating that aquatic plants and their fillers had a better removal effect on TP in sewage.

From the perspective of the concentration values, after 32 days of the experiment, the order of TP removal rate from high to low was CK2 > CK1 > H > J + S > Y > S > J > Y + S. It can be seen from Figure 2 that the TN concentration of water under all treatments showed a downward trend with time in this month. The decreasing trend of TN concentration in CK2 and CK1 was not obvious, and the decreasing trend of TP concentration in other treatments with aquatic plants was obvious. After aquatic plants treatment, the TP concentration decreased significantly in the first 22 days and tended to be moderate in the next 10 days, but it still showed a downward trend, in which TP concentration occasionally increased. Compared with CK2, the final TP concentrations of CK1, J, S, Y, J + S, Y + S and H were 89.84%, 45.21%, 46.63%, 47.74%, 49.02%, 40.20% and 54.73% of CK2, respectively. Compared with CK1, the final TP concentrations of J, S, Y, J + S, Y + S and H were 50.28%, 51.85%, 53.09%, 54.51%, 44.70% and 60.86% of CK1, respectively. It can be seen that aquatic plants and aeration treatment played an important role in the reduction of TP concentration, but the reduction of TP concentration was relatively lower than that of TN.

From the perspective of TP removal rate, the initial concentration of each treatment was taken as the reference, the order of TP removal rate from high to low was Y + S (68.87%) > S (63.92% > J (62.78%) > Y (62.09%) > J + S (60.75%) > H (57.08%) > CK1 (28.71%) > CK2 (19.87%), and the TP removal rate by Y + S was the highest. It can be inferred that Y + S had a better treatment effect for sewage with low phosphorus concentration, and H had a good effect for sewage with large phosphorus concentration, which was not the best for sewage treatment with low phosphorus concentration. In addition, both emergent plants and floating plants used in this experiment had broken stones in flowerpots as substrate, and roots grew in the substrate. However, submerged plant H did not have the substrate, and the removal effect of phosphorus was not as good as that of S, Y, and J, which might also be related to the absence of substrate. Nevertheless, the removal rate of TP by H in one month was still about 60%, and the effect was also good. The significant difference analysis of SPSS showed that there was no significant difference in TP removal rates between CK1 and CK2, but there were significant differences between the treatments with and without aquatic plants. In the aquatic plant group, except for S and H, there was no significant difference among other plants. It showed that aquatic plant treatment had a significant effect on the removal of TP in sewage, and S had a good effect on TP removal. Similar to the change of TP concentration with time, the removal rate of TP was larger before 22 days, and then the removal rate increased slightly, indicating that aquatic plants and aeration treatment had a better effect in the early stage.

Table 4. Microbial quantity and water quality after each treatment cycle.

Treatment	Y	J	S	H	Y+S	J+S	CK1	CK2
Number of microorganisms (amounts)	7.8×10^8	1.6×10^6	1.6×10^8	8.2×10^8	1.8×10^9	2.5×10^8	9.1×10^5	6.2×10^4
TN (mg/L)	0.879	0.922	0.553	1.470	0.996	0.930	2.393	3.712
TP (mg/L)	0.242	0.229	0.236	0.277	0.203	0.248	0.455	0.506

3.3 *Number of environmental microorganisms*

The role of each part in the aquatic ecological restoration was analyzed from the number of microorganisms in root, stone, and water. There were a large number of microorganisms in stone fillers, some of which had a large number of bacteria, and some had a large number of actinomycetes. The number of actinomycetes per unit mass in J and J + S treatments even exceeded that of bacteria, indicating that the filler matrix also played an important role in sewage treatment. Relatively speaking, the number of microorganisms in water was very small. The number of microorganisms per unit mass in water was very small regardless of whether there were measures or not in all treatments, and the number of actinomycetes was only a little more, indicating that in the aquatic ecological restoration, there were not enough microorganisms generated solely by the self-purification ability of water body, and the ability to treat sewage was not enough. It can be seen from the slightly poor removal ability of nitrogen and phosphorus by H without fillers. The number of microorganisms in the roots was the largest, and the microorganisms per unit mass in the roots treated with Y, J, S, H, Y + S, and J + S accounted for 94%, 60%, 99%, 99%, 99%, and 89%, respectively. It can be seen that plant roots can play an important role in sewage treatment. In terms of the total amount of microorganisms per unit mass, Y + S was the largest, and the removal effect of TN and TP was also the best. The number of microorganisms in CK1 and CK2 treatments without plant measures was very small, even 104 times different. It can also be seen from the diagram that there was a certain correlation between the number of microorganisms and TN removal rate, TN concentration, TP removal rate, and TP concentration. The more the number of microorganisms, the better the removal effect.

4 CONCLUSIONS

(1) From the perspective of nitrogen and phosphorus removal efficiency, the removal rates of TN and TP in each treatment, whether single plant or plant combination, were more than 75% and 57%, respectively. Y + S had the best treatment effect, with the TN removal rate of 89.4% and TP removal rate of 68.87%, indicating that the combination of landscape plants had a good sewage treatment effect.
(2) The plant communities formed by different aquatic plant combinations have higher removal rates of nitrogen and phosphorus than single plants, and the purification effect is also more stable. The water ecological restoration technology using the combination of aquatic plants can be applied in the construction of ecologically clean small watersheds and beautiful villages.

ACKNOWLEDGMENTS

This work was financially supported by the Provincial Key Research and Development Program of Jiangxi Province (20192BBGL70043), the National Natural Science Foundation of China (42067020 41761065) and the Project of Jiangxi Provincial Water Resources Bureau (201821ZDKT16, KT201718).

REFERENCES

Cai, P.-y., Liu, A.-q., Hou. X.-l., (2011) Study on effects of Seven Hydrophytes on Nitrogen and Phosphorys Removal from Domestic Sewage[J]. *Chinese Journal of Environmental Engineering*, 5(5):1067–1070.

Chen, Ch.-j., Zhao, T.-Ch., Liu R.-L., et al. (2017) Performance of five plant species in the removal of nitrogen and phosphorus from an experimental phytoremediation system in the Ningxia irrigation area. [J]. *Environmental monitoring and assessment*, 189(10):497.1–497.13

Fu, X.-y., He, X.-y., (2014) Comparison of Nitrogen and Phosphorus Removal Capability of Five Aquatic Plants[J]. *Journal Northwest Forestry University*, 29(3):79–82.

Hu, M.-h., Yuan, J.-h., Xiang, L.-c., et al. (2011) Influence of different nitrogen: phosphorus on growth characteristic of the perennial aquatic plant[J]. *Chinese Journal of Environmental Engineering*, 5(11):2487–2493.

Lei, H., Zhai, J.-p., Jiang, X.-y., et al. (2005) Experimental Study of Decontamination Ability of Three Hydrophytes in Different Seasons[J]. *Environmental Protection Science*, (3):44–47.

Li, X.-x., Rong, X.-m., Xie, G.-x., et al. (2018) Difference in Absorbability of Different Aquatic Plants on N and P in Surface Water and Its Mechanism[J]. *Journal of Soil and Water Conservation*, 32(1):259–263.

Pei, L., Liang, J., Zhou, Ch., (2014) Variation rule of pollutant in the treatment process of rural domestic sewage by soil filter system[J]. *Journal of Water Resources & Water Engineering*, 25(2):54–56.

Tang, A.-p., Wang, J.-b., Li, S., (2014) Rural Non-point Pollution Controlled Based on BMPs in Small Poyang Lake Basin[J]. *Resources and Environment in the Yangtze Basin*, 23(7):1019–1026.

Tian, R.-n., Zhu, M., Song, X.-x., et al. (2011) Nitrogen and phosphorus removal effects of different hydrophyte combinations under simulated eutrophic conditions[J]. *Journal of Beijing Forestry University*, 33(6):191–195.

Wang, C., Zheng, S.-S., Wang, P.-f., et al. (2014) Effects of vegetations on the removal of contaminants in aquatic environments: A review[J]. *Journal of Hydrodynamics*, 26(4): 497–511.

Energy Revolution and Chemical Research – Kok-Keong Chong and Zhongliang Liu (Eds)
© 2023 The Authors, ISBN: 978-1-032-36554-1

Research on the application of interception engineering in the treatment of blue algae in South Taihu Lake

Yaolan Zhang*
Zhenjiang Design Institute of Water Conservancy & Hydro-electric Power, Hangzhou, Zhejiang Province, China

Yuyu Ji
Zhejiang Institute of Hydraulics & Estuary, Hangzhou, Zhejiang Province, China

ABSTRACT: Taking the South Taihu cyanobacteria interception project as the research object, combined with satellite imagery and on-site water quality monitoring, the water quality inside and outside the enclosure project and the cyanobacteria blocking effect are analyzed. The results show that the interception project has a significant effect on the interception of cyanobacteria. The average chlorophyll-a concentration inside the enclosure is 56.6% lower than the average chlorophyll-a concentration outside the enclosure. The satellite images of the same period also confirm that the cyanobacteria are effectively intercepted outside the enclosure; the wave-eliminating piles in the interception project keep the water in the enclosure in a relatively calm state, which is of positive significance for the development of water ecology in the enclosure. In general, the interception project will help improve the water landscape in the enclosed area, and the implementation of the project will have greater economic and social benefits for lakeside cities.

1 INTRODUCTION

Taihu Lake is located in the lower reaches of the Yangtze River in China, with a water area of 2,425 km^2, which is the third-largest freshwater lake in China. With the rapid economic and social development of cities around Taihu Lake, industrial and domestic sewage, as well as non-point source pollution caused by agriculture and aquaculture, are imported into Taihu Lake through the river (Bian 2017; Rajeshkumar 2018), gradually causing the self-purification capacity of Taihu Lake's body of water to be insufficient to deal with serious excess nutrients (Yang 2010). The high salt level leads to the eutrophication of Taihu Lake.

The direct consequence of the eutrophication of Lake Tai is algae blooms (Qin 2016). Algae blooms are generally defined as the rapid reproduction of phytoplankton and clusters on the water surface. Algae blooms will cause structural disorder and functional decline in the aquatic ecosystem, and the ecological balance will be disrupted. Specifically, the transparency of the water body is reduced, the dissolved oxygen is reduced, the quality of the water body is deteriorating, and the lake ecosystem is degraded. Algae blooms, which are the consequences of lake eutrophication, further aggravate lake eutrophication and form a vicious circle of algae blooms and eutrophication (Zhang 2012). Taking 2007 as an example, an unprecedented cyanobacteria

*Corresponding Author: 25192482@qq.com

 DOI 10.1201/9781003332657-5

bloom occurred in Taihu Lake (Fan 2018; Zhang 2011). A huge amount of cyanobacteria accumulated in Gonghu Bay, decomposed and entered the water inlet of the Gonghu Water Plant in Wuxi City. The drinking water of 2 million people was polluted, causing the most serious water pollution incident in the history of Taihu Lake. Therefore, how to control cyanobacteria has become an urgent need to solve the water environmental safety problem of the cities along Taihu Lake.

In response to the water environmental safety problems caused by cyanobacteria in Taihu Lake, the coast of Taihu Lake has gradually explored and carried out enclosure projects and lake ecological management projects (Liu 2011; Qin 2007, 2013), but there is still a lack of research on the application effects of the project after implementation. Based on this, this study takes the South Tai Lake blue-green algae interception project (including soft enclosures and wave-eliminating piles) as the research object, analyzes the improvement of the water quality along the coast of Taihu Lake and the interception effect of the blue-green algae after the completion of the interception project. We provide references and guidance for the partition project and the ecological restoration project in the enclosed area.

2 MATERIALS AND METHODS

2.1 *Study Area*

The study area is located in the coastal zone of the South Taihu New District, Huzhou City, China. The entrances of Changdou Port and Xiaomei Port in the area are the main entrances of the Tiaoxi Water System of Taihu Lake, and there is an interceptor to intercept the invasion of blue algae in South Taihu Lake outside the entrance of the lake. The project includes 4216-meter wave-eliminating piles and a soft enclosure project. Its location is shown in Figure 1.

Figure 1. Location of South Taihu Lake.

2.2 *Sample Collection*

This study was conducted on-site sampling and testing of the water quality at 7 points inside and outside the interception project on October 31, 2019. The layout of the points is shown in Figure 2 below, where Points 1, 2, and 3 are outside the enclosure; Points 4, 5, and 6 are located on the inner side of the enclosure; and Point 7 is located in the Xiaomei Port of the river into the lake. The appearance of the collected water is shown in Figure 3.

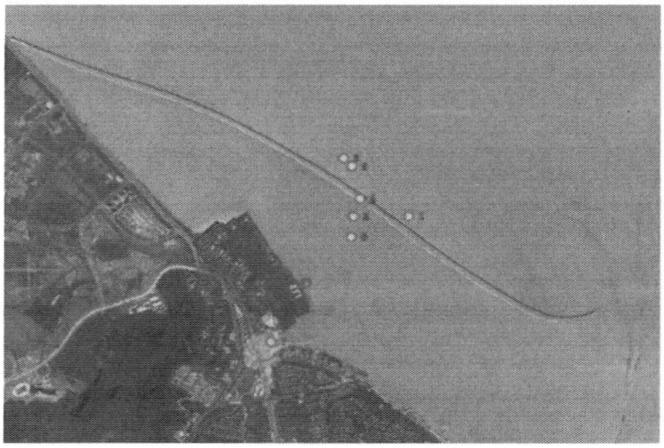

Figure 2. Location of sampling points.

Figure 3. Water quality at sampling points.

2.3 *Sample Analysis*

For lake water samples, the portable multi-parameter water quality analyzer was used to measure the water temperature (WT) and dissolved oxygen (DO) of the surface of the water on site, and 250 mL of 50 cm surface water samples were collected on site at each point and brought back to the laboratory for determination of the water body chlorophyll concentration. The concentration of chlorophyll-a is determined with reference to Chen Yuwei's improved hot ethanol extraction and spectrophotometry, with colorimetric wavelengths of 650 nm and 750 nm (Shimadzu UV-2450 spectrophotometer) (Wu 2021). When measuring chlorophyll-a, filter the phytoplankton on the membrane with a 47 mm diameter GF/F glass fiber filter membrane (Whatman company, pore size is about 0.7 μm), place the filter membrane in a 10 ml conical centrifuge tube, and cover it. Then, store it in the dark and in the freezer for more than 24 hours. Before the measurement, in a dark light environment, use 90% hot ethanol to grind and extract the pigment on the filter membrane, and then filter the extract with GF/C glass fiber filter membrane (Whatman company, pore size is about 1.2 μm), spectrophotometric method of determination.

3 RESULTS AND ANALYSIS

3.1 *Status of Cyanobacteria in Taihu Lake*

The prevention and control of cyanobacteria in Taihu Lake are under great pressure. From 2010 to 2017, the average density of cyanobacteria in Taihu Lake and South Taihu Lake showed a gradual increase, especially in 2016 and 2017. The average concentration of cyanobacteria in Taihu Lake in 2017 and 2018 was 82.82 million, respectively. The average concentration of cyanobacteria/L and 117.66 million/L in South Taihu Lake in 2017 and 2018 were 55.48 million/L and 10.07 million/L, respectively. Under the influence of Taihu Lake's backflow, the prevention and control pressure of cyanobacteria in surrounding cities is relatively high. The number of cyanobacteria in Taihu Lake is shown in Table 1.

Table 1. Outbreaks of cyanobacteria in Taihu Lake.

Time	Average of Taihu Lake Nutritional index	The time of the largest water bloom area in Taihu Lake	The largest water bloom area in Taihu Lake (km^2)	The proportion of cyanobacteria occurring in the total area of Taihu Lake (%)
2010	61.5	8.15	983.8	42.1%
2011	60.8	7.21	997.5	42.7%
2012	61.0	8.29	991.4	42.4%
2013	62.1	11.19	1092.4	46.7%
2014	61.5	11.26	854.0	36.5%
2015	61.0	11.27	1091.4	46.7%
2016	62.3	11.2	936.4	40.0%
2017	61.6	5.10	1403.0	60.0%

Although the pressure on the prevention and control of cyanobacteria in Lake Taihu is relatively high, the interception effect of cyanobacteria in South Taihu Lake is significant. Before 2011, the average density of cyanobacteria in South Taihu Lake was the same as that of the entire Taihu Lake. After 2011, the average density of cyanobacteria in both South Taihu Lake and the entire Taihu Lake increased simultaneously. However, the average concentration of cyanobacteria in South Taihu Lake is lower than that of the entire Taihu Lake, indicating that a series of cyanobacteria enclosure projects implemented in South Taihu Lake has a significant interception effect. The cyanobacteria trend is shown in Figure 4.

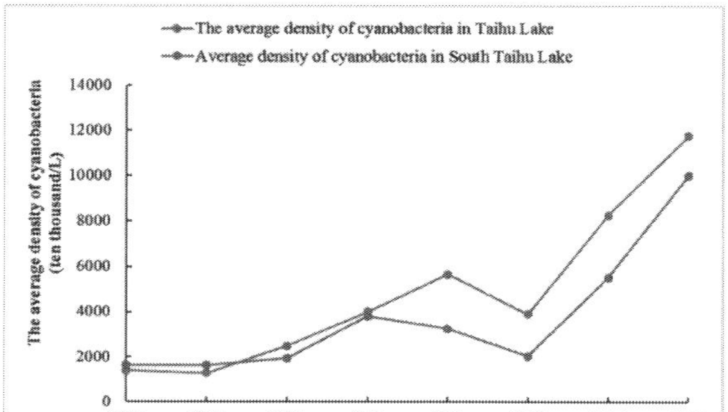

Figure 4. The average density of cyanobacteria in Taihu Lake and South Taihu Lake.

3.2 *Enclosure and Interception Effect*

1. On-site water quality monitoring results

The water quality test results showed that the concentration of chlorophyll a from the outside of the enclosure to the inside of the enclosure showed a decreasing trend. The average concentration of chlorophyll a on the outside of the enclosure was 11.7 μg/L, which belonged to the area of heavy cyanobacteria accumulation. The average concentration of chlorophyll a inside the enclosure was Compared with the average concentration of chlorophyll a on the outer side of the enclosure, the concentration of chlorophyll a decreased by 56.6%. The average concentration of chlorophyll a in the river channel into the lake was 0.11 μg/L, which was 97.8% lower than the average concentration of chlorophyll a on the inner side of the enclosure. Compared with the average concentration of chlorophyll a outside the enclosure, it decreased by 99.1%, as shown in 5 below.

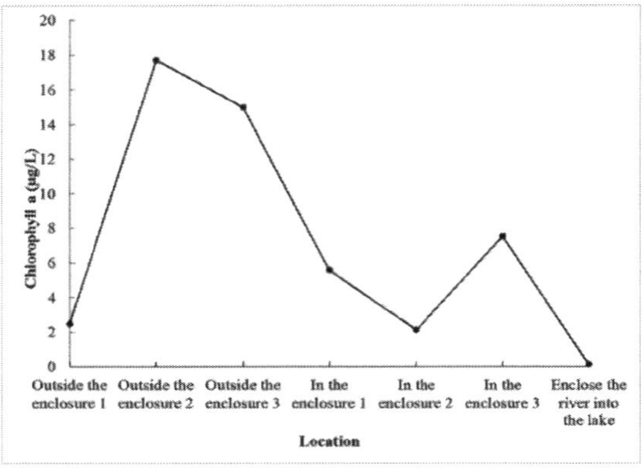

Figure 5. The concentration of chlorophyll a on the outside and inside of the interception project.

The dissolved oxygen concentration in the water from the outside to the inside of the enclosure also showed a decreasing trend. The average dissolved oxygen concentration in the water outside the enclosure was 10.51 mg/L, and the average dissolved oxygen concentration in the water inside the enclosure was 8.97 mg/L. Compared with the average dissolved oxygen concentration in the water outside the enclosure, the dissolved oxygen concentration in the water entering the lake channel inside the enclosure was 9.58 mg/L, which was 6.80% higher than the average concentration of chlorophyll a inside the enclosure. The average concentration of chlorophyll a on the lateral side of the septum decreased by 8.85%, as shown in Figure 6 below.

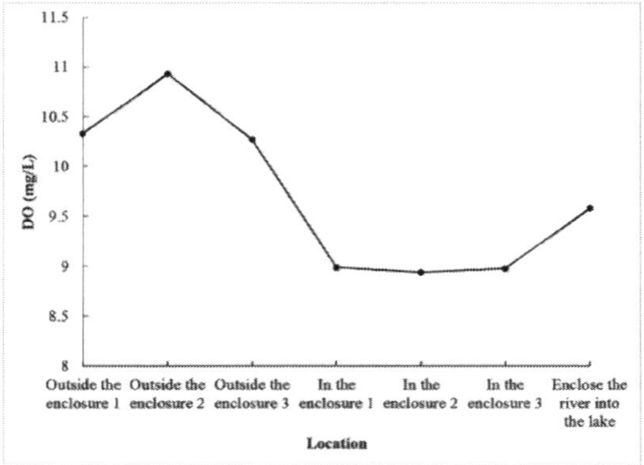

Figure 6. Dissolved oxygen concentration on the outside and inside of the interception project.

The water temperature on the outer side of the enclosure and the inner side of the enclosure were relatively close, 19.34°C and 19.48°C, respectively, while the water temperature of the river into the lake was relatively low, 18.58°C, as shown in Figure 7 below.

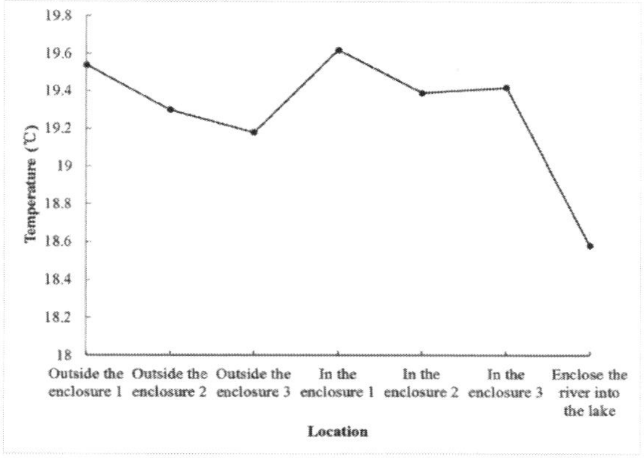

Figure 7. Water temperature outside and inside the interception project.

2. Satellite image analysis results

The enclosure provides a physical line of defense to prevent the cyanobacteria blooms that have been produced from flowing into the lakeside waters with the lake, reducing the density of algae in the water body in the enclosure, reducing the depth pressure on cyanobacteria treatment by urban water plants, ensuring the safety of water sources and improving the lakeside surroundings. The landscape plays an important role. The satellite imagery is used for analysis this time. Take the satellite image of September 8, 2019 as an example, Area 1 is a closed area, and the cyanobacteria are effectively intercepted by the soft enclosure after entering the wave-eliminating pile; take the satellite image of September 15, 2018 as an example, there is an interception project in Area 2, and the cyanobacteria are effectively blocked on the outside of the enclosure.

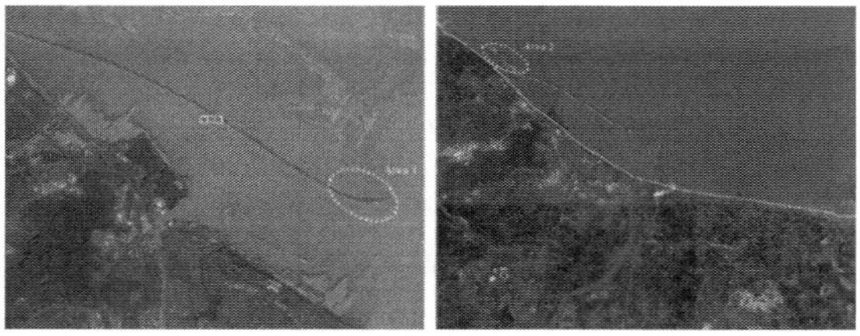

Figure 8. Satellite image of cyanobacteria interception near the interception project.

4 DISCUSSION

Based on the results of the water quality inspection on the inside and outside of the enclosure on October 31, 2019, it is obvious that due to the physical isolation of the enclosure, the chlorophyll a concentration values inside and outside the enclosure are significantly different, and the average chlorophyll a concentration inside the enclosure is higher than that of the enclosure. The average concentration of chlorophyll a on the outer side of the barrier decreased by 56.6%, which better blocked the cyanobacteria that drifted from the northwest side of Taihu Lake by the Taihu Lake current. In addition, the cyanobacteria on the outer side of the barrier were also well blocked from flowing back into the river channel of the lake. The average concentration of chlorophyll a in the river into the lake is only 0.11 μg/L. While blocking the cyanobacteria, the enclosure has a greater impact on the dissolved oxygen in the waters inside and outside the enclosure. Under the action of the lake breeze on the outside of the enclosure, the reoxygenation process continues to occur in the water while the inside of the enclosure is less affected by the lake breeze. The medium oxygen supplement ability is weak, and the dissolved oxygen concentration is relatively low. The water quality monitoring results also show that the dissolved oxygen concentration in the water inside the enclosure is 14.70% lower than the dissolved oxygen concentration in the water outside the enclosure.

Through satellite impact analysis in 2018 and 2019, exogenous floating cyanobacteria in Taihu Lake gathered near the outside of the soft enclosure and were blocked on the outside, while the inner cyanobacteria and Changdougang channel cyanobacteria were less affected (Zhai 2004). In addition, comparing the situation of cyanobacteria in Changdou Port and Yangjiapu Port on the west side at the same time, it was found that Changdou Port was affected by the interception project, and the number of cyanobacteria in the inland watercourse was relatively small. The concentration of cyanobacteria in the river is significantly higher than that of Changdou Port. Based on the comparison of the above two aspects, it can be seen that the Taihu cyanobacteria interception project has a significant effect on the cyanobacteria interception in the project area.

Wave-elimination piles are usually installed next to the enclosure project (Zhao 2020). The existence of the wave-elimination piles also reduces the impact of wind and waves outside the enclosure on the shoreline of the inner lake, keeping the water in the enclosure in a relatively calm state. At the same time, the enclosure serves as the South Taihu Lake. The first line of defense for intercepting cyanobacteria in the new area has played a better blocking effect on the outer cyanobacteria, ensuring that the water body in the enclosure has good light transmittance, and reducing the work pressure of cyanobacteria salvaging on the inner side of the enclosure. Therefore, the existence of wave-eliminating piles and enclosures can provide better environmental conditions for aquatic plants and other water ecological restoration work (such as reed planting, etc.) in the enclosure, and play a positive role in improving the water quality of the enclosure.

In summary, in the absence of enclosure facilities, patches of algae accumulate in the mouth of the lake, and the ecosystem in the restoration area may suffer a greater blow. Sheets of cyanobacteria cover the surface of the water. Floating leaves and submerged plants are easy to die and decompose. The cyanobacteria secrete algae toxins while consuming dissolved oxygen in the waters, threatening the safety of the animals and plants in the waters, and the algae blooms can emit more Odor seriously affects the aquatic landscape. The enclosure project effectively intercepted the blue algae in South Taihu Lake and has become the first physical protection barrier for the inland rivers in Huzhou City along the South Tai Lake, which has played a positive role in improving the lakeside water ecosystem around the South Tai Lake New Area (Wang 2019; Yin 2021). It also makes it possible for the enclosed area to become a multi-functional compound space integrating tourism, leisure, and hydrophilicity. The economic, ecological, and social benefits of the project are significant.

5 CONCLUSIONS

(1) The interception project has achieved remarkable results in the interception of cyanobacteria. The on-site water quality monitoring results show that the average chlorophyll a concentration inside the enclosure is 56.6% lower than the average chlorophyll a concentration outside the enclosure, and the interception effect of the enclosure is significant, based on 2018–2019 satellite images. It was also found that the cyanobacteria were effectively intercepted on the outside of the enclosure, reducing the number of cyanobacteria inside the enclosure, and improving the water environment in the enclosure.

(2) The interception project is of positive significance for the development of water ecological management in the enclosed area. The interception project includes wave-eliminating piles and soft enclosures. The existence of wave-eliminating piles reduces the impact of wind and waves

outside enclosures on the inner lake shore zone and allows the body of water in the enclosure to be kept in a relatively calm state. At the same time, the enclosure serves as the first line of defense to intercept cyanobacteria outside the port of South Taihu Lake. The luminosity reduces the pressure of cyanobacteria salvage on the inner side of the enclosure, thereby providing better environmental conditions for the restoration of aquatic plants and other water ecology in the enclosure.

(3) The interception project helps to improve the water landscape in the enclosed area. The interception project effectively intercepts the blue algae that invades the inland rivers and has become the first physical protection barrier for the rivers in Huzhou City along the South Tai Lake. The lakeside water ecosystem around the new area plays a positive role in improving and provides the possibility for the enclosed area to become a multi-functional complex space integrating tourism, leisure, and water treatment.

REFERENCES

Bian B, Yan Z, Qin Z. (2017). Pollution Characteristics and Risk Assessment of Heavy Metals from River Network Sediment in Western Area of Taihu Lake [J]. *Environmental Science*, 38(4): 1442–1450.

Fan Yamin, Jiang Weili, Liu Baogui et al. (2018). *Temporal microcystin dynamics of the source water and finished water in a waterworks of Lake Taihu*, 30(1): 25–33.

Liu X, Lu XH, Chen YW. (2011). The effects of temperature and nutrient ratios on Microcystis blooms in Lake Taihu, China: An 11-year investigation. *Harmful Algae*, 10(3): 337–343.

Qin B Q. (2013). A large-scale biological control experiment to improve water quality in eutrophic Lake Taihu, China. *Lake Reservoir Manage*, 29: 33–46.

Qin Boqiang, Hu Weiping, Liu Zhengwen. (2007). Ecological engineering experiment on water purification in drinking water source in Meiliang Bay, Lake Taihu [J]. *Acta Scientiae Circumstantiae*, 2007, 27(1): 5–12.

Qin BQ, Yang GJ, Ma J R et al. (2016). Dynamics of variability and mechanism of harmful cyanobacteria bloom in Lake Taihu, China [J]. *Chin Sci Bull*, 61(7): 759–770.

Rajeshkumar S, Liu Y, Zhang X Y, et al. (2018). Studies on seasonal pollution of heavy metals in water, sediment, fish, and oyster from the Meiliang Bay of Taihu Lake in China [J]. *Chemosphere*, 191(4): 626–638.

Wang SS, Gao YN, Li Q et al. (2019). Long-term and inter-monthly dynamics of aquatic vegetation and its relation with environmental factors in Taihu Lake, China [J]. *Science of the Total Environment*, 651: 367–380.

Wu Donghao, Jia Genghua, Wu Haoyun, et al. (2021). Chlorophyll-a concentration variation characteristics of the algae-dominant and macro-phyte-dominant areas in Lake Taihu and its driving factors, 2007-2019 [J]. 2021, 33(5): 1364–1375.

Yang GS, Ma R H, Zhang L et al. (2010). Lake status, major problems, and protection strategy in China [J]. *J Lake Sci*, 22(6): 799–810.

Yin Xueyan, Yan Guanghan, Wang Xing. (2021). Research on the integration and application of aquatic vegetation restoration technology in the lakeshore zone of Taihu Lake [J]. *Journal of East China Normal University* (Natural Science), 4: 26–38.

Zhai S H, Han T, Chen F. (2004). Self-purification capacity of nitrogen and phosphorus of Lake Taihu on the basis of mass Balance [J]. *Journal of Lake Sciences*, 26(2): 185–190.

Zhang R Q, Wu F C, Li H X, et al. (2012). Deriving aquatic water quality criteria for inorganic in China by species sensitivity distributions [J]. *Acta Scientiae Circumstantiae*, 32(2): 186–195.

Zhang Y, Lin S, Qian X et al. (2011). Temporal and spatial variability of chlorophyll a concentration in Lake Taihu using MODIS time-series data. *Hydrobiologia*, 661(1): 235–250.

Zhao Jixnxiao, Xu Shikai, Ding Wenhao. (2020). Relation between typical wind and water level distribution in Taihu lake [J]. *Jiangxi Hydraulic Science & Technology*. 2020, 46(6): 413–421.

Energy Revolution and Chemical Research – Kok-Keong Chong and Zhongliang Liu (Eds)

Research on evaluation of green energy transformation in China

Xinli Xiao* & Junshu Feng
State Grid Energy Research Institute Co, Ltd, Beijing, China

ABSTRACT: Reducing greenhouse gas emissions has become a consensus reached by countries all over the world. All countries have adopted various measures to promote their own carbon reduction processes. In all fields, carbon emissions brought by energy consumption account for a large proportion. Therefore, the energy transformation indicators can reflect the potential and effectiveness of national energy transformation to a certain extent. This paper provides a comprehensive evaluation index system of the effectiveness of China's energy transformation from the three dimensions of energy supply, energy consumption, and sustainable development; then introduces an analytic hierarchy process (AHP) to evaluate the effectiveness of China's energy transformation; and finally puts forward targeted suggestions.

1 INTRODUCTION

Carbon dioxide emitted by human activities is the main source of greenhouse gases, while fossil fuel combustion is the main activity mode of carbon dioxide emitted by human activities. In 2020, energy-related carbon dioxide emissions have accounted for about 87% of the total global carbon dioxide emissions. Differences in resource endowments and technological advantages determine the different paths of low-carbon transformation in countries all over the world, but the overall situation is 'reducing coal, stabilizing oil, increasing gas and vigorously developing renewable energy (Wang 2022). Taking China as an example, as the world's largest energy producer, consumer, and carbon dioxide emitter, nearly 88% of carbon dioxide emissions come from the energy system. In order to achieve the goal of carbon neutralization, China needs to achieve its rapid and in-depth transformation through measures such as improving energy efficiency, reducing coal use, and significantly increasing the proportion of clean energy. To a certain extent, by evaluating the effectiveness of energy transformation, the overall effectiveness of 'carbon peaking and carbon neutralization' can be reflected.

2 CONSTRUCTION OF INDEX SYSTEM

Through literature research and expert interviews, this paper comprehensively combs and summarizes the evaluation of the effectiveness of national energy transformation (Ma 2020; Pei 2017), and then constructs the evaluation index set of energy transformation from the three aspects of energy supply, energy consumption, and sustainable development to form the evaluation index system of the energy transformation process. The specific index system is shown in Table 1.

*Corresponding Author: 912727784@qq.com

 DOI 10.1201/9781003332657-6

Table 1. Evaluation index system of energy transformation.

Primary indicators	Secondary index	Tertiary indicators
Effectiveness of energy transformation	Energy supply	total primary energy supply; the proportion of non-fossil energy in total energy supply; total installed power generation capacity; the proportion of non-fossil energy; the proportion of renewable energy; total power generation; the proportion of non-fossil energy; and proportion of renewable energy
	Energy consumption	total primary energy consumption; total power consumption; total coal consumption; the proportion of electric energy in energy consumption; the proportion of coal in energy consumption; per capita energy consumption; per capita power consumption; energy consumption intensity; power consumption intensity; and energy consumption of transportation industry
	Sustainable development	total carbon dioxide emissions; per capita carbon dioxide emissions; energy carbon intensity; power carbon intensity; power price; and energy self-sufficiency rate

3 COMPREHENSIVE EVALUATION METHOD

According to the construction of the national energy transformation effect index system, the index has a typical hierarchical structure. Therefore, this paper uses the AHP method to comprehensively evaluate the effect of China's green transformation (Fei 2013). AHP is a decision analysis method combining qualitative and quantitative methods to solve multi-objective complex problems. It was proposed by T.L. Saaty, a professor at the University of Pittsburgh in the 1970s. This method is an effective multi-criteria decision-making method to express and deal with people's subjective judgments in quantitative form. It is the most common method of subjective weighting, and it is also an effective method for people to objectively describe subjective judgment. This method is an effective multi-criteria decision-making method to express and deal with people's subjective judgments in quantitative form. It is the most common method of subjective weighting, and it is also an effective method for people to objectively describe subjective judgment.

The AHP is a multi-objective decision-making technology, which can effectively make up for the shortcomings of traditional modeling methods and better deal with qualitative factors. The AHP decomposes the complex problem into various constituent factors and then groups these factors according to the dominant relationship to form a hierarchical structure. The relative importance of various factors in the hierarchy is determined by pairwise comparison. Then, the overall ranking of the relative importance of the decision-making scheme is determined by synthesizing the judgment of the decision-maker. The whole process embodies the basic characteristics of people's decision-making thinking, that is, decomposition and comprehensive judgment. The AHP is a combination of quantitative and qualitative, so it greatly improves the effectiveness, reliability, and feasibility of evaluation (Wu 2020).

4 ANALYSIS OF EVALUATION RESULTS

4.1 *Index weight*

Using the index weight calculation process, we determine the index weight level of energy transformation, as shown in Table 2. This will enable a comprehensive evaluation of the energy transformation in the next stage.

Table 2. Evaluation index system of energy transformation.

Primary indicators	Secondary index	Tertiary indicators	Weights
Effectiveness of energy transformation	Energy supply	total primary energy supply	0.04
		the proportion of non-fossil energy in the total energy supply	0.06
		total installed power generation capacity	0.04
		the proportion of non-fossil energy	0.04
		the proportion of renewable energy	0.06
		total power generation	0.04
		the proportion of power generation of non-fossil energy	0.04
		the proportion of power generation of renewable energy	0.06
	Energy consumption	total primary energy consumption	0.03
		total power consumption	0.04
		total coal consumption	0.04
		the proportion of electric energy in energy consumption	0.05
		the proportion of coal in energy consumption	0.05
		per capita energy consumption	0.03
		per capita power consumption	0.05
		energy consumption intensity	0.03
		power consumption intensity	0.04
		energy consumption of the transportation industry	0.04
	Sustainable development	total carbon dioxide emissions	0.04
		per capita carbon dioxide emissions	0.03
		energy carbon intensity	0.05
		power carbon intensity	0.05
		power price	0.02
		energy self-sufficiency rate	0.03

4.2 *Evaluation results*

The indicators of energy production, consumption and sustainable development are shown in Figures 1–3.

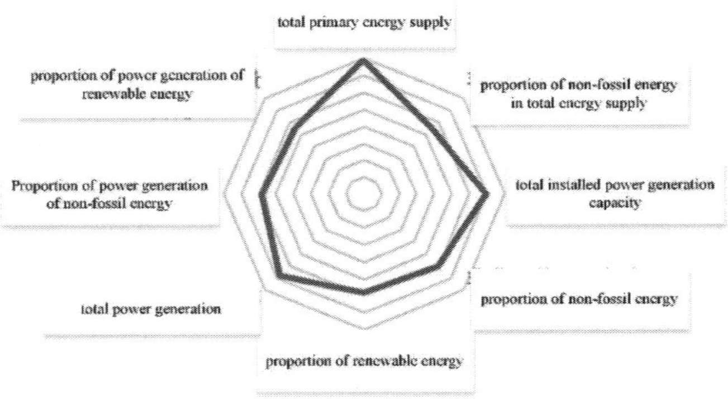

Figure 1. Overview of energy production index score.

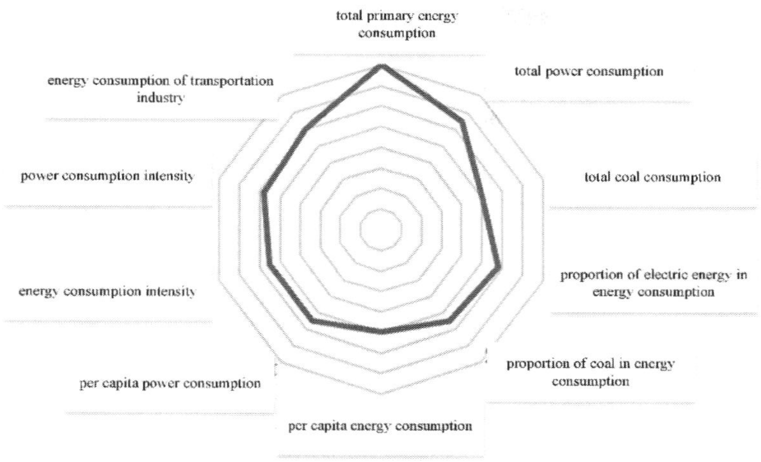

Figure 2. Overview of energy consumption index score.

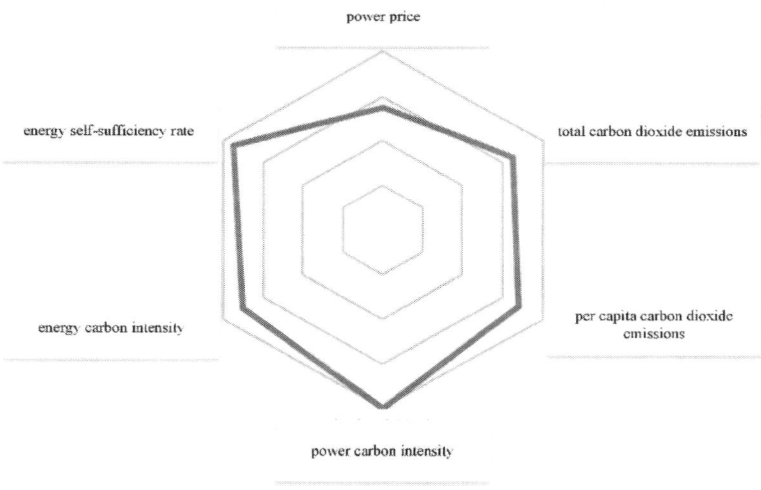

Figure 3. Overview of sustainable development index scores.

In terms of energy supply, with the continuous strengthening of China's development and production capacity, the indicators of energy supply capacity, such as primary energy supply, the proportion of total power generation installed capacity, and total power generation, continue to improve, which is consistent with the strategic policy of China's energy supply security. However, although structural indicators such as the proportion of non-fossil energy and the proportion of renewable energy production have shown positive results, these indicators will need to be developed in the future, which will be influenced primarily by China's energy endowment and its historical structure.

In terms of energy consumption, electric substitution is an important means for China to adjust the demand side structure in the future, and the proportion of electric energy in China's terminal consumption still needs to be improved. There is still a certain gap in indicators such as per capita power consumption and energy consumption per unit of GDP compared with international developed countries. China is in the transition period from high-speed development to high-quality development, and the level of energy consumption and energy efficiency will be further improved.

In terms of sustainable development, the carbon intensity per unit of power has reached its peak, but there is still a certain gap between the ultimate goal of low-carbon power and environmental protection. Recently, China's energy system reform has developed at a fast pace, exploring the restoration of the market value of energy factors, highlighting the role of marketization, accelerating the transformation of energy, and inaugurating a scientific and reasonable market mechanism in the future.

5 CONCLUSIONS

In this paper, we take China as an example and introduce the AHP method to conduct a comprehensive evaluation of the effectiveness of energy transformation. From the evaluation results, the green energy supply capacity has been continuously strengthened, and the proportion of renewable energy still has room to improve. Power substitution has achieved remarkable results, but there is still a certain gap with developed countries; the carbon reduction process will continue to advance, and the market will play a more prominent role.

ACKNOWLEDGMENTS

This paper is supported by the Science and Technology Project of State Grid Corporation of China (Grant No. 1400-201957296A-0-0-00): Research and Development of Key Decision Support Models for Energy Internet Enterprises (Phase II construction of think tank research platform).

REFERENCES

Guangqiang, Pei. (2017). *Investigation on the process of energy transformation in major western countries since modern times—centered on the four countries of Britain*, Netherlands, and virtue. Collected Papers of History Studies (04): 75–88.

Hong, Wu. & Shaohui, Chen. (2020). Research on the relationship between the expenditure structure of state-owned assets management budget and welfare effect-based on AHP. *Journal of Technical Economics & Management* (06): 74–78.

Limei, Ma. & Yifei, Dong. (2020). Innovation policy driven and national energy transformation: the case of Denmark. *Modernization of Management* 40(04): 23–28.

Xiao, Wang. & Nan, Li. (2022). China's energy transformation situation and policy suggestions. *Oil & Gas Storage and Transportation*: 1–7.

Yuejun, Fei. & Lili, Xu. (2013). Research on risk assessment model of shipwreck based on AHP. *Marine Forecasts* 30(06): 67–72.

Energy Revolution and Chemical Research – Kok-Keong Chong and Zhongliang Liu (Eds)

Mathematical modeling and capacity configuration of heating system in high altitude area

Meixiu Ma, Wenzhuo Yao* & Gaoqun Zhang
State Key Laboratory of Advanced Transmission Technology (State Grid Smart Grid Research Institute Co., Ltd.), Beijing, China

Bo Qu
Department of Energy Consumption, China Electric Power Research Institute, Beijing, China

Wenqiang Zhao
State Grid Electric Power Company of Qinghai Province Electric Power Research Institute, Qinghai, China

Linhai Cai & Wei Kang
State Key Laboratory of Advanced Transmission Technology (State Grid Smart Grid Research Institute Co., Ltd.), Beijing, China

Guobin Lei
State Grid Electric Power Company of Qinghai Province Electric Power Research Institute, Qinghai, China

ABSTRACT: Qinghai is located at a high altitude and has abundant wind power resources, but it's urgent to solve the problem of heating. A heating system based on phase change heat storage is established. The load aggregator is introduced into the system so that the load aggregator can interact with the power grid, the wind power, and the heat user. Finally, the optimal configuration model of the heat storage device is established to manage the capacity of the heat storage boiler. The results showed that the net profit of the aggregator could reach 327,274 Yuan in one week. Taking 100 school buildings in Maduo, Qinghai as an example, the optimal capacity of a heat storage device is 90 MWh.

1 INTRODUCTION

Qinghai province is located at a high altitude and has abundant wind power resources. However, since the temperature difference between day and night is large and the climate is cold, it's urgent to solve the problem of heating. Renewable energy has intermittent drawbacks. The intermittent output of wind power causes more challenges to the safe and stable operation of the power grid. If the integrated energy development model is widely used, which is making full use of renewable energy, that will bring the complementarity of the power grid and new energy into play and greatly improve the reliability and sustainability of power supply on the load side.

Scholars have done a lot of research on wind abandonment and load-side resources of power systems. Hughes et al. (Hughes 2019) proposed that the use of distributed electric boilers and renewable energy heating technology in electric heat storage systems could effectively consume more renewable energy. Mazidi et al. (Mazidi 2014) used Latin hypercube sampling and the k-mean algorithm to generate typical scenarios of uncertainty in wind power output. The results showed that stochastic optimization algorithms could make system standby smaller based on wind power elimination when the demand side participates in system peak-shaking and standby simultaneously.

*Corresponding Author: ywzh1412@163.com

DOI 10.1201/9781003332657-7

Falsafi et al. (Falsafi 2014) established the response model of power prices and proposed a two-stage stochastic programming model, in which the units and loads make the system dispatch more flexible, and provide more peak-shaking capacity for wind power access, minimize system operating costs, and pollute the least. Hu Landan et al. (Hu 2016) discussed the collaborative control strategy between multi-energy complementary systems and flexible load demand response. The combination of the demand-side and energy supply side made the scheduling more flexible. Wang Beibei et al. (Wang 2017) constructed a wind power dispatch model with stochastic and adjustable robust optimization considering demand response according to the different prediction accuracy characteristics of wind power in different time scales. The example showed that this model can significantly reduce the overall cost of dispatch compared with a single optimization model.

Böttger et al. (Böttger 2014) put forward the system where electric boilers and thermal power plants supply heat to the heating network together to solve the problem of wind abandonment. The results showed that the distribution of electric boilers in the regional heating system could effectively reduce the phenomenon of wind abandonment. Lv Quan (Lv 2013) put forward the idea of decoupling thermoelectric units to actively participate in peak regulation of wind power under the constraints of heat and power fixing. Yue Yunli et al. (Yue 2020) built a cooperative game model among power grid companies, wind power enterprises, and heating enterprises. Meanwhile, taking the Zhangjiakou area as an example, they analyzed the cost benefits of all parties. The results showed that the cooperative mechanism could effectively promote the implementation of wind power heating, and all participants could obtain a certain income.

In this paper, a model of a heating supply system is established which combines the characteristics of a high-temperature phase change heat storage device. Load aggregators are introduced into the heating systems to enable them to interact with power grids, wind power, and thermal users as intermediaries in the market. Finally, the optimal configuration model of the heat storage unit is established, and the capacity of the heat storage unit is rationally planned.

2 MODEL INTRODUCTION

2.1 *The overall design of the model*

To make full use of abundant wind power resources, this paper constructs a four-element information interaction model of "wind power-power grid-load aggregator-user". In the actual process of cold and heat supply, aggregators, as the core, rely on information interaction with all parties to maximize load regulation within the region, realize the time transfer of grid load, and increase the potential of wind power consumption to complete the overall benefit maximization control. At moment i, the information interaction structure among the four groups is shown in Figure 1.

2.2 *Load aggregator – information interaction between four agents*

2.2.1 *Information interaction between the power grid and load aggregator*

A load aggregator motivates users according to the subsidy price of peak clipping provided by power grid enterprises while at the same time controlling the heat release of phase change heat storage devices at high temperatures to aid peak clipping. The information conveyed by the load aggregator to the power grid enterprise is the total load reduction in the area managed by the load aggregator:

$$L_{A,i} = L_{X,i} + L_{DC,i} \tag{1}$$

where $L_{A,i}$ (kWh) is total load reduction, $L_{X,i}$ (kWh) is the total amount of load that users proactively cut, and $L_{DC,i}$ (kWh) is the heat released by the heat storage.

2.2.2 *Information interaction between wind power plants and load aggregators*

Wind power generation has the characteristic of reverse peak regulation, and the period of low power consumption is also the period of high wind power generation. Therefore, wind power plants

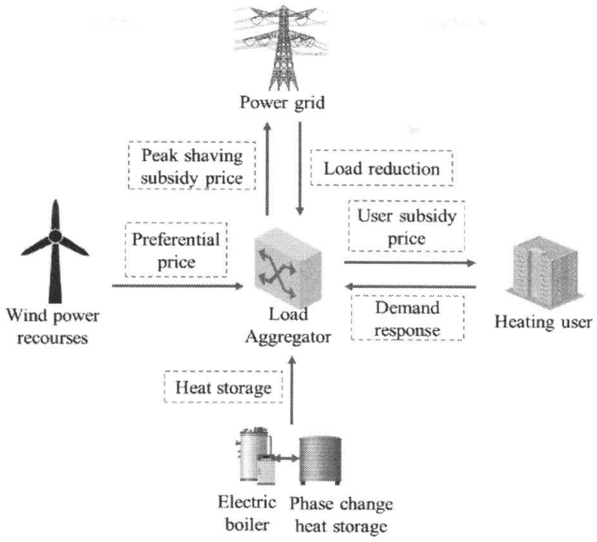

Figure 1. Structure of information interaction among four parties.

can sign electricity sales contracts with load aggregators and sell electricity to aggregators at a lower price for a specified period. Aggregators can use preferential rates for heating and for storing heat in heat storage devices during that time.

2.2.3 *Information interaction between load aggregators and users*

To further encourage users to participate in load response, the load aggregator will issue additional user subsidies to users according to the response ratio of the user side. Meanwhile, at the same time, it will also pass the cost of typical users in the current cycle to users to encourage more users to participate in load response. The user transmits the load and load reduction at this time to the load aggregator.

The relationship between the total load reduction by users and the active load reduction by users is as follows:

$$L_{X,i} = \sum_{m=1}^{M} N_{m,i} L_{X,m,i} \tag{2}$$

where $N_{m,i}$ is the number of users participating in the response, and $L_{X,m,i}$ (kWh) is active load reduction per user.

2.3 *Multi-income model of multi-market players*

Based on the framework of the information interaction model to be constructed, to participate in the four corners of the high-temperature phase-change thermal storage subject to obtaining corresponding income, load aggregators require a combination of three parties to transfer information at a reasonable user subsidy to build a win-win benefit mechanism.

The new revenue for the wind power plant comes from the electricity consumed by the load aggregator during the contract period and from the heat stored in the heat storage body. The income of the wind power plant at moment i is as follows:

$$E_{W,i} = P_{W,i} \cdot L_{C,i} \tag{3}$$

where $L_{C,i}$ (kWh) is the stored heat all the time.

The revenue of load aggregators comes from the subsidy award of the power grid and the heating income of the heat storage regenerator, and the expenditure part is the subsidy given to users and the heat storage cost of the heat storage regenerator. The returns of the load aggregator at moment i are as follows:

$$E_{LA,i} = P_{R,i} \cdot L_{A,i} + P_i \cdot L_{DC,i} - P_{B,i} \cdot L_{X,i} - P_{W,i} \cdot L_{C,i} \tag{4}$$

where $P_{w,i}$ is preferential wind power price and $P_{w,i}$ is consumer subsidy price.

Part of the revenue of users comes from the subsidies given by aggregators and the reduced load and saved heat consumption. The revenue of users at moment i is as follows:

$$E_{U,i} = P_{B,i} \cdot L_{X,i} + P_i \cdot \sum\nolimits_{m=1}^{M} L_{X,m,i} \tag{5}$$

where P_i is the peak electricity price.

The income of power grids mainly comes from the load reduction of users. Thus, the income is usually measured by the load reduction in peak hours. Here, the income is measured by the load reduction ratio, as shown in Equation (6).

$$\theta = \frac{L_{A,i}}{\sum\nolimits_{m=1}^{M} N_{m,i} L_{m,i}} \tag{6}$$

3 TEST RESULTS AND DISCUSSIONS

In this paper, 100 school buildings in Maduo County in Qinghai Province are taken as demand-side heat users to establish a quadrilateral interaction model. According to the revenue model of the load aggregator, the net revenue of the load aggregator can be calculated every week, as well as the income and expenditure details of the load aggregator (including heating revenue, heat storage expenditure, peak shaving revenue, and user subsidy expenditure). The change in weekly income in the heating cycle is shown in Figure 2. The net revenue curve is mainly caused by changes in heating revenue, which depends on user load and is easily affected by the weather. According to the calculation, the net profit of the aggregator for this week is 327,274 Yuan.

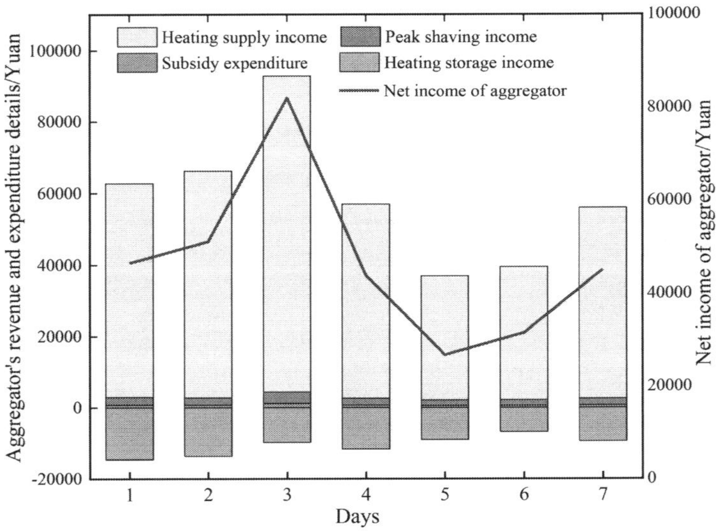

Figure 2. Aggregator's weekly net income and various revenue and expenditure details.

The capacity allocation of the heat storage device is very important for the economy and environmental protection of the system. During the operation of the heating system, it will be affected by the income of the aggregator, the power cost of the wind power plant, the cost of the power plant, etc. Figure 3 shows the income of aggregators, wind power costs, and power grid costs under different heat storage device capacities.

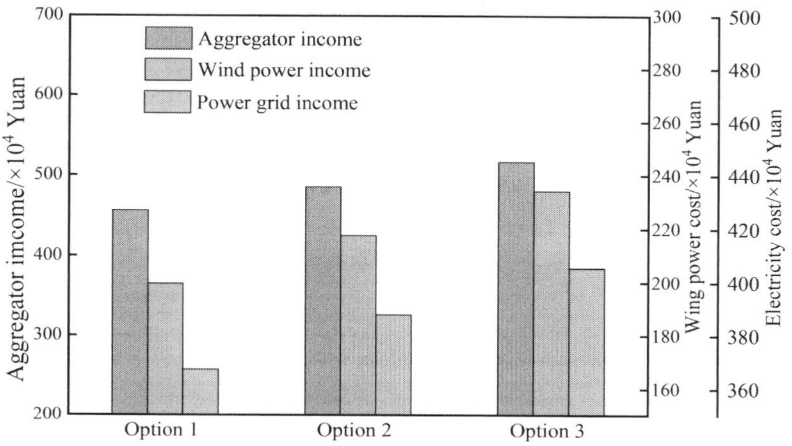

Figure 3. System income and cost under different capacity configurations.

Therefore, a capacity allocation model of the heat storage device is formulated, and an investment payback period index is introduced to plan and analyze the heat storage device whilst maintaining the normal operation of the system. The expression of dynamic investment payback period n_p is:

$$\sum_{t=0}^{n_p} \frac{f_{in} - f_{out}}{(1+i)^t} = 0 \tag{7}$$

where f_{in} is annual revenue; f_{out} is annual expenditure; and I is the discount rate.

(1) The equipment investment cost can be described as:

$$C_{inv} = \sum_{i=1}^{N_D} P_i \cdot V_i \tag{8}$$

where N_D is the number of types of equipment in the system, P_i is the capacity cost of class i equipment, and V_i is the equipment capacity of class i equipment.

(2) Operation and maintenance cost

Maintenance and operation costs refer to the dynamic capital spent to ensure the equipment's normal operation during its service life, which can be expressed as:

$$C_{\mathrm{om}} = \sum_{t=1}^{8760} \sum_{a=1} \lambda_a^{om} P_{a,t} \Delta t \tag{9}$$

where λ_a^{om} is the unit power operation and maintenance cost of equipment a, and $P_{a,t}$ is the output of equipment a in period t.

(3) User compensation cost

The system generally signs a load reduction contract with users when participating in load demand response and can uniformly control users with similar demand response strategies and compensation standards. The load after the thermal/electrical load is reduced in period t can be expressed as:

$$L_{b,t}^{cut} = (1 - \nu_{b,t}\zeta)L_{b,t}^{cut^*} \tag{10}$$

$$C_{com} = \sum_{t=1}^{8760}\sum_{b=1} \lambda_b^{cut}\upsilon_{b,t}\zeta L_{b,t}^{cut^*}\Delta t \tag{11}$$

where $\nu_{b,t}$ is the 0–1 state variable of the b-th reducible load in period t; $\nu_{b,t} = 1$ is the reducible load is reduced in period t; $\nu_{b,t} = 0$ indicates that no reduction occurs; ζ is the load reduction factor; $L_{b,t}^{cut}$ are the energy load in period t before and after the b-th load reduction scheduling respectively; λ_b^{cut} is the unit compensation price of b-th that can reduce the load.

(4) Heating income

The heating fee shall be charged according to the building area.

$$C_{hot} = A \cdot S_{hot} \tag{12}$$

where C_{hot} is the heating income during the heating period; A is the heating area (m^2); S_{hot} is the heating income (Yuan/m^2).

The calculation results are shown in the following table:

Table 1. Dynamic investment payback period under different schemes.

	Option A	Option B	Option C	Option D
Device capacity	70 MWh	80 MWh	90 MWh	100 MWh
Dynamic payback period	7.6 years	7.9 years	7.1 years	8.5 years

4 CONCLUSIONS

Qinghai province has abundant wind power resources, but it's urgent to solve the problem of heating. In this paper, a heating supply model based on high-temperature phase change heat storage is established. The load aggregator is introduced into the heating system so that the load aggregator can interact with the power grid, the wind power, and the heat user as the market intermediary. Finally, the optimal configuration model of the heat storage device is established to plan the capacity of the heat storage device reasonably. Based on our analysis, we obtain the following conclusions:

(1) Load aggregators act as intermediaries in the market to interact with the power grid, wind power, and heat users. Through the aggregator model, aggregators can make a net profit of 327,274 Yuan in one week.
(2) The capacity allocation model of heat storage devices is established, and the dynamic investment payback period indicator is introduced to study the heat storage device under the condition of ensuring the normal operation of the system. Taking 100 school buildings in Maduo County in Qinghai Province as an example, the optimal capacity of heat storage devices is 90 MWh.

ACKNOWLEDGMENTS

This study was funded by the Research and Development Project of State Grid Corporation of China (Project title: Study on heat storage technologies for renewable energy consumption in high altitude and cold regions; Grant No. 5419-202134244A-0-0-00).

REFERENCES

Böttger, D. 2015. Control power provision with power-to-heat plants in systems with high shares of renewable energy sources—An illustrative analysis for Germany based on the use of electric boilers in district heating grids. *Energy* 82.

Falsafi, H. 2014. The role of demand response in single and multi-objective wind-thermal generation scheduling: A stochastic programming. *Energy* 64.

Hu, L, D. 2016. CCHP multipurpose complementary optimization strategy considering demand response. *China Southern Power Grid* 10: 75–81.

Hughes, L. 2009. Meeting residential space heating demand with wind-generated electricity. *Renewable Energy* 35.

Lv, Q. 2013. Review and prospect on methods of participating in wind power peak regulation of thermal power plants. *Electric Power* 46: 129–136+141.

Mazidi, M. 2014. Integrated scheduling of renewable generation and demand response programs in a microgrid. *Energy Conversion and Management* 86.

Wang, B, B. 2017. Stochastic-adjustable robust hybrid day-ahead scheduling model with demand response participating in wind power dissipation. *Proceedings of the CSEE* 37:6339–6346.

Yue, Y, L. 2020. Research on the multi-cooperative game of electric heating considering wind power dissipation. *Zhejiang Electric Power* 39:95–102.

Energy Revolution and Chemical Research – Kok-Keong Chong and Zhongliang Liu (Eds)
© 2023 The Authors, ISBN: 978-1-032-36554-1

Study on power grid maintenance and operation cost based on combined weight TOPSIS model

Li Huang
State Grid Fujian Electric Power Co., Ltd., Fujian, Fuzhou, China

Zebang Yu
State Grid Economic and Technological Research Institute Co., Ltd., Beijing, China

Lijian Zheng & Shaohong Lin
State Grid Fujian Electric Power Co., Ltd., Fujian, Fuzhou, China

Jing Yu*
Power Economic Technology Research Institute of State Grid Fujian Electric Power Co., Ltd., Fujian, Fuzhou, China

Shaojun Jin
State Grid Zhejiang Electric Power Co., Ltd., Zhejiang, Hangzhou, China

ABSTRACT: At present, the country has deepened the reform of the system and mechanism of the power industry, promoted several policies represented by the supervision and examination of power transmission and distribution costs, electricity price verification, and the continuous liberalization of power sales side marketization, strengthened the supervision and supervision of the operation and development of power grid enterprises, and the lean production and operation management requirements of power grid enterprises are continuously improved. This paper analyzes the maintenance cost input characteristics, constructs the index system, determines the TOPSIS weight by hierarchical analysis and entropy method, verifies the effectiveness of the model, and provides the cost input decision in different regions. This can effectively improve the configuration efficiency and economy to the economic benefit of power grid enterprises.

1 INTRODUCTION

In recent years, research on the comprehensive evaluation of the input and output of power grid enterprises has been rich. Guo (2021) built a power grid value creation system from three aspects: social, user, and operation value; and it constructs the power grid development efficiency and efficiency index system from the level of strategic results and operational process. He (2020) established a comprehensive input-output evaluation system, calculates the comprehensive efficiency of an asset group based on a cloud element model, forms a six-order path map of the dynamic comprehensive efficiency of an asset group, carries out the traceability analysis of the comprehensive efficiency of an asset group, and forms a comprehensive overall input-output efficiency evaluation of a macro asset group. Li (2021) used the data envelopment analysis (DEA) method to analyze the investment and construction benefits of the Yunnan power transmission and distribution network industry, which has a guiding role in the Yunnan power transmission and distribution network construction planning and investment. Based on the power grid monitoring and operation center. Liu (2021) proposed the use of the hierarchical analysis method (HAM) combined

*Corresponding Author: 345332300@qq.com

 DOI 10.1201/9781003332657-8

with the variable weight method to select appropriate benefit evaluation indicators and data. The evaluation framework for high-quality grid development may give method support for high-quality grid development by providing index comprehensive evaluation method and assessment rules from the equipment, technology, management, service, operation, and design evaluation index systems.

To summarize, current investment results and their comprehensive evaluation results may offer some reference for the index system and evaluation method. However, there is little research focused on measuring maintenance efficiency and comparing the efficiency of various regions, so this paper provides some novelty in combining the input-output characteristics of maintenance and operation cost and providing a basis for decision-making in maintenance and operation cost.

2 CONSTRUCTION OF EVALUATION INDEX SYSTEM FOR POWER GRID MAINTENANCE AND OPERATION COSTS

In this section, we fully integrate the analysis of the input-output characteristics of grid maintenance and operation costs, taking existing evaluation index systems of grid enterprises as a reference, and finally, construct an efficiency evaluation index system based on the combing of power grid operation and experts' opinions.

Table 1. Diagnostic and evaluation index system for the business and development of power grid enterprises.

Number	First-level indicator	Secondary indicators
1	Power supply capacity index	Household distribution capacity
2	Safety and reliability index	Qualified rate of comprehensive voltage
3	Social Security Index	Tax contribution rate
4	Service Competition Index	Get the power index
5	Asset operation index	Income from ten thousand yuan
6		The comprehensive line loss rate
7	Cost management index	KWH cost
8	PI	Profit margin per kilowatt hour
9	Investment motivation index	The growth rate of investment in power sales of the unit power grid

3 CONSTRUCTION OF THE COST AND EFFICIENCY EVALUATION MODEL OF POWER GRID MAINTENANCE AND OPERATION

3.1 *Combined weight calculation method*

(1) The basic principles and steps of AHP

The hierarchical analysis (AHP) method first divides the decision problem into different target layers. Each target layer represents an important part of the decision problem, and then, according to the rating criteria and evaluation specification, combined with the method of solving the judgment matrix, calculates the weight of each target layer element, and finally uses the weighted sum method to calculate the final weight of the total target, where the largest weight is the optimal scheme. The steps are as follows.

1) First, the hierarchical structure is constructed, and the decision problems are refined and decomposed, combed, and structured into a hierarchy from top to bottom. Complex problems are decomposed into multiple core elements.
2) Construct the judgment matrix. To reflect the weight correspondence between each element, it is necessary to construct a judgment matrix, generally using the 1–9 scale method, and using the number 1–9 and its reciprocal as the scale, to evaluate the correspondence relationship between the elements.

3) Take a single-ranking consistency test. Generally, the consistency index CI is used to test whether the judgment matrix is reasonable and whether there are logical errors. It is usually believed that the judgment matrix is reasonable when CI is less than 0.10.
4) Take a total ranking consistency test. When the single-ranking consistency meets the requirements, the consistency test of the total ranking is also required. If the test passes, the current weight ranking results can be used as the final decision basis.

(2) Basic principles and steps of entropy weight method

The calculation steps of the entropy weight method are as follows:

1) Use n evaluation indicators to evaluate m plans to be selected.
   In Equation (1), x_{ik} refers to the estimation of the evaluation index i of the scheme k to be selected; x_i^* is the ideal value of the evaluation index i. The value varies according to the characteristics of the evaluation index. The higher the yield index, the better; for the loss index (reverse index), the smaller the better (it can be converted to a positive index first).
2) Define the proximity D_{ik} of x_{ik} to x_i^*:

$$D_{ik} = \begin{cases} \dfrac{x_{ik}}{x_i^*} & x_i^* = \max\{x_{ik}\} \\ \dfrac{x_{ik}}{x_i^*} & x_i^* = \min\{x_{ik}\} \end{cases} \tag{1}$$

3) Normalization processing of D_{ik}:

$$d_{ik} = D_{ik} \Big/ \sqrt{\sum_{i=1}^{n}\sum_{k=1}^{m} D_{ik}^2} \tag{2}$$

4) Overall entropy: the entropy E of m schemes to be selected with n evaluation indicators can be described as:

$$E = -\sum_{i=1}^{n}\sum_{k=1}^{m} d_{ik} \ln d_{ik} \tag{3}$$

5) Calculate overall entropy when the index is independent of the scheme:
   If the relative importance of the evaluation index is independent of the scheme to be selected, the entropy is calculated by the following formulas:

$$E = -\sum_{i=1}^{m} d_{ik} \ln d_i \tag{4}$$

$$d_i = \sum_{k=1}^{n} d_{ik} \tag{5}$$

   In this way, the uncertainty of the relative importance of the evaluation index i for the evaluation of selection decisions can be determined by the following conditional entropy.
6) Evaluate the conditional entropy of the index i:

$$E_i = -\sum_{k=1}^{m} \frac{d_{ik}}{d_i} \ln \frac{d_{ik}}{d_i} \tag{6}$$

Judging from the extremities of the entropy (k= 1–m), namely $d_{i1}, d_{i2}, \ldots, d_{ik}$, the closer the equal, the greater the conditional entropy, and the greater the uncertainty of the evaluation index to the evaluation scheme.

7) Normalize the above formula to obtain the entropy value of the evaluation decision importance of the evaluation index i.

$$e(d_i) = -\frac{1}{\ln m}\sum_{k=1}^{m}\frac{d_{ik}}{d_i}\ln\frac{d_{ik}}{d_i} \tag{7}$$

(3) Combined weight calculation process

The AHP is determined by the subjective weight, which is determined through the experience of the experts, and the index weight is determined through the combination of qualitative and quantitative methods, which has a certain subjectivity. The Entropy weight method is the objective weight determination, and the weight of the index of each index has a certain objective but ignores the influence of the subjective factors on the relationship between the accuracy of the weight determination and the same meets the characteristics of practical work.

3.2 *Fundamental principles of the TOPSIS theory*

The main principles of the TOPSIS method are as follows:

There are m evaluation objects and n evaluation indicators. The attribute value of the j-th indicators of the i-th evaluation object is $(i = 1, 2, \ldots, m; j = 1, 2, \ldots, n)$, constituting the raw data matrix P.

$$\mathrm{p} = \begin{bmatrix} \mathrm{p}_{11} & \mathrm{p}_{12} & \cdots & \mathrm{p}_{1n} \\ \mathrm{p}_{21} & \mathrm{p}_{22} & \cdots & \mathrm{p}_{2n} \\ \cdots & \cdots & \cdots & \cdots \\ \mathrm{p}_{m1} & \mathrm{p}_{12} & \cdots & \mathrm{p}_{mn} \end{bmatrix} \tag{8}$$

A scientific method is used to normalize the index data and conduct dimensionless calculations so that the analysis results do not depend on the magnitude and units of the indexes. Then, we can obtain the normalized data matrix L.

$$\mathrm{L} = \begin{bmatrix} \mathrm{L}_{11} & \mathrm{L}_{12} & \cdots & \mathrm{L}_{1n} \\ \mathrm{L}_{21} & \mathrm{L}_{22} & \cdots & \mathrm{L}_{2n} \\ \cdots & \cdots & \cdots & \cdots \\ \mathrm{L}_{m1} & \mathrm{L}_{12} & \cdots & \mathrm{L}_{mn} \end{bmatrix} \tag{9}$$

$$L_{\mathrm{ij}} = \frac{P_{\mathrm{ij}}}{\sqrt{\sum_{\mathrm{i=1}}^{\mathrm{m}} P_{\mathrm{ij}}^2}} \tag{10}$$

Combined with the combined weight value in Step 4, the positive and negative ideal solutions of each index in the evaluation object are determined respectively.

The European distance from the optimal worst vectors is calculated.

The positive ideal distance is:

$$D_{\mathrm{i}}^{+} = \sqrt{\sum_{\mathrm{j=1}}^{\mathrm{m}} (L_{\mathrm{ij}} - \mathrm{F}_{\mathrm{j}}^{+})^2} \tag{11}$$

The negative ideal distance is:

$$D_{\mathrm{i}}^{-} = \sqrt{\sum_{\mathrm{j=1}}^{\mathrm{m}} (L_{\mathrm{ij}} - \mathrm{F}_{\mathrm{j}}^{-})^2} \tag{12}$$

Then, we can get the evaluation index of the evaluation object i:

$$C_{\mathrm{i}} = \frac{D_{\mathrm{i}}^{-}}{D_{\mathrm{i}}^{-} + D_{\mathrm{i}}^{+}} \tag{13}$$

Finally, the closeness value of each index can be obtained.

4 EMPIRICAL ANALYSIS

According to the construction results of the index system, nine power supply enterprises under F Electric Power Company were selected as the research objects to evaluate and analyze the efficiency of different regions, combined with the method and model principles, to realize the scientific evaluation of the efficiency. The original data is shown in Table 2.

Table 2. The post-normalized data matrix.

Number	Evaluation index	A	B	C	D	E	F	G	H	I	J
1	Household distribution capacity	0.67	0.72	1.00	0.67	0.74	0.65	0.80	0.80	0.77	0.60
2	Qualified rate of comprehensive voltage	1.00	1.00	1.00	0.75	1.00	1.00	1.00	1.00	1.00	1.00
3	Tax contribution rate	0.56	0.73	0.63	0.52	0.58	0.52	0.66	0.72	0.60	1.00
4	Get the power index	0.97	0.97	1.00	0.97	0.96	0.97	1.00	0.97	0.96	0.97
5	Income from ten thousand yuan	0.47	0.52	0.88	0.68	0.59	0.81	0.82	0.96	0.87	1.00
6	Comprehensive line loss rate	0.53	0.57	1.00	0.51	0.54	0.57	0.80	0.85	0.74	0.80
7	KWH cost	0.75	0.80	1.00	0.88	0.77	0.88	0.93	0.93	0.74	0.91
8	Profit margin per kilowatt hour	0.60	0.60	1.00	0.61	0.60	0.75	0.80	0.90	0.60	0.65
9	Growth rate of investment in power sales of the unit power grid	0.88	0.96	0.60	0.60	1.00	0.60	0.70	0.97	0.97	0.61

The comprehensive weights were calculated by hierarchical analysis and entropy method to obtain the following results:

Table 3. Comprehensive weight calculation results.

Number	Evaluation index	Subjective weight	Objective weight	Comprehensive weight
1	Household distribution capacity	0.08	0.08	0.08
2	Qualified rate of comprehensive voltage	0.02	0.08	0.07
3	Tax contribution rate	0.15	0.12	0.13
4	Get the power index	0.01	0.14	0.11
5	Income from ten thousand yuan	0.20	0.10	0.12
6	Comprehensive line loss rate	0.19	0.10	0.12
7	KWH cost	0.03	0.12	0.10
8	Profit margin per kilowatt hour	0.13	0.16	0.15
9	The growth rate of investment in power sales of the unit power grid	0.19	0.1	0.12

The final evaluation results are calculated according to the relative paste schedule.

Table 4. Relative paste schedule calculation.

Project	A	B	C	D	E	F	G	H	I	J
Relative paste progress	0.2364	0.3392	0.6145	0.1947	0.3229	0.3293	0.4970	0.6933	0.4484	0.5449

According to the method principle, the ranking according to the size of the paste schedule is as follows: the larger the close value, the better the scheme; the smaller the close value, the worse the scheme. The area with a relatively large paste schedule is the area with the best maintenance and operation cost efficiency. Therefore, the H company is the most effective.

5 CONCLUSION

This paper proposes the grid maintenance operation cost efficiency evaluation model based on the combined weight TOPSIS theory, which strengthens the input and output efficiency evaluation and improves the development quality and efficiency of lean management of power grid enterprises. This research comprehensively promotes the whole process of lean resource allocation and has positive guiding significance and practical value for improving grid operation efficiency, optimizing resource allocation, and improving the level of fine investment management.

ACKNOWLEDGMENT

This study was supported by the Research on Operation Cost Allocation Decisions and Efficiency Evaluation Technology of Power Grid Maintenance Based on Fine Operation (Grant No. 52130N21000L).

REFERENCES

Guo Li, Fang Xiang, Wu Shuang. Research on the Benefit and Efficiency Evaluation of Power Grid Development Based on Value Maximize [J]. *Electrical appliances and Energy efficiency management Technology*, 2021(01): 90–97.

He Optics, Xue Xiaolin, Xu Qing, Li Chen, Mu Huaitao. *Comprehensive Efficiency Measurement of Asset Group Based on Cloud Material Yuan Model* [C] / / Management Innovation Practice of China Electric Power Enterprises (2019), 2020: 139–141.

Li Shuo. Input-output performance of Yunnan power transmission and distribution network based on DEA efficiency evaluation model [J]. *Yunnan Hydropower*, 2021, 37(11): 243–246.

Liu Wenjing, Fu Xianlan, Wu Jiekang, Wu Feng, Liu Shujian. Analysis on comprehensive benefits of grid enterprises based on DEA-RBFNN [J]. *Shandong Electric Power Technology*, 2021, 48(10): 33–39+59.

Yu Na, You Weiyang. Build the evaluation framework system for high-quality development of power grid [J]. *Management of China Electric Power Enterprises*, 2019(30): 46–47.

Energy Revolution and Chemical Research – Kok-Keong Chong and Zhongliang Liu (Eds)
© 2023 The Authors, ISBN: 978-1-032-36554-1

Analysis of runoff characteristics of non-point source pollutants of different land use types in Yinma River Basin

Chun Zhe Li
College of Mechanical Engineering, Jilin Engineering Normal University, Changchun, P.R. China

Mei Qing Fan*
Measurement Biotechnique Research Center, College of Food Engineering, Jilin Engineering Normal University, Changchun, P.R. China

ABSTRACT: Nitrogen and phosphorus loss caused by surface runoff and soil erosion is the main cause of agricultural non-point source pollution and surface water eutrophication in rivers and lakes. In this paper, four typical land types are selected to establish a runoff plot. The runoff characteristics of non-point source pollutants in different land use types with surface runoff under natural rainfall conditions are analyzed by field and laboratory tests. The main conclusions are as follows: (1) Given the same rainfall, intensity, and terrain gradient, the runoff yield of grassland and forest land is small and slow, whereas that of dry land and paddy fields is large and fast. (2) In the early rainfall period, the maximum value of total nitrogen and soluble nitrogen appeared before the maximum value of runoff, and pollutant concentration decreased as the runoff value increased. (3) The nitrogen loss in dry land and paddy fields is higher than that in forest land and grassland, but the change of total nitrogen and soluble nitrogen concentration with runoff is relatively weak. (4) Soluble nitrogen loss accounts for a greater proportion of nitrogen loss in dry land and paddy fields, and the soluble nitrogen content is also higher. The initial concentration of total phosphorus changed the most with runoff. Finally, with the continuous erosion and leaching of runoff, the pollutant concentration gradually decreases.

1 INTRODUCTION

The occurrence of non-point source pollution is influenced by many factors such as land use, topography, and slope, but it is generally believed that runoff generated during rainfall is the main reason for non-point source pollution (Duan et al. 2020; Hu et al. 2018; Maguffin et al. 2020; Patricio et al. 2016), which is known as the original driving force of non-point source pollution (Bin 2021). The loss patterns of non-point source pollutants with surface runoff are mainly two kinds (Landscape 2015), which are dissolved state and sediment combined state (granular state). Soils are the interface between rainfall and runoff. With the loss of soils and water, a large amount of surface runoff and sediment is taken away, along with nitrogen, phosphorus, and other nutrient elements in the soil. Their loss is not only a waste of fertilizer but also the main means of non-point source pollution into the environment of the basin (Coastal 2015). Surface runoff accumulation interacts with various pollutants and soil particles, and the interaction process is the key to the loss of non-point source pollutants. Therefore, to analyze the condition of rainfall and the migration mechanism of non-point source pollutants, many scholars have examined the rainfall conditions under the loss of pollution sources and the characteristics of the pollutants. This lays a foundation for studying the migration mechanism of non-point source pollutants in different land use types.

*Corresponding Author: fanmeiqing_011022@126.com

 DOI 10.1201/9781003332657-9

2 RESEARCH METHODS

2.1 *Selection principle of runoff plot*

Xinyang Township in Nongan County in the Yinma River Basin was selected as the sampling area because it has four types of land use. Four runoff plots of land types were selected. The situation of runoff plots is shown in Table 1.

Table 1. The situation of runoff plots.

No.	Land type	Area Slope	Geographical	Location
1	Paddy field	1.58	15.7	Nong'an County
2	Dry land	1.33	16.9	Nong'an County
3	Grassland	1.79	13.2	Nong'an County
4	Woodland	1.55	14.6	Nong'an County

The flow rate can be expressed as $Q = V \times A$, where Q is the flow rate (m^3/s); V is the velocity of cross section (m/s); A is the area of section (m^2). Section area and field measurements are taken simultaneously, using a ruler for measuring the water level and a buoy to measure the flow rate. Water samples are collected in the center of the section and on both sides of the shore, and these samples are then mixed into a bottle. Water samples are stored at 4°C with acid in the dark and sent to the laboratory within 24 hours for analysis. Among the most important components are total nitrogen, soluble nitrogen, total phosphorus, and soluble phosphorus.

2.2 *Test method of samples*

(1) Determination of total nitrogen

The testing and analysis of samples shall be carried out according to the China GB standard (Standard code: GB 11849-89). Total nitrogen was collected by UV-visible (UV-vis) spectrum with alkaline potassium persulfate oxidation, by adding sodium hydroxide to neutralize the hydrogen ions in the sample. When using potassium sulfate as an oxidant, at high temperatures (approximately 120–124°C) and in alkaline medium conditions, the nitrite and ammonia nitrogen in the water sample can be effectively oxidized to a nitrate; on the other hand, organic nitrogen compounds in the water sample can also be oxidized to nitrate. Next, the UV-vis spectrum is used to measure the absorbance at the wavelengths of 275 nm and 220 nm, and the absorbance of nitrate nitrogen is calculated to measure the total nitrogen content. Usually, the molar absorption coefficient is 1.47×10^3 L/(mol·cm). Water samples are filtered through a 0.45 μm micron microporous membrane for the determination of soluble nitrogen, and the process is the same as that used for total nitrogen determination.

(2) Determination of total phosphorus

The test method analysis of total phosphorus is in accordance with the China GB standard (Standard code: GB11893-89) using the potassium persulfate digestion molybdenum-antimony photometric method. The standard stipulates that nitric acid-perchloric acid or potassium persulfate must be used as an oxidizing agent, ammonium molybdate spectrophotometry to measure the total phosphorus. Total phosphorus mainly includes inorganic phosphorus, organic phosphorus, particulate phosphorus, and dissolved phosphorus. This standard is suitable for the detection of industrial wastewater, domestic sewage, and surface water. Under neutral conditions, nitric acid-perchloric acid or potassium sulfate is used to digest the sample, and all the total phosphorus in the sample is oxidized into orthophosphate. In addition, in acidic media, ammonium molybdate reacts with orthophosphate and reacts with antimony salt to form molybdate heteropoly acid, which is rapidly reduced by ascorbic acid to form a blue complex.

For the determination of soluble phosphorus, the water sample was filtered by 0.45 μm microporous membrane, which was the same as the determination of total phosphorus.

2.3 *Monitoring of rainfall runoff*

In rainfall events, the surface runoff at the outlet of the runoff plot is monitored synchronously. According to the monitoring results, the variation of pollutant loss concentration and runoff with rainfall time is recorded under different land use patterns, and the variation relationship and relationship characteristics are analyzed. In the whole experiment process, the key analysis of typical years is carried out first, and then a general conclusion is drawn based on the comprehensive analysis of experimental data.

3 ANALYSIS OF RUNOFF CHARACTERISTICS OF NON-POINT SOURCE POLLUTANTS

Samples are collected at the exit of the runoff plot of four land use types, and the contents of total nitrogen, total phosphorus, soluble nitrogen, and soluble phosphorus in each runoff sample were tested in the laboratory. The experimental results are shown in Tables 2 and 3.

Table 2. The concentration range of different land use types in total nitrogen and dissolved nitrogen.

	Total nitrogen		Dissolved nitrogen	
Land type	Average	Range	Average	Range
Dry land	1.575	0.437–2.449	0.900	0.079–1.654
Paddy field	1.227	0.245–1.908	0.765	0.065–1.523
Woodland	1.148	0.105–1.738	0.725	0.026–1.428
Grassland	0.910	0.165–1.528	0.714	0.069–1.158

Table 3. The concentration range of different land use types in total phosphorus and dissolved phosphorus.

	Total phosphorus		Dissolved phosphorus	
Land type	Average	Range	Average	Range
Dry land	0.0848	0.113–0.0628	0.0265	0.0108–0.428
Paddy field	0.0844	0.0628–0.135	0.0258	0.0102–0.0578
Woodland	0.0715	0.0539–0.0924	0.0165	0.0106–0.0328
Grassland	0.0608	0.0477–0.0895	0.0154	0.0057–0.0267

3.1 *Characteristics analysis of rainfall runoff and nitrogen loss over time for different land use types*

According to the experimental data, the loss load of total nitrogen, soluble nitrogen, and runoff in different land types follow the following rules:

(1) Under the condition that the rainfall amount, rainfall intensity, and terrain gradient selected for the experiment are the same, the runoff generated by grassland and forestland is small and slow, while the runoff generated by dry and paddy fields is large and the speed of runoff is fast. Mainly due to different vegetation coverage on the soil surface, afforestation and lawn planting can effectively reduce soil erosion and total nitrogen loss (Boers 1996).

(2) In all land use types, at the beginning of the precipitation, the concentration of total nitrogen, and soluble nitrogen in the runoff before the peak appeared to arrive, with the occurrence of rainfall, the runoff value increases gradually, and the concentration of the pollutants is smaller (Zhang 2014).
(3) Among the four land use types, the loss of nitrogen in upland and paddy fields was higher (Napoli et al. 2017), while the loss concentration of total nitrogen and soluble nitrogen in forest and grassland had a weaker change pattern with runoff.
(4) The amount of soluble nitrogen loss in dry and paddy fields accounts for a large proportion of nitrogen loss (Pierzynski et al. 1994), and soluble nitrogen content is also high (LIANG et al. 2014).
(5) Among the four land use types, dry land had the highest average concentration of total nitrogen loss (1.58 mg/L). Woodland was the smallest, and the concentration was 0.910 mg/L.

3.2 *Characteristics of rainfall runoff and PHOSPHORUS loss over time in different land use types*

According to the experimental data, the loss load of total nitrogen, soluble nitrogen, and runoff in different land types follow the following rules:

(1) In all land use types, at the initial stage of rainfall, the concentration of total phosphorus reaches before the maximum runoff value (Qian et al. 2010), which is similar to that of total nitrogen. On the contrary, at the end of rainfall, the concentration of total phosphorus loss increases in reverse, presenting a U-shaped trend, which may be mainly due to the accumulation of fertilizer nutrients in the soil which was washed by rain before, and the leaching of salt in the soil phosphorus at the later stage (Yu et al. 2020).
(2) In terms of phosphorus loss of the four types, the amount of soluble phosphorus loss accounts for a small proportion of phosphorus loss, and the concentration of total phosphorus is far greater than that of soluble phosphorus (Delgado et al. 1995).
(3) The variation range of soluble phosphorus is small, indicating that there is a certain limit to the concentration range that soluble components can reach under certain water environmental conditions, and its solubility may be affected by water environmental conditions more than the influence of flow changes on it (Jiang et al. 2019; Nazarbakhsh et al. 2020).
(4) It can be seen that the concentration of total phosphorus loss in upland and forestland varies greatly in the process of rainfall and runoff, while the concentration of total phosphorus loss in paddy fields and grassland is relatively stable and the variation range is relatively small. Secondly, the average loss concentrations of total phosphorus in a dry field, paddy field, and paddy field were almost the same, which were 0.0849 mg/L and 0.0834 mg/L, respectively. The average loss concentrations of total phosphorus in forestland and grassland were 0.0715 mg/L and 0.0609 mg/L, respectively.

4 CONCLUSIONS

Four representative land types were selected to establish runoff plots. The runoff characteristics of non-point source pollutants of different land use types with surface runoff under natural rainfall were analyzed by field and laboratory tests. The main conclusions are as follows:

(1) Under the condition that the rainfall amount, rainfall intensity, and terrain gradient selected for the experiment are the same, the runoff generated by grassland and forestland is small and slow, while the runoff generated by dry and paddy fields is large and the speed of runoff is fast.
(2) In all land use types, the maximum concentrations of total nitrogen and soluble nitrogen appeared before the maximum runoff value in the early rainfall, and the pollutant concentration decreased with the increase of runoff value.

(3) Among the four land use types, the loss of nitrogen in upland and paddy fields was higher, while the loss concentration of total nitrogen and soluble nitrogen in forestland and grassland was weaker with the change of runoff.
(4) From the form of nitrogen loss, the amount of soluble nitrogen loss in dry and paddy fields accounted for a larger proportion of nitrogen loss, and the content of soluble nitrogen was also high. Therefore, after the surface runoff was generated by rainfall, nitrogen mainly followed surface runoff in the form of dissolved nitrogen. The variation range of total nitrogen loss concentration in upland and paddy fields was large, while the variation range of total nitrogen loss concentration in forestland and grassland was relatively stable. Secondly, the average concentration of total nitrogen loss in four land use types was upland > paddy > grassland > woodland.
(5) The initial pollutant concentration of total phosphorus was the highest with runoff. Finally, with the continuous erosion and leaching of runoff, the concentration of pollutants also gradually decreased. On the contrary, at the end of rainfall, the loss concentration of total phosphorus increased in reverse. In terms of the phosphorus loss form of the four land use types, the concentration of total phosphorus is far greater than that of soluble phosphorus, indicating that the amount of sediment bonded phosphorus loss is relatively high. That is, most of the runoff loss is granular phosphorus combined with sediment, that is, the loss state of phosphorus is mainly sediment combined phosphorus. It can be seen that after the surface runoff is formed after the heavy rain, phosphorus is mainly lost through the formation of sediment binding form along with the surface runoff. Under the environmental conditions of runoff generated by rainfall, soluble phosphorus components have a certain limit, and the change of external flow has little influence on soluble phosphorus. It can be seen that the variation range of total phosphorus loss concentration in upland and forestland is large in the process of rainfall and runoff, In the run-up to rainfall, it is important to reduce and avoid farming activities such as soil ploughing and loosening that can exacerbate phosphorus loss and soil erosion, while the variation range of total phosphorus loss concentration in paddy field and grassland is relatively stable.

REFERENCES

Bin Xu. (2021). Assessment of a Gauge-Radar-Satellite Merged Hourly Precipitation Product for Accurately Monitoring the Characteristics of the Super-Strong Meiyu Precipitation over the Yangtze River Basin in 2020. *J. Remote Sensing*, 123(1): 96–107.

Delgado AN, Periago EL, Viqueira FD. (1995). Vegetated filter strips for water purification: A review. *J. Bioresource Technology*, 51: 13–22.

Duan, J., Liu, Y.J., Yang, J., Tang, C.J., & Shi, Z.H. (2020). Role of groundcover management in controlling soil erosion under extreme rainfall in citrus orchards of southern China. *J. Journal of Hydrology*, 582.

Hu, F., Liu, J., Xu, C., Wang, Z., Liu, G., Li, H., & Zhao, S. (2018). Soil internal forces initiate aggregate breakdown and splash erosion. *J. Geoderma*, 320: 43–51.

Jiang, C., Liu, J., Zhang, H., Zhang, Z., & Wang, D. (2019). China's progress towards sustainable land degradation control: Insights from the northwest arid regions. *Ecological Engineering*, 127: 75–87.

JMI Dahl, Coastal. (2015). Ecological Impacts on the Nearshore Bathing Areas. J. Journal of Environmental Protection and Ecology, 16(2): 424–433.

Liang Y., Yi Y.Y., Sun Q.F. (2014). The Impact of Migration on Fertility under China's Underlying Restrictions: A Comparative Study Between Permanent and Temporary Migrants. *J. Social Indicators Research*, 116(1): 307.

Maguffin S.C., Abu-Ali L., Tappero R.V., Pena J., Reid M.C. (2020). Influence of Manganese Abundances on Iron and Arsenic Solubility in Rice Paddy Soils. *J. Geochimica Et Cosmochimica Acta*, 406(1–2): 129–135.

Napoli, M., Marta, A.D., Zanchi, C.A., & Orlandini, S. (2017). Assessment of soil and nutrient losses by runoff under different soil management practices in an Italian hilly vineyard. *Soil and Tillage Research*, 168: 71–80.

Nazarbakhsh, M., Ireson, A. M., & Barr, A. G. (2020). Controls on evapotranspiration from jack pine forests in the Boreal Plains Ecozone. *Hydrological Processes*, 34(4): 927–940.

Patricio, Cid., Helena, Gómez-Macpherson., Hakim, Boulal., Luciano Mateos. (2016). Catchment scale hydrology of an irrigated cropping system under soil conservation practices. *J. Hydrological Processes*. 30(24).

PCM Boers. (1996). UKGrsHP: a UK high-resolution gauge–radar–satellite merged hourly precipitation analysis dataset. *J. Water Sci Technol*, 33(4): 183–189.

Pierzynski G.M., Sim J.T., Vance G.F. (1994). *Soil and environmental quality*. M. Boca Raton: Lewis Publisher, CRC Press.

Qian Hong, Hong Yu, Jun Fengniu. (2010). Parameter uncertainty analysis of non-point source pollution from different land use types. *J. Science of the Total Environment*, 40(8): 1971–1978.

Yu, J., Li, X.F., Lewis, E. (2020). UKGrsHP: a UK high-resolution gauge–radar–satellite merged hourly precipitation analysis data set. *J. Climate Dynamics*, 64(5): 169–180.

ZH B, EH M. Landscape and Geochemical Features of Man-made Pollution Zones of Aktobe Agglomerations. *J. Oxid Commun*, 2015, 38(2): 852–856.

Zhang Z. (2014). *Simulation and prediction of non-point source pollution in Yitong River Basin*. Doctoral dissertation of Jilin University.

Energy Revolution and Chemical Research – Kok-Keong Chong and Zhongliang Liu (Eds)
© 2023 The Authors, ISBN: 978-1-032-36554-1

Analysis and discussion of "zero discharge" technology for high saline wastewater

Zhan Liu*
Institute of Energy Sources, Hebei Academy of Science, Shijiazhuang Hebei, China
Hebei Engineering Research Center for Water Saving in Industry, Shijiazhuang Hebei, China

Na Li
Hebei Sun-water Treatment Co., Ltd., Shijiazhuang, Hebei, China

Mei-fang Yan, Yu-hua Gao & Hai-hua Li
Institute of Energy Sources, Hebei Academy of Science, Shijiazhuang Hebei, China
Hebei Engineering Research Center for Water Saving in Industry, Shijiazhuang, Hebei, China

ABSTRACT: The shortage of water resources and the shortage of capacity of the freshwater environment is an important bottleneck restricting the development of modern industry in China. One of the major challenges of industrial development is the treatment and discharge of high-concentration brine. The zero-discharge solution of wastewater is an important way to resolve the contradiction between the modern chemical industry and water resources and the environment. In this paper, the utilization of high-salt wastewater and its treatment technology are discussed. Hopefully, this will provide theoretical guidance for the treatment of high salt wastewater. The highly saline water can eventually be treated by the salt separation process system to obtain a single crystalline salt and condensate that meet industrial standards. The crystalline salt and condensate are used resourcefully and the recovery rate of the highly saline wastewater reaches 95%, achieving near-zero discharge and greatly reducing the burden on the enterprise and the environment.

1 INTRODUCTION

China's industrial production produces a large amount of tailwater, including coal chemical companies, steel companies, dairy products processing industries, and pesticide and pharmaceutical firms. It is called high salt water when the salt content in the discharged water exceeds 1000 mg/L. This kind of high-salinity wastewater contains a large amount of soluble inorganic salts, such as Cl^-, SO_4^{2-}, Na^+, Ca^{2+}. At the same time, the wastewater also contains some organic pollutants. Therefore, it is difficult to treat the high-salt wastewater produced by production (Bian et al. 2019; Deng et al. 2014).

If the untreated high-content salt water is directly discharged into natural water bodies, it will cause pollution of river water and groundwater bodies, which will pose a certain threat to the organisms in the water bodies and humans. The presence of many inorganic salt ions in the wastewater increases the density of the water body. The activated sludge in the water will float up due to the low relative density, causing the loss of activated sludge. If human beings drink brackish water for a long time, it will cause endocrine disorders, and the gastrointestinal function will be disordered, which will endanger the health of the human body. Therefore, how to efficiently treat high-content brine, which is one of the difficult-to-degrade wastewaters, is not only beneficial to the recycling and

*Corresponding Author: liuzhan1216@126.com

 DOI 10.1201/9781003332657-10

the energy saving, emission reduction of water resources, but also directly affects the sustainable development of the ecological environment and the health of human beings (Tian et al. 2016).

Moreover, as early as 2012, the State Council issued *Opinions on Implementing the Strictest Water Resources Management System*, and at that time it had put forward the highest total water consumption in the country from 2012 to 2030. This means that not only many water-saving measures such as air cooling need to be applied in industrial projects but also all the wastewater produced in production should be reused as much as possible. It can be said that the sustainable development of enterprises needs the support of "zero discharge" of wastewater (Duan et al. 2019; Wang 2016).

Therefore, this requires enterprises to deal with the high-salinity wastewater according to the quality of the wastewater produced during the production process, and have the corresponding "zero discharge" engineering process, and make all the water produced after the treatment be recycled. In this way, the goal of "zero discharge" of wastewater can be achieved.

2 TRADITIONAL "ZERO-EMISSION" PROCESS

The traditionally applied high-salt wastewater "zero discharge" technology has achieved zero discharge of water. The waste high-concentration brine is directly processed to obtain crystalline inorganic salt. The obtained solid salt is generally composed of solid mixed salt (sodium chloride and sodium sulfate account for the majority, which may also contain a small number of heavy metal ions and organic matter). The salt is landfilled or stored, and the produced water is recycled.

At present, the most traditional high-salt wastewater "zero discharge" technologies are evaporation pond treatment technology, thermal evaporation treatment technology, and membrane treatment technology.

The main categories and characteristics of traditional "zero-emission" processes are shown in Table 1.

Table 1. Traditional "zero-emission" processes.

Treatment	Principle	Device	Products	Product handling
Evaporation pond treatment	Natural evaporation	Evaporation ponds	Mixed salt	Landfill
Multi-effect evaporation	Heating/pressurizing	Multi-effect evaporation units	Mixed salt	Landfill
MVR	Heating/pressurizing	Steam compressors and evaporators	Mixed salt	Landfill
Membrane distillation	Vapor pressure difference	Membrane units	Mixed salt	Landfill

2.1 *Evaporation pond treatment process*

The principle of the evaporation pond is to use natural evaporation technology to treat high salt water. This method needs to consider the climatic conditions of the region. Under suitable climatic conditions, the high salt water is naturally evaporated, and finally, mineral salt crystals can form at the bottom of the evaporation pond. The process flow of the evaporation pond is to discharge the high salt water into the evaporation pond, and the concentrated water in the pool is sprayed into the air by a pump and a waterscape nozzle to form an umbrella-shaped water shape, which can increase the surface area of concentrated water evaporation. At the same time, there will be steam in the evaporation pond for heat exchange to increase the water temperature and increase the evaporation capacity. The solid salt is finally obtained, and the salt is taken out from the evaporation pond and sent to the dry salt pond for landfill disposal (Sun et al. 2015).

2.2 *Thermal evaporation process*

Thermal evaporation technology mainly includes three treatment technologies, such as multi-effect evaporation, mechanical compression evaporation, and membrane distillation.

(1) Multi-effect evaporation
The function of multi-effect evaporation is mainly realized through the application of evaporators. High-salt water is introduced into the system of multiple evaporators, and the evaporation process is completed through a certain process so that the treatment of high salt water can be realized (Cui et al. 2017).

(2) Mechanical compression evaporation
The mechanical compression evaporation technology uses a vapor compressor to compress the secondary vapor, and the compressed secondary vapor is introduced into the thermal cycle of the original system to achieve the treatment of high brine content (Wang et al. 2015).

(3) Membrane distillation
Membrane distillation technology is a technology that uses the principle of vapor pressure difference to achieve high brine treatment. Although this technology has high application efficiency, it has the disadvantage of large resource consumption.

2.3 *Membrane treatment technology*

Membrane treatment technology refers to membrane separation technology. Its core technology is the selective permeability of the membrane. By using this selective permeability of the membrane, the target mixture can be separated, purified, and concentrated.

The product of the traditional high-salt water treatment process currently used is solid waste, and the generated solids are sludge and solid mixed salt. The solid mixed salt is easily soluble in water. If there is rainwater, the solid mixed salt will dissolve and penetrate the groundwater, thereby causing secondary pollution. Even if the solid mixed salt is stored properly, it cannot be recycled and reused. The wastewater produced in the industry contains trace amounts of heavy metal ions and residual organic matter. The traditional high-salt water treatment process does not extract them, but crystallizes these substances together with the inorganic salts in the wastewater and enters the solid mixed salt. The presence of heavy metal ions and residual organic matter may cause the solids obtained after treatment to be extremely dangerous (Wu et al 2017; Zuo et al. 2018). Therefore, the crystalline miscellaneous salt obtained by the traditional high-salt wastewater "zero discharge" technology is called hazardous waste and must be disposed of in strict accordance with relevant regulations. This has brought a great economic burden not only to enterprises but also to the environment (Wu 2019).

By reforming and renewing the traditional "zero discharge" technology to improve the treatment process of high-concentration brine, the product solid mixed salt is developed into elemental crystalline salt with reusable value, and the product water after the crystalline salt is extracted is recycled and reduced which is the realization of "zero discharge" of high-salt wastewater and "salt separation of elemental substances". Two high-purity inorganic salt products, sodium sulfate, and sodium chloride are obtained. The water quality of the product water can meet the water quality requirements of the primary reclaimed water. Both the crystal salt and the product water can be recycled. This will be the inevitable development trend of "zero discharge" of high-salt wastewater (Wan 2018).

3 SALT SEPARATION PROCESS

At present, two main salt separation processes have been initially applied, namely, thermal method salt separation and nanofiltration salt separation.

The principles and characteristics of the salt separation process are shown in Table 2.

Table 2. Salt separation process.

Treatment	Principle	Device	Products	Product handling
Thermal salt separation	Differences in solubility of inorganic salts	Multi-effect evaporation units	High purity crystalline salt	Reuse
Nanofiltration fractionated salt	Selective permeability of nanofiltration membranes	Membrane units	High purity crystalline salt	Reuse

3.1 *Thermal method of salt separation*

The core of the thermal method of salt is variable temperature crystallization, The variable temperature crystallization technology is specially set up according to the different solubility of the two main inorganic salts (Na_2SO_4 and NaCl) contained in the high-salt wastewater. When the temperature is lower than 40°C, the solubility of Na_2SO_4 will increase obviously with the increase of temperature, but when the temperature is above 40°C, the solubility will decrease with the increase of temperature. Although the solubility of NaCl increases slightly with the increase in temperature, it is not greatly affected by temperature. Therefore, when recovering sodium sulfate, first evaporate and concentrate the high-salt wastewater, and then precipitate Na_2SO_4 crystals at a relatively high temperature. By controlling the concentration of the evaporating Na_2SO_4 endpoint, it is ensured that the concentration of the two inorganic salts falls in Na_2SO_4 when the evaporation end point crystallization area (where there is no NaCl precipitation), thereby obtaining Na_2SO_4 with qualified purity. The solid-liquid mixture obtained at the end of the evaporation is centrifuged, and the mother liquor is cooled to precipitate mixed salts, which can further remove the residual Na_2SO_4 in the liquid. The remaining mother liquor after solid-liquid separation contains only a small amount of Na_2SO_4, most of which is NaCl. After evaporation, NaCl with qualified purity can be obtained. By controlling the end-point concentration of the evaporation for the second time, the end-point concentration of the evaporation can fall in the NaCl crystallization area (where there is no Na_2SO_4 precipitation). By controlling the concentration of the evaporation end point twice, high-purity crystalline salt Na_2SO_4 and NaCl can be obtained, and the product water evaporated can also meet the water quality requirements of reclaimed water (Zhang 2015).

3.2 *Nanofiltration of salt*

The process of salt separation by nanofiltration uses the selectivity and retention characteristics of nanofiltration membranes for divalent salts to realize the separation of monovalent salt NaCl and divalent salt Na_2SO_4 in high brine. The permeate of high-salt water passing through the nanofiltration membrane mainly contains NaCl, while Na_2SO_4 mainly exists in the concentrated nanofiltration water, and both of the filtrates have been concentrated by the nanofiltration membrane. Then, the nanofiltration membrane permeate and nanofiltration concentrated water are respectively subjected to crystallization treatment. Finally, the recovery of sodium chloride and sodium sulfate crystalline salt can be achieved (Li et al. 2016; Xiong et al 2018).

The quality of the crystalline salt obtained after the treatment of the two salt separation systems can reach industrial purity, and the water quality of the condensate meets the standard for reuse.

Sodium sulfate: conform to GB/T 6009-2014 industrial sodium sulfate II qualified products, sodium chloride products implement GB/T 5462-2015 "industrial salt" in the industrial dry salt secondary product standards, condensate water to meet the circulating water supplement water quality indicators. The specific indicators are shown in Table 3 (% indicates mass fraction in the table) and Table 4.

Table 3. Quality standards for crystalline salt.

Item	Na_2SO_4	NaCl
Na_2SO_4(%)	≥ 97	
NaCl (%)		≥ 97.5
H_2O	≤ 1	≤ 0.80
Water insoluble matter (%)	≤ 0.2	≤ 0.2
Ca^{2+}, Mg^{2+}(%)	≤ 0.4	≤ 0.60
Cl^- (%)	≤ 0.90	
SO_4^{2-} (%)		≤ 0.90
Appearance		White crystal or slightly yellow, greenish-white

Table 4. Condensate water quality standards.

Indicators	Unit	Requirements
COD	mg/L	≤ 30
TOC	mg/L	–
TDS	mg/L	≤ 600
NH_3-N	mg/L	≤ 10
Cl^-	mg/L	≤ 250
Ca^{2+}	mg/L	≤ 175
Mg^{2+}	mg/L	≤ 50
SiO_2	mg/L	–
TP	mg/L	–
TSS	mg/L	≤ 1
Na^+	mg/L	–

4 CONCLUSION

Both the thermal method and the nanofiltration method for extracting salt from simple substances have been initially applied, and two crystalline salts with qualified purity can be successfully obtained, namely Na_2SO_4 and NaCl. Moreover, the water quality of the final product water can meet the water quality requirements of primary reclaimed water.

However, the nanofiltration membrane is prone to contamination, blockage, and corrosion. Once the nanofiltration membrane is contaminated or corroded, its retention efficiency will be greatly reduced; thermal separation has a relatively wider range of applications and better desalination performance, but thermal separation requires higher water quality, and a larger footprint for the whole system, and higher energy consumption. and sulfate concentrations are only suitable for use where there is a large difference. The next step in the research is to optimize the system for the disadvantages of both processes and to rationalize the treatment units to achieve higher performance, higher recovery rate, and a wider range of applications for the treatment of high brine in the future.

REFERENCES

Bian Xiaotong, Huang Yongming, Guo Rutao, et al. Research progress on elemental salt separation and resource utilization of high-salt wastewater [J]. *Inorganic Salt Industry*, 2019, 51(08): 7–12.

Cui Fengxia, Li Rong, Chen Weina. Progress in zero-discharge evaporative crystallization technology for high-salt wastewater [J]. *Guangzhou Chemical Industry*, 2017, 45(1): 21–23.

Deng Dandan, Shen Ying, Zhu Weina, et al. *Proceedings of the Annual Conference of the Chinese Society of Environmental Sciences* (Chapter 5), 2014, 4136–4140.
Duan Xiaobing, Mao Songlin, Xu Guofeng. Technical transformation of secondary desalination project for high salt water content [J]. *Metallurgical Power*, 2019(05): 68–70.
Li Kun, Wang Jianxing, Wei Yuansong. The application status and prospect of nanofiltration in water treatment and reuse [J]. *Acta Scientiae Circumstantiae*, 2016, 36(8): 2714–2729.
Sun Jia, Zhang Guohui, Wu Chunming. Preliminary study on the treatment of high salt water by Evaporated Tang in Ulan Chabu City [J]. *Environment and Development*, 2015, 4(12): 67–70.
Tian Caiyun, Li Zongshuo, Wang Chunli, et al. Research progress of forward osmosis separation technology and its application in high salt water treatment [J]. *Xinjiang Environmental Protection*, 2016, 38(04): 30–35+52.
Wan Lei. "Zero-discharge and qualitative crystallization" process technology of high brine in coal chemical industry [J]. *Ju She*, 2018(02):187–188.
Wang Hang. Progress of the "Zero Emission" Project of Great Wall Energy Chemical Co., Ltd. with high salt water content [J]. *Coal Processing and Comprehensive Utilization*, 2016(04): 34–38+71.
Wang Jian, Guo Tianjiao, Feng Ming, et al. Current status and research progress of high-salt industrial wastewater treatment technology [J]. *Coal Chemical Industry*, 2015, 43(3): 18–21.
Wu Yanfang, Zhang Junling, et al. Research on the operation effect of zero-discharge technology of coal chemical industry's high-salt wastewater resource utilization [J]. *Coal Processing and Comprehensive Utilization*, 2017(06): 32–35.
Wu Yanjun. Operation problems and countermeasures of concentrated brine resource utilization device [J]. *Large Nitrogen Fertilizer*, 2019, 42(02): 134–137.
Xiong Rihua, He Can, Ma Rui, et al. The salt crystallization process of high-salt wastewater and its technical and economic analysis [J]. *Coal Science and Technology*, 2018, 46(09): 37–43.
Zhang Jijun, *A method of recycling high-salt wastewater*: China, 105036222A [P]. 2015-8-19.
Zuo Wu, Ge Shifu, Zhu Huajun, et al. Research progress in heat treatment technology of high-salt organic waste liquid [J]. *Environmental Engineering*, 2018, 36(4): 47–51.

Energy Revolution and Chemical Research – Kok-Keong Chong and Zhongliang Liu (Eds)

Research on the implementation path of building a green and low-carbon port

Zhongfei Kang*, Wei Du & Teng Tian
Tianjin Research Institute for Water Transport Engineering, M.O.T., Tianjin, China

ABSTRACT: Many ports at home and abroad have done a lot of work to support the construction of green and low-carbon ports, and jointly promote the construction of green and low-carbon ports by improving the production energy consumption of main equipment, operation technology, and auxiliary facilities, strengthening the top-level design, strengthening the construction of concept and system, and gradually establishing the company's low-carbon management methods, assessment, reward and punishment methods. By studying and exploring the advantages of ports in green and low-carbon construction, we can identify a construction path that can be popularized, promote the construction of domestic green and low-carbon ports, and demonstrate the effectiveness of the transportation industry.

1 INTRODUCTION

Climate change has become a major issue affecting the development of human society and the global political and economic pattern. To mitigate global climate change and reduce its disastrous impact on human society, the United Nations requires all countries to clarify their greenhouse gas emission reduction tasks and targets. As one of the contracting countries and a responsible large developing country, China submitted the national independent contribution document on climate change "strengthening action on climate change—China's national independent contribution" on June 30, 2015. The goal of independent action by 2030 is determined: carbon dioxide emissions will peak around 2030 and strive to reach the peak as soon as possible; carbon dioxide emissions per unit of GDP decreased by 60%–65% compared with 2005 (He 2021).

2 CONSTRUCTION OF ZERO-CARBON PORTS AT HOME AND ABROAD

Developed countries such as the European Union and the United States pay more attention to energy-saving and low-carbon development. They have reached the peak of carbon as early as 2016, and have studied carbon neutralization and zero-carbon for many years.

Port of Rotterdam, the Netherlands, launched the "regiment air quality action project", carried out cleaning projects, and encouraged port enterprises to use clean energy. The carbon dioxide capture and storage plan will capture 500,000 tons of carbon dioxide per year from 2019 to 2025. Amsterdam port, the Netherlands, will carry out "clean shipping" (Liu 2019), make rational use of land and environment, control and reduce carbon dioxide emissions, and reduce emissions by 49% by 2030 and 90% by 2050.

The Ports of Los Angeles and Long Beach in the United States have launched the "clean air plan for the port group of San Perot Bay", and for heavy trucks, they will transition from clean trucks to

*Corresponding Author: 642456933@qq.com

 DOI 10.1201/9781003332657-11

zero-emission trucks; for port machinery, zero-emission will be achieved and idling will be limited in 2030.

At present, Ningbo port and Shanghai port have begun to study the action plan for carbon peak. The research on the zero-carbon wharf of domestic ports is mainly reflected in the utilization of new and renewable energy, including solar photovoltaic, wind power, hydrogen energy, etc.

Solar energy: the warehouse roof of bulk cargo wharf in the Dalian port area (40,000 m^2), with a total distributed photovoltaic power generation of 2.45 million kWh; the container terminals of Quanzhou port and Qingdao port will build photovoltaic power stations on maintenance sites and office buildings.

Hydrogen energy: at present, three hydrogen energy collection trucks in Qingdao port have been put into live test operation in Qianwan container terminal; the first hydrogen-powered automatic rail crane, with 60 kW fuel cell, provides energy power for the port "rail crane" equipment.

Wind energy: seven wind turbines have been put into operation in Jiangyin port, Jiangsu Province. Since the first one was connected to the grid in May 2018, distributed wind power has generated 27 million kWh, and the wind power substitution rates of the two-port areas have reached 48% and 60% respectively.

In 2018, Meishan, Ningbo launched the construction of an "international demonstration area of near-zero carbon emission" (Liu 2021). As the core of "port industry city" in Meishan District, Meishan port area of Zhoushan port, Ningbo, plans to achieve 80% electrification of port energy consumption equipment, 90% renewable energy power consumption, 100% renewable energy power consumption in 2050, and finally achieve net-zero carbon emission of greenhouse gases.

3 EXPLORATION OF THE IMPLEMENTATION PATH OF BUILDING A GREEN AND LOW-CARBON PORT

Many domestic ports have done a lot of work to support the construction of green and low-carbon ports, and jointly promote the construction of green and low-carbon ports by improving the production energy consumption of main equipment, operation technology, and auxiliary facilities, strengthening the top-level design, strengthening the construction of concept and system, and gradually establishing the company's low-carbon management methods, assessment, reward, and punishment methods.

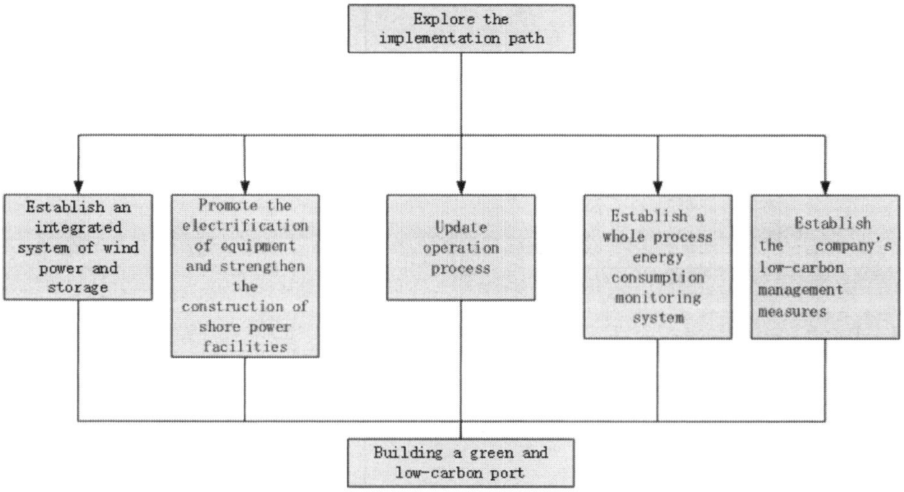

Figure 1. Path system diagram of building a green and low-carbon port.

3.1 *Establish a "scenery storage integration" system*

Developing distributed photovoltaic wind energy can reduce port energy costs, reduce pollutant emissions, optimize energy consumption structure, and promote the reform process of national green ports.

At the same time, supporting construction can establish a "scenery storage integration" system, take multiple measures, and realize all green supply of energy demand. Through the construction of a wind turbine generator and photovoltaic energy generation system, in conjunction with the establishment of a corresponding energy storage system, it is possible to reduce the power consumption of the wharf.

Distributed photovoltaic wind power covers a small area and is feasible in the port. On the premise of good site selection, it will not affect the production and operation of the port. It can optimize the energy consumption structure and improve pollution prevention and green management.

3.2 *Promote the electrification of equipment and strengthen the construction of shore power facilities*

All direct carbon emissions from ports are generated by fossil energy, and there are a large number of horizontal vehicles and non-road mobile machinery using fuel. By promoting the construction of green ports, the consumption and proportion of fossil energy in the whole port area can be reduced, which plays an important role in reducing the emission of pollutants in the process of port operation and production and in the process of energy production and transportation.

To enhance the top-level design of the port and emphasize the use of clean energy, the wharf should be powered by electricity in loading and unloading equipment, horizontal transportation equipment, and production auxiliary equipment, to reduce the number of fossil fuels utilized by the port. During the process of mechanical renewal of the old oil port, priority should be given to the acquisition of electrified equipment, as well as the trial implementation of large-scale electrified mobile machines.

Facilities and equipment with low energy consumption and high efficiency are preferred. Modern electromechanical conversion equipment is utilized with high efficiency. By analyzing and calculating the appropriate illuminance, we can do a reasonable arrangement of LED lamps for operation lighting, as well as apply segmented control, automatic dimming, and other technologies.

Other methods include encouraging the construction of shore power facilities and to establish a real-time monitoring system for shore power to allow ships calling at a port to use fewer fossil fuels, thereby reducing the overall carbon emission intensity of the port.

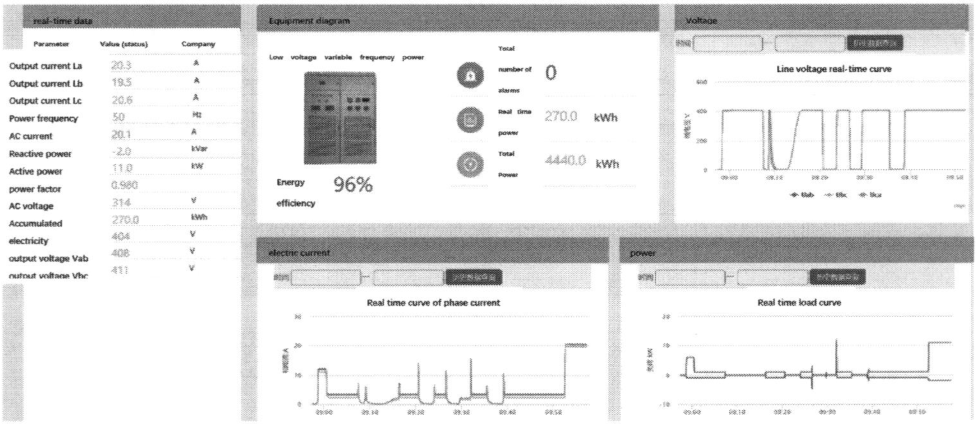

Figure 2. Schematic diagram of shore power monitoring system.

3.3 *Update operation process*

This process may include:

1) Utilize modern communication, sensing, information, and artificial intelligence technologies to perform automatic detection, intelligent monitoring, intelligent control, and intelligent scheduling of terminal operation processes using 4G/5G communication components, cameras, millimeter waves, lasers, ultrasonics, differential GNSS, and other hardware components;
2) Maximize agile production and flexible production to improve resource utilization, thereby reducing operating costs and energy consumption;
3) Strengthen the automation level of the operation process and improve the stability of operation.

3.4 *Establish a whole process energy consumption monitoring system*

The wharf should utilize all the advantages of information technology and digital management, make full use of the implementation, refinement, and intelligence of IoT, organically integrate energy management and production, and construct an intelligent green energy management and control system covering the entire system. The system integrates energy consumption management, environmental monitoring, lighting control, intelligent power consumption monitoring, shore power monitoring (Wang 2017), cold box monitoring, HVAC, and electric valve control systems to enable the efficient management of all elements of energy and the environment. Through in-depth analysis of energy consumption data and operation data, this paper explores the implementation path of saving energy, reducing consumption, and improving production efficiency from the perspective of energy management (Li 2021).

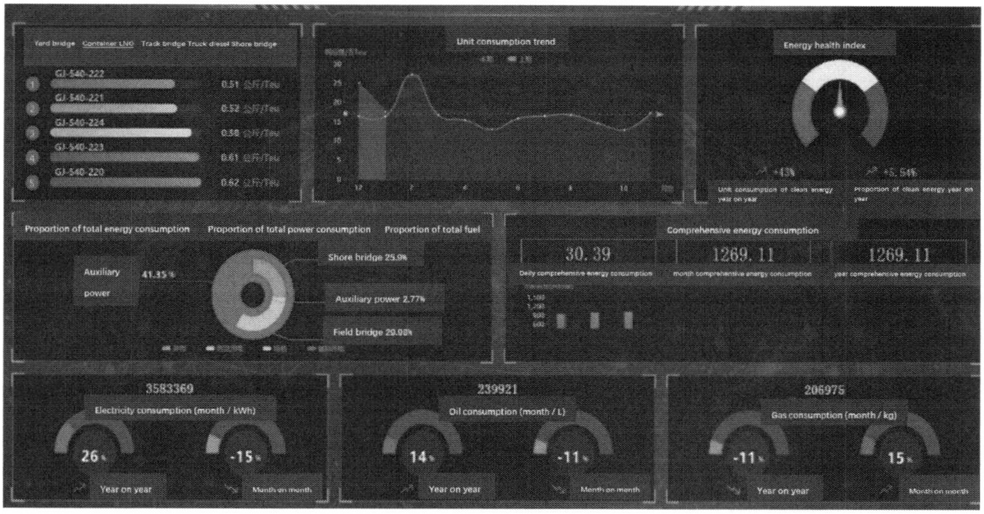

Figure 3. Diagram of energy consumption monitoring system.

3.5 *Establish the company's low-carbon management measures*

The company should enhance employees' enthusiasm for energy conservation and environmental protection, strengthen the construction of the concept and system, and adopt energy-efficient and low-carbon technologies for production and energy consumption such as main equipment, process, and auxiliary facilities. Further, the company should also develop a zero-carbon energy consumption management system as well as a carbon trading management system, such as the company's zero-carbon management measures, assessment, reward and punishment systems, and its internal and external carbon trading management systems.

4 CONCLUSIONS

The term "world-class green port" refers to a modern port that is geared toward "safety, convenience, efficiency, sustainability, economy, and innovation". Among the characteristics of "green development" are that it is deeply rooted in the hearts of the people, that the efficiency and environmental management systems are sound, that the equipment is eco-friendly, that the production organization is intelligent and efficient, that the resource elements are seamlessly interconnected, that the environmental pollution is controlled, and that the opening and integration are guided by science and technology.

The construction of a green low-carbon port meets the needs of China's socialist ecological civilization construction and high-quality development at this stage. It is an important part of building a world-class green port and helps the port achieve high-quality development. It is the need for the port industry and even the transportation industry to take the lead in achieving the goal of carbon peak and carbon neutralization.

Furthermore, under China's dual carbon targets, enterprises in the port sector can consider management of the energy varieties and emissions levels of collection and distribution vehicles entering and leaving the wharf, in conjunction with deployment and requirements of carbon peak in the field of transportation. This can be accomplished by utilizing information platforms such as production scheduling systems and gate management systems. Also, they can promote the access of clean energy-driven vehicles to the wharf for operations with preferential collection and distribution of ports, so that the whole supply chain of loading and unloading production and logistics is continuously promoted as low-carbon.

ACKNOWLEDGMENTS

This paper was financially supported by the National Key R&D Program of China (Grant No. 2021YFB2601600; Subject No. 2021YFB2601604); Central-level Nonprofit Scientific Research Institutes Basic Research Project—Special Funds (Grant No. TKS20200205); and Tianjin Transportation Science and Technology Development Plan Project (Grant No. 2019B-10.).

REFERENCES

He, L.L. & Jiao, Y.Q. & Jia, R. & Liang, Y. (2021). Review on the Research Status of Air Pollutant Emission in Port Area in the Development of Green Port J. *Journal of Chongqing Jiaotong University* (Natural Sciences Edition). 40(8), 78–87.

Li, N. (2021). Development of Power Quality Monitoring Technology J. *Journal of Dalian Maritime University* (Social Science Edition). 20(3), 71–78.

Liu, L.L. & Yang, R. & Kang, Z.F. (2021). Research on the Construction Path of Zero Carbon Port J. *Transport Energy Conservation & Environmental Protection*. 17(5), 14–17, 26.

Liu, L.L., & Guo, X. & Ding, J. &, Liu, Z.H. (2019). Study on methodology of voluntary emission reduction of greenhouse gases from shore power projects J, *Journal of Waterway and Harbor.* 40(4), 445–449.

Wang, Y.T. & Tang, G.L. & Yu, J.J. & Yu, X.H. & Zhang Y. (2017). Application of shore power system for ship in container terminals J. *Port & Waterway Engineering*. (9), 103–107.

Energy Revolution and Chemical Research – Kok-Keong Chong and Zhongliang Liu (Eds)
© 2023 The Authors, ISBN: 978-1-032-36554-1

A brief review of thermal management technologies for new energy vehicles

Jiansheng Guo*, Liang Qi & Jiao Suo
Automotive Data of China Co., Ltd., Tianjin, China

ABSTRACT: To optimize resource allocation and reduce the burden of conventional energy, new energy vehicles have been developing rapidly in recent years. However, there are still bottlenecks in the development of new energy vehicles, such as safety concerns, and mileage anxiety. The thermal management of new energy vehicle is not only related to the performance of auto parts but also affect the service life. Given the key issues in the construction of thermal management systems for electric vehicles, conventional thermal management requirements and technologies are sorted in this work. Firstly, the thermal requirements of the battery, crew cabin, and motor drive system are outlined. Secondly, the existing thermal management methods of battery, passenger compartment, and air conditioning system are systematically summarized, and the thermal management control methods of new energy vehicles battery are focused on analysis. Finally, the shortcomings of current research are summarized and a research outlook is carried out, and it is pointed out that low-carbon, integrated and intelligent thermal management technology is the future direction of development.

1 INTRODUCTION

Due to the shortage of fossil energy and the increasingly serious problem of environmental pollution, new energy vehicles, which become the main direction of future development, are becoming more and more popular. What is more, with the promotion of government policies, the development of the new energy vehicle industry has been promoted. Electric motors are seen as a solution to replace internal combustion engine technology, and the power source is mainly provided by the onboard battery pack. However, due to the limited space for loading batteries on the vehicle, the number of batteries required for normal operation is also larger. For the Joule effect, the amount of waste heat will generate with different heat generation rates during the battery's discharge at different rates. If the electric vehicle battery pack fails to dissipate heat in time, the temperature of the battery pack system will be too high or unevenly distributed. As a result, the battery charge-discharge cycle efficiency will be reduced, and the power and energy of the battery will be affected. In serious cases, it will also lead to thermal runaway, affect the service life of the battery pack, and affect the safety and reliability of the system. Hence, EVs and HEVs need to develop an effective technology of effective cooling and heating system to control the temperature of the battery pack system.

Nowadays, considerable research efforts have been devoted to developing an advanced battery thermal management (BTM) system, which can be summarized as an active or passive technology, air or liquid cooling, phase change materials (PCM), and composite technology. Different technologies have different advantages and shortcomings. The selection of a suitable BTM system depends on many factors, including but not limited to volume limitation, installation cost, and work efficiency. Furthermore, the thermal management system of the vehicle is closely related to the comfort of the passengers through the air conditioning system. Passenger comfort is closely related and is also an important influence on consumers' choice basis and experience.

*Corresponding Author: guojiansheng@catarc.ac.cn

DOI 10.1201/9781003332657-12

In this work, the focus is on talking about the effect of temperature on battery performance and further discussing the current status and development direction of thermal management techniques. The last section concludes and provides some suggestions for the development of thermal management systems for new energy vehicles.

2 COMPONENTS THERMAL REQUIRES

2.1 *Battery*

The performance and service life of the battery are closely related to thermal management, as shown in Figure 1. The essence of battery charge and discharge is the complex chemical reaction between positive and negative materials and electrolytes, the ambient temperature will seriously affect the reaction rate of the chemical reaction. Especially, battery aging and capacity degradation in high-temperature environments is an irreversible process.

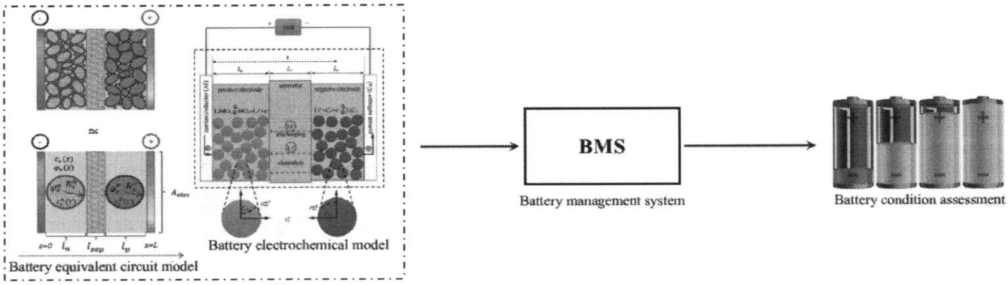

Figure 1. Battery thermal management is closely related to battery performance and service life.

In addition, as a manned tool, safety performance is considered the primary factor. At high temperatures, the electrode materials will arise a variety of exothermic reactions, and the continuous build-up of heat may cause a battery fire. Through the summary of previous studies, it is found that the battery fire is mainly caused by the combustion of alkane, olefin gas, and electrolyte steam generated by the internal reaction of the battery. These alkane and olefin gases are mainly formed by the reaction of electrolyte with embedded lithium and its decomposition reaction at high temperatures. The power cells used in new energy vehicles, which usually made up of hundreds or thousands of batteries, and the increased number will significantly increase the risk of battery failure. Therefore, it is necessary to control the battery temperature reasonably even if the waste heat is removed, so that the battery can operate within the appropriate temperature range as far as possible. This is where low-temperature thermal management comes in.

2.2 *Crew cabin*

Research shows that when people are in a thermal comfort environment, their thinking and reaction capacity are in the best condition. Therefore, the thermal comfort of the crew cabin has always been one of the focuses of vehicle thermal management. Usually, the heat load of new energy vehicles mainly comes from five aspects: fresh air heat load, envelope heat load, personnel heat load, solar radiation heat load, and equipment heat load (Liu 2014). However, for the non-uniform environment in the car, it is necessary to measure not only the thermal sensation of the whole body but also the local thermal sensation of the body. Thus, based on the heat balance equation, the heat balance equation of the human comfortable state is shown as follows, which is not considered the external work done by the human body:

$$M = Q_t = Q_{res} + E_s + Q \tag{1}$$

$$Q_{res} = 1.7 \times 10^{-5} M\,(5867 - P_a) + 0.0014M(34 - t_a) \tag{2}$$

$$E_s = 3.05 \times 10^{-3} (5733 - 6.99M - P_a) + 0.42(M - 58.15) \quad (3)$$

where M is human metabolic heat production, Q_t is total heat exchange between human and environment, E_s is the heat exchange of human skin evaporation, Q is the convective radiative heat exchange between human body and environment, and P_a is the partial pressure of ambient water vapor.

The relationship between the average temperature of human skin (t_{sk}) and the total heat dissipation in the comfort condition (Q_t) is related as follows:

$$t_{sk} = 35.77 - 0.028Q_t \quad (4)$$

$$t_{sk} = 36.4 - 0.054Q \quad (5)$$

The equivalent space temperature can be calculated according to the heating power and surface temperature of each area of the body heat, and finally, the comfort of each part can be evaluated by the equivalent space temperature.

3 CURRENT SITUATION OF THERMAL MANAGEMENT TECHNOLOGIES

3.1 *Heat dissipation of power battery*

Generally, the research about heat dissipation of power batteries could be divided into air cooling, liquid cooling, solid phase change materials heat dissipation, and heat pipe heat dissipation. The first three techniques are now commonly used in vehicle thermal management of power batteries.

3.1.1 *Air cooling*

Air cooling is one of the earliest methods, which can be divided into natural convection heat transfer and forced convection heat transfer. The traditional method is to increase the heat exchange area by increasing the number of heat exchange pipes. However, this method will increase the cost and the volume of the heat exchanger. Thus, to effectively improve the heat exchange per unit area, fins are often installed on heat exchanger pipes. Part of the research focuses on geometric parameters such as pipe diameter, number of tube rows, and fin spacing. More attention was paid to the heat transfer and flow mechanism of the heat exchanger. Saboya and Sparrow used mass transfer technology to determine the local mass transfer coefficient of one row, two rows, and three rows of coils. The results illustrate the distribution of the local heat transfer coefficient in the form of heat transfer coefficient through analogy. Mahamud and Park adopted reciprocating air flow by using a flip gate valve and unique pipeline to improve the temperature distribution of traditional unidirectional flow. Two dimensional CFD model and lumped capacitance thermal model were used in the simulation (Hong 2018). Park et al. designed an air-cooling system with a conical manifold to meet the cooling performance of prismatic batteries without changing the layout of the existing battery system (Lu 2018). Wang et al. analyzed the thermal performance of the battery pack under different air-cooling positions and concluded that when the fans sit on the top of the battery pack, it owns the best heat dissipation performance.

3.1.2 *Liquid cooling*

Liquid cooling is to use fluid material with high thermal conductivity to directly or indirectly contact the battery for heat dissipation. Compared with air cooling, liquid cooling owns the advantages of high convective heat transfer coefficient and low velocity. Commonly used liquid coolants are deionized water, a mixture of water and ethylene glycol, and nano-fluids. Direct contact liquid cooling requires the coolant to directly contact the battery, so the coolant used is required to be insulated and have high thermal conductivity, commonly including silicon-based oil, mineral oil, etc. However, direct contact liquid cooling may have problems such as poor contact between the coolant and the battery, which could not solve the heat dissipation well. Indirect contact liquid cooling has no insulation requirements and often uses deionized water or a mixture of water and glycol as coolant. The shortcoming of indirect contact liquid cooling is very obvious: the weight of the coolant and condenser pipe increases the load of the electric vehicle, the heat dissipation system is more complex, and the maintenance becomes more difficult.

3.1.3 *Phase change heat transfer*

The common phase change material could release or absorb a large amount of latent heat during phase change, which can effectively absorb the heat of the battery pack. The PCM for battery thermal management is paraffin. Paraffin, which is a common product of the petrochemical industry, owns the advantage of low price and easy access and has stable chemical properties in the range of hundreds of degrees Celsius. However, the thermal conductivity of paraffin is not high, so it is not suitable for working conditions requiring rapid heat dissipation. Thus, light, and high thermal conductivity materials such as foam aluminum and foam copper are often selected to improve thermal conductivity. Compared with liquid cooling heat transfer, phase change heat transfer usually immerses the battery pack in phase change materials such as nano-fluids. It could meet the cooling requirements of various shapes and types of batteries. Ling et al. (Ling 2014) studied the performance of PCM based thermal management system, and the simulation results are in good agreement with the experimental data.

3.2 *Thermal comfort of the crew cabin*

Thermal comfort has been defined by Fanger as 'The condition of mind which expresses satisfaction with the thermal environment'. Most studies concerning the thermal radiation comfort of passengers in vehicles can be divided into environmental conditions inside the vehicle that affect the human thermal comfort and the human's response to and perception of its interaction with the environment, the evaluation for the perception of human thermal comfort as shown in Figure 2. To explain human interaction with the environment, many thermoregulatory models and manikins have been developed, which could be divided into physiological or psychological models.

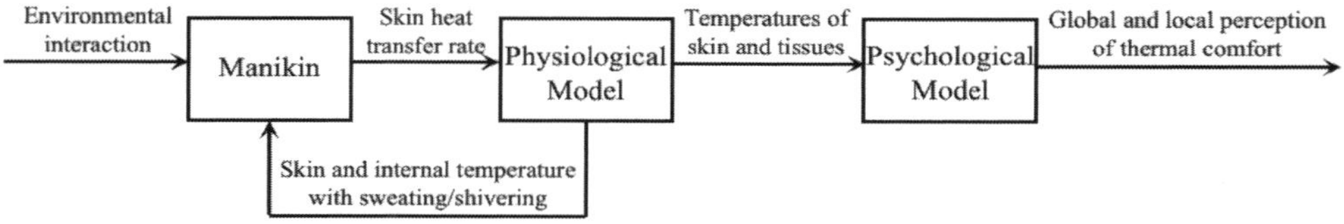

Figure 2. Block diagram for evaluating the perception of human thermal comfort.

The thermal comfort psychological models could be summarized as two methods. The first one is a direct method that analyzed the environmental data and human thermal comfort responses to the environment through a prediction equation, and the second method is to model the physiological responses of humans to the environmental conditions and then connected with thermal sensation. Thermal sensation depends on heat source temperature, heat transfer coefficient, the rate of heat transfer of the body, and human responses. The predicted mean vote (PMV) developed by Fanger and the humid operative temperature developed by Gagge are the most well-known human thermal psychological models, which could apply in uniform steady-state environments. PMV represents the average of the cold and hot feelings of most people in the same environment, and its value can be in the range of -3 to $+3$, and a zero value stands for the optimum thermal comfort situation. The detailed index is calculated according to the heat balance of the human body, which need to consider the metabolic rate of human activities, the thermal insulation value of clothing, and environmental factor, such as ambient temperature, ambient humidity, and air convection velocity, etc. However, PMV has high accuracy under steady-state conditions and has limitations for an unsteady environment. The Gagge model regards the human body as a cylinder with a two-layer structure, the outer layer stands for the skin structure, and the inner layer stands for internal organs. Though Gagge's model introduces the concept of skin moisture, which fully considers the heat taken away by sweat evaporation. However, the model is only used in the environment of low human activity, light clothing, and natural convection. Thus, lots of studies have been devoted to refining the model through theoretical, experimental, and simulation studies. Considering the differences in metabolism, blood flow rate, and heat transfer coefficient, parts of the studies focused on the relationship between thermal sensation and different parts of the human trunk. Huang et al found that the head and foot have the greatest impact on the overall thermal comfort. Further, the thermal comfort of different age groups and gender were studied and found that older people have lower thermal sensitivity, poorer thermoregulation, and higher thermal expectation temperature, due to the decline in body functions, slow metabolic generation, and other reasons. Due to skeletal muscle, body fat percentage differences, and metabolic rate differences, the average skin temperature is 0.2–0.4 degrees Celsius higher in men than in women. At the same activity level, men produce more heat and tend to feel hotter. Women are more likely to accumulate heat, resulting in faster skin warming. What is more, from a psychological viewpoint, women are more likely to be dissatisfied with the hot environment. Based on human activity, clothing thermal resistance, and major meteorological parameters, physiological equivalent temperature (PET) parameters are proposed. The temperature and humidity in the driver's cabin were analyzed by numerical simulation method in different seasons, extreme weather, and different clothing thickness conditions.

3.3 *System control of thermal management*

The main components of the thermal management control system of new energy vehicles are sensors, actuators, and controllers. Most of the sensors are composed of temperature sensors and pressure sensors, while the actuator takes the electric compressor and electronic expansion valve as the core, and also includes HVAC (heating ventilation air conditioning) blower, cooling fan, electronic water pump, and other peripheral parts. Most of the sensors are composed of temperature sensors and pressure sensors, while the actuator takes the electric compressor and electronic expansion valve as the core, and also includes HVAC (heating ventilation air conditioning) blower, cooling fan, electronic water pump, and other peripheral parts.

Compared with conventional fuel vehicles, new energy electric vehicles should not only focus on comfort but also on energy efficiency. Common feedback control methods for thermal management of new energy vehicles include start-stop control, PID (proportion integration differentiation) control, model predictive control (MPC), global MPC control, and control combined with intelligent algorithms. The start-stop control is relatively simple, which is directly controlled through the start and stop of the system. However, its control action is not continuous, which could cause system vibration easily and be difficult to achieve energy consumption control. Thus, directly controlled will not be described in detail.

3.3.1 *PID control*

PID control, one of the earliest developed control strategies, has the advantage of algorithms simply, good robust, and reliably, which has been used in the field of automotive thermal management and control systems for a long time. The principle of the PID controller is that the control system will switch the heat pump circuit to heating or cooling depending on the outside temperature, monitor the battery pack, cabin temperature, or other modules, and feedback to the PID controller according to the difference with the preset temperature. Compared with the start-stop control method, PID control could change accurately when the working condition changes linearly. However, the thermal management of new energy vehicles is a nonlinear thermodynamic system, PID control will have the problem of control instability such as oscillation during the variable working conditions. To deal with this problem, considering that the thermodynamic properties of the different thermal management sub-modules are coupled with each other, the integral scaling parameters of the PID control also need to cooperate with each other, but the PID control system will become extremely complex. Thus, PID control is further combined with intelligent algorithms such as fuzzy control and neural networks.

3.3.2 *Model predictive control*

Model predictive control (MPC) is a forward control method based on model prediction. The basic control process mainly consists of model building, prediction development, control command, feedback regulation control command, and feedback adjustment. Compared with PID control, MPC could easily handle multi-input and multi-output systems and has the characteristics of high stability, fast response, and strong optimization ability.

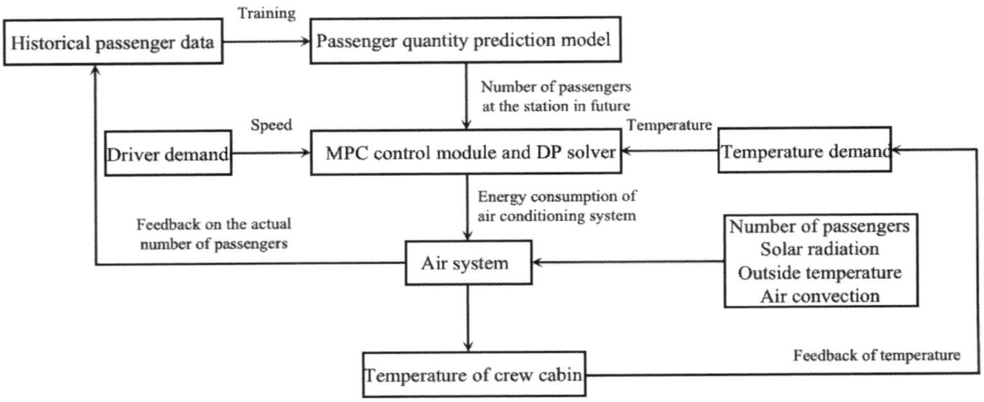

Figure 3. Model predictive control block diagram, taking temperature control as an example.

MPC has been applied in the thermal management field of new energy electric vehicles. but MPC control is mainly local application or combined with intelligent algorithms because the thermal management system of new energy vehicles is complex and has many model variables and MPC control relies on the model. Xia et al. used the MPC control strategy for local control of battery heating, and the constraints of model predictive control are the accuracy of battery temperature control and its temperature range under different operating conditions. The economical objective of the optimization is the power consumption of PTC heating, and a relaxation factor is introduced to establish a soft constraint to prevent the phenomenon of no feasible solution due to a fixed constraint. Mohammad et al. (Panchal 2017) proposed a control scheme for a smart car thermal management system with brake braking energy recovery. The test results show that compared with the simple feedback control model, the MPC control model and the global optimal control model can save 33.69% and 35.58% of system energy consumption by using brake braking energy and coordinating water pump and fan speed, respectively. The test results show that compared with the simple feedback control model, the MPC control model and the global optimal control model can save 33.69% and 35.58% of system energy consumption by using brake braking energy and coordinating water pump and fan speed, respectively. This energy-saving effect increases significantly with the increase in vehicle weight. For new energy vehicles, thermal management systems that need to consider more complex global variable objective multivariable control, especially simulation objectives for cabin temperature, battery, motor, and electronic control temperature need to involve road information, personal information, external loads, and other factors, offline optimization and other means are needed to reduce the model predictive control on the computational volume and dependence of computational efficiency. Combining with a fuzzy neural network to build the model off-line could avoid the failure of predictive control caused by too much calculation of global optimization. However, simple off-line optimization will cause deviations between the model and the actual situation. Therefore, the update frequency of the prediction model, the selection of the prediction domain, and the trade-off between MPC control and the thermal inertia of the thermal system itself are the problems that need to be further solved in the global application of MPC control in the thermal management system of new energy vehicles.

3.3.3 *Control based on intelligent algorithm*

In addition to the two classical control methods, the thermal management of new energy vehicles is paying more and more attention to comfort and energy management, such as machine-learning algorithms based on user characteristics and habits, etc. Based on PID feedback control and MPC predictive control, more control methods that combine specific intelligent algorithms have been derived. The fuzzy PID control enables self-learning intelligent control strategies based on different user habit characteristics.

The user's demand for car comfort is getting higher and higher, according to the user's characteristics, weather conditions, operating conditions, etc., will gradually be incorporated into the design of thermal management control algorithms, and the thermal management of new energy vehicles has the characteristics of highly non-linear and strong system coupling. Intelligent control methods combined with intelligent algorithms are still the key direction for future research and development.

4 CONCLUSIONS

With the in-depth development of the automotive industry, electrification, intelligence, and network connectivity will become the main direction of the development of the automotive industry. In addition to the need to improve battery performance and increase driving range, the thermal management of new energy vehicles also should be more energy-saving and low-carbon, intelligent, and function centralized.

4.1 *Low carbon*

Green and low-carbon automobile thermal management is not only the general trend of the development of the automobile industry but also the demand for China's national strategy of emission peak and carbon neutrality. Next-generation refrigeration technology should, on the one hand, comply with national standards, regulations, and specific policies on environmental pollution and climate warming; on the other hand, it should also meet the evolution of the inherent demand characteristics of new energy vehicles and the diversity of functions required in different regions, such as the new energy vehicle air conditioning system should improve the refrigerant adjustment range, precision, could be personalized according to user habits.

4.2 *Modularization*

The structure of the traditional car is relatively simple, while the complexity of the thermal management function requirements leads to a significant increase in the number of components and excuses of the thermal management system of the new energy vehicle, the decentralized installation of parts will cause the structure to be redundant, the cost will also increase the possibility of vibration, noise, and other problems. Thus, functional modularity and device integration are urgent needs driven by the rapid development of new energy vehicles and batch industrialization of thermal management.

4.3 *Intelligent*

The future thermal management of new energy vehicles tends to be integrated and intelligent, and the amount of control to be designed and the amount of data to be processed increases significantly, so relying on traditional control methods will increase costs and make it difficult to achieve high accuracy. Therefore, the new energy vehicle thermal management which is based on the MPC control method should be combined with intelligent algorithms to meet the needs of individual users and diversified road conditions, and also should have the ability of fast response, high precision, and diversified predictions.

REFERENCES

A, M.K., et al., Reducing auxiliary energy consumption of heavy trucks by onboard prediction and real-time optimization-ScienceDirect. *Applied Energy*, 2017. 188: p. 652–671.

Hong, S., et al., Design of flow configuration for parallel air-cooled battery thermal management system with secondary vent. *International Journal of Heat and Mass Transfer*, 2018. 116(JAN.): p. 1204–1212.

Li, L., Design and Experiment of a Heat Pump Air-conditioning System for Electric Vehicles. *Journal of Refrigeration*, 2013. 34(3): p. 60–63.

Ling, Z., et al., Experimental and numerical investigation of the application of phase change materials in a simulative power batteries thermal management system. *Applied Energy*, 2014. 121: p. 104–113.

Liu, Z., et al., Shortcut computation for the thermal management of a large air-cooled battery pack. *Applied Thermal Engineering*, 2014. 66(1–2): p. 445–452.

Lu, Z., et al., Parametric study of forced air-cooling strategy for lithium-ion battery pack with staggered arrangement. *Applied Thermal Engineering*, 2018. 136: p. 28–40.

Panchal, S., et al., Thermal design and simulation of mini-channel cold plate for water-cooled large-sized prismatic lithium-ion battery. *Applied Thermal Engineering*, 2017. 122: p. 80–90.

Peng, F., et al., Effects of different ambient temperatures on performance of electric vehicles' heat pump air conditioning. *Journal of Beijing University of Aeronautics and Astronautics*, 2014. 40(12): p. 1741–1746.

Taniguchi, Y. and K.P. Labs, *Topics Simulation Models for Predicting Car Occupants' Thermal Comfort*. 2001.

Yin H. *PMSM loss calculation and temperature field analysis* (in Chinese). Master Dissertation. Harbin: Harbin Institute of Technology, 2015.

Energy Revolution and Chemical Research – Kok-Keong Chong and Zhongliang Liu (Eds)
© 2023 The Authors, ISBN: 978-1-032-36554-1

Simulation study on minimum spatial relation of electric field for sensor size

Yudan Wang & Li Zhang*
Department of Mechanical and Electrical Engineering, Wenhua College, Wuhan, China

Guo Zhao & Haoran Song
School of Electrical and Electronic Engineering, Hubei University of Technology, Wuhan, China

ABSTRACT: As the power frequency electric field is measured in a power field environment that is a narrow space, such as a balcony, greenhouse, or grape trellis, the distortion caused by placing the electric field measurement probe in the measured space will alter the charge distribution on the nearby metal frame, resulting in an inaccurate measurement. In this paper, we study the influence of different sensor sizes on electric field distortion. ANSYS is used to simulate the relationship between the sensor size and the minimum space size suitable for measurement, and the influence of the sensor size on the electric field distortion is analyzed. We also get the expression for the quantitative relationship between the sensor size and the minimum spatial size.

1 INTRODUCTION

The measurement of the electric field strength has always been the focus of electromagnetic environment monitoring of power transmission and transformation projects (Wang 2019; Xue 2019; Zhao 2018; Zhou 2019). Many experts and scholars have analyzed and studied the influence of different factors such as effector placement (Wei 2019), ambient temperature and humidity (Zhai 2020), and the human body (Yu 2013) on electric field strength measurement results. In addition, many experts have proposed different design schemes for electric field strength measurement systems, which have been applied to a certain extent (Chen 2016; Deng 2020; Liu 2021; Wang 2020).

Usually, the measurement of electric field intensity is in the line corridor and nearby areas. These areas are relatively open and there is no metal structure nearby. After the electric field probe is put in, the electric field distortion caused by the probe will not affect other metal structures, so it will not change the original value of the measured electric field. The electric field measurement results are reliable. In the measurement of the balcony, greenhouse, and grape frame, due to the narrow space, the distortion caused by the insertion of the electric field probe will affect the charge distribution on the nearby metal frame, so the size of the original electric field in the measured area is changed, resulting in large measurement error.

At present, due to the distortion of the electrode to the measured field, to solve the coupling effect of the electric field distortion caused by the detection of electrodes with different components, the current applied electric field sensor probe has a large size and complex system. The distortion range caused by the probe is proportional to the probe size. The smaller the probe is, the smaller the distortion range is. In addition, the smaller the probe electrode is, the smaller the electric field difference within the electrode-enclosed area, and the lower the degree of nonuniformity, leading to a more accurate measurement of the nonuniform electric field. Usually, the error of engineering measurement should be controlled by $\pm 10\%$, and the current conventional probe size is more than

*Corresponding Author: 371017343@qq.com

DOI 10.1201/9781003332657-13

10 cm. When measuring in narrow spaces such as balconies, vegetable greenhouses, and grape frames, the error exceeds the engineering requirements.

Based on the above background, this paper uses ANSYS simulation software to quantitatively analyze the relationship between the size of the sensor probe and the minimum space of the measured electric field and obtains the corresponding mathematical relationship expression, which can provide a basis for the size requirements of the electric field measurement probe with different space sizes.

2 CALCULATION AND MEASUREMENT OF ELECTRIC FIELD STRENGTH

2.1 *Calculation of electric field intensity*

(1) Calculation of electric field intensity of vacuum midpoint charge

The electric field intensity of the point charge can be calculated according to Equation (1).

$$E = kQ/r^2 \tag{1}$$

where k is the electrostatic constant, $k = 9.0 \times 10^9 \mathrm{N} \cdot \mathrm{m}^2/\mathrm{C}^2$, Q is the electric quantity of the charge, and r is the distance to the charge.

(2) Calculation of uniform electric field strength

The calculation formula of uniform electric field intensity is shown in Equation (2).

$$E = U/d \tag{2}$$

where U is the potential difference between two points, and d is the distance along the direction of the electric field.

(3) Calculation of electric field strength between parallel plate capacitors

Let the charge density of each side of the parallel plate A and B be $+\sigma$ and $-\sigma$, the dielectric constant between plates is $\varepsilon_0\varepsilon_r$, and the distance is d. E_A and E_B are electric field strength of Plate A and Plate B, respectively.

1) The electric field intensity outside the two plates is:

$$E_\mathrm{A} + E_\mathrm{B} = 0 \tag{3}$$

2) The electric field intensity between the two poles is:

$$\begin{cases} E_\mathrm{A} + E_\mathrm{B} = E \\ E_\mathrm{A} = E_\mathrm{B} = \sigma/\varepsilon_0\varepsilon_r \\ E = 2\sigma/2\varepsilon_0\varepsilon_r = \sigma/\varepsilon_0\varepsilon_r \end{cases} \tag{4}$$

2.2 *Electric field intensity measurement*

The principle of the electric field sensor is that when the conductor is in an electric field (electrostatic field or alternating electric field), there will be induced charges on the surface of the conductor, which is in an electrostatic equilibrium state with the external field strength. And the larger the external field strength, the more induced charges on the conductor surface. The induced charge is added at both ends of the capacitor, and the potential difference is formed at both ends of the capacitor to produce the induced voltage. By measuring the voltage at both ends of the capacitor, the electric field intensity can be determined based on the relationship between the voltage and the electric field intensity.

Figure 1 shows the structure diagram of a one-dimensional spherical sensor, which consists of two hemispherical metal electrodes, an internal capacitor, and a connecting wire. The resulting wire is used to access the conditioning circuit of the measuring instrument. In space, the induced charges appear on the two metal electrodes in the electric field of the electric field intensity E, and

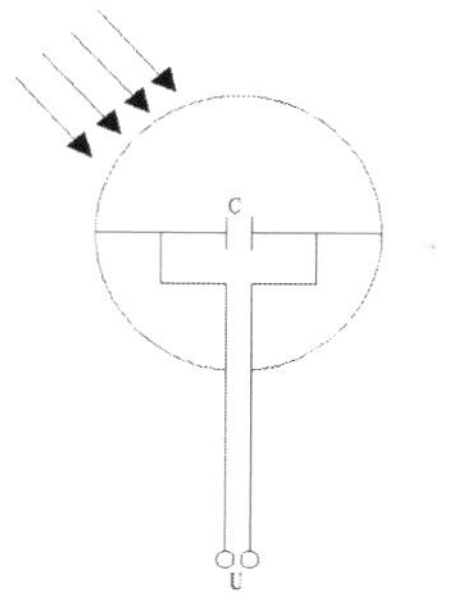

Figure 1. Spherical electric field sensor.

the two electrodes are connected by the capacitor C so that the two ends of the C will produce the potential difference E. E can be used as a measurement signal. The relationship between E and U is shown in Equation (5).

$$U = \frac{-3\pi \varepsilon R^2}{C} E \tag{5}$$

where R is the radius of the ball on the spherical sensor and ε is the dielectric constant.

3 SIMULATION METHOD AND MODEL

To study the influence of different sensor sizes on electric field distortion, this paper uses ANSYS to simulate the relationship between sensor size and the minimum space size suitable for measurement, analyzes the influence of sensor size on electric field distortion, and determines the relationship between sensor size and the minimum space size suitable for measurement. The simulation model is shown in Figure 2.

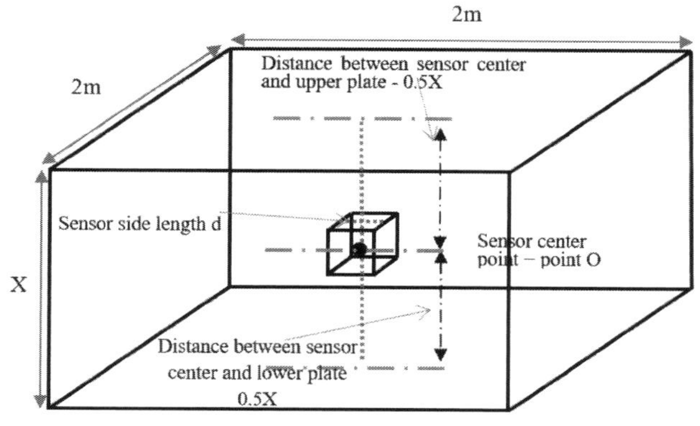

Figure 2. Sensor simulation model.

The model shown in Figure 2. shows the electric field distribution of the metal cube placed in the center of a rectangular air area (the center of the cube is located at 1/2 of the length, width, and height of the rectangular area). The edge length of the metal cube sensor is d. The rectangular area is a square of 2 m * 2 m above and below, and the height is X (in m). Assuming that the sensor center is zero O, the distance between the sensor center and the upper plate is −0.5 X.

Simulation conditions: identify the left and right sides of the cuboid, set data, simulate in the symmetry plane, and do not apply any further boundary conditions. The upper surface voltage is X*1000 V and the lower surface voltage is 0 V (that is, the electric field in the control area is 1000 V/m). All nodes of the metal body are voltage coupled to form a suspended metal body.

4 SIMULATION RESULT

(1) Sensor side length d is 0.02 m

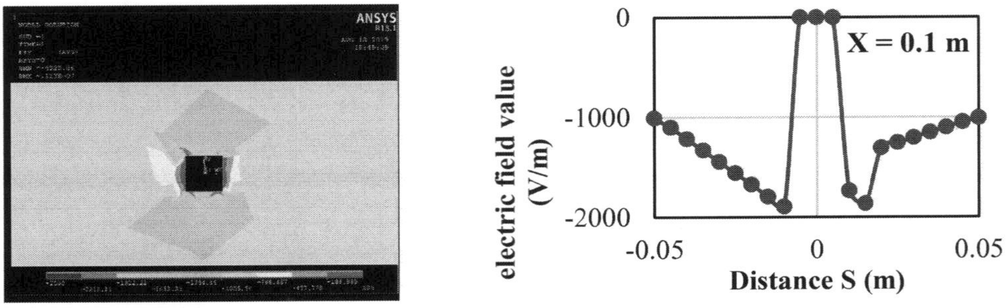

Figure 3. d = 0.02 m, X = 0.1 m.

It can be seen from Figure 3 that the maximum distortion value of the electric field caused by the introduction of the sensor is 1.89 kV/m, the position is 0.01 m away from the upper of the sensor center, and the electric field value is 1.008 kV/m at the position of the upper and lower plates.

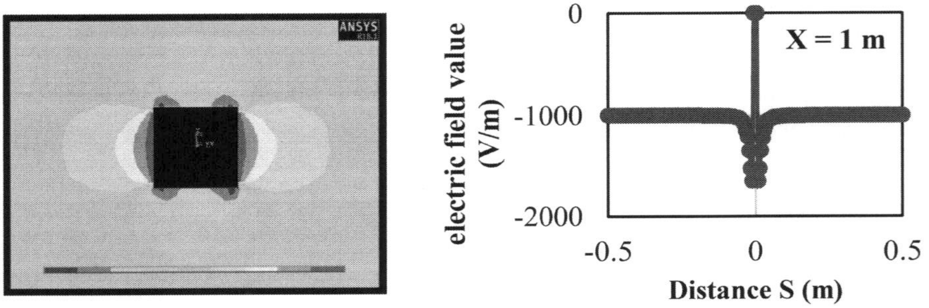

Figure 4. d = 0.02 m, X = 1 m.

Figure 4 shows that the maximum distortion of the electric field caused by the introduction of the sensor is 1.658 kV/m, the position is 0.01 m from the center of the sensor to the lower plate, and the electric field value is 1 kV/m at the position of the upper and lower plates.

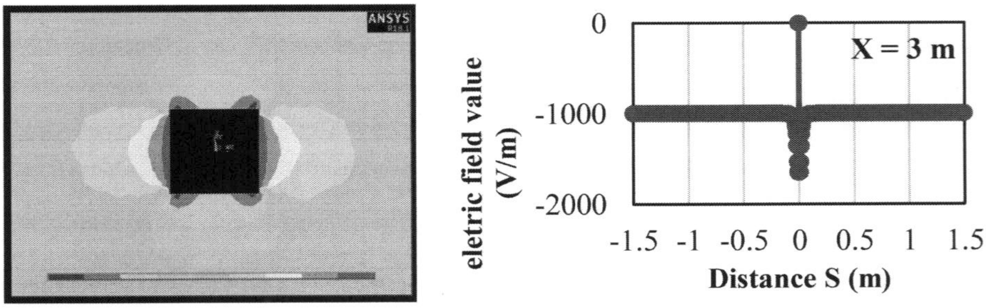

Figure 5. d = 0.02 m, X = 3 m.

Figure 5 shows that the maximum distortion value of the electric field caused by the introduction of the sensor is 1.654 kV/m, the position is 0.01 m from the center point of the sensor to the upper and lower plates, and the electric field value is 1 kV/m at the position of the upper and lower plates.

(2) Sensor side length d is 0.09 m

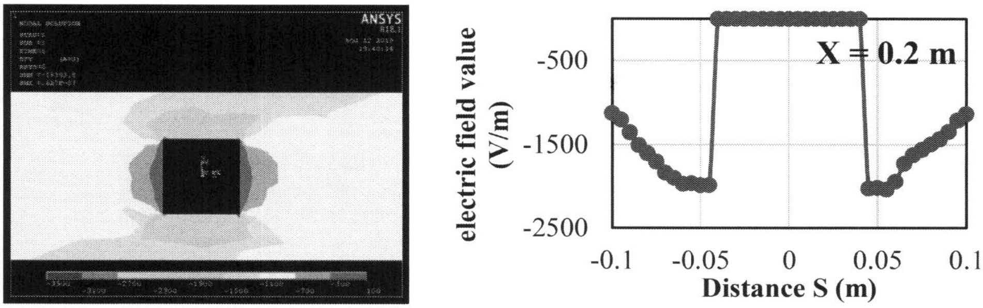

Figure 6. d = 0.09 m, X = 0.2 m.

It can be seen from Figure 6 that the maximum electric field distortion caused by the introduction of the sensor is 2.025 kV/m, and the position is 0.045 m from the center point of the sensor to the lower plate. The electric field value at the position of the lower plate is 1.138 kV/m, and the upper plate is 1.129 kV/m.

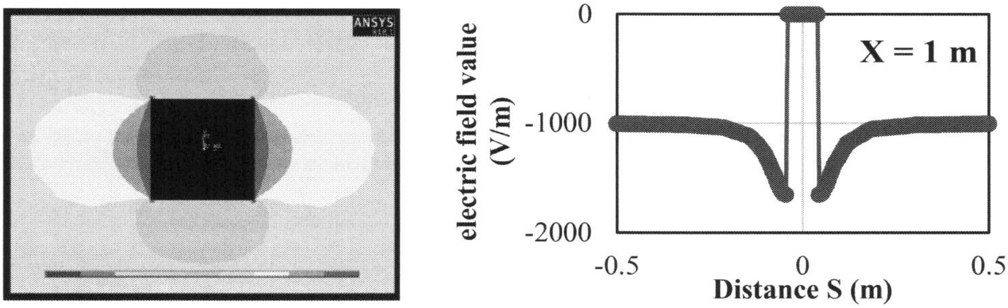

Figure 7. d = 0.09 m, X = 1 m.

It can be seen from Figure 7 that the maximum distortion value of the electric field caused by the introduction of the sensor is 1.655 kV/m. The position is 0.045 m from the center point of the sensor to the lower plate, and the electric field value is 1.007 kV/m at the position of the upper and lower plates.

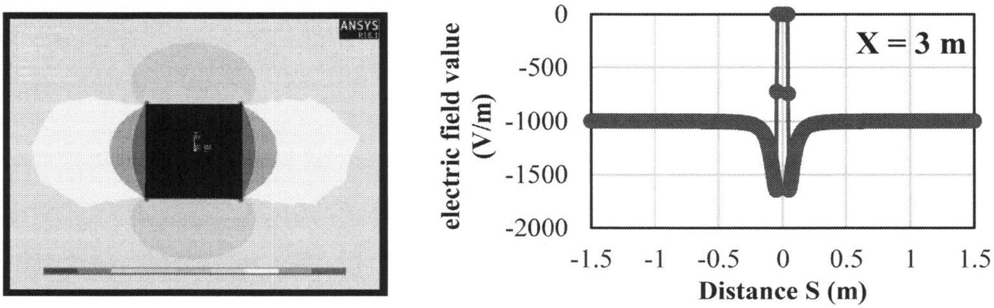

Figure 8. d = 0.09 m, X = 3 m.

It can be seen from Figure 8 that the maximum distortion value of the electric field caused by the introduction of the sensor is 1.646 kV/m, the position is 0.05 m from the center point of the sensor to the upper plate, and the electric field value is 1 kV/m at the position of the lower plate.

5 ANALYSIS OF SIMULATION RESULTS

Through the simulation analysis of different sensor side lengths and cuboid height combinations, the maximum electric field distortion value caused by the introduction of the sensor and the electric field value at the position of the upper and lower plates are obtained. The two indexes below are used to analyze and judge whether the size of the sensor affects the distortion of the electric field:

(1) Through analysis and comparison, the maximum distortion value of the electric field of the three size sensors is 1.65 kV/m when the height of the cuboid is 2 m. Considering that the height of the cuboid of each size is 2 m as the calibration size, the percentage of the maximum distortion value of the other sizes is compared and analyzed, which is expressed by S. When the percentage is less than 5%, it is not considered to affect the electric field distribution on the surface of the probe:

$$S = \frac{\text{MDEF-SDEF}}{SDEF} * 100\% \leq 5\% \tag{6}$$

where MDEF is the maximum distorted electric field value and SDEF is the 2 m standard distorted electric field value.

(2) Considering the error effect of the simulation model, it is considered that the electric field value at the upper and lower plates is not effected within the error range of 5%. The error between the maximum value of the upper and lower plates and the standard value is expressed as P:

$$P = \frac{\text{MEF-SEF}}{SEF} * 100\% \leq 5\% \tag{7}$$

where MEF is the maximum electric field on the surface of the plate and SEF is the standard electric field between the plates.

Table 1 shows the results of the combination calculation of different sensor side lengths and cuboid heights.

Table 1. The relationship between the side length of the sensor and the height of the cuboid.

Sensor side length D (m)	Height of cuboid X (m)	Maximum distortion value (kV/m)	S (%)	Maximum value of upper and lower plates (kV/m)	P (%)
0.02	0.1	1.89	14.13	1.008	0.8
	0.2	1.654	−0.12	1.001	0.1
	0.4	1.664	0.48	1.001	0.1
	0.6	1.675	1.15	1	0
	1	1.658	0.12	1	0
	3	1.654	−0.12	1	0
0.05	0.1	2.236	35.27	1.098	9.8
	0.2	1.748	5.75	1.023	2.3
	0.4	1.668	0.91	1.021	2.1
	0.6	1.658	0.3	1.006	0.6
	1	1.654	0.06	1.001	0.1
	3	1.653	0	1	0
0.09	0.2	2.025	22.43	1.138	13.8
	0.4	1.688	2.06	1.118	11.8
	0.6	1.666	0.73	1.034	3.4
	1	1.655	0.06	1.007	0.7
	3	1.646	−0.004	1	0

The analysis of Table 1 shows that the sensor size and the minimum space size suitable for measurement are within 5% of S and P parameters. The results are as follows:

Table 2. The relationship between the side length of the sensor and the height of the smallest cuboid.

Sensor side length d (m)	Height of cuboid X (m)
0.02	0.2
0.05	0.4
0.09	0.6

The relationship curve shown in Figure 9 is obtained by analyzing Table 2.

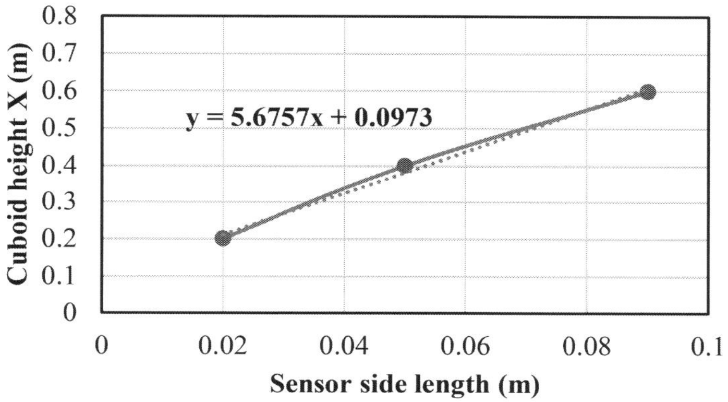

Figure 9. The relationship between the sensor size and the smallest accurate measurement space size.

Through trend simulation, we can get the relationship between the sensor side length and the minimum cuboid, as shown in Equation (8).

$$y = 5.6757x + 0.0973 \quad (8)$$

6 CONCLUSIONS

In this paper, the influence of different sensor sizes on the electric field distortion is studied. ANSYS software is used to simulate the relationship between the sensor size and the minimum spatial size suitable for measurement.

(1) The maximum distortion percentage S and the error P between the maximum value of the upper and lower plates and the standard value are defined to determine whether the sensor size affects the distortion of the electric field. If S is less than 5%, it is considered that the electric field distribution on the probe surface is not affected; if P is within 5%, it is considered that the electric field value at the upper and lower plates is not affected.

(2) Through the ANSYS simulation of the spatial height of the measured electric field under different sensor sizes, the corresponding electric field distribution nephogram and electric field value distribution map along the vertical section of the sensor are obtained. The simulation results are analyzed and the trend fitting is carried out, and the quantitative relationship between the sensor size shown in Equation (8) and the minimum spatial size suitable for measurement is obtained.

REFERENCES

Aiming, Chen. Jing, Zhou. Lin, Zhou. Xisheng, Li. & Zhen, Lv. (2016). Research on the wearable gauge for the power-frequency electric field. *J. Electrical Measurement & Instrumentation*. 53(16), 124–128.

Bin, Zhai. Jian, Ning. Jianhui, Li. Yigang, Ma. Wei, Wang. & Zhibin, Zhao. (2020). Influence of Environmental Humidity on the Measurement Results of Space Electric Field Under the Transmission Line. *J. Southern power system technology*. 14(03), 17–22.

Daojing, Wang. Xiaohu, Zhu. Min, Yin. Wei, Yan. & Tong, Liu. (2019). Research on measurement method of space electric field in cross transmission area of AC transmission lines. *J. Electronic Design Engineering*. 27(20), 79–83.

Guanghui, Wei. Kaifu, Ji. & Lingyuan, Meng. (2019). The Effect of EUT on Environment Electromagnetic Radiation E–Field Test. *J. Transactions of Beijing Institute of Technology*. 39(11), 1173–1179.

Luxing, Zhao. Xiang, Cui. Jiayu, Liu et al. (2018). Discussion on Measurement Method of Total Electric Field for DC Transmission Lines. *J. Proceedings of the CSEE*. 38(2), 644–652.

Mengting, Yu. Jingang, Wang. & Jian, Li. (2013). The Experimental Study on the Human Body Influence on Measurement of High Voltage Power Frequency Electric Field. *J. Electrical Measurement & Instrumentation*. 50(06), 24–27+48.

Xiaocui, Xue. (2019). Study on the measurement and calculation method of the intensity of induced electric field. *J. Audio Engineering*. 43(11), 76–77.

Yifei, Liu. Liang, Ma. Yinhui, Cheng. Wei, Wu. Jinghai, Guo. & Mo, Zhao. (2021). Pulse Electric Field Measurement System with Sensitivity Self-calibration Based on Optical Fiber Transmission. *J. High Voltage Engineering*. 47(04), 1478–1484.

Yifei, Wang. Mingxiang, Huang. Hongzhi, Bian. Dongyang, Yang. & Zhengcai, Fu. (2020). Design of power frequency electric field measurement system based on Wi-Fi communication. *J. Industrial Instrumentation &amp*. 2, 22–27.

Yongqiang, Deng. Jing, Chen. Cong, Zheng. & Chunhua, Peng. (2020). Optimization Design of Electric Field Measuring Instrument Holder Based on Finite Element Simulation and Differential Evolution Algorithm . *J. Smart Measurement*. 48(2), 71–77.

Zishu, Zhou. (2019) *Research on Measurement Method of Synthetic Electric Field in UHV Transmission Engineering*. D. North China Electric Power University.

Energy Revolution and Chemical Research – Kok-Keong Chong and Zhongliang Liu (Eds)
© 2023 The Authors, ISBN: 978-1-032-36554-1

Analysis and research on the influencing factors and influencing mechanism of environmental protection facilities in power transmission and transformation project

Xiaoyan Yu, Min Yu, Xiaoyong Yang & Fuyan Liu*
Economic and Technological Research Institute, State Grid Zhejiang Power Co., Ltd., Hangzhou, China

ABSTRACT: There is presently a lack of good relationships between power grid enterprises' transmission and transformation projects' environmental assessment reports, environmental protection investment saving plans, and engineering and economic documents, and the accounting of environmental protection investment in transmission and transformation projects is unclear, which hinders enterprises from having a full understanding of environmental protection and energy saving investments. Based on the analysis of the cost of environmental protection facilities in substation projects, this paper proposes an analysis method of factor influence mechanism based on the Pearson method and principal component analysis method, which provides a reference for power grid companies to improve the efficiency of environmental protection costs in power transmission and transformation projects.

1 INTRODUCTION

With the rapid development of power grid construction of power grid enterprises and the increasingly strict supervision policies of environmental protection and water conservation, strengthening the investment efficiency of environmental protection facilities has become a new requirement in the environmental protection management of power transmission and transformation projects.

Wu et al. (Wu et al. 2020) starts from the analysis of laws and regulations and the path selection of power transmission and transformation projects, and summarizes the influence of environmental and water conservation constraints on the site selection and line selection of power transmission and transformation projects. It is recommended to fully use satellite remote sensing monitoring, UAV aerial photography, and three-dimensional technologies such as digital models, especially the application of a map of environmental protection for engineering design and management, are used to prevent the occurrence of malignant events of environmental protection and water conservation. He (He 2021) advocates environmental protection and soil and water conservation as an indispensable part of engineering construction, so as to rationally develop natural resources, reduce the impact on the ecological environment, and avoid man-made soil erosion. Based on mobile Internet, machine learning, UAV oblique photography, and other technologies, and following the supervision idea of "air and space integration", Li et al. (Li et al. 2021, 2020; Zhao et al. 2020) explores high-definition satellite images to assist in large-scale monitoring of environmental protection and water conservation, and oblique photography technology to assist environmental protection. This protection method can ensure the technical solutions for detailed inspection, strengthen the supervision of the design, construction, and completion acceptance process, and timely discover and supervise the rectification of environmental and water conservation problems.

As a result, the current relevant scholars are more concerned with the effect of water protection on the construction of power grid engineering and have presented management technologies and

*Corresponding Author: ngxy20213@163.com

DOI 10.1201/9781003332657-14

methods from the perspectives of scheme design and construction process detection thus helping to improve the supervision efficiency of circular water protection of power transmission and transformation projects. There is, however, insufficient data on the influencing factors and the methodology used for calculating the cost of environmental water conservation measures.

2 A FISHBONE DIAGRAM ANALYSIS OF THE INFLUENCING FACTORS OF THE COST OF ENVIRONMENTAL PROTECTION FACILITIES IN TRANSMISSION AND TRANSFORMATION PROJECTS

2.1 *Theoretical principles of fishbone diagram*

Fishbone diagrams, as their name suggests, depict the skeleton of a fish, with the head and tail connected by thick lines in the manner of vertebrae. The tail of the fish represents the problem or status quo, the head represents the goal, and the spine represents all the steps and factors involved in that process. When people come up with a factor, use a fishbone to represent this factor, and mark all the relevant items with different fishbones. Then it will be refined and each factor will be analyzed. The fishbone branch will be used to represent the related elements of each main cause. Additionally, people can find the relevant elements through the three-level and four-level bifurcations. With repeated contemplation, a fishbone diagram develops a general framework. The answers should be provided for each branch or fork. Finally, categorize the remaining issues, tasks, and work needed. In this way, it is easy to find out which are the major causes of present concerns, how they can be resolved, which can be addressed immediately, and what resources need to be mobilized. Fishbone diagrams may appear similar to tree diagrams. They are tools that aid in analyzing and thinking, clarifying ideas, and identifying problems. Consultants often use the fishbone diagram analysis method to conduct causal analyses.

2.2 *Analysis of the influencing factors*

Tracing the cost of environmental protection facilities from the perspective of driving factors, and the results are as follows:

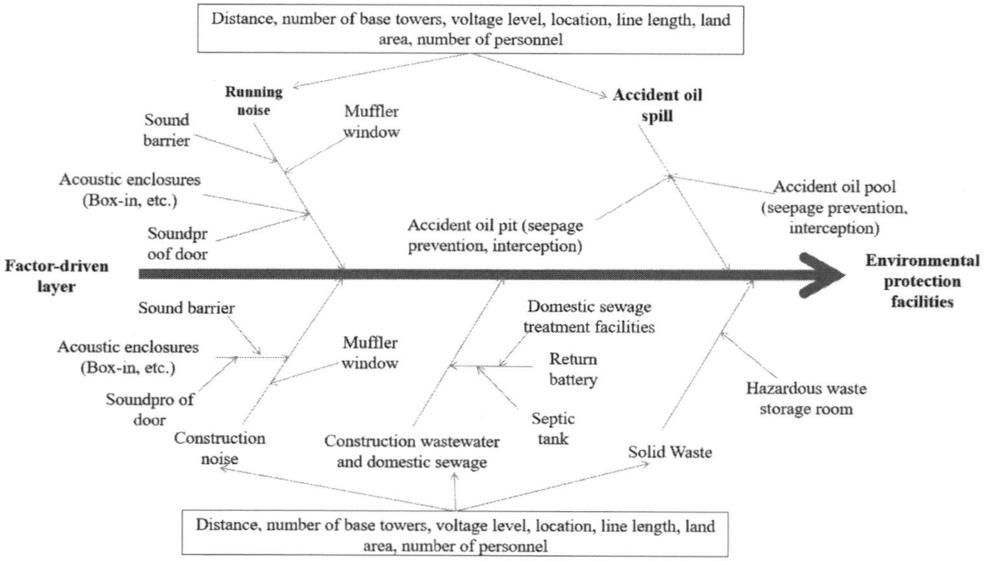

Figure 1. Analysis of influencing factors of environmental protection facilities for power transmission and transformation projects based on driving layer.

From the above fishbone diagram, it can be found that from the perspective of driving factors, traceability analysis is carried out to systematically sort out the cost of environmental protection facilities in the early stage of construction of power transmission and transformation projects, from construction to completion acceptance, and the cost of environmental protection facilities in power transmission and transformation projects is mainly affected by construction. Noise, operation noise, construction wastewater, domestic sewage, accidental oil leakage, and solid waste.

3 STUDY ON THE INFLUENCE MECHANISM OF ENVIRONMENTAL PROTECTION FACILITIES IN POWER TRANSMISSION AND TRANSFORMATION PROJECTS BASED ON PRINCIPAL COMPONENT ANALYSIS

3.1 *Basic principles of the principal component analysis*

Principal Component Analysis (PCA) was first proposed by Hotelling in 1933. The basic principle is to use the idea of dimension reduction to make a linear combination of multiple relevant indicators under the loss of little information into several comprehensive indicators, called the main component. The PCA was very similar to the factor analysis. The difference is that factor analysis extracts factors first and then decompose variables into a linear combination of factors; while PCA combines variables linearly to obtain the principal component. Meanwhile, PCA is one of the methods of factor extraction in factor analysis, and it can also be implemented with SPSS software, and the operation steps are the same as the factor analysis. Finally, the results are interpreted from different perspectives.

When identifying the influencing factors of the cost of power transmission and transformation projects, people often consider as many factors as possible, leading to the overlap of information and increasing the complexity of the research problem. Therefore, it is hoped that the principal components are extracted from these factors for study through a linear combination.

With p variables $x_1, x_2, \ldots, x_p$, in principle, p principal components can be extracted, and the mathematical model is:

$$\begin{cases} Y_1 = u_{11}x_1 + u_{11}x_1 + \cdots + u_{1p}x_p \\ Y_2 = u_{21}x_1 + u_{22}x_1 + \cdots + u_{2p}x_p \\ \vdots \\ Y_\mathrm{p} = u_{p1}x_1 + u_{p2}x_1 + \cdots + u_{pp}x_p \end{cases} \text{或} Y = UX \tag{1}$$

If all the p principal components are extracted, the meaning of simplifying the data will be lost. Therefore, the first few principal components that already contain more than 80%–90% of the information are extracted. The principal component is a linear combination of influencing factors, and the coefficient of each factor represents its contribution to the principal component. The steps of principal component extraction are as follows:

a. Factors were identified, and sample data were obtained.
b. Data standardization was performed to eliminate the effects of different dimensions.
c. Conduct correlation analysis of the identified influencing factors to determine whether they are suitable for principal component analysis. The more factors, the higher the correlation between the factors, and the better the effect of the principal component analysis.
d. Principal components were extracted. Extraction principle: the principal component eigenvalue is >1, and usually the cumulative contribution rate is $>80\%$.
e. Analysis component matrix (factor analysis is analysis rotating component matrix) to obtain the load of factors on the principal component and judge the influence degree of factors.

3.2 *Analysis of cost factors based on the Pearson method*

In statistics, Pearson's correlation coefficient (Pearson correlation coefficient), also known as Pearson's correlation coefficient (PPMCC or PCCs), is used to measure the correlation (linear correlation) between two variables X and Y (linear correlation). In natural science, the Pearson correlation coefficient is widely used to measure the correlation between two variables, with values ranging between-1 and 1. It evolved from Carl Pearson from a similar but slightly different idea proposed by Francis Galton in the 1880s. This correlation coefficient is also called the "Pearson's product-moment correlation coefficient".

The Pearson correlation coefficient between two variables is defined as the quotient of the covariance and standard deviation between two variables:

$$\rho_{X,Y} = \frac{\operatorname{cov}(X, Y)}{\sigma_X \sigma_Y} = \frac{E[(X - \mu_X)(Y - \mu_X)]}{\sigma_X \sigma_Y} \tag{2}$$

The above formula defines the overall correlation coefficient, and the Greek lowercase letter ρ is often used as a representative symbol. By estimating the covariance and standard deviation of the sample, the Pearson correlation coefficient can be obtained, which is usually represented by the English lowercase letter γ:

$$\gamma = \frac{\sum_{i=1}^{n} (X_i - \overline{X})(Y_i - \overline{Y})}{\sqrt{\sum_{i=1}^{n} (X_i - \overline{X})^2} \sqrt{\sum_{i=1}^{n} (Y_i - \overline{Y})^2}} \tag{3}$$

γ can also be estimated from the mean of the standard scores of the sample points of (X_i, Y_i), and an expression equivalent to the above formula can be obtained:It can also be estimated from the standard score mean of the sample point to obtain an expression equivalent to the above formula:

$$\gamma = \frac{1}{n-1} \sum_{i=1}^{n} \left(\frac{X_i - \overline{X}}{\sigma_X} \right) \left(\frac{Y_i - \overline{Y}}{\sigma_Y} \right) \tag{4}$$

Where $\frac{X_i - \overline{X}}{\sigma_X}$, $\overline{X}$ and X_i are the standard score, sample mean and sample standard deviation, respectively.

The absolute values of the population and sample Pearson coefficients are less than or equal to 1. If the sample data points fall exactly on the line (in the case where the sample Pearson coefficient is calculated), or if the bivariate distribution is exactly on the line, the correlation coefficient is equal to 1 or -1.

The Pearson correlation coefficient possesses an important mathematical property in that the position or scale of the two variables does not affect the coefficient, i.e., the invariant of the change (determined by the symbol). As a result, if we move X to and Y to where a, b, c, and d are constants, the correlation coefficients of both variables remain the same (which holds true for both the population and sample Pearson correlation coefficients). In a more general linear transformation, the correlation coefficient is altered as follows:

Since $\mu_x = E(X)$, $\sigma_x^2 = E[(X - E(X)^2)] = E[X^2] - E^2(X)$, Y are also similar, and $E[(X - E(X)(Y - E(Y)] = E(XY) - E(X)E(Y)$, the correlation coefficient can also be expressed as:

$$\rho_{X,Y} = \frac{E(XY) - E(X)E(Y)}{\sqrt{E(X^2) - (E(X)^2)} \sqrt{E(Y^2) - (E(Y)^2)}} \tag{5}$$

For the sample, the Pearson correlation coefficient:

$$\gamma_{xy} = \frac{\sum x_i y_i - n\overline{xy}}{(n-1) s_x s_y} = \frac{n \sum x_i y_i - \sum x_i \sum y_i}{\sqrt{n \sum x_i^2 - (\sum x_i)^2} \sqrt{n \sum y_i^2 - (\sum y_i)^2}} \tag{6}$$

The above equation gives a simple single-process algorithm for calculating the sample Pearson correlation coefficient, but it relies on the data involved, and sometimes it can be numerically unstable.

3.3 *Empirical analysis*

The empirical analysis is based on influence factors library, select distance, base quantity, voltage level, area, line length, area, personnel number sequence and power transmission and transformation engineering ring water conservation special cost publicity and education training, vegetation protection, animal protection, water environment protection, fee, atmospheric environment protection, soil environmental protection, solid waste disposal, demolition site protection fee sequence. Here, the sequence of atmospheric environment protection fees is taken as an example, and the correlation coefficient with the sequence of each factor is analyzed, as shown below.

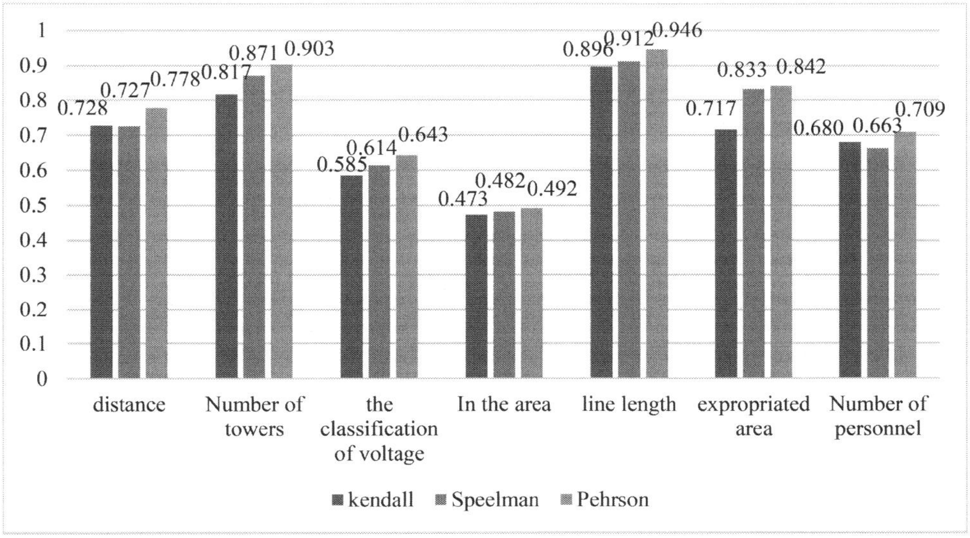

Figure 2. The correlation analysis results of environmental protection fee.

The influencing factors are line length, number, base, distance, land area and number of personnel. Principal components analysis is performed using SPSS to extract the principal components.

Table 1. Variance explained in the table.

	Initial eigenvalue			Extract the square sum to load		
Element	amount to	% of the variance	accumulate (%)	amount to	% of the variance	accumulate (%)
1	.3506	88.428	88.428	.3506	88.428	88.428
2	0.384	6.412	94.839			
3	0.198	3.317	98.156			
4	0.108	1.801	99.957			
5	0.002	.043	100.000			

In this section, the atmospheric environment protection fee is taken as an example, for which the main influencing factors are analyzed. It can be seen from the table that only by calculating the first principal component (line length), the cumulative variance contribution rate has reached 88.428%, more than 85%. Therefore, one principal component was selected for subsequent calculation, which ensured the data integrity while determining the main influencing factors. For the atmospheric environment protection cost, the line length is the first main component affecting its cost.

4 CONCLUSION

In order to improve the control level of power transmission and transformation project, the paper firstly analyzes the cost of the principal component analysis method, and verifies the effectiveness of the actual cases. The research method proposed in this paper can provide reference for power grid enterprises to scientifically analyze the influencing factors of environmental protection facilities costs of power transmission and transformation projects, and improve the input efficiency of engineering environmental protection costs.

ACKNOWLEDGMENT

This study was supported by the Science and Technology Project of the State Grid Zhejiang Electric Power Co., Ltd (Grant No. B311JY21000C).

REFERENCES

He Lefeng. Study on Construction Environment Protection and Soil and Water Conservation Measures of Plateau Transmission Line Engineering [J]. *Water Conservancy and Hydropower Technology* (Chinese and English), 2021, 52(S2): 154–156. DOI:10.13928/j.cnki.wrahe.2021.S2.040.

Li Wanzhi, Li Yang, Tan Rongrong, Li Yang, Wei Tao. Research on the Integrated Application of Water Protection and Space Protection in Power Grid Engineering [J]. *Electric Power Survey and Design*, 2020(S1): 194–199. DOI: 10.13500/j.dlkcsj.issn1671-9913.2020.S1.036.

Li Zhibin, Li Yang, Tan Rongrong. Research and application of key technologies of integrated supervision of ring water, air and sky conservation in UHV Line Engineering [J]. *Beijing Surveying and Mapping*, 2021, 35(12): 1512–1517. DOI: 10.19580/j.cnki.1007-3000.2021.12.004.

Wu Jian, Bai Xiaochun, Li Rui, Lu Lin, Zhang Dianmao, Huang Changyu. Influence of restriction factors on site selection and solution of power transmission and transformation project [J]. *China Soil and Water Conservation*, 2020(07): 21–23. DOI:10.14123/j.cnki.swcc.2020.0159.

Zhao Guozhong, Li Peimin, Yang Zhenghe, Wang Qian, Chen Siyuan, Wang Tao.Research and Practice of Space and Earth Integration in the Intelligent Management of Qinghai Power Grid Infrastructure [J]. *Science and Technology Innovation*, 2020(26): 123–125.

Energy Revolution and Chemical Research – Kok-Keong Chong and Zhongliang Liu (Eds)
© 2023 The Authors, ISBN: 978-1-032-36554-1

Research on the prediction of power energy consumption of campus buildings with ARMA model

X.X. Hong*
Wuhan University of Technology Design and Research Institute Co., Ltd, Wuhan, Hubei Province, China

W.Y. Wu
School of Civil Engineering and Architecture, Wuhan University of Technology, Wuhan, Hubei Province, China

ABSTRACT: In the context of the global energy crisis and economic crisis, energy conservation and emission reduction has become a key social goal. Campus buildings are a building group with full energy-saving potential, and reasonable campus energy management policies can fully tap the energy-saving potential of campus buildings. Based on the collected daily power consumption of four campus buildings, this study adopted the time-series prediction model, ARMA to forecast the daily power consumption in 30 days. To evaluate the accuracy of the predicted model, MAE, RMSE, and MAPE are adopted. Through verification, it is found that the prediction performance of the model is good, and the accuracy can meet the needs of campus building energy management.

1 INTRODUCTION

In response to the global energy crisis and environmental problems, China has made a clear action plan on energy conservation and carbon emission reduction. Currently, building energy consumption presents an increasing trend with urbanization promotion. Among them, college campus accounts for a large proportion of total energy consumption. There are nearly 2000 universities in China, accounting for 8% of the social energy consumption. The per capita energy consumption index of campus buildings is four times of residential buildings. The large difference between campus buildings and residential buildings indicates the energy-saving potential of the campus. However, due to the low level of energy management and the lack of energy data, the sustainable development of the campus is restricted.

Therefore, it is significant to carry out reasonable energy management and control on the campus. While the prediction of building energy consumption can provide valuable reference values for energy management and control. A novel prediction method of the campus building energy consumption considering occupant behavior was developed (Zhang et al. 2021). The relationship between occupant behavior and electricity consumption is established. Compared with the prediction models based on environmental parameters, the proposed model can reflect energy use laws with higher accuracy. Meanwhile, it can be applied online without complex algorithms and software simulation. The impact of occupancy rates and local environmental conditions, such as temperature, humidity ratio, solar radiation, and wind speed, on the actual electric energy consumption of a campus building for both working and non-working days, was investigated (Kim & Srebric 2020). A predicted model of electric energy consumption based on an ANN method with the LM-BP algorithm was proposed, which can provide reliable predictions of energy consumption compared with the linear regression modeling (Tang & Röllin 2018). The campus electricity consumption

*Corresponding Author: 674800161@qq.com

DOI 10.1201/9781003332657-15

was divided into two categories: the basic and the variable, and establishes a two-part building electricity forecasting model based on human behavior (Ding et al. 2019). The basic electricity consumption is related to the building area, while the variable electricity consumption is related to the building occupancy. The model is verified with the actual situation, and the error is found to be less than 5%.

In this study, based on the historical daily power consumption of four campus buildings in a university, the time-series prediction method of the ARMA model is used. With using the data of the first 135 days as the training data, the daily power consumption of 30 days is forecasted as the verification. Finally, it is found that the error is within the allowable range, which can meet the needs of the school. The established prediction is capable to give useful reference for the optimized operation of the campus energy system.

2 PREDICTION METHOD

2.1 *ARMA*

ARMA model is an integrated model based on the autoregressive model (AR) and moving average model (MA). It is a common prediction method for time series prediction. The basic principle is that the data sequence formed by the prediction index over time is regarded as a random sequence. The dependency of this group of random variables reflects the continuity of the original data in time. Meanwhile, the effect of influencing factors has its own change law. Assuming that the influencing factors are $x_1, x_2, \ldots, x_K$, it is analyzed by regression analysis (eq1)

$$Y_t = \beta_1 x_1 + \beta_2 x_2 + \cdots + \beta_p x_p + Z \tag{1}$$

where Y_t is the observed value of the prediction object and Z is the error. As the prediction object, Y_t is affected by its own changes, and its law can be reflected by the following formula (Eq.2).

$$Y_t = \beta_1 x_{t-1} + \beta_2 x_{t-2} + \cdots + \beta_p x_{t-p} + Z_t \tag{2}$$

The error term is a time-dependent variable, which is expressed by the following formula (Eq.3).

$$Z_t = \varepsilon_t + \alpha_1 \varepsilon_{t-1} + \alpha_2 \varepsilon_{t-2} + \cdots + \alpha_q \varepsilon_{t-q} \tag{3}$$

Thus, the ARMA model expression can be obtained (Eq.4):

$$Y_t = \beta_0 + \beta_1 Y_{t-1} + \beta_2 Y_{t-2} + \cdots + \beta_p Y_{t-p} + \varepsilon_t + \alpha_1 \varepsilon_{t-1} + \alpha_2 \varepsilon_{t-2} + \cdots + \alpha_q \varepsilon_{t-q} \tag{4}$$

2.2 *Order determination of ARMA model*

In the ARMA model, where P is the number of autoregressive terms and Q is the number of moving average terms. ACF method and PACF method are used to confirm P and Q. The autocorrect function (ACF) and the paracord function (PCF) are used in MATLAB to get the order of autoregressive terms and moving average terms respectively.

As shown in Figure 1 and Figure 2, for the first building, the values of P and Q are determined as $P = 1$ and $Q = 20$ according to the ACF method and PACF method. Similarly, the value of P and Q for the second building can be determined according to Figure 3 and Figure 4, which present that P is 1 and Q is 3. And the order of autoregressive terms for the third building and the fourth building is the same ($P = 1$) as presented in Figure 5 and Figure 6. While for the order of the moving average terms for the two buildings is $Q = 1$ and $Q = 13$ respectively, as illustrated in Figure 7 and Figure 8.

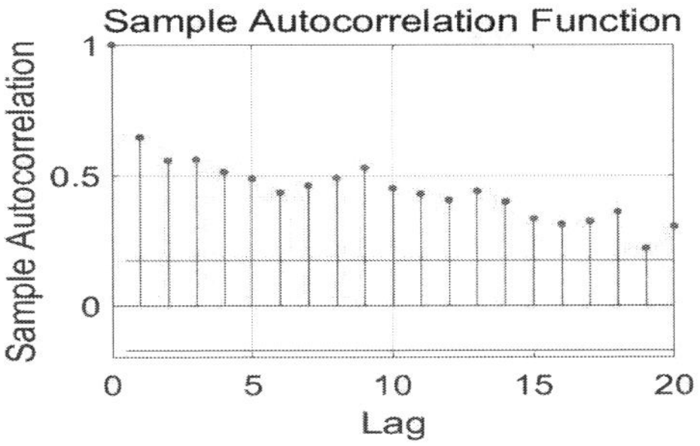

Figure 1. ACF for the first building.

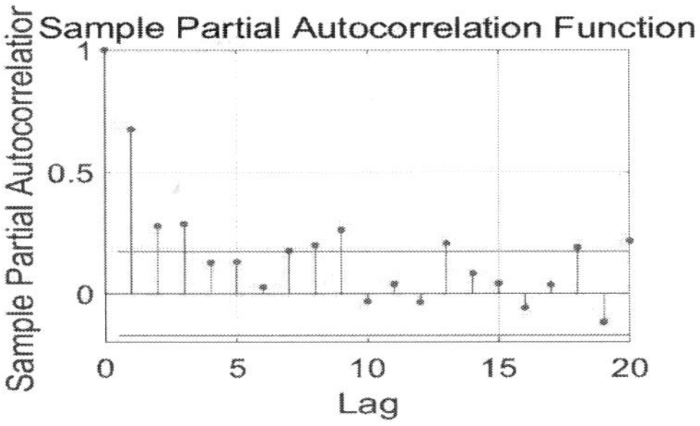

Figure 2. PACF for the first building.

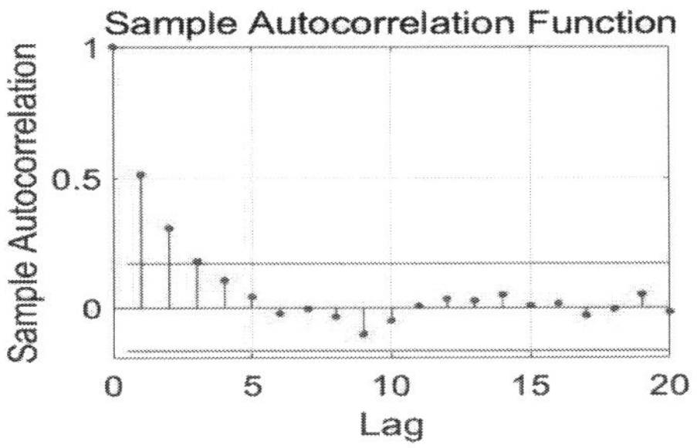

Figure 3. ACF for the second building.

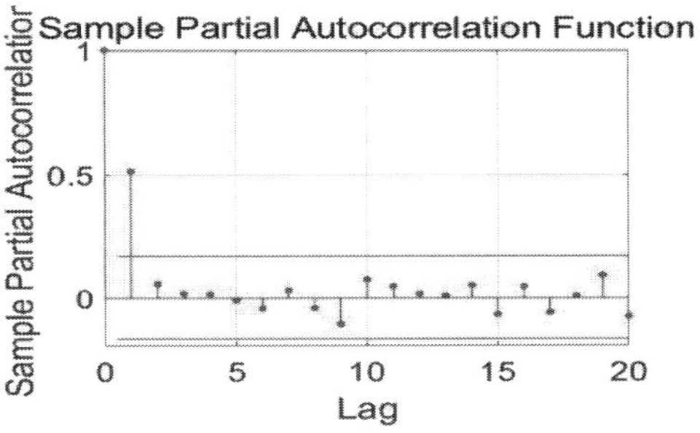

Figure 4. PACF for the second building.

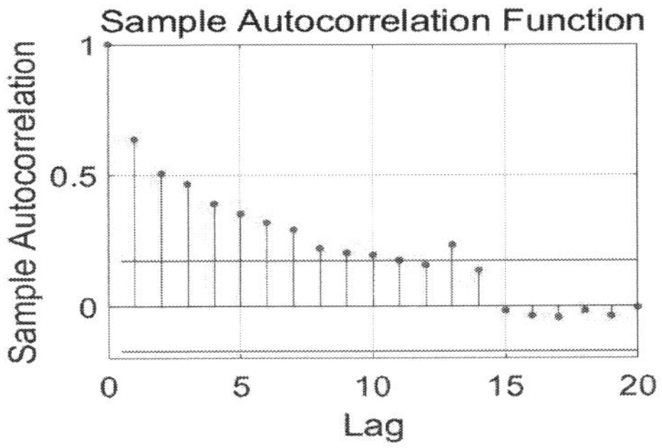

Figure 5. ACF for the third building.

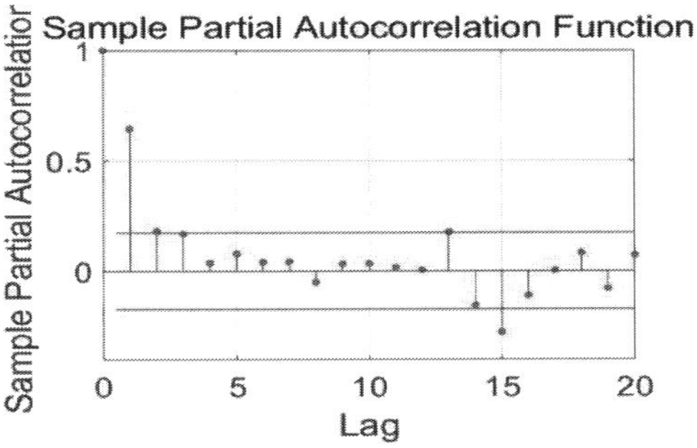

Figure 6. PACF for the third building.

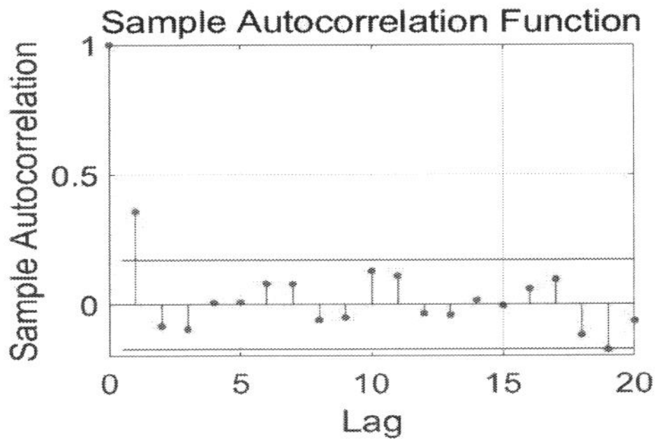

Figure 7. ACF for the fourth building.

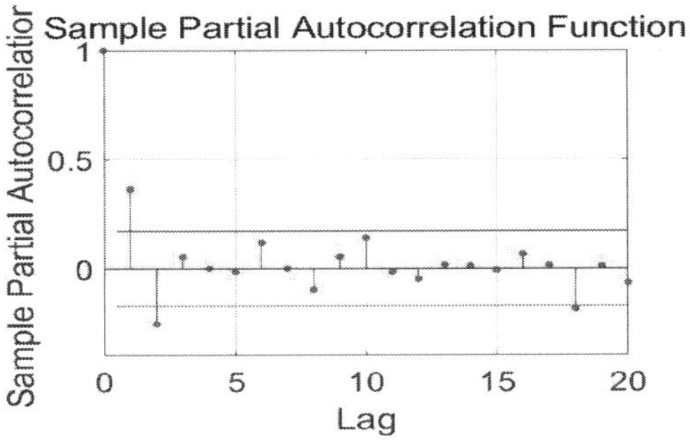

Figure 8. PACF for the fourth building.

2.3 *Prediction results*

This paper collects the historical daily power consumption of four buildings in a university from Jan 1st, 2020 to Dec 31st, 2021. Due to the epidemic and holidays, this paper removes some useless data, and finally obtains 165 days of effective historical data. In this paper, the first 135 days are used as training data, and the daily power consumption after 30 days is predicted as verification data. Based on the ARMA prediction model, the daily electricity consumption is forecasted for the selected four buildings. Figure 9 shows the prediction results of the first building. Compared with the actual collected data, it is found that the predicted result is slightly higher than the actual data, which is the result of the original time series in the training set with an increasing trend. However, the discrepancy between the predicted data and the real data is small, which is within the acceptable precited error range. Figure 10 shows the predicted results of the second building. It can be seen that the prediction result is relatively stable since P and Q have smaller values than in the first building. Compared with the actual data, except for several suddenly expanded points, the overall prediction performance of the model is good, and the changing trend in the next 30 days is correctly predicted.

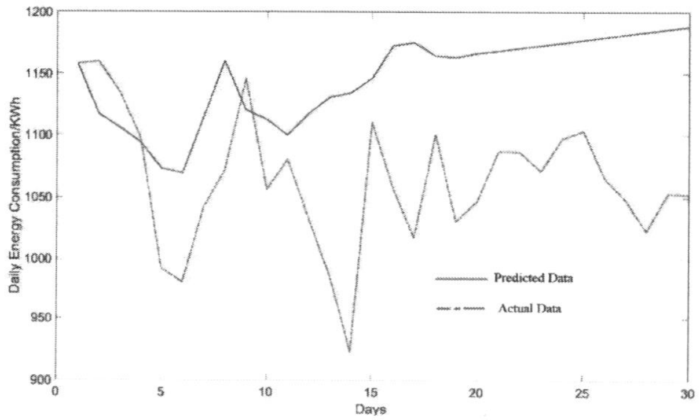

Figure 9. Comparison of the predicted results and real data of the first building.

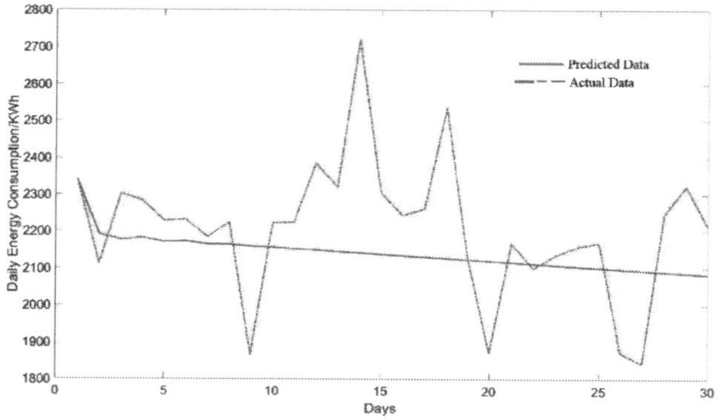

Figure 10. Comparison of the predicted results and real data of the second building.

Figure 11 shows the prediction results of the third building. The predicted results have a larger discrepancy with the actual electricity data compared with the other three buildings, and the predicted accuracy is the worst. It can also be indicted from the evaluation indicators in the next part. But in general, the prediction error is still within the acceptable range. Figure 12 shows the prediction results of the fourth building. The line connecting the highest point of the predicted result

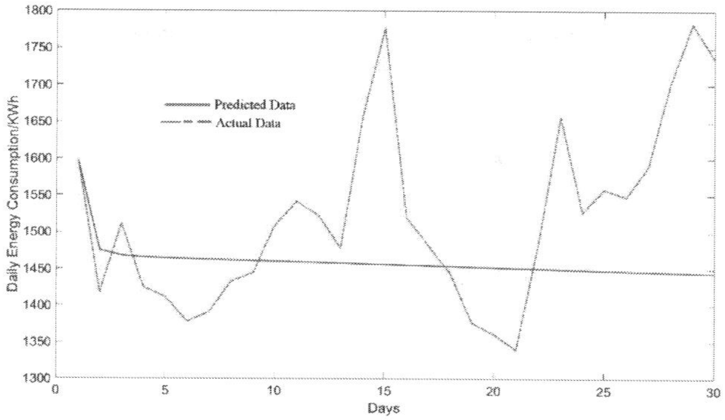

Figure 11. Comparison of the predicted results and real data of the third building.

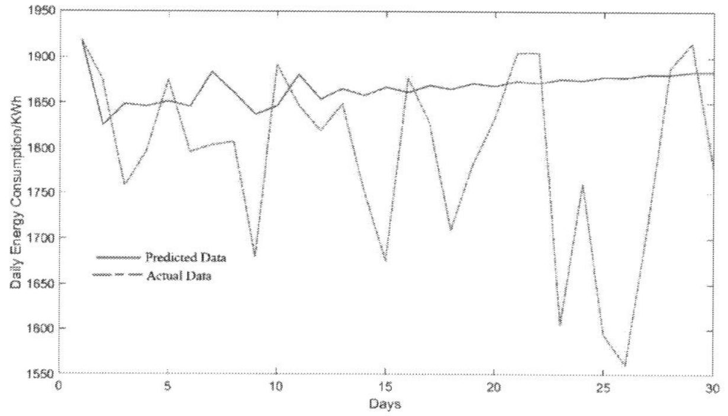

Figure 12. Comparison of the predicted results and real data of the fourth building.

curve and the actual result curve coincides, which also shows that the prediction model has a good performance in high-load transportation. But on the whole, the prediction error is also within the acceptable range.

2.4 *Prediction and evaluation*

In order to evaluate the predicted accuracy of the model, mean absolute error (MAE), root mean square error (RMSE), and mean absolute percentage error (MAPE) are selected as the evaluation indices. The mathematical expression is as follows (Eqs.5, 6, and 7).

$$MAE = \frac{1}{\mathrm{n}}\sum_{t=1}^{n}\left|\hat{y}_i - y_i\right| \tag{5}$$

$$RMSE = \sqrt{\frac{1}{\mathrm{n}}\sum_{t=1}^{n}(\hat{y}_i - y_i)^2} \tag{6}$$

$$MAPE = \frac{1}{\mathrm{n}}\sum_{t=1}^{n}\left|\frac{y_i - \hat{y}_i}{y_i}\right| \tag{7}$$

MAE represents the difference between the average daily predicted daily power consumption and the real daily power consumption. RMSE also represents the comparison between the predicted daily power consumption and the real daily power consumption. MAPE indicates that the error of the predicted value relative to the real value is equivalent to the percentage of the real value. Compared with MAE and RMSE, MAPE is more direct and can reflect the error rate.

The evaluation indices of four buildings are listed in Table 1. The average error of the four buildings is about 100kwh, the RMSE is close to 10, and the error rate is less than 10%. Therefore, on the whole, the preliminary prediction of the campus building energy consumption using ARMA model meets the requirements.

Table 1. The values of MAE, RMSE, and MAPE for the four buildings.

Category	MAE/kWh	RMSE	MAPE
The first building	89.0170	9.4349	8.5744
The second building	140.296	11.8447	6.3117
The third building	132.3234	11.5032	8.9966
The fourth building	89.6171	9.4666	5.2878

3 CONCLUSION

This paper collected the daily power consumption of four campus buildings in a university from 2020 to 2021. Through the data preprocessing, a total of 165 days of electricity data are filtered as the effective one. To forecast the daily electricity consumption of campus buildings, the time-series prediction model ARMA is used. And 135 days are selected as the training set, and a total of 30 days is set as the test set. At last, three indices, MAE, RMSE, and MAPE, are used to evaluate the predicted accuracy of the prediction model, it is found that the prediction results of the three models meet the requirements of campus building prediction. The error rate ranges from 5.28% to 8.57%, which means the predicted model can provide an effective reference for the operation of campus building energy systems.

ACKNOWLEDGMENT

This work was supported by the Fundamental Research Funds for the Central Universities (WUT: 2021IVA038).

REFERENCES

Ding Y, Wang Q.C., Wang Z.X., Han S.X. & Zhu N (2019). An occupancy-based model for building electricity consumption prediction: A case study of three campus buildings in Tianjin. *Energy Build.*, 202, 109412.

Kim M.K, Kim Y.S. & Srebric J (2020). Predictions of electricity consumption in a campus building using occupant rates and weather elements with sensitivity analysis: Artificial neural network vs. linear regression. *Sustain. Cities Soc.*, 62, 102385.

Tang W.H. & Röllin A. (2018). Model identification for ARMA time series through convolutional neural networks. *Decis. Support Syst.*, 146, 113544.

Zhang C.Y., Zhao T.Y. & Li K.S. (2021). Quantitative correlation models between electricity consumption and behaviors about lighting, sockets and others for electricity consumption prediction in typical campus buildings, *Energy Build.*, 253, 111510.

Energy Revolution and Chemical Research – Kok-Keong Chong and Zhongliang Liu (Eds)
© 2023 The Authors, ISBN: 978-1-032-36554-1

Asteroid mining predictions based on logit model

Yudan Guo*, Meixi Lin & Maolin Cheng
College of Computer and Network Security, Chengdu University of Technology, Chengdu, China

ABSTRACT: With the development of the world aerospace industry, human access to space is becoming more common, meaning that the rich mineral resources of asteroids could be exploited by humans soon. To measure the impact of asteroid mining on world equity, we first choose nine indicators to model a country's combined national power and quantify world equity by measuring a country's combined national power. we determine the future vision of asteroid mining and analyze its impact on world equity, and we propose some hypothetical conditions for future asteroid mining. The obtained trend of the asteroid mining industry is applied Logit Model. We take the stable period which is a longer existence period to quantify the asteroid mining industry specifically, and we find that the development of the asteroid mining industry can effectively reduce the global inequity.

1 INTRODUCTION

Asteroids are bodies in the solar system that move around the Sun like planets, but are much smaller in size and mass than planets, hence the name asteroids (Li 2012).

As the mineral resources of the earth gradually cannot meet the growing human demand, space mining has received more and more attention, and in recent years with the support of governments, private enterprises, research institutes, etc., space mining has been enjoying unprecedented development opportunities (Zhang et al. 2020). And because mineral resources are an important material basis for the survival of human society, they are an important guarantee for national security and economic development (Xu & Wang 2011). However, major developed countries and large multinational mining companies already control most of the strategic mineral resources on Earth (Rong et al. 2015), so the development of asteroid mining can help alleviate the current mineral resource scarcity problem.

This paper aims to:

- Integrate various evaluation methods of comprehensive national power, ignore secondary factors, build a model, select nine indicators to quantify comprehensive national power, and judge world inequity through the variance of comprehensive national power.
- Imagine the conditions of asteroid mining, quantify and measure the mining capacity and value of a country, and analyze its impact on world equity.

2 GLOBAL EQUITY

2.1 *Definition of global equity*

In today's society, there are many definitions of fairness in the world, some measure fairness in terms of respect for human rights, and some consider the equal distribution of wealth as fairness. However, we believe that the most intuitive means of reflecting equity in the world is the level of development of a country. If countries have the same level of development, then the resources and rights available per capital are similar, and then equity can be achieved both materially and

*Corresponding Author: guoyudan@stu.cdut.edu.cn

DOI 10.1201/9781003332657-16

spiritually. In this paper, we use 9 factors shown in Figure 1 as indicators to determine world equity. The model in this paper adopts the calculation method analogous to Klein's formula and selects several countries in each of the four categories of developed countries, upper-middle-income countries, lower-middle-income countries, and developing countries, and calculates the variance using the above-mentioned indicators, so as to compare the development level of each country and then judge the world equity.

2.2 *A model for measuring global equity*

2.2.1 *Theoretical basis of the model*

The model in this paper calculates the comprehensive national power and then applies the variance to measure world equity. Systematic quantitative comparison and analysis of comprehensive national power are conducted, and it is considered that comprehensive national power is divided into two parts: material power and spiritual power. These two components can interact with each other to contribute to the improvement of a country's comprehensive national power. We define NP as National Power, R_i as a share of certain resources in the world total, a_i as the weight of a resource. The equation for measuring comprehensive national power is:

$$\mathrm{NP} = \sum_{i=1}^{N} (a_i * R_i) \tag{1}$$

On the one hand, this paper uses the dimensionless weight method to calculate the share of each country's major resources in the world total. Firstly, the world equity studied in this paper refers to relative national power, and we are more interested in whether a country's comprehensive national power or strategic resources are relatively increasing or decreasing relative to another country. Secondly, when calculating the comprehensive national power using nine indicators, each indicator unit is different and cannot be added together. Using the specific gravity method and data normalization, the above problem is avoided and has additivity, i.e., different indicator units are transformed into a uniform unit (percentage), which constitutes the comprehensive national power, and also has comparability, i.e., international and historical comparisons are made.

Finally, we define x_i as combined national power of a country in a given year, p_i as assuming that each country has the same share of combined national power, E(X) as an average combined national power, and D(X) as comprehensive national power variance. So, world fairness was determined by calculating the variance according to the following formula.

$$D(X) = \sum_{i=1}^{n} (x_i - E(X))^2 p_i \tag{2}$$

2.2.2 *The Process of model implementation*

In calculating the combined national power, we followed the different weights as shown in Figure 1 and obtained Table 1.

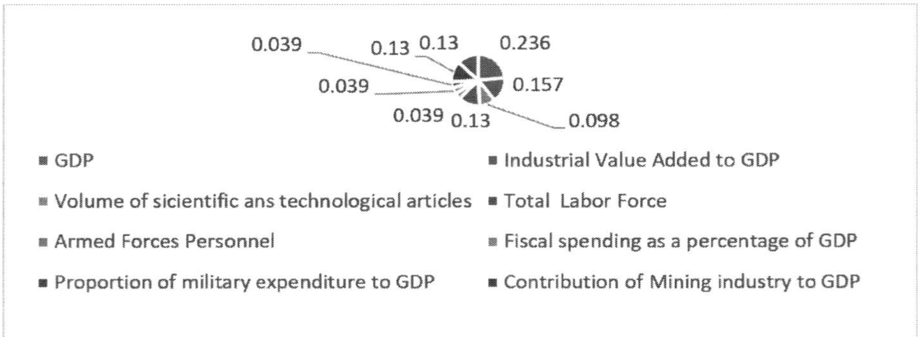

Figure 1. Nine indicators that determine comprehensive national power.

Table 1. Comprehensive national power of different countries.

Country Name	2014	2015	2016	2017	2018
Australia	0.558846566	0.601842167	0.592994784	0.566318643	0.572828708
azerbaijan	0.26286374	0.268289693	0.264626936	0.267544679	0.253525616
Burkina Faso	0.203114708	0.242367534	0.269910765	0.237979493	0.218072013
Bulgaria	0.227832332	0.246139338	0.235834118	0.229134145	0.226457841
Belarus	0.21023799	0.221090682	0.209516883	0.205009498	0.195418377
Brazil	0.966299764	0.980287724	0.987634627	0.980725301	0.986859907
Central African Republic	0.045594706	0.0594915	0.056818248	0.048518379	0.045095782
Canada	0.406628838	0.438170471	0.428665751	0.41950534	0.43636239
Colombia	0.517346314	0.535063637	0.539103393	0.522531896	0.516412861
Costa Rica	0.130937219	0.145701459	0.149286481	0.145086028	0.152522373
Ethiopia	0.256054363	0.27256713	0.304298753	0.308175251	0.314915743
Jordan	0.211189922	0.246816379	0.23163926	0.208054486	0.191596545
Sri Lanka	0.287473576	0.326751582	0.350371679	0.346845172	0.326734126
Nicaragua	0.135416224	0.137373885	0.141255522	0.124555817	0.121065095
Nepal	0.168053794	0.166372743	0.17293247	0.150655577	0.151858774
The Philippines	0.327510931	0.349511107	0.345864276	0.331285958	0.327725248
Poland	0.451781359	0.489775717	0.490588636	0.480065398	0.488237655
Portugal	0.245031561	0.253722262	0.259624016	0.239204589	0.241447244
Singapore	0.209550102	0.231447082	0.231120075	0.187683615	0.183048975
Uganda	0.198095799	0.221334421	0.222730558	0.215524973	0.212131339
In Zambia	0.290433949	0.318889002	0.291793564	0.318781072	0.298722186

The weights we use for the nine indicators in the comprehensive national power equation are derived from the reference weights given by the Klein equation and the dynamic comprehensive power equation (Zhu & Xiao 1999).

In Figure 1, we can see that the two factors with the largest weight are GDP and Industrial Value Added to GDP. These two factors will also be mentioned below.

Compared with Klein's equation, the factors we consider add the influence of different social institutions, etc. To a certain extent, it can better reflect the objective establishment of the results obtained by the model we have established.

2.2.3 *The result of the model*

Through the above experiment, we obtained the combined national power variance, which is an indicator quantity to measure world equity.

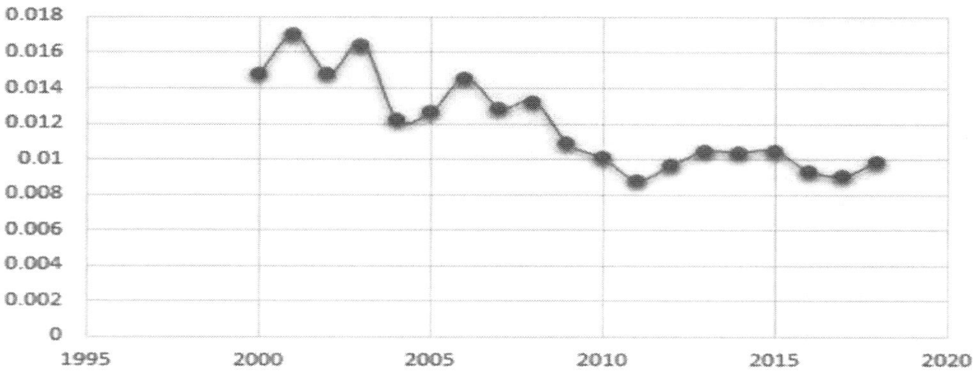

Figure 2. Comprehensive National Power Variance.

2.3 *Validation of the model*

In validating the model, we calculated the variance of the HDI obtained in Figure 4. It is clear that the variance shows a decreasing trend, indicating that the stability is increasing year by year. And the results obtained by our proposed model are also more volatile in the first period and more stable in the later period, but also show a decreasing trend overall. The results of our model are about the same as those obtained according to HDI, thus verifying the validity of our model.

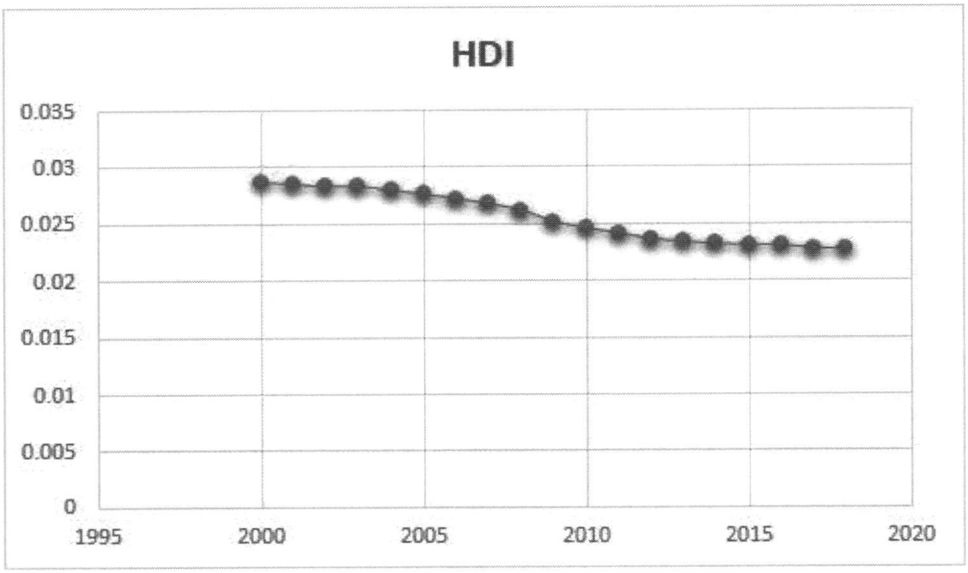

Figure 3. HDI Variance.

3 ASTEROID MINING

3.1 *The future of asteroid mining*

We assume the following conditions for future mining of asteroids, in the future of asteroid mining, most countries and private mining companies around the world will have the ability to mine asteroids profitably, and the mineral resources will not flow singularly to one or a few countries and regions. As human technology gradually advances, when rare metals also become readily available, humans can stop collecting resources on Earth and the polluted and damaged nature will be restored. The UN grants mining rights to the spacecraft to ensure they acquire a certain period to mine, which facilitates the management and is not the same as granting ownership of an asteroid, preventing multiple miners from competing for a single asteroid mining site. Although people cannot claim ownership of the moon or asteroids, they can take possession of the resources you extract from them.

3.2 *The impact of mining on global equity*

We believe that asteroid mining is similar to high seas fishing. A similar provision is made in the 1958 Convention on the High Seas. No state is allowed to exercise rights over a ship of another state on the high seas. Even warships are not allowed to board foreign ships on the high seas, except for special exceptions, such as where there is sufficient evidence of suspected piracy or the slave trade. The principle of freedom of the high seas is upheld by the international community's desire to protect the common good. This is the freedom of navigation, freedom of fishing, and later, with the development of aircraft, people proposed the freedom of overflight in the second half of the 19th century, and the freedom to lay submarine cables, etc. These are collectively known as freedom of use.

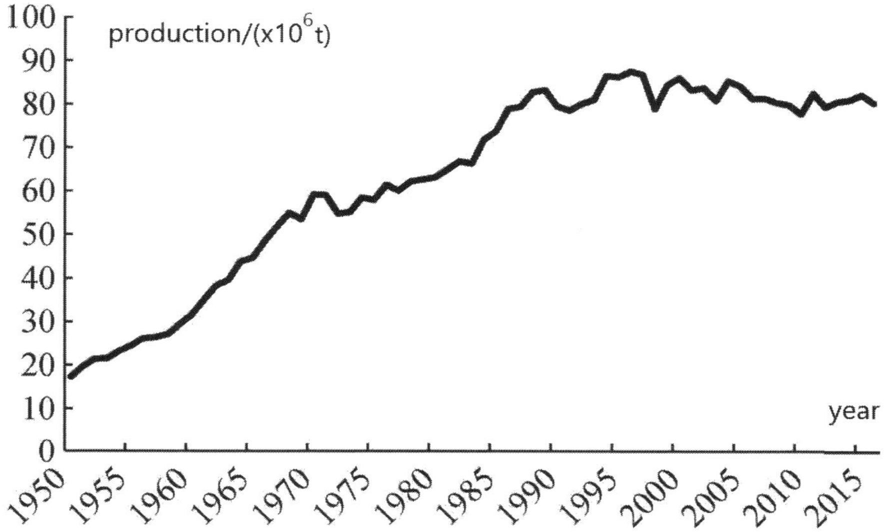

Figure 4. Global Fishing Production.

In Figure 5, the year-to-year output of the world's fisheries conforms to the basic trend of the S-curve, which is the Logit model. We believe that the development of the asteroid mining industry also satisfies the trend of the S-curve due to the multiple similarities between high seas fishing and space mining.

3.3 *The result of the model*

In considering the impact of asteroid mining on global equality, firstly, we use the number of spacecraft and the number of satellites as a measure of a country's aerospace technology, which measures that country's mining capacity. We define $SP_{country}$ as number of national spacecrafts, SA_{country} as number of national satellites, SC as national space capability, k_1 and k_2 as constants.

$$SP_{country} * k_1 + SA_{\text{country}} * k_2 = \text{SC} \tag{3}$$

Secondly, we quantify mining capacity by assuming that 1 unit of mining capacity can extract minerals valued Vd. Therefore, we consider that mining from asteroids affects a country's GDP and the contribution of the mining industry to GDP.

In Figure 6, we compare the variance between the combined national power obtained by successive calculations and find that the combined national power gap becomes smaller, thus indicating an increase in global equality after mining from asteroids. For the reason for this situation, we take Singapore as an example and analyze it as follows. Since Singapore's mineral resources are underdeveloped, our analysis is conducted after assuming that countries have mastered space and aviation technology. Therefore, for countries with backward economies or underdeveloped mineral resources, the comprehensive national power becomes stronger after asteroid mining activities, and thus world equality is strengthened.

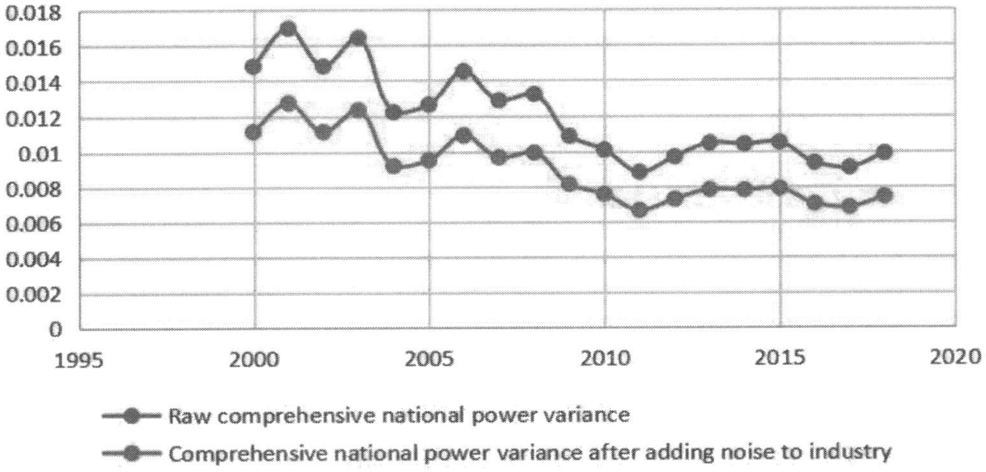

Figure 5. Comparison of Comprehensive National Power Variance.

4 CONCLUSION

In this paper, we first design a model to measure global equity, and finally, use historical data to validate our model. And then we propose some hypothetical conditions for future asteroid mining. The trend of the asteroid mining industry is obtained using Logit Model. We take the stable period which is a longer existence period to quantify the asteroid mining industry specifically, and we find that the development of the asteroid mining industry can effectively reduce the global inequity.

REFERENCES

Bing Li. Mining on asteroids[J]. *Nature and Technology*, 2012(06): 22–26.

Dongmei Rong, Haixu Gu, Li Na. An analysis of the competitive situation of mineral resources in the "Belt and Road" region[J]. *Contemporary Economics*, 2015(06): 42–44.

Kefei Zhang, Huaizhan Li, Yunjia Wang, Kazhong Deng, Changgui Li, Qianxin Wang, Xinhua Liu, Yaoshe Xie, Yabo Duan, Yang Yang. An investigation of space mining-current status, opportunities and challenges[J]. *Journal of China University of Mining & Technology*, 2020, 49(06): 1025–1034. DOI:10.13247/j.cnki.jcumt.001165.

Xian Zhu, Lazhen Xiao. Study on the method of comparing comprehensive national power[J]. *Statistics and Decision Making*. 1999(1).

Yingjie Xu, Lei Wang. The importance of mineral resources[J]. *Science and Technology Communication*, 2011(19): 28+49.

Energy Revolution and Chemical Research – Kok-Keong Chong and Zhongliang Liu (Eds)
© 2023 The Authors, ISBN: 978-1-032-36554-1

Shale gas development area optimization model based on order relation – TOPSIS method

Haipeng Gu & Tie Yan
School of Mathematics and Statistics, Northeast Petroleum University, Heilongjiang Province, China

Yuan Yuan
The Eighth Operation Area of the NO. 1 Production Plant, China

Xiaowei Li
School of Mathematics and Statistics, Northeast Petroleum University, Heilongjiang Province, China

Shihui Sun
School of Petroleum Engineering, Northeast Petroleum University, Heilongjiang Province, China

Yang Cao
The Eighth Operation Area of the NO. 4 Production Plant, China

ABSTRACT: The optimization of shale gas development areas is an important guarantee for the realization of the shale gas production target, which can provide a decision-making basis for the optimization and adjustment measures of exploration and development in the next step. Based on the collection and collation of a large number of literature about shale gas development and the reference of existing research, five representative and universal key indicators are analyzed and screened out from the geological factors, geochemical factors, and productivity factors that affect the selection of shale gas development. The order relation analysis method and entropy weight method are used to determine the combined weight of each index, it not only reflects the preference of the experts in the subjective weight but also fully reflects the influence degree of the information implied by the objective index data itself so that the weight value achieves the unity of the subjective and objective. The weighted TOPSIS method (approximate ideal solution) was used to establish the optimization model of shale gas development areas, by calculating the relative closeness between each evaluation area and the ideal area, we select the development potential area. Finally, the model is used to optimize some specific shale gas development areas in China, and the optimization results are consistent with the actual exploration and development information, indicating that the shale gas development area optimization model based on the order relation analysis -TOPSIS method is feasible and effective.

1 INTRODUCTION

After more than 10 years of shale gas exploration and development, China's shale gas output broke through 200×10^8 m^3 in 2020, and China is the second-largest shale gas producer after the United States. Caineng Zou, Shengxiang Long et al pointed out in the 14th Five-year shale gas development plan that China's shale gas output is expected to reach in 2025. Shale gas will be an important component of the future growth of natural gas production and the realization of carbon peak and carbon neutrality goals (Long et al. 2021; Zou et al. 2021). China is rich in shale gas resources and the effective development technology is increasingly mature. However, compared with North America, China still faces difficulties and challenges in large-scale commercial development. Shengxiang Long et al. proposed that to guarantee the realization of shale gas production goals in China, energy enterprises need to further optimize exploration and development

DOI 10.1201/9781003332657-17

goals and implementation plans. At present, the evaluation methods of shale gas development selection are mainly qualitative description methods and a combination of qualitative description and quantitative calculation. The qualitative description method evaluated the development block by the geological characteristics and parameters. Quantitative evaluation methods mainly include the statistical learning method, fuzzy mathematics method, and multi-objective decision analysis method. Statistical learning methods, such as the multiple regression linear model, are used to optimize favorable areas for shale gas development. However, multiple linear regression models often require a large number of indicators and a large amount of data, so that they can fit well. The amount of data in the shale gas development area often does not meet the requirements of the multiple linear regression model. The data are relatively fragmented. Therefore, the multiple linear regression model has certain limitations and errors in the selection evaluation (Zhou et al.2021). Fuzzy mathematical methods mainly include fuzzy optimization method, fuzzy matter-element method, and fuzzy comprehensive evaluation method. The core of evaluating shale gas development favorable areas by fuzzy mathematics method is to establish a fuzzy comprehensive evaluation model, which can quantify the non-quantitative indexes with fuzzy mathematical language. In this way, non-quantitative indexes can participate in a quantitative calculation, However, quantitative assignment of non-quantitative indexes is the main subjective assignment, and evaluation results may be different due to the result of the assignment. Moreover, a fuzzy membership matrix is not easy to establish (Li et al. 2013; Ren et al. 2012; Xie et al. 2016). Multi-objective decision analysis method mainly uses the analytic hierarchy process and grey relational analysis method to evaluate the target area. When using the above methods to evaluate the selection, most of the methods to determine the weight of indicators are analytic hierarchy processes. The analytic hierarchy process is a subjective method for determining weights, which needs to establish the consistency evaluation matrix of indicators. When there are many evaluation indexes, the data statistics are relatively large, and it is difficult to establish the judgment matrix, and the calculation of eigenvalues and eigenvectors is also relatively complex (Guo et al. 2015; Liang et al. 2014; Lu et al. 2021). In the evaluation of the shale gas development constituency, the index standard system will be slightly different according to the block characteristics of different regions. To study a set of universally applicable shale gas development block optimization models, which is not restricted by the index standard system, this paper establishes a selection evaluation model for shale gas development based on the TOPSIS method, which is widely used in economics, sociology, and engineering. The order relation analysis method without a consistency test is used to get the subjective weight of each index reflecting the expert preference. At the same time, the entropy weight method which makes full use of the data is used to get the objective weight of each index. The comprehensive weight reflecting subjective and objective weight can be obtained. Finally, the weighted TOPSIS method is used to optimize the development block. The model has no requirements for the number of indicators and does not need to establish evaluation standards in advance. The calculation is simple and easy to operate. An example is given to verify that the method and model are reasonable and effective.

2 ESTABLISHMENT OF EVALUATION INDEX

We investigated basic geological data and analyzed the studies of relevant scholars. The paper combs and summarizes the evaluation indexes affecting the shale gas development selection area, and selects the general key parameters as the evaluation indexes. This has a certain reference value for the optimization of most shale gas development areas. Li et al. (2013) established a selection index standard system for shale gas core development blocks from the aspects of geological conditions, economic benefits, and development environment, which contains 10 evaluation indexes. Xie et al. (2016) selected 13 evaluation indexes affecting shale gas development selection based on geochemistry, shale reservoir, gas-bearing property, and recoverable and preservation conditions. Based on the influence of different geological parameters on shale gas enrichment, Liang et al. (2014) established an index standard system of seven indexes. Guo et al. (2015) selected 12 evaluation indexes that affect shale gas development selection from underground geological factors of shale gas accumulation, shale gas development, and surface factors. Wang et al. (2013) discussed the key evaluation parameters and their value criteria that affect the scale and productivity of shale gas reservoirs and discussed 12 geological impact indicators and 12 engineering impact indicators. Taking the above literature as a reference, the author selected one or several representative indicators from the geological, geochemical, and productivity aspects affecting shale gas development, to optimize shale gas development blocks with fewer evaluation indicators and simpler evaluation models.

Total organic carbon: The proportion of organic carbon content in the pay zone is directly proportional to shale gas content. The content of organic carbon in the pay zone is high, and its shale gas content is also high. Studies in North America suggest that the organic carbon content of shale gas reservoirs with commercial value is usually between 2% and 10% (Jiang et al. 2012). Shale beds with higher organic carbon content generally have greater resource potential. The higher the organic carbon content is, the stronger the gas generation capacity is, and it is beneficial to the development of organic pores in shale.

Porosity: Shale pore structure characterization is an important indicator to evaluate shale reservoir performance, and porosity is a key parameter to reflect reservoir physical properties (Editorial board of shale gas geology and exploration and development practice series. 2011. Research progress of shale gas geology in China. Beijing: Petroleum Industry Press). There are adsorption, dissociation, and dissolution of shale gas in the reservoir. Adsorbed mainly occurs on the surface of kerogen or minerals, The free gas and dissolved gas of shale gas are mainly stored in the pores or fractures of the production layer, Porosity directly controls the content of free shale gas. The greater the porosity, the stronger the gas storage capacity, and the average porosity of shale reservoirs in four typical shale gas fields successfully developed in China is between 3% and 5%. The porosity of major shale gas-producing formations in the United States ranges between 4% and 6%. Through the comparison of relevant parameters of typical shale gas fields at home and abroad, the proportion of free gas is large, usually between 40% and 80%. The proportion of free gas in shale gas fields in South Sichuan and Fuling is between 60% and 80%. It shows that the content of free gas in shale gas is relatively high (Jiang et al. 2012).

Thermal maturity: Organic matter must be matured to form oil and gas. The maturity of organic matter determines the hydrocarbon-generating capacity of source rocks, and the maturity reflects whether the organic matter has entered the stage of gaseous hydrocarbon generation (gas generation window) (Guo et al. 2015). Generally, the maturity of organic matter is positively correlated with shale gas content. shale gas of RO in North America is between 1.0% and 3.1%. The thermal maturity of shale gas fields in southern Sichuan and Fuling in China is also mainly between 1.7% and 3.1%, which is in the mature high mature stage.

Net pay thickness: Effective thickness is the thickness of organic-rich shale. A certain thickness of a gas-bearing shale reservoir is the guarantee of shale accumulation, which reflects the enrichment degree of the shale gas reservoir and is also an important factor to measure the amount of shale gas resources. The greater the effective thickness of shale, the greater the exploration potential. The average thickness of shale gas reservoirs in the six major shale gas development areas in the United States is more than 30 m. In general, the effective thickness of shale gas reservoirs should be greater than 15 m (Li et al.2011; Zhao et al. 2011).

Gas content: Gas content is not only an important index for evaluating shale gas resource potential but also an important basis for judging whether shale gas has commercial exploitation value. Shale gas mainly occurs in reservoirs in the form of free gas and adsorbed gas. Generally, the higher the gas content, the better the gas content of the reservoir (Jiang et al. 2012; Li et al. 2011; Wang et al. 2013; Zhao et al. 2011).

Organic carbon content and thermal maturity are important factors affecting hydrocarbon generation conditions. Porosity is an important factor affecting the occurrence conditions of shale gas.Net pay thickness and gas content are important factors affecting resource development potential and commercial value. This paper aims to optimize shale gas development blocks by using these five less typical indicators, to find a kind of optimization model with high universality and simple operation.

3 THEORETICAL METHOD AND MODEL

3.1 *Entropy weight method*

The entropy weight method is an objective weighting method based on the degree of dispersion between decision information data to measure the impact of index data on decision objects (Li et al. 2013; Zhang et al. 2022). This method fully mines the hidden information in the data and reflects the difference in mapping behind the data through entropy value. If the information entropy value of an index is smaller, the difference of an index will be larger, the uncertainty of the index value will be smaller, the information provided by the data will be more important, and the weight of the index will be greater in decision-making (comprehensive evaluation). This method makes full use of data and is objective and easy to calculate. The calculation steps of the entropy weight method are as follows:m is the number of blocks (number of decision targets)

(1) Calculate the proportion of j index value in i block: p_{ij}

$$p_{ij} = x_{ij} \Big/ \sum_{i=1}^{m} x_{ij} \tag{1}$$

Where x_{ij} is the i-th index value of the j block.

(2) Calculate the entropy of the index:

$$e_j = -\frac{1}{\ln m} \sum_{i=1}^{m} p_{ij} \cdot \ln p_{ij} \tag{2}$$

(3) Calculate the entropy weight of the j indicator: w_j

$$w_j = 1 - e_j \Big/ \sum_{j=1}^{n} (1 - e_j) \tag{3}$$

Where n is the number of indicators.

3.2 *Rank correlation analysis*

In the evaluation of the shale gas development selection area, it is necessary to screen and make decisions on various indicators, and the influence degree of each indicator in the evaluation system is different. With the change in decision-making problems, the weight of various indicators will also change, and the Rank Correlation analysis method can meet the needs of such changes (Gong et al. 2020; Ma et al. 2017; Xu et al. 2014). The Rank Correlation analysis method to determine the weight is clear and simple, which can fully reflect the experience and preference of experts without a consistency test. The steps of the rank correlation analysis method to calculate the index weight are as follows:

(1) Determine the Rank relationship.

The evaluation index has the following order relation with a certain evaluation criterion: $X_i \succ X_j$ (X_i is more or less important than X_j)

(2) Determine the relative importance of adjacent indicators.

Experts' rational judgment on the ratio of relative importance between evaluation indexes X_{i-1} and X_i is as follows:

$$r_i = w_{i-1} \big/ w_i \tag{4}$$

where w_i is the weight of i-th indicator; $i = n, n-1, \ldots, 3, 2$. Refer to table 1 for the assignment of r_i.

Table 1. Reference table of values of r_i

r_i	instructions
1.0	Index X_{i-1} is as important as indicator X_i
1.2	Index X_{i-1} is slightly more important than index X_{i-1}
1.4	index X_{i-1} is more important than index X_{i-1}
1.6	Index X_{i-1} is strongly more important than indicator X_{i-1}
1.8	Index X_{i-1} is extremely more important than indicator X_{i-1}
1.1,1.3,1.5,1.7	In the middle of all situations

(3) Calculation of index weight:

$$w_n = \left(1 + \sum_{k=2}^{n} \prod_{i=k}^{n} r_i\right)^{-1} \tag{5}$$

where n is the number of indicators; $w_{i-1} = r_i w_i$, $i = n, n-1, n-2, \ldots, 3, 2$. The specific derivation process of the formula can be found in reference.

3.3 *TOPSIS method*

TOPSIS method is a commonly used and effective method in multi-objective decision analysis. It is a ranking method based on the approximation degree between the evaluation target and the ideal target. This method can objectively and comprehensively reflect the dynamic changes in shale gas development blocks. It is a dynamic model based on data. By defining a measure in the target space. In this paper, the European measure is adopted to calculate the degree that which each target is close to the positive ideal solution and far away from the negative ideal solution to optimize the shale gas development block (Li et al. 2013; Zhang et al. 2022).

Calculation steps:

(1) The decision matrix is constructed based on the decision objective

$$\overline{X} = \begin{pmatrix} x_{11} & x_{12} & \cdots & x_{1n} \\ x_{21} & x_{22} & \cdots & x_{2n} \\ \cdots & \cdots & \cdots & \cdots \\ x_{i1} & x_{i2} & \cdots & x_{in} \\ \cdots & \cdots & \cdots & \cdots \\ x_{m1} & x_{m2} & \cdots & x_{mn} \end{pmatrix} \tag{6}$$

Where x_{ij} represents the i-th attribute value of the j-th target; $i = 1, 2, \ldots, m, j = 1, 2 \ldots, n$.

(2) The indicators are the same as the trend. Usually, the indicators are positive, that is, the cost indicators are transformed into benefit indicators. The reciprocal method and range method are generally used to construct the forward matrix.

$$Y = \begin{pmatrix} y_{11} & y_{12} & \cdots & y_{1n} \\ y_{21} & y_{22} & \cdots & y_{2n} \\ \cdots & \cdots & \cdots & \cdots \\ y_{i1} & y_{i2} & \cdots & y_{in} \\ \cdots & \cdots & \cdots & \cdots \\ y_{m1} & y_{m2} & \cdots & y_{mn} \end{pmatrix}, \text{ where } y_{ij} = \begin{cases} x_{ij} & \textit{Benefit index} \\ 1/x_{ij} & \textit{Cost index} \end{cases} \tag{7}$$

(3) Normalize and forward the matrix.

$$Z = \left[z_{ij}\right]_{m \times n}, \text{ where } z_{ij} = \frac{y_{ij}}{\sqrt{\sum_{i=1}^{m} y_{ij}^2}} \tag{8}$$

(4) Construct positive and negative ideal goals.

$$\text{Positive ideal goal: } Z^+ = \left\{ \max_{1 \le i \le m, 1 \le j \le n} z_{ij} \middle| i = 1, 2, \ldots, m \right\} = \left\{ z_1^+, z_2^+, \ldots, z_n^+ \right\} \tag{9}$$

$$\text{Negative ideal goal: } Z^- = \left\{ \max_{1 \le i \le m, 1 \le j \le n} z_{ij} \middle| i = 1, 2, \ldots, m \right\} = \left\{ z_1^+, z_2^+, \ldots, z_n^+ \right\} \tag{10}$$

(5) Calculate the weighted Euclidean distance between the target to be decided, the ideal target, and the negative ideal target.

$$D_i^+ = \sqrt{\sum_{j=1}^{n} w_j \left(z_{ij} - z_j^+ \right)^2}, D_i^- = \sqrt{\sum_{j=1}^{n} w_j \left(z_{ij} - z_j^- \right)^2}, i = 1, 2, \ldots, m \tag{11}$$

Where w_j is the comprehensive weight ($w_j = \alpha_j \beta_j / \sum_{j=1}^{n} \alpha_j \beta_j$); α_j is the weight determined by the entropy weight method; β_j is the weight determined by the order relation method.

(6) Calculate the relative closeness.

$$D_i = \frac{D_i^-}{D_i^+ + D_i^-}, 0 \le D_i \le 1, i = 1, 2, \ldots, m, \tag{12}$$

The larger D_i is, the closer the evaluated target is to the positive ideal target, that is, the optimal target sought.

4 CASE APPLICATION AND ANALYSIS

10 blocks of a basin in Southwest China cited in reference (Li et al. 2013) are optimized. Li et al. used 11 evaluation indexes to optimize 10 blocks. In this paper, five universally applicable indicators are used to reflect the physical properties of shale reservoirs, which are effective thickness/*m*, organic carbon content/%, maturity/%, porosity/%, and gas content (m^3t^{-1}). Based on these five indicators, an optimization model for shale gas development block is established by using the sequential relations-TOPSIS method. Firstly, the decision matrix is constructed:

$\overline{X} =$ (13)

	net pay thickness	*Total organic carbon*	*Thermal maturity*	*Porosity*	*Gas content*
*block*1	9.6	1.5	2.5	2.3	3.7
*block*2	26	0.34	1	1	1.2
*block*3	38	1.2	1.2	1.7	2
*block*4	21.7	4.5	3	6.5	5
*block*5	22.8	2.8	1.7	4.6	4
*block*6	42	2	1.5	2.8	3.9
*block*7	28	2.4	1.6	4.5	3.6
*block*8	16.5	1.7	1.5	3.4	4.3
*block*9	34.6	2.5	1.2	2	1.5
*block*10	25.8	3	3.8	5	5.3

Note: some data in reference (Li et al. 2013) are cited.

Analyzing whether all the five selected indicators are positive. If there are negative indicators, namely cost indicators, it is necessary to apply Formula (7) to forward the negative indicators, and then construct the forward matrix and standardize. After analysis, all the indexes selected in this paper are positive. The positive index standardization matrix is established through formula (8).

$$Z=\begin{pmatrix} & \begin{matrix}\textit{Net pay}\\ \textit{thickness}\end{matrix} & \begin{matrix}\textit{Total organic}\\ \textit{carbon}\end{matrix} & \begin{matrix}\textit{Thermal}\\ \textit{maturity}\end{matrix} & \textit{Porosity} & \begin{matrix}\textit{Gas}\\ \textit{content}\end{matrix} \\ Block1 & 0.10810 & 0.19400 & 0.37896 & 0.19353 & 0.31596 \\ Block2 & 0.29278 & 0.04397 & 0.15158 & 0.08414 & 0.10247 \\ Block3 & 0.42791 & 0.15518 & 0.18190 & 0.14304 & 0.17079 \\ Block4 & 0.24436 & 0.58194 & 0.45475 & 0.54693 & 0.42698 \\ Block5 & 0.25675 & 0.36210 & 0.25769 & 0.38706 & 0.34158 \\ Block6 & 0.47296 & 0.25864 & 0.22738 & 0.23560 & 0.33304 \\ Block7 & 0.31531 & 0.31037 & 0.24254 & 0.37865 & 0.30742 \\ Block8 & 0.18580 & 0.21984 & 0.22738 & 0.28601 & 0.36720 \\ Block9 & 0.38962 & 0.32330 & 0.18190 & 0.16829 & 0.12809 \\ Block10 & 0.29053 & 0.38800 & 0.57602 & 0.42072 & 0.45259 \end{pmatrix} \tag{14}$$

Formula (9) is applied to determine the ideal block vector:

$$Z^{+}=(0.47296, 0.58194, 0.57602, 0.54693, 0.45259) \tag{15}$$

Formula (10) is applied to determine the negative ideal block vector:

$$Z^{-}=(0.10810, 0.04397, 0.15158, 0.08414, 0.10247) \tag{16}$$

The entropy weight method was used to determine the entropy weight of effective thickness, organic carbon content, maturity, porosity, and gas content as follows:

$$\alpha=(0.1423, 0.1498, 0.3231, 0.2029, 0.1819) \tag{17}$$

The order relation method was used to determine the order relation weights of the above indicators:

$$w=(0.1418, 0.1707, 0.2995, 0.1923, 0.1958) \tag{18}$$

Formula (12) is applied to calculate the relative closeness:

Table 2. Relative closeness value.

	D_i^+	D_i^-	D_i
Block 1	0.289585985	0.174778752	0.376382481
Block 2	0.416076362	0.069548997	0.143215333
Block 3	0.353263755	0.135912471	0.277839485
Block 4	0.109287645	0.375952087	0.774775977
Block 5	0.22924111	0.229370453	0.500141016
Block 6	0.275086435	0.208044823	0.430617601
Block 7	0.242874126	0.213483707	0.467798932
Block 8	0.29151383	0.171512875	0.37041681
Block 9	0.327261454	0.162235231	0.331432747
Block 10	0.119179242	0.353048741	0.747623508

Ten blocks are optimized by relative closeness degree, and the results are as follows:

Table 3. Optimization results of weighted TOPSIS method.

Block 4	Block 10	Block 5	Block 7	Block 6	Block 1	Block 8	Block 9	Block 3	Block 2
0.77478	0.74762	0.50014	0.46780	0.43062	0.37638	0.37042	0.33143	0.27784	0.14322

Li et al. (2013) analyzed 11 indicators and applied fuzzy optimization analysis to optimize the results:

Table 4. Optimization results of fuzzy optimization method.

Block 4	Block 10	Block 5	Block 7	Block 6	Block 9	Block 3	Block 8	Block 1	Block 2
0.9466	0.9321	0.5633	0.4826	0.4219	0.3182	0.3173	0.2727	0.2437	0.0546

Note: The data in Table 4 are from references (Li et al. 2013)

By comparing the optimization results of the model in this paper with those in reference (Li et al. 2013), it is found that the optimization results of the two models are consistent. The results show that the order relation-TOPSIS model is reasonable and effective. The comparison shows that it is feasible to optimize the exploitation of the development block based on five typical evaluation indexes: effective thickness, organic carbon content, maturity, porosity, and gas content.

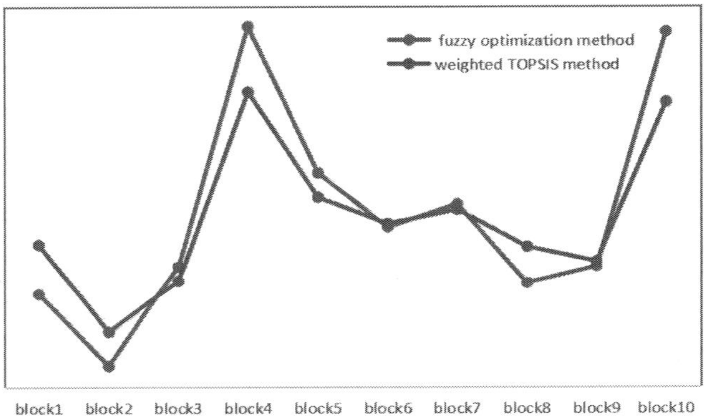

Figure 1. Comparison between weighted -TOPSIS method and fuzzy optimization method.

5 CONCLUSIONS

In this paper, the Rank Correlation Analysis method and the TOPSIS method are adopted to study the optimization of favorable areas for shale gas development, The main conclusions can be summarized as follows:

(1) Based on the comprehensive analysis of the influencing factors of shale gas development area and referring to the existing research, five universal typical indexes are selected from the multiple indexes of geological factors, geochemical factors, and productivity factors, which are effective thickness, organic carbon content, maturity, porosity, and gas content. The model calculation shows that these five indicators can optimize shale gas development blocks, and these five indicators apply to most shale gas development blocks. The model has good universality.

(2) The subjective weight and objective weight of each index affecting the selection area of shale gas development are determined by the order relation method and entropy weight method. Finally, the comprehensive weight reflecting the subjective and objective weight is determined by the combination weight formula. The comprehensive weight not only reflects the experience of experts but also fully follows the data law of each index, to effectively avoid the influence deviation caused by a single weight calculation method.

(3) The weighted TOPSIS optimization model ranks the selected shale gas development blocks by calculating the relative closeness between the selected shale gas development blocks and the

ideal development blocks, and the ranking results are consistent with the actual situation. The model makes full use of the original index data of each block, is not restricted by the number of samples and standards, is simple to operate, and has strong feasibility and universality.

In the future work, we should combine the evaluation of the "sweet spot" area and the favorable area of shale gas reservoir, and establish a general mathematical model, to better serve the exploration and development of shale gas. We can try to establish a TOPSIS model based on interval numbers to systematically evaluate shale gas blocks with complex geological parameter characteristics whose evaluation index value is interval number.

ACKNOWLEDGMENT

This paper was supported by the National Natural Science Foundation of China (Grant No. 52174060) and the Research on the construction technology of Gulong shale oil big data analysis system of Heilongjiang Province.

REFERENCES

Bing Liang, Yuanyuan Dai, Tianyu Chen, Weiji Sun, Bing Qin. Grey correlation optimization for shale gas exploration and development areas of complicated geological parameter features[J]. *Journal of china coal society*, 2014, 39(003):524–530.

Caineng Zou, Qun Zhao, Liantao Cong, HongYan Wang, Zhensheng Shi, et al. Development Progress, potential and prospect of shale gas in China[J]. *Natural Gas Industry*, 2021,41(01):1–14.

Can Li, Fengrong Zhang, Taifeng Zhu, Ting Feng, Pingli An. Evaluation and correlation analysis of land use performance based on entropy-weight TOPSIS method[J]. *Transactions of the Chinese Society of Agricultural Engineering*, 2013,29(05):217–227.

Chengju Gong, Weiwei Li, Yajun Guo. Rank Correlation Analysis Method In Group Evaluation[J]. *Operations research and management science*, 2020,29(11):152–156.

Dong Zhang, Bin Zhang, Rui Zhang, Zhoujian An. Study of the optimal allocation of multi-energy complementary cogeneration system based on entropy weight-TOPSIS[J/OL]. *Journal of Huazhong University of Science and Technology* (Natural Science Edition): 1-7[2022-01-20].DOI:10.13245/j.hust.220204.

Editorial board of shale gas geology and exploration and development practice series. *Research progress of shale gas geology in China* [M]. Beijing: Petroleum Industry Press, 2011.

Fujie Jiang, Weiqi Pang, Xuecheng OuYang, Jigang Guo, Zhipeng Huo, et al. The main progress and problems of shale gas study and the potential prediction of shale gas exploration[J]. *Earth Science Frontiers*, 2012,19(02):198–211.

Guogen Xie, Rupeng Zhang, Jinhua Yang. Shale Gas Favorable Area Evaluation Model Based on the Fuzzy Matter Element Analysis[J]. *Geological Science and Technology Information*, 2016,35(06):98–102+111.

Ji Ma, Xizhe Liu. Evaluation of health status of low-voltage distribution network based on order relation-entropy weight method[J]. *Power System Protection and Control*, 2017,45(06):87–93.

Jian Xu, Zhendong Du, Hongxiao Lin, Na Yuan, Hao Zhang. Confirmation of Index Weight of Water-saving Social Construction Based on Order Relation Analysis Method[J]. *Water Resources and Power*, 2014,32(10):132–13.

Jingzhou Zhao, Chaoqiang Fang, Jie Zhang, Li Wang, Xinxin Zhang. Evaluation of China shale gas from the exploration and development of North America shale gas[J]. *Journal of Xi'an Shiyou University* (Natural Science Edition), 2011,26(02):1–7+110+117.

Lei Ren, Bin Dou, Guoliang Liu, Shanshan Zhu. Geological and Regional Selection of Shale gas in Lower Jurassic of Sichuan Basin Based on Comprehensive Fuzzy Evaluation[J]. *Journal of Oil and Gas Technology*, 2012,34(09):177–180+

Shengxiang Long, Ting Lu, Qianwen Li, Guoqiao Yang, Donghui Li. Discussion on China's shale gas development ideas and goals during the 14th Five-Year Plan[J]. *Natural Gas Industry*, 2021,41(10):1–10.

Shiqian Wang, Shuyan Wang, Ling Man, Dazhong Dong, Yuman Wang. Appraisal method and key parameters for screening shale gas play[J]. *Jounal of Chengdu University of Technology* (Science &Technology Edition) [J]. 2013,40(06):609–620.

Weifeng Wang, Peng Liu, Chen Chen, Huili Wang, Shuai Jiang, Zhichao Zhang. The study of shale gas reservoir theory and resources evaluation[J]. *Natural Gas Geoscience*, 2013,24(03):429–438.

Wuguang Li, Shenglai Yang, Zhenzhen Wang, Qian Dong, Keliu Wu, Haiyang Wang [J]. *Journal of china coal society*, 2013,38(02):264–270.

Xiuying Guo, Yicai Chen, Jian Zhang, Ling Man, Haiqiao Zheng, Xiaojun Tong, et al. Assessment index selection and weight determination of shale gas plays: A Case study of marine shale in the Sichuan Basin[J]. *Natural Gas Industry*, 2015,35(10):57–64.

Yanjun LI, Huan Liu JiaXia Liu, Lichun Cao, Xuecheng Jia. Geological regional selection and an evaluation method of resource potential of shale gas[J]. *Journal of Southwest Petroleum University* (Science & Technology Edition), 2011,33(02):28–34+8–9.

Yaqiu Lu, Jin Wang, Mengxi Cao. Evaluation method of shale gas development area selection based on improved analytic hierarchy process[J]. *Reservoir evaluation and development*, 2021,11(02):70–77.

Yexin Zhou, Ankun Zhao, Qian Yu, Di Zhang, Qian Zhang, Zihui.Lei A new method for evaluating favorable shale gas exploration areas based on multi-linear Wufeng-Longmaxi Formations, Upper Yangtze Region[J]. *Sedimentary Geology and Tethyan Geology*, 2021,41(03):387–397.

Energy Revolution and Chemical Research – Kok-Keong Chong and Zhongliang Liu (Eds)
© 2023 The Authors, ISBN: 978-1-032-36554-1

Design of clearing mechanism for renewable energy in spot market

Fan Zhang* & Zechen Wu
State Grid Energy Research Institute, Beijing, China

ABSTRACT: The major function of the electricity market trading in the new power system is new energy. We must create a fair and sensible method for new energy to engage in the market, ensuring that all types of energy market participants benefit from victory. The win's impact on the grid's safe and dependable functioning. The paper proposes a design principle for new energy to participate in the spot market and designs the spot market mechanism as well as a market declaration, liquidation, and settlement compensation mechanisms to promote the consumption of new energy, taking into account the drastic increase in the proportion of new energy in the future. With the secondary energy and the secondary cleansing method, the result of new energy usage, the settling cost, and the simulation of the market user's income are all high, as are the result of new energy consumption, the price of new energy in the future, and the calculation of market participant's income. Based on the conduct of market players, cost allocation is fairer and more appropriate.

1 INSTRUCTIONS

Global warming and energy scarcity have become increasingly significant in recent years. Environmental and energy issues have become two important issues that governments all around the world are grappling with. Wind power and solar power generation have exploded in popularity across the world. By the end of 2019, China's total installed wind and PV capacity had surpassed 414 GW, accounting for 20.6 percent of the country's total installed capacity. By 2025, the installed capacity of new energy will account for more than 20% of primary energy, and new energy power generation will account for more than 60% of total power generation, according to the 2050 high percentage new energy development scenario and route study report (Xu 2019).

Many nations are now contemplating new energy involved in the spot market. Electricity spot markets have developed in countries such as the United States, the United Kingdom, and Germany. The market rules are designed differently, but the starting premise is essentially the same. Many national incentive programs have been implemented, and new energy development and consumption have been effective (Sun 2007).

China is now in the spot market exploration stage. Several trial payment operations have been conducted in the initial batch of eight-spot pilot locations, but the new energy techniques that are engaging in the market early on are cautious. Fujian uses a bid price proportional system to decide the quantity of power to enter the market bid based on the bid rate coefficient, allowing the transition from planning to market mode to be completed (Han 2020). The notion of trade agents is proposed in the text (Di 2020). Trading agents function as price receivers in the spot market in the early stages, not quoting volume and altering government-approved contract power according to market manipulation before progressively transitioning to the market model. Zhejiang New Energy has not joined the spot market, and preferential power generation and redemption acquisition are the market entrance strategies. Shandong has chosen a mechanism in which the majority of the electrical energy produced by new energy units is sold as set outputs, with the remainder being quoted as quoted quantities (SHI 2020). We presented and examined Sichuan Electric Power's spot

*Corresponding Author: zhangfansgeri@163.com.cn

DOI 10.1201/9781003332657-18

market mechanism, as well as a high and low double market model to fit the Sichuan power grid's operational characteristics (Lu 2021). To aggressively restrict the output of thermal power plants and encourage them to participate in deep peak shaving auxiliary services, Yamanishi's power spot market has implemented a joint optimum operating method for the power energy market and the deep peak shaving market. Optimize the distribution of important auxiliary resources. When fresh energy is abandoned, a new energy consumption process is initiated to ensure secondary liquidation, according to a previous study (Chen 2021).

The problem of China's renewable energy consumption is closely related to the country's energy resources and the characteristics of the energy mix. At the same time, this problem is aggravated by the large scale of centralized development and excessive local growth in the renewable energy-enriched areas. In addition, in recent years, the increasing rate of China's demand for electricity is slowing down, and the construction of the power grid and the policy support for renewable energy are lagging. Therefore, if we do not pay enough attention to it and solve it as soon as possible, then problems in renewable energy consumption will be more prominent (Xu 2019). The reasons for the above problems are as follows:

First, we organized the design principles of new energy that will reach the spot market ahead of Japan, as well as the design philosophy of Japan's market model. Then, in the future, install a large percentage of new energy. To assure the consumption of new energy, the new energy unit for secondary clearance is re-quoted when it abandoned energy in the market the day before, the spot market's operating method is developed, and the payment mechanism is improved. Finally, the usefulness of the suggested technique is demonstrated using an example.

2 DESIGN OF SPOT MARKET MECHANISM

2.1 *Market participation*

With the rise of the scale of wind and light new energy, the technology is maturing, and parity Internet connection is increasingly becoming a reality. As with other power sources, new energy power generating firms engage in the day ahead and real-time market in the form of volume quotation, as well as market bidding, market clearing, and dispatching operations. To gain priority power generating chances, rely on the benefit that its variable cost is lower than that of other forms of power generation firms.

2.2 *Market operation process*

Take the day-ahead market as an example to describe the organization process of the spot market. This process can be applied to the day or real-time market organization in the same way, and will not be repeated.

1) One-time clearing

   First, new energy generators and traditional power generators declare the volume price curve, and the load side submits the load demand curve based on load projection information in the day ahead market stage. The market operating organization clears through safe unit combination and optimum economic dispatching (SCUC + sced) and calculates the clearing results, which include the plan for unit output and power generation, as well as the node marginal electricity price. The formal clearing outcome is the one-time cleaning result if there is no energy abandonment at this moment.

2) Secondary clearing

   When new energy units are abandoned in the market, the right to a secondary quotation is granted to new energy units, but the quotation of generation units stays constant. The initial clearing's winning offer price will be made public. Because the first new energy unit's quotation is higher than the clearing price and there is no bid winner output, the new energy unit's quotation approach in the second quote will be somewhat conservative. To win the bid, it will not be more than the winning bid price of the first clearing.

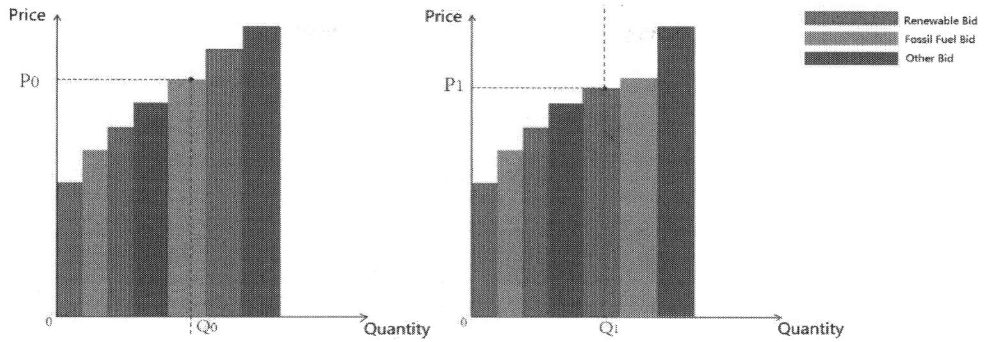

Figure 1. Quotation sequence of primary clearing and secondary clearing.

P0 is the initial clearing bid price, and new energy unit B failed to clear since the quote was greater than the bid price, as illustrated in Figure 1. The secondary quote for new energy units is expected to be lower than the clearing price. To satisfy the load at the time, the secondary clearing just won the bid, but the thermal power unit with the initial winning output failed to win the offer, equating to a power generating replacement right between the discarded unit and the heating power unit.

3 CLEARING MODEL

3.1 *The day-ahead market-clearing model*

The market's first clearing, which took place a few days ago, was identical to the customary clearing. The day-ahead market clearing optimization model's goal is to reduce the total power purchase costs as well as unit starting and shutdown costs. The restrictions primarily include system balance, system standby, power flow, new energy unit and thermal power unit power limit constraints, climbing and landslide limits, and so on.

The new energy units will be re-quoted, and the decision variables in the clearing model will change, when the primary clearing of the day ahead market produces abandoned energy and the second clearing is carried out. The entire power purchase cost is the objective function in the day-ahead market, and the unit starting and shutdown costs must also be considered. Thus, the clearing model is as follows:

$$\min\left(\sum_{i=1}^{N}\sum_{t=1}^{T}\sum_{s=1}^{M} C_{i,s}P_{i,s,t} + \sum_{i=1}^{N}\sum_{t=1}^{T}\left(1 - I_{i,t-1}\right)I_{i,t}\lambda_{i,t}^{\mathrm{th}}\right) \tag{1}$$

where N is the number of generator sets; M is the total number of quoted segments of the unit; T is the total number of periods; $P_{i,s,t}$ is the bid winning power of unit I in segment s of period T after secondary quotation; $C_{i,s}$ is the price corresponding to segment s declared by unit I after the second quotation; $\lambda_{i,t}^{\mathrm{th}}$ is the startup and shutdown cost of the thermal power unit; $I_{i,t}$ is the running state of the thermal motor unit I at time t (0–1 variable).

Load balance constraint

In each trading period, the balance between supply and demand of system load shall be ensured:

$$\sum_{i=1}^{N}\sum_{s=1}^{M} P_{i,s,t} = D_t \tag{2}$$

where D_t is the system load demand of time t.

Unit operation constraints

The output of the generator set shall meet its maximum and minimum output:

$$I_{i,t}P_{i,\min} \leq \sum_{s=1}^{M} P_{i,s,t} \leq I_{i,t}P_{i,\max} \tag{3}$$

where: $P_{i,\min}$ and $P_{i,\max}$ are the minimum and maximum technical output of unit i respectively;

3.2 *Real-time market*

The objective function of the real-time market is to minimize the total power purchase cost, and the mathematical model is described as

$$\min \sum_{i=1}^{N}\sum_{t=1}^{T}\sum_{s=1}^{M} C_{i,s}P_{i,s,t} \tag{4}$$

The constraints are similar to the previous clearing model.

4 CONCLUSION

This paper presents the design principles for new energy participants in the spot market, designs the spot market mechanism and market declaration, clearing, and settlement compensation mechanisms to promote new energy usage, and eventually simulates and calculates the consumption volume, clearing cost, and income of new electricity market subject areas in the cases of a high percentage of new energy.

Under the framework of a unified national market, China's renewable energy can mainly realize large-scale optimal allocation of resources through medium and long-term transactions across provinces. Specific means include participating in the electricity market and eliminating the problem of peak regulation and frequency modulation brought by the fluctuation of renewable energy through a flexible short-term transaction. Thus, gradually transforming it into a complete "long-term spot" market system. The results show that, compared to normal clearing, using the secondary clearing mechanism can promote new energy consumption and reduce system operation costs, and the corresponding cost distribution is fairer and more reasonable based on market subjects' behavior, and that it has achieved good results. The goal of this paper is to propose an effective spot mechanism for promoting new energy consumption in the market form of full electricity bidding and a high proportion of new energy, to provide a reference value for the design of China's power spot market mechanism aimed at promoting new energy consumption.

ACKNOWLEDGMENT

This project was funded by the State Grid Corporation of China headquarters management science and technology project (Grant No.SGHBJY00NYJS2000020).

REFERENCES

Chen Qixin, Liu Xue, Fang Xichen (2021). Power market clearing mechanism considering renewable energy security consumption [J]. *Power system and automation*, 6 (45): 26–33.

Dai jiefen, Chen Huwei (2020). Research on the transition mechanism of new energy participating in electricity spot market [J]. *Zhejiang electric power*, 39 (12): 78–84

Han Bin, Yan Jinghua, Sun Zhen (2020). Analysis on the initial mode of Fujian electric power spot market [J]. *Power system automation*, 3 (29), 1–6

Shi Xinhong (2020). Mode design of new energy participating in provincial spot market [J]. *Global energy Internet*, 3 (5): 451–460
Sun Yuanzhang, Wu Jun, Li Guojie (2007). Influence of wind power generation on power system [J]. *Power grid technology*, 31 (20): 55–62
Xu Tanghai, Lu Zongxiang, Qiao Ying (2019). High proportion renewable energy power planning with multiple types of flexible resource coordination [J]. *Global energy Internet*, 2 (1): 27–34.

Energy Revolution and Chemical Research – Kok-Keong Chong and Zhongliang Liu (Eds)
© 2023 The Authors, ISBN: 978-1-032-36554-1

A rapid calculation method of the cost of temporary power supply project based on modular technology: A case study of the 2022 Beijing Winter Olympic Games

Yanfeng Guo
Beijing Electric Power Economic and Technology Research Institute, Beijing, China

ABSTRACT: Temporary power supply project is characterized by a tight schedule, heavy tasks, and a short construction period, which makes estimated investment more difficult. Considering major events, epidemic prevention and control, rescue and disaster relief, and other situations, the demand for temporary power supply engineering construction is growing based on the need for emergency management. To improve the work efficiency of temporary power supply project, this paper takes the Winter Olympic Games power supply project as an example and proposes a rapid calculation method of the cost of a temporary power supply project based on modular technology, which reflects the project cost level more quickly and precisely and provides a reference for estimated investment more accurately. The application in actual project management further verifies the reasonableness and applicability of the method, which can meet the emergency management needs of rapid decision making, capital arrangement, speeding up construction, and guaranteeing operation.

1 INTRODUCTION

Electric power engineering is an important project related to the people's livelihoods of the country. With energy transition and the gradual increase of renewable energy penetration, the security and reliability of power supply in the grid are more important (Wang et al. 2022). To guarantee a safe power supply to venues such as temporary large-scale events or emergency infrastructure and reduce power supply accidents, emergency control during real-time operation of the power grid is imperative (Li et al. 2022).

The temporary power supply project provides power supply for the operation of temporary facilities from the low-voltage outlet line of the main and sub-distribution room of the venue to the terminal distribution box, including low-voltage cable outlet line, cable protection facilities, and terminal distribution box; if the main and sub-distribution room and other formal power distribution facilities in the venue do not have the conditions to lead the load, the new temporary box transformer and its external power supply project are also included in the scope of temporary power supply. The organization and design, equipment selection, and cost control of a temporary power supply project are not taken seriously in practice and cannot be considered a permanent power supply, but once the power failure will cause irreparable losses.

Temporary power supply projects require higher quality and reliability of power supply than ordinary power supply projects, and at the same time, the economy of the project should be considered. However, after receiving the notice of a temporary power supply project, there is generally no corresponding time for the preliminary feasibility study, preparation of preliminary design and accounting of engineering quantity, etc., which brings certain difficulties to the cost calculation of temporary power supply project (Islam et al. 2021).

Therefore, it is necessary to conduct in-depth research on the cost of a temporary power supply project and choose a suitable cost calculation method, so that the total investment of a temporary power supply project can be measured quickly. It will help more effectively to better meet the

 DOI 10.1201/9781003332657-19

needs of temporary power supply project construction and more accurately provide a reference for estimated investment.

The temporary power supply project for the Winter Olympic Games is a temporary power supply service purchased by the Winter Olympic Organizing Committee during the competition. The temporary power supply project of the Winter Olympic Games has higher requirements on the construction schedule, construction quality and power supply reliability, and more accurate and timely requirements on cost control level. According to the general management process of temporary power supply project cannot meet the needs of the Winter Olympics. Therefore, it is urgent to quickly and accurately calculate the modular design of the temporary power supply projects for the Winter Olympic Games.

Therefore, this paper takes the 2022 Beijing Winter Olympic Games as an example and proposes a rapid calculation method of the cost of a temporary power supply project based on modular technology. Firstly, it analyzes the composition of the construction cost of the temporary power supply projects for the Winter Olympic Games. Then, based on the modular technology, it divides the actual engineering quantity into unit modules according to the main technical conditions of the power supply scheme and compiles the "unit cost" modules. Finally, it takes the unit modules as the typical design basis to form a "package" typical design scheme based on the actual load demand of the temporary power supply project. The constructed method can be used to rapidly calculate the cost of a temporary power supply project.

2 ANALYSIS OF THE COST COMPONENTS OF TEMPORARY POWER SUPPLY PROJECTS FOR THE WINTER OLYMPIC GAMES

The total cost of temporary power supply projects for the Winter Olympic Games includes civil and erection costs, original equipment costs, and basic reserve fees.

(1) Civil and erection costs

Civil and erection cost includes direct cost, indirect cost, profit, and tax. Direct cost includes direct project cost and measure cost. Indirect cost includes the enterprise management fee and stipulated fee.

(2) Original equipment costs

The composition and determination of original equipment cost are divided into two cases: one is that the new equipment is leased to the Olympic Organizing Committee by State Grid Beijing Electric Power Company after the purchase of it, and the other is that the existing equipment of Beijing Jingdian Power Grid Maintenance Group Co., Ltd. is directly leased to the Olympic Organizing Committee.

(3) Other costs

Other costs refer to other relevant expenses necessary to complete the construction of the project other than civil and erection costs, original equipment costs, and basic reserve fees. Other costs include construction site acquisition and cleanup fee, project construction management fee, project construction technical service fee, and other relevant costs. The construction site acquisition and cleanup fee includes the transportation fee for returning the demolished materials to storage, which is calculated at 1.23% of the original equipment cost. In addition, other costs also include security costs and equipment storage costs. Security cost is 500 yuan per person per day according to the principle of one person for each piece of equipment. Equipment storage cost is calculated according to the rental store, with the low-voltage equipment at 4.7 yuan per day and the high voltage equipment at 37.3 yuan per day. The rental period is 270 days.

(4) Basic reserve fee

Basic reserve fee refers to the construction funds reserved for the increased costs due to design changes, possible losses caused by general natural disasters, and the costs of temporary measures taken to prevent natural disasters, as well as possible losses caused by other uncertainties.

3 RAPID CALCULATION METHOD OF THE COST OF TEMPORARY POWER SUPPLY PROJECT BASED ON MODULAR TECHNOLOGY

Modularization is a method to simplify complex product systems (Shao et al. 2021). Modular systems provide logistical flexibility in investment size and timing that can be strategically exploited to mitigate risk (Cao et al. 2011). The typical design scheme cost of temporary power supply projects for the Winter Olympic Games implements the modular design concept. It compiles the "unit cost" modules, and on this basis, a "package" typical design scheme is constructed. The method can reduce the repetitive design of the temporary power supply project, standardize the design, equipment, material, process, and construction, to avoid cost differences, and make the cost calculation of the temporary power supply project more accurate and faster.

3.1 *Compilation of the "unit cost" modules based on modular technology*

Referring to the temporary power supply engineering scheme of other important projects. this paper splits the actual quantity of work according to the actual load demand of the temporary power supply project based on the engineering drawings. Then composes unit modules close to the actual project according to the selection of equipment, main materials, and bill of quantities of the temporary power supply project.

In the actual project, to ensure the reliability of the power supply, SSTS box, and ATS box are used as low-voltage equipment; pluggable low-voltage cables are used, and aviation plugs are used as the connection method, which can be recycled repeatedly so that it cannot only shorten the construction period but also effectively control the project cost. In this paper, 20 "unit cost" modules with a high application ratio in actual projects are selected, considering the characteristics of temporary power supply projects for the Winter Olympic Games. The content and cost of each module are shown in Table 1.

Table 1. The "unit cost" modules of temporary power supply projects for the Winter Olympic Games.

Name	Model and specification	Unit	Integrated unit price (yuan)	Of which: civil and erection costs (yuan)	Of which: original equipment cost (yuan)	Of which: other costs (yuan)	Remark
shutter	Two-input and four-output	Stand	428785	201826	140000	86959	Amortized over five times
Box transformer	630 kVA	Set	465210	251094	121993	92123	Amortized over five times
Cable branch box	One-input and four-output	Set	110532	39569	40000	30963	Amortized over three times
Low-voltage terminal box	One-input and eight-output	Set	68912	32433	11667	24811	Amortized over three times
ATS box	400 A	Set	174579	43623	95000	35956	Amortized over three times
SSTS box	400 A	Set	583659	83808	400000	99850	Amortized over three times
Security Box	One-input and three-output	Set	29893	9928	1000	18965	Amortized over three times
UPS	200 kVA	Set	465388	65388	400000	/	Monthly lease
UPS	120 kVA	Set	285756	45756	240000	/	Monthly lease

(*continued*)

Table 1. Continued.

Name	Model and specification	Unit	Integrated unit price (yuan)	Of which: civil and erection costs (yuan)	Of which: original equipment cost (yuan)	Of which: other costs (yuan)	Remark
UPS	80 kVA	Set	199156	39156	160000	/	Monthly lease
UPS	30 kVA	Set	78693	18693	60,000	/	Monthly lease
Diesel generators	800 kW	Set	188614	/	/	/	Weekly lease
High voltage cables	ZC-YJY22-8.7/ 15kV-3×150 (direct burial through the pipe)	Km	1165481	995779	/	169702	Including 1 km of ϕ150 hot dipped steel pipe, 2 sets of intermediate joints, and 10 sets of terminal heads
Low-voltage cables	ZRC-REF-0.6/ 1 kV-1×150 (50 m per section)	Section	11253	7606	2008.1	1639	Amortized over ten times
Low-voltage cables	ZRC-REF-0.6/ 1 kV-1×150 (25 m per section)	Section	6494	4214	1333.1	947	Amortized over ten times
Low-voltage cables	ZRC-REF-0.6/ 1 kV-5×50 (50 m per section)	Section	10376	6251	2613.8	1511	Amortized over ten times
Low-voltage cables	ZRC-REF-0.6/ 1 kV-5×50 (25 m per section)	Section	5930	3525	1541.3	864	Amortized over ten times
Low-voltage cables	ZC-ERF-0.6/ 1 kV-5×35 (25 m per section)	Section	4752	2832	1226.3	693	Amortized over ten times
Low-voltage cables	ZC-ERF-0.6/ 1 kV-3×10 (25 m per section)	Section	3151	2393	299.2	459	Amortized over ten times
Rubber raceway	Two-hole	Piece	96.04	53.27	28.8	13.97	Amortized over ten times

3.2 *Construction of the "package" typical design scheme based on modular technology*

Based on the "unit cost" modules and taking into account the actual situation of temporary power supply projects for the Winter Olympic Games, this paper constructs a "package" typical design scheme for it, which is divided into 3 basic modules and 8 optional modules, According to different temporary power demand, the "package" basic modules and optional modules can be chosen selectively to form a complete typical design scheme and calculate the estimated investment of temporary power supply projects.

(1) Basic modules

This paper considers the actual situation of temporary power supply projects for the 2022 Beijing Winter Olympic Games. Firstly, three "package" basic modules are proposed according to different power supply radiuses, including "50 m power supply radius", "100 m power supply radius" and "150 m power supply radius". The load capacity that can be carried and the composition of equipment and materials of each basic module are shown in Table 2.

Table 2. Overview of the "package" basic modules.

Module name	Load capacity that can be carried	Composition of equipment and materials
50 m power supply radius	230 kVA	1 low-voltage terminal box (one-input and eight-output, containing communication module), 5 low-voltage cables (ZRC-REF-0.6/1 kV-1 × 150, 50 m per section), 300 insulated rubber raceway (two-hole).
100 m power supply radius	230 kVA	1 cable branch box (one-input and four-output, containing communication module), 3 low-voltage terminal boxes (one-input and eight-output, containing communication module), 5 low-voltage cables (ZRC-REF-0.6/1 kV-1 × 150, 50 m per section), 3 low-voltage cables (ZRC-REF-0.6/1kV-5 × 50, 50 m per section), 600 insulated rubber raceway (two-hole).
150 m power supply radius	2 × 230 kVA	2 cable branch boxes (one-input and four-output, containing communication module), 6 low-voltage terminal boxes (one-input and eight-output, containing communication module), 10 low-voltage cables (ZRC-REF-0.6/1 kV-1 × 150, 50 m per section), 6 low-voltage cables (ZRC-REF-0.6/1kV-5 × 50, 50 m per section), 1200 insulated rubber raceway (two-hole).

(2) Optional modules

At the same time, according to the principles of power supply for Winter Olympic venues and the principles of power supply reliability guarantee configuration for distribution equipment of Beijing Winter Olympic and Winter Paralympic Games, combined with the characteristics of power supply for equipment, special needs of venues and other factors, this paper proposes eight optional modules with different functions for temporary power supply projects, including 400 A ATS box, 400 A SSTS box, 200 KVA UPS, 120 KVA UPS, 80 KVA UPS, 30 kVA UPS, 800 kW diesel generator and security box. The load capacity that can be carried and the composition of equipment and materials of each optional module are shown in Table 3.

For important low-voltage loads, ATS boxes or SSTS boxes are used to supply dual utility power. For particularly important low-voltage loads and sensitive low-voltage loads, ATS boxes or SSTS boxes are used to supply dual utility power, while uninterruptible power supply (UPS) devices are added to ensure continuous power supply (Rahmat et al. 2017). For security loads such as perimeter cameras, a security box is used to supply power, and the superstructure power is supplied by dual utility power.

Using the cost of the full-cost modules above, the estimated investment of each "package" basic and optional module can be quickly calculated, as shown in Table 4. According to the actual demand for power supply for the Winter Olympic venues, the cost of the typical design scheme of the temporary power supply projects for the Winter Olympic Games can be obtained quickly, flexibly, and conveniently by selecting the appropriate "package" basic and optional modules and summing their integrated unit price.

Table 3. Overview of the "package" optional modules.

Module name	Load capacity that can be carried	Composition of equipment and materials
ATS box (400 A)	230 kVA	1 ATS box (400A, two-input and four-output, containing communication module), 5 low-voltage cables (ZRC-REF-0.6/1 kV-1 × 150, 25 m per section), 120 insulated rubber raceway (two-hole).
SSTS box (400 A)	230 kVA	1 SSTS box (400 A, two-input and four-output, containing communication module), 5 low-voltage cables (ZRC-REF-0.6/1 kV-1 × 150, 25 m per section), 120 insulated rubber raceway (two-hole).
UPS (200 kVA)	160 kVA	1 UPS (200kVA, for 20 min power supply), 5 low-voltage cables (ZRC-REF-0.6/1 kV-1 × 150, 25 m per section), 120 insulated rubber raceway (two-hole).
UPS (120 kVA)	96 kVA	1 UPS (120 kVA, for 20 min power supply), 5 low-voltage cables (ZRC-REF-0.6/1 kV-1 × 150, 25 m per section), 120 insulated rubber raceway (two-hole).
UPS (80 kVA)	64 kVA	1 UPS (80 kVA, for 20 min power supply), 1 low-voltage cable (ZRC-REF-0.6/1 kV-5 × 50, 25 m per section), 40 insulated rubber raceway (two-hole).
UPS (30 kVA)	24 kVA	1 UPS (30 kVA, for 20 min power supply), 1 low-voltage cable (ZRC-REF-0.6/1 kV-5 × 35, 25 m per section), 40 insulated rubber raceway (two-hole).
Diesel generator (800 kW)	/	1 diesel generator (800 kW, with quick-access plugs and cables).
Security box (One-input and three-output)	6 kVA	1 security box (one-input and three-output), 1 low-voltage cable (ZRC-REF-0.6/1 kV-3 × 10, 25m per section), 40 insulated rubber raceway (two-hole).

Table 4. Costs of modules of the "package" typical design scheme.

Module Type	Module Name	Unit	Quantity	Integrated unit price (yuan)
Basic Modules	50 m power supply radius	Set	1	153988
	100 m power supply radius	Set	1	662207
	150 m power supply radius	Set	1	924570
Optional Modules	ATS box (400 A)	Set	1	218573
	SSTS box (400 A)	Set	1	627653
	UPS (200 kVA)	Set	1	509382
	UPS (120 kVA)	Set	1	329750
	UPS (80 kVA)	Set	1	208928
	UPS (30 kVA)	Set	1	87286
	Diesel generator (800 kW)	Set	1	188614
	Security box (One-input and three-output)	Set	1	36885

4 A CASE STUDY OF RAPID COST CALCULATION OF TEMPORARY POWER SUPPLY PROJECT BASED ON MODULAR TECHNOLOGY

This paper takes the temporary power supply of the BRS broadcast commentators' workshop of the Winter Olympic Games test event as an example to verify the applicability of the rapid cost calculation method proposed in this paper. The BRS broadcast commentators' workshop is located on the upper floor of the spectator stands of the test event venue, about 100 meters away from the PowerPoint, and the main power equipment is computers, which is a particularly important low-voltage load. The total electrical load is about 60 kW.

To meet the demand for uninterrupted power supply and high-power quality of the load in the above case, the power supply method should be determined as a dual utility (one main and one backup). According to the load type, power quality demand, and power supply radius, this paper firstly selects two sets of "100 m power supply radius" from the "package" basic modules, and then selects one set of "SSTS box (400 A)", one set of "UPS (80 kVA)" and one set of "diesel generator (800 kW)" from the "package" optional modules, so that they can be combined to form a "package" typical design scheme for temporary power supply project of this case.

Using the above power supply scheme to calculate the estimated investment of it, the cost is 462,285 yuan × 2 (two sets of "100 m power supply radius") + 627,653 yuan (one set of "SSTS box (400 A)") + 208,928 yuan (one set of "UPS (80 kVA)") + 188,614 yuan (one set of "diesel generator (800 kW)") = 1,949,765 yuan. In a similar way can the cost be calculated by using the "unit cost" modules. The calculated cost is the same.

Through the above case, it can be seen that the rapid calculation method of the cost of a temporary power supply project based on modular technology constructed in this paper can adapt to the characteristics of a temporary power supply project such as tight time and urgent task, etc. The "unit cost" modules eliminate the compilation of engineering cost software. Based on it, the modules of the "package" typical design scheme can be combined flexibly to simplify the design scheme and provide the estimated investment at the first time, which provides effective and accurate costing support to simplify the design of temporary power supply project and lays a solid foundation for the construction of it.

5 CONCLUSION

Modular technology is widely used to promote the development of engineering technology and management mode. In this paper, modular technology is introduced into the field of temporary power supply engineering cost, and the proposed rapid cost calculation method is closer to the actual project and more convenient to use. Firstly, based on the main technical conditions of the power supply scheme, this paper divides the actual quantity of work into unit modules and the cost of the full-cost modules is compiled. Then, according to the actual power supply radius, the "package" typical design scheme is constructed. Finally, the costs of the "package" basic modules and optional modules are obtained by summing the cost of the corresponding unit modules, and the calculation method of the typical design scheme cost of a temporary power supply project is established to provide help for the rapid calculation of temporary power supply project cost in the future.

REFERENCES

Cao, H., Zhang, H., and Liu, M. (2011): Global modular production network: From system perspective. *Procedia engineering* 23, 786–791.

Islam, M. S., Nepal, M. P., Skitmore, M., and Drogemuller, R. (2021): Risk induced contingency cost modeling for power plant projects. *Automation in Construction* 123, 103519.

Li, X., Wang, X., Zheng, X., Dai, Y., Yu, Z., Zhang, J. J., Bu, G., and Wang, F. (2022): Supervised assisted deep reinforcement learning for emergency voltage control of power systems. *Neurocomputing* 475, 69–79.

Rahmat, M. K., Karim, A. Z. A., and Salleh, M. N. M. (2017): *Sensitivity analysis of the AC uninterruptible power supply (UPS) reliability.* 2017 International Conference on Engineering Technology and Technopreneurship (ICE2T), pp. 1–6.

Shao, Y., Hu, Y., and Zavala, V. M. (2021): Mitigating investment risk using modular technologies. *Computers & Chemical Engineering* 153, 107424.

Wang, Y., Sun, Y., Zhang, Y., Chen, X., Shen, H., Liu, Y., Zhang, X., and Zhang, Y. (2022): Optimal modeling and analysis of microgrid lithium iron phosphate battery energy storage system under different power supply states. *Journal of Power Sources* 521, 230931.

Energy Revolution and Chemical Research – Kok-Keong Chong and Zhongliang Liu (Eds)
 ISBN: 978-1-032-36554-1

Analysis of the distribution characteristics of audible noise of UHV AC transmission lines

Li Zhang & Yudan Wang*
Department of Mechanical and Electrical Engineering, Wenhua College, Wuhan, China

Guo Zhao & Shulin Li
School of Electrical and Electronic Engineering, Hubei University of Technology, Wuhan, China

ABSTRACT: To provide corresponding technical support for solving electromagnetic environmental disputes reasonably, the distribution characteristics of audible noise of UHV AC lines are studied. By measuring the background noise of a 1000 kV UHV transmission line in rainy weather, the frequency distribution characteristics of audible noise are analyzed by using an A-weighted network, and the noise is detected by transverse and longitudinal angles to analyze the spatial distribution of noise. The analysis results show that the high-frequency noise of the audible noise is obvious at 4 kHz, and the attenuation is large at 16 kHz. The noise distribution in the low-frequency band is mainly concentrated at 200 Hz and below, and the noise decreases with the increase of the distance from the line center.

1 INTRODUCTION

While improving the transmission capacity and reducing the occupation of transmission line corridors, the UHV power grid has attracted wide attention due to the electromagnetic environment problems caused by high voltage and large currents (Liu 2013). As an important index of the electromagnetic environment, the level of audible noise directly affects the selection and arrangement of transmission lines (Sun 2015), it has become a decisive factor in the selection of conductors for some UHV transmission lines (Li 2022). Audible noise of UHV AC transmission lines has been studied at home and abroad. Existing research mainly focuses on line segment test, corona cage test, prediction model and calculation of audible noise level, and determination of audible noise limit (Zhang 2008). Since there is no actual UHV AC transmission project in foreign countries except the former Soviet Union, there is a lack of measured data on audible noise of actual UHV long lines in China and abroad. The transmission line of 1,000 kV Jindongnan-Nanyang-Jingmen UHV AC pilot demonstration project was officially put into commercial operation in 2009, which made China have good technical conditions to carry out the measured research on the electromagnetic environment of actual UHV transmission lines. But the existing measured data of audible noise of UHV transmission lines are mainly in the test results of no rain, snow, and lightning weather required by environmental protection (Li 2019). Absence of bad weather, and accumulated statistical data, while audible noise is greatly affected by environmental climate (Luo 2017), residents' complaints about audible noise are mainly concentrated in the rain, snow, fog, and other bad weather. At the same time, the calculation results of the audible noise prediction formula are mostly statistical values (Xie 2016). Therefore, it is necessary to carry out the long-term test of the audible noise of UHV transmission lines and conduct a statistical analysis of the measured results (Chen 2012). Based on the actual measurement of the audible noise of the UHV transmission line in rainy

*Corresponding Author: 243291474@qq.com

DOI 10.1201/9781003332657-20

weather, this paper analyzes its propagation law and provides corresponding technical support for solving the problem of electromagnetic environment disputes.

2 AUDIBLE NOISE EVALUATION CRITERIA AND CONTROL VALUES

Under sunny weather conditions, the audible noise generated by transmission line corona is not large, but under rainy weather conditions, the water droplets below the conductor increase the corona discharge intensity, and the audible noise value is about 15–20 dB larger than that under sunny weather. Therefore, the monitoring and control value of audible noise should focus on rainy days for AC transmission lines. Transmission line audible noise mainly includes two parts: 1) Broadband noise generated by wind and rain in rainy weather; 2) 50 Hz and its multiple frequencies.

China has promulgated the standard "DL/T1187-2012 1000 kV overhead transmission line electromagnetic environment control value", which stipulates the control value of audible noise. The audible noise near the environmental sensitive point should meet the requirements of GB 3096 corresponding functional zoning. The sound environmental quality standard GB 3096 – 2008 specifies the audible noise of different types of places in different periods as shown in Table 1. The standard stipulates the classification requirements of 0–4 categories of sound environmental functional areas.

Table 1. Noise standard in China [unit: dB(A)].

Category	Daytime	Night
0	50	40
1	55	45
2	60	50
3	65	55
4a	70	55
4b	70	60

3 HUM MEASUREMENT

Weather conditions for 1000 kV AC transmission line background noise test: humidity 68%, temperature 28°C, wind speed 3 m/s. The measurement sites were all surrounded by farmland without houses. The tower is sheep-shaped. The minimum height of the conductor at the tower is 32.5 m, and the distance between the edge conductor and the centerline is 26 m. Starting below the projection of the middle phase wire, the data is measured every 1 m to a horizontal distance of 15 meters below the middle phase wire. B & K 3050 noise spectrum analyzer and B & K 4231 calibration is adopted; the accuracy of the sound level meter is level I; the frequency range is from 20 Hz to 20 kHz; the noise range is from 20 dB to 140 dB.

The background noise is measured at the same measuring point as much as possible when the line is a power cut. The ambient noise is constantly changing, resulting in the measured background noise being different at different times. To reduce this effect, the measurement of background noise is carried out at the same time as the measurement of transmission line noise, and the selected measuring points are kept as similar as the natural and social environment around the transmission line, to reduce the error caused by the measurement of background noise. The background noise is measured about 150 meters away from the line side wire. Figure 1 and Figure 2 are background noise FFT and CPB spectrum distribution.

It can be seen from the FFT and CPB diagrams of the background noise that the background noise is mainly concentrated in the region below 200 Hz at low frequency, and the sound pressure level at 4 kHz and above high frequency is low.

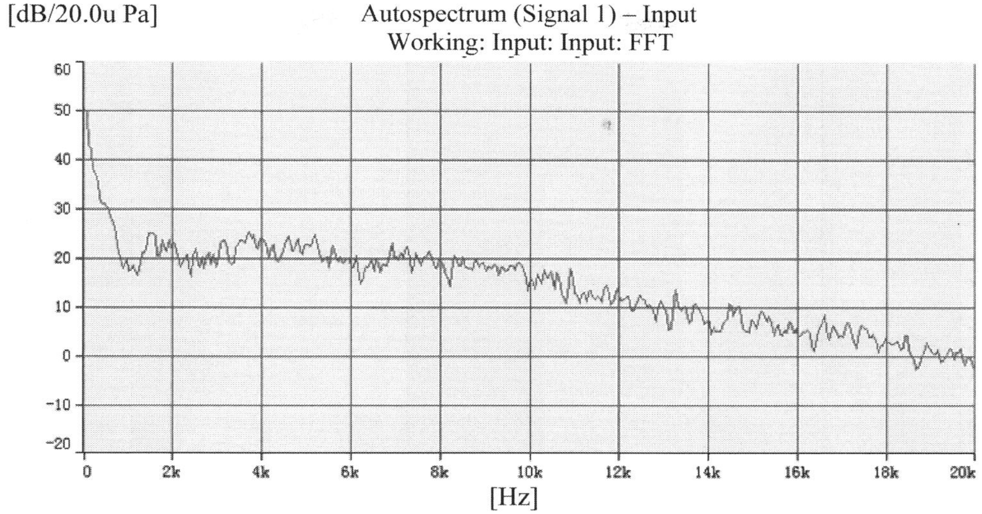

Figure 1. FFT Spectrum Distribution of Background Noise.

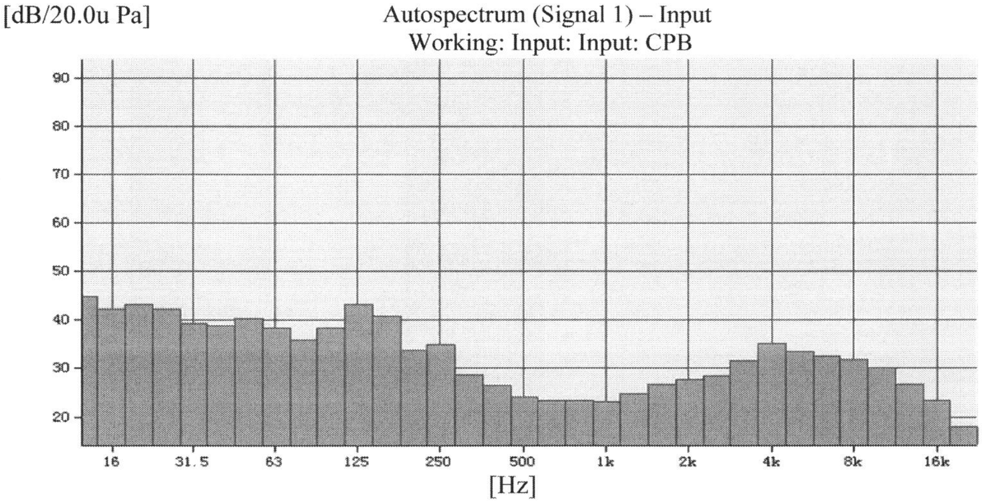

Figure 2. Spectrum distribution of background noise CPB.

4 ANALYSIS OF PROPAGATION LAW OF AUDIBLE NOISE

4.1 *Frequency distribution characteristics of audible noise*

The frequency distribution characteristics of audible noise are illustrated by taking the distribution map of some measured audible noise as an example. Figure 3 shows the spectrum of audible noise of typical AC transmission lines (four-bundle and eight-bundle conductors). Four measuring points are selected to measure the audible noise level below the line. The measurement point information here is consistent with the above electromagnetic field measurement point information. Figures 4 to 7 are the distribution map of field-measured audible noise. It can be seen that the corona noise of the transmission line begins to decay when the frequency is very high, and the level of environmental

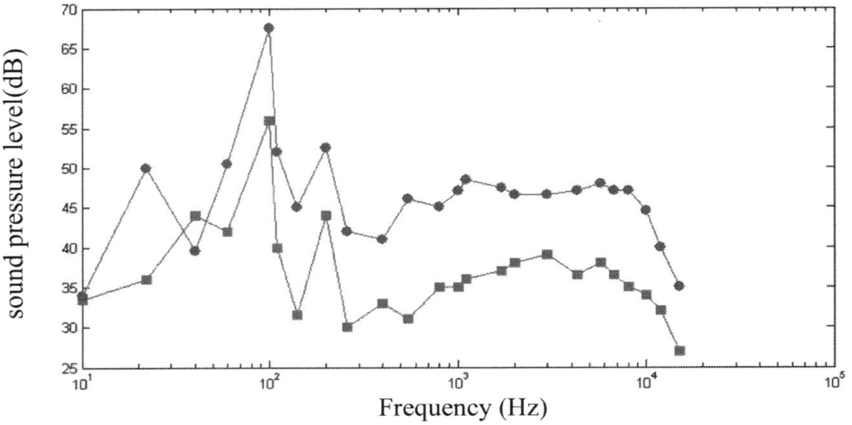

Figure 3. Spectrum of typical AC transmission line audible noise.

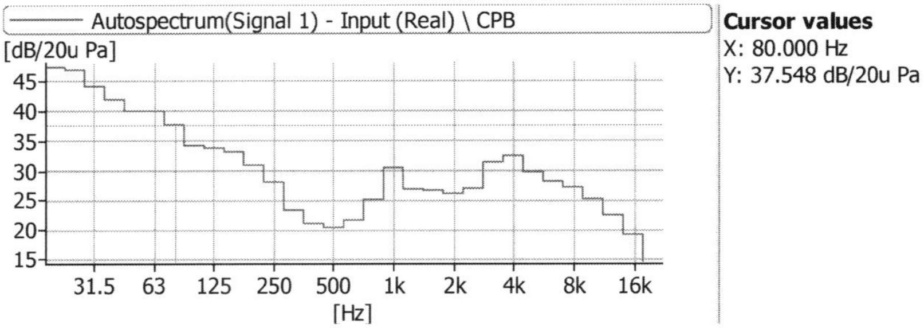

Figure 4. Spectrum of audible noise at measuring point 1.

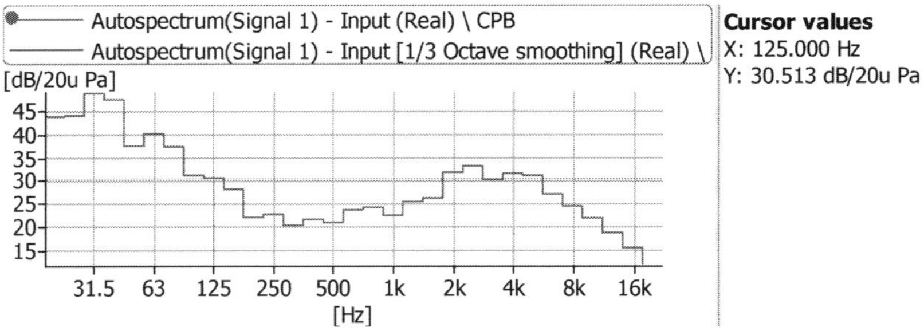

Figure 5. Spectrum of audible noise at measuring point 2.

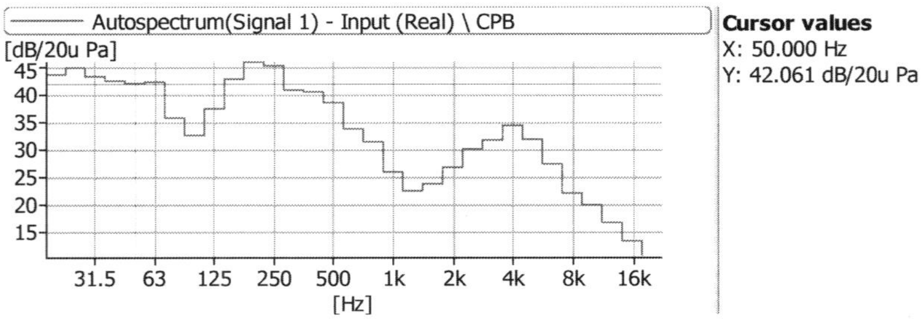

Figure 6. Spectrum of audible noise at measuring point 3.

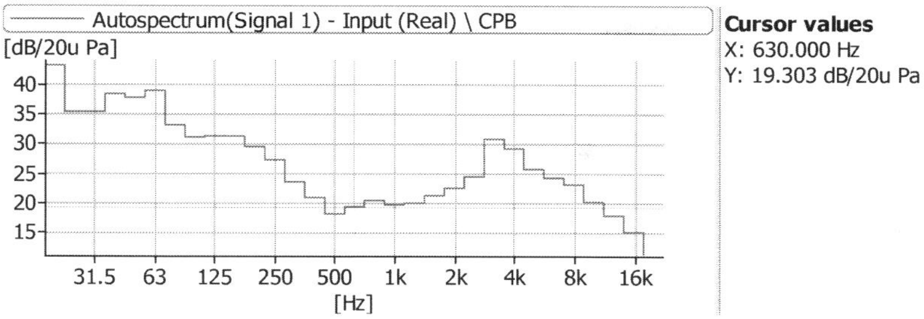

Figure 7. Spectrum of audible noise at measuring point 4.

noise decreases obviously with the increase of frequency after 100 Hz. In this way, in the case of low environmental noise, the high-frequency noise generated by the transmission line is easy to be distinguished. It is precise because of this characteristic that the corona noise of the transmission line gives a sense of hearing abnormality. Therefore, the use of an A-weighted network in the measurement of corona noise is more realistic. The measurement results reflect the law of high-frequency random noise, and the influence of the main energy concentrated in the low-frequency area can be ignored.

Figures 4–7 show that the variation of audible noise is the same, high-frequency noise is more obvious at 4 kHz to 16 kHz attenuation. Low-frequency noise distribution is mainly concentrated at 200 Hz and below.

4.2 *Spatial distribution characteristics of audible noise*

Two measuring points with different ground heights are selected for measurement, and the transverse and longitudinal distribution characteristics of audible noise are analyzed.

1) Transverse distribution characteristic

Audible noise values are measured from 0 m to 20 m below the line from the center point and then measured at 40 m, 60 m, 80 m, and 100 m respectively. The measured results of audible noise from 1 m to 20 m from the center of the line are shown in Table 2, and the trend is shown in Figure 8.

Table 2. Measurements of Audible Noise from 1 m to 20 m from Line Center.

Distance from the center point (m)	Audibility noise (dB(A))	Distance from the center point (m)	Audibility noise (dB(A))
1	32.7	11	39.5
2	34.8	12	39
3	36.9	13	38.8
4	37.7	14	40.7
5	39.3	15	40.1
6	39.6	16	39.5
7	39.4	17	39.5
8	39.2	18	39.4
9	39.5	19	40.3
10	39.5	20	40.1

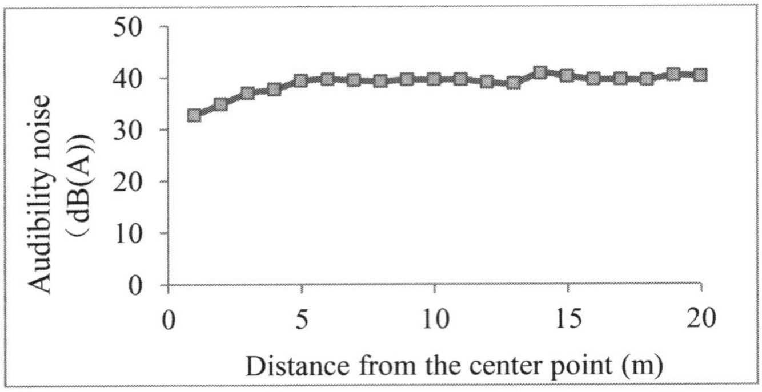

Figure 8. Transverse distribution characteristic curve of audible noise.

Obviously, from the above figure, there is no attenuation of audible noise near the wire.

To study the attenuation law of audible noise, the measuring point is extended to 100 m, in which the height of the first measuring point is 36.5 m and the height of the second measuring point is 33.6 m. The measurement results are shown in Table 3.

Table 3. Measurements of Audible Noise from Line Center to 100 m.

	Audibility noise (dB(A))	
Distance from line center	Measure point one	Measure point two
0	33.8	38.1
10	39.5	36.5
20	40.1	35.7
40	35.2	34.3
60	33.9	32.8
80	28.7	27.6
100	25	23.4

2) Longitudinal distribution characteristics

Due to the different measurement locations of the two groups of data, the wire-to-ground height, environment, and other factors are slightly different, so the data are slightly different. In general, the audible noise levels are between 20 dB and 50 dB. Figure 9 below shows the trends of two sets

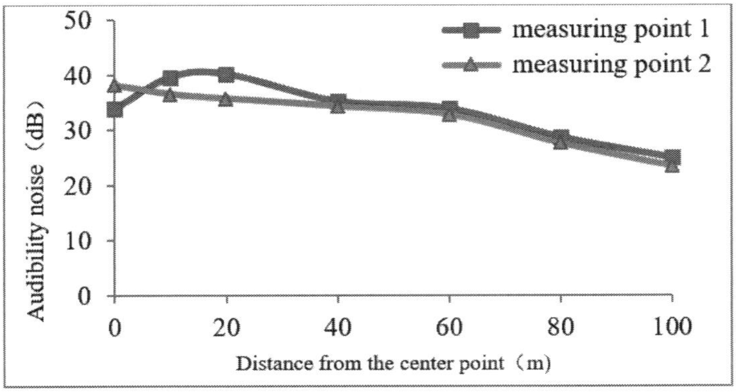

Figure 9. Trend of Audible Noise from Line Center to 100 m.

of data with distance. And the audible noise value of measuring point 1 is slightly larger than that of Measuring Point 2. The higher the height of wire to the ground is, the lower the audible noise is.

It can be seen from Figure 9 that the audible noise gradually decreases with the increase of the distance from the line center point, and the attenuation is obvious when the distance is beyond 80 m.

5 CONCLUSIONS

In this section, the measurement points below the typical UHV AC transmission lines are selected to measure and analyze the audible noise, and the variation law of each measurement value with distance is obtained. The conclusions are as follows:

1) The high-frequency noise of audible noise is obvious at 4 kHz, and the attenuation is large at 16 kHz. Low-frequency noise distribution is mainly concentrated in the 200 Hz and below frequency band; audible noise decreases with the increase of distance from the line center point, and the attenuation is obvious when the distance is beyond 80 m. The higher the wire-to-ground height is, the lower the audible noise is.
2) The line noise level meets the limits specified in the electric power industry standard "DL/T 1187-2012 1000 kV overhead transmission line electromagnetic environment control value".
3) Increasing the number of split wires can effectively reduce the level of audible noise around the wire, which can be prioritized in engineering design.

REFERENCES

Chao Luo, Shubo Fang, Zhiming Zha, Qing Han, Weifang Yao, Xueying Hua Zhang Xiaoqin. Research on Distribution Characteristics of Electromagnetic Field and Attenuation Law of Pure Sound for UHV AC Transmission Line [J]. *High Voltage Apparatus*, 2017, 53(08): 94–99.

Guangzhou Zhang, Yao Zhang, Baoquan Wan, et al. Optimization of conductor configuration for UHV double-circuit transmission lines[J]. *High Voltage Engineering*, 2008, 34(9): 1039–1843 (in Chinese).

Hui Li. Discussion on Influencing Factors of Audible Noise of 1000 kV AC Transmission Line [J]. *Technology Innovation and Application*, 2022, 12(04): 64–66.

Huichun Xie, CUI Xiang, LIU Huagang, CHEN Yuchao. Corona Characteristic Comparison of (8+2) Bundled Wire and Normal Wire Applying in UHV AC Double-circuit Transmission Line on the Same Tower [J]. *High Voltage Engineering*, 2016, 42(3): 966–972.

Pengfei Li, ZHANG Yao Zhang, Tong Zhang, Kejie Dai, Yifan Liu, Zhimin Zhao. Weather Factors and Prediction Method for "100 Hz" Pure Tone of UHV AC Transmission Lines [J]. *High Voltage Engineering*, 2019, 45(09): 2971–2979.

Xin Sun, Weijiang Chen, Jiayu Lu, et al. *Environmental impact assessment of AC electric power transmission and transformation project* [M]. Beijing: Science Press, 2015: 189–192 (in Chinese).

Yuchao Chen, Xie Hunchun, Zhang Yao, et al. Audible noise prediction of UHVAC transmission lines based on corona cage[J]. *High Voltage Engineering*, 2012, 38(9): 2189–2194 (in Chinese).

Zhenya Liu. *Ultra-high voltage AC & DC grid* [M]. Beijing: China Electric Power Press,2 013: 233–261 (in Chinese).

Energy Revolution and Chemical Research – Kok-Keong Chong and Zhongliang Liu (Eds)
© 2023 The Authors, ISBN: 978-1-032-36554-1

Research on energy efficiency evaluation model in typical energy system scenario

Liang Song, Xiaolong Xu, Xiaoguang Tang, Tong Liu, Peng Gao & Lin Liu
Zaozhuang Power Supply Company of State Grid Shandong Electric Power Company, Zaozhuang, Shandong, China

Hao Liu*
School of Economics and Management, North China Electric Power University, Beijing, China

ABSTRACT: In order to adapt to the trend of China's energy development and coordinate the relationship between energy security and low-carbon transition, integrated energy system services came into being. Integrated energy system service provides users with cold, heat, electricity, and gas comprehensive service with high system comprehensive energy efficiency through coordinated optimization and integration of resources and technologies so that various energy forms in the system interact with each other and form unique functions of the system as a whole (Cheng 2017). However, there are still many problems to be solved in the evaluation of energy efficiency of the comprehensive energy system in China: there is no clear energy efficiency evaluation system; there is a lack of understanding of energy evaluation; the development of energy efficiency requires a large amount of investment in energy efficiency improvement, which affects the economic benefit target (Hu 2018). Scenarios based on this, this article selects industrial park energy system, the selection of indicators and build the model, evaluation of energy efficiency of industrial park construction scheme, by building an integrated consideration of electricity, gas, cold and hot pluripotent complementary arrangement in use, and through the energy production, transportation, consumption, storage of evaluation system, so as to realize the comprehensive efficiency of scientific evaluation to the entire society. Thus, we can adjust the energy use structure of enterprises, create a new growth pole of the company, improve energy use efficiency, promote industrial transformation and upgrading, and improve the overall energy efficiency level of the society.

1 INTRODUCTION

In terms of energy transmission, the park has built heat and cold transmission network on the basis of transmission and distribution lines; in terms of energy use, the main energy users are mainly industrial enterprises, so there is a certain relationship between the industrial structure and composition (NEL AJH 2018). This paper develops a systemic energy efficiency index system with the park as a whole, from the source-grid-load perspective of an integrated energy system (Xu 2020). The energy efficiency evaluation index system for industrial parks is shown in the table below

*Corresponding Author: 1332892714@qq.com

 DOI 10.1201/9781003332657-21

Table 1. Energy efficiency evaluation index system for typical scenarios in industrial parks.

Target	Primary index	Secondary index
Energy efficiency evaluation index system for typical scenarios in industrial parks	Supply-side energy efficiency	Total energy supply Energy equipment acquisition costs Operating costs Initial investment cost
	Transmission-side energy efficiency	Supply and demand balance for hot, cold, and electricity Average voltage Average load factor Power Factor Current imbalance rate The cumulative time of voltage failure Energy efficiency ratio of network transmission Loss of network transmission volume
	Demand-side energy efficiency	Energy demand Energy consumption intensity Cost of energy use Renewable Energy Utilisation Energy use efficiency Equipment utilization efficiency The proportion of clean energy installed Equipment energy efficiency ratio

2 ENERGY EFFICIENCY EVALUATION MODELS FOR ENERGY SYSTEMS

2.1 *Determination of weights based on the AHP method*

The steps of hierarchical analysis modelling include: building a structural model of the hierarchical order; constructing a judgment matrix; hierarchical single ordering and consistency testing. The specific implementation is as follows.

2.1.1 *Modelling the structure of the hierarchy*

When applying AHP analysis to an assessment problem, the problem is first systematized by hierarchizing the factors and constructing a model of a hierarchy of factor progressions. Under this model, complex problems are broken down into components that form several levels of elements by attributes and relationships. The factors at the upper-level act as guidelines for the factors at the lower level, and the lower-level factors are refinements of the upper-level factors (Li 2020).

2.1.2 *Constructing the judgment matrix*

Once the recursive hierarchy model has been established, the affiliation of the upper and lower-level factors is determined. In order to determine the importance of the elements of each level relative to the objectives of the previous level, a judgment matrix needs to be constructed. The Saaty 1–9 scale is used in this report to indicate the relative importance between factors using a specific numerical scale (1–9).

Table 2. Judgment matrix scales and meanings.

Scale	Meaning
1	Two factors are of equal importance
3	Slightly more important
5	Obviously important
7	Strongly Important
9	Extremely important
2,4,6,8	Median of the above two adjacent judgments

2.1.3 *Hierarchical single sort*

Hierarchical ranking refers to the ranking of the importance of the factors at the same level of the hierarchical analysis in relation to the factors of the indicators at the previous level. This is generally determined by calculating the eigenvectors of the judgment matrix corresponding to the largest eigenvalues.

The judgment matrix $A=(a_{ij})_{n\times n}$ is normalized by columns to obtain the matrix $\overline{A}=(\overline{a}_{ij})_{n\times n}$, as:

$$\overline{a}_{ij}=a_{ij}/\sum_{j=1}^{n}a_{ij},\ (i,j=1,2,\ldots,n) \tag{1}$$

The average of the sum of the elements of each row in the matrix is calculated as:

$$w_i=\frac{1}{n}\sum_{j=1}^{n}\overline{a}_{ij} \tag{2}$$

It is calculated to obtain $w=[w_1,w_2,\ldots,w_n]^T$, which is the requested eigenvector.

The maximum characteristic root of the judgment matrix $\lambda_{\max}$ is calculated.

$$\lambda_{\max}=\frac{1}{n}\sum_{i=1}^{n}\frac{(A\omega)_i}{\omega_i} \tag{3}$$

2.2 *Energy efficiency evaluation model based on fuzzy integrated evaluation method*

The AHP-based fuzzy comprehensive evaluation method is a comprehensive evaluation method that takes full account of the complementary nature of AHP and fuzzy comprehensive evaluation methods in terms of strengths and weaknesses, as well as the fuzzy nature of certain uncertain problems (Tian 2019). The advantage of the method lies in the elimination of uncertainty in the indicators with a quantitative approach. In the process of constructing the indicators, a combination of qualitative and quantitative indicators is used, which can more appropriately describe the evaluation object and is conducive to improving the accuracy of decision-making. The AHP-based fuzzy integrated evaluation model is as follows.

2.2.1 *Determinants set and rubric set*

The factor set represents the combination of evaluation factors of the evaluated object. If there are n factors, it can be expressed as $U = (u_1, u_2, \ldots, u_n)$, A rubric set is a finite collection (e.g. excellent, good, fair, poor) of fuzzy ratings. The rubric can be divided into m levels depending on the actual situation, as $V = (v_1, v_2, \ldots, v_m)$.

2.2.2 *Determining weight allocation*

Each factor in the factor set has a different degree of importance, so the AHP method was used to assign weight to them.

2.2.3 *Constructing the affiliation matrix*

The degree of affiliation r_{ij} indicates how likely the evaluation subject is to v_j evaluate the evaluated object under the indicator u_i. The affiliation vector is $R_i = (r_{i1}, \underset{\sim}{r}_{i2}, \ldots, r_{im}), i = 1, 2, \ldots, n$, and meets $\sum_{j=1}^{m} = 1$. The affiliation matrix of the affiliation vector is $\widetilde{R} = (R_1, R_2, \ldots, R_n)^T = (r_{ij})_{n\times m}$.

2.2.4 *Calculating the overall evaluation results*

Based on the determination of the weight vector w and the affiliation matrix $\widetilde{R}$, the results of the overall evaluation $\widetilde{B}$ are calculated.

$$\widetilde{B} = w \cdot \widetilde{R} = (a_1, a_2, \ldots, a_n) \tag{4}$$

2.2.5 *Normalization process*

In order to make the results of the comprehensive evaluation comparable, this paper normalizes the results of the comprehensive evaluation.

$$b_i = \frac{\widetilde{b}_i}{\sum_{j=1}^{m} \widetilde{b}_j} \tag{5}$$

3 CASE STUDIES OF ENERGY EFFICIENCY EVALUATION AND ENERGY EFFICIENCY ANALYSIS UNDER TYPICAL SCENARIOS

3.1 *Calculation of subjective weights for the AHP method*

Experts are asked to score each indicator of each type of industrial park, and the relationship of the judgment matrix of the first-level indicators can be obtained. The feature vector can be calculated by the judgment matrix, and the feature vector is the weight between each secondary index.

$$w = \begin{bmatrix} 0.2526 & 0.4660 & 0.8468 \end{bmatrix}^T$$

Normalizing the weights yields:

$$w = \begin{bmatrix} 0.1635 & 0.3015 & 0.5480 \end{bmatrix}^T$$

Similarly, the judgment matrix can be formed for each secondary index and tertiary index, and the feature vectors and weights can be calculated. The weight ratio of different second-level indicators to first-level indicators can be derived, the weight ratio of different third-level indicators to second-level indicators can be obtained, and finally, the weight ratio of 20 third-level indicators to first-level indicators of the parking industry can be obtained by calculation. The specific index weights are shown in the following table.

Table 3. High and low levels of importance for each indicator.

Second-level indicators		Third-level indicators		Weight to the first level indicator
Factor	Weights	Factor	Weights	
Supply-side energy efficiency	0.1635	Total energy supply	0.0737	0.0120
		Energy equipment acquisition costs	0.2383	0.0389
		Operating Costs	0.1985	0.0324
		Initial investment cost	0.4895	0.0800
Transmission-side energy efficiency	0.3015	Hot and cold electric process balance	0.0926	0.0279
		Average voltage	0.1444	0.0435
		Average load factor	0.0972	0.0293
		Power Factor	0.1613	0.0486
		Current imbalance rate	0.1478	0.4456
		The cumulative time of voltage failure	0.0898	0.0271
		Network transmission energy efficiency ratio	0.1247	0.0038
		Network transmission volume loss	0.1422	0.0429
Demand-side energy efficiency	0.5480	Energy demand	0.0877	0.0480
		Energy consumption intensity	0.1003	0.0549
		Cost of energy use	0.1252	0.0686
		Renewable Energy Utilization	0.1545	0.0847
		Energy use efficiency	0.1513	0.0829
		Equipment utilization efficiency	0.0524	0.0287
		The proportion of clean energy installations	0.0563	0.0309
		Equipment energy efficiency ratio	0.1060	0.0581

3.2 *Comprehensive evaluation calculation based on fuzzy comprehensive evaluation method*

Due to the limited data samples, this paper uses the fuzzy comprehensive evaluation method based on the AHP-entropy power method to obtain the data that cannot be collected directly by fitting and research on the basis of keeping the authenticity of the data as much as possible, and if the data are missing and difficult to obtain, they are filled according to the way of empirical values. This comprehensive evaluation selected an industrial park before and after as the evaluation object, through the collection of detailed data of each indicator and the classification of each indicator level in the area. After data processing to obtain the comprehensive evaluation index values and affiliation parameters, the specific parameters are shown in the following table.

Table 4. Comprehensive evaluation index values and affiliation parameters.

Third-level indicators	Before	After	parameter 1	parameter 2	parameter 3	parameter 4
Total energy supply	31	40	40	35	30	25
Energy equipment acquisition costs	23	27	30	26	22	18
Operating Costs	56	70	90	80	70	50
Initial investment cost	58	70	90	80	50	40
Hot and cold electric process balance	25	18	0	10	25	50
Average voltage	20	21	90	84	70	60
Average load factor	70	82	80	70	60	50
Power Factor	60	86	90	75	60	80
Current imbalance rate	22	13	0	0.876	5.02	87.6
The cumulative time of voltage failure	30	5	0	0.876	5.02	87.6
Network transmission energy efficiency ratio	8	5	1	5	7	9

(continued)

Table 4. Continued.

Third-level indicators	Before	After	parameter 1	parameter 2	parameter 3	parameter 4
Network transmission volume loss	10	6	2	5	3	6
Energy demand	90	92	0	10	25	50
Energy consumption intensity	88	90	1	5	7	9
Cost of energy use	0.8	0.6	0.35	0.5	0.7	0.9
Renewable Energy Utilization	56	60	90	80	70	55
Energy use efficiency	72	85	90	80	70	60
Equipment utilization efficiency	25	18	0	10	25	50
The proportion of clean energy installations	36	45	75	65	50	20
Equipment energy efficiency ratio	4.5	0.55	0.45	0.5	0.57	66

The affiliation of each evaluation index to the evaluation set before and after the construction of the project is shown in the following table.

Table 5. Affiliation of each evaluation index to the evaluation set for pre/post-construction projects.

Indicators	Excellent	Good	Medium	Poor
Total energy supply	0/0	0.400/0.750	0.600/0.250	0/0
Energy equipment acquisition costs	0/0.750	0.400/0.250	0.600/0	0/0
Operating Costs	0/0.500	0/0.500	0.700/0	0.300/0
Initial investment cost	0/0.667	0.733/0.333	0.267/0	0/0
Hot and cold electric process balance	0/0	0.9/0	0.100/1	0/0
Average voltage	0/0.889	0/0.111	0.600/0	0.400/0
Average load factor	0/0.6	0/0.4	0.900/0	0.100/0
Power Factor	0/0	0/0.995	0.302/0.005	0.698/0
Current imbalance rate	0/0.067	0.200/0.933	0.800/0	0/0
The cumulative time of voltage failure	0/0.500	0.400/0.500	0.600/0	0/0
Network transmission energy efficiency ratio	0/1	0/0	1/0	0/0
Network transmission volume loss	0/0	0/0.533	1/0.476	0/0
Energy demand	0/0	0/1	0.500/0	0.500/0
Energy consumption intensity	0/0	0/0.500	0.500/0.500	0.500/0
Cost of energy use	0/0	0/0	0.800/0.667	0.200/0.333
Renewable Energy Utilization	0/0.500	0.800/0.500	0.200/0	0/0
Energy use efficiency	0.400/0.100	0.600/0.900	0/0	0/0
Equipment utilization efficiency	0/0	0/0	0.467/0.167	0.533/0.833
The proportion of clean energy installations	0/0	0/0.750	0/0.250	1/0
Equipment energy efficiency ratio	0/0	0/0.714	0.060/0.286	0.940/0

From the results of the weighting of the tertiary indicators under each secondary indicator, based on this, the overall different levels of affiliation values are calculated as:

Table 6. Evaluation results.

Second-level indicators	Comprehensive evaluation results of secondary indicators				Secondary indicator level
	Excellent	Good	Medium	Poor	
Supply-side energy efficiency	0.0000	0.0964	0.0956	0.0086	Good
Transmission-side energy efficiency	0.0000	0.0863	0.6229	0.0665	Medium
Demand-side energy efficiency	0.0320	0.0704	0.0953	0.1153	Poor
Comprehensive evaluation results	0.0320	0.2531	0.8138	0.1904	Medium

From the above table, it can be seen that before construction, the supply-side energy efficiency aspect performs well, the other stages perform average, and the demand-side energy efficiency aspect is very poor. In a comprehensive view, the pre-construction evaluation of the region is mostly at a moderate to low level, and the overall evaluation is moderate and needs improvement.

Based on the results of the weights of the third-level indicators under each second-level indicator, the comprehensive evaluation results of the second-level indicators are calculated, and the affiliation level of each subsystem of comprehensive energy efficiency can be obtained, as shown in the following table.

Table 7. Evaluation results.

Second-level indicators	Comprehensive evaluation results of secondary indicators				Secondary indicator level
	Excellent	Good	Medium	Poor	
Supply-side energy efficiency	0.0928	0.0933	0.0145	0.0000	Good
Transmission-side energy efficiency	0.1258	0.3568	0.2970	0.0000	Good
Demand-side energy efficiency	0.0220	0.1871	0.0623	0.0415	Good
Comprehensive evaluation results	0.2406	0.6372	0.3738	0.0415	Good

4 CONCLUSION

From the above table, we can see that after the construction, the performance of all indicators is good. In a comprehensive view, the area is mostly evaluated at a good level, the overall evaluation is also good, and the industrial park can play a positive and good role.

ACKNOWLEDGMENTS

This paper was supported by the State Grid science and technology projects "Research and Application of Energy Efficiency Service Technology under the Framework of Energy Internet-Research on Energy Efficiency Evaluation System and Energy Efficiency Analysis Method in Typical Scenarios of Topic 2 (SGSDZZ00YXJS2001196)".

REFERENCES

Cheng Lin, Zhang Jing, Huang Renle, et al. Case analysis of multi-scenario planning based on multi-energy complementation for integrated energy system[J]. *Electric Power Automation Equipment*, 2017, 37(6): 282–287.

Hu Biao, Sun Xue. Energy Efficiency Evaluation Analysis of the Three Urban Agglomerations in China[J]. *IOP Conference Series: Earth and Environmental Science*, 2018, 189(6).

NEL AJH, VOSLOO JC, MATHEWS M J. Financial model for energy efficiency projects in the mining industry[J]. *Energy*, 2018, 163:546–554.

Tao Xu, Jianxin You, Hui Li, Luning Shao. Energy Efficiency Evaluation Based on Data Envelopment Analysis: A Literature Review[J]. *Energies*, 2020, 13(14).

Yuanyuan Li, Zhenning Zhao, Tongrui Cheng, Jinjing Li, Li ng Bai. Method for Energy Efficiency Evaluation of Coal-fired Unit Based on Environmental Protection and Reliability[J]. *E3S Web of Conferences*, 2020, 204.

Yuan Tian, Zicong Yu, Kang Liu, Yongqiang Zhu, Ruihua Xia. The research on the index system of extended energy efficiency evaluation[J]. *IOP Conference Series: Earth and Environmental Science*, 2019, 227(4).

Energy Revolution and Chemical Research – Kok-Keong Chong and Zhongliang Liu (Eds)
© 2023 The Authors, ISBN: 978-1-032-36554-1

Analysis and countermeasures of safety management of crude oil transfer in Ningbo Port

Majing Lan & Pan Shao
China Waterborne Transport Research Institute, Beijing, China

ABSTRACT: In order to build a safety management model for lightering operations of crude oil in Ningbo Port, and promote the continuous improvement of the safety management level of it, this paper analyzed the reasons for the safety of lightering operations of crude oil in Ningbo Port and put forward the main countermeasures, based on the background of lightering operations of crude oil of Ningbo Port. Meanwhile, this paper adopted the method of combining literature summarization and safety management research to promote Ningbo Port to establish a long-term mechanism for safety management of lightering operations of crude oil, standardize lightering operations of crude oil in Ningbo Port, and improve its safety management level.

1 INTRODUCTION

To better implement the national energy development strategy, China has deployed a refinery with a crude oil processing capacity of 10 million tons per year and a crude oil storage depot with a reserve capacity of 10 million tons per year in Ningbo Technology Development Zone, Guangxi Province. However, the 300,000-ton crude oil terminal and the 300,000-ton waterway of the supporting refinery were still under construction, which has brought tremendous pressure and risks to the normal production of the 10 million tons refinery project of Petro China. Relying on the two 100,000-ton crude oil berths that have been put into production to directly berth foreign ships or transfer them from other ports in China to unload oil can't meet the continuous oil refining needs. Once the supply is cut off, the risk of the refinery is extremely high. Therefore, it is urgent to choose another channel to solve the problem of crude oil transferring. After comprehensive consideration of many aspects, as well as factors such as risk, time, efficiency, and cost, it is better to choose the method of anchorage lightering operations of crude oil in Ningbo Port, so as to solve the problem of crude oil transferring.

2 SAFETY MANAGEMENT OF LIGHTERING OPERATIONS OF CRUDE OIL IN NINGBO PORT

Lightering operations of crude oil is recognized as a high-risk offshore operation in the world. Once an accident of crude oil transfer occurs, huge oil spills may occur, leading to unimaginable consequences and huge and serious environmental pollution incidents, impacting port production, and bringing extremely bad social influence (Liu 2013). Therefore, in order to effectively deal with the safety issues of lightering operations of crude oil that have emerged under the new situation and effectively improve the level of safety management, the transportation industry is actively exploring corresponding management models. Ningbo Port, which has never used anchorages to transfer crude oil, faces severe challenges both in terms of business technology and security risk management and control. Based on the principle of law, the personnel of Ningbo Port have strengthened their management concept, combined with the actual situation, carefully analyzed various problems of offshore crude oil transfer at Ningbo Port, and explored countermeasures for the safety management of lightering operations of crude oil under the new situation.

DOI 10.1201/9781003332657-22

3 EXISTING PROBLEMS AND COUNTERMEASURES

3.1 *Emphasis on the main problems, moving forward the risk, and controlling threshold*

Lightering operations of an oil tanker are divided into ongoing lightering and anchor lightering (Li 2013). Due to the higher technical requirements for ongoing lightering, Ningbo Port has considered using anchorage to anchor the aboard ship, using the lighter ship (barge) to single-string berth, unberth and barge, as shown in Figure 1.

Figure 1. Lightering operations of crude oil.

3.1.1 *Selecting anchorage*

The appropriate choice of anchorage is a compulsory condition for the safe anchors of oil tankers. For the choice of anchorage, it is necessary to comprehensively consider the impact of the tonnage of the anchored ship, the bottom quality of the anchorage, wind, and waves, climate, and the navigation environment of the surrounding waters. In addition, sufficient waters must be set aside for the use of aboard ships, daughter ships, tugboats, and anti-fouling ships on duty in accordance with the port's general plan. It is required to invite a design unit with relevant qualifications to design and demonstrate by experts. Meanwhile, the navigation safety demonstration and anchorage safety demonstration are needed to pass, and the opinions of marine management departments should be consulted, so as to achieve legal, compliance, scientific and safe selection of anchorage. Therefore, Ningbo Port has selected three anchorages of 200,000 tons and 300,000 tons with water depth and bottom quality suitable for anchoring as crude oil transfer anchorages, after comprehensive consideration, evaluation and demonstration.

3.1.2 *Company qualifications and talents of lightering operations of crude oil*

Only a few domestic ports such as Guangzhou, Ningbo, Zhanjiang, and Maoming have carried out large-scale crude oil ships transfer at sea with small volumes, and most of them are considered auxiliary operations for load shedding. There are a few ports in Japan, Singapore, Australia, and the United States that have engaged in similar transfers abroad. There is not much experience for oil tankers to learn in anchorage lightering operations, and not any specific and complete operating standards. If Ningbo Port establishes a new crude oil transfer company and team, not only will it fail to meet the time-critical refinery production, but there will also be risks during the operation. Therefore, Ningbo Port introduces other companies' aid. After evaluation, it introduces a special team formed by a shipping company with crude oil transfer experience directly under the leadership of Ningbo Port Group and establishes Crude Oil Handling Co., Ltd. to take charge of loading and unloading crude oil in Ningbo Port.

3.1.3 *Selecting crude oil barge ship*

The lightering operation of crude oil requires long ship age and high performance due to high operational risks. Ningbo Port is expected to transfer 4-5 crude oil aboard ships every month on average. And the lighter ships need to frequently use the port for entry and exit. In order to ensure the lightering operation of crude oil can be carried out smoothly and without interruption, it is necessary to use newer ships with stable and good performance as far as possible. Therefore, the lightering operation of crude oil at Ningbo Port selects two 70,000-ton oil tankers with relatively short age and stable performance to carry out the operation, which can not only meet the needs of accelerating the transfer speed but also satisfy the requirements of risk prevention and emergency replacement.

3.1.4 *Leakage risk of anchorage outside the port*

The waters of Ningbo are close to the waters of Donghai. Once the crude oil leaks accidentally, it may not only cause pollution in the waters of Ningbo and Donghai but also lead to related diplomatic troubles. Therefore, the outer anchorage of Ningbo Port has chosen waters far away from the boundary line with the Vietnamese sea. At the same time, the newly established crude oil leakage logic tree was used for evaluation and analysis, so as to find shortcomings, supplement the corresponding equipment and facilities, and strengthen the safety management and prevention of crude oil transfer. The crude oil leakage logic tree is shown in Figure 2.

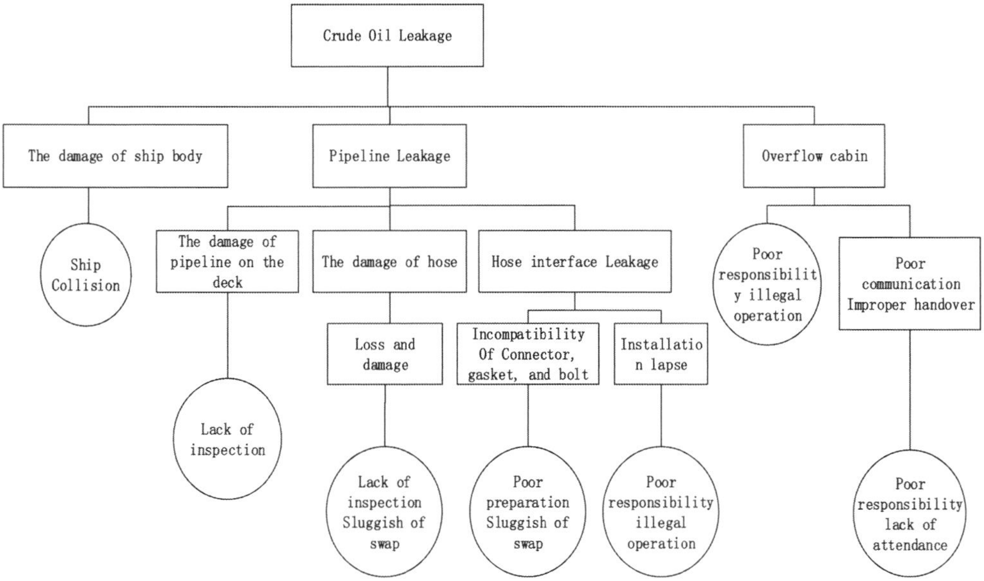

Figure 2. Crude oil leakage logical tree.

3.2 *Building a safety management system for tanker transfers and crude oil transfers*

The safety management of the lightering operation of crude oil involves many issues such as fire and explosion accidents, shipwrecks, sea crashes, oil pollution, etc. All kinds of problems is dependent on the standardized management of the system. For example, if the safety management system is not comprehensively constructed, there will be confusion in emergency command and hidden dangers in safety management, which will lead to major accidents after emergencies not being dealt with in time. Due to the subjective and objective factors such as high rents of large crude oil vessels, the limited stay in the port for crude oil transfers, the one-way approach channel at Ningbo Port and wind, weather and climate, as well as the variety of crude oil required by CNPC's refineries needing to be in place in time, a three-level safety management coordination mechanism for the lightering operation of crude oil has been initially established at Ningbo Port, that is to say, the joint cooperation and support of the three major working links of the Port Coordination Committee, the Maritime Safety Assessment Committee and the Transfer Site Coordination Committee.

The first one is the coordination committee of the lightering operation of crude oil. The Coordination Committee for lightering operation of crude oil is established by the port management department and the local port management office, including the member of the port management department, the port office, the port joint inspection unit, the transfer company, the marine anti-pollution company, and the terminal (factory) owner. They organize a coordination meeting at the end of each month, so as to summarize the operation work of this month and coordinate and arrange the work for the next month. As for the Production Coordination Committee, its main task is to coordinate customs, inspection and quarantine, maritime affairs, border inspection, and other port management departments to efficiently support the relevant procedures for the lightering operation, to supervise the port dispatch department and the pilotage department to rationally arrange the pilotage dispatch and the entry and exit of crude oil sub-ships, and to check the implementation of safety assessment and safety regulations by the transfer company.

The second one is the maritime safety assessment meeting. The maritime department regularly organized a safety assessment meeting, with the port management department, the maritime search and rescue command center (or maritime ship management agency), the pilotage department, the maritime anti-pollution company, and the transfer company. They will hold the meeting in the middle of every month in principle. The meeting mainly studies and evaluates the recent and current weather and wind current characteristics, and conducts a safety assessment of the lightering operation of the month, especially the impact of seasonal swells and windy days on the transfer of crude oil and the impact of crude oil containing a high concentration of hydrogen sulfide on the safety of transfer personnel. And preventive measures will be continuously evaluated and controlled. Meanwhile, it will check the implementation of emergency drills such as cable breaks, collisions, leaks, terrorist attacks, surges, and gale attacks.

The third one is the coordination meeting of the on-site lightering operation. A pre-barge coordination meeting, led by the transfer company is established to make preparations for coordination and inspection before leaving and opening the pump. An on-site coordination meeting must be held before the lighter ship approaches and departs from the aboard ship. The staff of the lighter ship, the transfer company, the tugboat, and the pilots need to attend the meeting. The pre-barge meeting will mainly study the specific berthing and departure plans and methods for the specific size of crude oil aboard the ship, weather, and tidal conditions, arrange staff to inspect and check the specific transfer subjects and the onboard watchkeeping matters, and coordinate and communicate with the aboard ship and notify the specific transfers that require cooperation and precautions.

Through the establishment of the above three-level consultation system, the parties involved in port administration, ports, transfers, pilotage, and pollution prevention will be coordinated to achieve the mechanism of "government-led, inter-departmental linkage, close collaboration, and gradual progress" so as to build a safety management network of crude oil transfer.

3.3 *Strengthening the training and education of crude oil transfer and consolidating the safe foundation*

Although the personnel of the crude oil transfer company can engage in the lightering operation of crude oil, the waters, wind, and waves conditions of Ningbo Port are quite different from those of Ningbo Port. For example, Ningbo Port doesn't have the southwest swells in summer and the strong winds in autumn and winter. In addition, the foggy conditions of Ningbo Port and Ningbo Port are inconsistent. Due to insufficient estimation of the hazards of the southwest swell and insufficient experience of the operators, three cables were disconnected continuously in the anchorage transfer. Therefore, it is imperative to strengthen safety education and training in practice.

The training of personnel focuses on technical safety clarification and on-site comments in the coordination meeting, so as to improve their business level (Wu 2010). At the same time, employees are organized to accept collective training, consolidate their original knowledge and skills, and improve their safety awareness of the transfer site in their spare time.

3.4 *Strengthening guarantee and emergency management*

The supply of crude oil to the PetroChina refinery needs to be on time. Once the supply is cut off, the risk of the refinery is extremely high. If an accident occurs, it will cause heavy casualties. In order to ensure the safe and timely supply of crude oil from the PetroChina Refinery, the crude oil transfer agency company has designated skilled personnel to properly resolve the procedures and logistics organizations involved in the various processes of the crude oil transfer, with the support of the port administration and port management departments. In this way, it guarantees the preparations for various organizations related to the crude oil transfer and creates reliable conditions for the aboard ship to start lightering operation immediately when it arrives at the port.

The lightering operation of crude oil is of high risk, and the consequences should be disastrous once the leakage occurs (Wang 2018). Therefore, safety precautions and emergencies must be prepared in advance. Ningbo Port has established rules and regulations to ensure the simultaneous implementation of safety management and emergency response and constructed a rear shore-based support and front emergency command system. As for rear shore-based support, the maritime emergency command center and the port production dispatching department implement 24-hour duty monitoring, timely grasp the situation of the transfer, and coordinate the port emergency tug, emergency oil fence, and other emergency equipment and facilities to rescue at any time. As for the front emergency command, it is clearly required that the transfer company must have at least one company-level leader to lead the lightering operation at the site. The leader will serve as the person in charge of on-site safety management and the emergency commander. Meanwhile, he or she is responsible for organizing and coordinating the safety management and emergency command of on-site lightering operations, and coordinating the first-class pilots of anti-pollution ships and aboard ships on-site to perform guard and emergency response.

4 CONCLUSION

At present, the volume of lightering operation of crude oil at Ningbo Port reaches 20 million tons. Ningbo Port has also developed into the largest domestic crude oil transfer port (Zhang 2014). The lightering operation of crude oil at Ningbo Port has initially formed a long-term safety management mechanism for the lightering operation of crude oil in Ningbo Port by controlling risks, building a safety management system, strengthening education and training, and emergency management. However, with regard to the continuous expansion of oil refinery production and the increase of the reserves of the national reserve oil depot, the risk of increased lightering volume, the requirements for personnel, equipment, and facilities, and the improvement of mechanisms will still be the key research topics for future lightering. Therefore, the joint risk control by government-led, departmental supervision, multi-party collaboration, and front-to-back linkage will be the future of the safety management system for the lightering operation of crude oil at Ningbo Port.

REFERENCES

Li Yeqing, Discussion on Pilotage Operation of Crude Oil Transfer [J], *Port Economics*, 2013, No.113(02): 53–55.

Liu Zhuo, *Study on the Safety Evaluation and Strategy of Crude Oil Transfer from Ship to Ship at Anchorage* [D], Dalian Maritime University, 2013.

Wang Jun, *Research on Safety Production Management of G Company Crude Oil Transfer* [D], Dalian Maritime University, 2018.

Wu Wanqing, Offshore Crude Oil Transfer Management [J], *China Maritime Affairs*, 2010, No.59(06):9–12,11.

Zhang Zhi Cong, Safety Management of Offshore Crude Oil Transfer with High Hydrogen Sulfide [J], *World Shipping*, 2014, v. 37; No. 231(09): 53–55, 53.

Energy Revolution and Chemical Research – Kok-Keong Chong and Zhongliang Liu (Eds)
© 2023 The Authors, ISBN: 978-1-032-36554-1

Cost assessment of clean energy in a country in Northeast Asia based on LOCE model

Jing Chen*
China Energy Investment Corporation (China Energy), Dongcheng District, Beijing, China

ABSTRACT: In order to cope with global climate change and reduce carbon emissions, countries around the world are gradually controlling the use of fossil energy. In order to ensure energy security, regional energy cooperation is becoming more and more important. Based on the LOCE model, this paper conducts an energy cost assessment on a country rich in clean energy resources in Northeast Asia and provides data reference for future energy cooperation among Northeast Asian countries. Research shows that the cost of wind power and photovoltaic power generation in a country in Northeast Asia will continue to decrease, the cost of wind power development projects is lower than that of photovoltaic projects, and greater economic benefits can be obtained by investing in wind farms. In addition, in the process of regional energy cooperation in Northeast Asia, the comprehensive benefits created in the follow-up are also huge, and the clean energy cooperation in Northeast Asia should be gradually promoted.

1 INTRODUCTION

In the context of tackling global climate change and effectively preventing atmospheric pollution, more and more countries have elevated "carbon peaking" and "carbon neutrality" to national strategies and put forward the vision of a carbon-free future. Clean energy, mainly renewable energy and nuclear energy, is gradually becoming the main source of energy resources. The Northeast Asian region is rich in clean energy resources, but the cost analysis of transnational energy cooperation is still unclear and there is an urgent need to conduct a cost assessment on energy cooperation. China emphasizes the importance of "gradually forming a new development pattern in which the domestic cycle is the mainstay and the domestic and international cycles promote each other". The North-East Asia region is rich in clean energy resources, and extensive cooperation will facilitate the optimal allocation of resources among the countries in North-East Asia. At the same time, it is also China's initiative to respond to the requirements of the "new development pattern".

Currently, numerous scholars have conducted studies on the levelised cost of electricity. The literature (Mulligan et al. 2015) estimated the levelised cost of electricity for organic photovoltaic (OPV) and verified that the levelised cost of electricity for OPV is more competitive with coal-fired generation. In (Pillai & Naser 2018), the economic performance of grid-connected photovoltaic (PV) systems was analyzed in terms of the levelised cost of electricity. The literature (D. L. T et al. 2015) provides a cost analysis of Spanish High Concentration Photovoltaic (HCPV) technology using levelised cost of electricity. The literature (Keck et al. 2019) analyses the impact of electrical energy storage on renewable energy grids using factors such as levelised electricity costs. The paper (Qingyou et al. 2021) investigates the economics of rooftop grid-connected PV based on the levelised cost of the electricity model. In the paper (Xiaoping & Yuting 2018), an LCOE model was constructed to provide a basis for the economic evaluation of wind power projects in China, taking into account the actual situation of wind power projects. The paper (Yanhong et al. 2016) analyses the optimal allocation of microgrid power based on the levelised cost of electricity.

*Corresponding Author: 120202206276@ncepu.edu.cn

DOI 10.1201/9781003332657-23

2 LEVELISED COST OF ELECTRICITY (LCOE) MODEL

2.1 *Pre-project financing LCOE model cost analysis*

The Levelised Cost of Electricity (LCOE) is the cost per kWh of building and operating a project over its entire life cycle. Considering the whole life cycle of wind and photovoltaic projects, the pre-financing levelised cost of electricity (LCOE) consists of two main components: the static initial investment in the project and the operation and maintenance costs. The formula for calculating the LCOE of a project is as follows.

$$LCOE = \frac{CAPEX + \sum_{N}^{n=1} \frac{OPEX_u}{(1+r)^n}}{\sum_{N}^{n=1} \frac{AEP_n}{(1+r)^n}} \tag{1}$$

where CAPEX is the static initial investment of the project, including equipment and installation costs, construction costs, other costs, etc.; OPEX is the operating costs of the project, including wages and benefits, repair and maintenance costs, material costs, insurance costs and other costs, etc.; AEP is the annual power generation capacity of the project; N is the operating life of the project; n is the year of operation of the project (1,2,3, …, n); and r is the discount rate.

Wind power and photovoltaic projects in certain countries in North-East Asia are eligible for preferential fixed asset VAT credits, and the impact of fixed asset VAT credits needs to be considered in the LCOE modelling. The impact of the salvage value of recovered assets also needs to be considered, and the LCOE formula can be amended to:

$$LCOE = \frac{CAPEX - \sum_{T_{vat}}^{n=1} \frac{VAT_n}{(1+r)^2} + \sum_{N}^{n=1} \frac{OPEX_n}{(1+r)^2} - \frac{V}{(1+r)^N}}{\sum_{N}^{n=1} \frac{AP_n}{(1+r)^n}} \tag{2}$$

In Equation (2), VAT is the VAT credit on fixed assets; V is the VAT credit on fixed assets over the life of the asset; V is the salvage value of the recovered asset. In Equation (2), the portion of the VAT credit on fixed assets is deducted. In addition, the portion recovered from the salvage value of the recovered assets is also considered.

2.2 *LCOE model cost analysis after project financing*

At present, the common financing scheme for wind power and photovoltaic projects in certain countries in Northeast Asia is debt financing. After considering the financing, the construction of the LCOE model needs to take into account the financing costs: that is, the initial investment in the project should take into account the interest during the construction period, and the finance costs need to be taken into account in the operating period pay-as-you-go costs, and the formula for calculating LCOE should be amended as follows:

$$LCOE = \frac{CAPEX - \sum_{T_{vat}}^{n=1} \frac{VAT_n}{(1+r)^n} + \sum_{N}^{n=1} \frac{OPEX_n}{(1+r)^n} - \frac{V}{(1+r)^N}}{\sum_{N}^{n=1} \frac{AEP_n}{(1+r)^n}} \tag{3}$$

In Equation (3), CAPEX is the initial investment of the project, including the cost of equipment and installation works, construction works, other costs, and interest during the construction period; OPEX is the pay-as-you-go cost of the project during the operation period, including wages and benefits, repair and maintenance costs, material costs, insurance costs, financial costs, and other costs; AEP is the annual power generation capacity of the wind power project; N is the operational life of the wind power project. n is the year of operation of the project (1,2,3,..., n); r is the discount rate; VAT is the VAT credit on fixed assets; V is the VAT credit on fixed assets; V is the residual value of the recovered assets. Most wind and photovoltaic projects require debt financing due to the large initial investment amount of the project. In the analysis in this paper, the appropriate LCOE model will be selected for the analysis depending on the situation.

3 CLEAN ENERGY LOCE ASSESSMENT FOR A COUNTRY IN NORTH-EAST ASIA

We have selected actual data from a country in North-East Asia and measured the LCOE levels for wind and PV power projects in areas with high-quality resources only, taking a discount rate of 8% and 5%, with the assessment conditions assuming a 25-year operating life of the renewable energy power plant based on the year of commencement of the project.

3.1 *LOCE analysis for wind and photovoltaic sites*

The initial investment cost of the wind farm consists of the cost of the turbine equipment and its installation, the cost of the turbine infrastructure, and other costs; the O&M cost consists of depreciation, labor costs, insurance, and taxes. In this case, the years 2020, 2026, and 2030 are based on Vestas' three turbine models: 2 MW-Ø110 m, 3.5 MW-Ø125 m, and 5 MW-Ø145 m respectively (Ø indicates diameter), where turbines with larger rotor diameters generally have a lower wind load factor in strong wind conditions.

A region of the country with an average solar irradiance of 1,700 kW-h/m^2 or 1,750 kW-h/m^2 was chosen to measure the potential for slightly higher conversion efficiencies of the selected polysilicon - double-glass PV modules by 2036, if double-glass PV modules are commercially available on a large scale. According to the start-up year of the project, the PV module conversion efficiencies are predicted to be 17%, 19%, and 23% respectively based on PV industry trends.

The initial investment costs and operating costs of wind and photovoltaic power generation for different models are shown in Table 1.

Table 1. Wind power, photovoltaic power project costs.

Year	Investment cost of wind turbine equipment	Wind turbine construction cost	Other Fees	Operating cost/ ($/kW/year)	Photovoltaic equipment investment costs	Photovoltaic construction costs	Other Fees	Operating cost/ ($/kW/year)
2020	1300	250	250	60	410	400	70	26
2026	1050	230	220	50	330	330	60	20
2036	720	200	230	45	250	280	50	15

3.2 *Analysis of LOCE measurement results*

Based on the above analysis, combined with the LCOE measurement model, LCOE assessments were conducted for wind and PV projects under different resource conditions, discount rates, and commissioning costs, and the results are shown in Tables 3 and 4. As shown in Tables 2 and 3, the average LCOE of wind power development projects in 2020 is $47.52/MW-h, while the average LCOE of PV development projects is $62.57/MW-h. The LCOE of wind power development projects is lower compared to that of PV.

Table 2. Wind power LCOE measurement results.

Year	Average wind speed	Wind load factor	LCOE $/MW·h-8%	LCOE $/MW·h-5%
2020	8 m/s	47.3%	55.9	45.8
2026		47.9%	45.6	37.3
2036		45.6%	38.2	31.9
2020	9 m/s	54.2%	48.6	39.8
2026		55.9%	39.8	32.6
2036		52.9%	32.1	29.6

Table 3. PV LCOE measurement results.

Year	Solar radiation intensity	Load factor	LCOE $/MW·h-8%	LCOE $/MW·h-5%
2020	1700 kW·h/m^2	17.8%	69.9	56.9
2026		19.9%	50.4	40.9
2036		23.1%	34.5	27.9
2020	1750 kW·h/m^2	18.3%	68.1	55.4
2026		20.4%	49.2	39.9
2036		23.7%	33.7	27.2

As can be seen from the LCOE trend in the graph below, in the short term, wind farm development projects in a country in North-East Asia are still more economical than ground-based solar farm development projects for wind resources with average wind speeds greater than 8 m/s. In the long term, as the efficiency of PV modules improves and the cost of PV decreases, it can be expected that the LCOE of wind and solar projects will fall to the same level as wind power.

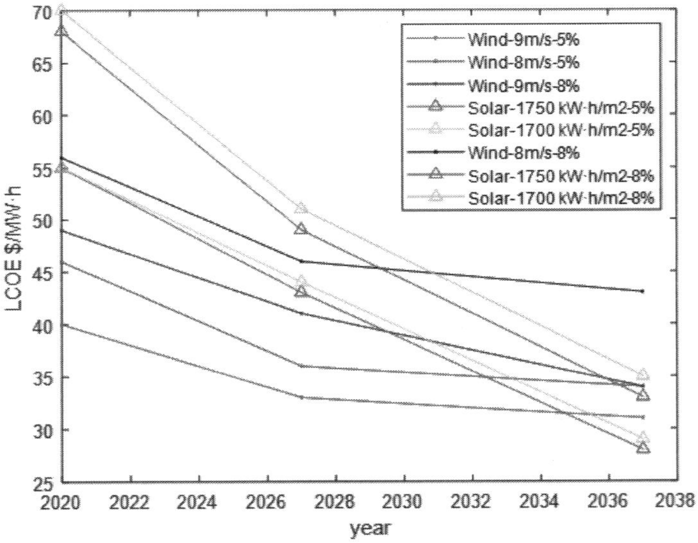

Figure 1. The trend in LCOE for wind and PV projects.

4 CONCLUSIONS

The above information shows that the cost of wind and photovoltaic power generation in one of the Northeast Asian countries will continue to decrease and that wind power development projects are more affordable than photovoltaic projects. From a cost perspective alone, if China were to consider investing in a clean energy base outside of China, it should invest in wind power stations first to gain greater economic benefits. In addition, clean energy projects have significant environmental benefits, and because of the huge potential for clean energy development in Northeast Asia, the benefits of regional energy cooperation in reducing air pollutants are enormous, helping Northeast Asian countries to reach their 'carbon neutrality' targets as soon as possible. Finally, the political role of energy cooperation in North-East Asia can provide a strong guarantee for long-term economic benefits. Therefore, from a comprehensive perspective, energy cooperation in Northeast Asia should be carried out gradually with long-term benefits as the goal. As the main body to promote the construction of the energy internet in Northeast Asia, the relevant power companies in Northeast Asia need to work actively.

REFERENCES

A, D. L. T., P. Pérez-Higueras a, J.A. Ruíz-Arias b, & E. F. F ernández c a. (2015). Levelised cost of electricity in high concentrated photovoltaic grid-connected systems: a spatial analysis of Spain. *Applied Energy*, 151, 49–59.

Keck, F., Lenzen, M., Vassallo, A., & Li, M. (2019). The impact of battery energy storage for renewable energy power grids in Australia. *Energy*, 173(APR.15), 647–657.

Mulligan, C. J., Bilen, C., Zhou, X., Belcher, W. J., & Dastoor, P. C. (2015). Levelised cost of electricity for organic photovoltaics. *Solar Energy Materials and Solar Cells*, 133, 26–31.

Pillai, G., & Naser, H. (2018). Techno-economic potential of largescale photovoltaics in Bahrain. *Sustainable Energy Technologies & Assessments*, 27, 40–45.

Qingyou Y, Qifeng W & Guangyu Q. (2021). An empirical study on the economics of rooftop grid-connected photovoltaics based on the equalization power cost model. *Science and Technology and Industry* (10),73–79.

Xiaoping M & Yuting W. (2018). Research on power cost of Wind Power project leveling in China. *Wind power* (12), 42–44.

Yanhong L, Yongqiang Z & Xin W. (2016). Optimization of power supply configuration in microgrid based on power cost analysis. *China Southern Power Grid Technology* (02),56–61.

Energy Revolution and Chemical Research – Kok-Keong Chong and Zhongliang Liu (Eds)
© 2023 The Authors, ISBN: 978-1-032-36554-1

Research and practice of tank risk management based on tank overhaul data statistics

Xixiang Zhang*, Yufeng Yang, Wuxi Bi, Qiang Zhang & Shuo Liu
National Pipe Network Research Institute, National Engineering Laboratory for Pipeline Safety, Langfang, Hebei, China

ABSTRACT: Storage tanks are important equipment and facilities in petroleum exploitation, production, storage and transportation, petrochemical production, and other systems. They store a large number and many kinds of chemicals, and most of them are flammable, explosive, toxic, and harmful. Once an accident occurs, it will seriously threaten the safety of people's lives and property. To effectively reduce the operation risk of storage tanks, the members of the research group sorted out the overhaul data of more than 180 storage tanks under the jurisdiction of the company. A storage tank risk evaluation system is established based on the RBI principle. The research provides a guarantee for the safe operation of storage tanks of the company.

1 INTRODUCTION

As an important production facility in the field of the petrochemical industry, the atmospheric storage tank is used to store crude oil, product oil, various dangerous chemicals, and chemical raw materials. With the development trend of large-scale, large-scale, continuous, and automatic petrochemical production, the volume of the largest oil tank in the world has reached 240000 m^3, and the volume of the largest storage tank built and used in China is 150000 m^3, of which 100000 m^3 storage tank has been common (Li & Xing 2018; Xiao et al. 2021; Zhao et al. 2021). Once a safety accident happens to such large storage tanks, it will lead to serious economic losses, environmental pollution, and casualties. Explosions, fires, and environmental pollution incidents caused by leakage of various storage tanks at home and abroad occur from time to time (Shui et al. 2009; Yang & Wang 2006). In January 1988, the diesel storage tank of an oil company in Pennsylvania cracked vertically, and all 1480 m^3 diesel leaked. Although there were no casualties, it caused serious environmental pollution. On June 27, 1997, an explosion and fire accident occurred in the storage tank area of a chemical plant, resulting in 48 casualties and a direct economic loss of 170 million yuan. The occurrence of tank fires, explosions, and other accidents has caused serious casualties, environmental pollution, and huge economic losses. The safe operation and management of tanks are facing great pressure. Especially under the background of the normalization of low oil prices, asset owners in the petroleum and petrochemical industry are actively seeking new ways to reduce costs and increase efficiency in the operation and maintenance of equipment and facilities, and carrying out tank risk assessment has gradually become the consensus of the tank industry (Shi & Shuai 2013).

At present, the management of atmospheric pressure storage tanks in China's petrochemical enterprises has not been brought into the track of legal management, and the number of storage

*Corresponding Author: zhangxx11@pipechina.com.cn

 DOI 10.1201/9781003332657-24

tanks is large, so it is difficult to realize the inspection one by one. In order not to affect production, all enterprises adopt the method of sampling inspection. However, this sampling inspection method often relies on manual experience and does not have a reasonable classification and hierarchical management method, which will cause two disadvantages: on the one hand, the result of sampling inspection is that most of the storage tanks have no defects. If the tank opening inspection is frequent, it will cause serious economic losses; on the other hand, the storage tanks with great potential risks cannot be inspected on schedule, and the aging and extended service of storage tanks are serious, resulting in many potential safety hazards.

To effectively reduce the operation risk of storage tanks, the members of the research group sorted out the overhaul data of more than 180 storage tanks under the jurisdiction of the company. The members sorted and analyzed the tank overhaul data to obtain the importance of various failure factors of the tank. At the same time, a tank risk assessment system is established based on the RBI principle. The purpose is to provide some ideas for tank safety management.

2 TANK RISK FACTOR ANALYSIS BASED ON TANK OVERHAUL DATA STATISTICS

After sorting out the overhaul data of more than 180 storage tanks under the jurisdiction of the company, it is found that the average overhaul time of storage tanks is 136.25 days, and the specific time distribution is shown in Figure 1. The top causes of tank opening overhaul are: the overhaul period has expired, aging seal, failure of rain shield, failure of the heating coil, damage of floating ladder, failure of central drainage pipe, failure of auxiliary drainage pipe, and failure of the steam dewaxing pipeline. See Figure 2 for the distribution of specific causes of tank opening overhaul. From the perspective of maintenance subjects. The most repaired part is the floating cabin. The maintenance contents of the floating cabin include failure of the floating cabin, cleaning of accumulated water, grinding and de-rusting and anti-corrosion painting, etc. Among them, the failure of floating cabins accounts for more than half. The distribution of each factor is shown in Figure 3.

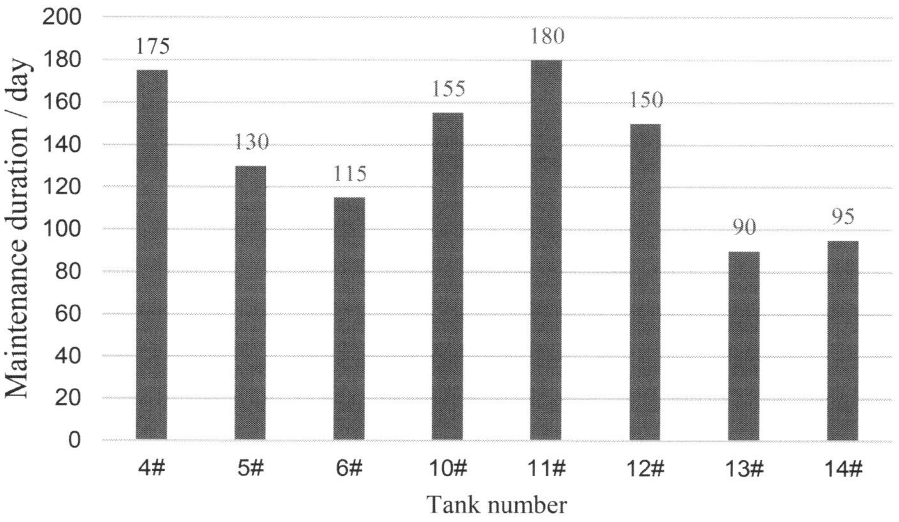

Figure 1. Statistics on the duration of an overhaul in a reservoir area.

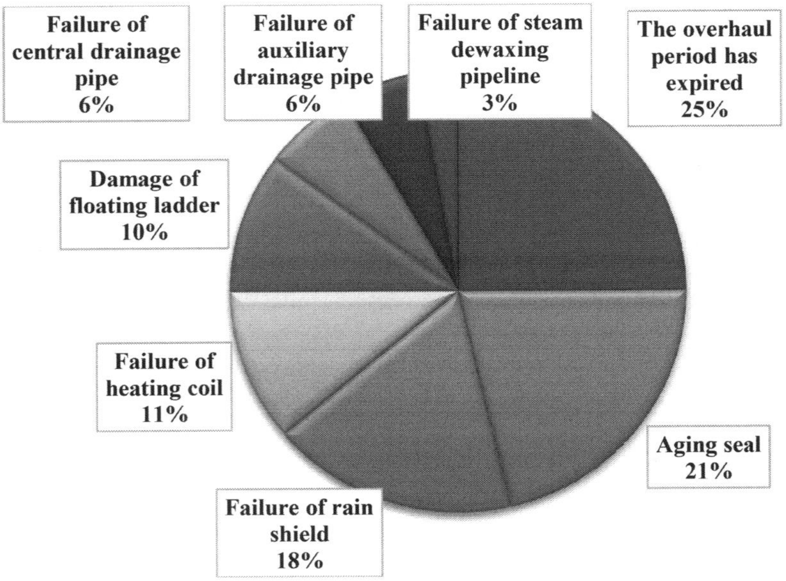

Figure 2. Causes of tank opening overhaul.

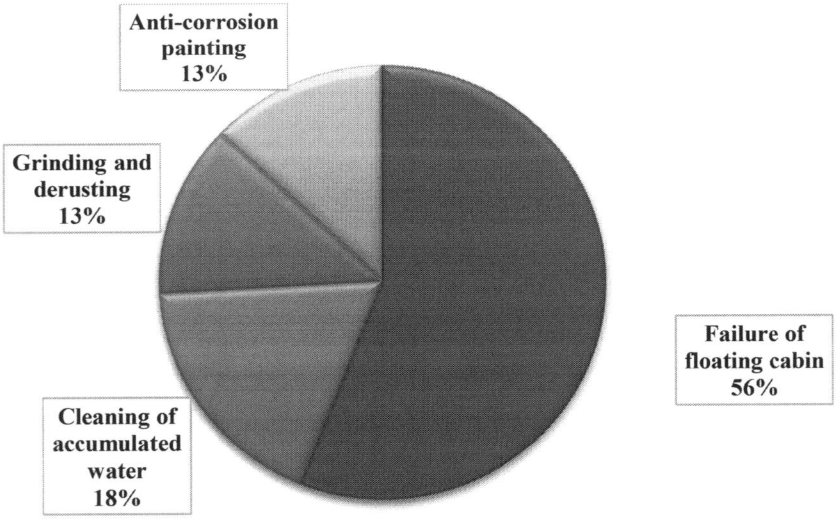

Figure 3. Causes of floating tank failure.

3 TANK RISK ASSESSMENT BASED ON TANK OVERHAUL DATA STATISTICS

Risk-based inspection (RBI) technology is a management method based on risk assessment and analysis to evaluate the defects or hazards existing in equipment or system, and then formulate an inspection and maintenance strategy (API Publ 353 Managing systems integrity of terminal and tank facilities 2006; Chen & Wang 2012). Risk-based inspection (RBI) technology was first proposed in the field of nuclear power in the 1970s to deal with extreme safety accidents with low possibility and high consequences. Then it gradually extended to the petrochemical industry. In the early 1990s, organizations such as the American Society of Mechanical Engineering (ASME) and the American Petroleum Association (API) began to devote themselves to the promotion and application of RBI technology and the formulation of standards, and developed risk assessment methods suitable for the petrochemical industry. In the middle and late 1990s, European and American developed countries tried to apply RBI Technology in the inspection and maintenance of pressure equipment, and achieved good results. They formulated a series of inspection specifications and evaluation standards, including the relevant standards of API 580 and API 58139 risk assessment, as well as the relevant industrial standards of inspection, inspection, and maintenance such as API 750, API 510, and API 750. After the successful application of RBI Technology in China, relevant articles of association and regulations have been issued, including supervision regulations on safety technology of fixed pressure vessels (TSGR 0004-2009), supervision regulations on safety technology of pressure pipelines – industrial pipelines (TSG D0001-2009) and rules for periodic inspection of pressure vessels (TSGR 7001-2013). These documents specify the relevant specifications for the application of RBI technology. In addition, relevant standards such as guidelines for the implementation of risk-based inspection of pressure equipment systems (GBT26610. 4-2014) and risk-based inspection of atmospheric storage tanks (GBT30578-2014) have been formulated according to api581 to guide the RBI calculation and evaluation of pressure equipment and atmospheric storage tanks. At present, RBI technology has been widely used in various fields, including nuclear power, electric power, chemical industry, bridge construction, and so on.

In the process of RBI, the key is to establish a scientific and reasonable risk assessment model. The assessment of risk level is based on the two aspects of failure possibility and failure consequence. The calculation results are represented by a 5×5 risk matrix, as shown in Figure 4. The colors red, orange, yellow, and green correspond to high risk, high risk, medium risk, and low risk respectively. The inspection strategy is formulated by comprehensively considering the damage mechanism and other factors.

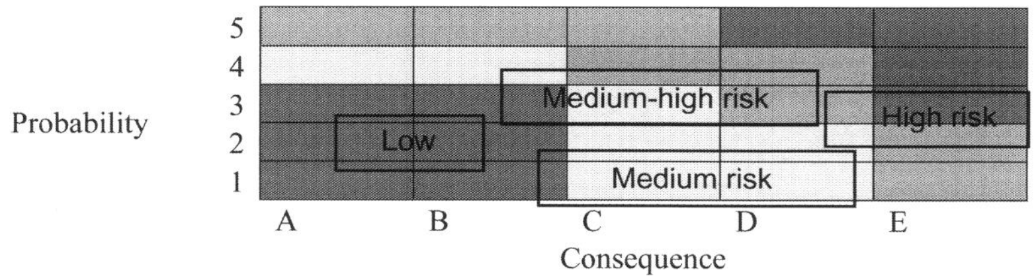

Figure 4. Tank risk matrix.

The risk of the atmospheric storage tank is calculated according to formula (1):

$$R(t) = F(t) * C(t) \tag{1}$$

Where:

F (t) indicates the possibility of failure;

C (t) indicates failure consequence;

T indicates that the risk is a function related to time;

In terms of failure possibility, according to the sorting and analysis of tank overhaul data, the relevant factors of failure possibility include the time from the last overhaul, seal aging, flashing failure, heating coil failure, floating ladder damage, central drainage pipe failure, auxiliary drainage pipe failure, and steam dewaxing pipeline failure.

Failure consequences, including environmental consequences, refer to the impact on soil, water, and ecology; Public consequences refer to the impact on the safety, health, and facilities of the station and surrounding personnel; Commercial consequences refer to economic losses and reputation losses. The main influencing factors of failure consequences are type and quantity of leaked oil products; Scope and object of leakage influence; Soil conditions, recyclability of leaked oil products; Damage to the environment, society, equipment, and operation and the duration of the damage; Specific environment affecting the consequences of leakage, including surrounding ecology, population density, and facility configuration.

The data of the company's storage tanks are entered into the storage tank risk assessment system, as shown in Figure 5, to obtain the risk classification of the whole company's storage tanks. According to the risk classification of storage tanks, reasonably arrange the tank overhaul and maintenance strategy.

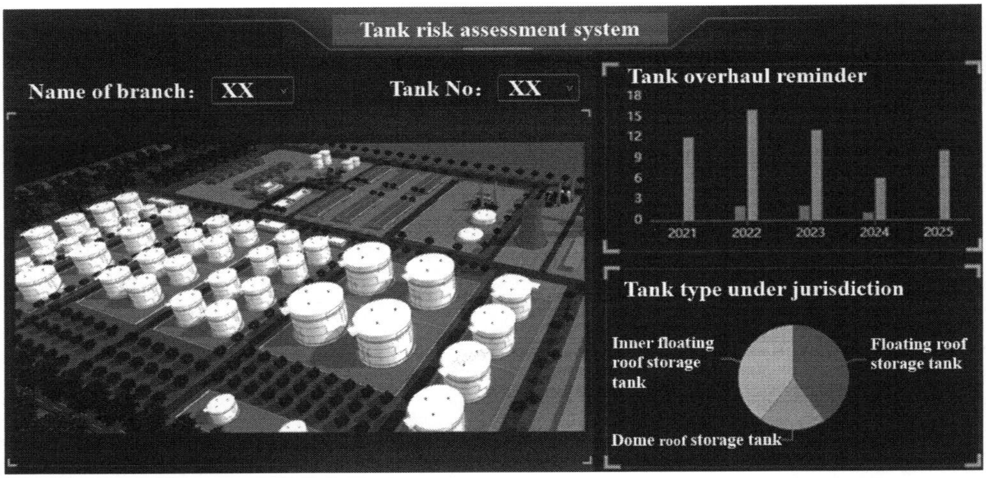

Figure 5. Tank risk assessment system.

4 MITIGATION MEASURES

According to the results of the tank risk assessment system, take risk mitigation measures to reduce the probability or consequences of specific failure events(Li 2021; Wang, 2021; Xing et al. 2021). Four factors need to be considered in selecting risk mitigation measures: (1) environmental and public factors, including population distribution, land use, location and slope of adjacent groundwater and aquifer, adjacent waters, site topography, and drainage system, the permeability of local soil and backfill, toxicological factors, epidemiological factors, physical and chemical properties of products (toxicity, flammability, solubility, volatility, and viscosity); (2) Operational factors, including remaining service life of the pipeline, storage tank or system, effectiveness of preventive measures, station facilities and storage capacity. Product type, inventory turnover rate (storage time); (3) Company experience, including previous control measures, historical maintenance records, operator qualifications and experience, existing training, maintenance, inspection, and operation methods; (4) Site factors, including training plan, site experience, management feasibility, leakage event frequency, device design, environmental or safety problems of special devices.

5 CONCLUSION

In this paper, research based on tank overhaul data statistics is adopted to study tank risk management. The main conclusions can be summarized as follows: (1) The tank risk assessment system is based on B / S architecture. Users can query and modify data through the network to facilitate data maintenance. This study improves the company's tank risk management level. (2) Based on the graphical management platform of GIS, the tank risk management system provides a more intuitive risk early warning function for enterprises. In terms of future work, big data technology and intelligent technology should be carried out to enhance the Tank risk management level.

REFERENCES

API Publ 353 *Managing systems integrity of terminal and tank facilities* [S].2006.

Chen Qingjuan, Wang Sanming Application research and improvement of RBI Technology in Chinese enterprises [J] *China work safety science and technology*, 2012, 08 (6): 191–196.

Li Bo Practice and application of integrity management of atmospheric storage tank [J] *Equipment management and maintenance*, 2021 (13): 4–6 DOI:10.16621/j.cnki.issn1001-0599.2021.07.03.

Li Jianhong, Xing Shu Research and application of integrity management system of atmospheric storage tank [J] *Petrochemical equipment technology*, 2018 (1).

Shi Lei, Shuai Jian Research on integrity management system of large crude oil storage tank [J] *Chinese Journal of safety science*, 2013,23 (6): 151–157.

Shui Biyuan, AI Muyang, Feng Qingshan Technical ideas for integrity management of oil and gas stations [J] *Oil and gas storage and transportation*, 2009, 28 (7): 11–14.

Wang Jian Safety control in overhaul of large crude oil storage tanks [J] *Equipment management and maintenance*, 2021 (24): 12–14 DOI:10.16621/j.cnki.issn1001-0599.2021.12D.

Xiao Zhuyun, Liu Guozhi, Wu Dong Integrity management method and application of atmospheric storage tank [J] *Chemical management*, 2021(29): 179–180+183 DOI:10.19900/j.cnki.ISSN1008-4800.2021.29.086.

Xing Shu, Wang Jing, Wang Shi, Yan He Application and Research on risk-based inspection of large atmospheric crude oil storage tank [J] *China special equipment safety*, 2021, 37 (11): 62–67.

Yang Zupei, Wang Weibin Research progress of oil and gas pipeline integrity management system [J] *Oil and gas storage and transportation*, 2006, 25 (8): 7–11.

Zhao Yanze, Chen Yanyan, Wang Yanxiu Integrity management technology and application of in-service atmospheric storage tank [J] *Oil and gas field surface engineering*, 2021, 40 (03): 70–75.

Energy Revolution and Chemical Research – Kok-Keong Chong and Zhongliang Liu (Eds)

Research on the development of environmental energy efficiency in Liaoning province under the goal of carbon neutralization

Gao Huan & Sun Bo*
Dalian Polytechnic University, Dalian, Liaoning Province, China

ABSTRACT: In the global energy consumption environment, China's coal consumption accounts for about 70% of the total energy consumption, and it is one of the few countries whose energy consumption is dominated by coal. The total amount of industrial wastewater, waste gas, and solid pollutants discharged by the heavy industry remains high. In particular, the total amount of carbon dioxide emitted by the massive use of coal ranks at the top in the world. Although a large part of the total amount of greenhouse gases was produced in developed countries during the period of rapid industrial development in the past, developed countries have basically completed the process of industrialization, with a high level of clean energy use and less pressure to reduce emissions. However, China is currently in the stage of a developing country, and its economic development largely depends on non-renewable energy. It is still a long way to go to complete the emission reduction task on the basis of maintaining sustainable economic development.

1 INTRODUCTION

Global warming has caused wide public concern at present, low carbon life people often mention the words, and carbon neutrality is an extension to the concept of low carbon development. It requires low, even lower human movement of carbon emissions, through afforestation and other ways to absorb the carbon dioxide emissions, to achieve the goal of zero carbon, so as to realize carbon neutrality. Many studies have shown that a country's economic system, environment, energy, research and development, and economic growth have interactional and interactional endogenous relations, rather than a one-way relationship. At present, there are relatively many research achievements on environmental pollution, energy consumption, and economic growth, which lays a good theoretical and empirical foundation for the study of this paper. Energy and environment are the two ends of economic growth constraint, in which environmental regulation on energy consumption and economic growth as well as energy control on economic growth double force mechanism will promote energy productivity and economic development mode transformation. The use of energy will inevitably bring pollution, which will inevitably affect the total utility of society. The way to improve energy efficiency through the forced mechanism of environmental regulation is the application of new technology and upgrading of industrial structure, which also promotes the transformation of China's economic development mode.

2 CARBON NEUTRALITY AND ENVIRONMENTAL ENERGY PROFILE

2.1 *The implications of carbon neutrality*

IPCC issued the "Special Report on Global warming 1.5°C″ and pointed out that carbon neutrality refers to an organization within a year of carbon dioxide emissions through carbon dioxide

*Corresponding Author: sunbo_0709@126.com

DOI 10.1201/9781003332657-25

elimination technology to achieve balance, or called net-zero carbon dioxide emissions. The carbon neutrality goal is to reduce global $CO2$ emissions by about 45 percent by 2030 compared to 2010 levels and to achieve net-zero $CO2$ emissions by 2050. The priority of carbon neutrality is to limit global warming to 1.5°C by the end of the century. Carbon neutrality is not only to control climate change, but also a fundamental measure to protect the ecological environment, helping to protect biodiversity and ecosystems and avoid more species extinction. Carbon neutrality accelerates the low-carbon and green transformation of the energy system and creates new economic growth points for the world. According to the Energy Transition 2050 report released by the International Renewable Energy Agency, carbon neutrality will increase global GDP by 2.4% and create 7,106 additional jobs in the energy sector.

Carbon neutrality involves energy science. Based on the evolution of the Earth system, carbon neutrality studies the formation and distribution, evaluation selection, development and utilization, orderly replacement and development prospect of energy from time scale and space scale, which focuses on the interaction and co-evolution of the earth, energy, and human beings. Carbon neutrality includes three core contents of energy science: (1) The formation of energy under the background of the Earth system and the feedback of energy consumption on the earth's climate and environment reflect the relationship between the earth and energy; (2) The earth's environment breeds human evolution and human behavior changes the earth's environment, which reflects the relationship between the earth and human beings. (3) Human beings use technology to develop energy and energy to drive the progress of human society, which reflects the relationship between human beings and energy. Carbon neutrality is also a social science, involving the harmonious and sustainable development of man and nature, human energy utilization, and the dynamic balance of the earth's ecosystem. It is not only an important research topic in the field of humanities and social sciences, but also the combination, expansion, and innovation of social science and carbon neutrality with the times. Carbon neutralization focuses on the fundamental changes in the international energy economy, energy policies, and energy laws and regulations under the goal of carbon neutralization, which involves major theoretical research issues in the field of social science. At present, the theoretical research on carbon neutrality has gradually become the forefront of international science and technology and humanities research. The technical connotation of carbon neutrality includes the whole process of carbon dioxide emission, capture, utilization, storage, and removal caused by human production and life and the related technical system. First of all, we should study and formulate economic and industrial policies from the source to guide human low-carbon life and consumption and reduce carbon emissions produced and lived by human beings. Secondly, with the goal of reducing and lowering carbon emissions, we should research relevant technologies to reduce carbon emissions while meeting the energy and material needs of human production and life. Finally, we should develop carbon capture, sequestration, and carbon removal technologies to effectively reduce $CO2$ concentration in the earth's atmospheric system. The technical connotation of carbon neutrality involves the following four aspects: (1) It is attempted to utilize carbon emission reduction technologies such as clean utilization of fossil energy, clean energy substitution, resource recycling, and utilization, energy conservation, and efficiency improvement; (2) It is advocated to use renewable energy such as wind energy, solar energy, ocean energy, and geothermal energy, as well as zero-carbon technologies such as hydrogen energy, new material storage, smart energy, nuclear energy and controlled nuclear fusion; (3) The focus is on carbon dioxide capture, utilization, storage, conversion and forestry, marine, soil carbon sequestration, and other negative carbon technology; (4) IT is centered on the carbon tax system, carbon trading system, composite carbon emission trading system, carbon economy and carbon industry policy, carbon fiscal subsidies, and other carbon economy technology.

2.2 *Environmental energy under low carbon*

With the gradual deepening of reform and opening-up, China has experienced more than 30 years of high-intensity development and made remarkable achievements in economic development. However, with the continuous improvement of China's economic scale, the deep-seated contradiction between economic development and resources and the environment has become increasingly prominent, especially the environmental problems caused by the use of fossil energy such as coal and oil, which are increasingly recognized by people. In this context, the coordinated development of resources, environment, and economic growth has become one of the central issues of sustainable development. To explore the sustainable development of the economy under the constraints of resources and environment, it is necessary to analyze and verify the evolution path between resources and environment and economic development from both theoretical and empirical aspects, so as to provide a beneficial reference for the formulation and implementation of subsequent environmental and economic policies.

The energy issue is an overall and strategic issue concerning the development of human society and the international political and economic pattern, as well as a global issue affecting climate change and environmental pollution. China is the world's largest energy producer and consumer, as well as the world's largest emitter of carbon dioxide. Reducing energy intensity and carbon emissions is an important goal of China's energy development strategy. Energy problems are ultimately energy utilization problems, and improving energy efficiency is the key. The connotation of energy efficiency lies in the contribution of energy consumption to maintaining and promoting human sustainable development. Ecological efficiency takes into account both the ecological and economic benefits of economic activities, and it is required that the intensity of environmental impact and resource utilization be adapted to the earth's carrying capacity level, which is the concentrated embodiment of sustainable development goals. At present, the research on ecological efficiency involves enterprises, industries, regions, and countries. The core idea of both energy efficiency and ecological efficiency is to create higher social value with less resource consumption and less environmental impact. The study of energy ecological efficiency is not only an evaluation of ecological efficiency focusing on energy resources, but also an evaluation of energy efficiency based on environmental factors. Improving energy ecological efficiency is the essential requirement of developing a low-carbon economy, and the key to realizing economic development from high energy consumption, high carbon, and high pollution to low energy consumption, low carbon, and low pollution.

3 CORRELATION ANALYSIS BETWEEN CARBON NEUTRALITY AND ENVIRONMENTAL ENERGY ISSUES

3.1 *Carbon use in agricultural areas of Liaoning Province*

Due to the unreasonable use of fertilizers and pesticides, serious agricultural pollution has been caused. The agricultural environment problem has become an important issue in our country in recent years. With the ecological civilization construction proposed for the first time at the 17th National Congress, an industrial structure, growth mode and consumption pattern based on energy resources and protection of the ecological environment have basically been formed, and it is pointed out that resource-saving, environment-friendly and ecological civilization is an effective way to meet people's needs ecologically under the framework of sustainable development. Therefore, the key to realizing a resource-saving and environment-friendly "dual-oriented society" in economic and social development is to improve ecological efficiency. In this context, ecological efficiency indicators have been widely used in agricultural development in recent years, and improving agricultural ecological efficiency has become a key factor for China's agricultural economic growth and the common development of resources and the environment (Renewable 2017).

It can be seen that the living energy consumption in rural areas of Liaoning Province is gradually increasing. With the continuous progress of the economy, technology, and science and technology, the people's living standard is also gradually improving. Since 2009, the rural domestic energy consumption in Liaoning Province has increased from 2.2127 million tons of standard coal to 4.9164 million tons of standard coal in 2019. Rural energy supply and demand show a diversified trend, and

Table 1. Rural domestic energy consumption in Liaoning Province from 2009 to 2019.

code	year	area	quantity	unit
1	2009	Rural areas of Liaoning Province	212.27	10000 tons of standard coal
2	2010	Rural areas of Liaoning Province	252.72	10000 tons of standard coal
3	2011	Rural areas of Liaoning Province	286.02	10000 tons of standard coal
4	2012	Rural areas of Liaoning Province	319.60	10000 tons of standard coal
5	2013	Rural areas of Liaoning Province	290.71	10000 tons of standard coal
6	2014	Rural areas of Liaoning Province	310.79	10000 tons of standard coal
7	2015	Rural areas of Liaoning Province	411.32	10000 tons of standard coal
8	2016	Rural areas of Liaoning Province	438.83	10000 tons of standard coal
9	2017	Rural areas of Liaoning Province	453.27	10000 tons of standard coal
10	2018	Rural areas of Liaoning Province	469.94	10000 tons of standard coal
11	2019	Rural areas of Liaoning Province	491.64	10000 tons of standard coal

renewable energy develops rapidly. At present, the rural consumption structure has evolved from "coal, straw, and firewood" to "electricity, coal and biogas", supplemented by liquefied petroleum gas and natural gas, and the proportion of commercial energy has been increasing. With the acceleration of urbanization, the rural population is decreasing, but the economic development and the acceleration of scientific and technological processes still increase the total rural energy consumption, which also reflects the continuous increase of the total rural domestic energy consumption in Liaoning Province.

3.2 *Current situation of energy consumption in rural production*

Energy consumption for rural production includes energy consumption for planting, breeding, and primary processing of agricultural products. Rural energy supply includes commercial energy input outside rural areas and energy development inside rural areas (Tian 2012). The rural energy production mentioned in this study only refers to the development of energy inside rural areas. Energy development in rural areas includes not only the development of various renewable energy sources, such as new biomass energy such as fuel ethanol, biodiesel, and briquette, and renewable energy such as hydropower, wind energy, solar energy, and geothermal energy, but also the development of traditional biomass energy such as direct combustion of firewood and straw (Zhu 2016). Rural internal energy is also different from commercial energy and non-commercial energy. Even the same energy may exist as both commercial energy and non-commercial energy (He 2017).

Table 2. Terminal consumption of primary industry in Liaoning Province from 2009 to 2019.

code	year	area	quantity	unit
1	2009	Rural areas of Liaoning Province	249.85	10000 tons of standard coal
2	2010	Rural areas of Liaoning Province	266.73	10000 tons of standard coal
3	2011	Rural areas of Liaoning Province	284.55	10000 tons of standard coal
4	2012	Rural areas of Liaoning Province	287.59	10000 tons of standard coal
5	2013	Rural areas of Liaoning Province	283.99	10000 tons of standard coal
6	2014	Rural areas of Liaoning Province	287.54	10000 tons of standard coal
7	2015	Rural areas of Liaoning Province	288.08	10000 tons of standard coal
8	2016	Rural areas of Liaoning Province	285.70	10000 tons of standard coal
9	2017	Rural areas of Liaoning Province	291.68	10000 tons of standard coal
10	2018	Rural areas of Liaoning Province	292.53	10000 tons of standard coal
11	2019	Rural areas of Liaoning Province	302.12	10000 tons of standard coal

It can be seen that the energy consumption of the primary industry in Liaoning Province also increased from 2.4985 million tons of standard coal in 2009 to 3.0212 million tons of standard coal in 2019. In recent years, rural energy in Liaoning Province has generally shown a good development trend, and the industrial technology of all parties has made great progress. With the continuous popularization and development of the Internet, farmers will also use the dissemination of the Internet to understand the development of agricultural resources, to understand the news and current events, make rational use of resources, and take measures to protect the environment and save resources.

4 CONCLUSIONS AND COUNTERMEASURES

Reducing global carbon dioxide emissions is an important way to achieve the sustainable development of the world economy. As a link between environmental energy issues and economic development, the carbon trading market has become a hot research topic in the world today.

(1) Dependence on the market itself for improvement. Some experts point out that the development of the domestic carbon trading market will be an absolute trend in the future. Therefore, it is necessary to improve the domestic carbon trading system as soon as possible to make the market fully active. In terms of supply and demand, environmental pollution in China is heavy and energy efficiency is low. The greenhouse gas emissions will increase, and the demand for carbon dioxide emission permits in Liaoning province will increase.

(2) Government policy support encourages enterprises to develop emission reduction technologies. First, it is necessary to establish and improve the legal system for environmental protection and strengthen supervision (Bai 2016). Efforts should be doubled to increase the penalties for environmental pollution, or assign greenhouse gas emission quotas to each enterprise. If an enterprise exceeds the emission quota, it will have to buy from other enterprises that have the remaining emission quotas, which is now known as initial carbon allocation. Second, we will endeavor to increase investment in research and development projects of energy conservation and emission reduction technologies, and adopt fiscal support policies of government spending, thereby attracting foreign investment, striving for financing, and fundamentally solving the problem of energy conservation, emission reduction, and environmental protection technology.

REFERENCES

Bai Yu. *Biomass energy is still the "backbone" of rural household energy* [n] China Electric Power News, 2016-05-21

He Xiangsheng. Analysis on the important impact and role of rural energy work on ecological environment construction [J] *Green technology*, 2017 (4): 109–110

Renewable energy may become the main force of rural energy in China [J] *Energy and Environment*, 2017 (1): 38 [2017-08-26].

Tian Yishui. Energy supply in the process of building a new socialist countryside [J]. *Journal of agricultural engineering*, 2012 (1): 35

Tian Yishui, Zhao Lixin, Meng Haibo, et al. Technical and economic evaluation of biomass energy utilization in rural China [J] *Journal of agricultural engineering*, 2011, 27 (13): 1–5

Zhu Yueying. *Biomass energy: the most potential renewable energy* [n] China Petroleum News, February 22, 2016 (004)

Energy Revolution and Chemical Research – Kok-Keong Chong and Zhongliang Liu (Eds)
© 2023 The Authors, ISBN: 978-1-032-36554-1

Practice of underbalanced drilling technology in oilfield

Chunlai Chang & Jielei Cui
Petrochina Bohai Drilling Mud Technical Service Company, Tianjin Binhai New Area, China

Wen Li, Yubo Sun, Fu Tao & Huijun Zhang
Petrochina Bohai Drilling Fifth Drilling Company, Hejian Hebei, China

Graciela Daniels*
Central Arizona College, Coolidge, AZ, USA

Shiela Kitchen
University of New England, Armidale, NSW, Australia

ABSTRACT: Because conventional drilling cannot be used in the formation of fractures and caves, underbalanced drilling research is carried out. Through the study of the advantages and key technologies of underbalanced drilling, the principle and characteristics of underbalanced drilling are clarified. The results show that underbalanced drilling has the advantages of protecting oil and gas layers and reducing well loss complexity, and its key technology lies in the underbalanced drilling system and the determination of drilling fluid density. Finally, the underbalanced drilling cases of YB103 well in Sichuan Basin and X73 well in Huabei Oilfield are analyzed. The success of the study is conducive to the application of underbalanced drilling.

1 GENERAL INSTRUCTIONS

With the continuous development of oil and gas exploitation, oil fields gradually enter the ranks of low porosity and low permeability, and the exploration effect is not ideal all the time (LI 2020). Every year, a large number of human and material resources are invested to deal with the above complex situation, but the effect has been poor. Given the above situation, as a new drilling technology, underbalanced drilling can effectively reduce drilling time, reduce drilling accidents, improve single well production, and contribute to the fine reservoir description and optimization of operation and management (Y 2010).

As early as the 1950s, the Sichuan Basin first began to use underbalanced drilling, and then Xinjiang oilfield, Tuha oilfield, Changqing oilfield, Yumen oilfield, and so on gradually began to use (Z 2013). Underbalanced drilling techniques commonly used at present include blowout completion, two-stage drilling completion, downhole blowout preventer completion, downhole suction valve completion, and balanced wellbore pressure completion. The key to underbalanced drilling technology lies in well control, drilling fluid, program design, and special tools(Melamed 2000). At present, underbalanced drilling equipment is mainly needed in China, but some researchers have been carried out in China. For example, Sichuan Administration Bureau has developed a 10.5 MPa, 17.5 MPa rotary control head, ZQF-1400 drilling fluid gas separator, true dehydrator, automatic collection system of underbalanced drilling data, underbalanced coring tool, etc., which effectively promoted the development of underbalanced drilling technology in China (L 2017).

*Corresponding Author: graciela_daniels@stu.centralaz.edu

DOI 10.1201/9781003332657-26

2 ADVANTAGES AND DISADVANTAGES OF UNDERBALANCED DRILLING

2.1 *Advantages of underbalanced drilling*

Maximum protection of oil and gas layers: underbalanced drilling can make the drilling fluid column pressure is less than the formation pressure, the drilling fluid will basically not pollute the oil and gas layers, which is conducive to the discovery and protection of oil and gas layers, improve the productivity of oil and gas Wells; To minimize the loss of well and other complex working conditions: for the formation with low pressure, fractures and caves, the underbalance drilling can prevent loss of well, especially the effective use of aerated drilling fluid and foam drilling fluid to reduce the pressure of drilling fluid column, reduce the loss of well and sticking accidents; Minimize the hydration of water-sensitive shale: underbalanced drilling can greatly reduce the drilling fluid, especially the filtrate of drilling fluid into the formation, effectively control the hydration expansion and collapse of mud shale, so that the well hole is more stable; maximize drilling efficiency: the use of underbalanced drilling, because the drilling fluid column pressure is less than the bottom hole formation pressure, is conducive to rock breaking, improve the rate of penetration.

2.2 *Disadvantages of underbalanced drilling*

Directly increase drilling costs: the reasons include: (1) increased equipment; (2) the area of the well site increases; ④ input equipment and management costs are also increased. Take nitrogen drilling as an example, that is, nitrogen is added to the drilling fluid. If nitrogen cannot be specially provided and can only be produced on-site, for remote areas, the cost of under-balanced drilling will increase due to the high cost of nitrogen production equipment. In addition, if nitrogen, natural gas, and other injection drilling fluid are used as a mitigation agent in this way, it requires manufacturing equipment, compression equipment, filling equipment, etc. And in the completion of the forced tripping equipment and so on.

There are some unsafe factors; When drilling with air, improper handling will cause an underground explosion, fire, blowout, well collapse, and drilling tool rot. Underbalanced drilling is a continuous process in which it is impossible to maintain underbalanced drilling in the wellbore for the entire operation period. Because the drilling fluid cannot form a filter cake on the borehole wall under negative pressure, if the underbalanced state is not maintained, the drilling fluid filtrate and the harmful solid phase will take the opportunity to invade the reservoir. If cementing is carried out at this point, it can cause more serious damage to the pay zone. With the development and progress of technology and process, these problems will be gradually solved and the negative impact will be gradually reduced.

3 KEY TECHNOLOGIES OF UNDERBALANCED DRILLING

3.1 *Underbalanced drilling fluid*

The type of drilling fluid should be selected according to drilling geological conditions, well structure, rock type, fluid pressure, and physical and chemical properties of drilling fluid.

Due to the inaccuracy of pre-drilling prediction of formation pressure, the reference value is usually given, which cannot achieve real-time and effective pre-drilling pressure prediction. Therefore, the drilling fluid density should be timely adjusted to keep it lower than the formation pore pressure.

Wellbore stability of drilling fluid is directly related to the success of drilling. When drilling in paste rock and clay mineral-rich areas, wellbore instability and formation collapse are easy to occur, and complex accidents such as stuck drilling are easy to occur. Therefore, the wellbore stability of drilling fluid is highly required. Only when the drilling fluid column pressure is higher than the collapse pressure of the formation can the wellbore stability be ensured.

For underbalanced drilling, how to carry rock debris out effectively is the key to ensuring the safety of drilling. Compared with balanced drilling, an underbalanced drilling machine has a faster drilling speed, so it requires more rock carrying capacity for underbalanced drilling.

Especially for areas with high sulfur content and high humidity, to effectively protect the casing, the drilling fluid is required to have anti-corrosion ability. Materials such as deoxidizer and hydrogen sulfide are usually added.

3.2 *Underbalanced drilling fluid density*

Formation pore pressure and annular pressure loss should be calculated before underbalanced drilling fluid density is determined. Formation pore pressure calculation includes pre-drilling pressure prediction, while drilling monitoring and post-drilling evaluation. Eaton (1972) established the expression of the relationship between formation pore pressure and logging acoustic time difference based on the experience and theoretical analysis in the Gulf of Mexico and other areas. This relationship does not change with the change in lithology or depth. The principle is that the overburden pressure gradient determines the relationship between the ratio of the actual value and the normal trend value of the observed parameters of compaction and the formation pore pressure (Figure 1).

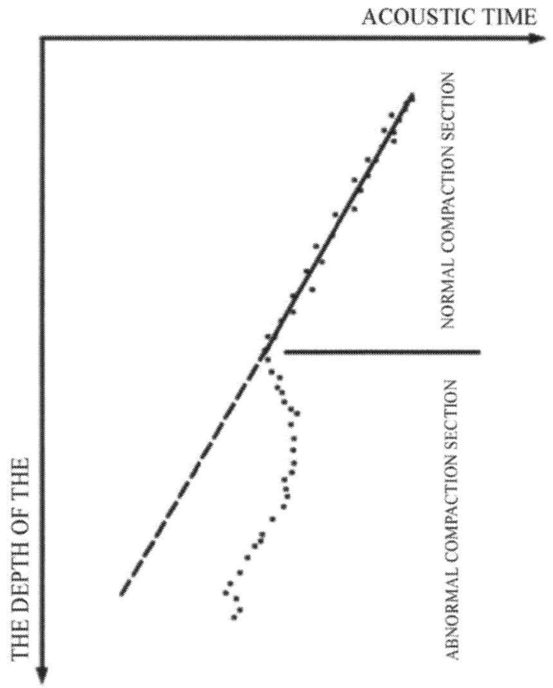

Figure 1. Schematic diagram of Eaton method.

The DC-index method is commonly used to calculate while drilling monitoring. D (DC) index method is based on the principle of compaction law of mud shale and the increase of mechanical drilling rate of under pressure formation and the influence of pressure difference on mechanical drilling rate. Meanwhile, the influence of drilling parameters on mechanical drilling rate is considered to monitor formation pressure. Due to the influence of various factors on pre-drilling pressure prediction, the accuracy of calculation results is not high, so it is necessary to carry out monitoring while drilling to timely understand the pressure changes in drilling, to effectively avoid the occurrence of special situations.

Annular pressure loss is directly related to the design of drilling fluid density. According to the principle of maximum bit water power or impact force in jet drilling, the pump displacement of conventional over-balanced and near-balanced drilling fields is designed to ensure that the circulation state of drilling fluid reaches turbulence. However, compared with underbalanced drilling, the standard is to reduce the load on the equipment, such as the wellhead rotating blowout preventer and choke manifold, to increase the reliability of well control, and to avoid the scouring effect of turbulence on the open hole oil and gas layer, to meet the safety requirements. Therefore, the displacement of underbalanced drilling fluid should not be too large, mainly laminar flow.

3.3 *Calculation of bottom hole pressure control in underbalanced drilling*

For underbalanced drilling, the key to its success lies in the control of bottom hole pressure. If the bottom-hole pressure is not properly controlled, it will lead to overkill, which will pollute the oil and gas layer, affect the productivity of oil and gas well, and cause a blowout and other complicated situations. Bottom-hole pressure is affected by formation parameters, wellbore geometry, pump rate, and gas injection rate, gas-liquid production while drilling, surface control procedures, gas injection methods, and other factors. To effectively reduce the influence of these factors, it is necessary to take targeted measures. For example, MWD calculations are used to effectively obtain bottom-hole pressure and determine underbalanced pressure based on changes in bottom-hole pressure.

3.4 *Well control in underbalanced drilling*

The problem of well control in underbalanced drilling is complicated for many reasons: underbalanced drilling requires negative pressure difference in the bottom hole; The wellhead uses a rotating blowout preventer, which is always rotating during drilling. Bottom-hole pressure is complicated by the intrusion of local zone fluids into the drilling fluid, which can cause changes in bottom-hole pressure and, if not properly controlled, can lead to blowouts.

(1) Underbalanced drilling fluid. Underbalanced drilling fluid should be selected by multiple factors such as underbalanced drilling type, well structure, and formation pore pressure.
(2) Underbalanced drilling can be realized only when a negative pressure difference occurs at the bottom of the hole. Therefore, the drilling fluid type should be adjusted at any time according to the formation pressure change.
(3) In Underbalanced drilling well control equipment, the key is to rotate the BOP, and choke manifold two choices. Compared with a conventional wellhead, an underbalanced drilling wellhead is equipped with a rotary control head, so sufficient space should be reserved during installation. If the well category belongs to the high production gas well or the high-pressure oil and gas well, we should choose the higher-grade combination form; Throttle manifold should be equipped with the same level of a pressure gauge; In addition, throttle lines should be equipped with two – or three-wing throttle lines.

4 KEY EXAMPLES OF UNDERBALANCED DRILLING

4.1 *Well YB103H, Sichuan Basin*

Well YB103H is located in Yuanba area of Sichuan Basin. From bottom to top, it encounters Changxing Formation, Feixianguan Formation, Jialingjiang Formation, Leikoupo Formation, Xujiahe Formation, Ziliujing Formation, Qianfoya Formation, Shaximiao Formation, Suining Formation, Penglaizhen Formation, and Jianmenguan Formation, with stable stratigraphic deposition. However, the process of oil and gas exploration and development in Yuanba area is affected by the high ground pressure gradient, frequent gas layers, and low drilling rate by conventional drilling technology, especially when drilling into the middle to lower Jurassic and Upper Triassic strata. As a new drilling technology, liquid underbalanced drilling technology has been applied in this area.

YB103H was started in 3267.00 m into the underbalanced drilling process, predicting formation pressure are shown in Table 1, because the yuan is now in Shaxi temple, thousand-buddha cliff formation is generally not oil gas, to ensure the speed-up effect, can be lower than the formation pressure coefficient of drilling fluid density, but because of the uncertainty of the formation pressure is very big, the initial reasonable drilling fluid density cannot be sure.

Table 1. YB103H underbalanced formation pressure coefficient.

Depth/m	Predicts fluid pressure coefficient
3267–3380	1.0–1.25
3380–3630	1.0–1.25
3630–3880	1.2–1.4
3880–4330	1.3–1.5
4330–4865	15–1.8

Two special situations occurred in the well drilling process:

(1) flame burning occurred at 3951 m; (2) 4336 m drilling encounters conglomerate.

The total hydrocarbon value increased from 0.8% to 1.01%, C1 increased from 0.4% to 1.76%, circulation pool drilling fluid increased to zero, the outlet drilling fluid density remained unchanged, viscosity rose from 48s to 50s, hydrogen sulfide concentration was zero, ignition cylinder began to burn, orange flame, flame height 1–3 m. While maintaining well control safety, the underbalanced three-way drilling was continued, with the flame never extinguished and the height kept below 3 meters. The effective liquid-gas separator oscillates badly after drilling, and the flame height exceeds 15 m.

4336 m drilling encountered conglomerate layer, to avoid the occurrence of bit accident, drilling to 4338 m drilling, out of the bit tooth loss, a tooth broken more than 60, during drilling jump serious. Six drill bits were used to drill through the conglomerate section, and the wear of the tooth tire body and the hand was serious. Therefore, it was necessary to choose a bit with special diameter retaining teeth to improve the wear resistance of the cone, prolong the service life of the bit and improve the drilling time.

4.2 *Well X73, Huabei Oilfield*

Well X73 is located in the structure-lithologic trap north of Well Wen31 in Nanmazhuang structural belt, Raoyang Depression, Jizhong Depression. Shahejie Formation, Dongying Formation, Guantao Formation, Minghuazhen formation, and Pingyuan Formation are developed successively from bottom to top. The target layer is Es2, and the sedimentary environment is delta-lacustrine deposition, which is developed from the northeast delta. The drilling depth was 3386.0 m. Three open design well depth structures: D339.7 mm×157.66 m/D444.5 mm×160 mm+D177.8 mm×3026.57 m/D241.3 mm×3028 m+D127 mm×3413 m/D152.4 mm×3417.23 m, among which, the underbalanced drilling section is 3028–3386 m.

The wellhead casing pressure was 1.3 mpa after the second well opening and the bottom hole underpressure value was calculated to be between 1.3–1.7 MPa. The wellhead pressure value was kept at about 1.2 MPa during the drill-out process. The drill-out process went smoothly and a small area was found on the ground, indicating that the bottom hole was under pressure.

5 CONCLUSIONS

With the increasing difficulty of drilling, underbalanced drilling has a huge advantage in areas with developed pores and fractures. It can maximize the protection of oil and gas layers and minimize the probability of complex accidents such as well loss. However, it also has disadvantages such as increasing drilling costs and increasing unsafe factors. The key to underbalanced drilling lies in drilling fluid optimization, drilling fluid density determination, and well control. Only by accurately determining these three key parameters can underbalanced drilling proceed safely and smoothly.

REFERENCES

Gensheng, L.I., Xianzhi, S.O.N.G. & Shouceng, T.I.A.N. (2020). Intelligent drilling technology research status and development trends. *Oil Drilling & Production Technology*, 48(1), 1–8.

Hu, Y., Wei, Z., Lisu, L., Baozhong, Z.H.A.O., Penggao, Z.H.O.U. & Shouming, Z.H.O.N.G. (2010). Drilling technology for fractured volcanic rocks in Ludong area, Junggar Basin. *Oil Drilling & Production Technology*, 32(4), 22–25.

Kuanliang, Z. (2013). Research and practice of underbalanced horizontal well dri1ling technology in deep buried-hill reservoirs in Nanpu Oilifeld. *Oil Drilling & Production Technology*, 35(4), 17–21.

Melamed, Y., Kiselev, A., Gelfgat, M., Dreesen, D. & Blacic, J. (2000). Hydraulic hammer drilling technology: developments and capabilities. *J. Energy Resour. Technol*, 122(1), 1–7.

Wanjun, L., Haiqiu, Z., Junfeng, W., Zhongxiang, C., Jiangbo, L. & Yixin, G. (2017). Application of optimized and fast drilling technology to the first long horizontalsection well in north Tluwa oilfield. *China Petroleum Exploration*, 22(3), 113.

Energy Revolution and Chemical Research – Kok-Keong Chong and Zhongliang Liu (Eds)
© 2023 The Authors, ISBN: 978-1-032-36554-1

Hydrocarbon accumulation environment and geological resource potential in jizhong depression

Yuqiong Li
The No. 2 Oil Production Plant of Huabei Oilfield Company, CNPC, Langfang Hebei, China

Yuxi Yang
Erlian Oil Production Plant of Huabei Oilfield Company, CNPC, Xilihot, China

Fengming Zhang, Ting Wang & Wenhui Cai
The No. 2 Oil Production Plant of Huabei Oilfield Company, CNPC, Langfang Hebei, China

Siqi He
Exploration and Development Research Institute of Huabei Oilfield Company, CNPC, Renqiu Hebei, China

Graciela Daniels*
Central Arizona College, Coolidge, AZ, USA

Qiang Qin
COSL-EXPRO Testing Services (Tianjin) Co., Ltd, Tianjin, China

ABSTRACT: The oil and gas exploration practice shows that the slope belt of the Central Hebei depression has good conditions for oil and gas accumulation, but it is difficult to explore oil and gas in this area because of the relatively single slope structure style, diverse sedimentary types, multiple change periods, and complex relationship between various types of traps and accumulation configuration. Based on sedimentology and tectonic geology, the geological environment and main controlling factors of oil and gas accumulation in the slope belt of Jizhong depression are studied in this paper. The results show that there are three types of buried-hill hydrocarbon accumulation modes in Jizhung depression, which are fault-step zone, slope zone, and central uplift zone. The central uplift zone is still the main hydrocarbon accumulation mode in this area and the step-fault zone and slope zone will be the focus of future exploration in this area.

1 GENERAL INSTRUCTIONS

The oil-rich sag of Bohai Bay Basin is rich in oil and gas resources. The discovered oil reserves are mainly distributed in the forward structural belt of the sag, while the sag area, which accounts for about 70% of the sag area, has a low exploration degree and great exploration potential. Jizhong depression is a first-order tectonic unit in the northwest of Bohai Bay Basin. It starts from Yanshan uplift in the north, reaches Xingheng uplift in the south, is adjacent to Taihang Mountain uplift in the west, and reaches Cangxian uplift in the east. The exploration area is about 32,000 km^2. Since the Paleozoic, Jizhong Depression has experienced multiple periods of tectonic movement, especially the late Yanshan tectonic movement, which uplifted the North China platform and caused long-term weathering and denudation, resulting in regional buried hill top unconformity. Due to the influence of palaeotopography and denudation, the buried hill strata are exposed from Archaean to Middle and upper Proterozoic to Mesozoic (ZOU 2015).

*Corresponding Author: graciela_daniels@stu.centralaz.edu

DOI 10.1201/9781003332657-27

At the end of the Yanshan Movement, the Gaoyang fault and Dabaiqi isotensional fault formed in the south of the Lixian slope, cutting the Paleozoic base. At the same time, the activity of the basin margin fault caused the slight tilt in the south of the slope, which showed the characteristics of the slope. Although Renxi fault continues to be active, its influence is very limited in the south, so the fault in the south ceased to be active during the period of weak tectonic activity, such as the late Member of Sha-3 and the First member of Sha-3. After the deposition of Sha-4 Kongdian and Sha-3, thousands of meters of strata have been deposited, and the low-lying areas on the paleotopography have been filled up (Z 2012). During this period, the slope was gentle and there was no big topographic relief (J 2012). At the end of Dongying, the regional uplift intensified and the Xishan Act ii movement began. The episodic movement of the structure led to the activation of Gaoyang and Dabaichi ancient faults, as well as the generation of some small faults with small fault spacing.

Buried-hill oil and gas reservoirs are the main types of oil and gas reservoirs in Jizhong Depression. According to their formation mechanism, they can be divided into three types and five subtypes: residual hill type, fold type (anticline, fault anticline), and fault-block type (single fault type, fault step type, fault horst type). The oil and gas in buried hills of Jizhong Depression are mainly derived from overlying Paleogene source rocks, followed by carboniferous-Permian coal measure source rocks, and accumulated in middle-upper Proterozoic and Lower Paleozoic carbonate reservoirs, which is a typical "new paleoreservoir" type oil and gas accumulation assemblage (JIAO 2020). The buried hill oil and gas are mainly distributed near the oil source, rich in the east and poor in the west, oil in the south and gas in the north, and enriched in the buried hill traps near the fault zone. The fault and unconformity are well developed in the area due to the multi-stage tectonic movement, which constitutes the transport systems of various types of buried hill oil and gas reservoirs. Hebei central depression buried hill reservoirs, the author of this paper, the use of seismic, drilling and analysis of test data, combined with the oil and gas exploration practice, analyzing the characteristic of the buried hill reservoirs of the petroleum system, discusses the patterns of buried hill reservoir, expounds that the law of buried hill hydrocarbon accumulation, for the prediction of favorable exploration direction, and provides the basis for looking for scale reserve zone (D. L 2009).

2 ACCUMULATION MODE

2.1 *Characteristics of stratigraphic sequence*

The study area is mainly composed of Kongdian formation, Shahejie Formation, and Dongying Formation in Paleogene, and Guantao Formation and Minghuazhen Formation in Neogene.

Hole store group. The formation was formed in the early stage of the division and filling of the Paleogene faulted lake basin. The lithology of the formation is mainly coarse clastic such as heterogeneous conglomerate and breccia with a red sandy mudstone layer. In the middle and upper part of the formation, there are dark mudstone and gypsum sections, which are mainly alluvial fan and fluvial facies deposits.

Sha He Street Section. The formation is widely distributed in the sag and is the main oil-bearing reservoir and oil-generating reservoir of the Paleogene in Raoyang Sag. It can be divided into four sections: (1) The fourth Member of Shahejie Formation (Es4): the thickness of the formation varies widely, generally ranging from 0 to 1000 m, and there are no deposits in Gaoyang, Yanling, Renqiu, and other areas. It was formed in the early stage of fault depression division and filling, and it has two third-order normal cycles with thick layers at the bottom and thin layers at the top, and two reservoir development zones are developed. The fourth member of Shahejie formation is in unconformity contact with the underlying Kongdian Formation, and its bottom boundary is equivalent to seismic reflection layer T7, which can only be identified in the local area of Raoyang sag. (2) Es3 Member: After the rapid filling of Kongdian Formation and Es4 Member, Es3 members developed in the stable sedimentary stage when the lake basin expanded and sank deeply, and the strata were widely distributed. It covers most of Jizhong Depression, and the stratum thickness is generally 150~800 m. The Es3 Member can be divided into three sub-sections: Es3 Lower (Es3 lower), Es3 Middle (Es3 Middle), and Es3 Upper (Es3 upper). The third member of Shahejie formation is a deep lake environment, mainly composed of purplish-red mudstone and gray sandstone. The bottom boundary of the third member of The Shahejie formation corresponds to the seismic reflection layer T6, and the top is bounded by the "mud neck" marker layer and the second member of the overlying Shahejie Formation. (3) Es2 Member of The Shahejie Formation (Es2): Formed in the late uplift of the faulted lake basin, with relatively thin thickness ranging from 53.5 m to 213 m. Regionally it is a set of brownish-red mudstone interbedded with light gray sandstone. The bottom of the second member of Shahejie formation corresponds to the T5 seismic reflector, which is in unconformity contact with the underlying strata. At the top, the purplish-red mudstone section is separated from the bottom conglomerate of the First member of Shahejie formation. The top interface corresponds to T4 seismic reflection layer, and the first member of Shahejie formation is in a regional unconformity contact relationship. (4) Es1 member: with an apparent thickness of 185–789 M, it can be divided into lower Shahejie sub-member and upper Shahejie sub-member. The sedimentary period of the lower sub-member of Sha-1 was mainly lacustrine deposition during the extension period of lacustrine fault and depression. Its bottom develops a set of gray sandstone, commonly known as "tail sandstone"; There is a set of dark gray-black mudstone, brown oil shale mixed with gray thin sandstone, gray dolomite, and oolitic limestone in the upper part of the formation, commonly known as the "special lithologic section", usually as a regional marker. During the sedimentary period of the upper sub-member of The First Shahejie formation, the lacustrine area began to shrink, and the lacustrine basin entered the period of faulting, depression, and extinction. At the top, the purplish-red and gray-green mudstone thickens, and there is carbonaceous mudstone locally, and gray sandstone.

2.2 *Accumulation model of central uplift zone*

The central uplift belt accumulation model is the main accumulation of buried hills in Hebei central depression type, its characteristic is the central uplift belt in fault developed complex buried hill monadnock system and buried hill in multiple hydrocarbon generation subsags lithostatic, subsags generated oil and gas along by the side of the fault on one side of the steep slope and gentle slope unconformity of buried hill trap for hydrocarbon, two-way can form to the top of buried hill reservoirs and hidden hill reservoirs. The discovered Renqiu buried-hill reservoir belongs to this type of accumulation model, which has found oil layers in Wumishan formation, Cambrian, and Ordovician reservoirs under the Paleogene unconformity.

2.3 *The pattern of accumulation of slope zone*

According to the distance between the source rock and trap and the difference between the oil and gas transport system, the slope zone of Jizhung depression can be divided into the inner zone, middle zone, and outer zone from the center of the depression, and the buried hill traps such as residual hills, fault blocks, and fault horsts can be developed. Because the slope zone is relatively shallow buried, weathered, and leached to a large extent, its internal reservoirs are well developed, and it is in the direction of oil and gas migration for a long time, so it has good reservoir forming conditions. The oil and gas mainly migrated laterally along unconformity, during which the oil and gas migrated step by step into buried hill traps in different parts and accumulated into reservoirs through the vertical regulation of faults.

2.4 *Fault step zone accumulation model*

Buried hill controlled the limitation of concave fault, since to the buried hill hydrocarbon generation sub-sags are successively lifted off order structure, rise in buried hill formation by an extrusion process, tilted and wrinkles, and weathering, denudation and leaching effect, so the top of the buried hill in this class can develop monadnock and block buried hill trap, such as the southern Ma Zhuang buried hill, Hejian buried deep-buried hill, north and west buried hill, etc. Oil and gas mainly enter the high buried hill reservoir through the transport system composed of faults and unconformity in a stepped manner. This model is characterized by unidirectional hydrocarbon supply from buried hills, and hydrocarbon migration is mainly vertical along faults with unconformity lateral transport and short migration distance.

3 MAIN CONTROLLING FACTORS OF HYDROCARBON ACCUMULATION

3.1 *Source rock conditions*

There are three sets of source rocks in the study area: one is the "special lithology section" of the lower Shahejie formation, one is the dark mudstone source rock of the third Shahejie formation, and one is the dark mudstone source rock of the ES4-EK section of the north trough of Beidian Lake. The lower member of Shahejie Formation is the most widely distributed source rock and is the most important source rock in the study area, while the third member of Shahejie Formation and ES4-EK source rock are far away from the slope and their oil supply range is relatively limited, so they are the secondary source rock of the slope oil supply.

Es1x source rocks are widely distributed in the slope, covering almost the entire slope, generally over 26 m and up to 72 m. They are also developed on the Northern Slope of the slope, with thicknesses ranging from 10–25 m. The organic carbon content of source rocks is generally higher in the lower carboniferous area, with the high-value area in the northeast of the slope, and the average organic carbon content is 3.3.5%. It gradually decreases to the south, and the middle and outer zone in the south of the slope is the low-value area, with TOC less than 0.5%, indicating that the source rocks are non-source rocks. The average content of soluble organic bitumen "A" in the source rocks of Shahaizhi basin is 0.47%, and the average total hydrocarbon content is 2,693 PPm. The abundance of organic matter in most Wells is relatively low.

Microscopic observation and scanning electron microscopy showed that kerogen in the lower Shahe-submember of the hydrocarbon source layer was amorphous, and granular, with no clear outline and regular edges. The color was yellow to brown, and shell bodies were rare. Cellulose, lignin, and vitrinite were observed, which were typical type 1 parent materials (Figure 1). In addition, the saturated hydrocarbon chromatographic results of the source rocks show that phytane has a high Ph content and a preponderance of phytane, indicating that the parent sedimentary environment is a strong reductive sedimentary environment. Almost all the source rocks in the first Shaxia area have entered the stage of immature and low mature oil generation and have the conditions to form a large amount of immature and low mature oil.

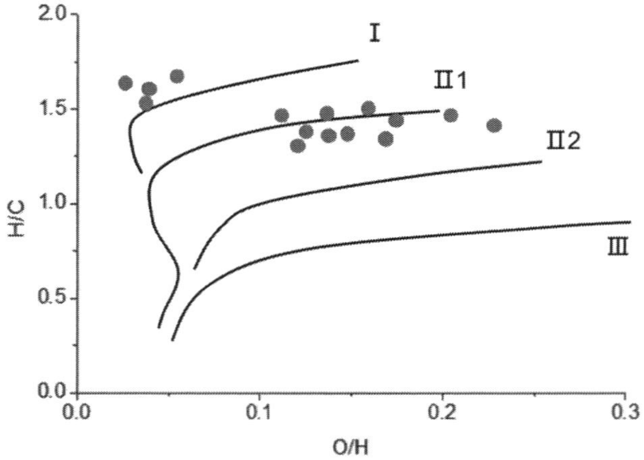

Figure 1. The slope kerogen element composition paradigm diagram in the study area.

The source rocks of the third Member of Shahejie formation are the main source rocks of mature crude oil in the study area, and the source rocks that contribute to the enrichment of oil and gas in the slope are confined to the northern part of the inner slope zone and the far trough. The organic carbon content of the dark mudstone in ES3 members is medium, with an average of 0.77%. The soluble organic matter content of chloroform bitumen "A" and total hydrocarbon is relatively high, with an average of 0.23% and 1692 pmm. The ratio of hydrocarbon to TOC is as high as 19.2–23.3%. The organic matter type of source rocks in the third member of Shahejie Formation is mainly the mixed type (ii 1) of metsaprosaic mud, and the IH of all samples is less than 410 mg/g, mainly the mixed type (ii 2-iii). In the mature stage, the hydrocarbon source of Shahe-3 member has a burial depth of 2800–3800 m, a temperature of 98°C–128°C, and is in the stage of massive oil generation. The vitrinite reflectance is 0.55–0.85%.

Es4-ek source rocks are mainly distributed in the Dianbei trough at the northern end of the slope, and the distribution direction of the set of source rocks is NW. The average thickness is generally 50～120 m, local paste. The average content of organic carbon is 0.84%, the average content of bitumen "A" is 0.042%, the main type of organic matter is type ii 2, and the Ro distribution is between 0.62 and 1.17, which is below the maturity threshold. It is A set of medium hydrocarbon source rocks and the main hydrocarbon source rocks in Baiyangdian area. The migration distance of oil and gas is affected by the range of effective source rock, and the influence factors of source rock on reservoir forming conditions of slope in the study area are mainly reflected in the distance between trap and oil source. If the trap is located in the center of abundance factor set to 1, since in center abundance factor increases gradually, when outer overlap trap outer edge and the mature hydrocarbon source rock, oil and gas abundance factor to maximum, then decreased with increasing distance trap from a hydrocarbon source rock, if sub-sag, not a round shape, the abundance of the different parts of the trap factor change is bigger, but the overall trend remains the same.

According to the specific situation of the slope in the study area, the effective influence range of source rocks in the lower Shahejie formation should be 20–40 km. As oil shale in the lower Shahejie formation is ubiquitous in the whole slope area, Es2 oil reservoir in the whole slope is abundant, mostly self-generated and self-stored oil reservoir, and even forms the feature of "full slope oil-bearing". A hydrocarbon source rocks at the bottom of the sand three sections and its main RaoYang sag formed in the near the southern slope and near the lake in the north of the north sag, slope distance is far, to source outside secondary lateral migration of oil-gas reservoir, so the whole slope Es3 reservoir is Es2 paragraphs reservoir development situation is bad, and outside sources of the secondary hydrocarbon reservoirs are due to the lateral migration is given priority to. In most cases, a unilateral reservoir style is formed, that is, one side is blocked by faults and the other side is lithology as boundary conditions, so it is easier to form oil and gas traps on the side of coexisting large faults.

3.2 *Sedimentary facies environment*

Through the observation and analysis of the core, the vertical evolution and lateral variation characteristics of the core in the slope study area are obtained.

(1) Core characteristics of Sha 3 Member. The sandstone in this section has a fine grain size and is mostly silty sand rock. Various sedimentary structures are developed, such as trough cross-bedding tabular cross-bedding, and parallel bedding. Furthermore, the development of wormholes and the occasional thin coal interlayer indicate that the hydrodynamic conditions in this area are strong and the unidirectional traction flow is dominant, and it belongs to the sedimentary environment of the water-land transition zone. Combined with the analysis of the structural position of the coring well, the third member of the Shahe formation in the coring area should belong to the environment of the delta plain and delta front.
(2) Core characteristics of the second Member of Shahejie Formation. From the observation of the core, the sandstone of the second member of Shahejie Formation is coarser than that of the third member of Shahejie Formation. The lithology is mostly medium-fine sandstone and siltstone, but there is also a small amount of carbonate breccia. The mudstone is mostly variegated,

indicating that the sedimentary environment is sometimes above water and sometimes under water. The watercourse characteristics of this section are very similar to those of Sha 3 section. From the observation of the core of this section, we also found the cutting structure, deformation structure, fault dislocation, and ball-pillow structure, which reflects that the terrain has a certain slope and strong hydrodynamic force. Combined with the tectonic position of the core well, it is considered that the sedimentary environment of the middle and north members of the Second Member of Shahejie formation is in fluvial facies or distributary channels of the delta plain.

3.3 *Hydrocarbon migration conditions*

The observation of the core shows that the reservoir has strong heterogeneity, and there is usually no oil near the scour surface of the core, while the plate-like cross-bedding and parallel bedding above the scour surface and trough cross-bedding are mostly oil-bearing parts. The lithology of the oil-bearing rocks is generally fine sandstone, which is relatively porous and has good physical properties. The occurrence of oil in the top and bottom and no oil in the middle may be caused by the deterioration of the physical property of the sand body upward, the separation of gray-green and black muddy bands in the upper part, or the upper sandstone is caused by calcareous cementation.

The source rocks are widely distributed on the whole slope of the study area, and the hydrocarbon accumulation is mainly nearby, so there is no high requirement for the transport channels. In general, there are three types of hydrocarbon migration channels in the slope region: faults, unconformities, and connected sand bodies. The two regions developed by the slope are unintegrated into the zone of pore fracture, which can become the lateral migration channel of oil and gas. There are two sets of unconformity: one is the unconformity between member 1 and Member 2, and the other is the unconformity between member 2 and Member 3. These two sets of unconformity cooperate with the upper and lower connected sand bodies, which can form a good transport system for oil and gas lateral migration. The sandstones at the top of the upper member of Sha-3 and the tail of the lower member of Sha-3 were developed in a high system tract or ascending semi-cycle system tract, at which time the provenance was sufficient and the space could be small. The sand bodies were superposed vertically, connected horizontally, and had good lateral connectivity, which provided the necessary conditions for oil and gas migration channels.

4 CONCLUSIONS

An effective transport system is key to hydrocarbon accumulation in trough areas. Subsags area often cut development in hydrocarbon source rock, and continued activity in Ming town, formation sedimentary period of oil source fault, fault fall plate usually large-scale sedimentary sand body development, is advantageous to the hydrocarbon source rock of lateral oil and gas gathering, oil source faults and hydrocarbon source rock internal sand body combination, oil and gas formation migration upwards, and gathered in the favorable reservoir accumulation. The hydrocarbon enrichment degree above the source in the trough area depends on the vertical transport capacity of the transport system outside the source and the hydrocarbon generation capacity of the fault cutting through the source rock. Generally, the stronger the hydrocarbon generation capacity of the source rock is, the better the relationship between the oil-source fault activity and the configuration of the accumulation period is, and the higher the hydrocarbon enrichment degree above the source in the trough area is.

REFERENCES

Caineng, Z. O. U., Guangming, Z. H. A. I., Zhang, G., Hongjun, W., Zhang, G., Jianzhong, L. & Liang, K. (2015). Formation, distribution, potential and prediction of global conventional and unconventional hydrocarbon resources. *Petroleum Exploration and Development*, 42(1), 14–28.

Caineng, Z., Zhi, Y., Shizhen, T., Wei, L., Songtao, W., Lianhua, H. & Qiulin, G. (2012). Nano-hydrocarbon and the accumulation in coexisting source and reservoir. *Petroleum Exploration and Development*, 39(1), 15–32.

Chengzao, J., Zheng, M. & Zhang, Y. (2012). Unconventional hydrocarbon resources in China and the prospect of exploration and development. *Petroleum Exploration and Development*, 39(2), 139–146.

Fangzheng, J. I. A. O., Caineng, Z. O. U. & Zhi, Y. A. N. G. (2020). Geological theory and exploration & development practice of hydrocarbon accumulation inside continental source kitchens. *Petroleum Exploration and Development*, 47(6), 1147–1159.

Gautier, D. L., Bird, K. J., Charpentier, R. R., Grantz, A., Houseknecht, D. W., Klett, T. R. & Wandrey, C. J. (2009). Assessment of undiscovered oil and gas in the Arctic. *Science*, 324(5931), 1175–1179.

Energy Revolution and Chemical Research – Kok-Keong Chong and Zhongliang Liu (Eds)
© 2023 The Authors, ISBN: 978-1-032-36554-1

Research on environmental impact assessment and ecological protection in hydropower planning

DaZhuang Wang
Beijing Zhonghuan Geyi Technology Consulting Co Ltd, Beijing, China

HengYu Chen
Beijing Zhonghuan Geyi Technology Consulting Co Ltd, Chengdu, Sichuan, China

Na Cao*
Environmental Engineering Assessment Centre, Ministry of Ecology and Environment, Energy Assessment Department, Beijing, China

ABSTRACT: Integrated river basin planning is a tool used to guide the development and use of water resources, the management of the water environment, and the optimal allocation of water resources in a river basin. Firstly, under the guidance of comprehensive environmental assessment theory, this paper constructs a set of index systems for environmental impact assessment in the process of basin planning. Then, the expert scoring method is used to construct the judgment matrix of the index system, and the hierarchical analysis method is used to calculate the weight value of each index. Finally, the environmental impact of the planning area is evaluated and analyzed through the comprehensive analysis method. The study analyzes, predicts, and evaluates the environmental impacts arising from the implementation of the integrated watershed planning, and proposes measures to prevent or reduce environmental impacts, having far-reaching significance in promoting the sustainable development of the watershed environment.

1 INTRODUCTION

The difficulty of environmental impact assessment of comprehensive river basin plans is increased by the wide scope of the assessment, the many influencing factors, the large planning time span, the cumulative nature, the holistic nature, the spatial specificity, and the uncertainty. In the environmental impact assessment of comprehensive river basin plans, relevant qualitative and quantitative indicators need to be selected to reduce the complexity of the environmental assessment, and these indicators constitute a scientific, complete, and reasonable indicator system (Wang 2017). The establishment of an environmental impact assessment index system for watershed planning is to combine the planning objectives and clarify the content of the assessment, so that the evaluation of the degree of environmental impact after the implementation of the comprehensive watershed plan can be more specific and systematic, and the index system can describe the state of the environment within the evaluation area and provide a scientific basis for decision-makers (Kader 2014). Given the immaturity of environmental impact evaluation of comprehensive watershed planning in China in terms of theory and evaluation technology, by analyzing the characteristics of environmental impact evaluation of comprehensive watershed planning and combining the practical experience of environmental impact evaluation of comprehensive watershed planning in China, we try to build a unified index system for environmental impact evaluation of comprehensive watershed planning, and apply this index system to the environmental impact evaluation of a comprehensive watershed management plan for Case verification.

*Corresponding Author: 40656020@qq.com

 DOI 10.1201/9781003332657-28

2 SELECTION OF EVALUATION INDICATORS SYSTEM

2.1 *Ecological indicators*

The main land-use types within the planned river section are bare rocky gravel land and Gobi, so the impact on vegetation is minimal. However, during the construction period, the large number of construction workers will cause some damage to the vegetation and environment.

Level of impact on aquatic ecosystems: The plan will also have an impact on aquatic life in the upper reaches of River A for the following reasons: ① The implementation of the plan will change the hydrological situation of the river, thus affecting the dynamic balance of the natural river, e.g. the river in the reservoir area will change from fast-flowing to slow-flowing or still water after reservoir storage. ② The construction of power stations interferes with the natural sediment transfer process, e.g. sedimentation in reservoirs. ③ The planned power station interrupts the circulation of nutrients, which are deposited in the reservoir area. ④ The dams blocking the rivers lead to a decline in biodiversity.

Landscape changes in the basin: Before the implementation of the plan, there was little human disturbance in the planned basin. With the implementation of the plan, artificial power stations will appear on the natural river channels and link roads will be established between the power stations to form corridors, so the landscape pattern will change, landscape heterogeneity will decrease and landscape fragmentation will increase. In this paper, seven concepts such as patch area and patch perimeter, landscape boundary density, separation index (Gusman 2014), fragmentation index, diversity index, uniformity index, and dominance index are used to illustrate the changes in landscape pattern and landscape spatial heterogeneity and diversity.

2.2 *Water environment indicators*

Degree of hydrological change: The main stream of river A is a reservoir-level hydropower terrace development scheme. After the completion of the step power station, a series of reservoirs will be formed in the river section above a certain reservoir, which will increase the water surface area and widen the water surface in the river valley, which will lead to the cumulative effect on the water environment.

Water quality impact: The impact of the construction and operation of the planned river stage hydropower station on the regional water quality mainly comes from domestic sewage and production wastewater during the construction period, with the main sources of pollution being the drainage of people's daily lives and the sediment, suspended matter and petroleum generated by the construction.

Groundwater cycle: According to the geological analysis of the basin, the basins in the middle and lower reaches of River A are relatively complete hydrogeological units, with their independent conditions of recharge, runoff, and discharge (Chen 2012). The "river and aquifer" water resource system. The large quantity and repeated cyclic transformation of water resources are conducive to the use of water resources and the improvement of their utilization, which is an important feature of the "river and aquifer" water cycle system in the arid zone.

2.3 *Water environment indicators*

Reservoir-induced earthquakes: The conditions for reservoir-induced earthquakes are generally twofold: firstly, geological and tectonic conditions, often with high initial stresses and active fractures in the reservoir seismic zone; and secondly, reservoir infiltration and storage conditions. Fractures and fissures are common in limestone areas, which are prone to deep seepage, as well as in swollen, easily softened rocks, which can easily induce reservoir earthquakes.

Reservoir bank stability: The construction of reservoirs in the upper reaches of rivers changes the original geological environment of the reservoir area, which may lead to environmental geological problems such as bank collapse, landslides, etc., which will cause different degrees of harm to local land resources, engineering efficiency, and water quality. Bank collapse generally occurs in plain and hilly reservoirs, while slumping and landslides are more common in mountainous ravine reservoirs.

3 DETERMINATION OF THE WEIGHTING OF EVALUATION INDICATORS

3.1 *Hierarchical analysis process*

Building a hierarchical structure: When applying AHP to a decision problem, it is important to first structure the problem hierarchically and to construct a hierarchical institutional model. These levels can be divided into three categories: the top level (the objective level), the middle level (the criterion level), and the bottom level (the solution level). The number of levels in the hierarchy is related to the complexity of the problem and the level of detail required for the analysis, but generally, there is no limit to the number of levels, with no more than nine elements dominating each level.

Constructing a comparative judgment matrix: Based on the structural system established by the DPSIR model, a comparative judgment matrix is constructed, which represents the relative importance of each indicator in a certain factor layer. See Table 1, Cij refers to the quantitative value obtained by comparing Ci and Cj in the indicator layer, and specifically refers to the importance of the factor layer B. The value of Cij is assigned using a 9-level scale (see Table 2 for details, using numbers and their reciprocals to express the importance of the two comparisons between factors, while using a certain actual representative ratio as the basis for achieving quantitative evaluation.

Table 1. Matrix of judgements at the indicator level.

C	C_1	C_2	C_3	...	C_j	...	C_n
C_1	C_{11}	C_{12}	C_{13}	...	C_{1j}	...	C_{1n}
C_2	C_{21}	C_{22}	C_{23}	...	C_{2j}	...	C_{2n}
...	...	...	...	...	...	...	...
C_i	C_{i1}	C_{i2}	C_{i3}	...	C_{ij}	...	C_{in}
...	...	...	...	...	...	...	...
C_n	C_{n1}	C_{n2}	C_{n3}	...	C_{nj}	...	C_{nn}

Table 2. The nine-level scale method and what it means.

Scale	Degree	Description
1	Same	indicates that Ci and Cj are equally important compared to each other
3	Slightly	indicates that Ci is slightly more important than Cj when compared to Ci and Cj
5	noticeable	Indicates that Ci is significantly more important than Cj when compared to Ci and Cj
7	Strong	Indicates that Ci is more strongly important than Cj when compared to Ci and Cj
9	Extremely	indicates that Ci is more extremely important than Cj when compared to Ci and Cj
2, 4, 6, 8	Intermediate	A comparative value between 2 adjacent scales
Countdown	–	If the ratio of the importance of the two factors Ci and Cj is Cij, then the ratio of the importance of Cj to Ci is Cji = 1/Cij

Calculation of weights: The four main methods are geometric mean, arithmetic mean, eigenvector, and lowermost squares.

The consistency of the judgment matrix and its test: The relationship between the judgment indicators in the judgment matrix is the result of human judgment, and the importance between the indicators is sometimes not entirely logical, especially when there are more indicators, there may be errors, so the calculated weight coefficients need to be tested logically.

3.2 *Ecological indicators*

Table 3 shows the constructed comparative judgment matrix of ecological and environmental indicators. The ecological environmental indicators are the criterion layer (B), the degree of impact of terrestrial ecosystems, the degree of impact of aquatic ecosystems, changes in watershed landscapes, changes in ecosystems in water-reducing river sections, and soil and water flows are the indicator layers (C). The scaling method proposed by A. L. Satty for integers between 1–9 and their reciprocal scales was used to calculate the weights of C1–C5.

Table 3. Ecological Indicators Judgement Matrix.

B_1	C_1	C_2	C_3	C_4	C_5
C_1	1	1	1/2	1	3
C_2	1	1	1/2	1	3
C_3	2	2	1	2	6
C_4	1	1	1/2	1	3
C_5	1/3	1/3	1/6	1/3	1

3.3 *Water environment indicators*

Table 4 shows the comparative judgment matrix of the constructed water environment indicators. The water environment indicators are the criterion layer (B), and the hydrological variability, water quality impact, water eutrophication, and groundwater circulation are the indicator layers (C). The scaling method proposed by A. L. Satty was used to scale the integers between 1–9 and their reciprocal scales, and the weights of C6–C9 were calculated.

Table 4. Water Environment Indicators Judgement Matrix.

B_2	C_6	C_7	C_8	C_9
C_6	1	1/2	1/2	1/3
C_7	2	1	1	1/2
C_8	2	1	1	1/2
C_9	3	2	2	1

3.4 *Environmental geological indicators*

Table 5 shows the comparative judgment matrix of the environmental geological indicators of the structure. The environmental geological indicators are the criterion layer (B), and the reservoir-induced seismic potential and reservoir bank stability are the indicator layer (C). The scaling method proposed by A. L. Satty for integers between 1–9 and their reciprocal scaling was used to calculate the weights of C10–C11.

Table 5. Environmental Geological Indicators Judgement Matrix.

B_3	C_{10}	C_{11}
C_{10}	1	1
C_{11}	1	1

3.5 *Social environment indicators*

Table 6 shows the comparative judgment matrix constructed for the socio-environmental indicators. The socio-environmental indicators are the criterion layer (B), and the socio-economic development,

Table 6. Social Environment Indicators Judgement Matrix.

B_4	C_{12}	C_{13}	C_{14}
C_{12}	1	1/3	1
C_{13}	3	1	3
C_{14}	1	1/3	1

population health, and energy use are the indicator layer (C). The scaling method proposed by A. L. Satty for integers between 1 and 9 and their reciprocal scaling was used to calculate the weights of C12–C14.

3.6 *Analysis of the weight of the criterion layer*

The criterion layer includes ecological environment indicators, water environment indicators, environmental geology indicators, and social environment indicators.

Table 7. Criteria level judgment matrix.

A	B_1	B_2	B_3	B_4
B_1	1	2	3	5
B_2	1/2	1	2	4
B_3	1/3	1/2	1	2
B_4	1/5	1/4	1/2	1

Hierarchical total ranking and consistency check: Hierarchical total ranking refers to the calculation of the weights of each sub-indicator under ecological environment indicators, water environment indicators, environmental geology indicators, and social environment indicators relative to the target layer. From the results, it can be seen that the results of the hierarchical total ranking have satisfactory consistency and accept the results of the analysis. The largest weight of the hierarchical total ranking is watershed landscape change (C3) with a weight of 0.1790, followed by reservoir-induced seismic potential (C10) with a weight of 0.1282, and the third is groundwater circulation (C9) with a weight of 0.1220.

3.7 *Assessment of indicators*

From the above analysis, it can be seen that the implementation of the water-energy plan for the section above a reservoir in the mainstream of River A in Gansu Province will have a great impact on the ecological environment, water environment, environmental geology, and social environment of the river basin, and at the same time make full use of water resources to generate great economic benefits and develop the regional economy.

Through the analysis of ecological environment indicators, water environment indicators, environmental geological indicators, and social environment indicators and their sub-indicators, it can be concluded that, in the evaluation of the environmental impact of a section of the main river A above a reservoir, according to the weight analysis, the most important ecological environment indicators are landscape changes in the basin water environment indicators are groundwater circulation environmental geological indicators are the possibility of earthquakes induced by the reservoir social environment indicators are the population Health. According to the total hierarchical ranking analysis, the most important indicators are, in order of importance, the change in the watershed landscape, the possibility of reservoir-induced earthquakes, and the groundwater cycle.

The above-mentioned indicators, which have been filtered and given greater weight, can best reflect the degree of impact on the ecological environment, water environment, environmental geology, and social environment as well as the rationality of the plan after the implementation of the water-energy plan for a section of the mainstream of the river A above a reservoir.

4 ECOLOGICAL PROTECTION MEASURES

4.1 *Reduction measures for ecological impacts*

The relationship between the implementation schedule, investment, and the regional landscape ecosystem is taken into account during the implementation of the plan, so that the construction works are coordinated with ecological protection. To reduce the impact of plan implementation and operation on ecology (Li 2003), flora and fauna, the following measures need to be taken.

- To reduce the impact of plan implementation personnel on plants and animals, the boundaries of plan implementation should be delineated and the arrangement of plan implementation projects beyond the boundaries should be strictly prohibited. Planning and implementation personnel should be prohibited from entering other areas and carrying out fishing or other activities that hinder ecological protection in the river section where the plan is implemented.
- During the construction period, public announcements and brochures will be distributed to enhance ecological protection publicity and education for planning and implementation personnel and nearby residents, and through the development of a strict system, unauthorized felling of trees and illegal hunting by planning and implementation personnel will be strictly prohibited to mitigate the impact of planning and implementation personnel on local terrestrial flora and fauna.
- The operation mode and season of the planning and implementation machinery should be reasonably arranged, and the blasting regulations for the implementation period should be concentrated (Duan 1993). As much as possible, ropeways should be used to solve the problems of transportation and slag out of branch caverns, and to reduce the damage to vegetation caused by the construction of temporary planning and implementation of transport facilities.

4.2 *Restoration measures for ecological impacts*

Some of these impacts are temporary and will disappear with the end of the project, while others can be eliminated through ecological restoration. The main elements of ecological restoration include: determining the ecological restoration plan, including the restoration objectives, location, scope, area, budget, etc., in conjunction with the layout of the project and the implementation of the plan, and conducting a cohesive economic and ecological assessment of the restoration plan.

5 CONCLUSION

In this paper, we have studied the development of environmental impact assessment (EIA), the basic theory of the evaluation index system, and the method of determining the weight of planning EIA indicators. It summarizes the main environmental impacts of river basin hydropower development planning, establishes an environmental impact assessment index system for hydropower planning, assigns values to them, determines the weights of each criterion and index layer using the hierarchical analysis method, and compares the case studies, the following conclusions are drawn: environmental impact assessment is an effective means of implementing sustainable development strategies; the development of hydropower basin gradients, while making full use of hydropower resources and creating huge economic and social benefits, may also have some negative impacts on the basin environment.

REFERENCES

Chen Yue. (2012). Exploring ecological and environmental impact assessment of water conservancy projects [J]. *Talent*. (28):29.

Duan Kaijia. (1993). Environmental assessment of hydraulic and hydropower project construction [J]. *Sichuan hydropower*. (03):92–93+96.
Khanizullah Gusman. (2014). Water ecological environment evaluation of Guanmenshan reservoir water conservancy scenic area[J]. *Jilin Water Resources*, 2014(06):47–49. DOI:10.15920/j.cnki.22-1179/tv.06.006.
Li Xiaokai. (2003). Environmental assessment of water conservancy projects[J]. *Northeast Water Conservancy and Hydropower*, 2003(10):4–7+23–59. DOI:10.14124/j.cnki.dbslsd22-1097.10.002.
Mukdeth Kader. (2014). Exploration of remote sensing feature parameter set in ecological environment evaluation of water conservancy projects[J]. *Information Technology and Informatization*. (12):84–86.
Wang Xingguo. (2017). A brief discussion on strategic environmental impact assessment in water resources planning and design [J]. *Heilongjiang science and technology information*. (11):197.

Energy Revolution and Chemical Research – Kok-Keong Chong and Zhongliang Liu (Eds)
© 2023 The Authors, ISBN: 978-1-032-36554-1

Evaluation of national rainwater development and utilization potential based on Geoclimatic zoning

Ying Wang, Junde Wang, Zhenjuan Su & Li Qin*
Gansu Academy for Water Conservancy, Lanzhou, China

ABSTRACT: In response to the fact that we are short of systematic evaluation of rainwater utilization potential, and lack the results of rainwater utilization potential for the whole country and different levels of climate zones, this paper defines and divides suitable areas for rainwater harvesting and utilization based on the differences of geography, climate and geological conditions in China, and also describes the rainwater development and utilization potential from the perspective of "appropriate" development and utilization of unconventional water resources and constructs a calculation model to evaluate the rainwater development and utilization potential in China based on geological and climatic dimensions. The results show that China's rainwater harvesting and utilization area is divided into core area and general area, where the core area is divided into Loess Plateau hilly area and the Karst area. In 2020, China's rainwater development and utilization potential are 111.59 billion m^3, while the rainwater development and utilization rate is only 0.7%, so there is a large potential for the development and utilization of rainwater resources in the future, and we can vigorously develop and utilize the rainwater in places where there is demand.

1 INTRODUCTION

With the global worsening situation of drought and the increasing scarcity of water resources, rainwater has become a new way to solve the water crisis in water-scarce countries and regions. Rainwater is the most fundamental source of water resources and rainwater resource potential is the basis for ensuring sustainable economic and social development and guaranteeing that rainwater resources can be exploited sustainably (Wang et al. 2007). In recent years, the evaluation of rainwater resources utilization potential is a popular topic of research by scholars at home and abroad, and Li proposed a model for calculating the target potential of rainwater resources utilization in watersheds based on the definition of theoretical (Li et al. 2005), realistic and target potential of rainwater resources utilization. Cai studied the potential of rainwater resource utilization in the mountainous areas in the south of Ningxia and redistributed precipitation (Cai & Zhang 2004). Zhao used GIS technology to establish a quantitative evaluation model of rainwater resource utilization potential (Zhao et al. 2007) and evaluated the potential of rainwater resource utilization in the region of the Loess Plateau area. Sara analyzed the endogenous resource potential of urban rainwater runoff, using a campus in Barcelona as an example (Sara & Anna 2017). Meng introduced the SWAT model to calculate the rainwater resources utilization potential and analyzed the temporal and spatial distribution patterns by taking the Songnen Plain in Heilongjiang Province as an example (Meng et al. 2020). Meanwhile, the application of the SCS-CN method, set-pair analysis method, and gray correlation theory (Wang et al. 2005, Xu et al. 2005) has broadened the diversity of evaluation methods for rainwater resources utilization potential. Summarizing the previous studies, the rainwater resource utilization potential has gradually changed from the initial theoretical research and qualitative analysis to quantitative research, and some achievements have been

*Corresponding Author: irrigation@126.com

DOI 10.1201/9781003332657-29

made. However, the evaluation of rainwater resource potential is concentrated in certain regions and drainage basins, the evaluation of rainwater resources utilization potential is not systematic, and the understanding of the driving and constraining mechanisms of rainwater development and utilization and its regional differences is not deep enough and comprehensive, and the results of rainwater resources utilization potential of the whole country and different levels of climate zones are lacking. Therefore, it seems very necessary to explore the evaluation method for evaluating the rainwater resource potential in China. This paper defines the suitable area for rainwater utilization from the geological, geographical and climatic scale, defines the rainwater development and utilization potential from the perspective of appropriateness, evaluates the national rainwater development and utilization potential by ArcGIS and Thiessen Polygons, innovates the evaluation theory and method of "appropriateness" development and utilization of unconventional water resources, makes the development and utilization of unconventional water resources in China more scientific and reliable at the macro and micro levels, providing technical support for the macro decision making and planning design of unconventional water resources control and allocation in China.

2 PRINCIPLE AND FRAMEWORK OF RAINWATER HARVESTING AND UTILIZATION AREA DIVISION

2.1 *Principle of project construction suitability*

Rainwater harvesting and utilization projects are generally suitable for construction in areas with precipitation above 250 mm, and they can be implemented in areas with precipitation around 200 mm when special needs arise (Li & Li 2010). In this paper, the areas with precipitation lower than 200 mm are first excluded, and Beijing, Hebei Province, Shanxi Province, Inner Mongolia Autonomous Region, Anhui Province, Jiangxi Province, Shandong Province, Henan Province, Hubei Province, Guangdong Province, Guangxi Zhuang Autonomous Region, Chongqing Municipality, Sichuan Province, Guizhou Province, Yunnan Province, Shaanxi Province, Gansu Province, Qinghai Province and Ningxia Hui Autonomous Region, where rainwater utilization projects are concentrated, are adopted as the basic research areas.

2.2 *Principle of rainwater resource demand*

China's geography and geological conditions vary greatly from east to west and from north to south, and the geographical characteristics of rainwater harvesting and utilization are obvious, which are mainly concentrated in the hilly areas of the Loess Plateau and Karst areas where water use is difficult in mountain areas. The Loess plateau area is one of the poorest areas with the most serious soil erosion, more prominent ecological and environmental problems, and backward socio-economic development in China (Sun et al. 2013). With an arid climate, scarce precipitation with uneven distribution within the year, and low utilization of rainfall, 60%-70% of rainfall lost as surface runoff and ineffective evaporation (Zhang 2008), this region is the most water-stressed area in China, with extreme difficulties in domestic water use and harsh conditions for agricultural production. Although the Karst area has abundant rainfall, less evaporation capacity and rich surface and groundwater resources, surface water is not easily stored and groundwater is buried deep and not easily exploited, thus forming an arid water-scarce area under a humid climate (Wang & Shi 2006), resulting in a prominent contradiction between supply and demand of water resources in local mountains and difficulties in water supply for domestic production and agricultural irrigation. Rainwater resources are the most important water resources in the Loess Plateau area and Karst area.

Therefore, considering the different natural geology, topography and geomorphology characteristics, the Loess Plateau hilly areas with serious resource-based water shortage and the Karst areas with serious seasonal water shortage are classified as the core area of rainwater harvesting and utilization, and other areas as the general area of rainwater harvesting and utilization according to local conditions.

2.3 *Principle of climate zoning*

Climate conditions are the key factor to determine rainwater resources, different climatic zones have different rainfall and different degree of rainwater utilization. Therefore, the core area and general area of rainwater harvesting and utilization are overlaid with Agro-climatic layers and divided into smaller climatic zones to calculate rainwater development and utilization potential more reasonably and accurately.

The dividing framework of rainwater harvesting and utilization area is shown in Figure 1.

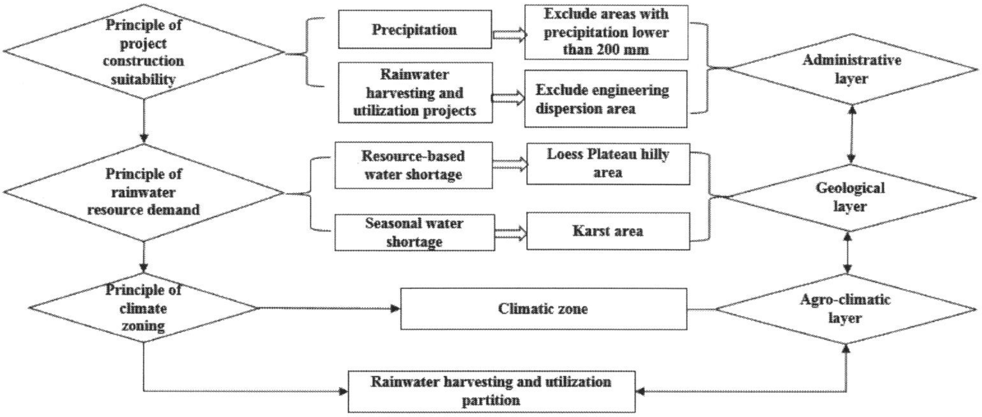

Figure 1. The dividing framework of rainwater harvesting and utilization area.

3 BASIC CONCEPT AND POTENTIAL EVALUATION METHOD FOR RAINWATER DEVELOPMENT AND UTILIZATION

3.1 *Basic concept of rainwater development and utilization*

Rainwater harvesting and utilization is a form of rainwater utilization and also a form of water resource development. All forms of water resources on Earth come from rainwater. In this sense, rainwater utilization is the utilization of rainwater in its original state or the utilization of rainwater at the initial transformation stage, and as for other forms of water resources formed by rainwater after multiple transformations, such as the utilization of runoff in rivers with large drainage basin areas and groundwater extraction, they should not fall into the category of rainwater utilization (Zhu & Li 2004). Through such means as changing the microscopic shape of the ground surface, regulating the infiltration capacity of the soil, the rainwater harvesting and utilization could effectively change the distribution of rainwater on the ground surface and the way of surface runoff collection, extend the time of surface runoff, or change the path of surface runoff movement, thus achieving the local collection of runoff and serving the purpose of rainwater harvesting and utilization. China's rainwater harvesting and utilization mainly include on-site utilization, off-site utilization and overlapping utilization methods.

Rainwater resources utilization is a new concept proposed in recent years, and the basic connotation of rainwater resources utilization lies in a process of emphasizing the evaluation and utilization of rainwater as a resource, and thorough planning and design, taking corresponding engineering measures to transform rainwater into a usable water source (Zhang 2008), and finally producing its value. The rainwater resources utilization potential refers to the maximum capacity of developing and utilizing rainwater resources under certain technical and economic conditions, in a specific area and within a certain period (Niu et al. 2005; Zhao & Wu 2007). Rainwater utilization potential reflects the connotation of appropriateness, so we define rainwater resources from the perspective of "appropriateness"

The theoretical utilization of rainwater resources is the total amount of precipitation in the drainage basin or region. The amount of rainwater resources that can be utilized under the constraints of the current economic conditions and the level of science and technology is the number of rainwater resources available. Under the current engineering conditions of rainwater collection, storage and utilization, the amount of rainwater resources that have been realized is the current utilization amount of rainwater resources. Rainwater development and utilization potential is the difference between the number of rainwater resources available and the current utilization amount of rainwater resources.

3.2 *Evaluation method of rainwater development and utilization potential*

3.2.1 *Rainwater development and utilization potential model construction*

Based on the definition of rainwater development and utilization potential, Equations (1), (2), and (3) are constructed:

$$W_t = 10^3 \sum_{i=1}^{n} PA_i \tag{1}$$

Where: W_t is the theoretical amount of rainwater resources, in m^3; P is the annual precipitation under the designed frequency, in mm; A_i is the area of different catchment surfaces, in km^2.

$$W_e = 10^3 \sum_{i=1}^{n} \lambda_i PA_i \tag{2}$$

Where: W_e is the available amount of rainwater resources, in m^3; λ_i is the catchment coefficient of different catchment surfaces; P is the annual precipitation under the designed frequency, in mm; A_i is the area of different catchment surfaces, in km^2.

$$\mathrm{W} = W_e - W_y \tag{3}$$

Where: W is rainwater development and utilization potential, in m^3; W_e is the amount of rainwater resources available, in m^3; W_y is the current amount of rainwater resources utilization, in m^3.

3.2.2 *Calculation method of annual design precipitation under climate zoning scale*

Factors such as uneven temporal and spatial distribution of precipitation across the country and large seasonal fluctuations have caused a large number of rainwater resources to be lost, and the interception and storage of runoff resources is the main way to solve the water shortage problem. Large-scale evaluation of rainwater development and utilization potential requires the division of precipitation units based on geomorphological zoning. The data of precipitation come from the actual measurement data of weather stations, and the precipitation observed by weather stations can only represent the rainfall in a smaller area around that weather station, and the distribution is uneven. Therefore, this paper uses the Thiessen Polygons of ArcGIS to calculate the designed annual rainfall after excluding the invalid precipitation in a certain area. The calculation of regional average precipitation is below:

$$\bar{x} = \frac{f_1 x_1 + f_2 x_2 + \cdots + f_n x_n}{f_1 + f_2 + \cdots + f_n} = \frac{1}{F} \sum_{i=1}^{n} f_i x_i = \sum_{i-1}^{n} A_i x_i \tag{4}$$

Where: x_i is the amount of precipitation at the precipitation observation point, in mm; f_i is the area of the Thiessen polygon, in km^2; n is the number of precipitation observation points or Thiessen polygons in the region; F is the total area of the region, in km^2; A_i is the Weight coefficient of the precipitation station.

3.2.3 *Statistical method of underlying surface area for rainwater harvesting and utilization*

In terms of the current level of rainwater harvesting and storage technology, the only way to use rainwater resources is off-site utilization. Through the construction of small catchment fields and a certain number of water storage facilities, the precipitation in a certain area will be gathered, stored and used for other areas of rainwater utilization. So the rainwater development and utilization potential is the potential of rainwater utilization under the off-site utilization model. At present, urban land, rural settlements, rural roads, highway land, other construction lands, bare land, and bare rocky land can collect, store and utilize rainwater by building small catchment fields or small water storage facilities, and such area is classified as off-site utilization area.

3.2.4 *Method of determining annual rainfall catchment coefficient*

By analyzing the rainfall process line of multi-year average precipitation at typical stations, the annual rainfall catchment of the underlying surface is calculated from the single-site precipitation and the number of rainfall sites, and the relationship between the annual precipitation and the catchment coefficient is deduced, and then the rainfall catchment coefficient is obtained.

The catchment coefficient of single-site rainfall is the area occupied by the rainfall meter that should be taken when calculating the unit precipitation. It should be calculated according to the actual effective area based on the following formula:

$$\lambda_0 = \frac{P_0 * 10^5}{A_0 * P} \tag{5}$$

Where: λ_0 is the single catchment coefficient; P_0 is the actual net precipitation; A_0 is the effective catchment area; P is the single precipitation amount.

The catchment coefficient for the whole year is calculated by the following formula:

$$\lambda = \sum P_i * \lambda_i / \sum P_i \tag{6}$$

Where: λ is the year-round catchment coefficient; P_i is the precipitation amount for each rainfall event during the year; λ_i is the catchment coefficient of each rainfall site, which is calculated according to the rainfall and rain intensity of each rainfall site.

The values of catchment coefficients are different for different catchment surfaces, and the catchment coefficients are different for different climatic zones. The catchment coefficient considers the catchment discount due to topography, infiltration, and the retention effect of surface soil and water conservation measures (Jin et al. 2017).

3.3 *Data source*

Table 1. Data source.

S/N	Data category	Data source
1	Design annual precipitation	Precipitation was observed at key weather stations nationwide from 1984 to 2015, with invalid rainfall excluded.
2	The underlying surface area of rainwater harvesting and utilization	Land use data were interpreted from remote sensing data in 2015, where the road data of the underlying surface is obtained from the national road distribution data in 2016.
3	Catchment coefficient	Technical Code for Rainwater Collection, Storage and Utilization (GB/T 50596-2010) and Theory, Technology and Practice of Rural Rainwater Harvesting and Utilization
4	Loess Plateau hilly area	Source from vector data of Resource and Environmental Science and Data Center
5	Karst area	Source from the Karst Science Data Center, obtained by deciphering from a 1: 500,000 geological map.
6	Agricultural climatic zoning	The Source comes from the Resource and Environmental Science and Data Center, which divides China into 38 agricultural natural regions based on temperature zones and humidity.

4 RESULTS OF RAINWATER HARVESTING AND UTILIZATION ZONING AND POTENTIAL EVALUATION

4.1 *Rainwater harvesting and utilization area division*

Per the principle of rainwater harvesting and utilization area division, China's rainwater harvesting and utilization areas are divided into core areas and general areas considering the different natural geography, geological geomorphology and climate characteristics. The core area is divided into the Loess Plateau area and Karst area, and each area is subdivided into seventeen climatic zones according to Agro-climatic zones, as shown in Figure 2.

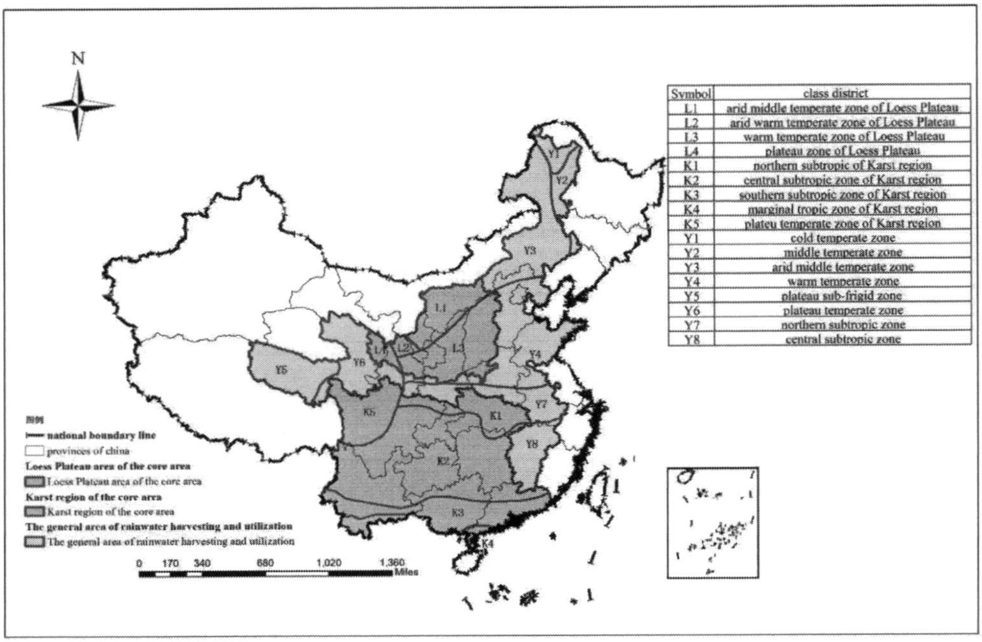

Figure 2. National rainwater harvesting and utilization zoning map.

Loess Plateau area includes all of Shanxi Province, northern Shaanxi, central and eastern Gansu, all of Ningxia, eastern Qinghai, southern Inner Mongolia and the northwest part of Henan. The region is divided into four climatic zones according to the climatic region: plateau temperate zone, arid middle temperate zone, arid warm temperate zone, and warm temperate zone. The region involves a total area of 649,300 km^2, accounting for 14% of the entire rainwater harvesting and utilization project distribution area.

Karst region includes Sichuan, Chongqing, Yunnan, Guizhou, Hubei, Hunan, Guangxi and Guangdong areas, which can be divided into five climatic regions based on climate zones: northern subtropics, central subtropics, southern subtropics, marginal tropics and plateau temperate zones. This region involves an area of about 1,935,100 km^2, accounting for 43% of the entire rainwater harvesting and utilization project distribution area.

The general area of rainwater harvesting and utilization includes most parts of Qinghai, southern Shaanxi, Hebei, southeastern Inner Mongolia, Beijing, Shandong, and most parts of Henan, Anhui and Jiangxi. Based on the climate zones, this area can be divided into eight climate zones: plateau sub-frigid zone, plateau temperate zone, cold temperate zone, middle temperate zone, arid middle temperate zone, warm temperate zone, northern subtropic zone, and central subtropic zone.

This region involves an area of about 1,950,500 km^2, accounting for 43% of the entire rainwater harvesting and storage project distribution area.

4.2 *Potential for rainwater development and utilization in different climatic zones*

According to the rainwater development and utilization potential function model, the rainwater development and utilization potential of different climatic zones are calculated, as shown in Figure 3.

The climate of the Loess Plateau hilly area gradually warms from north to south, and the precipitation gradually increases from north to south. When the design frequency is 50%, the size of rainfall is ranked as warm temperate zone > plateau temperate zone > arid middle temperate zone > arid warm temperate zone. And another factor that determines the potential of rainwater development and utilization is the underlying surface area for rainwater harvesting and utilization, which is the largest in the warm temperate zone of the Loess Plateau hilly area, with suitable climate, dense population, urban land, rural residential land and highway land, and the integrated underlying surface area of seven types of land utilization types: warm temperate zone > arid middle temperate zone > plateau temperate zone > arid warm temperate zone. The size of the catchment coefficient is changing with the size of precipitation, therefore, combining the three factors, the warm temperate zone has the largest potential for rainwater development and utilization, which is 3.890 billion m^3 The arid warm temperate zone is the smallest, which is 0.158 billion m^3

The precipitation in the Karst area gradually increases from north to south, and when the design frequency is 50%, the size of precipitation is ranked as marginal tropics > southern subtropics > central subtropics > northern subtropics > plateau temperate zone. Combining the areas of the seven types of land utilization types in the underlying surface, in the Karst zone, the area of the underlying surface: central subtropical zone > plateau temperate zone > southern subtropical zone > northern subtropical zone > marginal tropical zone. Whereas the rainwater development and utilization potential are the largest in the subtropics, which is 15.919 billion m^3. The marginal tropical potential is the smallest, which is 2.479 billion m^3

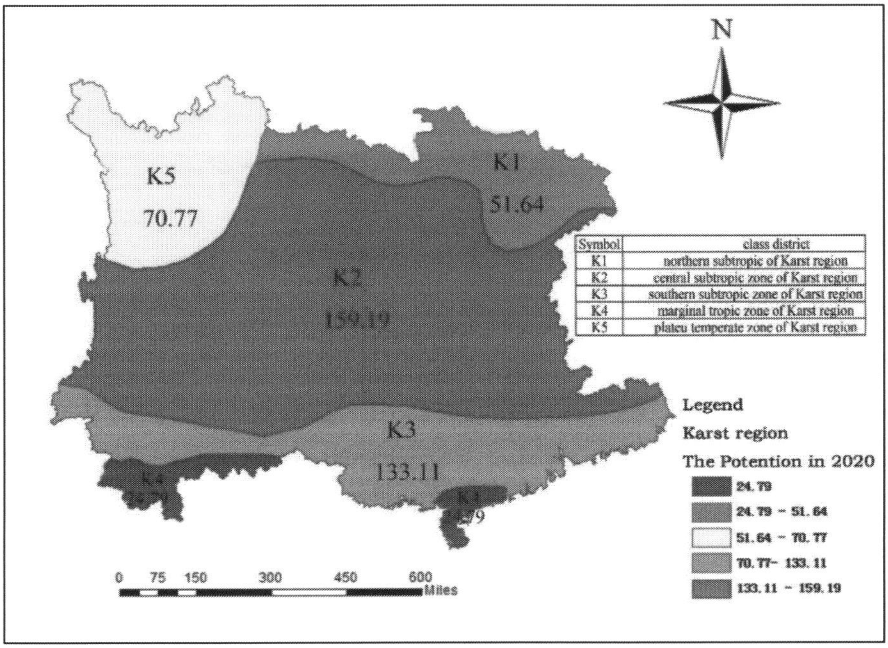

Figure 3. The chart of rainwater development and utilization potential in different climatic zones.

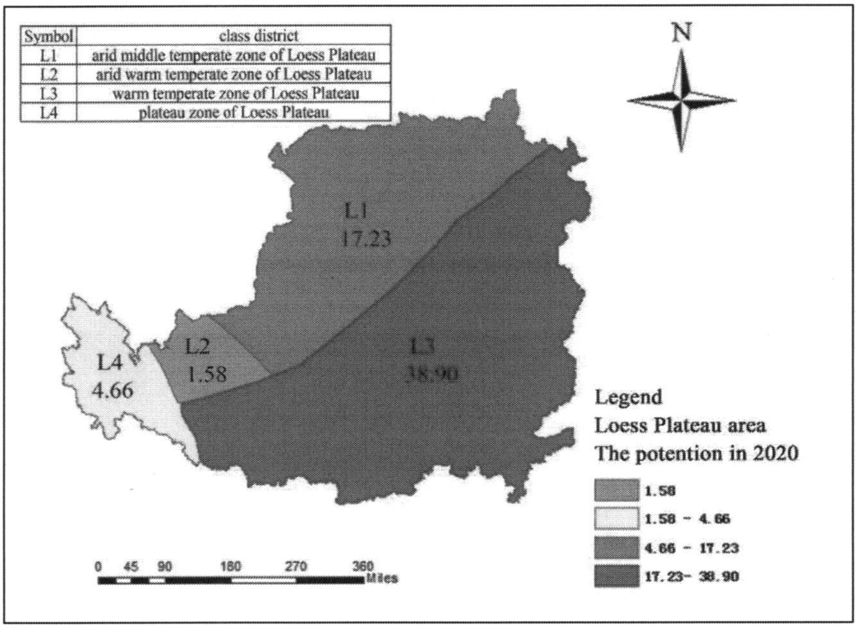

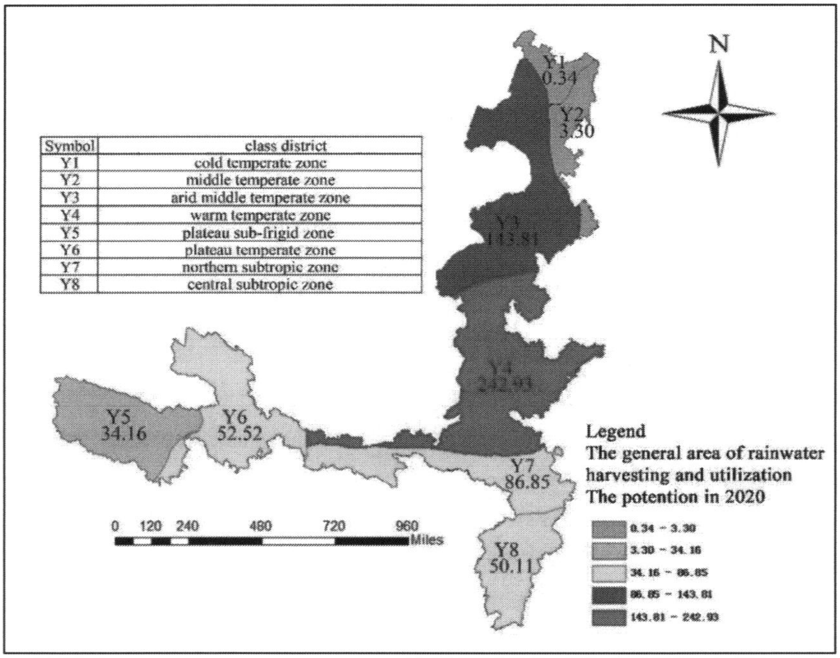

Figure 3. Continued.

4.3 *Evaluation of national rainwater development and utilization potential*

According to the rainwater resource development and utilization potential function model, the national rainwater resource development and utilization potential are calculated with the climate overlaying administrative division as the unit, as shown in Figure 4 below.

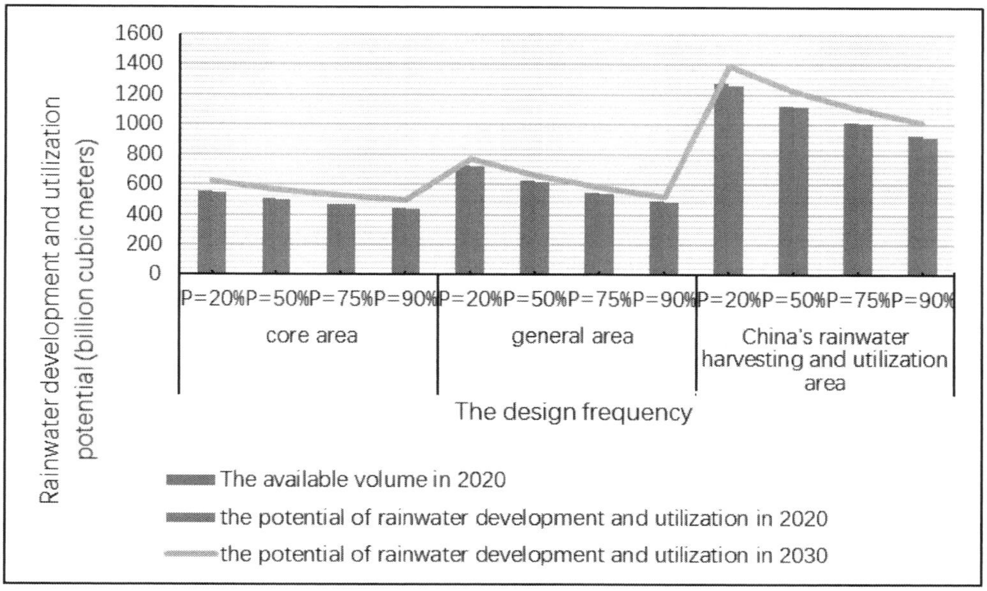

Figure 4. National rainwater development and utilization potential in 2020 and 2030.

After calculation and analysis, when the design frequency is 50%, the potential of rainwater development and utilization in China in 2020 is 111.590 billion m^3, and the potential of rainwater development and utilization in 2030 is 122.7 billion m^3. while the utilization amount of rainwater harvesting projects in China in 2020 is only 0.794 billion m^3, accounting for about 0.7% of the available amount in 2020, it can be seen that at this stage, China's use of rainwater resources is very low, and the scale is small, there is a lot of space for the use of rainwater resources in the future, and it can be developed and utilized vigorously in places where there is demand.

5 CONCLUSIONS AND RECOMMENDATIONS

China's geographic, climatic and geological conditions vary greatly from east to west and from south to north, and rainwater harvesting and utilization have obvious geographical characteristics. Through the principle of suitable construction of rainwater harvesting and utilization project, the principle of demand for rainwater resources, and the climatic conditions of rainwater resources, the national rainwater utilization area is defined and divided. The large scope is divided into rainwater harvesting and utilization core area and general area, the core area is divided into Loess Plateau hilly area and Karst area according to the geological structure characteristics, and each area is further divided into different climatic zones according to the Agro-climatic characteristics, such a hierarchical division breaks through the difficulty of evaluating the rainwater resource potential in a large scope, making the evaluation results more scientific and reliable.

The previous evaluation of rainwater resource potential focuses on small regions, small drainage basins, and other scopes, and lacks the results of rainwater development and utilization potential of the whole country and different levels of climatic zones. Based on the definition of rainwater harvesting and utilization zoning, this paper defines the rainwater development and utilization potential from the perspective of appropriateness by analyzing the form of rainwater resource transformation, proposes the theoretical amount of rainwater resources, the available amount of rainwater resources, and the current amount of rainwater resources utilization, and analyzes and evaluates the national rainwater development and utilization potential by constructing the evaluation method of rainwater development and utilization potential. The evaluation results show that the current degree of utilization of rainwater resources in China is low and the scale is small, and there is a lot of space for future rainwater utilization, and it is recommended to increase the development of rainwater resources in the Loess Plateau area and Karst area in the core area of rainwater harvesting and utilization.

ACKNOWLEDGMENT

This study was supported by the Gansu Provincial Scientific Research Program for Water Conservancy (No. 2021-71-15; 2021-71-09), Gansu Provincial Science and Technology Program (Grant No. 21JR7RA766), and National Key R&D Program of China (Grant No. 2017YFC0403504).

REFERENCES

Cai, J., Zhang, Y. (2004) Study on rainfall source potential and rainfall redistribution on hillside fields of mountain area in the south of Ningxia. *Research of Soil and Water Conservation*, 11(3):257–259.

Guo, W., Xu, J., Lu, S. (2005) Application of set pair analysis method to comprehensive evaluation of region water resources exploration potentiality. *Journal of Irrigation and Drainage*. 24 (4), 66–68.

Jin, Y., Zhou, L., Tang, X. (2017) *Theory technology and practice of rainwater collection and storage utilization in rural areas*. China Water & Power Press.

Li, H., Cao, J., Zhang, W. (2005) Model to calculate the target potentials of rainwater conversion into water resources in catchments. *Agricultural Research in the Arid Areas*, 23(2):159–63.

Li, Q., Li, Y. (2010) *Technical code for rainwater collection, storage and utilization*. Beijing: China Water & Power Press.

Meng, F., Li, T., Fu, Q. (2020) Study on regional rainwater resource potential calculation and spatial and temporal distribution law. *Journal of Hydraulic Engineering*, 51(8):1–13.

Niu, W., Wu, P., Feng, Hao. (2005) Calculation method and utilization planning evaluation on regional rainwater resources potential. *Science of Soil and Water Conservation*. 3(3), 40–44.

Sara, A., Anna, P. (2017) Urban rainwater runoff quantity and quality-A potential endogenous resource in cities. *Journal of Environmental Management*. 189:14–21.

Sun, D., Li, Y., Jin, Y. (2013) Comprehensive evaluation of utilization of urban rainwater resource in loess plateau of Gansu province. *Journal of Irrigation and Drainage*. 32 (1), 13–17

Wang, Y., Cai, J., Zhang, Y. (2007) The utilization potential of rainwater resources in semiarid loess hilly area of southern Ningxia. *Journal of Arid Land Resources and Environment*. 21(3):44–49.

Wang, H., Wang, X. (2012) SCS-CN-based approach for estimating collectable rainwater in watershed-scale. *Transactions of the Chinese Society of Agricultural Engineering*. 28(12), 86–91.

Wang, L., Shi, L. (2006) *The Availability of Rainwater Resources at Karst Mountains in Southwest China*. GUI ZHOU Science.

Xu, J., Guo, W., Lu, S. (2005) Grey relational analysis and evaluation on exploration potentiality of regional rainwater resource. *Journal of Irrigation and Drainage*. 24(3), 50–52.

Zhao, X., Wu, P., Wang, W. (2007) Regional rainwater harvesting potential assessment model based on GIS. *Transactions of the CSAE*, 2(23):6–10.

Zhang, B. (2008) Discussion on basic characteristics and developmental status of rainfall-harvesting technique in arid areas. *Journal of Irrigation and Drainage*. 27(2), 119–122.

Zhu, Q., Li, Y. (2004) On the theoretical and practical significant of rainwater harvesting and utilization. *Journal of Hydraulic Engineering*. (03), 60–65.

Zhang, H. (2008) General situation and zoning of utilization for rainwater resource in China. *Journal of Hydraulic Engineering*. 27(5), 125–127.

Zhao, X., Wu, P. (2007).Regional rainwater harvesting potential assessment model based on GIS. *Transactions of the CSAE*. 2(23), 6–10.

Energy Revolution and Chemical Research – Kok-Keong Chong and Zhongliang Liu (Eds)
© 2023 The Authors, ISBN: 978-1-032-36554-1

Research on calculation method of construction management cost based on improved technique for order preference by similarity to an ideal solution

Min Yin
Economic and Technological Research Institute of State Grid Anhui Electric Power Co., Ltd., Hefei, Anhui, China

Hui He
State Grid Anhui Electric Power Co., Ltd., Hefei, Anhui, China

Tianrui Fang
Economic and Technological Research Institute of State Grid Anhui Electric Power Co., Ltd., Hefei, Anhui, China

ABSTRACT: The construction cost of the power grid project is high and the cost items involved are complex. The costs related to the implementation of construction management are affected by many factors and cannot be reasonably calculated so the actual costs cannot match the work content. As China's power construction enters the period of structural adjustment and transformation, it has become a new trend of power development to adopt scientific project management methods to improve the cost management level of power grid projects. Accordingly, this paper studies the construction management cost calculation method, which is based on the traditional cost control method, to improve the availability of the collected power grid project sample data, and the established model has good robustness. The application in actual project management further verifies the effectiveness of the model.

1 INTRODUCTION

With the formulation and implementation of China's double carbon goal, the transformation process of the power industry has been accelerated, and the scale of new energy power grid connections has increased. Under this background, the investment scale of power grid infrastructure projects has continued to increase. State Grid Corporation of China actively responds to the national development strategy, aiming at accelerating the construction of an international leading energy Internet enterprise with Chinese characteristics, continues to promote the innovation of power grid capital construction management mechanism, and ensures the overall improvement of safety, quality, cost, and efficiency of power grid engineering construction.

The power grid infrastructure project has a high investment amount, long construction period, great construction difficulty, and many participants. In the process of project implementation, all participants need close contact and cooperation. All participants, including the owner, designer, construction contractor, equipment, and material suppliers, form the power grid infrastructure project management structure (Chen et al. 2000). The project construction management fee is the cost incurred by the construction unit in organizing, managing, coordinating, and supervising the project. The reasonable compliance of the calculation, use, and settlement of the project construction management fee greatly affect the quality and efficiency of the project management work of the construction unit (Iromuanya et al. 2013), which is the basis for ensuring the normal and efficient development of the project management work during the project construction. At the same time, it is also one of the bases of project construction cost accounting.

DOI 10.1201/9781003332657-30

The construction management fee is characterized by a large number of cost items, a diverse cost formation process, a broad range of methods of determining the cost quota, as well as a long use period. Due to these characteristics, calculating, utilizing, and settling the project construction management fee can present certain difficulties.

With the diversity of project construction management mode and market transaction mode in the construction market, part of the project management in the power grid construction project needs to be entrusted by the construction unit to a third party, and even needs the joint participation of multiple third parties (Li et al. 2014). In this case, different entities need to use the same cost, that is, the work content corresponding to a certain cost is shared by multiple participants, and the current cost calculation standard lacks the corresponding cost calculation method, which affects the reasonable calculation and accurate control of the project construction management cost, and may have an indirect impact on the project construction quality and progress (Odeck et al. 2014).

Therefore, it is necessary to carry out in-depth research on the power grid construction management cost, especially the cost calculation method, use way and calculation basis of the use cost of multiple subjects, to improve the standardized expenditure level of power grid construction management cost, ensure the quality and efficiency of the project management of the project construction unit, and improve the level of accurate cost control of power grid engineering projects.

Therefore, this paper first identifies the costs used by multiple subjects in the project construction management fee, then designs the applicability evaluation method of the existing cost calculation method, and finally improves the TOPSIS method to weight the samples collected in the actual project management work and improve the availability of the samples. The research results show that the proposed method can effectively measure the work of all participants in project construction management.

2 ANALYSIS OF PARTICIPANTS IN PROJECT CONSTRUCTION MANAGEMENT

The construction management fee is an important reflection of the management behavior of the construction unit. Strengthening the standardized management of project construction management fees will help to realize the continuous and accurate control of project costs. However, the construction management fee has the characteristics of many cost items and many cost subjects. With the diversification of the construction market management mode and the increasingly fierce market competition, the phenomenon of multiple subjects using the same cost often occurs in practical work. Moreover, there is no clear allocation standard between various expenses and expenditure subjects in the current construction management fee, resulting in uneven distribution among various expense users, resulting in great flexibility in the disbursement of power grid project construction management fees, and difficult management and control. Through extensive analysis, the participants in construction management are determined, as shown in Table 1.

Table 1. Analysis results of expense user.

Expense name		Expense user type	Users
Project legal person management fee	Relevant application fees	Multi-subject	Owner, consulting company, and construction unit
	Construction or rental expenses of temporary office space for project management personnel	Multi-subject	Owner and construction unit
	Purchase or lease necessary office furniture	Single-subject	Owner
	Staff salaries	Single-subject	Owner

(*continued*)

Table 1. Continued.

Expense name			Expense user type	Users
	Heating and heatstroke prevention fees, etc		Single-subject	Owner
	Use fees of fixed assets, tools and appliances, water, and electricity		Single-subject	Owner
	Project archives management fee		Multi-subject	The owner, the designer, the constructor, the supervisor, and the third-party file manager
	Contract signing and notarization fees, etc	Contract signing and notarization fee, legal consultant fee, and project audit fee	Single-subject	Notary office/legal personnel/consulting company/audit company
		Consulting fee	Multi-subject	Several consulting companies
		Project informatization management fee	Multi-subject	Owner and constructor
	Engineering conference fee, etc	Engineering conference fee	Multi-subject	Owner, constructor, and designer
		Business reception fee	No subject	/
	Fire and public security expenses		Multi-subject	Owner and constructor
	Expediting and inspection fees for equipment and materials		Single-subject	Provincial material company
	Stamp duty, house property tax, vehicle and vessel tax, vehicle insurance premium		Single-subject	Owner
	Labor safety acceptance and evaluation fee of a construction project, etc		Single-subject	Third-party acceptance evaluation organization/clearing company
Bidding fee			Multi-subject	Designer, consultant, and bidding agency
Project supervision fee			Multi-subject	Engineering supervision company, environmental supervision organization, and water and soil conservation supervision organization
Equipment and material supervision cost			Multi-subject	Owner, equipment supervision organization, and material construction organization
Cost consultation and completion settlement audit fee during construction			Multi-subject	A consulting company undertaking corresponding work
Project insurance premium			Single-subject	Finance department

3 APPLICABILITY ANALYSIS OF EXISTING COST CALCULATION METHODS

3.1 *Analysis purpose*

To further analyze the applicability of the existing cost calculation methods in the calculation, use, and settlement of project construction management costs, the integrity of cost composition, compliance of use methods, the applicability of calculation methods, degree of settlement dispute, and completeness of regulations/standards of various multi-party subjects' use costs are investigated based on Delphi method. Experts are invited to judge the five applicability indicators of various expenses in combination with their own experience and the nature and characteristics of various expenses.

3.2 *Design of Likert five-level scale*

Based on Likert's five-level scale (Tian et al. 2013), the applicability analysis scale of the cost calculation method is designed to provide necessary tools for expert investigation. A total of 30 valid questionnaires are collected in this investigation.

In the applicability analysis scale, according to the principle of Likert's five-level scale, the cost composition integrity, use way compliance, calculation method applicability, settlement dispute degree, and specification/standard integrity of various multi-party subject use fees are divided into five levels for evaluation. Five stars indicate the highest applicability and one star indicates the lowest applicability. During the investigation, experts gave objective and real feedback according to their actual management experience.

3.3 *Result of Likert five-level scale*

Assign values to the degree of applicability respectively, and one to five stars correspond to one to five points respectively, which can convert the evaluation results of each expert into corresponding scores.

Make statistics of the results recovered from the scale, convert the expert's rating results on the applicability of the existing cost calculation methods into corresponding scores, and analyze the total and average scores of the integrity of the cost composition, the compliance of the way of use, the applicability of the calculation methods, the degree of settlement disputes, and the integrity of regulations/standards.

To further analyze the comprehensive applicability of the cost calculation method, this study analyzes the importance of various applicability evaluation indicators. Considering the impact of cost composition on cost calculation, use and settlement, the applicability indicators of cost composition are given the highest weight. After collecting expert opinions, the weight of each indicator is finally determined according to the integrity of cost composition The compliance of use route, the applicability of calculation method, degree of settlement dispute and integrity of regulations/standards are 0.4, 0.2, 0.1, 0.1 and 0.2, respectively. After the weighted summation of the average applicability score, the comprehensive applicability score is obtained, and finally, the average score and comprehensive applicability score are obtained, as shown in Table 2.

Table 2. Comprehensive applicability scoring results.

Expense name	Cost composition integrity score	Use route compliance score	Calculation method applicability score	Settlement dispute score	Regulation/ standard integrity score
Relevant application fees	1.80	3.73	4.20	2.13	3.60
Construction or rental expenses of temporary office space	3.33	4.53	2.87	4.13	2.67
Project archives management fee	1.67	4.07	2.53	3.47	3.73
Consulting fee	3.53	4.27	2.33	4.20	3.87

4 COST CALCULATION METHOD BASED ON TOPSIS METHOD

To calculate the cost proportion of each entity in the cost used by multi-party entities, based on the principle that the fundamental motivation of cost formation is the activity itself, this paper proposes a calculation method based on the improved TOPSIS method. It applies to the calculation of the cost used by all multi-party entities with labor cost as its primary constituent. The method is divided into two parts: the calculation model of management personnel's labor consumption and labor unit price and the calculation method of cost proportion based on management work content.

4.1 *Calculation of labor consumption and labor unit price based on TOPSIS method*

The approximate ideal solution ranking method (Weber et al. 2014) is a classical method for multi-objective decision-making of finite schemes. This method was first proposed by Hwang and Yong in 1981. It was further improved and developed by Chen and Hwang in 1992. The TOPSIS method is based on two basic concepts, one is a positive ideal solution and the other is a negative ideal solution. The so-called positive ideal solution refers to a virtual optimal scheme, in which all attribute values are the optimal values of each attribute in the evaluated scheme; the negative ideal solution is just the opposite. It refers to a virtual worst scheme, and all attribute values are the worst values of each attribute in the evaluated scheme. The specific method of TOPSIS is to set the positive ideal solution and negative ideal solution according to the original matrix after standardization and calculate the distance between the evaluated scheme and the positive ideal solution and the negative ideal solution. If one scheme is closest to the positive ideal solution and furthest away from the negative ideal solution, it is the best scheme among all the evaluated schemes. At the same time, other schemes can also be ranked by their proximity to the ideal solution. The advantage of the TOPSIS method is that it makes full use of the original data information, and it also has good applicability in the case of a relatively small number of samples.

The steps of the TOPSIS method are as follows:

(1) Establish raw data matrices A and B. Matrix A is the labor consumption matrix, and matrix B is the labor unit price matrix, where a_{ij} represents the total labor consumption of the j-th type of management personnel in the i-th engineering sample of certain work content; b_{ij} represents the labor unit price of the j-th type management personnel in the i-th project sample of certain work content. Firstly, take calculating the labor consumption of various managers of a certain work content as an example.

(2) Determine the positive ideal solution z^+ and negative ideal solution z_j^-, which can be expressed as:

$$A^+ = (a_1^+, a_2^+, \dots, a_m^+), \quad A^- = (a_1^-, a_2^-, \dots, a_m^-) \tag{1}$$

Among them,

$$A_j^+ = \max\{a_{ij}\} \quad i = 1, 2, \dots, n; j = 1, 2, \dots, m \tag{2}$$

$$A_j^- = \min\{a_{ij}\} \quad i = 1, 2, \dots, n; j = 1, 2, \dots, m \tag{3}$$

(3) The Euclidean distance y_i^+ from each engineering sample $A_i = (a_{i1}, a_{i2}, \dots, a_{im})$ to the positive ideal solution z^+ and the Euclidean distance y_i^- to the negative ideal solution z_j^- are calculated respectively. Their expressions are:

$$y_i^+ = \sqrt{\sum_{j=1}^{m} (a_{ij} - a_j^+)^2}, i = 1, 2, \dots, n \tag{4}$$

$$y_i^- = \sqrt{\sum_{j=1}^{m} (a_{ij} - a_j^-)^2}, i = 1, 2, \dots, n \tag{5}$$

(4) Calculate the relative closeness T_i between each engineering sample and the ideal solution. In this paper, the closeness to the rational solution shall prevail. T_i can be described as:

$$T_i = \frac{y_i^-}{y_i^+ + y_i^-} \tag{6}$$

(5) Calculate the engineering weight ω_i of each sample based on T_i. The greater the T_i is, the closer the construction technology level of the sample project is to the average level, and the greater the sample weight is; conversely, the smaller the sample weight. The calculation formula is as follows:

$$\omega_i = \frac{\frac{1}{y_i}}{\sum_1^n \frac{1}{y_i}} \tag{7}$$

(6) Finally, according to the obtained weight, calculate the weighted average of various managers to obtain the final labor consumption of various managers of the work content. The formula is as follows:

$$a_j = \frac{\sum_{i=1}^n \omega_i * a_{ij}}{\sum_{i=1}^n \omega_i} \tag{8}$$

Then, using the same method, the labor unit price of various managers of the work can be calculated.

Taking a certain expense used by multiple entities as an example, the labor consumption and labor unit price results of various managers of certain work content of the expense are shown in Table 3.

Table 3. Calculation results of labor consumption and unit price of various management personnel in work content.

Class A Managers		Class B Managers		Class C Managers		Class D Managers	
Labor consumption	Labor unit price	Labor consumption	Labor unit price	Labor consumption	Labor unit price	Labor consumption	Labor unit price
a	b	c	d	e	f	g	h

4.2 *Result of labor consumption and labor unit price based on TOPSIS method*

Based on the calculation of management personnel labor cost, this paper gives the cost calculation method according to the principle of matching with the work content and completes the determination of the cost-sharing proportion when multiple entities use the same cost.

Firstly, the cost proportion of each work content needs to be calculated based on the labor cost of managers. The formula is as follows:

$$\varepsilon_k = \frac{Q_k}{R} * \lambda_k = \frac{\sum_{j=1}^m (a_{kj} * b_{kj})}{R} * \lambda_k \tag{9}$$

where,

Q_k is the comprehensive labor cost of the management for item k;

R is the total amount of a certain expense used by multiple entities, which is calculated according to the provisions of the pre-regulations;

a_{kj} refers to the labor consumption of j-th type management personnel required for k-th management work;

b_{kj} is the labor unit price of j-th management personnel required for item k management work;

λ_k is the proportion coefficient of the labor cost of item k management work; $\lambda_k = \frac{P_k}{Q_k} + 1$, where P_k refers to other expenses except for the labor cost of the k-th management work.

Based on the above formula, the cost proportion corresponding to each work content of a certain cost used by multiple entities can be calculated. Then, when the use expenses of multiple entities occur, the proportion of the expenses corresponding to the work content undertaken by one entity can be accumulated to obtain the proportion of the total expenses that should be calculated by the entity. The formula is as follows:

$$\varphi = \sum_{i=1}^{n} \varepsilon_i \tag{10}$$

Where,

φ is the proportion of expenses that should be accrued by an entity;

ε_i is the proportion of expenses corresponding to the t-th management work undertaken by an entity.

Using the above formula, the cost-sharing proportion can be determined when multiple entities use the same cost. The formula is as follows:

$$C = R * \varphi \tag{11}$$

where C is the expense amount that an entity should calculate.

5 CONCLUSION

The TOPSIS method has a good effect on weighting. The conventional TOPSIS method is mostly used in weighing indicators. This paper applies the concept of TOPSIS to weighting samples and to more reasonably quantify the impact of sample data on actual analysis. In view of the problems existing in the management of power grid construction management cost, firstly, the actual management status is analyzed, and the applicability of the index is then analyzed based on the Likert five-level scale. Finally, a cost calculation method matching the actual work content of managers is established, and the application suggestions of deepening the research results of multi-agent use cost are put forward. The research results show that the proposed method can serve as a guide for the preparation of relevant cost standards or accounting guidance in the future.

REFERENCES

Chen, C. T. (2000). Extensions of the TOPSIS for group decision-making under fuzzy environment. *Fuzzy Sets & Systems*, 114(1), 1–9.

Iromuanya, C., Hargiss, K. M., & Howard, C. (2013). *Critical Risk Path Method: A Risk and Contingency-Driven Model for Construction Procurement in Complex and Dynamic Projects*. IGI Global.

Li, Z. H., Gao, Q. X., & Hui, J. (2014). Risk assessment on fuzzy fault tree analysis for power transmission line field maintenance work. *Advanced Materials Research*, 1044-1045, 412–416.

Odeck, J. (2014). Do reforms reduce the magnitudes of cost overruns in road projects? statistical evidence from norway. *Transportation Research Part A Policy and Practice*, 65(JUL.), 68–79.

Tian, G. S., Liu, J. C., Zhou, J. J., Liu, J. L., & Li, B. (2013). Efficiency evaluation model of power transmission & transformation project investment based on ahp-fuzzy method. *Advanced Materials Research*, 732–733, 1303–1307.

Weber, C. K., Miglioranza, M. H., Moraes, M. A., RT Sant'Anna, Rover, M. M., & Kalil, R. A., et al. (2014). The five-point Likert scale for dyspnea can properly assess the degree of pulmonary congestion and predict adverse events in heart failure outpatients. *Clinics*, 69(5), 341–346.

Energy Revolution and Chemical Research – Kok-Keong Chong and Zhongliang Liu (Eds)

Development of an egg incubator with PID closed-loop control

BinHao Luo* & JiYi Liang
Information Engineering College, Jiangmen Polytechnic, Jiangmen, China

ABSTRACT: A PID closed-loop control poultry egg incubator is proposed, which is made of composite polyimide. Taking advantage of its wide temperature, chemical corrosion resistance, and high strength, it absorbs the heat energy discharged by the outdoor unit of a central air conditioner or the water pump of a high-rise building as the heat source for hatching poultry eggs in the incubator. Through the incubation experiment of egg-laying animals, the upper and lower limit data of temperature and humidity of the best hatching rate of all kinds of poultry eggs are obtained. Combined with the PID closed-loop control programming technology of MCU interrupt and pointer, the closed-loop automatic control of temperature and humidity in the incubator is achieved to realize the optimal hatching of poultry eggs.

1 INTRODUCTION

In China, the egg incubator in the farm is generally heated by an electric heating carrier to achieve the temperature control effect. This method makes the temperature in the box unbalanced and wastes electric energy because the temperature rise coefficient of the heater is a physical quantity. When long-term constant temperature incubation is required, it can only be achieved by frequently switching the heater on and off, increasing the failure rate of the incubator. After multiple investigations, at this stage, domestic incubators have not yet added humidity control. Experiments have shown that the influence of temperature and humidity has an important relationship with the hatchability of eggs. To this end, it is necessary to develop a PID closed-loop control egg incubator with a remote monitoring function, long-term work, and high reliability, and at the same time, a mobile phone can be used to monitor the temperature and humidity in the incubator. Li et al. (Li et al. 2021) collected real-time temperature and humidity data through the temperature and humidity sensor DHT11 module and input it into the MCU in the form of digital signals. The preset data exchange temperature and humidity data in the egg temperature and humidity interrupt service program. The selected category realizes the automatic adjustment of temperature and humidity PID and satisfy the ideal temperature and humidity incubation environment, thereby improving the hatching rate of eggs and the survival rate of pups.

2 BASIC IDEA OF THE PROJECT

Li et al. took STC12C5608 as the main control chip, and input the temperature and humidity incubation target data of the selected type of eggs through the temperature and humidity setting module, and the temperature and humidity acquisition module (Li et al. 2021) DHT11. They collected the current temperature and humidity data and compared the selected target data, using PID. The temperature and humidity control algorithm (Gao & Liu 2017) adjusts the temperature in the incubator within the target temperature and humidity range: the temperature and humidity acquisition module collect multiple differentials of the temperature and humidity of the incubator

*Corresponding author: 1010790157@qq.com

 DOI 10.1201/9781003332657-31

before and the integral sample data for a period and the current temperature and humidity and the target temperature and humidity. The comparison item is obtained from the quotient, and the compensation amount is calculated according to the proportional item, integral item, and differential item obtained by the calculation of the data collected by the single-chip microcomputer. The compensation amount corresponds to the humidity of the incubator. The main control chip (MCU) has its own PWM (Peng et al. 2021) pin to control the intermittent rotation of the fan by changing the duty cycle input drive module so that the hot air flows in the incubator. The MCU can automatically adjust and control the Fans 1 and 2 through the PID algorithm closed-loop. Timely switch or alternate operation to achieve the effect of constant temperature and humidity. Fan 1 is a suction motor, and Fan 2 is a blower motor. The MCU drives the LCD to display the incubation time, expected end time, temperature, and humidity in the incubator area, and feeds back to the intelligent remote-control terminal GRM532NW-C. The intelligent remote-control terminal GRM532NW-C receives the data and sends it to wireless monitoring equipment such as mobile phones. Wireless monitoring equipment such as mobile phones can also remotely control the intelligent remote-control terminal GRM532NW-C to change the temperature and humidity. When there is vibration in the incubator or when the data is abnormal, the incubator cannot work normally, the MCU will feed back to the wireless monitoring equipment such as mobile phones through the intelligent remote-control terminal GRM532NW-C for real-time monitoring, and the alarm module will also send out an alarm signal. The laptop is connected to the remote IO GRM532NW-C through the wireless router, so that the laptop and the remote IO are in the same network segment, and the data dictionary is written with the configuration software GRMDev5 of the giant control remote IO, and the lower computer PLC is connected and downloaded to the remote IO device. Open the drawing software GRMWebGUIDeveloper on the mobile phone interface and enter the device name, device ID, and password, and select the network refresh for the data dictionary, so that the data dictionary written by the configuration software GRMDev5 can be synchronized to the drawing software on the mobile phone interface. It is only reflected in the data dictionary of the configuration software, and the user-defined variable name is synchronized to the data dictionary of the mobile phone drawing software after hanging the point. The drawing component can select the binding variable through the component input attribute, and call it from the data dictionary. When the layout of the drawing components is completed, you can choose to compress the package and upload the project, so that the mobile phone can enter the mobile phone interface through the PLC cloud client to control and monitor operations. After the laptop is connected to the remote server through the connection software, the PLC remote download and remote monitoring can be performed, as shown in Figure 1.

2.1 *Working principle*

The egg input module is composed of chicken, goose, duck, turtle and other egg-laying category buttons and the input interrupt service routine of the single-chip microcomputer. The egg incubator project is composed of a single chip STC12C5608, temperature and humidity sensor DHT11 (Han 2017), intelligent remote-control terminal GRM532NW-C, hot air valve, fan drive circuit, LCD panel, humidifier, hot air collector, etc. The hot air received in the high-rise water pump or the central air-conditioning hot air collector is sent to the incubator through the hot air valve, and the temperature and humidity collected by the temperature and humidity sensor are sent to the microcontroller STC12C5608 in the form of digital signals for data processing, subroutine flow chart, and main Program flow chart (Figure 2 and Figure 3).

By comparing the historical experience data of temperature and humidity with a high hatching rate of eggs for quantitative conversion (Du 2018), the single-chip microcomputer adopts the method of pointer programming (Du 2018), the temperature and humidity control closed-loop system (Figure 4), and the PID control algorithm is adopted. Among them, the temperature and humidity control system are connected in parallel with the differential control and the integral control, and the PID regulation closed-loop control circuit block diagram when the proportional gain control is connected in series is shown in the figure. The associated parameters are as follows: the integral time constant range $T = 2.5$ ms $\sim$ 25 s; the inertial constant of the differential link $\varepsilon = 0.02$; the maximum lead angle of the circuit $\varphi c = 74°C$, thus the differential control can be

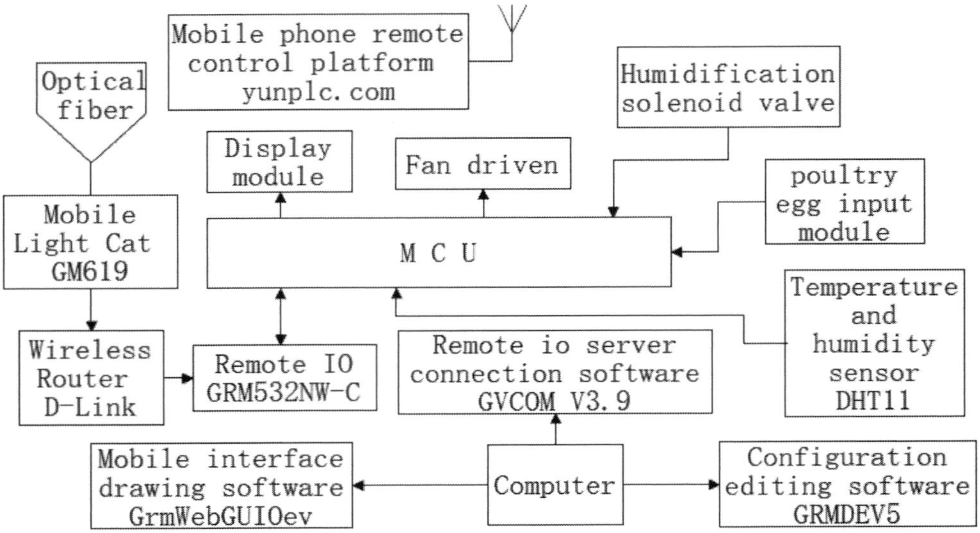

Figure 1. Diagram of the structure of remote IO control.

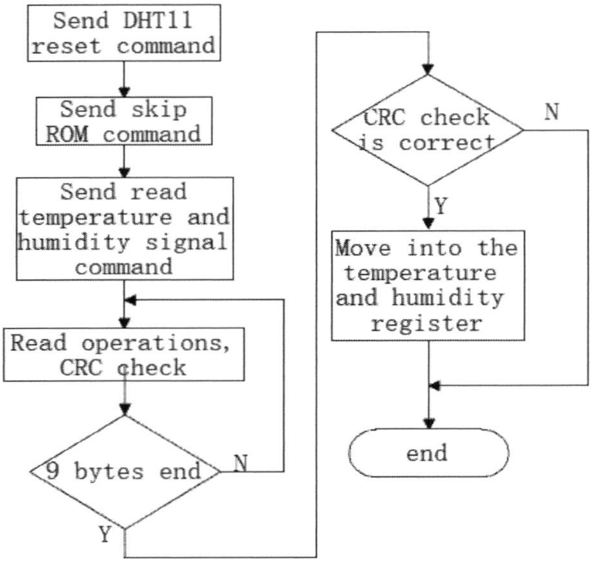

Figure 2. Flowchart of the subroutine for reading out temperature and humidity.

realized, the system response speed is fast, and the differential time constant is Tc = 0.5442 ms. And because STC12C5608 has the output function of 4 PWM pins, it can realize intelligent temperature and humidity control according to PID temperature and humidity control (Sun & Wei 2020) module. At the same time, the timing switch module is controlled to switch Fan 1 and Fan 2 to work in turn. When the humidity is not enough, the MCU controls the humidifier to humidify. When the temperature is too high, the MCU controls the hot air valve to turn off and the temperature drops. The user only needs to select the goose egg mode; duck egg mode; egg mode; turtle egg mode and

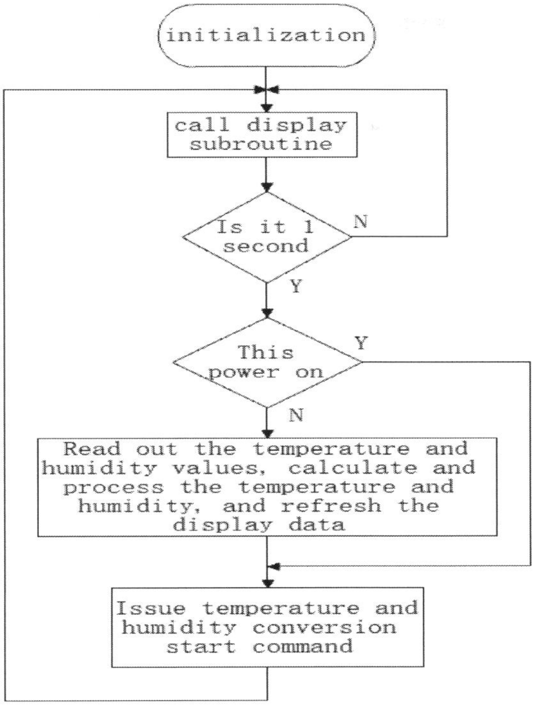

Figure 3. Main program flow chart.

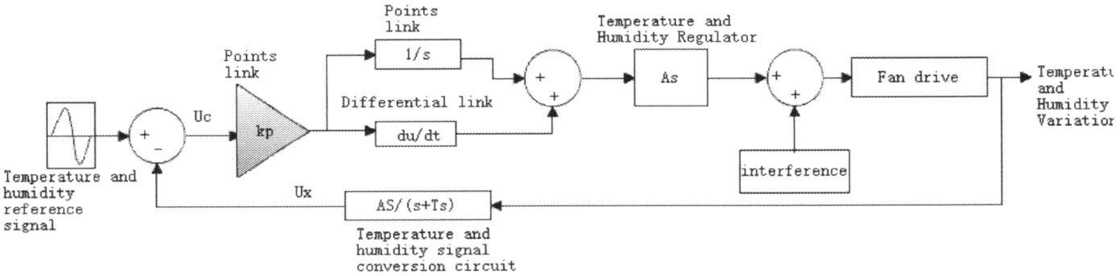

Figure 4. Block diagram of closed-loop control circuit.

can send it to the MCU interrupt (Qiu 2019) to execute the selected interrupt service routine. When the incubation is over, it will alarm. Since the camera is on the outside of the incubator, it can also be transmitted to the mobile phone through the smart terminal through the camera (Li et al. 2017), or the incubation situation can be monitored directly on the mobile phone through the 360 remote monitoring camera (Li & Li 2021).

2.2 *Fan speed control circuit*

The rotation speed of fans MG1 and MG2 is controlled by the bidirectional thyristor. The single-chip microcomputer sends the trigger signal to the thyristor through the optocoupler to control the conduction angle of the thyristor, thereby controlling the rotation speed of Fan 1 and Fan 2. In order to reliably turn off the fan power supply in the state of shutdown and over-temperature protection, a relay is added to the circuit to control the fan power supply. The fuse connected in series to the relay coil circuit is a thermal fuse of 45°C. When the temperature exceeds 45°C, the thermal fuse will be blown, preventing the fan from continuing to work, and sucking in the hot air and causing the temperature of the box to rise. The LED light-emitting tube connected in parallel with the fan is used to indicate the working status of the fan (Figure 5).

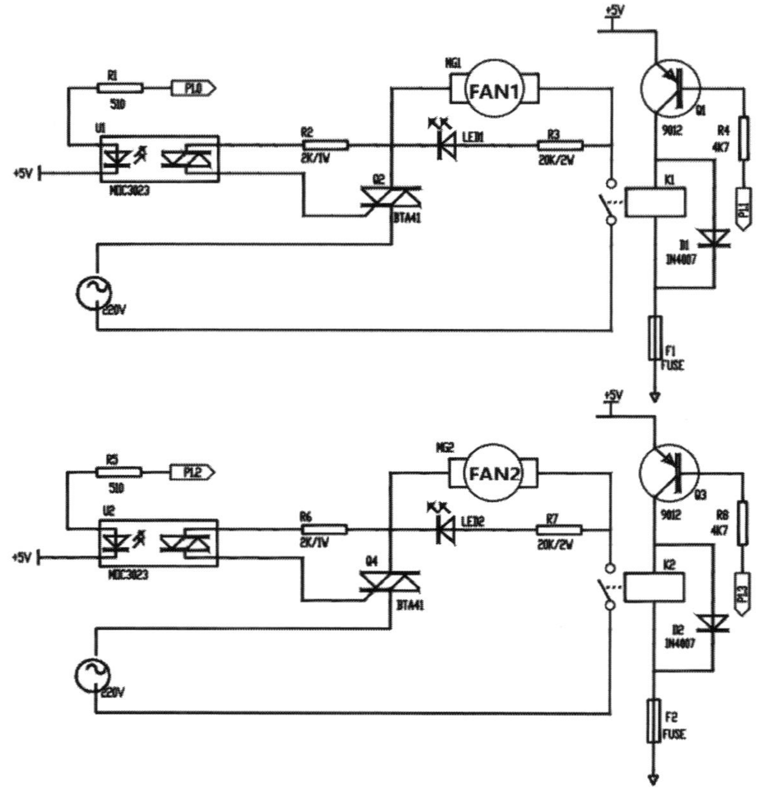

Figure 5. Fan control circuit diagram.

In the thyristor trigger signal, it is necessary to carry out zero-crossing detection of the mains (Wei et al. 2020) to realize the phase delay of the trigger pulse. In this circuit, the triode 8050 and a “NO” gate is used to realize the zero-crossing detection (Figure 6).

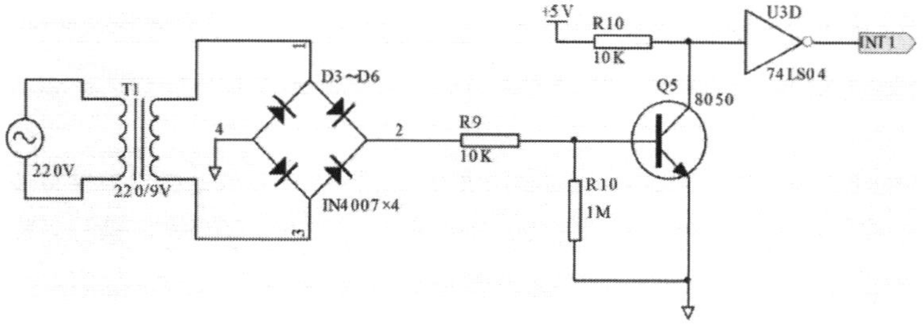

Figure 6. Trend chart of hatching of various types of eggs.

2.3 *Calculation of temperature and humidity subroutine*

The subroutine for calculating temperature and humidity converts the read value in RAM to BCD code, and judges the normal hatching value of temperature and humidity. The calculation process of temperature and humidity subroutines is shown in Figure 7.

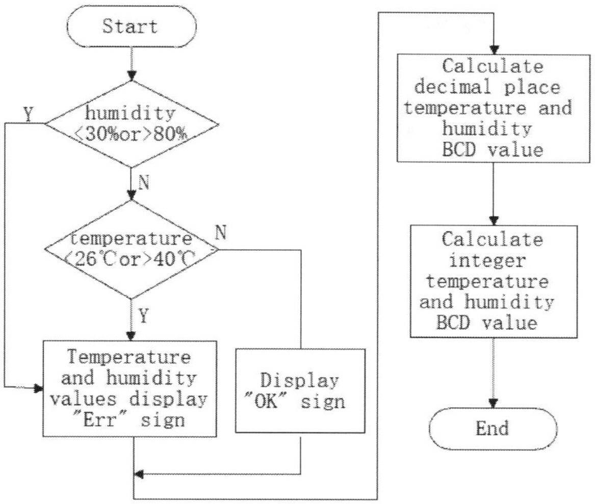

Figure 7. Flowchart of the subroutine for calculating temperature and humidity.

2.4 *Display data refresh subroutine*

The display data refresh subroutine is mainly to refresh the display data in the display buffer. When the highest data display bit is 0, the symbol display will be shifted to the next bit. The flowchart of the data refresh subroutine is shown in Figure 8.

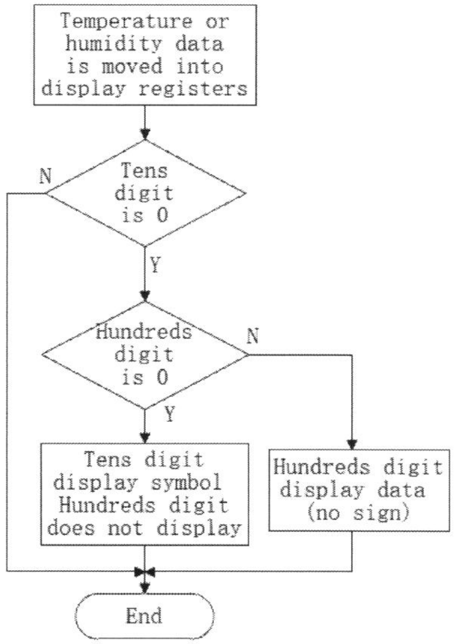

Figure 8. Flow chart of the data refresh subroutine.

2.5 *3D model structure design drawing*

The heat collection device includes a heat collector and a heat collection pipe, one end of the heat collection pipe extends to the incubator, the other end is connected to the heat collector, and the heat collection pipe extends into the incubator; the tower fan includes an air outlet hood and the grille (Chen et al. 2021); the tower fan is arranged in the incubator; the heat collection pipe is connected with the air outlet hood; the outside of the air outlet hood is provided with a grille. The incubator is a cabinet-type structure, and a supporting plate and a supporting block are arranged in the incubating box. The supporting plate is supported and arranged on the supporting block by the supporting block, and the supporting block is fixedly installed on the inner side of the incubator. The incubator is provided with a box door and a humidifier. An observation window is arranged on the box door. A temperature and humidity sensor are arranged in the incubator. An alarm is arranged outside the incubator. A liquid crystal display is arranged outside the incubator. And a liquid crystal display is arranged outside the incubator. It is electrically connected to the temperature and humidity sensor. The heat collector is connected with a fixing rod, and the materials such as the box body and the supporting plate are made of polyimide (Figures 9 to 12).

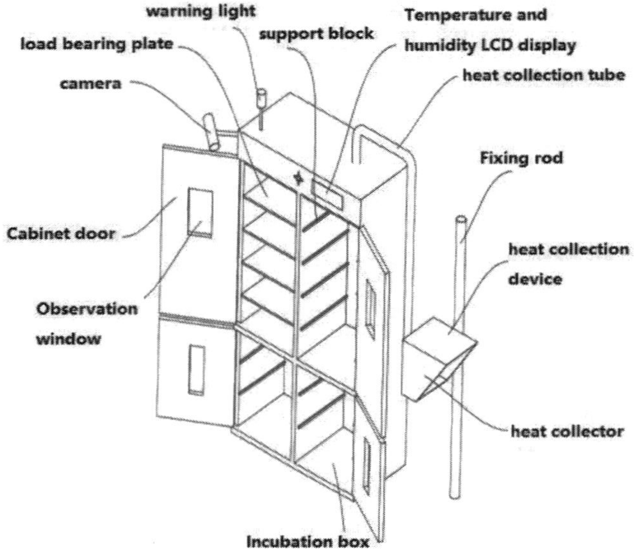

Figure 9. Structure diagram of incubator.

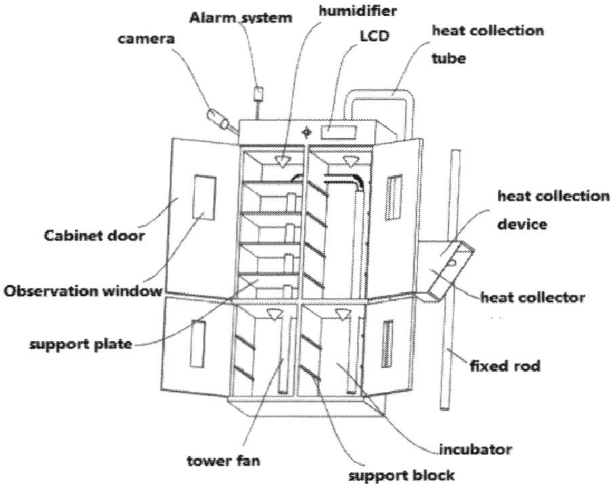

Figure 10. Schematic diagram of the inside of the incubator.

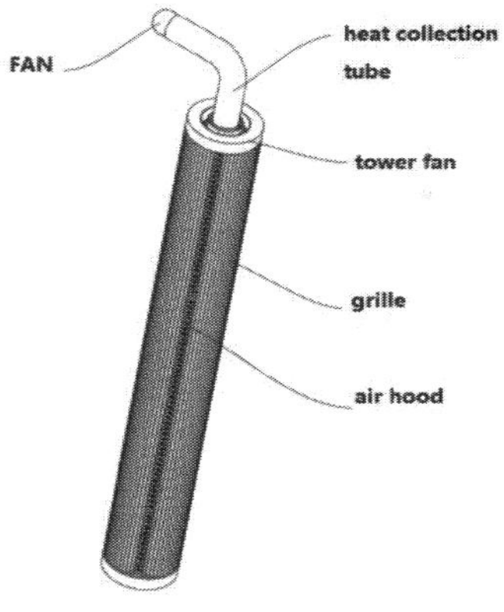

Figure 11. Structure diagram of heat collection device.

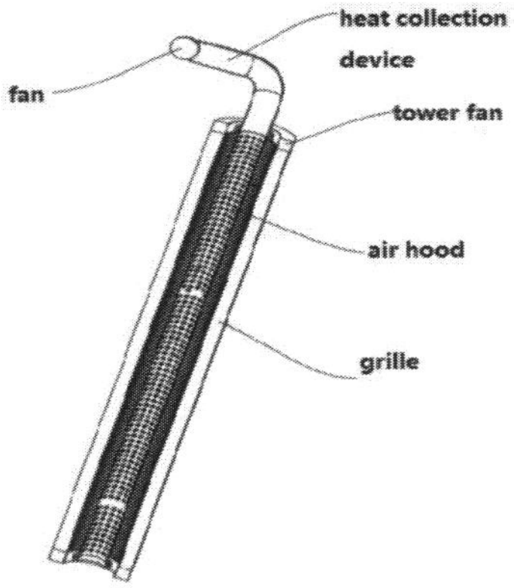

Figure 12. Schematic diagram of the interior of the heat collection device.

3 RESEARCH ON THE CONTROL ALGORITHM OF HUMIDITY AND TEMPERATURE REGULATION

The temperature and humidity of the incubator needs to be controlled near the range of the target parameters. In order to realize this function, the research on the realization of the temperature control algorithm has been carried out. The single-chip microcomputer uses PID control algorithm to adjust the temperature in the incubator within the target temperature and humidity range: the temperature and humidity acquisition module collects multiple differentials of the current incubator temperature and humidity and integral sample data for a period of time, and the current temperature and humidity are negotiated with the target temperature and humidity. The comparison item is obtained, and the compensation amount is calculated according to the proportional item, integral item and differential item obtained by the calculation of the data collected by the single chip microcomputer. Stop humidification for a period of time corresponds to the compensation amount corresponding to the humidity of the incubator. The PID algorithm formula is as follows:

$$U(k) = K_P\{e(k) + \frac{T}{T_i}\sum_{n=0}^{k} e(n) + \frac{T_d}{T}(e(k) - e(k-1))\}$$

The first term in parentheses is the proportional term, the second term is the integral term, the third term is the differential term, and the front is just a coefficient. In many cases, it only needs to be used when the temperature and humidity in the incubator are discrete. The working method of prolonging the life of the fan motor and the redundancy design research can ensure the uninterrupted operation of the incubator (Figure 4 PWM output waveform). The alternate operation of the fan can avoid the fan from working for too long, which will make the fan's motor coil enameled wire heat up and cause the motor coil insulating oil to melt, and reduce the service life of the motor. The brushless motor with long service life of the fan is driven by a PWM signal, and a 1 kHz rectangular wave is intended to be used to drive the motor.

4 SCIENTIFIC AND APPLICATION PROSPECTS

Compared with the traditional incubator, the smart incubator in the project adds a humidity control function to ensure that the hatchlings of poultry can live in a suitable humidity environment, and avoid the death of the young due to excessive evaporation of water in the body when the humidity is too low. The remote IO incubator has a remote monitoring function, which can remotely monitor the temperature and humidity data of the incubator throughout the day, which can effectively prevent failures or special environments that cause excessive temperature and reduce the success rate of hatching. The incubation temperature can be flexibly set through the panel, and different optimal incubation temperature and humidity can be given to different breeds of eggs for incubation. The upper and lower limit data of real-time incubation temperature and humidity can be obtained from the historical experience data of temperature and humidity of hatching eggs, so as to achieve the effect of optimal hatching rate. The incubation heat source absorbs heat energy from the central air-conditioning outdoor unit, and uses a unique fan tower fan blade design to keep the incubation environment ventilated, saving energy and environmental protection. The intelligent terminal module is adopted to make monitoring more convenient. The smart terminal GRM532NW-C has the function of remotely modifying temperature and humidity. The incubator box is made of polyimide foam material, which has the advantages of good thermal insulation effect, light weight, flame retardant and good durability. For traditional poultry farms that need to improve the hatchability of young poultry, smart incubators have the prospect of replacing outdated hatching equipment. The remote IO incubator camera has the function of remote monitoring, so the idle space of the hotel (roof or idle warehouse) can be used to collect heat from the outdoor unit of the central air conditioner or the water supply pump of the high-rise building to realize the energy-saving incubation of the egg hatching industry. Through this project, the enthusiasm of the hotel's idle staff can be aroused, and the egg hatching (Fang et al. 2017) of energy saving and emission reduction can be realized to generate income. It is also possible to make women at home in high-rise buildings participate in this work and publicize it. As an entrepreneurial project, this paper establishes a cooperative for egg hatching to increase income and have great prospects.

5 CONCLUSION

The remote IO incubator (Chen 2021) has the functions of remote monitoring, abnormal work alarm, simultaneous incubation of different types of eggs, etc., and can achieve the effect of energy saving and emission reduction. The project works also integrate the advantages of intelligent terminal, single-chip microcomputer, PID constant temperature technology (Wei et al. 2021), reliability redundant design, statistical method to optimize the incubation temperature and humidity data, and have the two-way control function of humidity and temperature, which can even make the incubator have the ability to incubate. Rare animals such as parrots, turtles, crocodiles and other related species can be cultivated through the incubator, and the practical ability of incubation scientific research can be improved. It can also be promoted to the residents of large hotels and high-rise buildings, making full use of the heat energy emitted by machinery and equipment as renewable energy, saving energy and environmental protection, and having huge market potential.

ACKNOWLEDGMENT

This work was supported by 2020 Guangdong Province Colleges and Universities Natural Science Characteristic Innovation Project "Research and Development of an Internet + Intelligent Temperature Adjusting Water Purifier" (Project No.: 2020KTSCX340); and 2021 Jiangmen City Fundamental and Theoretical Scientific Research Science and Technology Program "Egg Incubator with Automatic Temperature and Humidity Adjustment by Remote IO" (Project No.: Jiang Ke [2021] No. 87).

REFERENCES

Chen Juan. *Research and Design of Poultry Incubation Environment Monitoring System Based on Internet of Things* [D]. Guangxi University, 2021.DOI:10.27034

Chen Shiren, Huang Yong, Xiao Xi, Yang Fan. Development of active air intake grille strategy based on economy and thermal balance (Continued 1) [J]. *Automotive Engineer*, 2021(11):19–20.

Du Jiehui. *Research on temperature compensation and closed-loop system optimization of high-precision micromachined capacitive accelerometer* [D]. Zhejiang University, 2018.

Fang Ting, Ye Ming, Zhou Ping. Design of temperature and humidity control system for egg hatching [J]. *Anhui Agricultural Sciences*, 2017, 45(33): 223–226.DOI:10.13989/j.cnki.0517-6611.2017.33.071.

Gao Yan, Liu Hongxia. A crop soil moisture control algorithm based on PID neural network [J]. *Computer and Modernization*, 2017(06):122–126.

Han Tuanjun. Design of high-precision laser engraving platform control system based on STC12C5608AD [J]. *Machine Tools and Hydraulics*, 2017, 45 (20):126–129.

Li Bingtao, Li Shuqiao. Using the ESP8266 module to read the current data of the RS485 energy meter and realize remote monitoring on the mobile phone [J]. *Electronic Production*, 2021(21):39–42.DOI:10.16589/j.cnki. cn11-3571/tn.2021.21.011.

Li Du, Huang Zhangwei, Cheng Xianghui, Chen Xiai. Design of remote monitoring system based on STM32 camera power supply [J]. *Fujian Computer*, 2017, 33(09):123–124.DOI:10.16707/j.cnki.fjpc.2017.09 .070.

Li Zhiwei, Dong Wei, Huang Shuangcheng. Design of agricultural greenhouse temperature and humidity monitoring system based on DHT11 [J]. *Industrial Instrumentation and Automation Device*, 2021(01): 39–43.

Peng Haijun, Li Bin, Dangzheng. Design of fast PWM closed-loop control based on FPGA [J]. *Power Electronics Technology*, 2021, 55(11): 101–103+111.

Qiu Xuemin. "Overview of Microcontroller Interruption" Micro-Lesson Teaching Design [J]. *Digital World*, 2019(12):181.

Sun Guofa, Wei Wei. A Class of Strict Feedback System Variable Proportional Gain Accurate Differential Compensation Control [J]. *Control and Decision*, 2020, 35(06): 1490–1496. DOI: 10.13195/j.kzyjc.2018.1153.

Wang Yi. *Design of intelligent remote monitoring and control terminal for electric vehicles* [D]. Xihua University, 2019. DOI: 10.27411/d.cnki.gscgc.2019.000321.

Wei Haibo, Liu Jie, Fang Shengli, Wang Long, Mei Jianwei. Research and Design of AC Zero-Crossing Detection Circuit [J]. *Journal of Hubei Institute of Automotive Industry*, 2020, 34(02):64–66+71.

Wei Hongli, Zhou Jianbo, Wang Qingyue, Wei Xiu. Design of temperature control system based on fuzzy PID [J]. *Foreign Electronic Measurement Technology*, 2021, 40(09):111–116.DOI:10.19652/j.cnki.femt.2102790.

Energy Revolution and Chemical Research – Kok-Keong Chong and Zhongliang Liu (Eds)
© 2023 The Authors, ISBN: 978-1-032-36554-1

Annual publication trend and reference co-citation analyses of hydrogen safety-related publications

Wenwen Duan
Soochow College, Soochow University, Suzhou, Jiangsu, P.R. China

Shuo Wang
College of Mechanical and Electrical Engineering, ShiHezi University, ShiHezi, Xinjiang, P.R. China

Ruichao Wei*
Research Institute of New Energy Vehicle Technology, Shenzhen Polytechnic, Shenzhen, Guangdong, P.R. China
School of Automobile and Transportation, Shenzhen Polytechnic, Shenzhen, Guangdong, P.R. China

ABSTRACT: Hydrogen energy is a secondary energy source that uses hydrogen gas or hydrogen-containing substances as intermediate substances. Hydrogen energy can be used as an alternative to fossil energy, effectively reducing greenhouse gas emissions such as carbon dioxide. Hydrogen energy is also accompanied by safety issues during its use. Based on the Web of Science Core Collection database and bibliometrics, this article provides an in-depth analysis of publication trends and reference co-citation for publications related to hydrogen safety. The results show that 4842 articles published from 1992 to 2021 were mainly focused on the field of energy fuels, and the number of publications increased exponentially; the three stages of the research in order are as follows: laboratory mechanistic studies, hydrogen application safety, and hydrogen-related battery safety. This study provides a macro-overview of published articles in areas related to hydrogen safety.

1 INTRODUCTION

The safety of hydrogen as an energy carrier has always been a major concern for the engineering community (Wei et al. 2022). The range of hydrogen combustion and explosion concentration in air is particularly wide, up to 4%~75% (volume ratio), which is far more than natural gas. Hydrogen ignition energy is particularly low at 0.02 mJ, one tenth of that of gasoline. (Hansen 2020) These unfavorable safety intrinsic characteristics add to the difficulty of using hydrogen safely. In recent years, several hydrogen safety accidents have occurred worldwide, resulting in multiple injuries and property damage (Mao 2020). Therefore, the safety of hydrogen must be taken seriously in order to ensure the effective use of hydrogen energy.

In recent years, a large number of publications related to hydrogen safety have been published in the Web of Science Core Collection (WoS CC) (Abohamzeh et al. 2021; Ustolin et al. 2020; Yang et al 2021). However, there are no studies that provide a macroscopic overview of hydrogen safety-related publications. This study quantified the publication quantities and co-citation of references related to hydrogen safety based on WoS CC and the bibliometric software CiteSpace. The results of the study can provide an overview of information for external investors and researchers new to the field of hydrogen safety.

*Corresponding author: richardwei@szpt.edu.cn

DOI 10.1201/9781003332657-32

2 DATA AND METHODS

2.1 *Data collection*

The information on an annual number of publications and reference co-citation in this study was obtained from the WoS CC, which is a worldwide renowned citation indexing database widely used for scientific research and evaluation because of its ground-breaking content, high-quality data, and long history (Li et al. 2019). The editions in WoS CC in this study include Science Citation Index Expanded (SCI-EXPANDED), Social Sciences Citation Index (SSCI), and Current Chemical Reactions (CCR-EXPANDED), and Index Chemicus (IC). The retrieval code set according to the retrieval format is TS=(hydrogen) AND TS=(safety)) AND ((DT==("ARTICLE" OR "REVIEW")) NOT (PY==("2022")). The index date was set from 1900-01-01 to 2021-10-01. The reason for choosing 1900 as the starting time of the index date is because the WoS database starts in 1900. As a result, 4,842 records were obtained from 106 countries/regions, 4,266 research institutions, 1,375 journals and 19,053 authors. It is worth noting that as the WoS CC continues to be updated, the results of the same retrieval setup will be slightly different (Liu 2013).

2.2 *Bibliometric methods and visualization tools*

CiteSpace was employed for visualizing and analyzing trends in co-citation and frontier of reference. It is designed as a tool for progressive knowledge domain visualization (Chen 2004). Combining the concepts of research front (De Solla Price 1965) and intellectual base (Persson 1994), Chen et al. (Chen 2006) defined a research front as the state of the art of a specialty, which can be conceptualized as a time-variant mapping Φ(t) from its research front Ψ(t) to its intellectual base Ω(t).

$$\Phi(t) : \Psi(t) \rightarrow \Omega(t) \tag{1}$$

$\Psi(t)$ is a set of research front terms that gradually emerge or suddenly change as time moves forward. It is expressed as:

$$\Psi(t) = \left\{ term | term \in S_{title} \cup S_{abstract} \cup S_{descriptor} \cup S_{identifier} \wedge IsH\mathrm{ot}Topic(term, t) \right\} \tag{2}$$

where S_{title} denotes a set of terms; *IsHotTopic*(*term*, *t*) denotes a Boolean function.

$\Omega(t)$ consists of groups of articles cited by articles in which research-front terms were found, which is expressed as:

$$\Omega(t) = \{ article | term \in \Psi(t) \wedge term \in article_0 \wedge article_0 \rightarrow article \} \tag{3}$$

where $article_0 \rightarrow article$ denotes that $article_0$ cites *article*. $\Psi(t)$ and $\Omega(t)$ indicate the internal and external impact on the research-front terms, respectively.

Centrality BC_i is a parameter that indicates the importance of nodes in the network, which is calculated as:

$$BC_i = \sum_{s \neq i \neq t} \frac{n_{st}^i}{g_{st}} \tag{4}$$

where g_{st} is the number of shortest paths from node s to node t, and n_{st}^i is the number of shortest paths through node i. The analysis in Section 3.6 of this study was performed with the assistance of CiteSpace.

3 RESULTS AND DISCUSSION

3.1 *Annual publication trend*

Table 1 lists the annual number of hydrogen safety-related publications worldwide and in major active countries. The data in Table 3 are plotted in figure 1. As can be seen in Table 1, the first

Table 1. Annual publications on hydrogen safety published globally and in major active countries (as of October 2021).

Total	China	USA	Germany	Japan	Korea
1	0	0	0	1	0
1	0	1	0	0	0
3	0	0	0	0	0
1	1	0	0	0	0
3	0	1	0	0	0
2	0	1	0	0	0
169	16	36	16	8	7
240	27	43	30	31	12
220	25	48	17	21	5
254	33	53	19	14	19
283	52	59	23	25	14
309	53	67	27	24	19
348	82	68	27	30	18
454	116	102	50	37	22
495	158	80	41	27	34
606	243	87	32	26	29
752	298	110	50	40	44
701	285	92	42	27	49

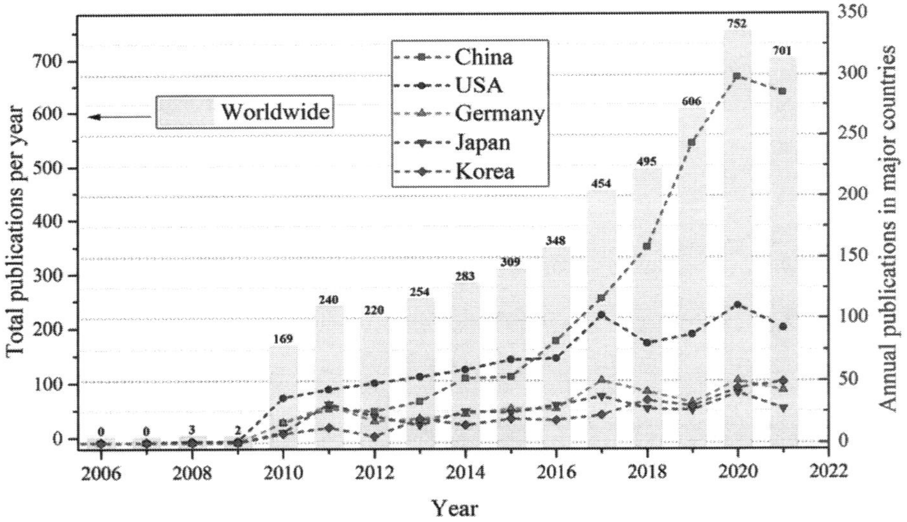

Figure 1. Trends in the increase of publications globally and in major active countries. (as of October 2021).

documented article, which belongs to Chemistry Medical in WoS, was published by a Japanese researcher in 1992. In this article, Suzuki et al. (Suzuki et al. 1992) investigated the safety boundary of hydrogen in a novel bronchodilator. From 1992 to 2009 of the 11 publications on hydrogen safety research included in the WoS CC publication, six belong to the category of Chemistry Medical, three in Chemistry Organic, and three in Chemistry Multidisciplinary.

In 2008, the Hydrogen Research Advisory Committee published a document with concerns about hydrogen explosion, fire prevention, and detection (Zalosh & Barilo 2009). Concerns about hydrogen safety awareness were also published in a white paper by the International Energy Agency in the same year (Tchouvelev 2008). In 2009, the NFPA 2 issued a draft on hydrogen fire and explosion safety requirements (Pratt et al. 2015). Moreover, EU HYPER published an Installation Permitting Guide on hydrogen fuel cell safety requirements and hydrogen safety standards (Pasman & Rogers 2012). As the requirements for hydrogen safety were highlighted by key institutions in 2018 and 2019, the number of article reviews on hydrogen safety included in the WoS CC proliferated in 2010, as shown in Table 1 and Figure 1. Trends in the increase of publications globally and in major active countries (as of October 2021). Of the 169 publications in 2010, the top four categories were Energy Fuels, Chemistry Physical, Electrochemistry, and Nuclear Science Technology. The corresponding numbers of publications were 43, 40, 39, and 21, respectively.

As seen in Figure 1, the number of publications increased annually and exponentially from 2012 to 2020. This is due to the frequent occurrence of accidents in hydrogen-related environments and the increased interest in related fields (Yang et al. 2021). In addition, the country that published the highest number of publications in the field of hydrogen safety from 2010 to 2015 was the USA, while it became China after 2016.

3.2 *Reference co-citation analysis*

If a publication cites two references at the same time, these two references constitute a co-citation relationship. A co-citation analysis can help identify key points in the research development process and the interrelationships between publications (Zou et al. 2018). The CiteSpace calculation parameter "look back years" is set to eight years in this study, that is, if a publication was issued in 2000, its citations before 2008 will be considered. Figure 2 presents the co-citation network of the top 11 clusters after clustering the obtained 265 references. The information of the top 11 clusters sorted by the number of citations is listed in Table 2. The 11 red labels in Figure 2 represent the themes of the 11 clusters.

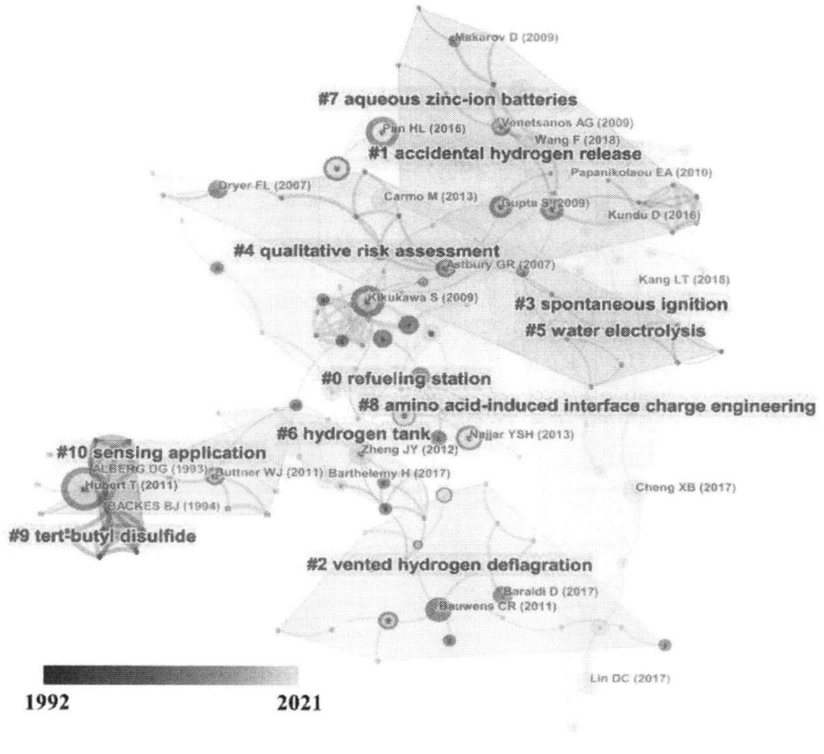

Figure 2. Trends in the increase of publications globally and in major active countries. (as of October 2021).

Table 2. Cluster information in the analysis of reference co-citation.

Cluster	Size	Silhouette	Mean (year)	LLR	Label
0	20	0.937	2014	70.66	refueling station
1	19	0.986	2008	52.82	accidental hydrogen release
2	19	1	2012	341.74	vented hydrogen deflagration
3	18	0.963	2007	91.98	spontaneous ignition
4	17	0.892	2011	79.35	qualitative risk assessment
5	16	0.982	2017	47.33	water electrolysis
6	12	0.99	2013	150.04	hydrogen tank
7	12	1	2016	75.57	aqueous zinc-ion batteries
8	11	0.855	2018	26.88	amino acid-induced interface charge engineering
9	10	1	1994	13.79	tert-butyl disulfide
10	10	1	2010	73.34	sensing application

Combining the mean years of citations (MYC) and labels in Figure 2 and Table 2, the references co-citation analysis can be divided into three periods. The MYC of the references for the first stage is before 2012, which contains clusters #1 "accidental hydrogen release," #2 "vented hydrogen deflagration," #3 "spontaneous ignition," and #9 "tert-butyl disulfide." This period focused on laboratory mechanism studies on the safety of hydrogen. The experimental and simulation studies on the combustion behaviors of hydrogen-air mixtures were carried out during this stage. The MYC for period two is from 2010 to 2014 and the clusters include #0 refueling station, #4 qualitative risk assessment, #6 hydrogen tank, and #10 sensing application. The second period focused on the safety of hydrogen applications, including the application of sensors, the safety of refueling stations and storage tanks, and the risk assessment for relevant hydrogen application sites. The MYC of the third period is from 2016 to 2018. Its research on hydrogen safety focused on hydrogen-related battery safety, specifically hydrogen evolution in zinc and lithium metal anode batteries.

4 CONCLUSIONS

The annual publication trend and reference co-citation of hydrogen safety-related publications were analyzed based on the WoS CC database and bibliometric methods in this study. The following conclusions can be drawn from this study.

(1) Research related to hydrogen safety began in 1992 and proliferated in 2010. China, the USA, Germany, Japan, and South Korea are the five most active countries in this field of research. After 2015, China maintained its first place in the world in terms of the quantity of publications in the field.
(2) Based on the reference co-citation analysis, the three stages of hydrogen safety research are "laboratory mechanism studies," "the safety of hydrogen applications," and "the safety issues raised by hydrogen evolution in zinc and lithium metal anode batteries."

This study provides a macro-overview of publication trends and reference co-citation for hydrogen safety. The next step would be an in-depth study on the correlation between hydrogen safety and the hydrogen economy.

ACKNOWLEDGMENTS

This work was supported by the Scientific Research Startup Fund for Shenzhen High-Caliber Personnel of SZPT (No. 6022310043K) and the Postdoctoral Later-stage Foundation Project of Shenzhen Polytechnic (6021271019K).

REFERENCES

Abohamzeh, E., Salehi, F., Sheikholeslami, M., Abbassi, R., Khan, F.(2021) Review of hydrogen safety during storage, transmission, and applications processes. *J Loss Prev Process Ind.*,72:104569.

Chen, C. (2004)Searching for intellectual turning points: Progressive knowledge domain visualization. *Proc Natl Acad Sci USA.*, 101:5303–10.

Chen, C. (2006)CiteSpace II: Detecting and visualizing emerging trends and transient patterns in scientific literature. *J Am Soc Inf Sci Technol.*, 57:359–77.

De Solla Price, D., J.(1965) Networks of Scientific Papers. *Science*, 149:510–5.

Hansen, O.R. (2020)Hydrogen infrastructure—Efficient risk assessment and design optimization approach to ensure safe and practical solutions. *Process Saf Environ Prot.*, 143:164–76.

Li, J., Goerlandt, F., Li, K.,W. (2019)Slip and fall incidents at work: A visual analytics analysis of the research domain. *Int J Environ Res Public Health*, 16:1–20.

Liu, X. (2013)Full-Text Citation Analysis?: A New Method to Enhance. *J Am Soc Inf Sci Technol.*, 64:1852–63.

Mao, Z.(2020) *Hydrogen safety* (in Chinese). Chemical Industries Press.

Persson, O.(1994) The intellectual base and research fronts of JASIS 1986–1990. *J Am Soc Inf Sci.*, 45:31–8.

Pratt, J., Terlip, D., Ainscough, C., Kurtz, J., Elgowainy, A. (2015) *H2FIRST Reference Station Design Task: Project Deliverable 2-2*. The United States.

Pasman, H., J., Rogers, W., J. (2010)Safety challenges in view of the upcoming hydrogen economy: An overview. *J Loss Prev Process Ind.*, 23:697–704.

Suzuki, F., Kuroda, T., Nakasato, Y., Ichikawa, S. (1992) *New bronchodilators.1.1,5-Substituted 1H-imidazo[4,5-c]quinolin-4(5H)-ones*, 4:4045–53.

Tchouvelev, A., V. (2008) *Knowledge gaps in hydrogen energy*. A white paper. Revision 1. IEA.

Ustolin, F., Paltrinieri, N., Berto, F.(2020) Loss of integrity of hydrogen technologies: A critical review. *Int J Hydrogen Energy*, 45:23809–40.

Wei, R., Lan, J., Lian, L., Huang, S., Zhao, C., Dong, Z., et al.(2022) A bibliometric study on research trends in hydrogen safety. *Process Saf Environ Prot.*, 159:1064–81.

Yang, F., Wang, T., Deng, X., Dang, J., Huang, Z., Hu, S., et al. (2021)Review on hydrogen safety issues: Incident statistics, hydrogen diffusion, and detonation process. *Int J Hydrogen Energy*, 46:31467–88.

Zalosh, R., Barilo, N. (2009)Wide area and distributed hydrogen sensors. *Proc Int Conf Hydrog Saf.*, 16–8.

Zou, X., Yue, W., L., Vu, H., Le. (2018)Visualization and analysis of mapping knowledge domain of road safety studies. *Accid Anal Prev.*, 118:131–45.

Energy Revolution and Chemical Research – Kok-Keong Chong and Zhongliang Liu (Eds)
© 2023 The Authors, ISBN: 978-1-032-36554-1

Investigations and considerations of the current energy development in Zhejiang province

Kang Zhang* & Gaoyan Han
State Grid Zhejiang Electric Power Research Institute, Hangzhou, P.R. China

Wei Wang
Hangzhou E.Energy Technology Co., Ltd., Hangzhou, P.R. China

Lijian Wang
Zhejiang Institute of Metrology, Hangzhou, P.R. China

Liwei Ding & Hongkun Lv
State Grid Zhejiang Electric Power Research Institute, Hangzhou, P.R. China

ABSTRACT: To achieve the carbon peak goal by 2030 and carbon neutrality by 2060, Zhejiang devotes itself to the construction of a high-level national demonstration province of clean energy and proposes the policy of "double control" of energy. A clean, low-carbon, safe and efficient energy system should be built to satisfy future energy development. This study investigates the current energy supply and consumption in Zhejiang. The power generation installation and structure are evaluated. Moreover, energy consumption per unit of GDP and carbon emission are compared. To satisfy the growing electricity demand, the import ratio of non-fossil energy power can be improved. The annual utilization hours of gas power and nuclear power units can be increased while the exploitation of photovoltaic and offshore wind resources can also reduce the electric power carbon emissions.

1 INTRODUCTION

The energy security strategy of "Four Revolutions and One Cooperation" has given a new direction for China's energy development and pointed out a new path for China's energy development. In September 2020, China proposed "achieve carbon peak by 2030 and carbon neutral by 2060", which further clarified China's energy development goals in the new era and forced China to accelerate the construction of a clean, low-carbon, safe and efficient energy system (BP. Statistical Review of World Energy 2019).

Zhejiang faithfully implements the "Eight-eight strategy", strives to create an "important window", tries to be a pioneer province of socialist modernization, devotes itself to the construction of a high-level national demonstration province of clean energy, and proposes the policy of "double control" of energy (Yuan et al. 2020; Zhang et al. 2017).

In this study, the energy supply and consumption are analyzed while the power generation installation and structure are evaluated. Moreover, energy consumption per unit of GDP and carbon emission are compared and effective actions are suggested. The results are significant to the construction of a clean-energy demonstration province.

*Corresponding Author: zhangkang@zju.edu.cn

 DOI 10.1201/9781003332657-33

2 CURRENT ENERGY SITUATION

2.1 *Energy supply and consumption*

The coal, oil and natural gas resources are scarce in Zhejiang Province. By 2019, there are almost no detected reserves of oil and natural gas resources, while the detected coal reserve is only 93 million tons. The raw coal, crude oil, and natural gas consumed in Zhejiang province almost all depend on external input (Zhang et al. 2019).

In 2019, the total energy consumption in Zhejiang province was 224 million tons of standard coal. The coal consumption was about 137 million tons, equivalent to about 100 million tons of standard coal, accounting for 45.3% of the total primary energy consumption (the national average is 57.6%), as seen in Figure 1. The petroleum consumption was 25.49 million tons and natural gas consumption was 14.8 billion cubic meters, accounting for 16.8% and 8.0% of the total primary energy consumption respectively (the national average was 19.7% and 7.8%). Besides, the non-fossil energy consumption accounted for 19.8% (the national average was 14.9%) while the external thermal power and other sources accounted for 10.1%, showing strong dependence on external electricity.

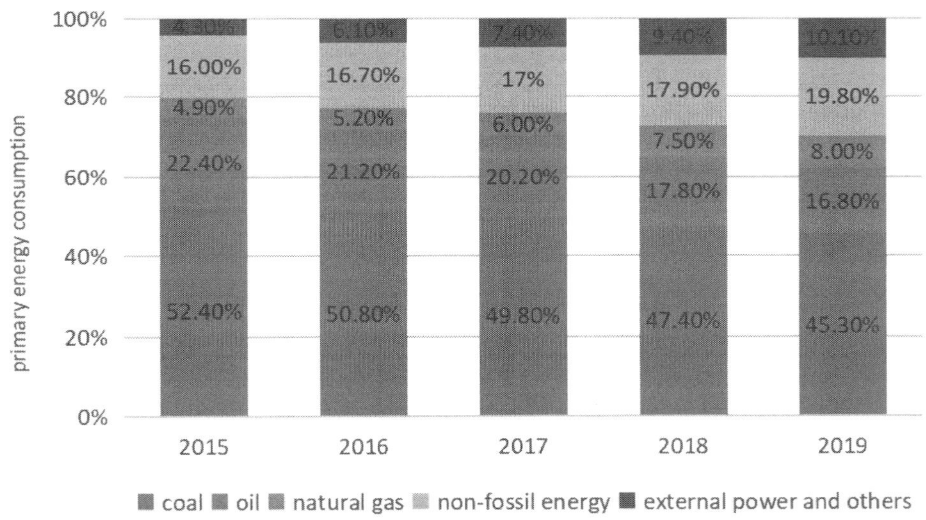

Figure 1. Primary energy consumption of Zhejiang.

On the whole, the energy demand in Zhejiang is strong while the energy resources are scarce. It has a low energy self-sufficiency rate and strong external energy dependence.

2.2 *Power generation installation and structure*

By 2019, the total installed capacity in Zhejiang province was 97.89 million kW. Figure 2 shows the power generation installation of Zhejiang in 2019. It can be seen that the installed coal power capacity was 46.49 million kW, accounting for 47.5% (the national installed capacity of coal power is 51.8%). The installed gas power capacity was 12.61 million kW, accounting for 12.9% (7.4% nationwide). The installed nuclear power capacity was 9.08 million kW, accounting for 9.3% (2.4% nationwide). The installed hydropower capacity was 11.7 million kW, accounting for 12.0% (17.8% nationwide). The non-water renewable energy installed capacity was 16.9 million kW, accounting for 17.3% (20.6% nationwide).

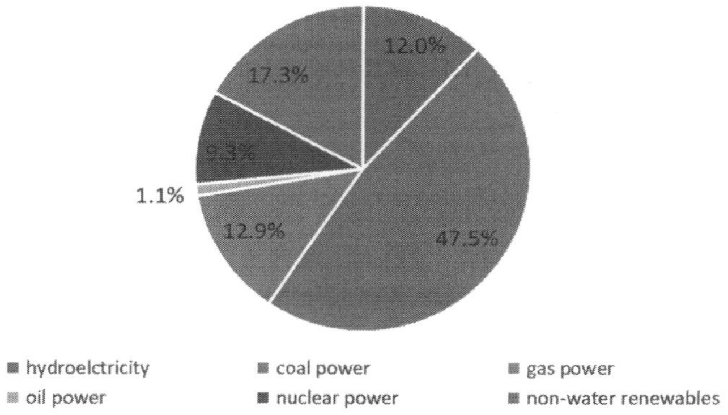

Figure 2. Power generation installation of Zhejiang in 2019.

In 2019, the total electricity consumption in Zhejiang was 470.6 billion kWh while more than one-third was purchased. The provincial power generation in 2019 was 354.4 billion kWh while figure 3 shows that coal power generation accounted for 61.9%, slightly lower than the national average of 64.7%. The gas power generation accounted for 4.2%, lower than the national average of 6.7% (Tong et al. 2018). Nuclear power generation accounted for 17.7%, much higher than the national average of 4.8%. The hydropower generation (including pumping storage) accounted for 7.3% (4.7%), lower than the national level of 17.8% (only 0.4%). Moreover, the non-water renewable energy power generation accounted for 7.3%, lower than the national average of 8.6%.

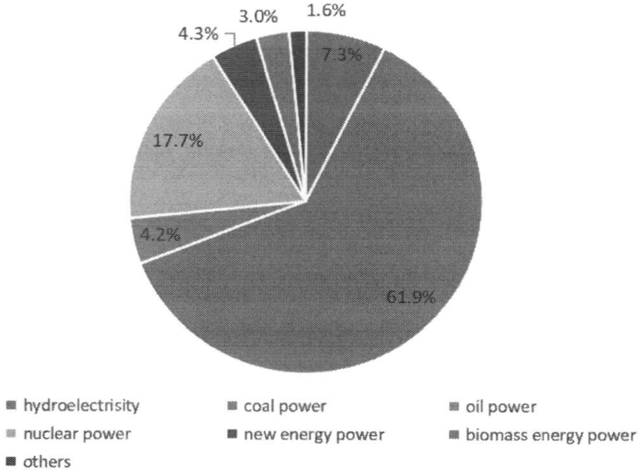

Figure 3. Power generation structure of Zhejiang in 2019.

On the whole, the imported electricity in Zhejiang province is very high, with a maximum import electricity load of 30.3 million kW. Although the installed gas power capacity in Zhejiang province is high in our country, its utilization hours are fairly low, which results in a small proportion of low-carbon energy generation.

Besides, wind and solar energy resources are scarce in Zhejiang. This makes the installed capacity and power generation of non-water renewable energy lower than the national average level. Moreover, the nuclear power generation in Zhejiang is high, which accounts for a higher ratio of zero-carbon energy power generation than the national average level.

Therefore, the power generation structure in Zhejiang is better than the national average level. To satisfy the growing electricity demand in Zhejiang, the import ratio of non-fossil energy power should be improved. The annual utilization hours of gas power and nuclear power units can be increased while the exploitation of photovoltaic and offshore wind resources can also reduce the electric power carbon emissions in Zhejiang.

2.3 *Energy consumption per unit of GDP and carbon emission*

As estimated, the total carbon emission of Zhejiang in 2019 was about 420 million tons while the total primary energy consumption was about 220 million tons of standard coal, and the annual GDP reached about 903.85 billion US dollars (Ye et al. 2018).

Table 1 indicates the energy consumption per unit of GDP and carbon emission in 2019. The energy consumption and carbon emission per unit GDP of Zhejiang province are about 2.5 tons of standard coal per \$10000 and 4.6 tons of carbon dioxide per \$10000, lower than the national average level of 0.9 tons of standard coal per \$10000 and 2.3 tons of carbon dioxide per \$10000. The energy efficiency and low carbon level are among the top in China. However, it is still 0.2 tons of standard coal per \$10,000 and 0.7 tons of carbon dioxide per \$10,000 higher than the world average level.

Table 1. Energy consumption per unit of GDP and carbon emission in 2019.

Content	Zhejiang	national average	world average
energy consumption per unit GDP (Tons of standard coal per \$10000)	2.5	3.4	2.7
carbon emission per unit GDP (Tons of CO_2 per \$10000)	4.6	6.9	3.9
carbon emission per unit of primary energy (Tons of CO_2 per ton of standard coal)	1.8	2.0	1.7

The carbon emission per unit of primary energy in Zhejiang is about 1.8 tons of carbon dioxide per ton of standard coal, better than the national average level of 2.0 tons of carbon dioxide per ton of standard coal, but higher than the world average level of 1.7 tons of carbon dioxide per ton standard coal per unit of primary energy.

3 ENERGY CHALLENGES

3.1 *Limited energy resources*

The coal, oil, and gas energy resources are quite scarce in Zhejiang and these resources rely mostly on imports. The solar energy resource in Zhejiang is also relatively poor. Its total annual horizontal radiation is 1050–1400 kWh/m^2, lower than the national average total horizontal radiation of 1470.9 kWh/m^2. Moreover, the land resource in Zhejiang is also scarce and the total amount of available terrestrial solar energy is relatively limited. The wind energy in Zhejiang has obvious seasonal differences and the land wind energy is scarce while only offshore wind power has some exploitation potential.

On the whole, fossil energy and renewable energy resources are both scarce in Zhejiang. This situation will be more severe with the increasing energy consumption and the restriction of the "carbon peak by 2030 and carbon neutrality by 2060" target.

3.2 *The conflict between energy supply and demand*

The total energy demand in Zhejiang is expected to reach 303 million tons of standard coal in 2025, while the renewable energy exploitation resource is limited and the total carbon emission of fossil energy is restricted. This makes the conflicts between energy supply and energy demand more severe in Zhejiang. Moreover, with the large-scale access to renewable energy, and the increase in peak valley and seasonal differences, the balance of energy supply and demand becomes more prominent. On the other hand, the energy facility and supply capacity cannot fully meet the growing demand for high-quality energy due to the multiple constraints of lacking power and gas transmission channels, decreasing renewable energy subsidies, and limited space resources.

4 CONCLUSIONS

This study investigates the current energy supply and consumption in Zhejiang. The power generation installation and structure are evaluated. Moreover, energy consumption per unit of GDP and carbon emission are compared. The conclusions and suggestions are as follows.

(1) Zhejiang has a low energy self-sufficiency rate and strong external energy dependence. The energy consumption and carbon emission per unit GDP of Zhejiang are lower than the national average level but higher than the world average level.
(2) The energy development in Zhejiang has challenges in limited energy resources and high energy demand. To satisfy the growing electricity demand, the import ratio of non-fossil energy power can be improved. The annual utilization hours of gas power and nuclear power units can be increased while the exploitation of photovoltaic and offshore wind resources can also reduce the electric power carbon emissions.

ACKNOWLEDGMENTS

This work was financially supported by the State Grid Zhejiang Electric Power Co. Ltd. under the Project "Research on Combined Cooling and Heating Technology for Multi-energy Transformation of High Elastic Power Grid in Industrial Cluster Area Based on High-Temperature Heat Storage" (Grant No. 5211DS20007N).

REFERENCES

BP. *Statistical Review of World Energy* 2019 [J].

Tong J.L., Lv H.K., Cai J.C., et al. (2018) Review on development and application prospect of domestic natural gas distributed energy resource. *Zhejiang Electric Power* 37(12): 1–7.

Ye Q.C., Lou K.W., Zhang B., et al. (2018) Design and optimization of multi-energy complementary integrated energy system. *Zhejiang Electric Power* 37(7): 5–12.

Yuan K., Feng C., and Yang J. (2020) How does China manage its energy market? A perspective of policy evolution. *Energy Policy* 147: 111898.

Zhang D.H., Wang J.Q., Lin Y.G., et al. (2017) Present situation and prospect of renewable energy in China. *Renewable and Sustainable Energy Reviews* 76: 865–871.

Zhang Z.X., Lu Q., and Zhang S.X. (2019) Research on development trends and strategies of integrated energy services in China. *Zhejiang Electric Power* 38(2):1–6.

Energy Revolution and Chemical Research – Kok-Keong Chong and Zhongliang Liu (Eds)

Variation characteristics of meteorological drought in Taihu Lake Basin from 1981 to 2016

Shouwei Shang
State Key Laboratory of Hydrology, Water Resources and Hydraulic Engineering & Science, Nanjing Hydraulic Research Institute, Nanjing, China
State Key Laboratory of Hydrology, Water Resources and Hydraulic Engineering& College of Hydrology and Water Resources, Hohai University, Nanjing, China

Xiting Li, Yintang Wang, Tingting Cui*, Leizhi Wang & Jiandong Chen
State Key Laboratory of Hydrology, Water Resources and Hydraulic Engineering & Science, Nanjing Hydraulic Research Institute, Nanjing, China

ABSTRACT: Based on the meteorological data of 10 national surface meteorological observation stations in the Taihu Lake Basin from 1981 to 2016, the K-index method is used to calculate the seasonal and annual drought indices of the Taihu Lake Basin in the past 36 years, and analyze the various characteristics of drought on various time scales in the Taihu Lake Basin. Results indicate that the drought in the Taihu Lake Basin is mainly light drought and moderate drought, and autumn is the most frequent drought season. The areas with high drought frequency are mainly concentrated in the northern part of the Taihu Lake Basin, and HX has the highest drought frequency. From 1981 to 2016, the drought intensity and the number of drought stations in the Taihu Lake Basin showed an increasing trend except for summer and winter, indicating that the level and scope of drought increased. In terms of the persistence of the drought trend, the future drought trend on each time scale is the same as that in the past 36 years, and the persistence of the spring and autumn trend is strong. In terms of the affected area of drought, the drought in the Taihu Lake Basin is dominated by global drought, and the frequency of global drought has increased after 2000.

1 INTRODUCTION

Drought is one of the most common meteorological disasters in the world. Drought is usually caused by a region where the amount of precipitation in a certain period is lower than that of a normal year (Dai 2011; Sun 1994). Under the background of changes in global climate conditions, extreme drought events occur frequently and their duration tends to increase (Yang et al. 2011). The average annual drought-affected area in China is as high as 22 million hm^2, accounting for more than 40% of the area affected by various disasters (Wang 2007). In the next 40 years, China's climate will generally show a trend of warming and drying. The frequency and duration of extreme droughts will be the largest. Drought will still be the main meteorological disaster that China will face in the future (Liu et al. 2016; Zhao et al. 2010).

The drought index is an important means to study drought. Zhang Haoqiang et al. (Zhang et al. 2021) used the percentage of precipitation anomaly, K index, Z index, standardized precipitation index, and standardized precipitation evapotranspiration index to analyze the drought situation of typical sites in the Huaibei area. Li Jun et al. (Li et al. 2021) used the CSDI index to analyze the drought in the Pearl River Basin and the future trend of drought. Huang Wanhua et al. (Huang et al.

*Corresponding Author: ttcui@nhri.cn

DOI 10.1201/9781003332657-34

2010) used the standardized precipitation index and the percentage of precipitation anomalies as drought indicators, conducted a detailed analysis of the evolution characteristics of seasonal drought in southern China, and concluded that the degree of drought in southern China showed a deepening trend. The Taihu Lake Basin is densely populated and economically developed. Although the area is rich in water resources, the annual precipitation distribution is uneven and the inter-annual variation is large, and seasonal droughts occur frequently in the basin. However, the current drought research in this region mainly focuses on the whole middle and lower reaches of the Yangtze River, and there are few studies on the Taihu Lake Basin. According to the characteristics of the Taihu Lake Basin, in this paper, the drought K index considering evaporation factors is used, combined with the ratio, intensity, and frequency of drought stations, to analyze the temporal and spatial variation characteristics of drought on the annual and seasonal scales in the Taihu Lake Basin from 1981 to 2016. To provide a scientific basis for water resources planning and disaster prevention and mitigation in Taihu Lake Basin.

2 STUDY AREA AND DATA

2.1 *Study area*

The Taihu Lake Basin is located in the southern wing of the Yangtze River Delta, bordering the East China Sea in the east, Hangzhou Bay in the south, Tianmu and Maoshan Mountain in the west, and the Yangtze River in the north. The geographical location is between 119°08E∼′121°55′E and 30°05′N∼32°08′N. It spans the four provinces of Jiangsu, Zhejiang, Shanghai, and Anhui, with a basin area of 36,895 km^2. According to the topography and water system characteristics, the whole basin is divided into seven water conservancy districts: HX, ZX, HQ, HJH, WCXY, YCDM, and PDPX (Figure 1).

2.2 *Data*

This paper collects the daily meteorological data from 1981 to 2016 from 10 national surface meteorological observation stations in the Taihu Lake Basin. The meteorological data comes from the National Meteorological Science Data Center (http://data.cma.cn/). There are 10 sites in the study area. The data of each meteorological station has good continuity and has been quality-controlled, and is consistent, reliable, and representative. The specific distribution is shown in Figure 1.

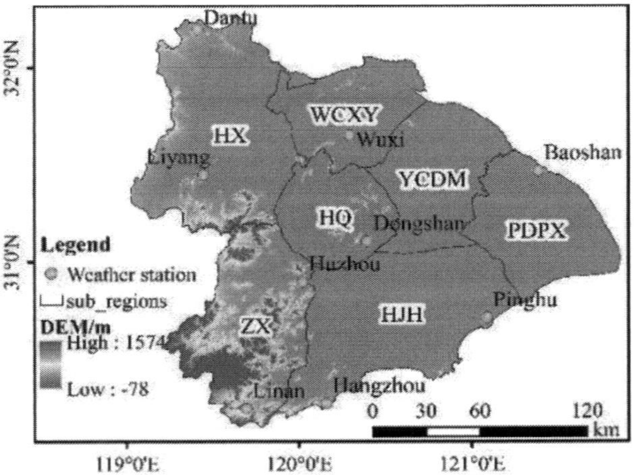

Figure 1. Distribution of meteorological stations in Taihu Lake Basin.

3 METHOD

3.1 *Reference crop evapotranspiration calculation method*

The Penman-Montes equation is a semi-empirical and semi-theoretical formula proposed by British H.L. Penman to calculate the reference evaporation. The specific Penman-Montieth formula is as follows:

$$ET_0 = \frac{0.408\Delta(R_n - G) + 900\Upsilon/(T + 273.15)u_2(e_s - e_a)}{\Delta + \Upsilon(1 + 0.34u_2)} \tag{1}$$

where G and R_n are the surface heat capacity and crop canopy net radiation, both in $MJ/(m^2 \cdot d)$; T is the average temperature, °C; u_2 is the wind speed at 2 m on the surface, m/s; e_s and e_a are saturation water vapor pressure and actual water vapor pressure, kPa; Δ, γ are the slope of the saturated water pressure curve and the hygrometer constant, kPa/°C.

3.2 *K index*

The K index reflects the drought situation by the ratio of the relative variability of precipitation to the relative variability of evaporation and is often used to characterize atmospheric and soil drought. The calculation formula is:

$$K = P'/E' \tag{2}$$

where P′ is the relative variability of precipitation: $P' = P/\overline{P}$; P is the rainfall in a certain period; $\overline{P}$ is the annual average rainfall in a certain period; E′ is the relative variability of evaporation: $E' = E/\overline{E}$, E is the evaporation in a certain period; $\overline{E}$ is the multi-year average evaporation in a certain period.

3.3 *Drought characteristic assessment methods*

The formula $P_i = \frac{n}{N} \times 100\%$ is used to evaluate the frequency of drought in a meteorological observatory in a year with data, where N is the total number of years for which a meteorological observatory has precipitation data, and n is the number of years in which drought occurs at the meteorological observatory. The subscript i denotes the different meteorological observation stations.

Use the formula $S_{ij} = \frac{1}{m}\sum_{i=1}^{m}(1 - K_i)$ to calculate the drought intensity of the index K, and use the formula $P_j = \frac{m}{M} \times 100\%$ to calculate the ratio of drought stations to measure the size of the affected area of drought. Where m is the number of meteorological stations where drought occurs, M is the total number of meteorological stations in the target area; i represents different meteorological stations, and j represents different years. Its drought intensity grade distribution is shown in Table 1. The definition of the affected area of drought is: when $P_j \geq 50\%$, it is a global drought; when $50\% > P_j \geq 25\%$, it is a regional drought; when $25\% > P_j \geq 10\%$, it is a local drought; when $P_j < 10\%$, there is no obvious drought.

Table 1. Classification of drought intensity.

Drought rating	S	Type of drought
1	≤ 0	no drought
2	0~0.2	light drought
3	0.2~0.5	moderate drought
4	0.5~0.8	severe drought
5	>0.8	very dry

The Mann-Kendall method is used to judge the variation trend characteristics of drought (Ali et al. 2019; Liu et al. 2007), and the Hurst method (Jiang & Deng 2004; Yan et al. 2018) is used to judge the persistence of the drought variation trend, and the Hurst index is between 0 and 1. When the Hurst exponent is $0 < H < 0.5$, the time series is inversely persistent, that is, the future trend is opposite to the past one, and the closer it is to 0, the stronger the inverse persistence; when $H = 0.5$, the time series is random, which means that the changing trend cannot be judged; when $0.5 < H < 1$, the time series is persistent, that is, the future changing trend is consistent with the past changing trend, and the closer the H value is to 1, the stronger the continuity.

4 RESULTS DISCUSSION AND ANALYSIS

4.1 *Spatial distribution characteristics of drought frequency*

The IDW method is used to interpolate the drought frequency at different time scales of each site in the Taihu Lake Basin, and the spatial distribution of the drought frequency in the Taihu Lake Basin is further analyzed, as shown in Figures 2 and 3.

Figure 2 shows the annual scale drought frequency distribution in the Taihu Lake Basin. Figure 2(a) shows that the 36-year drought frequency at each site in the Taihu Lake Basin is light drought > moderate drought > severe drought. No extreme drought events occurred, among which only Dantu Station has experienced severe drought with a frequency of 2.78%. Figure 2(b) shows that the spatial distribution of drought frequency in the Taihu Lake Basin shows a high-low distribution trend from northwest to southeast. The drought frequency is the highest in HX, WCXY, YCDM, and PDPX, with a drought frequency of more than 55%. The area with the lowest drought frequency is HJH, where the drought frequency is below 45%.

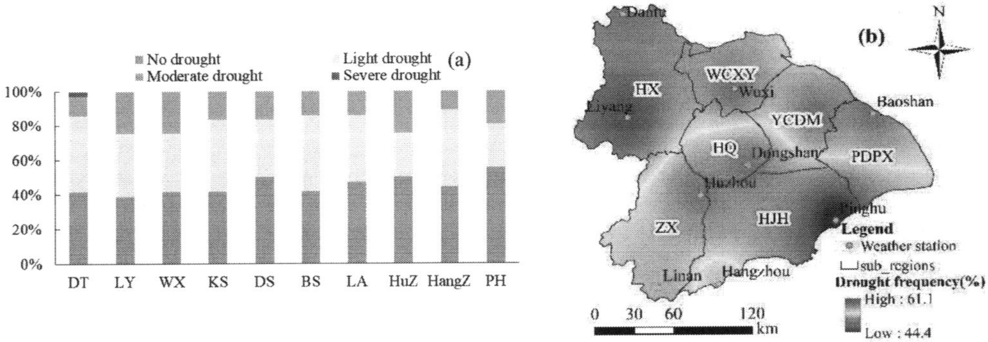

Figure 2. Drought frequency of different grades and spatial distribution of drought frequency at each site on an annual scale in the Taihu Lake Basin.

Figure 3 shows the spatial distribution of drought frequency at the seasonal scale in the Taihu Lake Basin. It can be seen from the figure that in spring, the drought frequency in the Taihu Lake Basin ranges from 44.4% to 55.6%, with an average of 50.6%. The drought frequency is high in the west and low in the east. The drought frequency in PDPX and YCDM is the lowest, and the drought frequency in HX is higher. In summer, the drought frequency ranges from 44.4% to 63.9%, with an average of 57.1%. The drought frequency is higher in the north and lowers in the south. The drought frequency in ZX is the lowest, and the drought frequency in YCDM is the highest. In autumn, the frequency of drought ranges from 55.5% to 72.2%, with an average of 60.3%. The drought frequency is low in the west and high in the east. The area with the lowest drought frequency is ZX, with a drought frequency of 55.5%; the area with the highest drought frequency is PDPX, and the drought frequency reached 72.2%. In winter, the drought frequency ranges from 41.7% to 52.8%, with an average of 49.1%. The drought frequency is high around the middle and low in the middle. The HQ has the lowest drought frequency at 41.7%, and Kunshan, Dantu, and Lin'an have the highest drought frequency, reaching 52.8%. In general, the frequency of droughts in each season in the Taihu Lake Basin is roughly the same, and autumn is the most frequent drought season in the Taihu Lake Basin.

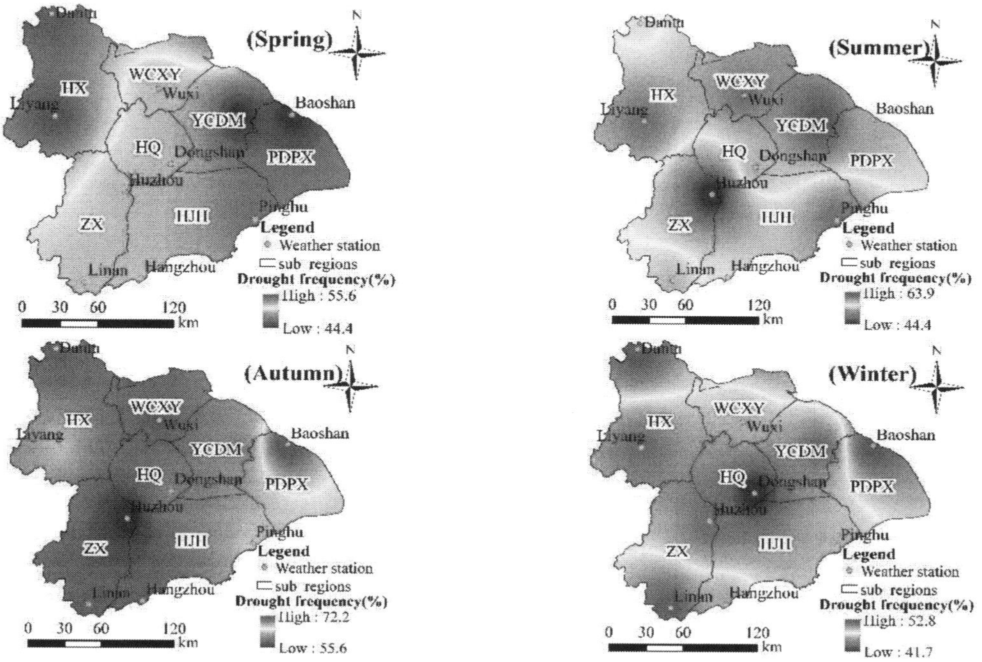

Figure 3. Seasonal drought frequency distribution in Taihu Lake Basin.

4.2 *Drought intensity and range variation characteristics*

4.2.1 *Annual scale*

Figure 4 shows the changing process of the drought intensity and station-time ratio in the Taihu Lake Basin on an annual scale from 1981 to 2016. Figure 4 shows that the variation range of the drought intensity in the Taihu Lake Basin is 0–0.29, and the average drought intensity is 0.12. The annual scale drought in the Taihu Lake Basin is mainly light droughts are the mainstay, with 23 light

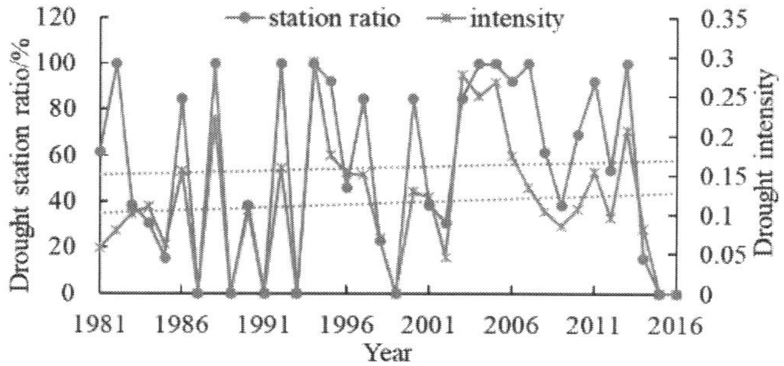

Figure 4. The variation characteristics of the annual scale drought intensity and the ratio of drought stations in the Taihu Lake Basin from 1981 to 2016.

droughts occurring in 36 years. At the same time, the moderate drought was mainly distributed from 2000 to 2010, of which 2003 to 2005 were three consecutive years of moderate drought, and the intensity of drought in 2003 was the highest at 0.28. From 1981 to 2016, there are 19 years of global drought, 7 years of regional drought, and 3 years of local drought, among which the global drought occurred once every two years on average.

4.2.2 *Seasonal scale*

Figure 5 shows the changing process of drought intensity and station-time ratio in the Taihu Lake Basin on a seasonal scale from 1981 to 2016. It can be seen from the figure that:

(1) Spring drought. From the scatter plot of the change of drought intensity grades in spring, the variation range of drought intensity in spring is 0 to 0.64, with an average drought intensity of 0.16, mainly light drought and moderate drought, with a total of 29 occurrences, including 12 moderate droughts. The moderate drought mainly occurred after 2000, and the light drought occurred 16 times, concentrated before 2000.

    There are 19 global spring droughts in the Taihu Lake Basin from 1981 to 2016, mainly after 2000. Among them, 2013 to 2015 was a three-year global drought. From 2005 to 2009, the drought station-time ratio reached 100%, and drought has the widest impact. The influence of the spring drought in the Taihu Lake Basin is steadily expanding.

(2) Summer drought. The summer drought intensity in the Taihu Lake Basin varies from 0 to 0.44, with an average drought intensity of 0.21. Moderate droughts are dominant, with a total of 21 occurrences, occurring every two years on average. 12 light droughts, and occur every three years on average. The summer drought in 1994 was the most severe, with a drought intensity of 0.44, the most severe summer drought since 1980.

    A total of 20 global summer droughts occurred in the Taihu Lake Basin from 1981 to 2016, and 9 occurred after 2000, accounting for 45% of the global summer droughts. The summer droughts in 2004 and 2005 were the most affected in the Taihu Lake Basin since the 21st century, and the drought station-time ratio reaches 100%. There are 7 regional droughts and 10 local droughts. To sum up, the summer drought in the Taihu Lake Basin is dominated by global drought.

(3) Autumn drought. The variation range of drought intensity in Taihu Lake Basin in autumn is 0–0.58, with an average drought intensity of 0.22. Autumn droughts are mainly light and moderate droughts, with a total of 26 occurrences, 12 light droughts, and 14 moderate droughts in autumn. From 1981 to 2016, a severe drought occurred three times in autumn in the Taihu Lake Basin, in 1995, 2001, and 2011, respectively. The drought intensity reached more than 0.55.

    There are 23 global autumn droughts in the Taihu Lake Basin from 1981 to 2016. Among them, the autumn droughts from 1994 to 1995, 2001 to 2004, and 2011 had the largest impact, with the station-time ratio reaching 100%. The global autumn drought is mainly concentrated in the 14 years from 2001 to 2014. Regional droughts occurred 3 times, and local droughts occurred 2 times. In general, the autumn drought in the Taihu Lake Basin is dominated by global drought, and the number of global droughts has increased after 2000.

(4) Winter drought. The winter drought intensity in the Taihu Lake Basin varies from 0 to 0.68, with an average drought intensity of 0.19. There are three severe droughts in 1986, 1999, and 2011. The drought intensity in 1986 reached 0.68, the most severe from 1981 to 2016. 11 times of moderate drought and 18 times of light drought. In general, the Taihu Lake Basin is dominated by light drought and moderate drought in winter.

    From 1981 to 2016, there are 15 global winter droughts in the Taihu Lake Basin, mainly between 1981 and 2000, there are 11 occurrences in total, accounting for 73.3% of the global drought years, and the ratio of drought stations was 100%. Regional drought and local droughts are 5 and 7 times, respectively. In general, the winter droughts are mostly global, and after 2000, the number of global winter droughts decreased and the affected area has been significantly reduced.

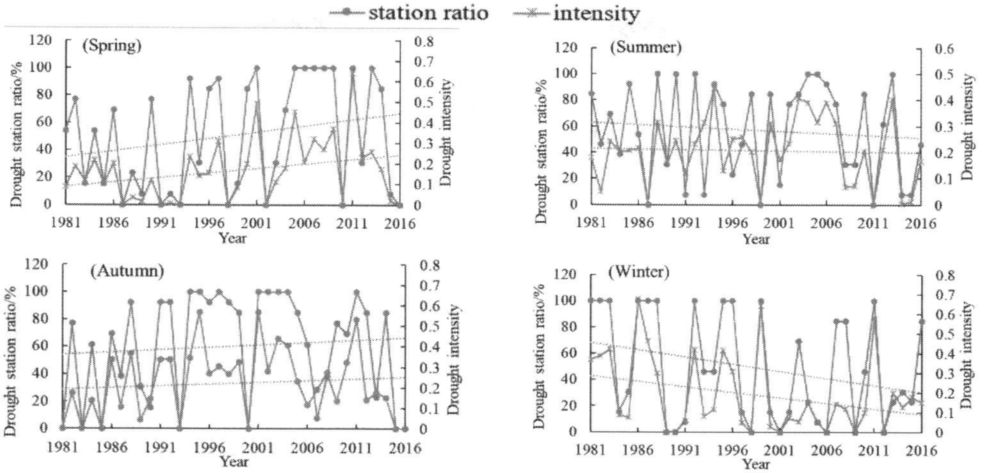

Figure 5. Seasonal-scale drought intensity and station-time ratio variation characteristics in Taihu Lake Basin from 1981 to 2016.

4.3 *Drought trends*

The Mann-Kendall and Spearman methods are used to calculate the drought trend from 1981 to 2016, and the Hurst method is used to study the persistence of the drought trend in the Taihu Lake Basin. The results are shown in Table 2.

Table 2. Drought intensity and station ratio trend test results at different time scales in the Taihu Lake Basin.

Time scale	Type	Trend	Z	Q	H	Future trends
Year	Drought intensity	↑	0.20	0.59	0.73	↑
	Station ratio	↑	0.13	1.01	0.71	↑
Spring	Drought intensity	↑	1.53	1.71	0.81	↑
	Station ratio	↑ *	1.73	2.08	0.77	↑ *
Summer	Drought intensity	↓	−0.75	−0.65	0.55	↓
	Station ratio	↓	−0.31	−0.24	0.54	↓
Autumn	Drought intensity	↑	0.23	0.78	0.80	↑
	Station ratio	↑	0.27	1.17	0.88	↑
Winter	Drought intensity	↓	−0.01	−1.05	0.53	↓
	Station ratio	↓ *	−1.70	−0.94	0.65	↓ *

(Note: "↑" indicates an upward trend; "↓" indicates a downward trend; "*" indicates a significant trend)

From Table 2, it can be seen that from 1981 to 2016, the annual-scale drought intensity and station-time ratio in the Taihu Lake Basin showed an upward trend, but they do not pass the $\alpha = 90\%$ confidence level test, and the upward trend is not significant. On the seasonal scale, from 1981 to 2016, the drought intensity and station ratio in spring and autumn showed an upward trend, and in summer and winter showed a downward trend. Among them, spring and winter passed the $\alpha = 90\%$ confidence level test. The trend is significant. The above analysis shows that in the annual scale of the Taihu Lake Basin, the probability of drought in spring and autumn will increase, the intensity of drought will further increase, and the scope of influence will further expand.

From the Hurst indices of drought intensity and the station-time ratio at different time scales in the Taihu Lake Basin from 1981 to 2016 in Table 2, the inter-annual and seasonal Hurst indices are all greater than 0.5. Therefore, the future drought trends on each time scale are the same as those in the past. The 36-year changing trends are consistent, among which the drought intensity in spring and autumn and the changing trends of the station-to-number ratio are strongly persistent.

5 CONCLUSIONS

(1) On the annual scale of Taihu Lake Basin from 1981 to 2016, the drought frequency showed a high-low spatial distribution from northwest to southeast. The drought frequency in HX, WCXY, YCDM, and PDPX is relatively high, with a drought frequency of 55%. The area with the lowest drought frequency is HJH, where the drought frequency is below 45%. On the seasonal scale, autumn is the most frequent drought season.
(2) On the annual and seasonal scales, the drought grades in the Taihu Lake Basin are dominated by light drought and moderate drought. In terms of the affected areas of droughts, all of them are dominated by global droughts, and on average, global droughts occur every two years. At the same time, the affected areas of drought in different seasons present obvious stage characteristics. Since 2000, the influence range of spring drought and autumn drought has been expanding, and the influence range of winter drought has shown a decreasing trend.
(3) In terms of drought trends, the drought intensity and the ratio of drought stations in the Taihu Lake Basin from 1981 to 2016 show an increasing trend, but the overall trend is not significant. On the seasonal scale, the drought station ratio increased significantly in spring and decreased significantly in winter, and the changing trend of drought intensity and station ratio in other seasons is not obvious.
(4) In terms of the persistence of the drought trend, according to the Hurst index, the future drought trend on each time scale is the same as that of the past 36 years. The drought in summer and winter shows a weakening trend, and the other scales show an increasing trend. The changing trend in spring and autumn has strong continuity.

ACKNOWLEDGMENT

The research was funded by the National Key R&D Program of China (Grant No. 2019YFC0408903), and the special funded project for basic scientific research operation expenses of central public welfare scientific research institutes of China (Grant No. Y521007; Y521022)

REFERENCES

Dai Aiguo. (2011) Drought under global warming: A review. *J. WIREs Climatic Change*, 2:45–65.
Huang W H, Yang X G, Li M S, et al. (2010) Evolution characteristics of seasonal drought in the south of China during the past 58 years based on standardized precipitation index. *J. Transactions of the CSAE*, 26(7): 50–59.
Jiang T H, Deng L T. (2004) Some Problems in Estimating a Hurst Exponent-A Case Study of Applications to Climatic Change. *J. Scientia Geographica Sinica*, (02):177–182.
Li J, Wu X S, Wang Z L, et al. (2021) Research on the characteristics of future drought changes in the Pearl River Basin based on a new comprehensive drought index. *Journal of Hydraulic Engineering*, 52(04):486–497.
Liu P, Guo S L, Xiao Y, et al. (2007) Trend and Change Point Analysis of Hydrological Time Series Based on Resampling Methods. *Journal of China Hydrology*, 27(2):49–53.
Liu Q, Yan C R, He W Q. (2016) Drought Variation and Its Sensitivity Coefficients to Climatic Factors in the Yellow River Basin. *Chinese Journal of Agrometeorology*, 37(06):623–632.
Rawshan Ali, Alban Kuriqi, Shadan Abubaker, Ozgur Kisi. (2019) Long-Term Trends and Seasonality Detection of the Observed Flow in Yangtze River Using Mann-Kendall and Sen's Innovative Trend Method. *J. Water*, 11(9).
Sun R Q. (1994) A review of drought definition and its indicators. *Journal of Catastrophology*, 1:17–21.
Wang C Y. (2007) *Research progress on major agro-meteorological disasters*. China Meteorological Press, Beijing.

Yan J J, Zhang J, Lei Y, et al. (2018) Analysis of changing trend of grassland NDVI in the Ili valley of Xinjiang during 2000–2016. *J. Acta Agrestia Sinica*, 26(4): 859–868.

Yang T, Lu G H, Li H H, et al. (2011) Advances in the study of projection of climate change impacts on hydrological extremes. *J. Advances in Water Science*, 22(2):279–286.

Zhang H Q, Zhu Y H, Lv H S, et al. (2021) Analysis of five drought indices at different time scales in Huaibei Plain based on typical station. *J. Water Resources and Power*, 39(04):15–19.

Zhao J F, Guo J P, Xu J W, et al. (2010) Trends of Chinese dry-wet condition based on wetness index. *J. Transactions of the CSAE*, 26(8): 18–24.

Energy Revolution and Chemical Research – Kok-Keong Chong and Zhongliang Liu (Eds)
© 2023 The Authors, ISBN: 978-1-032-36554-1

Evaluation index system and evaluation method for multi-energy systems

Long Zhao, Donglei Sun, Dong Liu, Bingke Shi, Liang Feng & Rui Liu*
Economic & Technology Research Institute of State Grid Shandong Electric Power Company, Shandong, China

ABSTRACT: With the rapid development of multi-energy systems, comprehensive and effective evaluation index systems and evaluation methods are becoming more and more important for the development of multi-energy systems. To this end, this paper constructs an evaluation index system of multi-energy systems from multiple dimensions of economy, environmental protection, energy efficiency, reliability and technology. Considering the subjectivity and objectivity of empowerment and the uncertainty of index weight, an evaluation method of multi-energy systems based on analytic network progress (ANP), entropy weight and interval TOPSIS is proposed. The case study compares different scenarios and verifies the scientificity and effectiveness of the proposed evaluation index system and evaluation method.

1 INTRODUCTION

Traditional energy systems are separated from each other and have limited coordination, which is difficult to adapt to the rapid development of renewable energy. Multi-energy systems can break the original mode of separate planning, design, and operation of traditional energy systems, and coordinate different energy systems organically, to improve the security, flexibility and reliability of various energy systems.

At present, the engineering construction of multi-energy systems is still immature. Engineering projects of Multi-energy systems require comprehensive evaluation before construction. However, the current evaluation index system for them is relatively scarce, and the most studies only put forward specific indexes of multi-energy systems from aspects of the economy (Wu et al. 2018), energy efficiency (Liu et al. 2019; Tang et al. 2017), reliability (Liu et al. 2019) and so on to complete the evaluation of a certain performance of multi-energy systems. In terms of the evaluation method, only a single subjectivity or objectivity has been achieved (Tang et al. 2017), and relatively little consideration has been given to the uncertainties in various aspects. In the process of engineering construction and development for multi-energy systems, whether the evaluation link is perfect or not is related to the effectiveness of the feedback link in multi-energy systems. If the feedback link cannot play a great role, it will affect the good development mechanism of engineering construction for multi-energy systems. Therefore, it is particularly important to construct a comprehensive and effective evaluation index system and evaluation method, which is both subjective and objective, for the development of multi-energy systems.

This paper focuses on the development effect of multi-energy systems and proposes a multi-dimensional evaluation index system suitable for multi-energy systems from five aspects include economy, environmental protection, energy efficiency, reliability and technology. Considering the combined weighting model which combines the advantages of subjective and the uncertainty of weighting, an evaluation method of multi-energy systems based on ANP-entropy weight and interval TOPSIS is proposed.

*Corresponding Author: liu_rui@zju.edu.cn

 DOI 10.1201/9781003332657-35

2 EVALUATION INDEX SYSTEM OF MULTI-ENERGY SYSTEMS

Considering the development characteristics of multi-energy systems comprehensively, the primary indexes are determined from five dimensions economy, environmental protection, energy efficiency, reliability and technology. The economy index should reflect the investment and construction level of an engineering project for multi-energy systems and the development value of multi-energy systems directly. The environmental protection index should reflect the environmental friendliness of energy supply and consumption mode for multi-energy systems. The energy efficiency index should reflect the utilization efficiency of different kinds of energy in multi-energy systems. The reliability index should reflect the risk level of energy outage in multi-energy systems. The technology index should reflect the utilization and loss of diverse equipment in multi-energy systems.

The secondary indexes of the economy index include operating income, construction cost, operating expenses, net present value, internal rate of return and payback period for dynamic investment. The secondary indexes of the environmental protection index include proportion of renewable energy, environmental pollution emission level, environmental pollution emission reduction level and environmental protection benefit; the secondary indexes of the energy efficiency index include primary energy utilization rate, primary energy saving rate, comprehensive energy utilization efficiency and energy cost; the secondary indexes of reliability index include energy shortage rate, the average time of energy shortage, outage ratio and reliability income. The secondary indexes of the technology index include equipment utilization rate, equipment operation efficiency, energy loss and energy supply quality. The constructed multi-dimensional evaluation index system of multi-energy systems is shown in Figure 1.

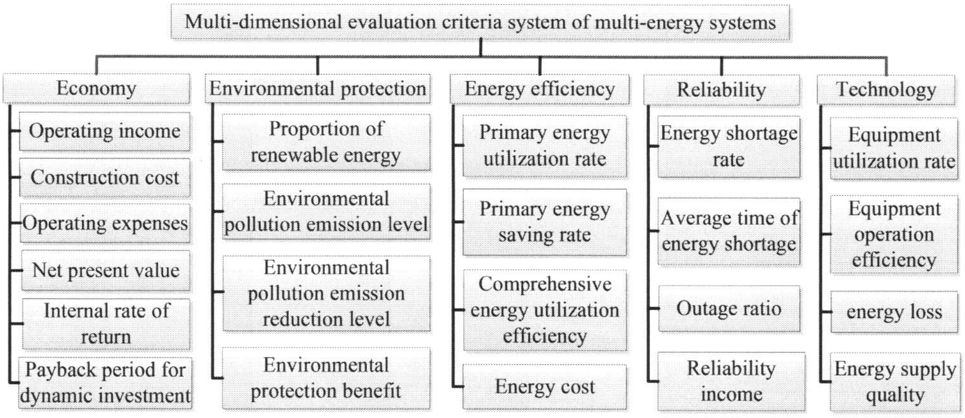

Figure 1. Multi-dimensional evaluation index system of multi-energy systems.

3 EVALUATION METHOD FOR MULTI-ENERGY SYSTEMS

3.1 *Combined weighting model based on ANP-entropy weight*

In the aspect of weight determination for secondary index, combined weighting model based on ANP-entropy weight is adopted to determine the weight of secondary index of multi-energy systems for evaluation index system. The steps are as follows:

Step1. Weights calculation by ANP method:

1) Establish a judgment matrix. The expert method is used to judge the relative importance of each index, and the judgment matrix is constructed according to the level nine scaling method. Assume that the scheme layer under the criterion layer of ANP has element group $C_1, \ldots, C_n$,

where C_i contains element $e_{ik}(i=1,\ldots,N,\ k=1,2,\ldots,n_j)$, the indirect dominance of each element of element group C_i is compared and analyzed according to its influence on e_{ik}, that is, construct a judgment matrix.

2) Establish supermatrix. The maximum eigenvalue and eigenvector of the judgment matrix are calculated and the priority weight vector is obtained by the roots method. Then the consistency test is carried out. If the consistency test is passed, the local weight vector-matrix W_{ij} can be obtained according to the rule that the column vector represents the importance of element C_i to C_j. The priority weight vectors of all mutually affecting elements of the scheme layer are combined to obtain the supermatrix W.

   Then, W is normalized to get the weighted supermatrix $\bar{W}=(\bar{W})_{n\times n}$, $\bar{W}_{ij}=a_{ij}W_{ij}(a_{ij}$ is the weighting factor).

3) Compute limit supermatrix. The limit relative priority weight vectors of each supermatrix are calculated as follows:

$$\lim_{k\to\infty}\frac{1}{N}\sum_{k=1}^{N}\bar{W}^k \tag{1}$$

The value of the corresponding row of the original matrix is the weight value of each evaluation index.

Step2. Weights calculation by entropy weight method:

1) Determine the target sequence. The data of each index of benefit index and cost index are normalized respectively, then the data of each index is standardized and the normalized decision matrix $E=(e_{ij})_{m\times n}$ is obtained.
2) Calculate weights. Information entropy Y_j of the j-th index is calculated, and then the weight of each index is calculated.

Step3. Establish a combined weighting model:

Assuming that the weights of the secondary indicators calculated by ANP method and entropy weight method are $w'=(w'_1,w'_2,\ldots,w'_n)^T$ and $w''=(w''_1,w''_2,\ldots,w''_n)^T$, the final weight can be obtained:

$$w_j=\alpha w'_j+\beta w''_j \tag{2}$$

where α and β are the proportions of weight calculated by the ANP method and entropy weight method to final weight ($\alpha>0,\beta>0,\alpha+\beta=1$). When $0\le\alpha+\beta\le 1$, the interval weight can be obtained:

$$[w_j]=[\min\{w'_j,w''_j\},\max\{w'_j,w''_j\}]=[\underline{w_j},\overline{w_j}] \tag{3}$$

where $\underline{w_j}$ and $\overline{w_j}$ are the upper and lower limits of $[w_j]$ respectively.

So, the evaluation result of the secondary index of the i-th evaluation scheme $[Z_i]$ is:

$$[z_i]=\sum_{j=1}^{n}[w_j]e_{ij} \tag{4}$$

The obtained evaluation result is an interval number.

3.2 *Interval TOPSIS evaluation method*

Because there are different preferences among decision-makers and different development periods of multi-energy systems, the preference degree of each primary index may fluctuate within a certain range, so the weight of primary indexes in the design evaluation process is an uncertain value. Therefore, this paper adopts the interval TOPSIS method (Jahanshahloo et al. 2006) to analyze the evaluation model of multi-energy systems whose weight values are interval numbers. The steps are as follows:

1) Establish an interval evaluation matrix. According to the interval attribute values of different indexes of different scenarios, the interval evaluation matrix is constructed.

2) Standardized interval evaluation matrix. The benefit index and cost index are respectively standardized for the interval evaluation matrix.
3) Calculate Euclidean distance. The positive and negative ideal interval numbers of the index weight P_w^+ and N_w^- are calculated, then the Euclidean distance between $[w_j]$ and $d([w_j],[P_w^+])$, $[w_j]$ and $d([w_j],[P_w^+])$ can be gotten.
4) Calculate the relative similarity of interval weights and normalize them. The calculation formula is:

$$c([w_j]) = \frac{d([w_j],[N_w^-])}{d([w_j],[P_w^+]) + d([w_j],[N_w^-])} \tag{5}$$

$$w_j = \frac{c([w_j])}{\sum_{j=1}^{n} c([w_j])} \tag{6}$$

where $c([w_j])$ is the relative similarity of interval weight, and w_j is its normalization result.
5) Calculate the weighted normalized decision matrix. Assuming that each element $[v_{ij}] = w_j[e_{ij}]$, $[E]$ is the set of $[e_{ij}]$, and w is the set of w_j, then the weighted normalized decision matrix $[V]$ is:

$$[V] = w[E] \tag{7}$$

6) Calculate the relative similarity of each scenario. Similar to 3), calculate the Euclidean distance between the ith scenario and the ideal interval number of positive and negative; Then, similar to 4), the relative similarity $c([v_i])$ of the ith scenario can be calculated.
7) Evaluate each scenario. Sort $c([v_i])$ obtained in 6) from large to small, and the larger its value is, the better scenario is.

4 CASE STUDY

4.1 *Experimental settings*

Taking a multi-energy system as the study object, the average annual radiation intensity is 1500 kWh/m^2, and the average wind speed is 5 m/s. The demand for power, heating, and cooling is about 3.70 MW, 2.90 MW, and 2.40 MW respectively. The annual power consumption of this system is about 4271.95 MWh. The purchase price from the power grid is 0.64 CNY/kWh, the network loss rate is 5%, and the energy efficiency ratio is 3. The hybrid energy storage system has $LiFePO_4$ Li-ion batteries, supercapacitor, and lead-acid batteries, their capacity is 100 kW·2 h, 100 kW·10 s, and 150 kW·1 h respectively. CCHP system is mainly composed of a 600-kW gas engine, 300 kW gas turbine, and 400-kW absorption chiller. The cost of each piece of equipment is shown in Table 1.

Table 1. The cost of each equipment.

Equipment	Investment cost (CNY/kW)	Fixed maintenance cost (CNY/kW·a)	Variable maintenance cost (CNY/kW·a)	Annual average utilization hours(h)	Gas cost (CNY/kWh)
Gas engine	2340	172	0.0021	2500	0.9395
Gas turbine	7800	771	00043	2500	0.1620
Absorption Chiller	1200	200	0.0080	1200	/
Photovoltaic	12600	90	0.0042	1250	/
Wind	3500	37	0.0050	1400	/
lead acid batteries	1200	30	0.0015	/	/

There are three scenarios. Scenario1: power network, 1.6 MW CCHP system, and energy storage system. Scenario2: power network, 2.4 MW photovoltaic, 1.3 MW CCHP system, and energy storage system. Scenario3: power network, 2.0 MW photovoltaic, 0.3 MW wind, 1.3 MW CCHP system, and energy storage system. Index values of each scenario are shown in Tables 2–6.

Table 2. Economy index.

Secondary index	Scenario1	Scenario2	Scenario3
Operating income (CNY)	665200	779800	827200
Construction cost (CNY)	1005000	3149800	2832300
Operating expenses (CNY)	496600	392400	393000
Net present value (CNY)	783300	149200	1044600
Internal rate of return(%)	19.48	8.58	12.38
Payback period for dynamic investment (year)	8.1	19.3	12.4

Table 3. Environmental protection index.

Secondary index	Scenario1	Scenario2	Scenario3
Proportion of renewable energy (%)	0	42.13	41.01
Environmental pollution emission level (t)	1001.2	747.7	751.9
Environmental pollution emission reduction level (t)	0	167.5	163.0
Environmental protection benefit(CNY)	0	89500	87100

Table 4. Energy efficiency index.

Secondary index	Scenario1	Scenario2	Scenario3
Primary energy utilization rate (%)	87.12	67.08	69.75
Primary energy saving rate (%)	0	23.53	22.91
Comprehensive energy utilization efficiency (%)	68.3	81.9	78.4
Energy cost (CNY/MWh)	2500	2100	2200

Table 5. Reliability index.

Secondary index	Scenario1	Scenario2	Scenario3
Energy shortage rate (%)	114.79	86.59	86.93
Average time of energy shortage (h)	6.04	5.85	5.73
Outage ratio (%)	0.5879	0.5334	0.5319
Reliability income (CNY)	26575200	25857700	26116200

Table 6. Technology index.

Secondary index	Scenario1	Scenario2	Scenario3
Equipment utilization rate (%)	28.54	14.27	14.79
Equipment operation efficiency (%)	52.89	25.74	27.38
energy loss (MWh)	493.27	415.77	418.17
Energy supply quality	87	95	91

4.2 *Experiment result*

The ANP method and the entropy weight method are used to calculate the weights of secondary indexes, and the results are combined to calculate the weights of each index. The evaluation results of calculated secondary indicators are shown in Tables 7 to 8. It can be seen that scenario 1 has the best result in the secondary indexes of equipment index, scenario 2 has the best performance in the two secondary indexes of environmental protection and energy efficiency indexes, and scenario 3 has the best performance in the two secondary indexes of economy and reliability indexes.

Table 7. The values of secondary index.

	Secondary index				
Scenario	Economy	Environmental protection	Energy efficiency	Reliability	Technology
Scenario1	0.3542	0	0.2412	0.1662	0.6650
Scenario2	0.2722	0.9014	0.7212	0.6157	0.4400
Scenario3	0.4689	0.8797	0.6414	0.6815	0.4043

Table 8. The evaluation results of the secondary index.

Index \ Ranking	1	2	3
Economy	Scenario3	Scenario1	Scenario2
Environmental protection	Scenario2	Scenario3	Scenario1
Energy efficiency	Scenario2	Scenario3	Scenario1
Reliability	Scenario3	Scenario2	Scenario1
Technology	Scenario1	Scenario2	Scenario3

Interval weight values of the five first-level indicators of economy, environment, reliability, energy efficiency and equipment are respectively set as: $[w_1]=[0.24, 0.26]$, $[w_2]=[0.26, 0.34]$, $[w_3]=[0.24, 0.26]$, $[w_5]=[0.09, 0.11]$, so that the positive ideal interval number and the negative ideal interval number of interval weight value can be calculated as [0.34,1] and [0,0.09] respectively. Then, the relative similarities of interval weight calculated are $c([w_1])=0.2826$, $c([w_2])=0.3517$, $c([w_3])=0.2826$, $c([w_4])=0.0907$ and $c([w_5])=0.0907$ respectively, the normalized result is $w_1=0.2573$, $w_2=0.3202$, $w_3=0.2573$, $w_4=0.0826$ and $w_5=0.0826$. Finally, the relative similarity of each scenario can be obtained and sorted, as shown in Table 9.

Table 9. The relative similarity of each scenario.

Ranking	Scenario3	Scenario2	Scenario1
Relative similarity	0.6570	0.6314	0.2219

As can be seen from Table 9, scenario 3 has the best performance, followed by scenario 2 and scenario 1 has the worst performance.

To sum up, the evaluation results are completely consistent with the actual situation, and the evaluation results of primary and secondary indexes are consistent with the numerical analysis results of secondary indexes of the actual scenario, indicating the effectiveness of the proposed evaluation method.

5 CONCLUSION

This paper puts forward a multi-dimensional evaluation index system of multi-energy systems from five aspects of the economy, environmental protection, reliability, energy efficiency, and technology. In addition, considering subjectivity, objectivity, and the uncertainty of evaluation weight, this paper proposes an evaluation method for multi-energy systems based on ANP, entropy weight, and interval TOPSIS. The effectiveness of the proposed evaluation method is verified by analyzing the evaluation results of three scenarios of a multi-energy system.

ACKNOWLEDGMENTS

This work was financially supported by the Science and Technology Project of SGCC (Grant No. SGSDJY00GPJS2000248).

REFERENCES

Jahanshahloo G.R., Lotfi F. Hosseinzadeh, Izadikha M. (2006) An algorithmic method to extend TOPSIS for decision-making problems with interval data. *Applied Mathematics and Computation*, 175: 1375–1384.

Liu Hong, Li Jifeng, Ge Shaoyun, et al. (2019) Impact Evaluation of Operation Strategies of Multiple Energy Storage Systems on Reliability of Multi-energy Microgrid. *Automation of Electric Power Systems*, 43(10): 36–45.

Liu Hong, Zhao Yue, Liu Xiaoou, et al. (2019) Comprehensive Energy Efficiency Assessment of Park-level Multi-energy System Considering Difference of Energy Grade. *Power System Technology*, 43(8): 2835–2843.

Tang Yanmei, He Guixiong, Liu Kaicheng, et al. (2017) *Study on method of comprehensive energy efficiency evaluation for distributed energy system*. In: IEEE Conference on Energy Internet and Energy System Integration, Beijing. pp. 1–5.

Wu Shengyu, Xu Bo, Lu Gang, et al. (2018) *Coordinated Development Evaluation on Integrated Energy Systems in China*. In: IEEE Conference on Energy Internet and Energy System Integration, Beijing. pp. 1–6.

Energy Revolution and Chemical Research – Kok-Keong Chong and Zhongliang Liu (Eds)
© 2023 The Authors, ISBN: 978-1-032-36554-1

Application of horizontal subsurface flow constructed wetland in initial rainwater treatment

Yucheng Ding*, Wei Ding, Yuechen Wei & Yang Wang
CCCC Hehai Engineering Co., Ltd., Nanjing, P.R. China
CCCC First Harbor Engineering Company Ltd., Tianjin, P.R. China

ABSTRACT: Initial rainwater treatment is an important concept of urban water environment management in China at present. According to the actual situation of a city in the south of China, this paper designed a wetland park for initial rainwater treatment. After preliminary treatment, the initial rainwater or polluted river water entered the horizontal subsurface flow constructed wetland and then was discharged into the natural water body after being treated again. Through the trial operation results, it can be seen that the wetland effluent meets the Class II standard of surface water, far exceeding the design requirements, and can be directly discharged into natural water bodies. After biochemical treatment, there is no significant difference between indexes of the initial rainwater and river water. After treatment, four indexes of wetland effluent have been significantly reduced, among which SS has the largest reduction rate, exceeding 90%, followed by NH3-N, TP and CODcr, with a minimum reduction rate of 33.3%, which indicates that the wetland has a very significant effect on sewage purification. The results show that the horizontal subsurface flow constructed wetland can be used to purify the initial rain and polluted river water, enhance the urban landscape, conform to the concept of "sponge city", and is an effective measure of urban water environment control.

1 INTRODUCTION

The initial rain refers to the rain in the initial period of rainfall. According to the Code for design of environmental protection of chemical industry projects (GB 50483-2019) (Code for design of environmental protection of chemical industry projects 2019), the initial rain refers to the initial rainfall of 15-30 min or the initial rain of 20-30 mm in thickness. At the beginning of rainfall, the rainwater had dissolved lots of polluting gases such as acid gas, automobile exhaust gas and industrial waste gas. After the rain falls to the ground, it scours asphalt linoleum roofs, asphalt concrete roads, and construction sites, etc. As a result, the initial rain contains a large number of pollutants such as organic matter, pathogens, heavy metals, grease, suspended solids, etc. So the pollution degree of the initial rain is relatively high, which usually exceeds that of ordinary urban sewage.

As early as the 1970s and 1980s, Europe and the United States and other developed countries have focused their attention on non-point source pollution such as urban rainwater runoff. At present, the treatment of initial rainwater in China is still in its infancy. If initial rainwater is directly discharged into the natural water body, it will cause very serious pollution to the water body, so the initial rainwater must be abandoned. Rain and pollution switching devices can be set up to divert the initial rainwater to sewage pipes. Or collect the initial rainwater for centralized treatment and then discharge it into the natural water body.

*Corresponding Author: dingyc_ouc@163.com

DOI 10.1201/9781003332657-36

Wetlands are rich in plants and microorganisms, which have a strong purification effect on sewage, and have been widely used in the purification of domestic sewage, tail water from sewage treatment plants and sewage from rivers and lakes. However, its application in initial rainwater treatment is still very rare. This paper will discuss the application of constructed wetlands in initial rainwater treatment based on a wetland park in the south.

2 CONSTRUCTED WETLAND

Constructed wetland is a technology in which sewage and sludge are distributed on artificially constructed and controlled ground similar to the marsh. In the process of flowing in a certain direction, sewage and sludge are treated mainly by the physical, chemical and biological synergy of soil, artificial medium, plants and microorganisms.

The substrate is the carrier for the growth of plants and microorganisms in the constructed wetland. According to the needs, the materials with different particle sizes are laid on the constructed wetland bed with a certain thickness. Soil, sand and gravel are the traditional substrates of constructed wetlands. In recent years, zeolite, limestone, shale, alum (Knox et al. 2010), plastics and ceramics are widely used.

There are many kinds of plants that can be planted in constructed wetlands, which can generally be divided into floating plants (eichhornia crassipes, duckweed, azolla, sophora japonica, rhododendron manshuriensis, etc.), emergent plants (reed, cattail, acorus tatarinowii, etc.) and submerged plants (elodea japonica, myriophylla, etc.) (Lu & Jiang 2009; Ren et al. 2011). The principles of screening plants mainly include adaptability of plants, strong growth ability of plants, landscape, diversity, economy and easy management (Niu et al. 2004).

The overall flow mode of constructed wetland system can be roughly divided into a surface flow and subsurface flow. In most studies, subsurface flow is further divided into horizontal subsurface flow and vertical flow (Vymazal 2007). A surface flow wetland is a rectangular structure where sewage flows over the surface of the wetland. It has a simple structure and low engineering costs. However, because of the overflow of sewage on the surface of the packing, it is easy to breed mosquitoes and flies, which will have a bad effect on the surrounding environment, and its treatment efficiency is low. Horizontal subsurface flow wetland, sewage seepage in the gap among the filler, can make full use of the surface of the filler and plant roots on the biofilm and other effects of sewage treatment, effluent quality is good. In vertical flow wetland, sewage flows along the vertical direction, has a strong oxygen supply capacity, full nitrification, covers a small area, and can achieve a large hydraulic load in long-term operation. The sewage in the horizontal subsurface flow wetland seeps through the gaps among the filler, which can make full use of biofilms on the surface of fillers and plant roots and other functions to treat the sewage, and the effluent water quality is good. The vertical flow wetland has a strong oxygen supply capacity, sufficient nitrification and a small area, which can realize long-term operation with a large hydraulic load.

3 PROJECT OVERVIEW

During the treatment of black and odorous water bodies in a city in the south of China, the technical route of "three-water separation, decentralized storage, treatment and reuse" was adopted, and a city-wide initial rainwater treatment system was established. For the initial rainwater treatment of 73 ha catchment area, the three elements of "water treatment, sponge city construction and landscape" were considered comprehensively, and the existing wetland park was reconstructed according to local conditions for initial rainwater treatment. Due to years of disrepair, the biochemical reaction pond in the existing wetland park has been unable to operate, and the equipment has been seriously damaged. The park is littered with vegetation, green algae, and black and smelly water. After the renovation, not only the initial rainwater treatment can be completed, but also the existing wetlands can be renewed and the image of the city can be enhanced.

4 DESIGN AND IMPLEMENTATION OF WETLANDS

In this area, the scale of initial rainwater is 2,846 m^3, and the rainwater collection volume is 3,000 m^3 at the initial stage of the design. The process of "pretreatment + secondary treatment + wetland" is adopted, and the effluent quality standard is not lower than quasi-class IV (Environmental quality standards for surface water 2002). The pretreatment system includes an inlet cross well, mechanical grille, lifting pump room and storage tank. The secondary treatment system includes a biological reaction tank and inclined tube sedimentation tank. After treatment, the initial rainwater enters the constructed wetland for further treatment and is discharged into the nearby river after treatment. The processing flow was shown in Figure 1.

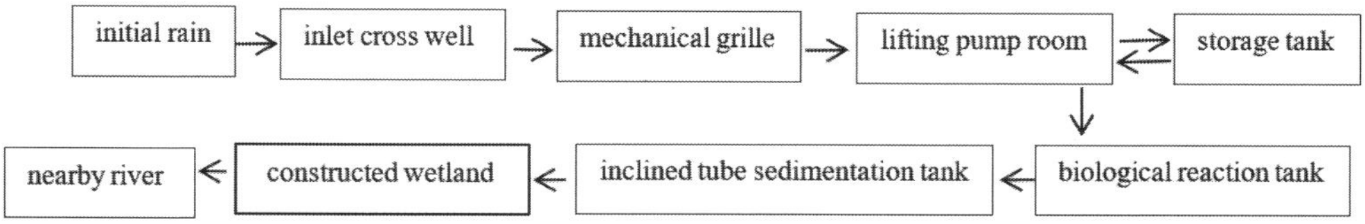

Figure 1. Initial rainwater treatment process.

In order to meet the operation requirements of the treatment system, when the initial rainwater is not enough, the standby river water should be introduced. When the pollutants in the water are not enough to meet the survival needs of microorganisms in the treatment system, municipal sewage should be introduced. The operating conditions of the wetland park are divided into two types. The first condition corresponds to rainy days. The treatment object was the initial rainwater, with a water volume of 3000 m^3/d and a hydraulic load of 0.24 $m^3/(m^2 \cdot d)$. The second condition corresponds to sunny days. The treatment object is river water (the water quality should meet the requirements of wetland water intake), the water volume of the treatment scale is 10,000 m^3/d, and the hydraulic load is 0.80 $m^3/(m^2 \cdot d)$. The sludge production is small, and the sludge is concentrated and transported for centralized treatment and disposal. The sludge produced is of small quality, and the sludge is concentrated and transported for centralized treatment and disposal.

The main control indexes of wetland effluent include BOD_5, COD_{cr}, NH_3-N and TP. The design indexes of inlet and outlet water are shown in Table 1.

Table 1. Design inlet and outlet water quality index (unit: mg/L).

Design water index	COD_{cr}	BOD_5	NH_3-N	TP
Inlet water	<100	<20	<25	<3.0
Outlet water	<30	<6	<1.5	<0.3

According to the actual working conditions of the project, the horizontal subsurface flow constructed wetland was adopted. The wetland area was 12,710 m^2, and the hydraulic load of the wetland was 0.24 $m^3/(m^2 \cdot d)$. The wetland was divided into 5 zones, which were arranged in parallel. To ensure the uniformity of water distribution, partition walls are added in each zone to form sub-units, as shown in Figure 2. The water is distributed in the pipes on both sides of each zone and purified by the substrate and the root zone, and then enters the middle catchment channel for outlet water.

The structure of the wetland is composed of 300 mm compacted clay, composite geomeme, 700 mm filler and 300 mm planting soil from bottom to top. The size of the filler from inlet to outlet is coarse to fine. They are gravel with a particle size of 25 mm to 80 mm, gravel with a particle size of 10 mm to 25 mm, and zeolite with a particle size of 10 mm to 25 mm.

Considering the water purification effect and landscape effect comprehensively, papyrus, reed, canna, iris, lythrum, calamus and other plants were mainly planted in the wetland to form a unified landscape effect. The real scene of the wetland park after construction is shown in Figure 3.

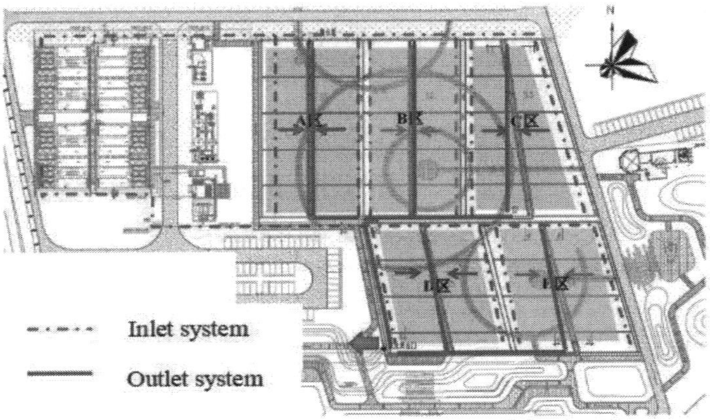

Figure 2. Plan design of constructed wetland.

Figure 3. Real scene of the wetland park.

5 ANALYSIS OF WETLAND WATER PURIFICATION EFFECT

At present, the project has been completed and has been in trial operation for 3 months. The operation results are good. The water quality of wetland inlet and outlet water was tested once a day. Some water quality test data are shown in Table 2 and Table 3.

By comparing Table 2, Table 3, Table 1 and 4, the following conclusions can be drawn.

(1) The initial rainwater and river water after biochemical treatment can meet the requirements of wetland water intake and ensure the safe operation of the wetland.
(2) The wetland effluent reaches the surface water class II standard, far exceeding the design requirements, and can directly discharge the natural water body.
(3) After biochemical treatment, the initial rainwater and river water had no significant difference in the values of the five indexes.
(4) After treatment, in addition to BOD_5, the water outlet reduction rate is not very significant because the water inlet value is very low. Four indicators of initial rainwater and river water are significantly reduced, among which SS reduction rate is the largest, more than 90%, and then NH3-N, TP and COD_{cr}, successively reduced, the minimum reduction rate is 33.3%. The results show that the purification effect of wetlands on sewage is very significant.

Table 2. Test results of wetland water quality during initial rainwater treatment (unit: mg/L).

Water quality indicators		Test results								Average value	Reduction rate (%)
COD_{cr}	Inlet	15	9	13	19	17	16	14	14	14.6	46.2
	Outlet	9	7	<2	8	10	9	10	10	7.9	
BOD_5	Inlet	<2	<2	<2	<2	<2	<2	<2	2.33	/	/
	Outlet	<2	<2	<2	<2	<2	<2	<2	<2	/	
NH_3-N	Inlet	0.69	0.79	1.09	0.53	0.55	0.25	0.50	0.27	0.584	73.4
	Outlet	0.15	0.32	0.29	0.07	0.12	0.09	0.03	0.17	0.155	
TP	Inlet	0.10	0.22	0.11	0.20	0.16	0.11	0.12	0.14	0.145	47.4
	Outlet	0.05	0.08	0.10	0.07	0.05	0.08	0.10	0.08	0.076	
SS	Inlet	33	63	72	37	31	42	37	61	47.0	92.3
	Outlet	3	2	2	2	5	5	5	5	3.6	

Table 3. Test results of water quality in and out of wetland during river treatment (unit: mg/L).

Water quality indicators		Test results								Average value	Reduction rate (%)
COD_{cr}	Inlet	15	15	25	17	12	14	15	15	16.0	33.3
	Outlet	11	11	13	17	9	3	10	10	10.5	
BOD_5	Inlet	<2	<2	<2	<2	<2	<2	<2	2.35	/	/
	Outlet	<2	<2	<2	<2	<2	<2	<2	<2	/	
NH_3-N	Inlet	0.79	0.75	0.79	0.45	0.32	0.37	0.25	0.61	0.541	77.0
	Outlet	0.09	0.12	0.16	0.35	0.08	0.22	0.12	0.14	0.160	
TP	Inlet	0.16	0.18	0.14	0.07	0.11	0.13	0.10	0.16	0.131	43.8
	Outlet	0.07	0.09	0.09	0.06	0.07	0.03	0.07	0.09	0.071	
SS	Inlet	29	42	31	10	19	26	21	33	26.4	93.9
	Outlet	1	3	5	3	2	6	4	2	3.3	

Table 4. Standard limits of surface water environmental quality standards (unit: mg/L).

Index	COD_{Cr}	BOD_5	NH_3-N	TP
Class I	$\leq$15	$\leq$3	$\leq$0.15	$\leq$0.02
Class II	$\leq$15	$\leq$3	$\leq$0.5	$\leq$0.1
Class III	$\leq$20	$\leq$4	$\leq$1.0	$\leq$0.2

6 CONCLUSIONS

Through the research on the application and purification effect of horizontal subsurface constructed wetland in the treatment of initial rainwater, the conclusions are obtained as below:

(1) The horizontal subsurface constructed wetland can be used to purify the initial rainwater and polluted river water, enhance the urban landscape, conform to the concept of "sponge city", and be an effective measure of urban water environment control.
(2) The wetland effluent reaches the surface water class II standard, far exceeding the design requirements, and can directly discharge the natural water body.

(3) After biochemical treatment, the initial rainwater and river water had no significant difference in the values of the five indexes.
(4) After treatment, except for BOD_5, which has a low inflow time and a low outflow rate, the other four indexes of the initial rainwater and river water have been significantly reduced. Among them, the reduction rate of SS is the largest, exceeding 90%, followed by NH_3-N, TP and CODcr, with the lowest reduction rate of 33.3%. It indicates that the wetland has a very significant effect on sewage purification.

REFERENCES

Code for design of environmental protection of chemical industry projects (GB 50483-2019).

Environmental quality standards for surface water (GB 3838-2002).

Knox A S, Nelson E A, Halverson N V, et al. (2010) Long-Term Performance of a Constructed Wetland for Metal Removal. *Soil and sediment Contamination*, 19(6): 667–685.

Lu Wei-Wei, Jiang Ming. (2009) Simulation of calamagrostis wetland's purification for phosphorus in sewage. *Wetland Science*, 7(1): 5–9.

Niu Xiao-Yin, Chang Jie, Ge Ying, et al. (2004) The role of lolium perenne in constructed wetland ecological engineering for eutrophicate water purification. *Wetland Science*, 2(3): 202–207.

Ren Jun, Fu Zhao-Wen, TAO Ling, YANG Qian. (2011) Accumulation Effect of Phragmites australis, Acorus calamus and Scirpus tabernaemontani on Zn2+ in Water Body. *Wetland Science*, 9(4): 322–327.

Vymazal J. (2007) Removal of nutrients in various types of constructed wetlands. *Science of the Total Environment*, 380(1): 48–65.

Energy Revolution and Chemical Research – Kok-Keong Chong and Zhongliang Liu (Eds)

Numerical analysis of water environment improvement of river and lake based on two-dimensional flow dynamics and numerical simulation

Yanfen Yu
Zhejiang Province Qiantang River Basin Center, Zhejiang, China

Pingping Mao*
Zhejiang Qiantang River Seawall Property Management Co., Ltd., Zhejiang, China

Lili Li, Fuqing Bai, Yishan Chen, Feng Xie, Le Yang & Mei Chen
Key Laboratory for Technology in Rural Water Management of Zhejiang Province, Zhejiang University of Water Resources and Electric Power, Hangzhou, China

ABSTRACT: Shaoxing coastal plain has dense river and lake water networks, poor river network fluidity, deteriorating water quality, and frequent eutrophication of river and lake water bodies. Water diversion outside the region and connection of river and lake water systems are important measures to improve the water environment of plain rivers and lakes. To demonstrate the effect of water diversion on the improvement of river and lake water environment, the plane two-dimensional water environment mathematical model is used to calculate the water environment of Baima Lake before and after water diversion. From the perspective of water quality indexes BOD and DO, the water quality after water diversion is improved from Class V water to Class III water, the improvement effect of water quality is obvious, and the connection plan is feasible.

1 INTRODUCTION

The comprehensive renovation project of six lakes in Yudong plain is a key water conservancy project in Zhejiang Province. Among them, Baima Lake has a long and narrow terrain, and there are many scattered sandbars and islands in its hinterland. See Figure 1. Due to the poor circulation of the water body in Baima Lake, the local water body is highly eutrophic, and the water quality reaches class V water. To improve the water environment of Baima Lake, combined with the local water resources, the connection between water diversion outside the region and the river lake water system is a more feasible measure to improve the water environment of Baima Lake (Gao et al. 2015; Liu et al. 2014; Tong et al. 2016; Yang et al. 2018). However, the effect of water diversion improvement still needs to be demonstrated by the feasibility study. Using a two-dimensional mathematical model of the water environment is an important tool and means to improve the water environment (Wu & He 2019). By establishing a two-dimensional numerical model, this paper divides Baima Lake into eight regions, and determines whether the connection plan is feasible by studying the changes of water environment, indexes BOD and DO in each region after transferring water (Chen et al. 2015, 2012; Chen & Ying 2016; Cui et al. 2017).

*Corresponding Author: maopingping@hzsteel.com

DOI 10.1201/9781003332657-37

Figure 1. Location map of Baima Lake.

2 ESTABLISHMENT AND VERIFICATION OF PLANE 2D WATER ENVIRONMENT MATHEMATICAL MODEL

2.1 *The basic equation of plane two-dimensional water environment mathematical model*

Water quality equation:

$$\frac{\partial hC}{\partial t} + \frac{\partial uhC}{\partial x} + \frac{\partial vhC}{\partial y} = \frac{\partial}{\partial y}\left(E_x h\frac{\partial C}{\partial x}\right) + \frac{\partial}{\partial y}\left(E_x h\frac{\partial C}{\partial y}\right) + S + F\left(C\right)$$

Where C is the pollutant concentration (mg/L); u, v is the velocity component in X and Y directions respectively; E_x and E_y are diffusion coefficients in X and Y directions respectively (m/s^2); S is the source and sink item ($g/m^2/s$); F(C) is biochemical reaction term.

2.2 *Establishment of model*

2.2.1 *Model scope*

The river network and water system are a whole, so we need to take the whole river network in Yudong, Shaoxing as a model. Because the six lakes are located in the middle of the Shangyu river network, there are too many uncertain factors in the boundary conditions of inflow and outflow of its model. Taking the river network of Shangyu plain as a whole system can determine the correct boundary conditions. In this way, the rainfall in the southern mountainous area, the diversion of Shangpu sluice, the diversion of Cao'e River, the diversion of Siming lake, as well as the No. 2 sluice and Xindong Inlet sluice can be taken as the boundary conditions, so that the model is established based on correct boundary conditions. The research area, grid division, and boundary conditions of the model are shown in Figure 2. The water area is divided into 140,552 grids and 112,641 grid nodes.

2.2.2 *Model validation*

In this verification calculation, "20131006" Typhoon Fitow with relatively recent flood and waterlogging, typical rainfall and complete measured data is used for verification calculation. The

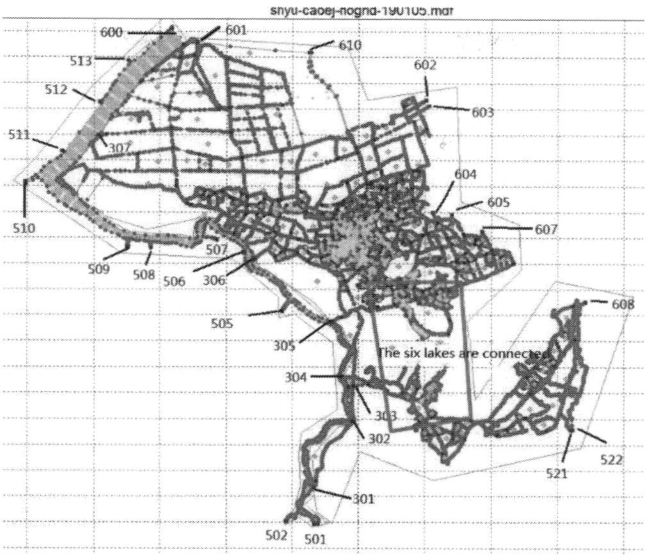

Figure 2. Study area, mesh generation and boundary conditions of the model.

verification calculation of the maximum water level and water level process of each representative station is relatively consistent with the actual situation, so it is considered that the calculation model is reliable and can be used for the calculation of various design conditions.

3 IMPACT OF RIVER LAKE CONNECTION PROJECT ON WATER QUALITY OF BAIMA LAKE

Because Baima Lake is too narrow and long, and the lake body is widely distributed, in order to facilitate the analysis of water flow and water environment improvement in each region, Baima Lake is divided into 8 regions, all of which are marked by a to h, as shown in Figure 3.

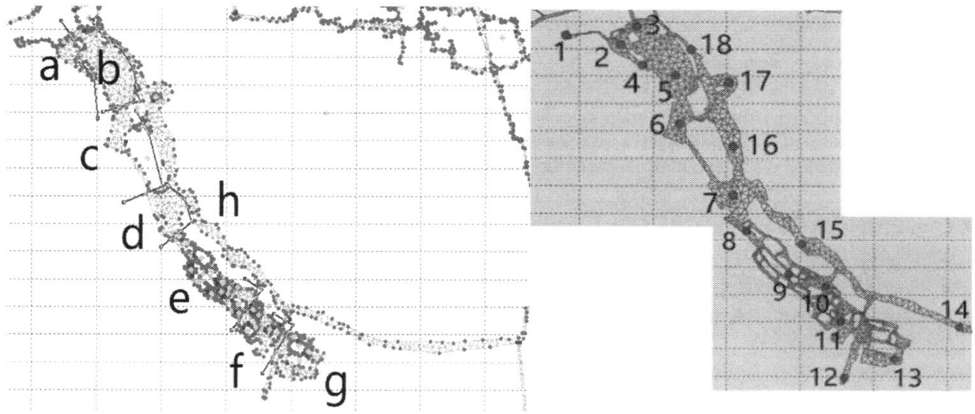

Figure 3. Baima Lake zoning map and analysis location map.

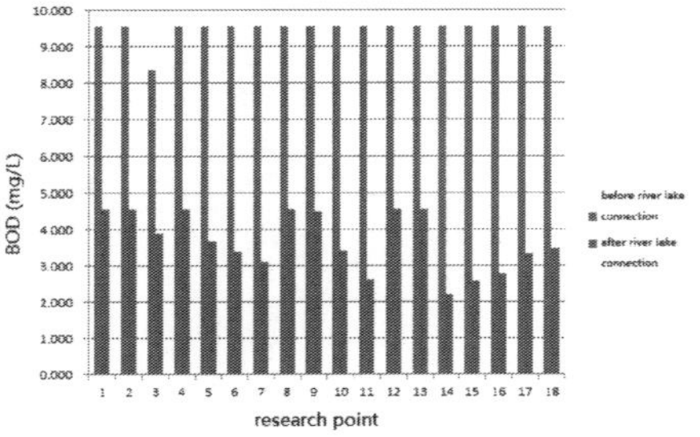

Figure 4. Comparison of BOD value before and after river lake connection.

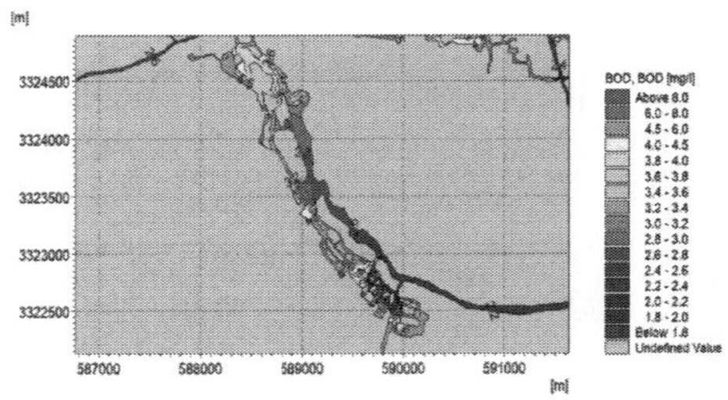

Figure 5. BOD value contour map after river lake connection.

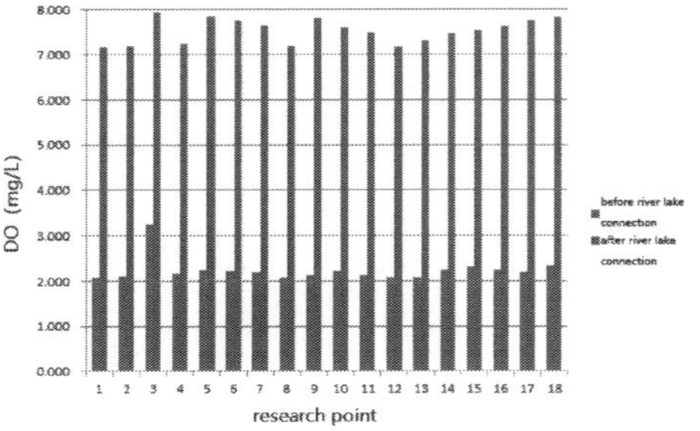

Figure 6. Comparison of DO value of research points before and after connection.

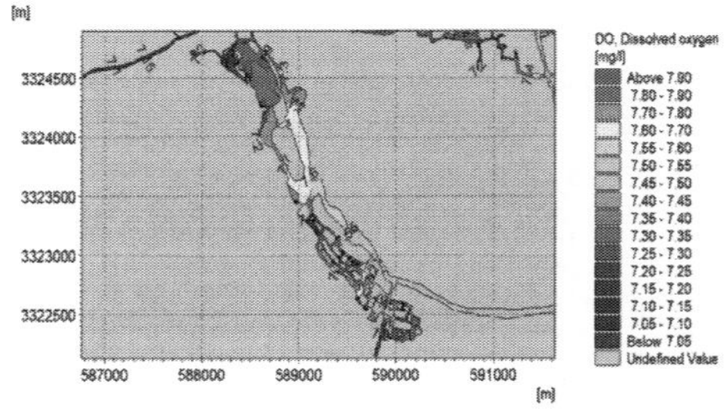

Figure 7. DO contour map of Baima Lake after river lake connection.

BOD (Biochemical Oxygen Demand) concentration is set to be 10 mg/L and DO (Dispersed Oxygen) concentration is 2 mg/L in the initial conditions option of the eco lab module.

3.1 *BOD analysis*

Figure 4. shows the comparison of BOD value of the research points before and after the river lake connection. Figure 5 shows the BOD value distribution of Baima Lake in step 444 after the river lake connection. As can be seen from Figure 4 and Figure 5, after Baima Lake is connected with the river and lake, the BOD value decreases significantly and the water quality improves significantly. The BOD value is all reduced to less than 5 mg/L and the lowest BOD is 2.2 mg/L. However, the BOD value at various locations is also quite different and the BOD value is unevenly distributed.

3.2 *Do analysis*

Figure 6 shows the comparison of DO value of research points before and after the river lake connection, and Figure 7 shows the distribution of the DO value of Baima Lake after the river lake connection. It can be seen from Figure 6 and Figure 7 that after Baima Lake is connected with the river and lake, the DO value increases significantly, the DO of the basin is generally greater than 7 mg/L, and the water quality is improved significantly, but the DO value at various locations is also quite different, and the distribution of DO value is uneven. Combined with the data of BOD and DO of the research points, the conclusions are summarized as follows.

In Region a, the water quality has improved significantly, and the BOD value and DO value are relatively evenly distributed in the whole basin. The BOD value decreased from about 9.57 mg/L to about 4.57 mg/L, and the DO value increased from about 2.1 mg/L to about 7.2 mg/L. According to the indexes analysis, the water quality has been improved to Class IV water. However, compared with other watersheds, the water quality is still relatively poor and needs to be further improved.

In Region b, the water quality has improved significantly. The BOD value is relatively evenly distributed in the whole basin, and the BOD value is reduced from about 9.56 mg/L to 3.5-4 mg/L. The DO value rises from about 2.2 mg/L to 7.8-8.0 mg/L. In the whole basin, the upper half of the basin is better and the lower half of the basin is lower. Combined with the indexes analysis, the water quality has been improved for Class III water.

In Region c, the water quality has improved significantly. The BOD value is low at the water inlet. The farther away from the water inlet, the higher the BOD value. The BOD value decreases from about 9.56 mg/L to 3-3.7 mg/L. The DO value rises from about 2.2 mg/L to 7.7-7.8 mg/L. The distribution of DO value in the basin is relatively uniform. Combined with the indexes analysis, the water quality has been improved for Class III water.

In Region d, the water quality has improved significantly and the BOD distribution is uneven. The BOD value near the H basin (right bank) is small, which is 2.17 mg/L. The farther away from the H basin, the greater the BOD value, which increases to 4.55 mg/L. The DO value increases from about 2.2 mg/L to 7.45-7.7 mg/L, and the distribution of DO value is relatively uniform.

In Region e, water quality has improved significantly, but BOD value and DO value are unevenly distributed in the whole basin. At the water inlet connected with the H basin, BOD value is about 3.7 mg/L, the DO value is about 7.6 mg/L, and the water quality is good. Because there are many hinterlands of the e basin and the water fluidity is not strong, the water quality in the hinterland is worse. The BOD value is as high as 4.57 mg/L and the DO value is as low as 7.1 mg/L. The water quality still needs to be improved.

In Region f, the water quality has improved significantly, but the BOD value is not evenly distributed in the whole basin. At the water inlet connected with the H basin, the BOD value is about 3 mg/L, the DO value is about 7.6 mg/L, and the water quality is good. The basin closest to the left bank has the largest BOD value, which is 4.6 mg/L and the DO value is 7.4 mg/L. The water quality of the basin closest to the left bank needs to be continuously improved.

In Region g, water quality has improved significantly, but BOD value and DO value are unevenly distributed in the whole basin. At the water inlet connected with the H basin, the BOD value is about 2.25 mg/L, the DO value is about 7.5 mg/L, and the water quality is good. As there are many hinterlands in the G basin and the fluidity of the water body is not strong, the water quality in the hinterlands is worse. The BOD value is as high as 4.57 mg/L and the DO value is as low as 7.1mg/L. The water quality still needs to be improved.

In Region h, the water quality is significantly improved. The water quality is the best near the Zaoli Lake and Baima Lake diversion tunnel. The BOD value is as low as 15 mg/L. As it flows northward, the water quality changes more and more, and the water quality becomes worse and worse. The BOD value increases to 3.5 mg/L, while the DO value also decreases along the way, from 7.47 mg/L to 7.82 mg/L.

4 CONCLUSION

This paper establishes and verifies the plane two-dimensional water environment mathematical model, and calculates the water environment of Baima Lake before and after water diversion. From the water quality indexes BOD and DO, the water quality after water diversion is improved from Class V water to Class III water. The research results show that the improvement effect of water quality is obvious, and the connection plan is feasible.

ACKNOWLEDGMENTS

This research was supported by the Joint Funds of the Zhejiang Provincial Natural Science Foundation of China (No. LZJWZ22C030001; No. LZJWZ22E090004); the Funds of Water Resources of Science and Technology of Zhejiang Provincial Water Resources Department, China (No.

RB2115); the National Key Research and Development Program of China (No. 2016YFC0402502); and the National Natural Science Foundation of China (No. 51979249).

REFERENCES

Chen, Z.H., Chen, X.H., Du, J. & Xiong, Y.J. (2012) Regulation and effect prediction of water environment diversion in river network area. *Water resources protection.*, 28(03): 16–21.
Chen, Z.T., Hua, L. & Jin, Q.N. (2015) Study on effect evaluation of water diversion to improve water quality of urban river network. *Journal of Changjiang Academy of Sciences.*, 32(07): 45–51.
Chen, F. & Ying, X.L. (2016) Research and application of river network model in Shaoxing Plain. *Journal of Zhejiang University of water resources and hydropower.*, 28(05): 43–47.
Cui, G.B., Chen, X., Xiang, L., Zhang, Q.C. & Xu, Q. (2017) Effect evaluation of water system connection on improving water environment in plain river network area. *Journal of water conservancy.*, 48(12): 1429–1437.
Gao, Q., Tang, Q.H. & Meng, Q.Q. (2015) Evaluation on improvement effect of connecting water environment of tidal river and lake water system. *People's Yangtze River.*, 46(15): 38–40 + 50.
Liu, B.J., Deng, Q.L. & Zou, C.W. (2014) Study on the necessity of river lake water system connection project. *People's Yangtze River.*, 45(16): 5–6+11.
Tong, Y.Y., Li, D.F. & Nie, H. (2016) Study on hydrodynamic and plane two-dimensional mathematical model of river network in Datian Plain. *Journal of Zhejiang University of water resources and hydropower.*, 28(01): 14–17.
Yang, W., Zhang, L.P., Li, Z.L., Zhang, Y.J., Xiao, Y. & Xia, J. (2018) Study on river lake connection scheme of urban lake group based on water environment improvement. *Journal of Geography.*, 73(01): 115–128.
Wu, Y. & He, L.Q. (2019) Discussion on constructing hydrodynamic model of Cao'e River flow area in Shaoxing City. *Zhejiang water conservancy science and technology.*, 47(03): 21–23 + 31.

Research on energy chemical properties and material structure

Energy Revolution and Chemical Research – Kok-Keong Chong and Zhongliang Liu (Eds)
© 2023 The Authors, ISBN: 978-1-032-36554-1

Study on detoxification property of H_2O_2 decontaminants against GD under subzero environment

Shaohua Wei, Hongpeng Zhang, Shaoxiong Wu, Haiyan Zhu*, Lianyuan Wang, Nan Xiang, Yuefeng Zhu & Zhenxing Cheng
Institute of Chemical Defence, Beijing, China

ABSTRACT: Compared with traditional decontaminants and decontamination technologies, the hydrogen peroxide decontaminant shows better decontamination effect and development prospects in the subzero environment. To improve the nucleophilic performance of H_2O_2decontaminants in decontamination against GD, the influence factors such as the addition of K_2MoO_4, pH, solvent types, and H_2O_2 concentration in the H_2O_2system had been investigated in this paper. The results showed that K_2MoO_4 had no obvious effect on the nucleophilic performance of the system. The decontamination reaction rate was increased significantly with the addition of a small amount of alkali. For the alkali-H_2O_2 decontaminant with a pH greater than 8, the decontamination ratio against GD was above 99.9% for 30 minutes. Replacing Ethylene glycol (EG)with propylene carbonate (PC) as the solvent, the decontamination reaction activity remained, and the freezing point of the solution was reduced to −35°C. Reducing H_2O_2 concentration had a negative effect on the reaction activity.

1 INTRODUCTION

Chemical warfare agents (CWAs) are highly toxic and have a great hazard to personnel. Although most stocks of CWAs have been destroyed, they still appeared in some local wars or terrorist attacks. Decontamination, as one of the important chemical defense technologies, should be improved to deal with various situations, and decontaminants play an extremely important role in the professional decontamination process (Ganesan et al. 2010; Kim et al. 2011; Richardt & Blum 2008). When troops are attacked by CWAs and the decontamination cannot be carried out in time under a subzero environment, the liquid or solid poisons attached to vehicles, equipment, or clothing will have a great risk of transfer, causing greater harm to personnel. Therefore, it is necessary to study the low-temperature decontamination performance of decontaminants.

GD (Soman, Pinacolyl methylphosphonofluoridate) is a lethal organophosphorus nerve agent that can inactivate acetylcholinesterase in the human nervous system (Young & Watson 2020). The main decontaminants used against GD include alkalis, hypochlorites, and peroxides. Water-based hypochlorite decontaminants were limited in the range of application at low temperatures due to their high freezing points. Alkali-alcohol-amine system decontaminants that can be used in subzero environments were highly corrosive with their strong alkalinity. Peroxide decontaminants were developed rapidly in recent years due to their high reactivity against CWAs through adding additives, activators, and low corrosivity to metals, rubbers, and plastics. Typical examples are DF-200 (Tucker 2009), and Decon Green developed in the United States, in which the main active ingredient is H_2O_2.

At present, activators of H_2O_2 mainly included metal ions, metal oxyacid salts, and NH_3, et al. Among them, metal oxyacid salts (Guidotti et al. 2012; Wagner & Yang 2001; Zhang et al.

*Corresponding Author: zhuhyuse@163.com

DOI 10.1201/9781003332657-38

2011; Zhao & XI 2018) were usually paid more attention to. Domestic research teams have also conducted systematic studies on the reaction mechanism of the activated H_2O_2solutions, but those were all carried out under room temperature and only from the level of reaction performance (Zhao et al. 2020, 2015, 2014). In this work, the influence regulation of H_2O_2 decontamination's nucleophilic performance against GD, such as the addition of K_2MoO_4, pH, solvent types, and H_2O_2 concentration was studied under a subzero environment.

2 EXPERIMENTAL

2.1 *Chemicals*

The toxic agents of GD had a purity >95%. It was handled only by well-trained personnel using appropriate safety procedures because of its high toxicity.

30% (wt%) Hydrogen peroxide (H_2O_2) and 28% (wt%) Ammonia (NH_3) were collected from Beijing Chemical Plant. The agents, purities, and sources are as follows: Propylene Carbonate (PC), Chemical Purity, Sinopharm Chemical Reagent Inc; Ethanolamine (MEA), Analytical Purity, West Asia Chemical Inc (Shandong, China); Sodium Sulfate (Na_2SO_3) and Sodium Hydroxide (NaOH), Analytical Purity, Tianjin Guangfu Fine Chemical Research Institute; Dichloromethane (CH_2Cl_2), Chromatographic Purity, Meryer Reagent Company.

2.2 *Droplet detoxification experiments*

The droplet detoxification experiments were done in test tubes of 25mL with a stopper. Firstly, 20 μL GD in 25mL test tubes and the pre-prepared decontaminants in conical flasks were put in a low and constant temperature stirring reaction bath (DHJF-8002) at a preset temperature for 30 minutes. Secondly, a 1mL decontaminant was taken from the conical flask and added to the test tube. After reacting for a certain time under stirring at 800 r/min, 6 mL 15% Na_2SO_3 and 10 mL extractant CH_2Cl_2 were added, and shaken for 20 s immediately. After standing for 30 min, the extractant layer solution was taken for analysis by diluting the corresponding multiples.

2.3 *The detection of GD*

GD was detected by GC-FPD through the Internal standard Method (30uL triethyl phosphate as an internal standard in a 1mL sample). The GD residual concentration was quantitatively analyzed by Agilent HP 7890 equipped with an HP-5 capillary column (30 m$\times$0.32 mm$\times$0.25 μm), employing a temperature program (60-280°C at 15°C /min^{-1}) and 60.0 mL/min N_2 as carrier gas.

3 RESULTS AND DISCUSSION

3.1 *The effect of pH on decontamination against GD*

NH_3 (28wt%) with a low freezing point was chosen as the pH adjuster. By adding different amounts of NH_3 to H_2O_2 (30wt%), NH_3-H_2O_2 decontaminants with different pH (determined by HI2221 HANNA pH /ORP tester) were prepared. Compared with 20% ammonia decontaminant, their decontamination effects against GD at −20°C were shown in Table 1.

It can be seen from the results in Table 1 that when NH_3 and H_2O_2 were mixed in a certain volume ratio, the pH was lower than 20% $NH_3{\cdot}H_2O$, but their decontamination effects against GD had been significantly better than 20% $NH_3{\cdot}H_2O$. When the volume ratio of NH_3 to H_2O_2 decreased from 4:1 to 1:5, the pH dropped from 11.38 to 9.54, but the decontamination ratio for 5 minutes remained no less than 99.0% and even reached 99.9%. The pH of 20% $NH_3{\cdot}H_2O$ was about 12.47, but the decontamination ratio against GD was 90.4% for 30 min, and the reaction rate constant k was 0.08 min^{-1}, which was significantly lower than NH_3-H_2O_2 decontaminants. For the

Table 1. The decontamination effect of several NH_3-H_2O_2 decontaminants against gd and their freezing points.

Decontaminant V_{NH3}:V_{H2O2}	pH	Decontamination ratio (%) 5 min	30 min	Reaction rate constant k (min^{-1})	Freezing point (°C)
4:1	11.38	99.0	>99.9	>9.2×10^{-1}	−21.0~-22.0
1:1	10.80	>99.9	>99.9	>1.4	−20.0~-21.0
1:5	8.76	>99.9	>99.9	>1.4	−22.0~-23.0
20%$NH_3\cdot H_2O$	12.47	32.4	90.4	7.8×10^{-2}	−23.0~-24.0
$H_2O_2^a$	4.50	–	5.8	1.9×10^{-4}	<−25

(The reaction rate constant k was calculated according to the first-order reaction rate formula, a: H_2O_2 solution with PC, the volume ratio of H_2O_2 to PC was 6:4).

H_2O_2 decontaminant without NH_3, due to the low value of pH, the decontamination ratio against GD was poor at only 5.8% after decontaminating for 30 min. In NH_3-H_2O_2, the higher the content of NH_3 was, the greater the pH was, the stronger the corrosivity of the decontaminant was, and the easier decomposition of $H_2O_2$was. So, the amount of alkali should be reduced as far as possible without significantly reducing the detoxification property of H_2O_2 decontaminant against GD. The data in Table 2 also shows that the freezing points of H_2O_2 decontaminants without adding solvent were above −23°C. Therefore, to improve the stability of H_2O_2 decontaminants and reduce their corrosivity and freezing point, the decontamination effects against GD of H_2O_2 decontaminants (pH$\leq$9) that contain a certain amount of solvent and low alkali content (NH_3, MEA, and NaOH) varied with pH was investigated. Meanwhile, to suit the needs of universal decontamination, the decontamination effects of H_2O_2 decontaminants that contain K_2MoO_4 were also compared. The results were shown in Table 2 and Figure 1.

Table 2. The decontamination effect of several H_2O_2 decontaminants against GD and their freezing points.

Formulas System[a]	Amount of alkali (%)	pH	Decontamination ratio (%) 1 min	30 min	Reaction rate constant k (min^{-1})	Freezing Point (°C)
MEA-H_2O_2	0	4.50	–	5.8	2.0×10^{-3}	<-25
	0.5	6.80	25.2	82.5	2.9×10^{-1}	<-25
	1	8.25	89.2	>99.9	>2.2	<-25
	2	8.54	94.2	>99.9	>2.8	<-25
	4	8.89	95.9	>99.9	>3.2	<-25
MEA-MoO_4^{2-}-$H_2O_2^b$	0	6.40	10.0	46.8	1.0×10^{-1}	<-25
	1	8.25	82.8	>99.9	>1.8	<-25
	3	8.55	99.3	>99.9	>5.0	<-25
	4	9.00	98.3	>99.9	>4.1	<-25
NH_3-MoO_4^{2-}-$H_2O_2^b$	0	6.40	10.0	46.8	1.0×10^{-1}	<-25
	0.5	7.75	68.5	99.9	>1.2	<-25
	2	8.38	85.6	>99.9	>1.9	<-25
	4	8.65	99.2	>99.9	>4.8	<-25
	6	9.00	99.6	>99.9	>5.5	<-25
NaOH-MoO_4^{2-}-$H_2O_2^b$	4[c]	9.00	96.3	>99.9	>3.3	<-25

(The reaction rate constant k was calculated according to the first-order reaction rate formula, a: H_2O_2 solution with PC, the volume ratio of H_2O_2 to PC was 6:4, b: The added content of K_2MoO_4 was 0.04 M, c: NaOH solution was prepared to the concentration of 10M in advance).

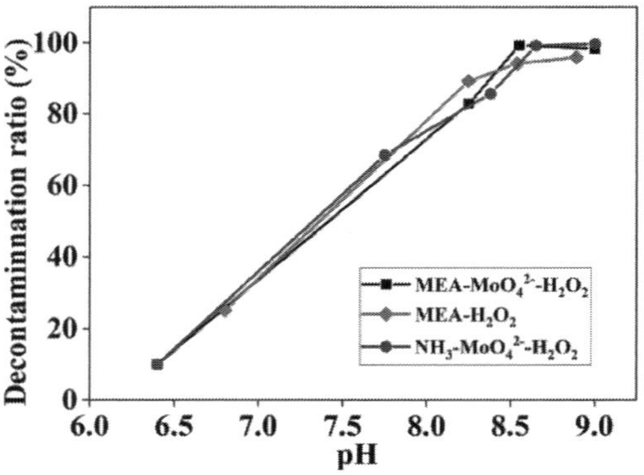

Figure 1. The decontamination effects of several H_2O_2 decontaminants against GD varied with pH. (PC as a solvent, the added content of K_2MoO_4 was 0.04M, a: An amount of alkali was added according to their pH values, $T = -20°C$, $t = 1$ min).

MEA was selected to adjust the pH in the H_2O_2decontaminants without catalyst. The results showed that the decontamination ratio was increased significantly by the addition of a small amount of alkali. For example, when 0.5% MEA was added, the pH increased from 4.5 to 6.8, and the reaction rate constant k was increased more than 100 times from 2.0×10^{-3} to 2.9×10^{-1}. The increase of pH in H_2O_2 decontaminant enhanced its decontamination effects against GD gradually. There was no residual GD detected (decontamination ratio was above 99.9%) when pH was greater than 8.25 after decontamination for 30 minutes.

The results above showed that the nucleophilic performance of the H_2O_2 solution was improved significantly with the addition of a small amount of alkali (such as 1% MEA), achieving the ideal detoxification property against GD relatively. Compared with MEA-H_2O_2 without catalyst, the decontamination activity of MEA-MoO_4^{2-}-H_2O_2 was unchanged significantly. For several alkali-modified activation formulations with a pH of 9, NH_3-MoO_4^{2-}-H_2O_2 showed a slightly better decontamination effect against GD for 1 min than MEA-MoO_4^{2-}-H_2O_2 and NaOH-MoO_4^{2-}-H_2O_2. It may be that the pH of decontaminants decreased during the nucleophilic reaction, while NH_3 had a better pH buffering effect than the other two alkalis on weakly alkali conditions. The decontamination ratios of three decontaminants with catalyst (pH=9) were all above 99.9% for 30 minutes.

3.2 *The effect of solvents on decontamination against GD*

The nucleophilic reaction activity against GD of three decontaminants with PC, EG, and no solvent were compared and the results were shown in Table 3.

Table 3. The effects of solvents on decontamination against GD in NH_3-H_2O_2 decontaminants and freezing points.

Decontaminants	pH	Decontamination ratio (%) 1 min	Decontamination ratio (%) 30 min	Reaction rate constant k (min^{-1})	Freezing Point (°C)
NH_3-H_2O_2-EG[a]	9.00	>99.9	>99.9	>6.9	−35
NH_3-H_2O_2-PC[a]	9.00	>99.9	>99.9	>6.9	−25
NH_3-H_2O_2[b]	9.00	>99.9	>99.9	>6.9	−22

(T=-20±1 °C, the reaction rate constant k was calculated according to the first-order reaction rate formula, a: The volume ratio of H_2O_2 to solvent (EG or PC) was 6:4, and the content of NH_3 was 10%, b: the content of NH_3 was 20%).

It can be seen in Table 4 that replacing the solvent PC with EG, the decontamination ratio against GD remained above 99.9% for 1 min at −20°C. It indicated that the decontamination reaction activity did not decrease obviously after the solvent was replaced by EG, and the freezing point had dropped from −25 °C to −35 °C. Even the 20% NH_3-H_2O_2 decontaminant (without solvent) showed an ideal decontamination effect for the decontamination ratio against GD reached above 99.9% for 1 min at −20 °C. But compared with the decontaminant containing 40% EG, its freezing point rose from −35 °C to −22 °C, which limited its range of application at low temperatures. Under the maintaining reaction activity, the addition of EG realized the reduction of the amount of ammonia and H_2O_2 added.

3.3 *The effect of H_2O_2 concentration on decontamination for GD*

30% H_2O_2 was replaced with 3% H_2O_2 diluted by water, and EG as a solvent, pH adjusting by NH_3 or MEA, the decontamination effects of decontaminants against GD were investigated at different pHs. The results were shown in Table 4. After diluting H_2O_2 concentration 10 times to 1 M, the pH increased significantly to 10 above, while the decontamination ratio against GD for 5 minutes decreased slightly only, remaining 90% above still, reaching 99.9% above for 30 minutes. With the addition of solvent EG, the freezing point all reached lower than −30°C.

Table 4. The decontamination effect against GD at different pH in diluted H_2O_2 decontaminants.

Decontaminant[a]	Initial Content of H_2O_2 (wt%)	pH	Decontamination ratio (%) 5 min	30 min	Freezing Reaction rate k (min^{-1})	constant Point (°C)
6% NH_3-H_2O_2	30	9.27	99.6	>99.9	>1.1	<-35
6% NH_3-H_2O_2	3	10.48	96.6	>99.9	>0.7	<-35
2% NH_3-H_2O_2		10.01	95.2	>99.9	>0.6	<-35
6% MEA-H_2O_2		10.82	98.7	>99.9	>0.9	−36
4% MEA-H_2O_2		10.71	99.2	>99.9	>1.0	−33
2% MEA-H_2O_2		10.57	93.4	>99.9	>0.5	−33

(The reaction rate constant k was calculated according to the first-order reaction rate formula, a: Taking EG as a solvent, the volume ratio of H_2O_2 to EG was 6:4).

4 CONCLUSION

This paper studied the influence factors such as K_2MoO_4, pH, solvent types, and H_2O_2 concentration on the nucleophilic reaction performance of H_2O_2 decontaminants in decontamination against GD. Results showed that alkali was the main factor that regulates the nucleophilic reaction rate. The detoxification property against GD increased significantly by adding a small amount of alkali in H_2O_2 decontaminants. Alkali-H_2O_2 decontaminants with pH>8 reached the decontamination ratio of 99.9% for 30 min at −20 °C. K_2MoO_4 had no obvious effect on the nucleophilic performance of H_2O_2 decontaminants. The solvent mainly lowered the freezing point and had no significant effect on nucleophilic reaction activity. Taking EG as a solvent instead of PC, the decontamination reaction activity improved slightly, and the freezing point was significantly reduced to -35 °C. But the reaction activity decreased when the content of H_2O_2 was reduced. When H_2O_2 concentration decreased to 1 M, the decontamination ratio was above 90% after decontaminating for 5 min.

REFERENCES

Ganesan K, Raza S K, Vijayaraghavan R. Chegemical warfare ants[J]. *Journal of pharmacy and bioallied sciences*, 2010, 2(3): 166–178.

Guidotti M, Rossodivita A, Ranghieri M C. *Nano-Structured Solids and Heterogeneous Catalysts: Powerful Tools for the Reduction of CBRN Threats*[J]. Springer Netherlands, 2012, (1): 278–284.

Kim K, Tsay O G, Atwood D A, et al. Destruction and detection of chemical warfare agents[J]. *Chemical reviews*, 2011, 111(9): 5345–5403.

Richardt A, Blum M M. *Decontamination of Warfare Agents: Enzymatic Methods for the Removal of B/C Weapons*[M]. 2008.

Tucker M D. DF-200: *A Broad Spectrum Decontaminant for Biological Warfare Agents, Biological Pathogens, and Other Toxic Materials*[C]. The Annual Meeting and Exhibition, 2009.

Wagner G W, Yang Y C. *Universal decontaminating solution for chemical warfare agents* [P]. US, 6245957. 2001.

Young R A, Watson A, *Handbook of Toxicology of Chemical Warfare Agents* (Third Edition)[M]. Academic Press, 2020, 97–126.

Zhang L, XI H L, Wang Q, et al. Kinetics and Mechanism of the Degradation Reaction of 2-chlororthyl Ethyl Sulfide by Sodium Percarbonate/Sodium Molybdate[J]. *Environmental Chemistry*, 2011, (10): 1695–1699.

Zhao S P, XI H L. Rapid activation of basic hydrogen peroxide by borate and efficient destruction of toxic industrial chemicals (TICs) and chemical warfare agents (CWAs)[J]. *Journal of Hazardous Materials*, 2018, 367(1): 91–98.

Zhao S P, Zhu Y B, XI H L, et al. Detoxification of mustard gas, nerve agents and simulants by peroxomolybdate in aqueous H_2O_2 solution: Reactive oxygen species and mechanisms[J]. *Journal of Environmental Chemical Engineering*, 2020, 8(5): 104221.

Zhao S P, XI H L, Zuo Y J, et al. Oxidation Kinetics and Products of Methyl Phenyl Sulfide, a Sulfur Mustard Simulant by Sodium Molybdate Catalyzed Hydrogen Peroxide Solution[J]. *Journal of Molecular Catalysis (China)*, 2015, 29(1): 45–51.

Zhao S P, XI H L, Wang Q. Two-Step Oxidation of Thioanisole as Yperite (HD) Simulant by Modified aqueous H2O2 Solution[J]. *Environmental Chemistry*, 2014, 33(9): 1546–1552.

Energy Revolution and Chemical Research – Kok-Keong Chong and Zhongliang Liu (Eds)

Experimental investigation on pyrolysis and combustion of sewage sludge: Pyrolysis and combustion characteristics, product characteristics

Chengxin Wang*, Shunyong Wu & Guanghui Zhang
Hefei General Machinery Research Institute Co. Ltd, Hefei, Anhui, China

ABSTRACT: In this paper, pyrolysis and combustion characteristics are investigated by Thermogravimetric-Infrared spectroscopy (TG-FTIR). And the characteristics of volatile products during pyrolysis and combustion were discussed in comparison with the characteristic functional groups and the mass-to-charge ratio of volatile products. The effects of different atmospheres were compared by combining the weight loss process and volatile evolution. The results show that there are two significant weight loss peaks in the process of weight loss and the characteristic temperature of combustion is earlier than that of pyrolysis. The presence of oxygen results in a much lower total volatile organic matter combustion than pyrolysis.

1 INTRODUCTION

Sewage sludge is a by-product of treating urban domestic sewage in the sewage treatment plant. With the intensification of urbanization, it is predicted that the production of sewage sludge with a water content of 80% in China will reach 60-90 million tons in 2020(Wang et al. 2019). Sewage sludge is not only easily putrefied, but also contains a lot of toxic substances including pathogenic microorganisms, heavy metals, non-biodegradable organics, dioxins, etc., which increased sludge production.

At present, the environmental pollution problems caused by sludge are mainly related to improper disposal methods. Thermochemical conversion is an efficient method for energy recovery and environmental protection (Cieślik et al. 2014). Based on volatile matter and calorific value, sludge can be considered an inferior fuel, and the goal of producing high-value-added products and recovering energy is achieved through thermochemical conversion techniques including pyrolysis, combustion, and co-firing (Magdziarz & Wilk 2013; Soria-Verdugo et al. 2013). Therefore, before developing various methods for sludge treatment, it is necessary to fully understand the weight loss characteristics, volatile evolution characteristics, and solid product properties of sludge itself as fuel when it is burned or pyrolyzed at different temperatures(Lin et al. 2017). Pyrolysis, as the initial stage of the combustion process, plays a decisive role in ignition, flame holding, product distribution, and burnout (Chen et al. 2015; Selcuk & Yuzbasi 2011).

For most comparative studies of combustion and pyrolysis, the effect of different atmospheres on thermal degradation cannot be explained. Therefore, the experimental methods of thermogravimetric-infrared spectroscopy-mass spectrometry and solid characterization were used to compare and study the weight loss characteristics, volatile evolution characteristics, and solid product properties during the combustion and pyrolysis of sludge. Theoretical guidance is provided for the thermochemical conversion and utilization of sludge.

*Corresponding Author: wcx2046@mail.ustc.edu.cn

DOI 10.1201/9781003332657-39

2 EXPERIMENTAL SAMPLES AND METHODS

The sewage sludge from Hefei Sewage Treatment Plant in Anhui Province was selected as the experimental raw material. The sewage sludge is dried, grounded, and selected to have a particle size of 75–150 μm. Before the experiment, the sludge samples were placed in a 105°C oven for 12 h to remove moisture. The industrial analysis and elemental analysis of the samples are shown in Table 1.

Table 1. Ultimate analyses and proximate analyses of the sewage sludge.

Proximate analyses (wt.%)					Ultimate analyses (wt.%)				LHV (MJ/kg)
C_{ad}	H_{ad}	O_{ad}	N_{ad}	S_{ad}	V_{ad}	FC_{ad}	A_{ad}	M_{ad}	
26.33	4.56	15.59	4.84	0.87	48.12	4.07	44.31	3.50	9.70

ad- air drying base; V- Volatile; A-Ash; FC- Fixed Carbon; M- Moisture; LHV-Lower Heat Value.

Thermogravimetric analyzer TG (TGA Q5000IR, TA instrument, American) was used to evaluate the weight loss characteristics of sludge pyrolysis and combustion. Before the start of each experiment, a blank experiment was performed to ensure a stable baseline. The thermogravimetric analyzer was fed with nitrogen (pyrolysis) and air (combustion) at a gas flow rate of 100 mL/min. The temperature program was set to increase from room temperature to 100°C at a rate of 20°C/min, holding for 30 minutes, and then heating up to 800°C at three heating rates of 10, 20, and 30°C/min. The purpose of holding is to remove the water vapor absorbed during the addition and removal of the sample.

The thermogravimetric analysis combined with Fourier transform infrared spectroscopy (FTIR, Frontier, Perkin Elmer, USA) and mass spectrometry MS (Clarus SQ 8T, Perkin Elmer, USA) was used to evaluate sludge pyrolysis and combustion volatilization Product characteristics. The volatiles generated in the thermogravimetric analyzer were sent to the gas cell of the FTIR before being purged by the carrier gas into the gas cell of the MS. The TG, FTIR, and MS were connected by a transfer line (TL-9000, Perkin Elmer, USA). Lines were kept at 280°C to prevent condensation of some volatiles. FTIR collects volatile functional groups at wavelengths from 4000 to 450 cm^{-1} every 11 s. The mass-to-charge ratio of the volatiles can be continuously measured online with the ion source of the mass spectrometer running at 70 eV. Through the combination of TG-FTIR-MS, the types and content of volatile gases released during the combustion or pyrolysis of sludge can be characterized, and the volatile emission can be linked with the mass loss.

A horizontal tube furnace is used to pyrolyze the sludge and produce coke. Weigh 1$\pm$0.01g of sludge sample and spread it on the bottom of the corundum ark, and inject 100mL/min of nitrogen with a purity of $\geq$99.99%. The temperature program was set to increase from room temperature to 100°C at a rate of 20°C/min, holding for 30min, and then heating to 300, 400, 500, 600, and 700°C at 10 °C/min, respectively. The coke was collected after cooling to room temperature. The coke components were characterized by the KBr pellet method Fourier transform infrared spectroscopy (Nicolet 8700, Thermo Nicolet, USA).

3 RESULT AND DISCUSSION

3.1 *Analysis of weight loss characteristics of sewage sludge*

Figure 1 shows the thermogravimetric curves (TG) and derivative thermogravimetric curves (DTG) of sludge pyrolysis and combustion at three heating rates. The characteristic parameters are listed in Table 2. It can be seen that as the heating rate increases, the TG curve and the DTG curve move to the high-temperature region, but the weight loss ratio corresponding to the termination temperature in the same atmosphere is almost the same, this is because the higher heating rate shortens the reaction time of the sample and increases the temperature gradient between the sample particles increases due to the increase of the heat transfer rate, so the reaction with the high heating rate at the same temperature cannot be completed, resulting in delayed decomposition [14].

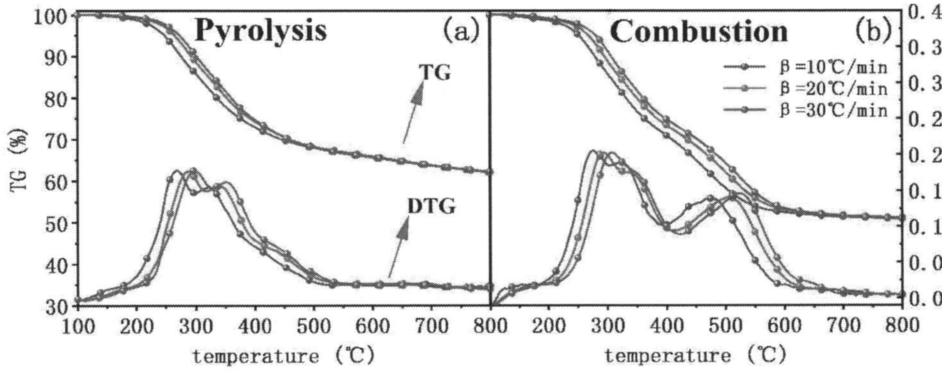

Figure 1. TG and DTG curves of sludge pyrolysis (a); combustion (b).

The thermogravimetric experiments were all kept at 100°C for 30 minutes, so there was no evaporation stage of water during the weight loss process. The weight loss process is described below with the characteristic parameters in the thermogravimetric curve at 20 °C/min.

Table 2. Sludge pyrolysis and combustion characteristic parameters.

	Heating rate (°C/min)	T_i^a(°C)	T_{p1}^b(°C)	DTG_{p1}^c(%/°C)	T_{p2}^b(°C)	DTG_{p2}^c(%/°C)	T_f^d(°C)
Pyrolysis	10	237	269	0.1779	322	0.1544	761
	20	242	286	0.1764	341	0.1559	773
	30	252	298	0.1777	352	0.1621	780
Combustion	10	236	275	0.2050	475	0.1388	653
	20	240	293	0.1935	505	0.1404	658
	30	249	307	0.2021	523	0.1458	670

T_i is the initial temperature, and the tangent line is drawn at the point with the largest weight loss rate on the TG curve, the horizontal line is drawn over the point where decomposition begins, and the temperature corresponding to the intersection of the two lines is the starting temperature [16]; T_{p1}, T_{p2}, and DTG curves show the temperature of the first and second peaks; DTG_{p1} and DTG_{p2} are the weight loss rate of the first and second peaks on the DTG curve; T_f is the termination temperature corresponding to the conversion rate of 98%;

For pyrolysis, the weight loss process of sludge can be roughly divided into two stages. The first stage is the devolatilization stage (242-520°C), which accounts for 85% of the total mass loss. The initial larger mass-loss rate in this stage is caused by the thermal decomposition of the biodegraded small organic matter in the sludge. Lead to. The mass-loss rate reaches its maximum at 286°C. After that, the macromolecular organic matter continues to depolymerize and decompose (including some biomass components), accompanied by devolatilization and coke generation. The second stage is the secondary decomposition stage of coke, which accounts for 15% of the total mass loss. In this stage, secondary decomposition of the coke formed in the first stage at a lower and almost constant mass-loss rate (0.015%/°C), and the product deposits in highly thermally stable residues up to 773°C. The residual mass was 62% at all three heating rates. There is no obvious demarcation between these two stages, and both have a certain quality loss. It shows that both devolatilization and carbonization exist in two stages. The first stage is dominated by devolatilization, and the second stage is dominated by secondary carbonization.

For combustion, the weight loss of sludge can also be divided into two stages: volatile combustion and coke oxidation stage, corresponding to the formation of two distinct peaks at 293°C and 505°C in the DTG curve. The second peak was formed when the DTG curve of the first stage did not drop to a smaller value, indicating that the devolatilization and coke oxidation stages overlapped in a large temperature range. For the volatile combustion and coke oxidation in the combustion process, the dominant role at this stage is lower than that of pyrolysis. Different from pyrolysis, the main feature of combustion is that oxidation participates in the weight loss process, which makes the characteristic temperature of combustion lower than that of pyrolysis, and the maximum mass loss rate is higher than that of pyrolysis. The volatiles was directly precipitated and burned around the sample, and the released heat provided a large amount of heat for the entire reaction system, thus significantly promoting the coke oxidation process. In the char oxidation stage, a significant weight loss peak (430°C-658°C) is formed, which indicates that the char formed by the polymerization of organic matter in the first stage is oxidized to form gaseous products (also called volatiles) in the second stage. The residual mass was 50% at all three heating rates.

3.2 *Volatile characteristics during pyrolysis and combustion*

Figure 2 is a three-dimensional 4000-450cm-1 infrared spectrum of volatile matter generated by sludge pyrolysis and combustion, where the X-axis is temperature, the Y-axis is the wavelength, and the Z-axis is the absorption intensity of the wavelength. The total amount of volatile matter is proportional to the absorption intensity of the infrared spectrum, so the pyrolysis process of sludge is more complex and has more volatile matter than the combustion process. Table 4 summarizes the possible volatile products and the functional groups possessed by the volatiles and their corresponding absorption bands. Figure 3 shows the spectral analysis of sludge pyrolysis and combustion at four selected temperature points.

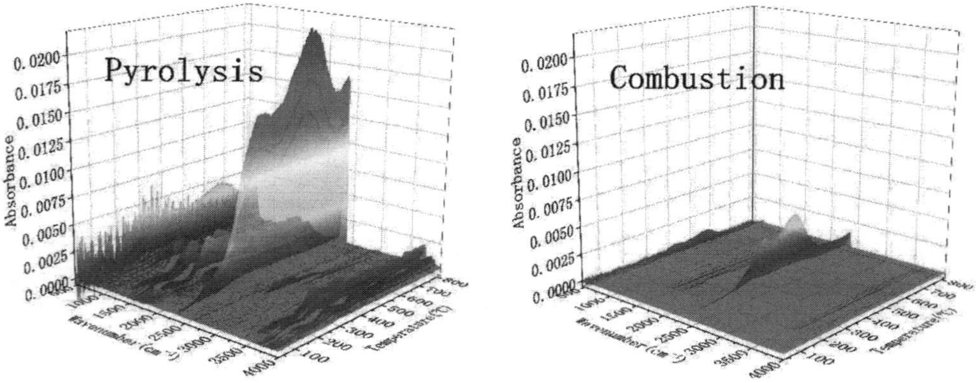

Figure 2. 3D infrared spectra of sludge pyrolysis and combustion.

As can be seen from Figure 2 and Figure 3, the combustion gas products detected online by infrared spectroscopy are almost only CO_2 with a wavelength of 2400-2240 cm-1. In the first weight-loss stage of the sludge, due to the obvious effect of high-temperature devolatilization, not only CO_2 is removed from the sample, but there are also many other organic substances. The volatiles needs to be purged by the carrier gas (air) through a long connecting line Entering the infrared spectroscopy chamber, this process allows the air to be thoroughly mixed with volatiles at 280°C and oxidized. The final test result is also almost only CO_2. Compared with the gaseous products of combustion, the gaseous products of pyrolysis are more abundant. Likewise, the predominant

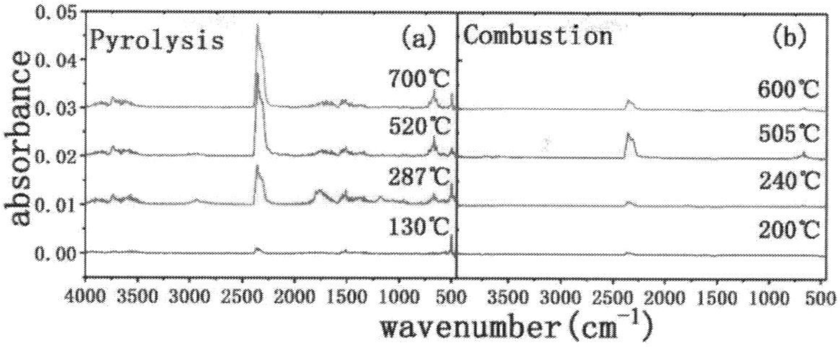

Figure 3. Infrared spectra of 4000-450cm-1 at 4 selected temperature.

pyrolysis gas product was detected as CO_2 from 2400-2240 cm-1 and 730-630 cm-1. Secondly, the tensile vibration of OH (4000-3500 cm-1) and the tensile vibration of C=O (1840-1950 cm-1) exist in the whole weight loss process and reach a maximum of around 340°C, which may be due to water and the formation of alcohols, phenols and ketones, aldehydes, esters, and carboxylic acids. Aromatic hydrocarbons (1590-1450 cm-1) are also present in the whole pyrolysis process starting from rapid weight loss similar to the formation stage of C=O, which may be caused by the aromatization of cellulose in sludge or benzene skeleton in lignin and the breakage of [17], since it is well known that lignin not only has a benzene ring structure but also decomposes in a wide temperature range [18]. For hydrocarbons with CH stretching vibrations (3100-2600 cm-1), especially the formation of CH_4 (3016 cm-1), its absorption intensity is not obvious and is only detected in the temperature range of 270-500°C, its sources are more extensive. However, C-O (1300-950cm-1) has a narrower temperature range (only 270-400°C) and lower spectral absorption intensity than C=O, which indicates that the pyrolysis product contains a large amount of aldehyde or ketone volatiles. In addition, weak CO was detected in the absorption band range of 2240-2060 cm-1, and the result of CO was not obvious in mass spectrometry, which was mainly caused by the secondary cracking of a small amount of volatiles and coke residues [19].

4 CONCLUSION

For the weight loss process, both pyrolysis and combustion are divided into two stages, but combustion has a more violent coke oxidation process. Compared with pyrolysis, the combustion phase has a lower characteristic temperature and d higher weight loss rate. The volatile products of pyrolysis and combustion are mainly CO_2, and the types of volatiles of pyrolysis are relatively rich, including alkanes, alcohols, phenols, ketones, aldehydes, esters, carboxylic acids, aromatic hydrocarbons, CO, etc.

ACKNOWLEDGMENT

This work is supported by Anhui Province Key Research and Development Program (No. 201904a07020039).

REFERENCES

Chen J, Mu L, Cai J, Yao P, Song X, Yin H, et al. Pyrolysis and oxy-fuel combustion characteristics and kinetics of petrochemical wastewater sludge using thermogravimetric analysis. *Bioresour Technol* 2015;198:115–23. https://doi.org/10.1016/j.biortech.2015.09.011.

Cieślik BM, Namieśnik J, Konieczka P. Review of sewage sludge management: Standards, regulations and analytical methods. *J Clean Prod* 2015;90:1–15. https://doi.org/10.1016/j.jclepro.2014.11.031.

Lin Y, Liao Y, Yu Z, Fang S, Ma X. A study on co-pyrolysis of bagasse and sewage sludge using TG-FTIR and Py- GC / MS. *Energy Convers Manag* 2017;151:190–8. https://doi.org/10.1016/j.enconman.2017.08.062.

Magdziarz A, Wilk M. Thermogravimetric study of biomass, sewage sludge and coal combustion. *Energy Convers Manag* 2013;75:425–30. https://doi.org/10.1016/j.enconman.2013.06.016.

Selcuk N, Yuzbasi NS. Combustion behaviour of Turkish lignite in O2/N2 and O2/CO2 mixtures by using TGA-FTIR. *J Anal Appl Pyrolysis* 2011. https://doi.org/10.1016/j.jaap.2010.11.003.

Soria-Verdugo A, Garcia-Hernando N, Garcia-Gutierrez LM, Ruiz-Rivas U. Analysis of biomass and sewage sludge devolatilization using the distributed activation energy model. *Energy Convers Manag* 2013. https://doi.org/10.1016/j.enconman.2012.08.017.

Wang T, Chen Y, Li J, Xue Y, Liu J, Mei M, et al. Co-pyrolysis behavior of sewage sludge and rice husk by TG-MS and residue analysis. *J Clean Prod* 2020. https://doi.org/10.1016/j.jclepro.2019.119557.

Energy Revolution and Chemical Research – Kok-Keong Chong and Zhongliang Liu (Eds)

Effect of different aeration levels on the wastewater treatment performance of an externally circulating electrically enhanced bioreactor

Bowen Wang, Hongguang Zhu & Kejia Wu
Bio-Energy Research Center, Institute of New Rural Development, Tongji University, Shanghai, China

ABSTRACT: In this paper, an external circulation electro-enhanced bioreactor (EC-EEB) is proposed for wastewater treatment based on the electro-enhanced bioreactor (EEB) proposed by Liu and Zhu. The DO value of the cathode is controlled by external circulation to enhance the denitrification effect of urban domestic wastewater. The wastewater treatment performance of the device was investigated at 20 ml/(min$_*$L), 27 ml/(min$_*$L), 33 ml/(min$_*$L), and 40 ml/(min$_*$L) of the reactor volume aeration under the conditions of applied voltage 0.7 V, HRT=24h and temperature 30°C, according to the external circulation 4 times/day. The results showed that EC-EEB can form a good aerobic and anaerobic environment. The volume aeration was positively correlated with the dissolved oxygen concentration. The best treatment effect of the device was achieved when the volume aeration was 33 ml/(min$_*$L). The average DO value of this anode chamber was 5.99 mg/L and the average DO value of the cathode chamber was 0.25 mg/L. The average COD and NH_4^+-N removal rates were 91.48% and 90.15%, respectively. At this time, the current density was (0.9844±0.5) A/m^2, and the average effluent NO_2^- and average effluent NO_3^- concentrations were 0.18 mg/L and 0.39 mg/L, respectively. Compared with the EEB system, the COD and NH_4^+-N removal rates are 3.58% and 6.84% higher respectively, both of which meet the Class A discharged standard, allowing for a more reliable effluent discharge.

1 INTRODUCTION

With the development of industrial technology and the increase in population, small towns are developing rapidly, while wastewater treatment facilities cannot keep up with the requirements of town development, and a large amount of untreated domestic wastewater is discharged into water bodies, resulting in a water circulation system that has been transformed into a nitrogen eutrophication system (Burow et al. 2010; Stuermer 2017). Biological treatment technologies based on electrochemical promotion have been widely used in urban domestic wastewater treatment in recent years (Huang et al. 2013; María et al. 2018; Sleutels et al. 2012; Tenca et al. 2013). One of the emerging technologies is microbial electrolysis cell (MEC), which provides a new way to treat the water environment. Xu (2014) constructed a single chamber MEC system with carbon cloth as an electrode to achieve 90% COD removal. Nguyen (2015) constructed a two-chamber MEC system using graphite felt electrodes with 79% nitrate-nitrogen removal. In the presence of denitrifying microorganisms, nitrate can be reduced to nitrogen gas by accepting electrons at the cathode and thus removed from the wastewater (Clauwaert et al. 2007). Liu and Zhu (2020) designed a novel microbial electrochemical denitrification reactor model: the electro-enhanced bioreactor (EEB) with carbon cloth bipolar plates and multiple compartments. The coupling of a microbial electrolytic cell with nitrification and denitrification was achieved in a single system. At an applied voltage of 0.7 V, 83.67% removal of ammonia nitrogen was achieved. The denitrification process of the EEB system provides the theoretical basis for MEC to treat high ammonia nitrogen

DOI 10.1201/9781003332657-40

wastewater. During the study, it was found that the EEB device could not effectively control the dissolved oxygen concentration in the cathode zone, which became the main factor affecting its denitrification efficiency. Lv (2011) held that elevating the dissolved oxygen concentration in the influent water in an electrode biofilm reactor also elevates the dissolved oxygen concentration in the cathode zone. To form a better anaerobic environment, the external circulation electrically enhanced bioreactor (EC-EEB) is proposed in this paper on this basis. It has two features: the use of carbon cloth bipolar plates and the adoption of external circulation. This design can increase the specific surface area of electrodes and provide better aerobic anaerobic partitioning to improve the degradation of organic matter and denitrification efficiency. Numerous studies have shown that the level of dissolved oxygen concentration is the main factor affecting the efficiency of microbial electrochemical denitrification (Microbiol Mol Biol. Rev, 1997). It is crucial to investigate the optimal dissolved oxygen content in the cathodic zone of the anode zone of the device. The volume aeration is positively correlated with the dissolved oxygen concentration within a certain range, and the dissolved oxygen concentration in the reactor is mainly controlled by changing the magnitude of the aeration in the laboratory stage. In this paper, the effect of different volume aeration on the performance of external circulation electro-enhanced bioreactor is investigated to enhance the denitrification efficiency of microbial electrochemical system and to achieve stable and reliable compliance with the discharge standards. It provides a reference for the current improvement of the EEB system to treat urban domestic wastewater.

2 MATERIALS AND METHODS

2.1 *EC-EEB experimental setup*

The external circulation electrically enhanced bioreactor (EC-EEB) consists of four parts: circulation unit, aeration unit, applied voltage unit, and reaction chamber, as shown in Figure 1. The circulation device is directly connected to the reaction chamber, and two circulation pumps are symmetrically distributed on both sides of the chamber. The carbon cloth bipolar plate is inserted into the reactor, a DC power supply is connected, and a real-time current detection device is connected in series into the circuit. The cathode and anode chambers are divided according to the connected positive and negative poles, and the aeration stone is connected to the anode chamber. It is needed to connect the power supply and run. The size of the reactor used in this study was 150 mm × 100 mm × 150 mm with an effective volume of 1.5 L (150 mm × 100 mm × 100 mm).

Figure 1. Experimental setup diagram.

2.2 *Measurement and analysis methods*

The assays for the experiments are shown in Table 1.

Table 1. Water quality testing methods.

Indicators	Testing Method
DO value	Portable Dissolved Oxygen Detector
COD	Fast-extinction spectrophotometric method
$NH_4{}^+$-N	Nascent reagent spectrophotometry
$NO_3{}^-$-N	UV spectrophotometry
$NO_2{}^-$-N	N-(1-naphthyl)-ethylenediamine spectrophotometric method

The real-time current data of the experiment is recorded in the memory card by a custom current acquisition device. The acquisition interval of current values was set to 10 min/time. The daily current data were exported between water changes to observe the current data anomalies and adjust the problems that occurred in the experiment in time. The effective area of the electrode is 0.015m^2 (150mm×100mm). The current density of the device is the current value divided by the effective area of the electrode, as shown in Equation 1.

$$J_A = \frac{I}{A} \tag{1}$$

where I (A) is the real-time current and A (m^2) is the surface area of the anode electrode.

2.3 *Start-up and operation of the device*

In this paper, nitrifying and denitrifying bacteria (model GANDEW-DEN) were used in the EC-EEB, and the start-up time of the device could be significantly reduced by using a direct culture of bacterial microorganisms. A white, reddish-brown flocculent microbial film was formed on the carbon cloth electrode through a 15-day incubation cycle at an applied voltage of 0.7 V, as shown in Figure 2.

Figure 2. Carbon cloth electrode plate hanging film effect.

The nature of most urban domestic wastewater is not very different, and the water does not contain heavy metals and toxic and harmful substances. However, due to living habits and other reasons, ammonia nitrogen and total nitrogen concentrations in small towns' domestic wastewater are higher than those in urban domestic wastewater. To better investigate the treatment effect of higher ammonia nitrogen wastewater, the experimental water was designed based on the following water quality (Ji et al. 2003): COD is 300 mg/L, $NH_4{}^+$-N is 50 mg/L, SS is 150 mg/L, TP is 4.1 mg/L, and pH is 7.5.

In this paper, four experimental groups with volumetric aeration rates of 20 ml/(min$_*$L), 27 ml/(min$_*$L), 33 ml/(min$_*$L) and 40 ml/(min$_*$L) were set. The experiments were conducted using a stable applied voltage of 0.7 V, a temperature of 30° C, a hydraulic residence time of 24 hours, and an external circulation rate of 4 times/day. Microorganisms were cultured in the device according to the start-up method described above, and the experiments were started after the device was stabilized. Continuous aeration and sequential batch water intake were used for operation. The effluent COD concentration, ammonia nitrogen concentration, nitrate-nitrogen concentration, nitrite nitrogen concentration, DO value, and current value were measured for 8 consecutive time points during the stable operation stage and analyzed and discussed.

3 RESULTS AND DISCUSSION

3.1 *COD removal effect*

As shown in Figure 3, the COD content of the effluent from the four experimental groups with different volume aeration amounts remained stable during the stable operation stage of the device, with the average effluent COD concentrations of 54.8 mg/L, 33.4 mg/L, 25.6 mg/L, and 21.3 mg/L, and the average COD removal rates of 81.75%, 88.88%, 91.48%, and 92.89%, respectively. The COD removal rate of the effluent from experimental group I had a decreasing trend, and the highest COD removal rate was 84.70%. The COD removal rate of experimental group II has been stable, and the highest COD removal rate was 90.30%. The COD removal rates of experimental groups three and four showed an increasing trend, with the highest COD removal rates of 93.20% and 94.23%, respectively. As the volume aeration rate of the experimental group increased from 20 ml/(min$_*$L) to 40 ml/(min$_*$L), the COD removal rate increased by 7.13%, 2.60%, and 1.41% in order, which indicated that higher volume aeration rate in a certain range provided better-dissolved oxygen concentration, which was beneficial to the utilization and removal of organic matter by aerobic microorganisms. The EC-EEB unit can effectively remove COD at a volumetric aeration rate of 40-60 ml/min, which meets the Class A emission standard. Considering the operating cost of the device, the volume aeration of 33 ml/(min$_*$L) is the best.

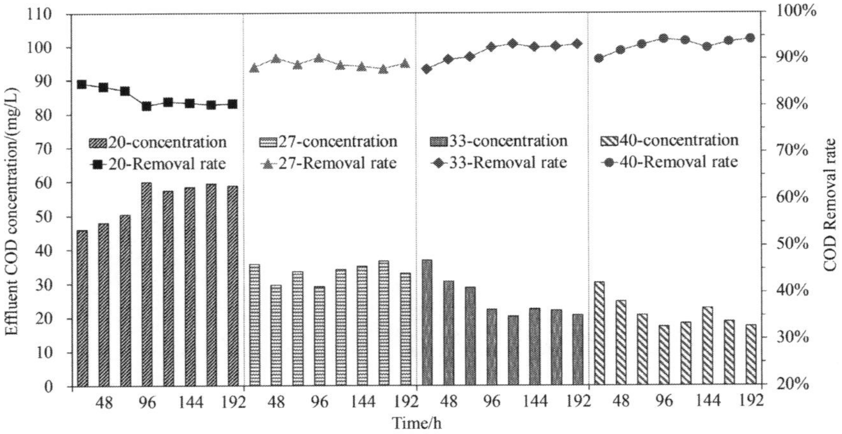

Figure 3. Effluent COD treatment results.

3.2 NH_4^+-*N removal effect*

As shown in Figure 4, with the increase in volumetric aeration, the NH_4^+-N content of the effluent water in the device decreased significantly and the removal rate increased significantly. The average effluent NH_4^+-N removal rates were 77.88%, 85.30%, 90.15% and 91.88%, respectively. The experimental results showed that the volume aeration of 20 ml/(min$_*$L) could not provide sufficient dissolved oxygen concentration to the device. It not only affected the organic removal effect but also could not provide a suitable environment for nitrifying bacteria to work. When volume aeration of 40 ml/(min$_*$L) was used, the phenomenon of NH_4^+-N accumulation occurred at the later stage when the NH_4^+-N concentration was reduced at the earlier stage. The reason for this may be that when the dissolved oxygen concentration in the reactor is high, the nitrification reaction proceeds well in the early stage of the reaction while the denitrification reaction is inhibited leading to the accumulation of large amounts of NO_x^--N, which will inhibit the nitrification reaction in the late stage of the reaction and further weaken the NH_4^+-N removal effect of the reactor (Mccarty 2018). The average effluent NH_4^+-N concentrations of the four experimental groups in the stabilization stage were 11.75 mg/L, 7.00 mg/L, 3.75 mg/L, and 3.28 mg/L. The effluent NH_4^+-N concentration of experimental group I reached the standard of Class I B, and the effluent ammonia-nitrogen concentrations of the remaining three groups reached the standard of Class I A. This indicates that the EC-EEB system has a good ability to carry out nitrification reactions in the volume aeration range of 27-40 ml/(min$_*$L). It can treat urban domestic wastewater with high ammonia nitrogen content well. Compared with the two-stage SBR process (Ji et al. 2003), EC-EEB improved NH_4^+-N removal by 9.15%-21.15% at the same HRT. In summary, the volume aeration rate of 33 ml/(min$_*$L) is the most suitable for the nitrification reaction of EC-EEB.

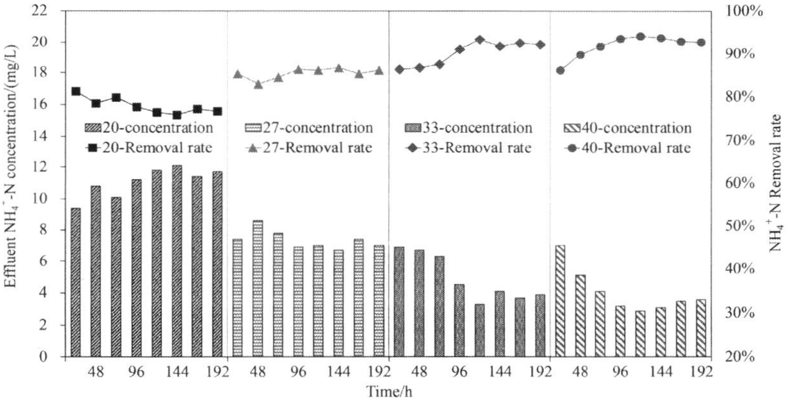

Figure 4. Effluent NH_4^+-N treatment results.

3.3 *NO_x^--N accumulation*

As shown in Figure 5, the trend of NO_3^-concentration in the effluent of the four experimental groups was smooth, and the average effluent NO_3^-concentration was 0.49 mg/L, 0.44 mg/L, 0.39 mg/L, and 0.40 mg/L, respectively. The variation trend of effluent NO_2^- concentration in the first three experimental groups was relatively smooth, and the effluent NO_2^- concentration in experimental group IV had a significant increasing trend, and the average effluent NO_2^- concentration was 0.15 mg/L, 0.17 mg/L, 0.18 mg/L, and 0.23 mg/L, respectively. The biological denitrification process requires an anaerobic environment and the dissolved oxygen concentration becomes the main factor affecting the denitrification efficiency of the device (Sengupta & Dick 2015). When the volume aeration was 20 ml/(min$_*$L), the dissolved oxygen content in the device was lower to facilitate the denitrification reaction, and the NO_2^- and NO_3^-concentrations in experimental group I were lower. When the volume aeration rate was 33 ml/(min$_*$L), the NO_3^-concentration in the device decreased significantly and NO_2^- did not accumulate significantly, which proved that the denitrification reaction in experimental group III was good. When the volumetric aeration rate was 40 ml/(min$_*$L), the NO_2^- concentration in the device increased in the later stages of the reaction, because increasing the dissolved oxygen concentration had almost no effect on the removal of NO_3^-concentration but caused a slight accumulation of NO_2^- (Lv et al. 2011). The microbial electrochemical denitrification process requires a good anaerobic environment, and a suitable dissolved oxygen concentration not only promotes the degree of nitrification reaction but also does not inhibit the denitrification reaction. The experimental results showed that when the volume aeration was 33 ml/(min$_*$L), the nitrification-denitrification reaction of the EC-EEB system was in good condition with the highest denitrification efficiency.

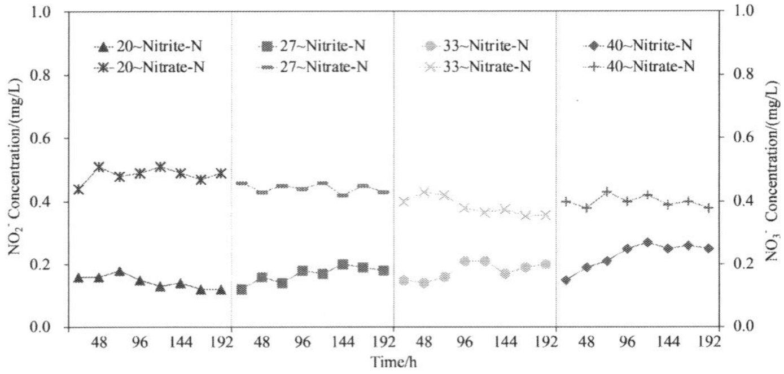

Figure 5. Accumulation of NO_x^--N in the effluent.

3.4 *Changes in dissolved oxygen*

As shown in Figure 6, the average DO values in the anode chamber of the four experimental groups increased with the increase of the volume aeration during the operation of the device, which was (3.23±0.41) mg/L, (4.56±0.39) mg/L, (5.99±0.35) mg/L, (7.51±0.46) mg/L, respectively; the average DO values in the cathode chamber were also well-controlled, which were (0.19±0.07) mg/L, (0.21±0.07) mg/L, (0.25±0.10) mg/L, and (0.55±0.45) mg/L, respectively. The dissolved oxygen concentration controlled by the volume aeration size mainly affects the denitrification reaction process in the microbial electrochemical system (Skeman & Macrae 1957). The high volume aeration in experimental group IV affected the activity of biological denitrification enzymes and inhibited the denitrification reaction in the later stages of the installation (Zumft 1997). The accumulation of nitrate and nitrite nitrogen in turn slowed down the reactivity of nitrifying bacteria. The lack of depletion of dissolved oxygen in the reactor leads to a corresponding increase in DO in the cathode zone during the cycle.

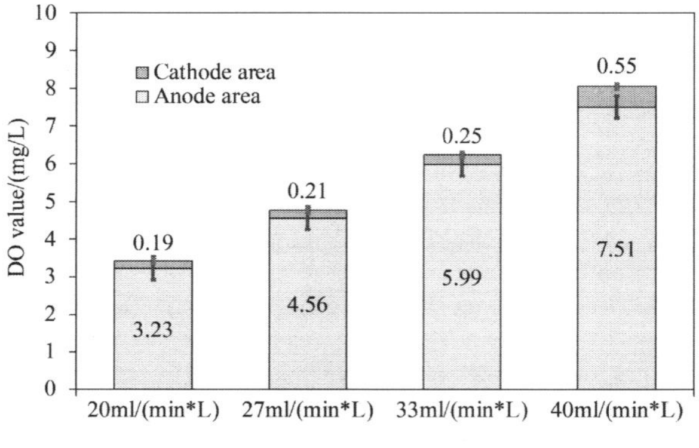

Figure 6. Variation of dissolved oxygen concentration.

The dissolved oxygen concentration in the anaerobic zone of the EEB system exceeded the upper DO limit of 2 mg/L for the denitrification reaction to proceed with the increase in volume aeration (Gillham & Cherry 1978). The cathodic zone of the EC-EEB system, on the other hand, could maintain a better anaerobic environment and achieved the purpose of establishing a better aerobic and anaerobic partition. In conclusion, 33 ml/(min$_*$L) volumetric aeration can provide better-dissolved oxygen concentrations, as well as good results for COD, ammonia nitrogen, and NOx–N removal.

3.5 *Electrochemical performance analysis*

As shown in Figure 7, the current density of experimental groups one, two, and three maintained a relatively stable trend throughout the experimental period, with the average current density of 0.8523 A/m^2, 0.8705 A/m^2, and 0.9844A/ m^2, respectively, and the device current density increased with the rise of volume aeration. The current density of experimental group IV remained stable before decreasing significantly, and the average current densities of the two stages were 1.0539A/ m^2 and 0.9274 A/m^2 respectively. The average current density of the whole experimental period was 0.9746 A/ m^2. During the experiment, reactors with sufficient volume aeration provide a good aerobic environment and a large number of electron donors are generated in the device to promote the extent of nitrification reactions. The COD removal in the reactor with insufficient volume aeration is inhibited leading to the current density being relatively at a lower level. In conjunction with section 3.2, the higher current density corresponds to higher ammonia nitrogen removal and the decreasing current density phase corresponds to a decrease in ammonia nitrogen removal, proving that current density has a greater effect on ammonia nitrogen removal (Li et al. 2012). The experiments in this section proved that the current density was within the range of (0.9844±0.5) A/m^2 at the volume aeration of 33 ml/(min$_*$L) when the reactor had a good and stable nitrogen removal effect.

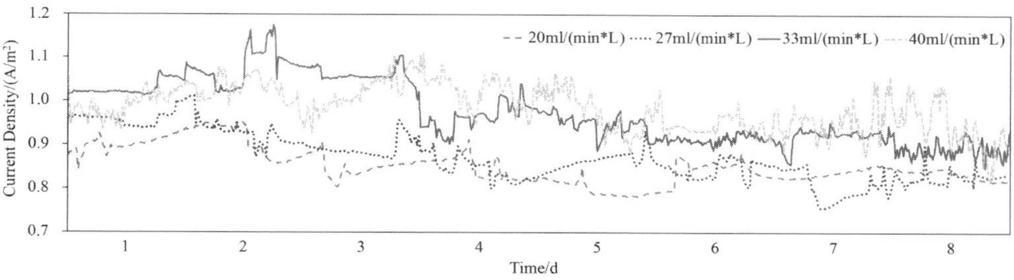

Figure 7. Current density variation.

4 CONCLUSION

In this paper, the effect of different aeration levels on the effectiveness of external circulation electro-enhanced bioreactors in treating urban domestic wastewater was investigated, and the results were as follows.

(1) The EC-EEB system adopts external circulation, which can provide a good aerobic environment and anaerobic environment. At the volume aeration rate of 33 ml/(min$_*$L), the average DO value of the anode chamber is 5.99 mg/L and the average DO value of the cathode chamber is 0.25 mg/L, which is most suitable for nitrogen removal in the CCP-EEB system.

(2) The best overall treatment effect of the EC-EEB unit was achieved at a volumetric aeration rate of 33 ml/(min$_*$L). The average COD removal rate and NH_4^+-N removal rate could reach 91.48% and 90.15%, respectively. The COD removal rate and ammonia removal rate were 3.58% and 6.84% higher than those of the EEB system, respectively. The effluent COD and NH_4^+-N concentrations both reached the Class A standard. The average NO_2^-concentration

and average NO_3^- concentration in the stable operation stage were 0.18 mg/L and 0.39 mg/L, respectively, and the current density was (0.9844±0.5) A/m^2 at this time.

(3) Volume aeration is positively correlated with dissolved oxygen concentration, too much volume aeration will mainly inhibit the denitrification process of the EC-EEB system, and the operation cost of the device, 33 ml/(min_*L) volume aeration is optimal considering the operation effect.

ACKNOWLEDGMENTS

This work is supported by the National Key Research and Development Program of China (Research and Development of Key Technologies and Special Equipment for Efficient Conversion of Intensive Farming Manure, No. 2018YFD0800103).

REFERENCES

Burow K R, Nolan B T, Rupert M G, et al. Nitrate in Groundwater of the United States, 1991-2003[J]. *Environ. Sci. Technol.*, 2010,44(13): 4988–4997.

Clauwaert P, Rabaey K, Aelterman P, et al. Biological denitrification in microbial fuel cells.[J]. *Environmental Science & Technology*, 2007, 41(9):3354–60.

Gillham RW, Cherry JA. Field evidence of denitrification in shallow groundwater flow systems[J]. *Water Pollut. ResCan*, 1978,13(1):53–71.

Huang B, Feng H, Wang M, et al. The effect of C/N ratio on nitrogen removal in a bioelectrochemical system[J]. *Bioresource Technology*, 2013, 132:91–98.

Ji J, Ge L, Wang Y. Study on two-stage SBR process for treating domestic wastewater in small towns[J]. *Journal of Yangzhou University* (Natural Science Edition), 2003(02):75–78. DOI:10.19411/j.1007-824x.2003.02.018.

Li Y., Wang CH R., He X W., Wang R B., Zhao N N., Huang L B., Qian Y. Removal of ammonia nitrogen from micropollutant water by electrochemical oxidation[J]. *Journal of Environmental Engineering*, 2012, 6(05):1553–1558.

Lv Jiangwei, Liu Jia, Shen Hong, et al.Effect of dissolved oxygen on the denitrification performance of electrode biofilm[J]. *Journal of Harbin Institute of Technology*, 2011, 43(2):6.

Mccarty P L. What is the Best Biological Process for Nitrogen Removal: When and Why?[J]. *Environmental Science & Technology*. 2018, 52(7):3835–3841.

María SM, et al. Pilot-scale bioelectrochemical system for simultaneous nitrogen and carbon removal in urban wastewater treatment plants[J]. *Journal of Bioscience and Bioengineering*, 2018, 126:S13891723 17308861-.

Nguyen V K, Hong S, Park Y, et al. Autotrophic denitrification performance and the bacterial community at biocathodes of bioelectrochemical systems with either abiotic or biotic anodes[J]. *Journal of Bioscience and Bioengineering*. 2015, 119(2):180–187.

Sengupta A, Dick W A. Bacterial Community Diversity in Soil Under two Tillage Practices as Determined by Pyrosequencing [J]. *Microbial Ecology*. 2015, 70(3):853–859.

Skeman VBD, Macrae IC. The influence of oxygen availability on the degree of nitrate reduction by pseudomonas denitrifications[J]. *Can J Microbiol.*, 1957, 3:505–530.

Sleutels THJA, Ter Heijne A, Buisman CJN, Hamelers HVM. Bioelectrochemical Systems: An Outlook for Practical Applications[J]. *Chemsuschem*, 2012, 5(6):1012–1019.

Stuermer M. Industrialization and the demand for mineral commodities[J]. *Journal of International Money and Finance*. 2017, 76:16–27.

Tenca A, Cusick R D, Schievano A, et al. Evaluation of low-cost cathode materials for treatment of industrial and food processing wastewater using microbial electrolysis cell[J]. *International Journal of Hydrogen Energy*, 2013.

Xu Y, Jiang Y, Chen Y, et al.Hydrogen Production and Wastewater Treatment a Microbial Electrolysis Cell with a Biocathode [J]. *Water Environment Research*. 2014, 86(7):649–653.

Xueyu Liu, Hongguang Zhu. Treatment of Low C/N Ratio Wastewater by a Carbon Cloth Bipolar Plate Multicompartment Electroenhanced Bioreactor (CBM-EEB) [J]. *ACS OMGEA*, 2020.

Zumft WG Cell biology and molecular basis of denitrification[J]. *Microbiol Mol Biol. Rev*, 1997, 61:533–616.

Energy Revolution and Chemical Research – Kok-Keong Chong and Zhongliang Liu (Eds)

Performance evaluation and optimization of methanol steam reforming reactor with waste heat recovery for hydrogen production

Min Zuo & Zhenzong He*
College of Energy and Power Engineering, Nanjing University of Aeronautics and Astronautics, Nanjing, China

ABSTRACT: Using engine waste heat to reform methanol for hydrogen generation can effectively recover waste heat from exhaust gas and enhance fuel usage. In this paper, computational fluid dynamics (CFD) and response surface methodology (RSM) are used in this paper to investigate the performance of a methanol reforming unit with waste heat recovery for hydrogen production. The impacts of different reactant and exhaust gas input characteristics on hydrogen generation performance were studied and optimized. These parameters include the exhaust gas inlet temperature and velocity, the reactant inlet temperature and velocity, and the steam-to-methanol ratio. The results indicated that the optimum inlet temperature of the reactants is 393 K, the velocity is 0.05 m/s, the steam to methanol ratio is 1.34, while the inlet temperature and velocity of the exhaust gas are 584.6 K and 1.35 m/s, respectively. A maximum methanol conversion of 83.73% and a minimum CO selectivity of 3.64E-6 can be obtained at this time.

1 INTRODUCTION

Economic development has led to dramatic increases in the number of motor vehicles (Costa & Grundstein 2016), which would then result in a large consumption of fossil fuels. There will inevitably be a shortage of fossil fuel energy and a rise in carbon emissions, and this could affect the productivity of society. Hence, it is very important to reduce engine emissions and to find alternative fuel sources.

The high calorific value and low greenhouse gas emissions of hydrogen make it a suitable substitute for conventional fuels (Prapinagsorn et al. 2018; Wang et al. 2018). However, the storage and transportation of hydrogen remain challenging. Researchers have conducted numerous studies on the development of new hydrogen storage methods. It remains unclear, however, whether hydrogen storage will be able to achieve the same volumetric energy density as gasoline and provide the desired vehicle range. Consequently, onboard hydrogen production technology has gained new attention, and one of the methods is to store hydrogen in liquid fuels (Santos Andrade et al. 2019). The liquid chemical methanol (CH_3OH) is considered to be suitable for storing hydrogen since it has a 40% higher hydrogen content per unit volume than liquid hydrogen. Li et al. (Li et al. 2021) reviewed the catalytic based methanol reforming organic hydrogen production process. The authors concluded that methanol's high hydrogen-to-carbon (H/C) ratio, flexibility, and sustainability make it an optimal choice for on-board hydrogen production. Pashchenko et al. (Pashchenko et al. 2020) used different fuels for hydrogen production and found that methanol steam reforming (MSR) has the largest heat recovery rate at 600 K temperature.

Utilizing MSR is an effective waste heat recovery technique, and it is particularly important to design the MSR reactor (Goldmann et al. 2018; Zhang et al. 2018). Kumar (Kumar et al. 2019) studied hydrogen production by MSR with waste gas heating. The results indicated that the methanol

*Corresponding Author: hezhenzong@nuaa.edu.cn

DOI 10.1201/9781003332657-41

conversion increased as the exhaust gas temperature increased. The volume fraction of hydrogen was approximately 42%. Using this method, hydrogen can be supplied to internal combustion engines on board, which increases the thermal efficiency of the system. Shu et al. (Shu et al. 2020) conducted a numerical study of the designed new MSR reactor. It was found that the flow rate of methanol had a significant effect on the heat flux, pressure loss and temperature distribution. Chen et al. (2020) conducted a numerical investigation of the thermal management design criteria for a methanol conversion reactor. It was found that temperature uniformity was the most critical factor that affected the performance of the reactor.

In conclusion, it can be seen that a large number of researchers have conducted experiments relating to the use of exhaust gas for on-board hydrogen reforming. However, there has been little research on the input parameters affecting the performance of the reformer and even less on optimizing the parameters of the reactant inlet and exhaust gas inlet. Therefore, following previous studies, the effects of reactant inlet temperature, velocity, steam to methanol (S/M) ratios, and the exhaust gas temperature and velocity on the performance of hydrogen production from the reformer are investigated in this paper. RSM is then used to optimize the input parameters to achieve maximum methanol conversion with the lowest CO selectivity. Our study findings can be applied as a reference for hydrogen reforming from exhaust gas waste heat.

2 MODEL DESCRIPTION

2.1 *Physical model*

We present in this paper a study of the use of engine exhaust gas waste heat to support the reforming of methanol to hydrogen. The aim was to determine the effect of varying parameters on the reactor performance, such as reactor flow parameters, S/M ratios, and exhaust gas temperatures. The reactor is a simple and commonly used small tubular reactor. A 3D model of the reactor is shown in Figure 1(a). The reforming reactor consists of concentric circular tubes with a length of 150 mm and a thickness of one mm. The methanol reforming reaction channel has a radius of 18 mm, while the heating channel has a radius of 36 mm. Reactants (methanol and water vapor) enter the inner tube, come in contact with the catalyst bed, react on its surface, and then exit from the outlet together with the products. During the reforming process, engine exhaust gas is drawn into the outer tube to provide heat. Here, the catalyst region is considered to be a porous medium, and the engine exhaust gas is simulated using hot air. The geometrical parameters of the reactor model are shown in Table 1.

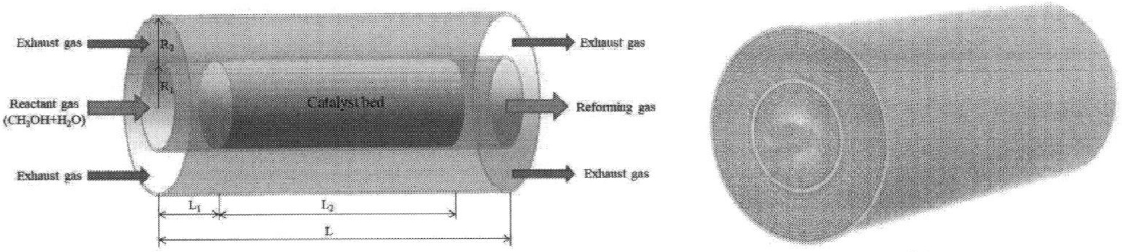

(a) (b)

Figure 1. Small tubular reforming reactor: (a) 3D model; (b) computational grid.

2.2 *Governing equations*

In this study, a 3D CFD model was used. To simplify the model, the following assumptions were made in the numerical simulation of the reforming reactor:

(1) The reactants are assumed to be incompressible, laminar, and stable gases.
(2) The outer reactor wall is considered adiabatic.

(3) The catalyst bed is isotropic and the chemical reactions occur only on the catalyst region.
(4) The effects of thermal radiation and gravity of the gas are neglected.
(5) The temperature and concentration differences between reactants and catalysts are ignored.

Considering the above assumptions, the governing equations of the model computational domain under the tri-stable condition are as follows.

The continuity equation is

$$\nabla \cdot \vec{u} = 0 \tag{1}$$

The momentum equation is

$$\varepsilon \left(\vec{u} \cdot \nabla \right) \vec{u} = -\frac{\varepsilon}{\rho} \nabla p + \frac{\varepsilon \mu}{\rho} \nabla^2 \vec{u} + S_m \tag{2}$$

Where S_m is the source term of the fluid induced by the porous catalyst, which arises only in the catalyst region and also causes a significant pressure drop. S_m can be expressed as

$$S_m = -\frac{\mu}{\rho \kappa} \vec{u} - \frac{\beta \vec{u}}{2} \left| \vec{u} \right| \tag{3}$$

Where β and κ can be expressed by the Ergun equation, which is shown as follows (Ergun 1952).

$$\kappa = \frac{D_p^2 \varepsilon^3}{150(1-\varepsilon)^2} \tag{4}$$

$$\beta = \frac{3.5(1-\varepsilon)}{D_p \varepsilon^3} \tag{5}$$

The component transport equation is

$$\varepsilon \left(\vec{u} \cdot \nabla c_i \right) = D_{eff} \nabla^2 c_i + \varepsilon \sum_{r=1}^{N} M_{w,i} R_{i,r} \tag{6}$$

Where D_{eff} is the effective diffusion coefficient of the gas, which can be corrected by the Stefan-Maxwell equation (Perng et al. 2013; Purnama et al. 2004) $D_{eff} = \varepsilon^{\tau} D_k$. The energy equation can be written as

$$(\rho c_p)(\vec{u} \cdot \nabla) T = K_{eff} \nabla^2 T + \varepsilon S_t \tag{7}$$

Where c_p and T denote the specific heat capacity and temperature of the fluid, respectively; K_{eff}($K_{eff} = \varepsilon\ K_f + (1\text{-}\varepsilon) K_s$) refers to the effective thermal conductivity, which characterizes the effect of the porous medium on the thermal conductivity; S_t is the source term caused by chemical reaction. S_t can be described as

$$S_t = -\sum_{i=1}^{N} \left(\frac{h_i^0}{M_{w,i}} + \int_{T_{ref}}^{T} C_{p,i} dT \right) R_i \tag{8}$$

The methanol water vapor reforming reaction mechanism is complex and varies with different catalysts. To effectively evaluate the methanol reforming reaction performance, the $CuO/Zn/Al_2O_3$dual rate reaction model proposed in the reference (Srivastava et al. 2021) is widely used and was employed in this study. The kinetic equations for the chemical reaction are

$$CH_3OH + H_2O \xrightarrow{k_1} CO_2 + 3H_2 \tag{9}$$

$$H_2 + CO_2 \xrightarrow{k_2} CO + H_2O; \quad CO + H_2O \xrightarrow{k_{-2}} H_2 + CO_2 \tag{10}$$

The rates of the above chemical reactions can be calculated using the Arrhenius model. The MSR rate and rWGS rate are expressed as follows.

$$R_{MSR} = k_1 C_{CH_3OH}^{0.6} C_{H_2O}^{0.4} exp\left(-\frac{E_{a1}}{RT}\right) \tag{11}$$

$$\mathrm{R_{rWGS} = k_2 C_{CO_2} C_{H_2} exp\left(-\frac{E_{a2}}{RT}\right) - k_{-2} C_{CO} C_{H_2O} exp\left(-\frac{E_{a2}}{RT}\right)} \tag{12}$$

The model geometry parameters and a series of other necessary kinetic parameters mentioned above in this study are detailed in Table 1.

2.3 *Boundary conditions*

The boundary conditions used in this study are as follows.
(1) Reactant inlet:

$$u = u_{in,R}, T = T_{in,R}, M_{CH3OH} : M_{H2O} = S/C \tag{13}$$

(2) Hot air inlet:

$$u = u_{in,H}, T = T_{in,H} \tag{14}$$

(3) Reformer outlet:

$$\frac{\partial \mathrm{u}}{\partial \mathrm{z}} = \frac{\partial T}{\partial z} = \frac{\partial C_i}{\partial z} = \frac{\partial P}{\partial z} = 0 \tag{15}$$

(4) Gas-solid interface:

$$u = 0, \lambda_m \frac{\partial T_m}{\partial z} = \lambda_c \frac{\partial T_c}{\partial z}, \frac{\partial M_i}{\partial z} = 0 \tag{16}$$

(5) Outer wall of the reformer:

$$\frac{\partial T_c}{\partial z} = 0 \tag{17}$$

Table 1. The input parameters.

Parameters	Values
Length of MSR, L (mm)	150
Distance between entrance and catalyst bed, L_1 (mm)	25
Length of Catalyst bed, L_2(mm)	100
Reforming reaction channel radius, R_1(mm)	18
Heating channel radius, R_2(mm)	36
Density of catalyst bed ($kg{\cdot}m^{-3}$) (Agrell et al. 2020)	1480
Permeability of catalyst bed, κ (m^2) (Karim et al. 2005)	2.379×10^{-12}
Mass diffusion coefficient, D_k ($m^2{\cdot}s^{-1}$) (Karim et al. 2005)	6.8×10^{-5}
Catalyst thermal conductivity (W m^{-1} K^{-1}) (Agrell et al. 2002)	1.0
Porosity of catalyst bed, ε	0.5
Activation energy for MSR (J mol^{-1}), E_{a1} (Agrell et al. 2002)	7.0×10^4
Pre-exponential factor for MSR, k_1(Agrell et al. 2002)	8.0×10^8
Activation energy for reverse WGS (J mol^{-1}), E_{a2} (Agrell et al. 2002)	1.0×10^5
Pre-exponential factor for reverse WGS, k_2 (Agrell et al. 2002)	4.0×10^8
Activation energy for backward WGS (J mol^{-1}), E_{a-2}(Agrell et al. 2002)	8.0×10^5
Pre-exponential factor for backward WGS, k_{-2} (Agrell et al. 2002)	4.0×10^8

3 NUMERICAL METHODS AND VALIDATION

3.1 *Numerical methods*

Ansys Fluent (v.19.1) is used to solve the control equations in this paper. SIMPLE-C is used to solve the pressure-velocity coupling. The pressure gradient is discretized using the PRESTO format, and the momentum, energy, and component transport equations are discretized using the second-order windward format. In the reformer, the inlet is governed by velocity-inlet boundary conditions, while the outlet is governed by pressure-outlet boundary conditions. Solutions are considered convergent when the residuals of all equations approach steady state and reach thermal and mass equilibrium, and when the residuals at convergence are less than 10^{-6}.

3.2 *Mesh independence verification*

Before the formal numerical simulation, the mesh independence was verified using a 3D model. The 3D model was meshed by ANSYS, and the 3D mesh is shown in Figure 1(c). According to Figure 2(a), methanol conversion and hydrogen (H_2) mole fraction are obtained for various mesh numbers for the following conditions: reactant inlet temperature of 433 K, the velocity of 0.15 m/s, the water-to-alcohol ratio of 1.5, inlet exhaust gas temperatures and velocity of 673 K and 2.5 m/s, respectively. It can be seen that when the grid number reaches 580,000, the methanol conversion and hydrogen molar fraction tend to be stable, so the grid number of 581,742 meets the requirement of calculation accuracy when subsequent calculations are performed.

3.3 *Model validation*

The CH3OH conversion and CO yield under the reformer proposed by Fukahori et al. (Fukahori et al. 2008) were calculated to demonstrate the accuracy of the numerical procedure in this paper, and the conditions used for the numerical simulations in this paper were identical to those of Fukahori et al. Figure 2(b) compares the current numerical results to the experimental data of Fukahori et al. The figure shows that the experimentally obtained methanol conversion and CO production are extremely close to the current modeling findings, indicating that the study is quite credible.

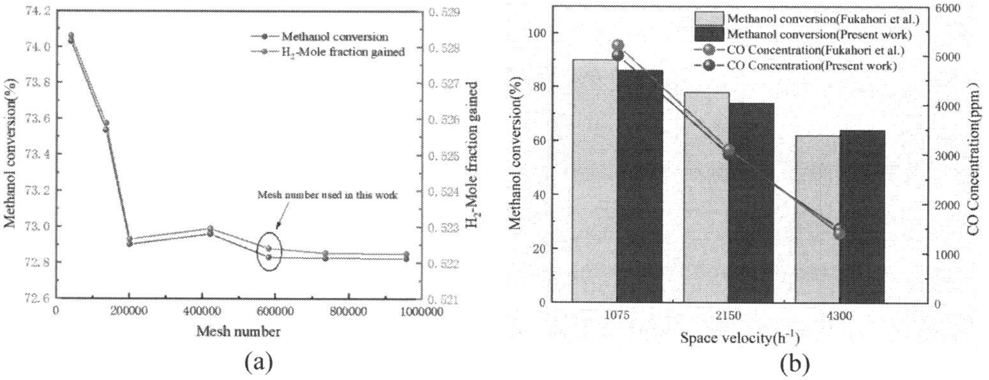

Figure 2. Validation: (a) grid-independent validation, (b) model validation.

4 RESULTS AND DISCUSSIONS

4.1 *Response surface analysis method*

In this paper, the reforming reactant inlet parameters and exhaust gas inlet parameters were optimized by CFD and RSM. RSM is an experimental method based on mathematics and statistics that helps to find out the relationship between independent factors and response, and applies to the problem of multivariate control system response (Pandey et al. 2018; Ralph 1989). This method allows us to obtain the best response efficiently using fewer experiments.

The Box-Behnken design approach was used in this investigation. The five influencing elements were intake temperature of reactants, velocity, water-alcohol ratio, input temperature, and velocity of hot air, and the range of values of the influencing factors is presented in Table 2. In addition, to assess the performance of the reformer reactor, methanol conversion and CO selectivity were used as indicators, as indicated below.

The methanol conversion is described as

$$X_{CH_3OH} = \frac{C_{CH_3OH,in} - C_{CH_3OH,out}}{X_{CH_3OH,in}} \times 100\% \tag{18}$$

The CO selectivity is described as

$$S_{CO_2} = \frac{C_{CO,out}}{C_{CO_2,out+C_{CO,out}}} \times 100\% \tag{19}$$

Table 2. Range of influence factors.

Factors	Range
Inlet temperature of reactant, $T_{in,reactant}$ (K)	393–473
Inlet temperature of hot air, $T_{in,air}$ (K)	523–673
Inlet velocity of reactant, $U_{in,reactant}$ (m/s)	0.05–0.25
Inlet velocity of hot air (m/s), $U_{in,air}$ (m/s)	0.5–2.5
Steam to carbon ratio, S/C	1.0–2.0

Methanol conversion and carbon monoxide selectivity are improved in this paper to increase methanol conversion while minimizing carbon monoxide selectivity. A total of 46 experiment sets were created. An ANOVA was performed on the model to test its correctness, and Table 3 displays the ANOVA findings for the two responses of methanol conversion and CO selectivity. The F-values of both responses, methanol conversion, and CO selectivity, are considerable in Table 3. The P-values are less than 0.0001, indicating that the fitted model is highly statistically significant. The R^2 and Adj. R^2 is close to 1 and the difference between Adj. R^2 and Pred. R^2 is less than 0.2, which proves that the model is desirable (Mosayebi et al. 2019; Batebi et al. 2020). Moreover, Figures 3(a) and 3(b) present the results of the comparison between the predicted and actual values of methanol conversion and CO selectivity models based on the SMR method and the CFD simulation data, respectively, which shows that the predicted and actual values are very similar, indicating that the model exhibits a high degree of accuracy in predicting the data.

Table 3. ANOVA for methanol conversion and CO selectivity.

Response	F-value	P-value	R^2	Adj. R^2	Pred. R^2
X_{CH_3OH} (%)	481.39	<0.0001	0.9974	0.9953	0.9876
S_{CO} (%)	245.99	<0.0001	0.9988	0.9948	0.9258

4.2 *Effect of exhaust gas inlet parameters*

To evaluate the effect of the exhaust gas on the hydrogen production performance of the reformer, the effects of the inlet temperature and velocity of the exhaust gas on the methanol conversion and

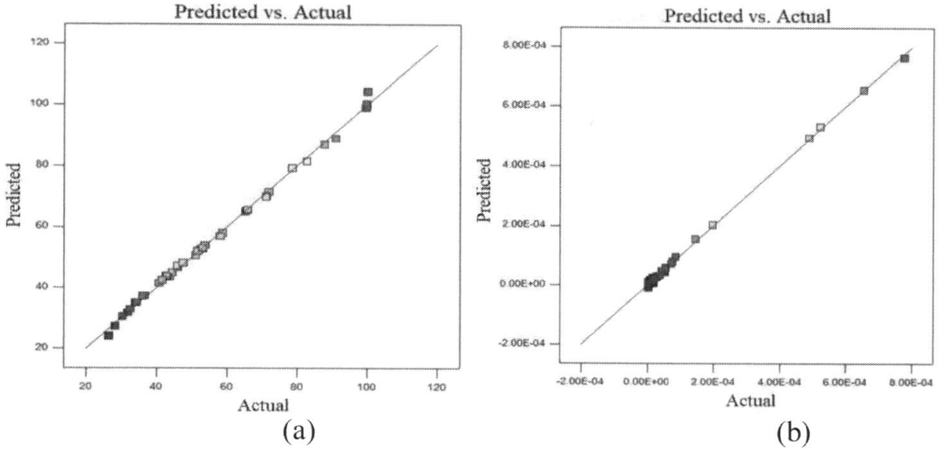

Figure 3. Comparison of actual values with model predictions: (a) Methanol conversion; (b) CO selectivity.

CO selectivity were investigated at the reactant inlet temperature of 433 K, the velocity of 0.15 m/s, and steam to methanol ratio of 1.5, as shown in Figures 4(a) and 4(b). Figure 4(a) shows that the methanol conversion steadily increases as the hot air input temperature and velocity increase. This is because MSR is a powerful endothermic reaction, and when the hot air input speed or temperature is sluggish or low, the hot air transfers less heat to the reactants per unit time and cannot deliver enough heat for the process. The heat delivered to the reactants rises as the hot air temperature and speed increase. The chemical reaction inside the catalyst develops quicker, the reforming reaction may run more smoothly, and the methanol conversion rate increases. The effect of hot air flow rate and temperature on CO selectivity is similar to that of methanol conversion, as shown in Figure 4(b). Because the rWGS reaction is also an endothermic reaction, temperature and flow rate rise, as does CO selectivity. The methanol conversion rose from 51.2% to 72.1% and the CO selectivity increased from 1.82 E-5 to 8.74 E-5 when the hot air temperature was 673 K and the speed increased from 0.5 m/s to 2.5 m/s. The methanol conversion of hot air increased from 43.8% to 72.1% when the hot air speed was 2.5 m/s and the air temperature was increased from 523 K to 673 K. Undoubtedly, the temperature of hot air had a significant impact on the performance of methanol reforming.

4.3 *Influence of reactant inlet conditions*

In this section, different inlet temperatures, velocities, and steam-to-methanol ratios of the reactants are discussed in detail with regard to their effects on methanol conversion and CO selectivity. Figure 5 shows how the temperature and velocity of the reactant inlet affect the rate of methanol conversion and CO selectivity. As the reactant temperature increases, methanol conversion and CO selectivity increase. As SMR and rWGS are endothermic reactions, the higher the reactant inlet temperature, in addition to the heat provided by the hot air, the better the reaction will be. At a constant inlet velocity (0.05 m/s), methanol conversion increased from 91.1% to 99.6% and CO selectivity increased from 1.46E-4 to 5.23E-4 as the temperature increased from 393 K to 473 K. Reactant velocity had the opposite effect of temperature on reformer performance, with methanol conversion and CO selectivity decreasing as velocity increased. A decrease in inlet velocity allows methanol vapor to remain in contact with the catalyst for a longer period of time, which enhances the chemical reaction. Meanwhile, the inlet velocity is reduced and the methanol vapor is able to absorb more heat from the exhaust gas during the same period, which also promotes the reforming reaction. The methanol conversion increased from 42.7% to 99.6% at constant temperature (473 K) when the

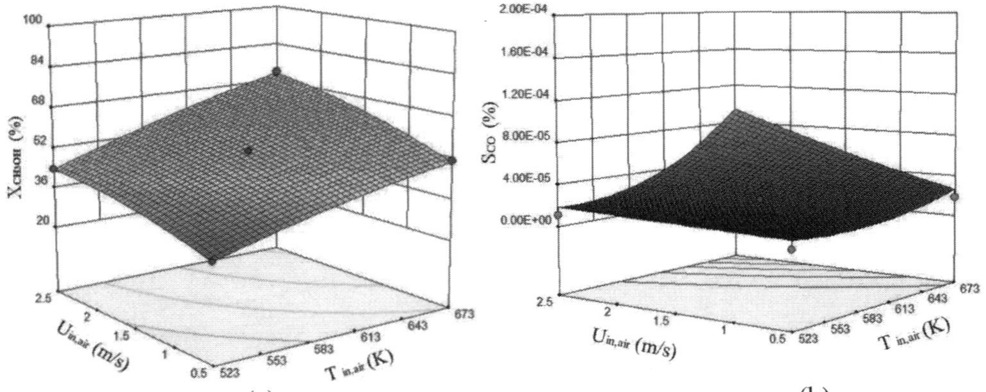

(a) (b)

Figure 4. Effect of exhaust gas inlet temperature and velocity on (a) methanol conversion, (b) CO selectivity.

inlet velocity was decreased from 0.25 m/s to 0.05 m/s, and the CO selectivity increased from 1.62 E-5 to 5.23 E-4 when the velocity was decreased from 0.25 m/s to 0.05 m/s. Furthermore, it can be seen that the influence of the reactant inlet velocity on the hydrogen production performance is much greater than that of the reaction temperature. As the inlet temperature of methanol vapor is raised and its velocity decreases, the methanol conversion can certainly increase. However, the CO yields also increase, and the catalyst is also more prone to sintering, which undoubtedly has an adverse effect on the reaction. It is therefore important to determine the optimum inlet temperature and speed.

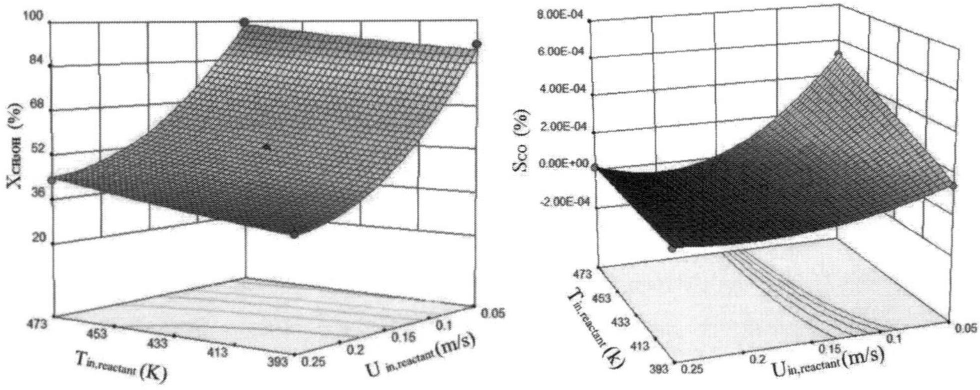

(a) (b)

Figure 5. Effect of reactant inlet temperature and velocity on (a) methanol conversion, (b) CO selectivity.

Figure 6 represents the effect of reactant inlet velocity and water-alcohol ratio on methanol conversion and CO selectivity when the reactant temperature was 433 K. As the water-alcohol ratio increased, methanol conversion and CO selectivity increased, but the increase was not very large, as shown in Figure 7. This indicates that increasing the water-to-alcohol ratio has little effect on methanol conversion, which is similar to the results of Srivastava et al. (Srivastava, Kumar, Dhar 201). Figure 5 shows that the methanol conversion and CO selectivity decreased significantly as the reactant inlet velocity increased, and the effect of reactant inlet velocity on the hydrogen production efficiency of the reformer was significant as well.

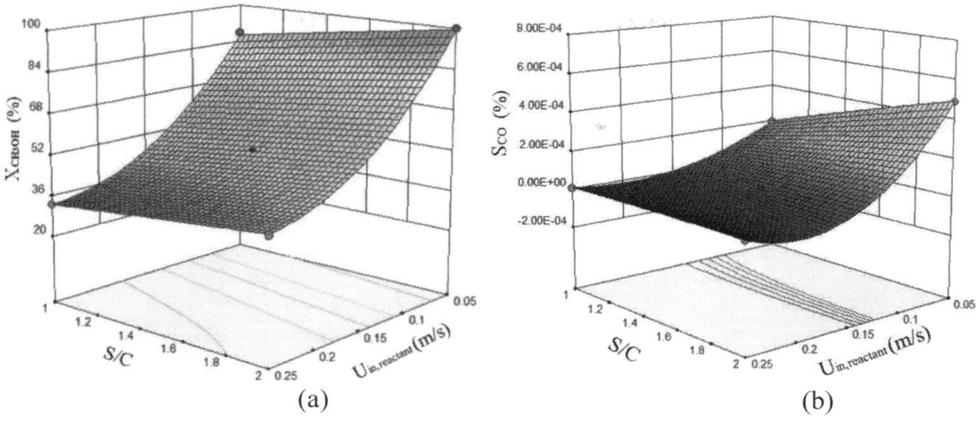

Figure 6. Effect of reactant inlet velocity and steam to methanol ratio on (a) methanol conversion, and (b) CO selectivity.

Figure 7 shows the effect of the water-to-alcohol ratio and inlet temperature on the hydrogen production performance of the reformer at a constant reactant inlet velocity (0.15 m/s). According to the results of the previous analysis, the methanol conversion and CO selectivity increased as the steam to methanol ratio and the inlet temperature increased, but the increase was not significant.

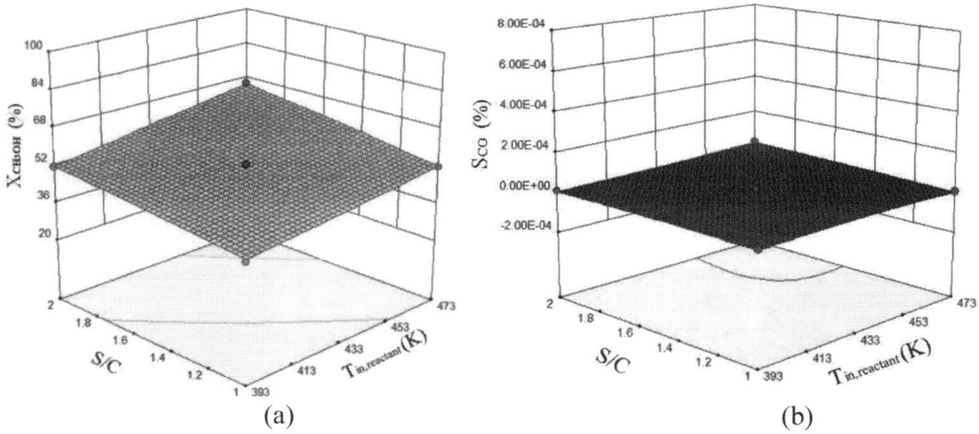

Figure 7. Effect of reactant inlet temperature and steam to methanol ratio on (a) methanol conversion, (b) CO selectivity.

4.4 *Analysis of optimization results*

Following the previous analysis, it is evident that the exhaust gas temperature, reactant inlet velocity, and steam to methanol ratio have an impact on the reformer hydrogen production performance. CFD numerical simulations and the response surface method have been used to optimize several parameters affecting the hydrogen production performance of the reformer. The methanol conversion and CO selectivity were selected as optimization targets to attain maximum methanol conversion and minimum CO selectivity. The optimized parameters are shown in Table 4. The optimized methanol conversion was 83.73% and CO selectivity was 3.64 E-6. A CFD simulation was conducted with the optimized reactant and hot air inlet temperatures, velocity, and steam to methanol ratio as input conditions. The optimized methanol conversion was 84.72 %, and the CO selectivity was 3.82 E-6. According to our results, the relative errors of methanol conversion and CO selectivity were 1.2% and 4.7%, respectively, which are within acceptable limits. The results also demonstrate the validity of the optimized model.

Table 4. Optimum values of variables obtained from the SMR performance.

Factors	Optimum value
The inlet temperature of reactant, $T_{in,reactant}$	393 K
The inlet temperature of hot air, $T_{in,air}$	584.6 K
Inlet velocity of reactant, $U_{in,reactant}$	0.05 m/s
Inlet velocity of hot air (m/s), $U_{in,air}$	1.35 m/s
Steam to carbon ratio, S/C	1.34
Response	Optimum value
Methanol conversion (CFD simulation value)	84.72%
Methanol conversion (RSM predicted value)	83.73%
CO selectivity (FD simulation value)	3.82 E-6
CO selectivity (RSM value)	3.64 E-6

5 CONCLUSIONS

Using CFD and RSM, this research evaluates and optimizes the methanol steam reforming reactor with waste heat recovery for hydrogen production. We investigated the influence of temperature, velocity, methanol/steam ratio, and exhaust gas inlet temperature and velocity on reactor performance using methanol conversion and CO selectivity as two responses, and our results are presented below.

(1) The performance of the reformer was enhanced as the methanol conversion and CO selectivity increased with the increase of the off-gas inlet temperature and velocity.
(2) The alcohol conversion and CO selectivity increase when the reactant inlet temperature rises and the inlet velocity falls, although the influence of inlet velocity on reformer performance is significantly larger than that of inlet temperature.
(3) Methanol conversion and CO selectivity increase slightly as the steam to methanol ratio increases, but the effect on reformer performance is smaller.
(4) After response surface optimization, the reactant input temperature was 393 K, the velocity was 0.05 m/s, the steam to methanol ratio was 1.34, and the exhaust gas inlet temperature and velocity were 584.6 K and 1.35 m/s, respectively. The CO selectivity was 3.64 E-6, and the methanol conversion was 83.73%.

The application of methanol reforming to onboard hydrogen production is a promising development, and the reformer plays an essential role in the methanol reforming process for hydrogen production. This study analyses and optimizes important parameters that influence the performance of the reformer, and the results could provide some useful recommendations for the use of engine waste heat and engine mixed hydrogen combustion in the future. However, the dynamic performance of vehicle exhaust gases is highly variable, and there are still many studies that need to be undertaken in order to fully apply this technology in practice. These include the optimization of exhaust gas utilization, the enhancement of heat transfer, and the improvement of reactor performance. Furthermore, it would be beneficial to combine the reactor and engine sections and conduct experiments on the reforming of engine waste heat for hydrogen production under different operating conditions in order to gain a better knowledge of the actual performance of the system.

REFERENCES

Agrell J, Birgersson H and Boutonnet M 2002 Steam reforming of methanol over a Cu/ZnO/Al2O3 catalyst: a kinetic analysis and strategies for suppression of CO formation *Journal of Power Sources* **106** 249–57.

Batebi D, Abedini R and Mosayebi A 2020 Combined steam and CO2 reforming of methane (CSCRM) over Ni–Pd/Al2O3 catalyst for syngas formation *International Journal of Hydrogen Energy* **45** 14293–310.

Chen J and Li L 2020 Thermal management of methanol reforming reactors for the portable production of hydrogen *International Journal of Hydrogen Energy* **45** 2527–45.

Costa D and Grundstein A 2016 An Analysis of Children Left Unattended in Parked Motor Vehicles in Brazil *International Journal of Environmental Research and Public Health* **13** 649.

Ergun S 1952 Fluid flow through packed columns *Fluid Flow Through Packed Columns* **48** 89–94.

Fukahori S, Koga H, Kitaoka T, Nakamura M and Wariishi H 2008 Steam reforming behavior of methanol using paper-structured catalysts: Experimental and computational fluid dynamic analysis *International Journal of Hydrogen Energy* **33** 1661–70.

Goldmann A, Sauter W, Oettinger M, Kluge T, Schröder U, Seume J R, Friedrichs J and Dinkelacker F 2018 A Study on Electrofuels in Aviation *Energies* **11** 392.

Karim A, Bravo J and Datye A 2005 Nonisothermality in packed bed reactors for steam reforming of methanol *Applied Catalysis A: General* **282** 101–9.

Kumar C, Rana K B and Tripathi B 2019 Effect of diesel-methanol-nitromethane blends combustion on VCR stationary CI engine performance and exhaust emissions *Environmental Science and Pollution Research* **26** 6517–31.

Li H, Ma C, Zou X, Li A, Huang Z and Zhu L 2021 On-board methanol catalytic reforming for hydrogen Production-A review *International Journal of Hydrogen Energy* **46** 22303–27.

Mosayebi A, Nasabi M and Abedini R 2019 Evaluation and modeling of Fischer-Tropsch synthesis in presence of a Co/ZrO2 catalyst *Petroleum Science and Technology* **37** 2338–49.

Pandey A, Belwal T, Sekar K C, Bhatt I D and Rawal R S 2018 Optimization of ultrasonic-assisted extraction (UAE) of phenolics and antioxidant compounds from rhizomes of Rheum moorcroftianum using response surface methodology (RSM) *Industrial Crops and Products* **119** 218–25.

Pashchenko D, Gnutikova M and Karpilov I 2020 Comparison study of thermochemical waste-heat recuperation by steam reforming of liquid biofuels *International Journal of Hydrogen Energy* **45** 4174–81.

Perng S-W, Horng R-F and Ku H-W 2013 Numerical predictions of design and operating parameters of reformer on the fuel conversion and CO production for the steam reforming of methanol *International Journal of Hydrogen Energy* **38** 840–52.

Prapinagsorn W, Sittijunda S and Reungsang A 2018 Co-Digestion of Napier Grass and Its Silage with Cow Dung for Bio-Hydrogen and Methane Production by Two-Stage Anaerobic Digestion Process *Energies* **11** 47.

Purnama H, Ressler T, Jentoft R E, Soerijanto H, Schlögl R and Schomäcker R 2004 CO formation/selectivity for steam reforming of methanol with a commercial CuO/ZnO/Al2O3 catalyst *Applied Catalysis A: General* **259** 83–94.

Ralph C S J 1989 Response Surfaces: Designs and Analyses *Technometrics*.

Santos Andrade T, Papagiannis I, Dracopoulos V, César Pereira M and Lianos P 2019 Visible-Light Activated Titania and Its Application to Photoelectrocatalytic Hydrogen Peroxide Production *Materials* **12** 4238.

Shu J, Fu J, Ren C, Liu J, Wang S and Feng S 2020 Numerical investigation on flow and heat transfer processes of novel methanol cracking device for internal combustion engine exhaust heat recovery *Energy* **195** 116954.

Srivastava A, Kumar P and Dhar A 2021 A numerical study on methanol steam reforming reactor utilizing engine exhaust heat for hydrogen generation *International Journal of Hydrogen Energy* **46** 38073–88.

Wang J, Wang H and Hu P 2018 Theoretical insight into methanol steam reforming on indium oxide with different coordination environments *Science China Chemistry* **61** 336–43.

Zhang Y, Li H, Han W, Bai W, Yang Y, Yao M and Wang Y 2018 Improved design of supercritical CO2 Brayton cycle for coal-fired power plant *Energy* **155** 1–14.

Energy Revolution and Chemical Research – Kok-Keong Chong and Zhongliang Liu (Eds)
© 2023 The Authors, ISBN: 978-1-032-36554-1

Study on pyrolysis characteristics of large-scale metabituminous coal in underground gasification based on thermogravimetric tests

Xiaojin Fu
North China Geological Exploration Bureau of Tianjin, Tianjin, China

Jing Wu*, Bowei Wang, Mingze Feng, Kaixuan Li & Jiaze Li
Key Laboratory of Mining Disaster Prevention and Control, Shandong University of Science and Technology, Qingdao, China
College of Safety and Environmental Engineering, Shandong University of Science and Technology, Qingdao, Shandong, China

ABSTRACT: During the underground gasification of coal, a large amount of coal pyrolysis occurs in the coal seam due to the special characteristics of the gasification channel. In this paper, the metabituminous coal from the Qianjiaying coal mine in Tianjin was used as the experimental object to carry out thermogravimetric simulations of pyrolysis of 15cm large-scale metabituminous coal under N_2 atmosphere and CO_2 atmosphere and to study the distribution pattern of pyrolysis products of large-scale metabituminous coal. It was found that in large-scale metabituminous coal pyrolysis, the activation energy of the pyrolysis reaction does not differ much when CO_2 and N_2 are used as the pyrolysis atmosphere, and the hysteresis of temperature transfer has a greater influence on the activation energy of metabituminous coal pyrolysis; the generation of CH_4 mainly occurs in the depolymerization stage, the escape of CH_4 starts to increase continuously from 300°C, and the CH_4 content starts to decrease around 700°C; while the generation of H_2 mainly occurs in the production in the polycondensation stage, the escape of H_2 from metabituminous coal pyrolysis starts from above 500°C, and the H_2 content continues to increase with the increase of temperature.

1 INTRODUCTION

China's energy has the characteristic of more coal and less oil, where coal resources are abundant and widely distributed, and the coal economy occupies a very important position in China's economy and social development (Li & Feng 2013; Peng et al. 2020). However, the traditional coal mining, transportation, and transformation processes generally suffer from low efficiency of coal resources utilization and serious environmental pollution, and about 50% of coal resources are abandoned underground due to the limitation of coal mining technology. Therefore, based on the spirit of the report of the 16th Party Congress, Academician Qian Minggao proposed the concept of green mining in coal mines (Xu 2019). The research and development of clean and efficient utilization technology of coal is an important foundation for the sustainable development of energy in China, serving as a necessary choice for the coordinated development of society and economy, and environment in China (Wang et al. 2019).

Underground coal gasification technology is a kind of mining method that low-carbonizes high-carbon resources and fluidizes solid resources (Krzysztof et al. 2012; Xu et al. 2020; Yang et al. 2008), which can turn coal into combustible gas under the action of high temperature in situ underground, and extract it for utilization, changing the way of traditional coal mining into chemical

*Corresponding Author: 3523684533@qq.com

DOI 10.1201/9781003332657-42

gas mining (Friedmann et al. 2009; Sajjad & Rasul 2015). UCG integrates the three processes of the well construction, coal mining, and surface gasification into one, involving geology, coal mining, rock mechanics, thermal engineering, coal chemistry, and other disciplines. It has the advantages of good safety, low investment, high efficiency, and low pollution (Xin et al. 2019). The coal gasification process is divided into two stages according to the different reactions: pyrolysis of coal and gasification of coke, and the different conditions of coal pyrolysis will have a great impact on the subsequent gasification reaction.

Experts and scholars mainly study the pyrolysis characteristics of small-scale coal blocks in the UCG process and find that the particle size of coal has a significant impact on the pyrolysis characteristics. The larger the particle size, the more loss in the pyrolysis process. The pyrolysis characteristics of large coal have directive significance for the pyrolysis of the underground coal gasification process. However, there is a lack of corresponding research on the change of organic matter in the pyrolysis process of large-scale coal samples from underground gasification, especially on the output law of characteristic products of large-scale metabituminous coal under different pyrolysis conditions. In this study, the pyrolysis conditions of metabituminous coal in the Qianjiaying Coal Mine were selected to simulate the underground gasification process. The large-scale metabituminous coal pyrolysis test was carried out in the N_2 atmosphere and CO_2 atmosphere, and the distribution law of 15cm large-scale metabituminous coal pyrolysis products under different temperature conditions was obtained.

2 MATERIALS AND METHODS

2.1 *Preparation of coal samples*

Using the metabituminous coal from Qianjiaying Coal Mine in Tianjin as the experimental object, firstly, the large metabituminous coal was selected, and cut into lump coal with a side length of 15cm with a cutting machine, holes were drilled in the middle part of the lump coal, so that the thermocouple can be inserted smoothly to measure the central temperature of the coal, and packed in a sealed bag for the thermogravimetric experiment of pyrolysis of large coal. An appropriate amount of metabituminous coal was taken for elemental analysis and industry analysis to understand the characteristic parameters of coal and provide the basis for pyrolysis experimental analysis. The measurement results of various parameters are shown in Table 1:

Table 1. Coal sample parameters of metabituminous coal in Qianjiaying Coal Mine.

	Industrial analysis						Elemental analysis				
Coal	Mad/%	Ad/%	Vd/%	Vdaf/%	Crc/%	Fcd/%	Cdaf/%	Hdaf/%	Ndaf/%	Odaf/%	St/%
Metabituminous coal	0.84	17.07	32.09	38.7	7	50.84	81.39	4.82	0.79	13	1.1

2.2 *Test method*

The large-scale metabituminous coal pyrolysis is carried out under a large-scale metabituminous coal pyrolysis thermogravimetric experimental apparatus. The metabituminous coal is heated by a resistance wire in the pyrolysis reactor, and the heating rate of the metabituminous coal is controlled using an electric current. The gravity and temperature sensors measure the weight and temperature of the system respectively and output the current signals, which are converted into gravity values as well as temperature values using Wonderware to output gravity and temperature values.

The calibration of the thermogravimetric system and the tare of the reactor are carried out before the large-scale metabituminous coal pyrolysis experiment. 30 KG standard weights are used to calibrate the thermogravimetric system and after the calibration is completed, the reactor is connected in the experimental condition, and tare is carried out on the gravity module. The prepared coals were then placed in the reactor, and temperature measuring thermocouples were inserted onto the surface of the coals and inside the coals to measure the internal and external temperature distribution. The pyrolysis gas is introduced 5 minutes in advance to replace the residual gas in the original reactor, the gas flow rate is adjusted to 0.5L/min, and the sealing performance of the whole system is tested with foam water. The current is adjusted to 10A to control the slow temperature rise of the system. Finally, the tail gas and tar produced during the pyrolysis process are collected and the power is cut off after the system has cooled down slowly.

3 RESULTS AND DISCUSSION

The pyrolysis experiment of 15cm large-scale metabituminous coal was carried out with N_2 and CO_2 as carrier gas respectively, and the pyrolysis TG curve was obtained, as shown in Figure 1:

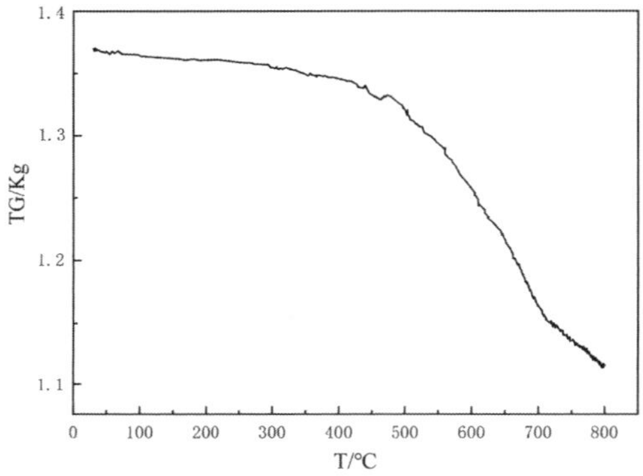

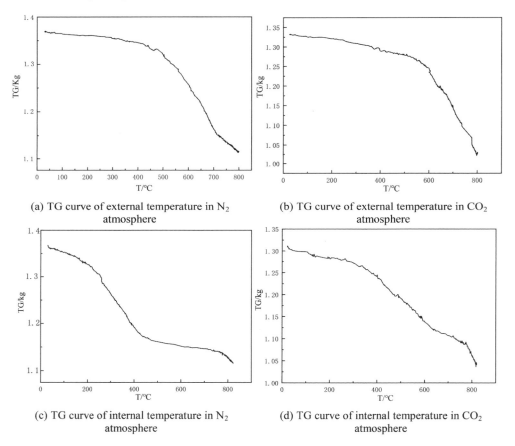

(a) TG curve of external temperature in N_2 atmosphere

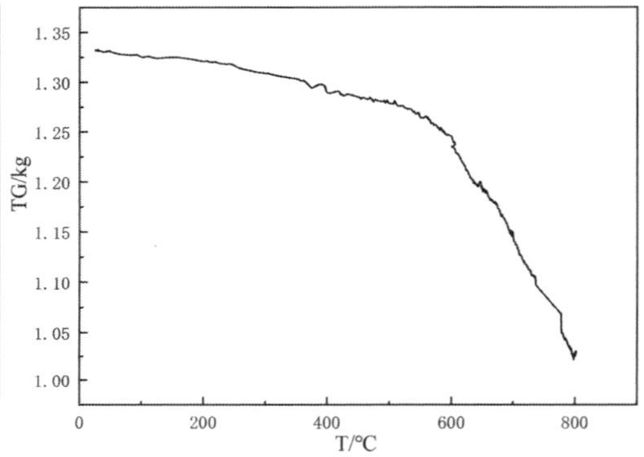

(b) TG curve of external temperature in CO_2 atmosphere

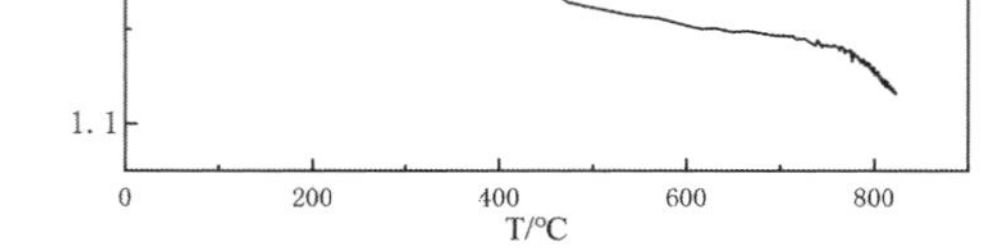

(c) TG curve of internal temperature in N_2 atmosphere

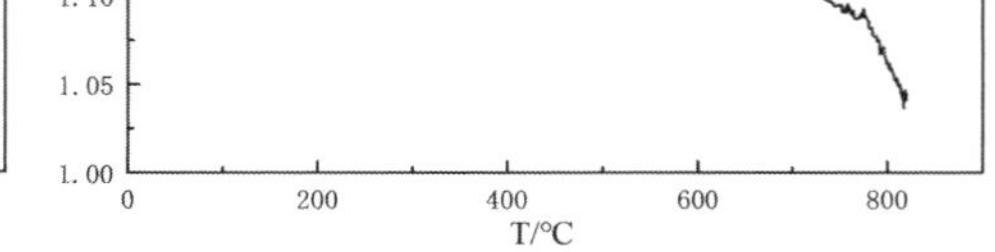

(d) TG curve of internal temperature in CO_2 atmosphere

Figure 1. Pyrolysis TG curve of 15cm large-scale metabituminous coal.

From the TG curve, it can be seen that the influence of the atmosphere on the TG curve during the pyrolysis of 15cm metabituminous coal is little. When 15cm metabituminous coal is pyrolysis in N_2 and CO_2 atmosphere, the temperature at which a large amount of loss of focus is reached is 500°C outside temperature and 300°C inside temperature respectively. Due to the lag of heat transfer, when the coal surface temperature reaches the metabituminous coal pyrolysis temperature, the metabituminous coal does not start a large number of pyrolysis, while when the internal temperature reaches the pyrolysis temperature, the coal starts a large number of pyrolysis, and the rapid decline point of weight loss curve appears. This shows that in the process of metabituminous coal pyrolysis, the coal surface temperature is not the main factor determining metabituminous coal pyrolysis, and the internal temperature plays a decisive role in metabituminous coal pyrolysis. Therefore, increasing the reaction residence time of metabituminous coal pyrolysis can fully transfer the temperature and increase the pyrolysis efficiency of metabituminous coal.

The heating rate and time curve of 15cm metabituminous coal pyrolysis is shown in Figure 2:

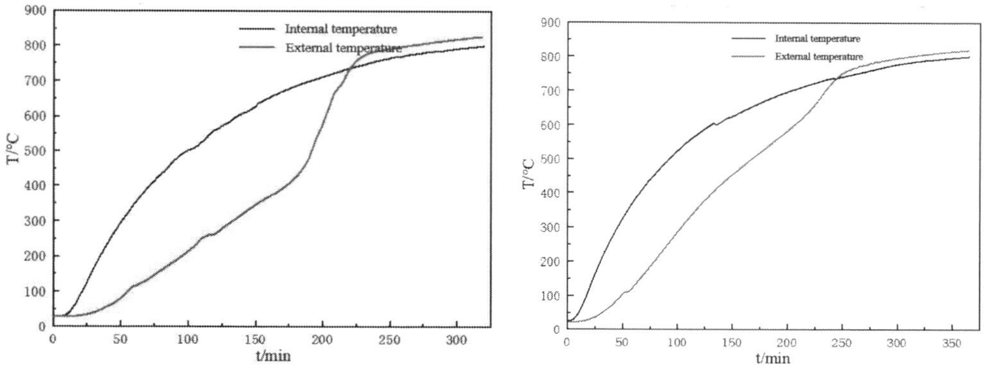

(a) N_2 atmosphere (b) CO_2 atmosphere

Figure 2. Temperature rise rate curve of 15cm metabituminous coal pyrolysis.

It can be seen from the heating rate curve that during the pyrolysis of metabituminous coal, the surface temperature of metabituminous coal basically remains unchanged when the heating rate is lower than 600°C and decreases when it exceeds 600°C. The internal temperature of metabituminous coal is lower than the surface temperature at the initial stage of the reaction. When the surface temperature is higher than 700°C and the internal temperature is higher than 450°C, the heating rate of the internal temperature is higher than the external temperature. This is because when the temperature is higher than 700°, the free radicals produced by the pyrolysis reaction of metabituminous coal combine with each other and produce an exothermic reaction. The surface of metabituminous coal is in the environment of the reactor, the gas in contact with it has heat transfer, and because the metabituminous coal is surrounded by metabituminous coal, the temperature of an exothermic reaction is not transmitted in time, so that the temperature rise rate of metabituminous coal is greater than that of metabituminous coal surface. At about 760°C, the inside and outside of metabituminous coal reach the same temperature and continue pyrolysis, and the internal temperature of metabituminous coal is higher than the external temperature.

The variation trend of pyrolysis component distribution of 15cm metabituminous coal with temperature is shown in Figure 3:

It can be seen from the component distribution diagram that the main gas components released during the pyrolysis of metabituminous coal are CH_4, H_2, and CO. Among them, CH_4 maintains a large escape concentration during the whole process, and the escape of H_2 is about 500°C. With the increase in temperature, the H_2 content continues to increase, the CH_4 content begins to decrease at about 700°C, and the H_2 content begins to decrease at about 750°C. This shows that in the process of coal pyrolysis, the content of hydrogen bond is higher than that of methyl, and the hydrogen bond breaking temperature needs to be higher than 500°C. During pyrolysis in N_2 and CO_2 atmosphere, the content of H_2 and CO in the CO_2 atmosphere is higher than that in the N_2 atmosphere, and the content of CH_4 in the N_2 atmosphere is higher than that in the CO_2 atmosphere, which is caused by the reforming reaction between CH_4 and CO_2 at high temperature. The reforming reaction is favorable for the underground gasification process aiming at the generation of syngas ($CO + H_2$).

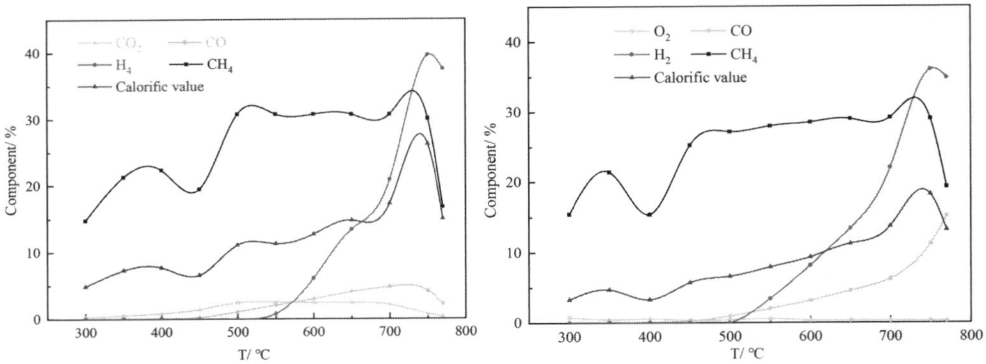

(a) N_2 atmosphere (b) CO_2 atmosphere

Figure 3. Pyrolysis composition temperature diagram of 15cm metabituminous coal.

The fitting curve of the 15cm metabituminous coal pyrolysis kinetic equation is shown in Figure 4:

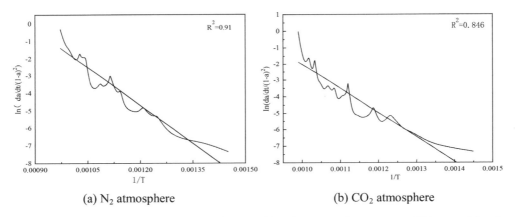

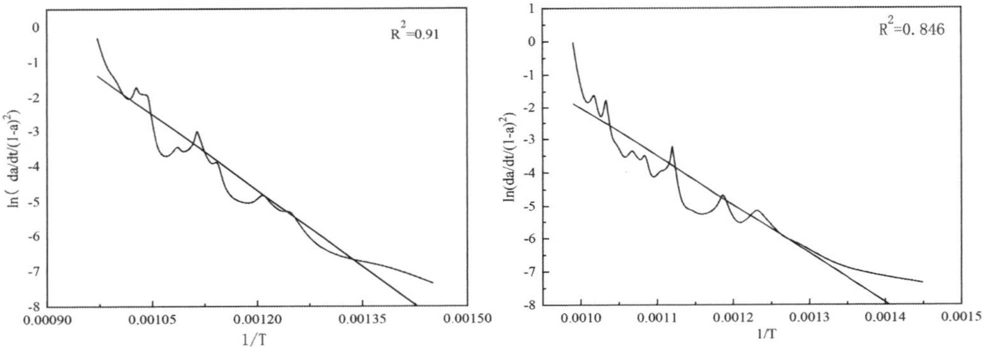

(a) N_2 atmosphere (b) CO_2 atmosphere

Figure 4. Kinetic fitting curve of metabituminous coal based on external temperature of 15cm metabituminous coal pyrolysis.

When the external temperature is used as the basis of the kinetic equation, the pyrolysis activation energies of 15cm metabituminous coal in the N_2 and CO_2 atmospheres are 120.76 KJ/mol and 122.6 KJ/mol respectively. The pyrolysis activation energies with CO_2 and N_2 as carrier gas are basically the same. It can be seen that the different atmosphere of metabituminous coal in the pyrolysis reaction process has little effect on the pyrolysis activation energy.

Taking the internal temperature of metabituminous coal as the kinetic equation, the fitting curve of the 15cm metabituminous coal pyrolysis kinetic equation is obtained, as shown in Figure 5:

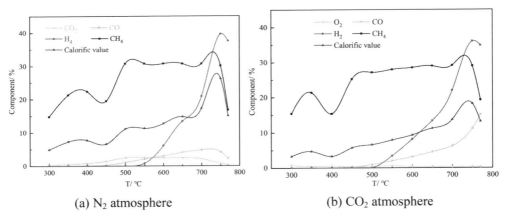

When the internal temperature of metabituminous coal is taken as the research object of the kinetic equation, the activation energy of pyrolysis of 15cm metabituminous coal in N_2 and CO_2 atmosphere is 47.77 KJ/mol and 36.82 KJ/mol respectively. In the process of metabituminous coal pyrolysis, the apparent activation energy of surface temperature is greater than that of internal temperature. According to the definition of activation energy, the energy required for pyrolysis at internal temperature is much lower than that at external temperature. This is also because in the pyrolysis process of large-scale metabituminous coal, due to the lag of temperature transmission,

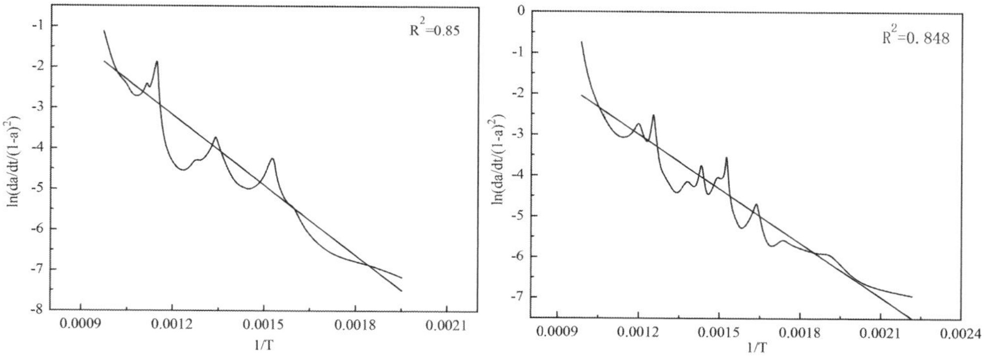

(a) N_2 atmosphere

(b) CO_2 atmosphere

Figure 5. Kinetic fitting curve of metabituminous coal based on the internal temperature of 15cm metabituminous coal pyrolysis.

when the internal temperature has not reached the minimum temperature required for pyrolysis, the external coal has begun pyrolysis. In this regard, the internal energy required is less than the external energy.

4 CONCLUSION

(1) For large-scale metabituminous coal pyrolysis, there is little difference in reaction activation energy when CO_2 and N_2 are used as pyrolysis atmosphere. When CO_2 is used in the pyrolysis atmosphere, the content of H_2 and CO in metabituminous coal pyrolysis gaseous material is significantly higher than that when N_2 is used in the pyrolysis atmosphere, which is beneficial to the underground gasification process with the main goal of producing syngas (CO + H_2).

(2) In the large-scale metabituminous coal pyrolysis process, quickly transferring heat to the interior is an important means to improve the pyrolysis efficiency of metabituminous coal. Increasing the residence time of the pyrolysis process can be used to fully transfer the temperature, so as to obtain more pyrolysis products.

(3) The escape of H_2 from metabituminous coal pyrolysis starts from above 500°C, and the escape of CH_4 continues to increase from 300°C. With the increase in temperature, the H_2 content continues to increase, the CH_4 content begins to decrease at about 700°C, and the H_2 content begins to decrease at about 750°C. This is because the generation of CH_4 mainly occurs in the depolymerization stage, while the generation of H_2 mainly occurs in the polycondensation stage. Therefore, the generation of H_2 requires a higher temperature.

ACKNOWLEDGMENT

This study was financially supported by the Shandong Natural Science Foundation (No. ZR2020ME084), the National Natural Science Foundation of China (No. 51504142), the Qingchuang Science and Technology Program of Shandong Province University (No. 2019KJG008), the Scientific Research Foundation of Shandong University of Science and Technology for Recruited Talents (No. 2017RCJJ013), the SDUST Research Fund (No. 2018TDJH102), and the Geological Survey Project of China Geological Survey (No. DD20190182).

REFERENCES

Friedmann SJ, Upadhye R, Kong FM. Prospects for underground coal gasification in the carbon-constrained world [J]. *Energy Procedia*, 2009,1(1):4551–4557.

Guofa Wang, Yongxiang Xu, Huaiwei Ren. Intelligent and ecological coal mining as well as clean utilization technology in China: Review and prospects[J]. *International Journal of Mining Science and Technology*, 2019, 29(02):161–169.

Jialin Xu. Strata control and scientific coal mining —A celebration of the academic thoughts and achievements of Academician Minggao Qian [J]. *Journal of Mining & Safety Engineering*, 2019, 36(01):1–6 [in Chinese].

Junlin Li Yongcheng Feng. A Preliminary Discussion on the Coupling Process of fluidized bed low-temperature pyrolysis and transport integrated gasification[J]. *Coal Chemical Industry*. 2013; 41:29–31 [in Chinese].

Krzysztof, S., et al. Experimental simulation of hard coal underground gasification for hydrogen production. *Fuel*, 91(2012) 40–50.

Lanhe Yang, Xing Zhang, Shuqin Liu, et al. Field test of large-scale hydrogen manufacturing from underground coal gasification (UCG). *International Journal of Hydrogen Energy*, 2008. 33(4): 1275–1285.

Lin Xin, Weimin Cheng, Jun Xie, et al. Theoretical research on heat transfer law during underground coal gasification channel extension process. *International Journal of Heat and Mass Transfer*. 2019,142:118409.

Min Xu, Lin Xin, Weitao Liu, et al. Study on the physical properties of coal pyrolysis in underground coal gasification channel[J]. *Powder Technology*, 376 (2020) 573–592.

Sajjad M, Rasul MG. Prospect of underground coal gasification in Bangladesh [J]. *Procedia Engineering*, 2015,105:537–548.

Zhengfu Peng, Xiaojun Ning, Guangwei Wang, et al. Structural characteristics and flammability of low-order coal pyrolysis semi-coke [J]. *Journal of the Energy Institute*, 2020,93(4):1341–1353.

Energy Revolution and Chemical Research – Kok-Keong Chong and Zhongliang Liu (Eds)
© 2023 The Authors, ISBN: 978-1-032-36554-1

Determination of iodine in food by sodium nitrite-iodine-starch system

Changqing Tu & Xinrong Wen*
College of Chemistry and Environment, Guangdong Provincial Key Laboratory of Conservation and Precision Utilization of Characteristic Agricultural Resources in Mountainous Areas
Jiaying University, Meizhou, Guangdong, P.R. China

ABSTRACT: A novel method for the determination of iodine in food by the sodium nitrite-iodine-starch system has been studied. In an acidic sodium chloride medium, potassium iodide can be quantitatively reduced by sodium nitrite to form I_2, then a blue complex with a maximum absorption wavelength of 580 nm is formed by I_2 and starch. A new method for the indirect determination of iodide content is established. The various effect factors on the determination of iodide by the sodium nitrite-iodine-starch system are investigated in detail. The results show that in the optimal conditions, a good linear relationship is obtained between the potassium iodide content and the absorbance of the blue complex within the potassium iodide concentration of 16.00~38.40μg/mL, and the linear regression equation is A=-0.4358+0.0293C (μg/mL). This proposed method had been successfully applied to determine iodine in kelp, and the recovery yields of the standard addition test are 96.1%~103.7%.

1 INTRODUCTION

Iodine is one of the essential microelements for humans. It can enhance basic metabolism and promote the growth and development of the human body. In daily life, the iodine needed by the human body mainly stems from food. Both iodine deficiency and iodine excess can cause diseases such as goiter and do harm to the human body. Thus, the study of the determination of iodine is of great importance and significance. On the basis of the literature report, the determination methods for iodine mainly included titration (Namekar S B 2017), spectrophotometry (Ajenesh C 2019; Araya M 2020), fluorescence spectroscopy (Hou 2018), electrochemical method (Espada-Bellido E 2017; Trésor K M 2008), ICP-MS (Hwang J B 2020; Márcia F M 2010; Ujang T 2013).

In this paper, a novel method for the determination of iodine in food by sodium nitrite-iodine-starch system is reported. In an acidic sodium chloride medium, I^-reacts with NO_2 to form I_2, then I_2 and starch form I_2-starch blue complex with a maximum absorption wavelength of 580 nm. Beer's law is obeyed between the potassium iodide content and the absorbance of the I_2-starch blue complex, the linear regression equation is A=-0.4358+0.0293C (μg/mL) within the potassium iodide concentration of 16.00~38.40 μg/mL. So, the content of I^- can be indirectly determined by measuring the absorbance of the I_2-starch blue complex. This proposed method has been applied to determine iodine in kelp with a satisfactory result.

2 EXPERIMENTAL

2.1 *Equipment and reagents*

UV-2401 UV-visible spectrophotometer (The Shimadzu Corporation, Japan); 723S spectrophotometer (Shanghai Precision & Scientific Instrument Co., Ltd).

*Corresponding Author: 198601012@jyu.edu.cn

DOI 10.1201/9781003332657-43

KI solution: 200.0 $\mu g \cdot mL^{-1}$. A stock of standard solution of 20.00 $mg \cdot mL^{-1}$ KI is prepared by dissolving 2.0000 g KI in 100 mL with distilled water. 1.00 mL 20.00 $mg \cdot mL^{-1}$ KI standard solution is transferred into another 100 mL brown volumetric flask, the solution is diluted to the mark with bidistilled water and mixed well, which is the 200.0 $\mu g \cdot mL^{-1}$ KI solution, shielding from light. $NaNO_2$ solution: 10.00 $mg \cdot mL^{-1}$. NaCl solution: 200.0 $g \cdot L^{-1}$. HCl solution: 1.0 $mol \cdot L^{-1}$. Starch solution:5.0 $g \cdot L^{-1}$. pH=6.86 buffer solution: KH_2PO_4-Na_2HPO_4, is prepared as references 10.

All reagents are of analytical reagent grade. Bidistilled water is used.

2.2 *Method*

A certain volume of KI solution or sample solution, NaCl solution 3.00 mL, HCl solution 0.50 mL, $NaNO_2$ solution 3.50 mL, starch solution 3.00 mL, pH=6.86 buffer solution 0.50 mL are added into a 25 mL volumetric flask. The solution is diluted to the mark with bidistilled water, mixed well, and placed at room temperature for 25 minutes in the dark. The absorbance of the I_2-starch blue complex is measured at 580 nm against the reagent blank.

3 RESULTS AND DISCUSSION

3.1 *Reaction mechanism*

(1) In NaCl medium, potassium iodide is quantitatively reduced by sodium nitrite to form I_2

$$I^- + NO_2 \rightarrow I_2$$

(2) A blue complex with a maximum absorption wavelength of 580 nm is formed by I_2 and starch.

$$I_2 + \text{starch} \rightarrow I_2 - \text{starch}$$

Beer's law is obeyed between the potassium iodide content and the absorbance of the I_2-starch blue complex. Therefore, the content of I^- can be indirectly determined by measuring the absorbance of the I_2-starch blue complex.

3.2 *Maximum absorption wavelength*

According to the experimental method, the determination solution of the I_2-starch blue complex formed from I^-, NO_2-, and starch in the system is prepared. At 500~650 nm, using a UV-2401 UV-visible spectrophotometer, the absorption spectrum of the I_2-starch blue complex is obtained (Figure 1). We can see from Figure 1 that the maximum absorption wavelength of the I_2-starch blue complex is 580 nm.

KI:3.00 mL; NaCl:5.00 mL; HCl:1.00 mL; $NaNO_2$:5.00 mL; starch:1.00 mL; pH=8.86 buffer solution:1.00 mL; reaction time:30 min.

3.3 *Reaction temperature*

The effect of the reaction temperature is seen in Table 1. When the reaction temperature is 25~40°C, the absorbance of the I_2-starch blue complex reaches its greatest and almost constant. Thereafter, the absorbance of the I_2-starch blue complex keeps decreasing with the increase of reaction temperature. Based on this result, the room temperature is used.

3.4 *Reaction time*

The effect of the reaction time is shown in Figure 2. It is found that the absorbance of the I_2-starch blue complex reaches greater and keeps constant when the reaction time is 25~40 min, and the

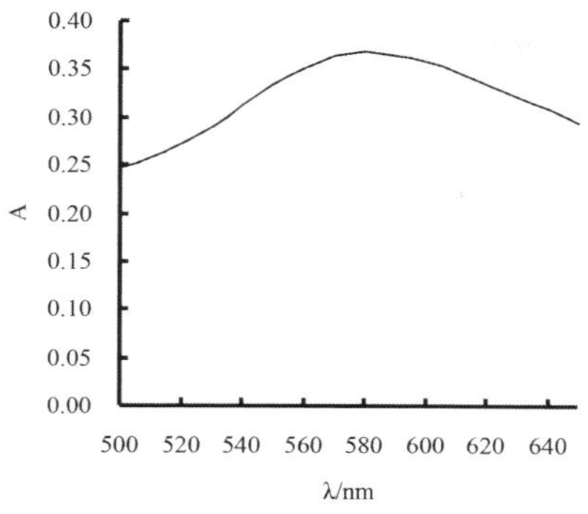

Figure 1. Absorption spectrum.

Table 1. The effect of reaction temperature on the absorbance.

Temperature /°C	25	30	35	40
Absorbance	0.378	0.377	0.378	0.376
Temperature /°C	45	50	55	60
Absorbance	0.367	0.361	0.354	0.342
Temperature /°C	65	70	75	80
Absorbance	0.330	0.325	0.320	0.312

Experimental conditions: KI:3.00 mL; NaCl:5.00 mL; HCl:1.00 mL; $NaNO_2$:5.00 mL; starch:1.00 mL; pH=8.86 buffer solution:1.00 mL; reaction time:30 min.

absorbance of the I_2-starch blue complex reaches the maximum value and remains substantially stable when the reaction time is 45~120 min. In terms of saving the measurement time, 25 minutes is selected as the reaction time.

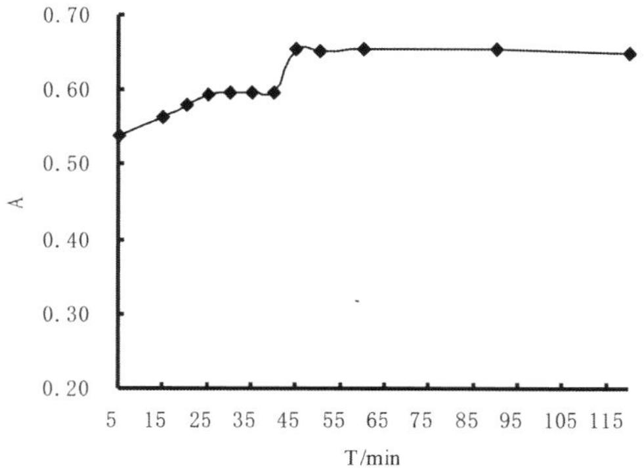

Figure 2. Effect of the reaction time.

KI:3.00 mL; NaCl:5.00 mL; HCl:1.00 mL; $NaNO_2$:5.00 mL; starch:1.00 mL; pH=8.86 buffer solution:1.00 mL.

3.5 *HCl solution dosage*

The effect of HCl solution dosage can be seen in table 2. We can see from Table 2 that the absorbance of the I_2-starch blue complex reaches the maximum value when the HCl solution dosage is 0.50 mL. So, 0.50 mL HCl solution is used.

Table 2. The effect of HCl solution dosage on the absorbance.

HCl solution dosage /mL	0.30	0.40	0.50	0.60
Absorbance	0.148	0.281	0.301	0.282
HCl solution dosage /mL	0.70	0.80	0.90	1.00
Absorbance	0.278	0.269	0.223	0.161

Experimental conditions: KI:3.00 mL; NaCl:5.00 mL; $NaNO_2$:5.00 mL; starch:1.00 mL; pH=8.86 buffer solution:1.00 mL; reaction time:25 min.

3.6 *NaCl solution dosage*

The NaCl solution acts as a medium to avoid the oxidation of I^- by O_2 in the air. The effect of NaCl solution dosage can be seen in Table 3. The results show that I_2-starch blue complex has a maximum absorption when the NaCl solution dosage is 4.00 mL. Therefore, the optimal dosage of NaCl is selected as 4.00 mL.

Table 3. The effect of NaCl solution dosage on the absorbance.

NaCl solution dosage /mL	1.00	2.00	3.00	4.00
Absorbance	0.351	0.323	0.385	0.397
NaCl solution dosage /mL	5.00	6.00	7.00	
Absorbance	0.372	0.362	0.336	

Experimental conditions: KI:3.00 mL; HCl:0.50 mL; $NaNO_2$:5.00 mL; starch:1.00 mL; pH=8.86 buffer solution:1.00 mL; reaction time:25 min.

3.7 *$NaNO_2$ solution dosage*

The effect of $NaNO_2$ solution dosage is seen in Figure 3.

The results show that the absorbance of the I_2-starch blue complex has the maximum value when the $NaNO_2$ solution dosage is 3.50 mL. Thus, 3.50 mL is employed.

KI:3.00 mL; NaCl:4.00 mL; HCl:0.50 mL; starch:1.00 mL; pH=8.86 buffer solution:1.00 mL; reaction time:25 min.

3.8 *Starch solution dosage*

The effect of starch solution dosage is shown in Table 4. We can see from Table 4 that the absorbance of the I_2-starch blue complex remains constant when the starch solution dosage is 2.50~4.00 mL. Hence, the starch solution dosage is chosen as 3.00 mL.

Experimental conditions: KI:3.00 mL; NaCl:4.00 mL; HCl:0.50 mL; $NaNO_2$:3.50 mL; pH=8.86 buffer solution:1.00 mL; reaction time:25 min.

3.9 *pH=6.86 buffer solution dosage*

KI:3.00 mL; NaCl:4.00 mL; HCl:0.50 mL; $NaNO_2$:3.50 mL; starch:3.00 mL; reaction time:25 min.

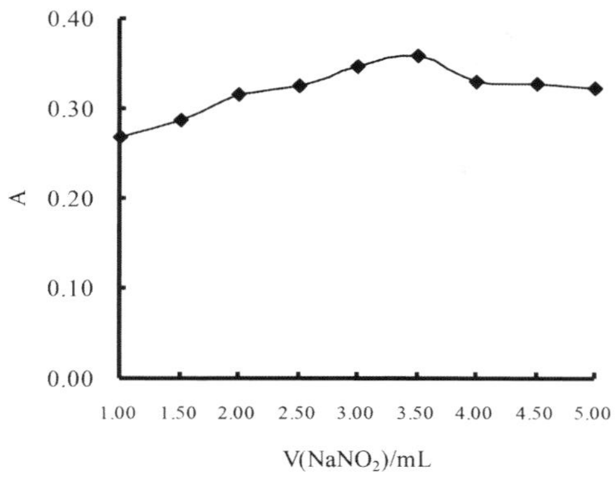

Figure 3. Effect of the $NaNO_2$ solution dosage.

Table 4. The effect of starch solution dosage on the absorbance.

Starch solution dosage /mL	0.50	1.00	1.50	2.00
Absorbance	0.231	0.322	0.344	0.362
Starch solution dosage /mL	2.50	3.00	3.50	4.00
Absorbance	0.348	0.340	0.342	0.341

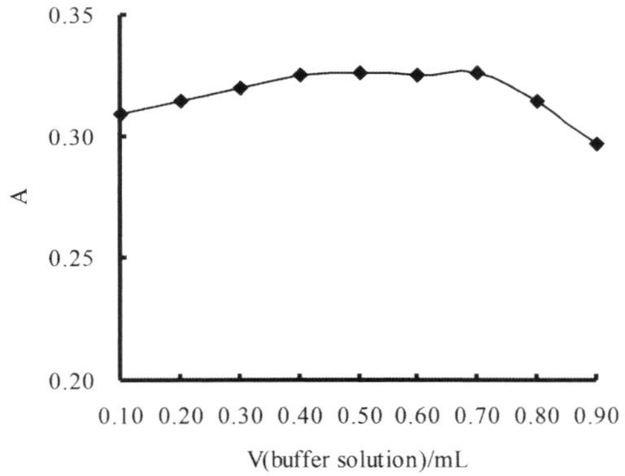

Figure 4. Effect of the pH=6.86 buffer solution dosage.

The effect of pH=6.86 buffer solution dosage is seen in Figure 4. Figure 4 shows that the absorbance of the I_2-starch blue complex absorbance is maximum and remains substantially unchanged when the pH=6.86 buffer solution dosage is 0.40~0.70 mL. Thus, 0.50 mL is applied.

3.10 *Calibration curve*

Under the optimum conditions, a series of determination solutions with different concentrations of KI are prepared and the absorbances of these solutions are measured at 580 nm against the reagent blank. The standard curve (Figure 5) is obtained using the concentration as the abscissa and the absorbance as the ordinate. Beer's law is obeyed between the concentration of KI and the absorbance in the range of 16.00~38.40 μg/mL KI concentration, the linear regression equation is $A=-0.4358+0.0293C(\mu g/mL)$ and the correlation coefficient is 0.9995.

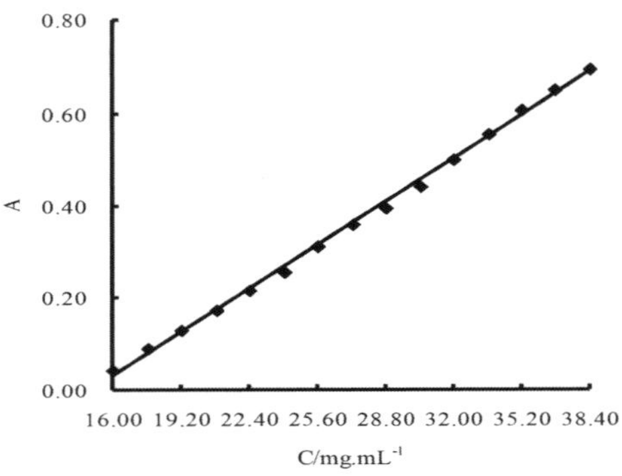

Figure 5. Calibration curve.

NaCl:4.00 mL; HCl:0.50 mL; $NaNO_2$:3.50 mL; starch:3.00mL; pH=6.86 buffer solution:0.50 mL; reaction time:25 min.

3.11 *Sample determination*

The kelp is dried in an oven at 105°C and crushed after the kelp sample is washed and cut into pieces. 50.00 g kelp sample is weighed and put in a crucible, then an appropriate amount of NaOH solution is added until the kelp sample is fully infiltrated. This kelp sample is carbonized on the electric stove until no smoke in the crucible, then it is put in the muffle furnace, heated to 600°C, and carbonized for 1.5 hours until the samples are all gray. 50.00 mL bidistilled water is added to the crucible and stirred. Then it is filtered and washed the slag with hot water, the filtrate is transferred to a 100 mL volumetric bottle, diluted to the mark with bidistilled water, and shaken well. This is the kelp sample solution.

According to the experimental method, 10.00 mL kelp sample solution is added, the absorbance of the I_2-starch blue complex is determined, and the content of the iodine is calculated. Meanwhile, the recovery tests of standard addition are performed. The results as shown in Table 5.

Table 5. The content of iodine in kelp.

Sample	Kelp
The content of iodine($\mu g \cdot g^{-1}$)	125.8
RSD (%)	0.2
Added ($\mu g \cdot mL^{-1}$)	2.400
	4.00
Recovered ($\mu g \cdot mL^{-1}$)	2.307
	4.149
Recovery (%)	96.1
	1037

Experimental conditions: NaCl:3.00 mL; HCl:0.50 mL; $NaNO_2$:3.50 mL; starch:3.00 mL; pH=8.86 buffer solution:0.50 mL; reaction time:25 min.

We can see from Table 5 that the content of iodine in kelp is 125.8$\mu g \cdot mL^{-1}$by this proposed method, and the recovery yields are 96.1%~103.7%.

4 CONCLUSION

A novel method for the determination of iodine by sodium nitrite-iodine-starch system has been established. This method has been successfully applied to the determination of iodine in kelp with satisfactory results. It is obvious that the determination of iodine by the sodium nitrite-iodine-starch system has certain practical significance and foreground of application.

REFERENCES

Ajenesh C, Matakite M, Surendra P. (2019) Determination of Iodine Content in Fijian Foods using Spectrophotometric Kinetic Method[J]. *Microchemical Journal.*, 148:475–479.

Araya M, Samantha G, Sebastián P. (2020) Iodine and Iodate Determination by A New Spectrophotometric Method using N, N-dimethyl-p-phenylenediamine, Validated in Veterinary Supplements and Table Salt [J]. *Analytical Methods.*, 12(2): 205–211.

Chang W B, Li K A. (1981) *Brief Handbook of Analytical Chemistry[M]*. Beijing: Beijing University Press, 262. (In Chinese)

Espada-Bellido E, Bi Z, Salaün, P, et al. (2017) Determination of Iodide and Total Iodine in Estuarine Waters by Cathodic Stripping Voltammetry Using a Vibrating Silver Amalgam Microwire Electrode[J]. *Talanta.*, 174(17): 165–170.

Hou W L, Chen Y, Lu Q J, et al. (2018) Silver Ions Enhanced AuNCs Fluorescence as A Turn-Off Nanoprobe for Ultrasensitive Detection of Iodide[J]. *Talanta.*, 180(17): 144–149.

Hwang J B, Su P J, Young K S, et al. (2020) Determination of Iodine in Foods by Inductively Coupled Plasma Mass Spectrometry After Tetramethylammonium Hydroxide Extraction[J]. *Journal of Analytical Chemistry.*, 75(11); 1399–1403.

Márcia F M, Paola A M, Cezar A B, et al. (2010) Iodine Determination in Food by Inductively Coupled Plasma Mass Spectrometry after Digestion by Microwave-Induced Combustion[J]. *Analytical and Bioanalytical Chemistry.*, 398(2):1125–1131.

Namekar S B, Lokhande S P, Jadhav G R, et al. (2017) Estimation of Iodine Content by Iodometric Titration and Spectrophotometric Evaluation Method in Commercially Available Salts [J]. *Indian Journal of Nutrition & Dietetics.*, 54(4): 465–476.

Trésor K M, Stéphanie P, Macours P, et al. (2008) Highly Sensitive Determination of Iodide by Ion Chromatography with Amperometric Detection at A Silver-Based Carbon Paste Electrode[J]. *Talanta.*, 76(3): 540–547.

Ujang T, Niikee S, Peter S W D, et al. (2013) Determination of Iodine in Selected Foods and Diets by Inductively Coupled Plasma-Mass Spectrometry[J]. *Pure and Applied Chemistry.*, 84 (2): 291–299.

Energy Revolution and Chemical Research – Kok-Keong Chong and Zhongliang Liu (Eds)
© 2023 The Authors, ISBN: 978-1-032-36554-1

Organic chemistry retrosynthesis strategies, research on basic rules of breaking bonds and fine tuning

Kaixiang Liang*
Beijing Bayi High School, Beijing, Beijing, China

Jinyu Wei
Wuhan No. 6 High School, Wuhan, Hubei, China

Song Wang
University of Delaware, Delaware, Newark, USA

Mohan Chen
Jinan Foreign Language Schools, Jinan, Shandong, China

Siyuan Zhang
Qingdao No. 58 Senior High School, Qingdao, Shandong, China

ABSTRACT: The study of organic chemistry is not an easy process. Recognizing and analyzing organic compounds is one of the biggest problems. After learning organic chemistry, students shall know the analysis of compound structure is one of the most important steps. This research project aims to introduce a common but useful method, to help students understand organic compound structures during their study. Retrosynthesis strategies belong to the one, which stands for a method of chemical which involves decomposing a specific molecule into the smallest structures that are easy to read and analyze, so as to assess the synthetic route. It is assumed that students get a specific organic but it is complex and hard to read and understand the properties. The process of retrosynthesis strategies involves disconnections and fine-tuning steps. In this way, works can get a successful and reasonable synthetic route from the smallest readable structure. With the structures, compounds should be able to be analyzed. After studying and understanding the retrosynthesis strategies, students are supposed to get a higher level of understanding of organic chemistry and can analyze specific complex structures by themselves.

1 INTRODUCTION

Organic synthesis was a creative field but a jungle that was entirely dark for chemists in the 19th. Scientists started research by finding valuable molecules for a disease that was rare in nature and tried to work on the synthesis. In 1950, Organic synthesis became an "art". During this stage, organic chemists synthesized many valuable products. For example, vitamin B12, a drug molecule essential for treating pernicious anemia and extracting from animals' livers before, was finally successfully produced by Robert Burns Woodward. His mean synthesized parts of vitamin B12 and then connected them. This approach used by Woodward was like the retrosynthesis method of another school of thought presented by Professor Elias James Corey, which later became the commonly accepted method for synthesizing all organic macromolecules.

*Corresponding Author: kaixiang.liang@bayims.cn

 DOI 10.1201/9781003332657-44

The theory of retrosynthesis analysis has promoted the rapid development of the whole field of organic synthesis since the 1970s, before professor E. J. Corey's theory was put forward, the total synthesis of organic compounds was less than 50, but by 2001, the number had exceeded 25,000, which fully proves the importance of retrosynthesis analysis. In 1991, this great professor of organic chemistry, E. J. Corey was awarded the Nobel Prize in Chemistry for his contributions to the field of organic chemistry. In the following paper, the authors will focus on retrosynthesis analysis and adopts several examples of disconnection principles and fine-tuning of different organic compounds, aiming to help students who are new to retrosynthesis analysis to have a preliminary understanding of this field. Both experienced scientists and beginners should start from the structure of the Target Molecule (TM.) This passage helps them to practice retrosynthesis analysis and use disconnection to cut bonds of the molecule to obtain Synthon (conceptual molecule obtained after cutting off chemical bond). The process results in simplified structural units that can be adjusted during synthesis and can be used to identify intermediates and total synthetic reaction routes of organic synthesis reactions quickly, reasonably, and effectively.

2 PRINCIPLES OF DISCONNECTIONS

The work summarized the basic principles of disconnections in this part of the paper. The proper cutting off always generally breaks where there are functional groups in the target molecule (TM); it should be cut on where there are branches, rather than the backbone of the molecule; the resulting synthesis route is reasonable, for example, the charge of the intermediate reactants is reasonable. A good disconnection also satisfies the right mechanism of forwarding reaction: the reaction process is as simplified as possible; all the intermediate reactants (synthons) should be recognizable materials. Also, a well-designed chart of the relationships between single functional groups is a good reference.

2.1 *Disconnections of simple olefins*

In the one-group disconnection of simple olefins, there are some cases for you to know about the mechanism. Olefins are more complicated than alcohols, and alcohol can produce them by dehydration. There are two ways for olefins to do disconnections, and a straightforward way is to use the property of the dehydration of alcohol (Warren 2013). As shown in the equation below (see Figure 1), in acidic conditions, alcohol can be dehydrated to form olefins. The disadvantage of doing so is that there is no way to control the carbon-carbon double bond's position and the formation route is too long.

$$Me_3C{-}OH \xrightarrow{H^+} Me_2C{=}CH_2 + H_2O$$

Figure 1. Single bond to double bond example (Warren 2013).

Compared to dehydration, the witting reaction eliminates the disadvantage mentioned above. The Wittig Reaction (see Figure 2), shown below, allows the preparation of an alkene by reacting an aldehyde or ketone with the ylide generated from a phosphonium salt. The geometry of the resulting alkene depends on the reactivity of the ylide. If R is an electron-withdrawing group, the ylide is stabilized and not as reactive as when R is alkyl. Stabilized ylides give E-alkenes predominantly, whereas non-stabilized ylides lead to Z-alkenes.

2.2 *Disconnections of aryl ketones*

Aryl ketone, the product of the reaction between aromatic hydrocarbon and an alkyl halide, could have the Friedel-Crafts alkylation reaction, which is considered to be one of the essential disconnections of aryl ketones. The main idea of this theory is shown in the picture below (H is replaced

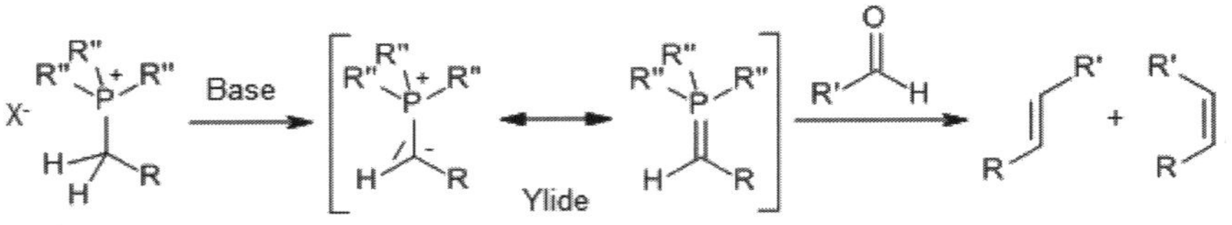

Figure 2. The Wittig Reaction (Warren 2013).

by an alkyl group and forms an aromatic hydrocarbon). Understand the Friedel- Crafts reaction principle, it is easy to find out the solution quickly (Tip: Benzene ring is the simplest aromatic hydrocarbon, which is used to study the principle of FC Rxn). The main idea of Friedel- Crafts reaction overall is shown below (Figure 3).

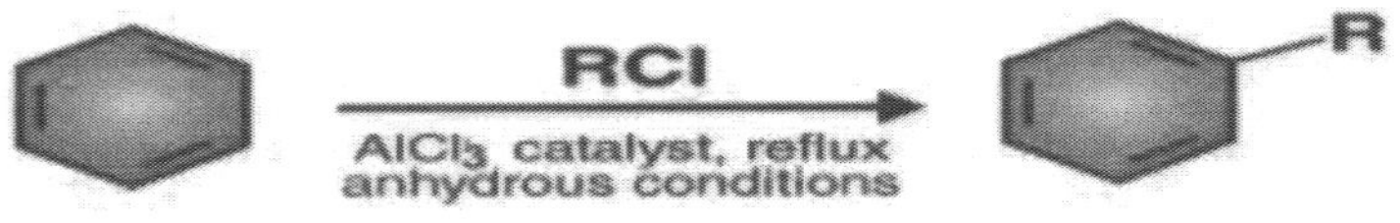

Figure 3. The big idea of Friedel-Crafts Alkylation reaction.

The disconnection of the following molecule can be easily estimated by the Friedel-Crafts reaction. There are two different answers because the carbonyl is located between two benzenes.

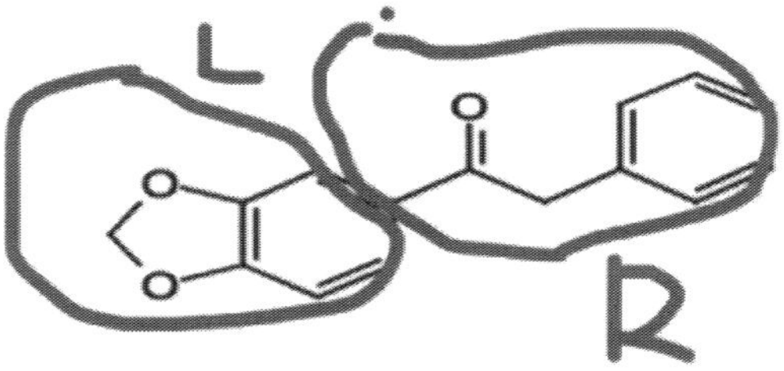

Figure 4. The final product of Friedel-Crafts reaction.

In Figure 4, if the 'left part' is considered in comparison to the carbonyl as an aromatic hydrocarbon, the right part must be R-Cl. According to the picture of the final product, the right part is the R group and it must contain chloride initially. It can be estimated that these two reactants must be phenylacetyl chloride and 1,3-benzodioxole (Figure 5).

Figure 5. The estimation of synthons.

Given that considering the left part as an aromatic hydrocarbon, the corresponding right part is R-Cl. Also, the catalyst will be $AlCl_3$. Cl ions will react with the AlCl3, and Cl attracts AlCl3 and makes Al- and Cl+. Then Cl will break the C-Cl bonds and lower the charge of Cl (make it become 0). At last, it will form $AlCl_4$ minus (-) and acylium cation, which will react with the 'left part,' aromatic hydrocarbon. Then the reaction between 1,3-benzodioxole and acylium cation will occur. To form a bond connecting two reactants and eliminating the charge, it must break the outer C=C double bonds and thus make them join while the H atom on the other side of this C=C bond will have a + charge. To eliminate the charge, it's necessary to lose the top H and render the C=C bond. Thus, Cl- is used (reference doesn't explain the reason for using Cl-, but it says that the AlCl4- formed before it is converted into $AlCl_3$ and Cl-, and since the catalyst will not participate in the reaction, this Cl- must stem from the breakdown of the $AlCl_{4-}$). Then, the Cl- is put into the reaction, and the negative charge will attract the H atom at the top of the C-C (C=C) bond. Thus, it's able to get HCl and the final product (Figure 6).

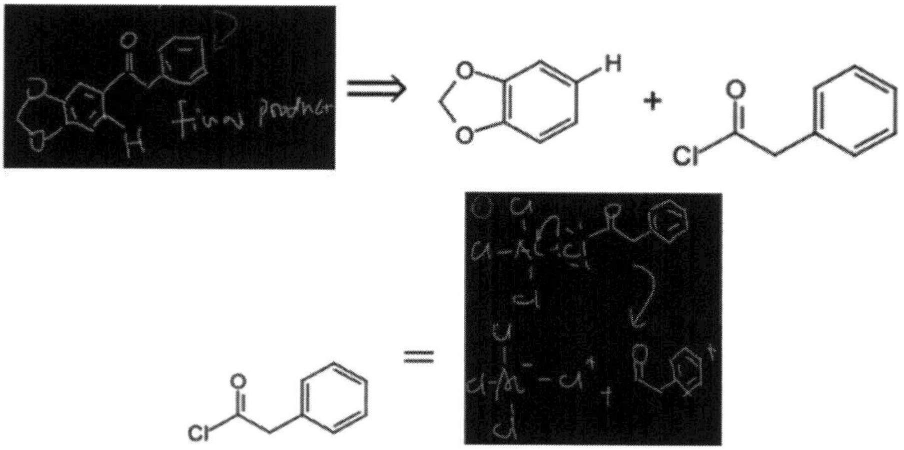

Figure 6. The whole process of synthesis.

Aromatics hydrocarbons react with alkylated halane, and the electrophilic substitution reaction takes place on an aromatic ring, and the hydrogen atom is replaced by alkyl to generate alkyl aromatics. Under the catalyst, a strong Lewis acid is formed. The most valuable and acknowledged catalyst is $AlCl_3$.

3 SPECIFIC AND DETAILED DISCONNECTIONS

3.1 *Specific and detailed disconnections: diels-alder reaction*

Diels-Alder reaction is one of the most important learning connections between alkene groups (Kloetzel 1948). The Discovery of this reaction was awarded the Nobel Prize in Chemistry in 1950. The Diels–Alder reaction provides a reliable way to form six-membered rings with good control over the regio- and stereochemical outcomes. It has served as a powerful and widely applied tool for the introduction of chemical complexity in the synthesis of natural products and new materials. It is also an important step in learning retrosynthesis. Here it is one of the easiest and most basic Diels-Alder reactions (Kloetzel 1948), which came from the reaction between Diene and Alkene to form Cyclohexene (Figure 7).

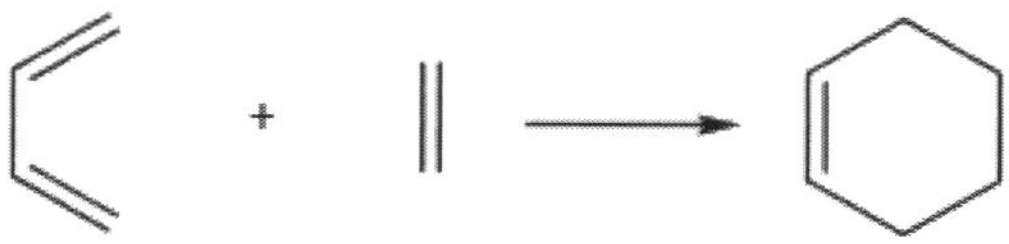

Figure 7. Forming of Cyclohexene.

In this reaction, it could be seen that the electrons' moving trail forming two connections between alkenes, also forms an alkene bond in a ring.

Fortunately, in the study of retrosynthesis, the Diels-Alder reaction is simple and easy to recognize. Here is another example to help understand the Diels-Alder reaction (Figure 8):

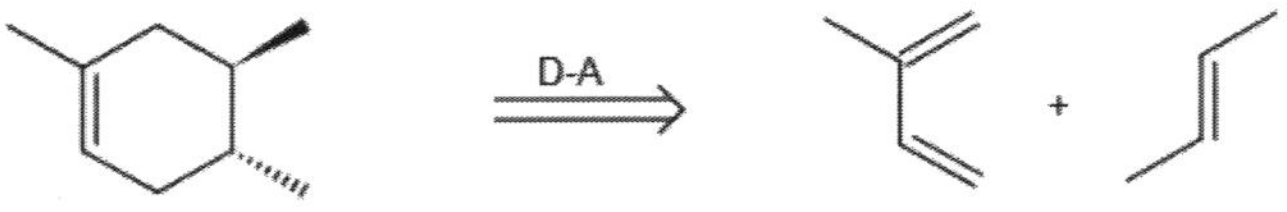

Figure 8. 1,4,5-Trimethylcyclohexene reverse.

In this reaction, if it is pondered with the retrosynthesis method, 1,4,5-trimethylcyclohexene could be considered as the formation of 2-methylbuta-1, 3-diene, and 2-Butene. And if the moving trail of electrons is checked, it is the same as the forming of cyclohexene (Nicolaou 2002). More complex compounds are shown in Figure 9:

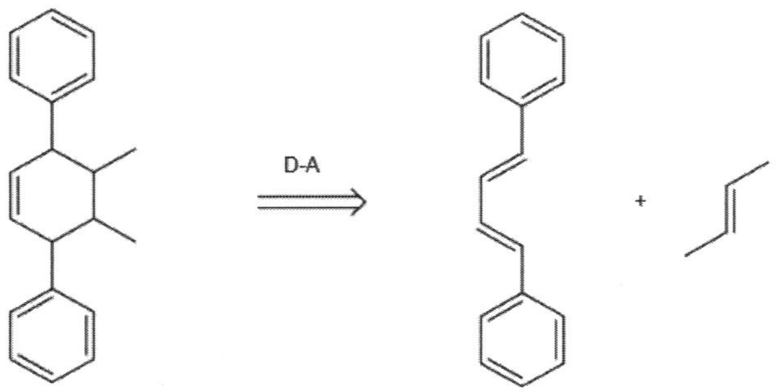

Figure 9. 3,6-diphenyl-4,5-dimethyl-1-cyclohexene formation reverse.

The intramolecular forces and reactions are the same. This device is advanced in retrosynthesis.

3.2 *Specific and detailed disconnections: diels-alder with alkyne diels-alder reaction*

On the other hand, the Diels-Alder reaction also works for alkyne compounds.

Figure 10. 1,4-Cyclohexadiene formation reverse.

Figure 10 shows the retrosynthesis idea of 1,4-Cyclohexadiene (Nicolaou 2002). Diene reacts with Acetylene to form it. To understand the reason, the method is to see the structure of acetylene. There are 2 pi electrons in alkyne so it perfectly fits the Diels-Alder reaction. So, for the triple bond in acetylene, it breaks one bond to form a new double bond in the ring, and that's why there are two alkenes in the production (Nicolaou 2002).

3.3 *Specific and detailed disconnections: heteroatoms and heterocyclic*

A heteroatom usually means atom O, N, or S. To do the disconnection, the bond that is not on the aromatic ring is used. Aromatic rings are a conjugated structure and their bonds are much stronger than usual. In the first example, with the heteroatoms, the C-O bond is firstly broken.

In this example, with the heteroatoms, it's easy to choose to break the C-N bonds because compared to the C=O bond, the C-N bond is obviously easier to break (Figures 11 and 12).

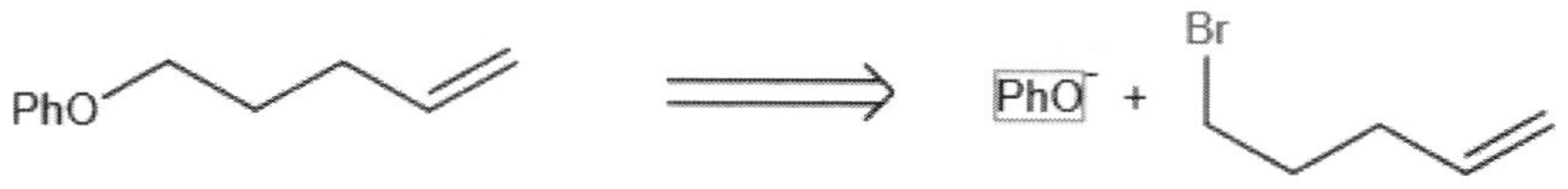

Figure 11. Breaking bonds at the end in heteroatom.

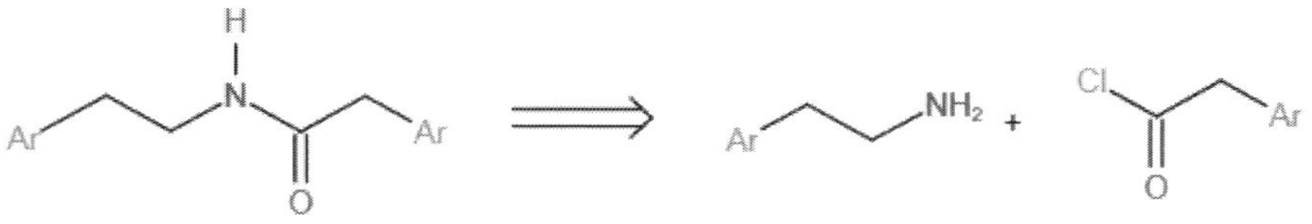

Figure 12. Breaking C-N bonds in heteroatoms.

Above all, when heteroatoms are in the carbon chain, breaking their bonds with carbon should be a great method to analyze the compound and think about in the retrosynthesis. Intramolecular reactions are faster and cleaner than intermolecular reactions. With respect to heterocyclic compounds, using retrosynthesis should be a great method. Here is an example to use retrosynthesis analysis a heterocyclic compound (Figure 13).

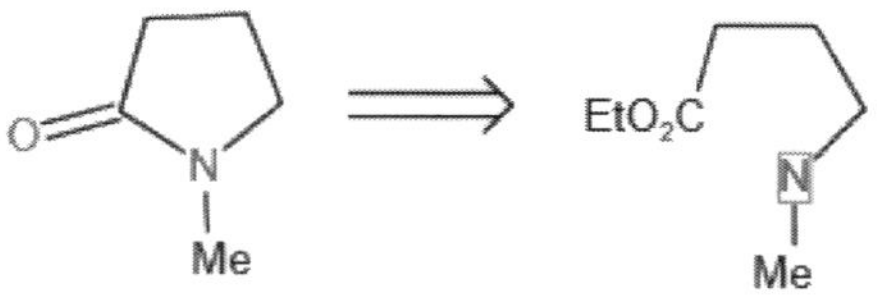

Figure 13. Disconnections of cyclic compounds.

In Figure 14, the example used retrosynthesis to analyze the compound, which is found to break the C-N bond first inside the heterocyclic compound.

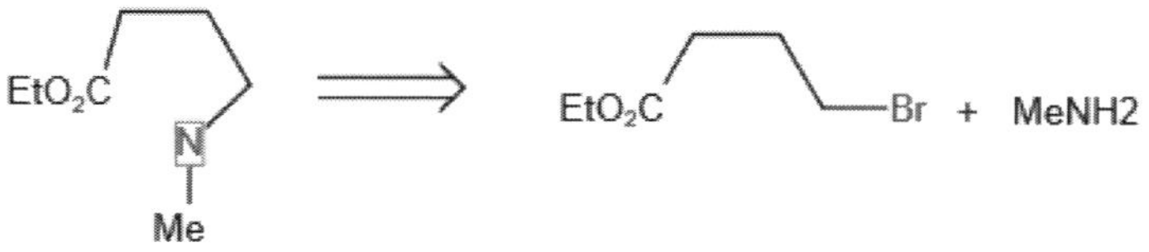

Figure 14. Breaking C-N bonds in the heterocyclic compound.

After breaking the first C-N bond and it is able to find that there is another C-N bond that could break. So, after two steps, a possible reactant group is found. So, compared to heteroatoms and heterocyclic compounds, the retrosynthesis method is used to break C-O, C-N, etc., bonds that would help students think about it effectively.

4 FINE TUNING

4.1 *Fine tuning: function group interconversion*

Functional group interconversion (FGI) is defined by Stuart Warren, due to "the process of converting one functional group into another by substitution, addition, elimination, oxidation or reduction, and the reverse process used in the (retrosynthetic) analysis (www.organic-chemistry.org 2021)". It is easy to think of examples of each of these categories. The FGI has multiple advantages for us to take. People could gain the molecules that are easier to gain with FGI as shown in the following reaction.

Many molecules of the ketone could be changed to alcohol in order to form synthons that are easier to process disconnections (Figure 15).

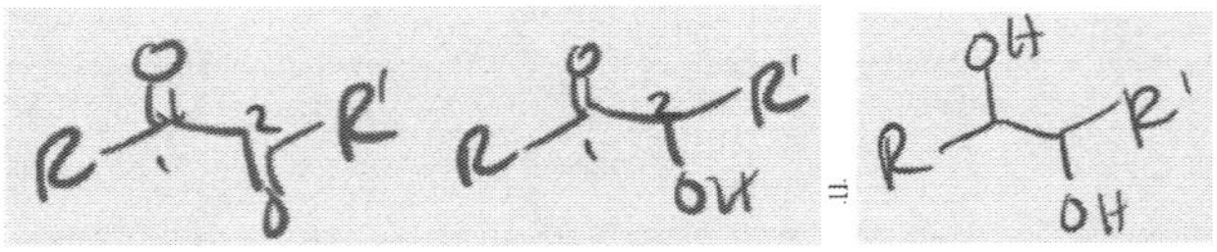

Figure 15. Transferring ketone to alcohol.

Moreover, it is possible to handle the unavailability of starting materials, adjust the synthons to more recognizable ones, and protect the reactive reactant by using this method (www.organic-chemistry.org 2021). There are more examples put in the Control Chapter.

4.2 *Fine tuning: function group addition*

During retrosynthetic analysis, additional functional groups are introduced. If necessary, it is needed to direct bond formation: Functional group addition (FGA) strategy in the retrosynthetic analysis involves introducing additional functional groups at strategic locations in a synthon. If necessary, it is necessary to guide further disconnections based on known powerful bond-making reactions. The addition of functional groups such as double bonds or carbonyl groups can serve to direct reactivity to specific locations of a molecule and allow disconnection based on dependable reactions. This can significantly simplify a synthesis (Figure 16) (Sykes 1986).

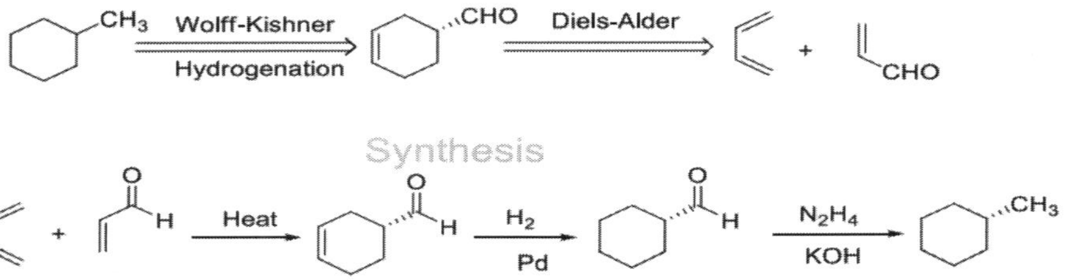

Figure 16. The example of using FGA to simplify the question.

For example, the introduction of a double bond in a cyclohexane target molecule may help key a disconnection based on a Diels Alder reaction, a powerful carbon-carbon bond-forming reaction (Sykes 1986).

4.3 *Fine tuning: function group removal*

During the analysis, the function group removal is defined that any special function group, like halogen and oxhydryl, being removed or replaced by hydrogen, which can be considered a reversible FGA reaction. It's an essential analytical way to have the retrosynthesis process. For example, the function group, amidogen, is removed and replaced with a double bond. This process is defined as the function group removal, FGR (Figure 17).

4.4 *Specific fine tuning: control*

The next thing to look at is something that happens before the carbonyl break is completed. Aspects of integration are controlled as a breakthrough from system analysis. Because some reagents will

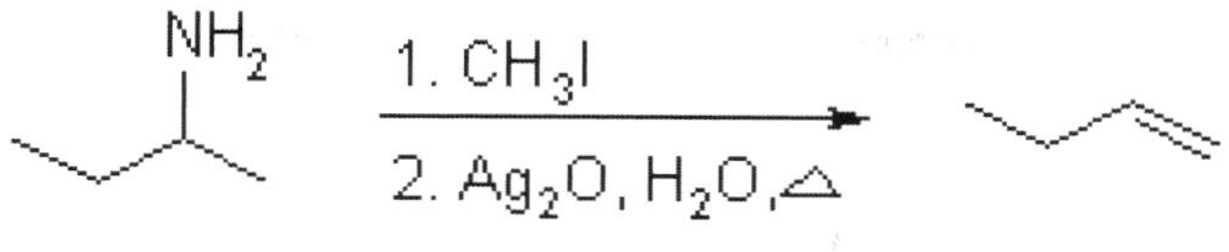

Figure 17. The removal of amidogen group.

form the wrong product in a reaction, a process is set up to protect the reagent is necessary. Any functional group can act as a protecting group providing that it can easily be added and removed and that it doesn't react with the reagent. To protect the functional group, rather than protect one part of a molecule, it is better to activate another. Taking the carbonyl group as an example, it can't protect the carbonyl group without stopping the reaction, thus, activating one position by adding a CO2Et group is possible by using the ester A below (see Figure 18).

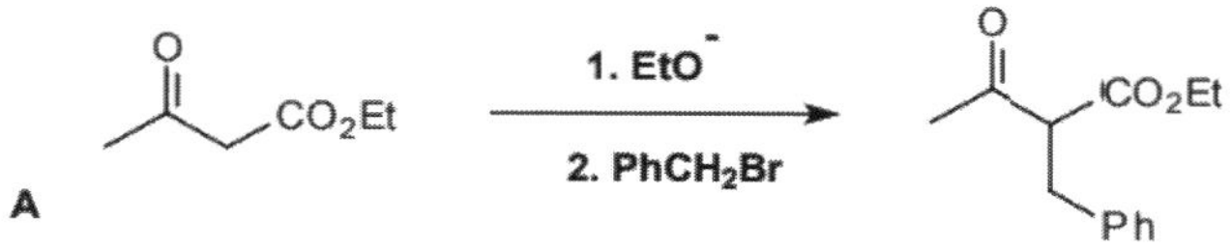

Figure 18. The protection of carbonyl group by adding CO2Et group.

And after this reaction, it is necessary to remove the activating group CO2Et, which can be removed by using hydrolysis and decarboxylation (see Figure 19 and Figure 20).

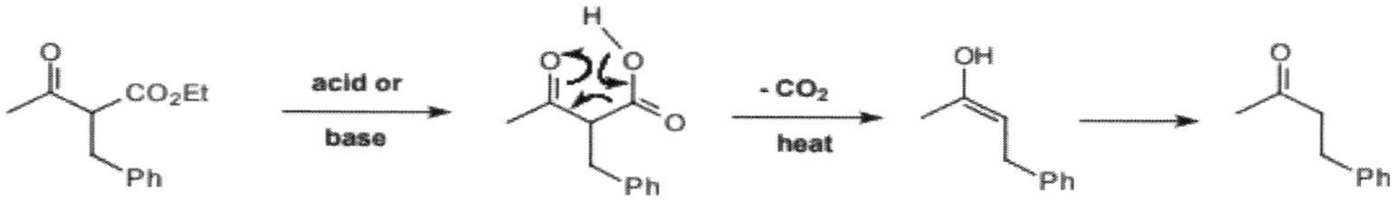

Figure 19. The protection of carbonyl group by adding CO2Et group.

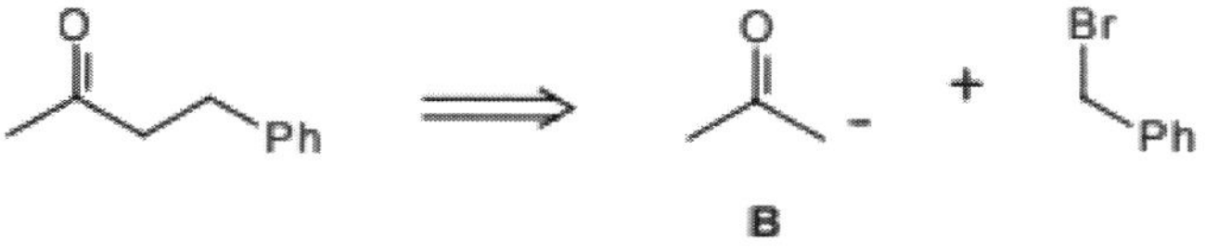

Figure 20. The general synthesis of ketones and disconnection for a synthon.

Protection and activation give us a reagent for the synthon - CH2CO2H. The acid is protected as an ester, and then another ester group is added as activation to obtain malonic ester: CH2(CO2Et)2 (see Figure 21).

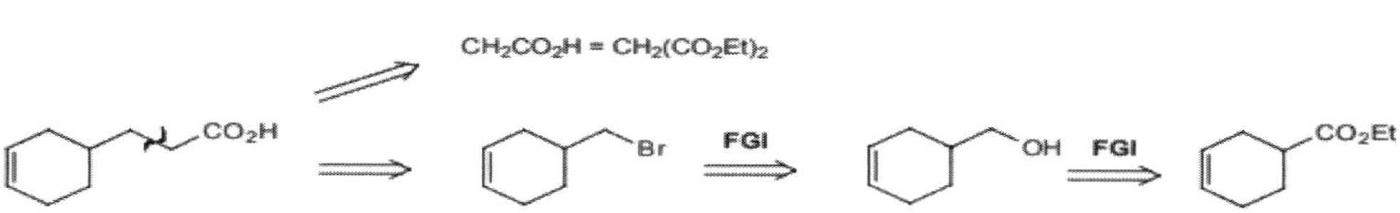

Figure 21. The protection of acid group with ester.

4.5 *Specific fine tuning: alcohol's functions*

To discuss the retrosynthesis in whole progress, alcohols are the simple way to analyze. They can be converted into a whole family of other functional groups. For reduction, alcohols can be transformed into alkane or olefins; for oxidation, they can be converted into aldehyde or ketone, eventually ending in carbon dioxide (www.vanderbilt.edu/AnS/Chemistry/Rizzo/chem223/FGI.pdf). In other words, almost every compound can be obtained by Functional Group Interconversions in alcohols (in Figure 22 shown below) (www.vanderbilt.edu/AnS/Chemistry/Rizzo/chem223/FGI.pdf). With some single bonds between carbon and oxygen, the compound can be considered to be equivalent to alcohol. To discuss the retrosynthesis of alcohol, it's easier for people to analyze the forward reaction which is how alcohol can be made.

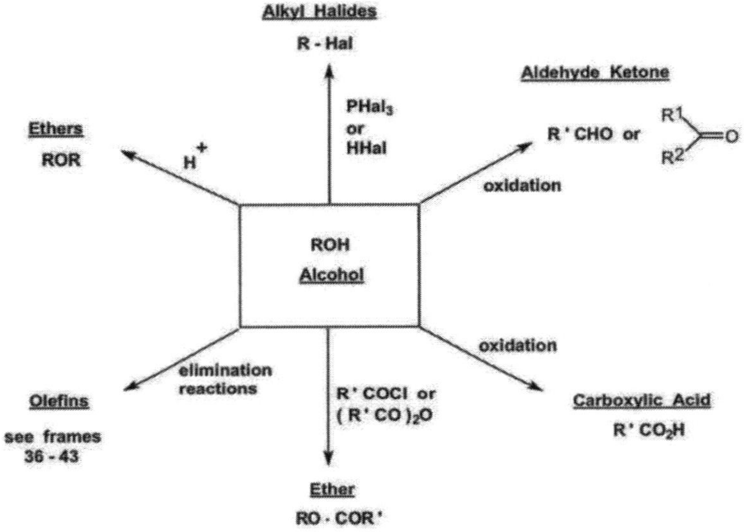

Figure 22. The FGI with alcohol.

First, the focus is on additional reactions and breaking the carbon-carbon double bond (Figure 23).

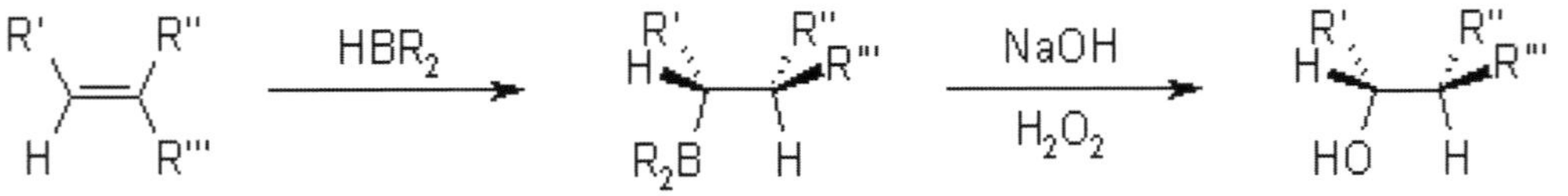

Figure 23. Brown Hydroboration.

The first step is based on the Markovnikov rule, in which the least hindered carbon is preferentially added to the boron atom. The target molecule with an alcohol functional group is fielded by the anti-Markovnikov rule.

In different catalysts, products are variable. Because the ozone has high oxidizability, it can make the disconnection in olefin by the reaction. In these cases, with a high oxidizable reagent, most hydrocarbon will get oxidation, transiting to alcohol (Figure 24).

The second is the oppenauer oxidation. The reaction is an equilibrium reaction, which means that to get the expected target, people can control the concentration of the whole compound by self-adding or removing it, based on Le Chatelier's principle. The reaction is shown below (Figure 25) (Xing 2016).

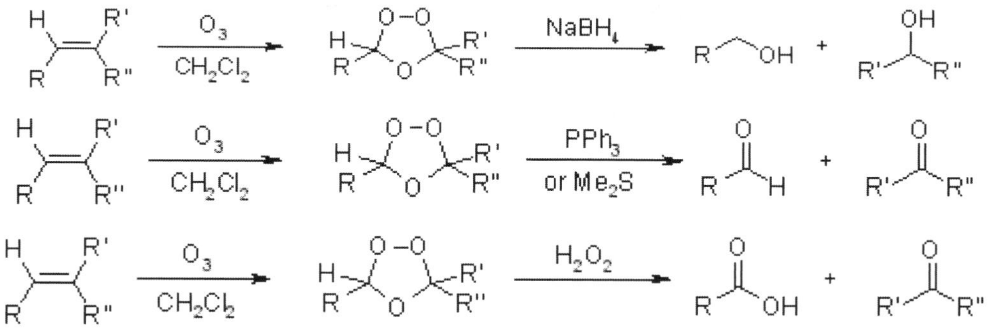

Figure 24. Converting the TS.

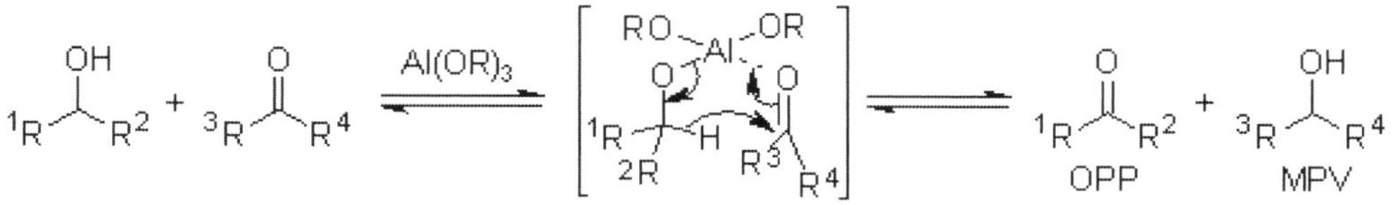

Figure 25. Using of aluminum and alcohol.

The aluminum-catalyzed hydride shift from the α-carbon of an alcohol component to the carbonyl carbon of a second component, which proceeds over a six-membered transition state, is named Meerwein-Ponndorf-Verley Reduction (MPV) or Oppenauer Oxidation (OPP) depending on the isolated product. If aldehydes or ketones are the desired products, the reaction is viewed as the Oppenauer Oxidation (Tedder 1987).

The third is the Heterolytic cleavage analysis (Grignard reagent). Within the ketone or aldehyde, the oxygen atom in carbonyl is most reactive, due to the property of the Lewis acid-base theory. The oxygen atom in carbonyl can provide lone electron pair, thus it can actively react with the cation (hydration in solution in most cases) or have a coordinate bond with some certain atom in the particular compound which nearly has less electronegativity (Jones 2010). In this case, the cation will attack and activate the central carbon into a carbon ion. Most of those kinds of experiments are operated in an acid environment in which plenty of hydration can exist.

The role of sodium cation is the same as that of the aluminum cation. Both make the coordinate bond with lone pair electron in oxygen atom, which makes the carbon positive to attract the hydride ion in borohydride or aluminum hydride. Thus, to neutralize the negative compound, the oxygen atom will attract the hydronium ion to form the alcohol functional group and water (Figure 26).

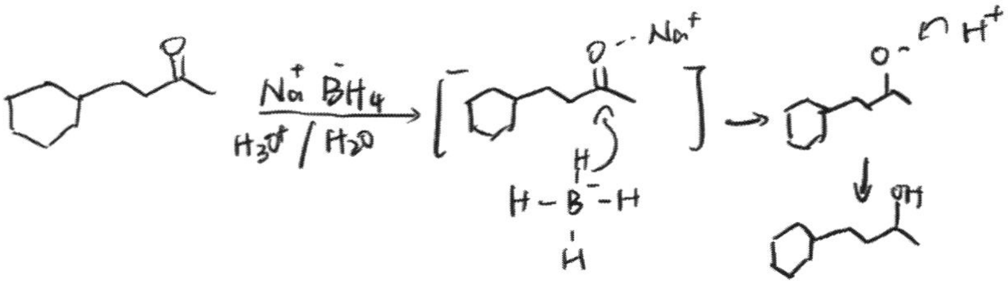

Figure 26. The Sodium borohydride/lithium Aluminum hydride.

The C-Mg bond is highly polarized and the C atom is partially negatively charged, making the Grignard reagent an excellent carbon nucleophile. It also attacks the oxygen atom in the carbonyl functional group to activate the central carbon into carbocation. The oxygen atom gets the hydrogen to form the alcohol (Figure 27) (Ayres 1930).

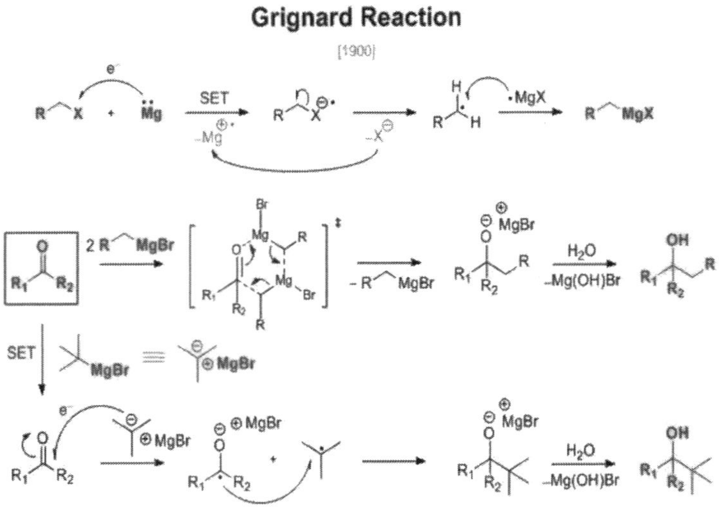

Figure 27. The reaction mechanism by using Grignard Reactant.

Conclusively, heterolysis cleavage is the most common way to analyze retrosynthesis. A pair of electrons can be attributed to one synthon with the negative; the other one will be considered positive. Forming the double bond can solve the synthon with a positive charge, and adding the hydrogen can solve the negative synthon, just like the Grignard reagent. The other direction to analyze the retrosynthesis is the redox reaction (Figure 28) (Fatahala 2017).

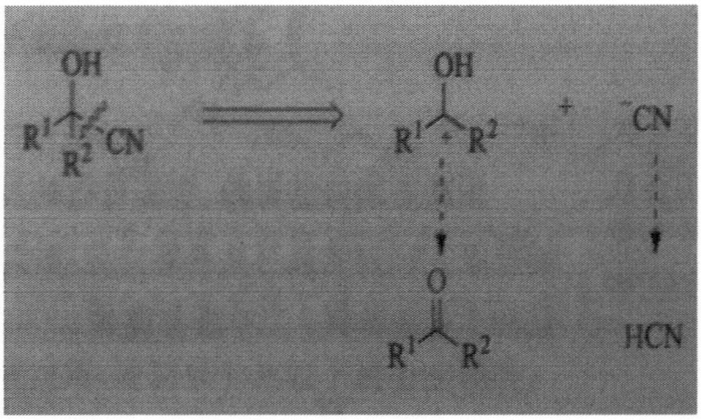

Figure 28. Grignard Reaction.

5 CONCLUSION

In modern days, traditional Chinese medicine is gaining a lot of attention around the world. Chemists are interested in the effects of different substances on the human body and want to create better clinical medicines based on historical data. Apart from a few plants, many of today's widely available drugs are synthesized by chemists in the laboratory. Therefore, some people believe that elemental medicine will be the trend of the future. The retrosynthesis reactions and methods this study discussed would help promote this process. The principles and examples mentioned above show how important and valuable the retrosynthesis method is in organic chemistry. The team studied the retrosynthesis method to help analyze more synthesizes of complex compounds. Using rules in disconnection to retrieve synthons and employing fine-tuning could creatively build target molecules (Morrow 2016).

In general, organic chemists have created several more efficient substances, such as synthetic polymers and dyes that are harder to fade and have a profound impact on lives. During this section, the fact of Disconnections tells students how the Retrosynthesis Strategies could work on many specific reactions such as the Diels-Alder reaction, Grignard reaction, and particular disconnection examples. The classical reaction that students learn how to recognize and identify is shown. Therefore, the research would be helpful for students to develop the method of drawing the ways of retrosynthesis, from the TM to the initial synthons step by step (Fatahala 2017). Above all, compounds and methods are presented to show how important and valuable the retrosynthesis method is in organic chemistry. Studying the retrosynthesis method helps students and chemists to analyze compounds more effectively and creatively. Students who get in touch with the retrosynthesis newly could study, for example, in this book, to find possible solutions for complex compounds.

REFERENCES

Ayres, F. D. (1930). Alpha pinene and the Grignard reagent.

Fatahala, Samar S. "Retrosynthesis Analysis; a Way to Design a Retrosynthesis Map for Pyridine AND PYRIMIDINE RING." *Annals of Advances in Chemistry*, 26 Sept. 2017, www.heighpubs.org/hjc/aac-aid1007.php.

Functional Group Interconversions. www.vanderbilt.edu/AnS/Chemistry/Rizzo/chem223/FGI.pdf.

Jones, M., & Fleming, S. A. (2010). *Organic chemistry, Chapter: a catalyzed addition to alkenes*. Norton.

Kloetzel, M. C. (1948). *"The Diels–Alder Reaction with Maleic Anhydride"*. Organic Reactions.

Morrow, Gary W. *"Organic Synthesis in the Laboratory." Bioorganic Synthesis*, 2016, doi:10.1093/oso/9780199860531.003.0011.

Nicolaou, K. C.; Snyder, S. A.; Montagnon, T.; Vassilikogiannakis, G. (2002). "The Diels-Alder Reaction in Total Synthesis". *Angewandte Chemie International Edition*.

"Organic Chemistry Oppenauer oxidation," "Ozonolysis criegee MECHANISM," "Brown hydroboration." Organic Chemistry, www.organic-chemistry.org/. Retrieved September 13, 2021

Sykes, G. (1986). A guidebook to mechanism in organic chemistry, Chapter: the addition or elimination reaction, *Longman Scientific & Technical*.

Tedder, J. M., & Nechvatal, A. (1987). *Basic organic chemistry: A mechanistic approach*. Wiley.

Warren, Stuart, and Paul Wyatt. *Organic Synthesis the Disconnection Approach*. Wiley, 2013.

Xing, Q. (2016). *Basic Organic Chemistry, the Graph about the cleavage of single Bond* (Vol. 1). Peking University Press.

Energy Revolution and Chemical Research – Kok-Keong Chong and Zhongliang Liu (Eds)
 ISBN: 978-1-032-36554-1

Research progress and application status of polylactic acid

Yuezhou Kang
Harold Vance Department of Petroleum Engineering, Texas A&M University, College Station, Texas, USA

Ruixiang Hou*
SINOPEC Beijing Research Institute of Chemical Industry, Beijing, P.R. China

ABSTRACT: Every year, a large number of abandoned plastic products are being discarded into the environment and not recycled and reused, which causes serious pollution to the natural ecological environment. In order to solve this problem, many countries have vigorously promoted plastic pollution control policies, and biodegradable plastics represented by polylactic acid (PLA) have gradually entered the perspective of researchers and the public. In this paper, the latest research progress on the synthesis and modification of PLA is introduced. The application and the future development direction of PLA are discussed, in the hope of offering opinions for the research on PLA.

1 INTRODUCTION

Since the birth of plastics, their excellent performance and low cost have been widely used in all aspects of production and life. From 1950 to 2015, 8.3 billion tons of plastic products have been produced by mankind, of which 4.9 billion tons are directly landfilled or abandoned, 800 million tons are directly burned, and 2.6 billion tons are still in use, of which 100 million tons are recycled plastic. The use of plastics will also show an increasing trend in the future, and it is expected that global plastic consumption will increase at an annual rate of 8%. The annual consumption of plastics will reach 700 million tons in 2030 (Geyer 2017; Jambeck 2017; Law 2017). It generally takes 200 – 700 years to degrade in the natural environment. The existence of a large number of waste plastics has greatly threatened the health of wild animals and human beings. Therefore, in the face of enormous white pollution, many countries actively promote plastic restriction, plastic prohibition, development, and the use of biodegradable plastics to reduce waste pollution.

PLA is a kind of polymer polymerized by lactic acid, the main raw material of which is abundant, mainly corn, wheat, and cassava containing starch. At the same time, the production process of PLA is pollution-free, because of its good biocompatibility, processing, mechanical properties, light transparency, and nontoxicity. So, it can be used in food, medicine, and other packaging materials. And PLA has good biodegradability, which can be completely degraded by microorganisms in nature, and eventually generate carbon dioxide and water without polluting the environment. Thus, PLA is an ideal green polymer material (mJarerat 2004; Tokiwa 2004). Based on this, in recent years, researchers have conducted a large number of studies on the synthesis, modification, and application of polylactic acid, and polylactic acid has gradually shown more attractive application value. Therefore, this paper made a detailed review of the synthesis, modification methods, and application scope of polylactic acid, hoping to provide some guidance for its continuous development of it.

*Corresponding Author: hourx76835.bjhy@sinopec.com

 DOI 10.1201/9781003332657-45

2 SYNTHESIS OF PLA

The synthetic routes of PLA can be divided into direct polycondensation, azeotropic dehydration polycondensation, and ring-opening polymerization of lactide. The former two routes are simple ways to obtain PLA, but industrially, high-molecular-weight PLA is mainly obtained by the third way. The synthesis and purification of lactide (Cargill 2001) have been the focus of a lot of research.

2.1 *Direct polycondensation*

Direct polycondensation is the direct condensation of the lactic acid monomer. In the presence of a dehydrator, the hydroxyl and carboxyl groups in lactic acid molecules were dehydrated by heating and directly condensed to synthesize oligomers. This process is simple, free from intermediate purification, and the processing cost is not high. But the problem is that the synthesized product has a low molecular weight, low strength, and easy decomposition. Direct polycondensation of lactic acid is usually carried out by distillation of condensed water with or without catalyst, during which the vacuum and temperature are gradually increased. Because it is difficult to completely remove water from the high viscosity reaction mixture, only tens of thousands of polymers can be obtained. Moreover, since the formation of polyester is a balanced and reversible reaction, the existence of by-product water makes the reaction difficult to generate polymers. The relative molecular weight of PLA is generally not high, the reaction time is long, the molecular weight distribution is wide, and the experimental reproducibility is poor. Therefore, how to improve the relative molecular weight of polymer products is the top priority of direct polycondensation. For example, Chen et al. (Chen 2006; Kim 2006; Yoon 2006) produced high molecular weight polymers by controlling the decompression and esterification rate. High molecular weight PLA (130000 Da) was prepared by direct bulk polycondensation with titanium butoxide as a catalyst at different polymerization times. Nagahata et al. (Nagahata 2007; Sano 2007; Suzuk 2007; Takeuchi2007) reported that the synthesis of PLA by microwave-assisted single-step direct polycondensation of LA could be effectively realized by the use of microwave irradiation. Compared with conventional polycondensation at the same temperature, the reaction time could be greatly shortened. At the same time, it was found that tin catalyst and decompression were very effective for obtaining polymers with molecular weight higher than 10000. Moreover, if the binary catalyst $SnCl_2/p-TsOH$ was used, PLA with Mw of 16000 was obtained within 30 min under decompression of about 30 mmHg. In addition, to aim at developing a practical and economic technology to produce PLLA effectively, Miyoshi et al. (Hashimoto 1996; Koyanagi 1996; Miyoshi 1996; Sakai 1996; Sumihiro 1996) carried out a continuous melt-polymerization experiment by using the combination of batch type stirred reactor and an intermeshed twin screw extruder. As a result, PLLA with Mw = 150000 was obtained from LA by the continuous melt polycondensation process.

2.2 *Azeotropic dehydration polycondensation*

Azeotropic polycondensation is a method for obtaining high chain length without using chain enhancers or additives and their related shortcomings. Azeotropic dehydration polycondensation developed by Mitsui Toatsu Chemicals (Mitsui) has made a breakthrough in increasing the molecular weight of LA polycondensation. In 1995, Ajioka et al. (Ajioka 1995; Enomoto 1995; Suzuki 1995; Yamaguchi 1995) used diphenyl ether as the solvent, tin powder as the catalyst, through 3A molecular sieve azeotropic reflux reaction for 40 h, the relative molecular weight of 300,000 PLA was obtained. The patents of the company were directly condensed by solvent method, and the condensed water was continuously removed by azeotropic distillation in the reaction process to obtain PLA with a relative molecular mass of 15,000 – 200,000 (Mitsui 1994). However, in order to achieve a sufficient reaction rate to obtain the required high concentration, this polymerization will produce a large number of catalyst residues. This can lead to many disadvantages during processing, such as degradation and hydrolysis. For most biomedical applications, catalyst toxicity is a highly sensitive issue. The solution is to deactivate the catalyst by adding phosphoric acid or to precipitate and filter the toxicity by adding strong acids such as sulfuric acid. In another patent, poly (lactic acid) with a relative molecular mass of 100,000 was obtained by direct polycondensation of diphenyl ether azeotropic dehydration. This method requires that the total molar content of impurities in raw lactic acid, such as methanol, ethanol, and acetic acid, should be less than 0.3% (Mitsui 1994). The bulk polymerization patents of Mitsui Toatsu Chemicals successively obtained high molecular weight PLA through the direct condensation process of liquid-phase polycondensation, granulation crystallization, and solid-phase polycondensation (Mitsui 2002). Moon et al. of Kyoto Institute of Technology (Kimura 2000; Lee 2000; Miyamoto 2000; Moon 2000) found that the two-component catalytic system composed of stannous chloride and p-toluene sulfonic acid had the best effect on the preparation of PLA by direct polycondensation. PLA with a relative molecular mass of 100,000 $\sim$ 200,000 can be obtained by polycondensation. After further crystallization-solid phase polycondensation, polymers with a relative molecular mass of 500,000 can be obtained (Kimura 2001; Lee 2001; Miyamoto 2001; Moon 2001; Taniguchi 2001).

2.3 *Ring-Opening Polymerization of Lactide*

Due to the low molecular weight of PLA obtained by the above two methods, it is difficult to meet the actual needs. Therefore, lactide ring-opening polymerization is often used to synthesize high-performance and high molecular weight PLA, which is also the mainstream way of large-scale industrial production. The method is divided into two steps. First, lactic acid is used as raw material, and then cyclic lactide is obtained by polycondensation and depolymerization. Finally, PLA is obtained by ring-opening polymerization of lactide.

However, in the production process of this method, the preparation and purification of lactide have certain technical difficulties. The first is the reactor material requirements. In the reaction process, the strong corrosion of high concentration lactic acid requires the application of special materials with corrosion resistance and high-temperature resistance in key units such as polycondensation and depolymerization reactors. High production durability has played a certain obstacle in the production cost. The second is the excessive viscosity of the reaction system. Under the polycondensation and depolymerization process, the decrease in flow rate leads to an increase in viscosity, which eventually leads to the volatilization of system products and inhibits the reaction in the positive direction. In addition, the selection of catalysts is also very important. At present, the mainstream metal catalysts are easy to remain in lactide, which is contrary to the current environmental protection concept. Therefore, metal residue treatment has become the focus of development.

According to the different initiator and reaction mechanisms, ring-opening polymerization of lactide can be divided into cationic ring-opening polymerization, anionic ring-opening polymerization, coordination insertion polymerization, and enzymatic ring-opening polymerization.

2.3.1 *Cationic ring-opening polymerization of lactide*

The mechanism of cationic ring-opening polymerization is that the catalyst cation first reacts with the oxygen atom in the monomer to form oxygen ions. Then, the alkoxy bond was broken, and the acyl cation was generated by the ring-opening reaction of a single molecule, which led to the chain growth, and finally, the PLA was prepared. The cationic ROP of lactones has been achieved using alkylating agents, acylating agents, Lewis's acids, and protic acids.

2.3.2 *Anionic ring-opening polymerization of lactide*

The mechanism of anionic ring-opening polymerization is that the carbonyl carbon in lactide breaks the acyl oxygen bond under the attack of the catalyst anion nucleophilic, forms the active center lactone anion, and then inserts into the main chain to initiate chain growth, and finally obtains PLA.

2.3.3 *Coordination insertion polymerization of lactide*

At present, coordination ring-opening polymerization is one of the most widely used and most studied synthesis methods of PLA. Most of the reaction mechanisms were that the metal active center coordinated with the oxygen atom on lactic acid, and then the acyloxy bond inserted the metal-oxygen coordination bond to promote the cleavage and opening of the acyloxy bond, thereby promoting the growth of the chain (Chen 1999; Cheng 1999; Fan 1999; L 1999; Nickol 1999; Wang 1999; Xie 1999).

2.3.4 *Enzymatic ring-opening polymerization of lactide*

In the past decade, the application of the enzyme in vitro polymer synthesis has been actively explored. Enzymatic polymerization is in vitro polymerization by separating enzyme-catalyzed abiotic synthesis pathways (Kobayashi 1999). This method is effective in the synthesis of natural polymers and has been extended to the synthesis of biodegradable polymers. Enzymes exhibit high stereo-, reaction-, and substrate specificity and come from renewable resources that can be easily recycled.

Table 1. Comparison of 4 ring-opening polymerization methods of lactide.

Method	Advantages	Disadvantages
Cationic ring-opening	High reaction velocity	Unstable, low optical purity
Anionic ring-opening	High activity catalyst for a solution or bulk polymerization	Easy racemization reaction, the low molecular weight of the product
Coordination insertion	Good reaction controllability, high molecular weight, and strength of the product	High cost
Enzymatic ring-opening	High stereo-, reaction-, and substrate specificity	Complex operation, low output, and high cost

3 MODIFICATION OF PLA

PLA is expected to be used as a substitute for the existed thermoplastics, and its wide range of applications can play a positive role in solving "white pollution" and "carbon neutralization" problems. However, there are still many shortcomings in the performance of the PLA, which need to be improved. For example, it is difficult to synthesize high molecular weight PLA by direct polycondensation method, because it is a hydrophobic material and is not soft and elastic enough to meet the requirements of tissue engineering and controlled release carrier of hydrophilic drugs when used as some medical materials. Due to different application conditions, PLA cannot simultaneously meet several applications in mechanical properties, hydrophilicity, and biological activity. The modification of PLA is focused, to find a method to improve molecular weight, reduce production costs, enhance its hydrophilicity and mechanical properties, and improve degradation performance. Therefore, it is necessary to improve the performance of the PLA in some aspects according to the demands. There are commonly three modification methods that can achieve this goal, namely blending, copolymerization, and composite modification.

3.1 *Blending modification*

Blending modification aims to improve the physical and mechanical properties of some defective polymers under the premise of maintaining the original excellent properties of the polymers while reducing the production cost. This modification method is to mix two or more kinds of polymers and achieve the purpose of modification by compounding the properties of each component of the polymer. In addition to the inherent excellent properties of each component, the blends exhibit new effects due to some synergistic effects between components. For example, since PLA is hard and brittle at room temperature, and its mechanical properties such as flexibility and elasticity are poor, blending modification is usually adopted to improve these properties. PLA was mixed with other monomers with flexibility and good compatibility, so that the mutual force between PLA molecules was reduced and the strength was decreased, so as to improve the flexibility and other properties. Therefore, blending modification is the most convenient and economical modification method at present.

Polyhydroxy butyrate polymer (PHB-di-rub) synthesized by Yeo et al. (He 2018; Kong 2018; Li 2018; Muiruri 2018; Tan 2018; Thitsartarn 2018; Yeo 2018; Zhang 2018) has a synergistic effect on PLA nucleation and toughening. Adding 10% PHB-di-rubble blends, the storage modulus increased by 32 %, the elongation increased by 128 times, the toughness increased by 84 times, and the strength and stiffness changed slightly. This degradable blend with high strength and high toughness is expected to replace the existing petroleum-based polymers and has broad application prospects in biomedical, automotive, and structural fields.

S. Alippilakkotte et al. (Alippilakkotte 2017; Sreejith 2017) introduced carboxyl and hydroxyl groups on the surface of PLA through alkaline hydrolysis of NaOH and then grafted gelatin onto the surface of PLA through coupling reaction. The combination of hydrophilic gelatin and hydrophobic PLA polymer can effectively improve tensile strength, elongation, flexibility, biocompatibility, and viscoelasticity. The SEM images of the grafted copolymer fibers showed that the surface of the fibers became smoother when crosslinked with gelatin. Compared with PLA / gelatin blend fiber, gelatin-g-PLA has higher hydrophilicity.

PLA / PCL is another kind of biodegradable PLA mixture system widely studied. PCL is a rubber polymer with low glass transition temperature, which can be degraded by hydrolysis or enzymatic hydrolysis. Kong et al. (Kong 2015; Yao 2015) prepared PLA / PCL composite foam sheet with a good performance by blending PLA and PCL, extrusion, and foaming. The results showed that the addition of PCL could improve the foaming effect of the material. When the PCL content was 10%, the PLA / PCL composite foam sheet had the best performance. The impact strength and compression strengths were 8.9 kJ / m2 and 31 MPa, respectively. The average cell diameter decreased from 2.8 mm to 1.1 mm, and the crystallinity increased from 12.68% to 17.95%.

3.2 *Copolymerization modification*

Copolymerization modification refers to the process of lactic acid polymerization, adding other monomers at the same time so that they are copolymerized. Adjusting the proportion of comonomers can change the properties of the polymer, such as crystallinity, hydrophobicity, and hydrophilicity. Copolymers can be prepared by condensation or ring-opening polymerization between lactic acid and these monomers. The copolymerization inserted the flexible segment into the PLA molecular chain, which reduced the regularity and crystallization ability of the whole molecular chain, and reduced the intermolecular interaction force. At the same time, the glass transition temperature and the melting point of PLA were improved, thereby enhancing the performance of PLA.

Gu Liangliang et al (Gu 2018; Li 2018; Macosko 2018; Nessim 2018) modified PLA with PEO-PPO-PEO triblock copolymer. Studies have shown that when the PEO-PPO-PEO content is 5%, the copolymer with high PPO content and large molecular weight forms droplets in PLA, which can significantly improve the ductility of PLA. Because the refractive index of PLA and PEO-PPO-PEO are basically the same, the blend is transparent, but the viscosity is lower than that of pure PLA.

J. Gug et al. (Gug 2016; Sobkowicz 2016) initiated an ester-amide exchange reaction by adding the catalyst to PLA and nylon 11 (PA11) to generate copolymer in situ. Studies have shown that the compatibility of the copolymerized two phases was improved after the reaction, and the mechanical properties of PLA were also improved. When 0.5% catalyst was added to the copolymer, the elongation at break increased to 116%. When the catalyst content is not more than 2% and the processing speed is maintained at about 250 r / min, the polymer can obtain better ductility.

M.A. Ghalia et al. (Dahman 2017; Ghalia 2017) used PEG to copolymerize and modify PLA and synthesized PEG-coPEG by melting polycondensation. It was found that when the volume ratio of PLA to PEG was 80:20 and 1.25% chain extender was added, the tensile strength and impact strength of the copolymer was increased to 70 MPa and 7.9 kJ / m2, respectively, and the elongation at break was increased by 17%.

3.3 *Composite modification*

Composite modification is a method to improve the overall performance of PLA by compounding PLA with various fibers, organic montmorillonite, and inorganic nanomaterials, and making use of some advantages of other materials to make up for the shortcomings of single PLA material (hydrophobicity, biocompatibility, thermal stability). The PLA material obtained by this method overcomes some defects of the original PLA and improves the performance of biological medicine, such as hydrophilicity, cell adhesion, strength, and degradation time as bone fixation materials. Nanomaterials have good volume effect, structure effect, and interface and surface effect, so they are widely used in the composite modification of PLA.

Zuo et al. (Chen 2020; He 2020; Li 2020; Wu 2020; Zuo 2020) constructed the compatible interface of PLA bamboo fiber / PLA composite by adding nano-SiO2. The results showed that the addition of nano-SiO2 was conducive to the crystallization and nucleation of PLA, and the initial decomposition temperature of the composite gradually increased with the increase of nano-SiO2. When the content of nano-SiO2 is 1.5%, the thermal decomposition temperature can reach 340 ° C, the crystallinity reaches the maximum, and the thermal stability is the best.

In order to improve the heat resistance and combustion performance of PLA, Hajibeygi et al. (Hajibeygi 2019; Shafiei-Navid 2019) used a new phosphorus-based organic additive (PDA) to modify hydroxyapatite (HA) nanoparticles in situ to prepare PLA nanocomposites. The results showed that PDA as a surface modifier could make HA uniformly dispersed in the PLA matrix. Compared with pure PLA, the initial decomposition temperature of PLA was increased by 20°C by adding 6% PDA and 2% HA nanoparticles. The peak heat release rate of pure PLA decreased from 566 W/g to 412 W/g.

Meng et al (Che 2018; Fu 2018; Jing 2018; Liu 2018; Xu 2018) modified the PLA surface with a traditional silane coupling agent (GF-S) and new graphene oxide (GF-GO). The results show that GF-S significantly improves the mechanical strength of the composites and maintains the original toughness of PLA. GF-GO has a good nucleation ability for PLA, and the crystallinity of PLA / GF-GO-10-9 and PLA / GF-GO-30-9 composites is 36.5%. The results show that the introduction of graphene oxide on the fiber surface can further enhance the crystallization and thermodynamic properties of the PLA matrix.

4 APPLICATION

PLA has good compatibility, degradability, tensile strength, and ductility, as well as good gloss and transparency. Therefore, it is generally widely used as a green polymer material, showing great application potential in many fields.

4.1 *Biomedicine*

The biomedical field is the earliest application field of PLA. PLA has high safety for the human body and can be absorbed by tissue, combined with its excellent physical and mechanical properties, so it can be used in the field of biomedicine. Such as disposable infusion tools, sutures, drug delivery microspheres, artificial fracture fixation materials, tissue repair materials, artificial skin, etc. The PLA surgical suture can be naturally degraded and discharged in the human body, avoiding the pain caused by the second operation of the patient. In vivo implantation of PLA instead of metal, instruments can avoid the stress shielding effect caused by metal instruments. Drug delivery microspheres use polyglycolic acid (PLGA), a copolymer of PLA, to make the active drug components (such as proteins or polypeptides) uniformly distributed in the biodegradable polymer microspheres matrix. After injection of drug delivery microspheres, the drug will be released into the organism. Cesur et al. (Alkaya 2019; Cesur 2019; Ege 2019; Ekren 2019; Erdemir 2019; Gunduz 2019; Kilic 2019; Kuruca 2019; Lin 2019; Oktar 2019; Seyhan 2019) added PLA, sodium alginate and nanofibers into oyster shell powder by electrospinning to prepare smooth surface nanofibers. The nanofibers have good compatibility with human bone tissue and no cytotoxicity, which can be used as human bone filler.

4.2 *Fiber textile*

PLA has excellent spinnability, and its fiber products have the advantages of safety, air permeability, moisture absorption, flame retardancy, and UV resistance. At the same time, PLA fibers can be made into single or compound fibers with a circular cross-section, polypropylene expanded filaments (BCFs, which can be used for weaving carpets and felts) with a triangular cross-section, crimp or non-crimp short fibers, bicomponent fibers, spun-bonded nonwovens, and melt-blown nonwovens. Therefore, PLA fibers are applied in the clothing market, household and decoration market, nonwovens market, bicomponent fibers, health, and medical fields.

The PLA fiber product Sorona developed by DuPont Company of America has good dyeing performance. The artificial leather made of Sorona is softer and more like real leather. It can be made into underwear, sportswear, imitation wool, medical supplies, household, and automobile decoration materials, and aerospace products. The sweat absorption of sportswear using Sorona is 3-4 times higher than that of cotton clothing. Now it has been applied to some Italian team clothing. PLA fiber Plas-tarch developed by Japan Coca-Colali Company can form a variety of composite fibers, which can be used in sports, uniforms, swear, nursing, decoration, and other aspects. In addition, it is widely used in agricultural materials, sanitary materials, aquatic materials, papermaking materials, etc.

4.3 *Packaging and coatings*

At present, various packaging materials are the largest and most potential application market of PLA. PLA has good water resistance, transparency, and printability, and its basic raw material lactic acid is one of the inherent physiological substances of the human body, which is nontoxic and harmless to the human body. Its applications in the packaging field can be mainly used as packaging belts, packaging film, agricultural film, foam plastic, tableware, horticultural film, and cold drink cups.

Wan et al. (Sun 2019; Wan 2019; Zhang 2019; Zhou 2019) used PLA, wood flour, and PMMA as raw materials to prepare PLA / wood flour / PMMA composites by melt blending method, and the degradation rate was evaluated. The results show that the hydrolysis rate of the prepared composite

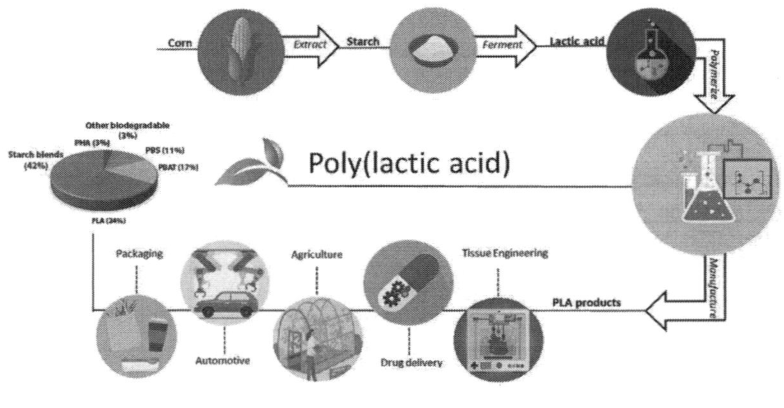

Figure 1. Applications of PLA.

is 8.56 times higher than that of pure PLA, and it has been widely used in the military controllable coating-filler model. In addition, PLA fiber also has the characteristics of moisture permeability, quick-drying, and skin affinity, which can be used for the manufacture of packaging wet paper towels, non-woven fabrics, space cups, and other products. Therefore, PLA can gradually replace the traditional polymer materials and show strong vitality in the application field of packaging coatings.

5 CONCLUSION

The research progress on the synthesis methods, the latest modification, and the applications of PLA is introduced in this paper. At present, there are many problems in the synthesis of PLA in industry, such as complex processes, long time-consuming, and high production costs. Moreover, the use of metal catalysts in the production process limits the application of PLA in biomedicine. Therefore, developing a green process for rapid and efficient synthesis of PLA is the focus of current research, including efficient non-metallic organic catalysts, enzyme catalysts, and the development of rapid synthesis technology.

The performance of PLA can be improved by modification. Although the physical modification method is simple, low cost, and easy to operate, the compatibility and dispersibility of materials need to be considered. The chemical modification method is to change the molecular structure of PLA, which makes PLA more stable. However, there are some defects such as high cost, incomplete decomposition of synthesis, environmental pollution, and complex operation. Therefore, it is necessary to consider how to reduce costs and make these methods environmentally friendly in the follow-up study.

PLA has been widely used in biomedicine, fiber textile, packaging coatings, and other fields. The production capacity is increasing and the cost is reducing, making PLA competitive in the plastic market, due to the enhancement of global environmental protection awareness and a profound understanding of the continuous deterioration of the living environment.

REFERENCES

Ajioka, M., Enomoto, K., Suzuki, K., & Yamaguchi, A. (1995). Basic properties of PLA produced by the direct condensation polymerization of lactic acid. *Bulletin of the Chemical Society of Japan*, 68(8), 2125–2131. https://doi.org/10.1246/bcsj.68.2125

Alippilakkotte, S., & Sreejith, L. (2017). Benign route for the modification and characterization of poly (lactic acid) (PLA) scaffolds for medicinal application. *Journal of Applied Polymer Science*, 135(13), 46056. https://doi.org/10.1002/app.46056

Cargill Incorporated. *Continuous process for the manufacture of lactide and lactide polymers*: US, 6326458[P]. 2001-12-04.

Cesur, S., Oktar, F. N., Ekren, N., Kilic, O., Alkaya, D. B., Seyhan, S. A., Ege, Z. R., Lin, C.-C., Kuruca, S. E., Erdemir, G., & Gunduz, O. (2019). Preparation and characterization of electrospun PLA/sodium alginate/orange oyster shell composite nanofiber for biomedical application. *Journal of the Australian Ceramic Society*, 56(2), 533–543. https://doi.org/10.1007/s41779-019-00363-1

Chen, G.-X., Kim, H.-S., Kim, E.-S., & Yoon, J.-S. (2006). Synthesis of high-molecular-weight poly (l-lactic acid) through the direct condensation polymerization of L-lactic acid in a bulk state. *European Polymer Journal*, 42(2), 468–472. https://doi.org/10.1016/j.eurpolymj.2005.07.022

Geyer R, Jambeck J R, Law K L.2017. Production, use, and the fate of all plastics ever made[J]. *Science advances*, 3(7), e1700782

Ghalia, M. A., & Dahman, Y. (2017). Investigating the effect Of MULTI-FUNCTIONAL Chain EXTENDERS on Pla/peg Copolymer properties. *International Journal of Biological Macromolecules*, 95, 494–504. https://doi.org/10.1016/j.ijbiomac.2016.11.003

Gu, L., Nessim, E. E., Li, T., & Macosko, C. W. (2018). Toughening poly (lactic acid) with poly (ethylene oxide)-poly (propylene oxide)-poly (ethylene oxide) Triblock Copolymers. *Polymer*, 156, 261–269. https://doi.org/10.1016/j.polymer.2018.09.027

Gug, J. I., & Sobkowicz, M. J. (2016). Improvement of the mechanical behavior of bioplastic poly (lactic acid)/polyamide blends by reactive compatibilization. *Journal of Applied Polymer Science*, 133(45). https://doi.org/10.1002/app.43350

Hajibeygi, M., & Shafiei-Navid, S. (2019). Design and preparation of poly (lactic acid) hydroxyapatite nanocomposites reinforced with phosphorus-based organic additive: Thermal, combustion, and mechanical properties studies. *Polymers for Advanced Technologies*, 30(9), 2233–2249. https://doi.org/10.1002/pat.4652

Jing, M., Che, J., Xu, S., Liu, Z., & Fu, Q. (2018). The effect of surface modification of glass fiber on the performance of poly (lactic acid) composites: Graphene oxide vs. Silane coupling agents. *Applied Surface Science*, 435, 1046–1056. https://doi.org/10.1016/j.apsusc.2017.11.134

Kobayashi, S. (1999). Enzymatic polymerization: A new method of polymer synthesis. *Journal of Polymer Science Part A: Polymer Chemistry*, 37(16), 3041–3056. https://doi.org/10.1002/(sici)1099-0518(19990815)37:16<3041: aid-pola1>3.0.co;2-v

Kong, Y., & Yao, Z. (2015). Preparation and Characterization of PLA / Polycaprolactone Foamed Composites. *Polymer Materials Science & Engineering*, 94–99.

Mitsui Toatsu Chemicals. (n.d.). *CN1130411C – Method for preparing polyhydroxy carboxylic acid. Google Patents*. https://patents.google.com/patent/CN1130411C/ko.

Mitsui Toatsu Chemicals, Incorporated. *Polyhydroxycarboxylic acid and preparation process thereof: US, 5310865*[P]. 1994-05-10.

Mitsui Toatsu Chemicals, Inc. *Process for the preparation of lactic acid polyesters: EP, 603889*[P]. 1994-06-29.

Mitsui Chemicals, Inc. *Process for preparing polyhydroxycarboxylic acid: US, 6429280*[P]. 2002-08-06.

Miyoshi, R., Hashimoto, N., Koyanagi, K., Sumihiro, Y., & Sakai, T. (1996). Biodegradable poly (lactic acid) with high molecular weight. *International Polymer Processing*, 11(4), 320–328. https://doi.org/10.3139/217.960320

Moon, S. I., Lee, C. W., Miyamoto, M., & Kimura, Y. (2000). Melt polycondensation OFL-lactic acid with sn(ii) catalysts activated by various proton acids: A direct manufacturing route to high molecular weight poly (l-lactic acid). *Journal of Polymer Science Part A: Polymer Chemistry*, 38 (9), 1673–1679. https://doi.org/10.1002/(sici)1099-0518 (20000501) 38:9<1673: aid-pola33>3.0.co;2-t

Moon, S.-I., Lee, C.-W., Taniguchi, I., Miyamoto, M., & Kimura, Y. (2001). Melt/solid polycondensation of L -lactic acid: An alternative route to poly (L -lactic acid) with high molecular weight. *Polymer*, 42(11), 5059–5062. https://doi.org/10.1016/s0032-3861(00)00889-2

Moon, S.-I., Taniguchi, I., Miyamoto, M., Kimura, Y., & Lee, C.-W. (2001). Synthesis and properties of high-molecular-weight poly (l-lactic acid) by melt/solid polycondensation under different reaction conditions. *High-Performance Polymers*, 13(2). https://doi.org/10.1088/0954-0083/13/2/317

Nagahata, R., Sano, D., Suzuki, H., & Takeuchi, K. (2007). Microwave-assisted single-step synthesis of poly (lactic acid) by direct polycondensation of lactic acid. *Macromolecular Rapid Communications*, 28(4), 437–442. https://doi.org/10.1002/marc.200600715

Tokiwa, Y., & Jarerat, A. (2004). Biodegradation of poly(l-lactide). *Biotechnology Letters*, 26(10), 771–777. https://doi.org/10.1023/b:bile.0000025927.31028.e3

University of Georgia. 2017. More than 8.3 billion tons of plastics made: Most have now been discarded. ScienceDaily. *Science Daily*. www.sciencedaily.com/releases/2017/07/170719140939.htm

Wan, L., Li, C., Sun, C., Zhou, S., & Zhang, Y. (2019). Conceiving a feasible degradation model of PLA-based composites through hydrolysis study to PLA/wood flour/polymethyl methacrylate. *Composites Science and Technology*, 181, 107675. https://doi.org/10.1016/j.compscitech.2019.06.002

Xie, W., Chen, D., Fan, X., Li, J., Wang, P. G., Cheng, H. N., & Nickol, R. G. (1999). Lithium chloride as a catalyst for the ring-opening polymerization of lactide in the presence of hydroxyl-containing compounds. *Journal of Polymer Science Part A: Polymer Chemistry*, 37(17), 3486–3491. https://doi.org/10.1002/(sici)1099-0518(19990901)37:17<3486: aid-pola6>3.0.co;2-2

Yeo, J. C., Muiruri, J. K., Tan, B. H., Thitsartarn, W., Kong, J., Zhang, X., Li, Z., & He, C. (2018). Biodegradable PHB-rubber copolymer toughened PLA Green Composites with Ultrahigh extensibility. *ACS Sustainable Chemistry & Engineering*, 6(11), 15517–15527. https://doi.org/10.1021/acssuschemeng.8b03978

Zuo, Y., Chen, K., Li, P., He, X., Li, W., & Wu, Y. (2020). Effect of nano-sio2 on the compatibility interface and properties of PLA-grafted bamboo fiber/PLA composite. *International Journal of Biological Macromolecules*, 157, 177–186. https://doi.org/10.1016/j.ijbiomac.2020.04.205

Energy Revolution and Chemical Research – Kok-Keong Chong and Zhongliang Liu (Eds)
 ISBN: 978-1-032-36554-1

Reactivity of $Ca_2Fe_2O_5$ oxygen carriers for chemical looping steam methane reforming

X.Y. Wang, M. Chen, Y.H. Hou, X.Y. Gao, Y.Z. Liu & Q.J. Guo*
College of Chemical Engineering, Qingdao University of Science and Technology, Qingdao, China

ABSTRACT: Chemical looping steam methane reforming (CL-SMR) can obtain high-quality syngas and pure hydrogen at the same time, with a prerequisite that oxygen carriers possess excellent reactivity. In this work, a composite oxygen carrier, i.e., $Ca_2Fe_2O_5$, was prepared by the sol-gel method and found to hold remarkable reactivity and syngas selectivity. At the optimal condition, the average CH_4 conversion, CO selectivity, and syngas yield were 36.33%, 78.79%, and 7.81 mmol/g_{OCs} respectively during the methane reforming stage, while hydrogen yield was 3.72 mmol/g during the water splitting stage. Ten redox-cycle experiments indicate that the syngas yield and H_2 yield were stable at 4.40 and 1.88 mmol/g_{OCs}. Characterization analysis of oxygen carriers indicates that Fe exsolution favors methane activation and decomposition, therefore enhancing the reaction efficiency.

1 INTRODUCTION

Hydrogen is considered the most promising energy source and carrier, in light of its potential renewability, high energy density, and environmental cleanliness. At present, the majority of industry hydrogen was produced via steam methane reforming (SMR) (Zhang 2021). However, a series of energy-intensive purification and separation devices increase the energy consumption and investment, together with emitting substantial CO_2. Therefore, how to prepare hydrogen more economically, efficiently, and cleanly has received extensive attention.

Chemical looping steam methane reforming (CL-SMR) uses an oxygen carrier to decouple methane steam reforming into methane reforming reaction and steam splitting reaction, and obtain high-quality syngas and pure hydrogen at the same time, which has attracted a lot of attention from international scholars and companies (Collins-Martinez 2021). In the reforming reactor, the partial oxidation reaction of methane can obtain a syngas (R1) with an H_2 to CO ratio of 2, which can be used for Fischer-Tropsch synthesis or synthesis of other chemicals; in the steam splitting reactor, the oxygen carrier after CH_4 reduction reacts with H_2O to obtain high purity hydrogen (R2).

$$yCH_4 + MeO_x \rightarrow MeO_{x-y} + yCO + 2yH_2 \tag{R1}$$

$$yH_2O + MeO_{x-y} \rightarrow MeO_x + yH_2 \tag{R2}$$

The preparation of highly selective, reactive, and stable oxygen carriers is a prerequisite for achieving CL-SMR. Among all candidate materials, perovskite oxides perform advantage in the CL-SMR process (Jiho 2017). However, it is also found that the carbon deposition phenomenon is more severe, and the CH_4 conversion is limited (Mihai 2012). Hence, scholars have carried out a lot of work on its modification to improve the anti-carbon deposition ability. Wang et al. (Wang 2020) found that the substitution of iron ions in $LaMnO_3$ can improve its reactivity and thermal stability, among them, $La_{0.85}MnFe_{0.15}O_3$ showed the highest syngas yield (3.78 mmol/g)

*Corresponding Author: qjguo@qust.edu.cn

DOI 10.1201/9781003332657-46

and hydrogen yield (1.76 mmol/g), and no carbon deposition occurred in 20 cycle experiments. However, perovskite oxide is confined by its structure, and it is difficult to be completely reduced it by CH_4. Therefore, affected by the reduction depth, its non-stoichiometric oxygen cannot be completely released, resulting in a relatively low yield of its target gas.

Iron-based oxides are considered ideal oxygen carrier materials due to their outstanding reactivity, low cost, and theoretical ability to completely release stoichiometric oxygen (Guo 2016). However, pure iron oxides are liable to sinter at high temperatures, resulting in serious deactivation. Moreover, the reduction of oxygen carriers by CH_4 is thermodynamically limited, and only a few of them can be oxidized back to the original valence state by H_2O steam. $Ca_2Fe_2O_5$ is among the oxygen carrier material that can theoretically release the stoichiometric oxygen of the active component completely. However, it has not been sufficiently studied in the CL-SMR hydrogen production process.

In the current work, $Ca_2Fe_2O_5$ oxygen carrier was prepared by the sol-gel method and employed for CL-SMR. Parameter including methane conversion, CO selectivity, H_2/CO molar ratio, syngas yield, H_2 yield, and redox stability was investigated. Combined with X-ray diffraction (XRD), scanning electron microscopy (SEM), and energy dispersive spectroscopy (EDS) techniques, the promotion mechanism of the structure was revealed.

2 EXPERIMENT

2.1 *Preparation of oxygen carrier*

$Ca(NO_3)_2{\cdot}4H_2O$ and $Fe(NO_3)_3{\cdot}9H_2O$ with 1:1 metal cations were weighed and dissolved in deionized water. As a complexing agent, citric acid with the molar ratio of citric acid to metal cations (1.2:1) was added. The viscous sols were obtained by heating and stirring at 80°C for 6 hrs, and then dried to obtain a solid sample. The solid samples were transferred to a ceramic crucible and placed in a muffle furnace. After being dried at 300°C for 1 h, the solid samples were calcined at 900°C for 6 hrs. The obtained samples were ground and collected for use.

2.2 *Experimental procedure*

The experiments were carried out in a self-built fixed-bed reactor system. 0.5 g oxygen carrier was placed in the middle of the reactor, being evaluated to 850°C with a heating rate of 20 K/min. During this period, high-purity Ar (50 ml/min) is used as a purge gas to ensure an inert atmosphere. As for the CH_4 reforming test, 5 ml/min of methane with 45 ml/min Ar was introduced. The reduction time was 18 min, and the reaction gas was collected every 3 min during the reduction process using a gas sampling bag and detected by a gas chromatography analyzer. As for the water-splitting test, the saturation water vapor at 75°C was carried by Ar gas into the reactor to react with the oxygen carrier. After the gas product was cooled and dried, the reaction gas was collected every 5 min using a gas sampling bag and sent to the gas chromatograph for detection. Until no hydrogen gas is produced the reaction is finished. The cycle stability performance of the oxygen carrier was investigated by a redox cycle test. The oxygen carrier is oxidized by H_2O vapor after the CH_4 reforming test. Between each cycle, the oxygen carrier is purged with Ar gas (50 ml/min) for 30 min to fully dry the oxygen carrier.

2.3 *Material characterization*

The composition of the oxygen carrier is characterized by an X-ray diffractometer (Rigaku D/max 2500 PC). The measurement conditions are as follows: Cu Kαlight source (λ = 0.15406 nm) with tube current 100 mA and voltage 40 kV. The scanning angle (2θ) ranged from 10°C to 85°C with a scanning rate of 0.02°C/s. To observe the microscopic morphology and elemental composition of the materials, a scanning electron microscope (Zeiss Sigma 300) equipped with an EDS spectrometer

(Oxford Xplore) was adopted. The samples need to be dried in advance to ensure that the samples do not contain moisture, after which heavy metal particles were sprayed on the surface of the samples to increase the electrical conductivity of the materials, and finally placed in the instrument for observation.

2.4 *Data analysis*

The evaluation indexes for the methane reforming stage include methane conversion: X_{CH_4}, CO selectivity: S_{CO}, Syngas yield: Ysyn, H_2/CO mole ratio: $m_{H_2/CO}$ as follows:

$$X_{CH_4} = \frac{n_{CH_{4,in}} - n_{CH_{4,out}}}{n_{CH_{4,in}}} \times 100\% \tag{Q1}$$

$$S_{CO} = \frac{n_{CO}}{n_{CO} + n_{CO_2}} \times 100\% \tag{Q2}$$

$$m_{H_2/CO} = \frac{n_{H_2}}{n_{CO}} \tag{Q3}$$

$$Y_{syn} = \frac{n_{CO} + n_{H_2}}{m_{OC}} \tag{Q4}$$

The evaluation indexes for the steam splitting stage are hydrogen yield (YH_2)'F

$$Y_{H_2} = \frac{n_{H_2}}{m_{OC}} \tag{Q5}$$

Where m_{OC} is the mass of oxygen carrier involved in the reaction, n_{H_2}, n_{CO} and n_{CO_2} are the molar numbers of H_2, CO and CO_2, respectively. $n_{CH_{4,in}}$ and $n_{CH_{4,out}}$ are the methane moles entering and leaving the reactor, respectively.

3 RESULTS AND DISCUSSION

3.1 *Characterization of oxygen carriers*

To investigate the structural changes of $Ca_2Fe_2O_5$ oxygen carriers before and after the reduction-oxidation reaction, the different state was characterized by XRD. As shown in Figure 1, a fresh sample corresponding to the characteristic peak of $Ca_2Fe_2O_5$ (JCPDS PDF #71-2264) maintains a single $Ca_2Fe_2O_5$ brownmillerite structure (Hu 2020). At 850°C, after the reaction with methane for 18 min, CaO and Fe characteristic peaks were also detected without other oxides containing Ca and Fe being formed, indicating that methane can reduce the $Ca_2Fe_2O_5$ oxygen carrier to CaO and Fe. After being oxidized by water vapor, the oxygen carrier returns to the original $Ca_2Fe_2O_5$ structure in one step, indicating that $Ca_2Fe_2O_5$ has excellent regeneration.

To investigate the microscopic morphological changes of oxygen carriers in the redox reaction, SEM and EDS characterization were performed. As shown in Figure 2(a), after the oxygen carrier undergoes the methane reduction reaction, the particles show irregular morphology, and there are similar spherical particles produced. The element distribution on the surface of the oxygen carrier is observed by EDS-mapping characterization. It can be seen that the signal of O elements on the surface of the material is weak and not easily observed, indicating that the oxygen carrier consumes a large amount of lattice oxygen after the reaction with methane. Partial oxygen elements observed come from the reduction product CaO and unreacted $Ca_2Fe_2O_5$. The Ca elements are evenly distributed, while the Fe elements are relatively aggregated. Combined with the XRD characterization, it can be inferred that the spherical particles may be Fe generated by the reduction of oxygen carriers by methane. Moreover, the Fe generated from $Ca_2Fe_2O_5$ oxygen carriers with methane is directly precipitated in the form of spherical particles, which may subsequently have a catalytic effect on methane. Figure 2(b) shows that the surface of the oxygen carrier after the reaction with water is regular and relatively dispersed between the particles with Ca, Fe, and O elements uniformly dispersed on the surface of the oxygen carrier. Meanwhile, lattice oxygen required could be replenished by H_2O and restored to the original $Ca_2Fe_2O_5$ structure, which verified the good regeneration ability of the oxygen carrier.

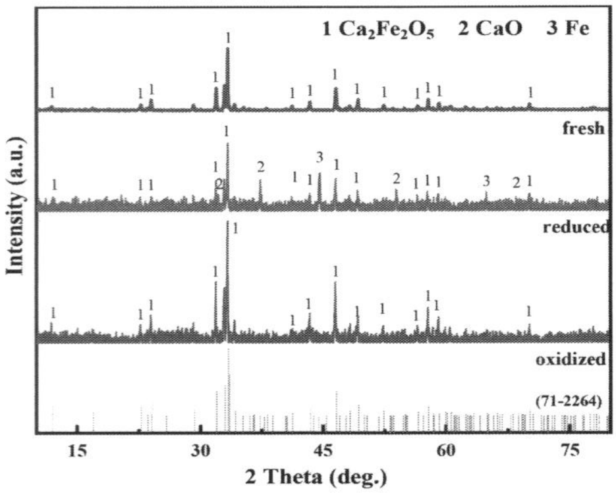

Figure 1. XRD patterns of $Ca_2Fe_2O_5$ oxygen carrier.

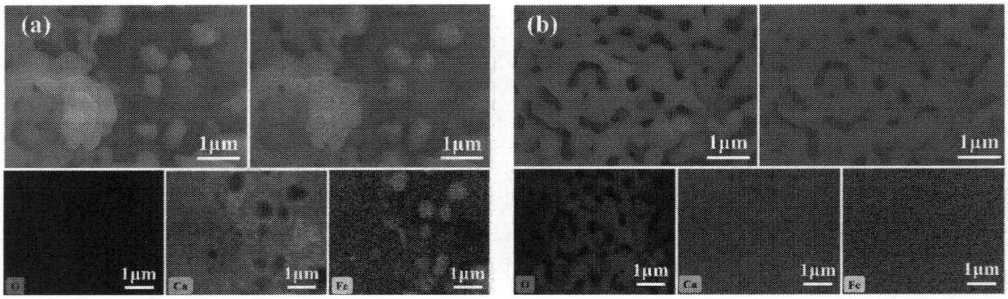

Figure 2. SEM and EDS characterization of $Ca_2Fe_2O_5$ oxygen carrier, (a) reduction, (b) oxidation.

3.2 *Reactivity of $Ca_2Fe_2O_5$ oxygen carrier*

The reaction performance of oxygen carriers during methane reforming and water splitting was depicted in Figure 3. As shown in Figure 3 (a), it is clear that the reaction between $Ca_2Fe_2O_5$oxygen carrier and methane was relatively gentle within 18 min, and the H_2 and CO contents gradually increase with the reaction time during the reaction, up to 9.15 % and 4.74 %, respectively at 18 minutes. It manifests that the lattice oxygen is released steadily without a large amount of methane reforming reaction. Combined with the previous characterization, it was found that the $Ca_2Fe_2O_5$ oxygen carrier was reduced to CaO and Fe during the methane reforming reaction. There is an exsolution of Fe from the oxygen carrier to catalyze the activation and decomposition of methane into active carbon deposits C* and H*. When lattice oxygen is sufficient, it can oxidize C* and H* to CO and H_2 without carbon deposits (Huang 2015). Furthermore, it can be speculated that this phenomenon can promote the reaction, which is the reason why the H_2 and CO content gradually increases with reaction time. During the reaction, the average methane conversion of $Ca_2Fe_2O_5$ oxygen was 36.33%, the CO selectivity was 78.79%, and the syngas yield was 7.81 mmol/g. Figure 3 (b) shows the H_2 production in the H_2O splitting reaction of the oxygen carrier, where the oxygen carrier reduced by methane reacts with water vapor to replenish the lattice oxygen and produce hydrogen at the same time. The gaseous product during this reaction is H_2 without C-containing oxides being detected, indicating that there is no carbon accumulation in the previous stage of the methane reforming reaction, and the hydrogen yield at this time is 3.72 mmol/g.

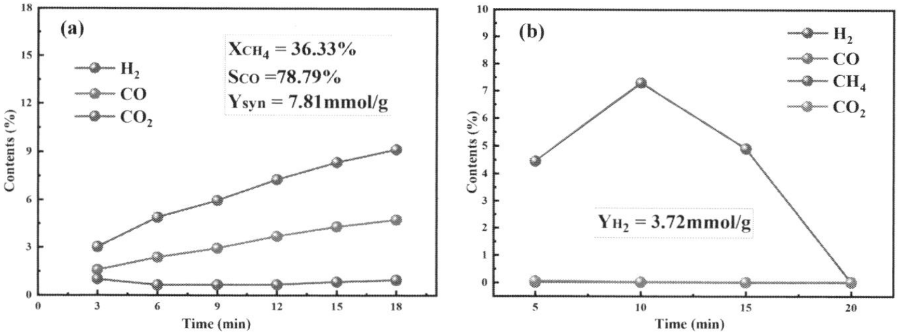

Figure 3. Reactivity of oxygen carrier during methane reforming (a) and water splitting (b) at 850°C.

3.3 *Cycle performance of $Ca_2Fe_2O_5$ oxygen carrier*

The feasibility of CL-SMR with $Ca_2Fe_2O_5$ oxygen carriers was verified in the previous section, and appreciable syngas and hydrogen yields were obtained during the reaction. Therefore, ten redox cycle experiments were conducted using $Ca_2Fe_2O_5$ oxygenate carriers to examine their cycle stability. Figure 4 shows the reaction performance of the $Ca_2Fe_2O_5$ oxygen carrier during the 10 redox cycle experiments. It can be seen that the methane conversion of the $Ca_2Fe_2O_5$ oxygen is 36.33% at the first cycle and maintained at about 21% during the subsequent cycles, while the CO

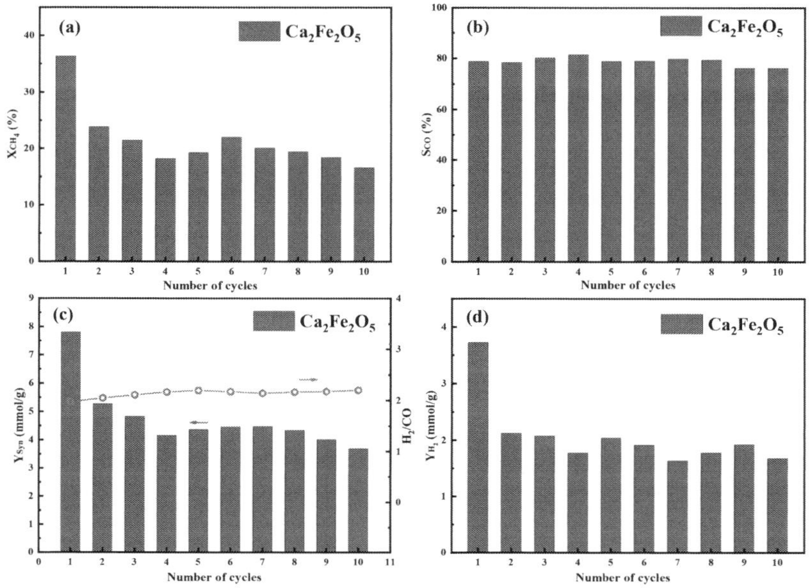

Figure 4. CH_4 conversion (a), CO selectivity (b), syngas yield, H_2/CO molar ratio (c), and H_2 yield (d) during 10 redox cycles at 850°C.

selectivity of the $Ca_2Fe_2O_5$ oxygen carrier remains relatively stable during the 10 redox cycles, averaging around 78.85%. As displayed in Figures 4 (c) and (d), the target gas of the oxygen carrier decreases significantly after the first cycle for both syngas and H_2 yields, and the trend is similar to that of the methane conversion of the oxygen carrier, and also decreases during the subsequent cycles. The H_2 yield gradually stabilized from 3.72 mmol/g_{OCs} in the first cycle to 1.88 mmol/g_{OCs}. It can be seen from the H_2/CO molar ratio that the $Ca_2Fe_2O_5$ oxygen carrier was maintained around the theoretical value of 2 in 10 cycles. It indicates that the side reaction of massive methane cracking did not occur during the cycle, which ensured the quality of the target gas product.

4 CONCLUSIONS

In this work, $Ca_2Fe_2O_5$ oxygen carriers were prepared by the sol-gel method and applied to the CL-SMR process. Based on our analysis, the following conclusions can be drawn:

(1) The CL-SMR process can be completed in two steps, CH_4 reforming and H_2O vapor oxidation, without additional oxidation steps. The oxygen carrier maintains the structure of $Ca_2Fe_2O_5$ after the redox reaction, which proves its ability to be recycled.
(2) During methane reforming, the average methane conversion and the selectivity of CO were 36.33 % and 78.79 % respectively, with the yield of syngas 7.81 mmol/g_{OCs}. As for water splitting, the yield of hydrogen was up to 3.72 mmol/g_{OCs}. The exsolution of elemental Fe is beneficial to catalytic activation of methane, therefore improving the reaction efficiency.
(3) During 10 redox experiments of the $Ca_2Fe_2O_5$ oxygen carrier, the syngas yield is stable at 4.40 mmol/g_{OCs}, and the H_2 yield is stable at 1.88 mmol/g_{OCs}.

ACKNOWLEDGMENT

This work was funded by the Joint Funds of the National Natural Science Foundation of China (Grant No. U20A20124), Shandong provincial natural science foundation (Grant No. ZR2020MB144), and the State Key Laboratory of High-efficiency Utilization of Coal and Green Chemical Engineering (Grant No. 2022-K3).

REFERENCES

Collins-Martinez V H, Cazares-Marroquin J F, Salinas-Gutierrez J M, et al. (2021). The thermodynamic evaluation and process simulation of the chemical looping steam methane reforming of mixed iron oxides. *J. RSC Advances*, 11(2): 684–699.

Guo Q, Yang M, Liu Y, et al. (2016). Multicycle investigation of a sol-gel-derived Fe_2O_3/ATP oxygen carrier for coal chemical looping combustion. *J. AIChE Journal*, 62(4): 996–1006.

Hu Q, Shen Y, Chew J W, et al. (2020). Chemical looping gasification of biomass with Fe_2O_3/CaO as the oxygen carrier for hydrogen-enriched syngas production. *J. Chemical Engineering Journal*, 379.

Huang L, Tang M, Fan M, et al. (2015). Density functional theory study on the reaction between hematite and methane during chemical looping process. *J. Applied Energy*, 159: 132–144.

Jiho, Yoo, Chung-Yul, et al. (2017). Determination of oxygen nonstoichiometry in $SrFeO_{3-\delta}$ by solid-state Coulometric titration. *Journal of the American Ceramic Society*, 100(6): 2690–2699.

Mihai O, Chen D, Holmen A. (2012). Chemical looping methane partial oxidation: The effect of the crystal size and O content of $LaFeO_3$. *Journal of Catalysis*, 293: 175–185.

Wang Y, Zheng Y, Wang Y, et al. (2020). Evaluation of Fe substitution in perovskite $LaMnO_3$ for the production of high purity syngas and hydrogen. *Journal of Power Sources*, 449(C): 227505.

Zhang H, Sun Z, Hu Y H. (2021). Steam reforming of methane: Current states of catalyst design and process upgrading. *J. Renewable and Sustainable Energy Reviews*, 149: 111330.

Energy Revolution and Chemical Research – Kok-Keong Chong and Zhongliang Liu (Eds)
© 2023 The Authors, ISBN: 978-1-032-36554-1

Research progress of $FASnI_3$ for tin-based perovskite solar cells

Jingbo Wang*
School of Physic, Northwest University, Xi'an, China

ABSTRACT: In the past decade years, Perovskite Solar cells (PSC) cells have flourished and have a very promising future. The highest reported conversion efficiency of Pb-based PSC has reached over 25% and is expected to replace silicon solar cells in the near future. However, the biggest problem of Pb-based PSC is that the Pb element is environmentally unfriendly, which hinders their commercialization, so in recent years, non-lead PSC, especially tin-based PSC, has gained widespread attention. The maximum conversion rate of tin-based PSCs has now increased from 6% to more than 12%. However, tin-based PSC has some problems of its own, which hinder its commercialization. This paper briefly introduces the development history of $FASnI_3$, one of the perovskite materials used in tin-based PSC, and several methods to improve the efficiency and stability of PSC. Finally, the development of tin-based PSC has prospected.

1 INTRODUCTION

1.1 *Structure of tin-based perovskite materials*

The general formula of the perovskite structure is ABX_3, where A is a monovalent organic or inorganic cation, which in tin-based perovskite materials is mainly MA^+, FA^+, Cs^+. B is Sn^{2+}, and X is mainly halogen element anions, such as Cl^-, Br^-, and I^-. Among them, BX_3^- ion is an octahedral structure, which is extended into a three-dimensional structure by angle sharing, and A ion is filled in it.

The energy band structure of tin-based perovskite depends mainly on the Sn and X-site ions, such as $CsSnI_3$, whose valence band top (VBM) is formed mainly by the coupling of the 5s orbital of Sn and the 5p orbital of I with antibonding features, while the conduction band bottom (CBM) is formed mainly by the coupling of the 5p of Sn and a very small amount of the 5p orbital of I with nonbonding features (Tao et al. 2019). When the I element is replaced by other halogen elements, the energy level is slightly changed by all three effects together, while when the A-site ion is changed, the electronic structure is affected by changing the lattice volume and introducing distortions, thus affecting the electronic structure, with a gradual increase in the degree of distortion from Cs to MA to FA, leading to a decrease in VBM (Huang & Lambrecht 2013).

1.2 *Problems faced by tin-based perovskite materials*

The tin-based perovskite energy band structure is a direct band gap and is narrower than the lead-based perovskite band gap, which is in the region of approximately 1.2 to 1.4 eV. Tin-based perovskite possesses a smaller exciton binding energy and higher carrier mobility (Tai et al. 2019). However, so far, the efficiency of tin-based PSC is much lower than that of Pb-based PSC for the following two reasons: (1) Sn^{2+} is highly susceptible to oxidation into Sn^{4+}, which is prone to self-doping effects (Kumar et al. 2014), resulting in a large number of defects in the lattice, and

*Corresponding Author: 1007439093@qq.com

DOI 10.1201/9781003332657-47

these traps can easily trap carriers when subjected to heat, leading to non-radiative recombination and seriously affecting the conversion efficiency (PCE) of perovskite cells; (2) the crystallization rate is too fast, making it difficult to form a uniform and dense perovskite films, and thus the fabrication process of Pb-based PSC cannot be directly applied to tin-based perovskite (Yokoyama et al. 2016).

2 FASNI$_3$ PSC RESEARCH PROGRESS

Currently, the most studied tin-based perovskite materials are $FASnI_3$, $MASnI_3$, $MASnBr_3$, $CsSnI_3$ and $CsSnBr_3$. In this paper, we summarize the research progress of $FASnI_3$ in recent years and list some strategies used in the development process.

2.1 *Compensators*

Due to the strong anti-bonding coupling between the 5s orbital of tin and the 5p orbital of iodine in perovskite, as well as the low formation energy of tin vacancies and the relative ease of defect formation, both theory and experiment show that adding tin compensators such as SnF_2, $SnCl_2$, and $SnBr_2$ to perovskite to create an atmosphere of tin excess can effectively enhance the chemical potential of Sn and increase the formation energy of tin vacancy defects, thus reducing surface defects (Xu et al. 2014).

The CBM position of $FASnI_3$ is −4.55 eV and the VBM position is −5.96 eV. $FASnI_3$ solar cells were first reported in 2015 by Koh et al. The authors used a precursor DMF solution containing 20% SnF_2 coated on porous TiO_2 by a one-step spin-coating method to obtain $FASnI_3$ solar cells with a PCE of 2.1%, and an open-circuit voltage (Voc) of 0.24 V (Koh et al. 2015). Unlike Pb-based perovskites, tin-based perovskites have many defects and the crystallization rate is too fast to directly follow the Pb-based perovskites preparation process (Yokoyama et al. 2016). To solve this problem, Lee et al. added 10% SnF_2 and pyrazine to the DMF/DMSO solution of $FASnI_3$ to optimize the film morphology with PCE up to 4.8% and Voc up to 0.32 V (Lee et al. 2016). By regulating the content of SnF_2 in the precursor solution and using ether as the antisolvent, Liao et al. reduced the cavity density to $10^{17}cm^{-3}$ and successfully suppressed carrier complexation to obtain a PCE of 6.22% and Voc of 0.47 V (Liao et al. 2016).

In 2021, Dai Z et al. (Dai et al. 2021) proposed a new strategy by choosing tin acetate instead of the conventional SnF_2 as the precursor additive. Compared with SnF_2, perovskite thin films prepared by the tin acetate additive have better crystallinity, higher stability, and lower defect density, and the cells can achieve a PCE of 9.93%, and run at maximum power point under standard AM1.5 G solar irradiation After 1000 h, the initial efficiency can still be maintained above 90%.

2.2 *Reducing agent*

In 2018, Gu et al. (Feidan et al. 2018) successfully reduced Sn^{4+} to Sn^{2+} by adding Sn powder to the $FASnI_3$ precursor solution prepared from 99% pure SnI_2. Through a series of controlled experiments, it was found that the presence of a small amount of Sn^{4+} in the precursor solution caused a sharp increase in the internal carrier concentration of the $FASnI_3$ device and significantly reduced the efficiency, and the introduction of Sn powder effectively reduced a large amount of Sn^{4+} and lowered the Sn^{4+} concentration, resulting in a lower concentration of defect states in the film and a significant reduction in the non-radiative composite process, with a final PCE of 6.75%.

In the same year, Kayesh et al. introduced solid hydrazine hydrochloride reducing additive to the $FASnI_3$ precursor, and by comparing the state of the $FASnI_3$ precursor solution with and without N_2H_5Cl in air, they found that hydrazine hydrochloride significantly inhibited the oxidation rate of Sn^{2+} and had a lower Sn^{4+} content at 2.5 mol% and a longer carrier lifetime with a PCE up to 5.4% (Kayesh et al. 2018). This led to the discovery of a reducing hydrazine group that could repair vacancy defects. Subsequently, Wang et al. (Wang et al. 2020) successfully prepared solar cells with a PCE of 11.4% by choosing milder phenylhydrazine hydrochloride (PHCl), and found that due to the doping of PHCl, the phenylhydrazine ions entered the lattice to swell the lattice, which was beneficial to improve the film morphology. Moreover, due to its large size, it can be a good barrier to water and air from outside. In the following year, Wang, Zhang, Gu, et al. (2020) synergistically introduced both halogen (Cl^- and Br^-) and phenylhydrazine ions. The introduction of Cl^- can regulate the nucleation and crystallization process without entering the perovskite lattice. Br^- in the lattice inhibited the migration of I^- and helped the phenylhydrazine cation to further reduce the film defect density. The PCE of the $FASnI_3$ cell was increased to 13.4% with a certified efficiency of 12.4%, indicating that the synergistic effect of the two can be very effective in passivating the defect state of the light-absorbing layer, while the PSC device also has excellent stability, and it was found that under normal conditions Br^- tends to aggregate under operating conditions, leading to phase separation, so the addition of Br in most literature reports did not significantly improve the stability of the devices, the experimental data showed that the introduction of phenylhydrazine cation effectively suppressed the phase separation and avoided this situation, and devices with better stability were obtained. The initial efficiency of 91% can be maintained in a glove box for 4800 hrs and 82% of the initial efficiency can still be maintained under continuous sunlight exposure for 330 hrs (Wang et al. 2021).

In addition to solid additives, in 2020, Meng et al. (Meng et al. 2020) introduced liquid formic acid (LFA) as a reducing solvent into the $FASnI_3$ perovskite precursor solution, and since the LFA solvent was volatile, there was no residual LFA in the $FASnI_3$ perovskite films. Finally, $FASnI_3$ perovskite films with high crystallinity, low Sn^{4+} content, low background doping, and low defect density were prepared, with PCE over 10%.

2.3 *Chelating agent*

Yang et al. (Yang 2020) introduced fluorinated (1,1,1,2,2,3,3,4,4-fluoro-substituted-6-dodecyl)-(F-PDI) into the grain boundaries of perovskites through a carbonyl group in F-PDI that provided lone electron chelation with uncoordinated Sn atoms in chalcogenides, thus passivating the internal defects. The F-PDI present in the grain boundaries of the film can also act as a charge transport medium to promote effective carrier transport at the grain boundaries and interfaces, and the final PCE obtained can reach 7.36%. Lin Z J et al. (Lin 2021) introduced 8-HQ with a bidentate ligand into $FASnI_3$, and the O and N atoms in 8-HQ were simultaneously coordinated with divalent Sn ions to form a stable chelate, which effectively inhibit the oxidation of Sn^{2+} to Sn^{4+}, and the final PCE obtained can reach 7.15%.

2.4 *Crystallization modulation*

Wu (Wu et al. 2020) et al. introduced high electron-density CDTA into $FASnI_3$ films to form a stable intermediate with the Sn-I backbone, which can regulate the crystallization rate of perovskite and form dense and uniform perovskite films. Also, the introduction of a π-conjugated system hinders the penetration of water, and the final obtained PCE can reach 10.1% and still maintain 90% of the initial value after 1000 hrs of visible light irradiation at 100 W/cm^2 in air. The team (Liu et al. 2020) also developed a templated growth method to reduce the $FASnI_3$ body defect concentration by pretreating with PAI before thermal annealing. The intermediate phase doped with PAI induced vertical growth of perovskite seeds along the (h00) plane through the spatial effect of alkyl chains, which reduced the defect density of $FASnI_3$ films from 2.89 × 1016 to 5.41 × 1015 cm^{-3}, extended the carrier lifetime from 4.1 ns to 6.88 ns, and increased Voc from 0.53 V to 0.73 V.

In 2021, Chang B et al. (2021) proposed a strategy to construct fluidic flexible polymer scaffolds by introducing environmentally friendly polyethylene glycol (PEG) polymers with many ethers bonding groups (C-O-C) into the precursor of $FASnI_3$ to modulate the nucleation and growth of perovskite particles and to fabricate uniform full-coverage perovskite films with low defect density. It is demonstrated that hydrogen bonding interactions between FA^+ and C-O-C and Lewis acid-base complexation between the uncoordinated Sn and C-O-C groups can effectively modulate the crystallization behavior of the films, improve the coverage of the perovskite films, and reduce the density of defect states. Polyethylene glycol acts as a fluid flexible polymer scaffold during the annealing process, playing a key role in cross-linking the chalcogenide particles and relieving internal stresses. With this approach, high-quality chalcogenide films with larger grain size, higher coverage, and higher carrier lifetime are obtained. The PCE of the optimized device can reach 7.53%, and the unencapsulated device maintains more than 90% of the initial PCE after aging for 720 hrs in a nitrogen glove box.

Liu G et al. (2021) achieved optimization of the film quality by modulating the crystallization rate of $FASnI_3$ perovskite by introducing ethylene-vinyl acetate copolymer (EVA) in the antisolvent. The strong Lewis acid-base complexation of the carbonyl group C=O in EVA with uncoordinated Sn atoms in the perovskite grains improved the grain size, optimized the grain orientation, and enhanced the PCE to 7.72%. The presence of EVA enabled the tin-based perovskite films to effectively prevent the penetration of moisture and oxygen into the perovskite grain boundaries and was able to inhibit their decomposition, and under a high humidity environment of 60% RH, the exposure After 48h, the initial efficiency of 62.4% can still be maintained.

In 2020, Cao K et al. (2020) regulated the crystal growth of $FASnI_3$ thin films by the seed growth method. First, a layer of tin-based perovskite was prepared by a typical solution method. Then, the deposition process was repeated in the first layer and the pre-coated layer could be used as a seed layer and affect the process of crystallization, as well as the quality of the perovskite film. By this method, high-quality tin-based perovskite films with compact crystals, large grain size, and small grain boundaries were obtained, and the resulting PSCs had a PCE of 7.32%.

Meng X Y et al. (2019) have done a series of studies in regard to crystallization kinetics. The crystallization rate of $FASnI_3$ can be slowed down by introducing hydrogen bonds. It was found that the addition of poly (vinyl alcohol) (PVA), and hydrogen bonding between O-PVA and $FASnI_3$ could introduce nucleation sites, slow down crystal growth, guide crystal orientation, and reduce trap states and inhibit iodine ion migration. $FASnI_3$-PVA PSC achieved higher PCE under reverse scan with Voc increased from 0.55 Voc to 0.63 V.

They (Meng et al. 2020) also found that the crystallization process of $FASnI_3$ thin films is surface controlled and the nucleation process of tin-based perovskite starts from the upper surface of the film. The crystallization quality of the crystals was improved by reducing the surface energy of the $FASnI_3$ films by adding organic cations containing fluorine (FOEI), and finally, conversion efficiency of 10.81% was obtained.

They (Meng et al. 2020) also regulated the crystallization kinetics of $FASnI_3$ by a non-classical nucleation mechanism based on pre-nuclear clusters (PNC). Piperazine dihydropyridine was introduced to tune the $FASnI_3$ perovskite precursor solution so that it can easily form stable clusters in the solution before nucleation. to lower the potential barrier for perovskite nucleation and form high-quality perovskite films with low defect density. The PCE of the prepared PSC can reach 11.39% efficiency.

2.5 *Low-dimensional structure*

In 2020, Li P et al. (2020) introduced 1,4-butanediamine (BEA) into $FASnI_3$ to develop a series of lead-free low-dimensional Dion-Jacobson phase perovskites, (BEA) $FA_{n-1}Sn_nI_{3n+1}$. The width of the $FA_2Sn_3I_{10}$ band gap is influenced by its high symmetry structural distortion. The introduction of the BEA ligand inhibits Sn^{2+} oxidation and stabilizes the one-dimensional chalcogenide structure (formation energy of about 106 j/mol). The compact (BEA) $FA_2Sn_3I_{10}$ dominated film enables a weakened carrier localization mechanism with a charge transfer the compact (BEA) $FA_2Sn_3I_{10}$ dominated film enables a weakened carrier localization mechanism with a charge transfer time of only 0.36 ps among the quantum wells. The carrier diffusion length of the electrons exceeds 450 nm and the aperture diameter exceeds 340 nm. 6.43% PCE can be achieved using $FA_2Sn_3I_{10}$ for solar cells with negligible hysteresis. These devices maintain more than 90% of the initial PCE after 1000 hrs and do not require encapsulation and preservation in a nitrogen atmosphere.

2.6 *Carrier transport layer*

The electron transport layer (ETL) and hole transport layer (HTL) contribute to the extraction and transport of electrons and holes, respectively, so the design of suitable ETL and HTL is crucial for the performance enhancement of PSC. TiO_2 and sprio-OMeTAD are the most used ETL and HTL materials, respectively, and in 2020 Gan Y et al. (2021) used simulation to study $MASnI_3$ perovskite cells with different ETL and HTL materials, C_{60}, CdS, $Cd_{0.5}Zn_{0.5}S$, IGZO, PCBM and ZnO were selected as ETL compared with TiO_2, and it was calculated that the order of Voc from high to low is: $Cd_{0.5}Zn_{0.5}S > C_{60} = TiO_2 > ZnO > IGZO > CdS$. $Cd_{0.5}Zn_{0.5}S$ was chosen as the ETL, and Cu_2O, CuI, CuSCN, $MASnBr_3$, NiO with sprio-OMeTAD as the HTL were calculated separately, and the Voc was ranked from high to low as $MASnBr_3$ > sprio-OMeTAD > Cu_2O > CuSCN > NiO > CuI. They concluded that when $Cd_{0.5}Zn_{0.5}S$ and $MASnBr_3$ obtained better open-circuit voltages than other materials when they were used as ETL and HTL, respectively. The optimal ETL and HTL materials are not necessarily the same for different chalcogenide cells, and the above method (Gan et al. 2021) can be used to study $FASnI_3$ and select more materials for comparison to find out the ETL and HTL materials suitable for $FASnI_3$. Further, we will verify the method through experiments.

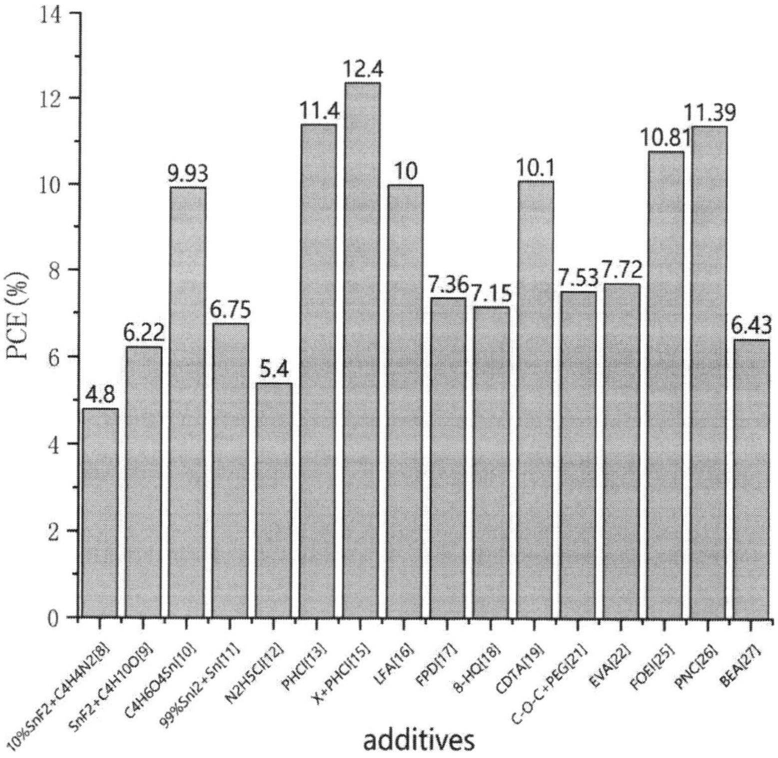

Figure 1. Conversion efficiency (PCE) of $FASnI_3$ of tin-based perovskite solar cells (PSC) with different additives.

3 CONCLUSION AND PROSPECT

In this paper, we present the progress of research to improve the performance of tin-based perovskite materials $FASnI_3$ using various methods, such as reducing additives, tin compensators, chelating agents, and crystallization kinetic modulation, and design of low-dimensional structures.

Although these methods help to improve the performance of tin-based PSC, with the maximum certified efficiency reaching 12.4%, their efficiency still lags behind that of Pb-based PSC. component modulation of A- and X-site ions can change the crystal structure of perovskite and prepare thermodynamically more stable perovskite phases, as well as adjust the energy level of perovskite and reduce the carrier complexation at the interface.

Thus, the follow-up work can be carried out in three aspects as follows:

1. Explore more suitable additives to improve the purity of the tin source, reduce the Sn^{4+} concentration, and achieve a controllable crystallization rate.
2. Use simulation to study $FASnI_3$ and identify ETL and HTL materials suitable for $FASnI_3$.
3. Conduct component modulation of A-site ions and X-site ions, and turn our attention to mixed cations and double-halide perovskites to find materials that are better matched with the transport layer.

REFERENCES

Cao K, Cheng Y, Chen J, et al. Regulated Crystallization of $FASnI_3$ Films through Seeded Growth Process for Efficient Tin Perovskite Solar Cells[J]. *ACS Applied Materials & Interfaces*, 2020.

Chang B, Li B, Pan L, et al. Polyethylene Glycol Polymer Scaffold Induced Intermolecular Interactions for Crystallization Regulation and Defect Passivation in FASnI 3 Films[J]. *ACS Applied Energy Materials*, 2021.

Dai Z, et al. Stable tin perovskite solar cells developed via additive engineering[J]. *Science China Materials*, 2021: 1–10.

Feidan G, Senyun Y, Ziran Z, et al. Improving Performance of Lead-Free Formamidinium Tin Triiodide Perovskite Solar Cells by Tin Source Purification[J]. *Solar RRL*, 2018: 1800136.

Gan Y J, Jiang Q B, Qin B Y, et al. Exploration of carrier transport layer for tin-based chalcogenide solar cells[J]. *Journal of Physics*, 2021, 70(3): 12.

Huang L, Lambrecht W R L. Electronic band structure, phonons, and exciton binding energies of halide perovskites $CsSnCl_3$, $CsSnBr_3$, and $CsSnI_3$, *Phys Rev B*, 2013, 88: 165203.

Kayesh M E, Chowdhury T H, Matsuishi K, et al. Enhanced photovoltaic performance of $FASnI_3$-based perovskite solar cells with hydrazinium chloride coadditive. *ACS Energy Lett*, 2018, 3: 1584-1589.

Koh t m, Krishnamoorthy t, Yantara n, et al. Formamidinium Tin-based Perovskite with Low Eg for Photovoltaic Applications[J]. *Journal of Materials Chemistry A*, 2015, 3(29):14996–15000.

Kumar, M H, Dharani S, Leong W L, et al. Lead-free halide perovskite solar cells with high photocurrents realized through vacancy modulation[J]. *Advanced Materials*. 2014, 26: 7122–7127.

Lee S J, Shin S S, Kim Y C, et al. Fabrication of Efficient Formamidinium Tin Iodide Perovskite Solar Cells through SnF_2–Pyrazine Complex[J]. *Journal of the American Chemical Society*, 2016, 138(12): 3974–3977.

Li, P., Liu, X., Zhang, Y., Liang, C., Chen, G., Li, F., Su, M., Xing, G., Tao, X., Song, Y. Low-Dimensional Dion-Jacobson-Phase Lead-Free Perovskites for High-Performance Photovoltaics with Improved Stability. *Angew. Chem., Int. Ed.* 2020, 59, 6909–6914.

Liao W Q, ZHAO D W, YU Y, et al. Lead-Free Inverted Planar Formamidinium Tin Triiodide Perovskite Solar Cells Achieving Power Conversion Efficiencies up to 6.22%[J]. *Advanced Materials*, 2016, 28 (42): 9333–9340.

Lin Z J. *Study on the morphology regulation and stability of tin-based chalcogenide solar cells* [D]. Nanchang: Nanchang University (2021).

Liu G L. *Crystallization regulation and stability study of chalcogenide solar cells* [D]. Nanchang: Nanchang University (2021).

Liu, X., Wu, T., Chen, J.Y., Meng, X., He, X., Noda, T., Chen, H., Yang, X., Segawa, H., Wang, Y., and Han, L. (2020). Templated growth of $FASnI_3$ crystals for efficient tin perovskite solar cells. *Energy Environ. Sci.* 13, 2896–2902.

Meng X, Wu T, Liu X, et al. Highly Reproducible and Efficient $FASnI_3$ Perovskite Solar Cells Fabricated with Volatilizable Reducing Solvent[J]. *Journal of Physical Chemistry Letters*, 2020, 11, 2965–2971.

Meng, Xiangyue; Li, Yunfei; Qu, Yizhi; Chen, Haining; Jiang, Nan; Li, Minghua; Xue, Ding-Jiang; Hu, Jin-Song; Huang, Hui; Yang, Shihe (2020). Crystallization Kinetics Modulation of $FASnI_3$ Films with Preânucleation Clusters for Efficient Leadâfree Perovskite Solar Cells. *Angewandte Chemie International Edition*, anie.202012280.

Tai, Qidong; Cao, Jiupeng; Wang, Tianyue; Yan, Feng (2019). Recent advances toward efficient and stable tin-based perovskite solar cells. *EcoMat*, (), eom2.12004.
Tao S, Schmidt I, Brocks G, et al. Absolute energy level positions in tin- and lead-based halide perovskites. *Nat Commun*, 2019, 10: 2560.
Wang C, Gu F, Zhao Z, et al. Self-repairing tin-based perovskite solar cells with a breakthrough efficiency over 11%. *Adv Mater*, 2020, 32: 1907623.
Wang C, Zhang Y, Gu F, et al. Illumination Durability and High-Efficiency Sn-Based Perovskite Solar Cell under Coordinated Control of Phenylhydrazine and Halogen Ions[J]. *Matter*, 2020.
Wang C, Zhang Y, Gu F, et al. Illumination durability and high-efficiency Sn-based perovskite solar cell under coordinated control of phenylhydrazine and halogen ions. *Matter*, 2021, 4: 709–721.
Wu T, Liu X, He X, et al. Efficient and stable tin-based perovskite solar cells by introducing π-conjugated Lewis base. *Sci China Chem*, 2020, 63: 107–115.
Xiangyue Meng, Jianbo Lin, Xiao Liu, Xin He, Tianhao Wu, Takeshi Noda, Xudong Yang, Liyuan Han. *Adv. Mater*. 2019, 31, 1903721.
Xiangyue Meng, Yanbo Wang, Jianbo Lin, Xiao Liu, Xin He, Tianhao Wu, Takeshi Noda, Xudong Yang, Liyuan Han. *Joule* 2020, 4, 902–912.
Xu P, Chen S, Xiang H J, et al. Influence of defects and synthesis conditions on the photovoltaic performance of perovskite semiconductor $CsSnI_3$. *Chem Mater*, 2014, 26: 6068–6072.
Yang J. *Stability study of high-efficiency chalcogenide solar cells* [D]. Nanchang: Nanchang University (2020).
Yokoyama T, Cao D H, Stoumpos C C, et al. Overcoming Short-Circuit in Lead-Free $CH_3NH_3SnI_3$ Perovskite Solar Cells via Kinetically Controlled Gas-Solid Reaction Film Fabrication Process[J]. *The Journal of Physical Chemistry Letters*, 2016, 7(5): 776–782.

Energy Revolution and Chemical Research – Kok-Keong Chong and Zhongliang Liu (Eds)
© 2023 The Authors, ISBN: 978-1-032-36554-1

Optimal allocation of heat storage device based on thermal load and distributed new energy power

Meixiu Ma
State Key Laboratory of Advanced Transmission Technology, State Grid Smart Grid Research Institute Co., Ltd., Chang-ping District, Beijing, China

Mingjun Jiang
State Grid Gansu Electric Power Company, Gansu, China

Mengdong Chen*
State Key Laboratory of Advanced Transmission Technology, State Grid Smart Grid Research Institute Co., Ltd., Chang-ping District, Beijing, China

Fan Luo & Lanlan Xu
State Grid Gansu Electric Power Company, Gansu, China

Kun Hou & Wei Kang
State Key Laboratory of Advanced Transmission Technology, State Grid Smart Grid Research Institute, Co., Ltd., Chang-ping District, Beijing, China

ABSTRACT: Distributed photovoltaic power has the problems of dispersion, flexibility, grid benchmarking, etc., which leads to the high light rejection rate of distributed photovoltaic power generation, which is an urgent problem to be solved for new energy power generation. The energy storage device can realize energy storage and time migration of energy, with high efficiency, good regulation performance, and strong flexibility, and can serve as an effective means to improve the photovoltaic power consumption. Aiming at the new energy power and heat load collaborative system, this paper establishes the minimum operating cost as the objective function, considers the heat balance and equipment operation constraints as the conditions, realizes the optimal allocation of heat storage device capacity, optimizes the power heating process, effectively avoids the peak value of conventional power consumption, and promotes the safe and stable operation of the power system.

1 INTRODUCTION

As environmental problems become increasingly serious, renewable energy is becoming more and more popular, and it is expected that by 2020, the capacity of renewable energy installations in China will reach 700–722 million kW, accounting for 35%–36.1% of the total installed capacity, including 60 million kW of decentralized PV, whose power generation capacity accounts for 1.3%–1.5% of the national power generation capacity (National Energy Administration 13th five-year plan for electric power development [EB/OL]. [2016-11-07]. http://www.nea.gov.cn.). As the scale of renewable energy access continues to increase, the abandoned wind and light rates are rising, with the abandoned wind rate reaching 16.5% and the abandoned light rate 9.2% in 2017 within the operation of the State Grid Corporation (Northwest regulatory bureau of the national energy administration. Notification on the grid-connected operation of new energy in Northwest China in 2016 [EB/OL]. [2017-01-18]. http://xbj.nea.gov.cn/website/Aastatic/news-176162.html.). Particularly serious among them is decentralized PV power.

*Corresponding Author: buaacmd@163.com

 DOI 10.1201/9781003332657-48

There is often a large degree of mismatch between decentralized PV power output and energy-using load demand in terms of time distribution. The simple way of dissipation, which only aims to guarantee source-load balance, essentially harms the economic interests of the grid and users, while the original intention of PV self-generation and self-consumption construction is also difficult to achieve. How to allocate energy storage equipment to achieve new energy consumption, improve the matching problem between source and load, and thus achieve flexibility and economy in the consumption of new energy generation is a problem that both plagues decentralized PV consumption and distribution network operators.

To solve these problems, this paper provides an optimal configuration of the capacity of new energy power storage devices to optimize the power supply process, reduce the peak-to-valley difference between electricity abandoned and electrical load, and effectively reduce the peak-to-valley variation between electricity abandoned and electrical load based on the lowest operating costs.

2 NEW ENERGY POWER AND THERMAL LOAD SYNERGY SYSTEM

2.1 *New energy power and thermal load synergy system*

The structure of the synergistic system of new energy power and heat load is shown in Figure 1, which mainly consists of a new energy power system, conventional power system, electric heat storage device, and heat load. The PV power in the decentralized new energy power produces heat directly, which is supplied to the heat load or sent to the electric heat storage device for storage, and the heat stored in the electric heat storage device is used to achieve heat supply through heat exchange equipment.

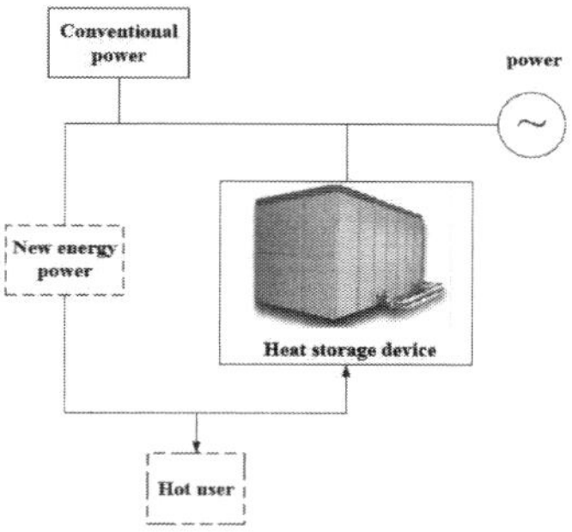

Figure 1. Structure diagram of new energy power and heat load coordination system.

2.2 *Features of thermal storage units*

The thermal storage device is a time-shiftable load, to achieve relatively independent operation and regulation of electrical and thermal loads, the thermal storage device system is shown in Figure 2. When new energy power fluctuations and grid access factors, new energy power can not be online when the heat storage, can be online when the heat will reduce the abandoned wind and light abandonment rate while reducing the cost of the load operation.

3 MATHEMATICAL MODEL OF A SYNERGISTIC SYSTEM OF NEW ENERGY POWER AND THERMAL LOADS

3.1 *Mathematical models for new energy power*

With new energy power as an example, the output power of a photovoltaic system is largely determined by the sunlight intensity, photovoltaic array area, and conversion efficiency of the photovoltaic generator. Since sunlight intensity satisfies the Beta distribution function, the power

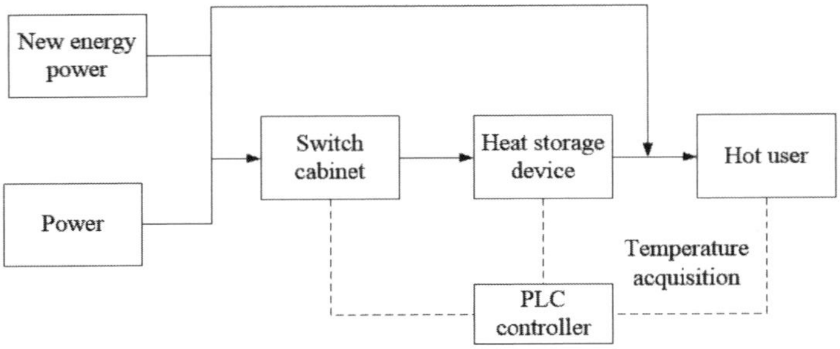

Figure 2. System diagram of heat storage device.

output probability density function of the photovoltaic generator is deemed to be the following: (Liao et al. 2017).

$$f(P_{PV})=\frac{\Gamma\,(a+b)}{\Gamma\,(a)\,\Gamma\,(b)}\left(\frac{P_{PV}}{P^{a-1}_{PVmax}\frac{P_{PV}}{P^{b-1}_{PVmax}}}\right) \tag{1}$$

$$Q_{PV}=P_{PV}\tan\phi \tag{2}$$

where Γ is the Gamma function; P_{PV} and Q_{PV} are the active and reactive power outputs of the PV system respectively; P_{PVmax} is the maximum power output of the PV system; a and b are the shape parameters of the Beta distribution; ϕ is the power factor angle.

3.2 *Mathematical model of an electric heat storage unit*

The electric heat storage device is a form of energy coupling unit, relying on electric heating elements energized to achieve the purpose of heat storage. It does not produce the chemical reaction of combustion, no black smoke, sulfur dioxide, carbon dioxide, and other waste emissions, the heat production efficiency of up to 95% or more, and a high degree of automation, safe and reliable operation. The heat storage device can achieve conventional electricity heat storage, also can consume new energy electricity heat storage, and at the same time can achieve customer-side heat supply.

$$H_{cr,t}=P_{cr,t}\eta_{cr}-(1-\mu)Q_{cr,t} \tag{3}$$

where $H_{cr,t}$ is the heat storage device time t heat storage capacity; $P_{cr,t}$ is the heat storage device time t heating power, η_{cr} is the heat storage device's electric heat conversion efficiency; μ is the heat storage device and heat exchange device heat loss rate; $Q_{cr,t}$ is the heat storage device time t of thermal power.

4 OPTIMISATION MODEL FOR CAPACITY ALLOCATION OF THERMAL STORAGE UNITS UNDER DECENTRALISED NEW ENERGY POWER

Based on the synergistic system of new energy power and thermal load, the policy guidance such as electricity market price mechanism and distributed new energy consumption incentive mechanism are considered, and the objective function of thermal storage device optimization is constructed from the economic cost.

4.1 *Target function*

To improve the consumption of decentralized new energy power, and to improve the clean heating of small clusters of residents. This paper proposes the configuration of thermal storage devices, which in turn reduces grid fluctuations caused by decentralized new energy power entering the grid and effectively reduces wind and light abandonment. The economic dispatching objective of this paper is to rationalize the new energy power or grid output situation to minimize the total operating cost while meeting the normal operating constraints. The overall objective function is as follows (Li et al. 2019; Lu et al. 2017).

$$\min A = C_{cr} + C_{gd} + C_{xd} \tag{4}$$

$$C_{cr} = H_{cr} V_{cr} \tag{5}$$

$$C_{gd} = \sum P_{gd,t} V_{gd,t} \tag{6}$$

$$C_{xd} = \sum P_{xd,t} V_{xd,t} \tag{7}$$

where A is the total system cost; C_{cr}, C_{gd}, and C_{xd} are the hourly equivalent operating cost of the thermal storage device, the conventional power operating cost, and the new energy power cost, respectively; H_{cr} is the design capacity of the thermal storage device; V_{cr} is the unit capacity cost of the thermal storage device; $P_{gd,t}$ and $P_{xd,t}$ are the conventional power and electric energy power operating power at time t respectively; $V_{gd,t}$ and $V_{xd,t}$ are the conventional power and electric energy power unit power cost at time t respectively.

4.2 *Binding conditions*

During the operation of the new energy power and heat load synergy system, the following power and heat energy balance relationships and thermal storage unit operating constraints need to be met.

4.2.1 *Energy balance constraints*

$$\sum M_{gd,t} + \sum M_{xd,t} = W_t \tag{8}$$

where $M_{gd,t}$ and $M_{xd,t}$ are the electrical and thermal loads of the system at moment t.

4.2.2 *Thermal storage unit operating constraints*

The operating constraints of the thermal storage unit are

$$0 \le P_{gd,t} \le P_{gd,max} \tag{9}$$

$$0 \le P_{xd,t} \le P_{xd,max} \tag{10}$$

$$H_{cr,T} = H_{cr,0} \tag{11}$$

$$H_{cr,min} \le H_{cr,t} \le H_{cr,max} \tag{12}$$

$$0 \le H_{cr\_in,t} \le H_{cr,nom} \tag{13}$$

$$0 \le H_{cr\_out,t} \le H_{cr,nom} \tag{14}$$

$$H_{cr\_out,t} H_{cr\_out,t} = 0 \tag{15}$$

where $H_{cr,T}$ and $H_{cr,0}$ are the termination capacity and initial capacity of the thermal storage unit respectively; $H_{cr,min}$ and $H_{cr,max}$ are the minimum and maximum capacities under stable operating conditions respectively; $H_{cr\_in,t}$ and $H_{cr\_out,t}$ are the input capacity and output capacity of the thermal storage unit respectively; $H_{cr,nom}$ is the rated capacity of the thermal storage unit.

5 EXAMPLE ANALYSIS

5.1 *Example of calculation*

Taking an area in Gansu Province as an example, the time-of-use tariff and decentralized PV feed-in subsidy price of Gansu Province's clean heating policy are shown in Figure 3. The PV power and heat load characteristics of the 84-kW decentralized small-scale PV power generation unit and the 1700 m^2 building are shown in Figure 4 as an example. At the same time, the system aims to have the lowest total operating cost and is equipped with thermal storage device capacity to enable it to fully consume decentralized new energy power.

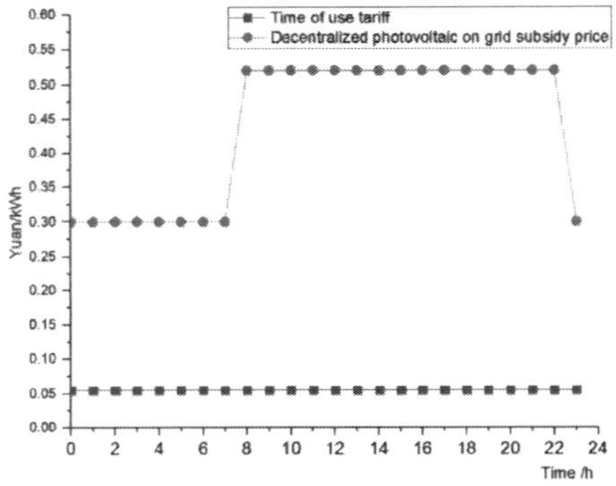

Figure 3. Time of use electricity price of clean heating and subsidy price of the distributed photovoltaic grid in Gansu Province.

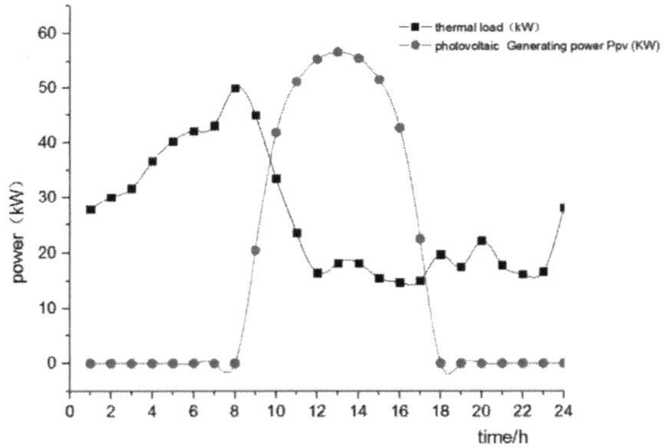

Figure 4. Output characteristic curve of photovoltaic power and thermal load.

5.2 *Genetic algorithm solving*

The previous paper constructs an optimization model for the capacity configuration of heat storage devices under decentralized new energy power with economic cost as the objective function, based on the energy balance constraint and the heat storage device operation constraint. The lower optimization model is to optimize the equivalent operating cost of the heat storage device when operating, the conventional power operation cost, and the new energy power cost, and the model is solved using a genetic algorithm. In the process of solving the optimal total system cost, a two-tier optimization model is used to cooperate, where the upper optimization model works to adjust the maximum power of the configured thermal storage device, and the lower optimization model treats the adjusted thermal storage device power value as a fixed value and then optimizes the conventional power operation cost, the new energy power cost and the electric and thermal loads of the large grid, and reasonably arranges the new energy power or grid The model then optimizes the cost of conventional power, the cost of new energy power and the electrical and thermal load of the grid, to minimize the total cost of microgrid operation. The upper optimization flow chart of the model is shown in Figure 5 (Wang et al. 2018).

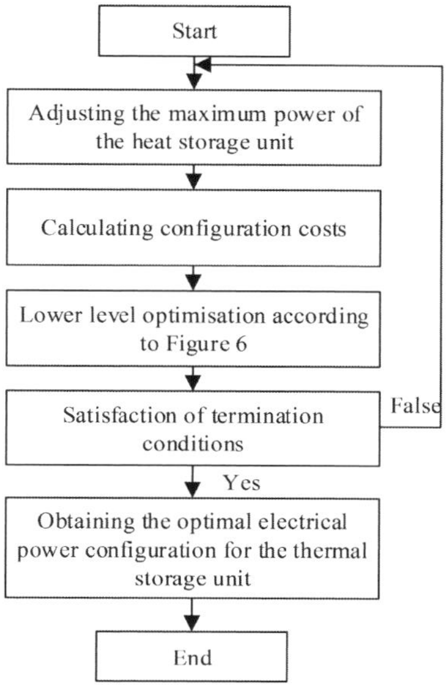

Figure 5. Upper layer optimization flow chart.

The steps are as follows.

(1) First, adjust the maximum value of the power of the heat storage device according to the selected range of maximum power values of the heat storage device, and pass this power value to the lower optimization model, when the power value of the heat storage device with the smallest target value of the upper optimization model is the final determining electric power value of the heat storage device configuration for the whole optimization process.
(2) Calculate the cost of the configuration at this power.
(3) Treat the adjusted power value as a fixed value and carry out lower-level optimization following Figure 6 to minimize the total cost of microgrid operation provided that the constraints are satisfied.
(4) If the termination condition is not satisfied at this point, return to Step 1 and recalculate, otherwise jump out of the loop.
(5) Calculate the sum of operating cost, electric power operating cost, and new energy power cost, and find out the smallest total micro-grid operating cost among all calculation results, and the corresponding optimal electric power configuration value of the heat storage device can be obtained.

The lower optimization flow chart of the model is shown in Figure 6. The steps are as follows:

(1) Input the original data and parameters, after the upper optimization model has adjusted the maximum power of the configured thermal storage device into the lower optimization model, the data, and parameters to be inputted are also the parameters of the genetic algorithm, the termination condition maximum number of iterations G.

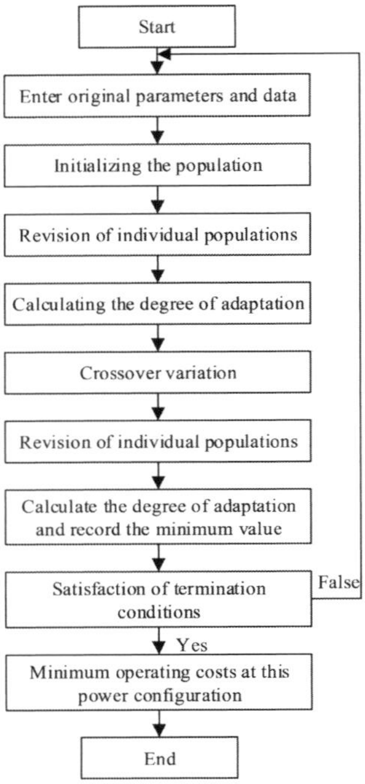

Figure 6. Lower layer optimization flow chart.

(2) The population is initialized and the number of iterations g = 1 is set.
(3) To meet the stable operation of the microgrid, the initialized population of individuals must meet the thermal load balance conditions, the electrical load balance conditions, and the operating constraints of each part of the model, so it is necessary to carry out the correction of the population individuals, the specific process is to first meet the thermal load balance conditions of the microgrid, then meet the operating constraints of each part, and finally meet the electrical load balance conditions, to complete the correction of the population individuals. The work of the individual population correction is completed.
(4) After the equilibrium conditions and constraints of the population individuals are satisfied, the adaptation degree of the individuals is calculated first.
(5) The genetic algorithm is then used to perform crossover operations and mutation operations.
(6) Since the individuals of the population after crossover and mutation may no longer satisfy the equilibrium and constraint conditions, the correction of the individuals of the population needs to be performed again, in the same way as in Step 3.
(7) Calculate the fitness of the individuals of the population that have undergone the crossover and mutation operations and record the minimum fitness and the corresponding individual value for the g-th generation.
(8) Determine whether the maximum number of iterations G has been reached at this point; if not, make g = g + 1 and return to Step 5 to re-run the genetic algorithm for the monarch programme, or jump out of the loop if it is satisfied.
(9) Find the minimum micro-grid operating cost in each generation of the recorded minimum adaptation and record the corresponding individual value, the minimum operating cost under this heat storage unit power configuration is obtained, and the output of each unit corresponding to it, and then receive the maximum power of the heat storage unit configuration adjusted by the upper optimization model and continue to repeat Steps 1 to 9 of the lower loop.

5.3 *Analysis of results*

According to the objective function and constraints, the relevant procedures are prepared and the results of the analysis of the operating costs of conventional power and new energy power with energy storage devices are measured, as shown in Figure 7.

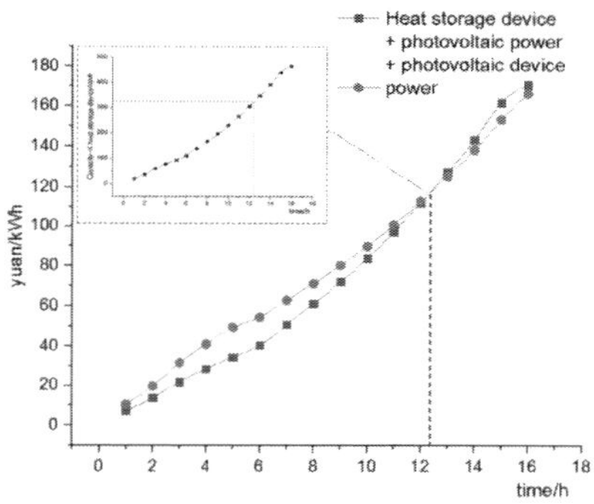

Figure 7. Operation cost comparison chart of conventional power and new energy power with the energy storage device.

As can be seen from Figure 7, with the increase in load heat supply, the cost of electricity heating increases and the cost of new energy electricity heating is significantly lower than the cost of conventional electric heating, but when the accumulated heating time reaches a certain value, the cost of new energy electricity heating exceeds the cost of conventional electric heating, i.e., the capacity of the heating device reaches 305 kWh. Heat storage devices can consume excess power and realize a shift of energy in time, but cannot increase their capacity infinitely, resulting in equipment selection on the large side, low equipment utilization, and high operating costs.

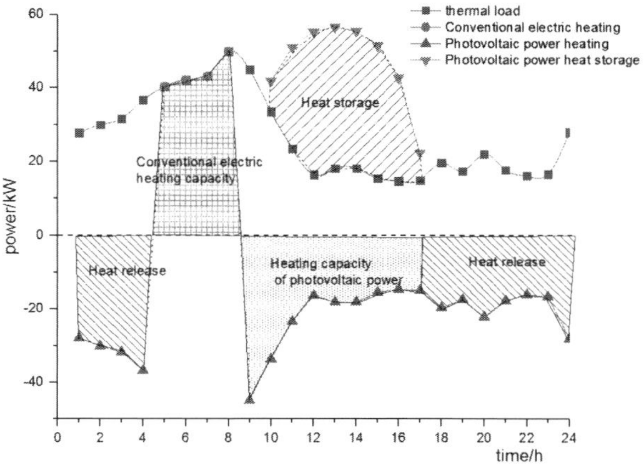

Figure 8. Heat load and optimal heating curve of distributed photovoltaic new energy.

Figure 8 shows the heat supply curve for optimizing the capacity of the thermal storage unit. As can be seen, when the new energy power is insufficient, priority is given to releasing the stored new energy power heat, while at the heat release end of the storage device, conventional power is used to supplement the heat supply. This not only avoids the 18:00 nighttime peak of regular electricity consumption but also makes up for the shortcomings of insufficient heat supply at the end of the heat storage device.

6 CONCLUSIONS

This paper takes the lowest cost of heating load as the target, considers heat balance and equipment operation constraints as the conditions, carries out the optimal configuration of the capacity of new energy power storage devices, and obtains the following conclusions.

(1) Compared with the decentralized direct heat supply method, the heat supply method equipped with heat storage devices can consume more new energy power. Not only does it substantially improve the regulation capacity and flexibility of the system, but it also has more room for adjustment through reasonable distribution control of electricity and heat, effectively improving the abandonment rate and reducing operating costs.

(2) By optimizing the comprehensive operation cost, the capacity of the heat storage device is obtained as 305 kWh, which accounts for about 50% of the total heat storage capacity of PV throughout the day, which not only reduces the overall investment of the heat storage device but also mentions the interactive response capability of conventional power and new energy power, which is conducive to the safe and stable operation of the power system.

ACKNOWLEDGMENT

This paper is funded by the research and development project of State Grid Corporation of China (Project name: 'Research and Application of Shared Power-heat Control Platform for Flexible Interactive Response'; Grant No. 5400-202033206A-0-0-00).

REFERENCES

Li Ruimin, Zhang Xinjing, Xu Yujie, Sun Wenwen. Research on optimal configuration of hybrid energy storage capacity for wind-solar generation system. *Energy Storage Science and Technology*, 2019, 5(3): 513–522.

Liao Qiuping, Lv Lin, Liu Youbo, et al. Reconfiguration based Model and algorithm of Voltage Regulating for Distribution Network with Renewable Energy [J]. *Automation of Electric Power Systems*, 2017, 41(18): 32–39.

Lu X J, Guo Q, Dong H Y. Multi objective optimization of hybrid energy storage micro grid based on CMOPSO algorithm [J]. *Acta Energiae Solaris Sinica*, 2017, 38(1): 279–286.

National Energy Administration 13th five-year plan for electric power development [EB/OL]. [2016-11-07]. http://www.nea.gov.cn.

Northwest regulatory bureau of the national energy administration. *Notification on grid connected operation of new energy in Northwest China in 2016 [EB/OL]*. [2017-01-18]. http://xbj.nea.gov.cn/website/Aastatic/news-176162.html.

Wang L M, Liu J C, Tian C G, et al. Capacity optimization of hybrid energy storage in microgrid based on statistic method [J]. *Power System Technology* 2018, 42(1): 187–194.

Energy Revolution and Chemical Research – Kok-Keong Chong and Zhongliang Liu (Eds)
© 2023 The Authors, ISBN: 978-1-032-36554-1

Effect of the reduced Ca-Fe oxygen carrier on products distribution during *Chlorella* chemical looping pyrolysis

M. Chen, X.Y. Wang, Y.H. Hou, J.J. Liu, Y.Z. Liu & Q.J. Guo*
College of Chemical Engineering, Qingdao University of Science and Technology, China

ABSTRACT: Chemical looping pyrolysis (CLPy) provides a novel strategy for biomass conversion by gaining upgraded bio-oils and clean syngas separately. The distribution of gas-liquid-solid products during *Chlorella* CLPy with reduced $Ca_2Fe_2O_5$ oxygen carrier (Re-CF) was investigated by using a fixed bed reactor. The results showed that compared with direct pyrolysis of *Chlorella*, CLPy significantly enhanced the hydrocarbon content and reduced the oxygen content of the bio-oils. The oxygen element migrates via two pathways: bulk ketonization of CaO and surface ketonization of metal oxides to ketones and hydrodeoxygenation (HDO) reactions of ketones, aldehydes, and carboxylic acids to alcohols. In the pyrolysis stage, the gas phase products were dominated by CO_2 and H_2, while the Re-CF was converted to $Ca_2Fe_2O_5$ and $CaCO_3$. In the gasification stage, CO-rich syngas was generated via the reactions between oxygen carrier and pyrolysis char. Furthermore, the oxygen carrier was reduced to its reduction state and the regeneration of the oxygen carrier was realized.

1 INTRODUCTION

Microalgae, as a third-generation biofuel, has become an excellent source of biomass energy due to its wide distribution, short growth cycle, and low arable land occupation compared to terrestrial biomass. Microalgae can be converted into bio-oils through pyrolysis reaction, which is one of the main methods of resource utilization of microalgae. However, compared with fossil fuels, bio-oils still have high oxygen content, which makes bio-oils have strong acidity and low calorific value. Although the quality of bio-oils can be improved by bio-oil upgrading, there is still a problem of harsh reaction conditions. Therefore, bio-oils are usually modified by catalytic pyrolysis to reduce oxygen content. Some researchers found that catalysts such as zeolite (Campanella 2012) and CaO (Veses 2014) can reduce the acidity and oxygen content of bio-oils and promote the formation of hydrocarbons. However, there is still a problem with catalyst deactivation during catalytic pyrolysis (Stanton 2018).

Syngas provides a new technology for the efficient and clean utilization of biomass. To solve the problem of catalyst deactivation, Liu et al. (Liu 2021) proposed a pyrolysis process based on the concept of chemical looping, i.e., chemical looping pyrolysis (CLPy), as shown in Figure 1. CLPy uses the reduced oxygen carriers to participate in the pyrolysis of biomass, which can significantly decrease the content of oxygenated and N-containing compounds in the bio-oils. At the same time, the gasification reaction of pyrolysis char with the oxidized oxygen carriers can produce rich H_2 or CO.

Although Liu et al. conducted a preliminary study on the CLPy process, there were few studies on the migration of oxygen elements during the pyrolysis process. Therefore, in this study, the effect of reduced $Ca_2Fe_2O_5$ oxygen carrier (Re-CF) on the product distribution of *Chlorella* was

*Corresponding Author: qjguo@qust.edu.cn

DOI 10.1201/9781003332657-49

studied in a fixed-bed reactor. Furthermore, the distribution of oxygen elements was investigated, to provide a theoretical basis for the preparation of high-quality bio-oils and chemicals by CLPy.

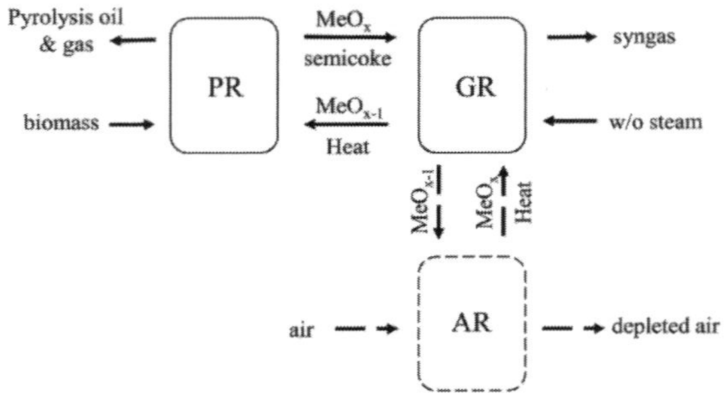

Figure 1. Schematic diagram of chemical looping pyrolysis.

2 MATERIALS AND METHODS

2.1 *Materials*

The results of proximate, elemental, and chemical analysis of *Chlorella* (Xi'an Ruiying Biotech Co., Ltd.) are shown in Table 1.

Table 1. Proximate, ultimate, and chemical analysis of *Chlorella*.

Proximate analysis, wt. %, ad		Ultimate analysis, wt./%, daf		Chemical analysis wt. %	
M	4.00	C	48.44	Protein	54.82
V	61.90	H	5.50	Lipid	17.89
A	5.72	N	9.82	Carbohydrate	14.43
FC	28.38	O	35.78	Others	12.86

$Ca_2Fe_2O_5$ oxygen carrier was prepared by sol-gel method. $Ca(NO_3)_3{\cdot}4H_2O$ $Fe(NO_3)_3{\cdot}9H_2O$ and citric acid were dissolved in deionized water according to the molar ratio of 1:1:2.4. The mixture was stirred to gel at 90 ° C. After drying completely, the precursor was transferred to a muffle furnace and combusted at 300 ° for 1 h, then calcined at 900 ° for 5 hrs. After being cooled to room temperature, the $Ca_2Fe_2O_5$ oxygen carrier was reduced with H_2 at 900 ° for 1 h to prepare the reduced $Ca_2Fe_2O_5$ oxygen carrier, which was labeled as Re-CF after grinding and screening.

2.2 *Experiment scheme*

The experimental device diagram of CLPy is shown in Figure 2. *Chlorella* and Re-CF oxygen carriers were mixed uniformly at the mass ratio of 1:1 (total 6.00 $\pm$ 0.02 g) and filled in the package. Before the experiment, the package was suspended at the top of the reactor, Ar was injected at 300 mL/min and lasted for 30 min. When the reactor was heated to 550 ° C, the package was quickly pushed into the reactor and lasted for 30 min. After the reaction, the package was quickly removed from the reactor. The bio-oils were collected by quantitative isopropanol, pyrolysis gas was collected by Tedlar bags every 4 min. After the pyrolysis experiment, the reactor was heated to 850 ° C and lasted for 40 min, for the gasification experiment of the pyrolysis char. The Ar flow rate is 300 ml/min. The gasification gas is collected by Tedlar bags every 4 min.

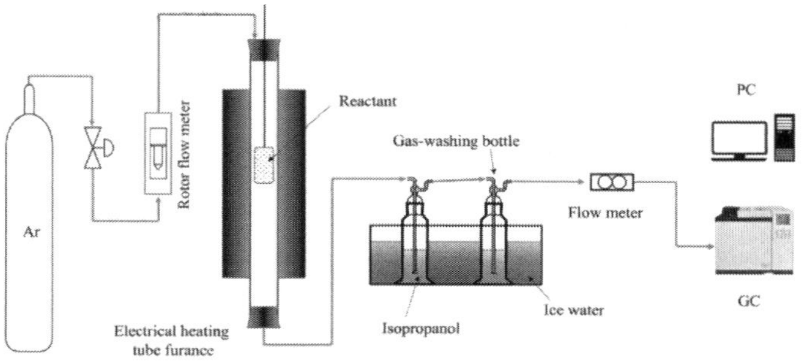

Figure 2. Experimental device diagram of *chlorella* chemical looping pyrolysis.

After being filtered by a PTFE filter, bio-oils were analyzed by GC-MS (Agilent 7890B-5977C) with HP-5MS capillary column (60 m × 0.25 mm × 0.25 μm). The composition of bio-oils was analyzed according to the NIST14 spectral library. The collected gas products were analyzed by GC (PE CLARUS 500). The composition of the oxygen carrier was characterized by XRD (Rigaku D/max 2500 PC).

3 RESULTS AND DISCUSSIONS

3.1 *Effect of Re-CF on the distribution of three-phase yield*

The distribution of three-phase yield during the direct pyrolysis and CLPy of *Chlorella* was shown in Figure 3. Compared with direct pyrolysis, the solid yield of *Chlorella* was greatly improved during CLPy, while the liquid yield and gas yield was reduced to a certain extent. It is attributed to the presence of Re-CF in the form of CaO and Fe, and reacts with oxygen-containing substances such as bio-oils, CO_2, and H_2O to produce $CaCO_3$ and $Ca_2Fe_2O_5$, thereby reducing the yield of gas products and causing an increase of solid yield.

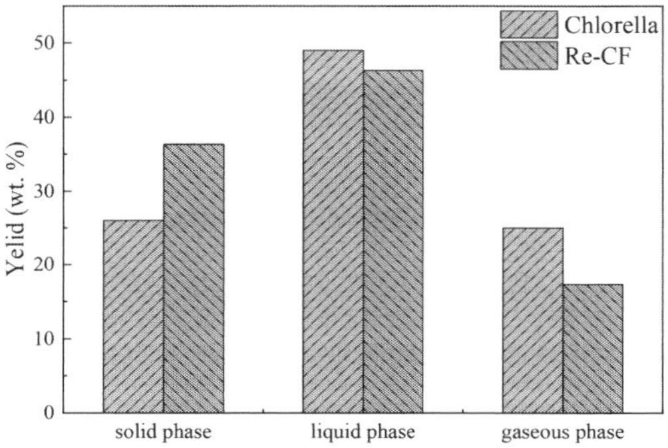

Figure 3. Distribution of three-phase yield of *Chlorella* pyrolysis.

3.2 *Effect of Re-CF on the composition of bio-oils*

The comparison of bio-oils in the direct pyrolysis and CLPy process of *Chlorella* was displayed in Figure 4a. It is clear that the content of N-containing compounds in the direct pyrolysis process is the highest, reaching more than 30% of the total bio-oil content, and is dominated by nitrile compounds (R-C≡N) and indole, both of which mainly come from the pyrolysis process of proteins. The main compounds of phenolic are phenol and p-cresol, which are mainly derived from the decomposition

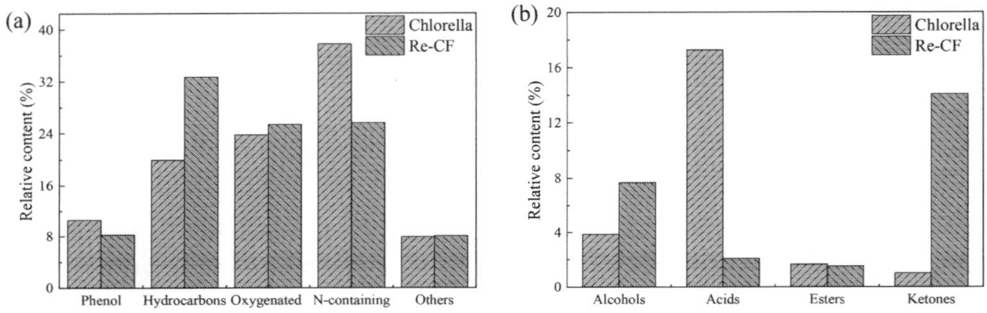

Figure 4. Relative content of the composition in bio-oils of *Chlorella* pyrolysis liquid compounds; (b) oxygenated compounds.

of tyrosine in proteins (Du 2013). In contrast, the content of N-containing compounds in CLPy reduced significantly, and the content of hydrocarbons increased. It may be explained by the reaction between Fe and CaO in the Re-CF oxygen carrier and N-containing compounds to generate FeNx and CaCNx compounds, which promoted the pyrolysis of nitrile and reduced the content of N-containing compounds (Yi 2017).

The composition of oxygenated compounds in the direct pyrolysis and CLPy of *Chlorella* was shown in Figure 4(b). The oxygenated compounds in the direct pyrolysis products of *Chlorella* are mainly carboxylic acids, which are mainly generated from the pyrolysis of lipids, and the content of alcohols and lipids is less. As for the CLPy process, the ketone content increases substantially and the carboxylic acid disappears almost completely. The reason may be that carboxylic acids are converted into ketones through ketonization, which involves the reaction of a-H atoms of two carboxylic acids, as shown in reaction R1 (Kumar 2018; Landoll 2011). Table 2 shows that the content and carbon chain length of ketones are consistent with carboxylic acids. For example, 3-octadecanone may be generated by the ketonization of hexadecanoic acid and propionic acid, resulting in an increase in the content of 3-octadecanone accompanied by a decrease in the content of hexadecanoic acid.

$$RCOOH + R'CH_2COOH = RCOCH_2R' + CO_2 + H_2O(R1)$$

In addition, the content of alcohol increased in the CLPy process. This phenomenon may be ascribed to the HDO reaction and catalytic hydrogenation of ketones, carboxylic acids, and other products in the bio-oils.

Element analysis of bio-oils is shown in Table 3. The contents of C and H elements in the CLPy process are higher than that in direct pyrolysis, which is consistent with the increasing hydrocarbon content in the above analysis. The content of O elements decreased significantly from 26.29 % to 15.36 %. Although the reduced Re-CF increases the content of oxygenated compounds, the number of O atoms is reduced by the ketonization of carboxylic acids and other substances, which realizes the removal of oxygen.

3.3 *Effect of Re-CF on the composition of pyrolysis gas*

The composition of pyrolysis gas from direct pyrolysis and CLPy of *Chlorella* are listed in Table 4. During the pyrolysis of *Chlorella*, CO and CO_2 mainly come from the pyrolysis of oxygen-containing functional groups, while CH_4 is mainly generated by the secondary pyrolysis of pyrolysis products. Compared with the direct pyrolysis process, H_2 content in the CLPy increases significantly, which is mainly because the CaO and Fe in the Re-CF react with H_2O in the pyrolysis gas to generate $Ca_2Fe_2O_5$ and a large amount of H_2 (Liu 2021). In contrast, the CO_2 yield decreased

Table 2. Bio-oils composition of direct pyrolysis and chemical looping pyrolysis of *Chlorella*.

				Peak area (%)	
NO.	RT (min)	Formula	Compound	*Chlorella*	Re-CF
1	5.648	C_6H_6O	Phenol	5.24	3.41
2	6.004	$C_8H_{18}O$	Octan-2-ol	1.07	2.57
3	6.142	$C_6H_{13}NO$	n-Hexanamide	1.83	0.77
4	6.64	C_8H_7N	Benzyl cyanide	–	0.87
5	6.745	$C_8H_{10}O$	p-Cresol	3.93	2.29
6	7.814	$C_6H_8O_2$	Ethenyl 2-methylprop-2-enoate	1.01	–
7	7.891	C_9H_{12}	Cumene	–	0.99
8	7.892	$C_9H_{12}O$	3-Isopropylphenol	1.39	–
9	8.117	C_9H_9N	3-Phenylpropanenitrile	3.39	2.95
10	8.34	$C_{10}H_{20}$	1-Decene	1.54	–
11	8.525	$C_{10}H_{12}O_2$	Ethyl 4-methylbenzoate	1.65	–
12	8.685	$C_{10}H_{22}O$	2,7-Dimethyl-1-octanol	–	2.57
13	8.931	C_8H_7N	1H-indole	14.33	12.22
14	9.207	$C_8H_{15}N$	Octanenitrile	1.11	–
15	9.826	$C_{13}H_{18}$	Ionene	1.36	1.49
16	9.901	$C_{12}H_{16}O$	hexanophenone	–	0.63
17	10.209	C_9H_9N	2-Methylindole	5.74	5.70
18	10.301	$C_{12}H_{26}$	Dodecane	–	1.67
19	11.448	$C_{10}H_{11}N$	2,3-Dimethylindole	–	1.09
20	11.495	$C_{12}H_{24}$	1-Dodecene	–	2.23
21	11.586	$C_{13}H_{28}$	3-Methyl-dodecan	2.06	3.97
22	13.957	$C_{16}H_{34}$	Hexadecane	7.82	7.23
23	15.437	$C_{20}H_{38}$	7,11,15-Trimethyl-3-methylidenehexadec-1-ene	7.08	9.02
24	15.882	$C_{20}H_{40}O$	3,7,11,15-Tetramethyl-2-hexadecen-1-ol	2.79	1.73
25	16.107	$C_{16}H_{31}N$	Hexadecanenitrile	4.05	–
26	16.109	$C_{18}H_{36}O$	3-Octadecanone	–	11.93
27	17.066	$C_{18}H_{36}O_2$	Ethyl Hexadecanoate	–	1.51
28	17.298	$C_{16}H_{32}O_2$	Palmitic acid	12.41	0.94
29	17.888	$C_{20}H_{42}$	n-Eicosane	–	3.34
30	18.758	$C_{16}H_{33}NO$	Palmitamide	2.90	2.01
31	18.853	$C_{18}H_{32}O_2$	Linoleic acid	2.63	–
32	19.139	$C_{18}H_{36}O_2$	Stearic acid	2.21	1.13
33	20.348	$C_{19}H_{38}O$	2-Oxononadecane	–	1.51

Table 3. Element analysis of bio-oils.

	Elemental analysis (wt. %)			
	C	H	N	O
Chlorella	56.87	7.11	9.73	26.29
Re-CF	66.72	8.72	9.20	15.36

during the CLPy, which mainly results from the CO_2 reacting with the Re-CF to generate $CaCO_3$ and $Ca_2Fe_2O_5$.

3.4 *Effect of Re-CF on the composition of syngas*

To investigate the effect of oxidized oxygen carriers on the gasification product of pyrolysis char, the gasification experiments of pyrolysis char in direct pyrolysis and CLPy were carried out. The

Table 4. Composition of pyrolysis gas from pyrolysis of *Chlorella*.

	Volume fraction (%)			
	H_2	CO	CH_4	CO_2
Chlorella	16.07	7.27	14.39	62.27
Re-CF	35.2	5.7	14.52	44.57

Table 5. Composition of syngas from gasification of pyrolysis char.

	Cumulative gas production (ml/g microalgae)			
	H_2	CO	CH_4	CO_2
Chlorella	37.98	7.13	2.17	3.91
Re-CF	55.36	91.99	2.82	6.46

cumulative gas production is shown in Table 5. H_2 is the main gasification product of pyrolysis char in direct pyrolysis, which is produced by the deep cracking of pyrolysis char. In contrast, the yield of CO was significantly increased during the CLPy process, mainly because the gasification reaction occurred between the pyrolysis char and the oxidized oxygen carrier. At the same time, $CaCO_3$ decomposes at high temperatures to produce CO_2, and then reacts with pyrolysis char to produce a large amount of CO, thereby increasing CO production.

3.5 *Characterization of the oxygen carrier*

The XRD patterns of Re-CF are shown in Figure 5. $Ca_2Fe_2O_5$ and $CaCO_3$ appeared after the pyrolysis reaction. Among them, $Ca_2Fe_2O_5$ mainly comes from the oxidation reaction of the oxygen carrier, which is formed by the reaction between Re-CF and the pyrolysis product, i.e., H_2O and CO_2. $CaCO_3$ is mainly derived from two aspects: one is the ketonization of CaO and carboxylic acid (Landoll 2011); the other is that CaO directly reacts with CO_2 in the pyrolysis gas. After the gasification reaction, the formation of CaO and Fe indicates that $Ca_2Fe_2O_5$ and $CaCO_3$ generated during pyrolysis are completely regenerated through gasification.

3.6 *Migration pathway of oxygen in CLPy*

Compared to the distribution of pyrolysis products in direct pyrolysis with that CLPy, the possible paths of oxygen elements during *Chlorella* CLPy are shown in Figure 6. Firstly, the phenolic compounds are mainly derived from the pyrolysis of tyrosine in proteins, and generate p-cresol and phenol through pathways a and b, respectively. Secondly, oxygenated compounds such as carboxylic acids and alcohols were formed by the direct pyrolysis of lipids and carbohydrates, and the reactions such as decarboxylation, HDO, and ketonization have occurred with Re-CF. There may be two kinds of ketonization paths in the process of CLPy: one is bulk ketonization of CaO (Landoll 2011), CaO in Re-CF oxygen carrier reacts with carboxylic acid to form $(RCOO)_2Ca$, and then decomposes to ketones and $CaCO_3$; the other one is surface ketonization of metal oxides (Kumar 2018), in the pyrolysis process, Re-CF experienced a rapid multi-step reduction/oxidation cycle, which combined with α-H of oxygen-containing functional groups in pyrolysis products, then oxidized by protons and hydroxyls in carboxylic acid molecules, and generated water and carbon dioxide. Finality, substances such as ketones, aldehydes, and carboxylic acids generate alcohol compounds through the HDO process, and some alcohols react with carboxylic acids to form lipids.

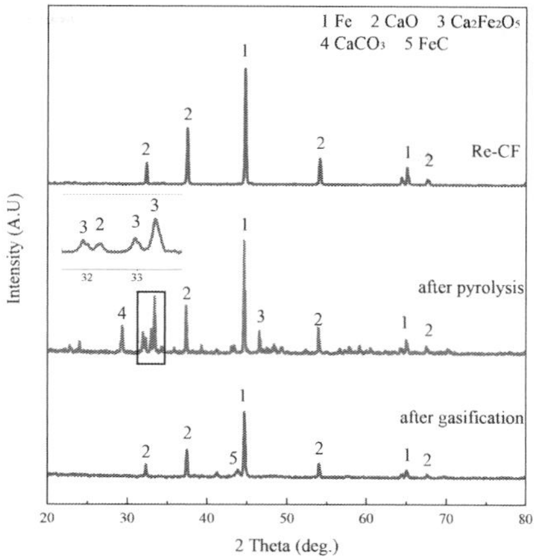

Figure 5. XRD spectra of Re-CF oxygen carrier after chemical looping pyrolysis.

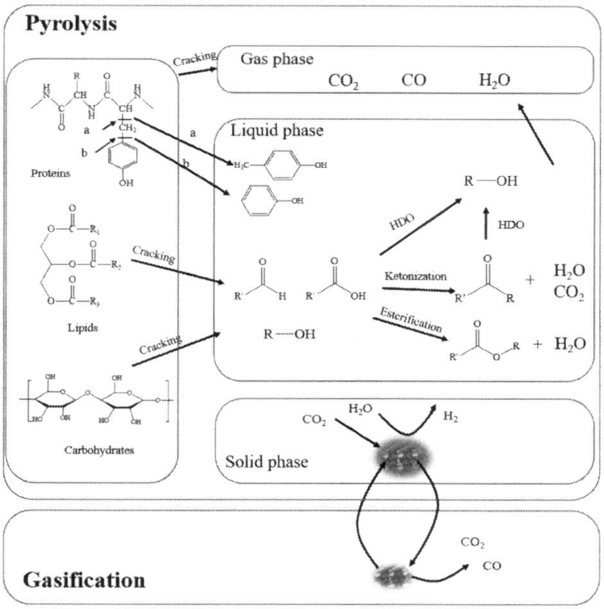

Figure 6. Oxygen migration pathway during chemical looping pyrolysis.

In addition, a large amount of CO_2 and H_2O were produced during the pyrolysis process and reacted with Re-CF to generate $Ca_2Fe_2O_5$ and $CaCO_3$, which promoted the transfer of oxygen to solid-phase products during CLPy. At the same time, H_2O reacts with Re-CF to generate H_2, which provides a sufficient hydrogen source for the HDO process. During the gasification of pyrolysis char, the oxidized oxygen carrier reacts with the char again, and oxygen in the oxygen carrier is released mainly in the form of CO, oxygen carrier is reduced to CaO and Fe. $CaCO_3$ generated in the pyrolysis process decomposes at high temperatures, and the released CO_2 generates CO through a gasification reaction with pyrolysis char, only a small amount is released in the form of CO_2.

4 CONCLUSIONS

CLPy can produce high-quality bio-oils and syngas simultaneously; it has the effect of Re-CF on the distribution of the pyrolyzed products of *Chlorella* CLPy and the possible migration pathway of oxygen element was investigated in this work. The following conclusions can be drawn: Compared with the direct pyrolysis, the CLPy process significantly improve the quality of bio-oils by the formation of hydrocarbons and reducing oxygenated content. In the pyrolysis stage, there are two main pathways, i.e., ketonization reaction and HDO process for the migration of oxygen during CLPy. In the gasification stage, the oxidized oxygen carrier reacts with the pyrolysis char to generate high-quality syngas, which realizes the oxygen transferring from the oxygen carrier to the gas phase.

ACKNOWLEDGEMENTS

The work was supported by the Joint Funds of the National Natural Science Foundation of China (Grant No. U20A20124), Shandong provincial natural science foundation (Grant No. ZR2020MB144), and the State Key Laboratory of High-efficiency Utilization of Coal and Green Chemical Engineering (Grant No. 2022-K3).

REFERENCES

Campanella A, Harold M P. (2012). Fast pyrolysis of microalgae in a falling solids reactor: effects of process variables and zeolite catalysts. *J. Biomass and Bioenergy* 46: 218–232.

Du Z, Hu B, Ma X, et al. (2013). Catalytic pyrolysis of microalgae and their three major components: carbohydrates, proteins, and lipids. *J. Bioresource technology* 130: 777–782.

Kumar R, Enjamuri N, Shah S, et al. (2018). Ketonization of oxygenated hydrocarbons on metal oxide-based catalysts. *J. Catalysis Today* 302: 16–49.

Landoll M P, Holtzapple M T. (2011). Thermal decomposition of mixed calcium carboxylate salts: Effects of lime on ketone yield. *J. Biomass and bioenergy* 35(8): 3592–3603.

Liu Y, Liu J, Wang T, et al. (2021). Co-production of upgraded bio-oils and H_2-rich gas from microalgae via chemical looping pyrolysis. *J. International Journal of Hydrogen Energy* 46(49): 24942–24955.

Stanton A R, Iisa K, Mukarakate C, et al. (2018). Role of biopolymers in the deactivation of ZSM-5 during catalytic fast pyrolysis of biomass. *J. ACS Sustainable Chemistry & Engineering* 6(8): 10030–10038.

Veses A, Aznar M, Martínez I, et al. (2014). Catalytic pyrolysis of wood biomass in an auger reactor using calcium-based catalysts. *J. Bioresource technology* 162: 250–258.

Yi L, Liu H, Lu G, et al. (2017). Effect of mixed Fe/Ca additives on nitrogen transformation during protein and amino acid pyrolysis. *J. Energy & Fuels* 31(9): 9484–9490.

Energy Revolution and Chemical Research – Kok-Keong Chong and Zhongliang Liu (Eds)
© 2023 The Authors, ISBN: 978-1-032-36554-1

Effects of ankaflavin on fatty acid content of muscle in obese mice

Huihui Li & Anni Su
College of Food Technology, Wuhan Business University, Wuhan, Hubei, China

Shenghong Zhou
Hubei Province Intangible Cultural Heritage Research Center, Wuhan, Hubei, China

ABSTRACT: Ankaflavin was a yellow pigment of monascus, which might have anti-obesity effects in recent studies. In this paper, the effect of ankaflavin on fatty acid content in muscle of obese mice was studied by the Gas Chromatography with Flame Ionization Detection (GC-FID) method. The results showed that ankaflavin did not significantly reduce the body weight and BMI of obese mice induced by a high-fat diet when the mice's daily effective dose was about 2.0 mg of ankaflavin per kilogram of body weight and the experiment was carried out for 12 weeks. However, ankaflavin could significantly reduce blood uric acid levels in obese mice. What is more, the results of GC-FID showed that ankaflavin could significantly reduce the contents of C14:1n5, C16:0, C18:0, and C18:1 in the muscles of obese mice, while increasing the contents of downstream metabolites C20:1n9 and C20:3n6. The above data provide a reference for an in-depth study of the functions of ankaflavin.

1 INTRODUCTION

According to a 2016 survey by the World Health Organization, more than 1.9 billion adults worldwide were abnormally overweight, of whom 650 million (13 percent of adults) were obese. Obesity had become the fifth leading risk factor for death and a major global public health problem. Accordingly, the Healthy China 2030 Planning Outline proposed to decrease the rate of overweight and obesity in China by 2030 through a sensible diet and healthy lifestyle.

Ankaflavin (AK) was a kind of yellow pigment of monascus, which belonged to lactone-ring azaphilones (Figure 1). It was a traditional fermented product with Chinese characteristics that could be used as food and medicine. As a pigment, ankaflavin was developing rapidly. In 2015, the *National Food Safety Standards for Food Additives Monascus Yellow Pigment* (Standard code: GB1886.66-2015) had been issued. At the same time, recent studies had shown that it also had many physiological activities, such as anti-cancer, anti-tumor, anti-diabetic, antioxidant, anti-obesity, and anti-inflammatory. Ankaflavin could be used as a functional factor in healthy food or medicine. Current studies had shown that ankaflavin not only had a strong effect on inhibiting fat variation, fat generation, and promoting lipolysis in preadipocyte models, but also could effectively reduce the total fat content in high-fat diet-induced obesity rats. To explore the possibility of ankaflavin as potential weight-loss food, the GC-FID method was used to study the effect of ankaflavin on the fatty acid content of muscle in obese mice.

2 MATERIALS AND METHODS

2.1 *Materials and instruments*

C57BL/6 male mice were purchased from *Hunan SJA Laboratory Animal Co., Ltd.* with the production license number SCXK (Xiang) 2019-0004. Animals were fed in the Hubei Food and Drug

DOI 10.1201/9781003332657-50

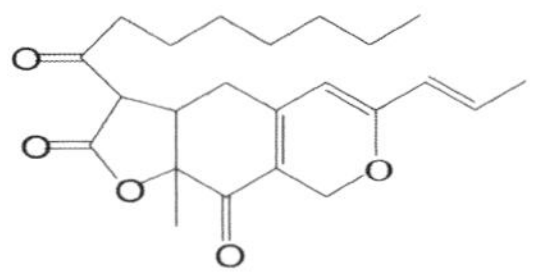

Figure 1. Diagram of the structure of ankaflavin.

Safety Evaluation Center. High-fat feed was purchased from *Trophic Animal Feed High-tech Co., Ltd*, China. Methyl heptahedron (CAS: 1731-92-6), Twenty-three carbon fatty acid methyl esters, 37 kinds of fatty acid methyl esters, and acetyl chloride were purchased from Sigma-Aldrich, USA. Methane, chloroform, n-hexane, potassium carbonate, 3,5-di-tert-butyl-4-Hydroxytoluene (BHT) were purchased from *Sinopharm Group Co. Ltd.* Ankaflavin (HPLC 95%) was purchased from *Chengdu Purify Technology Co., Ltd.* Liquid nitrogen, and glass centrifuge tubes of Pyrex were purchased from *Corning Inc.* GC8890-FID purchased from Agilent, USA.

2.2 *Experimental methods and measurement indicators*

2.2.1 *Animal experiments*

Animal experimental procedures were performed according to the National Guidelines for Experimental Animal Welfare (MOST of P. R. China 2006). Twenty male C57BL/6 mice were purchased at the age of four weeks and housed in a specific pathogen-free animal laboratory with a 12-h light/dark cycle at a constant temperature of 20–25°C and relative humidity of 40−60%. All animals had free access to water and normal food. After two weeks of acclimatization, the animals were randomly divided into two groups, namely, the FAT and AK groups.

After two weeks of acclimatization, the two groups were given a high-fat diet. Ankaflavin was mixed in drinking water in the AK group and the effective dose was about 2.0 mg of ankaflavin per kilogram of body weight per day, considering the photodegradation of ankaflavin. During the experiment, the body weight, body length, food intake, and water intake of mice were detected every week. After 12 weeks of the experiment, the mice were sacrificed by neck dislocation under isoflurane anesthesia, and the muscle tissue of the mice was taken and stored in liquid nitrogen for the extraction of fatty acids. The whole blood samples from three mice in each group were taken without anticoagulant. Two hours later, the whole blood was centrifuged (25°C, 1789 g, 10 min) to obtain serum, and the detection of blood biochemistry was completed within 24 hours.

2.2.2 *Detection of mouse blood biochemistry*

About 60 uL of serum was collected for the detection of nine indexes of serum biochemical routine on a Chemray-800 automated biochemistry analyzer according to standard experimental methods. Specific detection indicators included alanine aminotransferase (ALT), aspartate aminotransferase (AST), creatinine (CREA), uric acid (UA), total bile acid (TBA), glucose (GLC), triglyceride (TG), total cholesterol (CHO) and high-density lipoprotein (HDL). This part of the experiment was completed by *Wuhan Service Biotechnology Co., Ltd.*

2.2.3 *Fatty acid sample processing and detection*

The steps are as follows:

(1) Firstly, weigh about 25 mg of mouse muscle tissue, add 500 uL of a chloroform-methanol solution, and homogenize it on a tissue layer for 90 s at a frequency of 20 Hz;
(2) Then transfer 100 uL of the homogenate to a 10 mL Pyrex glass centrifuge tube, and add 20 uL of 1 mg/mL Methyl heptahedron internal standard and 2 mg/mL BHT and 1 mL methanol-hexane mixture;

(3) Under the condition of the liquid nitrogen ice bath, add 200 μL of acetyl chloride, and place it in a dark room at room temperature for 24 hrs until the methyl esterification reaction was completed;
(4) On the following day, add 2 mL of 6% K2CO3 to the glass centrifuge tube under ice bath conditions to neutralize the hydrochloric acid produced in the reaction. Then, extract it with 200 μL of n-hexane three times, and centrifuge at 800 g for 10 min, and the supernatants are combined;
(5) After the organic solvent was evaporated, extracts were redissolved in 200 μL hexane for GC-FID analysis.

Methylated fatty acids were measured on an Agilent 8890 gas chromatograph equipped with a flame ionization detector (FID). A DB-225 capillary GC column (10 m, 0.1 mm ID, 0.1 μm film thickness) from Agilent Technologies was employed with helium as a carrier and makeup gas. The sample injection volume was 2 μL with a splitter (1:10). the temperature of the injection port and the FID detector was 250°C. The column temperature was started from 115°C for 0.5 min and then increased to 205°C with a rate of 30°C/min. The temperature was kept at 205°C for 3 min then increased to 240°C (5°C/min) and kept at 240°C for 1.5 min.

The qualitative detection of fatty acids in samples is mainly determined by comparing the retention time of fatty acids in the standard. The fatty acid content was mainly quantified by the internal standard method (Methyl heptahedron).

2.2.4 *Statistics of experimental data*

The formula for calculating the fatty acid content was as follows: Cx (umol/g) = (100×1000×Sx)/(S17×Mx×mx), where Cx is the content of the fatty acid to be tested; Sx is the integral area of fatty acid methyl ester corresponding to the fatty acid to be measured; S17 is the integral area of internal standard methyl heptahedron; Mx is the molar mass of the fatty acid methyl ester while Mx is the mass of the sample (about 25 mg). In this experiment, eight samples were tested in each group, and the fatty acid content was expressed as mean ± SD. In addition, t-test of unrelated samples of SPSS19.0 was used to test the statistical differences between samples, and the significance level was set at 5%.

3 RESULTS AND ANALYSIS

3.1 *Effects of ankaflavin on body weight, BMI, and Lee's index of obese mice*

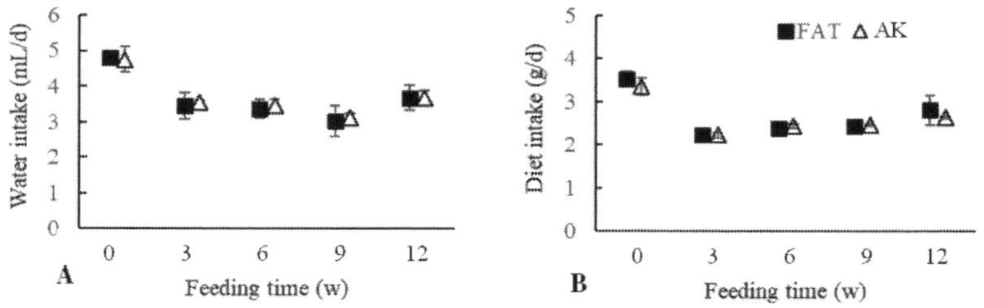

Figure 2. Effects of AK on water intake (A) and food intake (B) of obese mice.

After one week of adaptation, the mice were fed for 12 weeks. It could be seen from Figure 2 that except for the slightly higher water and food intake in the 0-th week, the daily water intake of the experimental mice was about 3.5 mL while the daily food intake was about 2.5 g. And at the same time point, there was no statistical difference in water intake and food intake between the obese group and the AK-fed obese mice. During the feeding period, the body weight of the mice increased steadily with age. From the sixth week then on, the weight of obese mice induced by a high-fat diet was significantly higher than that in the normal diet group (data were not shown), so a mouse obesity model was successfully constructed through a high-fat diet. However, there were no significant differences in body weight, BMI index, and lee's index between the obese and the obese ankaflavin groups throughout the whole feeding period (Figure 3), which were indicated that ankaflavin administration at a 2.2 mg/kg body weight per day for 12 weeks, did not significantly reduce the body weight of mice.

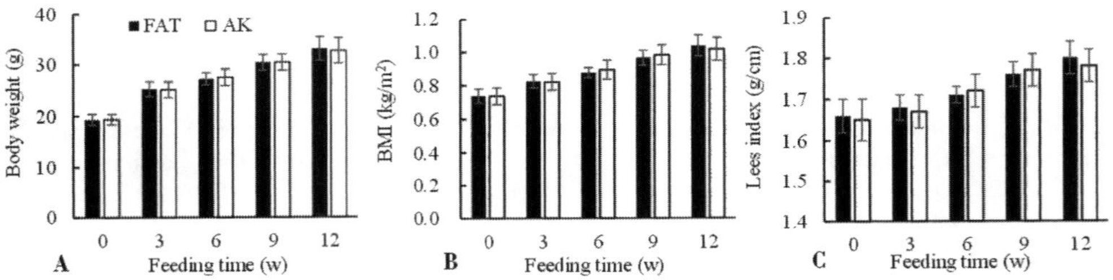

Figure 3. Effects of AK on the body weight (A), BMI (B), and Lee's index (C) of obese mice.

3.2 *The effect of ankaflavin on blood biochemistry in obese mice*

Table 1. Effects of AK on blood biochemistry in obese mice.

	FAT	AK
ALT (U/L)	25.06±3.3	24.82±6.44
AST (U/L)	103.89±15.75	117.07±14.42
CREA (μmol/L)	6.12±0.42	6.23±1.37
UA (μmol/L)	27.90±7.38	12.90±3.04 *
TBA (μmol/L)	3.15±0.73	3.97±0.93
GLU (μmol/L)	5.01±1.00	4.70±1.70
TG (μmol/L)	0.57±0.05	0.55±0.10
CHO (μmol/L)	3.83±0.22	3.64±0.30
HDL (μmol/L)	1.84±0.06	1.77±0.07

Note: * means $p< 0.05$.

The results of mice blood biochemistry (Table 1) showed that there were no significant differences in the contents of alanine aminotransferase (ALT), aspartate aminotransferase (AST), creatinine (CREA), total bile acid (TBA), glucose triglyceride (TG), total cholesterol (CHO) and high-density lipoprotein (HDL) between AK mice and obese group. The content of uric acid (UA) in the AK group was 12.90 μmol/L while the obese group was 27.90. The UA content in AK-fed obese mice was significantly lower than in the obese group. Uric acid was a purine metabolite. 80% of uric acid in the body was endogenous, formed by the synthesis of amino acids, phosphoribose, and other small molecular compounds in the body and the decomposition of nucleic acids. And the other 20% was derived from the diet. About two-thirds of uric acid was excreted in the urine through the kidneys and one-third through the gastrointestinal tract. Current studies had found that hyperuricemia was closely related to visceral obesity. In obese people with hyperuricemia, weight loss was an effective way to reduce uric acid levels. In this experiment, the administration of ankaflavin significantly reduced the blood uric acid level in the obese group, and its mechanism deserved further study.

3.3 *The effect of ankaflavin on muscle fatty acid in obese mice*

3.3.1 *GC-FID total ion current (TIC) diagram of fatty acid content in mice*

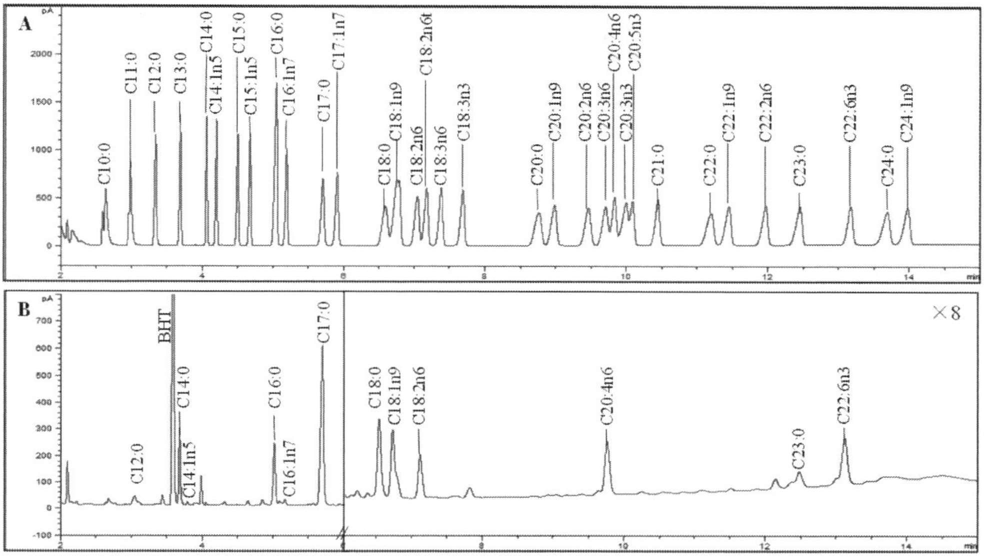

Figure 4. Total ion current diagram of fatty acid standards (A) and mice muscle fatty acid (B).

Disorders of lipid metabolism were the initiating link of obesity-related diseases. Lipid metabolism disorders could occur in many steps of lipid metabolism, including the synthesis, storage, mobilization, and decomposition of triglycerides, fatty acids, and cholesterol esters. In this paper, fatty acid composition in mouse muscle was determined by the GC-FID method. The qualitative detection of fatty acids in mouse muscle was determined by comparing the retention time of fatty acids in the standard substance. Figure 4 showed that 16 fatty acids were detected in mouse muscle, including six saturated fatty acids (C8:0, C10:0, C12:0, C14:0, C16:0, and C18:0), six monounsaturated fatty acids (C14:1n5, C16:1n7, C18:1n9, C18:1n7, C20:1n9, and C24:1n9), and four polyunsaturated fatty acids (C18:2n6, C20:3n6, C20:4n6, and C22:6n3). Among them, C18:1n9 and C18:1n7 in the mouse muscle could not be completely separated due to the close retention time of their peaks, so they were mixed in the calculation of fatty acid content in the following paper. In addition, it could be seen from the peak height and peak area that the content of long-chain fatty acids in mouse muscle was relatively high, of which C16:0 had the highest content while C18:0, C18:1, C18:2n6, C20:4n6, C22:6n3, and C24:1n9 were also higher.

3.3.2 *The effect of ankaflavin on muscle fatty acid in mice*

In this experiment, the content of fatty acids in mouse muscle was quantitatively calculated by the internal standard method, and the data were expressed as mean ± standard deviation. Due to the large individual differences in mice, the standard deviation of some fatty acids was also large. After statistical analysis, it was found that the contents of C14:1n5, C16:0, C18:0, and C18:1 in the muscles of mice were decreased after feeding ankaflavin (Table 2), while C20:1n9 and C20:3n6, which were downstream of their metabolism, were not detected in the obese group, but significantly increased after feeding ankaflavin. Among them, C20:3n6 belonged to the n6 series of polyunsaturated fatty acids, and it was generally believed that n3 and n6 fatty acids might have different effects on body fat through mechanisms such as adipogenesis, lipid homeostasis, and systemic inflammation. The n6 series of fatty acids could increase the content of intracellular triglycerides by increasing the permeability of the cell membrane, and at the same time had different

Table 2. The fatty acid content in muscle of obese and AK-fed obese mice.

Retention time (min)	Fatty acid	FAT (umol/g)	AK (umol/g)
1.98	C8:0	73.78±12.81	88.58±5.06
2.64	C10:0	39.94±10.70	48.04±8.05
3.34	C12:0	13.02±7.00	20.17±7.74
4.06	C14:0	36.10±11.72	26.79±4.45
4.20	C14:1n5	8.88±2.31	4.97±2.01*
5.05	C16:0	927.17±152.83	743.29±132.91*
5.20	C16:1n7	74.87±10.36	61.21±19.97
6.60	C18:0	481.65±88.53	350.28±52.86*
6.80	C18:1n9+C18:1n7	686.36±184.44	321.03±77.97*
7.18	C18:2n6	253.54±69.41	199.64±35.90
8.80	C20:1n9	ND	11.60±2.53*
9.72	C20:3n6	ND	331.85±46.22*
9.83	C20:4n6	441.64±194.84	308.95±63.53
13.18	C22:6n3	363.25±66.18	314.24±71.61
14.02	C24:1n9	347.56±93.46	442.63±20.51

Note: ND means not detected. * means $p< 0.05$.

regulatory effects on the inflammatory properties of their downstream eicosanoids, which ultimately affected preadipocyte differentiation and fat mass growth. The literature showed that ankaflavin acts as a natural PPAR-γ and AMPK activator. On the one hand, ankaflavin affected the differentiation of adipocytes by inhibiting C/EBPβ; on the other hand, it activated the activity of lipase and lipoprotein enzyme activity, to inhibit fat synthesis and accelerate the β-oxidation of fat. The two aspects jointly interfere with the occurrence and development of obesity.

4 CONCLUSION

To explore the weight loss function of ankaflavin, the effect of ankaflavin on fatty acid content in muscle of obese mice was studied by the GC-FID method. Firstly, a mouse obesity model was constructed through a high-fat diet which was called the FAT group while the AK group was administered ankaflavin through drinking water. The daily effective dose consumed by each mouse was about 2.0 mg of ankaflavin per kilogram of body weight, and the experiment was carried out for 12 weeks. Then according to the comparison of body weight, BMI index, and Lee's index of two groups, ankaflavin seemed to have no obvious weight loss effect, which may due to the low dosage of ankaflavin. However, it was found that ankaflavin could significantly reduce the blood uric acid level in obese mice through blood biochemical detection. What is more, the results of GC-FID showed that ankaflavin could significantly reduce the contents of C14:1n5, C16:0, C18:0, and C18:1 in the muscles of obese mice, while increasing the contents of downstream metabolites C20:1n9 and C20:3n6. The mechanism of ankaflavin changing fatty acid metabolites in mouse muscle was worthy of further study, which would help us to deeply understand the function of Chinese traditional fermented monascus product and fully explore its value in processing and utilization.

ACKNOWLEDGMENTS

This work is financially supported by the General Programs of Natural Science Foundation in the Science and Technology Department of Hubei Province (Grant No. 2019CFB764).

REFERENCES

Hsu, W. H. & Chen, T. H. & Lee, B. H. & Hsu, Y. W. & Pan, T. M. (2014). Monascin and ankaflavin act as natural AMPK activators with PPARα agonist activity to down-regulate nonalcoholic steatohepatitis in high-fat diet-fed C57BL/6 mice. *Food Chem. Toxi*. 64, 94–103.

Hu, X. H. & Zhang, L. M. (2019). Research progress of uric acid metabolic pathways. *J. Clin. Nephrol.* 19(12), 935–937.

Jou, P. C. & Ho, B. Y. & Hsu, Y. W. (2010). The Effect of Monascus Secondary Polyketide Metabolites, Monascin, and Ankaflavin, on Adipogenesis and Lipolysis Activity in 3T3-L1. *J. Agri. Food Chem.* (58), 12703–12709.

Lee, C. L. & Wen, J. Y. & Hsu, Y. W. & Pan, T. M. (2013). Monascus-fermented yellow pigments monascin and ankaflavin showed antiobesity effect via the suppression of differentiation and lipogenesis in obese rats fed a high-fat diet. *J. Agri. Food Chem.* 61 (7), 1493–1500.

Liu, Y. & Zeng, Y. (2016). Hyperuricemia and obesity. Chin. *J Cardiovasc Med.* 21(01): 11–13.

Zhuang, P. (2020). *Study on the role of polyunsaturated fatty acids in regulating glucose and lipid homeostasis in obesity based on the "lipid-gut-liver" axis*. Diss. Hangzhou, Zhejiang University.

Energy Revolution and Chemical Research – Kok-Keong Chong and Zhongliang Liu (Eds)
 ISBN: 978-1-032-36554-1

Visual analysis of bamboo furniture research based on CiteSpace

Sichao Li*, Jianhua Lyu, Ming Chen & Hongjia Guo
Forestry College, Sichuan Agricultural University, Chengdu, China

ABSTRACT: In this paper, we aim to investigate the research status of Chinese bamboo furniture, analyze its context, direction, and hotspots and predict its future development trend. In CNKI database, "subject = bamboo" and "subject = furniture" are used as the keywords for advanced retrieval. The time span is set as 2000–2021, and 675 literatures are obtained. The data is then imported into CiteSpace for visual comparative analysis. The results show that the number of articles published in relevant journals is low from 2000 to 2003, and the number of articles published has increased significantly from 2004 to 2016, reached a peak in 2016, and then gradually stabilized from 2017 to now. At present, a core author group with strong scientific research output capacity has not been formed. Hot research contents mainly include laminated bamboo timber, new manufacturing processes, bamboo culture, and the integrated use of bamboo-rattan and bamboo-wood. Based on the results, this paper proposes to strengthen the communication and cooperation among the core authors, jointly develop new bamboo manufacturing technology, and integrate the spiritual connotation of Chinese traditional bamboo culture into the design of bamboo furniture, so as to enhance the competitiveness of bamboo furniture in the market and promote the development of bamboo furniture.

1 INTRODUCTION

The knowledge map tool used in this study is CiteSpace software developed by Professor Chen Chaomei of Drexel University based on Java platform; the version is 5.8. R3 (Zhang & Zhu 2021). The imported data came from CNKI. Advanced retrieval was carried out with the keywords "theme = bamboo" and "theme = furniture". The time span was set to "2000–2021", and 675 documents were obtained. By drawing the network Atlas of keyword co-occurrence, author co-occurrence and hot research co-occurrence, this paper studies and analyzes the age distribution, author distribution and research hot spots in the research field of bamboo furniture, etc, so as to understand the development trends and research strength in this field.

2 ANALYSIS OF LITERATURE RESEARCH RESULTS OF BAMBOO FURNITURE

2.1 *Analysis of publication quantity*

The periodical literature data about bamboo furniture from 2000 to 2021 retrieved from CNKI database were counted and the literature time distribution map was drawn, as shown in Figure 1. It can be seen from the figure that the overall trend shows an upward trend before 2016, and the number of documents issued gradually maintains a stable trend after 2016. The literature on bamboo furniture can be divided into three development stages. The first stage is the embryonic stage (2000–2003). In this stage, the number of documents published was small, the number of

*Corresponding Author: ljh@sicau.edu.cn

 DOI 10.1201/9781003332657-51

documents had little change, and 12 documents were published annually. The second stage is the development stage (2004–2016). At this stage, the number of papers began to increase significantly, reached the peak of development in 2016, with an annual number of 57 papers. The third stage is the mature stage (from 2017 to now). In this stage, the number of documents issued is relatively stable, and the number of published papers still maintains a high level, with an average annual output of 42 documents, indicating that scholars are paying continuous attention to the research of bamboo furniture.

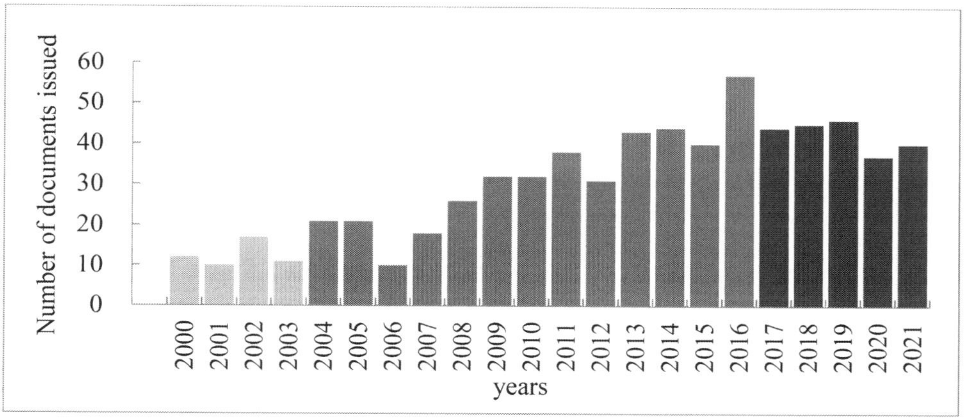

Figure 1. Chronological distribution of bamboo furniture research literature from 2000 to 2021.

2.2 *Core author analysis*

Based on the statistical analysis of 675 retrieved journal documents, the distribution map of authors is obtained, as shown in Figure 2. Then, the author information of the top 6 is screened according to the number of articles published by the author, as shown in Table 1. It can be seen that the author with the largest number of articles is Li Jiqing, with a total of 24 articles. According to price's law, the author with no less than 4 starting papers is the core author of research in this field. According to the statistics of this paper, there are 46 core authors, and a total of 282 papers have been published, accounting for 41.77% of the total number of papers, which is nearly 10 percentage points lower than the standard that the mass number of core authors should account for 50% of the total number of documents in price's law (Zong 2016). It can be seen that the core author group with strong scientific research output ability has not been formed in the research field of bamboo furniture in China. Figure 2 shows the core authors and their cooperative relationship. It can be seen that the core authors in this field are represented by Li Jiqing, Song Shasha and Wu Zhihui, etc. At present, the core authors in this research field are in a state of partial concentration and overall divergence. It can be seen that it is necessary to strengthen continuous research and cooperation in this area, and gradually form a stable and closely connected core author group to jointly promote the development of bamboo furniture research.

3 ANALYSIS ON RESEARCH HOTSPOTS OF BAMBOO FURNITURE

3.1 *Keyword co-occurrence map analysis*

The 675 retrieved documents are imported into CiteSpace, and the time division is 1 year. Pathfinder and pruning sliced networks are used for network pruning to highlight important structural features, and a keyword co-occurrence map is obtained, as shown in Figure 3. To preliminarily analyze the

Figure 2. Cooperation map of authors.

Table 1. Distribution of core authors.

Serial Number	Author	Post Volume
1	Li Jiqing	24
2	Song Shasha	15
3	Wu Zhihui	14
4	Chen Zujian	14
5	Zhang Qisheng	10
6	Zhang Zhongfeng	9

hot spots and frontiers of bamboo furniture development research, this paper summarizes and sorts the keyword frequency in the knowledge map, as shown in Table 2. The top ten high-frequency keywords mainly include bamboo furniture, furniture design, bamboo-based furniture, laminated bamboo timber, etc. Therefore, it can be inferred that the hot spots of bamboo furniture development in China are mainly concentrated in four aspects: (1) the application of different types of bamboo in furniture design; (2) the connotation of bamboo culture in the design of bamboo furniture; (3) the use of bamboo-rattan, bamboo-wood, and other materials in bamboo furniture design; (4) the research and development of new bamboo manufacturing processes such as laminated bamboo timber to improve the production efficiency of bamboo furniture.

3.2 *Keyword cluster map analysis*

On the basis of keyword co-occurrence, cluster analysis of keywords is carried out, as shown in Figure 4. The Q value is 0.7684 and the mean silhouette value is 0.9095, which means that the clustering result is reasonable. It can be seen from Figure 5 that the largest cluster is #0 bamboo furniture, which contains keywords such as bamboo products and manufacturing technology. The results show that the manufacturing technology of bamboo products such as bamboo furniture has attracted extensive attention from researchers. The low degree of mechanization of traditional original bamboo furniture leads to low production efficiency and quality assurance. However, the research and development of bamboo furniture with new manufacturing processes such as laminated bamboo timber can realize industrial production, so as to promote the development of bamboo furniture (Li & Wu, Zhang 2004).

Table 2. Key words of bamboo furniture research.

Counts	Centrality	Keywords
93	0.38	Bamboo Furniture
93	0.28	Furniture Design
41	0.28	Bamboo-Based Furniture
35	0.13	Laminated Bamboo Timber
31	0.09	Bamboo
22	0.16	Bamboo-Rattan Furniture
21	0.05	Innovative Design
18	0.05	Bamboo Culture
14	0.08	Original Bamboo Furniture
13	0.04	Bamboo-Wood Furniture

Figure 3. Keyword co-occurrence map.

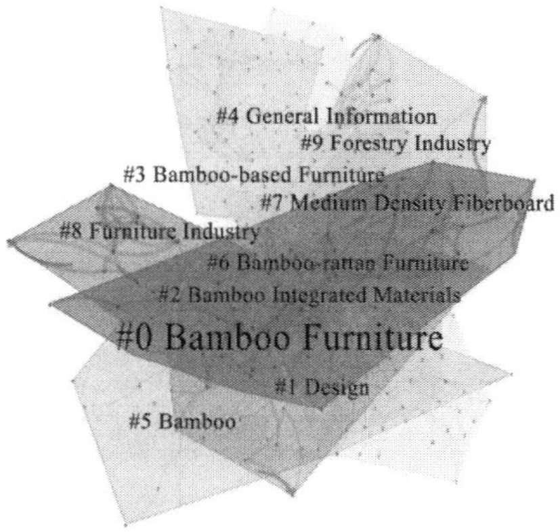

Figure 4. Keyword clustering map.

3.3 *Keyword emergence map analysis*

Emergent words are the key terms in this field formed by the significant increase in the use frequency of keywords in a certain period. Through the in-depth analysis of emergent words, we can clarify the phased frontier research field of bamboo furniture (Guo & Hao 2021). As shown in Figure 5, the emergent words are “forestry industry”, “reconsolidated bamboo”, “furniture design”, “innovative design”, “bamboo culture”, “bamboo flooring”, “imports and exports”, and “bamboo-rattan furniture”. Using the temporal changes of emergent words, we can understand the changes in research topics in this field over time. The research shows that the frontier issues and future trends of bamboo furniture research mainly include the following three aspects.

(1) Early research frontier. The emergence of bamboo-rattan furniture was in 2004, which was the frontier field concerned by scholars in the early stage.
(2) Mid-term research frontier. The emergent words at the forefront of the mid-term research are “bamboo flooring”, “forestry industry”, “imports and exports” and “reconsolidated bamboo” from 2009 to 2015. The emergent time range is 3–4 years, in which the appearance of “forestry industry” and “reconsolidated bamboo” is the highest, indicating that the research on bamboo in this period revolved around the import and export trade of bamboo flooring, centered on China’s forestry industry.

(3) The latest research frontier. The emerging terms in the latest research frontier are key terms that are still used frequently. Since 2013, the bamboo culture has been included in the research, the innovation in bamboo design since 2016, and the furniture design since 2019, which indicates that scholars have gradually turned their research to the innovative design of bamboo furniture and that the design research has been integrated with the bamboo culture in China's traditional culture.

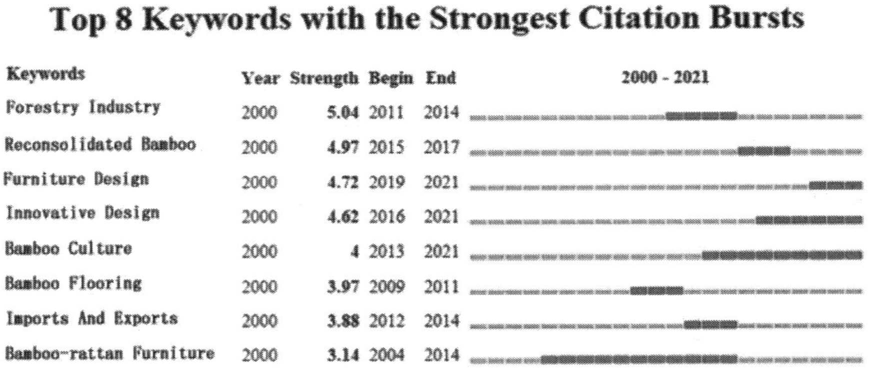

Figure 5. Clustering map of emergent words.

4 CONCLUSION AND OUTLOOK

4.1 *Conclusion*

In this paper, CiteSpace bibliometric software is used to visually analyze the literature in the research field of bamboo furniture in the CNKI database. The main conclusions are as follows:

(1) As evidenced by the time distribution of the literature, the number of articles published continued to increase after 2016, suggesting a constant interest on bamboo furniture in academic circles.
(2) According to the author distribution map, there are few exchanges and cooperation among the core author groups in the field of bamboo furniture research in China, which hinders the development of bamboo furniture research.
(3) It can be seen from the research hotspots that the current research on bamboo furniture focuses on materials, technology, and the embodiment of bamboo culture.

4.2 *Outlook*

Based on the above analysis results, the following prospects are put forward:

(1) According to the number of papers on bamboo furniture research, future research on bamboo furniture still belongs to the hot research field.
(2) There is a need to strengthen the cooperation and exchange among the core authors, to form a close group, and to constantly examine new research directions and hot spots in the process of communication and cooperation. Moreover, cross-regional and interdisciplinary cooperation can be achieved through the creation of a database of achievements by using communication technology and digital technology to promote the development of bamboo furniture research on all fronts (Fang & Yuan 2021).
(3) In the future, the focus of bamboo furniture research should continue to center on new bamboo manufacturing techniques, and the use of manufacturing and information industry technologies

to transform the traditional bamboo processing industry, so as to gradually realize mechanization and automation in the production process of bamboo processing, and to enhance the utilization of bamboo (Zhang 2003). In terms of material selection, Xiangfei bamboo (mottled bamboo), mottled mud bamboo, purple bamboo (black bamboo), and square bamboo are widely used in furniture production because of their unique decorative effect. Therefore, some new varieties with different colors and shapes can be specially cultivated for furniture production to enrich the decorative effect and meet the needs of different consumers (Li 2004). Furthermore, an innovative design of bamboo furniture should also incorporate the spiritual connotation of China's traditional bamboo culture, meet the demands of customers, increase added values of products, and enhance China's competitiveness in the international market for bamboo furniture (Li 2021) (Wang & Yang; Yuan 2020). Furthermore, the research and development of bamboo furniture should also follow the national policy and strategies.

REFERENCES

Fang Jingrong, Yuan Haiming. 2021. Knowledge graph analysis of cultural and creative industry research: Taking CiteSpace measurement tool as an example[J]. *Journal of Suzhou Arts and Crafts Vocational and Technical College*, (02): 14–20.

Guo Wei, Hao Ruinan. 2021. Research characteristics and trends of cultural and creative product design based on bibliometrics[J]. *Packaging Engineering*: 1–14.

Li Cisheng. 2004. Bamboo culture and bamboo furniture[J]. *Furniture and Interior Decoration*, (07): 54–55.

Li Jiqing, Wu Zhihui, Zhang Sheng. 2004. Research on the production technology of new laminated bamboo timber[J]. *Journal of Inner Mongolia Agricultural University (Natural Science Edition)*, 25(2): 95–99.

Li Zhensheng. 2021. Design and utilization methods and development strategies of bamboo resources under rural revitalization [J]. *Journal of Sanming University*, 38(02): 79–85.

Wang Lujia, Yang Ming, Yuan Zhe. 2020. A preliminary study of bamboo in furniture design[J]. *Western Leather*, 42(22): 28–29.

Zhang Kai, Zhu Bowei. 2021. Research progress, hot Spots and trends of inclusive design[J]. *Packaging Engineering*, 42(02): 64–69+103.

Zhang Qisheng. 2003. My country's bamboo processing and utilization should attach importance to science and innovation[J]. *Journal of Zhejiang Forestry University*, (01): 3–6.

Zong Shuping. 2016. Evaluation of core authors based on Price's law and comprehensive index method: Taking "China science and technology journal research" as an example[J]. *China Science and Technology Journal Research*, 27(12): 1310–1314.

Energy Revolution and Chemical Research – Kok-Keong Chong and Zhongliang Liu (Eds)
© 2023 The Authors, ISBN: 978-1-032-36554-1

Discussion on main occurrence characteristics of geothermal resources in Yutai sag of West Shandong

Peng Qin
Shandong Provincial Space Ecological Restoration Center, Jinan, China

Peng Yang
The 8th Bureau of Geological and Mineral Resource Prospecting of Shandong, Rizhao, China

Yiping Li
Shandong GEO-Surveying & Mapping Institute, Jinan, China

Hongliang Liu & Zhentao Wang
Shandong Provincial Space Ecological Restoration Center, Jinan, China

Jia Meng
Shandong Lunan Geological Engineering Survey Institute, Jining, China

Taitao Liang
Shandong Institute of Geological Sciences, Jinan, China

ABSTRACT: Geothermal energy is a type of green energy and it has been highly valued in recent years. According to the division of geothermal fields in Shandong Province, this area is a Yutai geothermal prospective area. However, few researchers conducted in-depth research in this area. This study is based on the project "Investigation and Evaluation of Geothermal Resources in Yutai Sag, Shandong Province". Through data collection, ground investigation, geothermal field measurement, geophysical exploration, geothermal well drilling, pumping test, water quality analysis, and other means, the occurrence conditions of geothermal resources, thermal reservoir characteristics, and geothermal fluid chemical characteristics in this area are analyzed and studied by using the relevant data of geothermal exploration wells. The depth of the constant temperature zone is 20 m, the temperature of the constant temperature layer is 16°C, the Cretaceous geothermal gradient is about 1.26°C/100 m, the geothermal gradient is about 2.94°C/100 m, and the Ordovician geothermal gradient is about 3.98°C/100 m. The limestone roof depth of the geothermal well is 2,089 m, the well depth is 2,309 m, the bottom-hole geophysical temperature measurement is 74.5°C, the wellhead temperature is 68.5°C, the maximum water inflow is 552 m^3/d, the radius of influence is 852 m, the water chemical type is Cl-Na type, and the salinity is 26454.54 mg/L. It is a hot water geothermal resource in low-temperature geothermal resources. Geothermal water contains fluorine, bromine, strontium, and iron concentrations that met the standards of named mineral water. The contents of lithium and boron can reach the mineral water concentration, with a high physiotherapy value. This study provides a geological basis for the exploration and utilization of geothermal resources in this area and related areas.

1 INTRODUCTION

Geothermal energy is a kind of green, low-carbon, recyclable, and renewable energy, which has the characteristics of large reserves, wide distribution, clean and environmental protection, good stability, and high utilization coefficient. In recent years, great advantages in energy supply and

 DOI 10.1201/9781003332657-52

energy conservation, and emission reduction have been highly valued. Yutai Sag is located in the middle and south of Yutai County, Jining City, Shandong Province. It belongs to the Jining stratigraphic area of the West Shandong stratigraphic division in the North China stratigraphic area. The western boundary is Jiaxiang Fault, the eastern boundary is the Sunshidian Fault, and the Fushan Fault is the northern boundary. According to the division of geothermal fields in Shandong Province, this area is a Yutai geothermal prospective area. It belongs to the blank area of geothermal research, and no one has conducted in-depth research. When Li Aijun and others studied the geothermal resources in southern Jining in 2015, they mentioned that Jiaxiang Fault and Sunshidian Fault was a gravity gradient zone and that the geothermal resources in the region were mainly affected by the conduction heat flow in the deep crust and upper mantle. The thermal reservoir was mainly Ordovician limestone, and the geothermal resources were calculated in the area where the burial depth of the Ordovician roof is less than 1,200 m (Li 2015). Therefore, there is no deep exploration data in Yutai County geothermal prospect area and the regional data are relatively scarce. The geothermal research data in Jining City are relatively abundant (Bi 2019; Cao 2008; Chen 2008).

In this paper, we evaluate the distribution of geothermal resources above a buried depth of 3,000 meters in Yutai Sag using the survey results from the project "Investigation and Evaluation of Geothermal Resources in Yutai Sag of Shandong Province" and the subsequent geothermal exploration data. It is concluded that about 31.8% of the heat produced by the decay of radioactive elements in this area comes from the deep crust, the rest of the heat comes from deep in the Earth's crust, the depth of the annual constant temperature zone in the area is 20 m, the temperature of the constant temperature layer is 16°C, and the geothermal gradient of the Cretaceous is about 1.26°C/100 m. After entering the coal strata downward, the geothermal gradient is about 2.94°C/100 m, and the geothermal gradient of the Ordovician is about 3.98°C/100 m. The limestone roof depth of a geothermal well is 2,089 m, the well depth is 2,309 m, the bottom-hole geophysical temperature measurement is 74.5°C, the wellhead temperature is 68.5°C, the maximum water inflow is 552 m^3/d, the radius of influence is 852 m, the hydrochemical type is Cl-Na type, and the salinity is 26454.54 mg/L. All these show that this area has abundant low-temperature geothermal resources.

2 GEOLOGICAL BACKGROUNDS

2.1 *Stratum*

The strata in the study area belong to the Jining stratigraphic area of the West Shandong stratigraphic division in the North China stratigraphic area. The strata in the area from old to new mainly develop Paleozoic Ordovician, Carboniferous-Permian, Mesozoic Jurassic, Cretaceous and Cenozoic Paleogene, Quaternary. Ordovician in this region is chiefly Majiagou group, with a regional stratigraphic thickness of 550–900 m, and it is in contact with the underlying unconformable parallel Jiulong group. The lower system is missing in the Yutai region Carboniferous sedimentary; only the upper system is developed. According to lithology, it is combined into the Yuemengou group, in which the roof depth is greater than 1,200 m, reaching a maximum depth of 6,000 m. The Jurassic strata in this region only develop the Middle-Upper System Zibo Group, and the thickness and burial depth change greatly. The thickness and burial depth in the center of the basin are large, while the burial depth in the southeast is shallow. Cretaceous strata in this region develop the lower system. According to lithology, it is combined into Laiyang group and Qingshan group, whose buried depth is about 1,000 m. Paleocene is developed in Paleogene, and the rock strata belong to the Guanzhuang Group. The thickness of the whole region is different, with a maximum thickness of 660 m. Neogene is distributed in the whole region because only under the Quaternary is it buried relatively shallow, about 60–150 m, toward the center of the basin sedimentary thickness. Quaternary is widely distributed throughout the region, with a thickness of 60–150 m and an average thickness of about 110 m; it gradually thickened from east to west.

2.2 *Structure*

The study area belongs to the North China plate (Level I), West Shandong uplift area (Level II), Southwest Shandong buried uplift area (Level III), Heze-Yanzhou buried fault uplift (Level IV), and Yutai buried depression (Level V), and the regional structure is dominated by faults. The structural pattern of the working area is affected by the activities of deep and large faults, forming a fault depression bounded by regional faults. The north-south and east-west faults intersect and cut. The faults near the north and south are the Jiaxiang fault, Jining fault, and Sunshidian fault. The faults near the east-west are mainly the Fushan fault, Yutai fault, and Shanxian fault.

2.3 *Hydrogeology*

The study area is located between the leading edge of the Yellow River alluvial plain and the piedmont alluvial plain. The geological structure of the study area is in a relatively declining area. It is mainly characterized by accumulation, low terrain, small ground slope, the large thickness of Quaternary and Neogene (Q+N), and Quaternary material mainly comes from the accumulation of the Yellow River in the west. It has the characteristics of fine aquifer particles (generally fine sand layer), weak groundwater horizontal runoff, uneven water abundance, and complex water quality. Aquifer pore development, groundwater has good storage space to accept atmospheric precipitation recharge. The depth of the aquifer bottom plate is about 20–40 m in the shallow loose rock pore water-bearing rock group. While the depth is about 120–140 m in the middle and deep loose rock pore water-bearing rock group. And there is a multi-layer thick and continuous distribution of sandy clay-based aquifer between the top and bottom plate. The buried depth of deep pore water aquifer bottom plate is 250–330 m, and its distribution is strictly controlled by sedimentary hydrodynamic conditions. Its burial characteristics, lithologic combination, and hydraulic properties have obvious regularity, which is the main drinking water source for residents in the area.

3 GEOTHERMAL GEOLOGICAL CONDITIONS

3.1 *Geothermal field characteristics*

3.1.1 *Horizontal direction*

Combining the geothermal contour map with the formation situation in the working area with buried depths of 30 m, 50 m, and 80 m (Figure 1), it is indicated that the southeast of the working area bedrock uplift area geothermal value is the highest; the ground temperature in the bedrock dumping area under the Quaternary sediments cover in the northwest of the working area is the lowest.

3.1.2 *Perpendicular direction*

The depth of the constant temperature zone roof in the working area is 10–20 m. The influencing factors of the depth of the bottom plate in the constant temperature zone are mainly the results of the comprehensive influence of multiple factors such as the heat flux intensity in the deep, the formation lithology, and its thermal conductivity, followed by the influence intensity of the external climate environment. The analysis of the depth position of the bottom plate of the constant temperature zone and the thermal properties of the rock and soil in this area indicates that the depth position of the bottom plate of the constant temperature zone is shallow in the area where ground temperatures are rapidly increasing, and the bedrock depth is shallow with good thermal conductivity in the lower warming zone, due to the greater influence of deep heat flow upwards.

3.1.3 *Geothermal heating rate of thermal storage cover*

It can be seen from the temperature measurement data and temperature measurement curve that the ground temperature shows a continuous upward trend below the constant temperature zone. At 2,100 m, the ground temperature fluctuates significantly, due to the sudden increase in temperature that is caused by entering the Ordovician Majiagou Group thermal storage aquifer. In addition, as the depth increases, the temperature of the ground also rises significantly (Figure 2).

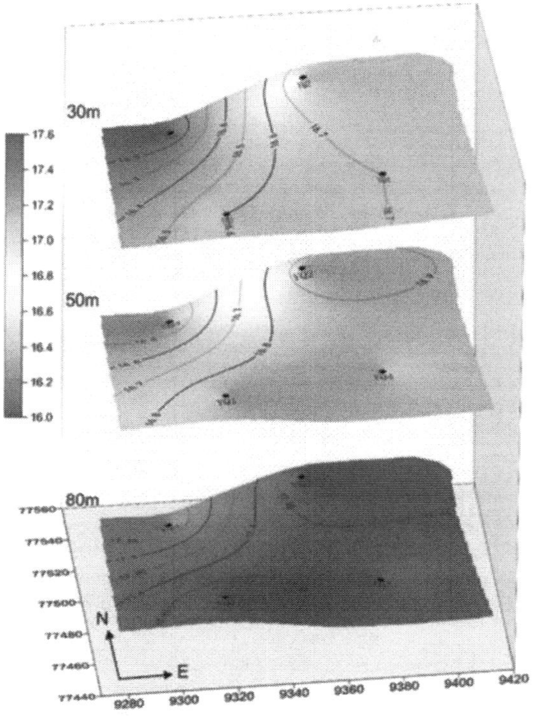

Figure 1. The contour map of ground temperature at the different depths.

The temperature measurements of the exploration well indicate that the Quaternary geothermal gradient is greater than 5°C below 100 m, which is abnormally high. This is because the water circulating within the well has not been fully cooled until logging. Geothermal fluid in the upper wellbore is greatly affected by temperature changes. The pore water runoff condition of the loose layer of the surrounding strata is good, and the cooling is rapid, resulting in a large difference in temperature vertically. On average, the Cretaceous geothermal gradient is about 1.26°C/100 m (the geothermal gradient of the upper basalt is about 0.36°C/100 m; the gradient of the lower mudstone and sandstone interbeds is about 1.38°C/100 m); in the coal strata, the gradient is about 2.94°C/100 m, and the gradient of the Ordovician is about 3.98°C/100 m. At a depth of 2100–2125 m, the geothermal gradient changed to –3.92°C/100 m, and there was a negative growth due to the strongly alternating activity of groundwater, and slurry leakage was observed when drilling to 2,118 meters. Logging data showed that the karst fracture section generally ranged from 2117.8 m to 2123.8 m and from 2215.5 m to 2190.4 m.

3.2 *Geothermal reservoir characteristics*

The geothermal reservoir type in this area belongs to the layered fissure geothermal field. The underground hot water occurs in the karst fissure of the Ordovician Majia Formation. The outlet water temperature of the geothermal well is 68.5°C, which belongs to the warm water geothermal resource in the low-temperature geothermal resources. The salinity of hot water is 26364.89 mg/L, and the hydrochemical type is Cl-Na.

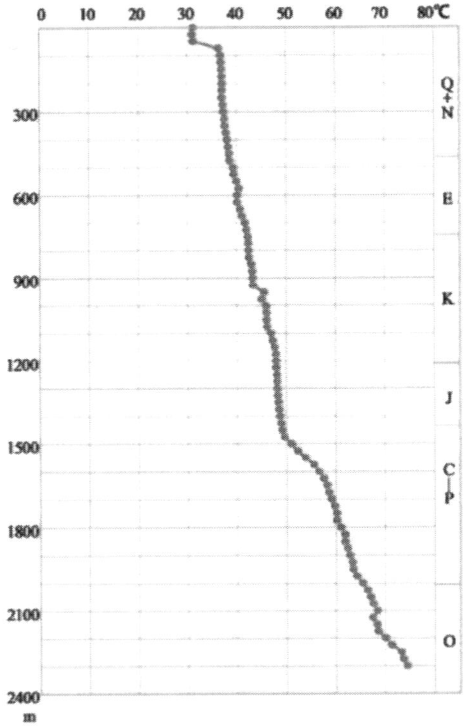

Figure 2. Well temperature and geothermal gradient map.

3.2.1 *Cover*

The depth of this exploration and research is 2,309 m, and the depth of the Ordovician thermal storage roof is 2,089 m. The Carboniferous-Permian, Cretaceous, Neogene, and Quaternary jointly constitute the thermal storage cover. The lithology of the cover is composed of multi-layer clay soil, sandy soil, sand layer, clay layer, and mudstone. It has good thermal insulation performance and is a good thermal storage insulation layer.

3.3 *Hot reservoir*

Ordovician limestone experienced tectonic movement, combined with weathering erosion and dissolution, forming fractures, fissures, and fracture zone. In the Ordovician strata, pore fractures are developed, which are the channels of deep geothermal water circulation and migration. The geothermal well is located in the Ordovician Majiagou group. The roof depth is 2,089 m, the exposed thickness is 220 m, the bottom hole geophysical temperature measurement is 74.5°C, the maximum water inflow is 552 m^3/d, and the wellhead temperature is 68.5°C. The karst fracture section has two main sections. One is 2117.8–2123.8 m, with a layer thickness of 6 m and a porosity of 12.4%, consistent with the drilling slurry leakage position; the other section is 2161.5 –2190.4 m, with a layer thickness of 28.9 m and a porosity of 9.0%, serving as the storage space for geothermal water. The limestone thermal reservoir of the Ordovician Majiagou Formation has deep burial, good water quantity, clear water quality, and higher water temperature in the working area, which has high development and utilization value.

3.4 *Heat source*

Heat flow is the movement of heat from the interior of Earth to the surface. The terrestrial heat flow measured in the shallow crust is the most direct reflection of the heat loss in the earth, which is composed of two parts: one is the heat generated by the decay of radioactive elements (U, TH, 40 K) in rocks shallow to the upper mantle, namely, the crustal heat flow; the other part comes from the heat of the upper mantle-the deep heat source. For a region, the crustal heat flow is different due to the difference in the content of radioactive elements and the corresponding thickness of strata, but the deep heat flow is relatively stable.

The thermal conductivity of limestone is 2.010 W/(m·°C). According to the BIRCH formula, the deep crustal temperature in this area is calculated. The working area is located in the North China Plate (Level I), West Shandong uplift area (Level II), West Shandong buried uplift area (Level III), Heze-Yanzhou buried fault uplift (Level IV), and Yutai buried depression (Level V). The deep earth heat source, magmatic-hydrothermal activity, and radioactive element decay are important heat sources in this area. The heat generated by the decay of radioactive elements accounts for about 31.8 %, and its heat comes from the deep crust. The new tectonic movement makes the heat source in the deep crust continue to pass up.

This is in agreement with Chen Moxiang's finding that the ratio of crustal heat flux variation to surface heat flux variation is mainly due to the heat flux of the silicon-aluminum layer in the crust, while radioactive element decay heat accounts for 37.8–46.1%. In 1981, Tianjin Geological Bureau conducted an investigation of geothermal resources and found that 62% of the heat flux in the North China Plain originates from the deep crust and 38% from the mantle.

3.5 *Conditions of heat conduction and water conduction channel*

The concealed fault structures are well developed in the region, which not only communicates the upwelling of deep heat sources, but also are the main channels for deep groundwater circulation. Jiaxiang fault is a tensile-torsional fault, which is the boundary fault between Yutai sag and Jinxiang sag. Drilling data confirm that the western plate of the fault is a relative upward plate, and the Yutai sag of the eastern plate is a relative downward plate. The vertical fracture distance of the fault is greater than 1,500 m. Jiaxiang fault (marked as "FJ1" in Figure 3) is distributed along Hanlou to Yanglou to Zhouzhuang to Wangfuzhuang. Its strike is northwest, its dip direction is northeast, and its dip angle is approximately 55° to 70°. In the apparent resistivity section, the vertical distribution has a "y" shape. The local distribution has a "step" shape along with the dip, and the development scale is large. Figure 3 shows obvious low-resistivity U-shape anomalies, which are reflected by obvious fault anomaly characteristics.

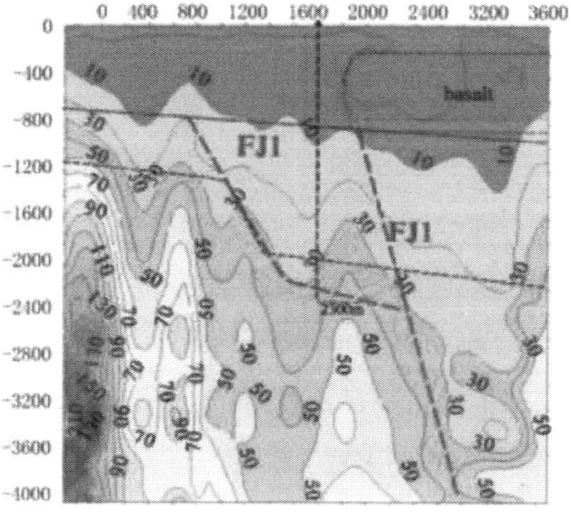

Figure 3. Magnetotelluric apparent resistivity contour map.

4 GEOCHEMICAL CHARACTERISTICS OF GEOTHERMAL FLUID

4.1 *Chemical composition characteristics of geothermal fluid*

The samples of geothermal wells in the area were collected and analyzed. The salinity of geothermal water in geothermal exploration wells was 26454.54 mg/L, the PH value was 7.28, the total hardness was 5900.67 mg/L, the main cation Na+ content was 7614.15 mg/L, the K^+ content was 194.54 mg/L, the Ca^{2+} content was 1849.03 mg/L, and the Mg^{2+} content was 311.58 mg/L. The main anion Cl^-content was 15060.28 mg/L, $SO_4{}^{2-}$ content was 1132.71 mg/L, $HCO_3{}^-$ content was 169.09 mg/L, and the hydrochemical type was Cl-Na.

In the chemical components of groundwater, there is a symbiotic relationship between the contents of many chemical components; that is, the ratio between the contents of some chemical elements tends to be fixed. This fixed relationship can analyze and judge the causes and environment of groundwater, which is called the element proportional coefficient method. The causes and formation environment of underground hot water can be analyzed by comparing the proportion coefficient of elements in geothermal exploration holes with the average proportion coefficient of elements in seawater (Hou 1987, Feng 2016, Ma 2018). Exploration hole: $\gamma Na/\gamma C1 \approx 0.78$, $Cl/Br \approx 319.1$, $Br/I \approx 47.9$; seawater: $\gamma Na/\gamma C1 \approx 0.85$, $Cl/Br \approx 300$, $Br/I \approx 1300$. In the process of geological history, Na^+ in marine sedimentary water generates cation exchange with the exchangeable calcium ions in the formation, so the Na^+ content decreases, and $\gamma Na/\gamma C1$ is less than 0.85. The value of Cl/Br is about 1.34 times that of seawater, which is judged as the dissolved water in the rock salt formation; Br/I content is only 1.15% of seawater, which is the characteristic of marine sedimentary water; Ca/Sr=27.0 can also determine that geothermal water is sedimentary water-related to seawater. The calculation of the proportion of the above elements can reflect that the underground hot water of the Ordovician Majiagou Formation in the area has the characteristics of dissolved and filtered water of marine salt-bearing sedimentary rocks.

The formation process and environment of underground hot water in the area are complex. A certain amount of water (including a part of seawater) will always be preserved during the diagenesis of the sediments in the sedimentary basin. Then, when the sediments are exposed to the surface during the geological period, the water preserved in the sediments begins to infiltrate and expel for atmospheric precipitation or surface water, and water alternation occurs. This interaction in the geological period and geological environment forms the characteristics of seawater, sedimentary water, and dissolved and filtered water.

Trace elements such as bromine, fluorine, strontium, iron, lithium, and metaborate in karst geothermal water in the working area are enriched to varying degrees. Compared with the Quality Standard of Medical Thermo-mineral Water, the contents of bromine, fluorine, strontium, and iron in geothermal water reached the designated mineral water concentration. The enrichment of trace elements in geothermal water is the result of the combined action of surrounding rock composition, poor underground hot water circulation alternating conditions, and favorable geochemical environmental conditions. The trace elements in the strata are the basis of the formation of the material composition in the groundwater. Under a favorable geochemical environment, the dissolution and continuous circulation of groundwater are the main reasons for the enrichment of trace elements. Therefore, the enrichment of various trace elements in the underground hot water in this area is mainly due to the continuous dissolution of the material composition in the surrounding rock during the slow movement of geothermal water, which makes the accumulation of various trace elements in hot water and gradually forms its high concentration.

4.2 *Isotope geochemistry*

4.2.1 *Stable isotope*

In this work, the hydrogen and oxygen isotopes of geothermal water from exploration wells were tested, with a $\delta^{18}O$ value of -8.86 and a δD value of -67.49. The hydrogen and oxygen isotope test data were put on the $\delta D - \delta^{18}O$ diagram (Figure 4). It can be seen that the geothermal water points fall near the global atmospheric precipitation line and the atmospheric precipitation line in China, indicating that the geothermal water originates from atmospheric precipitation. This karst geothermal water point falls below the precipitation line, and the obvious oxygen isotope drifts to the right, mainly due to the exchange of isotopes between water and rock. The retention time of hot water in the underground is long, and the interaction with surrounding rock is sufficient. The more ^{18}O obtained from the rock, the greater the oxygen drift (Zhang 2014, Liu 2009, Chai 2010).

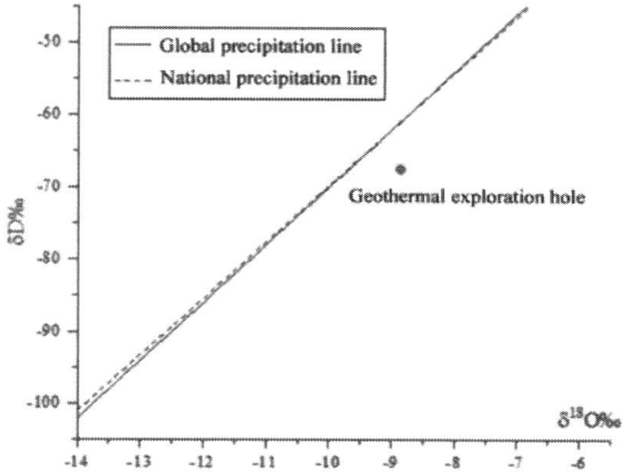

Figure 4. Relationship diagram of karst hot water δD-δ18O in the working area.

Water-rock reaction is one of the main controlling factors for hydrogen and oxygen isotope exchange of geothermal fluid. When the water-rock reaction occurs between geothermal water and rock, isotope exchange is limited to ^{18}O or ^{2}H. Because the content of $\delta^{18}O$ in hot water is much less than that of oxygen-containing minerals (for example, the δ18O of carbonate minerals is 29‰ VSMOW, and the $\delta^{18}O$ of silicate minerals is 8‰–29‰ VSMOW), the results of isotope exchange often increase the $\delta^{18}O$ in water, that is, the $\delta^{18}O$ oxygen drift occurs. However, the content of ^{2}H in rock-forming minerals is very low, within only a small amount of hydrogen-containing rock-forming minerals such as biotite and hornblende, so the isotope exchange of ^{2}H is not obvious in the evaluation area. When $\delta^{18}O$ is enriched due to a water-rock exchange reaction, the ^{2}H value is unchanged.

The isotopic composition of water and rock is always in an unbalanced state. In the geothermal system, higher temperatures always drive isotopic exchange between water and rock to reach isotopic equilibrium gradually. The result of isotope exchange makes the oxygen isotope in geothermal water relatively enriched, while the rock in contact with geothermal water becomes relatively scarce. The apparent age of geothermal water measured by ^{14}C is greater than 43,500 BP.

4.2.2 *Radioisotope*

The content of radioactive elements in the geothermal fluid of the geothermal exploration hole is higher, including total α 325.4 mBq·L^{-1}, total β 6662.5 mBq·L^{-1}, ^{226}Ra207.3 mBq·L^{-1}, ^{222}Rn1741.0 mBq·L^{-1}. The formation of hot water in the Ordovician Majiagou Formation is mainly due to the flow and accumulation of underground hot water in a sedimentary rock environment, and good contact conditions with rocks containing radioactive substances, thus forming high radioactive water.

5 CONCLUSIONS

This is our first study to examine the characteristics of geothermal resources in Yutai County depression. Our study adopts geothermal well drilling and water quality analysis methods to study the source of heat and chemical composition of geothermal water at the depth of the Ordovician roof, which is over 2000 meters deep. Our main conclusions are as follows:

(1) The Ordovician Majiagou Formation limestone is the main thermal reservoir in Yutai Sag, which belongs to the layered fractured thermal reservoir. The work area belongs to the North China plate (Level I), West Shandong uplift area (Level II), West Shandongnan buried uplift area (Level III), Heze-Yanzhou buried fault uplift (Level IV), and Yutai buried depression (Level V).
(2) The depth of the constant temperature zone is 20 m, the temperature of the constant temperature layer is 16°C, the geothermal gradient of Cretaceous is about 1.26°C/100 m (the geothermal gradient of the upper basalt section is about 0.36°C/100 m, the geothermal gradient of lower mudstone and sandstone interbed section is about 1.38°C/100 m), the geothermal gradient is about 2.94°C/100 m, and the geothermal gradient of Ordovician is about 3.98°C/100 m.
(3) The heat produced by the decay of radioactive elements accounts for 31.8%, and the rest of the heat comes from the deep crust. The neotectonic movement causes the heat source in the deep crust to transfer continuously upward.
(4) The limestone roof depth of the Majiagou Formation is about 2,089 m, and the average water temperature at the outlet of the geothermal well is 68.5°C. The hydrochemical type is Cl-Na type, and the salinity is 26454.54 mg/L. The contents of fluorine, bromine, strontium, and iron in geothermal water reach the standard of named mineral water concentration. The contents of lithium and boron reach the mineral water concentration and have high physiotherapy value.
(5) Geothermal water in the region is a hot-water geothermal resource in low-temperature geothermal resources, which can be used in heating, bathing, physiotherapy and health care, domestic hot water, and other fields. It cannot be directly used for fishery breeding and agricultural irrigation. This could serve as a helpful reference for future research.

In terms of future work, the recharge experiment should be carried out to further ascertain the geothermal recharge runoff.

REFERENCES

Bi S. K., *Research on the utilization of deep geothermal resources in Tangkou Coal Mine* [D], China University of Mining and Technology, 2019.

Cao H. S., Lu G. M., Shi J., Remote sensing geothermal anomalies and geological characteristics of geothermal fields in urban areas of Jining [J], *Shandong Land Resources*, 2008 (04).

Chai X. P., *Study on the genesis of helium-rich natural gas associated with geothermal water in Xi'an*, [D], Chang'an University, 2010.

Chen H. N., Li Z. Q., Dong H. J., Distribution and development prospects of geothermal resources in urban areas of Jining City [J], *Groundwater*, 2008 (03).

Feng M. Y., Song H. Z., Yang Q., et al., Gas composition and trace element content characteristics of geothermal water in part of Jiangsu Province and their indicative significance [J], *Hydrogeological Engineering Geology*, 2016 (01).

Hou D. Y., Underground hot water gas composition and its geochemical significance [J], *Hydrogeological Engineering Geology*, 1987 (01).

Li A. J., Zhang F., Wang P., Geothermal resources analysis in southern Jining [J], *Shandong Land Resources*, 2015 (06).

Liu J. C., Li R. X., Wei G. F., Causes and Sources of Geothermal Water-Soluble Helium in Weihe Basin [J], *Geological Science and Technology Information*, 2009, 28 (03).

Ma R., *Study on the genesis and water-rock interaction of medium-low temperature hot water of carbonate thermal storage type-taking Taiyuan*, Shanxi Province as an example [D], China University of Geosciences, 2018.

Zhang X., Liu J. C., Li R. X., et al., Distribution of geothermal water dissolved gas resources in Weihe Basin [J], *Geological Prospecting Series*, 2014 (03).

Energy Revolution and Chemical Research – Kok-Keong Chong and Zhongliang Liu (Eds)
© 2023 The Authors, ISBN: 978-1-032-36554-1

Synthesis and electrochemical performances of $LiNi_{0.5}Mn_{1.5}O_4$ materials prepared by a coprecipitation-hydrothermal method

Junhua Fan
College of Environmental and Chemical Engineering, Heilongjiang University of Science and Technology, Harbin, P.R. China

Yang Zhou
Energy & Environmental Research Institute of Heilongjiang Province, Harbin, P.R. China

Zhiyuan Xu, Jiazhi Yang, Hao Wang, Guang Liu & Guojiang Zhou
College of Environmental and Chemical Engineering, Heilongjiang University of Science and Technology, Harbin, P.R. China

ABSTRACT: Spinel $LiNi_{0.5}Mn_{1.5}O_4$ (LNMO) is considered to be a potential cathode material for lithium-ion batteries due to its low manufacturing cost, environmentally friendly, and three-dimensional lithium-ion channel. However, the presence of large amounts of Mn^{3+}leads to a decrease in electrical performance. In this paper, LNMO was prepared by coprecipitation-hydrothermal and co-precipitation methods. The characterization of materials was tested by X-ray diffraction, scanning electron microscopy, X-ray photoelectron spectroscopy, and electrochemical measurements. The content of Mn^{3+}, $Li_xNi_{1-x}O$ impurities and the particle size were reduced, and the crystallinity of LNMO was enhanced, which can improve the charge/discharge performance and rate capability of the materials. It was the superiority of LNMO synthesized by a coprecipitation-hydrothermal method. The discharge-specific capacity reached 127.98 $mAh{\cdot}g^{-1}$ at 1 C. Moreover, it increased to 114.43 $mAh{\cdot}g^{-1}$ and 85.48 $mAh{\cdot}g^{-1}$ at 5 C and 10 C, respectively. After 200 cycles, the capacity retention remained 84.06 % at 1 C.

1 INTRODUCTION

With the emergence of electric vehicles and hybrid vehicles, people greatly enhance the demand for energy storage materials (Hai 2013). Lithium-ion batteries (LIBs) have attracted much attention because of their light weight, large capacity, no memory effect, long cycle life, high-rate capability, and good safety (Chen 2019; Yi 2017). The cathode material is one of the important components of LIBs. Therefore, it has an important influence on electrochemical performance (Zong 2020). Conventional cathode materials such as $LiFePO_4$, $LiMn_2O_4$, and $LiCoO_2$ are all limited by low energy density (Li 2019, 2017). Therefore, the development of cathode materials with high energy density has attracted the attention of many researchers.

Spinel $LiNi_{0.5}Mn_{1.5}O_4$ (LNMO) is expected to be the most promising LIB cathode material due to its high operation voltage (~4.75 V), and high energy density (650 Wh/kg), and abundant resources (Wang 2017; Zong 2020). However, a large amount of Mn^{3+}, which exists in LNMO, is prone to disproportionation reactions to generate soluble Mn^{2+} and is eroded by the electrolyte. These lead to structural damage and capacity fading during cycling, especially at high rates (Dauth 2012; Liang 2021, 2018; Yin 2019). To avoid the disproportion reaction affecting the properties of the material, different synthesis methods can be used to control the content of Mn^{3+} in LNMO (Yin 2019). There are many synthetic methods such as the solid-state method, co-precipitation method, sol-gel method, hydrothermal method, etc, which are commonly used (Li 2017). The solid-state method is simple, but not uniform, so there will be a lot of impurities and non-uniform particles will appear in the sample (Siqin 2021). The co-precipitation method realizes the reaction at a molecular level, but the samples have low crystallinity and contain a large amount of Mn^{3+}, $Li_xNi_{1-x}O$ impurities (Feng 2013; Pan 2019; Xue 2014). The high-temperature and high-pressure hydrothermal method can guarantee the precise stoichiometric ratio of LNMO and control the size, and shape of the particle. This method will improve the crystallinity of the sample, reduce the generation of Mn^{3+} and $Li_xNi_{1-x}O$ impurities, and increase the rate capability and cyclic performance (Cheng 2016; Miao 2021; Xue 2014; Zhao 2017).

In this paper, the LNMOs were synthesized by the coprecipitation-hydrothermal method and co-precipitation method, and the samples were characterized and tested for their electrical properties. The purpose was to compare the effects of different preparation methods on the Mn^{3+} content, $Li_xNi_{1-x}O$ impurity amount, crystallinity, particle size, and uniformity on the electrical properties in the two samples.

2 EXPERIMENTAL

2.1 *Material Preparation*

LNMO was synthesized by the coprecipitation-hydrothermal method and segmented sintering reaction. 0.005 moL $NiSO_4$ and 0.015 moL $MnSO_4$ were dissolved in 30 mL deionized water, and then 0.024 moL Na_2CO_3 was dissolved in 30 mL deionized water with stirring for 0.5 h. The Na_2CO_3 solution was slowly added to the mixed metal salt solution, and 10 mL of ethanol was added dropwise. After agitating for 6 h, they were poured into a 100 mL Teflon-lined stainless-steel autoclave and heated at 180° for 10 h. The Ni-Mn carbonate precursor powder was obtained, when the samples were naturally cooled to room temperature, washed with deionized water and ethanol to neutral in turn, and dried in an oven at 120° for 8 h. The precursor was calcined at 500° for 5 h, 0.003 moL Li_2CO_3 was added for grinding, and then calcined at 800° for 8 h. Finally, the H-LNMO material was prepared. For comparison, keep them at room temperature for 10 h without the hydrothermal method, the G-LNMO was synthesized by coprecipitation method and segmented sintering reaction.

2.2 *Sample Characterization*

The crystal structure of the samples was analyzed by X-ray diffractometer (XRD, Brunker D8 Advance, Germany) with a Cu-Kα radiation source and a 2θ scan range of 10° to 80° with a step size of 0.02°. The microscopic morphology of the samples was studied using a scanning electron microscope (SEM, Phenom Pro X, The Netherlands) with an accelerating voltage of 10 kV. The surface chemistry of the samples was studied by an X-ray photoelectron spectrometer (XPS, AXIS ULTRA DLD, UK) with an X-ray source of monochromatic Al target, a voltage of 15 kV and a power 150 W, with a C 1s peak for binding energy correction (284.6 eV).

2.3 *Electrochemical Tests*

Electrochemical performance measurements of samples through installed CR2032 coin cell batteries. The prepared active material (LNMO), polyvinylidene fluoride (PVDF), and super-P were mixed in the appropriate amount of N-methyl pyrrolidone (NMP) solvent at a mass ratio of 8:1:1. They were stirred into a paste and coated on a current collector. After being dried in a vacuum oven and cropped, the electrode sheets with 1.5~2 mg active material were prepared. The electrode sheets were used as the cathode. and the lithium foil as a counter electrode. The lithium-ion battery is assembled in a vacuum glove box (O_2, H_2O less than 1 ppm) under high purity argon gas. The electrolyte was composed of 1 moL/L $LiPF_6$ solute and 1:1:1 solvent of ethylene carbonate (EC), methyl ethyl carbonate (EMC), and dimethyl carbonate (DMC) by volume. The batteries were tested for charge/discharge, rate, and cycling using a Sunway CT-3008 electrochemical test system (5 V/5 mA). Cyclic voltammetry (CV) test was at voltages from 3.5 V to 5.1 V with a scan rate of 0.1 mV s^{-1}, and electrochemical impedance spectroscopy tests (EIS) were at open-circuit voltages, in the frequency range from 0.01 Hz to 100 kHz with an amplitude of 5 mA. They were all performed by a CHI660E electrochemical workstation from Shanghai Chenhua.

3 RESULTS AND DISCUSSION

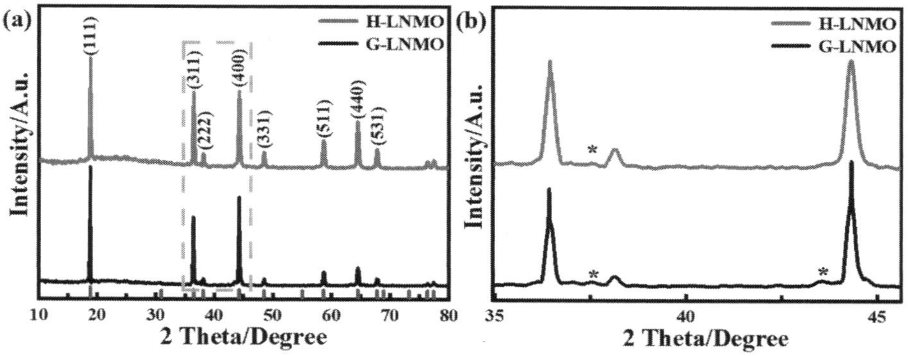

Figure 1. (a) XRD pattern, (b) local XRD pattern.

LNMO has two spatial structures, one is the $P4_332$ spatial structure with the ordered distribution of Mn^{4+} and Ni^{2+}, while the other is the Fd3m spatial structure with a disordered distribution of Mn^{4+} and Ni^{2+} (Liu 2019; Wang 2020). The Fd3m spatial structure possesses excellent rate properties (Hai 2013; Liu 2017). Figure 1(a) shows the X-ray diffractometer (XRD) of H-LNMO and G-LNMO. It can be seen that the H-LNMO and G-LNMO samples have no diffraction peaks at 15°. However, they all have strong intensity attribute to (111), (311), (400), (511), and (440) lattice planes, which is matched to the Fd3m spatial configuration of the LNMO (JCPDS: 80-2162) (Li 2020; Xue 2014). H-LNMO with stronger diffraction peaks has better crystallinity than G-LNMO. In Figure 1(a), the $I_{(311)}/I_{(400)}$ ratios of H-LNMO and G-LNMO are 0.93 and 0.79 respectively. The ratio of relatively larger $I_{(311)}/I_{(400)}$ implied that H-LNMO has good structural stability (Yi 2017). In addition, as shown in Figure 1b, H-LNMO has a smaller impurity peak at 37.5°, while G-LNMO has stronger impurity peaks at both 37.5° and 43.7°. The comparison shows that H-LNMO contains less $Li_xNi_{1-x}O$ impurities than G-LNMO (Liu 2017).

The microscopic morphology of the H-LNMO and G-LNMO were characterized by Scanning electron microscopy (SEM) shown in Figure 2. The two samples are all sphere morphology. In Figure 2(a), the particle size with 1~2 μm of H-LNMO is uniform and cohesionless which is conductive to sufficient contact between the electrolyte and the active materials. Therefore, the transport capacity of lithium ions will be improved, resulting in a better rate of performance (Xu 2014; Yi 2017). In Figure 2(b), the particle size distribution is uneven, some of the particles are 2.4 μm, and others are less than 1 μm. By comparison, the coprecipitation-hydrothermal method is easier to control the size of the sample than the coprecipitation method (Qin 2019). The size of LNMO is large to reduce the contact area between the active material and the electrolyte. It will lengthen the diffusion pathway of lithium ions, thereby affecting the rate capability of the sample. However, the size is small to increase the contact area, so the Mn^{3+} in the material is easily dissolved or reacted with the electrolyte, thus reducing the cycling performance (Xue 2014; Zeng 2021). According to the relationship between particle radius (r) and solvent dielectric constant (ε), $1/r=A+B/\varepsilon$, (A and B are constants), the radius is proportional to the dielectric constant. Therefore,

Figure 2. (a) SEM images of H-LNMO, (b) SEM images of G-LNMO.

it is further proved that H-LNMO with property radius possesses better electrochemical properties (Xue 2014).

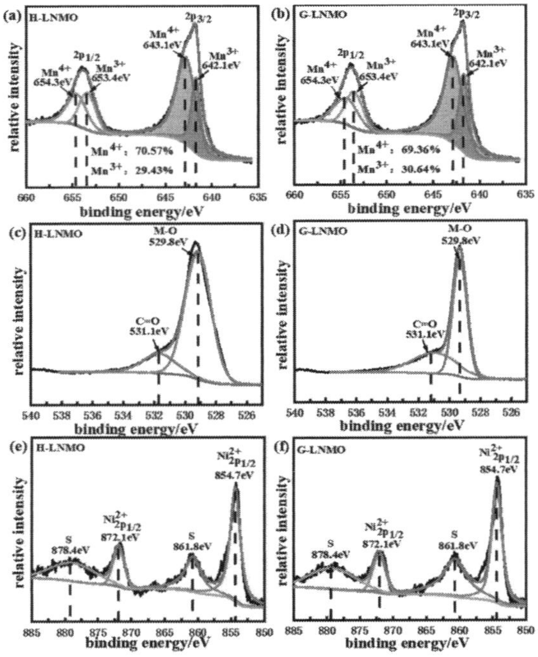

Figure 3. (a-b) Mn 2p, (c-d) O 1s and (e-f) Ni 2p.

The valence of elements in compounds was studied by an X-ray photoelectron spectrometer (XPS). The manganese XPS peaks of the two samples are consistent. The Mn^{4+} $2p_{1/2}$ and $2p_{3/2}$ are located at 654.3 eV and 643.1 eV, respectively. Moreover, The Mn^{3+} $2p_{1/2}$ and $2p_{3/2}$ are located at 653.4 eV and 642.1 eV. Therefore, the manganese element in LNMO has both trivalent and tetravalent (Li 2019). The relative contents of Mn^{3+} and Mn^{4+} of the two samples were calculated by the area ratio of Mn $2p_{3/2}$. It can be seen that the content of Mn^{3+} of H-LNMO is 40.83 % and that of Mn^{4+} is 59.17 %. However, the contents of Mn^{3+} and $Mn^{4+}$of G-LNMO are 46.09 % and 53.91 %, respectively. The content of Mn^{3+} in the H-LNMO is little than in the G-LNMO. The

Mn^{3+} on the surface tends to undergo a disproportionation reaction ($2Mn^{3+} = Mn^{2+} + Mn^{4+}$) to generate easily soluble Mn^{2+}, which damages the structure (Liang 2020; Qin 2019). Therefore, H-LNMO has good cycling stability. There are two main peaks of O 1s in Figure 3(c-d). One broad peak of 531.1 eV is oxygen in carbonate species (Li_2CO_3) and the other narrow peak of 529.8 eV is oxygen in M-O (M: metal ion) (Dauth 2012; Li 2019). The observation of carbonate species revealed the existence of Li_2CO_3, and a small amount of Li_2CO_3 could cover the LNMO to suppress the dissolution of Mn^{3+}. It enhanced the cycling stability of LNMO (Yin 2019). In addition, the C=O peak intensity of H-LNMO is significantly lower than that of G-LNMO, which indicates that H-LNMO has an advantageous performance (Wei 2020).

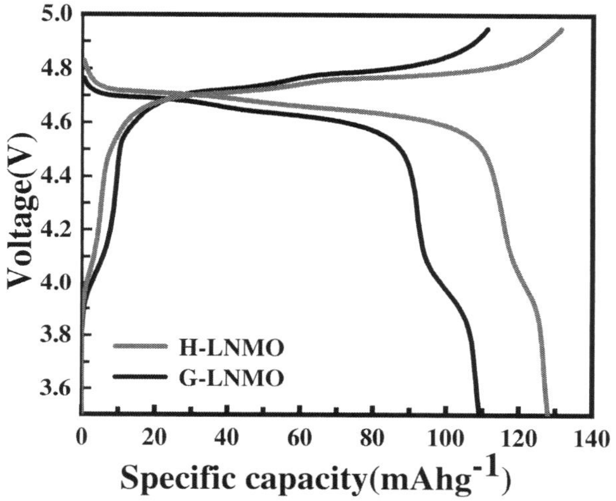

Figure 4. Charge/discharge curves of H-LNMO and G-LNMO.

The charge/discharge curve of the LNMO sample at 1 C shows in Figure 4. The two samples have a dominant discharge plateau at ~4.7 V, which is generated by the redox couples of Ni^{2+}/Ni^{3+} and Ni^{3+}/Ni^{4+} (Xu 2020). The discharge-specific capacities of H-LNMO and G-LNMO are 127.98 $mAh{\cdot}g^{-1}$ and 109.1 $mAh{\cdot}g^{-1}$ respectively. H-LNMO has a longer discharge plateau (~4.7 V) and higher discharge-specific capacity compared to G-LNMO. This is due to the high crystallinity and fewer $Li_xNi_{1-x}O$ impurities of H-LNMO. A smaller discharge plateau at ~4.0 V is generated by the redox pair of Mn^{3+}/Mn^{4+} (Gao 2020). The ~4.0 V plateaus show that the samples all contain Mn^{3+} and that there is a spatial structure of Fd3m, which is consistent with the XRD results (Li 2019; Piao 2018; Zhu 2014). In addition, the length of the discharge plateau at ~4.0 V reveals an estimate of the amount of Mn^{3+} in the sample. H-LNMO has a shorter discharge plateau compared to G-LNMO, and it contains less Mn^{3+}.

The LNMO samples were tested for 200 cycles at 1 C. The results (Figure 5) show that the cyclic capacity retention rates of the H-LNMO and G-LNMO samples were 85.04% and 81.37% respectively. The high cyclic capacity retention rate of the H-LNMO sample is attributed to its appropriate sphere morphology and small amounts of $Li_xNi_{1-x}O$ impurities corresponding to the $I_{(311)}/I_{(400)}$ ratio in XRD, which are further proved that H-LNMO possesses excellent structural stability.

Figure 6 shows the rate performance of the samples. It can be seen that H-LNMO exhibits better rate performance and higher discharge specific capacity than G-LNMO at rates of 1 C, 2 C, 5 C, and 10 C. H-LNMO sample maintains high discharge specific capacities of 114.43 and 85.48 $mAh{\cdot}g^{-1}$ at 5 C and 10 C, respectively. The main reason is that it has a suitable particle size which shortens the lithium ions diffusion pathway than the G-LNMO sample during operation.

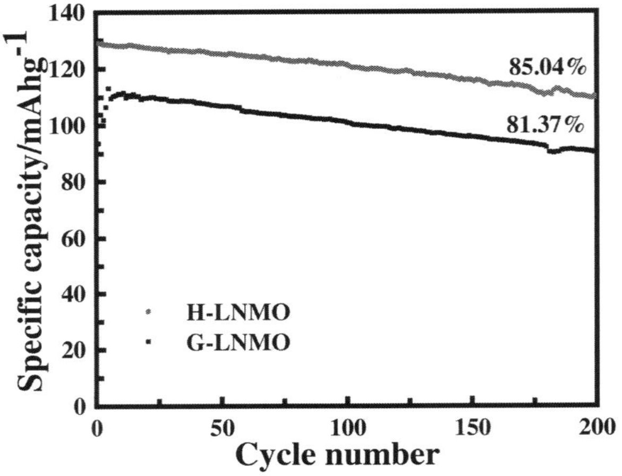

Figure 5. The cycle performance curve of H-LNMO and G-LNMO.

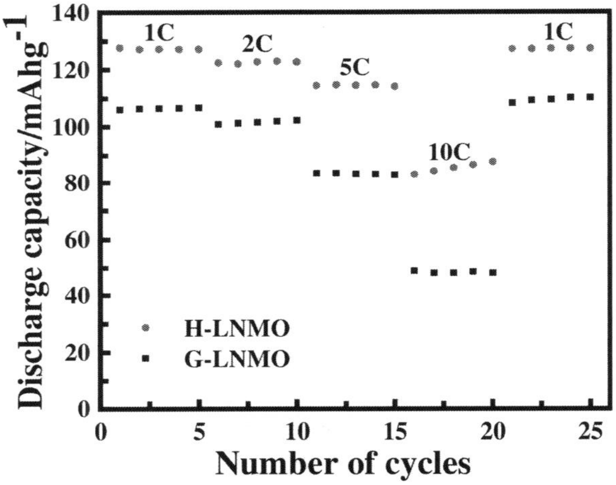

Figure 6. rate curves of H-LNMO and G-LNMO.

Figure 7 shows the cyclic voltammetry curves of the two samples. It compares the redox ability and lithium-ion insertion/extraction capacity. H-LNMO has a dominant peak at ~4.7V, which originated from the redox couples (Ni^{2+}/Ni^{3+} and Ni^{3+}/Ni^{4+}). Especially, the appearance of the peak splitting phenomenon, which is a typical feature of Fd3m spatial structure (Zong 2020). It can be seen from Figure 7 that the H-LNMO sample has narrower clearance and higher strength than the G-LNMO sample at ~4.7 V, indicating that the H-LNMO sample has better kinetics of lithium-ion insertion/extraction and good rate performance. It is consistent with the rate performance test results (Figure 6) (Zhao 2017). A small peak at ~4.0V (inset of Figure 7) can be attributed to the Mn^{3+}/Mn^{4+} redox couple. The content of Mn^{3+} of the samples can be seen by the area of the peak at ~4.0V. The peak area at ~4.0V of the H-LNMO sample is smaller than that of the G-LNMO sample, indicating that the content of Mn^{3+} in the H-LNMO sample is less, which corresponds to XPS characterization.

Electrochemical impedance spectroscopy tests (EIS) are used to further investigate the electrochemical properties of samples. Figure 8 shows that the EIS curve consists of a semicircle in

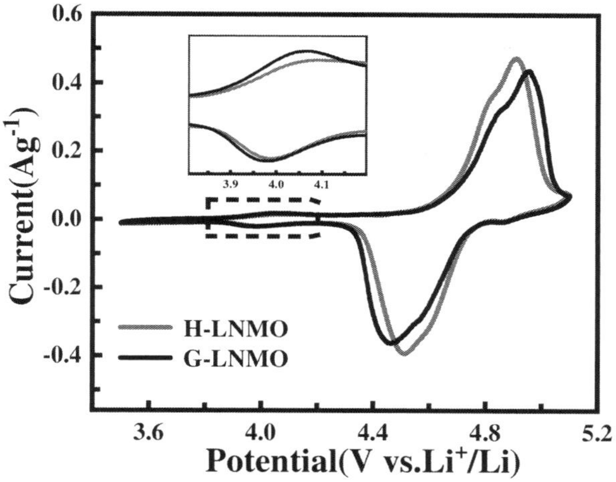

Figure 7. CV curves of H-LNMO and G-LNMO.

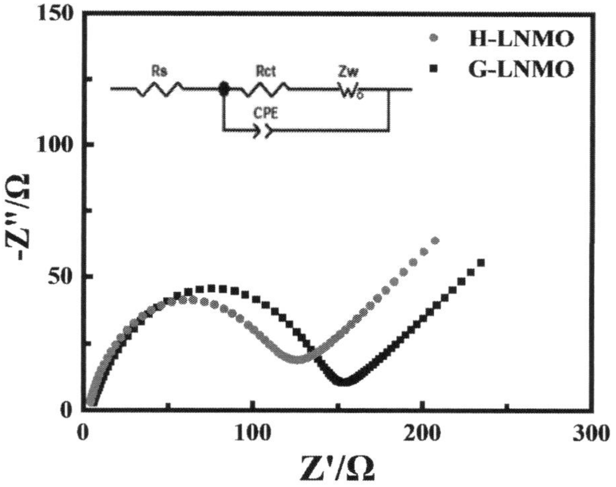

Figure 8. EIS curves of H-LNMO and G-LNMO.

the high-frequency region and a diagonal line in the low-frequency region. The semicircle in the high-frequency region represents the charge transfer resistance (R_{ct}) of the sample which is used to study the interfacial layer resistance between the electrode material and the electrolyte. If the radius of the semicircle is smaller, the impedance of the battery is lower. The diagonal line in the low-frequency region stands for the Warburg impedance (Zw). It is used to evaluate the rate of diffusion of lithium ions in the sample (Mo 2016; Wu 2017). Circuit fitting shows that H-LNMO (127.75 Ω) has a lower interfacial layer resistance than G-LNMO (150.63 Ω). The H-LNMO sample has an excellent interfacial impedance because it has a uniform, small particle size, and loose morphology. Therefore, it can contact the electrolyte fully. In addition, the low content of $Li_xNi_{1-x}O$ impurity in the sample can reduce the interface impedance and make the charge transfer easier (Zhou 2020). During the charge/discharge process, H-LNMO has good rate performance and high discharge-specific capacity due to its small internal resistance. It corresponds to the charge/discharge and the rate curve (Figure 5 and Figure 6).

4 CONCLUSION

In this paper, the coprecipitation-hydrothermal method is applied to study LNMO. The main conclusions can be summarized as follows: (1) H-LNMO exhibited excellent rate performance due to the obvious loose morphology with a uniform sphere and a particle size ranging from 1 $\sim$ 2 μm. (2) The content of Mn^{3+} and $Li_xNi_{1-x}O$ were reduced and crystallinity was increased. They improve the cycling performance and discharge-specific capacity of H-LNMO. (3) The discharge-specific capacity is 127.98 mAh·g^{-1} at 1 C. Moreover, it increased to 114.43 mAh·g^{-1} and 85.48 mAh·g^{-1} at 5 C and 10 C, respectively. After 200 cycles, the capacity retention remains 84.06 % at 1 C. In future work, more simple and low-cost synthetic methods should be considered to improve the electrochemical performance of LNMO materials.

REFERENCES

Chang Z, Dai D, Tang H, et al. (2010) Effects of precursor treatment with reductant or oxidant on the structure and electrochemical properties of LiNi0.5Mn1.5O4 [J]. *Electrochimica Acta*, 55(19): 5506–10.

Chen Z, Wang X, Tian X, et al. (2019) Synthesis of ordered $LiNi_{0.5}Mn_{1.5}O_4$ nanoplates with exposed {100} and {110} crystal planes and its electrochemical performance for lithium ions batteries [J]. *Solid State Ionics*, 333: 50–6.

Cheng J, Li X, Wang Z, et al (2016) Hydrothermal synthesis of $LiNi_{0.5}Mn_{1.5}O_4$sphere and its performance as high-voltage cathode material for lithium-ion batteries [J]. *Ceramics International*, 42(2): 3715–9.

Dauth A, Love J A. (2012). Synthesis and reactivity of 2-azametallacyclobutanes [J]. *Dalton Trans*, 41(26): 7782–91.

Feng J, Huang Z, Guo C, et al. (2013). An organic coprecipitation route to synthesize high voltage $LiNi_{0.5}Mn_{1.5}O_4$ [J]. *ACS Appl Mater Interfaces*, 5(20): 10227–32.

Gao C, Liu H, Bi S, et al. (2020). Insight into the effect of graphene coating on cycling stability of $LiNi_{0.5}Mn_{1.5}O_4$: Integration of structure-stability and surface-stability [J]. *Journal of Materiomics*, 6(4): 712–22.

Hai B, Shukla A K, Duncan H, et al. (2013). The effect of particle surface facets on the kinetic properties of $LiMn_{1.5}Ni_{0.5}O_4$ cathode materials [J]. *J Mater Chem A*, 1(3): 759–69.

Li L, Sui J, Chen J, et al. (2019). $LiNi_{0.5}Mn_{1.5}O_4$ microrod with ultrahigh Mn3+ content: A high performance cathode material for lithium-ion battery [J]. *Electrochimica Acta*, 305: 433–42.

Li S, Liang W, Xie J, et al. (2020). Synthesis of Hollow Microspheres $LiNi_{0.5}Mn_{1.5}O_4$ Coated with Al_2O_3 and Characterization of the Electrochemical Capabilities [J]. *Journal of Electrochemical Energy Conversion and Storage*, 17(3).

Li S, Yang Y, Xie M, et al. (2017). Synthesis and electrochemical performances of high-voltage $LiNi_{0.5}Mn_{1.5}O_4$ cathode materials prepared by hydroxide co-precipitation method [J]. *Rare Metals*, 36(4): 277–83.

Liang C, Li L, Fang H. (2018). New molten salt synthesis of $LiNi_{0.5}Mn_{1.5}O_4$ cathode material [J]. *Materials Research Express*, 5(7).

Liang G, Peterson V K, See K W, et al. (2020). Developing high-voltage spinel $LiNi_{0.5}Mn_{1.5}O_4$ cathodes for high-energy-density lithium-ion batteries: current achievements and future prospects [J]. *Journal of Materials Chemistry A*, 8(31): 15373–98.

Liang W, Wang P, Ding H, et al. (2021). Granularity control enables high stability and elevated-temperature properties of micron-sized single-crystal $LiNi_{0.5}Mn_{1.5}O_4$ cathodes at high voltage [J]. *Journal of Materiomics*, 7(5): 1049–60.

Liu G, Zhang J, Zhang X, et al. (2017). Study on oxygen deficiency in spinel $LiNi_{0.5}Mn_{1.5}O_4$ and its Fe and Cr-doped compounds [J]. *Journal of Alloys and Compounds*, 725: 580–6.

Liu H, Liang G, Gao C, et al. (2019). Insight into the improved cycling stability of sphere-nanorod-like micro-nanostructured high voltage spinel cathode for lithium-ion batteries [J]. *Nano Energy*, 66.

Liu Y, Lu Z, Deng C, et al. (2017). A novel LiCoPO4-coated core–shell structure for spinel $LiNi_{0.5}Mn_{1.5}O_4$ as a high-performance cathode material for lithium-ion batteries [J]. *Journal of Materials Chemistry A*, 5(3): 996–1004.

Miao X, Qin X, Huang S, et al. (2021). Hollow spherical $LiNi_{0.5}Mn_{1.5}O_4$ synthesized by a glucose-assisted hydrothermal method [J]. *Materials Letters*, 289.

Mo M, Chen H, Hong X, et al. (2016). Hydrothermal synthesis of reduced graphene oxide-$LiNi_{0.5}Mn_{1.5}O_4$ composites as 5 V cathode materials for Li-ion batteries [J]. *Journal of Materials Science*, 52(5): 2858–67.

Pan J-J, Chen B, Xie Y, et al. (2019). V_2O_5 modified $LiNi_{0.5}Mn_{1.5}O_4$ as cathode material for high-performance Li-ion battery [J]. *Materials Letters*, 253: 136–9.
Piao J-Y, Sun Y-G, Duan S-Y, et al. (2018). Stabilizing Cathode Materials of Lithium-Ion Batteries by Controlling Interstitial Sites on the Surface [J]. *Chem*, 4(7): 1685–95.
Qin X, Gong J, Guo J, et al. (2019). Synthesis and performance of $LiNi_{0.5}Mn_{1.5}O_4$ cathode materials with different particle morphologies and sizes for lithium-ion battery [J]. *Journal of Alloys and Compounds*, 786: 240–9.
Siqin G, Qilu, Tian W. (2021). Scalable synthesis of high-voltage $LiNi_{0.5}Mn_{1.5}O_4$ with high electrochemical performances by a modified solid-state method for lithium-ion batteries [J]. *Inorganic Chemistry Communications*, 134.
Wang J-F, Chen D, Wu W, et al. (2017). Effects of Na^+ doping on crystalline structure and electrochemical performances of $LiNi_{0.5}Mn_{1.5}O_4$ cathode material [J]. *Transactions of Nonferrous Metals Society of China*, 27(10): 2239–48.
Wang W-N, Meng D, Qian G, et al. (2020). Controlling Particle Size and Phase Purity of "Single-Crystal" $LiNi_{0.5}Mn_{1.5}O_4$ in Molten-Salt-Assisted Synthesis [J]. *The Journal of Physical Chemistry C*, 124(51): 27937–45.
Wei L, Tao J, Yang Y, et al. (2020). Surface sulfidization of spinel $LiNi_{0.5}Mn_{1.5}O_4$ cathode material for enhanced electrochemical performance in lithium-ion batteries [J]. *Chemical Engineering Journal*, 384.
Wu W, Qin X, Guo J, et al. (2017). Influence of cerium doping on structure and electrochemical properties of $LiNi_{0.5}Mn_{1.5}O_4$ cathode materials [J]. *Journal of Rare Earths*, 35(9): 887–95.
Xu D, Yang F, Liu Z, et al. (2020). Effects of Co doping sites on the electrochemical performance of $LiNi_{0.5}Mn_{1.5}O_4$ as a cathode material [J]. *Ionics*, 26(8): 3777–83.
Xu L, Lin X, Shang Y, et al. (2014). Modified KCl Molten Salt Method Synthesis of Spinel $LiNi_{0.5}Mn_{1.5}O_4$ with Loose Structure as Cathodes for Li-ion Batteries[J]. *International journal of electrochemical science*, 9(12):7253–7265.
Xue Y, Wang Z, Yu F, et al. (2014). Ethanol-assisted hydrothermal synthesis of $LiNi_{0.5}Mn_{1.5}O_4$ with excellent long-term cyclability at high rate for lithium-ion batteries [J]. *J Mater Chem A*, 2(12): 4185–91.
Yi T-F, Han X, Chen B, et al. (2017). Porous sphere-like $LiNi_{0.5}Mn_{1.5}O_4$-CeO_2 composite with high cycling stability as cathode material for lithium-ion battery [J]. *Journal of Alloys and Compounds*, 703: 103–13.
Yin C, Bao Z, Tan H, et al. (2019). Metal-organic framework-mediated synthesis of $LiNi_{0.5}Mn_{1.5}O_4$: Tuning the Mn^{3+} content and electrochemical performance by organic ligands [J]. *Chemical Engineering Journal*, 372: 408–19.
Zeng F, Zhang Y, Shao Z, et al. (2021). The influence of different calcination temperatures and times on the chemical performance of $LiNi_{0.5}Mn_{1.5}O_4$ cathode materials [J]. *Ionics*, 27(9): 3739–48.
Zhao E, Wei L, Guo Y, et al. (2017). Rapid hydrothermal and post-calcination synthesis of well-shaped $LiNi_{0.5}Mn_{1.5}O_4$ cathode materials for lithium-ion batteries [J]. *Journal of Alloys and Compounds*, 695: 3393–401.
Zhou M, Lang Y, Deng Z, et al. (2020). Effect of Presintering Atmosphere on Structure and Electrochemical Properties of LiNi0.5Mn1.5O4 Cathode Materials for Lithium-Ion Batteries [J]. *Chemistry Select*, 5(8): 2535–44.
Zhu X, Li X, Zhu Y, et al. (2014). $LiNi_{0.5}Mn_{1.5}O_4$ nanostructures with two-phase intergrowth as enhanced cathodes for lithium-ion batteries [J]. *Electrochimica Acta*, 121: 253–7.
Zong B, Deng Z, Yan S, et al. (2020). Effects of Si doping on structural and electrochemical performance of $LiNi_{0.5}Mn_{1.5}O_4$ cathode materials for lithium-ion batteries [J]. *Powder Technology*, 364: 725-37.

Energy Revolution and Chemical Research – Kok-Keong Chong and Zhongliang Liu (Eds)
 ISBN: 978-1-032-36554-1

Study on electrochemical performance of aluminum anode in seawater and intertidal environment of Shengli oilfield

Chao Yang, De-sheng Chen & Qing Han
Technology Inspection Center of Shengli Oilfield, SINOPEC, China

Peng Sun
Ocean Oil Production Plant of Shengli Oilfield, SINOPEC, China

Songxi Li
Engineering Technology Management Center of Shengli Oilfield, SINOPEC, China

Feng Wang*
China Special Equipment Inspection and Research Institute, China

ABSTRACT: In this paper, the electrochemical properties of aluminum anodes in intertidal and seawater environments were studied based on a self-designed soil dry-wet alternating experiment device. The results showed that the general corrosion rate and pitting rate of the aluminum anode in the intertidal environment was much lower than that in the seawater environment. With regard to Al anode in an intertidal environment, the working potential was more positive, the corrosion current density was lower, while the polarization resistance and charge transfer resistance were larger, indicating that the electrochemical activity of aluminum anode in the intertidal environment was lower than that in seawater environment.

1 INTRODUCTION

With the extensive use of offshore oil and gas, intertidal pipelines, as a key component connecting onshore and offshore, are subject to severe external corrosion in the intertidal environment. Cathodic protection technology is a key means of pipeline corrosion protection.

In marine environments, aluminum anodes have higher capacitance and lower protection rates compared to zinc anodes, providing longer-lasting cathodic protection. Currently, scholars in the relevant field have carried out research on the electrochemical performance of aluminum anodes in the intertidal environment (Lu 2022). However, there are no relevant reports on the electrochemical performance of sacrificial anodes in the intertidal environment (Cheng 2017). Therefore, the applicability of traditional sacrificial anodes in the intertidal environment of beach soil is debatable.

To this end, based on the self-designed intertidal experimental device, this paper simulates the intertidal environment in the Shengli oilfield and studies the electrochemical performance of the national standard I-type aluminum anode (Al anode) in seawater and intertidal environments through general corrosion rate, corrosion images, and electrochemical experiments.

*Corresponding Author: ht4s@163.com

 DOI 10.1201/9781003332657-54

2 EXPERIMENTAL SETTINGS

2.1 *Soil experiment device*

The experimental box has a size of 50 × 50 × 20 cm^3, wherein the height of the experimental soil is 15 cm, and the height of the liquid level in the wet condition of the corrosive environment is 2 cm. Four soil moisture content probes are set in the experimental soil, the height is 8 cm, and the distance from the wall of the box is 10 cm. The specimen is buried in the middle of the experimental soil. A water outlet hole with a diameter of 5 cm is opened directly below the specimen, the water outlet pipe and vacuum pump are connected below, and an intelligent control valve is installed to discharge the water in the experimental box through the vacuum pump.

Specific operation steps are as follows. The experimental intertidal time is set to 6:6 through the computer. In the wet condition, the liquid level gauge monitors the liquid level in the device. When the liquid level in the device is insufficient, the system controls the water inlet system to replenish the experimental solution. And in the dry condition, the soil moisture content is monitored through the soil moisture probe, the intelligent control valve and vacuum pump are controlled to extract the soil moisture, and the soil moisture content is controlled to be 24±1% in the dry condition.

2.2 *Specimen and corrosive environment*

According to GB/T 4948-2002 *Sacrificial Anode of Al-Zn-In series alloy*, I-type aluminum-based sacrificial anode (Al anode) is selected as the research object, in which the mass fraction of Zn is 3.2%, In is 0.031%, and the density was 2.84 g/cm^3.

According to the test results of the properties of the soil sampled on site, the ion content (mass fraction) of the intertidal solution in this paper is determined: total salt 0.503%, Mg^{2+} 0.03%, Ca^{2+} 0.02%, Cl^- 0.314%, $SO_4{}^{2-}$ 0.06%, and $HCO_3{}^-$ 0.05%. And the ion content (mass fraction) in seawater environment is total salt 2.971%, Mg^{2+} 0.52%, Ca^{2+} 0.116%, Cl^- 1.667%, $SO_4{}^{2-}$ 0.2%, and $HCO_3{}^-$ 0.1%.

Acetone is used to remove the protective oil film on the surface of the test piece, then deionized water is used to clean it, and finally, the length, width, thickness, and original weight of the samples are measured after dehydration and drying with absolute ethanol for 24 hours. The corrosion rate test is carried out in the intertidal environment and the seawater environment. The experimental temperature is set to 25°C. After the completion of the experiment, the Al anode is put into 68% concentrated nitric acid for 5–10 minutes and then washed with deionized water. The general corrosion rate is calculated after weighing.

Electrochemical experiments are carried out using Al anodes with different immersion times as the working electrode, wherein the reference electrode is a saturated calomel electrode, and the auxiliary electrode is a platinum electrode. The workstation used for the electrochemical test is PARSTAT 2273. After the whole system is stable (the open circuit potential of the test piece changes less than ±10 mV within 300 s), the polarization curve and EIS test are carried out, for which the scan rate of the polarization curve is set to 0.3 mV/s, the scanning potential range is ±250 mV (vs. open circuit potential), and the electrochemical impedance scanning frequency range is 10^5–10^{-2}Hz. The data processing adopts the "PowerSuite" software and "ZSimpWin" software that comes with the system.

3 RESULTS AND DISCUSSION

3.1 *Corrosion rate analysis*

Figure 1 shows the variation curve of the general corrosion rate and pitting rate of the national standard I-type Al anode with time at 25°C. It can be seen from Figure 1(1) that during the long-term immersion experiment, the general corrosion rates of Al anode in the intertidal environment and the seawater environment are basically unchanged, which are 0.074 mm/a and 0.224 mm/a, respectively. As a passive sacrificial anode, the Al anode is mostly used for cathodic protection of metals in seawater environment or crude oil storage tank sludge environment. The chloride ion concentration in seawater is high, which can effectively destroy the dense corrosion product film on the surface of Al anode. To ensure that the corrosion product film can be evenly peeled off, the continuous activation of Al anode should be ensured. In the intertidal environment, due to the relatively low content of active chloride ions in the soil, especially in the dry condition, the surface of the aluminum-based anode forms a hard corrosion product layer ($Al \rightarrow Al^{3+}+3e^-$; $O_2+2H_2O+4e^- \rightarrow 4OH^-$), which hinders the activation process of Al anode, so the general corrosion rate is low (Cheng 2017; Ma 2010; Zhang 2014).

Focusing on the pitting rate in Figure 1(2), the chloride ion concentration is higher in the seawater environment, which can effectively destroy the corrosion products on the surface of Al anode and promote the continuous activation of Al anode (0.787 mm/a). While the dense corrosion product film in the intertidal environment protects the Al anode well, the pitting rate is small (0.125 mm/a).

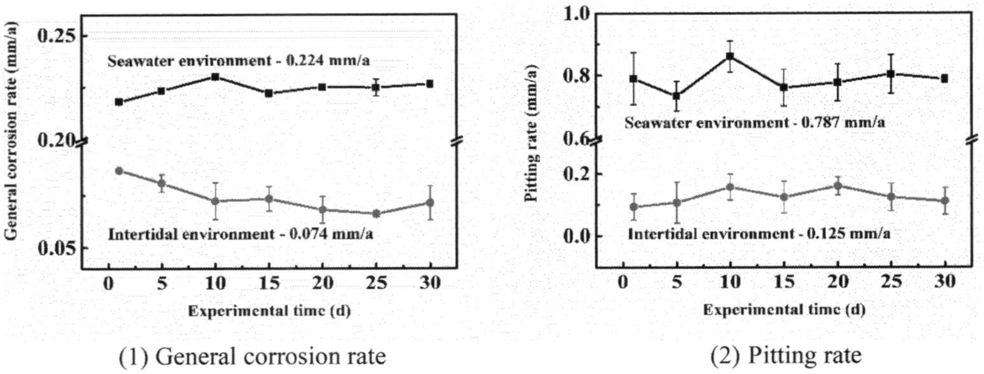

Figure 1. Variation curves of general corrosion rate and pitting rate of Al anode with time.

3.2 *Study on corrosion kinetics*

3.2.1 *Open circuit potential analysis*

Figure 2 shows the open circuit potential (E_{OCP}) of the I-type Al anode in different environments at 25°C. In the seawater environment, the E_{OCP} of Al anode is –1.149 V, which meets the requirements of GB/T 4948-2002 (the E_{OCP} is from –1.18 to –1.12 V), and in the intertidal environment, the E_{OCP} of Al anode is only –0.702 V. Therefore, it can be demonstrated that the Al anode cannot meet the requirements of on-site cathodic protection in the intertidal environment (Wen 2011).

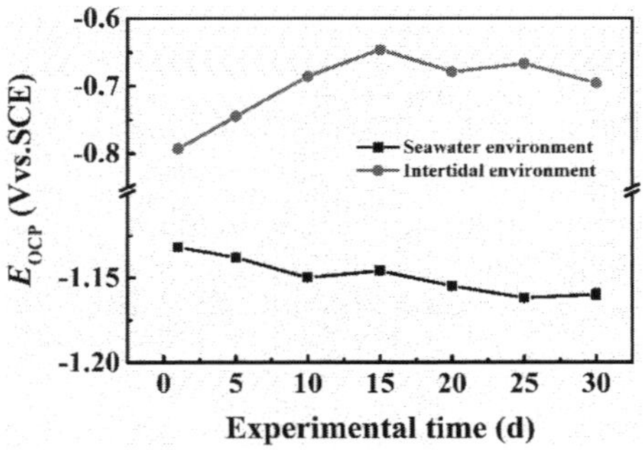

Figure 2. Open circuit potential (E_{OCP}) of Al anode in different environments.

3.2.2 *Polarization Curve Analysis*

Figure 3 shows the polarization curves of the I-type Al anodes in different environments at 25°C with time. It can be noted from the figure that in the seawater environment and the intertidal environment, the Al anode is in an activated corrosion state. In the seawater environment, the corrosion potential (E_{corr}) of Al anode and the corrosion current densities (I_{corr}) are -1.127 V and 96.24 mA/cm^2, respectively, while they are only -0.679 V and 10.05 mA/cm^2 in the intertidal environment, indicating that the activation performance of Al anode is lower in the intertidal environment (Kunst 2021; Yu 2018). From the corrosion current density results, the cathodic protection potential that can be provided is only 1/10 of that in the seawater environment.

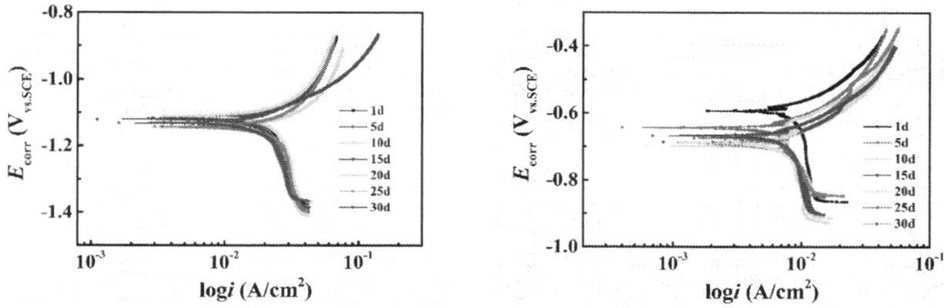

(1) Seawater environment (2) Intertidal environment

Figure 3. Polarization curves of Al anodes in different environments with time.

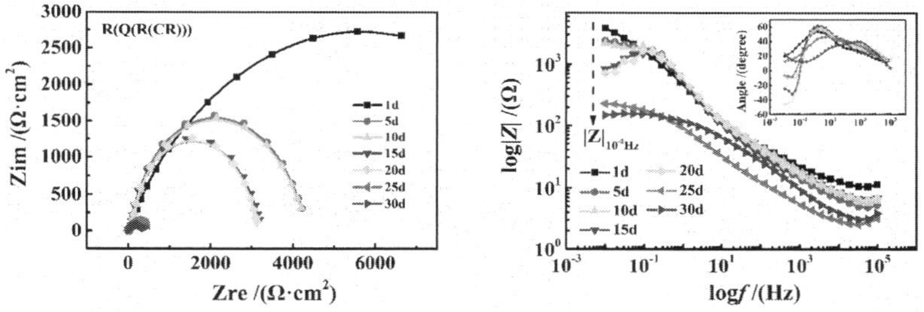

(1) Seawater environment

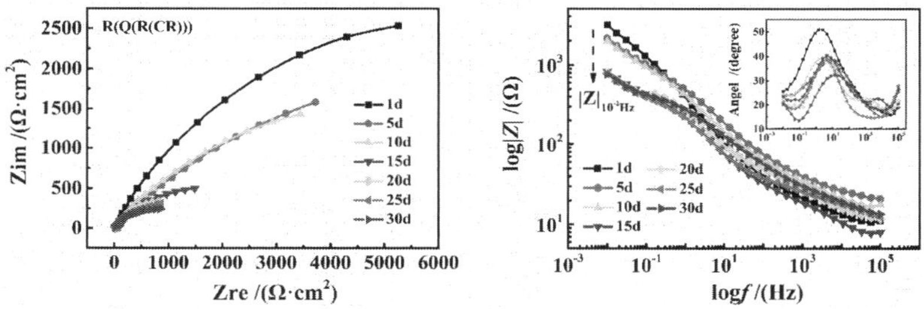

(2) Intertidal environment

Figure 4. Electrochemical impedance curves of Al anodes in different environments.

3.2.3 *Electrochemical Impedance Analysis*

Figure 4 shows the electrochemical impedance curves of the I-type Al anode in different environments at 25°C, and Table 1 displays the fitting results of the EIS curves. It can be seen from the figure that in different environments, the Nyquist diagrams of Al anodes all show double capacitive reactance characteristics, so the equivalent circuit is selected as $R_s(Q(R_p(C_{dl}R_{ct})))$, in which R_s is solution resistance, Q is constant phase angle element, R_p is surface resistance, C_{dl} is double-layer capacitance, and R_{ct} is charger transfer resistance. The radius of capacitive arc resistance decreased with experimental time, indicating that the corrosion process is relatively more likely to occur in different environments with longer experimental time. Comparing the R_p and R_{ct} of Al anodes in different environments, it can be seen that the R_p in the seawater environment is smaller than that in the intertidal environment, indicating that the corrosion product layer on the surface of Al anode is relatively loose and easy to fall off in the seawater environment. In the meantime, the R_{ct} in the seawater environment is smaller than that in the intertidal soil environment, indicating that the metal activation process of Al anode in the intertidal environment is relatively difficult. It proved that the Al anodes are difficult to provide continuous and effective cathodic protection in intertidal environments.

Table 1. Fitting results of EIS curves of Al anode in different environments.

	Seawater environment		Intertidal environment	
Time/(d)	$R_p/(\Omega\cdot cm^2)$	$R_{ct}/(\Omega\cdot cm^2)$	$R_p/(\Omega\cdot cm^2)$	$R_{ct}/(\Omega\cdot cm^2)$
1	2.79E+02	1.14E+04	2.10E+02	1.20E+04
5	1.65E+02	4.21E+03	2.52E+02	1.24E+04
10	2.13E+02	4.16E+03	3.97E+02	9.02E+03
15	9.70E+01	3.15E+03	1.64E+02	3.14E+03
20	1.49E+02	3.02E+03	2.69E+02	1.90E+03
25	1.43E+04	3.57E+06	1.41E+02	2.96E+03
30	2.31E-04	4.35E+02	2.11E+02	1.54E+03

4 CONCLUSIONS

In this paper, the electrochemical performance of Al anode in the intertidal environment and seawater environment was studied, and the following conclusions can be drawn:

(1) During the long-term immersion experiment, the general corrosion rate and pitting corrosion rate of Al anodes in the intertidal environment and seawater environment changed minimally, which was due to the relatively low content of active chloride ions in the beach soil, which hinders the activation process of Al anode, so the corrosion rate is low.
(2) The open circuit potential of Al anode in the seawater environment is –1.149 V, but it is only -0.702 V in the intertidal environment, which cannot meet the requirements of on-site cathodic protection. At the same time, the corrosion current density of Al anode in the intertidal environment is only 1/10 of that in the seawater environment, and the activation performance is low.
(3) The polarization resistance (R_p) of Al anode surface in the seawater environment is smaller than that in the intertidal environment, indicating that the corrosion product layer on the Al anode surface is loose and easy to fall off in the seawater environment. At the same time, the charge transfer resistance (R_{ct}) in the seawater environment is smaller than that of the intertidal environment, indicating that the metal activation process of Al anode is relatively difficult in the intertidal environment. It is also proved that the Al anodes are difficult to provide continuous and effective cathodic protection in the intertidal environment.

REFERENCES

Cheng, C. Q., Klinkenberg, L. I., Ise, Y., Zhao, J., Tada, E. & Nishikata, A. (2017). Pitting corrosion of sensitised type 304 stainless steel under wet–dry cycling condition. *Corros. Sci.* 118, 217–226.

Kunst, S. R., Bianchin, A. C. V., Mueller, L. T., Santana, J. A., Volkmer, T. M., Morisso, F. D. P., Carone, C. L. P., Ferreira, J. L., Mueller, I. L. & Oliveira, C. T. (2021). Model of anodized layers formation in Zn-Al (Zamak) aiming to corrosion resistance. *J. Mater. Res. Technol.* 12, 831–847.
Lu, Y., Wang, R. Q., Han, Q. H., Yu, X. L. & Yu, Z. C. (2022). Experimental investigation on the corrosion and corrosion fatigue behavior of butt weld with G20Mn5QT cast steel and Q355D steel under dry–wet cycle. *Eng. Fail. Anal.* 134, 105977.
Ma, J. L. & Wen, J. B. (2010). Corrosion analysis of Al-Zn-In-Mg-Ti-Mn sacrificial anode alloy. *J. Alloy. Compd.* 496, 110–115.
Wen, J. B., He, J. G. & Lu, X. W. (2011). Influence of silicon on the corrosion behaviour of Al-Zn-In-Mg-Ti sacrificial anode. *Corros. Sci.* 53, 3861–3865.
Yu, M., Zu, H., Zhao, K., Liu, J.H. & Li, S. M. (2018). Effects of dry/wet ratio and pre-immersion on stress corrosion cracking of 7050-T7451 aluminum alloy under wet-dry cyclic conditions. *Chinese J. Aeronaut.* 31, 2176–2184.
Zhang, X., Yang, S. W., Zhang, W. H., Guo, H. & He, X. L. (2014). Influence of outer rust layers on corrosion of carbon steel and weathering steel during wet–dry cycles. *Corros. Sci.* 82, 165–172.

Energy Revolution and Chemical Research – Kok-Keong Chong and Zhongliang Liu (Eds)
© 2023 The Authors, ISBN: 978-1-032-36554-1

Synthesis and properties of cross-linked starch-based flocculant

H.H. Li, Y.H. Gao & L.H. Zhang
Institute of Energy Sources, Hebei Academy of Sciences, Shijiazhuang Hebei, China

Z.F. Liu
Hebei Technological Innovation Center for Water Saving in Industry, Shijiazhuang Hebei, China

ABSTRACT: Cationic modified cross-linked starch-based flocculant IStAD was synthesized by graft copolymerization with cross-linked starch (ISt), acrylamide (AM), and methacryloyloxyethyl trimethylammonium chloride (DMC) as raw materials and ceric ammonium nitrate (CAN) as initiator. The effects of the amount of initiator and graft monomer on the flocculation performance of the product were studied, and the structure of the product was characterized by ATR-FTIR and SEM. The results showed that the optimal synthetic conditions of IStAD were m(ISt):m(CAN)=1:0.3,m(ISt):m(AM):m(DMC)=1:0.3:0.7, the synthesis temperature 60 ^{o}C and the synthesis time 4 h. When IStAD was used to treat 2 wt% kaolin suspension, the optimal dosage was 14 mg/L and the light transmittance of supernatant reached 85.7%.

1 INSTRUCTION

Starch-based flocculant is mainly synthesized by grafting groups with strong flocculation ability onto starch molecular chain through a chemical reaction, which can be modified according to actual needs (Hu 2020; Lapointe 2019). Compared with inorganic flocculants, starch-based flocculants have stronger stability, less dosage and better flocculation effect. And compared with synthetic polymer flocculant, starch has the advantages of rich sources, economic and environmental protection, easy degradation and so on. However, starch-based flocculant is composed of natural substances, which has certain limitations in storage time.

Cross-linked starch has lower water solubility, improved acid resistance and acid resistance, and more stable chemical properties compared with starch. It is widely used in food, pharmaceutical, papermaking, textile and other fields (Gurler 2020; Mao 2020). In the field of sewage treatment, cross-linked starch is often used as raw material for the synthesis of starch-based adsorbents, such as the synthesis of cross-linked carboxymethyl starch (Haq 2021) and insoluble starch xanthate (Feng 2017). In addition to synthetic adsorbents, cross-linked starch is less used in the field of sewage treatment. In recent years, people began to research and develop cross-linked starch-based flocculants, trying to use the special network structure of cross-linked starch to improve the net trapping and sweeping effect on impurities or colloidal particles in wastewater (Chang 2015), so as to improve the flocculation performance of products, but this research is still in its infancy.

In this paper, cross-linked starch-based flocculant ISt-g-PAM-co-PDMC (abbreviated as IStAD) was prepared with cross-linked starch (ISt) as raw material, acrylamide (AM) and methacryloyloxyethyl dimethyl ammonium chloride (DMC) as graft monomers and ceric ammonium nitrate (CAN) as initiator, and the flocculation performance of the product on kaolin simulated water samples were tested. It will provide a reference for expanding the application of cross-linked starch in the field of sewage treatment.

 DOI 10.1201/9781003332657-55

2 METHODS AND MATERIALS

2.1 *Reagents*

Cross-linked starch was self-made (Li 2020). Acrylamide (AM), analytical pure, was purchased from Tianjin Yongda Chemical Reagent Co., Ltd; China. Methacryloyloxyethyl trimethylammonium chloride (DMC, 75wt%) and ceric ammonium nitrate (CAN), analytical pure, were purchased from Shanghai Aladdin Chemical Co., Ltd., China. HNO_3, analytical purity, was got from Tianjin Kemio Chemical Reagent Co., Ltd., China. Kaolin, chemically pure, was purchased from Tianjin Fuchen Chemical Reagent Factory, China.

2.2 *Synthesis of cross-linked starch-based flocculant (IStAD)*

5 g of cross-linked starch and 130 ml of deionized water were put into a four-mouth flask and beat well. The system was vacuumed, and then nitrogen was introduced to make the subsequent reaction proceed under the protection of nitrogen. The starch suspension was gelatinized at 85 °C for 0.5 h, and then the temperature was reduced to 60 °C. 1.5 g of CAN was dissolved in 10 mL of HNO_3 solution (1 mol/L) as an initiator, then the solution of CAN was quickly injected into the starch solution with a syringe, and initiated for 5 min. The solution of graft monomers including 1.5 g AM and 4.7 g DMC (dissolved in 10 mL water) was injected into the system with an injection pump (TYD01, Baoding Leifu Fluid Technology Co., Ltd., China) within 40 min. The system reacted at a constant temperature for 4 h to obtain a kind of light-yellow solution. This solution was precipitated with ethanol, and the precipitate was washed and filtered with ethanol many times until there was no chloride ion in the filtrate. The filter cake was dried at 50 °C in a vacuum drier and obtained a light-yellow powder, namely IStAD.

2.3 *Characterization of IStAD*

Infrared spectra of IStAD were measured via a Frontier Fourier-transform infrared (FTIR) spectrometer (PerkinElmer, USA.) with the ATR attachment. The wave number range was 4000-650 cm-1. SEM images were observed on an Inspect S50 scanning microscope (FEI, USA). The acceleration voltage was 5 kV and the magnification was 2000 times. The content of nitrogen of IStAD was determined by a vario mICRO cube element analyzer (Elementar, Germany).

2.4 *Flocculation experiment*

1 g/L IStAD aqueous solution was prepared with deionized water for standby. Taking kaolin suspension as an experimental water sample, the flocculation performance of modified starch was tested. The experimental steps were as follows: 2 g kaolin was put into 100 mL deionized water and stirred at 300 r/min for 1 min, then a certain amount of IStAD solution was dosed in the suspension and continued stirring for 1 min. After stirring, the simulated water sample stood for 1 h, and then the supernatant was taken to measure the light transmittance with a UV-1500 UV spectrophotometer (Macy, China). The higher the light transmittance of the supernatant, the better the flocculation performance of the product.

3 RESULTS AND DISCUSSION

3.1 *Synthetic conditions of IStAD*

3.1.1 *Effect of dosage of CAN.*

The effect of the dosage of CAN on the flocculation was studied with the other conditions of *m*(ISt): *m*(AM): *m*(DMC) = 1:0.3:0.7, the reaction temperature 60°C, and the reaction time 4 h. As seen in Table 1, the amount of initiator was an important factor affecting the flocculation performance of the product. With the increase in the amount of initiator, the flocculation performance of the product was greatly improved. When m(CAN): m(ISt) was increased from 0.2 to 0.3, the optimal amount of the product was reduced from 50 mg/L to 14 mg/L. while when m(CAN): m(ISt)>0.3, the zeta potential still increased significantly, but the flocculation performance was not improved greatly, and the flocculation window became narrower, which is not conducive to practical application. Considering the cost and application, m(ISt): m(CAN) = 1:0.3 was selected.

Table 1. Effect of dosage of CAN on flocculation properties of IStAD.

m(ISt): m(CAN)	1:0.2	1:0.3	1:0.4
$N\%$	1.54	3.57	4.43
Zeta potential/mV	13.7	23.8	34.6
Optimal dosage/(mg·L^{-1})	50	14	13
Transmittance/%	80.5	85.7	83.5
Flocculation window/(mg·L^{-1})	30-60	10-22	10-16

3.1.2 *Effect of monomer dosage.*

Under the conditions of m(ISt): m(CAN) = 1:0.3, the reaction temperature 60°C and the reaction time 4 h, the effect of dosage of monomer on the flocculation was studied. It can be seen from Table 2 that the nitrogen content ($N\%$) of the product increased gradually with the increase of AM content in the graft monomer, but the zeta potential of the product increased first and then decreased. This is mainly because the reaction activity of AM is higher than that of DMC. When a small amount of AM is contained in the grafted monomer, more DMC monomers are introduced into the side chain connected to the starch skeleton while introducing PAM. However, when the mass ratio of AM to DMC exceeds 0.3:0.7, the amount of DMC introduced due to AM activity cannot offset the decrease of DMC in the monomer. Although the nitrogen content of the product has been increasing, the charge density of the product has decreased, which leads to a decrease in the flocculation performance of the product.

Table 2. Effect of monomer ratio on flocculation performance of IStAD.

m(ISt): m(AM): m(DMC)	1:0.1:0.9	1:0.2:0.8	1:0.3:0.7	1:0.4:0.6	1:0.5:0.5
$N\%$	2.36	2.65	3.57	3.62	3.73
Zeta potential/mV	18.2	22.1	23.8	20.4	19.2
Optimal dosage/(mg·L^{-1})	20	18	14	30	30
Transmittance/%	83.8	84.6	85.7	82.5	84.4
Flocculation window/(mg·L^{-1})	12-26	12-26	10-26	20-35	20-40

In conclusion, the optimal synthesis conditions of IStAD were: m(ISt): m(CAN) = 1:0.3, m(ISt): m(AM): m(DMC) = 1:0.3:0.7, reaction temperature 60 °C and reaction time 4 h.

3.2 *Characterization of IStAD*

3.2.1 *ATR-FTIR analysis*

ATR-FTIR analysis was carried out on corn starch, cross-linked starch, and IStAD, and the results are shown in Figure 1. Compared with that of the raw corn starch, the infrared spectrum of cross-linked starch did not change significantly, mainly because any characteristic functional groups did not be introduced into the molecular chains of the starch after cross-linking. While in the infrared spectrum of IStAD, in addition to maintaining the characteristic absorption peaks of starch, new absorption peaks appeared at 1727 cm^{-1}, 1665 cm^{-1}, 1476 cm^{-1}, and 1019 cm^{-1}. The absorption peaks at 1727 cm^{-1} and 1476 cm^{-1} were the stretching vibration absorption peak of C=O and the

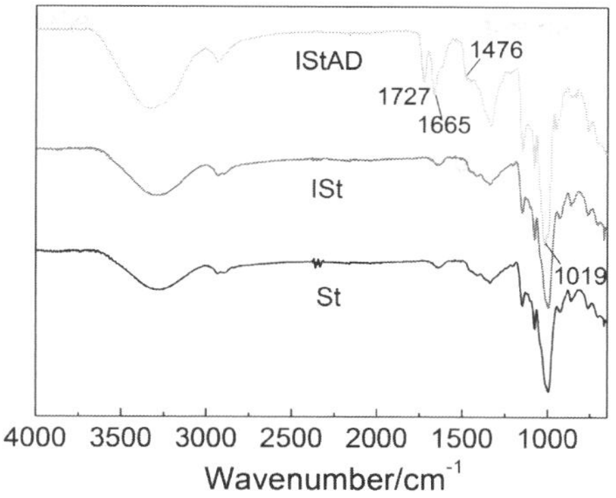

Figure 1. Infrared spectra of corn starch and modified starch.

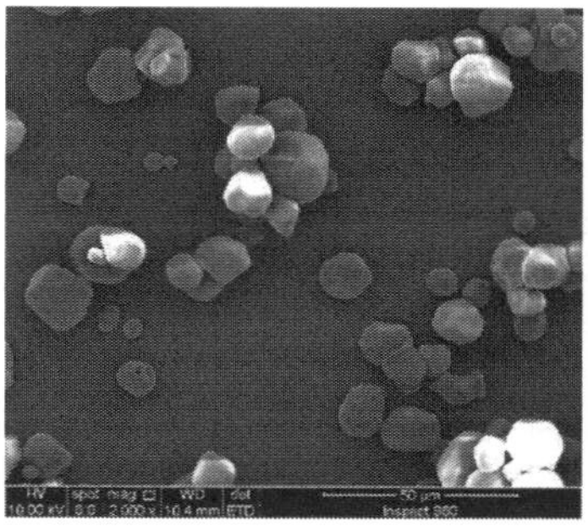

(a) Corn starch

(b) Cross-linked starch

(c) ISAtD

Figure 2. SEM photos of corn starch and modified starch.

vibration absorption peak of methyl (-$N^+(CH_3)_3$) in the quaternary amine group, respectively. The peak at 1665 cm^{-1} was caused by the overlap of the C=O stretching vibration absorption band and -NH_2 deformation vibration absorption band, and the intensity of this peak increased with the increase of AM content in the graft monomer, indicating that the primary amide group was connected to the starch molecule. The peak at 1019 cm^{-1} was the stretching vibration absorption peak of C1-O-C4. When the amide group was introduced into the starch molecule, a new hydrogen bond was formed between C1-O-C4 and the amide band, which significantly enhanced this peak.

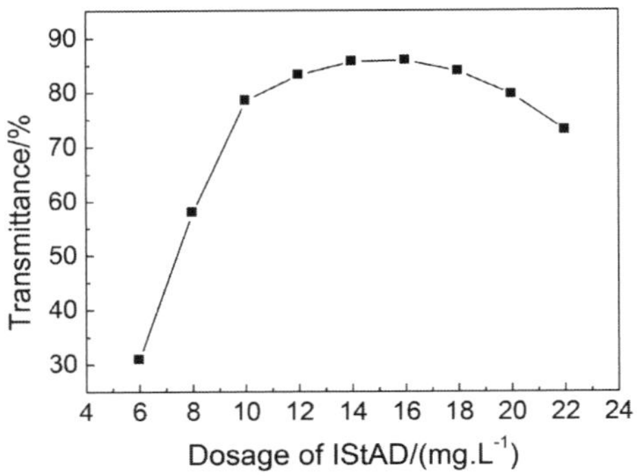

Figure 3. Effect of dosage of IStAD on flocculation.

3.2.2 *SEM analysis*

SEM photos of corn starch, cross-linked starch and IStAD are shown in Figure 2. As seen in Figure 2(a), the surfaces of corn starch granules were smooth, and the shapes were polygonal or spherical. This is due to the interweaving of straight and branched chains in starch molecules so that starch particles can maintain certain integrity. After cross-linking of corn starch, as shown in Figure 2(b), the particle shape did not change much, the particle volume increased slightly, and some particles had holes. After the grafting reaction of cross-linked starch, the starch particles expanded and the shell of some particles was broken, as shown in Figure 2(c).

3.2.3 *Flocculation performance of IStAD on kaolin*

The influence of IStAD dosage on the flocculation effect was tested, and the results are shown in Figure 3. As can be seen from Figure 3, with the increase in dosage, the light transmittance of supernatant also increased rapidly. The optimal dosage was 14 mg/L, and the light transmittance was 85.7%. When the dosage exceeded the optimal dosage, the supernatant could still maintain a certain light transmittance, and the flocculation window was 10-20 mg/L. When the dosage was greater than 20 mg/L, the light transmittance decreased, which is mainly because too many flocculant molecules adsorb on kaolin particles, resulting in the complete reversal of the surface charge of kaolin particles. The electrostatic repulsion between particles increases and the flocs are re-dispersed in water, resulting in a decrease in light transmittance.

4 CONCLUSION

Cross-linked starch-based flocculant IStAD was synthesized with cross-linked starch, acrylamide and methacryloyloxyethyl trimethylammonium chloride as raw materials and ceric ammonium nitrate as initiator. The optimal synthesis conditions of IStAD were as follows: m(ISt): m(CAN) = 1:0.3, m(ISt): m(AM): m(DMC) = 1:0.3:0.7, reaction temperature 60°C and reaction time 4 h.

The flocculation performance of IStAD was tested by using 2 wt% kaolin as the simulated water sample. The results showed that the optimal dosage of IStAD was 14 mg/L, the light transmittance was 85.7%, and the flocculation window was 10-20 mg/L.

ACKNOWLEDGMENTS

The research was sponsored by the Science and Technology Project of the Hebei Academy of Sciences (Grant No. 22706).

REFERENCES

Chang Q. 2015. New research area of flocculation in water treatment-macromolecule flocculant with the function of trapping heavy metal. *Acta Scienctiae Circumstantiae* 35(1): 1–11.

Feng K. & Wen G. H. 2017. Absorbed Pb^{2+} and Cd^{2+} ions in water by cross-linked starch xanthate. *International Journal of Polymer Science* 2017: 1–9.

Gurler, N., Pasa, S., Alma M. H., et al. 2020. The fabrication of bilayer polylactic acid films from cross-linked starch as eco-friendly biodegradable materials: synthesis, characterization, mechanical and physical properties. *European Polymer Journal* 127: 109588.

Ha F., Yu H. J., Wang L., et al. 2021. Synthesis of succinylated carboxymethyl starches and their role as adsorbents for the removal of phenol. *Colloid and Polymer Science* 299(11): 1833–1841.

Hu P., Xi Z. H., Li Y., et al. 2020. Evaluation of the structural factors for the flocculation performance of a co-graft cationic starch-based flocculant. *Chemosphere* 240(C): 124866.

Lapointe M. & Barbeau B. 2019. Substituting polyacrylamide with an activated starch polymer during ballasted flocculation. *Journal of Water Process Engineering* 28: 129–134.

Li H. H., Zhang L. H., Gao Y. H., et al. 2020. Synthesis and adsorption properties of insoluble starch xanthate (ISX). *IOP Conference Series: Materials Science and Engineering* 735: 012026.

Mao Y. X., Pan M. M., Yang H. L., et al. 2020. Injectable hydrogel wound dressing based on strontium ion cross-linked starch. *Frontiers of Materials Science* 14(2): 1–10.

Energy Revolution and Chemical Research – Kok-Keong Chong and Zhongliang Liu (Eds)
© 2023 The Authors, ISBN: 978-1-032-36554-1

Study on solving plasma equilibrium composition model with HLMA algorithm

Zhongyuan Chi, Weijun Zhang & Qiangda Yang
School of Metallurgy, Northeastern University, Shenyang, China

ABSTRACT: The Homotopy Levenberg-Marquardt Algorithm is used to solve singular nonlinear equations that are constructed by the plasma equilibrium composition model according to mass action law and settle a matter that is sensitive to the initial iteration values. The homotopy equation, $\boldsymbol{H}(\boldsymbol{x},t) = t\boldsymbol{F}(\boldsymbol{x})+(1\text{-}t)\boldsymbol{G}(\boldsymbol{x})$, is established by introducing auxiliary equation $\boldsymbol{G}(\boldsymbol{x})$ and homotopy factor sequence $[t_k]$. The auxiliary equation is plasma equilibrium composition at high temperature and assumes that it contains only electron and the highest atomic cation of atoms, such as T_{max}= 30000 K. The homotopy factor sequence $[t_k]$ is arranged in geometric, and the change of difference value of $t_k - t_{k-1}$ from large to small guarantees the continuity of the whole process. Finally, the Mg-CO mixture plasma for melting magnesium oxide crystals is taken as an example to verify the feasibility, the solution accuracy is less than 1×10^{-15}, and the particle distribution at atmospheric pressure from 300 to 30,000 K was calculated.

1 INTRODUCTION

Plasma is an important substance in modern industry, such as plasma cutting (Gani et al. 2021), welding (Murphy et al. 2010), plasma metallurgy (Bermudez et al. 1999), plasma display (Meunier et al. 1995), plasma surface modification (Trinh et al. 2019) and also exists in 60–100 km altitude. Determining what plasma is and how it is distributed is the guarantee of studying the application of plasma, which requires constructing and solving the plasma equilibrium composition model. The model is essentially a nonlinear equation of a singular Jacobian matrix, and its solution is very difficult. Existing iterative methods are very sensitive to the initial value of iteration and require close approximation to the real solution, which is not easy to achieve. In this paper, aiming at the particularity of plasma at high temperature, the Homotopy Levenberg-Marquardt Algorithm (HLMA) is used to solve the plasma equilibrium composition model at arbitrary temperature and pressure.

The plasma equilibrium composition model is made up of the Saha and Guldberg-Waage equation, Dalton's partial pressure, conservation of mass, and charge. Gleizes and Chervy (Gleizes et al. 1999) applied this method to construct SF_6 plasma model, and solved this model in the range of 300–20,000 K temperature and 0.1–1.6 MPa pressure. Wang, Rong *et al.* (Wang et al. 2011, 2012) studied and solved the model of carbon–argon, carbon–helium and carbon-water plasma at high temperatures. Murphy *et al.* (1994) respectively calculated the equilibrium composition, thermodynamic and transport properties of Ar, N_2, O_2, N_2-Ar and O_2-Ar plasma from 300 K to 30,000 K at 0.1 MPa. The above-mentioned scholars have made a detailed description of the model construction, but they have not yet defined a general and effective method for solving this model (Chi et al. 2021). A variety of existing iteration methods, such as the Newton iteration method (Chen 2016; Fang et al. 2018), Levenberg-Marquardt Algorithm (LMA) (Chen 2016), secant method (Sidi 2008) and so on, can theoretically solve nonlinear equations. But they are all sensitive to the initial value, and this is also the difficulty of solving nonlinear equations.

 DOI 10.1201/9781003332657-56

Seeking the method that solves nonlinear equations independent of the initial values is one of the keys to the study of the plasma equilibrium composition model (Gleizes et al. 2005; Smith & Missen 1982), which is also the focus of this paper. The homotopy algorithm has become an ideal method for solving equations because it has no strict limitation on initial values (Levenberg 1944). However, for a nonlinear problem with a singular Jacobian matrix, the construction of auxiliary equations $\boldsymbol{G}(\boldsymbol{x})$ and homotopy factor sequence $[t_k]$ are still the focus and difficulty. LMA (Chen 2016) is an iterative method for solving nonlinear equations with adaptive step size, which is suitable for singular and non-singular equations and has the advantage of fast calculation speed. In addition, LMA is an algorithm that has a Gauss-Newton algorithm or fastest descent method advantage by automatically controlling the adaptive step size. By constantly adjusting step size, the probability of jumping out of the local optimum can be increased. Combining the advantages of the above two algorithms, the Homotopy Levenberg-Marquardt Algorithm (HLMA) is put forward for solving the plasma equilibrium composition model in this paper. A series of equations were constructed by the homotopy method and solved by the LMA algorithm.

The main research contents of this paper are as follows: 1) build the plasma equilibrium composition model according to the law of mass action; 2) Describe the basic principles and procedure of HLMA in detail, including the construction and solution of auxiliary equations $\boldsymbol{G}(\boldsymbol{x})$, the rule of confirming homotopy factor sequence $[t_k]$ and solution procedure; 3) Based on the Mg50%-CO50% plasma, the particle number densities at 25000K and 24000K were calculated respectively, focusing on the calculation process and accuracy of HLMA algorithm; 4) the plasma composition from 300K to 30000K at atmospheric pressure was calculated and analyzed.

2 EQUILIBRIUM COMPOSITION MODEL

Plasma is a special substance obtained by neutral particles after a series of chemical reactions (Gleizes et al. 2005). Under the condition of chemical equilibrium, the chemical reactions of plasma can be divided into two types: ionization and dissociation. The relationship between the number density of the particles involved in the chemical reactions is expressed by Saha and Guldbeng-Waage equations. Equation (1) shows the plasma equilibrium composition.

$$\begin{cases} f_1(n_1, n_2, \cdots, n_N) = \sum\limits_i n_i z_i = 0 \\ f_m(n_1, n_2, \cdots, n_N) = \sum\limits_i n_i l_{m,i} - const \sum\limits_i n_i l_{1,i} = 0 \\ f_{N_e+1}(n_1, n_2, \cdots, n_N) = \sum\limits_i n_i k_B T - p = 0 \\ \vdots \\ f_{\dots}(n_1, n_2, \cdots, n_N) = n_e n_{r+1} - \frac{2Q_{r+1}}{Q_r}\left(\frac{2m_e \pi k_B T}{h^2}\right)^{3/2} \exp\left(-\frac{E_{I,r+1}}{k_B T}\right) n_r = 0 \\ f_{\dots}(n_1, n_2, \cdots, n_N) = n_A n_B - \frac{Q_A Q_B}{Q_{AB}}\left(\frac{2\pi m_A m_B k_B T}{m_{AB} h^2}\right)^{3/2} \exp\left(-\frac{E_d}{k_B T}\right) n_{AB} = 0 \end{cases} \tag{1}$$

Where m is molecular mass, the subscript e stands for electron; n_i and z_i are particle density and charge number of i−th molecule; p is pressure, T is temperature and k_B are Boltzmann constant. Q_r and Q_{r+1} is the partition function of non-electronic molecules in ionization reaction, and Q_{AB}, Q_A and Q_B is the partition function of reactant and products in dissociation reaction. In these equations, N_e is the number of atomic elements. The first equation states charge conservation; The second is conservation of mass, where $l_{m,i}$ represents the number of the m−th element in the i−th molecule. If the plasma contains only one element (N_e=1), the equation does not exist; otherwise, there will be N_e-1 equations. The third equation is the partial pressure equation; the fifth and sixth equations are the Saha equation and the Guldbeng-Waage equation respectively, and each equation contains three kinds of molecules. The coefficient difference between different reactions at the same temperature is as high as 5–7 orders of magnitude, and this is also the main reason for the singularity of the equations.

3 HOMOTOPY LEVENBERG-MARQUARDT ALGORITHM (HLMA)

Homotopy is a process in which some homotopy equation is constructed and solved by introducing an auxiliary equation $\boldsymbol{G}(\boldsymbol{x})$ with the solution and homotopy factor sequence $[t_k]$, and the whole process does not need to consider the initial value-sensitive problem. The elementary form of homotopy equation is,

$$\boldsymbol{H}(\boldsymbol{x}, t) = t\boldsymbol{F}(\boldsymbol{x}) + (1-t)\,\boldsymbol{G}(\boldsymbol{x}) \tag{2}$$

where $\boldsymbol{H}(\boldsymbol{x},t)$ is the homotopy equation, $\boldsymbol{F}(\boldsymbol{x})$ is the original equation to be solved, $\boldsymbol{G}(\boldsymbol{x})$ is the auxiliary equation and t is the homotopy factor. If $t=0$, homotopy equation is equivalent to the auxiliary equation; if $t=1$, it is original equation.

3.1 *Auxiliary equation*

The auxiliary equation is an important part of the homotopy equation, which requires the same dimension as the original equation and has a more accurate solution. In plasma, temperature and pressure are objective factors affecting composition distribution. It's common sense to know that the higher temperature of plasma, the higher the number density of electrons is at the same pressure. When the temperature is high enough, it can be assumed that the plasma contains only the highest valence atomic cations and electrons, other particles are converted into corresponding cations and electrons through a series of chemical reactions.

If the above assumptions are true at high temperature and taking it as the initial iterative value of the plasma equilibrium composition model, particle number densities can be accurately calculated. Similarly, the homotopy method is also satisfied if the plasma model has a more accurate solution and the temperature is higher than the solution temperature.

3.2 *Homotopy factor sequence*

Another important constituent part of the homotopy method is to determine the factor sequence $[t_k]$, $0 = t_0 < t_1 < t_k < \ldots < t_N = 1$, which not only affects the time of the homotopy equation solution but also the indicator of whether the calculation can be completed. Equation (3) is the expression of k-th homotopy equation, T and T_{max} represent the solution and the auxiliary temperature, respectively.

$$\boldsymbol{H}_{\mathrm{T}}(\boldsymbol{x}, t_k) = t_k\boldsymbol{F}_{\mathrm{T}}(\boldsymbol{d}) + (1-t_k)\,\boldsymbol{F}_{T_{\max}}(\boldsymbol{x}) = 0, k = 0, 1, 2, \cdots, N \tag{3}$$

It has been introduced in Section 2 that the singularity of the system of equations is caused by the coefficient difference between Saha and Guldbeng-Waage equations of different chemical reactions. The maximum coefficient C_{max} in the auxiliary equation and C in the original equation are selected as references and used to determine calculation steps N. In principle, the larger the homotopy step N is, the better. The factor sequence $[t_k]$ is carried out by geometric sequence. The expression of proportion coefficient q is Equation (4) and Equation (5) is applied to get homotopy coefficient t_k.

$$q = \left(\frac{C_{\max}}{C}\right)^{1/N} \tag{4}$$

$$t_k = \begin{cases} 0 & k=0 \\ \frac{C_{\max} - q^{N-k}}{C_{\max}} & 0 < k < N \\ 1 & k = N \end{cases} \tag{5}$$

3.3 *Computational process*

The process of solving plasma equilibrium composition model by HLMA algorithm is mainly divided into three steps: 1) determining the auxiliary equations, the condition is that the temperature is higher than the solution temperature and has a precise solution; 2) the homotopy factor sequence $[t_k]$ is determined to ensure the continuity of homotopy advance; 3) construct and solve each homotopy equation by LMA. The calculation steps are as follows and Figure 1 is the flow chart of HLMA calculation.

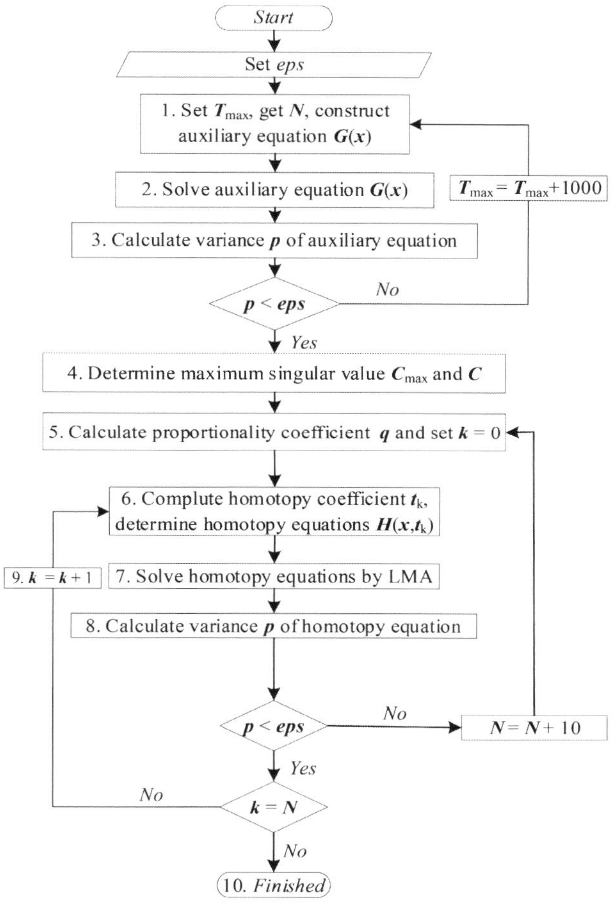

Figure 1. The calculation flow chart of HLMA.

1) Determine auxiliary working conditions T_{max}, homotopy steps N, and set calculation accuracy eps.
2) If the auxiliary equation has no solution, the number density of electron and highest valence atomic cations is determined according to the law of conservation of mass and charge, and take it as the initial value of iteration, use LMA to solve the auxiliary equation.
3) Calculate the variance of the auxiliary equation p and judge $p < eps$, yes, turn to 4; otherwise $T_{max} = T_{max}+1000$, turn to 1;
4) Seek out maximum singular values $\boldsymbol{C}_{max}$ and $\boldsymbol{C}$;
5) Calculate the $\boldsymbol{q}$ by equation (4) and set calculating cursor $\boldsymbol{k} = 0$;
6) Calculate the homotopy factor t_k and build new homotopy equation $\boldsymbol{H}(\boldsymbol{x},t_k)$;
7) Solve equation $\boldsymbol{H}(\boldsymbol{x},t_k)$ by LMA and calculate the variance p of this equation;
8) If $p < eps$, turn to 9; otherwise, calculate $N_{max} = N_{max} + 10$, and turn to Step 5;

9) If $k = N_{max}$, go to 10; if $k = k+1$, go to 6;
10) Finished.

4 CASE ANALYSIS

In this paper, the familiar Mg50%-CO50% mixture plasma in magnesium oxide crystal smelting is studied as an example. Taking 30,000K as the auxiliary working condition, the plasma particle number densities at 25,000 K and atmospheric pressure are calculated. On this basis, calculate the particle distribution at 24,000 K. Finally, the particle numerical densities of plasma from 300 K to 30,000 K at atmospheric pressure are calculated and analyzed.

In addition to electron, in Mg50%-CO50% plasma, also contains CO_2, CO, O_2, C_2, O, C, Mg and MgO eight neutral particles, Mg^+, Mg^{++}, O^+, O^{++}, O^{+++}, C^+, C^{++}, C^{+++}, O^{2+} and CO^+ ten ions. The calculation of partition functions for these particles has been described in detail in our earlier studies (Chi et al. 2022). There are 15 kinds of chemical reactions involved, including dissociation reactions of polyatomic molecules, such as $CO_2 \rightleftharpoons CO + O$, $CO \rightleftharpoons C + O$, $MgO \rightleftharpoons Mg + O$, $C_2 \rightleftharpoons C + C$ and $O_2 \rightleftharpoons O + O$, partial molecular ionization reaction and multistage atomic ionization reaction.

The homotopy calculation process and accuracy are shown in Figure 2. It is not difficult to find that the accuracy of solving the plasma equilibrium composition model by HLMA is less than 1×10^{-15}, the particle number density changes continuously with iteration, and the solution accuracy of each homotopy equation is good, less than 1×10^{-12}. It can also be seen that homotopy factor t increases exponentially with the homotopy step, can well adjust the difference between two adjacent homotopy equations and ensure the continuity of the calculation process. In addition, the difference between the auxiliary and the solution temperature is proportional to the homotopy step, so it is suggested to disintegrate a single calculation into multiple calculations when the difference is larger.

As can be seen from Figure 3, when T= 30,000 K, the proportion of the number density of electrons and the highest valence atomic cations to total density is about 95%, which is a more appropriate initial value for the iteration of the equilibrium composition model. The chemical property of metal Mg is very active, it immediately oxidizes when the O is decomposed at about 1,300 K. The first ionization temperature of Mg is 4,100 K, while the values of C and O are about

(a)

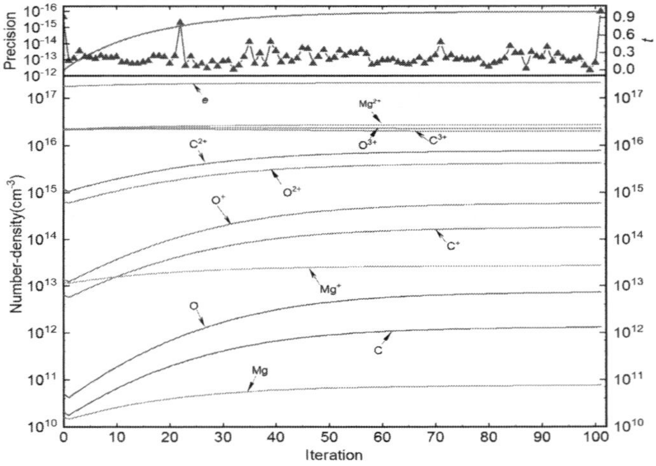

Figure 2. Variations of homotopy factor t, precision eps, and particle number density n_i with homotopy steps for CO-Mg plasma. (a) From 30,000 K to 25,000 K; (b) from 25,000 K to 24,000 K.

(b)

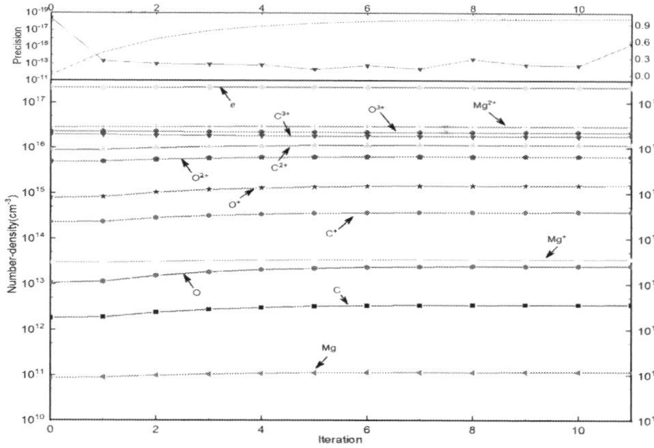

Figure 2. Continued.

(a)

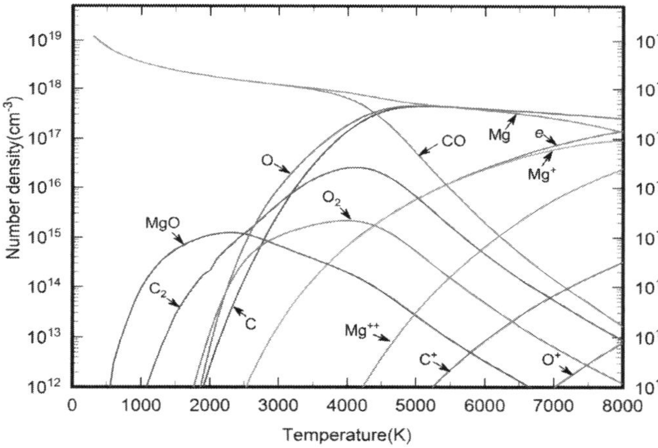

(b)

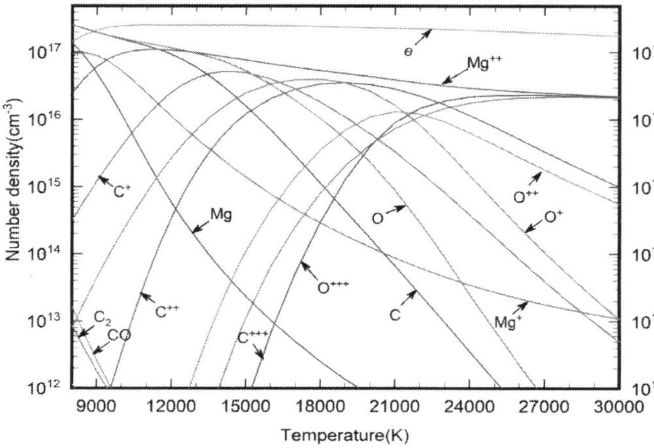

Figure 3. Equilibrium composition of Mg-CO plasma from 300 K to 30,000 K at atmospheric pressure. (a) range: 300–8,000 K, (b) range: 8,000–30,000 K.

5,100 K and 7,000 K, respectively. In addition, all particles are not shown in Figure 3, such as O^{2+} and CO_2. This is because the ionization energy required for O_2 is much higher than the dissociation, and the dissociation reaction takes place preferentially under the same conditions. The dissociation energy of CO_2 CO + O was greater than O_2 O + O, and O atomic supply of the latter led to a lower number density value of CO_2.

5 CONCLUSIONS

In this paper, the plasma composition model is established based on mass action law, combined with Dalton's partial pressure, mass conservation, and charge conservation. The Homotopy Levenberg-Marquardt Algorithm (HLMA) is put forward and used to solve this model. The results show that:

1) At high temperature, as T= 30,000 K, electron and atom highest valence positive atom is the main substance of plasma, and it is easy to obtain the number density distribution of particles with its initial iteration value. The homotopy factor sequence $[t_k]$ arranged by an equal ratio ensures the continuity of homotopy calculation.
2) HLMA algorithm can solve nonlinear equations of the singular Jacobian matrix if the auxiliary equations meet the requirements that have the same dimension as the solution equation and a more precise solution.
3) HLMA used for solving the fractional density of the plasma group has high accuracy and continuous convergence, and the calculation accuracy is less than 1×10^{-15}, which meets the requirements of calculation accuracy.

ACKNOWLEDGMENT

This study was supported by The National Key Research and Development Program of China (No. 2017YFA0700300) and the Fundamental Research Funds for the Central Universities (No. N2025032), and the Liaoning Provincial Natural Science Foundation (No. 2020-MS-362).

REFERENCES

Bermudez, Alfredo, M. Muñiz, Francisco Pena, and Javier Bullón. 1999. Numerical computation of the electromagnetic field in the electrodes of a three-phase arc furnace. *International Journal for Numerical Methods in Engineering* 46:649–658.

Chen, Liang. 2016. A high-order modified Levenberg–Marquardt method for systems of nonlinear equations with fourth-order convergence. *Applied Mathematics and Computation* 285:79–93.

Chi, Zhongyuan, Weijun Zhang, and Qiangda Yang. 2022. Study on Thermophysical Properties of Arc Plasma for Melting Magnesium Oxide Crystals at Atmospheric Pressure. *Energies* 15 (3).

Chi, Zhongyuan, Weijun Zhang, and Qiangda Yang. 2021. Thermophysical properties of pure gases and mixtures at temperatures of 300–30 000 K and atmospheric pressure: thermodynamic properties and solution of equilibrium compositions. *Plasma Science and Technology* 23 (12):125505.

Fang, Xiaowei, Qin Ni, and Meilan Zeng. 2018. A modified quasi-Newton method for nonlinear equations. *Journal of Computational and Applied Mathematics* 328:44–58.

Gani, Adel, William Ion, and Erfu Yang. 2021. Experimental Investigation of Plasma Cutting Two Separate Thin Steel Sheets Simultaneously and Parameters Optimisation Using Taguchi Approach. *Journal of Manufacturing Processes* 64:1013–1023.

Gleizes, A., B. Chervy, and J J Gonzalez. 1999. Calculation of a two-temperature plasma composition: bases and application to SF6. *Journal of Physics D Applied Physics* 32 (16):2060–2067.

Gleizes, A., J. J. Gonzalez, and P. Freton. 2005. Thermal plasma modelling. *Journal of Physics D: Applied Physics* 38 (9):R153–R183.

Levenberg, K. 1944. A Method for the Solution of Certain Non-Linear Problems in Least Squares. *The Quarterly of Applied Mathematics* 2:164–168.
Meunier, J., Ph Belenguer, and J. P. Boeuf. 1995. Numerical model of an ac plasma display panel cell in neon-xenon mixtures. *Journal of Applied Physics* 78 (2):731–745.
Murphy, A. B., C. J. Arundelli, and Plasma Processing. 1994. Transport coefficients of argon, nitrogen, oxygen, argon-nitrogen, and argon-oxygen plasmas. *Plasma Chemistry* 14 (4):451–490.
Murphy, Anthony B., Manabu Tanaka, Kentaro Yamamoto, Shinichi Tashiro, John J. Lowke, and Kostya Ostrikov. 2010. Modelling of arc welding: The importance of including the arc plasma in the computational domain. *Vacuum* 85 (5):579–584.
Sidi, Avram. 2008. Generalization Of The Secant Method For Nonlinear Equations. *Applied Mathematics E-Notes* [electronic only] 8.
Smith, W. R., and R. W. Missen. 1982. Chemical reaction equilibrium analysis: theory and algorithms: Wiley.
Trinh, Quang Hung, Duc Ba Nguyen, Md Mokter Hossain, and Young Sun Mok. 2019. Deposition of superhydrophobic coatings on glass substrates from hexamethyldisiloxane using a kHz-powered plasma jet. *Surface and Coatings Technology* 361:377–385.
Wang, Wei Zong, A. B. Murphy, J. D. Yan, et al. 2012. Thermophysical Properties of High-Temperature Reacting Mixtures of Carbon and Water in the Range 400–30,000 K and 0.1–10 atm. Part 1: Equilibrium Composition and Thermodynamic Properties. *Plasma Chemistry* 32 (1):75–96.
Wang, WeiZong, MingZhe Rong, Anthony B. Murphy, et al. 2011. Thermophysical properties of carbon–argon and carbon–helium plasmas. *Journal of Physics D: Applied Physics* 44 (35):355207.

Energy Revolution and Chemical Research – Kok-Keong Chong and Zhongliang Liu (Eds)

Influence of head variation on the stability of Kaplan turbine

Li Chen
Chongqing Datang International Wulong Hydro Power Development Co., Ltd., Wulong, China

ABSTRACT: This paper investigates the influence of stability under the different heads of the Kaplan turbine. Firstly, the stability tests of the Kaplan turbine under 15 m, 22 m, 26 m, and 32 m head are tested by a vibration inspection instrument. Then the experimental data such as vibration and pressure are analyzed. It is shown that with increasing head, the vibration area of the unit will increase. And as the head increases, the average pressure of the draft tube will decrease under the same load.

1 GENERAL INSTRUCTIONS

As the key core equipment of a hydropower plant, the safe and stable operation of the hydro-generator unit is very important to the production efficiency of the power plant. How to improve the operation stability of the hydro-generator unit is one of the current research hotspots. As we all know, hydraulic factors are one of the reasons that affect the vibration of the hydro-generator unit, and the operation stability of a hydro-generator unit with a wide range of head variation is also greatly affected by the operating head. At present, the influence of the head variation on the stability of the pump turbine has been studied, it is shown that the stability of the pump turbine can be improved by selecting an appropriate ratio of the maximum head of the pump to the minimum head of the turbine (Gong et al. 2020). Some literature studies the effect of head variation on the unbalanced capacity of units in pumped storage power stations, and shows that the capacity will be more unbalanced when the head amplitude increases (Gao 2018). Many researchers study the relationship between unit vibration and AGC under head variation, and the large fluctuations in load can be avoided by setting a reasonable AGC adjustment method (Huang & Huang 2016). Other researchers have analyzed the pressure pulsation in the draft tube of the Francis turbine under head variation, the results show that reasonable parameter selection is conducive to reducing the pressure pulsation in the draft tube (Wu et al. 2000). There are also research studying on the parameter selection of pumped storage power station units under ultra-wide hydraulic head variation, which provides a reference for the optimal design of the pump turbine (Chen et al. 2016, Wang et al. 2021).

The above literature show that a wider range of head variation will directly affect the operation stability of the turbine. However, most of the existing literature focuses on pump turbines and Francis turbines, while there are few studies on Kaplan turbines. In this paper, the Kaplan hydro-generating unit of a power station is taken as an example, the influence of the head variation on the stability of the Kaplan turbine is found by the stability test under multiple heads.

2 RESEARCH METHOD

2.1 *Unit parameters*

Among various types of hydraulic turbines, the axial flow turbine is suitable for hydropower stations with medium and low heads. The type of runner blade can be divided into fixed propeller type and

 DOI 10.1201/9781003332657-57

propeller type (Kaplan type). In contrast, the Kaplan turbine can adapt to a wider range of head variation and has a wider efficient and stable operation area, because its runner blade can rotate and adjust within a certain range. Table 1 shows the equipment parameters of the Kaplan turbine generator unit in a hydropower station. The ratio of the maximum head to the minimum head of the unit is 2.70.

Table 1. The equipment parameters of Kaplan turbine generator unit.

Type	ZZ-LH-860	Rated speed	83.3 r/min
Rated head	26.5 m	Runaway speed	212.1 r/min
Maximum head	35.12 m	Rated power	161.25 MW
Minimum head	13.0 m	Rated efficiency	93.1%
Design flow	680 m^3/s	Number of runner blades	6
Maximum output	167.86 MW	Number of guide vanes	28

2.2 *Test method*

The vibration inspection instrument is used to measure the vibration and pressure value of the unit which can analyze the amplitude and frequency spectrum of test signals. The vibration inspection instrument can also draw the corresponding test curve and print the analysis results graphically.

The test is set up with 17 vibration measured points (Table 2) and 1 pressure test point. According to the requirements of IEC 60994, the stability data are measured for the variable load conditions of the unit under 15 m, 22 m, 26 m and 32 m head respectively. During the test, the unit load is adjusted in increments of 10 MW on average (increments of 5 MW at 15 m head), until the load reaches the maximum load under the current head. The data is collected for about 2 minutes at each steady-state operating condition, and the upstream and downstream water levels of the hydropower station are recorded at the same time to calculate the head. The data analysis method is a peak-to-peak measurement at 8 cycles, with a 97% confidence level.

Table 2. The arrangement of measured points.

Positions	Horizontal vibration	Vertical vibration
Upper guide shaft	x and y directions	/
Lower guide shaft	x and y directions	/
Turbine shaft	x and y directions	/
Upper bracket	x and y directions	x-direction
Lower bracket	x and y directions	x-direction
Thrust bearing support	x and y directions	x-direction
Head cover	x direction	x direction

3 RESULT ANALYSIS

3.1 *Influence on vibration*

The vibration test results under the different heads are indicated in Figure 1 to Figure 6. Due to space limitations, this paper only presents the results of the measured points in the x-direction of each part.

As can be seen from the figures, except for the vertical vibration of the upper bracket (Figure 1b), lower bracket (Figure 2b) and thrust bearing bracket (Figure 3b), the vibration amplitude of the rest part of the unit tends to decrease with the increase of the load. The vibration amplitude decreases sharply when the load reaches a certain value and then remains stable. According to the vibration level of each part of the unit, we divide the operating area of the unit into the vibration area and the non-vibration area (Table 3). The boundary points of the vibration area under 15 m, 22 m, 26 m and 32 m heads are 20.7 MW, 42.1 MW, 51.2 MW and 58.6 MW, respectively. It shows that the vibration area increases as the head increases.

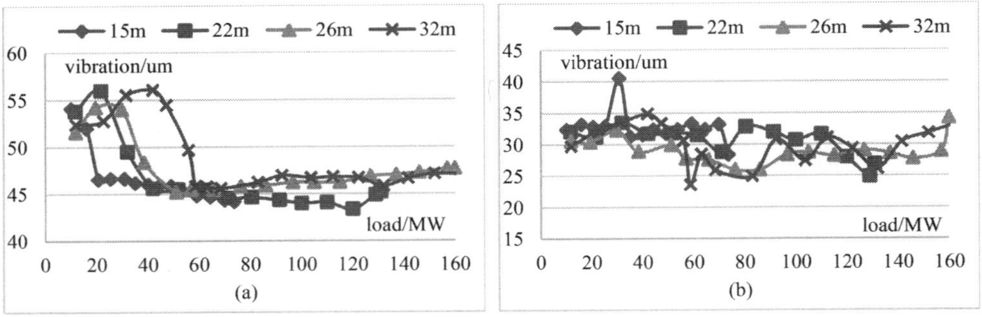

Figure 1. Vibration of upper bracket: (a) Horizontal vibration; (b) Vertical vibration.

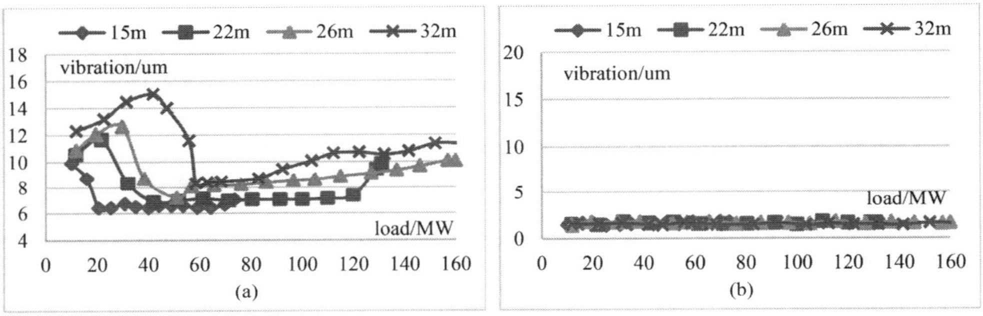

Figure 2. Vibration of lower bracket: (a) Horizontal vibration; (b) Vertical vibration.

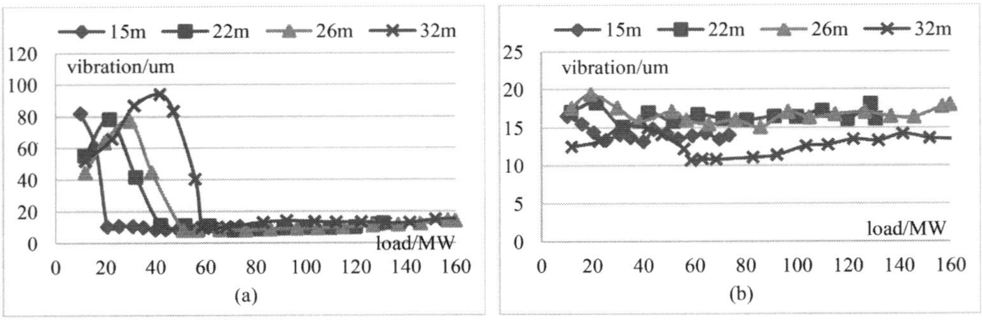

Figure 3. Vibration of thrust bearing support: (a) Horizontal vibration; (b) Vertical vibration.

3.2 *Influence on pressure*

The pressure pulsation results under different heads are indicated in Figure 7. The results of pressure pulsation are also affected by the amplitude of head variation, the boundary point between the

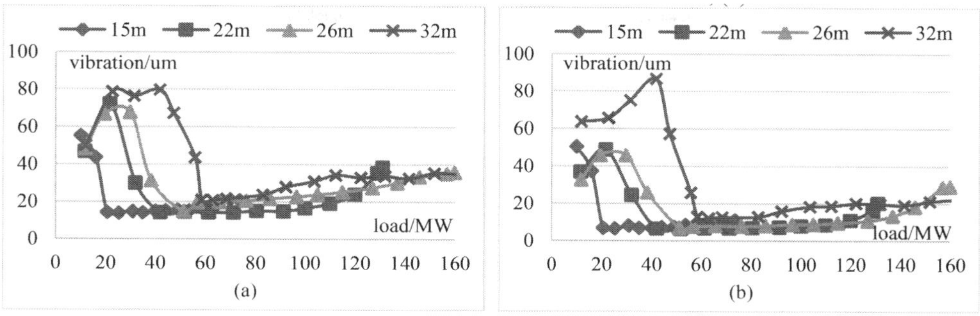

Figure 4. Vibration of head cover: (a) Horizontal vibration; (b) Vertical vibration.

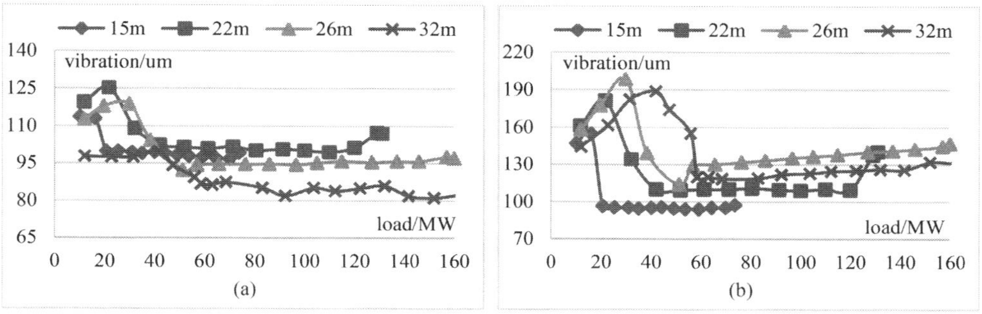

Figure 5. Vibration of guide shaft: (a) upper guide shaft; (b) lower guide shaft.

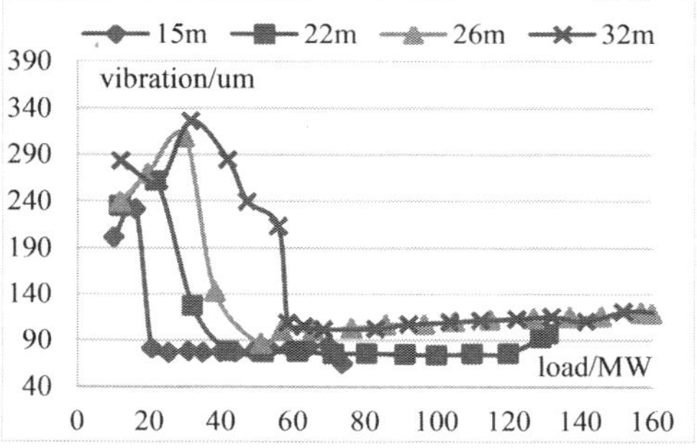

Figure 6. Vibration of turbine shaft.

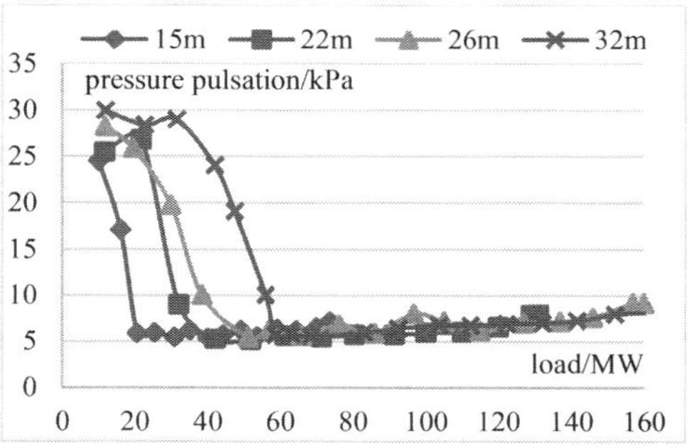

Figure 7. Pressure pulsation of draft tube.

vibration area and non-vibration area of the pressure pulsation under 15 m, 22 m, 26 m and 32 m heads are also 20.7 MW, 42.1 MW, 51.2 MW and 58.6 MW.

In addition, the average pressure of the draft tube under different heads and loads are listed in Table 4. As shown in Table 4, the pressure value under a single head decreases as the load increases, and the pressure at the same load under different heads also decreases with the increase of the head. For example, when the load is about 10 MW, the average pressure of the draft tube under 15 m, 22 m, 26 m and 32 m heads are 287.7 kPa, 216.8 kPa, 190.8 kPa and 128.3 kPa.

Table 3. The vibration area and non-vibration area of the unit under different head.

Head	Vibration area	Non-vibration area
m	MW	MW
15	0~20.7	20.7~73.7
22	0~42.1	42.1~131
26	0~51.2	51.2~160
32	0~58.6	58.6~161.3

Table 4. The average pressure of the draft tube.

15 m head		22 m head		26 m head		32 m head	
Load	Average pressure	Load	Average pressure	Load	Average pressure	Load	Average pressure
MW	kPa	MW	kPa	MW	kPa	MW	kPa
10.1	287.7	11.8	216.8	11.9	190.8	12.0	128.3
20.7	286.7	21.7	219.6	19.6	191.7	22.7	128.7
30.9	285.4	32.1	218.7	29.9	192.8	31.5	130.2
40.1	282.1	42.1	218.1	38.5	190.4	41.9	129.6
49.1	278.5	51.8	217.6	51.2	183.2	47.4	128.1
59.2	274.1	61.4	215.7	65.7	179.4	58.6	128.1
70.0	267.0	71.3	214.2	76.3	178.5	68.6	129.6
73.7	263.9	80.5	211.2	85.9	176.0	83.0	129.5
		91.4	207.9	96.7	174.0	92.4	128.3
		100.0	204.0	105.0	172.6	103.7	126.1
		110.0	200.1	115.0	169.4	112.3	124.9
		120.0	194.5	127.0	163.8	122.3	123.1
		131.0	185.9	137.0	160.4	132.1	121.1
				146.0	157.0	141.7	118.9
				157.0	152.7	152.0	116.6
				160.0	150.2	161.3	115.0

4 CONCLUSION

In this paper, the influence of stability in Kaplan turbine caused by the head variation is measured and obtained. The novelty of this paper primarily lies in two issues: 1) the vibration area of the unit will increase when the head increases and 2) the average pressure of the draft tube will decrease with the increase of the head under the same load.

REFERENCES

Chen, R. Tian, Y.J. Wang, X. & Peng, Z.N. (2016). Study on parameter selection and optimization design of pump-turbine with ultra-wide amplitude of head variation. J. *Water Resources and Hydropower Engineering* 47(1): 85–89.

Gao, S.W. (2018). Study on the unbalance of unit capacity of pumped storage power station under large head variation. J. *Zhi Huai* 2: 21–23.

Gong, R.Q. Chen, Y.L. & Qin, D.Q. (2020). High stability pump turbine with wide head variation. J. *Hydropower and Pumped Storage* 6(3): 34–36.

Huang, Z. & Huang, G.S. (2016). The research and implementation of the high range water head unit vibration cooperate with AGC control. J. *Guizhou Electric Power Technology* 19(1): 65–68.
Wang, X. Xue, P. & Peng, Z.N. (2021). Study on the turbine type selection for transnormal head variable amplitude hydro-power station. J. *Water Resources and Hydropower Engineering* 52(S1): 248–251.
Wu, G. Wei, C.X. Zhang, K.W. & Shong, L.R. (2000). Study of pressure fluctuations in draft tubes of Francis type hydraulic turbines under strong water head varying conditions. J. *Power Generation Equipment* 4: 22–27.

Energy Revolution and Chemical Research – Kok-Keong Chong and Zhongliang Liu (Eds)
© 2023 The Authors, ISBN: 978-1-032-36554-1

Electrochemical behavior and mechanism analysis of lithium rich materials $Li_{1.2}Mn_{0.54}Co_{0.13}Ni_{0.13}O_2$ coated with iron hydroxyphosphate

MoHan Wei & Xu Xiao
China Automotive Technology & Research Center Co. Ltd., Tianjin, China

Hui Sun*
China University of Petroleum, Beijing, China

TianYi Ma & WeiJian Hao
China Automotive Technology & Research Center Co. Ltd., Tianjin, China

ABSTRACT: The development of lithium-rich materials has always faced the key problems of voltage plateau attenuation and phase interface instability. In this study, $Li_{1.2}Mn_{0.54}Co_{0.13}Ni_{0.13}O_2$ coated with iron hydroxyphosphate (FeLLO) was prepared by the coprecipitation method, and its electrochemical properties were analyzed. The obtained material has a specific discharge capacity of 227.33 mAh/g in the first cycle, the capacity retention rate after 100 cycles is 82.3%, and the specific discharge capacity of 170.8 mAh/g can be maintained at a current density of up to 300 mA/g. Combining with the electrochemical characterization, in-situ Raman spectroscopy and potentiometric titration were used to analyze its optimization mechanism. It was found that iron hydroxyphosphate could form a stable lithium-ion channel during the first charging process, which effectively guaranteed the electrochemical stability of lithium-rich materials. This effective solution was proposed for the key problem of Li_2O precipitation in lithium-rich materials.

1 INTRODUCTION

Lithium-ion battery has been widely used in electric transportation and power grid energy storage. Its advantages such as long cycle life and no memory effect are considered to be the most suitable and potential chemical power source for the current social development (Tarascon 2001; Yoshino 2012). However, the rapid development of electric transportation technology puts forward higher requirements for the energy density of lithium-ion batteries (Etacheri 2011; Hesse 2017; Nitta 2015). In principle, the innovation of anode and cathode material in the electrochemical system is the fundamental way to improve the energy density (Marom 2011; Mukai 2014). Among the known cathode materials, the theoretical specific capacity of lithium-rich manganese-based cathode material (LLO) can be as high as 300 mAh/g, which is about twice the current commercial application of lithium iron phosphate and NCM materials. It is regarded as an ideal cathode material for the new generation of high-energy-density traction lithium-ion batteries (Ding 2018; Martha 2012; Peng 2019; Song 2013; Yin 2020). In addition, LLO materials have the advantages of being low cost, non-toxic and safe, which can meet the application requirements of lithium-ion batteries in the fields of electronic products, electric vehicles and energy storage (Li 2018; Redel 2019; Wang 2018).

*Corresponding Author: sunhui@cup.edu.cn

 DOI 10.1201/9781003332657-58

However, the industrialization and popularization of LLO still face many challenges, including voltage attenuation, irreversible capacity loss in the first cycle, and poor rate capability (Cui 2020; Hu 2020; Li 2015; Rana 2019). The main reasons for the instability of the electrochemical performance of LLO include unstable phase structure, poor conductivity and surface side reactions (Lu 2019; Lan 2019; Zheng 2018).

Surface structure optimization is an ideal way to solve the above problems. The surface optimization of LLO materials can improve their electrochemical performance from several aspects, including improving the lithium-ion conductivity, enhancing the conductivity of materials, inhibiting the surface phase transition of LLO materials, and establishing a more stable solid electrolyte interphase (SEI) (Lee 2018; Tomoya 2018). Metal oxides, phosphates and carbon-based materials are potential surface optimization materials; however, such materials have certain disadvantages of high cost and limited optimization effect (Pan 2018; Song 2013; Wang 2018). At present, there are few reports about a kind of material capable of solving the application problems of LLO materials in many aspects at the same time and have the advantages of being low cost and easily operated.

Iron hydroxyphosphate (FeHyPh) is widely used in the field of electrochemistry (Hautier 2011; Jugović 2009; Zhang 2011). It is an important precursor for the preparation of $LiFePO_4$ materials. It has ideal conductivity, ion diffusion rate and electrochemical stability(Deng 2014; Hassoun 2014). In this paper, LLO materials were prepared by a series of steps including coprecipitation, lithium addition and sintering. FeHyPh coated LLO materials (named FeLLO) were prepared by a simple and effective suspension coprecipitation method. Through the structural characterization and electrochemical analysis of FeLLO materials, it is proved that FeHyPh coating could significantly improve the electrochemical performance of LLO materials. Furthermore, the mechanism of FeHyPh coating optimizing the electrochemical performance of LLO was investigated by in-situ Raman spectroscopy and potentiometric titration. FeHyPh coating can stabilize the phase interface of LLO by establishing a lithium-ion channel to effectively improve its electrochemical performance, inhibit the irreversible phase transition reaction of LLO, and significantly improve the conductivity and lithium-ion diffusion of the coating composite. The results show that the first cycle discharge capacity of FeLLO material is 227.33 mAh/g, the capacity retention rate is 82.3% after 100 cycles, and the specific discharge capacity of 170.8 mAh/g can be maintained at a current density of 300 mA/g.

2 EXPERIMENT

2.1 *Preparation of LLO precursor*

The precursor material $Mn_{0.66}Co_{0.17}Ni_{0.17}O$ is prepared by the coprecipitation method. Specifically, $MnSO_4$, $CoSO_4$ and $NiSO_4$ are weighed at the molar ratio of 4:1:1, dissolved in deionized water, and then an equal molar amount of Na_2CO_3 is taken. Sodium citrate is added as a chelating agent, and ammonia is added to adjust pH to 11. Keep the reaction system under 80° and N_2 atmosphere, stir until coprecipitation is completed, and then keep aging for another 12 h. After centrifugation and washing, the precipitates were dried in vacuum at 110°C for 12 h, grounded into a fine powder and sintered at 500°C for 12 h at a heating rate of $1°C \cdot min^{-1}$ in a tubular furnace.

2.2 *Preparation of LLO*

The precursor is mixed with $LiOH \cdot H_2O$ in the molar ratio of 1:1.05, and then ground into fine powder. Next, the precursor is pre-burned at 550°C for 5h at a heating rate of $1°C \cdot min^{-1}$ in a tubular furnace and then heated to 900°C for 20 h. After sintering, $Li_{1.2}Mn_{0.54}Co_{0.13}Ni_{0.13}O_2$ (LLO) is obtained by grinding powder.

2.3 *Preparation of FeLLO*

LLO is mixed with $(NH_3)_2HPO_4$, deionized water and sodium citrate as chelating agents. Add the $FeCl_2$ solution slowly, keep the reaction system under 60°C and N_2 atmosphere, stir until coprecipitation is completed, and then keep aging in a nitrogen environment for 12 h. After centrifugation and washing, dry the product in vacuum at 110°C for 12 h. Finally, FeLLO [$Li_{1.2}Mn_{0.54}Co_{0.13}Ni_{0.13}O_2$ @ $Fe_4(PO4)_3(OH)_3$] can be obtained after being sintered at 500°C for 12 h at a heating rate of 1°C·min^{-1} in a tubular furnace.

3 RESULTS & DISCUSSION

As described above, FeHyPh coated $Li_{1.2}Mn_{0.54}Co_{0.13}Ni_{0.13}O_2$ material (FeLLO) was obtained by precursor preparation, lithium addition sintering and coating. To verify the crystal structure of the materials, X-ray diffraction (XRD) analysis was carried out for LLO and FeLLO with different FeHyPh ratios (1% FeLLO, 3% FeLLO and 5% FeLLO), as shown in Figure 1a. The diffraction peaks of LLO have an obvious α-$NaFeO_2$ structure. The splitting of (006)/(012) and (108)/(110) double peaks is obvious, which indicates a well-layered structure. $I_{(003)}/I_{(104)}$ at 1.367 reveals the low level of mixing degree of Li^+ and Ni^{2+}. Double peaks at 20~25° are the characteristic peaks of Li_2MnO_3, corresponding to the peaks of (020)/(110) crystal plane in Li_2MnO_3 exclusively, which are diffraction peaks of Li and Mn arranged orderly in the transition metal layer of $LiMn_6$ superlattice structure of Li_2MnO_3. Such results indicate that the Li_2MnO_3 structure is generated in the prepared LLO. There is no obvious difference in XRD peaks of FeLLO with different coating degrees. The (020)/(110) double peaks at 20~25° exist in all samples' XRD patterns, which indicates that the coating does not affect the layered structure and Li_2MnO_3 structure of LLO. The characteristic peak of FeHyPh was not observed in Figure 1 due to the low surface coating degree. To analyze whether the structure of FeHyPh in FeLLO remains stable, we prepared 20% wt. of FeLLO and carried out an XRD test. The characteristic peak of FeHyPh was found at about 12.8° which indicated that FeHyPh kept a stable structure during the coating reaction.

The particle morphology of the composite was observed by scanning electron microscope (SEM). As shown in Figure 2, the morphology of LLO is similar to that of 1% FeLLO, 3% FeLLO and 5% FeLLO particles, both of which are secondary particles formed by primary agglomeration of polygonal particles. The primary particle size is about 500nm, and the secondary particle size is about 3 μm. The similar morphology and particle size of FeLLO and LLO reveal that the coating process of FeHyPh has no effect on the morphology of LLO.

In principle, excellent core-shell materials need to reduce the thickness of the coating layer as much as possible on a complete layered structure. In order to make the electrochemical performance of LLO materials reach optimized status, at the same time find out the best ratio of FeHyPh and LLO in the composites, 1% FeLLO, 3% FeLLO and 5% FeLLO were characterized by high-resolution TEM to observe the state of the surface coating. Figures 3a and 3b are the detailed morphologic images of LLO materials. An obvious stripe pattern and clear edges can be seen in Figure 3b, which proves the layered structure of LLO. From Figure 3b, the lattice fringe spacing is 4.7 Å, corresponding to the (001) lattice plane of Li_2MnO_3, which further proves the existence of Li_2MnO_3in LLO. Figures 3c and 3d are detailed morphology images of 3% FeLLO. Under the higher resolution, it can be found that a coating layer with about 15 nm thickness is formed on the edge of the LLO, and the clear layered structure inside can be seen, which demonstrates that the regular coating of FeHyPh material on the surface of LLO. The ratio of raw materials used in the above experiments is appropriate and acceptable.

The electrochemical performance of prepared materials was tested as 2032 type coin half-cells using lithium metal as a counter electrode. Figure 4a is the attenuation curve of discharge capacity of LLO and FeLLO at constant current. The potential range is 2.0 ~ 4.8 V (vs. Li/Li^+), and the current density is controlled to 30 mA/g. The first cycle discharge capacity of LLO is 213.5 mAh/g. After 100 cycles, the specific capacity decreases to 96.5 mAh/g and the capacity retention

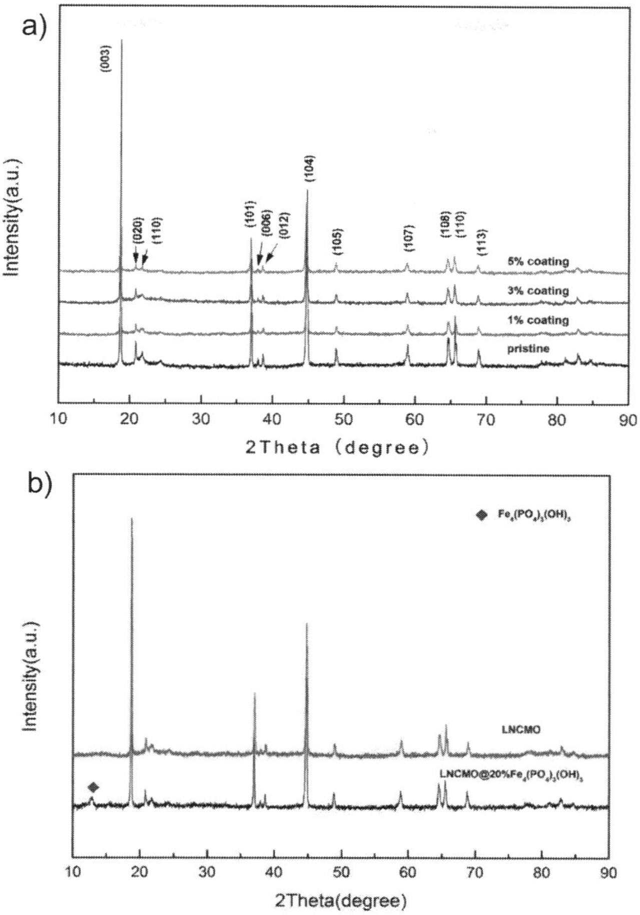

Figure 1. XRD patterns of the product. (a)LLO and 1%-5%FeLLO. (b)LLO and 20%FeLLO.

rate is 45.2%. From Figure 4a, it can be observed that the cycle stability of all FeLLO materials coated with different proportions is better than that of LLO materials. Specifically, after 100 cycles, discharge capacities of 1%, 3%, 5% FeLLO are 142.9 mAh/g, 187.1 mAh/g and 172.2 mAh/g, respectively, corresponding to capacity retention rates are 62.8%, 82.3% and 79.6%. Figure 4b shows the discharge capacity variation of LLO and FeLLO under different current density cycles. Among the materials, 3% FeLLO has the best performance. The first discharge capacity is 226.9 mAh/g, 214.8 mAh/g, 201.4 mAh/g and 170.8 mAh/g at the current density of 30 mA/g, 60 mA/g, 150 mA/g and 300 mA/g, respectively. According to the experimental results in Figure 4, 3% FeLLO shows better electrochemical performance and surface modification, which is consistent with the best coating structure of FeLLO in TEM morphology characterization.

The cycle stability of 3% FeLLO is about twice as high as that of LLO, and the discharge capacity at high current is increased by about 30%. To compare and analyze the electrochemical behavior of the two materials, we analyzed the electrochemical characteristic curves of the two materials with different cycle numbers at 30 mA/g current density. Figs.5a and 5b show the capacity curves of LLO and 3% FeLLO at different cycles. From the curves of the two materials, two electrochemical platforms were observed during the first cycle of charging. The 2~4.5V charging platform is caused by the release of Li^+ from $Li_2Mn_{1/3}Co_{1/3}Ni_{1/3}O_2$ in LLO, while the 4.5~4.8V platform

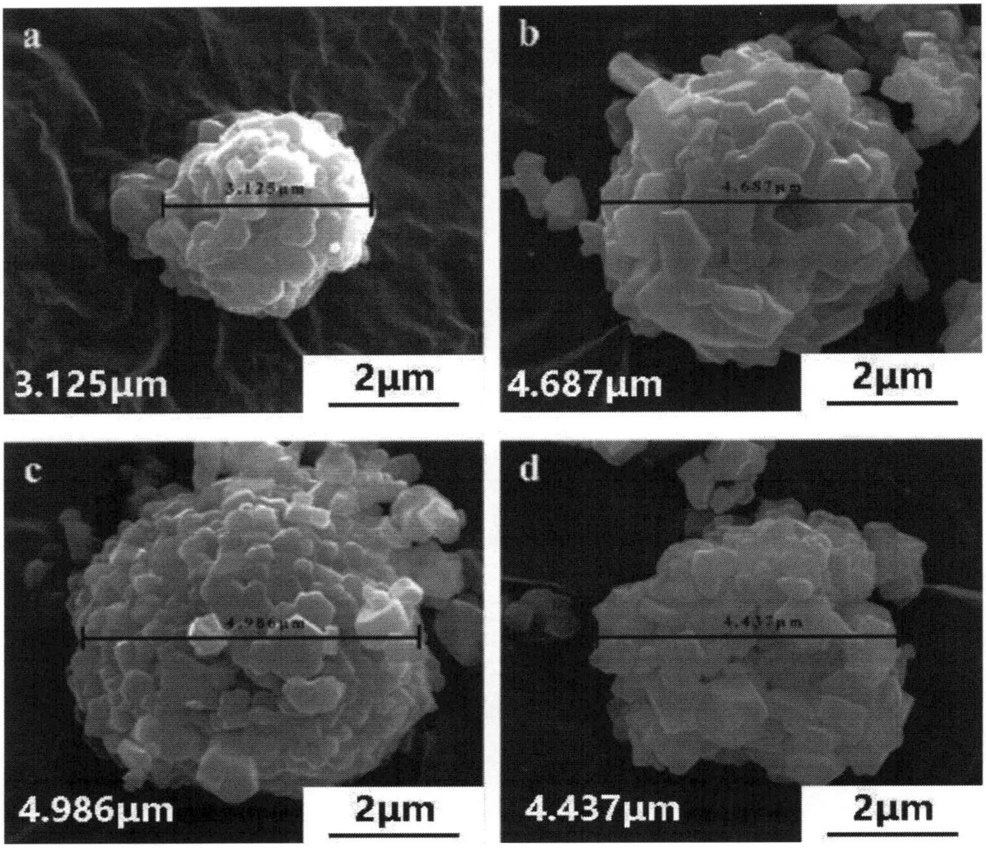

Figure 2. SEM images of the samples. (a)LLO, (b)1%FeLLO, (c)3%FeLLO, (d)5%FeLLO.

comes from the release of Li^+ from Li_2MnO_3 structure in LLO. The results show that the voltage plateau of LLO decreases from 3.7V to 3.0V, while that of FeLLO only decreases from 3.7V to 3.5V, which convinces that FeHyPh coating can effectively restrain the rapid voltage plateau degradation of LLO materials. In addition, it can be found that the Coulomb efficiency of FeLLO is 72.9%, which is 25.6% higher than that of LLO. We further analyzed the reason for this phenomenon of FeLLO in the first cycle by cyclic voltammetry (CV) test. Figs.5c and 5d show the CV curves of LLO and FeLLO, with the scanning rate at 0.1 mV s^{-1}. It can be observed that there are two oxidation peaks (4.01V and 4.65V) in the CV curves of LLO and FeLLO during the first cycle of charging. Among them, 4.01 V corresponds to the oxidation peaks of Ni^{2+} and Co^{3+}, and 4.65V corresponds to the oxygen loss of Li_2MnO_3, which does not appear in the charge curves of later cycles, indicating that the reaction of Li_2MnO_3 to Li_2O is irreversible. In the discharge curve, the peak at 3.7 V corresponds to the reduction process of Mn^{4+}, Ni^{3+} and Co^{4+}. It should be noted that the small peak of reduction potential corresponding to Mn^{4+}/Mn^{3+} of the two materials at 3.3V is not the same. Specifically, the reduction peak of LLO in the first cycle is not obvious, and the second and third cycles are hardly observed, while the reduction peak of 3.3V was observed in the first three cycles of FeLLO. This result shows that the coated structure of FeLLO can protect the internal Li_2MnO_3 structure more effectively and keep the electrochemical activity during the charge discharge cycle. On the other hand, the initial oxidation potential of LLO is about 3.65V, while that of FeLLO is about 3.75V. The increase in initial oxidation potential of FeLLO is due to

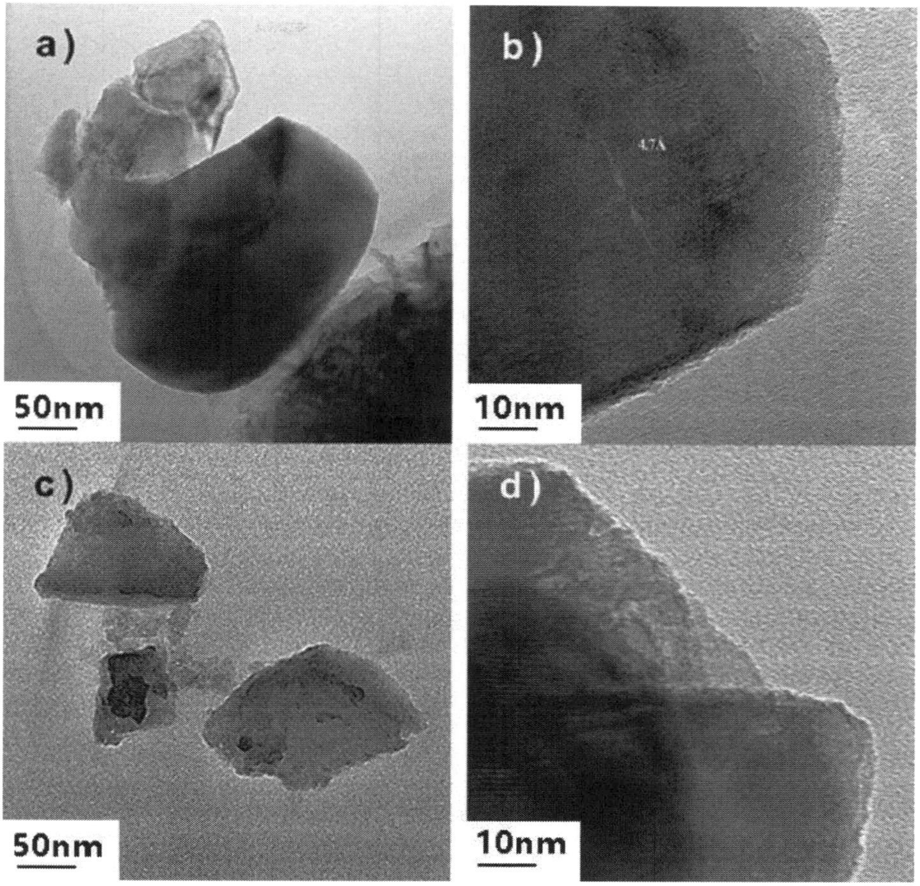

Figure 3. High resolution TEM images. (a, b) LLO, (c-d) 3%FeLLO.

the surface coating structure effectively preventing the lithium ions in LLO from escaping from the layered structure and directly entering the electrolyte. However, in the subsequent cycle, lithiation of FeLLO and lithium-ion channel render the initial oxidation potential the same as that of LLO.

The electrochemical process analysis revealed the difference in electrochemical reaction between the two materials. In order to further explore the electrochemical reaction mechanism of the two materials and the promotion effect of FeHyPh coating on the performance of LLO, we compared and analyzed the molecular structure changes of LLO and 3% FeLLO in the charge and discharge process by in-situ Raman spectroscopy. Figure 6a shows the Raman shift curves of the LLO half-cell at different potentials. The corresponding peaks of O-M-O (480 cm^{-1}) and MO_6(600 cm^{-1}) appear in the Raman spectra at the initial voltage of 2.8V. During the charging process after 4.0V, the peak intensity of MO_6/O-Mn-O gradually recedes to disappear, and the characteristic double peaks of Li_2CO_3 appear at 700 cm^{-1} ~ 800 cm^{-1} and increase gradually. This reveals that Li_2O will be produced when the potential is greater than 4V during the first cycle of charging. Figure 6b shows the Raman shift curve of 3%FeLLO half-cell under the same conditions. The characteristic peaks of O-M-O and MO_6 and the peak intensity ratio of MO_6/O-Mn-O decreased gradually with the charging process, but no Li_2O characteristic peak was observed. In order to show the results of in-situ Raman spectroscopy more intuitively, we plot the data as equipotential maps, as shown in Figs.6c and 6d. By comparing the equipotential diagrams of LLO and 3% FeLLO, the decreasing

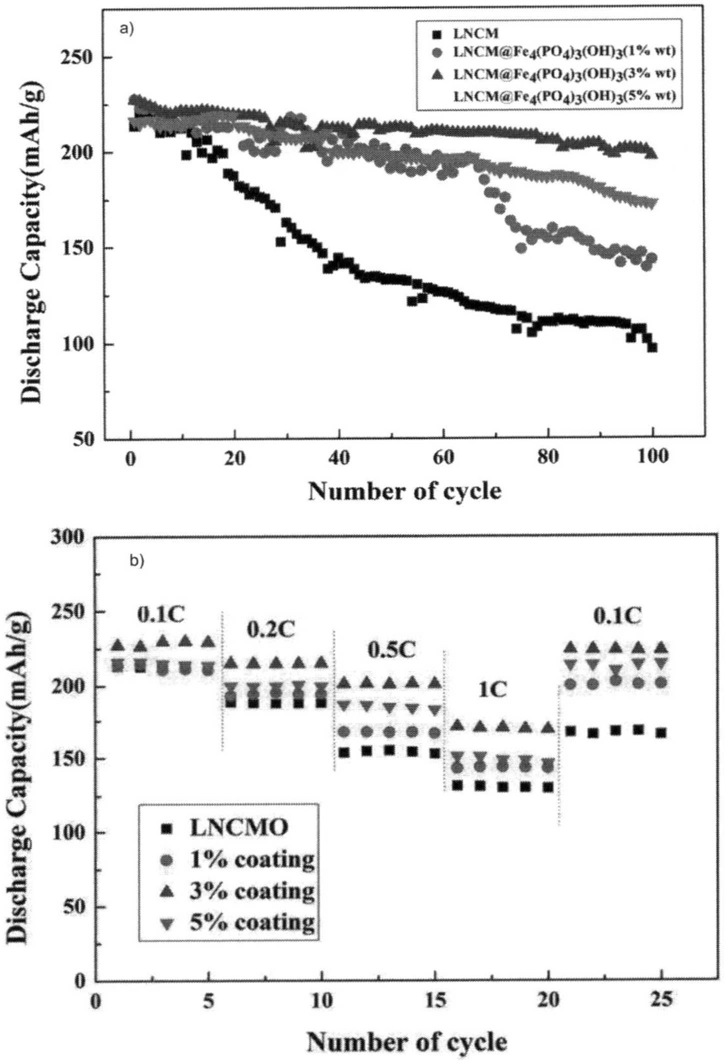

Figure 4. Electrochemical performance. (a) Cycle performance at constant current. (b) Discharge capacity under different current densities.

trend of MO_6/O-Mn-O peak intensity ratio in 3% FeLLO is significantly eased, and there is almost no characteristic peak of Li_2O. Combined with the results of in-situ Raman spectroscopy, it can be concluded that FeHyPh coating can alleviate the precipitation of Li_2O.

The result of in-situ Raman spectroscopy directly shows the evolution of the molecular structure of LLO and FeLLO during the charging process, and further reveals the mechanism of improving the electrochemical performance of LLO by FeHyPh coating. Combined with the results of electrochemical analysis, we use a schematic diagram to describe the different electrochemical processes of LLO and FeLLO. As shown in Figure 7, FeHyPh has excellent electrochemical stability and high conductivity. As a stable surface coating layer, FeHyPh can improve the cycle performance, rate performance and Coulomb efficiency of the composites. In addition, FeHyPh has a good pore structure and can be used as a "lithium container". During the first charging process, the lithium-ion

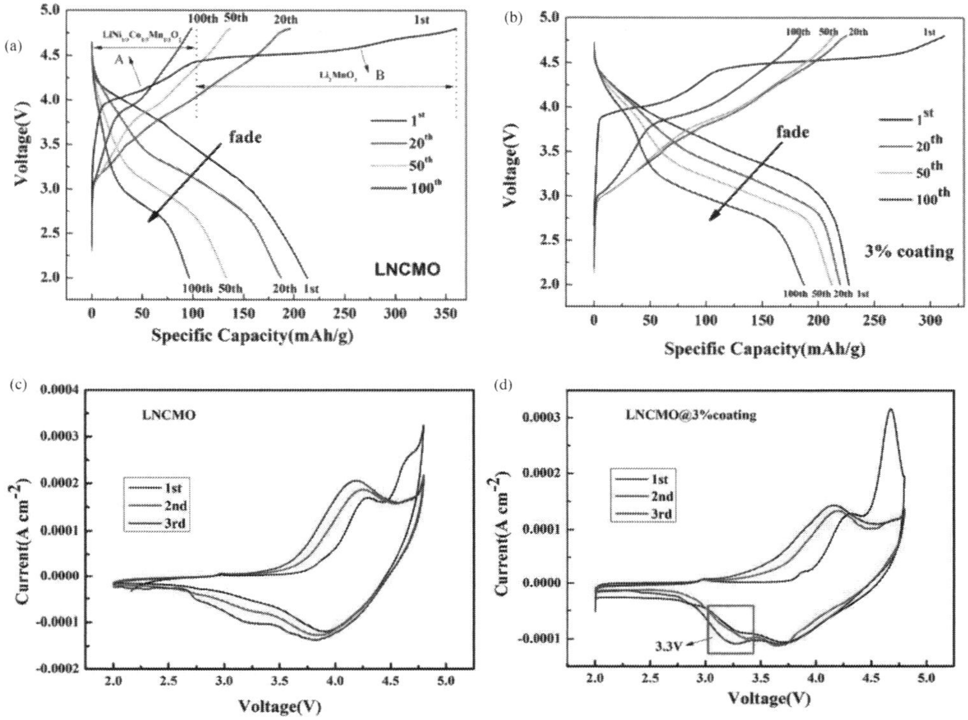

Figure 5. (a), (b) Charge-discharge profiles of LLO and FeLLO. (c), (d) Cyclic voltammetry curves of LLO and FeLLO.

channel between the LLO and the electrolyte is formed, which can effectively promote the stability of the internal LLO in the process of lithium insertion and extraction, and alleviate the irreversible phase transition of Li_2O precipitation.

To verify that the lithium-ion channel promoting the electrochemical stability of LLO was established during the first charging process, the diffusion coefficient of lithium ion in core-shell structure materials was analyzed by potentiometric titration (PITT). When the step potential ΔE is set at 0.2V, the diffusion coefficient of lithium ion in active substance is obtained by recording the change curve of current with time at different potentials and substituting the slope of the logarithmic curve of step current with time into the following formula for calculation:

$$D_{Li} = -\frac{d\ln(\mathrm{I})}{d\,\mathrm{t}}\frac{4L^2}{\pi^2} \tag{1}$$

Figure 8 shows the variation of lithium-ion diffusion coefficient with potential during the first charge and discharge of LLO and FeLLO. D_{Li} of LLO is higher than that of FeLLO in the potential range of 3.4 - 4V. When the voltage is higher than 4V, D_{Li} of FeLLO increases significantly and is always higher than that of LLO in the subsequent charging and discharging process. The results show that more lithium ions pass through and occupy the vacancies of FeHyPh layer at the first charge of 4V, and form a good lithiumion channel, which makes the D_{Li} of FeLLO increase significantly. In addition, in the subsequent electrochemical process, the lithiumion channel of FeHyPh layer is always stable, so it can always maintain a higher D_{Li}. This result is consistent with the previous electrochemical characterization and the mechanism of FeHyPh layer optimizing the electrochemical performance of LLO.

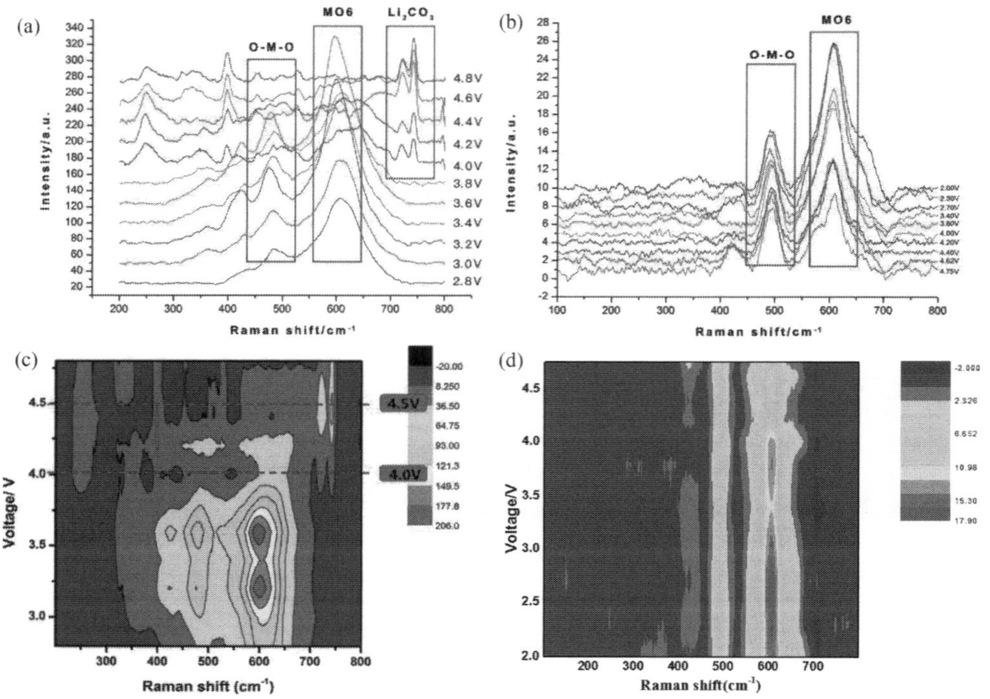

Figure 6. (a), (b) In situ Raman spectroscopy of LLO and FeLLO. (c), (d) Raman shift isoelectric diagrams of LLO and FeLLO.

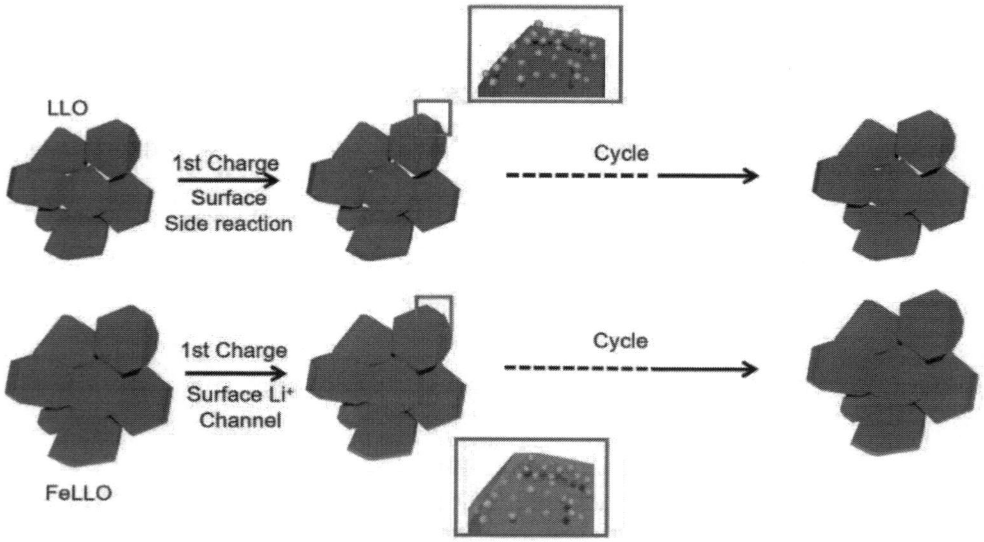

Figure 7. Schematic diagram of improving the electrochemical performance of FeLLO.

4 CONCLUSION

In this paper, $Li_{1.2}Mn_{0.54}Co_{0.13}Ni_{0.13}O_2$material was prepared by precursor preparation, lithium addition and sintering. The electrochemical performance of $Li_{1.2}Mn_{0.54}Co_{0.13}Ni_{0.13}O_2$ was

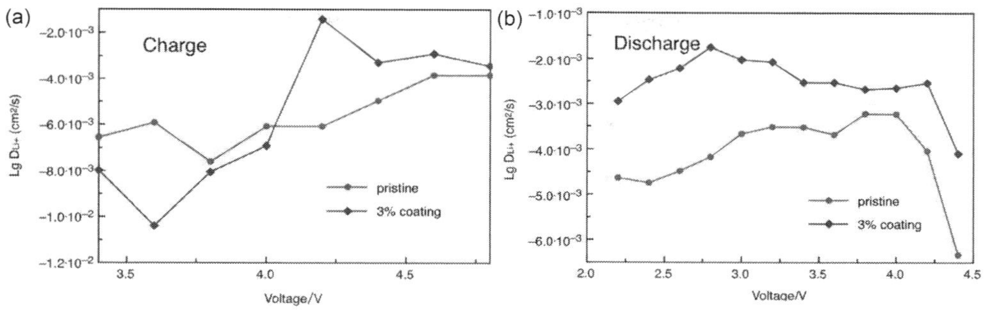

Figure 8. Variation of lithium-ion diffusion coefficient with potential during the first charge and discharge of LLO and FeLLO.

improved by simple liquid-phase coprecipitation and ferric hydroxyphosphate coating. The electrochemical comparison experiments show that the material coated with 3% FeHyPh can obtain the optimal electrochemical performance, with a discharge-specific capacity of 227.33 mAh/g in the first cycle, and the capacity retention rate is 82.3% after 100 cycles. Also, the discharge-specific capacity of 170.8 mAh/g can be maintained at a current density of up to 300 mA/g. Furthermore, we combined electrochemical characterization, in-situ Raman spectroscopy and potentiometric titration to reveal the mechanism by which FeHyPh can optimize the electrochemical performance of LLO: the appropriate pore structure and excellent electrochemical stability of FeHyPh make it form a lithium-ion channel between the LLO and the electrolyte during the first charging process, which effectively guarantees the stability of LLO in the process of lithium insertion and dilithium, and inhibits the Li_2O precipitation which is one of the most important problems in similar materials. Such this method is easily operated, low-cost, and can be applied to all kinds of lithium-rich manganesebased materials, which has a very high industrial application potential.

REFERENCES

Chen, J. 2018. Layered $Li[Li_{0.2}Ni_{0.133}Co_{0.133}Mn_{0.534}]O_2$ with porous sandwich structure as high-rate cathode materials for Li-ion batteries *Mater. Lett*, 217(2018) 284–287.

Cui, C. 2020. Structure and Interface Design Enable Stable Li-Rich Cathode *J. Am. Chem. Soc*, 142 (2020) 8918–8927.

Deng, S. 2014. Research Progress in Improving the Rate Performance of $LiFePO_4$ Cathode Materials *Nano-Micro Letters*, 6 (2014) 209–226.

Ding, X. 2018. Improving the electrochemical performance of Li-rich $Li_{1.2}Ni_{0.2}Mn_{0.6}O_2$ by using Ni-Mn oxide surface modification *J. Power Sources*, 390(2018) 13–19.

Etacheri, E. 2011. Challenges in the development of advanced Li-ion batteries: a review *Energy & Environmental Science*, 4 (2011) 3243–3262.

Hassoun, J. 2014. An advanced lithium-ion battery based on a graphene anode and a lithium iron phosphate cathode *Nano Lett*, 14 (2014) 4901–4906.

Hautier, G. 2011. Phosphates as lithium-ion battery cathodes: an evaluation based on high-throughput ab initio calculations *Chem. Mater*, 23 (2011) 3495–3508.

Hesse, H. 2017. Lithium-ion battery storage for the grid—A review of stationary battery storage system design tailored for applications in modern power grids *Energies*, 10 (2017) 2107–2149.

Hu, W. 2020. Mitigation of voltage decay in Li-rich layered oxides as cathode materials for lithium-ion batteries *Nano Research*, 13 (2020) 151–159.

Jugović, D. 2009. A review of recent developments in the synthesis procedures of lithium iron phosphate powders *J. Power Sources*, 190 (2009) 538–544.

Lan, X. 2019. Stabilizing Li-rich layered cathode materials by nanolayer-confined crystal growth for Li-ion batteries *Electrochim. Acta*, 333 (2019) 135466.

Lee, H. 2018. Characterization and Control of Irreversible Reaction in Li-Rich Cathode during the Initial Charge Process *ACS Appl. Mater. Interfaces*, 10 (2018) 10804–10818.

Li, J. 2015. Improve First-Cycle Efficiency and Rate Performance of Layered-Layered $Li_{1.2}Mn_{0.6}Ni_{0.2}O_2$ Using Oxygen Stabilizing Dopant *ACS Appl. Mater. Interfaces*, 7 (2015) 16040–16045.

Li, H. 2018. Improving rate capability and decelerating voltage decay of Li-rich layered oxide cathodes by chromium doping *Int. J. Hydrogen Energy*, 43 (2018) 11109–11119.

Lu, L. 2019. Revealing the Electrochemical Mechanism of Cationic/Anionic Redox on Li-Rich Layered Oxides via Controlling the Distribution of Primary Particle Size *ACS Appl. Mater. Interfaces*, 11 (2019) 25796–25803.

Martha, S.K. 2012. Surface Studies of High Voltage Lithium Rich Composition: $Li_{1.2}Mn_{0.525}Ni_{0.175}Co_{0.1}O_2$ *J. Power Sources*, 199 (2012) 220–226.

Marom, R. 2011. A review of advanced and practical lithium battery materials *J. Mater. Chem*, 21 (2011) 9938–9954.

Mukai, K. 2014. Factors Affecting the Volumetric Energy Density of Lithium-Ion Battery Materials: Particle Density Measurements and Cross-Sectional Observations of Layered $LiCo_{1-x}Ni_xO_2$ with $0 \leq x \leq 1$ *ACS Appl. Mater. Interfaces*, 6 (2014) 10583–10592.

Nitta, N. 2015. Li-ion battery materials: present and future *Materials today*, 18 (2015) 252–264.

Pan, H. 2018. Li- and Mn-rich layered oxide cathode materials for lithium-ion batteries: a review from fundamentals to research progress and applications *Molecular Systems Design & Engineering*, 3 (2018) 10.1039.C1038ME00025E.

Peng, Z. 2019. Enhanced electrochemical performance of layered Li-rich cathode materials for lithium-ion batteries via aluminum and boron dual-doping *Ceram. Int*, 45(2019) 4184–4192.

Rana, J. 2019. Structural Changes in a Li-Rich $0.5Li_2MnO_3$*$0.5LiMn_{0.4}Ni_{0.4}Co_{0.2}O_2$ Cathode Material for Li-Ion Batteries: *A Local Perspective J. Electrochem. Soc*, 163 (2019) A811–A820.

Redel, K. 2019. High-Performance Li-Rich Layered Transition Metal Oxide Cathode Materials for Li-Ion Batteries *J. Electrochem. Soc*, 166 (2019) A5333–A5342.

Song, B. 2013. Graphene-based surface modification on layered Li-rich cathode for high-performance Li-ion batteries *Journal of Materials Chemistry A*, 34(2013) 9954–9965.

Tarascon, J.M. 2001. Issues and challenges facing rechargeable lithium batteries *Nature*, 414 (2001) 359–367.

Tomoya, K. 2018. Strain-Induced Stabilization of Charged State in Li-Rich Layered Transition-Metal Oxide for Lithium-Ion Batteries *Journal of Physical Chemistry C Nanomaterials & Interfaces*, 122(2018) 19298–19308.

Wang, D. 2018. Integrated Surface Functionalization of Li-rich Cathode Materials for Li-ion Batteries *ACS Appl. Mater. Interfaces*, 10(2018) 41802–41813.

Yin, Z. 2020. High-capacity Li-rich Mn-based Cathodes for Lithium-ion Batteries *Chinese Journal of Structural Chemistry*, 303 (2020) 25–30.

Yoshino, A. 2012. The birth of the lithium-ion battery *Angewandte Chemie International Edition*, 51 (2012) 5798–5800.

Zhang, W.J. 2011. Structure and performance of $LiFePO_4$ cathode materials: A review *J. Power Sources*, 196 (2011) 2962–2970.

Zheng, J. 2018. Effect of calcination temperature on the electrochemical properties of nickel-rich $LiNi_{0.76}Mn_{0.14}Co_{0.10}O_2$ cathodes for lithium-ion batteries *Nano Energy*, (2018) 538–548.

Energy Revolution and Chemical Research – Kok-Keong Chong and Zhongliang Liu (Eds)
© 2023 The Authors, ISBN: 978-1-032-36554-1

Modeling and analysis of hydrogen-oxygen fuel cell

Benhai Chen, Dongchen Qin*, Tingting Wang, Jiangyi Chen & Ruikang Zhao
School of Mechanical and Power Engineering, Zhengzhou University, Zhengzhou, Henan

ABSTRACT: With the development of new energy vehicles, fuel cell vehicles are considered an ideal means of transportation in the future due to their use of clean energy. As the core component of fuel cell vehicles, proton exchange membrane fuel cell directly affects the performance of the whole vehicle, hence, the research on proton exchange membrane fuel cell is extremely significant. In this paper, the theoretical analysis of proton exchange membrane fuel cell is carried out, and the simulation model of MATLAB/Simulink is established. Through the fuel cell voltage and power output at a different temperature, different membrane water contents, and different gas partial pressures, the influences of temperature, membrane water contents and gas partial pressure on fuel cell performance were analyzed (Jixin 2019), and the relationship between different factors and fuel cell performance were investigated. It is concluded that with the increase in temperature, membrane water content and gas partial pressure, the fuel cell voltage and power also increase by different degrees.

1 INTRODUCTION

A fuel cell is a battery that converts chemical energy directly into electricity. Its fuel efficiency is very high, energy efficiency up to about 70%, nearly 40% higher than the general energy generation efficiency. It is a very ideal power generation device.

Due to the high complexity of the fuel cell internal system, there are different considerations for fuel cell modeling (Hannan 2014). Verbrugge and Bernardi only considered the gas transport in the Y direction. They developed a one-dimensional model of PEM fuel cell, which has few calculations and a simple structure. Springer et al. introduced the gas state equation to establish an isothermal, steady-state one-dimensional model. Du Xin et al. studied the law of gas diffusion and established a one-dimensional MODEL of PEMFC that could accurately reflect the influence of inlet gas partial pressure of anode and cathode and battery temperature on the voltage of a single battery. Mann et al. established a two-dimensional model considering the effects of inlet and outlet partial pressure (Azooz 2011), current density, cell temperature, film thickness and activation area on the output characteristics of fuel cells. Du Chunyu et al. established a THREE-DIMENSIONAL hydrodynamic PEMFC model considering the structure of the catalytic layer.

This paper mainly studies the influencing factors of fuel cell performance and explores the rules between different factors and fuel cell performance. The fuel cell simulation model is established according to the research requirements and the actual situation.

2 FUEL CELL MODELING

2.1 *Basic working principle of proton exchange membrane fuel cell*

The main reaction of a hydrogen-oxygen fuel cell is the electrochemical reaction of hydrogen and oxygen to produce water and heat.

*Corresponding Author: zzdxjxsz@163.com

DOI 10.1201/9781003332657-59

The reaction principle of the hydrogen-oxygen fuel cell is as follows:

The anode is: $H_2 \rightarrow 2H + 2e^-$ (1)

The cathode is: $\frac{1}{2}O_2 + 2H^+ + 2e^- \rightarrow H_2O$ (2)

The total chemical reaction can be expressed as:

$$H_2 + \frac{1}{2}O_2 \rightarrow H_2O + \text{Electricity} + \text{Heatenergy}$$

2.2 *Loss of fuel cell polarization voltage*

In fuel cells, the irreversible loss that affects the open-circuit voltage drop is polarization loss potential. The polarization voltage loss includes three types: active polarization, ohm polarization and concentration polarization. The actual voltage of the fuel cell is the ideal voltage minus the voltage loss due to polarization. The polarization characteristic curve of the fuel cell is shown in Figure 1.

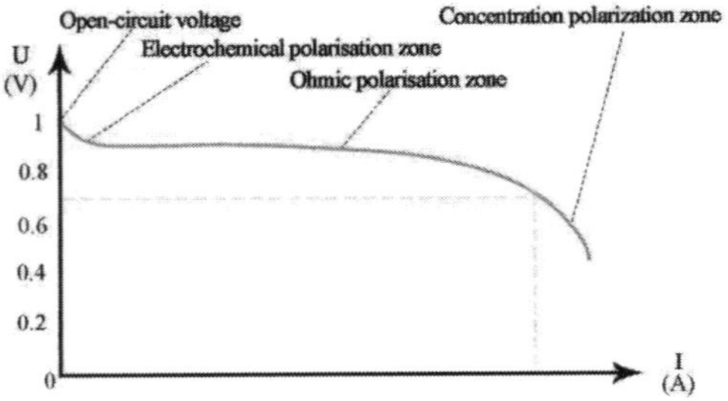

Figure 1. Polarization characteristic curve.

The voltage output of a single cell of hydrogen and oxygen fuel cell is

$$V_{cell} = E_{Nernst} - V_{act} - V_{ohm} - V_{con} \tag{3}$$

Where:
V_{cell} —Actual voltage of a single cell in a hydrogen-oxygen fuel cell;
E_{Nernst} —Thermodynamic electromotive force;
V_{act} —Loss of activation polarization voltage;
V_{ohm} —ohm polarization voltage loss;
V_{con} —Loss of concentration polarization voltage.

It can be seen from Figure 1 that on the polarization characteristic curve of H-O fuel cell, the electrochemical polarization region is activated polarization. With the increase of current density, the activation polarization loss occurs first, then the ohm polarization loss occurs, and finally the concentration polarization loss occurs in the concentration polarization region.

2.3 *Thermodynamic electromotive force of fuel cell*

The thermodynamic electromotive force can be expressed by the Nernst equation:

$$E_{Nernst} = \frac{\Delta G}{2F} + \frac{\Delta S}{2F}(T - T_{ref}) + \frac{RT}{2F}\left[\ln\left(P_{H_2}\right) + \frac{1}{2}\ln\left(P_{O_2}\right)\right] \tag{4}$$

Where:
E_{Nernst} ——Thermodynamic electromotive force;

ΔG ——The cell responds to changes in Gibbs free energy;
F ——Faraday constant;
ΔS ——The change in entropy of the battery reaction;
R ——Gas constant;
P_{H_2} ——Partial pressure of hydrogen gas;
P_{O_2} ——Partial pressure of oxygen gas;
T ——Battery temperature;
T_{ref} ——Reference temperature.

Substituting ΔG, ΔS and T_{ref} at normal temperature and standard atmospheric pressure into Equation (2), then we can get:

$$E_{Nernst} = 1.229 - 8.5 \times 10^{-4} \times (T - 298.15) + 4.308 \times 10^{-5} \times T \times \left[\ln\left(P_{H_2}\right) + \frac{1}{2}\ln\left(P_{O_2}\right)\right] \tag{5}$$

2.3.1 *Activation polarization of hydrogen-oxygen fuel cells*

Active polarization is caused by the movement of electrons between the anode and cathode and the destruction and formation of chemical bonds caused by chemical reactions at the anode and cathode of the h-O fuel cell.

The process of activation reaction is expressed by Tafel equation:

$$V_{act} = V_0 + V_a\left(1 - e^{\frac{-I_{fc}}{C_a}}\right) \tag{6}$$

Where:
V_0 —Active polarization voltage drop when current density is 0;
V_a —The coefficient;
I_{fc} —Current density;
C_a —The coefficient.

The values of V_0 and V_a can be processed by nonlinear regression fitting.

$$\begin{aligned} V_0 = {} & 0.28 - 8.5 \times 10^{-4}\left(T_{fc} - 298.15\right) + 4.3 \\ & \times 10^{-5} T_{fc}\left[\ln\left(\frac{P_{acthode} - P_{sat}}{1.01325}\right) + \frac{1}{2}\ln\left(\frac{0.12\left(P_{acthode} - P_{sat}\right)}{1.01325}\right)\right] \end{aligned} \tag{7}$$

$$\begin{aligned} V_a = {} & \left(-1.62 \times 10^{-5} T_{fc} + 1.62 \times 10^{-2}\right)\left(\frac{P_{O_2}}{0.1173} + P_{sat}\right)^2 \\ & + \left(1.8 \times 10^{-4} T_{fc} - 0.17\right)\left(\frac{P_{O_2}}{0.1173} + P_{sat}\right) + \left(-5.8 \times 10^{-4} T_{fc} + 0.5736\right) \end{aligned} \tag{8}$$

Where:
$P_{acthode}$ —Total cathode input gas pressure;
P_{sat} —Vapor saturation pressure
P_{sat} is related to the temperature of the battery, and the formula is:

$$P_{sat} = \exp\left[\frac{s_1}{T_{fc}} + s_2 + s_3 T_{fc} + s_4 T_{fc}^2 + s_5 T_{fc}^3 + s_6 \lg\left(T_{fc}\right)\right] \tag{9}$$

Where $s_1 \rightarrow s_6$ is the fitting coefficient.

2.3.2 *Ohm polarization of hydrogen-oxygen fuel cells*

Ohm polarization is caused by the obstruction of hydrogen ions through the proton exchange membrane and electrolyte of the hydrogen-oxygen fuel cell and the obstruction of electric charges through the cell material. Ohm polarization is mainly divided into two parts, one is the resistance that prevents hydrogen ions from crossing the proton membrane as they move between the poles, and another is the resistance that prevents the charge generated on the electrode from moving to the cathode.

Its expression is as follows:

$$V_{ohm} = i \times A_{cell} \times R_{ohm} = i \times A_{cell} \times \left(\frac{\delta_t}{\sigma A_{cell}}\right) = i \times \left(\frac{\delta_t}{\sigma}\right) \tag{10}$$

Where:

V_{ohm} —Sum of ionic resistance and material resistance;

σ —Electrical conductivity;

A_{cell} —Activation area of fuel cell;

n_c —Charge carrier number;

δ_t —Thickness of electrolyte layer.

2.3.3 *Concentration polarization of hydrogen-oxygen fuel cells*

Concentration polarization refers to the phenomenon that the concentration of reactants or products in the reaction zone of fuel cell electrode changes due to the migration movement including convection, electromigration, molecular diffusion and pure chemical transformation. Eventually, the potential of the electrode changes.

$$V_{con} = I_{fc}(C_1 \frac{I_{fc}}{I_{max}})^{C_2} \tag{11}$$

When $\left(\frac{P_{O_2}}{0.1173} + P_{sat}\right) < 2atm$,

$$C_1 = (7.16 \times 10^{-4}T - 0.622)\left(\frac{P_{O_2}}{0.1173} + P_{sat}\right) + (-1.45 \times 10^{-3}T + 1.68) \tag{12}$$

When $\left(\frac{P_{O_2}}{0.1173} + P_{sat}\right) \geq 2\text{atm}$:

$$C_1 = (8.66 \times 10^{-5}T - 0.068)\left(\frac{P_{O_2}}{0.1173} + P_{sat}\right) + (-1.6 \times 10^{-4}T + 0.54) \tag{13}$$

Where I_{max} refers to the limiting current density.

2.4 *Establishment and analysis of fuel cell model*

2.4.1 *The establishment of fuel cell model*

Through the theoretical analysis in Section 1.2, simulation models of fuel cell open circuit voltage, activation loss voltage, ohm loss voltage and concentration loss voltage are established. The models established in MATLAB/Simulink are shown in Figures 2 to 5.

The reversible voltage of the fuel cell is about 1.22V, which is calculated by the theoretical model of Nergst voltage.

Through the theoretical analysis and simulation modeling of Nergst equation and polarization voltage loss of hydrogen and oxygen fuel cell, the fuel cell monomer was modeled in MATLAB/Simulink software.

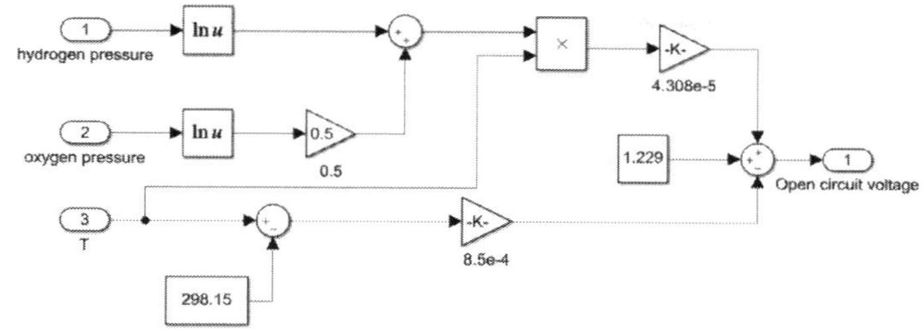

Figure 2. Open circuit voltage model.

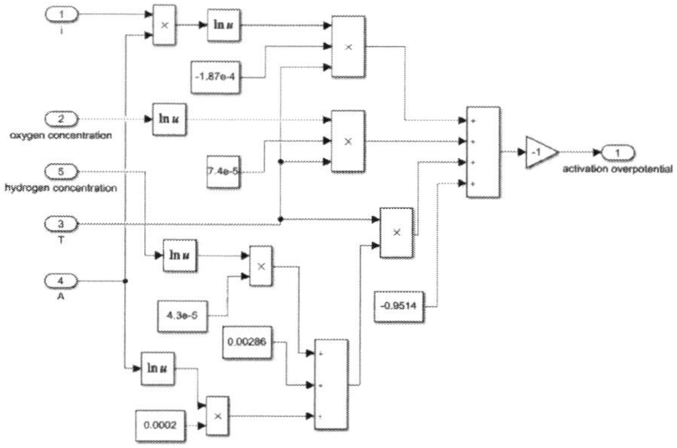

Figure 3. Activated loss voltage model.

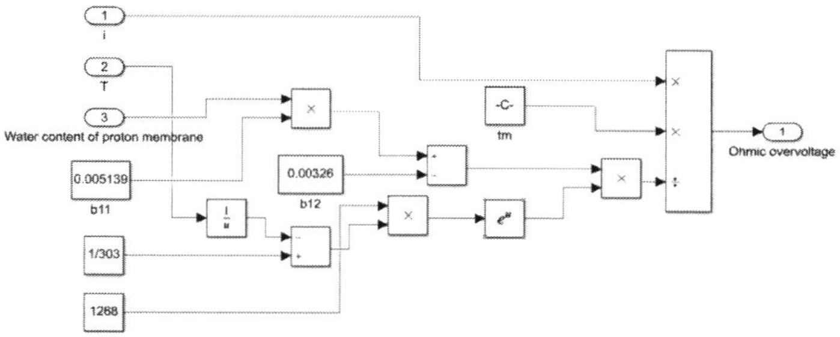

Figure 4. Ohm loss voltage model.

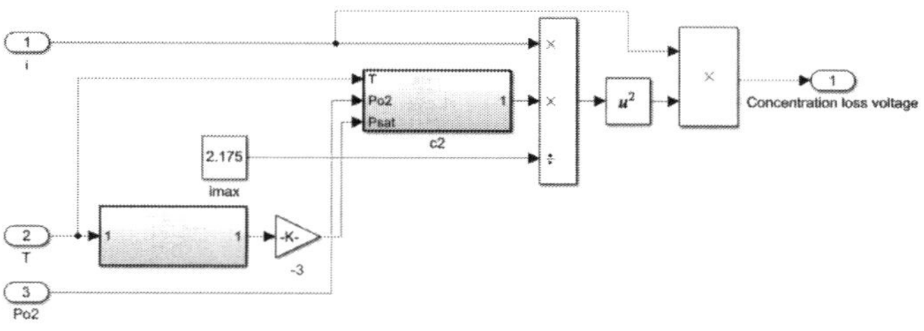

Figure 5. Concentration loss voltage model.

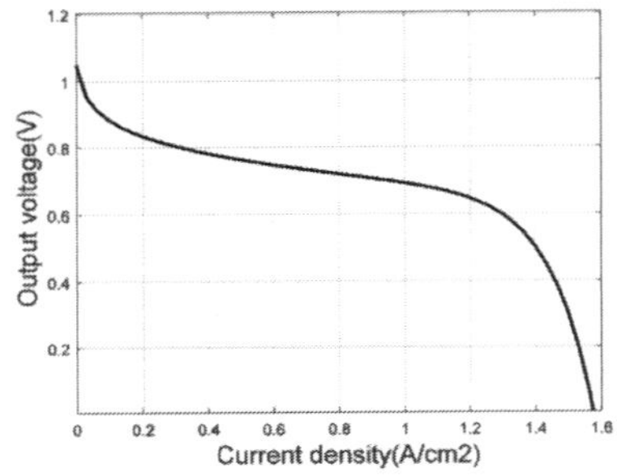

Figure 6. Variation curve of polarization voltage.

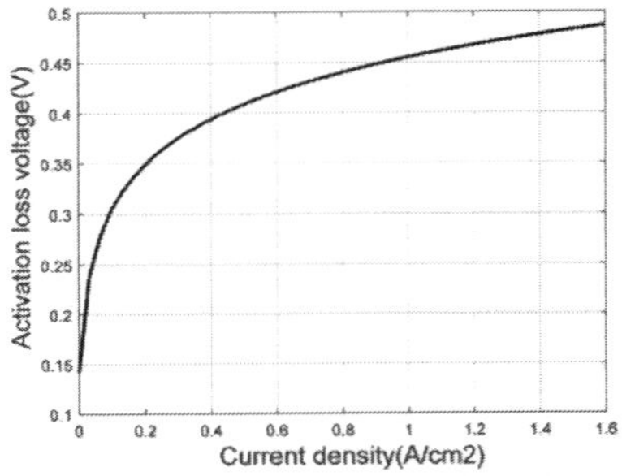

Figure 7. Curve of activation loss voltage.

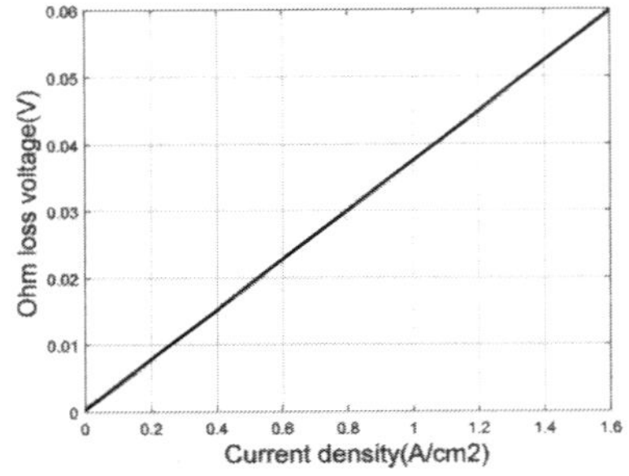

Figure 8. Variation curve of ohm loss voltage.

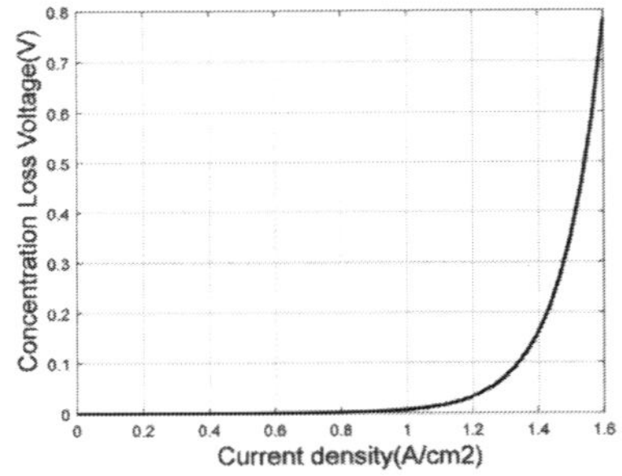

Figure 9. Variation curve of concentration.

2.4.2 *Analysis of fuel cell simulation results*

Through the fuel cell simulation models established in this paper, the following simulation results are obtained.

The reliability of the model was verified by comparing the simulation results with the theoretical results and experimental results as shown in Figure 6.

Figure 7 shows that the activation polarization voltage increases rapidly when the current density is low, the activation energy barrier accumulates rapidly, and the resistance increases with the current. When the current density $I > 0$, the increased rate of the activation polarization of the fuel cell begins to decrease.

Figure 8 shows that ohm loss voltage is directly proportional to current density in the hydrogen-oxygen fuel cell, and ohm polarization loss increases steadily in the process of current density increase compared with activation polarization loss and concentration polarization loss.

Figure 9 shows that in the hydrogen fuel cell, when the current density is less than 1, the concentration of loss-voltage of rate of change is very small, in this phase hydrogen fuel cell polarization is mainly for activation and ohm polarization, while with the increase of current density, when the current density is greater than 1, the concentration type loss voltage increases exponentially rapidly.

3 SIMULATION ANALYSIS OF FUEL CELL MODEL

3.1 *Influence of temperature on fuel cell performance*

According to the above theoretical analysis, the performance of fuel cells is affected by temperature. The temperature continues to exist in the whole operation of the fuel cell, and continues to affect the status of the fuel cell. Therefore, it is very important to explore the influence of temperature on the performance of fuel cells.

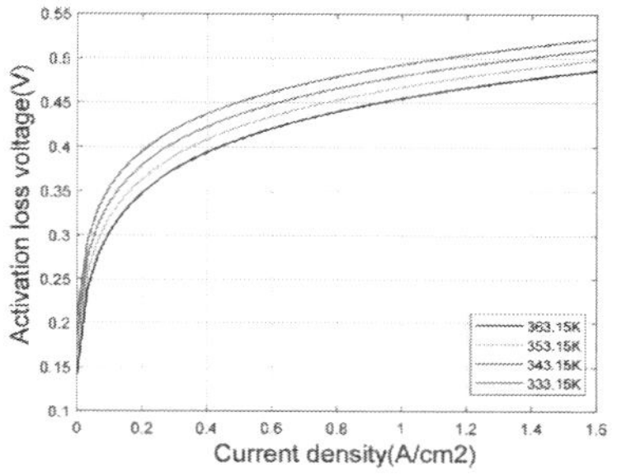

Figure 10. Activation loss voltage is affected by temperature.

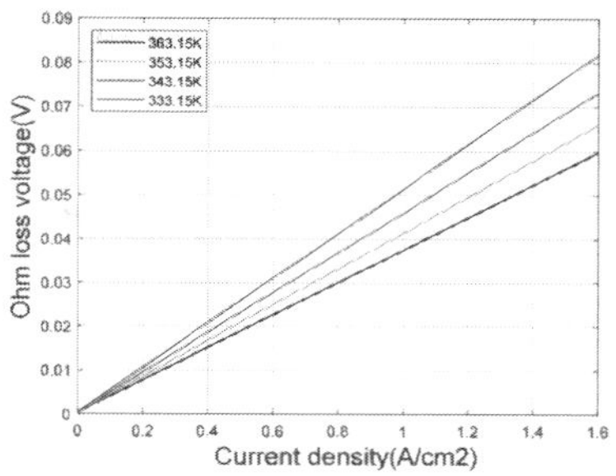

Figure 11. Ohm loss voltage is affected by temperature.

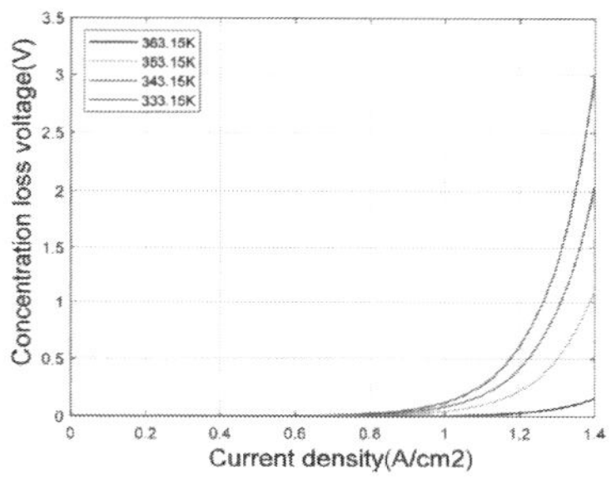

Figure 12. Concentration loss voltage is affected by the temperature.

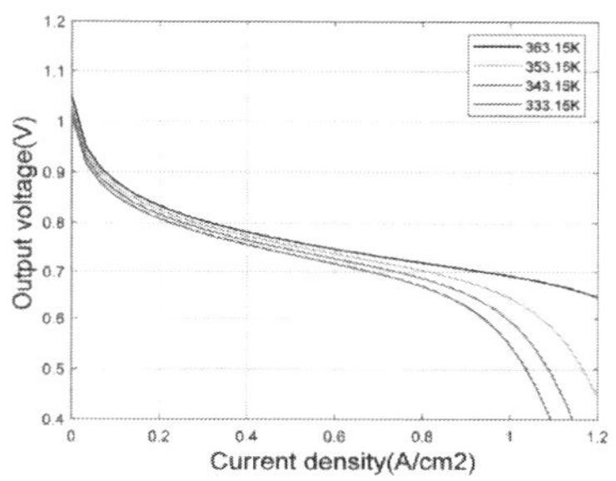

Figure 13. The polarization voltage of single fuel cell is affected by temperature.

Through simulation at 333.15 K, 343.15 K, 353.15 K, and 363.15 K, the activation loss voltage, ohm loss voltage, concentration loss voltage and polarization voltage of fuel cell monomer were obtained.

Figure 10 shows that the activation loss voltage of the fuel cell decreases with the increase of temperature, and the energy consumed by the chemical reaction in the fuel cell decreases in this process. The increase in temperature increases the molecular activity of the reactants and makes it easier for the chemical bonds of the reactants to be broken, thus reducing the energy required for the chemical reaction and thus reducing the activation loss voltage.

It can be seen from Figure 11 that the ohm loss voltage of the fuel cell decreases with the increase in temperature. In this process, the barrier of hydrogen ions through the proton exchange membrane, electrolyte and cell material decreases. The increase in temperature strengthens the molecular movement, making hydrogen ions more easily pass through the barrier, enhancing the fluidity of hydrogen ions, and reducing the ohm loss voltage.

Figure 12 shows that the concentration loss voltage of the fuel cell decreases with the increase in temperature. Moreover, the voltage change rate of concentration loss voltage slows down with the increase of temperature (Pukrushpan 2004), and the effect of gas diffusion is enhanced in this process, leading to the decrease of concentration loss voltage.

The influence of temperature on activation polarization, ohm polarization and concentration polarization of fuel cells shows that the increase in temperature will lead to the decrease of activation loss voltage, ohm loss voltage and concentration loss voltage of fuel cells. Therefore, it can be seen in Figure 13 that the polarization voltage of fuel cell increases with the rise in temperature.

Figure 14 shows the change of fuel cell power affected by temperature.

It can be seen from Figure 14 that the power output of curve A rises first and then declines. When the current density is less than 1.3, the power of fuel cell increases steadily with the increase

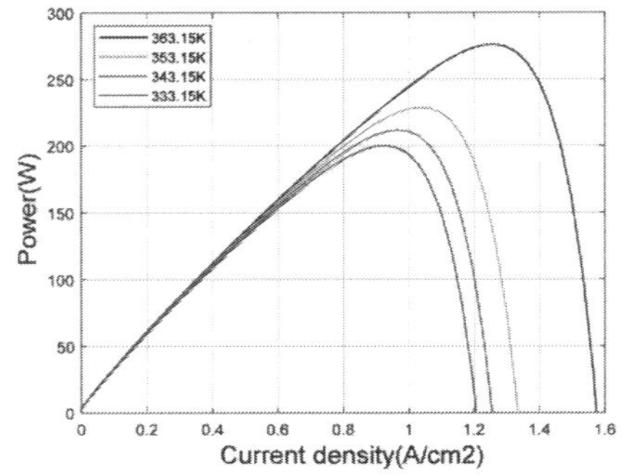

Figure 14. Power is affected by temperature.

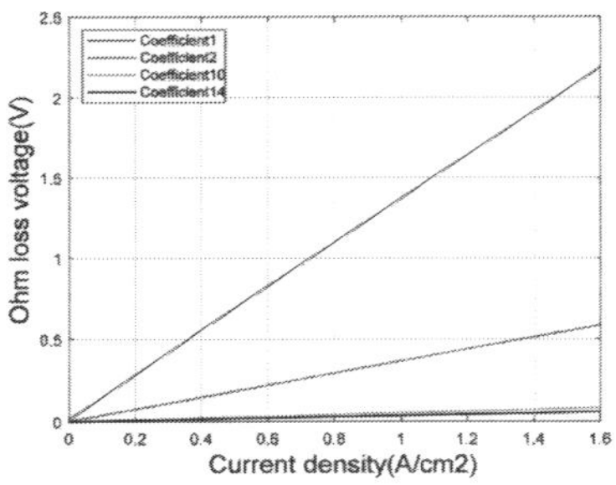

Figure 15. Ohm loss voltage is affected is affected by membrane water content.

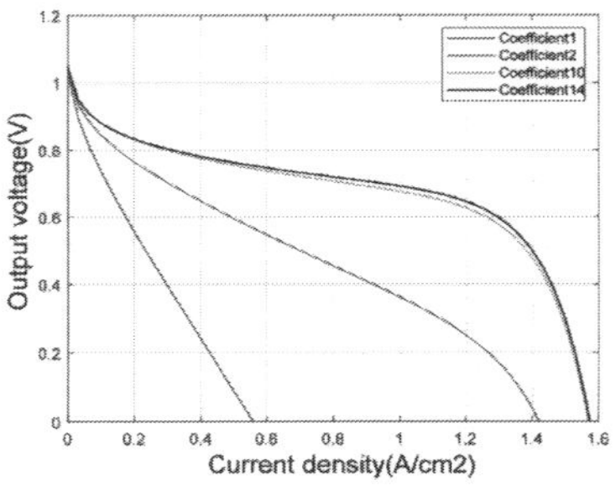

Figure 16. The polarization voltage by membrane water content.

of current density. When the current density is above 1.3, the power of the fuel cell decreases rapidly. At this time, the voltage of the fuel cell is dominated by concentration polarization. With the increase of temperature, the power output and maximum current density also increase. The highest point of output power gradually moves in the direction of greater current density.

3.2 *Influence of membrane moisture content on fuel cell performance*

The performance of the fuel cell is also affected by the water content of the proton membrane. Because the proton membrane water content mainly affects the transport of hydrogen ions between the two poles, the influence of proton membrane water content is mainly in the ohm polarization stage. The water content coefficient of the proton exchange membrane varies from 0 to 14 according to the relative humidity of the proton exchange membrane (0%-100%).

Figure 15 shows that the ohm loss voltage decreases as the membrane water content coefficient increases and the ohm loss voltage is maximal when the membrane water content coefficient is 1. The reason is that when the water content coefficient of the membrane is too low, the proton membrane cannot reach full wetting, resulting in the difficulty of hydrogen ion transfer. With the increase of membrane water content coefficient, it can be seen that the influence of membrane water content on ohm loss voltage gradually decreases.

Figures 15 and 16 show that the polarization curve of the fuel cell is greatly affected under the influence of ohm loss voltage. As the membrane water content increases, the ohm loss voltage decreases, and the voltage of the polarization curve decreases under the influence of the proton membrane water content.

It can be seen from Figures 15, 16, and 17 that when the water content of the proton membrane is low, it has a great influence on the polarization voltage, and the output power of the same fuel

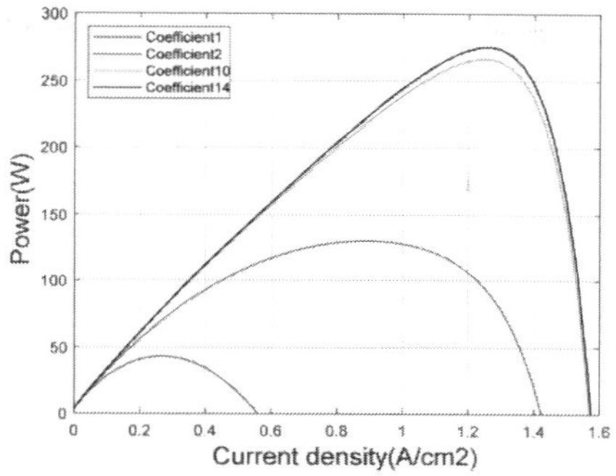

Figure 17. Output power is affected by membrane water content.

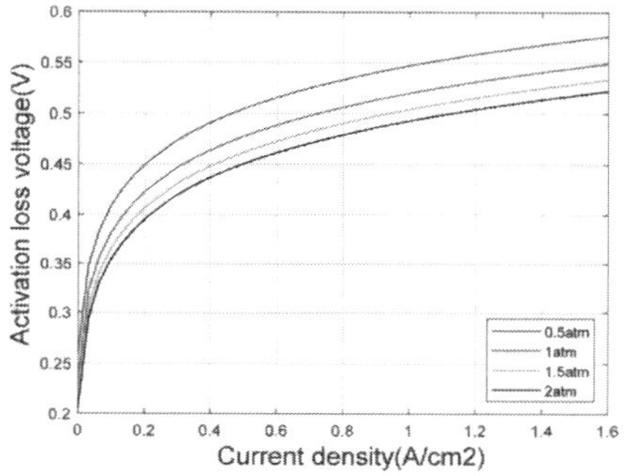

Figure 18. Activation loss voltage is affected by gas partial pressure.

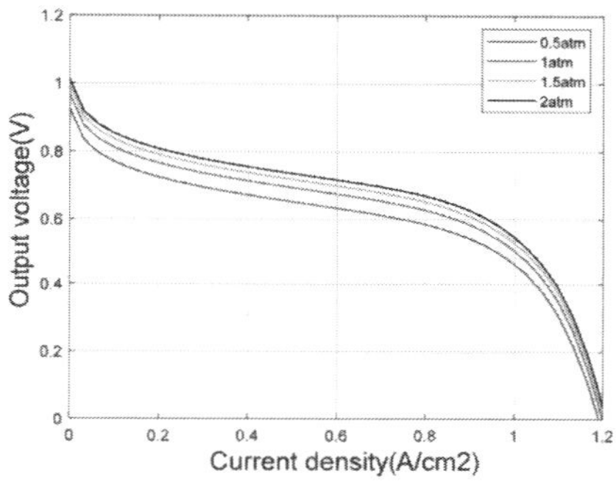

Figure 19. The polarization voltage is affected by the gas partial voltage.

cell is also greatly affected. It can be seen from Figure 17 that when the water content of proton membrane is low, the output power is low. With the increase in water content of proton membrane, the output power increases.

3.3 *Influence of gas partial pressure on fuel cell performance*

The gas partial pressure of fuel cell affects the gas reaction concentration. In this paper, the performance of fuel cell was studied at the gas partial pressure of 0.5 atm, 1 atm, 1.5 atm, and 2 atm at 333.15 K.

It can be seen from Figure 18 that the activation loss voltage decreases gradually as the partial pressure of the gas increases. When the pressure changes uniformly at 0.5 atm, the reduction of activation loss voltage is not uniform. With the increase of partial voltage, the amplitude of voltage drops gradually. This is because when the partial pressure increases, the gas concentration and the diffusion rate of the gas increase, resulting in the reduction of the reaction barrier and the reduction of activation loss voltage.

Figure 19 shows that the open-circuit voltage in the polarization voltage curve increases as the partial voltage increases, and the open-circuit voltage is affected by the gas reaction temperature. As the partial pressure of gas increases, the activation loss voltage decreases and the polarization voltage increases.

The fuel cell power variation affected by gas partial pressure is shown in Figure 20.

It can be seen from Figure 20 that the power of the fuel cell increases with the increase of gas partial pressure, and the maximum power point moves to the direction of high current with the increase of power.

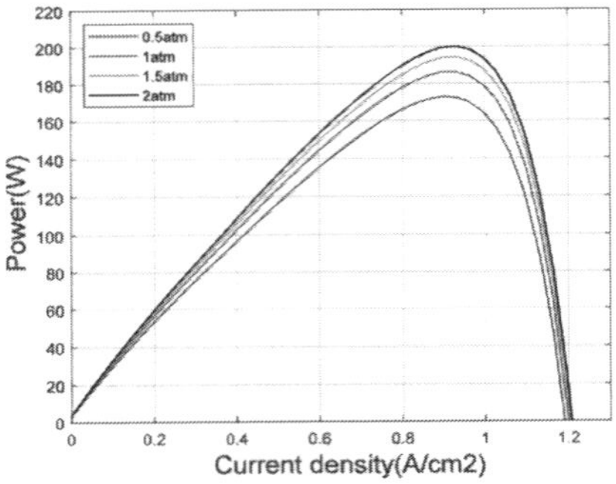

Figure 20. Power is affected by gas partial pressure.

4 CONCLUSION

In this paper, the fuel cell monomer model is established in MATLAB/Simulink by theoretical analysis of the energy equation, activation polarization, ohm polarization and concentration polarization of fuel cells (Das 2017). The simulated results of activation loss voltage, ohm loss voltage and concentration loss voltage are analyzed, and the accuracy of the simulation model is verified by comparing the simulated polarization voltage curve with the theoretical and experimental results.

The effects of temperature on activation loss voltage, ohm loss voltage, concentration loss voltage, polarization voltage and output power of fuel cell were investigated by simulation at different temperatures. It is concluded that the activation loss voltage, ohm loss voltage and concentration loss voltage of fuel cell decrease with the increase of temperature, while the polarization voltage and output power of fuel cell increase.

The influence of proton membrane water content on ohm loss voltage, polarization voltage and output power of fuel cell was also investigated by simulation under different proton membrane water content coefficients. It is concluded that the ohm loss voltage decreases rapidly with the increase of proton membrane water content when the water content is low and decreases slowly with the increase of proton membrane water content when the water content is high. The polarization voltage and output power of the fuel cell increase with the increase in water content of the proton membrane.

The influence of gas partial pressure on activation loss voltage, polarization voltage and output power of fuel cell was investigated by simulation under different gas partial pressure.

It is concluded that the activation loss voltage decreases with the increase of gas partial pressure, and the influence on voltage change decreases with the increase of gas partial pressure. The polarization voltage and output power of the fuel cell increase with the increase of partial voltage.

REFERENCES

Azooz R, Sayyah S M, Abd-Elrehem S S. Conducting Polymers. *LAP Lambert Academic Publishing*, 2011.

Das H S, Tan C W, Yatim A H M. Fuel cell hybrid electric vehicles: A review on power conditioning units and topologies[J]. *Renewable & Sustainable Energy Reviews*, 2017, 76(SEP.):268–291.

Hannan M A, Azidin F A, Mohamed A. *Hybrid electric vehicles and their challenges: A review*[J]. Elsevier Ltd, 2014, 29.

Jixin C, Mike V, J. Purewal, B. Hobein, S. Papasavva. Modeling a hydrogen pressure regulator in a fuel cell system with Joulee Thomson effect. *Int J Hydrogen Energy*, 2019 (44): 1272–1287.

Pukrushpan J.T, Stefanopoulou A G, Peng H. *Conrtrol of Fuel Cell Power System: Principles, Modeling, Analysis and Feedback Design*[M]. New York: Springer Science &Business Media, 2004.

Energy Revolution and Chemical Research – Kok-Keong Chong and Zhongliang Liu (Eds)

Strength tests and mechanical characteristics of asphalt concrete in pumped storage power plants

Qu Manli
Qu Manli, School of Civil and Architectural Engineering, Xi'an University of Technology, Xi'an, China

Lu Shiquan
Lu Shiquan, Drought Control Project Management Centers, Chongzuo, China

Ma Dong
Lu Shiquan, State Nuclear Electric Power Planning Design& Research Institute Co., Ltd., Beijing, China

ABSTRACT: Based on static triaxial tests, uniaxial compression, and tensile tests of asphalt concrete, the stress-strain characteristics of asphalt concrete specimens under different temperature and confining pressure conditions are analyzed by using the test method of different test temperatures at the same mixing ratio to make up for the shortcomings of the current theoretical research on asphalt concrete strength. The temperatures include 5°C, 10°C, 15°C, and 20 °C, and the confining pressures include 0.1 MPa, 0.2 MPa, 0.3 MPa, 0.4 MPa, 0.8 MPa, and 1.2 MPa. The results show that under the same temperature condition, with higher confining pressure, the stress-axial strain curve becomes moderate, and the failure stress and its corresponding axis become larger when reaching the peak, which indicates that the creep of asphalt concrete becomes more obvious with higher confining pressure. Under the same confining pressure, the stress-axial strain curve becomes moderate with the increase in temperature, the smaller the failure stress at the peak, the larger is the axial variation value, which shows that the creep of asphalt concrete becomes more obvious with the increase in temperature. The failure form of asphalt concrete is gradually transformed from brittle damage to plastic damage, and the transformation rate is higher, which indicates that asphalt concrete is very sensitive to the change in temperature. Static triaxial test parameters were determined based on the Mohr-Coulomb strength theory. The boundary point between the Mohr-Coulomb strength theory and the Mohr-Coulomb strength theory first increased and then decreased with the increase in temperature, and the peak value appeared when the temperature was close to 15°C. The parameters of the static triaxial test were determined based on the Mohr-Coulomb strength theory. The boundary point between the Mohr-Coulomb strength theory and the Mohr strength theory first increased and then decreased with the increase in temperature, and the peak value appeared when the temperature was close to 15°C. In this paper, with the static triaxial test as the main test, uniaxial compression and tensile test as the auxiliary test, the strength theory of asphalt concrete was improved by comparing and analyzing the results of the three tests, which can provide the theoretical basis for the strength analysis of asphalt concrete.

1 GENERAL INSTRUCTIONS

The application technology of hydraulic asphalt concrete started late in China. It was not until the 1970s that the technology was applied to the dam construction on a large scale, and the pouring asphalt concrete anti-seepage projects, such as Biliuhe in Liaoning and Niutou Mountain in Zhejiang, were successively built. Compared with foreign countries, the application of asphalt concrete as a seepage prevention body in the earth-rock dam started late in China. Because of the excellent characteristics of asphalt concrete, its anti-seepage technology has been rapidly developed. In the

DOI 10.1201/9781003332657-60

past 20 years, asphalt concrete has been widely used in the seepage prevention of earth-rock dams, concrete dams, masonry dams, and reservoirs, and anti-seepage lining of channels, reservoirs, and reservoirs.

More than 300 asphalt concrete panel projects, including storage ponds, have been completed worldwide. The highest asphalt concrete panel rock fill dam is the Osehenik reservoir in Austria, which was built in four phases in 1979 and has a maximum height of 106 m, followed by the Sabigawa Pumped Storage Power Station Upper Reservoir in Japan with an asphalt concrete panel rock fill dam of 90.5 m. The largest asphalt concrete panel dam with a seepage control area of 1.84 million m^2 is the Geeste Reservoir in Germany, followed by the upper reservoir of the LaMuelao pumped storage power station in Spain, with a seepage control area of 1.2 million m^2 of asphalt concrete. The first rock fill dam in China that completely used asphalt concrete panels for seepage control was the pilot dam of Zhengjia Reservoir built in 1976 with a height of 36 m. The asphalt concrete panel directional blast rock fill dam of Shijiayu Reservoir built in 1978 was 85 m high, with a crest length of 265 m and a crest width of 7.5 m.

Asphalt concrete is used more widely at present, but there is lack of systematic and in-depth research on the performance of asphalt concrete. Rahmani et al. (2013) characterized the nonlinear viscoelasticity of asphalt concrete materials that exhibit a nonlinear viscoelastic response at high stress/strain levels using cyclic uniaxial creep-recovery tests, taking into account the effects of the confinement pressure (Eisa et al. 2013). Seo et al. (2017) investigated the stress-strain properties of asphalt concrete by Unconfined Compression Strength test, Indirect Tensile Test, and Triaxial Test, and tested the performance of the Asphalt Core Rockfill Dam (Seo et al 2017). Khan et al. (2018) developed axisymmetric finite element models to investigate the nanoscale stress-strain distribution of the non-aggregated phase of asphalt concrete under-aged and unaged conditions. The nanoscale response of the non-aggregated phase of asphalt concrete under fixed and increasing amplitudes of cyclic loads was also investigated (Khan Zafrul et al. 2018). Tang et al. (2020) carried out uniaxial compression experiments using a hydraulic servo machine with four temperatures and four strain rates, analyzed the influence of temperature and strain rate on the failure modes, stress-strain curves, and mechanical characteristic parameters of hydraulic asphalt concrete, and proposed two relationship models of the coupling effect between temperature and strain rate (Tang et al. 2020).

It is found that temperature has a great influence on the strength and failure mode of asphalt concrete by many engineering applications. Previous experimental studies on asphalt concrete have been conducted for individual projects, where strength and failure forms were derived from tests based on specific temperature and mix ratio conditions. Taking Qiongzhong pumped storage power station in Hainan as the research object, the paper carried out the research on the strength test and stress-strain characteristics of asphalt concrete in pumped storage power station using the combination of outdoor sampling and indoor test based on the design ratio of asphalt concrete, to explore the effect of temperature on the strength and failure form of asphalt concrete. The indoor test uses a static triaxial test, uniaxial tensile, and compression test. By recording asphalt concrete under different temperatures and confining pressure conditions and stress-strain characteristics, this paper analyzes the influence of the stress and strain parameters of asphalt concrete under different test conditions and reveals the mechanical parameters of asphalt concrete, which can provide a reference for strength analysis and quality control of asphalt concrete.

2 METHODS AND MODEL

2.1 *Experimental setup*

In this paper, combining the asphalt concrete mix design of the Qiongzhong Pumped storage power station in Hainan and the materials of this project, pass aggregate is adopted with a density of 2.720 g/cm^3. The filler is mineral powder with a density of 2.722 g/cm^3. The asphalt is CNOOC NO.70 asphalt with a density of 1.035 g/cm^3. Design triaxial test specimens following the asphalt concrete test procedure. The specimen is a 100 mm diameter, 200 mm high tensile, and compressive specimen.

According to the requirements of the triaxial test of asphalt concrete, the confining pressures are σ_3= 0.1, 0.2, 0.3, 0.4, 0.8, and 1.2 MPa, respectively. The triaxial tests are carried out at 5°C, 10°C, 15°C, and 20°C, respectively. Two specimens are tested in each group, and the average of the two specimens is taken as the test results. There are 48 specimens in 24 groups. During the test, considering the different initial states of each group of specimens under different temperatures and confining pressures, the method of non-consolidation drainage was adopted. The loading speed was 0.2 mm/min. Uniaxial tensile and compressive tests were carried out under 5°C, 10°C, 15°C, and 20°C with the same confining pressures according to the tensile and compressive test requirements of hydraulic asphalt concrete. Each group tests three specimens. The test pieces of the static triaxial test, compressive test, and tensile test of asphalt concrete are shown in Figure 1. The test results are taken as the average value of the two data in each group, with a total of 24 specimens in 8 groups. The loading speed is 1% of the specimen height per minute.

2.2 *Methods*

2.2.1 *Static triaxial test*

(1) With the principal stress (σ) as the horizontal coordinate and the shear stress (τ) as the vertical coordinate, the stress circle under different confining pressure is drawn with (σ_1+σ_3)/2 as the center and (σ_1-σ_3)/2 as the radius in the horizontal coordinate. The envelope of the circles is made, the inclination of the envelope is the angle of internal friction φ, and the intercept of the envelope in the vertical coordinate is the bonding force c (Huang et al 2018).

(2) Plot the axial deflective stresses (σ_1-σ_3) under different confining pressures, axial strain relationship curve, and axial strain (ε_1) and volumetric strain (ε_v) relationship curve. Calculate the lateral strain (ε_3) from the volume strain according to the following formula.

$$\xi_3 = \frac{\xi_v - \xi_1}{2}$$

Where ε_3 is the lateral strain (%), ε_v is the volumetric strain (%), and ε_1 is the axial strain (%).

(3) According to the above curves, the nonlinear deformation modulus, Poisson's ratio, and other parameters of asphalt concrete under triaxial stress are obtained.

2.2.2 *Uniaxial compression test*

(1) The compressive strength of the specimen is calculated according to the following formula, being accurate to 0.01 MPa.

$$R_c = \frac{P_c}{A}$$

Where R_c is compressive strength (Mpa), P_c is the maximum load of the specimen under compression (N), and A is the cross-sectional area of the specimen (mm^2).

(2) The strain under the maximum stress of the specimen is calculated by the following formula.

$$\xi = \frac{\delta}{h} \times 100\%$$

Where ε is the strain of the specimen under maximum stress (%), δ is vertical deformation under maximum load (mm), and H is the height of the specimen (mm).

(3) Plot the stress-strain curve of the specimen. If σ-ε is a linear relationship, the slope of the linear segment is its modulus of deformation under compression. If σ-ε is a curvilinear relationship, the deformation modulus of the specimen is calculated by the following equation.

$$E_c = \frac{\sigma_{0.5P_c} - \sigma_{0.1P_c}}{\xi_{0.5P_c} - \xi_{0.1P_c}}$$

Figure 1. Asphalt concrete test piece (Left: static triaxial test; middle: compressive test; right: tensile test).

Where E_c is compressive deformation modulus (Mpa); $\sigma_{0.5Pc}$ and $\sigma_{0.1Pc}$ are corresponding to the compressive stress at $0.5P_c$, $0.1P_c$, respectively (Mpa); $\varepsilon_{0.5Pc}$ and $\varepsilon_{0.1Pc}$ are corresponding to the compressive strain at $0.5P_c$, $0.1P_c$, respectively (%).

2.2.3 *Tensile test*

(1) The tensile strength of asphalt concrete is calculated according to the following formula.

$$R_t = \frac{P_t}{A}$$

Where R_t is axial tensile strength (Mpa), P_t is axial maximum tensile load (N), and A is specimen cross-sectional area (mm^2).

(2) Tensile strain is calculated by the following formula.

$$\xi_t = \frac{\delta_t}{L} \times 100\%$$

Where $\varepsilon_t$is the axial tensile strain (%), δ_t is the axial tensile deformation (m), and L is the axial measurement distance (mm).

(3) Tensile deformation modulus is calculated by the following formula.

$$E_t = \frac{R_t}{\xi_t}$$

Where E_t is tensile deformation modulus (Mpa), R_t is a tensile stress (Mpa), and ε_t is a certain tensile strain (%).

For the determination of tensile deformation modulus, when the recorded load-deformation curve is not a straight line, the deformation modulus can be calculated by taking the slope of the cut line of 0.1–0.7 of the maximum load value P_t (Zheng et al 2019).

3 RESULTS AND DISCUSSION

3.1 *Static triaxial test results under different confining pressure and the same temperature*

Taking the temperature of 10°C as an example, the stress-strain curves of the static triaxial tests with different confining pressures under the same temperature conditions are shown in Figure 2. From Figure 2, it can be seen that the asphalt concrete stress-strain curve is nearly hyperbolic before reaching the failure point. The Duncan-zhang hyperbolic model was approximated and regressed by the E-μ model to obtain the nonlinear parameters C = 1.14 and $\varphi = 25.8°$.

With higher confining pressure, the stress-axis strain curve is more moderate, the peak failure stress and the corresponding axial variation are also increasing, the test time of asphalt concrete specimens is also getting longer, and the minimum value of the volume strain is getting smaller and smaller, and the maximum value of the volume strain is also getting smaller. When $\sigma_3 < 0.8$ MPa, the volume change is first compressed and then increased. When $\sigma_3 \geq 0.8$ MPa, the volume change is always in the compressed state. The tested asphalt concrete specimens are ductile damage with significant yielding due to slip along the 45° oblique section where the maximum shear stress is located. After the test, the asphalt concrete specimen slips along the 45° diagonal section of the maximum shear stress and presents an obvious yield phenomenon, which belongs to ductile failure.

3.2 *Static triaxial test results under different temperature conditions and the same confining pressure*

Taking 0.2 confining pressure as an example, the stress-strain curve of static triaxial test for asphalt concrete under the same confining pressure ($\sigma_3 = 0.2$ MPa) and different temperatures are shown in Figure 3.

Figure 3 shows that as the temperature increases, the stress-axis strain curve is more moderate, the failure stress at the peak becomes smaller, and the corresponding axis variation value becomes larger, the test time of asphalt concrete specimens is also getting longer, the minimum value of the volume strain is getting smaller, the maximum value of the volume strain is also getting smaller, and the volume change is first compressed and then increased. After the test, the asphalt concrete specimens did not have obvious plastic yielding at 5°C, which belonged to brittle failure. Asphalt concrete specimens at 10°C, 15°C, and 20°C were slipping along the 45° oblique section where the maximum tangential stress was located and an obvious yielding phenomenon occurred, which belonged to ductile damage.

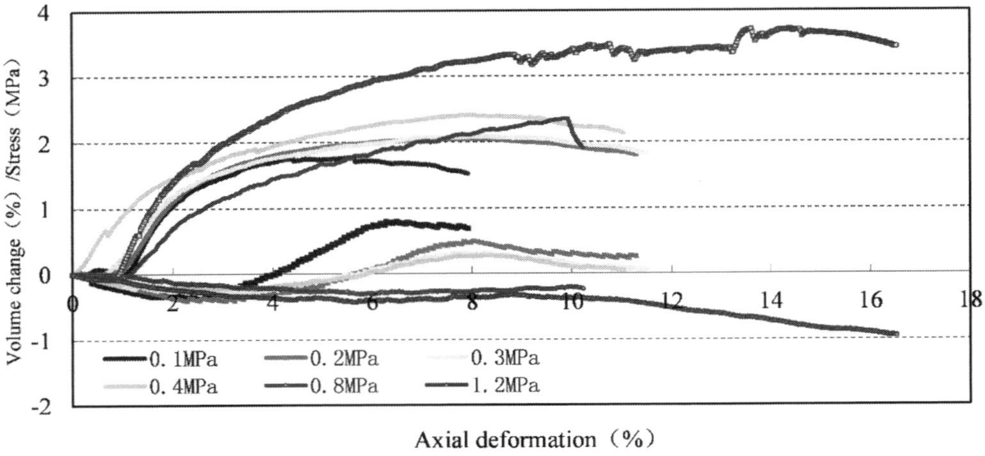

Figure 2. Static triaxial stress-strain curves of asphalt concrete under different confining pressures and the same temperature.

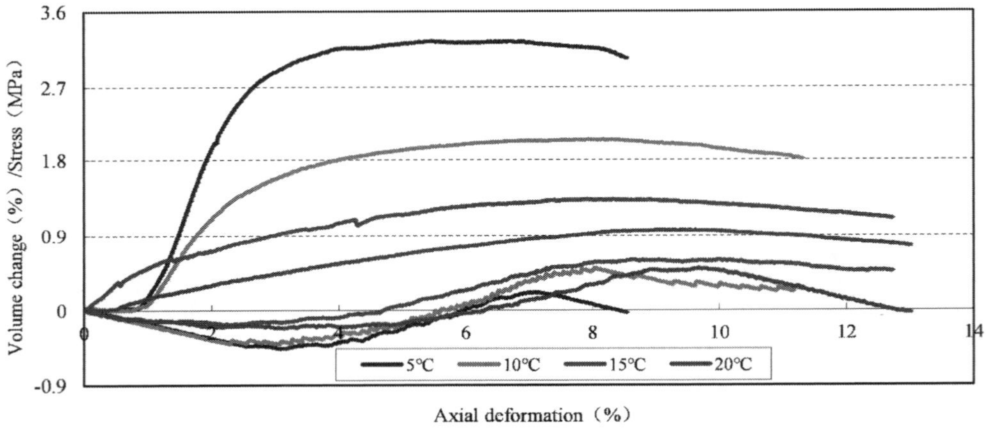

Figure 3. Static triaxial stress-strain curve of asphalt concrete under different temperatures and same pressure (σ_3= 0.2 MPa).

3.3 *Uniaxial compressive test results under different temperature*

Asphalt concrete was tested under 5°C, 10°C, 15°C, and 20°C for the compressive and tensile tests. Two specimens were made for each temperature, a total of 24 specimens were made, and the test results were averaged. By the uniform proportioning, the stress-strain curves of asphalt concrete at different test temperatures are shown in Figure 4.

Figure 4 shows that as the temperature increases, the stress-axis strain curve is more moderate, the failure stress at the peak becomes smaller, and the corresponding axis variation value becomes larger, the test time for asphalt concrete specimens is also gets longer. When the temperature is 5°C, 10°C, and 15°C, there is no obvious plastic yielding, which is brittle damage. When the temperature is 20°C, asphalt concrete specimens are along with the maximum tangential stress in the 45° oblique cross-section of the slip and the obvious yielding phenomenon, which is ductile damage.

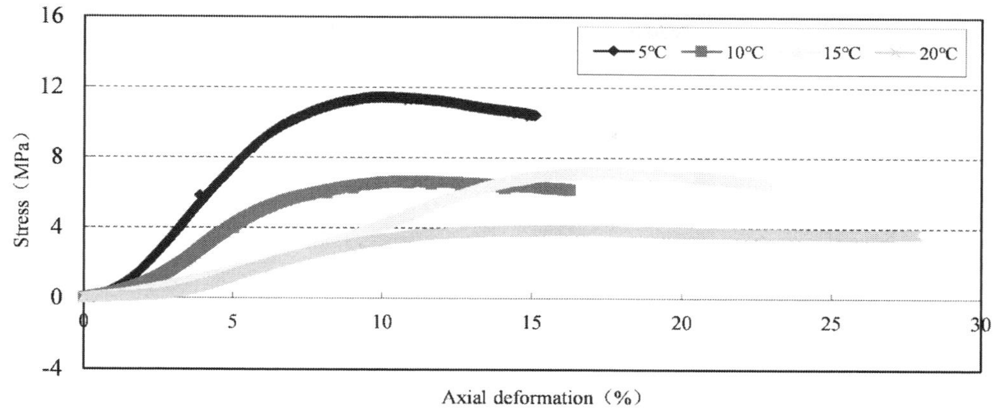

Figure 4. Uniaxial compressive stress-strain curve of asphalt concrete under different temperatures (5–20°C).

From the above we note that $\sin\theta = (x + y)z$ or:

$$K_t = \left(1 - \frac{R^2\tau}{c_a + \nu \tan\delta}\right)^4 k_1 \tag{1}$$

where c_a = interface adhesion; δ = friction angle at interface; and k_1 = shear stiffness number.

3.4 *Uniaxial tensile test results under different temperature*

By the uniform proportioning, the stress-strain curve of asphalt concrete under different test temperature conditions is shown in Figure 5.

Figure 5 shows that as the temperature rises, the stress-axis strain curve becomes more moderate, and the failure stress becomes smaller when it reaches the peak, and the corresponding axis variation value becomes larger. When the temperature is 5°C, 10°C, and 15°C the stress-strain curve reaches its peak and decreases rapidly. When the temperature is 20°C the stress-strain curve has no obvious peak and decreases insignificantly after reaching the peak. The test time for asphalt concrete specimens is also getting longer. When the temperature is 5°C, 10°C, and 15°C, there is no obvious plastic yield, belonging to brittle failure. When the temperature is 20°C, the asphalt concrete specimen belongs to a ductile failure, and the failure surface is at the connection between the specimen and the steel joint because there is a strong constraint here.

3.5 *Comparative analysis of different test results*

The asphalt concrete stress-strain curve is nearly hyperbolic before reaching the damage point from the static triaxial test curve. It approximates the Duncan-zhang hyperbolic model and regresses it to the E-μ model to obtain the nonlinear parameters. The asphalt concrete stress-strain curve is nearly hyperbolic before reaching the damage point from the uniaxial compressive and tensile test curves. The intercept C and the inclination Φ of the line tangent to the two stress circles by (σ_1) max and (σ_3) max are shown in Figure 6.

From Figure 6, larger values of C and smaller values of φ from the static triaxial experiment yield. Smaller C values and larger φ values from tensile and compression tests. The trend of C value decreases with the increase in temperature and the trend of φ value increases with the increase in temperature. The C and φ values of asphalt concrete determined by the tensile pressure test have a basic assumption that the internal properties of asphalt concrete remain the same when it is subjected to compression and tension. The three-axis test instrumentation is more complex, demanding, and

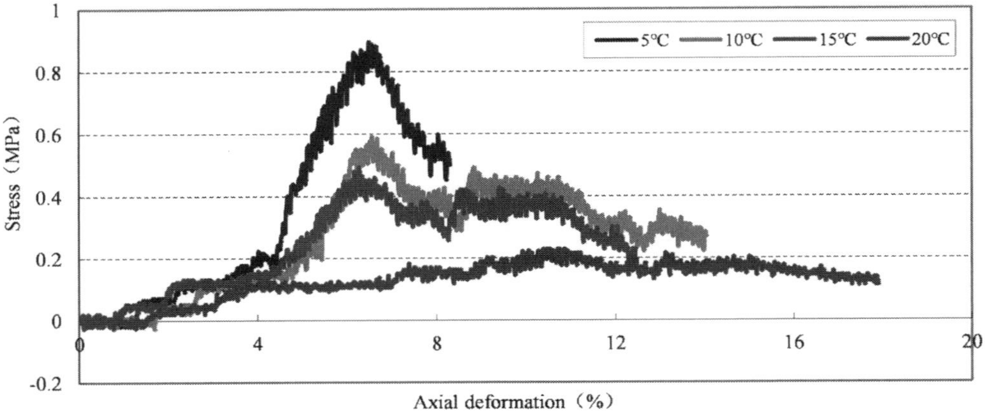

Figure 5. Uniaxial tensile stress-strain curve of asphalt concrete under different temperatures (5–20 °C).

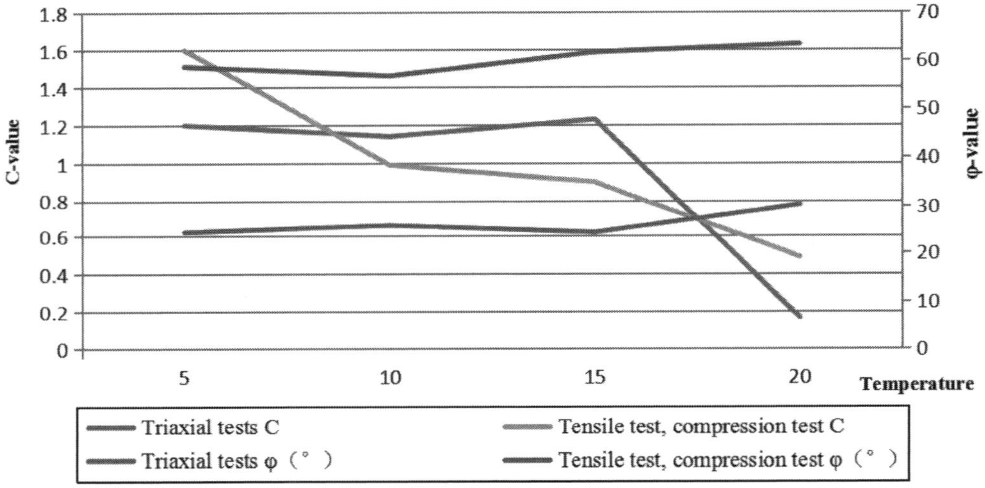

Figure 6. Variation of strength parameters of asphalt concrete at different temperatures and different tests.

difficult to operate. However, the triaxial test can simulate the real stress-strain state very well. Therefore, the final static triaxial test was used to determine the values of parameters C and φ for asphalt concrete.

4 CONCLUSIONS

(1) At the same temperature with higher confining pressure, the stress-axis strain curve becomes more moderate, and the failure stress and the corresponding axial variation at the peak also become larger. The testing time for asphalt concrete specimens also gets longer, which indicates that the higher the confining pressure the more obvious is the creep of asphalt concrete. The form of failure of asphalt concrete has a gradual transformation from brittle failure to plastic damage.

(2) The stress-axis strain curve becomes more moderate with the increase in temperature under the same confining pressure, the breaking stress becomes smaller when it reaches the peak, and the

corresponding axis variation value becomes larger. The test time of asphalt concrete specimens also gets longer, which indicates that the creep of asphalt concrete is getting more obvious. The failure form of asphalt concrete has brittle failure gradually transformed to plastic damage, and the transformation rate is faster, which indicates that asphalt concrete is very sensitive to changes in temperature.

(3) According to the Mohr-Coulomb strength theory, the Mohr-Coulomb envelope is formed by the calculated parameters C and φ derived from the static triaxial test. According to the Mohr strength theory, the stress circles are made by uniaxial compressive and tensile tests according to $(\sigma 1)$ max and $(\sigma 3)$ max, and the intercept C of the line tangent to these two stress circles forms a Mohr envelope with the value of the inclination Φ. The formed Mohr-Coulomb envelope and the Mohr envelope intersect at a point. Mohr-Coulomb intensity theory is used on the left side of the intersection, and Mohr intensity theory is used on the right side of the intersection. The cut-off point between the Mohr-Coulomb strength theory and the Mohr strength theory increases and then decreases with increasing temperature (between 15°C and 20°C), with a peak near 15°C.

REFERENCES

Huang Bin, et al. Static and Dynamic Properties and Temperature Sensitivity of Emulsified Asphalt Concrete[J]. *Advances in Materials Science & Engineering*, 2018, 2018:7067608.

Khan Zafrul H, et al. Evaluation of Nanomechanical Properties of Nonaggregate Phase of Asphalt Concrete Using Finite-Element Method[J]. *Journal of Materials in Civil Engineering*, 2018, 30(12):04018331.

Rahmani Eisa, et al. Effect of confinement pressure on the nonlinear-viscoelastic response of asphalt concrete at high temperatures[J]. *Construction and Building Materials*, 2013, 47:779–788.

Seo Jung-Woo, et al. Development of an asphalt concrete mixture for Asphalt Core Rockfill Dam[J]. *Construction and Building Materials*, 2017, 140:301–309.

Tang Rui, et al. Uniaxial Dynamic Compressive Behaviors of Hydraulic Asphalt Concrete Under the Coupling Effect between Temperature and Strain Rate[J]. *Materials*, 2020, 13(23):5348.

Zheng Mulian, et al. Fatigue character comparison between high modulus asphalt concrete and matrix asphalt concrete[J]. *Construction and Building Materials*, 2019, 206:655–664.

Energy Revolution and Chemical Research – Kok-Keong Chong and Zhongliang Liu (Eds)
© 2023 The Authors, ISBN: 978-1-032-36554-1

Contents of five heavy metals in agricultural land soil in Hengshui and ecological risk assessment

Zhongqiang Zhang
Department of Applied Chemistry, Hengshui University, Hengshui, Hebei, China

ABSTRACT: Twelve sampling points were set up in farmlands in Dengzhuang Town, Hengshui. An analysis was conducted on the contents of heavy metals Cr, Mn, Pb, Cu, and Cd in the samples. The pollution of heavy metals was evaluated with the geoaccumulation index (I_{geo}). Results showed that contents of Cr, Pb, Cu, and Cd at each sampling point were lower than the screening values of agricultural soil pollution risks in the *Soil Environmental Quality-risk Control Standard for Soil Contamination in Agricultural Land (Trial)* (GB15618-2018); Contents of heavy metals Cr, Mn, Pb and Cu were all lower than the background values of heavy metal contents in soil (BVSH) in Hebei Province, but the average content of Cd was slightly higher than BVSH. The average I_{geo} of Cr, Mn, Pb, and Cu contents was less than 0, and the I_{geo} at each sampling point was less than 0, based on which the soil can be regarded as pollution-free. The average I_{geo} of Cd was very close to 0, based on which the soil can be deemed to have slight pollution.

1 INTRODUCTION

Heavy metal atoms refer to 45 kinds of metallic elements with a density higher than 4.5 g/cm^3. Heavy metal pollution of soil is a long-term, concealed, accumulative and irreversible issue (Liao 2016), so it has become one of the focuses of current environmental pollution prevention and control. According to the National Soil Pollution Survey Report in 2014, the total over-standard rate of heavy metals in soil was 16.1% in China, among which the over-standard rate of heavy metals in cultivated soil was 19.4%, indicating an unoptimistic soil environment (Ministry of Environmental Protection, Ministry of Land and Resources 2014). Heavy metal pollution of soil will result in an increase of toxicity, significant change of the pH value, and structural damage, which will lead to the imbalance of nutrients, affect the activities of microorganisms in soil adversely, impair the transformation ability of substances and energy, decrease soil productivity and cause harm to agricultural production. Pollutants can be continuously spread and enriched through the food chain, posing a serious threat to food safety, human health, and ecological environment (Hua et al. 2021; Jia 2020; Liao 2016; Zhou et al. 2021).

Hengshui is located in the North China Plain. Dengzhuang Town is located in the eastern suburb of Hengshui, 3 km away from the urban area and 5km away from Hengshui Lake. With an area of 100 km^2, cultivated land of 100,000 mu (1 mu = 666.67 m^2), and a vegetable planting area of 51,000 mu, Dengzhuang Town is the Hometown of Cherry and Tomatoes in the North, a Direct Supply Base of Vegetable Basket Project in Beijing and a National Standardization Demonstration Base of Vegetables. Therefore, the quality of its farmland will directly affect the quality of agricultural products and thus the health of the general public including the people in the capital of China.

In this paper, points were set up in farmlands in Dengzhuang Town, and samples were taken in a targeted way through a field survey. The pollution of five heavy metals in farmland soil was systematically analyzed and evaluated with the geoaccumulation index, which provided a scientific basis for effective control of heavy metal pollution in the farmland environment of Hengshui, risk management, and ecological protection.

 DOI 10.1201/9781003332657-61

2 MATERIALS AND METHODS

2.1 *Instruments and drugs*

All the reagents used were superior pure or analytical pure reagents made in China, which were not further purified before use. The WFX-130A flame/graphite furnace atomic absorption spectrophotometer was bought from Beifen-Ruili Analytical Instrument Co., Ltd.; FA3204B electronic balance was bought from Shanghai Jinghai Instrument Co., Ltd., and the DHG-9145A electrothermal blowing dry box was bought from Shanghai Yiheng Scientific Instrument Co., Ltd.

2.2 *The setting of sampling points*

Twelve sampling points were set up in farmlands in Dengzhuang Town, Hengshui, located in115.6257E–115.9598E and 37.6903N–38.0128N, according to the distribution characteristics of farmlands in the town.

2.3 *Sample collection, treatment, and testing*

The latitude and longitude of sampling points were recorded in detail with GPS. Surface soil at the depth of 0~20 cm was collected at each sampling point with the plum blossom distribution method. 1kg soil samples were collected with the coning and quartering method after even mixing, kept in vinyl plastic bags, and transported to the laboratory. Samples were air-dried naturally indoors and impurities and gravel were removed. 100 g soil samples were taken with the coning and quartering method, ground and sieved with 100-mesh nylon screens, and placed in sealed plastic bags for later use. 0.1000 g soil samples were weighed with an analytical balance and transferred to a 23 mL stainless steel reactor. 7.5 mL nitric acid and 7.5 mL hydrofluoric acid were added respectively. The samples were then put into a blowing dry box. The temperature was kept at 160°C for 60 min. After being heated and cooled to room temperature, the samples were taken out of the reactor. The completely nitrated sample solution was transferred to a 100 mL volumetric flask for constant volume. The atomic absorption spectrophotometer was used to measure the contents of Mn, Cr, Cu, Pb, and Cd. First, standard solutions of Cr, Mn, Pb, Cu, and Cd were prepared accurately, and the standard curve R^2 was greater than 0.999. Then, the contents of the five heavy metals were measured 3 times to obtain the average value.

2.4 *Data processing and evaluation methods*

The geoaccumulation index proposed by German scholar Muller 1969 is a method widely used to evaluate the pollution degree of heavy metals in sediments and other substances.

The formula for calculating the geoaccumulation index is $I_{geo} = \log(C_i/kB_i)$ (1)

Where C_i is the content of the ith heavy metal in the sample (in $mg \cdot kg^{-1}$); B_i is the geochemical background value of the ith heavy metal. In this paper, BVSH (State Environmental Protection Administration, China National Environmental Monitoring Center, 1990) is used as the reference value; k is used to correct the difference between regional background values, which is usually 1.5. Table 1 shows the relation between the classification of I_{geo} and heavy metal pollution degree.

Table 1. Classification of I_{geo} and soil pollution degree.

I_{geo}	≦0	0~1	1~2	2~3	3~4	4~5	>5
Rating	0	1	2	3	4	5	6
Pollution degree	Pollution-free	Mild	Slightly moderate	Moderate	Slightly heavy	Heavy	Extremely serious

2.5 *Statistical analysis of data*

All experimental data were processed in Excel.

3 RESULTS AND DISCUSSIONS

3.1 *Distribution features of heavy metal contents in farmlands in Dengzhuang Town, Hengshui*

Table 2 shows heavy metal contents in farmlands in Dengzhuang Town, Hengshui. The pollution of five heavy metals in farmlands in Dengzhuang Town, Hengshui was evaluated based on BVSH.

Table 2. Heavy metal contents in farmlands in Dengzhuang Town, Hengshui ($mg \cdot kg^{-1}$).

Sampling point	Longitude	Latitude	Cr	Mn	Pb	Cu	Cd
1	115.8336	37.7109	0.260	9.722	1.224	0.149	0.131
2	115.9598	38.0128	0.199	9.693	1.158	0.133	0.173
3	115.7899	37.7173	0.264	11.811	1.910	1.220	0.192
4	115.8108	37.6957	0.262	11.385	1.417	0.263	0.146
5	115.7901	37.7183	0.296	9.174	1.319	0.132	0.174
6	115.8037	37.7222	0.195	10.906	0.995	0.182	0.165
7	115.8342	37.7121	0.300	10.623	1.352	0.149	0.133
8	115.7964	37.6903	0.261	13.657	0.866	1.976	0.152
9	115.6947	37.7761	0.264	10.201	1.266	0.276	0.118
10	115.6257	37.7169	0.265	10.818	1.174	0.350	0.180
11	115.7583	37.7339	0.199	9.582	1.219	1.260	0.142
12	115.7899	37.7178	0.328	11.728	1.129	0.238	0.160
Average value			0.258	10.775	1.252	0.527	0.156
Background value of heavy metal contents in the soil in Hebei Province			68.3	78.4	21.5	21.8	0.094

According to the test, the pH of farmland soil in Dengzhuang Town, Hengshui was 7.5.

According to Table 2, the average contents of Cr, Mn, Pb, and Cu and their contents at each sampling point were lower than BVSH, but the average content of Cd was slightly higher than BVSH and the content of Cd at only three sampling points was lower than BVSH, which may be related to the high background content of Cd in local soil. According to GB15618-2018(Ministry of Ecological Environment 2018), the screening values of agricultural soil pollution risks were 200 $mg \cdot kg^{-1}$, 120 $mg \cdot kg^{-1}$, 100 $mg \cdot kg^{-1}$, and 0.3 mg kg^{-1} for Cr, Pb, Cu, and Cd respectively, and there was no screening value for Mn. Average contents of heavy metals Cr, Pb, Cu, and Cd in farmlands in Dengzhuang Town, Hengshui were 0.258 $mg \cdot kg^{-1}$, 1.252 $mg \cdot kg^{-1}$, 0.527 $mg \cdot kg^{-1}$, and 0.156 $mg \cdot kg^{-1}$, respectively, which means that the concentrations of Cr, Pb, Cu and Cd at all sampling points were lower than the screening values of agricultural soil pollution risks in GB15618-2018. Moreover, the average contents of heavy metals Cr, Pb, and Cu in farmlands in the town and their contents at each sampling point were far lower than the screening values in GB15618-2018.

3.2 *Geoaccumulation index pollution evaluation*

Table 3 shows the I_{geo} results of heavy metal contents in farmlands in the town. According to table 3, the average I_{geo} of Cr, Mn, Pb, Cu, and Cd at each sampling point was -2.599313289, -1.037990043, -1.41092539, -1.792737137, and 0.043905486, respectively. That is, the average

$I_{geo}$of Cr, Mn, Pb, and Cu contents was less than 0, and the I_{geo} at each sampling point was less than 0, based on which the soil can be regarded as pollution-free.

The I_{geo} of Cd at three sampling points was less than 0, based on which the soil can be regarded as pollution-free. At all sampling points, the average I_{geo} of Cd was 0.043905486. Considering that the average I_{geo} of Cd was small and very close to 0, the soil can be regarded to be mild-polluted.

Table 3. I_{geo} of heavy metal contents in farmlands in Dengzhuang Town, Hengshui.

Sampling point	Cr	Mn	Pb	Cu	Cd
1	−2.595538615	−1.082651705	−1.420748301	−2.341361484	−0.031947817
2	−2.711658886	−1.083949109	−1.44482116	−2.390696112	0.08882699
3	−2.588908036	−0.998120652	−1.227496352	−1.428187922	0.134082116
4	−2.592210671	−1.014074287	−1.357159869	−2.094592004	0.015133743
5	−2.539220252	−1.107848586	−1.388284923	−2.393973821	0.091330136
6	−2.720477351	−1.032741828	−1.510706638	−2.254476365	0.068264832
7	−2.533390708	−1.04416014	−1.377553027	−2.341361484	−0.025367472
8	−2.593871455	−0.935052012	−1.571011827	−1.218760812	0.032624475
9	−2.588908036	−1.061764574	−1.406096013	−2.073638671	−0.077337105
10	−2.587266089	−1.036260345	−1.438861622	−1.970479708	0.106053392
11	−2.711658886	−1.088951155	−1.422526013	−1.414177208	0.003069232
12	−2.494638119	−1.001183364	−1.455835777	−2.137970796	0.05490087
Average value	−2.599313289	−1.037990043	−1.41092539	−1.792737137	0.043905486

4 CONCLUSIONS

(1) At 12 sampling points in farmlands in Dengzhuang Town, Hengshui, the contents of Cr, Mn, Pb, and Cu were all lower than BVSH, but the content of Cd was slightly higher than BVSH and at 9 sampling points was higher than BVSH, which may be related to the high background content of Cd in local soil.
(2) The average contents of heavy metals Cr, Pb, Cu, and Cd in farmlands in the town and their contents at each sampling point were lower than the screening values of agricultural soil pollution risks in GB15618-2018, but there was no screening value for Mn in GB15618-2018.
(3) The I_{geo} of Cr, Mn, Pb, and Cu contents in farmlands in the town was less than 0 and the I_{geo} at each sampling point was less than 0, based on which the soil can be regarded as pollution-free; the average I_{geo} of Cd was 0.043905486. Considering that the average I_{geo} of Cd was small and approached 0, the soil is determined to be subject to mild pollution.

To sum up, the contents of the five heavy metals Cr, Mn, Pb, Cu, and Cd in the farmland soil in Dengzhuang Town, Hengshui meet the standard for cultivated land, which indicates that the farmland soil in the town is generally at a safe level. In future work, we should strengthen the routine monitoring of heavy metals in the farmland soil in Hengshui, especially the detection of Cd, to ensure the healthy and sustainable utilization of soil.

ACKNOWLEDGMENTS

Scientific research project of institutions of higher learning in Hebei Province in 2019—Detection, Analysis, and Evaluation of Heavy Metal Contents in Agricultural Land in Hengshui (No. ZD2019310); Independent project of Hebei Provincial Key Laboratory of Wetland Ecology and Protection (to be established) in 2019—Investigation and Detection of Major Heavy Metal Pollutants in Sediments and Surrounding Soil of Hengshui Lake and Analysis and Study on Impact on Lake Water Quality (No. hklz201911); and Initial scientific research fund program of high-level talents in Hengshui University in 2018—Measurement and Analysis of Heavy Metal Contents in Agricultural Soil in Dengzhuang Town, Hengshui (No. 2018GC35).

REFERENCES

Hua Xiaozan, Cheng Bin, Zhao Ruifen, Huo Xiaolan, Wang Zhao, Wang Sen. (2021) Evaluation and Spatial Distribution Characteristics of Heavy Metal Pollution in Farmland Soil in Taiyuan City. *Journal of Irrigation and Drainage*, 40:101–109.

Jia Ning. (2020) Study on Restoration Technology of Heavy Metals in Farmland Soil at Different Pollution Levels. *Environmental Science and Management*, 45:85–89.

Liao Renmei. (2016) *Evaluation of Heavy Metal Pollution of Farmland Soil in Fengxian County, Shanxi Province*. Master's Thesis of Northwest A&F University, Xi'an.

Ministry of Ecological Environment. (2018) *Soil Environmental Quality-risk Control Standard for Soil Contamination In Agricultural Land* (Trial) (GB 15618–2018). China Environment Publishing Group, Beijing.

Ministry of Environmental Protection, Ministry of Land and Resources. (2014) *National Soil Pollution Survey Report*. http://www.gov.cn/foot /site1/20140417 /782bcb88840814ba158d01.pdf.

Muller G. (1969) Index of geoaccumulation in sediments of the Rhine River.*Geo Jounal*. 2: 109–118.

State Environmental Protection Administration, China National Environmental Monitoring Center. (1990) *Background Values of Soil Elements in China*. China Environmental Science Press, Beijing.

Zhou Yalong, Yang Zhibin, Wang Qiaolin, Wang Chengwen, Liu Fei, Song Yuntao and Guo Zhijuan. (2021) Potential Ecological Risk Assessment and Source Analysis of Heavy Metals in Farmland Soil-Crop System in Xiong'an New Area. *Environmental Science*, 42: 2003–2015.

Energy Revolution and Chemical Research – Kok-Keong Chong and Zhongliang Liu (Eds)
© 2023 The Authors, ISBN: 978-1-032-36554-1

Effect of YSZ Addition on Electrochemical Activity of Pt/YSZ Electrode

Jixin Wang & Jiandong Cui
GRINMAT State Key Laboratory of Advanced Materials for Smart Sensing, GRINM Group Co., Ltd., Beijing, China
GRIMAT Engineering Institute Co., Ltd., Beijing, China
General Research Institute for Nonferrous Metals, Beijing, China

Xiao Zhang
GRINMAT State Key Laboratory of Advanced Materials for Smart Sensing, GRINM Group Co., Ltd., Beijing, China
GRIMAT Engineering Institute Co., Ltd., Beijing, China

Wentao Tang & Changhui Mao*
GRINMAT State Key Laboratory of Advanced Materials for Smart Sensing, GRINM Group Co., Ltd., Beijing, China
GRIMAT Engineering Institute Co., Ltd., Beijing, China
General Research Institute for Nonferrous Metals, Beijing, China

ABSTRACT: The (Pt/YSZ)/YSZ sensor unit is the basic component of the NO_x sensor, which can detect the emission of nitrogen oxides in the exhaust gas and optimize the fuel combustion process. In this work, the effect of YSZ Addition on the electrochemical activity of the Pt/YSZ electrode was investigated. Pt/YSZ electrodes added with different contents of YSZ powders were prepared. The microstructure of the Pt/YSZ electrodes was observed by SEM. Chronoamperometry, linear scan voltammetry, and AC impedance were tested by the electrochemical workstation. The results show that increasing the addition amount of YSZ helps to improve the electrochemical activity of the Pt/YSZ electrode, which is benefited by the formation of the porous structure of the Pt/YSZ electrode. With the gradual increase of YSZ addition, the electrode conductivity first increased and then decreased. The activation energy reaches the minimum value (1.05 eV) when the YSZ content is 15 wt%.

1 INTRODUCTION

Nitrogen oxides (mainly NO and NO_2) bring acid rain and photochemical smog, which pose a great threat to human health and environmental safety (Killa S 2013). NO_x sensor is a key device to control this problem by monitoring the NO_x content in the exhaust gas and optimizing the fueling combustion process (Kato N 1996; Martin L P 2007; Ueda T 2010). At present, NO_x sensors are mainly divided into the following four types: potential type, mixed potential type, complex impedance type, and current type (Cai 2016; Kato N 1998; Park C O 2009; Shimizu Y 2018), of which the current type sensor is the only one commercially used until now. This type of NO_x sensor consists of two cavities and three oxygen-pumping cells with (Pt/YSZ) electrode/YSZ electrolyte sensor unit structure, which mainly consists of an oxygen-vacancy-rich YSZ electrolyte and two highly electrochemically active Pt/YSZ electrode. The detection process is shown in Figure 1. The main pump and auxiliary

*Corresponding Author: mao@grinm.com

DOI 10.1201/9781003332657-62

pump fully pump the O_2 in the exhaust gas in the two cavities, inducing the conversion of NO_2 into NO, and NO is finally decomposed into N_2 and O_2. When the decomposed O_2 is pumped away by the measuring pump oxygen cell, the concentration of decomposed O_2 can be obtained by measuring the corresponding pump current, and the NO_x concentration can be obtained after conversion. The electrochemical activity of the Pt/YSZ electrode seriously affects the performance of NO_x sensors. Moreover, the electrode composition, electrode thickness, sintering process, microstructure, and morphology are all key factors for the Pt/YSZ electrode (Jaccoud A 2007; Li, 2021; Sridhar S 1997). Jaccoud et al. (2007) found that the Pt electrode prepared by Pt slurry had better electrochemical performance than that prepared by sputtering. Nurhamizah (2017) found that the Pt electrode with a porous structure has better electrochemical performance. Chen (2011) and Liu (2017) found that a certain proportion of YSZ powder in the Pt slurry could improve the porous structure of the Pt electrode and improve the adhesion of the Pt electrode to the YSZ electrolyte. However, the addition of YSZ powder will reduce the conductivity of Pt. Excessive YSZ powder is obviously unfavorable to the electrochemical activity of the Pt/YSZ electrode. In this experiment, the effect of YSZ addition on the performance of the Pt/YSZ electrode was studied. The chronocurrent, linear scanning voltammetry, and AC impedance of the Pt/YSZ electrodes were tested by an electrochemical workstation. The microstructure of the Pt/YSZ electrodes was also observed by SEM.

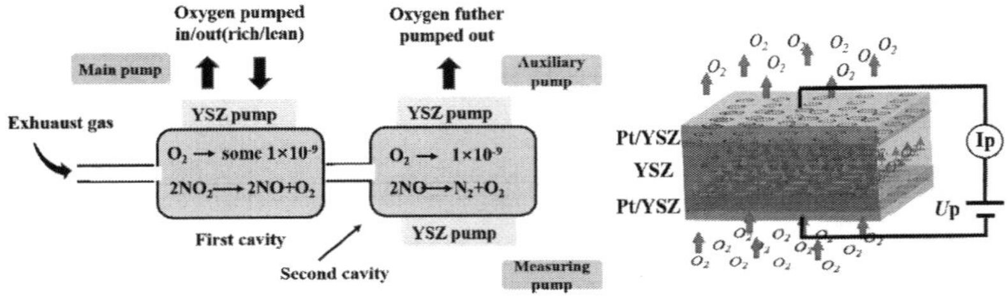

Figure 1. Schematic diagram of detection principle of current type NO_x sensor (left) NO_x concentration detection principle; (right) (Pt/YSZ)/YSZ sensor unit oxygen pumping principle.

2 EXPERIMENTAL PROCEDURE

The YSZ green tapes were prepared by the tape casting process. Different mass fractions of YSZ powder were added to the Pt slurry, and the additional amounts were 1 wt%, 5 wt%, 10 wt%, 15 wt%, and 20 wt%, respectively. The Pt/YSZ electrode slurry was printed on the YSZ green tapes, and the (Pt/YSZ)/YSZ sensor units were obtained by sintering at 1450°C under air condition. The sample schematic of the (Pt/YSZ)/YSZ sensor unit is shown in Figure 2. The microscopic morphology of the samples was observed by SEM (JSM-7610F, Japan). The electrical properties of the Pt/YSZ electrodes were tested by a CHI660D electrochemical workstation, the mixture of gases with 10 vol.%O_2 and 90 vol.%$N_2$was added during the test. Specific test parameters were as follows: (1) Chronoamperometric experiment: a fixed voltage of 600 mV was set between the two poles, the scanning time was 0-180 s, the test temperature was 750°C; (2) Linear scan voltammetry test: the scanning voltage was -0.6 V to 0.6 V, the test temperature was 750°C; (3) AC Impedance (EIS) Test: the frequency range was set to 0.001 Hz-10 MHz, the signal voltage was 500 mV, the test temperatures were 600°C, 650°C, 700°C, 750°C, and 800°C, respectively.

3 TEST RESULTS AND DISCUSSIONS

3.1 *Micromorphologies of Pt/YSZ electrodes*

The micromorphologies of the Pt/YSZ electrodes with different mass fractions of YSZ are shown in Figure 3. The pure Pt electrode is denser after sintering, and the Pt particles are sintered into

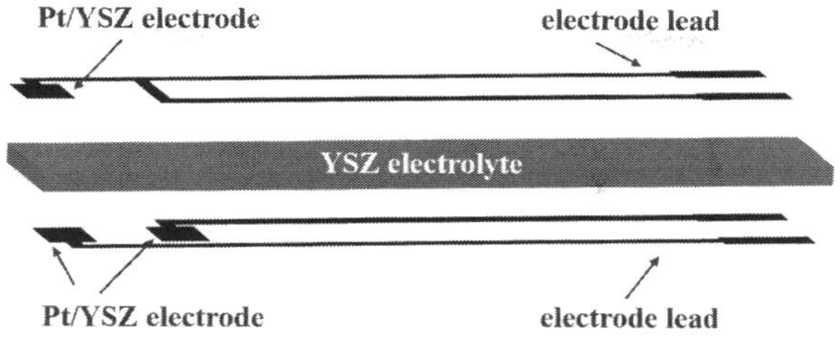

Figure 2. Sample schematic of the (Pt/YSZ)/YSZ sensor unit.

one piece to form a larger metal electrode structure with few holes. With the gradual increase of the amount of YSZ added, the porous structure of the Pt/YSZ electrode became more obvious, the number of pores increased, and it was not simply sintered into sheets, indicating that the YSZ powder had a certain inhibitory effect on the agglomeration of Pt. When the addition ratio of YSZ is 15 wt%, the large particles of Pt and the small particles of YSZ are uniformly mixed together to form a porous electrode structure after sintering, which is favorable for gas diffusion through the electrode. However, when the addition of YSZ continued to increase to 20 wt%, the densification and sintering trend of YSZ would destroy the porous structure of the electrode, resulting in fewer pores and a smaller diameter of pores. Comprehensive analysis shows that the Pt/YSZ electrode has the best porous structure when the addition ratio of YSZ is 15 wt%.

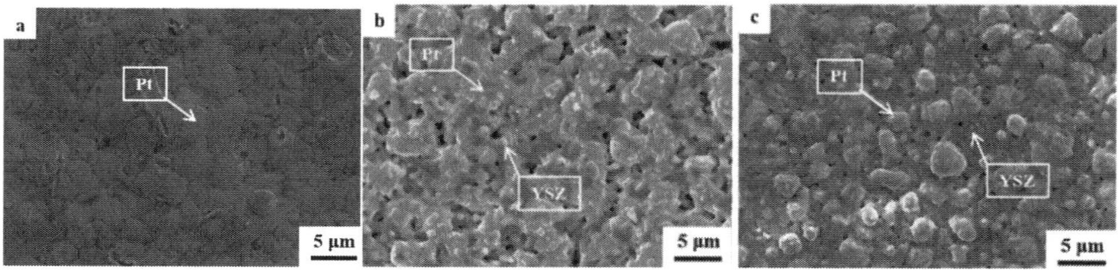

Figure 3. SEM morphologies of Pt/YSZ electrodes with different mass fractions of YSZ (a)0 wt%; (b)15 wt%; (c)20 wt%.

3.2 *Chronoamperometry*

A constant potential is applied to the Pt/YSZ electrodes to obtain a current-time curve, which is called chronoamperometry. The stability of the electrode can be investigated by observing the change in the current value, and the electrochemical activity can be analyzed by comparing the stable current value (Gyrgy Fóti 2009; Li 2021). The chronoamperometric curves of Pt/YSZ electrodes with different mass fractions of YSZ are shown in Figure 4, each curve shows the same trend of change, and the current reaches a stable value in a relatively short period of time. With the gradual increase of the amount of YSZ added, the current first increased and then decreased. When the YSZ content is 15 wt%, the current of the Pt/YSZ electrode reaches the maximum value.

3.3 *Linear scan voltammetry analysis*

In a certain potential range, apply a continuous triangular wave signal to the Pt/YSZ electrodes, scan from the cathode direction to the anode direction with a constant scanning rate, and record the curve of current versus voltage, which is called linear scan voltammetry (LSV) (Gyrgy Fóti 2009; Nicholson R S 1965). By analyzing the cathodic and anodic peaks of the linear voltammetry curve, the possible electrode reactions, the reversibility of the electrode reactions, and the source of

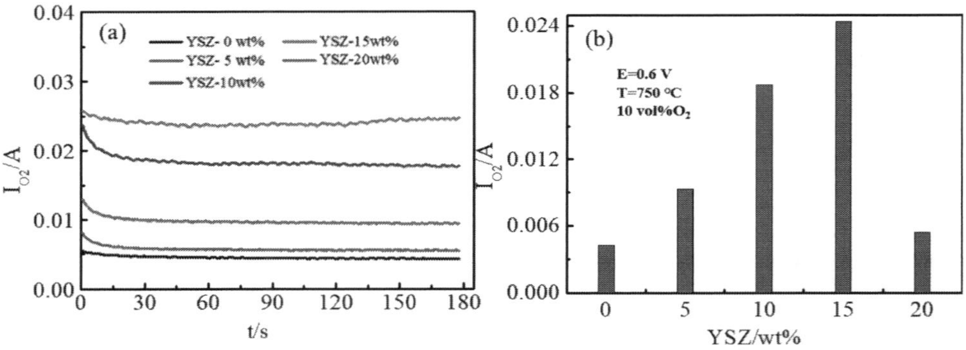

Figure 4. Chronoamperometric curves of Pt/YSZ electrodes with different mass fractions of YSZ (a) current versus time; (b) steady current value.

the reaction products can be studied. In the NO_x testing process, the cathodic and anodic reaction model of the Pt/YSZ electrode system can be expressed as (Jaccoud A 2007):

$$O^{2-} \overset{-2e^-}{\rightarrow} O_{atoms} \overset{+Pt_{atoms}}{\rightarrow} PtO \tag{1}$$

$$O^{2-} \overset{-2e^-}{\rightarrow} O_{atoms} \rightarrow \frac{1}{2} O_{2(gas)} \tag{2}$$

$$Pt\text{-}O + 2e^- \rightarrow Pt_{atoms} + O^{2-} \tag{3}$$

The speeds of the electrode reactions can be evaluated by comparing the slopes of the curves. The LSV curves of the Pt/YSZ electrodes are shown in Figure 5. The current changed drastically with increasing potential, approximately fitting the curve of the cathodic reaction to a straight line, and the fitting results are shown in Figure 5. It can be seen that the slopes of these curves gradually increase with the increase in the amount of YSZ added. The slope relationship of these curves is: 15 wt%>10 wt%>5 wt%>20 wt%>0 wt%. It can be inferred that the cathodic reaction rate of the Pt/YSZ electrode is the fastest when the YSZ content is 15 wt%, indicating the highest electrochemical catalytic activity.

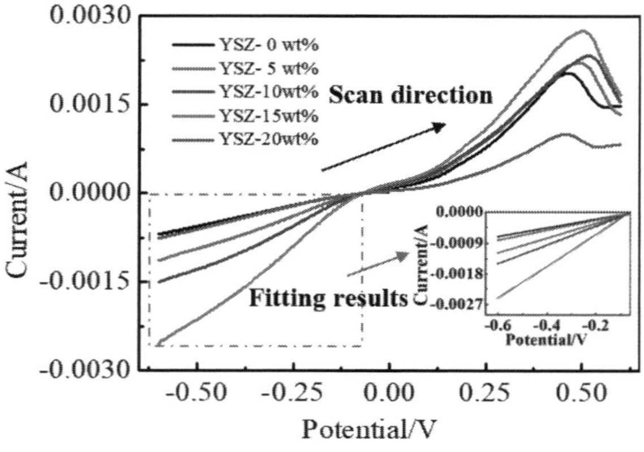

Figure 5. LSV curves of the Pt/YSZ electrodes with different mass fractions of YSZ.

3.4 *AC Impedance analysis*

Through the AC impedance, the conductivity of the electrolyte and electrode and the activation energy of the reaction can be calculated respectively. The greater the conductivity is, the easier the charge transfer process will become, and the smaller the reaction activation energy is, the higher the electrochemical activity of the electrode will be (Kovrova A I 2020; Shan 2017). The variation curves of Pt/YSZ electrode conductivity with different test temperatures (650°C, 700°C, 750°C, 800°C, and 850°C, respectively) under different YSZ additions are shown in Figure 6. With the gradual increase of YSZ addition, the electrode conductivity first increased and then decreased. When the YSZ content was 15 wt%, the Pt/YSZ electrode reached the maximum value. It is shown that the addition of YSZ to the Pt slurry increases the conductivity of the electrode, which can be explained by the Pt/YSZ electrode. The addition of YSZ improves the porous structure of the electrode, making the charge transfer easier.

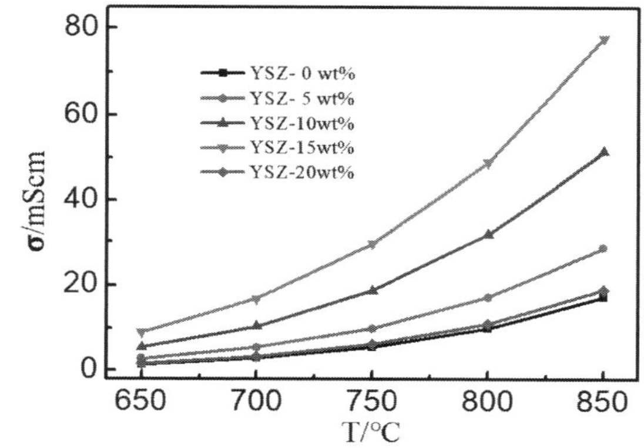

Figure 6. Variation curves of Pt/YSZ electrode conductivity under different YSZ additions.

The variation of the conductivity of the sample with the test temperature (600°C, 650°C, 700°C, 750°C, and 800°C, respectively) can be expressed as:

$$\sigma = \frac{A}{T} e^{-\frac{E}{kT}} \tag{4}$$

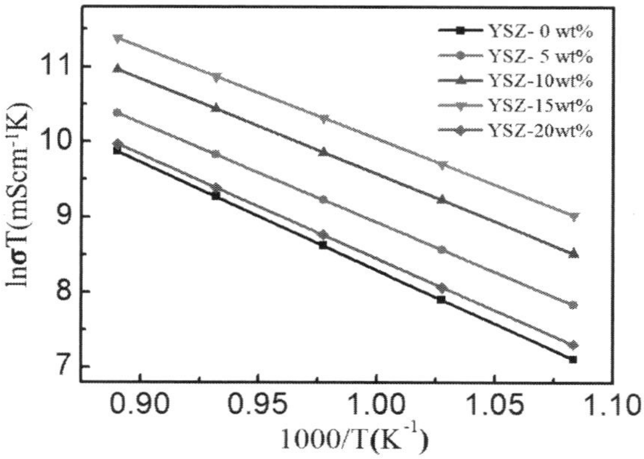

Figure 7. Arrhenius curves of Pt/YSZ electrodes under different YSZ additions.

Transforming it into another form:

$$\ln \sigma T = -\frac{E}{kT} + \ln A \tag{5}$$

Among them: σ is the conductivity(mScm^{-1}), T is the test temperature(K), k is the Boltzmann constant (1.38×10^{-23} J/K), E is the activation energy(eV), A is a constant. Therefore, taking 1000/T as the abscissa and $\ln \sigma T$ as the ordinate to draw a curve, according to the fitting slope of the obtained curve, the specific value of the diffusion activation energy can be calculated. The Arrhenius curves of Pt/YSZ electrodes under different YSZ additions are shown in Figure 7, and the activation energy of the electrode reaction can be obtained by linear fitting of each point, as shown in Table 1. With the increase of YSZ addition, the activation energy of the Pt/YSZ electrode reaction first decreased and then increased, and the activation energy of the Pt/YSZ electrode reaction reached the minimum value (1.05 eV) when the YSZ content is 15 wt%.

Table 1. The activation energy of electrode reaction.

YSZ additions/wt%	0	5	10	15	20
E lectrode activation energy/eV	1.23	1.13	1.09	1.05	1.19

4 CONCLUSION

Based on the results and discussions presented above, the conclusions are obtained as below:

(1) Increasing the YSZ additions ($\leq$15 wt%) helps to improve the electrochemical activity of the Pt/YSZ electrode, which is benefited by the porous structure of the Pt/YSZ electrode.
(2) With the gradual increase of YSZ addition, the electrode conductivity first increased and then decreased. The activation energy reaches the minimum value (1.05 eV) when the YSZ content is 15 wt%.

ACKNOWLEDGMENTS

This work was financially supported by the Shandong Province Key R&D Program (No.2020CXGC010203).

REFERENCES

Cai H, Sun R, Yang X, et al. (2016) "Mixed-potential type NO_x sensor using stabilized zirconia and MoO_3-In_2O_3 nanocomposites". *Ceramics International* 42(2016): 12503–12507.
Chen B X, Xin H E, Xiong X D, (2011) "Effect of YSZ Powder on Porous Platinum Electrode Activity". *Instrument Technique and Sensor* 12(2011): 6–9.
Gyrgy Fóti, Jaccoud A, Falgairette C, et al. (2009) "Charge storage at the Pt/YSZ interface". *Journal of Electroceramics* 23(2009): 175–179.
Jaccoud A, G Fóti, R Wüthrich, et al. (2007) "Effect of microstructure on the electrochemical behavior of Pt/YSZ electrodes". *Topics in Catalysis* 44(2007): 409–417.
Jaccoud A, Falgairette C, G Fóti, et al. (2007) "Charge storage in the O_2(g), Pt/YSZ system". *Electrochimica Acta*, 52(2007): 7927–7935.
Kato N, Nakagaki K, Ina N. (1996) "Thick Film ZrO_2 NO_x Sensor". proceedings of the International Congress & *Exposition*. 1996.
Kato N, Kurachi H, Hamada Y. (1998) "Thick film ZrO_2 NOx sensor for the measurement of low NO_x concentration". *SAE Transactions* 106(1998): 312–320.

Killa S, Cui L, Murray E P, Mainardi D S. (2013) "Kinetics of Nitric Oxide and Oxygen Gases on Porous Y-Stabilized ZrO_2-Based Sensors". *Molecules* 18(2013): 9901–9918.
Kovrova A I, Gorelov V P, Osinkin D A. (2020) "Impregnation of Pt|YSZ Oxygen Electrodes with Microquantities of Praseodymium Oxide". *Russian Journal of Electrochemistry* 56 (2020): 477–484.
Li C. (2021) "Study on the Preparation and Properties of Pt-YSZ Composite Electrode". *Chemical Management* 11(2021): 79–80.
Liu G, Xiong X, Xin H, et al. (2017) "Microstructure and Volt-Ampere Characteristics of Pt-Rh-YSZ Gas Sensing Electrodes". *Chinese Journal of Rare Metals* 41(2017): 1339–1346.
Martin L P, Woo L Y, Glass R S, (2007) "Impedancemetric NO_x sensing using YSZ electrolyte and YSZ/Cr_2O_3 composite electrodes". *Electrochem. Soc* 154(2007): 97–104.
Nicholson R S. (1965) "Theory and application of cyclic voltammetry for measurement of electrode reaction kinetics". *Analytical Chemistry* 37(1965): 1351–1355.
Nurhamizah A R, Ibrahim Z, Muhammad R, et al. (2017) "Effect of Annealing Temperature on Platinum/YSZ Thin Film Fabricated Using RF and DC Magnetron Sputtering." *Solid State Phenomena* 268(2017): 229–233.
Park C O, Fergus J W, Miura N, et al. (2009) "Solid-state electrochemical gas sensors". *Ionics* 15(2009): 261–284.
Shan K, Zhai F, Nan L I, et al. (2017) "Impedance Behavior of YSZ-$SrTi_{0.6}Fe_{0.4}O_{3-\delta}$ Ceramic Composite". *Journal of Ceramics* 38(2017): 908–912.
Shimizu Y, Nakano H, Takase S, et al. (2018) "Solid Electrolyte Impedance metric NOx Sensor Attached with Zeolite Receptor". *Sensors & Actuators B Chemical* 264(2018): 177–183.
Sridhar S, Stancovski V, Pal U B. (1997) "Effect of oxygen-containing species on the impedance of the Pt/YSZ interface". *Solid State Ionics* 100(1997):17–22.
Ueda T, Nagano T, Okawa H, Takahashi S (2010) "Amperometric-type NO_x sensor based on YSZ electrolyte and La-based perovskite-type oxide sensing electrode". *Journal of the Ceramic Society of Japan* 118(2010): 180–113.

Energy Revolution and Chemical Research – Kok-Keong Chong and Zhongliang Liu (Eds)
© 2023 The Authors, ISBN: 978-1-032-36554-1

Study on sensory quality of Gaoxiangyihong Congou black tea

Liangzi Zhang & Ziyue Zhang
College of Food Science and Technology, Wuhan Business University, Wuhan, China

Shenghong Zhou*
College of Food Science and Technology, Wuhan Business University, Wuhan, China
Chinese Tea Culture and Industry Research Institute, Hangzhou, China

ABSTRACT: Eight kinds of Congou black tea were studied through the traditional sensory evaluation, from the five aspects of tea shape, leaf bottom, tea color, aroma, and taste of the eight tea samples were measured with a colorimeter, and the derivative values were calculated; the total content of caffeine and free amino acids in the eight tea samples was detected by spectrophotometry, and the analysis was carried out in combination with the scores of the sensory evaluation. The aroma of tea soup was collected and detected by Fox-4400 electronic nose, and the data processing system of the electronic nose was used to analyze the detection results of the aroma. Through the above analysis, the correlation between artificial sensory evaluation and instrumental analysis methods is summarized, and a rapid detection method for sensory quality of Congou black tea is tried to be established. The results showed that: in terms of taste, the total amount of free amino acids in Gaoxiang Yihong Congou black tea was 0.035%, and the content of caffeine was 0.013%, which were negatively correlated with the total sensory score. In terms of aroma, the electronic nose can be obtained from the radar map and DFA map, which has advantages in the identification of tea varieties, but cannot directly reflect the quality of aroma.

1 INTRODUCTION

As a beverage product, black tea has the common characteristics of food. Color (Li 2005), aroma, taste, and shape are all important conditions for judging the sense of tea. In China, most tea leaves are habitually used the traditional sensory evaluation method (Li 2015) to evaluate the quality of tea (Huang 2017). Intelligent sensory analysis technologies such as the electronic nose and electronic tongue have the advantages of simple sample preparation and fast testing. The electronic tongue obtains information through sensors, and it extracts parameters from the sensor or the characteristic information and different classifications of samples. Statistical analysis methods were also used as the core evaluation of tea (Li 2017), but the cost was high, and it is difficult to popularize in many places. In addition to bionic technology, most detection institutions still use headspace adsorption and GC-MS analysis for the detection of tea aroma.

Based on traditional sensory evaluation, supplemented by modern science and technology, using modern scientific methods such as electronic nose (Ma 2019; Wang 2012) and colorimeter, the quality analysis of the tea soup (Lu 2002), aroma and taste of Gaoxiang Yihong Congou black tea was carried out (Qiu 2016), and the quality of tea soup, aroma, and taste of Gaoxiang Yihong Congou black tea was analyzed to find out the quality of its tea in the same category. In terms of its advantages and disadvantages compared with other Congou black teas, reasonable data analysis was carried out on the research on the quality of Gaoxiang Yihong Congou (Yuan 2018; Zhang

*Corresponding Author: 1021098393@qq.com

 DOI 10.1201/9781003332657-63

2019), in an attempt to provide a certain theoretical basis for finding a fast, efficient and accurate method for identifying the quality of Congou black tea.

2 MATERIALS AND METHODS

2.1 *Materials*

Basic lead acetate, concentrated hydrochloric acid, concentrated sulfuric acid, caffeine disodium hydrogen phosphate, ninhydrin hydrate, and glutamic acid were all of the analytical grades.

The CM-2600d colorimeter was bought from Konica Minolta Holdings, Japan; the SHZ-D (III) circulating water type multi-purpose vacuum pump was bought from Gongyi Ruide Instrument Equipment Co., Ltd.; the Fox-4400 electronic nose was bought from France Alpha M OS Company.

Select several of the most representative Congou black teas in China. Among all black tea samples, Lichuan Hong is the Gaoxiang Yihong Congou black tea among the traditional Congou black teas, Zhengshan Souchong and Yinghong No. 9 black tea are first-class teas. The remaining six are premium teas. Eight tea samples and their specific information are shown in Table 1.

Table 1. Black tea samples and origin.

Number	Black tea sample	Origin
1	Lichuanhong	Fujian Nanping Wuyi Mountain
2	Yihong Congou Black Tea	Yilin District, Yichang
3	Huang Guanyin	Maoling Reef Base of COSCO Ecological Tea Company in Wuyi Mount
4	Golden Peony 2017	Maoling Reef Base of COSCO Ecological Tea Company in Wuyi Mount
5	Golden Peony 2018	Maoling Reef Base of COSCO Ecological Tea Company in Wuyi Mount
6	Lapsang Souchong	Fujian Nanping Wuyi Mount
7	Ying Hong No. 9	Yingde City, Guangdong Province
8	Dianhong tea	Fengqing, Yunnan

2.2 *Experimental method*

2.2.1 *Sensory evaluation of black tea*

According to the evaluation methods of black tea in GB/T 23776-2018 and GB/T 14487-2018, the terms of appearance, aroma, soup color, taste, and Leaf bottom were described and scored respectively. The shape, taste, aroma, tea color, and Leaf bottom were scored by 9 professionally trained personnel. The scoring table is based on a percentage system, in which the shape of the tea leaves accounts for 25%, the taste accounts for 30%, the aroma accounts for 25%, the soup color accounts for 10%, and the bottom of the leaves accounts for 10%. We take 3 grams of tea samples respectively, pour 150 mL of boiled ultrapure water, cover and let stand for 5 minutes, and pour them into the judge's cup. After the judges observe and taste the scores, we get the average score, i.e., the final score. The sensory evaluation form is shown in Table 2.

2.2.2 *Detection of taste substances in tea soup*

For the detection of the total amount of free amino acids, refer to the national standard GB/T 8314-2013, and use spectrophotometry to determine its content at a specific wavelength.

The detection of caffeine in tea soup refers to the national standard GB/T 8312-2013, using the principle that caffeine in tea is easily soluble in water, after the sample tea soup is subjected to suction filtration and other processes, the interfering substances in the tea soup are removed, and its content is determined by a specific wavelength.

Table 2. Sensory evaluation table for tea.

Descriptive word	Quality traits	Score
Shape	The dried tea is compact and slender in shape, with tender stems and golden hairs (the outside of the dry tea looks covered with tiny fluff). The color of dry tea is dark and bright, the integrity of the whole tea leaves is high, and the content of tea stalks, tea pieces, and non-tea inclusions is low.	80~99
	The dried tea is compact and slender in shape, with tender stems and golden hairs (the outside of dried tea leaves looks covered with tiny fluff). The color of dry tea is dark and bright, the integrity of the whole tea leaves is high, and the content of tea stalks, tea pieces, and non-tea inclusions is low.	60~79
	The dried tea is compact in shape, with tender stems and golden hairs (the outside of dried tea leaves looks covered with tiny fluff). The color of dry tea is dark and bright, the integrity of the whole tea leaves is high, and the content of tea stalks, tea pieces, and non-tea inclusions is high.	40~59
Aroma	With fruity or floral aromas, the aroma is long, heavy, and sweet.	80~99
	With a fruity or floral aroma, the aroma is slightly lighter, persistent, and slightly sweet.	60~79
	The fruity or floral fragrance is weak or difficult to distinguish, slightly persistent and sweet	40~59
Taste	The tea soup is heavy, mellow, sweet, refreshing, and delicious.	80~99
	The tea soup has a slightly heavy taste, mellow, sweet, slightly bitter, and has a refreshing taste.	60~79
	The tea soup is bitter, astringent, slightly mellow, and slightly sweet.	40~59
Soup color	Bright and ruddy.	80~99
	Brighter, slightly reddish.	60~79
	Less bright or cloudy.	40~59
Leaf bottom	Soft and many buds, the color is red and bright, and the integrity is high.	80~99
	Tender and soft, with many buds, the color is still bright red.	60~79
	It is still tender, with many tendons (the tendons of tea leaves), and the color is still red and bright	40~59

2.2.3 *Detection of the aroma of tea soup*

The detection process is as follows: weigh 0.2 g of tea sample, put it into a special detection container for electronic nose, pour 10 mL of boiled ultrapure water into the container, seal it with plastic wrap and let it stand for 45 min, and collect the aroma for detection.

The carrier gas air generator pressure is 0.35 bar; the head air injection volume is 3 mL; the head air injection rate is 1 mL/s; the injection needle temperature is 60°C; the data acquisition time is 120 s, and the carrier gas flow rate is 150 mL/min.

2.2.4 *Analysis and processing of experimental data*

The experimental data were processed with Microsoft Office Excel 2007. The electronic nose data were analyzed and graphed by principal component with the electronic nose special software.

3 RESULTS AND ANALYSIS

3.1 *Sensory evaluation of black tea*

According to the calculation method of GB/T 23776-2018, the final scores of shape, aroma, taste, soup color, the bottom of the leaf, and total scores were obtained. The sensory evaluation scores are shown in Table 3.

Table 3. Sensory evaluation score.

Black tea sample	Shape	Aroma	Taste	Soup color	Leaf bottom	Score
Lichuanhong	69.85±18.5	77.22±11.69	69.81±14.43	74.78±13.15	75.41±14.81	72.73±8.30
Yihong Congou Black Tea	68.52±15.81	72.33±21.44	74.22±16,94	70.67±20.72	80.04±12.93	72.55±10.87
Huang Guanyin	67.59±16.61	86.41±10.38	80.44±16.38	77.11±13.63	61.81±14.91	76.53±7.78
Golden Peony 2017	71.26±19.15	67.37±15.75	65.67±15.69	76.00±12.73	66.22±16.22	68.58±7.53
Golden Peony 2018	68.04±13.45	59.70±13.14	62.81±15.68	69.93±16.27	75.00±11.45	65.27±7.23
Lapsang Souchong	75.96±17.10	75.11±14.81	74.52±15.75	76.44±13.12	75.07±14.44	75.39±13.73
Ying Hong No. 9	74.11±13.69	75.26±14.31	75.48±14.83	72.52±14.72	75.70±12.32	74.81±11.88
Dianhong tea	73.67±14.53	73.70±15.20	74.30±15.31	74.63±20.20	74.85±15.30	74.08±14.24

From the score of sensory evaluation and the data of visual analysis, it can be seen that in the shape of Lapsang souchong, the score is 75.96, most of them give a high degree of integrity, good clarity, tight lines, good color, and smooth, giving the best intuitive feeling, the Lichuan Red scored a below-average of 69.85, with most of the judges judging it to be tight and with a gold rating. In terms of aroma score, Huang Guanyin was the most popular among the judges, scoring 86.41. The Aroma was fruity and full-bodied, while Lichuan Red ranked second with a score of 77.22. The fruit aroma was sweet. In terms of taste score, Huang Guanyin ranked first with a high score of 80.4. Most people think that it tastes mellow and sweet, unusually refreshing (Yin 2020), while Lichuan Red scored 69.81 with a bitter taste and light taste. From the perspective of soup color, Huang Guanyin still scored the highest with red and bright color, while Lichuan Red scored 74.78 with red and bright color. From the bottom of the leaf score, Yihong Congou black tea scored the highest and Lichuan Red was still in the medium level. Finally, from the overall score, Huang Guanyin had the highest total score and the best overall sensory effect, while Lichuan Red was in the middle of the sensory evaluation among the eight kinds of tea samples.

3.2 *Detection and analysis of flavoring substances in tea soup*

According to Figures 1 and 2, and Table 4, the total free amino acid content and theine content were significantly negatively correlated with sensory total score and taste score, indicating that the lower the total free amino acid content and theine content (Wu 2020), the higher the score would be. Because theine is mainly bitter in people's taste sense and mainly reflected as sweet in tea, the less theine content. The more sweetness people taste, the more it fits the general taste criteria, and therefore the higher the score. The total content of free amino acids in these eight tea samples is greater than that of theine. Amino acids can relieve the bitter taste and improve the feeling of freshness, but the content is too small to be completely felt by the human senses. Although the content of caffeine in Lichuan red studied in this paper is moderate and the total content of free amino acids is high, the bitterness is prominent in people's senses, and the bitterness is not covered well. The overall taste score is also below medium. When studying Zunyi black tea, it was found that the more amino acid content, the higher people's sensory score of tea. However, in this experimental analysis, it was found that the total amount of free amino acids was negatively correlated with the total score, which may be due to the following two reasons: taste recognition threshold could not accurately taste the sweet taste of amino acids, and the tea was affected by moisture in the storage process, and the flavoring substances were affected to a certain extent (Zhao 2016).

3.3 *Electronic nose detection of the aroma of tea soup*

3.3.1 *Analysis of aroma response value of the sensor*

It can be observed from radar Figure 3 that in terms of aroma sensitivity, p30/1, PA/2, P30/2, T30/1, and other sensors have obvious differences in response values of tea sample aroma, while

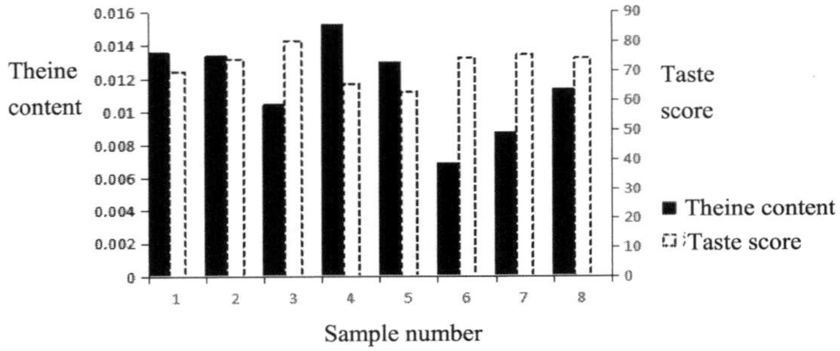

Figure 1. Theine content and taste score of eight kinds of tea samples.

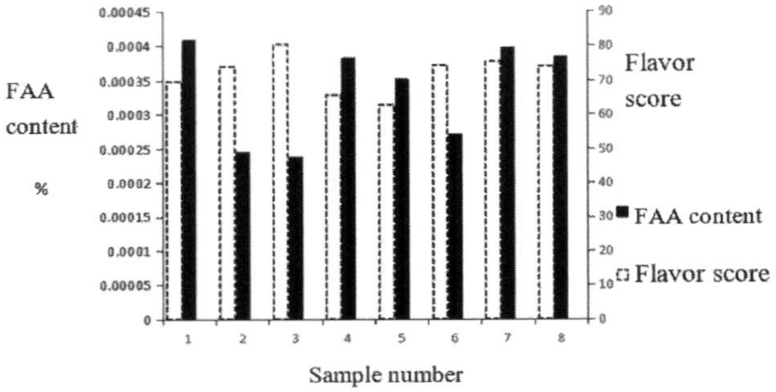

Figure 2. Total free amino acids (FAA) and flavor score of eight tea samples.

Table 4. Correlation coefficient r between total free amino acids and theine content on sensory results.

Item	Taste	Score
FAA	−0.51**	−0.34**
Theine content	−0.61**	−0.66**

(Note: "**" indicated that the p-value is less than 0.01, which is extremely correlated.)

for LY2/G, LY2/AA, and other sensors, the contribution percentage is very small, almost negligible. It can be well judged that the substances in tea aroma mainly come from p30/1, PA/2, P30/2, T30/1, and other sensitive aroma substances.

P30/1, 2, P30/PA/2, and T30/1 sensor to the tea aroma composition analysis of the contribution rate was higher more than 80%, the rest of the sensor was a response, but the contribution rate was low, according to the performance of their respective sensors showed that the sensor to the aromatic compounds, organic compounds and oxidation ability strong gas-sensitive (Zhou 2004). Therefore, in the subsequent research process, we adopt the numerical analysis of these sensors with large contribution rates.

3.3.2 *Comparison of DFA on aroma clustering of different black tea*

It can be seen from Figure 4 that the cumulative contribution of the electronic nose to discriminant factor analysis of eight tea samples reached 86.715%, in the DFA differentiation diagram, the position of the resulting diagram of repeated determination of each sample was similar, and the error of parallel determination was small. According to the above analysis, the electronic nose can

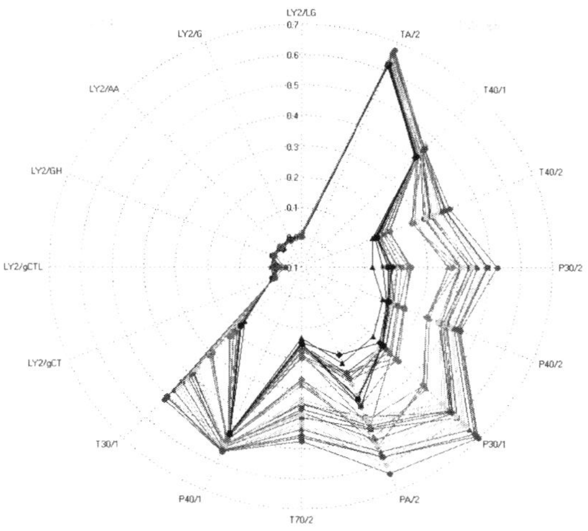

Figure 3. Aroma response values of 16 sensors of eight tea samples.

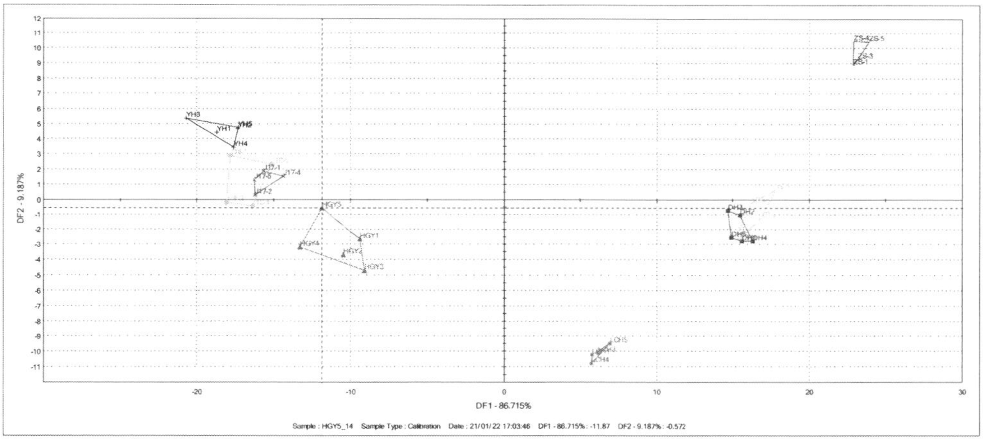

Figure 4. DFA analysis diagram of original data.

distinguish the aroma of eight kinds of tea well, and there are differences between each other (Zhou 2015).

In Figure 4, there are many overlapping areas between Golden Peony 2017 and Golden Peony 2018, which proves that the aroma components of the two have little difference. Lapsang souchong and Yinghong No. 9 also have a lot of overlap. Compared with them, the remaining four tea samples are all in their respective regions with no overlap and are far apart. Among them, Yihong Congou, Golden Peony 2018 and Golden Peony 2017 are in the same area as Huang Guanyin, while Lichuan Red, Yunnan Black Tea, Yinghong No.9, and Lapsang souchong are in the same area. Although the aroma components and degrees of the same tea are different, they are similar. Combine the aroma cluster distribution in Figure 4 with the sensory aroma score. In the left half, the score of Huang Guanyin was the highest, which was consistent with the phenomenon that the tea samples were far away from other tea samples in the figure. In addition, the score of tea samples in the right half was similar, among which the score difference between Lapsang Souchong and Yinghong No. 9 was small, which was consistent with the overlapping phenomenon in Figure 4. The size rule of the remaining tea samples was consistent with the distance. The aroma and sensory score of Lichuan black tea were second only to Huang Guanyin, which was close to Lapsang Souchong and Yinghong Nine, but the highest score of Huang Guanyin did not belong to the same region.

Table 5. Aroma sensory score.

Black tea sample	Lichuanhong	Yihong Congou Black Tea	Huang Guanyin	Golden Peony 2017	Golden Peony 2018	Lapsang Souchong	Ying Hong No. 9	Dianhong tea
Aroma score	77.22	72.33	86.41	67.37	59.70	75.11	75.26	73.70

Referring to Table 5 of the aroma score, we can find that Golden Peony 2017 and Golden Peony 2018 are black tea of the same variety and different years, and there are many overlapping areas. Huang Guanyin, Golden Peony 2017, and Golden Peony 2018 are black tea of the same region, and the clustering of aroma components is also similar. The aroma clustering substances of other tea samples are far apart and dispersed. It indicates that electronic nose DFA analysis can distinguish the varieties of tea samples through sensors. Other scholars also confirmed that DFA analysis is superior to PCA analysis. At the same time, the electronic nose had the function of distinguishing sample varieties.

4 CONCLUSION

The aroma of Gaoxiang Yihong Congou black tea is indeed superior to that of similar Congou black tea. There was a significant negative correlation between the total amount of free amino acids and theine content in taste. The electronic nose has great application prospects in the rapid detection of black tea varieties and main aroma distribution.

To sum up, the taste of Gaoxiang Yihong Congou black tea is not fresh and cool, but slightly bitter. In terms of aroma, it can be found to have advantages only from sensory evaluation. The brightness of the soup color is moderate, and the clarity needs to be improved. The color difference meter can accurately and quickly analyze the quality of the color of tea soup. The electronic nose has great advantages in distinguishing tea species and aroma, but it cannot determine the quality of aroma well, so it needs the assistance of other detection instruments.

ACKNOWLEDGMENT

This study was funded by the Guiding project of the Scientific Research Program of the Hubei Provincial Department of Education (Grant No. B2018265) and the Science and Technology Innovation Project of Wuhan Business University (Grant No. 2016KC06). The research findings were part of these projects.

REFERENCES

Huang Xiangshen. (2017). From Black Tea to Yichang Black Tea[J]. *Journal of Tongren University*, 2017, 19(11): 33–38.

Li Jianquan. (2017). Analysis of Aroma Components of Black Teas from Different Regions[J]. *Hunan Agricultural Sciences*, (8): 85–92, 97.

Li Lixiang, Mei Yu, Chang Shan, et al., (2005). Analysis of Liquor Color of Green Tea[J]. *Food and Fermentation Industries*, (10): 123–126.

Li Xiaoyuan. (2015). *Relationship between Sensory Character and Chemical Components of Congou's Flavor*[D]. Chinese Master's Theses Full-text Database.

Lu Jianliang, Liang Yuerong, Gong Shuying, Gu Zhilei, Zhang Lingyun, Xu Yuerong. (2002). Studies on Relationship between Liquor Chromaticity and Organoleptic Quality of Tea[J]. *Journal of Tea Science*, 2002(1): 57–61.

Ma Huijie, Jiang Bin, Pan Yulan, Li Yongmei, Ma Yan, Yang Guangrong. (2019). Identification of aroma characteristics of famous green tea and black tea from various regions by electronic nose[J]. *Food Science and Technology*, 44(1): 336–344.

Qiu Fangfang, Hao Qingqing, Deng Yuliang. (2016). Yichang Academy of Agricultural Sciences; Hangzhou Tea Research Institute of China Academy of Agricultural Sciences;. The Different Processing Method Effect on the Quality of Yihong Black Tea[J]. *Journal of Tea Communication*, 43(3): 23–26.

Wang Bei. (2012). The GC-olfactometric Analysis on the Discrimination of Different Milk Flavoring[J]. *Food and Fermentation Industries*, 38(3): 77–80.

Wu Jinchun, Wang Lanlan, Zhang Ji, Wang Ao, Li Jie, Xiang Liping. (2020). Inspection and Testing Institute for Product Quality of Zunyi; National Supervision & Inspection Center for Tea & Tea Product Quality (Guizhou);. The Relativity Analysis between Sensory Quality and Inner Components of Zunyi Black Tea[J]. *Modern Food*, (7): 198–201.

Yin Xia, Bao Xiaocun, Huang Jianan at al., (2020). Comparative Analysis of Sensory Quality of Black Tea in Hunan Province[J]. *Journal of Tea Communication*, 47(2): 297–302.

Yuan Bolan. (2018). Research on the Producing History and Value of "Yihong" Tea[J]. *China Three Gorges Tribune*, (2): 1–6.

Zhang Yijia .(2019). Research on the Social Life History of Yichang Black Tea in Southwest Hubei[D]. *Chinese Master's Theses Full-text Database*.

Zhao Dan. & Lu Cai you. (2016). College of Longrun Pu-erh Tea, Yunnan Agricultural University. Research Progress of the Aroma of Black Tea[J]. *Journal of Anhui Agricultural Sciences*, 44(23): 45–46, 83.

Zhou Yibin. & Wang Jun. (2004). The Development and Tendency on the New Technology Application in Tea Quality Evaluation[J]. *Journal of Tea Science*, (2): 82–85.

Zhou Ying, Liu Ren, Tan Ting, Huang Jianan. (2015). Methods evaluation on Gongfu black tea aroma clustering in different procession by the electronic nose analysis[J]. *Journal of Food Safety & Quality*, 6(5): 1611–1618.

Energy Revolution and Chemical Research – Kok-Keong Chong and Zhongliang Liu (Eds)
© 2023 The Authors, ISBN: 978-1-032-36554-1

Research of different scale materials for pesticide residue detection

Lu Xin*
Chemical and Environmental Engineering, University of Nottingham, Nottingham, Nottinghamshire, UK

Xinzheng Li
Yangtze Delta Region Institute of Tsinghua University, Jiaxing, China

ABSTRACT: The addition of pesticides is necessary, while excessive pesticides could be harmful to the well-being of humans down to the molecular level. Therefore, more precise detection methods for various pesticides are essential. This essay demonstrates the research of different scale materials for pesticide residue detection and the detection performance of these materials. Also, the advantages and disadvantages of materials with diverse sizes are compared, especially the single-atom materials. Conclusions on the extensive research for nanomaterials are drawn.

1 INTRODUCTION

1.1 *Pesticides residue*

There are about 3.5 million tonnes of pesticides used per year around the world (Sharma 2019). Consequently, the public presents a fear of chemistry due to the tremendous unrevealed side-effects used indirectly, among them, the rice fields consumed 80 percent of the global pesticides (Cabasan 2019). Moreover, with the latest technological advances, the threat pesticides cast on human health has dropped to the molecular level, which indicates that several molecules of these elements could impose negative effects on people. Thus, the technology to detect those components in molecular and atomic sizes should be focused on (Rawtani 2017). In other words, research on the detection of pesticide residue is urgently demanded (Meng 2021). Obviously, those new catalyst materials could broaden the range of detection and lower the detection limit to benefit the safety of humanity. This paper mainly mentioned nanomaterials and single-atom materials, more specifically, the detection performances for pesticides of these two materials were compared, and more accurate detection methods for various pesticides to reduce detection limits and ensure human food safety were explored. According to the standard of the European Commission, the MRL (maximum residue level) for those commonly used pesticides like malathion, chlorpyrifos, and parathion are 0.02, 0.01, and 0.05 mg/kg respectively (EU 2021).

1.2 *Nanomaterials*

Nanomaterial is a type of material with sizes ranging from 1 to 100 nanometers that possess superior connectivity, exceeds catalytic activity, and other features like nontoxic and clean. The synthesis and applications of this type of material have supported nanochemistry and material science for decades (Virginia Moreno 2018). Thus, it serves as an essential material for the preparation of biosensors (Krishnan 2019). Nanoparticles (NPs) possess excellent features, thus, it is necessary to study their applications and characteristics (Virginia Moreno 2018). Moreover, the demands for effective and sensitive nanoprobes for the detection of pesticides are one of the challenges nowadays, thus

*Corresponding Author: ssylx1@nottingham.edu.cn

DOI 10.1201/9781003332657-64

several formulae were used to test the performance of AuNPs as a colorimetric probe in pesticide detection (Satnami 2018). However, the research on using artificial nanozyme to detect pesticides is still little. The materials in existing research could be classified as monometallic oxide, bimetallic oxide, and various supported metal oxides including graphene and carbon nanotube (CNT). These nanocatalysts will be discussed in the following part. Despite all these advantages, a quite limitation of the coordination of unsaturated atoms in the nanoparticles can directly contact the reactants which means that the catalytic efficiency of metals in nanometer diameter is still deficient.

$$I\% = (i_0 - i_t)/i_0 \times 100 \tag{1}$$

$$R\% = (i_r - i_t)/(i_0 - i_t) \times 100 \tag{2}$$

1.3 *Single-atom materials*

The concept of 'single-atom catalyst' was first raised by academician Zhang Tao at the Dalian Institute of chemical physics, Chinese Academy of Sciences. Zhang reported a kind of iron oxide catalyst which contains single platinum atoms on its surface (Qiao 2011). Single-atom catalysts (SACs) should be an advanced version of catalysts to utilize for improved atomic efficiency (Li 2020). It has been stated by Zhang that the reduction of the size of the catalyst can effectively improve its surface free energy. Moreover, when the metal sizes drop to a single atom level, the catalytic activity of this catalyst would be significantly promoted. It is eligible to lower the cost of catalyst, maximize atomic utilization, achieve effective regulation of single-atom active sites, conducive to knowing the structure-activity relationship between the structure and performance of the catalyst (Single-atom nanozymes for the detection of organophosphorus pesticides). Additionally, the appearance of Density Functional Theory (DFT) has offered a strong tool for discovering the basic steps and mechanisms of heterogeneous catalysis on the atom scale for the last 20 years (Zhao 2015). This essay combined and compared the research results of the catalytic materials with different particle sizes including the nanomaterials and single-atom materials.

2 SINGLE-ATOM CATALYST FOR PESTICIDES RESIDUE DETECTION

2.1 *Carbon-based single-atom catalyst*

It can be discovered that the application of SACs in the energy field is considerable, while that for the application of biology is still infant which can be inferred from the fact that the catalysts being applied for detection of pesticides residue were reported until recent years around 2020 (Ge 2020). As mentioned above, to improve the atomic efficiency and the detection efficiency, Wu et al. synthesized a type of oxidase-like Fe-N-C single-atom catalyst through pyrolysis and this catalyst has higher oxidase activity compared with traditional nano-enzyme. The hierarchical porous structure was first characterized by TEM (transmission electron microscope) in Figure 1A down to the nano level. Then the figure was amplificated again by HRTEM (high-resolution transmission electron microscope) in Figure 1B and there were no nanoparticles observed. Eventually, HAADF-STEM (high-angle annular dark-field scanning transmission electron microscopy) exhibited some circle light spots which should be the Fe single-atoms. In Figures 1E to H, the corresponding EDS (energy dispersive spectroscopy) marked the distribution of C, N, and Fe respectively. XAFS (X-ray absorption fine structure) has also been used to confirm the formation of Fe-N-C single-atom catalyst. HAADF-STEM and XAFS are the most important ones as the insurance of single-atom catalysts because they can be used to illustrate that the Fe-based materials exist in the form of Fe-N with a coordination number of 2 which means there are only single atoms and no iron-iron bindings. Single-atom catalyst was successfully prepared. Furthermore, based on the discovery that the small molecules containing sulfhydryl groups could significantly inhibit the activity of Fe-N-C monatomic nanocrase, this single-atom catalyst was utilized for the sensitive organophosphorus pesticide detection. The linear range of detection was 0.1-25 mU mL^{-1}, and the detection limit was 0.014 mU mL^{-1} which is better than the previous detection methods as shown in Table 1.

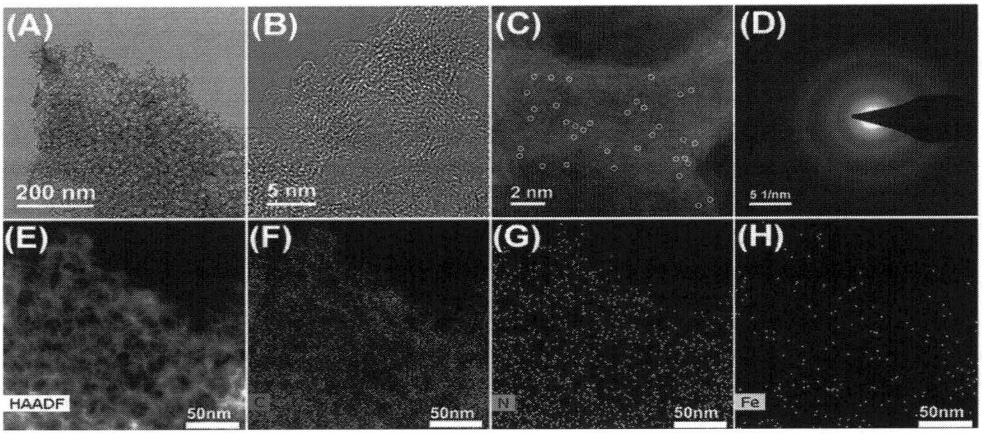

Figure 1. A) TEM image, B) HR-TEM image, C) HAADF-STEM image (Z-contrast), and D) SAED pattern of Fe-N-C SAzymes. E) HAADF-STEM image and F–H) the corresponding EDS elemental distribution of the C, N, and Fe atoms in the Fe-N-C SAzymes.

After the analysis of the Fe-based SACs, a problem about the detection performance increase after doping from other atoms such as boron has been raised. Jiao et al. analyzed a type of single-atom nanozyme which is called the boron-doped Fe-N-C. A similar characterization process to the previous research including HRTEM, XRD (X-ray diffraction analysis), HAADF-STEM, and XPS (X-ray photoelectron spectroscopy) stated that it was a type of Fe-B-N-C material without Fe nanoparticles and B atoms were successfully doped in the nanozyme. Additionally, HAADF-STEM, XAFS further proved that the Fe single atoms existed with a valence from 0 to +3. The charge transfer effects induced by boron are able to regulate the positive charge of the Fe atom in the center to increase the peroxidase-like activity by reducing the energy barrier for the formatting process of hydroxyl radical. The material has wide applications in the detection of small particles and enzyme activity. More specifically, the sensitive detection of pesticides relevant to the acetylcholinesterase activity because of the peroxidase-like activity of this catalyst would drop with the rising of the concentration of organophosphorus pesticides. As expected, the specific activities of Fe-B-N-C were 15.41 U/mg, much higher than that of Fe-N-C (4.09 U/mg). This catalyst can be used to detect OPs whose concentrations are from 8 to 1000 ng/mL and the detection limit for this pesticide is 2.19 ng mL^{-1}. Overall, this discovery opens up a new approach to the synthesis of single-atom nanozyme and demonstrates the gap between nanozyme and natural enzymes.

Table 1. Comparison of biosensors for the detection of OP using colorimetric.

Material	Linear range	LOD	Reference
Citrate-CeO_2	0–10 μM	<10 nM	(Cheng, 2016)
AChE-MnO_2-TMB	0.001–0.1 μg mL^{-1}	10 ng mL^{-1}	(Yan, 2017)
CDs/DNTB/ATCh/AChE	0.001–1.0 μg mL^{-1}	0.4 ng mL^{-1}	(Li, 2018)
Fe-N-C SAaymes	0.1–10 μg mL^{-1}	0.97 ng mL^{-1}	(Wu, 2019)
Fe-B-N-C SAC	0.008–1.0 μg mL^{-1}	2.19 ng mL^{-1}	(Jiao, 2020)

2.2 *Supported precious metal single-atom catalyst*

Ge et al (2020) utilized single-atom Pd-SA/TiO2 as a kind of platform of photocatalytic detection to detect highly sensitive and detective chlorpyrifos directly. It was synthesized by a method of situ photocatalytic reductions under the frozen state. The chemical characterization methods TEM, STEM, and HAADF-STEM were used to characterize the formation of Pd and they were uniformly separated on the surface of TiO2. XANES and EXAFS showed the presence of palladium oxides and a large amount of Pd-O bonds and a low number of Pd-Pd bonds to ensure the single atom structure of this material. According to the fact that the existence of chlorpyrifos would reduce the photocatalytic activity of Pd-SA/TiO2, this material could detect chlorpyrifos from 0.03 ng mL-1 to 10 μg mL-1, and the detection limit is 0.01 ng mL^{-1} which is much lower than the method reported before like Poly(3-hexylthiophene)/TiO2 Nanoparticle-Functionalized Electrodes with the detection range from 0.2 to 16 μM, the detection limit is 0.01 μM (Li 2011) and the Nitrogen functionalized graphene quantum dots/3D bismuth oxyiodine hybrid hollow microspheres with the detection limit of 0.03 ng mL^{-1} in the concentration of chlorpyrifos ranging from 0.1 to 50 ng mL^{-1} (Qian 2018). The U.S. Environmental Protection Agency banned the usage of this type of pesticide from 29/10/2021, which helps expand testing limits.

Basically, the chemical characterization process of a single-atom catalyst is more complicated, and the requirements are higher than that of nanoscale materials because it has to be proved that all the metal atoms are presenting in the form of a single-atom. In a contrast, the existing research relevant to the single-atom catalyst presents very promising results in detecting a certain type of pesticide. The obviously lower LOD and a wider range of detection compared with materials of a larger scale might promote the standard of detection to a new level.

3 NANOCATALYST FOR THE PESTICIDE RESIDUE DETECTION

3.1 *Nano metallic oxide catalyst*

The research of the nanoscale materials in the application of pesticide residue detection has become obviously more comprehensive and diverse than the single-atom materials. The one with the simplest structure is the nano metallic oxide catalyst. A type of gold nanoprobe was found by Satnami and his group (Satnami 2018) to detect the organophosphorus pesticides using the method based on the color change and redshift of the absorption band. The aggregation of gold nanoprobes exposed to the compounds containing mercaptan would cause a decrease in the absorbance during the synthesis of gold nanoprobe. The spherical Au nanoparticles size of around 13 nm was synthesized through the method reported. The characterization methods and the detection methods mainly include the TEM, UV–visible absorption spectrum, and Surface Plasmon Resonance (SPR). These results demonstrated that Au particles were well-dispersed, and they can be used as a common probe for the detection of acetylcholinesterase. This nanoprobe could be used for the detection of organophosphorus compounds from 0.30–17.30 ng mL^{-1} with the detection limit of 0.13 ng mL^{-1} (paraoxon), 0.37 ng mL^{-1} (parathion), 0.42 ng mL^{-1} (fenitrothion) and 0.20 ng mL^{-1} (diazinon). This method is easy, sensitive, and economical for the test of organophosphorus pesticides in real water. Moreover, the team of Tunesi (Tunesi 2018) used HR-SEM to characterize the nanostructure of CuO which had a diameter of 20 to 70 nm, and EDR to further prove the purity of this material. This material which can be synthesized under 100°C can replace the original material in an easy way. The detection performance of this catalyst with its method is compared with other systems in Table 2 sorted by the pesticides.

It can be seen from Table 2 that even if one certain material provides efficient detection for a certain type of pesticide, it does not work efficiently as well with other different pesticides. As a result, the research should be highly targeted.

Additionally, Fe-based nano metallic oxide materials seem to present a better ability in the recycling process due to their magnetic property. In 2017, Fan (2017) reported that magnetic Fe_3O_4 nanomaterial was combined with a new anion-exchange reagent to detect pyrethroid pesticides in

Table 2. Comparisons of the developed sensor with various other systems reported for organophosphate pesticides.

Pesticide	Sensor system	Linear range(μM)	LOD(M)	Reference
Chlorpyrifos (MRL of 0.01 mg/kg)	EITO-CuO	0.01–0.16	1.6×10^{-9}	(Tunesi 2018)
	AChE/Fc-F/GCE	0.005–1	3×10^{-9}	(Xia 2015)
	AChE/[BMIM][BF4]-MWCNT/CPE	0.01–1	4×10^{-9}	(Zamfir 2011)
Methyl parathion (MRL of 0.05 mg/kg)	EITO-CuO	0.01–0.16	6.7×10^{-9}	(Tunesi 2018)
	CNT	2–10	0.8×10^{-6}	(Deo 2005)
	Pralidoxime chloride (PAM-Cl)	0-19.3	0.215×10^{-6}	(Zheng 2017)

water. This material was characterized by SEM DLS and XRD that those particles were spherical with a diameter of less than 30 nm, and they were uniformly distributed. The detection limits of this catalyst on Cyhalothrin, Deltamethrin, Etofenprox, and Bifenthrin are relatively 0.21 $\mu g\ L^{-1}$, 0.19 $\mu g\ L^{-1}$, 0.16 $\mu g\ L^{-1}$ and 0.16 $\mu g\ L^{-1}$, and the linearity range is all 1-100 $\mu g\ L^{-1}$ (Fan 2017). This method has the advantages of a high recycling rate, well distribution condition, low pretreatment time, and low consumption of extracting solvent. Especially the nanomagnetic Fe_3O_4 allows the quick recycling by physical adsorption and electrostatic forces which simplified the process of centrifugation.

3.2 *Nano bimetallic oxide catalyst*

Compared with the single metal-based catalyst, bimetallic catalysts could represent some evolution in the detection performance due to the synergistic effect between the two kinds of metal elements. Wen and his members (Wen 2020) used an easy solvothermal method to synthesize a type of 2-dimension bimetallic Ni-Zn MOF NSs whose ratios of Ni/Zn could be adjusted. The MOF materials were successfully synthesized. Also, SEM and TEM revealed the flower-like cluster structure of this whole material which contains several nanosheets. The curled or blended edges further proved their flexibility. Other characterization methods such as the EDS and XPS showed the element composition of the bimetallic MOF. Phenol compounds in pesticides could be detected by such material from 0.08 to 58.2 μM with the detection limit of 6.5 nM which is much lower than the standards from the maximum amount of phenol compounds in freshwater set by the US Environment Protection Agency and the European Union (EU) respectively as 0.5 $\mu g\ L^{-1}$ and 0.1 $\mu g\ L^{-1}$.

Another pair of bimetallic PdAu was reported by de Barros and his members (Barros 2021) and it was characterized by XRD and TEM that there were Au particles in this material which means they were partially alloying with gold. TEM showed that Au-on-Pd nanoparticles were successfully synthesized and the diameter of 3.1∼5.9 nm spherical, and these were well-spread. Then, it was modified by the glassy carbon electrode (GCE) and finally formed the Nf/Au-on-Pd/GCE for pesticide detection. The molecular absorption spectrometry was used to prove the accuracy. It was tested to have an obvious current response to the 4-nitroaniline (4-NA) which can be found in the pesticide after degradation. The detection limit is 0.17 μM ranging from 0.5 to 50 μM. This catalyst can be used to detect 4-NA in the environment, especially in the application of management and in-loco monitoring. The comparison of the performance of this material and other works can be seen in Table 3.

3.3 *Supported nano metallic oxide catalyst*

3.3.1 *Graphene sheets supported metallic oxide catalyst*

Different from the previously mentioned Fe_3O_4 metallic catalyst, this part demonstrates the material with the feature of being supported by the graphene sheets. Boruah and his members (Boruah 2021)

Table 3. The comparison of t 4-NA determination results of NiZn-MOF NSs with other electrodes through the oxidization process.

Electrode architecture	Linear range/μM	LOD/μM	Reference
GCE	2-100	0.3	(Pfeifer 2016)
Yb_2O_3-ZnO/AgE	390-6250	390	(Umar 2019)
ZnO-CeO_2/AgE	250-5000	250	(Ahmad 2016)
ZnONRs/FTO	1-80	0.5	(A, R. A. 2017)
Nf/Au-on-Pd/GCE	0.5–50	0.17	(Barros 2021)

reported a kind of artificial nanozyme Fe_3O_4/rGO which can be used for the detection of simazine pesticide. The Fe_3O_4/rGO material was prepared by a simple method called co-precipitation. It was characterized by XRD and XPS for the content of elements and chemical bonds, HRTEM for the distribution, size, and shape of the materials, also Raman spectroscopy, AFM, TGA, and VSM. The average size of this material was 12 nm. Based on the fact that the hydrogen bond interaction between pesticide molecules and 3,3′,5,5′- tetramethylbenzidine (TMB) inhibits the catalytic activity of FDGs for TMB oxidation. This material can be used to detect the existence of water-base medium simazine pesticide and the detection limit is 2.24 μM from 0.01 to 50 μM. Compared with Fe_3O_4-TiO2/rGO in the same range which has a detection limit of 4.85 μM, this material was more effective. It can be used as a next-generation functional nanozyme that is highly sustainable and has the advanced ability for detection.

TiO2 is commonly used in nanocomposite catalysts for its excellent photocatalytic activity. Boruah and his members (Boruah 2020) synthesized a type of nanozyme Fe_3O_4-TiO2/rGO sized around 9 ± 0.2 nm through a one-step method called the hygrothermal method. The characterization tools include XRD, HRTEM, FESEM, and XPS. These results showed that the spherical Fe_3O_4 and tetragonal TiO2NPs are properly distributed on the graphene nanocomposite sheets. This catalyst can be applied for the detection of atrazine based on the colorimetric detection technique. In addition, the detection limit is 2.98 μg/L was obtained in the linear range of 2–20 μg/L which is suitable for detecting low concentration atrazine. This magnetically separable material is highly recyclable as a promising approach to the detection of pollution. These two cases show that the addition of another oxide can moderately change the detection direction.

3.3.2 *CNTs supported metallic oxide catalyst*

Except for the Graphene sheets, CNTs can also support the metallic oxide. In the research of Pathak and Gupta (Pathak 2020), Pd nanoparticles were synthesized by an easy method called the in-situ solvothermal method, and these particles were Inserted in a polypyrrole (PPy) shell covered with carbon nanotubes (CNT) and eventually formed PdNPs/PPy@CNTs. SEM and TEM were used to analyze the morphology of this material. Also, FTIR, SAED, and XRD support the determination of other factors like composition, structure, size, distribution, and phase uniformity. The detection was processed from the concentration of hydrazine from 0 to 1500 nM and the detection limit was 20 nM. The LOD is the ratio of three times the standard deviation to the possible ratio of the sensor near zero concentration. This essay demonstrated the availability of this material in the environment because of its facile synthesis method, convenient operation, excellent stability, and effectivity. Additionally, the limit of detection gave it promising practices in the future as well.

Despite the bimetallic catalyst, the synergistic effect would also affect the two loading substances. As a result, Moreno et al (Virginia Moreno 2018) reported a novel hybrid magnetic nanocomposite that decorated the surface of carbon nanotubes (CNTs) and octadecyl group-modified silica (nano SiO2C18) synthesized in a supercritical CO2 medium. This feature and the existence of nanocomposite were characterized by TEM, HRTEM, and XRD. This material can be reused at least 10 times and the detection limits for five neonicotinoids and four sulfonylurea pesticides in the water were

0.07 to 0.18 $\mu g\ mL^{-1}$. Synthesis processes of several magnetic nanocomposites were reported to gain new sorbents for magnetic solid-phase extraction of pesticides.

4 CONCLUSION

In this paper, the research of detection materials for pesticide residues was discussed. The paper started with the necessity of pesticide detection, then introduced the research on the materials relevant to pesticide residue detection. Through analyzing the correlated essays, the performance, merits, and demerits of nanocatalysts were comprehensively expounded. Compared with nanomaterials, single-atom catalysts have the highest atomic utilization rate, wider detection range, and lower limit of detection. Moreover, the principle and results of a single-atom catalyst can be predicted and explained by DFT. However, there are currently no systematic research and applications of SACs and the amount of relevant research is apparently smaller than nanomaterials because of the strict characterization standard and high cost. Also, the research corresponds to the combination of SACs with equipment remains to be discovered.

REFERENCES

Ahmad, N., Umar, A., Kumar, R., & Alam, M. (2016). Microwave-assisted synthesis of zno doped ceo2 nanoparticles as a potential scaffold for highly sensitive nitroaniline chemical sensor. *Ceramics International*, 11562–11567.

A, R. A., B, N. T., A, M. S. A., & A, Y. B. H. (2017). Development of highly-stable binder-free chemical sensor electrodes for p-nitroaniline detection. *Journal of Colloid and Interface Science*, 494, 300–306.

Barros, M., Winiarski, J. P., Elias, W. C., CEMD Campos, & Jost, C. L. (2021). Au-on-pd bimetallic nanoparticles applied to the voltammetric determination and monitoring of 4-nitroaniline in environmental samples. *Journal of Environmental Chemical Engineering*, 105821.

Boruah, P.K., G. Darabdhara, and M.R. Das. Polydopamine functionalized graphene sheets decorated with magnetic metal oxide nanoparticles as efficient nanozyme for the detection and degradation of harmful triazine pesticides. *Chemosphere*, 2021. 268: p. 129328.

Boruah, P. K., & Das, M. R. (2020). Dual responsive magnetic fe_3o_4-tio_2/graphene nanocomposite as an artificial nanozyme for the colorimetric detection and photodegradation of pesticide in an aqueous medium. *Journal of Hazardous Materials*, 385(Mar. 5), 121516.1–121516.17.

Cabasan, M. T. N., Tabora, J. A. G., Cabatac, N. N., Jumao-As, C. M., & Soberano, J. O. (2019). Economic and ecological perspectives of farmers on rice insect pest management. *Global Journal of Environmental Science and Management*, 5(1), 31–42.

Cheng, H., Lin, S., Muhammad, F., Lin, Y. W., & Wei, H. (2016). Rationally modulate the oxidase-like activity of nanoceria for self-regulated bioassays. *ACS Sensors, acssensors*. 6b00500.

Deo, R. P., Wang, J., Block, I., Mulchandani, A., Joshi, K. A., & Trojanowicz, M., et al. (2005). Determination of organophosphate pesticides at a carbon nanotube/organophosphorus hydrolase electrochemical biosensor. *Analytica Chimica Acta*.

EU, EU Pesticides database. 2021.

Fan, C., et al. In-situ ionic liquid dispersive liquid-liquid microextraction using a new anion-exchange reagent combined Fe_3O_4 magnetic nanoparticles for determination of pyrethroid pesticides in water samples. *Analytica Chimica Acta*, 2017. 975: p. 20–29.

Ge, X., Zhou, P., Zhang, Q., Xia, Z., Chen, S., & Gao, P., et al. (2020). Palladium single atoms on tio2 as a photocatalytic sensing platform for analyzing the organophosphorus pesticide chlorpyrifos. *Angewandte Chemie International Edition*.

Hongbo, Li, Jing, Li, Qin, & Xu, et al. (2011). Poly (3-hexylthiophene) /tio2nanoparticle-functionalized electrodes for visible light and low potential photoelectrochemical sensing of the organophosphorus pesticide chlorpyrifos. *Analytical Chemistry*.

Hongxia Li, X Yan, G Lu, & X Su]. (2018). Carbon dot-based bioplatform for dual colorimetric and fluorometric sensing of organophosphate pesticides. *Sensors and Actuators B: Chemical*.

Jiao, L., Xu, W., Zhang, Y., Wu, Y., & Guo, S. (2020). Boron-doped fe-n-c single-atom nanozymes specifically boost peroxidase-like activity. *Nano Today*, 35, 100971.

Krishnan, S. K., Singh, E., Singh, P., Meyyappan, M., & Nalwa, H. S. (2019). A review on graphene-based nanocomposites for electrochemical and fluorescent biosensors. *RSC Advances*, 9(16).

Li, L., Chang, X., Lin, X., Zhao, Z. J., & Gong, J. (2020). Theoretical insights into single-atom catalysts. *Chemical Society Reviews*.

Meng, S. G., Su, D. L., Yang, Y. F., Naqiuddin, M., & Ma, N. L. (2021). Omics technologies are used in pesticide residue detection and mitigation in the crop. *Journal of Hazardous Materials*, 420(1), 126624.

Pathak, A., & Gupta, B. D. (2020). Palladium nanoparticles embedded ppy shell coated cents towards a high-performance hydrazine detection through optical fiber plasmonic sensor. *Sensors and Actuators B: Chemical*, 326.

Pfeifer, Rene, Chaer, do, Nascimento, & Marco, et al. (2016). Differential pulse voltammetric determination of 4-nitroaniline using a glassy carbon electrode: a comparative study between cathodic and anodic quantification. *Monatshefte fur Chemie*, 147(1), 111–118.

Qian, L. A., Yy, B., Nan, H. B., Jing, Q. B., Ll, A., & Ty, A., et al. (2018). Nitrogen functionalized graphene quantum dots/3d bismuth oxyiodine hybrid hollow microspheres as remarkable photoelectrode for photoelectrochemical sensing of chlorpyrifos. *Sensors and Actuators B: Chemical*, 260, 1034–1042.

Qiao, B., Wang, A., Yang, X., Allard, L. F., Jiang, Z., & Cui, Y., et al. (2011). Single-atom catalysis of co-oxidation using pt1/feox. *Nature Chemistry*, 3(8), 634–41.

Rawtani, D., Khatri, N., Tyagi, S., & Pandey, G. (2017). Nanotechnology-based recent approaches for sensing and remediation of pesticides. *Journal of Environmental Management*, 206, 749–762.

Satnami, M. L., Korram, J., Nagwanshi, R., Vaishanav, S. K., Karbhal, I., & Dewangan, H. K., et al. (2018). Gold nanoprobe for inhibition and reactivation of acetylcholinesterase: an application to detection of organophosphorus pesticides. *Sensors and Actuators*, B267 (AUG.), 155–164.

Sharma, A., Kumar, V., Shahzad, B., Tanveer, M., & Thukral, A. K. (2019). Worldwide pesticide usage and its impacts on the ecosystem. *SN Applied Sciences*, 1(11), 1446.

Tunesi, M. M., Kalwar, N., Abbas, M. W. Karakus, S., Soomro, R. A., & Kilislioglu, A. , et al. (2018). Functionalized cuo nanostructures for the detection of organophosphorus pesticides: a non-enzymatic inhibition approach coupled with nano-scale electrode engineering to improve electrode sensitivity. *Sensors and Actuators*, B260(MAY), 480–489.

Umar, A., et al. Nitroaniline chemi-sensor based on bitter gourd shaped ytterbium oxide (Yb2O3) doped zinc oxide (ZnO) nanostructures. *Ceramics International*, 2019. 45(11): p. 13825–13831.

Virginia Moreno, EJ Llorent-Martínez, Zougagh, M. , & A Ríos]. (2018). Synthesis of hybrid magnetic carbon nanotubes – c18-modified nano sio2 under supercritical carbon dioxide media and their analytical potential for solid-phase extraction of pesticides. *The Journal of Supercritical Fluids*.

Wen, Y., Li, R., Liu, J., Zhang, X., & Sun, B. (2020). Promotion effect of zn on 2d bimetallic nizn metal-organic framework nanosheets for tyrosinase immobilization and ultrasensitive detection of phenol. *Analytica Chimica Acta*, 1127.

Wu, Y., et al., (2019). Oxidase-like fe-n-c single-atom nanozymes for the detection of acetylcholinesterase activity. *Small*, 15(43).

Xia, N., Zhang, Y., Chang, K., Gai, X., Jin, Y., & Li, S., et al. (2015). Ferrocene-phenylalanine hydrogels for immobilization of acetylcholinesterase and detection of chlorpyrifos. *Journal of Electroanalytical Chemistry*.

Yan, X., Song, Y., Wu, X., Zhu, C., Su, X. & Du, D., et al. (2017). Oxidase-mimicking activity of ultrathin mno2 nanosheets in colorimetric assay of acetylcholinesterase activity. *Nanoscale*, 9(6), 2317.

Zamfir, L. G., Rotariu, L. , & Bala, C. . (2011). A novel, sensitive, reusable and low potential acetylcholinesterase biosensor for chlorpyrifos based on 1-butyl-3-methylimidazolium tetrafluoroborate/multiwalled carbon nanotubes gel. *Biosensors & Bioelectronics*, 26(8), 3692–3695.

Zhao, Z. J. , Chiu, C. C. , & Gong, J. . (2015). Cheminform abstract: molecular understandings on the activation of light hydrocarbons over heterogeneous catalysts. *ChemInform*, 46(35), no-no.

Zheng, Q., Chen, Y., Fan, K., Wu, J., & Ying, Y. (2017). Exploring pralidoxime chloride as a universal electrochemical probe for organophosphorus pesticides detection. *Analytica Chimica Acta*, 78.

Energy Revolution and Chemical Research – Kok-Keong Chong and Zhongliang Liu (Eds)
© 2023 The Authors, ISBN: 978-1-032-36554-1

Research and application of comprehensive evaluation method for vehicle odor performance

Chen Cui, Shujie Xu, Xuefeng Liu & Boyang Tian
China Automotive Technology and Research Center Co., Ltd., Tianjin, China

ABSTRACT: Based on the odor subjective evaluation method in the automobile industry, the study establishes an odor comprehensive evaluation method, which includes five indexes: high-risk odor substances, odor objective intensity, odor concentration, odor description, and odor pleasantness. Through the practical application analysis of 53 vehicles from 2020 to 2021, we obtain comprehensive odor performance results. The above results can provide accurate data sources and method references for the rectification and improvement of vehicle interior odor in automobile enterprises, which will effectively promote the progress of odor control technology in the automotive industry.

1 INTRODUCTION

As an indispensable means of transportation for people's daily travel, the car has entered thousands of households. At the same time, the health and environmental performance of the car has also become increasingly important to consumers, which has become an important factor affecting the consumer's decision to purchase a car. Among them, the car odor as the first impression of consumers in the car cabin is the user's perception of the quality of automotive products key performance indicators, should be focused on control. In addition, according to the third-party platform information adopted by the State Administration of Market Supervision and Administration of Defective Products Recall Center, by the end of June 2021, the cumulative number of complaints about the odor in cars in China was as high as 27,728, involving independent, joint venture and imported brands. The above results show that the problem of odor in cars has become a common quality problem of automotive products, which should attract the attention of relevant automotive enterprises.

At present, there are several sets of odor evaluation standards established by enterprises, associations, and standardization organizations in the automotive industry, such as T/CMIF 13-2016, VDA 270, ISO 12219-7, etc. There is no unified odor evaluation standard at home and abroad. The existing odor evaluation standards in the automotive industry are based on the evaluation and control (Zhu et al. 2017) of in-vehicle odor through the subjective sniffing of odor evaluators and giving the odor intensity level, which is easily influenced by the subjectivity of odor evaluators and has the problems of poor consistency and low credibility; at the same time, the odor intensity level index is not enough to comprehensively characterize the in-vehicle odor status and has less guidance value for odor improvement. Based on the above industry situation, it is necessary to establish a set of comprehensive evaluation methods that can characterize the odor status of the vehicle and guide the improvement of the odor in the vehicle.

Based on the current subjective odor evaluation method, and on the basis of the domestic and foreign odor evaluation methods in the environmental protection, military, and medical industries, this study establishes a comprehensive evaluation method for odor performance in vehicles, including five indicators of high-risk odor substances, the objective intensity of odor, odor concentration, odor type, and odor pleasantness, and upgrades the original odor characterization method. Based on

 DOI 10.1201/9781003332657-65

this method, it can provide an accurate scientific method for automotive enterprises to rectify and improve in-vehicle odor, and strongly promote the progress of in-vehicle odor control technology in the automotive industry.

2 COMPREHENSIVE EVALUATION METHODS FOR IN-VEHICLE ODOR PERFORMANCE

2.1 *Experimental method*

According to the industry standard "method for the determination of volatile organic compounds and aldehydes and ketones in cabin interiors" (HJ/T 400-2007) and the international standard "Interior air of road vehicles - Part 1: Whole vehicle test chamber - Specification and method for the determination of volatile organic compounds in cabin interiors" (ISO 12219-1-2012) in ambient and high-temperature conditions, (ISO 12219-1-2021)respectively. The full spectrum analysis of volatile organic compounds (VOCs) was conducted under ambient and high-temperature conditions, and the high-risk odor substances were calculated based on the odor substance threshold database and the odor phase diagram analysis tool established by China Auto Data Co. At the same time, the coupled analysis of odor in the vehicle was conducted based on the "Evaluation of Odor in the Vehicle: Sensory and Photoionization Detector Coupling Analysis Method" (T/CAS 406-2020), and the objective intensity, odor concentration, odor type and odor pleasantness were detected and calculated.

2.2 *Performance indicators*

2.2.1 *High-risk odor substance indicators*

High-risk odor substances are gaseous organic substances present in the vehicle interior space emitted by interior materials, etc., with high content, high concern, and high contribution to the vehicle interior odor. The threshold dilution multiple is used as the only measure in screening high-risk odor substances.

The threshold dilution factor is the ratio (Yang 2016) of the substance concentration of an odorant substance to the olfactory threshold concentration of that substance. The significance is that in a multi-component gas system, the higher the threshold dilution factor of an odorant substance, the greater its contribution to the odor of the gas mixture, so the main substance causing the odor in the car is not the substance with the highest concentration content, but the substance with the high threshold dilution factor.

Based on the above definition, the calculation of high-risk odorant substances can be obtained.

(1) For a vehicle, the concentration of VOCs substance in the vehicle obtained by full-spectrum analysis of VOCs is a_i.
(2) The olfactory thresholds b_i for different in-vehicle VOCs substances is obtained through experimental tests, literature research, and model calculations, and the threshold dilution multiplier d_i for all detected in-vehicle VOCs substances is calculated.

$$_{i}d = a/b_{ii} \quad (1)$$

Where: d_i is the threshold dilution multiple of the ith in-vehicle VOCs substance, dimensionless. a_i is the concentration of the ith in-vehicle VOCs substance, $\mu g/m^3$; b_i is the smell threshold of the ith in-vehicle VOCs substance, $\mu g/m^3$.

(3) The VOCs substances in the vehicle with a threshold dilution factor $d_i \geq 1$ are the high-risk odorants of the vehicle.

2.2.2 *Objective intensity index of odor*

Odor intensity refers to the degree (Xie 2014) of irritation to the human olfactory organs when the odoriferous gases in the car are undiluted. In general, the odor intensity is expressed in the form of numbers, which can simply and intuitively reflect the pollution degree (Wang 2011) of odor substances. The odor intensity in the automotive industry is expressed in a six-level system, with numbers from 1 to 6, of which, level 1 represents no odor, and the stronger the odor, the larger the number.

Table 1. Description of odor intensity.

Grade	Odor intensity description
1	No odor
2	Perceptible, non-disturbing odor
3	Visibly detectable, non-disturbing odor
3.5	Between 3 and 4 levels
4	Clearly detectable with disturbing odor
5	Clearly detectable, with a strong disturbing odor
6	Unbearable odor

The objective intensity of an odor is based on the value measured by the sensor on the analyzer, and the intensity of the odor is calculated by the law or curve of the subject-objective coupling of the odor.

The objective intensity of odor is calculated as follows.

(1) Determining the odor intensity S_i of the gas sample at each dilution i in the order of the largest to smallest dilution.
(2) Specifying the dilution multiplier for S_i equal to 1.5 levels as the odor concentration $_iC/ODT$ of the original gas sample.
(3) Determining the volatile pollutant concentration C_1 of the original gas sample using a photoionization detector.
(4) Dividing C_1 by the odor concentration of the original gas sample to obtain the olfactory threshold ODT of the gas sample.
(5) Least-squares curve fitting of the logarithm of subjective odor intensity S_i and odor concentration ($_iC/ODT$) at different dilution multiples based on the Weber-Fechner subject-objective coupling equation (2) to obtain the values of the fitted curve equation constants k and b.

$$S = k \times \log_i({}_iC/ODT) + b \tag{2}$$

where: S_i is the subjective intensity of odor of gas sample at dilution multiple i, dimensionless; C_i is the concentration of volatile pollutant of the gas sample at dilution multiple i, ppm; ODT is the olfactory threshold of the gas sample, ppm. $_iC/ODT$ is the odor concentration of the gas sample, dimensionless; k is the equation constant, dimensionless; b is the equation constant, dimensionless.

(6) The volatile pollutant concentration C1 of the original gas sample is again determined using a photoionization detector, and the objective odor intensity S of the gas sample is calculated by substituting it into the equation (2) where ODT, k, and b are known.

2.2.3 *Odor concentration index*

Odor concentration is the dilution factor (Zhao et al. 2014) required to dilute a gas sample with clean air (odorless air) until the sample is odorless. In the field of in-vehicle odor evaluation, the odor concentration indicator represents the ease of odor dissipation in the vehicle. In general, the smaller the value of this indicator, the faster the odor dissipates in the car.

2.2.4 *Odor Type Indicator*

An odor type indicator is a descriptor or a set of descriptors expressing odor similarity, each descriptor corresponding to an odor reference sample.

2.2.5 *Odor Pleasantness Index*

The odor pleasantness index characterizes the degree of liking or disliking of an odor by the odor evaluator, which is a completely subjective evaluation index, and generally characterizes the pleasantness of a gas sample using a pleasantness scale defined by the following descriptors.

Table 2. Description of odor pleasantness.

Descriptors	Pleasure Level
Very pleasant	Level 3
Pleasure	Level 2
Mildly pleasurable	Level 1
Neutral	Level 0
Mild disgust	–1 level
Disgusted	–2 levels
Very pleasant	Level 3

2.3 *Experimental vehicles*

The experimental vehicle should be the M1 vehicle specified in GB/T 15089-2001, and in order to fully simulate the consumer's new car odor experience, the following conditions should be met.

(1) Within 3 months of going offline.
(2) The mileage is less than 100 km.
(3) The internal components' surface coverings (such as factory plastic film used to protect seats, carpets, etc.) are well preserved.
(4) The user has not used it and has not carried out the car odor treatment.

Based on the above conditions, 53 mainstream models in the market from 2020 to 2021 were selected to conduct the relevant experiments. The selected models include both car models and SUV models, as well as independent brand models, joint-venture brand models, and foreign brand models, which can basically represent the current level of in-vehicle odor performance in the automotive industry.

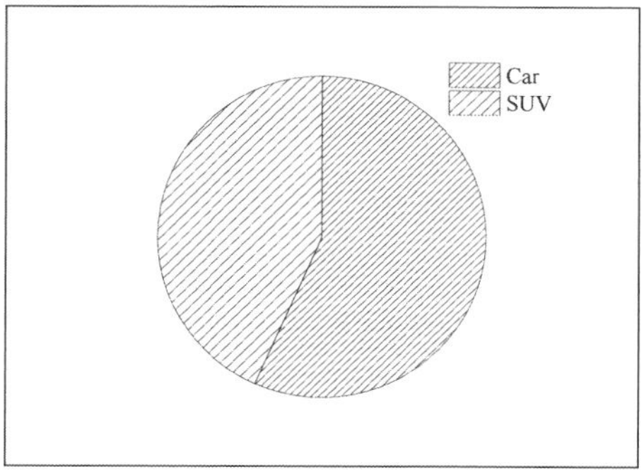

Figure 1. Distribution of different model categories.

As shown in Figures 1 and 2, from the model category, the selected models contain 30 sedans, accounting for 57%, and 23 SUV models, accounting for 43%. From the brand category, the selected models include 25 models of independent brands, accounting for 47%, and 28 models of non-independent brands, accounting for 53%.

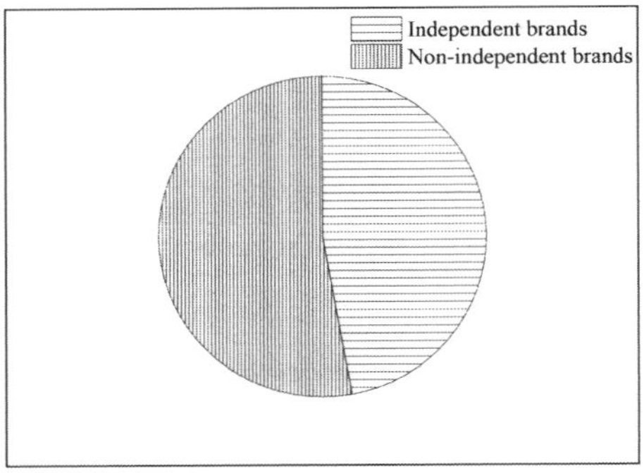

Figure 2. Distribution of different brand categories.

3 ANALYSIS OF DATA RESULTS

3.1 *Overall industry situation analysis*

Using the 53 vehicles mentioned above as research objects, we carried out in-depth research and analysis on five indicators: high-risk odor substances, the objective intensity of odor, odor concentration, odor type, and odor pleasantness.

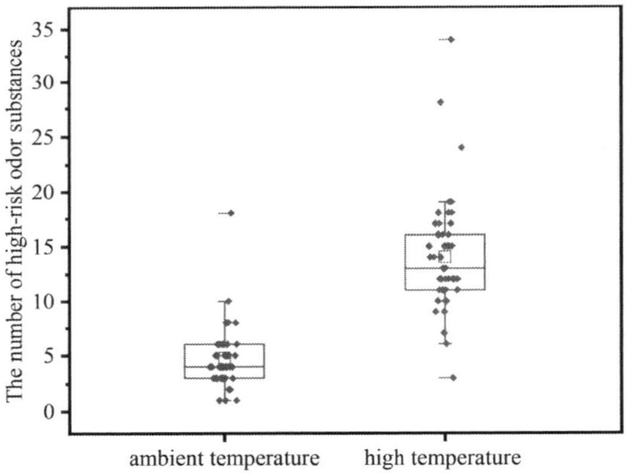

Figure 3. Distribution of the number of high-risk odor substances for 53 models.

As shown in Figure 3, the number of high-risk odor substances under ambient conditions was distributed between 1 and 18, with an average value of 5; the high-risk odor substances under high-temperature conditions were distributed between 3 and 34, with an average value of 14. The above results show that the number of key odor substances differs greatly from model to model and there are more types of odor substances emitted from automotive interior materials under high-temperature conditions, because the materials and processes used for interior trim differ greatly from model to model, which in turn causes differences in the substances emitted. At the same time, when the temperature rises, the automobile interior materials are more likely to volatilize odor substances.

As shown in Figure 4, the objective odor intensity results under room temperature conditions were mainly concentrated between 3.3 and 3.8 levels, with an average value of 3.5 levels; the objective odor intensity results under high-temperature conditions were distributed between 2.8 and 4.7 levels, mainly concentrated between 3.7 and 4.3 levels, with an average value of 4.0 levels.

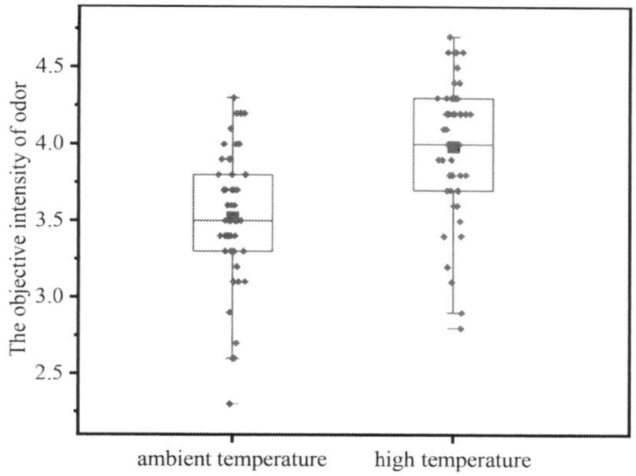

Figure 4. Distribution of the objective intensity of odor results for 53 models.

Under the same conditions, the difference between the maximum and minimum values of objective odor intensity was nearly 2.0 levels, indicating that the levels of odor intensity indicators of different models were uneven and differed significantly.

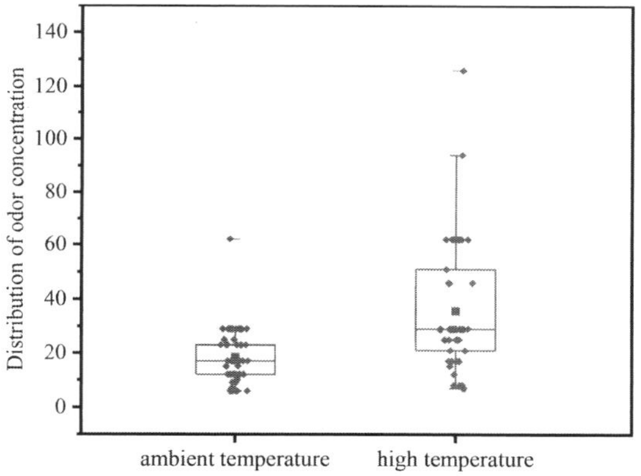

Figure 5. Distribution of odor concentration results for 53 models.

As shown in Figure 5, the odor concentration results are distributed between 6 and 62 under ambient conditions, with an average value of 18; the odor concentration results are distributed between 7 and 126 under high-temperature conditions, with an average value of 36. Thus, it can be seen that the odor concentration results of different models vary greatly, and the difference between the maximum and minimum values under the same conditions can be nearly 20 times; at the same time, the value of odor concentration under high-temperature conditions is significantly higher.

As shown in Figure 6, the top three odor types at room temperature and high temperature are leather, foam, and irritation, and these three odor types account for more than 50% of the odor, with special attention to the fact that there are models that show smoke, paint and fishy odor at room temperature and asphalt, smoke and paint at high temperature, which are odor types that are strongly objectionable to consumers, and relevant enterprises should pay attention to and strictly control.

As shown in Figure 7, the distribution of odor pleasantness at room temperature ranges from –2 to +1, with a mean value of –1, while the distribution of odor pleasantness at high temperature ranges from –3 to +2, with a mean value of –2. From the figure, it can be seen that consumers

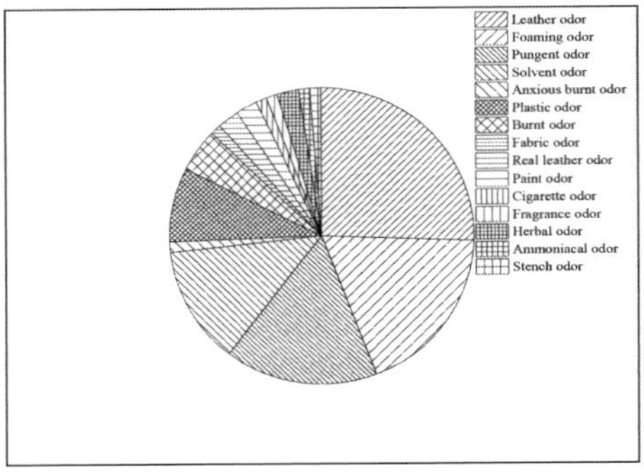

a) Ambient temperature conditions

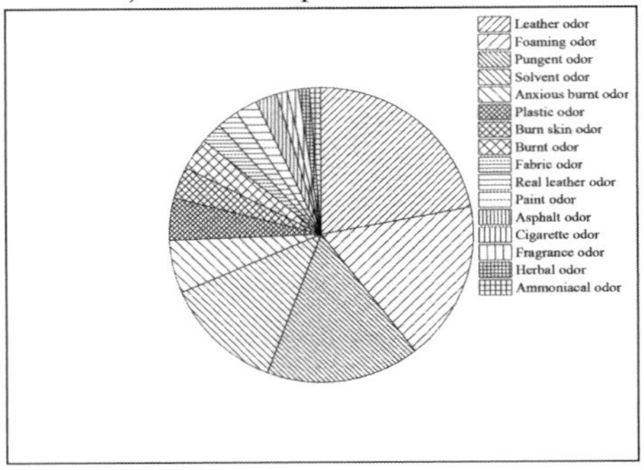

b) High-temperature conditions

Figure 6. Distribution of odor type results for 53 models.

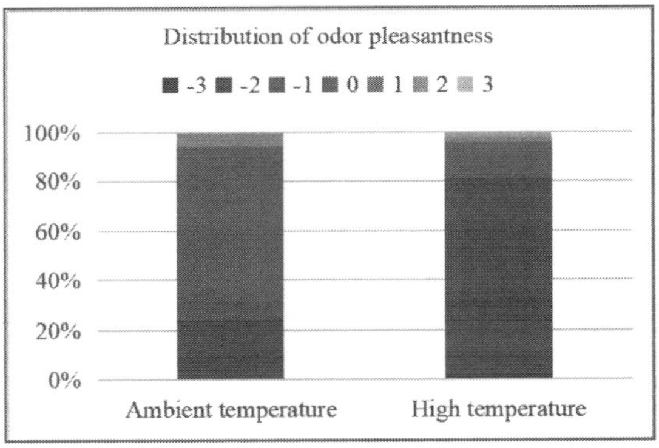

Figure 7. Distribution of odor pleasantness results for 53 models.

tend to feel more negative about car odor, and the higher the temperature, the stronger the negative feeling, and the worse the evaluation result of car odor.

3.2 *Comparative analysis of indicators of different model categories*

The above 53 vehicles are classified according to model categories, which are 30 sedan models and 23 SUV models. As the car and SUV models in the exterior size, interior materials, and other aspects of the difference, the interior odor performance of the two also has their own characteristics, and the following will be a comparative analysis of five indicators: high-risk odor substances, odor objective intensity, odor concentration, odor type, and odor pleasure.

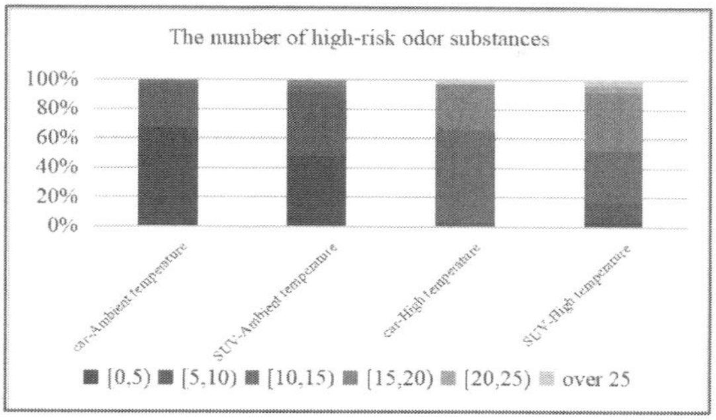

Figure 8. Distribution of the number of high-risk odor substances in different vehicle categories.

As can be seen from Figure 8, the number of high-risk odorant substances in sedan and SUV models under ambient conditions is mainly concentrated between 0 and 10, of which more than 60% of sedan models are distributed in the interval [0,5) and more than 80% of SUV models are distributed in the interval [0,10). The above results indicate that the number of high-risk odorant substances in sedan models is less than that in SUV models. And under high-temperature conditions, the number of high-risk odor substances in sedan models was mainly distributed in the interval [10,15) and SUV models were mainly distributed in the interval [15,20), exceeding the number of substances at room temperature by more than two times, and the number of substances in SUV models was still more than that in sedan models.

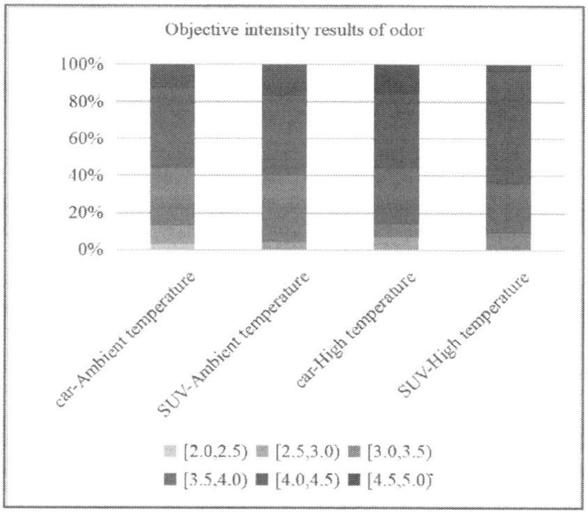

Figure 9. Distribution of objective intensity results of odor for different vehicle categories.

As can be seen from Figure 9, under normal temperature conditions, the objective intensity of odor of both car and SUV models is mainly concentrated at 3.0~4.0 level, the average value of car models is 3.5 level, the average value of SUV models is 3.6 level, and SUV models are slightly worse than car models. Under high-temperature conditions, the objective intensity of odor of car models is mainly concentrated in 3.5~4.5 levels, the objective intensity of odor of SUV models is mainly concentrated in 4.0~4.5 levels, and the average value of both car and SUV models is 4.0

levels. Although the distribution range of the two is different, the overall level of difference is not significant.

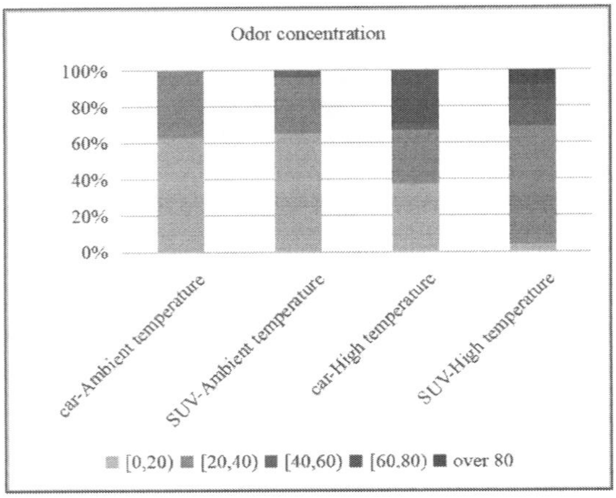

Figure 10. Distribution of odor concentration results for different vehicle categories.

As can be seen from Figure 10, under normal temperature conditions, the odor concentration of car and SUV models is mainly concentrated in the interval [10,20), with the average value of 17 for car models and 20 for SUV models; under high-temperature conditions, the distribution of car models is more different, mainly distributed in the interval [0,20), interval [20,40) and interval [60,80), and SUV models are mainly distributed in the interval. This shows that compared with the car models, the odor diffusion in SUV models is slower, and it takes a longer time to diffuse the odor under the same conditions.

As can be seen from Figure 11, whether at room temperature or high temperature, the top three odor types of cars and SUVs are leather, foam, and irritation, with the difference being that the proportion of odor types differs from model to model. At the same time, the car models have more types of odor performance, more than ten; while the SUV models have relatively single odor type performance characteristics, in addition to the common odor types, but also the solvent smell, plastic smell, leather smell, and paint smell.

As can be seen from Figure 12, the odor pleasantness of cars and SUVs is mainly concentrated at –1 under room temperature conditions and –2 under high-temperature conditions. The odor characteristics of the above two-car categories are relatively consistent, indicating that the odor pleasantness is not related to different car categories.

3.3 *Comparative analysis of indicators of different brand categories*

In order to examine the level of odor performance of different brand models, this paper classifies 53 vehicles according to brand categories: 25 models of independent brands and 28 models of non-independent brands. The following will be a comparative analysis of five indicators: high-risk odor substances, the objective intensity of odor, odor concentration, odor type, and odor pleasantness.

As can be seen from Figure 13, under normal temperature conditions, the number of high-risk odorant substances of non-autonomous brands was mainly concentrated in the interval [0,5) with an average value of 4, while the number of high-risk odorant substances of autonomous brands was mainly in the interval [5,10) with an average value of 6. Therefore, non-autonomous brands performed better; under high-temperature conditions, the difference between autonomous and non-autonomous brands was not significant, and the number of high-risk odorant substances was mainly distributed in the interval [10,20), and the average value was 14.

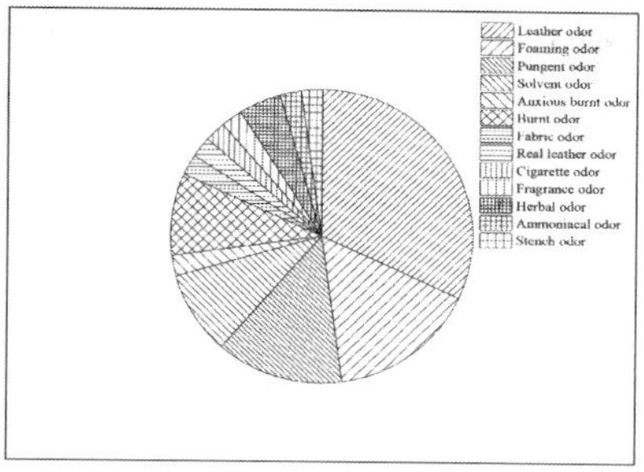

(a) Normal temperature conditions - car models

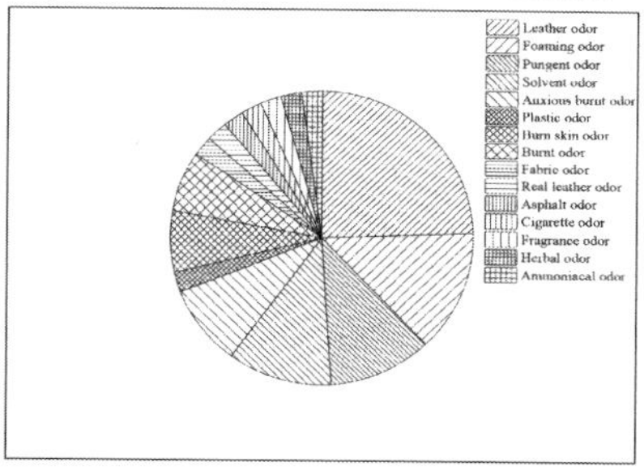

b) High-temperature conditions - car models

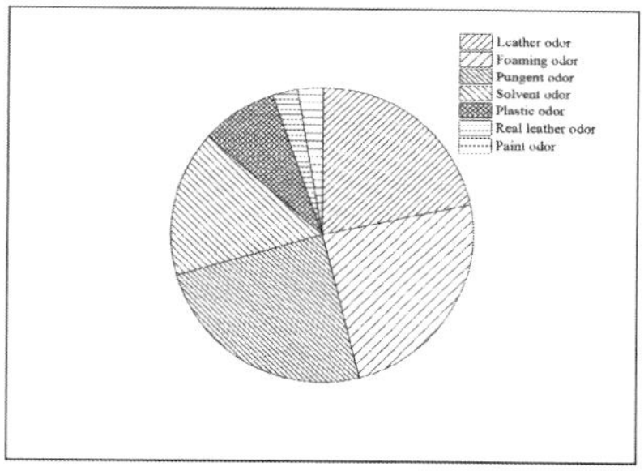

c) Ambient temperature conditions-SUV models

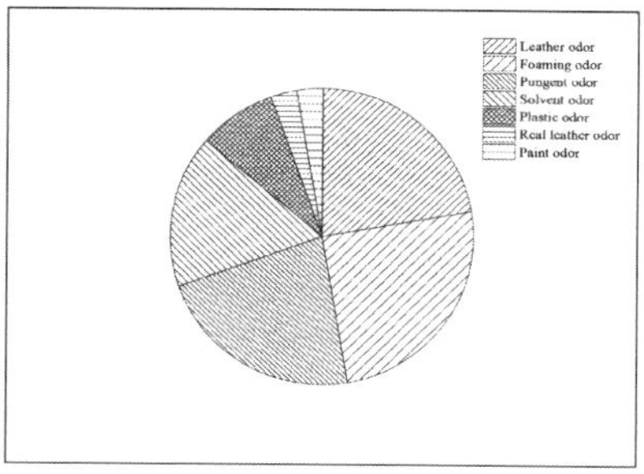

d) High-temperature conditions - SUV models

Figure 11. Distribution of odor type results for different vehicle categories.

As can be seen from Figure 14, under room temperature conditions, the objective intensity of the odor of autonomous brands is distributed between 2.5 and 4.5 levels, with an average value of 3.6; the objective intensity of the odor of non-autonomous brands is distributed between 2.0 and 4.5 levels, with an average value of 3.4. Therefore, the overall distribution of non-autonomous brands is better than that of autonomous brands, and the proportion of models with poor odor intensity

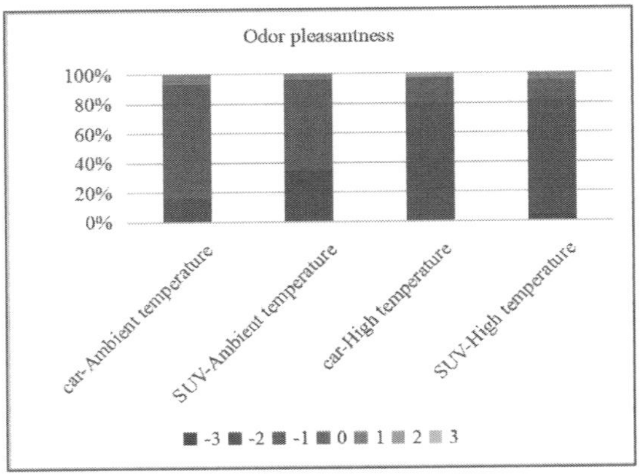

Figure 12. Distribution of odor pleasantness results for different model categories.

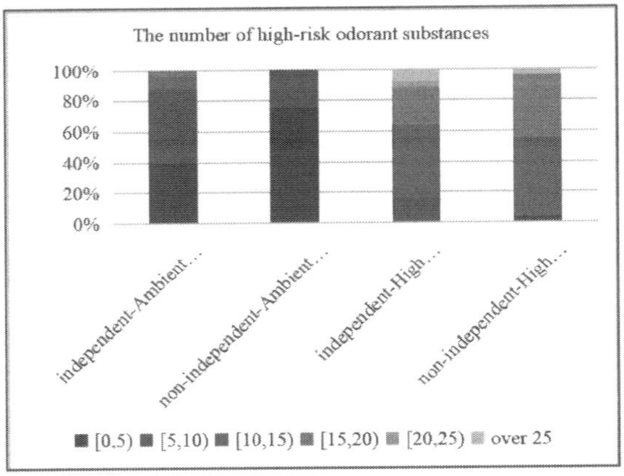

Figure 13. Distribution of the number of high-risk odorant substances in different brand categories.

(≥4.0 levels) is smaller. Under high-temperature conditions, the average value of the objective intensity of odor of both is 4.0 level, the level of odor intensity index of independent brand models is uneven, distributed in 2.0~5.0 level; the value of non-independent brand models is relatively concentrated, distributed in 3.0~5.0 level.

As can be seen in Figure 15, under normal temperature conditions, the independent brand model odor concentration values are more concentrated, distributed between 0 ~ 30, while the non-independent brand model odor concentration indicators level is uneven; high-temperature conditions are mainly concentrated between 20 ~ 40, with the average value of 36, and the difference between the two indicators is not significant.

As can be seen from Figure 16, under normal and high-temperature conditions, the odor types of autonomous and non-autonomous brands are mainly expressed as leather, foaming, and irritating odors, with different percentages of odor types for different brands; the difference is that certain autonomous brand models show fabric, smoke and ammonia odors, while certain non-autonomous brand models show asphalt, fishy and fragrant odors.

As can be seen from Figure 17, the odor pleasantness of independent and non-independent brands is mainly concentrated at -1 under room temperature conditions and -2 under high-temperature conditions. The odor characteristics of the above two brands are relatively consistent, indicating that the odor pleasantness has little relationship with the model brand.

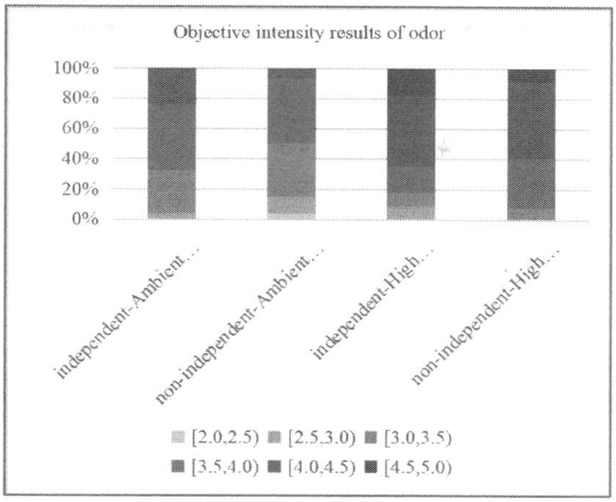

Figure 14. Distribution of the objective intensity of odor results for different brand categories.

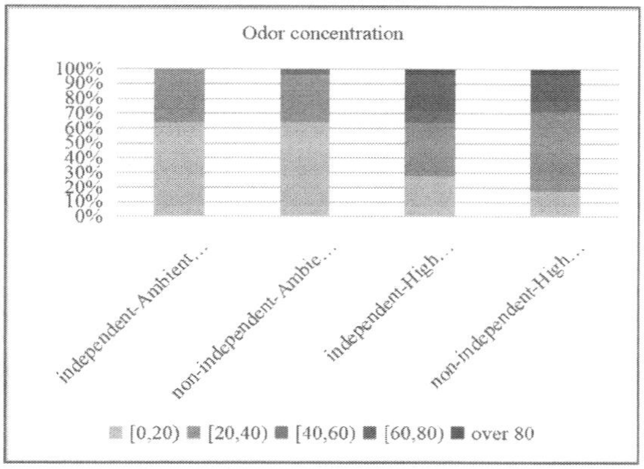

Figure 15. Distribution of odor concentration results for different brand categories.

4 CONCLUSION

In this paper, through a comparative analysis of five indicators of high-risk odor substances, the objective intensity of odor, odor concentration, odor type, and odor pleasantness, we basically figured out the overall level of the industry, different car categories, and different brand categories of in-car odor performance levels, and obtained the following conclusions.

(1) The number of key odor substances varies widely among different models. From the model category, the number of high-risk odor substances of car models is less than that of SUV models; from the brand category, the number of high-risk odor substances of non-autonomous brands performs better than that of autonomous brands. The reason for this is that the materials and processes used in the interior of different models differ greatly, which in turn causes differences in the substances emitted. Under high-temperature conditions, the number of high-risk odor substances of the same model is higher than under normal temperature conditions. The reason for this is that when the temperature rises, the interior materials of the car are more likely to evaporate odor substances.

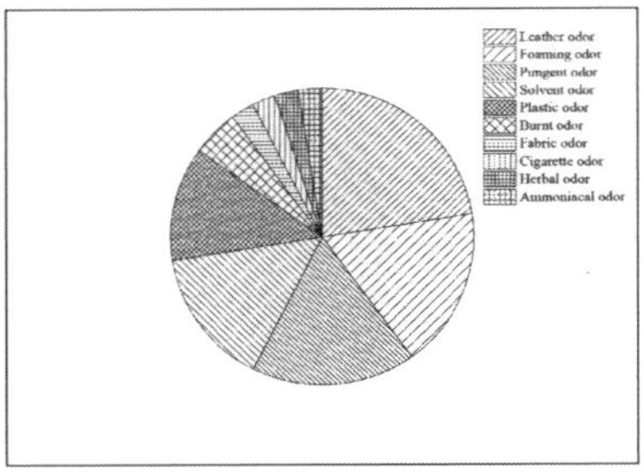

a) Ambient temperature conditions-own brand

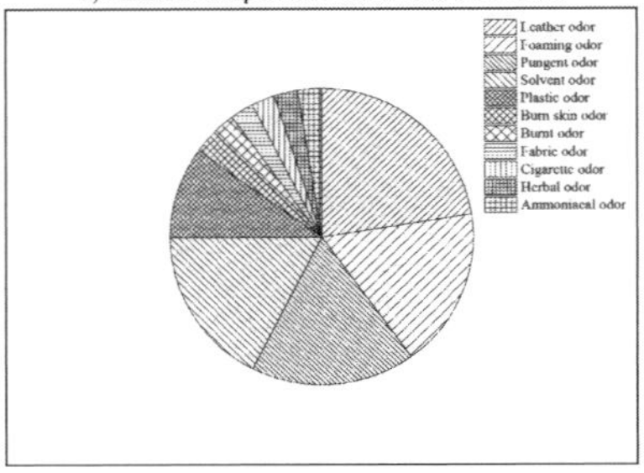

b) High-temperature conditions-own brand

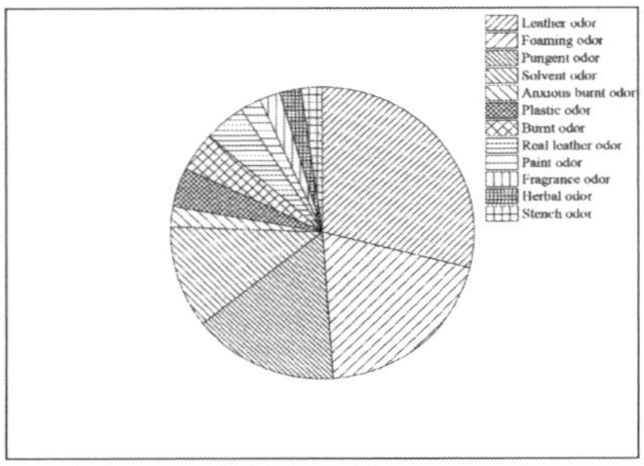

c) Ambient temperature conditions - non-owned brand

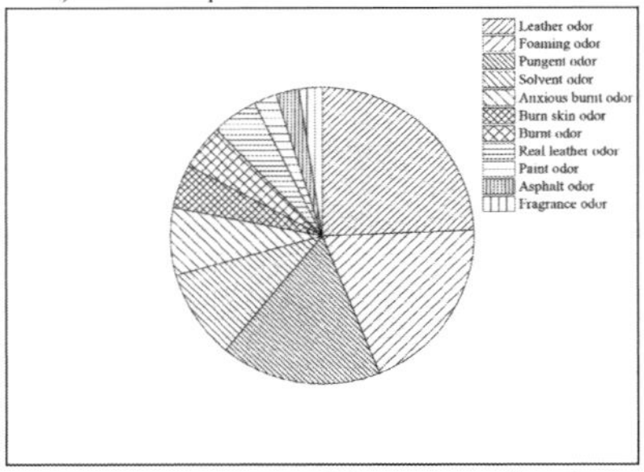

d) High-temperature conditions-non-owned brand

Figure 16. Distribution of odor type results for different brand categories.

(2) The average value of the objective intensity of odor of 53 models under ambient conditions is 3.5; among them, the average value of car models is 3.5, the average value of SUV models is 3.6, the average value of autonomous brand models is 3.6, and the average value of non-autonomous brand models is 3.4. Therefore, SUV models are slightly worse than car models, and non-autonomous brands are better than autonomous brands. Under high-temperature conditions, the average value of the objective intensity of odor of 53 models is 4.0 level, and the average value of

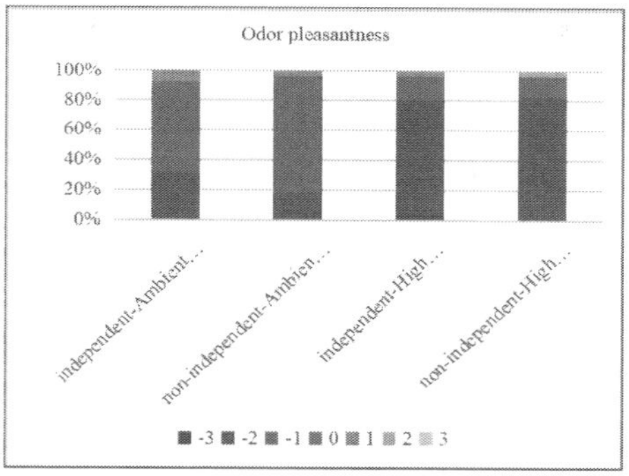

Figure 17. Distribution of odor pleasantness results for different brand categories.

car/SUV models and autonomous/non-autonomous brand models is consistent with the industry average, indicating that the level of different categories of models under high-temperature conditions is the same.

(3) The level of odor concentration varies widely by model category. In general, car models are better than the industry average, and SUV models are worse than the industry average, so the odor diffusion in SUV models is slower, and it takes longer to diffuse the odor in the car under the same conditions. In addition, the odor concentration indicators of different brand categories do not vary much, and the average level of different brands remains consistent with the overall industry average.

(4) The top three types of odor are leather, foam, and irritation, with special attention to the models at room temperature for smoke, paint, and fishy odor, and at high temperatures for asphalt, smoke, and paint. These odor types are strongly objectionable odor types to consumers, relevant enterprises should pay attention to and strictly control them.

(5) The odor pleasantness is not related to the model category and brand category, and the mean value of odor pleasantness is -1 under normal temperature and -2 under high temperature. Consumers' feelings about the car odor are more inclined to be negative, and the higher the temperature the stronger the negative feelings, and the worse the evaluation result of the car odor.

REFERENCES

HJ/T 400-2007, sampling and determination methods for volatile organic compounds and aldehydes and ketones in vehicles [S].

ISO 12219-1. *Whole vehicle test chamber - Specification and method for the determination of volatile organic compounds in cabin interiors* [S].

T/CAS 406-2020, *Evaluation of air odor in vehicles: A coupled sensory and photoionization detector analysis method* [S].

Wang Yuangang. *Research on Early Warning and Emergency Response System for Odor Pollution Accidents* [D]. Hebei University of Technology, 2011.

Xie Lei. *Research on Odor Pollution Source Analysis Technology and Traceability System*[D]. Hebei University of Technology, 2014.

Yang Weihua. *Emission characteristics and odor sensory assessment of typical odor pollution sources*[D]. Tianjin University of Technology, 2016.

Zhao Yan, Lu Wenjing, Wang Hongtao, et al. A study on odor pollution assessment index system for municipal solid waste treatment and disposal facilities[J]. *China Environmental Science*, 2014, 34(007):1804–1810.

Zhu Z.Y., Liu X.F., Liu W. Exploring ways to manage odor problems in vehicles[J]. *Environment and Sustainable Development*, 2017, 42(006):88–90.

Energy Revolution and Chemical Research – Kok-Keong Chong and Zhongliang Liu (Eds)
© 2023 The Authors, ISBN: 978-1-032-36554-1

A study on the magnetic powder enhanced flocculation of the hot rolling wastewater

Hong Wan
Wuhan University of Science and Technology, Wuhan, China
Wuhan REDSUN Chemical Co., Ltd., Huanggang, Hubei, China

Ao Chen*, Ya Hu* & Xiaoyu Yang
Wuhan University of Science and Technology, Wuhan, China
Academy of Green Manufacturing Engineering, Wuhan University of Science and Technology, Wuhan, China

Lin Wu
Wuhan University of Science and Technology, Wuhan, China
Wuhan REDSUN Chemical Co., Ltd., Huanggang, Hubei, China
Academy of Green Manufacturing Engineering, Wuhan University of Science and Technology, Wuhan, China

ABSTRACT: The magnetic powder, polyaluminum chloride (PAC) and polyacrylamide (PAM) composite flocculants were used to treat the hot rolling wastewater of Wuhan Iron and Steel Group. Changes in wastewater turbidity and oil content by changing the addition amount of each component of the composite flocculant are observed, and the treatment effect is discussed. The experimental results show that the optimal formulations of the composite flocculants are magnetic powder 5 mg/L, PAC 15 mg/L, and PAM 1.5 mg/L, respectively. By adding the optimized flocculant, the turbidity of the wastewater decreased from 82.6 NTU to 10.8 NTU and the oil content was decreased from 15.62 mg/L to 7.44 mg/L, respectively.

1 INTRODUCTION

Rolling is an important step in the production of iron and steel products. During this process, a lot of water, which is needed in device cooling, mechanical lubrication, and heat treating, turns into wastewater. There are two different kinds of the rolling process, hot rolling and cold rolling, and the quality of their wastewater is different too. The hot-rolling wastewater is generated by roll neck bearings and rolled pieces when water is used to cool the mill roll. There are two major pollutants (Wang et al. 2021; Yu et al. 2019). The first one is iron scales which are produced by steel plates during process-cycle, they go into the gutter with the cooling water and then reach the processing unit of turbid circulating water; the other one is oil, which is considered to be the lubricant and refrigerant in the hot-rolling process, including oil-water emulsion or palm oil. These oils also reach the processing unit of turbid circulating water when they are useless and turn into oily emulsion wastewater. Meanwhile, these two pollutants always stick together and become fat, which can easily affect the device safety by causing blocks to the tube or affecting the product quality by attaching to the product. These are important problems that need to be solved during the hot rolling turbid circulating water process (Lin et al. 2015; Xiong & Cang 2008).

So far, the process of hot rolling turbid circulating water has 3 steps, the swirl sedimentation tank (the first filtration), the horizontal flow sedimentation tank (the second filtration), and the oil filter (Gong 2012). The swirl sedimentation tank can wipe off big pieces of iron scale using the physical

*Corresponding Authors: chen_ao@wust.edu.cn and huya@wust.edu.cn

 DOI 10.1201/9781003332657-66

separation. The horizontal flow sedimentation tank, which can be used to remove smaller particles and oil material, usually needs some flocculant to enhance the effect. At last, the oil filter is set to ensure that the water quality meets the criterion, it intercepts pollutants that are not removed by the previous filtration. During the horizontal flow sedimentation, a composite flocculant, which includes the polyaluminium chloride (PAC) and the polyacrylamide (PAM), has always been used (Guo, Meng 2020).

A kind of magnetic powder is developed to enhance the hot rolling wastewater filtering effect of composite flocculant based on PAC+PAM composite flocculant. This magnetic powder is made by grinding and magnetizing iron and phosphor. When the magnetic powder mix with polyaluminium chloride (PAC) and polyacrylamide (PAM) to make the composite flocculant, it can stabilize the floc and adsorb some oil (Chen et al. 2011; Wang et al. 2014). By using the magnetic powder, the rate of PAC utilization is increased and the processing cost is reduced while the efficiency of processing is the same.

2 EXPERIMENTAL MATERIALS AND INSTRUMENTS

2.1 *Experimental materials*

The iron phosphor powder comes from a powder metallurgy company that belongs to an iron and steel enterprise. The powder is made by desiccation and ball-milling of the industrial iron phosphor. The turbidity of hot rolling wastewater is 82.6±5.0 NTU, the oil content is 15.62 mg/L. The polyaluminium chloride (PAC) and the polyacrylamide (PAM) are both pure chemical reagents.

2.2 *Instruments*

JJ-4 model type motor stirrer; HAC2100P model type nephelometer; OIL-6D model type automatic infrared measuring oil meter; stopwatch; N35 NdFeB permanent magnets.

3 METHODS

3.1 *The preparation of magnetic powder*

The process is as follows: put 20.0g iron phosphor powder into agate mortar and grind manually. Let the powder go through the 200-mesh sieve and collect the powder under the sieve, then repeat the step above on the powder which is left on the sieve until most powder can pass the sieve. Make the iron phosphor powder spread on a piece of paper with the density of 5 g/dm2, put several N35 NdFeB permanent magnets together and form a square, then put the paper along with the powder on the square and stay for 10 min. In the end, the powder would be magnetized.

3.2 *The experiment on flocculant*

Put the hot rolling wastewater 200ML into the 500ML conical flask, then add PAC or magnetic powder according to the volume. After stirring the mixture with the velocity of 300 r/min for 1 min, add a certain amount of PAM and stir for 30 s, then stir the mixture with the velocity of 60 r/min for 10 min. Then stay for 30 min, suck up the liquid 1cm below the liquid level and check the turbidity and oil content. The ratio of turbidity removal and oil removal is calculated according to Formulas 1 and 2.

$$\text{The ratio of turbidity removal} = \left(1 - \frac{Turb_m}{Turb_o}\right) \times 100\% \tag{1}$$

Where $Turb_m$ is the turbidity of liquid that is sucked up after processing and $Turb_o$ is the turbidity of original wastewater.

$$\text{The ratio of oil removal} = \left(1 - \frac{Oil_m}{Oil_o}\right) \times 100\% \tag{2}$$

Where Oil_m is the oil content of liquid that is sucked up after processing, and Oil_o is the oil content of original wastewater.

4 RESULT AND DISCUSSION

4.1 *The removal effect of turbidity and oil on different dosages of PAC*

Holding other conditions constant, we add different amount of PAC in different flocculation experiments (1 mg/L, 2mg/L, 4mg/L, 8mg/L, 12mg/L, 20mg/L, 30mg/L, and 40mg/L). Figure 1 and Figure 2 are the variety of the turbidity and the variety of the oil content of the hot rolling wastewater respectively.

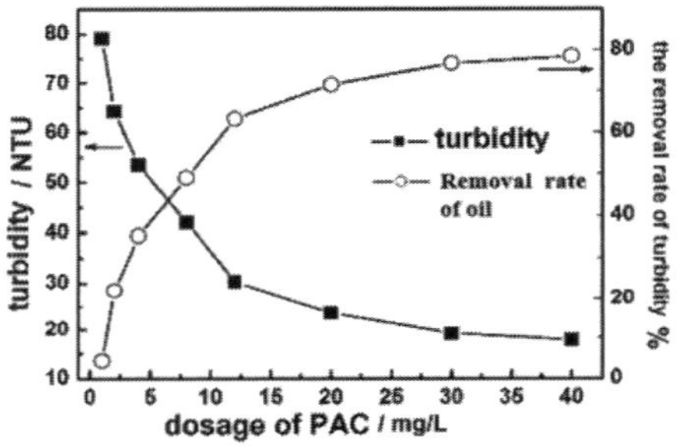

Figure 1. The variety of the turbidity of the hot rolling wastewater by adding PAC.

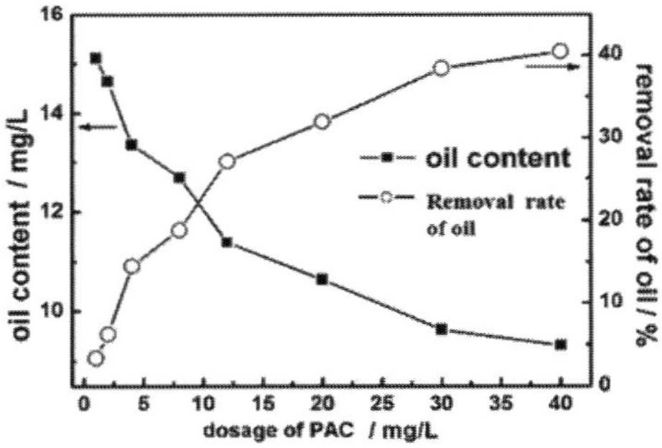

Figure 2. The variety of the oil content of the hot rolling wastewater by adding PAC.

Figures 1 and 2 indicate that the removal efficiency of the turbidity and the oil content increases as the amount of PAC is increased, while the increment rate of removal efficiency gets slower when the amount of PAC passes a certain value. In terms of turbidity, the efficiency of flocculation is close to saturation when the amount of PAC is more than 20 mg/L, it would be uneconomic to add more PAC. For the oil content, although the removal rate changes quicker than turbidity, it changes little when the amount of PAC is more than 30 mg/L. A reasonable amount of PAC should be given between 20 and 30 mg/L after the overall consideration. The difference between the change rate of oil content and the change rate of turbidity is mainly because the turbidity removal (the decrease of suspended particles) is a kind of flocculation, the availability factor of the Flocculation machine decreases when the number of suspended particles reduce to a certain value; on the other hand, oil removal is mainly a sort of absorption, so the curve is more linear (Chen 2020).

4.2 *The removal effect of turbidity and oil on different dosages of PAC+PAM.*

Under the condition of 20 mg/L PAC, we consider the effects of PAM dosage on turbidity and oil content of wastewater. The results are shown in Figure 3 and Figure 4.

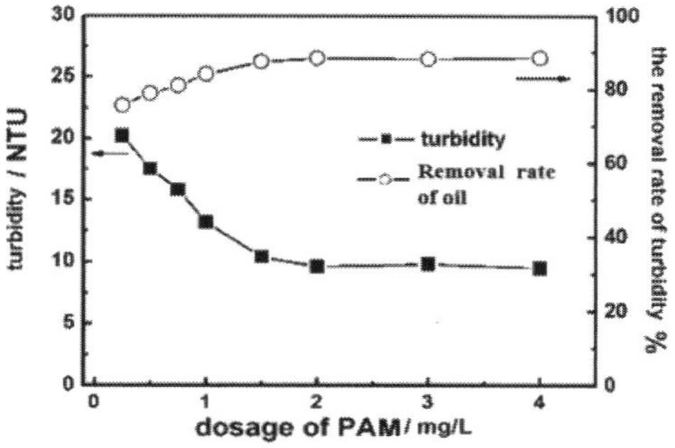

Figure 3. The variety of the turbidity of the hot rolling wastewater by adding PAC+PAM.

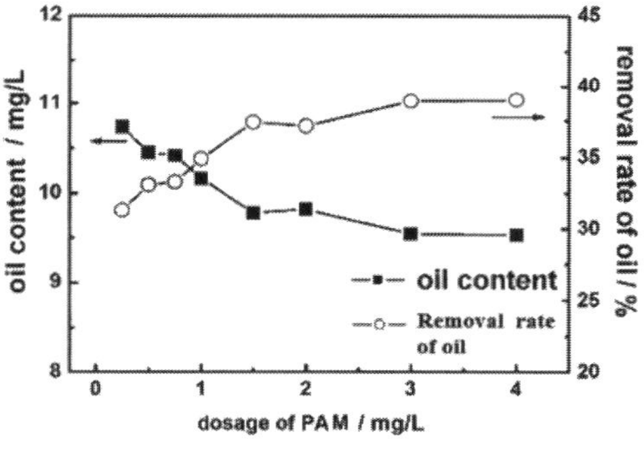

Figure 4. The variety of the oil content of the hot rolling wastewater by adding PAC+PAM.

Figure 3 and Figure 4 demonstrate that the turbidity can decrease to 10 NTU from 20 NTU before by mixing 0.25-4 mg/L PAM with 20 mg/L PAC. When the dosage of PAM is more than 1.5 mg/L, the accessorial effect of flocculation is close to saturation. This is mainly because PAM is a long chain macromolecule, which plays the role of bridgework, while PAC is mainly used for electric charge neutralization. They work like this: at first, the small particles in wastewater are neutralized and destabilized by PAC and gather together to be small clots, then PAM enhances the flocculation effect by connecting the small clots into big clots (Liu 2013). The bridging just needs a slight amount of PAM. When the input of PAM exceeds a certain amount, their increment effect of flocculation is not obvious. In terms of oil removal, the supplementary effect of PAM on oil removal is not significant (Liu & Zha 2015). The main reason is that the oil, which is neutral or nearly neutral, is not sensitive to the electric charge. Although oil cannot flocculate itself, some oil can be flocculated by the adhesion of particulate matters and so on. Therefore, the oil removing effect of slight PAM is limited.

4.3 *The removal effect of turbidity and oil on different dosages of PAC+PAM+magnetic powder*

The particulate matters (such as iron scale) in hot rolling wastewater usually are paramagnetic and can be attracted by the magnetic material. Hence, adding magnetic powder during the flocculation could aggregate and stabilize the floc in the water, thus enhancing the effect of flocculation. The efficiency of PAC can be improved through the rational use of magnetic powder, which can be a partial substitute for PAC. What's more, the processing cost can be reduced without affecting the water quality. In this experiment, the amount of PAM is fixed at 1.5 mg/L, and the sum of magnetic powder and PAC is fixed at 20 mg/L. We gradually increase the amount of magnetic powder and reduce the amount of PAC, and check the effect of this composite flocculant (PAC+ PAM+ magnetic powder) on turbidity and oil content. The results are shown in Figure 5 and Figure 6.

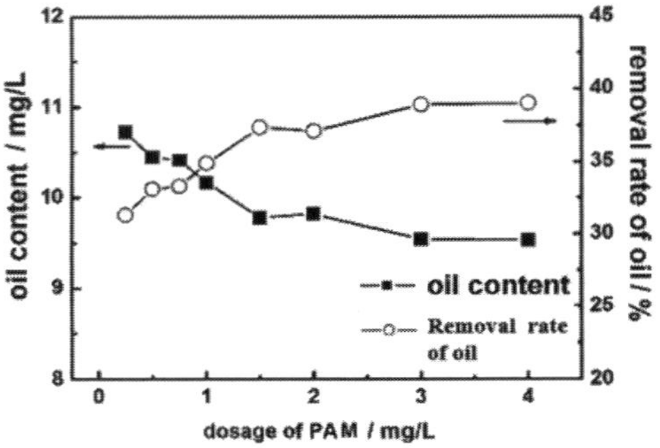

Figure 5. The variety of the turbidity of the hot rolling wastewater by adding PAC+PAM+magnetic powder.

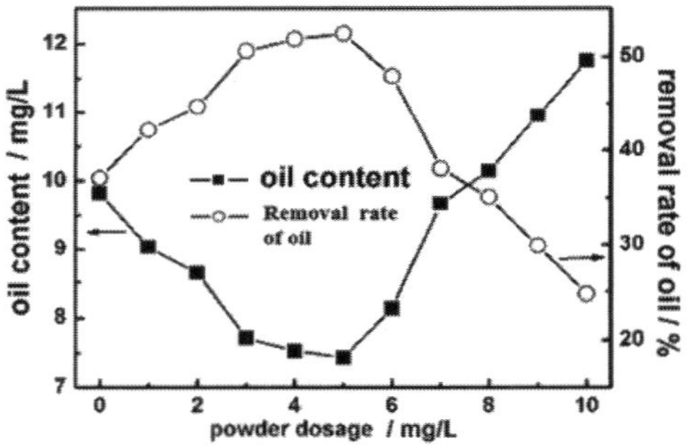

Figure 6. The variety of the oil content of the hot rolling wastewater by adding PAC+PAM+ magnetic powder.

Figure 5 shows that when the dosage of magnetic powder increases from 0 mg/L to 5 mg/L, and the dosage of PAC decreases from 20 mg/L to 15 mg/L, the turbidity after processing maintains at 10 NTU, which is the same as 20 mg/L PAC+1.5 mg/L PAM. This indicates that adding magnetic powder in this range can stabilize the floc and improve the efficiency of the PAC. When the dosage of the magnetic powder is more than 5 mg/L and the dosage of PAC is less than 15 mg/L, the turbidity increases with the increase of magnetic powder. This is mainly due to the reduced amount of PAC, which causes less neutralization for the charged particles in wastewater, while magnetic powder, which is neutral, cannot neutralize them. Therefore, the flocculating effect gradually decreases with the reducing amount of PAC. Figure 6 indicates that when the dosage of magnetic powder increases from 0 mg/L to 3 mg/L, and the dosage of PAC decreases from 20 mg/L to 17 mg/L, the oil content in wastewater is slightly reduced. When the dosage of magnetic powder increases to 5 mg/L, and the dosage of PAC decreases to 15 mg/L, the oil content in wastewater changes little. This is mainly because the magnetic powder can absorb some oil, and enhance the effect of absorbing oil when the amount of floc decreases a little. The removal effect of oil significantly decreases when the dosage of the magnetic powder is more than 5 mg/L, the main reason is also the reduced amount of PAC, which causes less amount of floc, the removal effect of oil would decrease due to lack of floc. To sum up, the best dosage of magnetic powder should be 5mg/L, which means about 25 percent of the PAC amount.

From the experimental results above, for the hot rolling wastewater with initial 82.6 NTU turbidity and 15.62 mg/L oil content, when using 20 mg/L PAC+1.5mg/L PAM as composite flocculant, the turbidity gets down to 10.3 NTU and the oil content is reduced to 9.81 mg/L. Meanwhile, the turbidity decreases to 10.5 NTU, and the oil content drop to 7.44 mg/L using 15 mg/L PAC+1.5 mg/L PAM+5 mg/L magnetic powder under the same condition. Thus, for 15 mg/L PAC+1.5 mg/L PAM+5 mg/L magnetic powder, compared with 20 mg/L PAC+1.5mg/L PAM, the removal effect of turbidity stays almost the same and the removal effect of oil content is enhanced. It has some advantages in efficiency. From economic consideration, the price of PAC is about 1,000 RMB/t, while the price of the magnetic powder is only 400 RMB/t. The cost can be reduced by 15 percent if one-fourth of PAC is replaced by magnetic powder.

5 CONCLUSIONS

(1) A new composite flocculant including PAC, PAM, and magnetic powder is developed by laboratory experiment. In addition, the specific content of proportion, which is proven to be the most reasonable in the laboratory so far, is 15 mg/L PAC, 1.5 mg/L PAM, and 5 mg/L magnetic powder.
(2) The magnetic powder can be made by further grinding and magnetizing industrial iron and phosphor powder. The manufacturing process is simple and easy to operate.
(3) The turbidity of hot rolling wastewater can decrease to 10.5 NTU and the oil content of the wastewater can drop to 7.44 mg/L using the new composite flocculant. Compared with the old composite flocculant (PAC+PAM), the new one is more efficient and reduces the cost by 15%. It has advantages in both efficiency and finance, we believe that it will have a broad market prospect.

ACKNOWLEDGMENTS

This work was financially supported by the Key Research and Development Program of Hubei Province (Grant No. 2021BAA063 and No. 2020BAB084), the National Natural Science Foundation of China (Grant No.61904130), and the Key Laboratory of Hubei Province for Coal Conversion and New Carbon Materials.

REFERENCES

Chen X. Q. (2020) *Study on the treatment of simulated dye wastewater by electroflocculation combined with electroflocculation and magnetic flocculation* [D]. Dalian Maritime University.

Chen Y., Li J., Chen X.L. (2011) The research on wastewater processing with flocculation[J] *Chinese water supply and drainage*, 27: 78–81.

Gong E.H. The Analysis of rolling wastewater processing[J] *Science and technology information* 2012,19: 289–290.

Guo J.W, Meng Y. (2020) Application of magnetic flocculation technology in advanced sewage treatment [J]. *China Comprehensive Utilization of Resources*, 38(07): 188–190.

Lin L., Li G.C., Yang T. (2015) Experimental study on the treatment of wastewater containing emulsified oil with composite flocculant PAC-PAM [J]. *Shandong Chemical Industry*, 44:3.

Liu H, Zha F.L. (2015) Experimental study on the treatment of hot rolling wastewater with composite flocculants (PAC+PAM+magnetic powder) [J]. *Journal of Wuhan University of Science and Technology*, 38:4.

Liu J. (2013) The research on the improvement of Hot Rolling turbid circulating water quality[J] *Guangdong chemistry*, 40: 223–224.
Wang J., Chen W., Liu C. (2014) The Research of the combination of PAM and magnetic powder on processing the drinking water[J] *Science technology and Engineering*, 14: 197–200.
Wang S.K, Cheng F., Guo X. F., et al. (2021) Experimental study on optimization of deep phosphorus removal from sewage by magnetic loading flocculation [J]. *Journal of Hebei Institute of Environmental Engineering*, 31(2):6.
Xiong G.H., Cang D.Q. (2008) The research of rolling turbid water with low turbidity or high oil content[J]. *Industrial water processing*, 28: 39-42
Yu F, Wang P.P., Zhao J.P., et al. (2019) Study on the influencing factors of composite flocculant PAC-PAM in the treatment of fluorine-containing wastewater [J]. *Organofluorine Industry*:4.

Energy Revolution and Chemical Research – Kok-Keong Chong and Zhongliang Liu (Eds)
© 2023 The Authors, ISBN: 978-1-032-36554-1

A control method of chemical composition based on fuzzy theory during recycling aluminum scrap process

Xiaohui Ao*
School of Mechanical Engineering, Beijing Institute of Technology, Beijing, China

JiaHong Zhang
Publishing House of Electronics Industry, Beijing, China

Yujun Bai
Beijing Satellite Manufacturing Plant Co., Ltd., Beijing, China

ABSTRACT: The stability of recycled castings' properties is seriously restricted by the complexity and variability of the chemical composition of Aluminum scrap. In this paper, a method based on the fuzzy theory used to evaluate the quality grade of the chemical composition of the alloy liquid after melting aluminum scrap together with a superiority control strategy of the chemical composition of the alloy liquid in the continuous feeding process were proposed, respectively. A quality evaluation model and a superiority control model of chemical composition were also established, and the corresponding evaluation principle was formulated. The verification example showed that the quality evaluation index could be used as the quality standard of alloy liquid, and the superiority control index could be used as the criterion for chemical composition adjustment. The application of this method could significantly improve the stability of chemical components and casting properties during the recycling of scrap aluminum.

1 INTRODUCTION

Due to its excellent mechanical properties and significant lightweight, the use of aluminum alloys is increasing year by year. The primary aluminum used to produce aluminum alloy castings can no longer meet the market demand, and recycled aluminum will gradually become an important raw material (Zhou et al. 2021; Zhu et al. 2021). Aluminum scrap is not only rich in raw materials but also can achieve significant energy savings and emission reduction (Ding et al. 2012). Therefore, the development of the aluminum scrap recycling industry has become an inevitable trend, especially the development of new equipment and advanced technology (Ighodalo et al. 2011; Verran & Kurzawa 2008). However, due to the complex composition of scrap aluminum, it is extremely difficult to accurately control the chemical composition in the recycling process of scrap aluminum. The large-scale aluminum scrap recycling melting furnace generally has a double chamber structure, the melting chamber is continuously fed by the automatic feeding machine for heating and melting, and the insulation chamber is used to stir melt and remove impurities. Due to the continuous addition of raw materials, the chemical composition of the alloy in the furnace is constantly changing. The fuzzy theory has a very good application effect on the chemical composition control in the metallurgical industry (Jahedsaravani et al. 2016a; Jahedsaravani et al. 2016b; Wicher et al. 2016), so the author tries to apply the fuzzy theory to the chemical composition control of recycling Aluminum scrap process. In this paper, in view of the lack of chemical composition

*Corresponding Author: xhao@bit.edu.cn

DOI 10.1201/9781003332657-67

quality evaluation, chemical composition fluctuations and difficulty to control, and other issues during the recycling of Aluminum scrap process, the quality characterization and evaluation of the chemical composition of the aluminum alloy were carried out, the quality grade and the superiority control strategy of the chemical composition of the alloy liquid were established, and the precise control of the chemical composition was finally realized.

2 QUALITY EVALUATION OF CHEMICAL COMPOSITION

2.1 *Evaluation strategy*

The quality evaluation of chemical composition based on the fuzzy theory is realized by normalizing the relevant indicators of the object to be evaluated according to the combination of fuzzy transformation and membership principle firstly, and then considering the influence of different factors on the evaluation object, the distribution weights were set, and finally, a reasonable quality evaluation conclusion was drawn. The Process of quality evaluation of chemical composition based on fuzzy theory is shown in Figure 1.

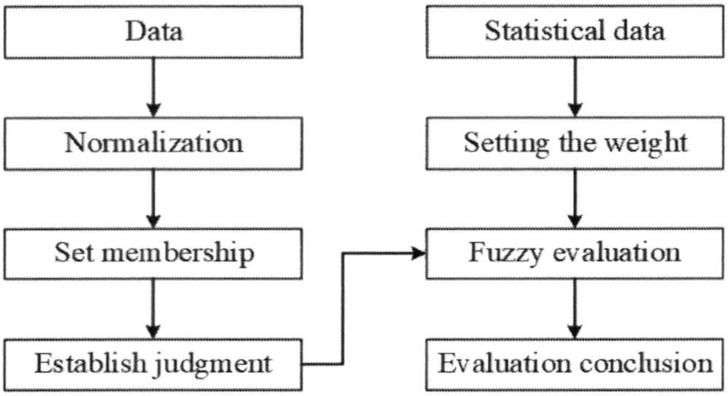

Figure 1. Process of quality evaluation of chemical composition based on fuzzy theory.

2.2 *Normalization*

The chemical composition of aluminum alloy is mainly divided into alloying elements and impurity elements, in which alloying elements have fixed content ranges and impurity elements only have the maximum content limits. The company will also develop different internal standards for each element based on casting technology, raw material cost, and castings' performance. Therefore, in order to conduct quality evaluation and analysis, it is necessary to normalize the data, that is, the content of each element should be converted to a specific value in the interval [0,1], where 0 is the best and 1 is the worst. The normalization of alloying elements and impurity elements can be calculated as:

$$e_i(\omega) = \begin{cases} 1 & \omega_{\min} < \omega \leq \omega_{\mathrm{imin}} \\ \dfrac{\omega_{\mathrm{opt}} - \omega}{\omega_{\mathrm{opt}} - \omega_{\mathrm{imin}}} & \omega_{\mathrm{imin}} < \omega < \omega_{\mathrm{opt}} \\ \dfrac{\omega - \omega_{\mathrm{opt}}}{\omega_{\mathrm{opt}} - \omega_{\mathrm{imax}}} & \omega_{\mathrm{opt}} \leq \omega < \omega_{\mathrm{imax}} \\ 1 & \omega_{\mathrm{imax}} \leq \omega < \omega_{\max} \end{cases} \tag{1}$$

$$e_i(\omega) = \begin{cases} \dfrac{\omega}{\omega_{i\max}} & 0 \le \omega \le \omega_{i\max} \\ 1 & \omega_{i\max} < \omega < \omega_{\max} \end{cases} \tag{2}$$

Where ω is measured content, $[\omega_{\min}, \omega_{\max}]$ is the national standard range, $[\omega_{imin}, \omega_{imax}]$ is the internal standard range, ω_{opt} is the optimum content.

2.3 *Fuzzy membership function*

Considering the characteristics of scrap composition and recycling process, the quality grades of chemical elements can be divided into four grades: 'excellent', 'good', 'qualified', and 'poor', which are expressed as S1, S2, S3, and S4 respectively. The 'poor' grade refers to meeting the national standard range but not meeting the internal standard, and the other three grades are further ratings for the internal standard. The membership function of each grade can be expressed as:

$$f_{s1}(e_i) = \begin{cases} 1 & 0 \le e_i \le \alpha \\ 1 - \sin\dfrac{\pi}{1-\alpha}(e_i - \alpha) & \alpha < e_i < \dfrac{1+\alpha}{2} \\ 0 & \dfrac{1+\alpha}{2} \le e_i \le 1 \end{cases} \tag{3}$$

$$f_{s2}(e_i) = \begin{cases} 0 & 0 \le e_i \le \alpha \\ \sin\dfrac{\pi}{1-\alpha}(e_i - \alpha) & \alpha < e_i \le 1 \end{cases} \tag{4}$$

$$f_{s3}(e_i) = \begin{cases} 0 & 0 \le e_i < \frac{1+\alpha}{2} \\ 1 - \sin\dfrac{\pi}{1-\alpha}(e_i - \alpha) & \dfrac{1+\alpha}{2} \le e_i < 1 \\ 0 & e_i = 1 \end{cases} \tag{5}$$

$$f_{s4}(e_i) = \begin{cases} 0 & 0 \le e_i < 1 \\ 1 & e_i = 1 \end{cases} \tag{6}$$

Through normalization and membership calculation, the membership matrix of n chemical elements can be obtained as follows:

$$R = \begin{bmatrix} f_{s1}(e_1) & f_{s2}(e_1) & f_{s3}(e_1) & f_{s4}(e_1) \\ f_{s1}(e_2) & f_{s2}(e_2) & f_{s3}(e_2) & f_{s4}(e_2) \\ \cdot & \cdot & \cdot & \cdot \\ f_{s1}(e_n) & f_{s2}(e_n) & f_{s3}(e_n) & f_{s4}(e_n) \end{bmatrix} \tag{7}$$

2.4 *The weight of each chemical composition*

This article is mainly aimed at the complex and changeable chemical composition of scrap aluminum, so the setting weights should be based on the historical statistics of the fixed furnace group. The equation for determining the weight of chemical element i according to furnace j historical data is:

$$W_i = \sum_{j=1}^{m}\sum_{r=1}^{j}\left|\omega_{i,j} - \omega_{i,r}\right| \Big/ \sum_{t=1}^{n} A_t \tag{8}$$

$$W = \left[\, W_1 \;\, W_2 \;\, \cdots \;\, W_n \,\right] \tag{9}$$

Where $\omega_{i,j}$ is the content of the chemical element i in the furnace j.

The advantage of setting the weights is that the greater the fluctuation, the greater the weight, which reflects the sensitivity to compositional changes.

2.5 *Quality assessment*

The membership matrix of quality evaluation of chemical composition can be obtained by:

$$T = W \bullet R = \left[\, T_1 \;\, T_2 \;\, T_3 \;\, T_4 \,\right] \tag{10}$$

Where, T_1, T_2, T_3, and T_4 are the membership of chemical element quality grades ('excellent', 'good', 'qualified', 'poor'). Therefore, a quality evaluation of the quality grade can be calculated. Here, the quality grade of chemical composition is evaluated by the principle of "membership nonzero of 'poor' grade" and "membership maximum". For example, when $T_4 \neq 0$, the chemical composition quality grade was recognized as a 'poor' grade. T4=0 means that all chemical components meet the internal control standards, and the grade with the maximum membership is regarded as the quality evaluation conclusion.

3 OPTIMIZATION CONTROL OF CHEMICAL COMPOSITION

3.1 *Control strategy*

Since the alloy liquid in the melting furnace is 70-90t, and the average feeding rate is 4t/h, the detection time can be set to once every half an hour. In the continuous smelting of scrap aluminum, the composition information of the feed includes the current composition and changing the content of the alloy liquid. The fuzzy quality evaluation process is shown in Figure 2.

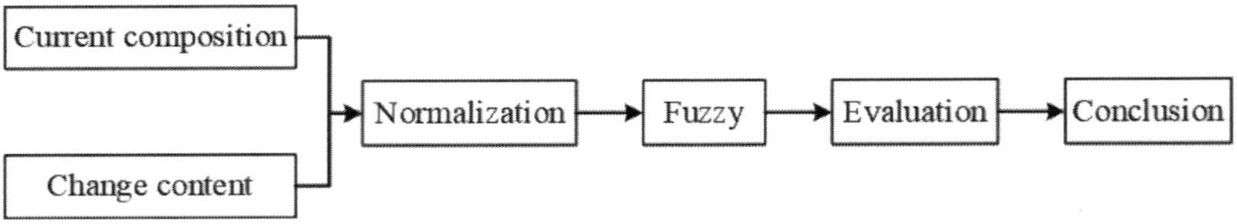

Figure 2. Fuzzy quality evaluation process.

3.2 *Normalization of control information*

In order to normalize the current component and the changing content, the value $e_i(\omega)$ is normalized by Equations (11). The normalization criterion: $e_i(\omega)$ increases with an increased tendency of the change content Δe; $e_i(\omega)$ reduces with a reduced tendency of the change content Δe, and the corrected criteria are shown in Figure 3.

$$E_i = \begin{cases} e_i\,(\omega)\,(1 + \Delta e) & \Delta e < 0 \\ e_i\,(\omega) + 0.5\Delta e & \Delta e \geq 0 \end{cases} \tag{11}$$

Where $\Delta e = e_i\,(\omega)_j - e_i\,(\omega)_{j-1}$.

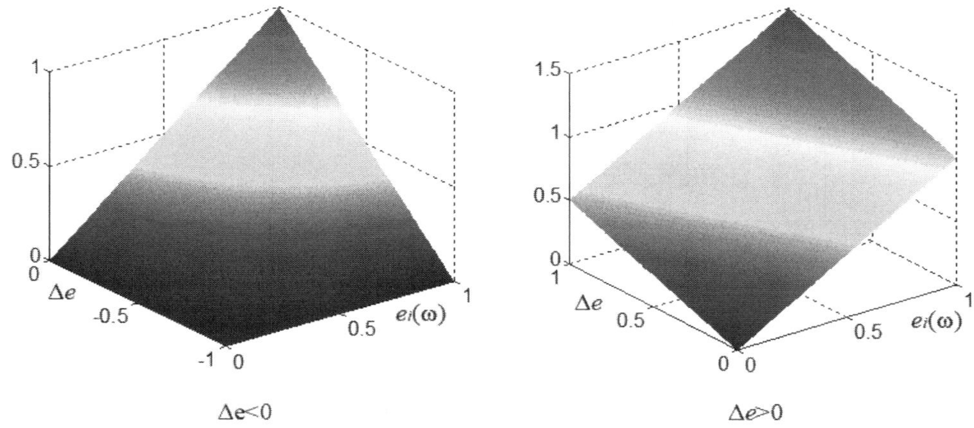

Figure 3. Correction rules of normalization.

4 VERIFICATION

Taking ADC12 as an example, the standard range of chemical composition of ADC12 is shown in Table 1. The chemical composition of the alloy liquid is measured every half an hour. Table 2 shows the monitoring data of the chemical composition of ADC12, and the corresponding quality membership matrix and superiority membership matrix can be calculated. The evaluation conclusion is shown in Table 3.

Table 1. Standard range of chemical composition of ADC12.

Element	Si	Fe	Mg	Cu	Zn	Mn
National	9.6-12.0	0.6-1.1	0.2-0.35	1.5-3.5	≦3.0	≦0.5
Internal	10.1-10.8	0.65-0.88	0.25-0.33	1.5-1.9	1.0-1.5	≦0.25

Table 2. Monitoring data of chemical composition of ADC12.

Time	Si	Fe	Mg	Cu	Zn	Mn
0.5	10.3	0.78	0.31	1.77	1.14	0.22
1	10.5	0.73	0.31	1.77	1.32	0.23
1.5	10.7	0.84	0.3	1.81	1.23	0.23
2	10.9	0.96	0.29	1.84	1.24	0.21
2.5	10.8	0.91	0.28	1.82	1.25	0.21
3	10.7	0.87	0.27	1.8	1.27	0.22

According to the quality evaluation of the 6 groups of measured data, it was found that after the 3rd group, the alloy liquid was evaluated as 'good', the 4th group skipped the 'qualified' and was directly evaluated as 'poor', and then the component was quickly adjusted, but the 5th group still staying at the 'poor' grade, until the 6th group was promoted to the 'good' grade. After the superiority control strategy was carried out, it was found that the 'poor' grade appeared in the 3rd group, indicating that although the chemical composition of the alloy liquid was 'good' at this time, the deterioration trend was obvious. If it is adjusted at this time, the 'poor' grade in the

Table 3. Evaluation conclusions of ADC12 for Scrap Aluminium.

	Quality membership					Superiority membership				
Time	S1	S2	S3	S4	QG	S1	S2	S3	S4	SG
0.5	0.6676	0.2913	0.0411	0	Excellent	–	–	–	–	–
1	0.4567	0.4841	0.0592	0	Good	0.3063	0.6057	0.088	0	Good
1.5	0.2191	0.6276	0.1533	0	Good	0.25	0.1996	0.1004	0.45	Poor
2	0.0839	0.3866	0.084	0.45	Poor	0.1747	0.2867	0.0886	0.45	Poor
2.5	0.0746	0.4277	0.0522	0.45	Poor	0.1956	0.66	0.1445	0	Good
3	0.0532	0.7224	0.2245	0	Good	0.0257	0.9058	0.0686	0	Good

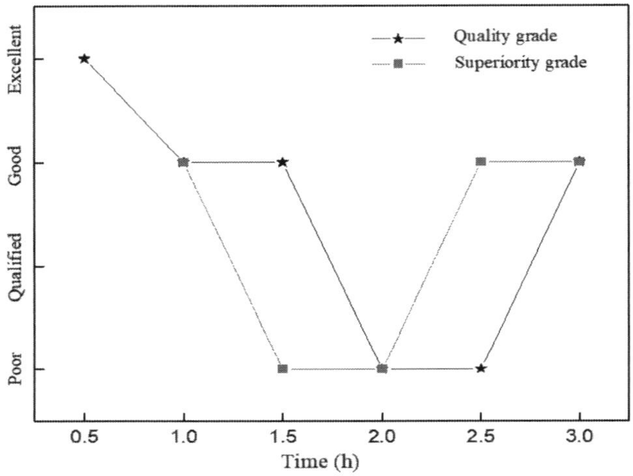

Figure 4. Trend of quality grade and superiority grade.

4th group can be avoided. The superiority grade of the 5th group was 'good', indicating that the effect of component adjustment was significant, which was proved by the 'good' grade in the 6th group. Figure 4 shows the trend between quality grade and superiority grade. It can be seen that the superiority grade trend and the quality grade trend have similar shapes, but the superiority grade arrives earlier, indicating that it has the ability to predict the quality grade of chemical components. In order to ensure the accuracy of chemical composition regulation, it is advisable to adopt a method combining quality evaluation and superiority control in the recycling process of scrap aluminum. The quality evaluation index is used as the product quality standard, and the superiority control index is used as the warning signal for the ingredient adjustment, and their combination can well realize the regulation of the chemical composition of recycling scrap aluminum.

5 CONCLUSION

In this paper, the quality evaluation method based on fuzzy theory was used to evaluate the chemical composition of the alloy liquid during the aluminum scrap recycling process. According to the characteristics of scrap composition and the recycling process, the chemical composition quality grades can be divided into four grades: 'excellent', 'good', 'qualified', and 'poor' respectively. The evaluation principle of "membership nonzero of 'poor' grade" and "membership maximum" was developed. The example showed that the fuzzy evaluation and superiority control strategy proposed in this paper can effectively control the chemical composition during the aluminum scrap recycling process.

ACKNOWLEDGMENTS

This work is supported by the National Natural Science Foundation of China (No. 52105504).

REFERENCES

Ding, N., Gao, F., Wang, Z., Gong, X., Nie, Z. (2012) Environment impact analysis of primary aluminum and recycled aluminum. *Procedia Eng.*, 27: 465–474.

Ighodalo, O. A., Akue, G., Enaboifo, E., Oyedoh, J. (2011) Performance evaluation of the local charcoal-fired furnace for recycling aluminum. *J. Emerging Trends Eng. Appl. Sci.*, 2(3): 448–450.

Jahedsaravani, A., Marhaban, M. H., Massinaei, M. (2016a) Application of statistical and intelligent techniques for modeling of the metallurgical performance of a batch flotation process. *Chem. Eng. Commun.*, 203(2): 151–160.

Jahedsaravani, A., Massinaei, M., Marhaban, M. H. (2016b) Application of image processing and adaptive neuro-fuzzy system for estimation of the metallurgical parameters of a flotation process. *Chem. Eng. Commun.*, 203(10): 1395–1402.

Verran, G. O., Kurzawa, U. (2008) An experimental study of aluminum can recycling using fusion in an induction furnace. *Resour., Conserv. Recycle.*, 52(5): 731–736.

Wicher, P., Zapletal, F., Lenort, R., Staš, D. (2016) Measuring the metallurgical supply chain resilience using fuzzy analytic network process. *Metalurgija*, 55(4): 783–786.

Zhou, B., Liu, B., Zhang, S., Lin, R., Jiang, Y., Lan, X. (2021) Microstructure evolution of recycled 7075 aluminum alloy and its mechanical and corrosion properties. *J. Alloys Compd.*, 879: 160407.

Zhu, Y., Chappuis, L. B., De Kleine, R., Kim, H. C., Wallington, T. J., Luckey, G., Cooper, D. R. (2021) The coming wave of aluminum sheet scrap from vehicle recycling in the United States. *Resour., Conserv. Recycle.*, 164: 105208.

Energy Revolution and Chemical Research – Kok-Keong Chong and Zhongliang Liu (Eds)
© 2023 The Authors, ISBN: 978-1-032-36554-1

The mechanism analysis of H radical effect on ignition of methane/air mixture

Yong Li*, Weizhuo Hua*, Conghui Huang*
Aviation Engineering School, Air Force Engineering University, Xi'an, Shaanxi, China

ABSTRACT: To demonstrate the physical characteristics of H radical, which influences the mixture gas ignition of methane and air by using GRI-Mech3.0, we conduct a numerical simulation of the methane ignition process by using a zero-dimensional, homogeneous, and completely mixed model. We also study active particles at different initial concentrations on methane ignition delay time. Through the analysis of the numerical result, when adding 0.5% H radicals, the ignition time was reduced by nearly 94.7%. Using the analysis of reaction path and sensitivity, the methane ignition upon adding H radical and the chemical reaction mechanism involved is revealed.

1 INTRODUCTION

In the future, the hypersonic engine has a pride place in technology. Until now, the hypersonic engine still faces various issues that need to be addressed, such as ignition, combustion, etc. In the hypersonic state, it is difficult to have enough time to ignite because of the very high flow rate and a short time. Through the experiment, it can be found that H has lower density energy than hydrocarbon. But the hydrocarbons need a longer igniting time than hydrogen. For instance, the fuel igniting time is approximately 5 ms to 10 ms (Segal et al. 1998; Tishkoff et al. 1997; Zakrzewski et al. 2004) within the condition of static pressure P = 50 kPa~100 kPa and static temperature T = 600 K~1000 K. Extensive research (Bocharov et al. 2004; Lenonov & Constantin 2008; Lan & HE 2009; Mceldowney et al. 2003; Packan et al. 2004) indicate that some plasma generated can enhance combustion efficiency, for the reaction process is influenced by some active particles. And those active particles can accelerate the reaction. But the physical characteristics of H active particles affect the ignition of CH_4 and the air mixture is non-obvious. This paper uses GRI-Mech3.0 and numerical simulation of the methane ignition process. It informs the influence of igniting mechanism of CH_4 and air mixture by using the analysis result of reaction path and sensitivity.

2 STUDY METHOD

2.1 *Equation*

The control equation of the temperature and reaction of kinds of fractions can be defined as (Kee et al. 1989):

$$\frac{dY_k}{dt} = \frac{\omega_k W_k}{\rho} \tag{1}$$

$$\frac{dT}{dt} = -\frac{1}{\rho c_p}\sum_{k=1}^{n} h_k \omega_k W_k, (k = 1, \ldots, n.) \tag{2}$$

*Corresponding Authors: lixuan_1984@126.com, hwz1991@sina.com and hch271@126.com.

 DOI 10.1201/9781003332657-68

Where Y_k corresponds to the k kinds of fraction and T corresponds to the temperature. T, ρ, c_p, h_k, ω_k, and W_k correspond to the time, density, ratio, and enthalpy. By the integral equation, we can have all kinds of processes of mole fraction and temperature.

The equation needs to input some fuel data of gas-phase chemical and thermodynamics, the gas-phase chemical data can be obtained through Equations 3 and 4, and the thermodynamics data can be obtained through Equation 5 (Kee et al. 1989; Zhang 2012):

$$\sum_{k=1}^{k} \nu'_{ki} \chi_k \Leftrightarrow \sum_{k=1}^{k} \nu''_{ki} \chi_k \tag{3}$$

$$k_{fi} = A_i T^{bi} \exp\left(-E_i/RT\right) \tag{4}$$

$$\frac{C^o_{pk}}{R} = a_{1k} + a_{2k} T_k + a_{3k} T_k^2 + a_{4k} T_k^3 + a_{5k} T_k^4 \tag{5}$$

2.2 *Sensitivity analysis*

The sensitivity analysis can calculate the influence of some parameters on the measure. The sensitivity defined as $(\tau_i\text{-}\tau_0)/\tau_0$. τ_i is the ignition time after increasing twice the reaction velocity. τ_0 is the ignition time by adopting the regular reaction velocity. When the $(\tau_i\text{-}\tau_0)/\tau_0<0$, it implies that the reaction of this can reduce ignition time. When the $(\tau_i\text{-}\tau_0)/\tau_0>0$, it implies that the reaction of this can increase ignition time.

Comparing with the sensitivity analysis, we can confirm that some reactions accelerated the reaction velocity and some decreased. Therefore, it is possible to increase or decrease the molar fraction of certain free radicals at the same temperature to accelerate certain chemical reactions or slow down some chemical reactions, to decrease the ignition time of H.

2.3 *Reaction path analysis*

To research the effect of the radical on the ignition of methane/air, we analyze different fractions that influence the reaction path. All kinds of reactions influence fractions are defined as (Zhang 2012):

$$\varphi_i = \frac{f_i}{\sum_{i=1}^{n} |f_i|} \tag{6}$$

Where f_i is the influence value of the i-th reaction; n is the number of different reactions. Normally, we can define the ignition process by using OH radicals and can analyze the effect of H radical concentration change.

3 RESULTS ANALYSIS

The condition of calculation is that the pressure is 1 atm and the temperature is 1200 K, with the composed of gas is the methane and air (CH_4:O_2: N_2=0.095:0.19:0.715).

Currently, research shows that the H radical can influence the ignition process. In the meantime, the survival time of H radical is longer in the gas discharge. Figure 1 shows the evolution in time of mole fractions and temperature changing under the condition of autoignition. Figure 2 shows the evolution in time of mole fractions and temperature change by adding 0.5% H radical. From both the figures, the ignition time can be reduced by adding 0.5% H radical. It can reduce ignition time by about 94.7% and the ignition time is upgraded from 3.8×10^{-2}s to 0.2×10^{-2}s.

Figure 3 shows the ignition delay time by adding various mole fractions H as the temperature changes. The ignition time has varying degrees of reduction by adding a fraction H radical. It

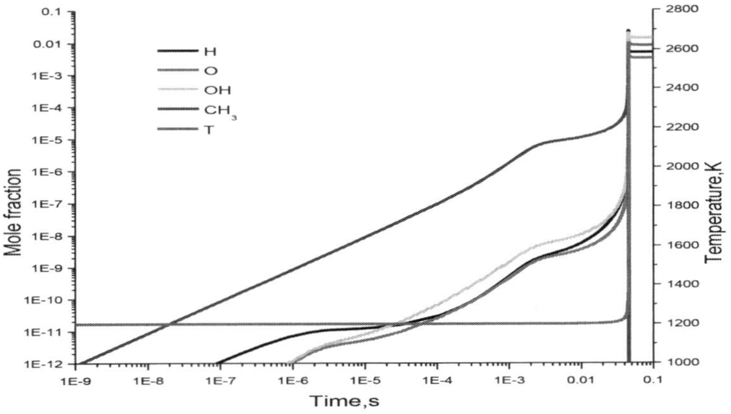

Figure 1. The evolution in time of mole fractions mole and temperature changing under autoignition.

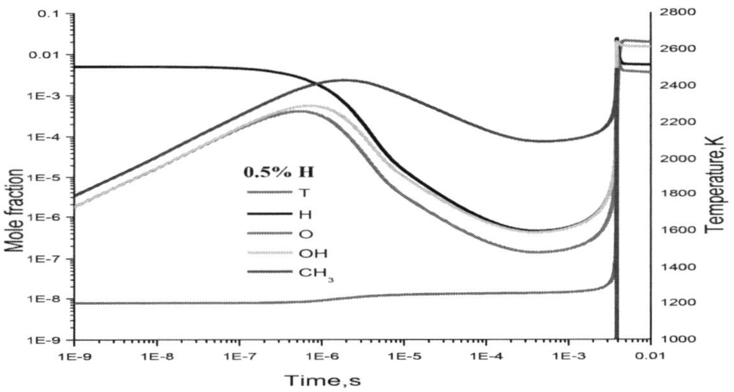

Figure 2. The evolution in time of fractions and temperature changing for adding 0.5% H.

also reduces different degrees with different temperatures. When the temperature is high, the H radical can impact ignition time slowly. Under the temperature of 1000 K, the ignition delay time is decreased by about 99.75% by adding 2% H radical, which is compared with autoignition. However, when the temperature goes up to 2400 K, the ignition delay time is only reduced by 8.3% from the same situation before.

The consequence of sensitivity analysis can be seen, which mixed 0.5% H or not in Figure 4. Sixteen significant reactions can be acquired from the experiment. S_k represents the influence of k variety of reactions. Comparing autoignition with mixed 0.5% H radical, the ignition time increased in reaction $CH_3(+M) <=> C_2H_6(+M)$. Through reactions R38, $H+O_2<=>O+OH$, R101, $OH+CH_2O<=>HCO+H_2O$, and $HO_2+CH_3<=>OH+CH_3O$ shorten the ignition time. Although when mixed with 0.5% H activity particles, some reactions extend ignition time in some links the ignition time of the final results is still decreased.

In Figure 5, it can be found that the analysis results depict the chemical reaction pathway of OH activity particles. There are 10 significant reactions. The main producing reactions are R38 and R119. The main consumption reactions are R98 and R97.

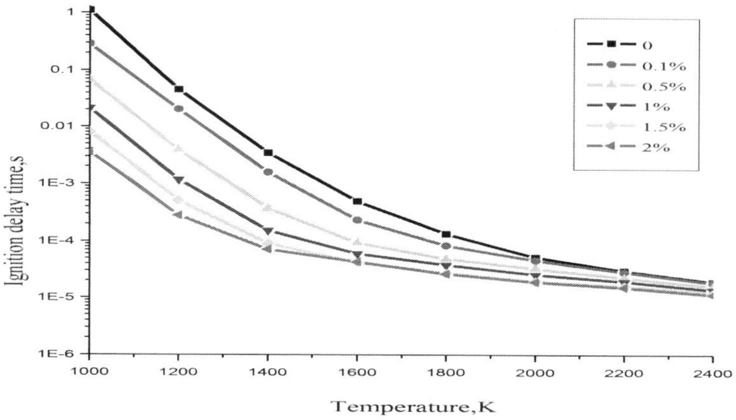

Figure 3. The ignition delay time by mixed different mole proportions H with different temperature.

Table 1. Main producing reaction.

Number	Reaction
R11	$O+CH_4$<=>$OH+CH_3$
R38	$H+O_2$<=>$O+OH$
R98	$OH+CH_4$<=>CH_3+H_2O
R119	HO_2+CH_3<=>$OH+CH_3O$
R156	CH_3+O_2<=>$OH+CH_2O$

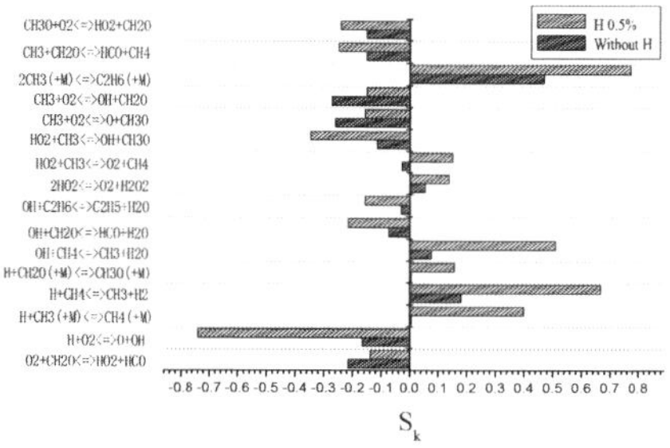

Figure 4. Results of sensitivity analysis without H and adding 0.5% H.

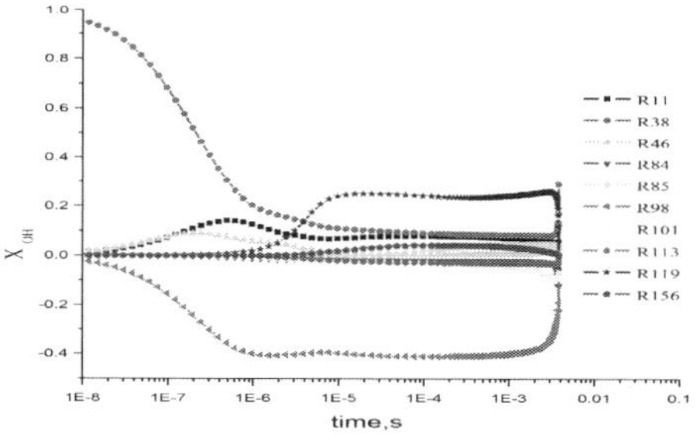

Figure 5. Chemistry reaction pathway results of OH activity particles.

4 CONCLUSION

The plasma has many productions of activity particles, such as H activity particles. Especially, the H radical can be maintained for a good while. Thus, this study analyzes that H activity particles affect the ignition. The article on H activity particles about the ignition process of CH_4 and air mixture indicates that H activity particles can shorten the ignition time by a mixed different proportion of H activity particles. Through analysis of sensitivity and chemistry reaction path, the ignition time is decreased by about 92.5% when mixed with 0.5% H activity particles. $H+O_2$<=>$O+OH$, $OH+CH_2O$<=>$HCO+H_2O$, and HO_2+CH_3<=>$OH+CH_3O$ reduced ignition delay time. While the temperature increased, the ignition time was reduced. Meanwhile, the effect of the H radical declined considerably.

REFERENCES

Bocharov, A. N. Bityurin, V. A. Filimonova, E. A. (2004). AIAA, Paper 2004–1017.

Kee, R. J. Rupley, F. M. Miller, J. A. (1989) *CHEMKIN-III: A Fortran Chemical Kinetics Package for the Analysis of Gas-Phase Chemical and Plasma Kinetics*. Sandia National Laboratory Report, SAND89-8009B.

Lan, Y. D. HE, L. M. (2009) Effects of plasma on the combustion of H2/Air mixture under different initial temperatures. *Journal of Propulsion Technology*, 30(6): 651–655.

Lenonov, S. B. Constantin, V. (2008) *Experiments on plasma-assisted combustion in M=2 hot test-bed pwt-50H*. AIAA, Paper 2008-1359.

Mceldowney, B. Meyer, R. Chintala, N. (2003) *Ignition of premixed hydrocarbon-air flows using a nonequilibium RF discharge*. AIAA, Paper 2003–3478.

Packan, D. Grisch, F. Attal-Tretout. (2004) *Study of plasma-enhanced combustion using optical diagnostics*. AIAA, Paper 2004–983.

Segal, C. Owens, M. G. Tehranian, S. (1998) *Flame holding configurations for kerosene combustion in a Mach 1.8 airflow*. AIAA, 97–2888.

Tishkoff, J. M. Drummond, J. P. Edwards, T. (1997) *Future direction of supersonic combustion research: Air Force/NASA workshop on supersonic combustion*. AIAA, 97–1017.

Zakrzewski, S. Milton, B. E. Pianthong, K. Behnia, M. (2004) *Supersonic Liquid Fuel Jets Injected Into Quiescent Air*. AIAA, 79–1238.

Zhang, P. (2012) *The research on ignition of methane/air mixture based on the simplified mode of plasma*. Beijing: Academy of Equipment, Beijing.

Energy Revolution and Chemical Research – Kok-Keong Chong and Zhongliang Liu (Eds)
© 2023 The Authors, ISBN: 978-1-032-36554-1

Application of low-temperature plasma and titanium dioxide improving new fouling resistant reverse osmosis membrane in water treatment

Shiyu Zhang*
College of Water Conservancy and Hydropower, Sichuan Agricultural University, Ya'an, China

ABSTRACT: With the increasingly serious problem of water pollution and the depletion of water resources, reverse osmosis membrane has become a hot spot in the field of water treatment. In this paper, vinylimidazole and titania were mixed and grafted on the reverse osmosis membrane with low-temperature plasma treatment, to enhance the flux, antibacterial, and fouling resistance of the improved membrane, and it is used in the fields of seawater and bitter alkaline water desalination, heavy metal sewage treatment, advanced treatment and reuse of urban sewage and rural livestock and poultry sewage, and drinking water safety.

1 INTRODUCTION

Polyamide (PA) membrane composite membrane (TFC) in reverse osmosis (RO) membrane is widely used in the fields of drinking water purification, seawater desalination, and sewage treatment because of its advantages of stable structure, excellent separation performance, and easy operation (Ang et al. 2016; Ang 2016). However, traditional RO membranes show some limitations in the process of sewage treatment, among which membrane fouling is the main factor limiting its popularization and application. In the RO membrane sewage treatment system, membrane fouling is caused by the interaction of pollutants, bacteria, and membrane surface, and membrane surface characteristics are one of the key factors affecting the formation of membrane fouling. If the RO membrane can be endowed with bacteriostasis, it can inhibit the growth and reproduction of adherent bacteria and reduce stubborn biological dirt in the middle and late stage of membrane pollution, to realize the long-term stable operation of the membrane system and further improve the application and promotion of RO membrane (Lyu et al. 2016).

Photocatalysis membrane can combine the physical separation of membrane filtration with the organic degradation and antibacterial properties achieved by photocatalysis. It shows great application potential in energy-saving, water purification, and wastewater treatment. Different from the traditional physical and biological water treatment processes, a photocatalytic membrane can produce hydroxyl radicals. Hydroxyl radical is the strongest oxidant in an aqueous solution, which is second only to fluorine base. It can degrade toxic and refractory pollutants into simple and harmless inorganic molecules without secondary waste. The generation of hydroxyl radicals can be initiated by catalysts such as titanium dioxide, zinc oxide, and Fenton's reagent. Among them, TiO2 has the advantages of low cost, non-toxicity, high chemical stability, organic degradation, and antibacterial properties. It is the most commonly used material for the preparation of photocatalytic films. In addition, nano-tio2 has many excellent characteristics in membrane improvement. The polymer membrane improved by TiO2 is usually more hydrophilic than the unmodified polymer membrane, and the membrane flux and salt rejection rate are significantly improved, and it has excellent pollution resistance, decontamination, and bacteriostatic properties (Leong et al. 2014). Therefore, the application of nano TiO2 in RO membrane improvement will effectively increase

*Corresponding Author: ReZsy03@163.com

DOI 10.1201/9781003332657-69

the hydrophilicity, membrane flux, salt rejection rate, fouling resistance, and bacteriostasis of improved RO membrane, and create a new type of high-performance improved RO membrane.

In addition, the main modification methods of reverse osmosis membrane are chemical immersion and graft coating plasma modification. Among them, low-temperature plasma modification refers to the polymerization, crosslinking, grafting, and other reactions on the surface of the material under the conditions of plasma excitation, to change the surface properties of the material. It has the advantages of being pollution-free, only acting on the surface of the material, and has no influence on the material body, and has become a hot spot for polymer material modification. In addition, among many membrane improvement materials, 1-vinyl imidazole is a heterocyclic compound. In recent research, vinyl imidazole gradually appears in the field of medicine and membrane treatment. Vinyl imidazole forms a new improved membrane through graft coating, which shows many excellent characteristics, including high hydrophilicity, high salt rejection rate, and high bacteriostasis. It can maintain the performance of molecular sieve in high concentration salt or ethanol solution (Meléndez-Ortiz et al. 2016), and the domestic research on the improvement of vinyl imidazole on reverse osmosis membrane is still in the blank stage, which has great significance of research and application prospect (Meléndez-Ortiz et al. 2016).

Combined with many excellent characteristics of nano-TiO2, low-temperature plasma, and vinyl imidazole, it will be applied to the improvement of RO membrane, which will create conditions for the preparation of a new RO membrane with excellent separation, permeability, bacteriostasis, and pollution resistance.

2 MATERIALS AND METHODS

2.1 *Preparation of new improved reverse osmosis membrane*

2.1.1 *Modification of bottom film by titanium dioxide (TiO2) and vinyl imidazole (PVI)*

Different concentration gradients of PVI and TiO2 were set to improve the membrane, and the optimal concentration of PVI and TiO2 in the improvement of the reverse osmosis membrane was studied. The preparation process is shown in Figure 1. Different volumes of vinylidene imidazole and titanium dioxide were extracted with a pipette gun and transferred to a test tube with a capacity of 10 ml. Then, DI water was added to the scale line of 10 ml to prepare titanium dioxide vinylidene imidazole mixed solutions with different concentration gradients for later use. Take two acrylic plate frames, place the dry bottom film between the two plate frames, clamp, and fix it. Pour the prepared titanium dioxide vinylimidazole mixed solution on the membrane surface, after the solution is evenly distributed, start timing for 10 min, then pour the remaining liquid on the membrane surface, and dry the membrane at 30°C.

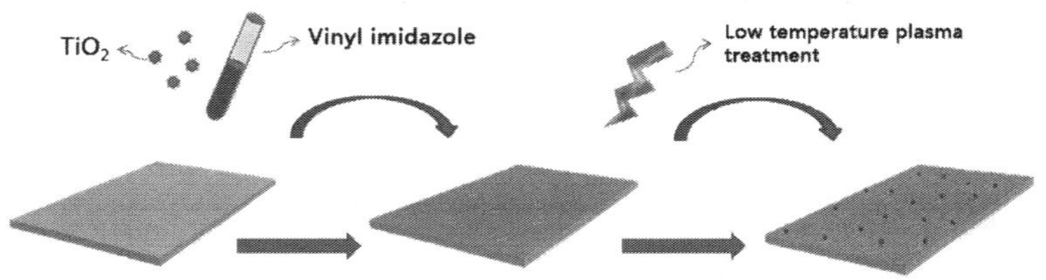

Figure 1. Process of preparing new improved RO membrane.

2.1.2 *Modification of base film by low-temperature plasma treatment (Plasma)*

The low-temperature plasma discharge equipment independently built by the water treatment and Membrane Technology Laboratory of Sichuan Agricultural University is used to modify the surface of the reverse osmosis membrane. Different plasma treatment times are set to study the best time for plasma treatment of reverse osmosis bottom membrane. The modification process is as follows: fix the treated improved film on the plate, open the computer control end, correct the moving zero point of the plasma generator, determine the treatment range, ensure that the film to be treated is in the working range, and set a series of low-temperature plasma treatment time; then turn on the low-temperature gas pump switch to deliver low-temperature gas to the plasma generator to ensure that the working temperature is in a low-temperature state; then, turn on the plasma generator and wait for its normal operation. Finally, run the control program to improve the reverse osmosis membrane through low-temperature plasma surface treatment.

2.2 *Performance test of new improved reverse osmosis membrane*

2.2.1 *Membrane hydrophilicity and surface charge*

The hydrophilicity of the membrane surface is characterized by the water contact angle. The smaller the contact angle is, the stronger the hydrophilicity is. The film contact angle is measured by a contact angle tester combined with image software. Before measurement, the sample was dried at room temperature for 6 hours, and ionic water was dropped from the microneedle with a stainless-steel needle on the smooth and flat film surface. To reduce the error, a reliable contact angle value is averagely obtained by at least 5 measurements on different film surfaces. Membrane surface charge is one of the important factors affecting membrane fouling behavior. In this experiment, surpass solid surface zeta potentiometer will be used to characterize membrane surface charge.

2.2.2 *Membrane flux detection*

Membrane permeation flux is mainly expressed by membrane water-specific flux and salt rejection rate. Membrane clear water-specific flux J refers to the amount of clear water passing through the unit membrane area under unit time and unit filtration pressure difference, which is an important indicator to characterize membrane permeability. Its calculation formula is as follows:

$$J' = \frac{V}{StP} \tag{1}$$

Where:
V—Permeate volume;
S—Effective area of membrane;
t—Filter time;
P—Filter pressure difference.

2.2.3 *Salt rejection rate Detection*

Retention rate R (%) is the ability of the membrane to prevent a component in the feed liquid from passing through or retaining a component, and its calculation formula is as follows:

$$R(\%) = \left(1 - \frac{C_p}{C_f}\right) \tag{2}$$

Where:
C_p —The concentration of each substance in the permeate;
C_f —The concentration of each substance of feed.

2.2.4 *Pollution resistance test*

The fouling resistance of the membrane can be measured by monitoring the interaction force between fouling and membrane surface and membrane fouling itself by atomic force microscope, and can also be calculated by monitoring the flux in the process of membrane filtration.

The fouling resistance of the membrane was evaluated by calculating flux recovery rate (FRR), total fouling rate (R), reversible fouling rate (R), and irreversible fouling rate (RIR).

$$FRR = \frac{J_{w1}}{J_{w2}} \times 100 \tag{3}$$

$$R_t = \left(1 - \frac{J_p}{J_{w1}}\right) \times 100\% \tag{4}$$

$$R_r = \left(\frac{J_{w2} - J_p}{J_{w1}}\right) \times 100\% \tag{5}$$

$$R_{ir} = \left(\frac{J_{w1} - J_{w2}}{J_{w1}}\right) \times 100\% \tag{6}$$

Where:
J_{w1} —Pure water flux, unit $L{\cdot}m^{-2}h^{-1}$;
J_{w2} —Water flux of cleaned membrane, unit $L{\cdot}m^{-2}h^{-1}$;
J_p —Flux after fouling, unit $L{\cdot}m^{-2}h^{-1}$.

In addition, a confocal microscope and electron microscope can be used to observe and compare the thickness of the fouling layer on the modified film and the original film at the same time, to visually compare and analyze the fouling resistance of the two.

3 DISCUSSION AND APPLICATION

As a hot research direction in the field of membrane improvement, the use of reverse osmosis membranes occupies an important position in the field of water treatment, attracting many researchers to explore and study.

In terms of research status and development trends of titanium dioxide, Rahinpour et al. found that the shorter the soaking time in TiO2 suspension, the better the fluidity of the membrane. However, the longer the soaking time, the more TiO2 nanoparticles coated on the membrane surface, which will block the membrane pores and reduce the fluidity (Rahimpour et al. 2008). You et al. used the theory that treated the commercial PVDF membrane with plasma and then purified the acrylic acid aqueous solution by immersing in different concentrations of nitrogen for graft polymerization to increase the hydrophilicity and obtained the modification of the polymer membrane. They tried to introduce more COOH functional groups, to increase the hydrophilicity of the membrane surface and deposit TiO2 nanoparticles. Hangzhou water treatment technology research and development center used TMC to modify the surface of hydrophilic nano-TiO2, and then added it to the polyamide layer of the composite reverse osmosis membrane to prepare the modified nano-TiO2 polyamide composite reverse osmosis membrane (You et al. 2012). The results showed that the surface of modified TiO2 was grafted with acyl chloride groups, and the dispersion in organic solvents was improved; SEM and AFM photos confirmed that TiO2 was evenly distributed on the membrane surface and the membrane surface roughness increased; the hydrophilicity of hybrid composite membrane was also improved to a certain extent; the results of membrane performance test confirmed that the water flux of composite membrane with TiO2 was higher than that of pure polyamide membrane, and the change of desalination rate was very small. When the addition amount of modified TiO2 was 0.05% (M/V), the water flux increased from 11.21 L/(M2·h) to 32.61 L/(M2·h), and the retention rate of NaCl reached 98.9%. The results show that the dispersion rate of polyamide membrane is improved and the desalting rate of polyamide membrane is improved. The above research shows that in the improvement experiment, the size of titanium dioxide, the concentration of titanium dioxide solution, and the interaction with other materials will affect the fluidity, permeability, hydrophilicity, bacteriostatic and pollution resistance, salt rejection rate, water flux and other properties of the membrane.

Research status and development of vinylidene imidazole. H. Ivan Melendez-Ortiz et al modified medical PVC with N-vinylimidazole to obtain bactericidal surfaces. Elizabeth Vazquez et al. prepared hydrophilic polysulfone membrane by γ irradiation grafting n-vinylimidazole, and obtained that the grafting process of VIM monomer could improve the hydrophilic property of the membrane, and the graft and irradiation did not affect the heat resistance and heat resistance of PSU membrane. Davari et al. also investigated grafting vinylimidazole and beet base groups onto polyamide membranes by γ -ray irradiation. As a result, grafting agents enhance resistance to contamination and resist nonspecific protein adsorption at neutral and alkaline pH, including inhibition of bacterial growth. The above research shows that vinylimidazole has good advantages in membrane improvement, which can greatly improve the bacteriostasis, stain resistance, durability and hydrophilicity of the membrane.

Research and achievements in low-temperature plasmas. Kim et al. investigated the influence of low-temperature plasma-modified polypropylene (PP) and polysulfone base membrane on the preparation of reverse osmosis composite membrane by interfacial polymerization, and found that plasma modification significantly improved the hydrophilicity of PP membrane, enabling PP membrane to be used for the preparation of reverse osmosis composite membrane. Kim et al. studied the influence of plasma-modified polyvinylidene fluoride (PVDF) membrane on the preparation of polyamide composite membrane, and found that the rejection rate of polyamide composite membrane prepared by PVDF membrane modified by oxygen and methane 1:1 mixed plasma glow discharge was significantly improved. Yang Dan et al. from Tianjin Polytechnic University solved the interface bonding problem of enhanced PVDF hollow fiber membrane by using low-temperature plasma to pretreat the surface of a knitted tube with PET as reinforcement and PVDF as casting liquid, and prepared the enhanced hollow fiber membrane by immersion-precipitation phase transformation method. The effect of plasma treatment time on the performance of the reinforced membrane was studied by measuring the pure water flux, porosity, average pore size, and mechanical properties of the membrane. The results show that plasma treatment has little effect on the pure water flux, porosity, and average pore size of the membrane, and has an obvious effect on improving the bond strength of the membrane interface. These studies indicate that low-temperature plasma has good properties in membrane improvement and can greatly improve the performance of RO membrane.

4 CONCLUSION

Facing the increasing requirements of the application environment, further research and application of new reverse osmosis membrane with high desalination rate, high throughput, low energy consumption, and pollution resistance has become an important research topic. More and more researchers have proposed the introduction of inorganic nanomaterials into organic polymer membranes. Organic-inorganic hybrid reverse osmosis membranes have become a new direction for future development. In this paper, vinyl imidazole was used as an organic phase solution and TiO2 inorganic phase solution to coat the polyamide (PA) film composite membrane, and the treated reverse osmosis membrane was treated by low-temperature plasma, It will greatly improve the comprehensive characterization of reverse osmosis and improve the characteristics of the composite membrane (including its permeability flux, salt rejection rate, hydrophilicity, electricity, roughness, pollution resistance, chlorine resistance, porosity, bacteriostasis, and durability). Find out the optimal proportion scheme of titanium dioxide concentration, vinyl imidazole concentration, and low-temperature plasma treatment time for preparing a high-performance reverse osmosis membrane, and apply it to seawater and bitter alkaline water desalination, heavy metal sewage treatment, advanced treatment, and reuse irrigation of urban sewage and Rural Livestock and poultry sewage. As the improved reverse osmosis membrane has the advantages of low energy consumption, high efficiency, and green environmental protection, to a certain extent, it can improve the current situation of water shortage and water pollution in China, realize the efficient transformation, clean utilization and intensive processing of water resources and other resources, effectively improve the utilization efficiency of China's resources and support the construction of ecological civilization.

REFERENCES

Ang W L, Mohammad A W, Benamor A, et al. Hybrid coagulation–NF membrane processes for brackish water treatment: Effect of pH and salt/calcium concentration[J]. *Desalination*, 2016, 390:25–32.

Ang, W.L., et al., Hybrid coagulation–NF membrane process for brackish water treatment: Effect of antiscalant on water characteristics and membrane fouling. *Desalination*, 2016. 393: p. 144–150.

Iván Meléndez-Ortiz, H. Carmen Alvarez-Lorenzo, Angel Concheiro, Víctor M. Jiménez-Páez,Emilio Bucio. Modification of medical grade PVC with N-vinylimidazole to obtain bactericidal surface[J]. *Radiation Physics and Chemistry*, 2016, 119.

Lyu, S.D., et al., Wastewater reclamation and reuse in China: Opportunities and challenges. *Journal of Environmental Sciences*, 2016. 39: p. 86–96.

Rahimpour, A., Madaeni, S.S., Taheri, A.H., Mansourpanah, Y. Coupling TiO2 nanoparticles with UV irradiation for modification of polyethersulfone ultra-filtration membranes, *J. Membr. Sci.* 313 (2008) 158–169.

Sookwan Leong, Amir Razmjou, Kun Wang, Karen Hapgood, Xiwang Zhang, Huanting Wang. TiO_2 based photocatalytic membranes: A review[J]. *Journal of Membrane Science*, 2014, 472.

You, S.-J., Semblante, G.U., Lu, S.-C., Damodar, R.A., Wei, T.-C. Evaluation of the antifouling and photocatalytic properties of poly(vinylidene fluoride) plasma-grafted poly(acrylic acid) membrane with self-assembled TiO_2, *J. Hazard.Mater.* 237–238 (2012) 10–19.

Energy Revolution and Chemical Research – Kok-Keong Chong and Zhongliang Liu (Eds)
© 2023 The Authors, ISBN: 978-1-032-36554-1

Comprehensive utilization of red mud from the perspective of circular economy

Han Lei
Wuhan University of Technology, Wuhan, China

ABSTRACT: As the solid waste in alumina production, the rate of red mud resource comprehensive utilization is low. With the deepening of the concept of a sustainable and low-carbon circular economic development system, the comprehensive use of red mud has become a research hotspot. By exploring the comprehensive use of red mud under the concept of circular economy, this article exposes the importance of innovating the path of red mud recycling for alumina production enterprises and related industries. This paper analyzes the current situation of the red mud industry, takes the problems as the starting point, and puts forward reasonable suggestions for the optimization of the way of comprehensive utilization of red mud under the concept of the circular economy. Through research, this paper finds that aluminum enterprises need to continuously enrich the use of red mud recycling, turn waste into treasure, and take the road of sustainable development based on introducing high and new technology and strengthening multi-party cooperation.

1 INTRODUCTION

Red mud is the bauxite residue after alumina leaching with strong alkalis. As it is well known, red mud is the largest solid waste discharged by the aluminum industry, and it is also one of the main areas of comprehensive resource use. By the end of 2020, the cumulative global red mud stock exceeded 5 billion tons. Currently, with the annual increase rate of 200 million tons, the environmental problems caused by the red mud stock are seriously threatening the sustainable development of the aluminum industry and are a common challenge faced by the global aluminum industries (Zhu 2022).

Since the second national environmental protection conference, China has actively defended the basic national environmental protection policy, carried out large-scale treatment of industrial pollutants, such as red mud and coal gangue, vigorously promoted energy efficiency, accelerated economic development, solidly promoted cleaner production, significantly reduced pollution emissions, and strengthened the comprehensive use of resources Measures such as continuous promotion of circular development promoted the global construction of "five in one" and achieved remarkable results in general, but the pollution of red mud and coal denim has not been significantly improved. According to statistics, currently, the use of red mud in China is less than 7%. The guidance opinions on the comprehensive use of red mud stipulate that support policies for the comprehensive use of red mud will be established and improved, central government financial support will be strengthened and a series of demonstration projects for the comprehensive application of red mud and promotion demonstration projects will be selected to give special financial support to the central government for cleaner production. At the same time, the full use of red mud will also be included in the support focus of the national special fund for technological transformation.

Through the research on the current situation of circular economy and red mud recycling, this paper brings the new concept of the circular economy into solid waste—the red mud recycling system. Under the background that the problem of red mud treatment continues to intensify and needs to be solved urgently, based on the technology research and development of existing enterprises and

DOI 10.1201/9781003332657-70

universities for the application of red mud in cement manufacturing, building bricks, composites, environmental protection materials, thermal insulation, and refractory materials, subgrade, and so on, this paper looks forward to a new way of green development of alumina enterprises.

2 CIRCULAR ECONOMY AND ITS ECONOMIC SYSTEM

"Circular economy," also known as "ecological economy" or "green economy," was first presented by economist Pierce in the blue book on green economy published in 1989; with the penetration and influence of ecological economics in the industrial economy, people also call the comprehensive development of industrial economy and ecological economy green industrial economy or collectively referred to as green economy, and believe that the green industrial economy is a new industrial economy with sustainable development (Yue 2009).

To guide and promote the accelerated development of the circular economy and achieve the objective of increasing the resource production rate by 15% proposed in the draft 12th Five-Year Plan, the state formulated the development strategy and the recent circular economy action plan in 2013 to elaborate a strategic plan for the development of the circular economy and take specific measures for future work.

The action plan sets out requirements for the comprehensive development and utilization of ores and associated tailings in the non-ferrous metal industry, energy conservation and consumption reduction, resource utilization of foundry waste, waste gases, liquid waste, and waste heat, the recycling of non-ferrous metal waste, and the construction of a circular economy industrial chain in the non-ferrous metal industry. All industrial companies and local governments constantly formulate circular economy development plans and implement circular economy policies according to their conditions.

Research on the circulation of the value of circular economy resources is mainly based on the basic logic of "circular economy is an economy that creates more value through the circulation of materials." The extent of engineering, management, economics, environmental science, and other disciplines of the circular economy determines the characteristics of cross-integration of research. At the same time, it integrates the basic theory of environmental accounting to further enrich the theoretical attributes, improve and form the fit of the data structure, systematically reveal the changing rules of circular economy resource value circulation, and build a system that is generally applicable to enterprises. The management standard and circular economy value transfer application guide system between enterprises and even society as a whole (Zhu 2021).

3 CURRENT SITUATION OF THE RED MUD CYCLIC UTILIZATION

3.1 *The rate of use of red mud is low and it is difficult for companies to deal with it*

Currently, most production plants in China produce 0.8 t~1.5 t red mud per 1 t alumina. Due to the huge amount of red mud produced, companies are not willing to spend a lot of money on treatment. In addition, due to the lack of corresponding high and new technology, companies choose to stack more and build a red mud warehouse to solve the problem of red mud. Therefore, environmental pollution problems caused by red mud leakage also occur frequently.

In recent years, the capacity scale of China's alumina industry expanded. In 2018, China's alumina capacity exceeded 80 million tons and production reached 72.5 million tons. The problem of red mud disposal caused by the expansion of the scale of the alumina industry has become increasingly prominent. According to incomplete statistics, the cumulative stock of red mud in China currently exceeds 1.1 billion tons. Due to the lack of economical and viable technology for comprehensive utilization of red mud resources, a large amount of red mud is stacked for a long time, which not only occupies a large amount of land, but also has the risks of red mud reservoir dam rupture, soil, and water pollution.

In September 2014, there was also a leak in the red mud reservoir of the Henan Branch of Chinalco, followed by a local rupture of the dam. A large amount of sewage from the red mud reservoir flows into nearby villages and farmland, and even approaches the Yellow River several kilometers away. In 2018, the red mud reservoir of Guangxi Xinfa Aluminum Power Co., Ltd. leaked many times, and its downstream villages were seriously affected. The pH value of the nearby land was alkaline, the cadmium content exceeded the standard, and the soil pollution risk screening value of super agricultural land (0.8 mg/kg) was 8.9 times, seriously polluting the environment. In May 2019, the filtrate leak from the alumina plant red mud dam occurred in Xiaoyi, Jiaokou, and other areas in Shanxi, resulting in an excessive concentration of sodium ions outside the reservoir, piping at any time, dam instability, and other safety issues, which seriously threatened the personal safety of nearby residents.

Currently, due to the lack of effective methods of treatment of large-scale industrialization at home and abroad, even in the past, most developed countries have been discharged into the sea, which has become a global problem of environmental protection. China has not yet developed a suitable technical and economic industrial path for the comprehensive large-scale use of red mud. In 2018, the total amount of comprehensive utilization of red mud in China was about 4.5 million tons, and the comprehensive utilization rate was about 4.29%, which was much lower than the average level of comprehensive utilization of bulk industrial solid waste in China. Figure 1 shows the tendency of the rate of red mud cyclic utilization in recent years. Although the rate is increasing since 2016–2018, the level is low.

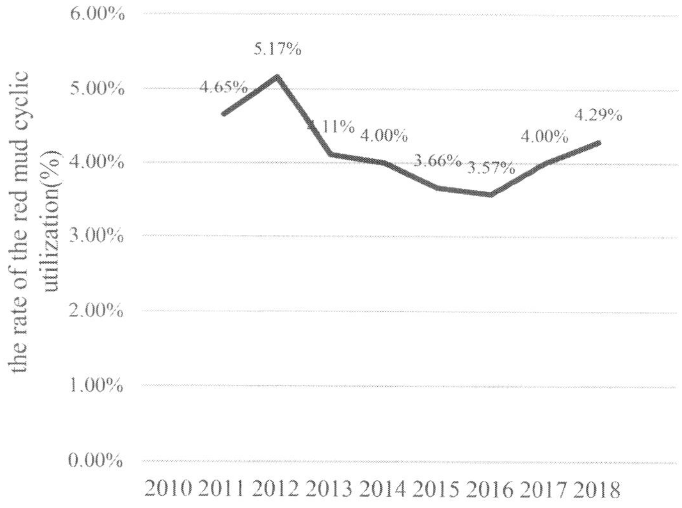

Figure 1. The rate of the red mud cyclic utilization in China in 2010–2018.

3.2 *Serious pollution caused by the accumulation of red mud*

Due to the production characteristics of the alumina industry, the pH value of the leaching solution of its discharge red mud is high, usually 12–14, which is much higher than the pH discharge standard for discharged liquid waste specified in the gb5058-85 standard for solid waste pollution control in the non-ferrous metal industry. If these strong alkaline solutions penetrate the soil, they will cause soil salinization and damage the environment. The content of associated metal elements in red mud increases with increased discharge of red mud, and is enriched in the storage of red mud, damaging the soil and polluting groundwater sources. In addition, there are even traces of radioactive elements (uranium, thorium, radio, etc.) in the red mud. Stacking for a long time will lead to the enrichment of radioactive elements and threaten the health of the surrounding residents.

Red mud particles have good adsorption on heavy metal ions, such as Cu^{2+}, Pb^{2+}, Zn^{2+}, Ni^{2+}, Cr^{6+} and Cd^{2+} in the soil. However, in the application of red mud, because the red mud itself contains bleach, some red mud also contains radioactive elements, which are directly harmful to human health and pollute the water.

3.3 *Increasingly rigorous control of solid waste*

Therefore, relevant state departments gove great importance to the environmental management of industrial solid waste, such as red mud, implement the law of the People's Republic of China on the prevention and control of environmental pollution by solid waste and other legal provisions, and actively promote the reduction, recycling, and harmless industrial solid waste, such as red mud, and vigorously realize the comprehensive use of red mud. Severe measures such as shutdown will be taken for subjective malicious leaks, falsification of monitoring data, repeated punishments and repeated offenders, and excessive sewage discharge during a severe pollution weather warning. In 2019, some alumina factories in Shanxi faced a forced shutdown, involving an annual production capacity of about 6.6 million tons. Therefore, strengthening the comprehensive use of red mud is an important measure to alleviate environmental pollution and potential safety risks caused by long-term storage of red mud and improve resource utilization efficiency. It is of great importance for the economy and resource-intensive use and promotes the green and high-quality development of the industry.

4 COMPREHENSIVE DEVELOPMENT STATUS OF RED MUD CYCLIC UTILIZATION UNDER CIRCULAR ECONOMY

Currently, the ways of using red mud resources include: the recovery of metallic elements such as iron from red mud, production of building materials such as bricks, tiles, cement, and ceramics, using red mud as the catalyst for chemical reaction, production of cement, fly ash brick, and red mud glazed brick with dealkalized red mud, using red mud as mine filler, and the application of red mud in plastic industry. Significant progress has been made in these areas. However, due to small amount of red mud and the problems of cost and technology, there is no effective method of large-scale industrial treatment at home and abroad. Recycling red mud in the market is mainly to treat it innocuously as filling material. Shandong Haiyi Transportation Technology Co., Ltd. innovatively dissolves red mud pollutants as subgrade fill application material. Shandong Weiqiao Aluminum Power Co., Ltd. realized the application of Bayer process red mud in the actual construction of highway engineering, and was applied to the reconstruction and expansion project of Jiqing expressway as a science and technology demonstration project of the Ministry of Communications. The application of red mud in Jiqing expressway reconstruction and expansion project is the first time in the world to apply red mud to expressway construction, which makes a new exploration of the technology of large-scale application of red mud and brings the comprehensive use of red mud to a new stage.

In terms of university research, many universities take the harmless treatment of red mud as the key field of their research. The research direction of Shandong University is the test of high content red mud-based subgrade filling material. Taiyuan University of Technology takes red mud as a subgrade material to study its dynamic performance. Wuhan University of Technology depends on its strong resistance to materials discipline and State Key Laboratory of Silicate Building Materials, the only national key laboratory in colleges and universities in the field of building materials, Taking the harmless integration of red mud as its research direction, it was carried out for the medium-term experimental phase. In addition, according to the chemical and mineral composition of red mud, some scholars may apply red mud to cement (He, Xin, Huang, Duan) to prepare slag cement, Portland cement, and sulfoaluminate cement. However, the chemical composition of red mud is unstable due to different sources of aluminum ore, which increases the difficulty of preparing raw cement flour from red mud. The content of iron oxide and alkali metals in the red mud is high, which is easy to lead to the melting of the prepared cement raw flour and increase the risk of ring formation and even blockage of the rotary furnace. Excessive alkali metal content in raw flour will lead to scaling of the preheater, which is easily leads to alkaline flashing of the hardened cement body and reduce the utilization rate of red mud in cement preparation. Du Yunpeng (Du, Tong, Xie, Zhang, Song, Fan, Cao) and other scholars have creatively proposed that valuable metals can be recovered from red mud through pyrometallugy, hydrometallurgy, alkaline melting process, bioleaching, physical method, reduction magnetic separation, and other methods to reduce the stacking of red mud and realize the comprehensive use of red mud.

5 COMPREHENSIVE PATHS OF USE OF RED MUD BASED ON CIRCULAR ECONOMY CONCEPT

5.1 *Strengthening cooperation between businesses and universities and relevant scientific research institutes*

Restricted by technology and production cost, most alumina production enterprises on the market do not have corresponding key technologies for the comprehensive application of red mud. Red sludge, such as alumina leaching residue from bauxite by strong alkalis, cannot be recycled due to a lack of technology, which will further aggravate resource waste and environmental pollution. Under the concepts of "sustainable development" and "green development," many national universities and research institutes have included red mud recycling in the key scope of research. Relevant alumina production and filling material production enterprises can strengthen joint research with universities and scientific research institutes, improve the comprehensive efficiency of using red mud, and reduce costs by introducing and utilizing high technology and relevant equipment. In the comprehensive use of red mud, companies can adopt corresponding processes to recover valuable metals, such as iron and gallium, in the alumina production process. Strengthen research and technological development and use red mud as raw material in the manufacture of cement, building bricks, composite materials, environmental protection materials, thermal insulation, and refractory materials, subgrade, etc.

5.2 *Valuing the main demonstration effect of leading companies*

Strict corporate responsibility continues to encourage enterprises to give full role to green transformation and promote high-quality enterprise development and formulate and implement green development strategies based on their reality. For example, promote core enterprises, such as Chinalco Group, strive to strengthen and expand energy-saving and environmental protection industries in Shanxi Province, improve the comprehensive utilization level of red mud, and improve the ability to supply green products and services.

5.3 *Extension of sustainable red mud cyclic utilization chain*

Alumina production companies need to make scientific and rational use of resources under the concept of circular economy and build an eco-friendly industrial cluster through an effective combination between industries. In the alumina production process, resource recycling is taken as the development direction to further extend the ecological chain of the papermaking industry. The alumina industry adheres to the goal of sustainable development. By optimizing production technology, production equipment, and production concept, it adopts the development way of reducing resource waste, improving pollutant emission standards, harmless pollutant treatment, and no environmental pollution, to turn the industry into an eco-friendly industry, as much as possible and make the necessary contributions to the construction of an environmentally friendly society and resource economy. Through the extension of upstream and downstream industries, red mud produced after alumina production can be rationalized to avoid environmental pollution and resource waste caused by accumulation.

6 CONCLUSION

The characteristics of the alumina industry determine that its development needs to consume certain energy and produce solid waste. As the most discharged solid residue in the aluminum industry, the issues with red mud need to be addressed urgently. The main task to solve the problem of red mud is to introduce high and new technology and equipment, reduce environmental pollution, and improve the comprehensive utilization level. If aluminum companies want to pursue development, they need to take the concept of circular economy as a guideline and take the road of sustainable development to ensure sustainable and stable development in industrial competition.

ACKNOWLEDGMENT

This paper was funded by the National Entrepreneurship Training Program of China.

REFERENCES

Du, Y., Tong, X., Xie, X., Zhang, W., Song, Q., Fan, P., Cao, Y. Present situation and Prospect of comprehensive utilization of rare earth scandium and other valuable metal resources in red mud. *Journal of the Chinese Society of Rare Earths. J.CSSCI.* in press.

He, Y., Xin, X., Huang, Y., Duan, G. Application and research progress of industrial solid waste red mud in cement preparation. *China Powder Science and Technology. J. CSCD.* in press.

Yue, Y. Discussion on product development based on green industrial economy. *Ecological Economy. J. CSSCI.* vol. 6, pp. 151–153, June 2009.

Zhu, P. Mechanism and analysis framework of value circulation of circular economy in cement enterprises. *Journal of Jishou University (Social Sciences). J. CSSCI.* vol. 3, pp. 112–119, March 2021.

Zhu, Y. Create a new situation of comprehensive utilization of red mud—the exchange and promotion meeting of comprehensive utilization of red mud will be held in 2021. *China Nonferrous Metals. J.* vol. 2, pp. 52–53, February 2022.

Energy Revolution and Chemical Research – Kok-Keong Chong and Zhongliang Liu (Eds)
 ISBN: 978-1-032-36554-1

Research status and prospect of demulsification technology

Liguang Xiao
Fushun Mineral Group Limited Liability Company Engineering Technology R&D Center, Fushun, China
Institute of Water Pollution Control in Liaoning River Institute Basin, Shenyang Jianzhu University, Shenyang, China

Maiyu Li, Yanping Zhang, Di Luo & Jinxiang Fu
Institute of Water Pollution Control in Liaoning River Institute Basin, Shenyang Jianzhu University, Shenyang, China

ABSTRACT: The current demulsification with good treatment effect and certain development prospects are described: physical demulsification (ultrasonic demulsification, microwave demulsification, electric demulsification, membrane demulsification), biological demulsification, and chemical demulsification. It summarizes the demulsification mechanism, influencing factors, application status, and advantages and disadvantages, makes a brief comparison of various technologies, and looks forward to the future development direction of demulsification.

1 INTRODUCTION

In the oil recovery process, the oil phase and the water phase are strongly stirred to form an emulsion, which is mainly divided into two types: one is O/W type, where oil is the dispersed phase, and water is the continuous phase of the oil-in-water emulsion; the other is W/O type, where water is the dispersed phase and oil is the continuous phase of an oil-in-water emulsion. However, due to excessive oil recovery, resulting in increased difficulty in oil recovery, primary oil recovery and secondary oil recovery technology cannot meet the use of demand, the popularity of tertiary oil recovery technology, especially the application of ternary composite drive (ASP) technology is more common, but it also leads to the increase of polymer and surfactant content in the recovery fluid, making the recovery fluid emulsion serious, multiple emulsions, such as oil-in-water-in-oil type, water-in-oil-in-water type, making it very difficult to separate oil and water. However, regardless of the type of emulsion, the demulsification techniques can be divided into physical, chemical, and biological demulsification, each with its applicable conditions. And with the complexity of the extracted fluid composition, the more typical technologies, such as sedimentation separation method, centrifugal method, flotation method, and treatment effect often fails to meet the requirements of the discharge or reinjection, research on more efficient emulsion demulsification technology is imminent. Therefore, we summarize the relatively new and promising emulsion-breaking technologies in recent years, which can be used as a reference for the development of emulsion demulsification technologies.

2 PHYSICAL DEMULSIFICATION

The existing physical demulsification techniques can be divided into two types (Gu 2015): first, the demulsification device applies field energy to the emulsion, including ultrasonic demulsification, microwave demulsification, electric demulsification, etc; second, the demulsification device

DOI 10.1201/9781003332657-71

applies external force to the emulsion, including centrifugal demulsification, gravity demulsification, membrane demulsification, etc. At present, ultrasonic, microwave, electric, and membrane demulsification are more effective in treating complex emulsions.

2.1 *Ultrasonic demulsification*

Ultrasonic action on fluids with different properties will produce displacement effects, water (oil) particles with the change of waveform, continuous accumulation, collision, and the formation of large droplets, under the action of gravity to achieve oil-water separation.

The influencing factors of ultrasonic demulsification are sound intensity, frequency, and radiation time, etc. Sun et al. (Sun 1999) treated W/O emulsion by ultrasound, and the results showed that sound intensity was the main controlling factor of ultrasonic demulsification, and the dewatering rate reached 98% at a frequency of 20 kHz, radiation time of 20 min, and the sound intensity of $0.31W/cm^2$. The dehydration rate increases with the increase of sound intensity and then decreases rapidly. There exists an optimal sound intensity to maximize the dewatering rate, and when this intensity is exceeded, ultrasonic transient cavitation becomes apparent, and the collapsed bubbles generate high-speed jets that crush the dispersed phase droplets, leading to secondary emulsification and a rapid decrease in the dewatering rate. In addition, Ali et al (2018) found that ultrasonic treatment of Iranian oil fields did not achieve effective oil-water separation without the addition of any demulsifier. Therefore, the ultrasonic process is not a complete substitute for chemical demulsification, and the effective combination of ultrasonic demulsification and chemical demulsification can improve demulsification efficiency and reduce the number of chemicals and energy consumption of ultrasonic equipment. For example, Xu et al (2019) used chemical, ultrasonic, and ultrasonic-chemical methods to treat SAGD water-bearing crude oil, respectively, at 60°C, 30 min of treatment time, and under the same ultrasonic conditions, the ultrasonic-chemical method resulted in 30% and 24% more dewatering than the chemical and ultrasonic methods respectively, demonstrating the synergistic effect of ultrasound and demulsifier.

Because ultrasound itself has good conductivity, so it can deal with many types of emulsions, but the amount of water that can be treated by the separate ultrasound method in engineering is small, the ultrasound generation equipment is complicated, and the sound intensity is not easy to control, and it is easy to produce secondary emulsification. Therefore, the development of ultrasonic demulsification should focus on the development of simple generation equipment, so that it can reduce energy consumption and drug consumption while maintaining processing efficiency and enhancing processing capacity.

2.2 *Microwave demulsification*

The current microwave demulsification mechanism is divided into two types, one is that the high-frequency magnetic field formed by microwaves can make the polar water molecules in the emulsion rotate at high speed, thus destroying the zeta potential of the oil-water interfacial film, the two phases of oil and water lose electrical equilibrium, water droplets (oil droplets) collide with each other to form large droplets, making the oil-water separation. Second, the thermal effect of microwaves makes the temperature gradient between the continuous phase to the dispersed phase, which can reduce the viscosity and interfacial tension, increasing the frequency of collision between droplets and reducing the resistance to agglomeration, thus promoting oil-water separation (Binner 2014).

Water content is one of the important factors affecting the effect of microwave emulsion breaking, Jiang et al (2005), through the treatment of crude oil from Oxheart Tuo in the Liaohe oilfield, compared the effect of microwave treatment of emulsions with different water content found that emulsions with high water content can reach stability after 10 min of microwave treatment and 30 min of standing, but the dehydration rate is less than 30% when the water content is below 40%. In addition, Igor N et al (2014) concluded from microwave demulsification experiments that the optimal water content of emulsions should be controlled at 18% to 56% to maximize cost-effectiveness.

The microwave method can also be used in combination with chemical methods to achieve better treatment results. Compared to conventional chemical demulsification, the microwave chemical method of demulsifier with less dosage and shorter heating time, especially for heavy oil-in-water emulsions, requires less demulsifier than conventional heating, and the separated water is clearer. For example, Yang (2006) treated emulsified crude oil by microwave chemical method, adding 50 ppm demulsifier, and the dewatering rate reached 94.9% when microwave radiation was used for 5 min. The water quality after microwave chemical dewatering was much better than that of the thermal chemical method.

Microwave demulsification technology is time-saving, efficient, and has a good effect on the treatment of difficult thick oil and compound drive oil, but there are many models of microwave ovens used in research, and the operating parameters vary greatly, making it difficult to control accurately, and the mechanism of action is not yet uniform. Therefore, the development of microwave demulsification should focus on the study of the mechanism of demulsification, followed by the development of a more stable microwave generator, so that the project can accurately control the parameters and reduce the energy consumption of the generator.

2.3 *Electric demulsification*

Electric demulsification is mainly used to treat W/O emulsions, where a high electric field is applied to the emulsion and the polarized droplets approach each other under the electric field, causing the interfacial film to break and agglomerate into large water droplets to achieve oil-water separation (Guo 2018). When the electric field strength reaches a certain threshold, the phenomenon of "electric dispersion" will occur and the droplets will break up to form smaller droplets, causing secondary emulsification and reducing the demulsification effect. Therefore, there is a critical electric field strength in the DC or AC demulsification process, beyond which electrical dispersion will occur. However, under the pulsed electric action, a suitable pulse width ratio is used and the electric field action time is short in one cycle, even though the electric field strength has far exceeded the critical field strength, the action time is short and the droplets cannot obtain enough energy for the electric dispersion phenomenon (Gong 2015).

Electric demulsification has high efficiency and relatively mature technology, but when the water content is too high, the electric field is unstable and likely to cause short circuits. Therefore, in practice, electric demulsification is more often used in the final stage of crude oil dewatering, for deep dewatering. Among the various types of electric fields, AC electric field is suitable for fast dewatering and DC electric field is suitable for deep dewatering, but both deal with W/O type emulsions with low water content, while pulsed electric field can also deal with O/W type emulsions at the same time. Akinori et al (2016) investigated the demulsification effect of different alternating electric fields on W/O emulsions, and the experimental results indicated that square waves could be more effective for oil droplet agglomeration, with a constant demulsification rate of 80% for square waves at frequencies greater than 8 Hz, compared to only 40% for sine waves. Boping et al (2018) investigated the kinematic aggregation behavior of oil droplets in O/W emulsions under the action of a bidirectional pulsed electric field (BPEF), and the results showed that the aggregation state of oil droplets was optimal when the BPEF voltage was 750 V, the frequency was 50 Hz, and duty cycle was 70%, and oil droplet chains could be formed at 10 s and large oil droplets were formed at 19 min.

With the complexity of the emulsion, the use of pulsed electric field demulsification will be an important direction for the development of electric demulsification, but the current electric demulsification experiments are mostly small static experiments, while industrialization will be faced with the dynamic situation of large flow rate, electrode corrosion off phenomenon will be more obvious, the long-term high-pressure operation at the site will also make the emulsion breaker safety factor decreased. Therefore, while emphasizing the development of pulsed electric demulsification, we should focus on the research and development of safety materials.

2.4 *Membrane demulsification*

Membrane demulsification technology uses the hydrophilic or lipophilic properties of membranes to trap emulsions, which pass through the membrane pores and rupture due to the uneven force on the droplet surface, and water or oil droplets are adsorbed on the membrane surface for aggregation, thus achieving the effect of oil-water separation. However, due to the limitation of membrane material development, the progress of membrane demulsification technology is relatively slow (Sun 2000).

Due to the variety of membrane materials, the conditions of application vary. Kocherginsky (2003) used flat nitrocellulose membranes to treat W/O emulsions to investigate the effects of transmembrane pressure, pore size, and thickness on the demulsification effect. Studies have shown that the filtration rate of emulsions is proportional to the transmembrane pressure, and the larger the pore size, the faster the rate increases and the less the effect of pressure, while the membrane thickness has almost no effect on the final water content. Xu et al (2020) proposed a chemical demulsification combined with membrane separation technique using an alcoholysis reaction to graft a hyperbranched phenamine resin block polyether demulsifier (AE2311) onto the surface of styrene-maleic anhydride (SMA) blended polyvinylidene fluoride (PVDF) membranes. Modified PVDF membranes with a grafting time of 9h have good underwater resistance to contamination and can be reused with long-term operational stability.

Membrane demulsification has the advantages of high efficiency, low energy consumption, and universality, but the membrane material with better performance is generally more expensive, and the membrane needs to be cleaned in time after a period of time, and the membrane flux and breaking rate will be reduced after cleaning, so the development of membrane demulsification should focus more on low cost and reusability.

3 BIOLOGICAL DEMULSIFICATION

Biological demulsification is the addition of biological demulsifier, i.e., biological products, and their fermentation cultures made from natural microbial organisms through biochemical processes such as screening, domestication and fermentation, thus dewatering the emulsion (Liu 2010). And Lu (2007) pointed out that microbial cells are also demulsification, and the cell surface has certain hydrophilic/hydrophobic properties, which can promote the coalescence of discontinuous phases to achieve the demulsification. For example, Cai et al (2019) reported a Halomonas sp. N3-2A, which could achieve 92.5% emulsion breakage within 24 h by the combined action of a biological demulsifier and a cell surface. However, most of the currently discovered biomilk-breaking bacteria act in conjunction with a biological demulsifier for demulsification.

The use of microbial surfactants for crude oil demulsification is generally slow and inefficient, but if the synergistic effect between surfactants is applied, combining microbial surfactants and conventional demulsifiers, it is possible to have an additive effect and thus produce a good demulsification effect. For example, Guo (2002) chose a microbial surfactant MS to be combined with three demulsifiers GE-189, XE-120, and XP-120, respectively, to dewater a simulated Brazilian P-20 crude oil emulsion, confirming this conclusion. Biological demulsification is simple and low cost, but the strain culture is difficult, the growth period is long, and the survival conditions of microorganisms are harsh in actual engineering, so the method is less applied. However, biological demulsifiers have low cost, low toxicity, diverse structures, and better environmental compatibility, so biological demulsification technology has good prospects for development, and further cultivation of strains with high viability and wide adaptability of the produced agents is the main driving force for the development of biological methods.

4 CHEMICAL DEMULSIFICATION

Chemical demulsification is the most common method of demulsification, the principle is to use a chemical demulsifier and the emulsifier on the oil-water interface film physical or chemical effects, changing the nature of the interface, so that the emulsion droplets flocculation, coalescence into large droplets and separation, and finally achieve the purpose of demulsification.

The type, structure, molecular weight, and dosage of the demulsifier will affect the effect of demulsification. For example, Alsabagh et al (2016) prepared ethylene oxide demulsifiers of 430, 650, and 1100 and investigated the relationship between their molecular weights and demulsification effect. The results showed that the larger the molecular weight of the demulsifier, the greater the interfacial activity, the smaller the interfacial tension gradient, the faster the rate of membrane drainage, and the better the demulsification effect. Yao et al (2013) treated O/W emulsion in an oil field in Daqing with a homemade demulsifier TS-24 to study the demulsification influencing factors. When the demulsifier concentration is too high, a new interfacial film will be formed at the oil-water interface, resulting in a decrease in the demulsification rate, so the demulsification rate will show a trend of increasing and then decreasing with the increase of dosage.

Chemical demulsification has the advantages of higher demulsification efficiency and lower cost, but because a large number of chemicals are added, the amount of sludge increases and even secondary pollution is produced, which increases the cost of treatment. As the composition of oilfield-produced water becomes more and more complex, the existing demulsifier gradually cannot meet the requirements of use, the future research and development of demulsifiers should be more inclined to be micro-efficient, universal, to reduce the sludge and secondary pollution generation.

5 CONCLUSION

Ultrasonic demulsification is non-polluting but lacks simple and low-cost industrialized equipment. The microwave demulsification is time-saving and efficient, but the engineering is difficult to control precisely. Electric demulsification does not cause environmental pollution but is costly, energy-intensive and unsuitable for treating crude oil emulsions with high water content. Membrane demulsification has low energy consumption and a wide application range, but high cost, poor economy, and more serious environmental pollution. It is not the technology that limits the development of physical demulsification, but the demulsification equipment. Most of the existing equipment is very bulky, complex, and low automation, so the development of easy-to-operate and low energy consumption devices will greatly promote the development of physical demulsification. Biological demulsification is low cost and pollution-free, but the biological strain is difficult to cultivate and the cycle is long; chemical demulsification is high efficiency and easy to apply, but it is easy to cause secondary pollution and the dosage is large. In practice, the use of physical and biological demulsification alone is relatively rare, and is usually used in combination with chemical demulsification, as a pretreatment or deep treatment process, thus reducing the amount of chemical demulsifier dosing.

In comprehensive comparison, chemical demulsification is the most widely used, but its environmental problems cannot be ignored, the development of small pollution and wide applicability of high-efficiency demulsifiers is the key to solving the problem; physical demulsification and biological demulsification with their non-polluting characteristics, has great prospects for development, and currently need to be combined with chemical demulsification to meet engineering needs. In future research, we should not only make simple, low energy consumption, high automation equipment and microorganisms with a short culture period and excellent effect, but also focus on the research of its demulsification mechanism, so that it can improve the treatment capacity and effect without adding chemicals, and make the demulsification green development.

REFERENCES

Akinori M, Yuichi H, Koichiro K, et al. Effects of organic solvent and ionic strength on continuous demulsification using an alternating electric field[J]. *Colloids and Surfaces A: Physicochemical and Engineering Aspects*, 2016, 506:228–233.

Ali K, Ali S, Payam P, et al. Experimental and theoretical study of crude oil pretreatment using low-frequency ultrasonic waves[J]. *Ultrasonics Sonochemistry*, 2018, 48.

Alsabagh A M, Hassan M E, Desouky S E M, et al. Demulsification of W/O emulsion at petroleum field and reservoir conditions using some demulsifiers based on polyethylene and propylene oxides[J]. *Egyptian Journal of Petroleum*, 2016, 25(4).

Binner E R, Robinson J P, Silvester S A, et al. Investigation into the mechanisms by which microwave heating enhances separation of water-in-oil emulsions[J]. *Fuel*, 2014, 116.

Boping R, Yong K. Aggregation of Oil Droplets and Demulsification Performance of Oil-in-water Emulsion in Bidirectional Pulsed Electric Field[J]. *Separation and Purification Technology*, 2018, 211.

Cai Q H, Zhu Z W, Chen B, et al. Oil-in-water emulsion breaking marine bacteria for demulsifying oily wastewater[J]. *Water Research*, 2019, 149.

Gong X, Zhang J, Tang J, et al. Research on electric demulsification [J]. *Energy and Environment*, 2015(02):16–17+26.

Gu G J, Liu G, Chen B, et al. Research progress of W/O type oil-water emulsion physical demulsification technology and device [J]. *Chemical Industry and Engineering Progress*, 2015, 34(02): 319–324.

Guo L P, Liu S. Progress in the analysis of the microscopic mechanism of electric demulsification and its influencing factors [J]. *Oilfield Chemistry*, 2018, 35(04):750–756.

Guo D H. Electrochemical methods to study the effect of microbial surfactants on demulsification of crude oil emulsions [J]. *Acta Petrolei Sinica(Petroleum Processing Section)*, 2002(04):27–32.

Igor N E, Aleksandr P L. Microwave treatment of crude oil emulsions: Effects of water content[J]. *Journal of Petroleum Science and Engineering*, 2014, 115.

Jiang H Y, Huang L, Wei A J. Experimental study of microwave dehydration of thick oil [J]. *Journal of Xi'an Shiyou University (Natural Science Edition)*, 2005(05):61–63+10.

Kocherginsky N M, Tan C L, Lu W F. Demulsification of water-in-oil emulsions via filtration through a hydrophilic polymer membrane[J]. *Journal of Membrane Science*, 2003, 220(1).

Liu W C, Yi S J. Advances in bio-demulsification of oilfield emulsions [J]. *Journal of Yangtze University (Natural Science Edition)*, 2010, 7(03):545–546+558.

Lu L J. *Study on biological demulsification bacteria, identification of demulsification active ingredients and their effects on oilfield produced water treatment system* [D]. Tongji University, 2007.

Sun B J, Yan D Q, Qiao W X. Ultrasonic dehydration study of emulsified crude oil [J]. *Acta Acustica*, 1999(03):327–331.

Sun N N, Jiang H Y, Wang Y L, et al. A Comparative Research of Microwave, Conventional-Heating, and Microwave/Chemical Demulsification of Tahe Heavy-Oil-in-Water Emulsion[J]. *SPE Production & Operations*, 2018, 33(02).

Sun Y, Luo G, Pu Y, et al. A new membrane demulsification technology [J]. *Modern Chemical Industry*, 2000(03):16–18.

Xu X Z, Cao D, Liu J, et al. Research on ultrasound-assisted demulsification/ dehydration for crude oil. [J]. *Ultrasonics sonochemistry*, 2019, 57.

Xu C, Yan F, Wang M X, et al. Fabrication of hyperbranched polyether demulsifier modified PVDF membrane for demulsification and separation of oil-in-water emulsion[J]. *Journal of Membrane Science*, 2020, 602.

Yang X G. Study of microwave radiation crude oil demulsification technology [D]. *Tianjin University*, 2006.

Yao L C, Liu W, Huo Y. Experimental study on the treatment of O/W emulsion wastewater by emulsion breaking-flocculation [J]. *Modern Chemical Industry*, 2013, 33(08):58–61.

Electrothermal chemical engineering and chemical preparation technology

Energy Revolution and Chemical Research – Kok-Keong Chong and Zhongliang Liu (Eds)
© 2023 The Authors, ISBN: 978-1-032-36554-1

Thermal behaviour of unsymmetrical dimethylhydrazine picrate

Xiaogang Mu*, Bo Liu & Jingjing Yang
Xi'an Research Institute of High Technology, Xi'an, China

ABSTRACT: The thermal decomposition behaviour of unsymmetrical dimethylhydrazine (UDMH) picrate was investigated via thermal analytic methodologies including differential scanning calorimetry (DSC) and thermogravimetry (TG). The exothermal decomposition of UDMH picrate is observed in the range of 175.8°C to 224.3°C, from which the peak decomposition temperature is determined to be 215.7°C. Kinetic parameters of decomposition reaction were calculated according to the Kissinger Equation, i.e., thermal decomposition activation energy E_a is 86.90 kJ·mol^{-1} and the pre-exponential factor ln$A$ is 16.86. The critical temperature of thermal detonation T_b was theoretically calculated to be 167.1°C based on DSC analysis.

1 INTRODUCTION

Picric acid (2,4,6-Trinitrophenol, PA) is a highly acidic nitrophenol compound that reacts with copper, potassium, lead, and barium to form picric acid metal salts (Feng 2008; Li et al. 2012; Wang et al. 2008), the corresponding salts (Dou et al. 2010; Jin et al. 2005; Li et al 2009; Li et al. 2009; Zhang et al. 2010) can also be synthesized with organic base compounds. As a new energetic material, picric acid energetic ion salt has been widely used in the field of propellant and explosive because of its good thermal stability, high density and energy, and low vulnerability. Therefore, the study of energetic ionic salts based on picric acid has attracted extensive attention in recent years.

Thermal analysis of energetic materials is essential not only to the safety technology of production, use, transportation and storage, but also to the basic theoretical research of combustion and detonation (Silva et al 2010). In this paper, the thermal decomposition kinetics of unsymmetrical dimethyl hydrazine (UDMH) picrate was studied using differential scanning calorimetry (DSC) and thermogravimetry (TG). The thermodynamic function values of the decomposition reaction were calculated by thermodynamic relations, and the thermal explosion critical point was estimated by non-isothermal DSC curves.

2 EXPERIMENTAL

2.1 *Synthesis of UDMH picrate*

UDMH and PA are weighed and then dissolved in ethanol. The UDMH solution was added to the ethanol-containing picric acid by continuous stirring at room temperature. The reaction terminates when the solution becomes cloudy and neutral. The bright yellow crystals were obtained by ethanol washing and drying at a constant temperature.

*Corresponding Author: muxg2001@163.com

DOI 10.1201/9781003332657-72

2.2 *Characterization*

TG/DTG analysis was carried out with Shimadzu TGA-50 thermogravimetric analyzer at a heating rate of 10°C/min and a mass of 0.5 mg.

DSC measurements were performed with NETZSCH DSC 204 at 2.5, 5, 10, and 15°C/min in the range of 50–300°C under the nitrogen flow.

3 RESULTS AND DISCUSSION

3.1 *TG/DTG and DSC analysis*

The thermal stability of UDMH picrate was investigated by TG/DTG experiments (see Figure 1). As can be seen from the diagram, the TG curve mass remains constant until 175.8°C, from which the weight begins to decline at an accelerated rate. The DTG results show that the thermal decomposition process mainly occurs between 175.8°C and 244.3°C. The peak of weight loss occurred at 215.7°C, and the proportion of weight loss reached 87.1 %.

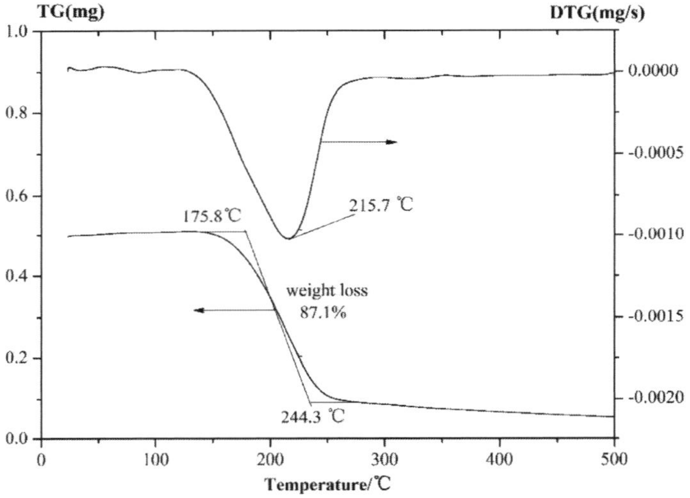

Figure 1. TG/DTG curve of UDMH picrate.

Figure 2 exhibits the DSC curves of UDMH picrate at various heating rates. The first endothermic peak is located near 150°C indicating the melting temperature of the picrate, and the exothermic peak of the picrate occurs during the decomposition process, which is in agreement with the results from TG/DTG. The enthalpy (ΔH), the initial decomposition temperature (T_0) and the peak decomposition temperature (T_p) can be calculated by the second peak integral, as shown in Table 1. Furthermore, it also indicates that T_p increases with the increment of heating rate, which is consistent with the general law of thermal decomposition.

3.2 *Thermal decomposition kinetics of UDMH picrate*

The kinetic model of thermal decomposition of solid materials is generally based on the thermal decomposition rate ($d\alpha/dt$), which is expressed as follows:

$$\frac{d\alpha}{dt} = kf(\alpha) \tag{1}$$

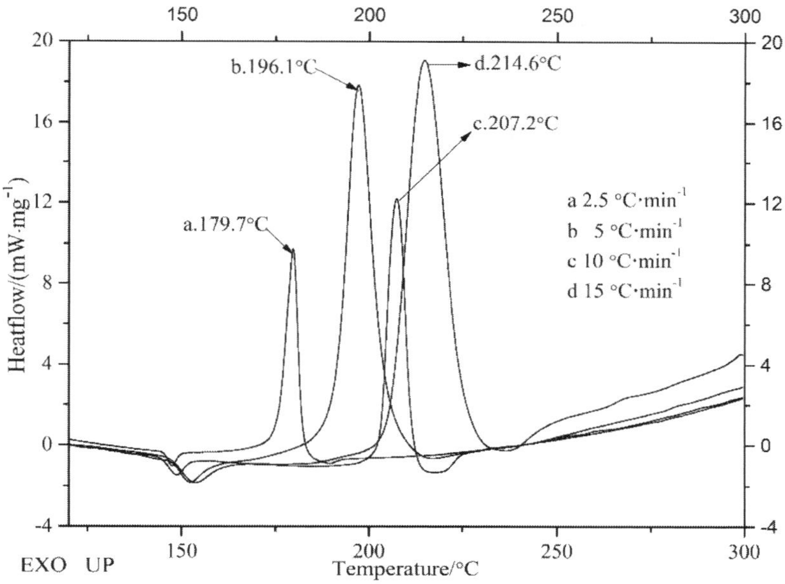

Figure 2. Non-isothermal DSC curves of UDMH picrate.

Table 1. Key parameters of non-isothermal DSC curves of UDMH picrate.

Heating rate (°C/min)	T_0(°C)	T_p (°C)	ΔH (J/g)
2.5	176.1	179.7	974.8
5	191.6	196.1	956.8
10	203.3	207.2	860.4
15	206.3	214.6	922.7

where:
$d\alpha/dt$: thermal conversion rate;
k: reaction rate constant;
$f(\alpha)$: kinetic function of thermal decomposition: $f(\alpha) = (1-\alpha)^n$

The kinetic parameter could be calculated by substituting the Arrhenius's equation (Equation (2)) in Equation (1), which is shown in Equation (3):

$$k = A\exp\left(-\frac{E_a}{RT}\right) \tag{2}$$

$$\frac{d\alpha}{dt} = A\exp\left(-\frac{E_a}{RT}\right)(1-\alpha)^n \tag{3}$$

where:
A: pre-exponential factor;
E_a: activation energy ($J{\cdot}mol^{-1}$);
R: universal constant of gases (8.314 $J{\cdot}K^{-1}{\cdot}mol^{-1}$);
T: absolute temperature (K).

Kissinger's method was chosen and applied in this study. The activation energy parameter (E_a) can be deduced from DSC data at different heating rates, with the maximum reaction rate occurring at the peak of the reaction temperature (T_p) and $\frac{d}{dt}\left(\frac{d\alpha_p}{dt}\right)=0$,
hence,

$$\frac{E_a}{RT_p^2} = \frac{An}{\beta}(1-\alpha_p)^{n-1}\exp\left(-\frac{E_a}{RT_p}\right) \tag{4}$$

Where,
β: heating rate (K/min);
T_p: peak temperature for specific heating rate (K);
When $n = 1$, convert Equation (4) to Equation (5):

$$\frac{E_a}{RT_p^2} = \frac{A}{\beta}\exp\left(-\frac{E_a}{RT_p}\right) \tag{5}$$

To get a logarithm on both sides of Equation (5), Equation (6) is obtained:

$$\ln\left(\frac{\beta}{T_p^2}\right) = \ln\left(\frac{RA}{E_a}\right) - \frac{E_a}{RT_p} \tag{6}$$

The data points for $\ln(\beta/T_p^2)$ and $1/T_p$ could be obtained from real DSC measurements and corresponding linear regression was drawn to fit whereby the activation energy was determined by the slope, i.e., E_a= 86.90 kJ·mol^{-1}, and the pre-exponential factor was calculated from the intercept, ln$A$= 16.86 (see Figure 3)

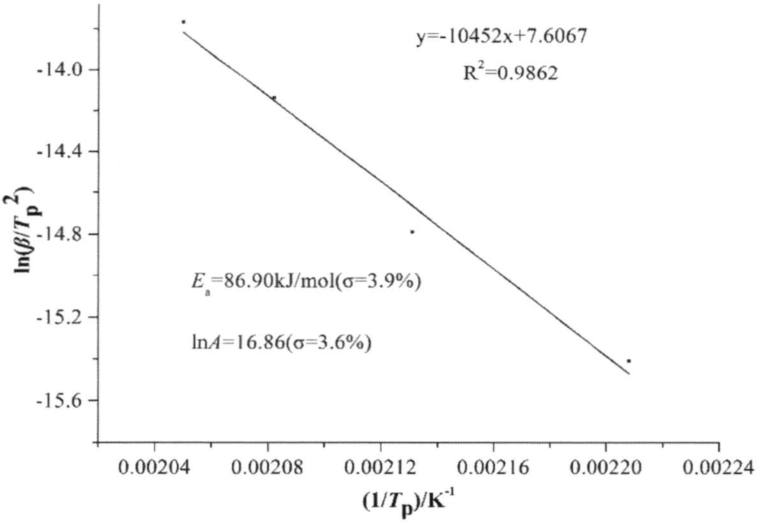

Figure 3. Kissinger's linear relationship of $\ln(\beta/T_p^2)$ and $1/T_p$.

3.3 *Critical temperature of thermal explosion of UDMH picrate*

Critical temperature point T_b is an essential parameter to characterize the thermal stability of energetic materials. Thermal explosive critical point T_b under non-isothermal conditions could be calculated by Equations (7) and (8) (Zhang et al. 1994).

$$T_{ei} = T_{e0} + b\beta_i + c\beta_i^2 + d\beta_i^3 \tag{7}$$

$$T_b = \frac{E - \sqrt{E^2 - 4ERT_{e0}}}{2R} \tag{8}$$

Where, T_{ei} is the extrapolated temperature and could be obtained from the non-isothermal DSC curve, and T_{e0} is the initializing value when $\beta = 0$. On the basis of the characteristic temperature under different heating rates, the least square regression analysis is applied to fit Equation (7):

$$T_{ei} = 148.6 + 14.30\beta_i - 1.232\beta_i^2 + 0.0369\beta_i^3$$

Hence, T_b is determined to be 167.1°C when $\beta = 0$ and $T_{e0} = 148.6$°C.

4 CONCLUSION

According to experimental measurements and a theoretical kinetic model of thermal decomposition, the following conclusions are obtained:

(1) The exothermic reaction of unsymmetrical dimethylhydrazine picrate mainly occurs in the range of 175.8°C–244.3°C.
(2) Combining the equation of Kissinger and DSC analytic tests, the activation energy of thermal decomposition E_a(86.90 kJ·mol^{-1}) was obtained.
(3) According to the non-isothermal DSC curve, the thermal detonation critical point $T_b = 167.1$°C.
(4) Unsymmetrical dimethylhydrazine picrate has good thermal stability and its energy properties will be further investigated in the next research plans.

REFERENCES

Dou, S., Zhang, J., Yang, X. et al. Preparation of ammonium picrate and its shock wave sensitivity (in Chinese). *Shanxi Chemical Industry*. 2010, 30 (4): 25–27.

Feng, J. Synthesis and characterization of a novel Salamo derivative copper (II) picrate complex (in Chinese). *Chinese Journal of Applied Chemistry*. 2008, 37 (6): 654–656.

Jin, M.C., Ye, F.C., Piekarski, C. et al. Mono and bridged azolium picrates as energetic salts. *European Journal of Inorganic Chemistry*, 2005, 18: 3760–3767.

Li, D., Fu, W., You, X. et al. Preparation and crystal structure of a novel complex [Cd(H2O)2(phen)2](PA)2 (in Chinese). *Chinese Journal of Inorganic Chemistry*. 2002, 18 (2): 209–213.

Li, D., Ren, Y., Zhao, F. et al. Preparation, structure and Quantum chemistry of C3N2H5+ C6N3O7H2- (in Chinese). *Chinese Journal of Explosives & Propellants*. 2009, 32 (6): 48–52.

Li, Z., Yan, Y., Ji, H. et al. Theoretical investigation into the structural, thermal and explosive properties of energetic ion salts of picric acid (in Chinese). *Chinese Journal of Explosives & Propellants*. 2009, 32 (6): 6–10.

Silva, G., Iha K., Cardoso, A. M. et al. Study of the thermal decomposition of 2, 2', 4, 4', 6, 6'-hexanitrostilbene[J]. *Journal of Aerospace Technology and Management*, 2010, 2: 41–46.

Wang, Z., Li, G., Jiang, X. et al. Effectiveness of potassium picrate on thermal behavior and ignition performance of RDX-containing ammonium nitrate propellants (in Chinese). *Chinese Journal of Explosives & Propellants*. 2008,31 (3): 29–31.

Zhang, G., Zhang, T., Liu, J. et al. Preparation of picric acid carbohydrazide energetic complexes and thermal influences on the decomposition of CL-20 (in Chinese). *Chinese Journal of Explosives & Propellants*. 2010, 33 (1): 38–42.

Zhang, T., Hu, R., Xie, Y. et al. The estimation of critical temperatures of thermal explosion for energetic materials using non-isothermal DSC. *Thermochimica Acta*. 1994, 244: 171–176.

Energy Revolution and Chemical Research – Kok-Keong Chong and Zhongliang Liu (Eds)
© 2023 The Authors, ISBN: 978-1-032-36554-1

Experimental study of MnOx-CeO_2/AC-catalyzed air oxidation for sulfide removal from wastewater

Ying Liu, Weixing Wang & Meng Du
Southwest Petroleum University College of Chemistry and Chemical Engineering, Chengdu, Sichuan, China

Jianhua Zhao
Southwest Petroleum University College of Engineering Nanchong, Sichuan, China

Chenyang Xu
Southwest Petroleum University College of Chemistry and Chemical Engineering, Chengdu, Sichuan, China

ABSTRACT: Air-catalytic oxidation is an environmentally friendly, energy-efficient, and low-cost method for the degradation and treatment of sulfur-containing wastewater, and the treatment effect of this method depends on the catalyst performance. Manganese oxide (MnO_x) is a commonly used catalyst in the air catalytic oxidation process, but its catalytic performance needs to be improved. This paper is to explore a method to improve the performance of MnO_x catalyst. Based on the principle of catalyst doping modification, the research idea of improving the catalytic performance of MnOx catalyst by doping it with rare earth element Ce was proposed. The experiment prepared the activated carbon-loaded MnOx catalyst (MnOxCeO_2/AC) with different Ce contents and evaluated its effect on catalytic air oxidation degradation of sulfur-containing wastewater, and relevant experimental results and conclusions were obtained: 1. Doping with rare earth element Ce can significantly improve the performance of MnO_x/AC catalyst, and the best performance conditions of the catalyst were optimized; 2. Single-factor experiments evaluated the effects of aeration, temperature, catalyst dosage, desulfurization rate, and initial sulfur content of wastewater on the desulfurization effect; 3. Mechanistic studies showed that the sulfide degradation in wastewater underwent a transformation from S2to S2O32 and the degradation of sulfide in wastewater underwent the process of conversion from S2-toS2O32- and finally toSO42-.

1 INTRODUCTION

Industrial wastewater containing hydrogen sulfide, sodium sulfide, sodium thiosulfate, and other sulfides is discharged in large quantities during the industrial production of industrial pharmaceuticals, tanneries, oil refining, coking, gas production, biofuels, petrochemicals, etc., which is referred to as sulfur-containing industrial wastewater (Li 2014). The industries that produce the most sulfur-containing industrial wastewater are the tanning industry and petroleum refining industry respectively. Sulfur-containing wastewater reacts easily with acid to produce hydrogen sulfide. And the sludge containing sulfide can also produce hydrogen sulfide gas in the absence of oxygen and volatilize into the air at the same time. Hydrogen sulfide gas not only pollutes the air but also seriously affects the quality of natural water resources. There is also a risk of poisoning if hydrogen sulfide is inhaled by humans. Industrial wastewater will have a negative impact on the production facilities in the plant, the reprocessing process, human beings, and the natural environment in which they live.

Most of the catalyst oxidation methods are used in China under the consideration of various factors, that is, using aeration to expose air into the wastewater for sulfur removal while adding an alkaline catalyst for catalytic reaction to improve the oxidation rate. Currently, the catalysts used

 DOI 10.1201/9781003332657-73

are mainly manganese, copper, iron, cobalt, and other heavy metal salts, as well as activated carbon (Jian 2016; Wang 2016).

2 PREPARATION OF CATALYST

In this experiment, a modified MnO_X-CeO/AC_2 was used as the catalyst, and the air was used as the oxidant to catalytically oxidize the removal of sulfide from the simulated wastewater under certain conditions. The simulated wastewater was a water sample prepared with sodium sulfide and distilled water in a certain ratio. The desulfurization efficiency was evaluated by measuring the ratio of sulfur ion content in the water sample before treatment to the sulfur ion content in the water sample after treatment (Kuang 2018).

The composite catalyst was made by the equal volume impregnation method, i.e., the volume and concentration of the solution required to prepare the composite catalyst were determined based on the water absorption of the activated carbon powder and the amount of catalyst host agent and additives to be loaded (Guan 2003; Su 2013; Wu 2014).

The cerium dioxide doped activated carbon loaded manganese-based catalyst was used as the catalyst for the catalytic air oxidation method in this experiment for several reasons.

In recent years, manganese oxides as adsorbents to adsorb pollutants in wastewater have become a research hotspot. With a large specific surface area and abundant surface hydroxyl groups, manganese oxides have a strong catalytic oxidation capacity and show excellent adsorption performance and catalytic activity for many pollutants.

Activated carbon materials are rich in microporous structure and surface functional groups, and studies have shown that catalysts loaded with activated carbon carriers exhibit better low-temperature SCR performance compared to Al_2O_3 and TiO_2 carriers (Huang 2018).

Among the numerous catalysts loaded with activated carbon (Fe, Co, Ni, V, Mn, Cu, Ce), only Ce and Mn showed the best low-temperature activity, which is related to the good redox properties of CeO_2 and the presence of multiple unstable valence states of manganese oxides, respectively (Huang 2018).

In order to combine the excellent performance of both, the effectiveness of activated carbon-loaded Ce-Mn-based catalysts for sulfide removal from wastewater will be investigated in this project.

3 EXPERIMENTAL RESULTS (INFLUENCING FACTORS)

3.1 *Doping ratio*

The following reaction results were obtained at a temperature of 20°C, a treatment time of 5 hours, a pH=12 before reaction, and an aeration rate of 4L/min.

From Figure 1-a, it is concluded that with the increase in reaction time, the desulfurization effect of catalysts with each doping ratio has a significant difference, in which, the catalyst with a doping ratio of 1:0.06 has the largest reduction of sulfur during the desulfurization reaction. The final desulfurization rate is also the highest, therefore, it can be concluded that the catalyst with a doping ratio of1:0.06 is the best catalyst for this experiment. From Figure 1-b it is observed that the addition or not of a catalyst has a decisive role in the desulfurization effect, and the difference between adding a catalyst and not adding a catalyst is obvious, on this basis, the catalyst doped with cerium dioxide also has a significant promotion effect on the improvement of desulfurization effect.

3.2 *Catalyst dosage and pH value*

From figure 2-a, it is noted that the sulfide removal rate increases continuously as the amount of metal-loaded catalyst used rises; when the catalyst dosage exceeds 10 g/L, i.e., the sulfur content

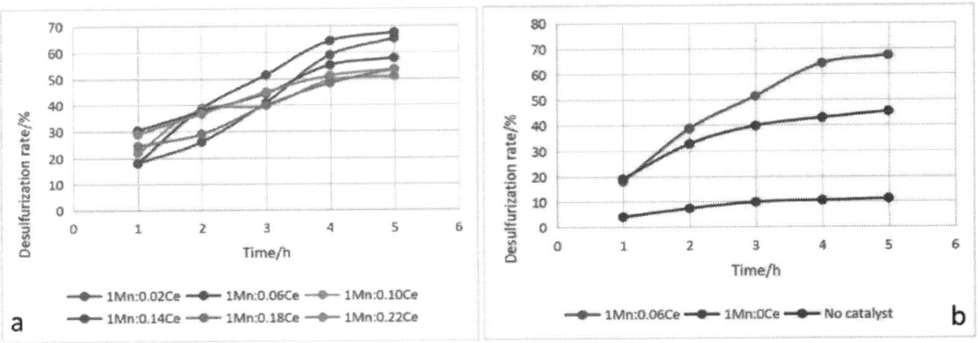

Figure 1. (a) Relationship between different doping ratios of catalysts and desulfurization rate; (b) Relationship between catalyst addition and doping and the desulfurization rate in the solution after the reaction.

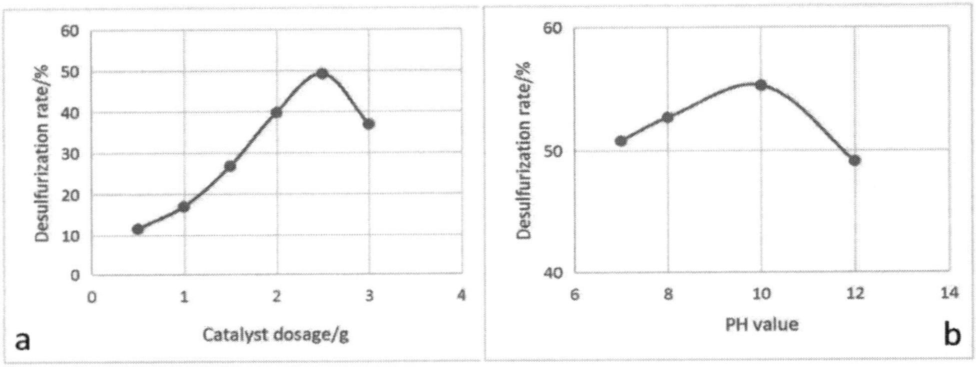

Figure 2. (a) Relationship between catalyst dosage and desulfurization rate; (b) Relationship between pH of solution before reaction and desulfurization rate.

in the wastewater sample is about 2.72 times the amount of manganese-based catalyst on activated carbon and about 90 times the amount of cerium dioxide, and the sulfide removal rate begins to decrease. It can be seen that the reaction reached the best desulfurization effect when the doping ratio was 1Mn:0.06Ce catalyst injection amount was 10g/L.

Figure 2-b shows that it is not the case that the higher the alkalinity, the better the desulfurization effect. Combined with the graphs of the experimental results, the maximum sulfide desulfurization rate occurs at pH=10, which shows the best desulfurization performance of the catalyst.

3.3 *Aeration volume and Temperature.*

The curves in Figure 3-a show that the aeration volume has a greater effect on the catalytic air oxidation process. With the increase of aeration, the gas-liquid ratio in the reaction increases, and the rate of sulfide content reduction in the simulated wastewater increases significantly, meanwhile, the rate of sulfide content removal is accelerated. The oxidation rate of sulfide ions is controlled by the transfer rate of oxygen in the solution, so in the actual industrial production process, improving the reaction rate can achieve the purpose, in addition to the need to increase the amount of air in the solution. The aeration device should also choose the model with high oxygen utilization rate.

From figure 3-b, the larger the temperature, the larger the total oxygen transfer coefficient kLa is, which can substantially enhance the oxygen transfer efficiency and increase the oxidation rate of sulfide. Moreover, the elevated temperature will reduce the saturated dissolved oxygen

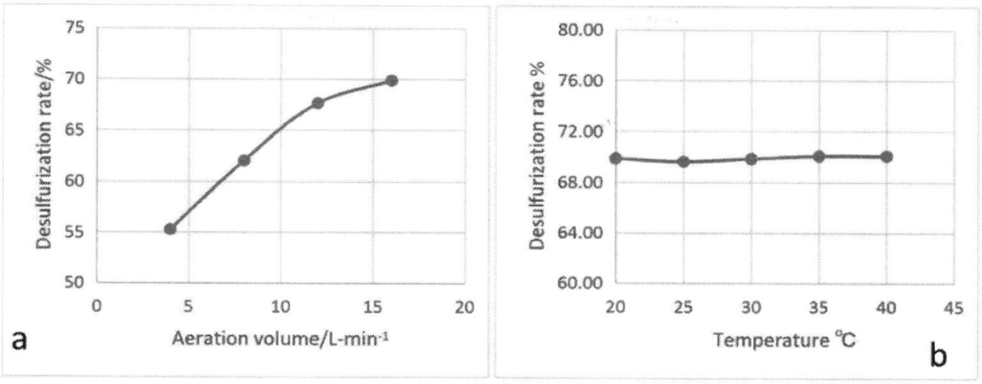

Figure 3. (a) Relationship between aeration and desulfurization rate; (b) Relationship between temperature and desulfurization rate.

concentration cs in water, which in turn weakens the oxygen diffusion drive and reduces the oxygen transfer rate, and thus reduces the oxidation rate of sulfide, one of which promotes desulfurization and the other inhibits it. It is known from the experimental results that the sulfide removal rate varies up and down by about 1% in the range of 20°C~40°C. Therefore, the effect of temperature on the desulfurization rate within this range is negligible.

3.4 *Initial sulfide content and Duration.*

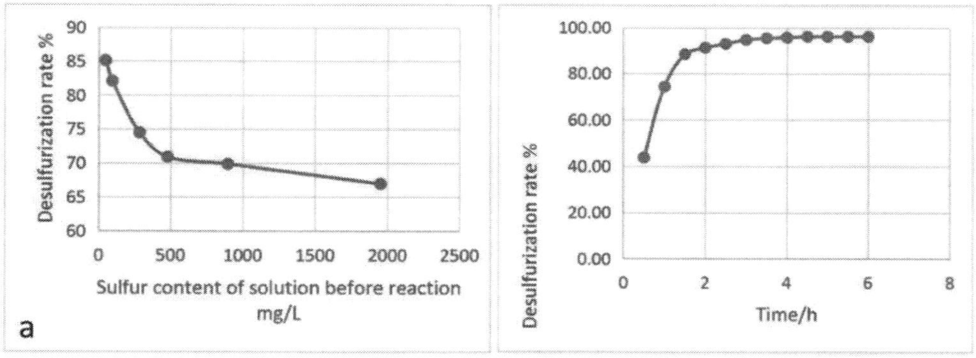

Figure 4. (a) Relationship between initial sulfide content and desulfurization rate; (b) Relationship between time and desulfurization rate.

From figure 4-a, it can be known that the initial sulfide content is negatively correlated with the efficiency of sulfide removal from the solution. This shows that this air-catalytic oxidation method is suitable for the removal of sulfur-containing wastewater with low concentrations of excess sulfide.

The slope of the curve in Figure 4-b indicates that the reaction rate is higher for treatment times within 3 hours than after 3 hours. The sulfide removal rate was up to 96.15% under optimal reaction conditions.

4 REACTION MECHANISM EXPLORATION

4.1 *Variation of sulfate content*

While the sulfide content in the solution after the reaction was measured in the above experiment where time was the influencing factor, the sulfate concentration in the solution was also measured, and the following Figures 5-a and 5-b were drawn from the experimental data (Liu 2007).

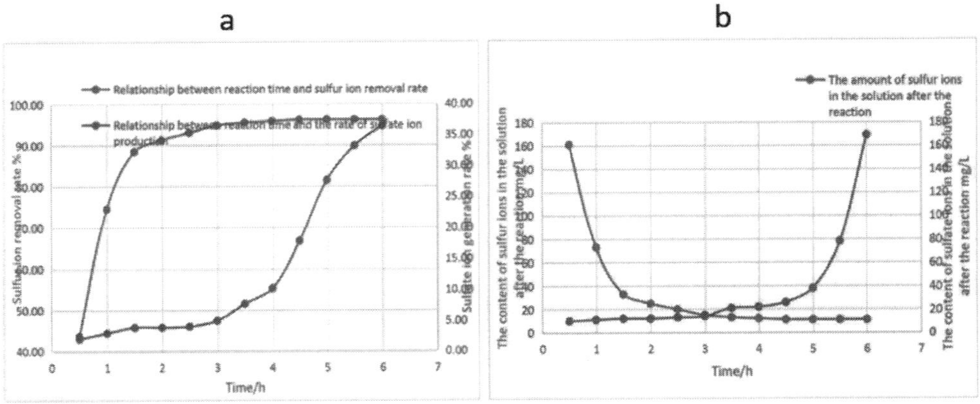

Figure 5. (a) Time versus sulfide and sulfate content in solution after reaction; (b) Time vs. sulfide removal rate and sulfate production rate.

With the increase in reaction time, Figure 5-a shows that the sulfide content in the solution gradually decreases while the sulfate content gradually increases; Figure 5-b shows that the removal rate of sulfide and the production rate of sulfate in the solution are increasing.

The rate of sulfide removal is larger and the rate of sulfate production is smaller within the reaction for 3 hours, while the rate of sulfide removal decreases and the rate of sulfate production increases after the reaction for 3 hours. This phenomenon indicates that the sulfur ions in the solution do not produce sulfate directly, but there are intermediate products. Through the combination of the literature review and the above analysis, it is assumed that the intermediate product is thiosulfate, and the mechanism of the catalytic air oxidation sulfide removal reaction is proposed as follows.

$$2Na_2S + 2O_2 + H_2O = Na_2S_2O_3 + 2NaOH \tag{1}$$

$$Na_2S_2O_3 + 2O_2 + 2NaOH = 2Na_2SO_4 + H_2O \tag{2}$$

4.2 *Characterization of the catalyst*

4.2.1 *x-ray diffraction and UV diffuse reflection*

Figure 6-a shows the XRD patterns of the catalysts loaded with MnO as well as different levels of cerium dioxide. The XRD results show that the activated carbon is successfully loaded with manganese oxide and cerium dioxide and the composites are polycrystalline in structure. The broad peaks in the range of 20–30° are attributed to the activated carbon in the hexagonal crystal system, and the peaks of cerium dioxide are smaller in the composites due to the low loading of cerium dioxide and overlap with the manganese oxides, with only smaller peaks.

The absorption bands at different positions of the diffuse reflectance UV-Vis spectra show the morphological and compositional variations of the different catalyst surfaces. The optical absorption bands are in the range of 200-250 nm due to the presence of $O_2^- \rightarrow Mn^{2+}$ and $O_2^- \rightarrow Mn^{3+}$ charge transfer in the black manganese ore structure. Activated carbon loaded with MnO or MnO-CeO_2has a strong visible light absorption with a maximum absorption wavelength of 400-500 nm

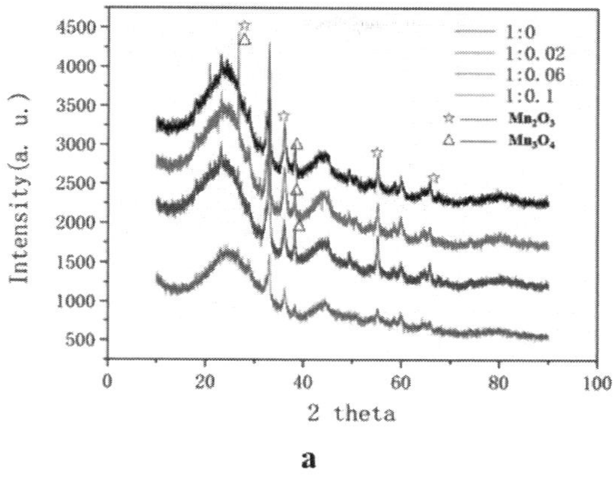

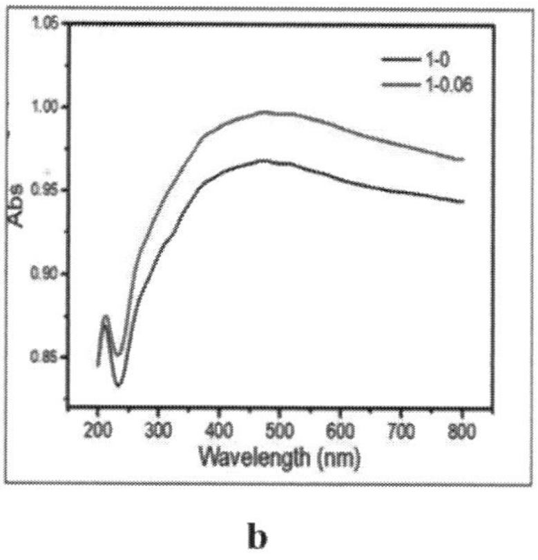

Figure 6. (a) Comparison of XRD patterns of catalysts with different doping ratios; (b) Comparison of UV absorption spectra of two doping ratio catalysts.

in the visible region. This maximum absorption wavelength suggests the successful loading of MnO, attributed to the d-d leap of $Mn^{3+}$and Mn^{4+}. The loading of cerium dioxide, on the other hand, improves the absorbance of the catalyst in the visible region, which may be due to the fact that cerium dioxide improves the dispersion uniformity of the catalyst and its own light absorption in the visible region to some extent. The principle is that visible photocatalysis takes advantage of the efficient charge separation efficiency of cerium dioxide or the absorption of visible light using manganese, and after forming a composite material, cerium dioxide in manganese can help manganese to carry out efficient charge separation, which will form holes and individual electrons, and generate free radicals, such as hydroxyl radicals or superoxide anions. The improved light absorption by cerium dioxide can be demonstrated using photocatalytic reactions.

4.2.2 *Potential*

Comparison of the potentials measured by the pH meter after the reaction for 4 hours in different cases. The results are shown in the figure.

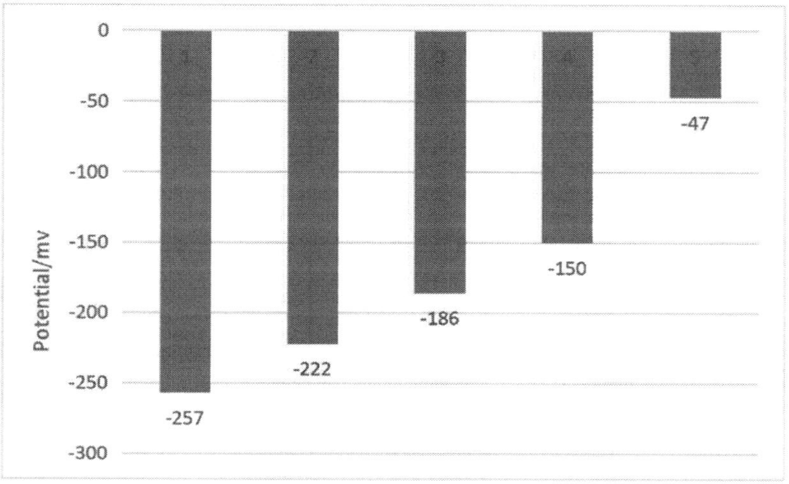

Figure 7. Comparison of potentials in different cases.

Figure 7 shows a comparison of potentials in different cases1 - Total potential of the original water sample solution 2- Total potential of the solution after reaction without catalyst 3- Total potential of the solution after reaction with 1:0 catalyst 4- Total potential of the solution after reaction with 1:0.06, and catalyst 5- Total potential of distilled water.

The original water sample solution was alkaline, so with the increase in desulfurization time, the solution alkalinity weakened, the concentration of reducing substances decreased, and the potential increased, so the change in the value of the potential can be judged according to the desulfurization effect. The results from the above graph show that when no catalyst was added, the potential value changed 35mV from the original water sample, while when the catalyst with only doped manganese oxide was added, the potential value changed 71mV, and when the catalyst with doping ratio of1:0.06 was added, the potential value changed 93mV. Thus, it can be seen that the catalyst was added to promote the desulfurization effect, and the doping of cerium dioxide will make the promotion effect of the desulfurization effect much higher.

5 CONCLUSIONS

The main experimental results and discussions of this topic are as follows.

(1) The optimal manganese-based activated carbon catalyst with a doping ratio of 1Mn:0.06Ce was studied, and the optimal reaction conditions when the air was used as the oxidant were: the catalyst dosage was 10 times the initial content of sulfide in solution, the temperature was 20°C, the aeration was 16 L/min, and the pH=10. Under this reaction condition, the sulfide was treated for 6 hours at a concentration of 286 mg/L simulated wastewater, and the conversion rate of sulfide could reach 96.15%.

(2) The effect of sulfide content in the solution before the reaction of simulated wastewater on the desulfurization effect was studied. The data showed that the higher the concentration of sulfide in the pre-reaction solution, the lower the removal rate, which shows that the catalytic reaction is suitable for the sulfur-containing wastewater with low concentration and over standard.

(3) The activated carbon surface was successfully loaded with manganese oxide and cerium dioxide, and the composite material was polycrystalline in structure. The doping of cerium dioxide improves the dispersion uniformity of the catalyst and its own light absorption in the visible region to a certain extent, which promotes the reaction.

ACKNOWLEDGMENTS

This article is one of the achievements of the Science and Technology Strategy School Cooperation Projects of the Nanchong City Science and Technology Bureau “Environmental protection sugar-base-anion Surfactant for Special Reservoir oil displacement” (no. SXHZ016).

REFERENCES

Guan H X; Qiu R C; Zhao X T. Treatment of High Strength Sulfide-Bearing Wastewater by Catalytic Air Oxidation Process [J]. *Journal of Lanzhou Railway University*, 2003, 22(06):21–23.

Huang L.H., Li X. Effect of MnO_x doping on CeO_2/AC catalyst for NH3 selective catalytic reduction of NOx [J]. *Journal of Functional Materials*, 2018, 49(06): 6124–6128.

Jain C K, Malik D S, Yadav A K. Applicability of plant-based biosorbents in the removal of heavy metals: a review[J]. *Environmental Processes*, 2016, 3 (2): 495–523.

Kuang H C. *Analytical Study of Sulfide in Soil under Hydrochloric Acid and Tin Dichloride Acidification System*, 2018, 43(09):144–147.

Li C.C. *Study on Sulfide Production, Influence of Sulfide on Biochemical System and Removal Technology in Sulfate Wastewater* [D]. Qingdao University of Science & Technology, 2014.

Liu X Q; Zhang J G. *Study on Methods to Detect Sulfate Ion of Groundwater* [C]. Proceedings of the second National Congress on Geotechnology and Engineering (Volume 2). Tianjin: Tianjin Municipal Engineering Design and Research Institute, 2007, 431–434.
Su D.H, Wang C.H; Zhou J.J; Lin W. Sulphide Removal from Wastewater with Attapulgite /Manganese Oxide Composites [J]. *China Leather*, 2013, 23: 24–28.
Wang P, Lv C X, Sheng Q, et al. Research development of uranium-containing wastewater treatment technologies[J]. *Xiandai Huagong/Modern Chemical Industry*, 2016, 36 (12): 23–27.
Wu N. *Research on caustic washing desulfurization from oil production and harmless treatment of sulfur wastewater*. [D]. Gansu: Lanzhou Jiaotong University, 2014.

Energy Revolution and Chemical Research – Kok-Keong Chong and Zhongliang Liu (Eds)
© 2023 The Authors, ISBN: 978-1-032-36554-1

Study on recovery of platinum from waste ternary catalyst

Yu Bo*
School of Automotive Engineering, Wuhan University of Technology, Wuhan, Hubei

ABSTRACT: In view of the existing industrial methods for recovering platinum metal, aqua regia is widely used. It has strong acidity and oxidation and high requirements for acid and corrosion resistance of industrial equipment. At the same time, a considerable number of polluting nitrogen and chlorine-containing gases are generated, which seriously pollute the environment in the production process. This paper provides a new method for dissolving platinum, which uses a mixture of hydrobromic acid, sodium chloride and hydrogen peroxide to dissolve platinum. Compared with the current industrial conventional methods, this method is more energy-saving and environment-friendly. Under the condition of ensuring a certain dissolution rate, it effectively reduces the generation of chlorine and nitrogen-containing gas. At the same time, the by-products are relatively easy to recover and deal with, which meets the requirements of green development advocated at present.

1 RESEARCH BACKGROUND

Platinum group metals include platinum (Pt), palladium (Pd), osmium (Os), iridium (Ir), ruthenium (Ru), and rhodium (Rh). Due to their high melting point, high-temperature oxidation resistance, corrosion resistance, and other characteristics, they are widely used in the automobile industry, jewelry and financial industry, high-end weapons, petrochemical industry, electronic industry, glass industry, medicine, health, energy, and environmental protection are indispensable materials in the national economy and national defense construction (Liu & Qin 2021). In particular, the use of fuel cell catalysts and exhaust three-way catalysts in the automotive industry is inseparable from the participation of platinum group metals. The reaction of the three-way catalyst is shown in the table below:

Table 1. Main chemical reactions catalyzed by three-way catalyst (Zhang et al. 2021).

Main chemical reaction	Chemical reaction formula
Oxidation reaction of CO and HC	$CO + O_2 \rightarrow CO_2$
	$H_2 + O_2 \rightarrow H_2O$
	$HC + O_2 \rightarrow CO_2 + H_2O$
No reduction reaction	$CO + NO \rightarrow CO_2 + N_2$
	$HC + NO \rightarrow CO_2 + N_2$
	$H_2 + NO \rightarrow H_2O + N_2$
steam reforming	$HC + H_2O \rightarrow CO + H_2$
Water-gas conversion reaction	$CO + H_2O \rightarrow CO_2 + H_2$

Although platinum group metals are important, their content in nature is only 69000t, and 91.3% of the reserves are in South Africa, 5.65% in Russia and less than 2% in other regions. (Zhu 2021) It

*Corresponding Author: jqxxyb@163.com

 DOI 10.1201/9781003332657-74

can be said that the reserves are rare and unevenly distributed. With the rapid development of China's platinum industry, China needs to import a large number of platinum metals from foreign countries (Li & Han 2021). At present, China has a large number of automobile reserves. 60% of platinum group metals are used for the manufacture of three-way catalytic converters for automobile exhaust every year. The content of platinum group metals in three-way catalysts for automobile exhaust is hundreds of times that in the earth's crust (Xie et al. 2020). And a large number of three-way catalytic converters are scrapped every year. Therefore, recycling is imminent. This paper compares the more common recycling methods and puts forward a new recycling method.

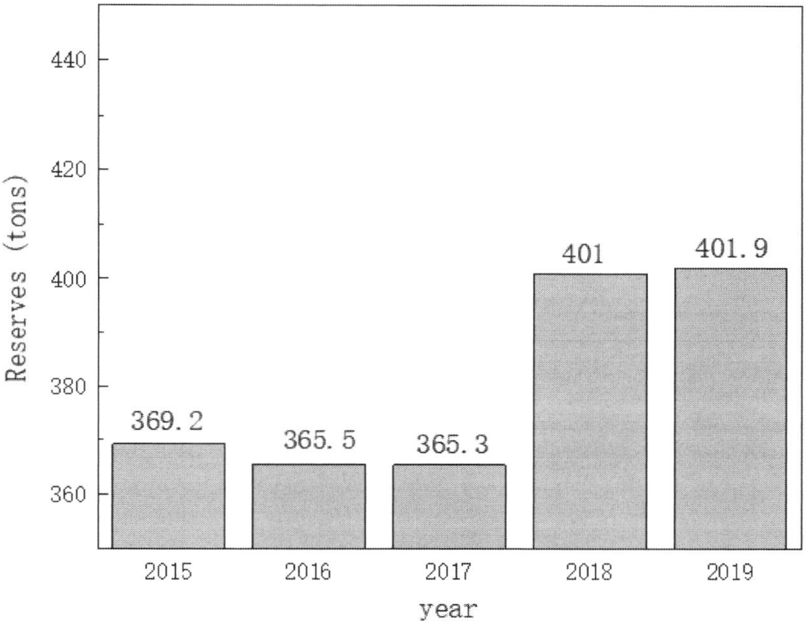

Figure 1. National bureau of statistics released the proved reserves of platinum group metals in China by 2019.

2 CURRENT RECYCLING METHODS AND THEIR DISADVANTAGES

At present, the common dissolution method of platinum is aqua regia dissolution. It is widely used in industry because of its simple process, high recovery rate, simple equipment, easy operation, and low cost. However, aqua regia has high requirements for equipment in the production process because of its strong acidity and strong oxidation. The equation for dissolving platinum is as follows:

$$3HNO_3 + 4HCl + 18Pt = 3NO_2 + 4H_2[PtCl_4] + 8H_2O,$$

in which the generated nitrogen oxide is volatile, which is the main reason for the formation of acid rain. At the same time, due to the active and interactive nitrogen oxide compounds, the generated pollutants and pollution effects are difficult to estimate.

Zhao Jishou (Zhao et al. 2008), etc. proposed the method of dissolving platinum with cyanide under pressure. The reaction equation is:

$$Pt + 8NaCN + O_2 + 2H_2O = 2Na_2[Pt(CN)_4] + 4NaOH$$

It is not easy to produce volatile substances in the reaction process. At the same time, the reaction avoids the environment of strong acid and strong oxidation and has low requirements for production equipment. However, the cyanide used is a highly toxic substance, which has great risks for production and has great hidden dangers for the safety of operators. At the same time, if raw materials or products leak into the environment, toxic pollution will be more serious.

In addition to the common wet dissolution process, Zhao Jiachun, etc. (Zhao et al 2018) used the copper capture method to recover platinum group metals from automotive catalysts. Under the conditions of CaO / SiO2 = 1.05, collector ratio of 35% ~ 40%, reductant ratio of 6%, the melting temperature of 1400°C, and melting time of 5 h. Japanese patent (Ezawa 1990) also studied an improved copper capture method. The main advantages of the copper capture method are as follows: good selectivity, high platinum content, and less loss in the melt. The melting temperature is lower than that of iron, and copper does little harm to the human body. And it has good recycling performance. However, the melting temperature is nearly 1500°C, the enrichment process is nearly 20h, and the energy consumption is very high. At the same time, smoke and dust will be generated in the smelting process, resulting in a certain degree of dust pollution.

To sum up, the existing methods are more or less polluted or difficult to deal with the generation of pollutants. Next, a method of dissolving platinum will be introduced to minimize the generation of pollutants and deal with the possible by-products at a low cost.

3 EXPERIMENTAL METHODS AND MATERIALS

3.1 *Experimental principle*

According to the principle of dissolving platinum in aqua regia,

$$3Pt + 4HNO_3 + 18HCl = 3H_2PtCl_6 + 4NO\uparrow + 8H2O$$

Based on the idea of chloride ion complexation and nitric acid as oxidant, hydrogen peroxide is used to replace oxidant, hydrobromic acid is used to replace hydrochloric acid to provide hydrogen ion, and sodium chloride is used to provide complex chloride ion. At the same time, because the oxidizability of bromine is lower than that of chlorine, it preferentially reacts with hydrogen peroxide in case of a side reaction to effectively inhibit the generation of chlorine. The generated bromine is liquid at room temperature and pressure, which is easy to recycle and reuse. The overall experiment is friendly to the environment. The main reaction equation is:

$$Pt + 4H^+ + 6Cl + 2H_2O_2 = PtCl_6^{2-} + 4H_2O$$

The main side reaction equation is:

$$2HBr + H_2O_2 = Br_2 + 2H_2O$$

The reaction does not produce volatile substances or substances that are difficult to treat later, which is more environmentally friendly.

3.2 *Reagents and instruments*

There are waste automobile three-way catalytic converter powder, hydrobromic acid (analytical purity, mass fraction $\geq$ 40%), sodium chloride (analytical purity, purity $\geq$ 99%), hydrogen peroxide (mass fraction $\geq$ 30%), sodium hydroxide (analytical purity, purity $\geq$ 99%), magnetic stirrer, muffle furnace (Bei Yike medium temperature furnace), and iron crucible.

Before the experiment, the obtained three-way catalytic converter powder was tested for all elements. It was measured that the main elements contained soluble aluminum is 14.87%, Er was 2.73%, Ce is 1.89%, and the content of platinum is only 0.03%. At the same time, there was a large amount of insoluble acid in the powderα-Al2O3 material and a large amount of silicon.

3.3 *Experiment procedure*

The ternary catalyst and matrix mixed powder 5g, mixed with NaOH with a mass ratio of 1:1.5, were evenly placed in a Muffle furnace and roasted at 650°C for 3h. After roasting, the powder appeared as a whole and was blue. After washing the roasted powder in 75°C water 2–3 times, the washed powder was put into 1mol/L hydrobromic acid for pickling 1–2 times to obtain white powder and dry it. According to the solid-liquid ratio of ternary catalyst and reaction solution 1:5, the mixture of NaCl, HBr, and 10%H_2O_2 was reacted, and the platinum content of the solution obtained was detected.

4 RESULTS AND DISCUSSION

4.1 *Pretreatment process*

During the experiment, the pretreatment process of the separation of precious metals and cordierite basal need time is longer, and the generated blue material in the process of the experiment is Na_2SiO_3. After washing and drying, the platinum content in the white powder reached 0.3%. As the precious metal in the catalyst used was platinum-palladium alloy, the content reached the proportion of platinum content in the catalyst, so it was considered that the substrate was completely removed. Due to the complexity of the structure of sodium silicate, not all sodium silicate generated in the roasting process will be removed in the pretreatment process. Therefore, the concentration and temperature of acid solution need to be strictly controlled in the pretreatment process to prevent the influence of sodium silicate on the whole experiment.

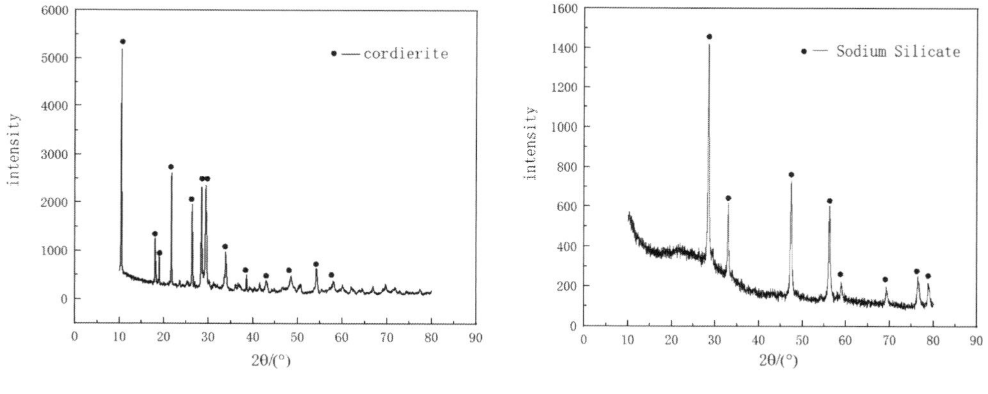

Figure 2. XRD analysis of ternary catalyst before and after pretreatment.

Because pretreatment cannot avoid the formation of refractory sodium silicate, if the conditions of the pickling process are not controlled, it will have a great impact on the whole pretreatment. As long as the specific reaction conditions are controlled, the impact of sodium silicate on the experiment can be avoided. To this end, we explored the effects of different acid concentrations on weight loss:

At room temperature, with the increase of acid concentration of pickling, it increases first and then decreases. When the concentration of hydrobromic acid is 3mol / L and the reaction temperature is 10°C, its solubility reaches 90%, but the overall change trend is small. With the continuous increase of temperature, the weight loss rate decreases significantly, and even increases under the conditions of high concentration acid and high temperature. During the experiment, when the pickling temperature is high, the solid after pickling is colloidal and difficult to dry, and it is sandy after drying. Therefore, with reference to the whole weight loss rate and from the perspective

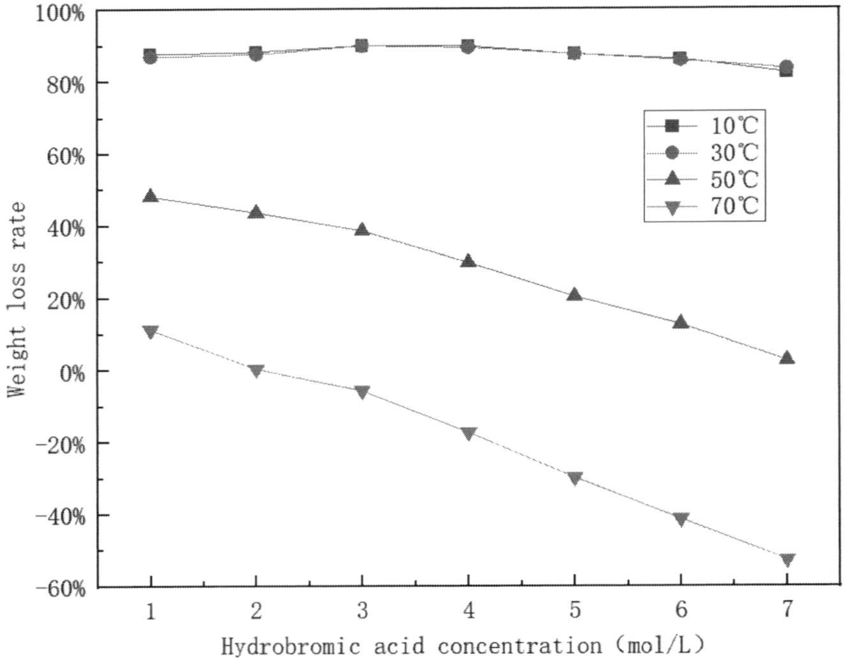

Figure 3. Weight loss rate of pickling under different temperatures and acid concentrations.

of economy, acid pickling with 1mol / L hydrobromic acid at room temperature can meet the needs of the acid pickling experiment.

4.2 *Effect of temperature and solid-liquid ratio*

Figure 4 is a broken line diagram of the solubility of the ternary catalyst under different experimental conditions, which discusses the dissolution rate of platinum at different temperatures and solid-liquid ratio under the conditions of 1mol/L hydrobromic acid, 5mol/L sodium chloride, and 10% hydrogen peroxide. The results showed that the dissolution rate reached 99.32% at the reaction temperature of 90 °C and the solid-liquid ratio of 1:5.

During the whole experiment, a small amount of gas was generated, but yellow or brownish yellow gas was rarely generated. After the reaction, the solution was orange as a whole, which was a side reaction caused by the reaction between hydrogen peroxide and hydrobromic acid under heating conditions.

$$H_2O_2 + 2HBr = Br_2 + 2H_2O + O_2\uparrow.$$

Due to the relatively low acidity and low bromine content in the reaction system, most of the generated bromine vapor is dissolved in the reaction system and does not overflow outward. Even if a small amount of bromine overflows through the condensation device at normal temperature, it can be used as a liquid reaction system. From the perspective of production cost and environmental protection, that is to save costs and reduce the damage to the environment.

Although the higher the reaction temperature is, the greater the reaction activation energy will be provided, which is more conducive to the forward progress of the reaction. However, due to the accelerated decomposition of hydrogen peroxide after 90°C, the H_2O_2 concentration will decrease rapidly with time in the reaction process with the increase of temperature, resulting in the rapid reduction of leaching rate (Study on enrichment mechanism and application of platinum group

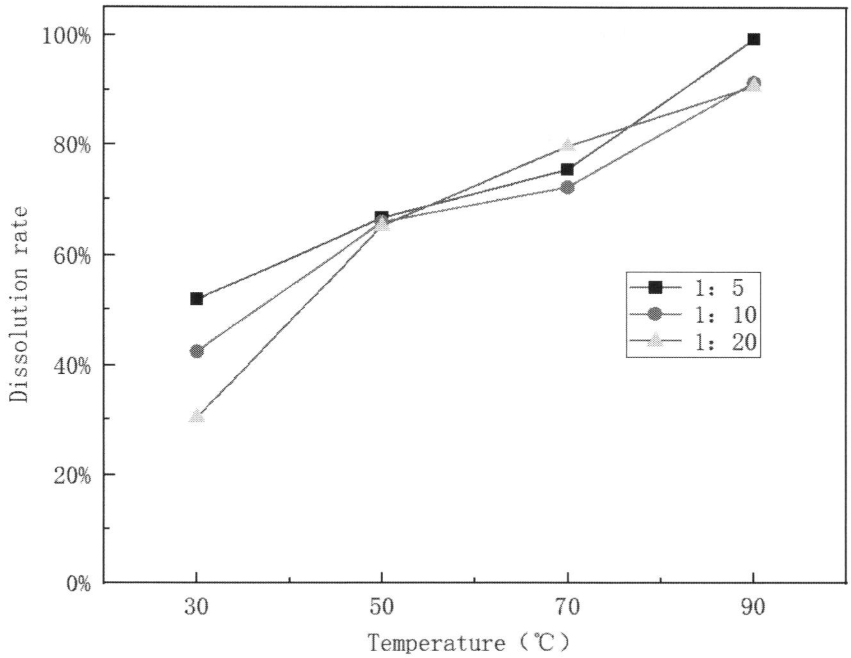

Figure 4. Leaching rate of platinum at different temperatures and solid-liquid ratio.

metals in Ding Yunji waste catalyst [D] Beijing: Institute of new materials technology, Beijing University of science and technology 2019), so the leaching effect after 90° is not discussed.

4.3 *Effect of HBr and H_2O_2 concentration*

Figure 5 shows the change of HBr concentration and H2O2 concentration in the reaction system under the conditions of 5mol/L sodium chloride, the solid-liquid ratio of 1:5, and a reaction temperature of 90°C. The effects of different concentrations on platinum dissolution are discussed

The experimental results show that the solubility increases first and then decreases with the increase of acid concentration. When the acid concentration is lower than 1mol/L, the acid concentration in the reaction system is too small, and the generated chloroplatinic acid may decompose, resulting in a decrease in the leaching rate. When the concentration of hydrobromic acid is higher than 1mol/L, as the temperature of the reaction system increases, a large amount of reddish-brown gas is generated in the reaction vessel, and the side reaction reacts violently, resulting in the rapid reduction of hydrogen ion concentration and hydrogen peroxide concentration in the system, which not only reduces the leaching effect, but also has strong oxidizing property. As equipment and operators have a lot of harm, they will cause environmental pollution if not properly handled. Similarly, when the concentration of hydrogen peroxide is high, the side reaction is violent and the main reaction is inhibited. When a low concentration system of oxidizing is insufficient, the dissolution rate will decrease.

$$H_2O_2 + 2HBr = Br_2 + 2H_2O + O_2 \uparrow$$

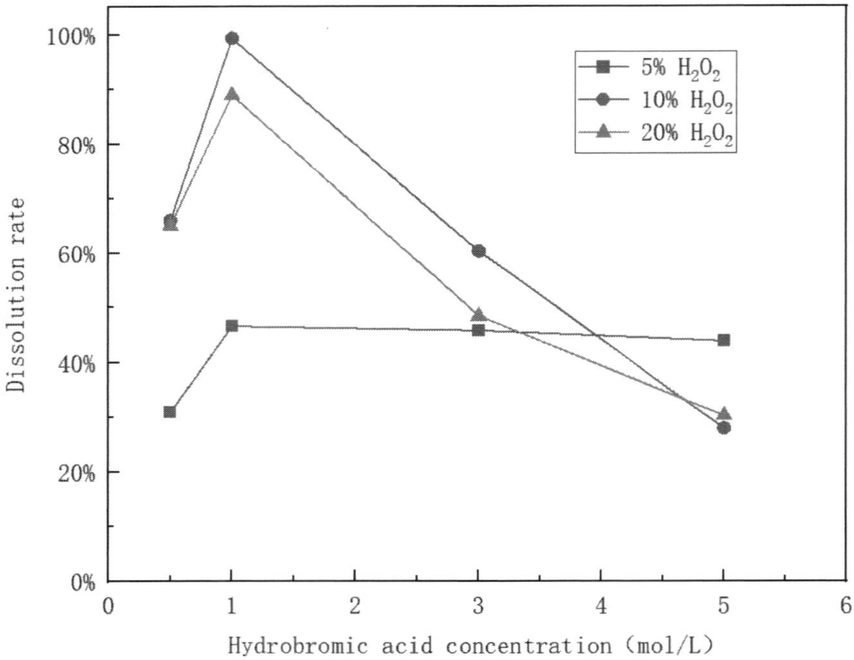

Figure 5. Leaching rate of platinum under different HBr and H2O2 concentrations.

5 CONCLUSION

To sum up, the following conclusions are obtained by dissolving hydrobromic acid, sodium chloride, and hydrogen peroxide under heating conditions:

(1) Based on the principle of aqua regia dissolution, when the concentration of hydrobromic acid is 1mol / L, the concentration of sodium chloride is 5mol / L, the mass fraction of hydrogen peroxide is 10%, the solid-liquid ratio is 1:5, and the reaction temperature is 90°C, the leaching rate reaches 99.32%.
(2) With the increase of acid concentration, leaching increases first and then decreases, which is due to the insufficient concentration of low concentration hydrogen ions and the instability of products. High concentration will lead to severe side reactions, which will affect the main reaction due to the low content of precious metals.
(3) No yellow-green gas was generated during the experiment. The generated gas was collected and detected after condensation. It was mainly oxygen, which was more environmentally friendly than the dissolution of aqua regia and other conventional methods. At the same time, the generated bromine gas can be used to prepare HBr with hydrogen after collection, and then generate reactants. The by-products can be reasonably utilized to reduce the generation cost.
(4) This paper innovatively proposes a green way to recycle platinum, which can provide a new idea for reducing cost and green production.

REFERENCES

Ezawa N. *Recovery of precious metals*: Japan, JP 92317423[P]. 1990-07-02.

Li Zhi, Han Zhimin, recovery and purification technology of platinum from domestic waste catalysts [J] *Tianjin chemical industry*, 2021, 35 (4), 71–75.

Liu Yanmin, Qin Bo. *Research progress of platinum purification and recovery process* [J] nonferrous metals, 2021, issue 3, 92–99.
Study on enrichment mechanism and application of platinum group metals in Ding Yunji waste catalyst [D] Beijing: Institute of new materials technology, Beijing University of science and technology, 2019.
Xie Xue, Qu Zhiping, Zhang Bangsheng, Liu Guiqing, Zhang Yifan, Zhang Baoming. Enrichment of platinum group metals in ineffective automobile exhaust purification catalysts [J] *Comprehensive utilization of resources in China*, 2020, 38 (5), 105–109.
Zhang Ruoran, Chen Qishen, Liu Qunyi, Yu Wenjia, Tan Huachuan. Global demand forecast and supply and demand situation of major platinum group metals Analysis [J] *Resource science*, 2015, 37 (05): 1018–1029.
Zhang Zhaoliang, He Hong, Zhao Zhen. Research and application of three-way catalyst for automobile exhaust for 40 years [J] *Environmental chemistry*, 2021, 40 (7): 1937–1944.
Zhao Jiachun, Cui Hao, Bao Simin, et al. Recovery of platinum, palladium and rhodium from spent automotive catalysts by copper capture [J], *precious metals*, 2018, 39 (1), 56–59 Fourteen.
Zhao Jishou, Huang Kun, Chen Jing. Experimental study on the dissolution kinetics of platinum in pressurized cyanidation [J] *Rare metals*, 2008, Vol. 32, No. 2, 211–215.
Zhu Qing. Analysis of the impact of the development of major international platinum group metal markets on China's economy [J]. *Comprehensive utilization of resources in China*, 2021, 39 (5): 94–98.

Energy Revolution and Chemical Research – Kok-Keong Chong and Zhongliang Liu (Eds)
© 2023 The Authors, ISBN: 978-1-032-36554-1

Research on flue gas denitrification technology of natural circulation boiler

Qiudong Hu
Shandong Huayu University of Technology, Dezhou, Shandong Province, China

ABSTRACT: The continuous progress of science and technology and the continuous development of the economy have led to the increasing demand for electric energy. At present, in addition to the use of clean energy sources such as nuclear energy, wind energy, and solar energy for power generation, traditional coal-fired power plants have always been the main source of power generation, compared with clean energy generation, coal-fired power plant boilers will emit a large number of environmentally harmful nitrogen, sulfur-containing flue gas. Based on the relevant parameters of a 2200T/h natural circulation boiler in a power plant in Dezhou, the design of the flue gas denitrification system is carried out, and based on the selective catalytic reduction (SCR) technology, the reduction effect of ammonia (NH_3) on the nitrogen oxide is utilized, under the action of catalyst, the NO_x in flue gas can be converted into non-toxic N_2 and H_2O, thus the NO_x emission can be reduced, and the goal of energy conservation and emission reduction can be achieved. The main design contents of this paper include the SCR reactor, spraying ammonia system, ash blowing system, and SCR-DCS control system, and the NO_x content is below 50mg/m^3, and the denitration rate is above 95%.

1 INTRODUCTION

Coal-fired boilers in power plants consume a large number of coal resources during the power generation process, and at the same time continuously emit nitrogen oxides (NO_x) into the atmosphere. As one of the main pollutants in the outdoor atmosphere, it will induce acid rain after entering the atmosphere, seriously endangering the safety of the biological environment. Therefore, it is necessary to reduce the emission of pollutants such as nitrogen oxides and sulfides in the flue gas of coal-fired power plants, and implement the green development concept of "lucid waters and lush mountains are invaluable assets". China pointed out in the "Air Pollutant Emission Standard for Thermal Power Plants" that the NO_x emission of ultra-low units should be lower than 50mg/m (Yu 2017). To meet the emission standards, various coal-fired power plants have designed and modified the boiler denitration system and adopted different denitration technologies, of which SCR denitration technology is the most widely used.

According to different generation mechanisms, NO_x generation pathways can be divided into three categories: thermal NO_x, fast NO_x, and fuel NO_x (Cao 2010; Lei 2010). Fuel NO_x accounts for about 80% of NO_x emitted by coal-fired power plants. There are two ways to control NO_x emissions: the first is to improve combustion technology to adopt low-nitrogen combustion or use natural gas, nuclear energy, and other clean energy to reduce the amount of NO_x generated from the source; the second is to reduce the amount of NO_x generated by coal-fired power plants. Denitration technologies such as SCR, SNCR, etc. can be used to reduce its emissions (Zhao 2012).

 DOI 10.1201/9781003332657-75

2 SELECTIVE CATALYTIC REDUCTION (SCR)

The SCR denitration method selected for the design is a technology that reduces its emissions by using ammonia to react with NO_x in the boiler flue gas. The reaction principle is to inject the diluted reducing agent ammonia gas (NH_3) into the smoke through the ammonia injection grille. In the gas, it is fully mixed with the NO_x in the flue gas and then evenly enters the SCR reactor. Under the "catalysis" effect of the layer-by-layer catalysts in the reactor, NH_3 "selectively" reacts with the NO_x in the flue gas to form non-toxic N_2 and H_2O (Yu et al. 2016). In addition to SCR denitrification technology, there are SNCR, liquid absorption methods, and microbial methods for flue gas denitrification of coal-fired boilers in power plants (Zhuang 2010).

3 MAIN FACTORS AFFECTING SCR DENITRATION EFFICIENCY

The denitration efficiency of SCR denitration equipment for coal-fired boilers in power plants is not only affected by the ammonia injection grille and catalyst, but also by the boiler flue gas temperature, flue gas dust content, sulfide content, escaped ammonia content, and other factors.

3.1 *Flue gas temperature*

At present, the flue gas temperature required by the SCR denitration equipment of conventional coal-fired power plant boilers is 280~420°C. If the flue gas temperature is higher than 420°C, the catalyst will be sintered under high-temperature operation, thereby reducing the reactivity of the catalyst (Wang 2017). If the flue gas temperature is less than 280°C, the operation at low temperature will slow down the reaction rate and reduce the denitration efficiency. Therefore, to meet the reaction temperature, the SCR denitration equipment adopts a high-ash type arrangement, which is arranged between the economizer and the air preheater.

3.2 *Dust content in flue gas*

The SCR denitration equipment adopts a high-ash type layout. Although it meets the reaction temperature requirements, the smoke and dust content in the flue gas is relatively large (measured at about 38700mg/m^3), and the impact on the SCR reactor catalyst is mainly reflected in three aspects:

(1) The soot will cause the catalyst in the denitration equipment to block and reduce the catalyst reaction efficiency;
(2) The alkaline substances such as K, Na, and Ca in the soot cover the surface of the catalyst, which will cause poisoning and failure of the active substances contained in the catalyst;
(3) The large particle ash in the soot will cause wear on the catalyst unit, greatly reduce the service life of the catalyst, increase the consumption of the catalyst, and increase the operating cost of the power plant.

Therefore, it is necessary to arrange the soot blowing device for the catalyst layer of the SCR denitration equipment, optimize the flue gas flow field, and reasonably select the catalyst spacing to reduce the adverse effect of the soot on the denitration equipment.

3.3 *Escaping NH_3 content*

The formation of fugitive ammonia is caused by the incomplete reaction of NO_x with the reducing agent NH_3 in the SCR reactor. Under ideal conditions, NO_x and NH_3 are evenly distributed on the flue section of the SCR reactor, NH_3 can fully participate in the reaction and be consumed, and the escaped ammonia content in the flue gas is zero. In the actual working condition, the NO_x concentration distribution in the flue gas is uneven, and the injected NH_3 cannot be guaranteed to be evenly distributed in the flue gas. For example, in the region where the NH_3 concentration is too high, it is likely that NH_3 cannot be completely reacted and consumed, thereby forming escaped ammonia. Sulfur trioxide (SO_3) in the flue gas will form $(NH_4)_2SO_4$ and NH_4HSO_4 with the escaped ammonia in the reactor. The strong water absorption of the two substances will cause the pores of the catalyst unit to block and corrode, greatly reducing the denitration efficiency of the catalyst. service life.

In addition, the escaped ammonia will not only cause harm to the catalyst of the denitration equipment, but also cause the blockage of the air preheater downstream of the denitration equipment. Therefore, the escaped NH_3 content in the flue gas should be controlled below 3%, which can be achieved by optimizing the design of the ammonia injection grille and the SCR-DCS ammonia injection control system.

3.4 *Uniformity of NO_x inlet distribution*

The NO_x content and distribution uniformity at the inlet of the SCR reactor directly affect the NO_x content at the outlet. When the flue gas enters the SCR reactor, the more uniform the distribution is, the more favorable the denitration reaction is. At the same time, the SCR-DCS system is optimized to monitor and analyze the NO_x content in the inlet and outlet flue gas of the denitrification equipment in real-time. When it is detected that the NO_x concentration in the outlet flue gas is high, the SCR-DCS system will comprehensively analyze the boiler load, inlet NO_x concentration, and other variables, adjust the dilution air volume in real-time, and reduce the NO_x concentration.

4 CATALYST SELECTION AND ARRANGEMENT OF SCR DENITRATION EQUIPMENT

4.1 *Catalyst selection and characteristic analysis*

Catalysts are the core of the SCR denitration system and can be classified into zeolite catalysts, precious metal catalysts, metal oxide catalysts, and activated carbon catalysts according to their material composition and operating temperature (Xing & Zhou 2016).

4.2 *Catalyst performance comparison*

SCR denitration system catalysts are divided into honeycomb type, flat plate type, and corrugated plate type according to different physical structures in addition to the material composition (Li 2016). According to different soot concentrations, the number of catalyst holes can be divided into 13×13 holes, 15×15 holes, 16×16 holes, 18×18 holes, and 20×20 holes. The specific characteristics are compared in Table 1.

Table 1. Performance comparison of catalysts with different pore numbers.

Serial number	Number of holes	Pitch (mm)	Applicable soot concentration (g/m^3)
1	20×20	7.6	≤35
2	18×18	8.2	≤45
3	16×16	9.2	≤55
4	15×15	10	≤60
5	13×13	11.9	≤100

4.3 *Catalyst module arrangement*

The cross-sectional dimension of the unit of the selected catalyst is 150 mm×150 mm, and each formed catalyst module is composed of 6×12 unit units, and its final size is 1910 mm×970 mm × 1200 mm. The cross-sectional dimension of the designed SCR reactor is 13890 mm × 13620 mm, and each layer of catalysts is arranged according to 7 × 14, that is, 98 catalyst modules are required for each layer, and a total of 392 catalyst modules are required for four layers.

5 DESIGN OF AMMONIA INJECTION SYSTEM FOR SCR DENITRATION EQUIPMENT

5.1 *Ammonia injection grid design*

The ammonia injection grid is arranged in the rising flue at the inlet of the SCR reactor and consists of several nozzles and nozzles. According to the consumption of reducing agent NH3, the dilution air volume, and the size of the SCR reactor, the main ammonia supply pipeline is divided into 7 groups of branch pipes on the outside of the SCR reactor, and each group of branch pipes consists of 4 nozzle pipes.

5.2 *Ammonia injection control*

There are multiple NO_x analyzers in the SCR denitration equipment, and the analysis data is connected to the ammonia injection control system. The system calculates the amount of ammonia injection required by the equipment in real time according to the detected NO_x concentration at the inlet and outlet of the SCR reactor and sends adjustment signals to the AIG and the dilution fan. Adjust the amount of ammonia injection and dilution air (Zhuo & Xu 2020), to control the NO_x emission concentration within a reasonable range.

5.3 *Ammonia and air mixing system*

After the ammonia and air mixing system receive the signal from the ammonia injection control system, by adjusting the ammonia flow and dilution air volume, a certain concentration of ammonia and air mixed gas is injected into the flue through the ammonia injection grill. This process requires the opening of the regulating valve to be accurate, and the amount of ammonia injected to match the actual required amount. If the ammonia injection is insufficient, the NO_x content in the flue gas will increase sharply, and the excessive injection of ammonia will lead to an increase in the ammonia that does not participate in the reaction in the flue gas. It causes the ammonia escape rate to rise and aggravates the generation of ammonium sulfate and ammonium hydrogen sulfate. Therefore, the ammonia and air mixing supply device not only needs a reasonable design of the ammonia injection grill, but also a set of precise ammonia and air mixing control systems.

The selected dilution fan specifications and models are: 9-19No.12.5D/75kW-4P-B3, in which 9-19 represents the national standard serial number, 12.5 represents the machine number, D represents the transmission mode (coupling transmission), 75kW-4P- B3 means that the matching motor is 75kW, the rated speed is 4 poles (1450r/min), the installation method is horizontal, and the motor brand is Guanglu Motor.

6 SELECTION AND LAYOUT OF SOOT BLOWING SYSTEM OF SCR DENITRATION EQUIPMENT

To meet the requirements of the working temperature of the catalyst, the equipment adopts a high-ash type layout. The flue gas contains a lot of dust without an electrostatic precipitator. The dust will accumulate on the surface of the catalyst after the long-term operation of the device, causing the catalyst to block and reduce the denitration effect. Therefore, to reduce the harm of the dust in the flue gas to the reactor and ensure the denitration efficiency, according to the size of the catalyst layer and the working conditions of the two soot blowers, 10 sets of Kecon IKT230GD/170 sonic soot blowers and 8 sets of PSAT were installed in each catalyst layer. Drake steam soot blower,

40 sonic soot blowers, and 32 steam soot blowers are arranged in four catalyst layers. Each layer of sonic soot blowers and steam soot blowers is installed on both sides of the inner wall of the reactor. The installation position does not affect catalyst replacement. The soot blowing medium is compressed steam from the boiler. The two sets of soot blowing systems are equipped with independent compressed gas storage tanks to stabilize the voltage.

6.1 *IKT230GD/170 Sonic Sootblower*

Kekang IKT230GD/170 sonic soot blower uses the compressed stream from the boiler to convert it into high-power sound waves. The generated sound waves have a certain amount of energy. Through the vibration of the sound waves, the dust in the flue gas cannot accumulate and form on the surface of the catalyst. shell.

6.2 *PSAT/D rake steam soot blower*

The soot blowing element of the PSAT/D rake steam soot blower is a soot blower, which consists of a central pipe and several branch pipes which are 90° with the central pipe and are equipped with special nozzles. The principle of soot blowing is to use boiler steam to eject air flow from the nozzle (the pressure is 0.8MPa) to clean the soot deposited on the surface of the catalyst.

7 CONCLUSIONS

Based on the relevant parameters of a 2200T/h natural circulation boiler in a power plant in Dezhou, the article designs the boiler flue gas denitrification system. Selective catalytic reduction technology (SCR) is used to utilize the reduction effect of ammonia on nitrogen oxides. The conversion of nitrogen oxides into non-toxic nitrogen and water is a widely used denitrification technology for coal-fired power plant boilers. Based on the flue gas parameters of the boiler, coal quality data, and catalyst characteristics, the formula is used to calculate the ammonia consumption of the reducing agent, the size of the reactor, the size of the catalyst, etc., and the design drawing is drawn. The following conclusions are drawn:

(1) It is calculated that when the SCR reactor meets the denitration requirements, the ammonia consumption of the reducing agent is 1292kg/h, and the total dilution air volume is 24548m3/h. A total of two dilution fans are arranged, one of which is a standby fan;
(2) A total of 4 layers of catalysts are arranged in the reactor, and the “3+1” arrangement is adopted, and the bottom catalyst is the spare catalyst layer;
(3) The reactor adopts a high-ash type layout, which is arranged between the economizer and the air preheater. There is a lot of soot in the flue gas, which will cause the catalyst to block. Therefore, steam and sonic soot blowers are arranged on each layer of catalyst;
(4) For the design of the ammonia injection system, which involves the calculation of the amount of ammonia injection, the calculation of the dilution air volume, etc., errors in each item will affect the index that the final denitrification rate is not less than 95%. For example, if the amount of ammonia injection is too large, it will cause excess ammonia in the flue gas and increase the generation of ammonium bisulfate and ammonium sulfate on the catalyst surface, which will not only cause the catalyst to block, but also increase the ammonia content in the air, endangering human health and environmental quality.

ACKNOWLEDGMENT

Thanks to the data support of the Dezhou Power Plant, this technology has been implemented and achieved practical results.

REFERENCES

Cao Ming. Analysis of industrial boiler SCR flue gas denitration technology [J]. *Internal combustion engine and accessories*, 2010, (11): 33–34.

Lei Jie. *Research on nitrogen emission flux from coal combustion in China* [D]. Nanchang: Nanchang University, 2010.

Li Na. *Design of flue gas denitrification control system by selective catalytic reduction method* [D]. Beijing: North China Electric Power University, 2016.

Wang Xin. *Research on optimization design and simulation of low-temperature SCR reactor* [D]. Beijing: North China Electric Power University, 2017.

Xing Jianmin, Zhou Rong. Research on the removal of harmful gases by high-temperature filter media [J]. *Journal of Chengdu Textile College*, 2016, 33(2): 18.

Yu Feixiang. *A comprehensive evaluation of flue gas pollution status and control technology of small and medium-sized industrial boilers in the Yangtze River Delta region* [D]. Zhejiang: Zhejiang University, 2017.

Yu Fengping, Li Qingyi, Zhao Jinlong. On-line measurement method of flue gas flow in selective catalytic reduction flue gas denitrification system [J]. *Thermal Power Generation*, 2016, 45(2): 101–104.

Yuqun Zhuo, Xuchang. The development of pollution control technology in coal combustion in China[J]. *Frontiers in Energy Volume*, 2020, 1(01): 9-15.

Zhao Baojiang. Overview of mercury removal technology in thermal power plants [J]. *Energy and Environmental Protection*, 2012, 27(6): 10–13.

Zhuang Jianhua. Application of SCR flue gas denitration technology [J]. *Power Generation Equipment*, 2010, 12(2): 142.

Energy Revolution and Chemical Research – Kok-Keong Chong and Zhongliang Liu (Eds)

Study on recovery of soman from decontaminant of zirconium hydroxide

ShaoXiong Wu, Haiyan Zhu, Lianyuan Wang*, Xiaoyu Liu* & Liang Ge
Institute of NBC Defence, PLA Army Beijing, China

ABSTRACT: Due to the active center of acid and alkali in zirconium hydroxide Zr(OH)4, it can react with chemical warfare agents (CWAs) such as mustard (HD), soman (GD), and VX. Studies on the decontamination of CWAs were often carried out by in-situ semi-quantitative analysis methods like Nuclear Magnetic Resonance (NMR) or Fourier Transform Infrared Spectroscopy (FT-IR), with problems of analysis distortion. In this paper, a way of accurate analysis with GC for residual GD in decon sample of GD with $Zr(OH)_4$ was tried to develop. It was found that a suitable neutralizer should be introduced for the effective recovery of GD. The effects of different kinds of neutralizers and extractants on recoveries of GD were investigated. By using 20 ml water as a neutralizer and 20 ml anhydrous ethanol as extractant, the recoveries of GD reached 36.39% and 82.37% for samples with low concentration (3 μL) and high concentration (30 μL), respectively. Rich active hydroxyl (-OH) group in neutralizer aqueous solution could destroy or inhibit the combination of hydroxyl groups and GD in $Zr(OH)_4$, accordingly, this promoted the extractant to separate and enrich GD from $Zr(OH)_4$, and improved the recovery of GD.

1 INTRODUCTION

Chemical warfare agents (CWAs) such as mustard (HD), Soman (GD), and VX, et al., bring a serious threat to personnel safety and the eco-environment in a century, due to their high toxicity, secrecy, and mass destruction (Munro et al 1999). Nerve agents such as Soman GD have multiple forms of killing and fast toxic effect, and it is difficult to treat (Davies et al 1996), so there are several possibilities to use them in future military action and terrorist attacks (Lane et al. 2020). Decontamination plays a significant role in eliminating the consequences of chemical attacks. Therefore, researchers are committed to developing various kinds of decontaminants, which are of high efficiency and environmentally friendly to eliminate the destruction of CWAs to personnel life and the environment (Kim et al 2011; Singh et al 2010; Yang et al. 1992).

George's team (Bandosz et al. 2012) reported that the solid zirconium hydroxide ($Zr(OH)_4$) has excellent degradation properties for CWAs in 2012. Since then, domestic and foreign research teams have taken different analytical methods and studied the adsorption and degradation properties of $Zr(OH)_4$ to CWAs and their simulants(Balow et al. 2017; Jang et al 2020; Jeon et al 2020, 2019; Long et al. 2020). Solid $Zr(OH)_4$ contains a large number of acidic and alkaline hydroxyl (-OH) groups and Lewis acid zirconium metal nodes, which can chemically absorb GD through hydrogen bonding or Lewis acid-base interaction(Balow et al 2017; Colon-Ortiz et al 2019; Jeon et al. 2020). Most works of literature reported that the dynamics characteristics of adsorption and degradation of nerve agents or their simulants by $Zr(OH)_4$, were detected by in-situ analysis methods such as Nuclear Magnetic Resonance (NMR)(Bandosz et al. 2012; Wagner et al. 2007) or Fourier Transform Infrared Spectroscopy (FT-IR)(Balow et al. 2017; Jeon et al. 2019, 2020; Long et al

*Corresponding Authors: wangly09@sina.com and 752792683@qq.com

 DOI 10.1201/9781003332657-76

2020). However, there is chemical adsorption between $Zr(OH)_4$ and nerve agents or their simulants, which widens the signal peaks of nerve agents and their degradation products, leading to the result that this in-situ analysis method has become semi-quantitative analysis.

Gas chromatography can quantitatively analyze the residual amount of toxic agent with high sensitivity after decontamination, however, the precondition to using this method is the enrichment of residual toxic agent by solvent extraction from the decontamination system. In this paper, neutralizers such as an aqueous solution are used to neutralize the activity of hydroxyl and Lewis acid groups from $Zr(OH)_4$, destroying the strong chemical adsorption between $Zr(OH)_4$ and GD, then the residual GD in solid-phase zirconium hydroxide decontamination system are extracted by solvent. High recovery of GD in the $Zr(OH)_4$ would be useful for the follow-up study of GD's degradation.

2 EXPERIMENT

2.1 *Chemicals and Solution*

The toxic agents of GD have a purity >95%. Because of their high toxicity. They were handled only by well-trained personnel using appropriate safety procedures.

The internal standard for soman is tributyl phosphate at a concentration of 98% obtained from Alfa Aesar (Shanghai) Chemical Co., Ltd.

Decontaminants: Nanometer $Zr(OH)_4$ (A.R.) was purchased from Shanghai Zhongye New Material Reagent Company. Ordinary zirconium hydroxide (A.R.) was purchased from Damao Chemical Reagent Factory (Tianjin). Activated clay was bought from Shanghai Macklin Biochemical Co., Ltd. Activated charcoal was bought from Jiangsu Xinghong Carbon Industry Technology Co., Ltd.

Extractants: The chromatographically pure dichloromethane was purchased from Puredil, Germany; Anhydrous ethanol and ethyl acetate (A.R.) was purchased from Beijing Chemical Works.

Neutralizers: Concentrated Hydrochloric acid (36%), Sodium tetraborate, sodium, and Hydrogen Phosphate (A.R.) obtained from Beijing Chemical Works. Add 10 μL concentrated hydrochloric acid into 100 mL deionized water to prepare hydrochloric acid neutralization solution, with the pH value of the solution measured by pH meter is 3.27; Weigh 1.906 g of sodium tetraborate accurately and dissolve it in 100μL deionized water to prepare borax neutralization solution of 0.05 mol/L; Weigh 0.071 g sodium dihydrogen phosphate solid accurately and dissolve it in 100 mL deionized water to prepare sodium dihydrogen phosphate neutralization solution of 0.005 mol/L.

Analytical reagents: Sodium Perborate and Benzidine Monohydrochloride (A.R.) were purchased from Sinopharm Chemical Reagent Co., Ltd. Acetone (A.R.) was purchased from Beijing Chemical Works.

2.2 *Apparatus*

The solution extracted by dichloromethane and ethyl acetate was detected by a gas chromatograph with a flame photometric detector (FPD) named N6890, bought from American Aligent Technologies, which was equipped with an HP-5 capillary column (30 m×0.32 mm×0.25 μm), employing a temperature program (60–280°C at 15°C/min), using triethyl phosphate as an internal standard of GD.

The UV-Visible Spectrophotometer named UV-6000PC was used for the detection of ethanol extraction solution after chemical coloration, bought from Shanghai Metash Instruments Co., Ltd.

A HI-2221 Specialty Laboratory pH/ORP/T Tester purchased from Italian Hanna Corporation was used to indicate the change of temperature and pH online.

The extracted mixture solution was centrifuged through a high-speed centrifuge that the model is HC-3018, bought from Anhui USTC Zonkia Scientific Instruments Co., Ltd.

Pipetting guns (10 μL ~ 10 mL) and conical flasks with a bottom area of about 30 m^2 were used for removing and laying out the soman agent, purchased from Ebend Life Science (Germany) and Beijing Glass Apparatus Factory, respectively.

2.3 *Preparation for the standard solution of GD*

Use the pipette gun to accurately take 30 μL or 3 μL of GD, and distribute it evenly into 50 to 100 droplets at the bottom (about 30 cm^2 in area) of the conical flasks with cover within 30 s, simulating the contamination density of 10 g/m^2 and the contamination density with decontamination rate of more than 90% respectively. Then add organic solvents such as dichloromethane, ethyl acetate, and absolute ethanol into the conical flasks above and shake for 1 min. After standing for 10 min, the extraction solution of dichloromethane and ethyl acetate is quantitatively analyzed by the GC-FPD internal standard method, and the extraction solution of absolute ethanol is quantitatively analyzed by spectrophotometer after its chemical coloration reaction through cross-linking method (Schneidman principle).

2.4 *Recovery of residual GD from decontamination samples*

Equably distribute 3 μL or 30 μL GD (completed within 30 s) in the conical flasks (the requirements are the same as 2.3 above) and add 1.0 g zirconium hydroxide powder into it, shake for 5 seconds to make the powder equably distributed at the bottom of the flask, then immediately add neutralizer and extractant, such as dichloromethane, ethyl acetate or absolute ethanol into conical flasks above and shake for 20 s, then centrifuge the organic phase for 1 minute after standing for 10 min, finally take the clear liquid for quantitative analysis. The analytical method is the same as 2.3 above.

2.5 *Calculation of GD's recovery rate*

Assuming that C_0 is the initial concentration of the GD in the disinfection system from GD standard solution, GD of concentration C_1 is the recovery concentration of GD in the extracted organic phase, then the recovery rate (R%) is calculated as:

$$R\% = \frac{C_1}{C_0} \times 100\%$$

3 RESULTS AND DISCUSSION

According to the literature, there are a large number of hydroxyl groups and Lewis acidic zirconium metal nodes contained in zirconium hydroxide. These groups can chemisorb GD by forming hydrogen bonds or coordination bonds with O atoms and F atoms of P=O and P-F bonds from GD(Jeon, Schweigert, Pehrsson, et al 2020). Then, the basic hydroxyl groups from $Zr(OH)_4$ or adsorbed water molecules nucleophilic attack GD adsorbed on $Zr(OH)_4$ and a hydrolysis reaction occurs. In this experiment, different organic solvents and neutralizers were selected to destroy the chemical adsorption between $Zr(OH)_4$ and GD, to make GD extracted from the solid-phase decontamination system of $Zr(OH)_4$ as much as possible to obtain a high recovery.

3.1 *Selection of extractants*

In the experiment, dichloromethane, ethyl acetate, and absolute ethanol were used as extractants to extract the residual GD from solid-phase $Zr(OH)_4$ decontaminant. The recoveries determined are shown in Table 1.

Table 1 shows that, among the three organic extractants, anhydrous ethanol has the highest recovery of GD from solid-phase decontaminant, followed by ethyl acetate, while the lowest is

Table 1. Recoveries of GD from solid $Zr(OH)_4$ by different extractants.

Extractant	Volume/ mL	3 μL-R/%	30 μL-R/%
Dichloromethane	40	.05	1.19
Ethyl acetate		6.10	33.60
Anhydrous ethanol		21.82	75.72

dichloromethane, and the recovery of high-dose GD (30 μL) is generally higher than that of low-dose GD (3 μL). For high dose contamination of GD, the molar amount of $Zr(OH)_4$ is more than 40 times. The contamination density of low-dose GD is lower, which makes the contact between $Zr(OH)_4$ and GD droplets more sufficient, and the adsorption and degradation are stronger, therefore, the recovery of low-dose GD is lower.

The recovery rate of GD in the decontamination system by the three solvents was consistent with the order of solvent polarity (ethanol > ethyl acetate > dichloromethane). It could be that because ethanol molecules have highly polar hydroxyl groups (-OH), which can occupy the adsorption and reaction active sites in $Zr(OH)_4$, weaken the chemical adsorption between $Zr(OH)_4$ and GD, then it is easy to separate GD from the solid-phase degradation system of $Zr(OH)_4$ to obtain a high recovery rate. The carbonyl group (C=O) of the ester group in ethyl acetate also has a certain polarity, but the polarity of the ester group is weaker than that of the hydroxyl group in ethanol, so the competition for the adsorption active sites of $Zr(OH)_4$ is weaker, and the molecular weight of ethyl acetate is larger, so its recovery rate of GD is lower. There is no active group in dichloromethane that can form a hydrogen bond with the active adsorption site of $Zr(OH)_4$, so it is quite difficult to weaken the chemical adsorption between $Zr(OH)_4$ and GD to obtain a lower recovery rate.

3.2 *Selection of neutralizers*

$Zr(OH)_4$ contains a large number of alkaline and acidic -OH groups with reactive activity. In this experiment, aqueous solutions of different pH values are used as neutralizers during the extraction process, to neutralize the active groups with adsorbability and reactivity from $Zr(OH)_4$, and the chemical adsorption between $Zr(OH)_4$ and GD was weakened, thereby obtaining a high recovery. The results are shown in Table 2.

In the experiment, the pH value of several mixtures of neutralizers and solid $Zr(OH)_4$ was determined. When 1 g nano $Zr(OH)_4$ was first mixed with 10 ml deionized water, its pH value was 9.80, which showed that there were free -OH groups in the nano $Zr(OH)_4$. Then water was removed and added to 10 ml water once more, pH values of this solution remained constant, which showed that this nano $Zr(OH)_4$ is rich in alkaline groups. As shown in table 2, the pH value of the mixture reaches higher than 10 after mixing an alkaline borax solution of sodium hydrogen phosphate solution with solid $Zr(OH)_4$. In theory, the enhancement of alkalinity will accelerate the degradation of GD, but the recovery of GD not only does not decrease but increases as result. The possible reason may be that solvent (water molecules) and solutes (borax and sodium hydrogen phosphate) all contain hydroxyl groups with strong activity, resulting in strong chemical competitive adsorption with GD molecules, which destroys or inhibits the combination between hydroxyl groups from $Zr(OH)_4$ and GD molecules, making it is easy for dichloromethane to extract GD from solid-phase $Zr(OH)_4$ decontaminant. When HCl solution was used as a neutralizer, the pH value of the mixture was neutral, which reduced the degradation rate of GD, and destroyed the alkaline -OH groups in $Zr(OH)_4$ meanwhile, the degradation ability of $Zr(OH)_4$ to GD reduced as a result.

Table 2 showed that when water is used as a neutralizer, the extraction effect of GD by dichloromethane and absolute ethanol had improvement in different degrees except for ethyl acetate, especially the promotion effect on the extraction of dichloromethane is more obvious. Other neutralization solutions, such as 0.05 M borax, 0.005 M Na_2HPO_4, and HCl with a pH value of 3.27, also significantly improved the recovery of GD by dichloromethane, especially for GD of high concentration (30 μL), the recovery reached more than 45%.

However, no matter what kind of neutralization was used, the recovery of GD is always higher when ethanol was used as extraction, especially when water was used as neutralization, the recovery was higher than that of dichloromethane, which reached 36.39% and 82.37% for the GD of low concentration (3 μL) and high concentration (30 μL) respectively, the reason may be that the ethanol-water system is rich in active hydroxyl groups and the low pH value during neutralization is not conducive to the hydrolysis of GD.

In conclusion, for solid-phase $Zr(OH)_4$, adding a large number of active hydroxyl groups through neutralizer and extractant, the combination of hydroxyl groups in $Zr(OH)_4$ and GD was destroyed or inhibited, to promote the separation and enrichment of GD from $Zr(OH)_4$ by extractant, it is an effective way to improve the recovery of the agent in this system.

Table 2. Recovery of GD from nano $Zr(OH)_4$ with different extractants and neutralizers.

Neutralization conditions				Extraction conditions		Recovery rate R/%	
		pH					
Neutralizer	Volume/ mL	Neutralization Solution	After neutralization[a]	Extractant	Volume/ mL	3μL	30μL
Water	10	7.09	9.80	Dichloromethane[b]	30	9.85	46.41
	10	7.09	9.80	Ethyl acetate[b]	30	Not detected	0.80
	20	7.09	9.80	Anhydrous ethanol[c]	20	36.39	82.37
0.05M Borax	10	9.61	10.23	Dichloromethane[b]	30	29.90	66.46
	20	9.61	10.23	Anhydrous ethanol[c]	20	26.80	63.47
0.005M Na_2HPO_4	10	9.04	10.60	Dichloromethane[b]	30	19.9	57.02
	20	9.04	10.60	Anhydrous ethanol[c]	20	21.60	62.35
HCl	10	3.27	7.69	Dichloromethane[b]	30	24.81	59.80

[a]: The pH value of solution mixed by 10 mL Neutralization solution and 1.0 g solid $Zr(OH)_4$.
[b]: Shake for 20 s during extraction; [c]: Centrifugated for 1minute after extraction shaking.

3.3 *Study on adsorption effect of GD by $Zr(OH)_4$*

To investigate the adsorption of zirconium hydroxide on GD, the solution of ethanol and water fixed at 1:1 was used as the extractant to detect the recovery rate of residual GD, which is from the solid-phase adsorption system or the low-dose GD degradation system by nano $Zr(OH)_4$, Industrial $Zr(OH)_4$, activated clay and activated carbon. Before the experiment, different detoxification (adsorption) agents were filtrated by a molecular sieve to ensure the particle size of each detoxification (adsorption) is uniform. The results are shown in Table 3.

Table 3. Recovery of GD from different adsorbents by ethanol-water solution as a neutralizer.

The volume of GD	Decontamination(adsorbent)	Recovery rate (R/ %)
3 μL	Nanometer $Zr(OH)_4$	30.10
	Industrial $Zr(OH)_4$	64.58
	Activated clay	90.14
	Activated charcoal	82.47

Table 3 shows It can be seen from Table 3 that the recovery rate of GD in the $Zr(OH)_4$ decontamination system is generally lower than that in activated clay and activated carbon adsorption or disinfection system. Activated carbon has no degradation effect on GD, mainly adsorbs the agent on its rich pore structure through physical adsorption. Therefore, the agent can be desorbed from the pore diameter of activated carbon through a highly polar solvent (ethanol: water = 1:1), to obtain a relatively high recovery rate.

It has been researched in the literature that the active clay mainly adsorbs toxic agents by physical adsorption (Bizzigotti et al. 2012). The interaction between metal oxides in active clay and toxic agents is mostly intermolecular forces, and the degradation rate of active clay on toxic agents is slower (half-life time is hours or more), therefore, a higher recovery can be obtained when the ethanol-water solution is used to extract GD in the solid-phase system of activated clay for the adsorption and degradation GD.

The recovery of GD from the two kinds of solid $Zr(OH)_4$ is lower than that from the adsorption or decontamination system of activated clay and activated carbon, this result can be explained from two aspects, first, the adsorption between $Zr(OH)_4$ and GD is not physical adsorption, but chemical adsorption is strong, so it is difficult to separate GD from $Zr(OH)_4$ by solvent extraction. Second, the degradation rate of GD by $Zr(OH)_4$ is very fast, and a large amount of GD has been degraded during the process of extraction of ethanol-water solution. George's team has studied the degradation of GD in solid-phase $Zr(OH)_4$ through analysis technologies such as NMR. Under the condition of 50% relative humidity, the half-life of the degradation of GD is 8.7min (Bandosz et al. 2012). In this view, the degradation rate of solid $Zr(OH)_4$ to GD is not instantaneous, the addition of an aqueous solution can reduce the degradability of solid $Zr(OH)_4$ for GD(Balow et al 2017; Jang et al 2020), according to this, the rapid degradation by $Zr(OH)_4$ in extraction process can be eliminated.

It is be showed in table 3 that the recovery of GD from solid-phase nano $Zr(OH)_4$ is lower than that from industrial $Zr(OH)_4$. In the experiment, the pH values of these two kinds of $Zr(OH)_4$ solutions were determined, the pH value of nano $Zr(OH)_4$ is 9.80, which is alkaline, while the pH value of industrial $Zr(OH)_4$ is 6.93, which is neutral. Alkaline conditions are more conducive to the degradation of GD, meanwhile, due to the smaller particle size of nano $Zr(OH)_4$, the adsorption of GD is stronger, and the recovery of nano $Zr(OH)_4$ is lower

By comparing the recovery rates of GD in different materials and combining them with literature analysis, it is further confirmed that solid $Zr(OH)_4$ has strong chemical adsorption on GD, and the adsorption is difficult to be destroyed by general solvents and neutralization solutions. Therefore, it is difficult to effectively recover GD in a solid-phase decontamination system of $Zr(OH)_4$ by the means of solvent enrichment after the solution is completely terminated, This may be the reason why most literature evaluates the detoxification property of $Zr(OH)_4$ by in-situ characterization, which can avoid the problem of low recovery in the pretreatment process

4 CONCLUSIONS

Literature investigation and experimental studies showed that solid $Zr(OH)_4$ has strong chemical adsorption and catalytic degradation for GD. In this paper, a large number of active hydroxyl groups from the neutralizer and extractant were added to destroy or inhibit the combination of hydroxyl groups and GD in $Zr(OH)_4$, accordingly, this promoted the extractant to separate and enrich GD from $Zr(OH)_4$, and improved the recovery of GD. When 20 ml water was used as a neutralizer and 20 mL ethanol was used as extractant, the recovery of GD with low concentration (3 μL) and high concentration GD (30 μL) from solid-phase $Zr(OH)_4$ reached 36.39% and 82.37% respectively. This pretreatment method is suitable for GC or the chemical coloration method to study the decontamination property of solid-phase $Zr(OH)_4$ for the adsorption and degradation of GD.

REFERENCES

Balow R B, Lundin J G, Daniels G C, et al. Environmental Effects on Zirconium Hydroxide Nanoparticles and Chemical Warfare Agent Decomposition: Implications of Atmospheric Water and Carbon Dioxide [J]. *ACS Appl Mater Interfaces*, 2017, 9(45): 39747–57.

Bandosz T J, Laskoski M, Mahle J, et al. Reactions of VX, GD, and HD with $Zr(OH)_4$: Near Instantaneous Decontamination of VX [J]. *The Journal of Physical Chemistry C*, 2012, 116(21): 11606–14.

Bizzigotti G O, Rhoads R P, Lee S J. *Handbook of chemical and biological warfare agent decontamination*[M]. ILM Publications, 2012.

Colon-Ortiz J, Landers J M, Gordon W O, et al. Disordered Mesoporous Zirconium (Hydr)oxides for Decomposition of Dimethyl Chlorophosphate [J]. *ACS Appl Mater Interfaces*, 2019, 11(19): 17931–9.

Davies H G, Richter R J, Keifer M, et al. The effect of the human serum paraoxonase polymorphism is reversed with diazoxon, *Soman and sarin* [J]. 1996, 14(3): 334–6.

Jang S, Ka D, Jung H, et al. Zr (OH) 4/GO Nanocomposite for the Degradation of Nerve Agent Soman (GD) in High-Humidity Environments[J]. *Materials*, 2020, 13(13): 2954.

Jeon S, Balow R B, Daniels G C, et al. Conformal Nanoscale Zirconium Hydroxide Films for Decomposing Chemical Warfare Agents [J]. *ACS Applied Nano Materials*, 2019, 2(4): 2295–307.

Jeon S, Schweigert I V, Pehrsson P E, et al. Kinetics of Dimethyl Methylphosphonate Adsorption and Decomposition on Zirconium Hydroxide Using Variable Temperature In Situ Attenuated Total Reflection Infrared Spectroscopy [J]. *ACS Appl Mater Interfaces*, 2020, 12(13): 14662–71.

Kim K, Tsay O G, Atwood D A, et al. Destruction and detection of chemical warfare agents [J]. *Chem Rev*, 2011, 111(9): 5345–403.

Lane M, Pescrille J D, Aracava Y, et al. *Oral Pretreatment with Galantamine Effectively Mitigates the Acute Toxicity of a Supralethal Dose of Soman in Cynomolgus Monkeys Posttreated with Conventional Antidotes* [J]. 2020, 375(1): 115–26.

Long J W, Chervin C N, Balow R B, et al. Zirconia-Based Aerogels for Sorption and Degradation of Dimethyl Methylphosphonate [J]. *Industrial & Engineering Chemistry Research*, 2020, 59(44): 19584–92.

Munro N B, Talmage S S, Griffin G D, et al. *The sources, fate, and toxicity of chemical warfare agent degradation products* [J]. 1999, 107(12): 933–74.

Singh B, Prasad G K, Pandey K, et al. *Decontamination of chemical warfare agents* [J]. 2010, 60(4): 428.

Wagner G W, Procell L R, Munavalli S. 27Al, 47, 49Ti, 31P, and 13C MAS NMR study of VX, GD, and HD reactions with nanosize Al2O3, conventional Al2O3 and TiO2, and aluminum and titanium metal[J]. *The Journal of Physical Chemistry C*, 2007, 111(47): 17564–17569.

Wang H, Mahle J J, Tovar T M, et al. Solid-Phase Detoxification of Chemical Warfare Agents using Zirconium-Based Metal Organic Frameworks and the Moisture Effects: Analyze via Digestion [J]. *ACS Appl Mater Interfaces*, 2019, 11(23): 21109–16.

Yang Y C, Baker J A, Ward J R J C R. *Decontamination of chemical warfare agents* [J]. 1992, 92(8): 1729–43.

Energy Revolution and Chemical Research – Kok-Keong Chong and Zhongliang Liu (Eds)
© 2023 The Authors, ISBN: 978-1-032-36554-1

Preparation of Fe_2O_3 supported N-doped carbon-based ORR catalyst by phthalocyanine polymer

Yanling Wu*, Ke Lu, Xianqi Mao, Liping Ma & Minghui Lu
School of Transportation and Civil Engineering, Shandong Jiaotong University, Ji'nan, China

ABSTRACT: Fuel cells can directly convert the chemical energy of the reaction between the oxidizer and reducing agent into electrical energy, presenting high energy conversion efficiency, so it is considered an important next-generation energy conversion device. The efficient ORR is an important pursuit of all fuel cells. Hence, Fe_2O_3/N-doped porous carbon (named γ-Fe_2O_3@NC) ORR catalyst *via* a pyrolysis process of iron phthalocyanine polymer material. The as-synthesized electrocatalyst of γ-Fe_2O_3@NC shows good electroactivity. The research provides a guide for the development of high-performance fuel cells.

1 INTRODUCTION

Electrochemical oxygen reduction reaction (ORR) determines the efficiency of fuel cells and rechargeable metal-air batteries. Because of its slow kinetics and large overpotential, it is very important to design an efficient ORR catalyst (Gopal et al. 2022).

Pure carbon-based materials are not ideal ORR catalysts. Recently, Fe_2O_3-doped carbon materials attract people's attention, but due to the easy agglomeration of Fe_2O_3 in the synthesis process, the poor conductivity limits its large-scale application (Gopal et al. 2022). Metal-phthalocyanine conjugated polymers have been widely studied for their effective anchoring of different nanoparticles (Xu et al. 2019).

Here, we present a simple synthesis procedure for Fe_2O_3-embedded porous N-doped carbon (named γ-Fe_2O_3@NC) by employing iron phthalocyanine polymer material, calcined at 800°. And the composition, morphology, and pore structure of γ-Fe_2O_3@NC catalyst were investigated using PXRD, SEM, and BET.

2 THE EXPERIMENT SECTION

The material preparation process is described below: The mixture of ferric chloride (0.9g), ammonium chloride (1.0g), ammonium molybdate (0.025g), urea (4.1g), and tetrahedral anhydride (2.1g) was thoroughly grounded, then put into 1000 mL open flask with a round bottom, heating with the temperature of oil bath to 240°C for 3h. The dark green product was dried and cooled. The above-described mixture was calcined at 550°C for 2h and then heated to 800°C for 2h during the whole process under a nitrogen atmosphere. Finally, the target catalyst was obtained (named γ-Fe_2O_3@NC) (Wu et al. 2022).

*Corresponding Author: wuyanling621@163.com

DOI 10.1201/9781003332657-77

3 RESULTS AND DISCUSSION

3.1 *XRD analysis*

From the PXRD in Figure 1, γ-Fe_2O_3@NC shows two peaks at 26 and 44°C, corresponding to the planes of graphite carbon (002 and 101, respectively). Meanwhile, additional diffraction peaks can be well indexed to Fe_2O_3 (JCPDS No.25-1402), suggesting the existence of Fe_2O_3 nanoparticles on the graphitic matrix of carbon (Qu et al. 2017).

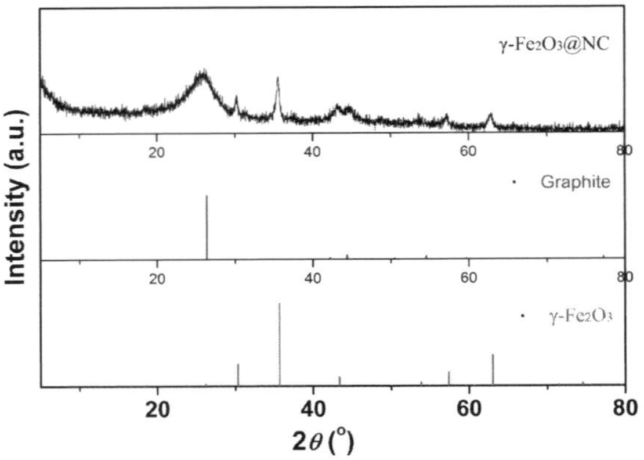

Figure 1. The PXRD pattern of γ-Fe_2O_3@NC.

3.2 *SEM and EDS analysis*

From the SEM in Figure 2, the γ-Fe_2O_3@NC shows a cluster structure. Notably, the formation of large crystals can be observed in the cluster-like structures, which are referred to as γ-Fe_2O_3. From the EDS in Figure 3, the γ-Fe_2O_3@NC includes C, O, Fe, and Mo. Otherwise, no N element was found in EDS. To further verify the presence of the N element, XPS survey spectra of the sample were tested. The elemental content of γ-Fe_2O_3@NC shows C (87.32%), N (6.03%), O (5.19%), Fe (1.3%) and Mo (0.159%).

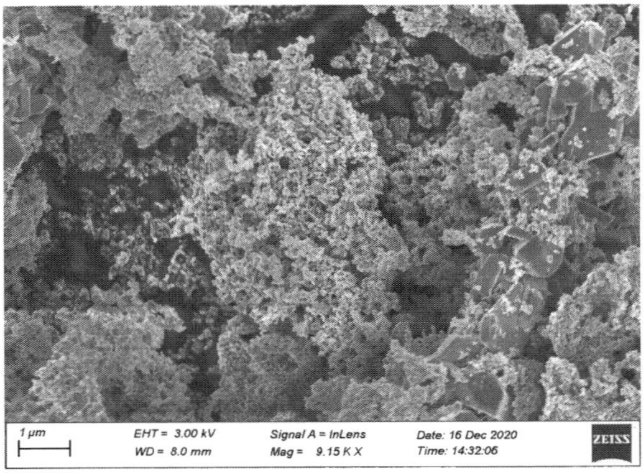

Figure 2. The SEM image of the γ-Fe_2O_3@NC catalyst.

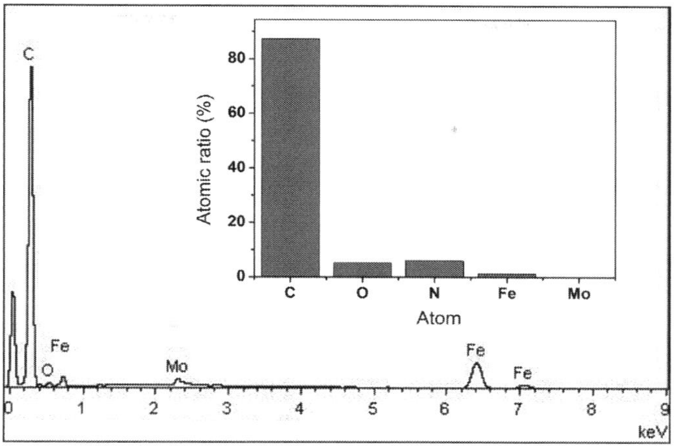

Figure 3. The EDS image of the γ-Fe_2O_3@NC catalyst (Insert the content of each element).

3.3 *Analysis of the specific surface area*

From Figure 4a and Table 1, we can see that γ-Fe_2O_3@NC catalyst with obvious hysteresis gyrus showed typical IV type characteristics, indicating the presence of mesoporous structure (Chen et al. 2022). The pore size of γ-Fe_2O_3@NC in Figure 4b showed a mesoporous nature.

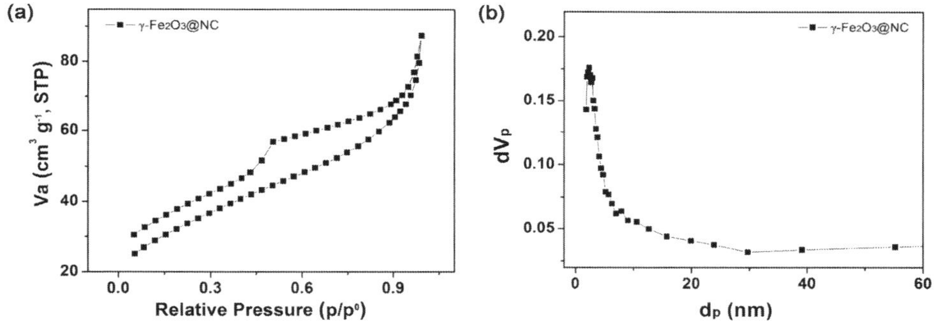

Figure 4. (a) The N_2 sorption isotherms and (b) corresponding BJH plots of γ-Fe_2O_3@NC catalyst.

Table 1. The parameters of the γ-Fe_2O_3@NC catalyst.

Catalyst	Surface area (m^2 g^{-1})	Average aperture (nm)	Pore volume (cm^3 g^{-1})
γ− Fe_2O_3@NC	114.77	2.34	0.13

3.4 *ORR catalytic properties*

The ORR activity of the γ-Fe_2O_3@NC catalyst was tested using CV in alkaline conditions, as shown in Figure 5a. Compared with γ-Fe_2O_3@NC under saturated N_2, a reduction peak at 0.79 V appears under saturated O_2. From LSV at different speeds in Figure 5b, the γ-Fe_2O_3@NC showed better catalytic activity (E_{onset}= 0.90 V, $E_{1/2}$ = 0.79 V, J_L= 5.05 mA·cm^{-2}). From the Koutecky-Levich (K-L) plots insert the Figure 5b, the γ-Fe_2O_3@NC showed good linearity. The electron transfer number for γ-Fe_2O_3@NC at the 0.50-0.60 V potential range was calculated to be 3.22, indicating that γ-Fe_2O_3@NC followed the 4-electron oxygen reduction path.

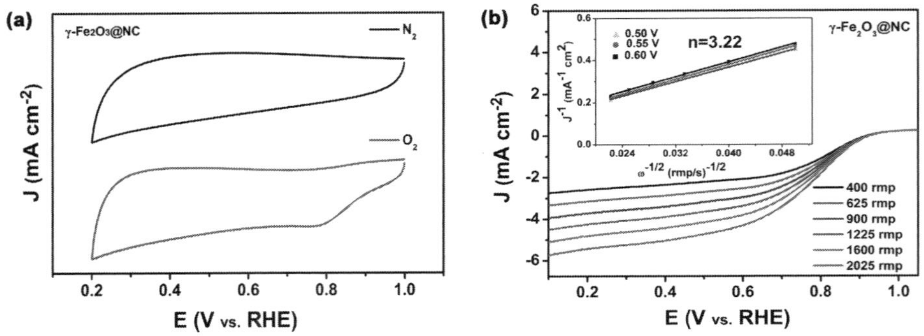

Figure 5. (a) CV curves of γ-Fe_2O_3@NC (O_2-saturated red line; N_2-saturated, black line); (b) LSV curves of γ-Fe_2O_3@NC at different rotational speeds (Inset: K-L plots of γ-Fe_2O_3@NC).

4 CONCLUSION

In this paper, a kind of iron phthalocyanine polymer was synthesized by a simple one-step method. Then a γ-Fe_2O_3@NC ORR catalyst with excellent performance was obtained after high-temperature calcination. This catalyst exhibits good ORR performance, which is attributed to the high amount of efficient Fe_2O_3 active sites and ordered porous structure. This study provides guidance for the development of non-platinum-based electrocatalysts for ORR.

ACKNOWLEDGMENTS

This project is financially supported by Shandong Jiaotong University (The Doctoral Scientific Research Foundation, BS50004952).

REFERENCES

Chen, F., Yao, C.C., Qian, J.C., et al. (2022) α-Fe_2O_3/alkalinized C_3N_4 heterostructure as efficient electrocatalyst for the oxygen reduction reaction. *J Mater Sci* 57: 2012–2020.

Gopal S.A., Poulose A.E., Sudakar C., et al. (2022) Kinetic insights into the mechanism of oxygen reduction reaction on Fe_2O_3/C composites, *ACS Appl. Mater. Interfaces*, 13: 44195–44206.

Qu, B., Sun, Y., Liu, L.L., et al. (2017) Ultrasmall Fe_2O_3 nanoparticles/MoS_2 nanosheets composite as high-performance anode material for lithium-ion batteries, *Sci. Rep.*, 7: 42772.

Wu, Y.L., Hou, Q.G., Qiu F.Y., et al. (2022) Co_2O_3/$Co_2N_{0.67}$ nanoparticles encased in honeycomb-like N, P, O-codoped carbon framework derived from corncob as efficient ORR electrocatalysts, *RSC Adv.*, 12: 207–215.

Xu, Z., Du, Y.C., Liu, D.W., et al. (2019) Pea-like Fe/Fe_3C nanoparticles embedded in nitrogen-doped garbon nanotubes with tunable dielectric/magnetic loss and efficient electromagnetic absorption, *ACS Appl. Mater. Interfaces*, 11: 4268–4277.

Energy Revolution and Chemical Research – Kok-Keong Chong and Zhongliang Liu (Eds)
© 2023 The Authors, ISBN: 978-1-032-36554-1

A review on characterization technologies for chemical features of activated carbon

Yue Wang*
The Institute of NBC Defense, Beijing, China

ABSTRACT: There is an increased global demand for activated carbon (AC) in the application of industry and agricultural production, such as chemical catalysis, water purification, medical engineering, and military defense. Researchers associate the chemical features of AC with its properties and thus its reactivity and price in the application. As a result, characterization methods and analysis play a significant role. This paper gives a brief review of characterization technologies for chemical features of AC, about the working mechanism, operation, and application examples of four technologies——FTIR, XPS, TPD, and TGA. To obtain information on elements composition, surface functional groups, electronic structures, and critical structures of AC, they are often used jointly as corroboration for each other.

1 INTRODUCTION

With various pore structures, large surface area, adjustable surface functional groups, special electron structures, and activated carbon (AC) is an excellent material for adsorbing and removing toxic and harmful molecules from either aqueous or gas phase (Sujata Mandal 2021). AC is also widely used as a catalyst carrier, electrode material, and hydrogen storage material. The quantity and type of surface chemical groups of AC determine its properties, such as surface charge, hydrophily(hydrophobicity), and reactivity (P. González-García 2018). The production process of AC includes pre-carbonization, carbonization, and activation, where special activators or methods like ultrasound and ionization can be used in these steps to adjust the type and quantity of chemical groups on the AC surface. There are many methods to characterize the chemical features of AC. This paper gives a brief introduction to the most commonly used methods.

2 CLASSIFICATION OF SURFACE FUNCTIONAL GROUPS

Surface chemical groups determine the surface properties of AC. Common functional groups are oxygen and nitrogen functional groups, classified by acidity and basicity properties as shown in Table 1 (Mao 2011).

Other functional groups may exist due to the additive agents in the production process, such as sulfur-containing groups and metal groups which have not been studied exhaustively about their bonding with carbon skeletons. As can be seen, the advance and further exploration of technologies for chemical characterization make a great difference.

*Corresponding Author: wy.1998.8@163.com

DOI 10.1201/9781003332657-78

Table 1. Common functional groups classified by acidity and basicity properties.

Acid functional group	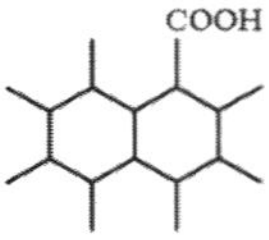 Carboxyl	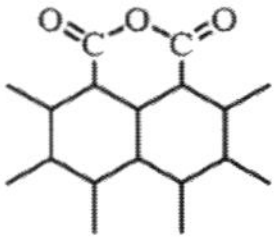 Carboxylic acid anhydrid	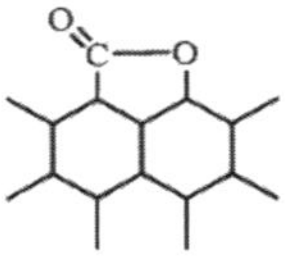 Lactone base	
Neutral functional group	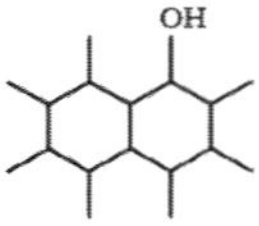 Oxhydryl	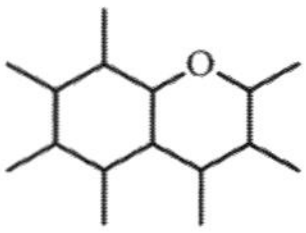 Ether group		
	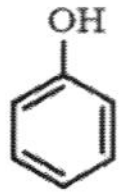 Phenolic hydroxyl	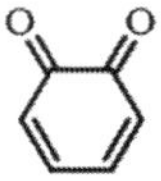 Benzoquinonyl		
Basic functional group	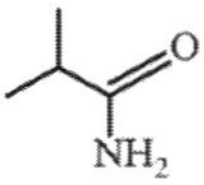 Amido	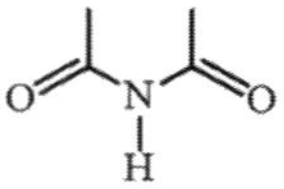 Acylamine group	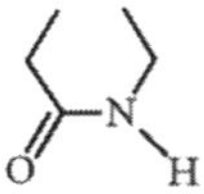 Lactam group	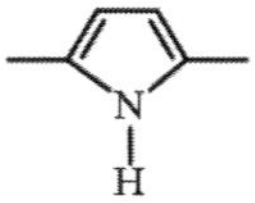 Acylamide group
	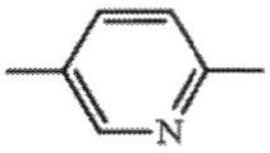 Pyridine-like groups	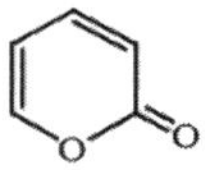 Pyranone group	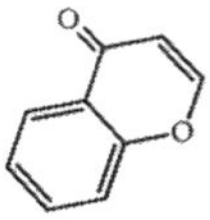 Benzopyranyl	

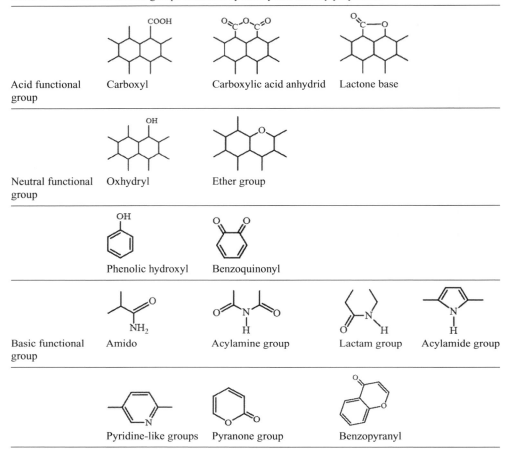

3 CHARACTERIZATION TECHNOLOGIES FOR CHEMICAL FEATURES

3.1 *Elemental analysis*

Elemental analysis can determine the content of each element in the sample. Despite the difference in structure and performance of analysis instruments, they are designed based on the spectrum, chromatography, and mass spectrum. The working principle is that under the action of the composite catalyst, the sample is combusted at a high temperature to generate nitrogen, nitrogen oxide, carbon dioxide, sulfur dioxide, and water. Then under the carrier gas, they are flushed into the separation and detection unit. Finally, the content of C, S, P, Si, and various metals can be measured quickly and accurately. This method is often used for the detection and analysis of elements in modified AC.

3.2 *Fourier-transform infrared spectroscopy (FTIR)*

Infrared spectroscopy is an instrument that can analyze the molecular structure and chemical composition of samples by using the absorption properties of infrared radiation at different wavelengths. It is commonly used to detect various chemical bonds and relative content contained in AC samples. The AC sample is black and has strong infrared absorption, which requires a larger proportion of potassium bromide ground into powder, mixed with the sample, and pressed into a plate, in order to dilute the sample for measurement.

Fourier-transform infrared spectrometer, known as the third-generation infrared spectrometer, is the most widely used currently. It uses the Michelson interferometer to make two polychromatic infrared lights varying at a certain speed interfere with each other and forms interference light, and then acts with the sample. The detector collects the infrared interferogram data which contains the sample information, and the data is then calculated with Fourier transformation. As a result, the infrared spectrograph of the sample will be obtained. FTIR is widely used because of its fast-scanning rate, high resolution, and stable reproducibility.

Infrared spectrum (IR) is often used with Raman spectrum, working as complementary and mutual verification.

Xiya Du (2018) et al. found that (as shown in Figure 1) all the coal-based AC samples showed absorption peaks at 1046, 1147, and 3436 cm^{-1}, and the peak at 3436 cm^{-1} enhanced after oxidation modification, which indicated more alcohol hydroxyl in samples. While the absorption peaks at 2917 and 2847 cm^{-1} were weakened, which indicated that the C-H bond was oxidized. Absorption peaks at 1636 and 1384 cm^{-1} belong to the C-O bond, which indicated the carboxyl group and pyrene group might exist. This shows that the acidic aerobic functional groups on the surface of the AC sample increase.

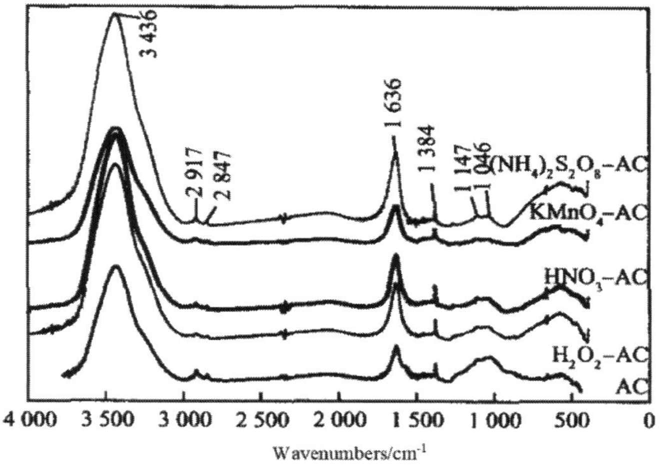

Figure 1. FTIR spectrogram of samples before and after oxidation.

3.3 *X-ray photoelectron spectroscopy (XPS)*

X-ray photoelectron spectroscopy (XPS) is an advanced analysis technology widely used in the microscopic analysis of electronic materials and components and is often used with Auger electron spectroscopy (AES) cooperatively. Since it can measure the inner layer electron binding energy of atoms and its chemical displacement accurately, it can provide information on molecular structure, atomic valence state, chemical state, molecular structures, and chemical bonds. Moreover, it can provide not only chemical information on the overall aspects but also information on the surface, micro-regions, and depth distribution. XPS can locate a single particle to get spectrum and images and obtain further information using argon ion sputtering etching for deep analysis.

Yuliang Hu (2018) et al. used PHI Quantera II XPS for full-spectrum scanning and narrow-spectrum scanning on C, O, and Cu elements, as well as scanning samples sputtered by Cu ion (named Cu-AC). Cu-AC was then further analyzed using Ar ion splash and XPS scanning after flash. Finally, the obtained XPS spectrogram was curve-fitted and provided with information on peak position, peak strength, and chemical displacement for analysis of the structure of surface functional groups. By consulting the XPS spectrogram of Cu in the NIST database, they found that the element Cu on the Cu-AC surface mainly existed as CuO, but as Cu_2O after sputtering.

Jie Liu et al. used Escalab MKI XPS (the British Vaccum Generator company) to test polyacrylonitrile (PAN) -based activated carbon fiber (ACF) at 1.3310^{-7} Pa vacuum. It was found that both the binding energy and displacement of N_{1s} and C_{1s} changed after being treated under different conditions, but neither of the O_{1s} was changed, indicating that the bond between carbon and nitrogen atoms changed under different conditions, while those with oxygen atoms did not change significantly. The data change is shown in Table 2.

Table 2. XPS results of samples before and after treatment.

	N_{1s}			O_{1s}			C_{1s}		
Sample	Binding energy E_B/eV	Displacement δ_E/eV	Content V/%	Binding energy E_B/eV	Displacement δ_E/eV	Content V/%	Binding energy E_B/eV	Displacement δ_E/eV	Content V/%
CF-A_0	398.40	–0.6	8.67	532.25	0.25	7.00	284.80	0.80	84.33
ACF-A_1	401.15	2.15	6.73	532.30	0.30	7.04	284.80	0.80	86.24
ACF-A_2	400.85	1.85	7.63	532.30	0.30	5.73	284.80	0.80	86.63
ACF-A_3	401.10	2.10	4.89	532.30	0.30	6.63	284.75	0.75	88.48

3.4 *Temperature programmed desorption (TPD)*

Temperature programmed analysis technology (TPAT) is a dynamically analyzing technology of the multiphase catalysis process. Specifically, it includes temperature-programmed desorption (TPD), temperature-programmed reduction (TPR), temperature-programmed oxidation (TPO), temperature-programmed vulcanization (TPS), temperature-programmed surface reaction (TPSR), etc. Among them, temperature-programmed desorption (TPD) is the most deeply studied, the most widely used, and relatively mature theory.

When using the TPD method, we heat samples in a programmed heating procedure (such as constant speed heating) to obtain the desorption quantity diagram related to temperature. In this process, the following phenomena may occur: (1) molecules desorb from the surface to the gas phase, then from the gas phase to the surface; (2) molecules spread from the surface to the second layer (subsurface), from the second layer to the surface, and (3) molecules diffuse in the inner pore.

The temperature programmed analysis technology (TPAT) can obtain the type, density, and energy distribution of adsorption centers on the sample's surface, the bonding energy and bonding state between adsorption centers and absorbents, as well as the dynamic behavior and reaction mechanism.

Yun Shu (2017) used TPD to characterize xylem-based AC and lignite-based AC which loaded potassium catalyst. They found that the desorption peak of CO on xylem-based AC was lower in temperature than on lignite-based, which means CO desorption on onxylem-based AC is easier. Whereas, CO desorption capacity is related to the reduction capacity of potassium catalysts, indicating that potassium catalyst reduction is easier on xylem-based AC. Thus, the catalyst dispersion on the xylem-based AC should have had a smaller particle size and a larger contact area with carbon. This result is also confirmed by the better selectivity of the xylem-based AC catalyst.

Xiaoyun Li (2008) et al used TPD to characterize the oxygenic functional groups on the surface of three AC: raw coconut shell-based AC, nitrate oxidated one, and the one treated in hydrogen at high temperature after nitrate oxidation. It can be seen from three desorption curves (as shown in Figure 2) that: the first AC has two CO_2 desorption peaks, located at 90 to 430°C and 550 to 660°C; and one CO desorption peak, located at 520 to 900°C. They respectively belong to carboxyl, lactone, and adventitia groups. The second AC shows a new desorption peak at 410°C, which can be attributed to the breakdown of the acid anhydride group. This is because when a large number of carboxyl groups are heated, some of the adjacent will be dehydrated to form anhydride groups, which are more stable in heat treatment. The total amount of CO_2 and CO of the second AC is larger than that of the first one, indicating that a large number of oxygenic functional groups were introduced into the carbon surface after nitrate acid treatment. While for the third AC, the desorption of CO_2 and CO both occur at a temperature above 500°, which indicates that after pretreatment at high temperature, some kinds of the thermally unstable groups are selectively removed, such as the carboxylic acid and acid anhydride groups. While thermally unstable groups, like enlactones, phenols, and carbonyl groups were reserved.

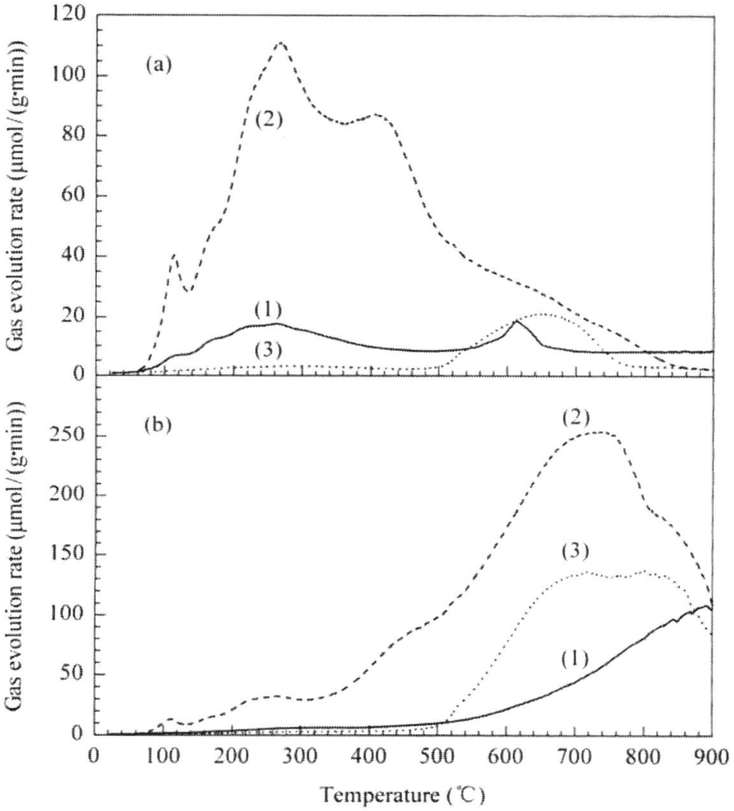

Figure 2. TPD profiles of CO_2 (a) and CO (b) from original and modified activated carbon materials (1) AC_1, (2) AC_2, (3) AC_3.

3.5 *Thermogravimetric analysis (TGA)*

Thermogravimetric analysis (TGA) is a method to measure the relationship between material mass and temperature or time under the programmed control temperature. By analyzing the thermal-gravimetric curve, we can know the information related to the mass, such as the composition, thermal stability, thermal decomposition of samples, and potential intermediate products.

Differential thermogravimetric analysis is derivated from the thermogravimetric analysis. It is a technique to record the first derivative of the TG curve (a shorter form of TGA) to temperature or time, namely the DTG curve. The vertical coordinate is the mass change rate, decreasing from top to bottom; the abscissa is the temperature or time, increasing from left to right.

The main advantage of TGA is its high quantification level and accuracy in measuring the mass change and rate of change. According to this merit, it can be said that as long as the material has mass changes during heating, it can be studied by TGA. Physical and chemical changes both can be detected by TGA because these processes such as sublimation, vaporization, adsorption, desorption, absorption, and gas-solid reaction, all have mass changes.

Yichen Cui (2005) et al. used TGA to analyze the reactivity changes of AC samples during the methane lysis reaction and obtain the reaction rate changes from the weight increasing rate curve, and the 'final weight increase'. It was found that the initial methane lysis rate on AC samples increased accordingly to the increase in the reaction temperature, but the reaction rate decreased faster. And the 'final weight' remains the same regardless of the temperature.

TGA is often used in combination with (DTG) to obtain the TG-DTG curve. TG curve can reflect the total loss of mass, while the peak of the DTG curve reflects whether there is a thermal cracking of the sample.

Xikai Liao (2015) et al. used the TG-DTG curve (as shown in Figure 3) to analyze the thermal decomposition process of AC prepared from waste filter bags. The TG curve obtained includes three weight loss stages. The first weight-loss stage occurred from 30 to 390°C, and the loss rate was only 0.44%. The corresponding DTG curve was mainly a straight line. This stage was mainly the volatilization of water. The second weight-loss stage occurred from 390 to 510°C, which was the major weight loss stage, and the weight loss rate was up to 78.53%. The DTG curve showed a distinct peak. The largest weight loss rate was at 448°C, reaching 33.5% min^{-1}. In this stage, the filter bag began to be pyrolyzed, with a violent reaction and a large number of volatile gases produced, resulting in a sharp increase in weight loss proportion. The third weight loss phase occurred from 510°C to 1000°C, with a small amount of volatile gas produced, but the weight loss rate is only 2.27%. There was no significant weight loss peak present in the DTG curve, indicating that the pyrolysis reaction was close to the end. The solid carbide left accounted for about 21.03%.

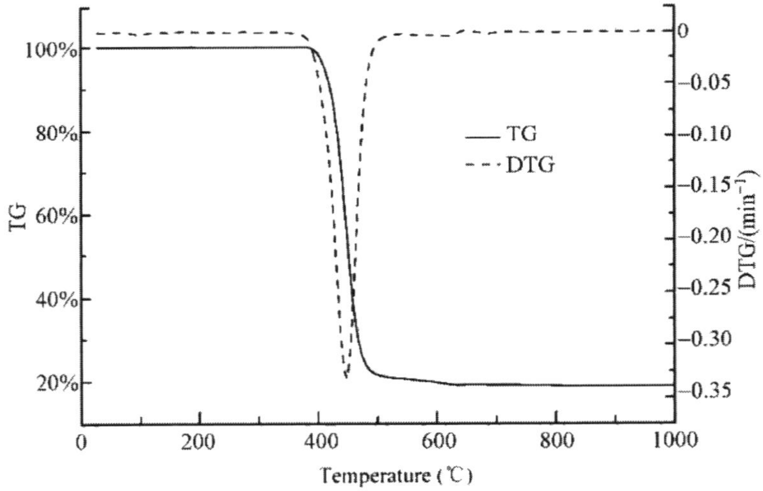

Figure 3. TG and DTG curves (in N_2) of AC prepared from waste filter bags.

4 CONCLUSION

This paper gives a brief review of the commonly used characterization methods for chemical features of activated carbon (AC), including elemental analysis, Fourier-transform infrared spectroscopy (FTIR), X-ray photoelectron spectroscopy (XPS), temperature-programmed desorption (TPD), and thermogravimetric analysis (TGA). FTIR gives information about the types and relative content of chemical bonds having dipole moment, which work as a reference for composition analysis. XPS gives information on electron structures, which are closely related to valence and crystal structure and provide desirable identification for certain components. TPD uses adsorption and desorption methods, which can reveal information on surface functional groups. TGA can be applied for all samples burnable, giving information on the proportion of components. These methods can be used jointly to obtain overall and complementary information, for a more precise analysis of the composition and as the basis for subsequent modification.

ACKNOWLEDGMENT

I would like to express my gratitude to my mentor Zhenxing Cheng professor, for his patient instruction and insightful guidance during my academic studies, which instructed and enlightened me every time I faced challenges. I also owe a debt of gratitude to the supervisors, Haiyan Zhu and Ting Miao, who have offered me valuable suggestions during the preparation and completion of this paper. Without their consistent and illuminating instruction, this paper could not have reached its present form.

REFERENCES

Aiqin Mao, Wang Hua, Tan Linghua, et al. Progress in the functional group characterization of the activated carbon surface [J]. *Applied Chemical Industry*, 2011, 40(7): 5.

González-García, P. Activated carbon from lignocellulosics precursors: A review of the synthesis methods, characterization techniques and applications[J]. *Chemical Engineering Journal*, 2018 (82) 1393–1414.

Sujata Mandal, Jose Calderon, Sreekar B. Marpu. Mesoporous activated carbon as a green adsorbent for the removal of heavy metals and Congo red: Characterization, adsorption kinetics, and isotherm studies[J]. *Journal of Contaminant Hydrology*, 2021 (243) 1–11.

Xiaoyun Li, Ding Ma, Xinhe Bao. Dispersal properties of Pt catalysts on different activated carbon species and their catalytic properties in methylcyclohexane dehydrogenation reactions [J]. *Catalytics Journal*, 2008, 29 (3): 259–263.

Xikai Liao, Xiaoyuan Man, Xunan Ning, etc. Effects of carbonization temperature on the properties of activated carbon and its characterization [J]. *Journal of Environmental Science*, 2015, 35 (11): 3775–3780.

Xiya Du, Xinzhou Chen, Bin Yang, et al. Performance of the catalytic removal of NO by oxidative-modified activated carbon [J]. *Chemical environmental protection*, 2018, 38 (1): 95–100.

Yichen Cui, Chang Yang, Ningsheng Cai, etc. Thermal weight analysis of the catalytic cleavage of methane on activated carbon [J]. *Combustion Science and Technology*, 2005, 11 (5): 480–485.

Yuliang Hu, Guohong Liu, Xiaodong Zhou, etc. XPS study of copper oxides in impregnated carbon [J]. *Guangzhou Chemical Industry*, 2018, 46 (7): 73–76.

Yun Shu, Fan Zhang, Fan Wang, etc. Biomass-based activated carbon-loaded potassium catalyst under oxygen-rich conditions [J]. *Fuel Chemistry*, 2017, 45 (6): 747–754.

Energy Revolution and Chemical Research – Kok-Keong Chong and Zhongliang Liu (Eds)
© 2023 The Authors, ISBN: 978-1-032-36554-1

The development, technology principles, and implementation of biofuels

Rui Wang
AP Program, The High School Affiliated to Beijing Normal University, Beijing, China

Yanlin Zhu*
Jinling High School, Nanjing, China

Binrui Huang
School of International Department, Beijing Bayi School, Beijing, China

ABSTRACT: This paper gives an overview and specific descriptions and explanations of biofuel, which will help people understand the development, technology, and implementation of biofuels. Although biofuels haven't been used as the main fuel source in nearly all the countries that are still using fossil fuels, the positive impacts it will have on solving the fossil crisis and greenhouse effect caused by CO_2 have encouraged many governments to highlight the necessity of using or reusing the biofuels. Scientists from different countries are dedicated to investigating enhancing biofuels' feasibility and increasing the biofuel's proportion in the market. Overall, the tendency of using more biofuels is inevitable, promising, and impendent.

1 INTRODUCTION

Due to the rapid development of technology and economy, which are used to stimulate society's growth, people have exploited a tremendous number of fuels such as coal, natural gas, fossil fuels, and petrochemical sources. The demand for fossil fuels is increasing, but non-renewable sources like fossil fuels will be exhausted in less than 100 years as predicted by the World Energy Forum (Debabrata Das 2019). Therefore, finding and using replaceable energy sources is an emergency, to deal with the global fuel crisis and the climate change that is caused by the CO_2 emissions from fossil fuels largely.

Biofuel is a practical solution, it is extracted and purified from the biomass. Biomass is easily accessible, environmentally friendly, and innoxious. There are three forms of biofuels: solid biofuels, gaseous or liquid biofuels, and feedstocks. The feedstocks are in four generations. The first generation relates to food crops; the second generation includes non-edible plant residues; the third generation involves food waste and algae, and the fourth generation corresponds to genetic modification. People had already practiced those sources to create values. Agricultural sustainability, social and ethical concerns, economic feasibility, and environmental pollution are impeding the usage of the first-to-fourth generation sources severally. The first-generation sources such as ethanol are practiced non-ideally; the algal oil, biodiesel, and aviation fuel from the third generation have weaknesses; and the fourth-generation sources require to be technological advancements, like metabolic engineering and synthetic biology. The second-generation biofuels are relatively more viable, because it comes from nonedible plant biomass, which is technologically simpler, and it produces much less greenhouse gas in their life cycle. Specifically, bioethanol, biobutanol, bio-oil,

*Corresponding Author: zhuyanlin666666@126.com

 DOI 10.1201/9781003332657-79

biohydrogen, and biomethane (Prakash Kumar Sarangi 2018). The first to fourth generations all demonstrate special benefits, shortcomings, and potential. They uniquely present special properties that help solve the energy depletion, address the social concerns, then ensure economic growth and environmental security. For now, all the generations require technological advancements, and second-generation biofuels are relatively more viable.

Since the earliest human civilization appeared, solid biofuels had been used for heating purposes. Then came the industrial revolution, and liquid biofuels were the main source of energy to boost automobile industries with various inventions. The first internal combustion engine was designed to run on a blend of ethanol and turpentine. The first-ever diesel engine, invented by German scientist Rudolph Diesel, was intended to run using vegetable oil (Webb 2013). But the fossil fuel then replaced biofuels because of the cheaper price and high efficiency. The petroleum shortage brought bioethanol back to be used during WWI. Moreover, the oil crisis in the 1970s prompted the reconsideration of applying biofuels. The U.S. and Brazil typically implemented the production and transportation through first-generation biofuels. According to the data in 2017, biofuels occupied 10.3% of the total global energy supply, there are 87% from the forestry section, 10% from agriculture, and 3% from waste energy (Debabrata Das 2019). Biodiesel and bioethanol are respectively enhancing the technological renewable rates by 6.5% and 3.0% (Peake S 2018). However, starvation and global warming situations, technological challenges, and economic issues are the problems that people are concerned about and resolving.

Scientists have been investigating and developing several kinds of biofuels in the first-to-fourth generation sources. The typical sources are bioethanol, biobutanol, bio-oil, biohydrogen, and biomethane from the second generation; the algal oil, biodiesel, and aviation fuel from the third generation. Their specific illustration and history of them are discussed in the following passages.

Bioethanol: The first internal combustion engines were designed by Samuel Morey in 1826, and Nicholas Otto designed the famous one in 1876. Henry Ford successfully produced pure ethanol cars in 1896, then led the ethanol or gasoline-ethanol hybrid power to be used in 1908. After World War I, the cheaper petroleum-based fuels diminished the ethanol demand. But, at present, ethanol has been used on a large scale, including in Brazil, European countries, and the United States. It is hopeful to be dominating the market in the next 20 years because it has a high-octane number, vast vaporized heat production, and great sustainability. However, ethanol has a lower efficiency compared to gasoline, the sources used for producing bioethanol are lignocellulosic biomass that is unacceptable to sacrifice for eating purposes. There are examples of producing bioethanol by the use of corn and beans in Figure 1. And the water diluting and transporting techniques tend to be challenging in using bioethanol as an effective biofuel. So, biobutanol comes in as an alternative.

Figure 1. The sources of bioethanol such as corn and bean products (Brittany Rycroft 2019).

Biobutanol: Louis Pasteur reported microbial fermentation butanol production in 1861 firstly. Then Albert Fitz tried to extract butanol from glycerol from two strains of bacteria. The researchers like Pringsheim, Bredemann Shardinger, and Beijerinck conducted research substantially on biobutanol, which was industrially synthesized in 1912-1914 by famous ABE fermentation. Traditionally, biobutanol has been used as a solvent in producing several kinds of products such as cosmetics, the intermediate in manufacturing methacrylate and butyl acrylate, and the extractant in synthesizing pharmaceutical products. Biobutanol has the advantages such as safe, higher heating value than ethanol, a comparable ratio of air to fuel and octane, and less hygroscopic peculiarity. Unfortunately, the biobutanol investigation was paused during WWI, due to the increase in food demand. Recently, Dupont and British Petroleum have declared to commercialize the production of biobutanol through ABE fermentation (Kumar & Gayen 2011). Biobutanol reacts with volatile organics, then stimulus the photochemical formation, which is flammable to human eyes, throat, and nose, but large-scale research could be done to solve it. It is still promising.

Bio-oil: It is a pretty new term. The bio-oil is derived from biomass pyrolysis. Bio-oil contains plenty of aromatic compounds, such as phenol derivatives, esters, and alkanes. It requires the oxygen or anaerobic biomass pyrolysis circumstance to be produced at high temperatures.

Despite the bright perspective of bio-oil, there is still a long route to letting bio-oil be used as a competitive product in the commercialization market.

Algal oil: Biofuels from algae fall under the third-generation biofuels, which demand relatively lower input. Since it requires no arable land or pesticides to grow; it is not dependent on seasonality, and it has higher biofuel productivity than terrestrial plants. However, finding the optimal algal oil production process is a challenge, facing the limitation of light penetration in cultured areas and thinking about removing high water content in it at a low cost. In the 1950s, Oswald and Goluke proposed to use microalgae for biofuel production. During the 1970s, the U.S., Australia, and Japan adopted relevant research programs on algal oil fuel. Then, the US Department of Energy made a series of shifts in the financial support of algal oil fuel production. Considering the "peak oil" circumstance and enormous CO_2 emissions, discovering the algal oil fuel is useful. For now, the US-based Algal Biomass Organization is working as the pioneer, and nations in Europe, Asia, and elsewhere are developing the algae as a biofuel. Figure 2 provides us with an image of how people study and grow algae in tanks.

Figure 2. The algae are growing in tanks (Doug Tribou 2018).

Biodiesel: The diesel engine was developed by Rudolph Diesel in the 1890s, it had become the first choice of engine worldwide because it was reliable and economic. He held the first patent for the compression ignition engine, issued in 1893. Diesel became known for his innovative engine which could use a variety of fuels. The French government and Dr. Diesel were involved in investigating vegetable oil fuels as the early experimenters who expected that pure vegetable oils could power early diesel engines in remote areas of the world. Petroleum was not available at the time. In the 1980s, the biodiesel industry was established in Europe. As of 2005, global biodiesel production reached 1.1 billion gallons, with occupied the market largely in the European Union (Pacific Biodiesel 2019). The following Figure 3 depicts a typical factory for producing biodiesel. The trend of developing the biodiesel industry is increasing.

Figure 3. A typical factory for producing biodiesel (Advanced Fuel Solutions 2019).

Biohydrogen: H_2 heralds the alternative to petroleum-based transportation fuels because its combustion yields exclusively pure water. In Figure 4, part of various processes of producing biohydrogen-utilizing waste as the substrate is shown. Biohydrogen can be used directly from solar energy through certain oxygenic and anoxygenic photosynthetic microorganisms (Mckinlay & Harwood 2010). In 1766, Cavendish used the drainage method in many experiments and finally found that the content of flammable gas in the air was less than 9.5 percent or more than 65 percent, which would burn but not explode when ignited. Lavoisier heard of this and repeated Cavendish's experiment, stating that water was a compound of hydrogen and oxygen. In 1787, Lavoisier officially identified hydrogen as an element, and he recognized hydrogen as "the maker of water". Recently, astronomical programs, like NASA's Space Shuttle or the International Space Station have used hydrogen as the source of the main engine, and it was used as the main component of coal gas during the 19th and 20th centuries. For now, the global-scale production and utilization of hydrogen are required, especially the biomass hydrogen production should be strategically planned.

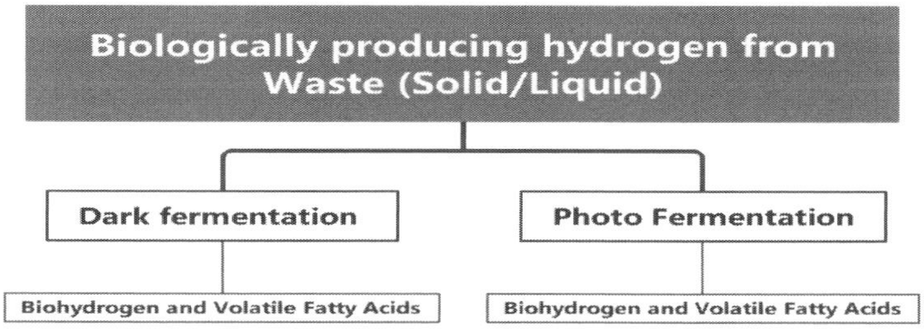

Figure 4. Part of various processes of producing biohydrogen-utilizing waste as substrate. (Naresh Kumar Amradi. 2016).

Biomethane: People discovered the flaming phenomenon in the marshall over the world in thousand years ago. Chinese have documented it in 1066-771 BC, the A. Volta from Italia ensured the components in it are carbon and hydrogen. In 1808, Dalton inaugurated the atomic and molecular theory, because he investigated methane's molecular structure. The methane intrigued the chemistry development based on the dedication came from Kekule et al. in the 19th 60s, Van Hoff et al. in the 19th 70s, and Pauling et al in the 20th 20s (Arthur Wellinger 2013). People produce biomethane from the high moisture-containing biogenic wastes, the biogas upgrading, and the specific agricultural feedstocks (Prakash Kumar Sarangi 2018). Producing it does not need to occupy a lot of arable lands. Some factories that are operating for producing biomethane, such as in Figure 5, there is a factory that is being built in Metheringham, UK. Biomethane production and upgrading technologies have gained significance in Europe, since the introduction of the EU landfill directive in 1999 (Dhamodharan Kondusamy 2021). Since biomethane has high greenhouse gas emissions, its properties are like fossil fuels, which results in causing environmental constraints overweight the benefits. Europe, the People's Republic of China, and the United States account for 90% of global production (IEA 2018). So, the efforts to ameliorate biomethane's environmental properties are highlighted.

Figure 5. The biomethane plant that is being built in Metheringham, UK. (Photo: Agraferm Technologies) (Sun and Wind Energy 2015).

Aviation fuel: Jet fuel and aviation gasoline are combined to produce aviation fuels. There are five airports that have regular biofuel distribution today (Bergen, Brisbane, Los Angeles, Oslo, and Stockholm), with others offering occasional supply (British Petroleum Company 2021). Figure 6 gives a picture of a real airplane that uses biofuel. The ITAKA group in Europe is developing Synthetic Paraffinic Kerosene (SPK) that is environmentally, economically, and socially feasible. In addition, Europe is attempting to boost the aviation industry by pre-processing waste cooking oil including esterification and thermal-catalytic processing. Aviation biofuels have been transformed from a dicey experiment to a certified sustainable alternative. It is promising though it undertakes commercial constraints and uncertainties, the limited availability of feedstock, and the lack of national and international policy support. The IEA anticipates biofuels reaching around 10% of aviation fuel demand by 2030, and close to 20% by 2040 (IEA 2018).

Figure 6. A picture of biofuel using an airplane (Jason Paur 2011).

2 BASIC PHYSICS & TECHNOLOGY PRINCIPLES

Biofuel has four generations, and each of them has very different production ways. However, most of them have similar usages, which is combustion.

2.1 *The first generation of biofuels and production*

The first generation of biofuels are made of food crops grown on arable lands, such as crop sugar, starch, and oil, these raw materials are converted into biodiesel or ethanol through transesterification or yeast fermentation.

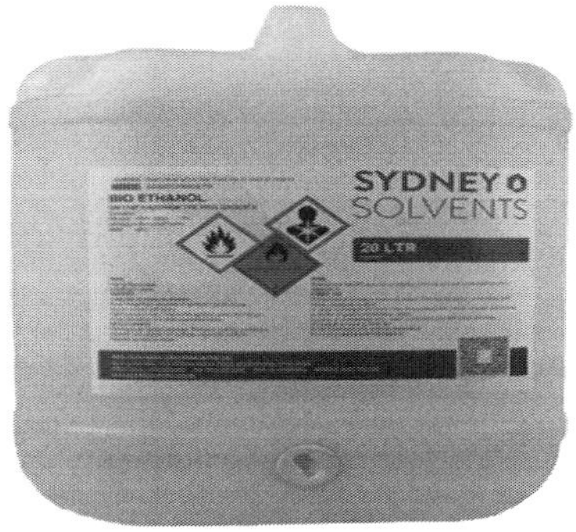

Figure 7. A product of bioethanol (Sydney Solvents 2021).

2.1.1 *Bioethanol production*

Figure 7 shows a product of bioethanol. Firstly, bioethanol can be produced from sugarcane, the sugarcane is washed, chopped, shredded, and crushed into juice in the end. Then the juice will be further filtered to kill micro-bacterial. Bagasse, which is the by-product, will be used as the energy resource of the factory, producing heat and steam. The juice will be separated into pure crystals and molasses, molasses will have a process of fermentation for 4 to 12 hours. Secondly, it can also be produced from cereal crops. Their molecules were long chain glucose polymers, which need to break down onto polymers a simple sugar through two kinds of processes, dry-milling or wet milling. Dry milling is converted from the raw material into flour and added water. Then add dextrose and Ammonia to convert them into simple sugar and control PH. After 40 to 50 hours of low-temperature fermenting, the ethanol can be produced after distillation. Wet milling is a process that mixes raw material with sulfuric acid, waits for 24 to 48 hours of decomposition, and the corn germ could be separated from the mixture. Then centrifugal separators are used to separate starch, then the rest process is like dry milling.

Figure 8. A display of biodiesel mixture (Jueves 2014).

2.1.2 *Biodiesel production*

In Figure 8, there is a display of a biodiesel mixture that is produced from glycerin from vegetable fats and oil through transesterification. Firstly, the phospholipids and water are removed in order not to produce a fatty acid, which is also called soap. Secondly, esterification can help excess fatty acid to convert into biodiesel and increase the yield. The oil is mixed with catalysts and dissolved in methanol, heated, and then converted into biodiesel, while the rest will be dehydrated. Thirdly, the mixture is added with KOH and transesterification process, and Biodiesel and glycerol would be produced. In the end, biodiesel should be recovered and refined.

2.2 *The second generation of biofuels and production*

Figure 9. The raw material of the production (Talha Akbar Kamal 2021).

The second generation of biofuel is produced from lignocellulosic or woody biomass or agricultural residues, and the raw material in Figure 9 gives a picture of residues. The product is mostly ethanol. It solved the problem of making ethanol by the use of the food of humans and increase the price of it, which makes poor people cannot afford it. Lignocellulosic materials are sent to pretreatment operation, through stream explosion, about 70 percent of hemicellulose would be hydrolyzed into pentoses. A filter is used to separate liquid and solid, and pentoses can be converted into ethanol through fermenting. The solid fraction would be sent to enzymatic hydrolysis and be separated into water-rich in glucose. The solid part, in the end, will be used for combustion and provide energy for the factory.

2.3 *The third generation of biofuels and production*

Figure 10. Seaweeds in Cummingston coast, Scotland (A. Alaswad 2015).

The third generation of biofuel is mainly produced from algae. It is a type of plant that grows in water and can be divided into two types, seaweeds, and microalgae. Figure 10 represents the image of seaweeds on Cummingston coast, Scotland. It depends on photosynthesis to survive, absorb carbon dioxide from the air and get energy from the sun. It has a significantly high rate of photosynthesis, about 3 percent to 8 percent, while some plants grown on land have only 0.5 percent. Seaweeds also have an advantage because of their short life circle, easiness to breed, no need for freshwater or skill to cultivate, and higher volumetric production rates which lead to greater biomass densities. Biogas is the most efficient production of seaweeds, anaerobic digestion, which is a process of bacteria in an environment without oxygen and producing carbon dioxide and oxygen. Firstly, hydrolysis is a process of organic polymers such as fats and carbohydrates breaking down into small molecules like monosaccharides, amino acids, and fatty acids. They will have a process Organism produces energy and methane with the nutrients. At the same time, a small number of ammonium salts, carbon dioxide, and hydrogen will be produced. In the end, anaerobic digestion occurs, methanogens create methane in the final product, here are the reactions:

$$CH_3COOH \rightarrow CH_4 + CO_2 \tag{1}$$

$$CO_2 + 4H_2 \rightarrow CH_4 + 2H_2O \tag{2}$$

Hydrogen is not likely to dissolve in water, there is another method to convert carbon dioxide into methane, when the raw material contains many lipids and acetogenesis reaction occurs, and reducing agents will be produced at the same time. It could also happen when there is sugar and produce acetic acid, pyruvic acid will be produced. The pyruvic acid will convert to acetic acid in acetogenic bacteria and produce carbon oxide at the same time.

2.3.1 *Biodiesel production*

The microalgae can be converted into biodiesel through a process called trans-esterification, triglyceride reacts with the alcohol with the impetus from catalysts such as sodium hydroxide or potassium hydroxide. Alcohol helps the fat or oil convert into ester radically. Fat reacts with alcohol and forms mono-alkyl ester. The product's glycerol is denser than biodiesel significantly, they can be separated by gravity or centrifuge. Figure 11 shows the mechanism of transesterification:

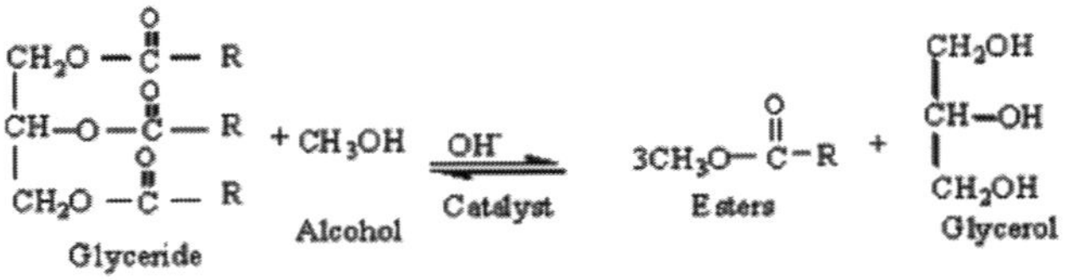

Figure 11. The mechanism of transesterification is shown (Big yard of science2018).

2.3.2 *The fourth generation of biofuels and production*

The fourth generation of biofuel is electrical fuels and solar fuels, which means producing fuels that store energy from the sun. For example, using the oxygenic photosynthesis of algae and changing the respiratory metabolism of certain algae, they will produce hydrogen at a fast rate. However, the technology is still not fully-fledged yet and needs more research to support it.

3 IMPLEMENTATION

Back in prehistoric times, humans have already used bioenergy and biofuels for domestic purposes. However, with the discovery of fossil fuels, humans replaced biofuels with these relatively high efficiencies and more accessible forms of energy. In recent years, with the worsening of environmental problems, the voice of environmental protection is getting louder and louder. Biofuel gradually returning to the public eye due to its sustainability and renewability. In the year 2000, 175 thousand barrels of biofuel were produced, however, they only equivalented to about 0.2% of the oil production in 2000 (N. Sönnichsen 2021). In the year 2019, the percentage increased to 1% (N. Sönnichsen 2021). The overall trend of biofuel production is increasing, largely due to people's increasing awareness of the environmental crisis, as the graph below shows.

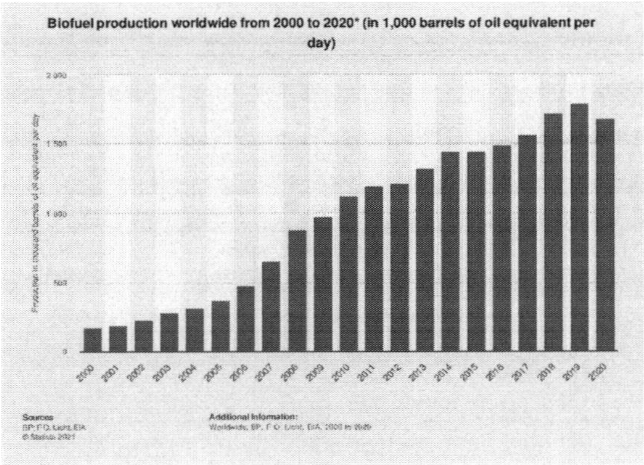

Figure 12. Biofuel production worldwide from 2000 to 2020*(in 1,000 barrels of oil equivalent per day) (N. Sönnichsen 2021).

In 2018, biofuel met around 3.4% of the transport fuel demand. Biofuels share 93% of the renewable energy used for transportation. In Figure 12, by the year 2024, biofuels are expected to increase 24% in transportation, which expands to 4.1% of the transport fuel demand (N. Sönnichsen 2021). The forecast for biofuels in transportation is shown below.

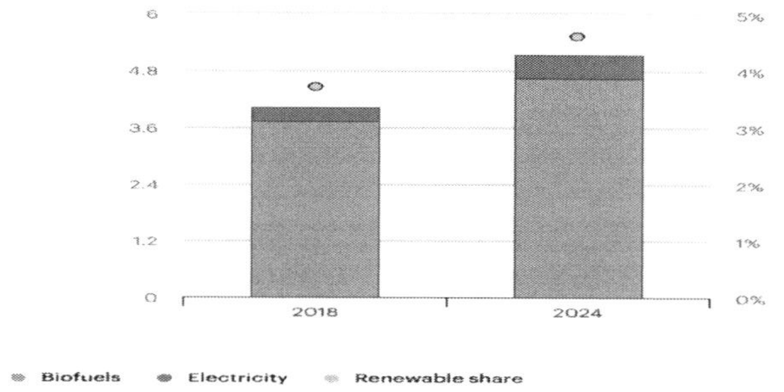

Figure 13. Renewable energy in the transportation of 2018 and 2024 (N. Sönnichsen 2021).

As shown in Figure 13, the use of biofuels has been presenting an increasing tendency. However, the phenomenon that biofuel is used more than electricity in transportation does not apply in developed countries. In developed counties like the United States, electricity is normally used more extensively. There are 30343 alternative fueling stations, 75% of which are electric fueling stations. Apparently that E85 (85% ethanol) fueling stations are more than electric in just a few states. Even though biofuels have the largest part in the transport fuel, they still have not been widely used in counties like the United States which held the largest production of biodiesel and ethanol. The United States produced as twice much as the production in Brazil, the second-largest ethanol production (Alternative Fuels Data Center. 2021). The graph below shows the top 5 ethanol production countries in the world.

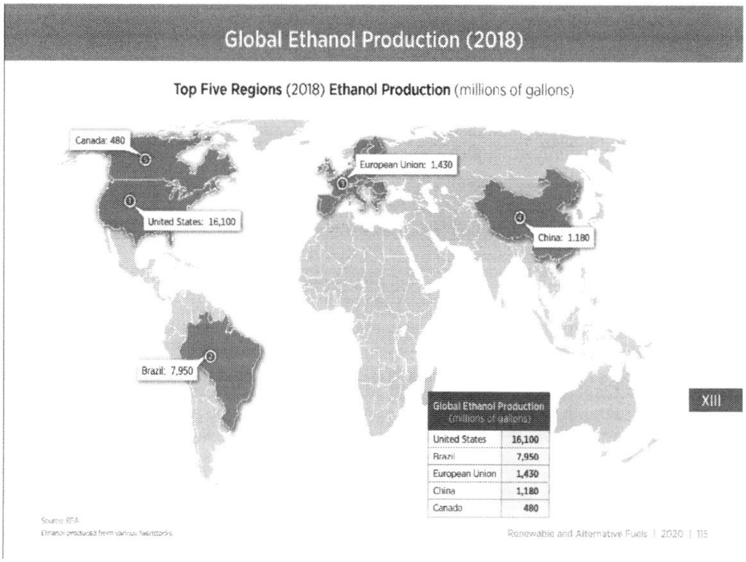

Figure 14. Top Five Regions (2018) Ethanol Production (millions of gallons) (Alternative Fuels Data Center. 2021).

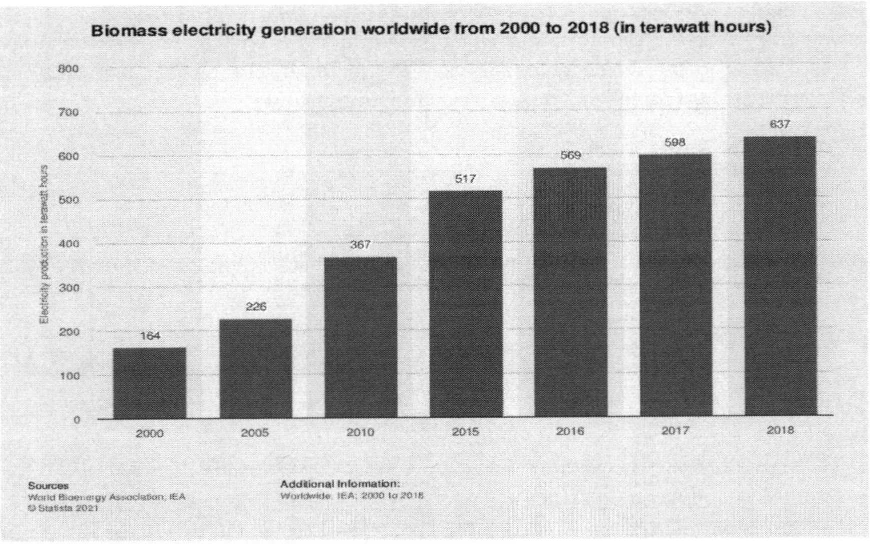

Figure 15. Biomass electricity generation worldwide from 2000 to 2018 (in terawatt-hours) (World Bioenergy 2020).

Facts show that the price of biofuels is higher than fossil fuels, which is one of the biggest reasons why biofuels are not widespread (Alternative Fuels Data Center 2021). As shown in Figure 14, five countries are using a relatively great amount of ethanol that is categorized as biofuel. There is still a long way to go for biofuels to replace fossil fuels in the future. Other than using biofuels as an alternative to fossil fuel, it has also been used to generate electricity. In 2018, 637.3 terawatt-hours of electricity were generated by biofuels, which hold about 2.2% of the global electricity generation. However, using biofuels to generate electricity is only a small part of renewable generation.

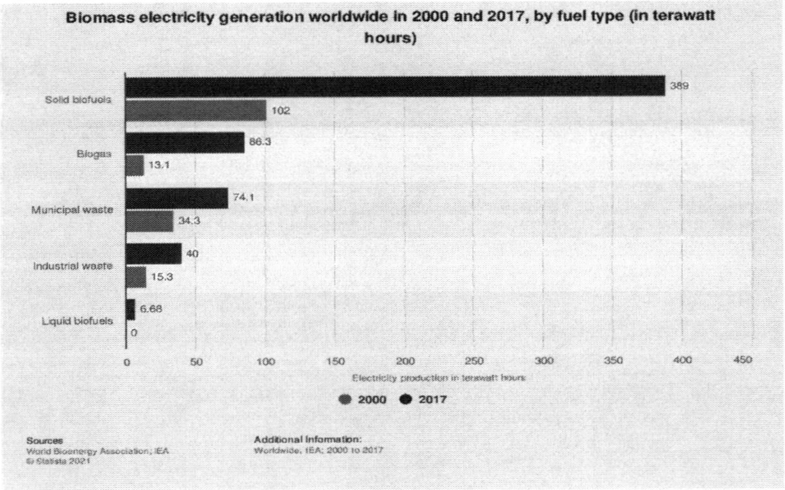

Figure 16. Biomass electricity generation worldwide in 2000 and 2017, by fuel type (in terawatt-hours) (World Bioenergy 2020).

Biofuels merely produced 8.3% of the renewable generation in 2018, which is the second last in renewable sources, slightly higher than solar PV (7.9%) (World Bioenergy 2020). Regardless of the small fraction that biofuels generated, the overall trend of biofuels' generation is increasing as the following graph shows.

In biomass electricity generation as illustrated in Figure 15, solid biofuels account for the largest proportion of various biofuels, biogas, and liquid biofuels come after. Biogas increased rapidly, it increased by almost 6 times from 2000 to 2017. Liquid biofuels haven't been used for electricity generation in 2000, and they held the smallest part in 2017 (World Bioenergy 2020). The Biomass electricity generation worldwide from 2000 to 2018 (in terawatt-hours) is shown below in Figure 16.

4 CONCLUSIONS

This paper uses a timeline to reveal the history of various types of biofuels, and then further introduces the technology within different generations of biofuel. The following part mentions the implementation of biofuels. This information views biofuel from a comprehensive perspective. Biofuel is clean and renewable energy that makes less pollution than fossil fuel. It holds tremendous potential since it has a net-zero carbon dioxide emission in theory, and it has been considered an alternative source of fossil fuel. If biofuels are widely used, the large number of greenhouse gases from fossil fuels will be reduced. However, the drawback of biofuel should be considered as well as seeing the bright future of biofuels. The biofuels produced by crops have the potential risk of intensifying soil erosion and worsening the famine. Crops may also change the biodiversity and affect the ecosystem. The waste and by-products cannot be ignored as they consumed many crops and food. In the future, new technology should be employed to solve the problems of using crops as raw materials. For instance, residue and waste are used to generate biofuel. Furthermore, new policy and technology should focus on lowering the price of biofuel, making it more available to the general public. It is believed that biofuel will take over the fuel market sooner or later.

REFERENCES

Advanced Fuel Solutions. (2019) *The why and how of blending biodiesel*. https://yourfuelsolution.com/the-why-and-how-of-blending-biodiesel-a-terminal-operators-guide/.

Alaswad, A., Michele Dassisti, T. Prescott, Abdul Ghani Olabi. (2015) *Technologies and developments of third-generation biofuel production*. https://www.researchgate.net/figure/Seaweeds-in-Cummingston-coast-Moray-Scotland_fig1_282301293.

Alternative Fuels Data Center. (2021) *EERE — Alternative Fuels Data Center, Alternative Fueling Station Counts by State*. http://www.afdc.energy.gov/afdc/fuels/stations_counts.html.

Alternative Fuels Data Center. (2021) *Alternative Fuel Price Report*. https://afdc.energy.gov/fuels/prices.html.

Arthur Wellinger, Jerry Murphy, David Baxter (Woodhead Publishing), 2013. *The biogas Handbook, Science, Production, andApplications*. https://www.sciencedirect.com/book/9780857094988/the-biogas-handbook.

Big yard of science (Sina Science and Technology), 2018. *Microchlorella: Eats carbon dioxide and squeezes out fuel*. http://tech.sina.com.cn/d/a/2018-08-14/doc-ihhtfwqq6628926.shtml.

Brittany Rycroft (Modern Ethanol Fireplaces). (2019) *What Is Bioethanol Fuel*? https://modernethanol fireplaces.com/blogs/news/what-is-bioethanol-fuel.

British Petroleum Company. (2021) *Statistical Review of World Energy*. https://www.bp.com/content/dam/bp/business-sites/en/global/corporate/pdfs/energy-economics/statistical-review/bp-stats-review-2021-full-report.pdf.

Debabrata Das, Jhansi L. Varanasi (Taylor & Francis Group), 2019. *Fundamentals of Biofuel Production Processes*. https://doi.org/10.1201/b22274.

Dhamodharan Kondusamy, MehakKaushalSaumya Ahlawat, Karthik Rajendran. (2021) *The role of techno-economic implications and governmental policies in accelerating the promotion of biomethane technologies*. https://www.sciencedirect.com/science/article/pii.

Doug Tribou (Michigan Radio), 2018. *Algae in the gas tank? U of M researchers study problems holding back algal biofuel*. https://www.michiganradio.org/news.

IEA. (2018) *An introduction to biogas and biomethane*. https://www.iea.org/. reports/outlook-for-biogas-and-biomethane-prospects-for-organic-growth.
Jason Paur. (2011) *High Cost Makes Aviation Biofuel Slow to Take Off.* https://www.wired.com/2011/11/aviation-biofuel.
Jueves. (2014) *Precio del biodiesel de Argentina*. https://www.rosariobioenergysa.com/novedades/precio-del-biodiesel-de-argentina.
Naresh Kumar Amradi. (2016) *Various processes of producing biohydrogen-utilizing waste as substrate*. https://www.researchgate.net/figure/Various-processes-of-producing-biohydrogen-utilizing-waste-as-substrate_fig6_295906955.
Pacific Biodiesel US. (2019) *Statistics of global biodiesel production*. https://www.biodiesel.com/ history-of-biodiesel-fuel.
Peake, S. (2018) Introducing Renewable Energy. In: Stephen, P., Bob, E. (Eds.), *Renewable Energy, Power for a Sustainable Future. Oxford University Press*, Oxford. pp. 22-55.
Prakash Kumar Sarangi, Sonil Nanda, Pravakar Mohanty (Springer), 2018. *Recent Advancements in Biofuels and Bioenergy Utilization*. https://link.springer.com/book/10.1007%2F978-981-13-1307-3.
Sönnichsen (Statista), N. 2021. *Biofuel production worldwide from 2000 to 2020*. https://www.statista.com/statistics/274163/global-biofuel-production-in-oil-equivalent/.
Sun and Wind Energy. (2015) *Biomethane plant in the UK inaugurated*. https://www.sunwindenergy.com/bioenergy/biomethane-plant-uk-inaugurated.
Sydney Solvents. (2021) *Eco Alcohol-Based 80% Sanitiser Liquid 20 Litre*. https://www.sydneysolvents.com.au/bio-ethanol-20-litre.
Talha Akbar Kamal. (2021) *Resource Base for Second-Generation Biofuels*. https://www.bioenergyconsult.com/second-generation-biofuels/.
World Bioenergy. (2020) *Global Bioenergy Statistics 2020*. https://worldbioenergy.org/uploads/201210%20WBA%20GBS%202020.pdf.

Energy Revolution and Chemical Research – Kok-Keong Chong and Zhongliang Liu (Eds)
© 2023 The Authors, ISBN: 978-1-032-36554-1

Experimental study on sludge combustion behavior and heavy metal migration

Chengxin Wang*, Shunyong Wu & Guanghui Zhang
Hefei General Machinery Research Institute Co. Ltd, Hefei, Anhui, China

ABSTRACT: In this paper, the combustion characteristics and heavy metal migration characteristics of sludge were studied by experimental methods, and the weight loss process of sludge under programmed temperature was evaluated by thermogravimetric analysis. The BCR continuous extraction method and the inductively coupled plasma mass spectrometry method were used to determine and compare the distribution of heavy metal forms in the original sludge and the sludge ash. The results show that the combustion process can be divided into two stages, devolatilization and volatile combustion. Combustion significantly reduces the proportion of heavy metals in the residue state and oxidizable state. The combustion process can reduce the environmental risk of heavy metals, but it is still high. The immobilization of heavy metals by combustion is limited.

1 INTRODUCTION

Sludge is a by-product of the sewage treatment plant in the process of sewage treatment. The sediment produced by multiple biological treatments and dosing treatment processes of sewage is sludge. According to reports, by the end of 2018, the annual output of urban domestic sewage sludge was 31 million tons, and the proportion of total sludge output increased year by year. Research forecasts show that China's annual sludge production will reach 60-90 million tons in 2020 (Liu 2020). Such an explosive increase in sludge production, if not properly disposed of, will bring severe pressure on the environment.

The biodegraded humus in the sludge organic matter contains a variety of pollutants, which are mainly organic compounds with complex functional groups, including hydroxyl, carboxyl, ester, aldehyde, carbonyl, and polycyclic aromatic hydrocarbons (PAHs). Some aromatic compounds combined with chlorine, such as polychlorinated biphenyls (PCBs), chlorophenols (PCs), and polychlorinated dibenzodioxins (PCDDs), can be volatilized at lower temperatures and can be volatile in pollutants (Sonoyama N 2011; Xu 2017). In the case of a large amount of mud accumulation, the water body and soil are seriously polluted, and it is easy to make people sick (Dai 2007). The sludge production process includes multiple biodegradations, which determine that the sludge contains a large number of microorganisms, including bacteria, viruses, and parasites. The storage method of long-term exposure to the air allows microorganisms to multiply in a large number of nutrients, which is easy to spread diseases.

Combustion is a promising sludge disposal method (Huang 2011). Through centralized combustion treatment, the requirements of sludge resource utilization, harmlessness, and reduction are achieved to the greatest extent (Vamvuka D 2019). Combustion can oxidize and decompose all the organic matter in the sludge, kill microorganisms, and realize the inertization of heavy metals. The calorific value of dried sludge is 1/3 to 1/4 of standard coal, and slightly lower than that of biomass, and it can be self-sustaining combustion without adding auxiliary fuel (Cai 2014; Xie 2018).

*Corresponding Author: wcx2046@mail.ustc.edu.cn

DOI 10.1201/9781003332657-80

2 EXPERIMENT

2.1 *Material*

The sewage sludge from Hefei Sewage Treatment Plant in Anhui Province was selected as the experimental raw material. The sewage sludge is dried, ground, and selected to have a particle size of 75-150 μm. Before the experiment, the sludge samples were placed in a 105°C oven for 12 h to remove moisture. The industrial analysis and elemental analysis of the samples are shown in Table 1.

Table 1. Ultimate analyses, proximate analyses of sewage sludge

Proximate analyses (wt.%)					Ultimate analyses (wt.%)				LHV (MJ/kg)
C_{ad}	H_{ad}	O_{ad}	N_{ad}	S_{ad}	V_{ad}	FC_{ad}	A_{ad}	M_{ad}	
26.33	4.56	15.59	4.84	0.87	48.12	4.07	44.31	3.50	9.70

ad- air drying base; V- Volatile; A-Ash; FC- Fixed Carbon; M- Moisture; LHV- Lower Heat Value.

The sludge ash was prepared according to the biomass ash production method of ASTM E1755, and the sludge ash composition was analyzed by X-ray fluorescence spectrometry (XRF). The results are shown in Table 2. Nearly half of the ash content is SiO_2. Due to the use of additives in the sewage treatment process, the content of Al_2O_3 and P_2O_5 in the ash is relatively high.

Table 2. Sludge ash composition analysis.

	SiO_2	Al_2O_3	P_2O_5	Fe_2O_3	CaO	MgO	K_2O	Na_2O	TiO_2	SO_3
(wt.%)	48.88	17.18	11.10	6.64	3.50	3.61	3.12	0.81	0.76	0.67

2.2 *Thermogravimetric experiment*

Sludge combustion weight loss characteristics experiments were carried out on a thermogravimetric analyzer TGA Q5000IR. The heating programs of both combustion and pyrolysis were raised from room temperature to 100°C for 10 min to avoid the influence of moisture absorption by the samples exposed to the air, and the temperature was then increased from 100°C to 10°C/min, 20°C/min, 30°C/min heating rate (β) increased to 800°C. Air was used in the combustion experiment, with a flow rate of 1 atm and 100 ml/min. The number of samples weighed in each group of experiments was 10 ± 0.5 mg. A blank experiment was performed before the start of each group of experiments to ensure a stable baseline. After the blank experiment, the TG experiment was repeated three times to ensure repeatability and consistency. When the deviation of each data point of the three repeated experiments is within 1%, the average of three sets of data is used as the experimental result, otherwise, the experiment is repeated. Background gas is purged for 5 min before starting the temperature program.

2.3 *Heavy metal chemical speciation experiment*

The chemical forms of heavy metals will change during the heat treatment process. The four forms obtained by the BCR sequential extraction method are called weak acid extraction state, reducible state, oxidizable state, and residue state. The five forms obtained by Tessier sequential extraction are called exchangeable state, carbonate bound state, iron, and manganese oxidation state-bound state, organically bound state, and residue state. The BCR method extracts four fractions: first, the acid-soluble/exchangeable fraction (F1), second, the reducible fraction (F2), third, the oxidizable fraction (F3), and finally, the residual fraction (F4). The weak acid extraction state is equivalent to the exchangeable state and the carbonate state, mainly due to the presence of heavy metals in the form of ionic bonds or carbonates, which are very sensitive to pH and easy to decompose, with the strongest toxicity. In the reducible state, heavy metals exist in the form of iron-manganese oxides, which are potentially toxic. When the heavy metal exists in the form of organic matter, the oxidation state is relatively stable in nature and does not cause high harm to the environment. In the residual state, heavy metals are sequestered in the form of stable minerals, which are difficult to release and have no toxicity. In the case of the same concentration of heavy metals, the different chemical forms distribution, and the harm to the environment is also different. This chapter uses the BCR four-step continuous extraction method to operate in accordance with the national standard GB/T 25282-2010, and the acid-soluble state, the reduced state, the oxidized state, and the residue state are recorded as F1, F2, F3, and F4, respectively.

3 RESULT AND DISCUSSION

3.1 *Analysis of sludge thermogravimetric*

Figure 1 shows the TG and DTG curves of sludge (SS) at a heating rate of 20°C/min. From the perspective of the overall weight loss process, combustion can be divided into two stages. The first stage, in the temperature range of 120–420°C, is the combustion stage of devolatilization and volatile matter and occurs at a position near 157°C. A smaller peak appears on a shoulder at 340°C. The second stage is the decomposition and combustion of macromolecular organic matter and the combustion of coke in the temperature range of 420-593°C. But when the temperature is above 610°C, the decomposition of inorganic substances such as calcium carbonate causes a slight decrease in the TG curve and fluctuations in the DTG curve. First of all, for the first stage, weight loss starts at 120°C, and a smaller peak appears at 157°C. This is because the sludge composition is more complex, and a large number of small molecular amino acids, short peptide chains, or other molecular fragments are decomposed by bio-degradation, which is stable. It has poor performance and will decompose in a concentrated manner before the ignition temperature. This is also the reason why the weight loss onset temperature is lower compared to other biomasses. The weight loss ratio of stage one is about 49%, which is consistent with the volatile content in industry analysis, and the weight loss peak appears at 290°C. The mass loss in this process is due to the decomposition of organic matter and biodegradation with a simple structure. Caused by the breakdown of organic matter such as amino acids, lipids, humic acids, and aliphatic compounds. This stage is also accompanied by the depolymerization and polymerization of macromolecules (proteins) and the gradual formation of coke. Repeated experiments found an insignificant shoulder at 340°C, possibly due to the decomposition of a lesser substance. For stage two, the coke remaining after stage one devolatilization and some organics that are difficult to decompose at low temperature are oxidized in the range of 420-593°C.

3.2 *Analysis of heavy metal form*

Figure 2 shows the distribution of heavy metal chemical forms in sludge and sludge ash after combustion at 700°C, and the effect of the combustion process on the chemical forms of heavy metals was studied.

It can be seen from Figure 4.7 that after the combustion process, the proportions of F1 and F2 of heavy metals decrease to varying degrees. The F3 ratio of different heavy metals has two situations: rising and falling. However, the F3+F4 ratio of all heavy metals is on the rise, which indicates that the environmental risk of heavy metals in the combustion ash is reduced during the combustion process.

The F1+F2 ratio of Cd, Zn, Mn, and As in SS is about 60%, and the F1+F2 ratio of Cd, Zn, Mn, and as in ASS-700 is also over 40%, indicating that although the combustion process can reduce

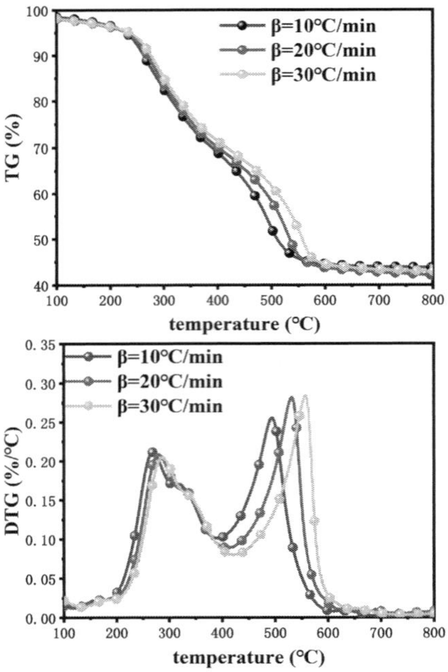

Figure 1. TG/DTG curves of sewage sludge.

the environmental risk of heavy metals, being still higher, and the immobilization of heavy metals by combustion is limited.

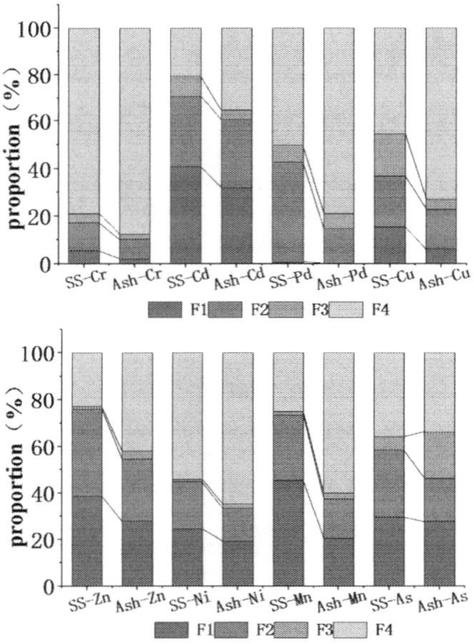

Figure 2. Heavy metal chemical forms in sludge and sludge ash after combustion at 700°C.

4 CONCLUSION

The combustion process can be divided into two stages, the combustion stage of devolatilization and volatile matter; combustion significantly reduces the proportion of heavy metals in the residue state and oxidizable state. The combustion process can reduce the environmental risk of heavy metals, but it is still high, and the immobilization of heavy metals by combustion is limited.

ACKNOWLEDGMENTS

This work is supported by Anhui Province Key Research and Development Program (No. 201904a07020039).

REFERENCES

Cai J, Wu W, Liu R. An overview of distributed activation energy model and its application in the pyrolysis of lignocellulosic biomass. *Renew Sustain Energy Rev* 2014; 36: 236–46. https://doi.org/10.1016/j.rser.2014.04.052.

Dai J, Xu M, Chen J, Yang X, Ke Z. PCDD/F, PAH and heavy metals in the sewage sludge from six wastewater treatment plants in Beijing, China. *Chemosphere* 2007; 66: 353–61. https://doi.org/10.1016/j.chemosphere.2006.04.072.

Huang M, Chen L, Chen D, Zhou S. Characteristics and aluminum reuse of textile sludge incineration residues after acidification. *J Environ Sci* 2011. https://doi.org/10.1016/S1001-0742(10)60662-6.

Liu Y, Ran C, Siddiqui AR, Siyal AA, Song Y, Dai J, et al. Characterization and analysis of sludge char prepared from bench-scale fluidized bed pyrolysis of sewage sludge. *Energy* 2020; 200:117398. https://doi.org/10.1016/j.energy.2020.117398.

Sonoyama N, Nobuta K, Kimura T, Hosokai S, Hayashi J. Production of chemicals by cracking pyrolytic tar from Loy Yang coal over iron oxide catalysts in a steam atmosphere. *Fuel Process Technol* 2011; 92: 771–5. https://doi.org/10.1016/j.fuproc.2010.09.036.

Vamvuka D, Sfakiotakis S, Pantelaki O. Evaluation of gaseous and solid products from the pyrolysis of waste biomass blends for energetic and environmental applications. *Fuel* 2019; 236: 574–82. https://doi.org/10.1016/j.fuel.2018.08.145.

Xie W, Wen S, Liu J, Xie W, Kuo J, Lu X, et al. Bioresource Technology Comparative thermogravimetric analyses of co-combustion of textile dyeing sludge and sugarcane bagasse in carbon dioxide/oxygen and nitrogen/oxygen atmospheres: Thermal conversion characteristics, kinetics, and thermodynamics. *Bioresour Technol* 2018; 255:88–95. https://doi.org/10.1016/j.biortech.2018.01.110.

Xu ZX, Liu P, Xu GS, Liu Q, He ZX, Wang Q. Bio-fuel oil characteristic from catalytic cracking of hydrogenated palm oil. *Energy* 2017; 133: 666–75. https://doi.org/10.1016/j.energy.2017.05.155.

Energy Revolution and Chemical Research – Kok-Keong Chong and Zhongliang Liu (Eds)
© 2023 The Authors, ISBN: 978-1-032-36554-1

Research progress on quick start-up of filter for manganese removal from groundwater

Bing Leng
CNNC (Shenyang)Water Services Co, Shenyang, China
Institute of Water Pollution Prevention and Control in Liaohe River Basin, Shenyang Jianzhu University, Shenyang, China

Sijia Zhang, Xing Jin, Xin Li & Yanping Zhang
Institute of Water Pollution Prevention and Control in Liaohe River Basin, Shenyang Jianzhu University, Shenyang, China

Zhiyuan Zheng
CNNC Fushun Environmental Protection Technology Co., Ltd, Fushun, China

Jinxiang Fu*
Institute of Water Pollution Prevention and Control in Liaohe River Basin, Shenyang Jianzhu University, Shenyang, China

ABSTRACT: The processes and mechanisms of natural oxidation, contact oxidation, biological demanganization and chemical oxidation demanganization were introduced. The common filter materials and their characteristics in waterworks were expounded, and the effect and research progress of oxidation by oxidant, adjustment of filter operation parameters, cultivation of high-quality manganese removal bacteria, and rapid start-up of the filter by artificial loading of manganese oxide were described. It is considered that it is of great significance to eliminate the mature period of filter media and prospect the future development direction.

1 INTRODUCTION

As an important water resource, groundwater has the advantages of wide distribution and good water quality. With the development of industry and the improvement of people's living standards, water pollution is becoming more and more serious. Manganese is a common inorganic pollutant in groundwater. If the content of manganese in water is too high, it will have a negative impact on people's production and life. Excessive manganese intake can lead to neurological impairment; in life, excessive manganese content can pollute living utensils, make sanitary ware attached to yellow-brown rust spots, and block water pipelines. Therefore, China's new "Sanitary Standard for Drinking Water" GB5749-2006 stipulates that the maximum content of manganese should not exceed 0.1 mg/L.

According to the different filter materials used in the water treatment, the maturity period of the manganese removal filter adopting the contact oxidation method is also different. The contact oxidation filter used at home and abroad basically adopts the natural fixation method, that is, the manganese active filter membrane is naturally formed by filtering the manganese-containing water. However, the maturity period of the natural method for forming the manganese active filter membrane is longer, and even one month is needed. The slow one needs half a year or even longer,

*Corresponding Author: fujinxiang@sina.com

 DOI 10.1201/9781003332657-81

so it is particularly important to study the rapid start-up of the contact oxidation filter for manganese removal.

2 MANGANESE REMOVAL METHOD

2.1 *Natural oxidation method*

Natural oxidation is the earliest manganese removal method in China, and the treatment process includes aeration, oxidation, precipitation, filtration, and so on (Li 2010). The process flow is long, the equipment is huge, the investment is large, the operation management is complex, and the manganese removal effect is not ideal (Li 2011). Under neutral conditions, manganese ions are oxidized by dissolved oxygen in the air, and the oxidation rate of manganese is very low. Only when the pH value is increased to above 9.5 can the oxidation rate of manganese in the air be accelerated. Usually alkaline substances are added to meet the requirements, but because the pH value of the treated water is high, exceeding the national drinking water quality standards, acid needs to be added to the water to meet the water quality standards. The cost is high, the processing flow is complex, and the process is tedious (Li 2016).

2.2 *Contact oxidation method*

Manganese removal by contact oxidation was proposed by Academician Li Guibai in the 1950s (Tian 2019). Its treatment process is to aerate the groundwater containing manganese and filter it through a filter, forming a manganese active filter membrane mainly composed of high-valent manganese on the surface of the filter material. This active filter membrane has a catalytic effect on the oxidation of divalent manganese, and the filter membrane is constantly updated so that the reaction of manganese removal continues. Due to the different chemical compositions of different filter materials (Tang 2018), the maturity of filter materials is different, and the adsorption capacity of pollutants is also different. Even if the same filter material is used in the test, the maturity of filter materials is also different. As reported, the shortest maturity period of manganese sand is 30 days, but some need 6 months to mature.

2.3 *Biological manganese removal method*

In the 1980s, biological manganese removal technology was studied in China, and it was found that there were a large number of microorganisms on the surface of the filter material, so there was a new idea for biological manganese removal, that is, biological catalytic oxidation manganese removal (Hou 2016). In the 1990s, academician Zhang Jie put forward the theory of "biological manganese fixation and removal" (Zhong 2019), which proposed that the oxidation of divalent manganese in the filter layer is biological oxidation under neutral pH conditions (Bai 2018). The process of biological manganese removal is divided into three stages: diffusion, adsorption, and oxidation (Zheng 2013). Firstly, the divalent manganese in the water is diffused to the surface of the filter material, and the divalent manganese diffused to the surface of the filter material is adsorbed by a manganese active filter membrane, iron, manganese bacteria, and manganese oxide, which is finally oxidized under the action of manganese-oxidizing bacteria and extracellular polymers of the manganese-oxidizing bacteria, so that the divalent Mn is oxidized into tetravalent Mn.

2.4 *Chemical oxidation method*

The chemical oxidation method is to add chemical agents with strong oxidizability to water (Hou 2016), oxidize the divalent manganese in water into high-valent insoluble manganese oxide through oxidation, and remove it through filtration or coagulation sedimentation process. The agents used for manganese removal mainly include ozone, potassium permanganate, chlorine dioxide, etc.

3 FILTER MATERIALS ARE COMMONLY USED IN WATER PLANTS

3.1 *Quartz sand*

Quartz sand is a very common non-metallic mineral, low price, hard, wear-resistant, strong impact resistance, used as a filter material for several years without replacement, cost savings, is a good natural filter material, is an ideal material for water treatment. Shao Yuezong et al. (Shao 2016) used quartz sand as raw material to investigate the effect of dissolved oxygen concentration on the synchronous removal of ammonia nitrogen, and manganese in groundwater by iron and manganese oxide film on the surface of quartz sand filter material. The results showed that dissolved oxygen concentration was the main factor affecting the synchronous removal of NH^{4+}-N and Mn^{2+}.

3.2 *Manganese sand*

Natural manganese sand contains a large number of high-valence manganese oxides, which have a certain adsorption effect on manganese in the water. Manganese removal depends on adsorption before manganese sand is mature, and manganese removal depends on oxidation after manganese sand filter material is mature. Zhang Liyan (Zhang 2012) compared the manganese removal effect of natural manganese sand, ordinary quartz sand, water plant ripe sand, and filter used sand. The results show that the natural manganese sand and the ripe sand from the water treatment plant have a better manganese removal effect, but the natural manganese sand has a slightly worse turbidity removal effect due to its larger particle size; the ordinary quartz sand has a worse manganese removal effect because it only has filtration effect; the used sand from the filter has a certain manganese removal ability.

3.3 *Zeolite*

Zeolite is a kind of framework aluminosilicate mineral (Tang 2018), which is a common natural ore with a low price and large specific surface area and can absorb and filter substances. Zhao Da (Zhao 2015) carried out iron-loaded zeolite and explored the performance of iron-loaded membrane zeolite in pollutant treatment. The experimental results showed that the natural zeolite modified by ferric chloride increased the removal effect of phosphate, sulfide, and hexavalent chromium.

4 QUICK STARTUP OF THE FILTER

4.1 *Oxidant oxidation quick starting*

4.1.1 *Ozone oxidation starts quickly*

Ozone is a strong oxidant, which can rapidly oxidize and remove Mn^{2+} in the absence of a catalyst or a low pH environment. The reaction equation is as follows: $2Mn^{2+}+2O_3+4H_2O=2MnO_2(OH)_2+2O_2+4H^+$.

Zhong Shuang et al. (2011) used cheap river sand as filler and adopted intermittent ozone aeration to rapidly oxidize Mn^{2+} in raw water into water-insoluble compounds attached to the surface of river sand, thus rapidly forming a manganese activated filter membrane with manganese removal capability. The manganese activated filter membrane can be formed in only 5 days so that the effluent is continuously stabilized below 0.1 mg/L, the maturity period of the manganese removal filter is greatly shortened, and the rapid start-up of the manganese removal filter is realized.

4.1.2 *Quick start of potassium permanganate oxidation*

Sun Chengchao (2019) studied the rapid startup of contact oxidation manganese removal filters from four aspects of oxidant, hardness, $KMnO_4$ dosage, and filter material. The quartz sand can only remove about 0.3 mg/L of Mn^{2+}in 120 days under the condition of natural oxidation, and the rapid start-up of the contact oxidation filter can be realized after the addition of $KMnO_4$ for 48 days.

4.1.3 *Chlorine dioxide oxidation method*

Chlorine dioxide is a strong oxidant with safe and efficient oxidation characteristics. Chlorine dioxide oxidation is a process in which free divalent manganese is oxidized to insoluble high-valent manganese by chlorine dioxide and then removed by filtration.

Liu Qingyuan (2010) investigated the effect of chlorine dioxide dosage, temperature, pH, pre-oxidation time and Mn^{2+}dosage on the removal of iron and manganese by chlorine dioxide pre-oxidation. The results showed that the dosage of chlorine dioxide was the main factor affecting the manganese removal effect, while the pre-oxidation time and temperature had little effect on the manganese removal effect.

4.2 *Adjust operation parameters and culture high-quality demanganization strains for rapid startup*

Zhang Jie (2018) studied the rapid startup and stable operation of the biofilter for groundwater with low temperature, high-speed railway, and high manganese content. For the groundwater with low temperature, high-speed railway, and high manganese content, the temperature is 5 ~ 6°C, the total iron concentration is 10 ~ 20 mg/L, and the total manganese concentration is 0.8 ~ 1.5mg/L, through the startup mode of inoculating mature filter sand, backwashing supernatant reflux and changing backwashing intensity, The rapid startup of the low-temperature high-iron-manganese biofilter is realized, and the startup time is only 60 days.

4.3 *Artificial preparation of load manganese oxide for fast start-up*

Manganese oxide is prepared by a specific chemical synthesis method and then is loaded on the surface of the filter material by embedding technology to eliminate the mature period of the filter material. Li Jincheng (2011) prepared a chemically loaded manganese oxide filter material of the same type as biological manganese oxide and then used a chemical oxidant instead of manganese-oxidizing bacteria to oxidize and regenerate the saturated filter layer. Manganese oxides were prepared in acidic medium and alkaline medium respectively, and it was found that the surface manganese content of manganese oxides prepared in acidic medium was much higher than that of manganese oxides prepared in alkaline medium, and the manganese contents were 4. 62 mg Mn/g and 0. 51 mg Mn/g, respectively. And the stability of the filter layer under acidic conditions is much higher than that under alkaline conditions. The chemically loaded manganese oxide filter material is not only less affected by the environment, but also eliminates the mature time of the filter material, so as to realize the rapid use of the manganese removal filter material.

Cong Lyu et al. (2017) studied the preparation of manganese oxide-loaded zeolite and its performance in removing divalent manganese from groundwater. In the filtration experiment, the removal effect of Mn^{2+} ions by loaded manganese oxide zeolite was better, and the removal rate of Mn^{2+} ions reached 98% -100% in a very short start-up time. In addition, during the filtration process, new flocculated manganese oxides with mixed-valence states of manganese were formed on the surface of the loaded manganese oxide zeolite, which further promoted the adsorption and oxidation of Mn^{2+} ions.

Anna Georgiadis et al. (2017) studied the adsorption of birnessite sand synthesized by reducing permanganate with lactic acid on quartz sand and prepared quartz sand loaded with manganese oxide for water treatment and adsorption of metal cations. The results show that the attachment and aggregation of the loaded manganese oxide quartz sand increase with the increase of pH value and the extension of reaction time.

5 CONCLUSION

At present, quartz sand and natural manganese sand are used as the filter material for manganese removal in most water plants, but it takes a long time for the surface of these filter materials to form a mature manganese activated filter membrane, so it is of great significance to shorten or even eliminate the maturity period of the filter material and maintain the stability of the treatment effect for the manganese removal of groundwater, the rational use of groundwater and the conservation of water resources.

REFERENCES

Anna Georgiadis and Thilo Rennert. (2017) A simple method to produce birnessite-coated quartz sand. *J. Plant Nutr. Soil Sci.* 1–5.

Chaochun Tang, Rongming Xu. (2018) Research Progress of Chemical Manganese Removal Technology. *Technology of Water Treatment.* 44(12):14–19.

Chengchao Sun. (2019) *Contact Oxidation Filter Enhanced by Potassium Permanganate for Manganese Removal: Quick Start and Performance of Treatment.* Harbin Institute of Technology.

Cong Lyu, Xuejiao Yang, Shengyu Zhang, Qihui Zhang&Xiaosi su. (2017) Preparation and performance of manganese oxide coated zeolite for removal of manganese contaminated in groundwater. *Environmental Technology.*

Da Zhao. (2015) *Synthesis of Iron Oxide Coated Zeolite (IOCZ) and Its Experimental Investigation on Controlling River Pollutants.* Harbin Institute of Technology.

Guibai Li, Xing Du, Huarong Yu, Fangshu Zhai, Heng Liang. (2016) Some thoughts on innovation and development of iron and manganese removal from groundwater. *Water supply and drainage.* 9–16.

Hexia Hou. (2016) *Study on the Construction of Autotrophic Biological Filter and the Effect of Manganese Removal.* The Qingdao University of Technology.

Jie Zhang, Yulin Wang, Dong Li, Hang Yang, Huiping Zeng. (2018) Experimental Research of Low-temperature, High Concentration Iron and Manganese in Groundwater by Bio-filter Purification. *Journal of Beijing University of Technology.* 44(9):1239–1246.

Jincheng Li. (2011) *Removal of high content manganese from groundwater with manganese oxide-coated filter.* The Ocean University of China.

Junbing Wu, Tinglin Huang, Ya Cheng, Jie Liu. (2017) Exploration of the factors for the rapid start-up of the chemical catalytic oxidation filters for the simultaneous removal of iron, manganese, and ammonia. *China Environmental Science.* 37(3)1007–1008.

Laisheng Zhu, Tinglin Huang, Ya Cheng, Yanfeng Yang, Nan Zhao. (2017) Effect of influent manganese concentration on startup of contact oxidation. *China in Groundwater Treatment.* 33(21):6–12.

Lin Zhong. (2019) *Study on Chemical Action and Biological Action of Manganese Sand on Manganese Removal from Groundwater.* Harbin Institute of Technology.

Liyan Zhang. (2012) *Study on Investigation and Removal of Iron and Manganese in Shenzhen Source Water.* Harbin Institute of Technology.

Na Zheng. (2013) *Mechanism of catalytic oxidation of Mn2+in groundwater by active filter membranes.* Xi'an University of Architecture & Technology.

Qingyuan Liu. (2010) *Research on Removal of Manganese by Chlorine Dioxide Pre-oxidation and Oxidation By-product Removal of Chlorite.* Nanjing University of Science&Technology.

Shuang Zhong, Cong Lu, Sijia Wang, Fengjun Zhang, Lei Zhu. (2011) Quick start-up of filter for removing manganese by contact oxidation method. *CIESC Journal.* 62(5):1435–1440.

Wei Li. (2010) *Study on Adsorption Removal of Mn(?) From Water Using Modified Materials.* Qingdao Technological University.

Xiaoli Bai. (2018) *Kinetics and Pilot Study of Simultaneous Removal of Ammonium and Manganese from Surface Water by Manganese Co-oxides Film.* Xi'an University of Architecture & Technology.

Xuan Tian. (2019) *Effects of ammonium and water quality on manganese removal by manganese co-oxide film in surface water.* Xi'an University of Architecture& Technology.

Yuezong Shao, Tinglin Huang, Xinxin Shi, Yang Wang, Ya Cheng, Hao Bu. (2016) Effect of DO concentration on the removal of ammonium and manganese ion in groundwater by Fe/Mn co-oxides film coating on quartz sands. *Chinese Journal of Environmental Engineering.* 10(11): 6159–6164.

Yu Zhang, Rui Sun, Hui-ping Zeng, Jie Zhang. (2018) Screening of Manganese Oxidation Bacteria in Biological Iron and Manganese Removal Filter. *China Water& Wastewater.* 34(3):68–76.

Energy Revolution and Chemical Research – Kok-Keong Chong and Zhongliang Liu (Eds)
© 2023 The Authors, ISBN: 978-1-032-36554-1

Dechlorination of high-chlorine waste biomass by hydrothermal treatment with additives assistance

Yousheng Lin*, Pengwei Chen & Qing He
Guangdong Provincial Key Laboratory of Distributed Energy Systems, School of Chemical Engineering and Energy Technology, Dongguan University of Technology, Dongguan, China

ABSTRACT: In order to further reduce the chlorine content of high-chlorine agroforestry waste biomass (wheat, Cl 9356 mg/kg) and avoid alkali metal problems caused by alkaline additives, this paper explored the hydrothermal dechlorination effect of an alkali-free dechlorinating agent γ-Al_2O_3. Four alkaline additives were also studied for comparison. The experiments were carried out at 235°C with a 4g: 50 ml solid-to-liquid ratio for 60 min. And the additive/biomass ratios were between 0.5% and 4%. Results showed that the dechlorination efficiency of 2% and 4% γ-Al_2O_3 increased from 84.65% to 95.09% and 96.08%, respectively. Followed by 4% Na_2CO_3 (95.51%) and 4% K_2CO_3 (95.00%). The Cl contents of these corresponding hydrochars were less than 0.1%. With additives assistance, the Cl ions dissolve, the nucleophilic substitution reaction, C–Cl bond cleavage in organic compounds, and the synergetic dechlorination and dealkalization processes were enhanced. Therefore, hydrothermal dechlorination was promoted. But employing additives assistance also reduced the hydrochar yields. The carbon contents of some hydrochars were reduced. The energy grade of hydrochars derived from dechlorination additives was lower than that of hydrochar without additives assistance. Through comprehensive consideration, 1%~2% γ-Al_2O_3 was a more suitable approach for deep dechlorination of high-chlorine waste biomass by hydrothermal treatment in this work. These findings provided some fundamental data to find some cheap but efficient additives for deep dechlorination of waste biomass.

1 INTRODUCTION

To limit global warming, the extraction and use of fossil fuels must be strictly limited (Welsby D 2021). As an abundant and nearly CO_2-neutral energy source, renewable biomass plays an important role in the carbon neutralization process (Qin 2022). The thermal conversion technologies, such as combustion, pyrolysis, and gasification, can convert biomass energy into electricity, biofuel, syngas, and other high value-added products. Due to the fertilization practices, the chlorine contents of agroforestry waste biomass are usually much higher than aquatic biomass and coal. Therefore, special attention should be paid to the disadvantages caused by high-chlorine waste biomass in such applications, including chlorine corrosion of heat exchangers and the formation of health hazardous polychlorinated dioxins and furans (Ren 2017).

As low chlorine biomass is preferred in the thermochemical conversion process, attention can be paid to literature reports showing that hydrothermal treatment (HTT) reduces the Cl content of high-chlorine solid fuel (Gandon-Ros G 2020; Huang 2019; Shen, 2020). The leaching effects, elimination reaction ($–CH_2–CHCl \rightarrow –CH{=}CH– + HCl$), and substitution reaction ($–CH_2–CHCl + [OH^-] \rightarrow –CH_2–CHOH– + [Cl]$) were enhanced under a hydrothermal environment, resulting in the conversion of insoluble organic Cl to soluble inorganic. Therefore, the process is also known as hydrothermal dechlorination. In our previous study, we employed this effect to significantly reduce

*Corresponding Author: linys@dgut.edu.cn

DOI 10.1201/9781003332657-82

the Cl content of three agroforestry biomass and the dechlorination efficiency reached approximately 90% at high HTC severity (Lin 2022). However, the Cl contents of hydrochars derived from wheat still exceeded 0.2%. How to further reduce the chlorine content in a milder environment is a topic worthy of in-depth study. Zhao et.al (2018) and Gandon-Ros et.al (2020) demonstrated that alkaline additives could promote hydrothermal dechlorination. Considering that alkaline additives may further aggravate the alkali metal problem in subsequent biomass utilization, this work explored the effects of using γ-Al_2O_3 and alkali-free dechlorinating agents on hydrothermal dechlorination. For comparison, four alkaline additives were also studied.

2 MATERIALS AND METHODS

Wheat, a typical agroforestry waste biomass, with 9356 mg/kg Cl content was used in this work. The wheat was collected from Hebei Province, China. A 100 ml Hastelloy autoclave reactor was employed in this study. The desired temperature was 235°C with a 4g: 50 ml solid-to-liquid ratio for 60 min. The γ-Al_2O_3, K_2CO_3, KOH, Na_2CO_3, and NaOH were provided by Shanghai Aladdin (Analytical pure). The additive/biomass ratios were between 0.5% and 4%. The hydrochar yield was calculated as follows:

$$Hydrochar\ yield = \frac{mass\ of\ hydrochar}{mass\ of\ raw\ sample} \tag{1}$$

The Cl content was measured by flask combustion ion chromatography method according to EN 14582: 2016. The dechlorination efficiency (DE) was obtained by the following equation:

$$DE = 1 - \frac{Cl\ content\ of\ hydrochar}{Cl\ of\ raw\ sample} \times hydrochar\ yield \tag{2}$$

It is noted that the DE was equal to the proportion of Cl migrated to the liquid phase. Surface morphologies of all samples were acquired by scanning electron microscopy (SEM, FEI Quanta FEG 250). More special information about the experimental process, sample detection, analysis methods, etc., could be found in our previous study (Lin 2022).

3 RESULTS AND DISCUSSION

3.1 *Dechlorination and chlorine redistribution of waste biomass during HTT with additives assistance*

Figure 1 showed the Cl content and DE of all samples during HTT with different additives assistance. As a high-Cl fuel, the Cl content of wheat declined significantly from 9356 mg/kg to 2723 mg/kg after HTT. With the additive's assistance, the Cl contents of hydrochars further declined. For example, the Cl content declined from 2140 mg/kg to 703 mg/kg with increasing the γ-Al_2O_3ratio from 0.5% to 4%. As presented in Figure 1(a), the Cl contents of hydrochars derived under 2%, 4% γ-Al_2O_3, and 4% K_2CO_3, KOH, and Na_2CO_3 conditions were all less than 0.1%. These solid hydrochars could be considered low Cl fuel and exhibited a low slagging tendency (Lin 2022).

Figure 1(b) also showed the DE of all hydrothermal dechlorination conditions. Except for without additives, 0.5% γ-Al_2O_3, and 2%NaOH conditions, the DEs of other conditions were all above 90%. The additive introduced more OH ions and promoted the hydrolysis of organic matter. Figure 3 exhibited the SEM images of the raw waste biomasses and hydrochars. The surface structures of hydrochars obtained with additives were damaged more seriously, which was conducive to releasing inorganic Cl ions and accelerating the migration of Cl ions. Therefore, the Cl ions dissolve, the nucleophilic substitution reaction, C–Cl bond cleavage in organic compounds, and the synergetic dechlorination and dealkalization process were all enhanced. Compared to alkaline additives, the alkali-free dechlorinating agents γ-Al_2O_3performed betteron hydrothermal dechlorination. As shown in Figure 1. (b), the DEs of 1%, 2%, and 4% γ-Al_2O_3 were 91.95%, 95.09%, and 96.08%, which increased by 8.62%, 12.33%, and 13.50% compared with the condition without additive, respectively. Compared to the above four alkaline additives, the molar mass of γ-Al_2O_3 combined with Cl ($AlCl_3$) was more under the same adding mass. The DEs of 4% Na_2CO_3 and 4% K_2CO_3 also reached 95.51% and 95.00%, respectively. Zhao et.al also found that the Na_2CO_3 showed higher DE than NaOH or KOH in their experimental conditions (Zhao 2018). They attribute the reason that the use of alkali caused a reduction of BET surface area of PVC. The dechlorination effect of the NaOH additive was the worst among the additives. The Cl content of hydrochar obtained from 4% NaOH was higher than 0.1% (1043 mg/kg). This might be related to the resorption effect of free chlorine in the liquid phase. As displayed in Figure 3, the porous structure of hydrochar derived from 4% NaOH was more obvious than other hydrochar which was beneficial to improving the adsorption capacity of inorganic ions (Lin 2022; Ma 2019). Thus, this resorption effect weakened the hydrothermal dechlorination effect of NaOH.

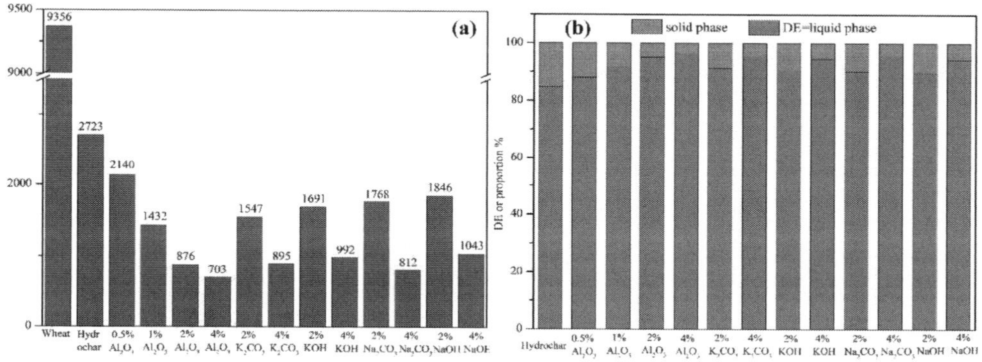

Figure 1. The Cl content (mg/kg) of wheat and hydrochars (a) and DE or Cl redistribution of hydrochars (b) during HTT with different additives assistance.

Interestingly, for the four alkaline additives, the DEs values were obviously improved when the addition ratio was increased from 2% to 4%. However, the improvement effect for γ-Al_2O_3 was relatively weak. The above analysis results suggested that the dechlorination effect of five additives was γ-Al_2O_3> Na_2CO_3> K_2CO_3>KOH>NaOH. In general, after comprehensive consideration of the dechlorination effect and use cost, deep dechlorination of high-chlorine waste biomass by hydrothermal treatment with 1%~2% γ-Al_2O_3 was a more suitable approach.

3.2 *Hydrochar yield and Fuel properties of hydrochars*

The above findings confirmed the facilitation effect of additives on hydrothermal dechlorination. However, the effects of additives assistance on the hydrothermal process and the fuel properties of hydrochar are rarely reported. Therefore, it was necessary to analyze these effects in order to evaluate the chlorination process of high-chlorine waste biomass by HTT with additives assistance more comprehensively.

The physicochemical properties (%) of all samples as dry basic and hydrochar yields (%) with different additives assistance were presented in Table 1. As listed in Table 1, all hydrochar yields of HTT with additives assistance conditions were below that of HTT without additives. The published paper reported that partial hemicelluloses and some amounts of amorphous cellulose contained in biomass can be removed by alkaline treatments (Sun 2022). With the additive's assistance, the chain break of the hydrogen and covalent bonds linking or cross-linking with hemicellulose, cellulose and lignin accelerated. This process would lead to further hydrolysis and degradation of cellulose and lignin.

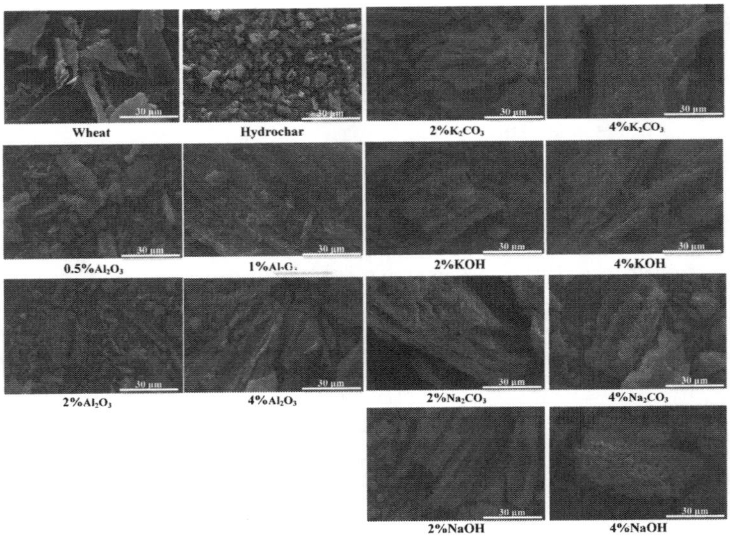

Figure 2. SEM images of wheat and hydrochars during HTT with different additives assistance.

It was worth noting that the C contents of some hydrochars were higher than that of hydrochar without additives (55.83%), such as 0.5%γ-Al_2O_3(56.75%), 2%Na_2CO_3 (57.26%), 2%KOH (56.9%) and 2%NaOH (56.73%), etc. However, the C contents of some hydrochars showed an unexpectedly decreased, such as 1%-4% Al_2O_3 and 2%-4%K_2CO_3, etc. The main reason for this phenomenon was that the massive decomposition of organic matter made more carbon transfer to the liquid phase under these conditions. The C content of the hydrochar obtained from 4%γ-Al_2O_3 was only 50.73%. The O element and ash content present similar tendencies. The O element and ash content of the hydrochar obtained from 4% γ-Al_2O_3 were 31.84% and 10.32%, respectively, which were higher among all samples. From this point of view, the adding ratio should not exceed 2% when using γ-Al_2O_3 as a hydrothermal dechlorination agent. This result was consistent with the above analysis.

Table 1. Physicochemical properties (%) of all samples as dry basic and hydrochar yield (%).

Dechlorinate additives	C	H	N	S	O	Cl	Ash	Hydrochar yield
Wheat	41.64	5.86	0.50	0.07	44.93	0.9356	6.07	–
Hydrochar	55.83	5.51	0.64	0.02	28.94	0.2723	8.79	52.73
0.5%γ-Al_2O_3	56.75	6.26	1.02	0	26.90	0.2140	8.86	52.71
1% γ-Al_2O_3	54.85	6.44	0.9	0	28.62	0.1432	9.05	52.62
2% γ-Al_2O_3	54.82	6.12	1.03	0	28.51	0.0876	9.43	52.46
4% γ-Al_2O_3	50.73	6.12	0.92	0	31.84	0.0703	10.32	52.15
2%K_2CO_3	54.36	6.45	1.09	0	28.99	0.1547	8.96	52.54
4%K_2CO_3	54.22	6.63	1.08	0	28.86	0.0895	9.12	52.23
2%Na_2CO_3	57.26	6.74	0.92	0	25.91	0.1768	8.99	51.78
4%Na_2CO_3	56.84	6.83	0.86	0	26.31	0.0812	9.08	51.70
2%KOH	56.9	6.66	0.98	0	26.27	0.1691	9.02	52.06
4%KOH	55.17	6.56	1.13	0	27.89	0.0992	9.15	51.89
2%NaOH	56.73	6.77	0.91	0	26.35	0.1846	9.06	52.05
4%NaOH	55.52	6.85	0.99	0	27.37	0.1043	9.17	51.57

A Van Krevelen diagram of raw wheat and hydrochars was described in Figure 4. Four typical coals, i.e., anthracite, bituminous, sub-bituminous, and lignite, were listed for comparison (Kim 2014). As it showed, the raw wheat was in the upper left corner which was away from the coal area.

This implied that wheat is a low-grade solid fuel. Due to the decarboxylation and dehydration reactions, the wheat was coalified by lowering the H/C and O/C atomic ratios. The process improved the coal ranks of hydrochar without dechlorination additives to a value comparable to lignite. It could also see from Figure 4 that although the O/C atomic ratio showed a downward trend, the H/C atomic ratio increased instead of decreasing due to the increase of H element content and the decrease of C element content (Table 1). This finding also suggested that dechlorination additives have a certain inhibitory effect on the decarboxylation and dehydration reactions. Compared to the hydrochar without dechlorination additives, the energy grade of hydrochars derived from dechlorination additives has been lowered. In addition, the hydrochar of 4% γ-Al_2O_3had the lowest energy grade of all hydrochars. Therefore, it could be said that the realization of deep dechlorination needs to sacrifice energy grade to a certain extent.

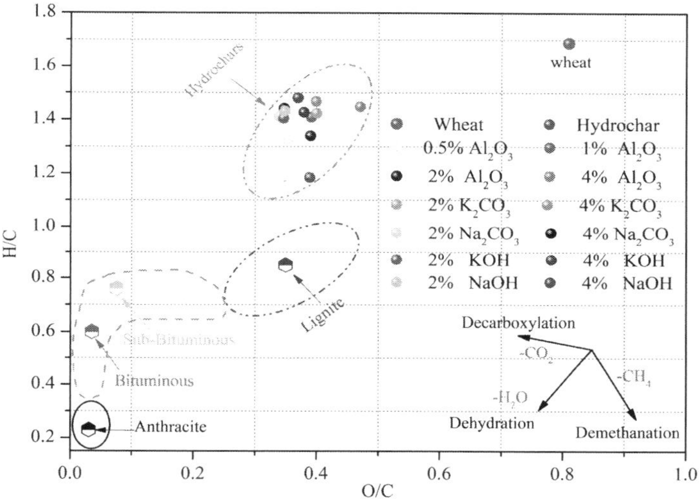

Figure 3. Van Krevelen diagram of wheat and hydrochars with different additives assistance.

4 CONCLUSION

In this work, alkali-free additives γ-Al_2O_3 and four alkaline additives (K_2CO_3, KOH, Na_2CO_3, and NaOH) were employed as dechlorinating agents for deep dechlorination of wheat by hydrothermal treatment. Results showed that the hydrothermal dechlorination effect was improved with additives assistance. At the same adding ratio, the effect of γ-Al_2O_3 was the best, followed by Na_2CO_3, K_2CO_3, KOH, and NaOH. However, a decrease in the C content was observed in some hydrochars with additives. And the energy grades of all hydrochars with additives were lower than that of hydrochar without additives. This finding suggested that the realization of deep dechlorination needs to sacrifice energy grade to a certain extent. From this point of view, the high adding ratio of dechlorinating agents is not recommended. For γ-Al_2O_3, the 1%~2% adding ratio was a more suitable approach in our experimental conditions.

ACKNOWLEDGMENTS

This work was financially supported by the Guangdong Basic and Applied Basic Research Foundation (2019A1515110271), the National Natural Science Foundation of China (11902075), and Guangdong Provincial Key Laboratory of Distributed Energy Systems (2020B1212060075).

REFERENCES

Gandon-Ros G., Soler A., Aracil I., Gómez-Rico M. F. Dechlorination of polyvinyl chloride electric wires by hydrothermal treatment using K_2CO_3 in subcritical water. *Waste Management*, 2020, **102**: 204–211.

Huang N., Zhao P., Ghosh S., Fedyukhin A. Co-hydrothermal carbonization of polyvinyl chloride and moist biomass to remove chlorine and inorganics for clean fuel production. *Applied Energy*, 2019, **240**: 882–892.

Kim D., Lee K., Park K. Y. Hydrothermal carbonization of anaerobically digested sludge for solid fuel production and energy recovery. *Fuel*, 2014, **130**: 120–125.

Lin Y., Ge Y., He Q., Chen P., Xiao H. The redistribution and migration mechanism of chlorine during hydrothermal carbonization of waste biomass and fuel properties of hydrochars. *Energy*, 2022, **244**: 122578.

Ma D., Feng Q., Chen B., Cheng X., Chen K., Li J. Insight into chlorine evolution during hydrothermal carbonization of the medical waste model. *Journal of Hazardous Materials*, 2019, **380**: 120847.

Qin F., Zhang C., Zeng G., Huang D., Tan X., Duan A. Lignocellulosic biomass carbonization for biochar production and characterization of biochar reactivity. *Renewable and Sustainable Energy Reviews*, 2022, **157**: 112056.

Ren X., Sun R., Chi H.-H., Meng X., Li Y., Levendis Y. A. Hydrogen chloride emissions from combustion of raw and torrefied biomass. *Fuel*, 2017, **200**: 37–46.

Shen Y. A review on hydrothermal carbonization of biomass and plastic wastes to energy products. *Biomass and Bioenergy*, 2020, **134**: 105479.

Sun S.-F., Yang H.-Y., Yang J., Shi Z.-J. The effect of alkaline extraction of hemicellulose on cocksfoot grass enzymatic hydrolysis recalcitrance. *Industrial Crops and Products*, 2022, **178**: 114654.

Welsby D., Price J., Pye S., Ekins P. Unextractable fossil fuels in a 1.5°C world. *Nature*, 2021, **597**(7875): 230–234.

Zhao P. T., Li T., Yan W. J., Yuan L. J. Dechlorination of PVC wastes by hydrothermal treatment using alkaline additives. *Environmental Technology*, 2018, **39**(8): 977–985.

Energy Revolution and Chemical Research – Kok-Keong Chong and Zhongliang Liu (Eds)
© 2023 The Authors, ISBN: 978-1-032-36554-1

Study on the process of preparing C4 olefin by catalytic coupling of ethanol

Xinkai Yuan, Hanqi Jiang* & Lan Zeng
School of Computer and Information Technology, Beijing Jiaotong University, Beijing, China

ABSTRACT: C4 olefins are one of the most important raw materials for chemical products and ethanol can be used as a feedstock for the production and preparation of C4 olefins. In the preparation process, catalyst combinations and reaction temperatures can affect both feedstock conversion and C4 olefin yield, and finding the optimal configuration required for the preparation process can effectively improve yield and efficiency and save costs. Based on the analysis of experimental data, this paper discusses the reasons affecting the magnitude of ethanol conversion, C4 olefin selectivity, and C4 olefin yield, and proposes a regression model and grid search-based approach to solving the optimal configuration of catalyst combination and reaction temperature.

1 INTRODUCTION

For the investigation of the effect of different catalyst combinations and temperatures on the preparation of C4 olefins, we proposed both univariate analyses based on control and multi-factor analysis based on the LightGBM regression model to jointly discuss the effect of catalyst combination and temperature on the conversion of ethanol and the magnitude of C4 olefin selectivity. We split the catalyst combination into six variables such as feed type and Co loading. The results showed that temperature was the most important factor affecting the above ethanol conversion and C4 olefin selectivity, and that increasing the temperature significantly enhanced ethanol conversion and C4 olefin selectivity. Subsequently, we developed a regression model for C4 olefin yield on catalyst combination and temperature based on the LightGBM regression model and solved the optimization problem by sampling the values of the variables using a grid search method.

2 INVESTIGATION OF THE EFFECT OF DIFFERENT CATALYST COMBINATIONS AND TEMPERATURES ON THE PREPARATION OF C4 OLEFINS

2.1 *Analysis of the problem*

Amid examining the effect of catalyst combinations on the conversion of ethanol and the magnitude of C4 olefin selectivity, we used single-factor and multi-factor analyses. We fixed the temperature to form a control and considered the effect of a single variable change on the ethylene preparation effect while ensuring that the other variables were the same. For the multi-factor analysis, we modeled the regression of conversion and selectivity on the catalyst combination and reaction temperature using the catalyst combination and reactant temperature as independent variables, while considering the effect of updates to the independent variables on the magnitude of the target values. Catalyst combinations can be split into six categories: packing method (Class I and II), Co loading, Co/SiO2 loading, Co/SiO2 and catalyst carrier loading ratio, ethanol concentration, and

*Corresponding Author: 20722010@bjtu.edu.cn

DOI 10.1201/9781003332657-83

whether or not catalyst carrier HAP is used. We will base our discussion on these six variables and build regression models from them.

2.2 *Model building*

2.2.1 *Control-based univariate analysis*

The key to conducting a one-way analysis (GJ Székely 2007) is to construct controls, and for the six variables mentioned in the problem analysis regarding catalyst combinations, a number of control experiments were designed as shown in Table 1. Each set of controls in Table 1 ensures that the variables remain the same except for the variables in the catalyst combination under consideration. Section 2.3.1 gives the results of the one-way analysis based on Section 2.3.1 gives the relationship between ethanol conversion and C4 olefin selectivity and catalyst combinations based on the single-factor analysis.

Table 1. Control experiments that can be used in the single-factor analysis.

Variables in the catalyst portfolio	No. of catalyst combinations that can be used for the control
Filling method	[A12, B1], [A9, B5]
Co load	[A1, A2, A4, A6], [A9, A10]
Co/SiO2 charge	[A3, A8], [B1, B2, B3, B4, B6]
Charge ratio	[A12, A13, A14]
Ethanol concentration	[A1, A3], [A2, A5], [A7, A8, A9], [B1, B5], [B2, B7]
Whether to use HAP	[A11, A12]

2.2.2 *Multi-factor analysis based on LightGBM regression*

The six variables of the catalyst combination (packing method, Co loading, total charge, Co/SiO2 and catalyst carrier charge ratio, ethanol concentration, and whether or not to use catalyst carrier HAP) and temperature were used as independent variables in the multifactorial model, with ethanol conversion and C4 olefin selectivity, where packing method and whether or not to use catalyst carrier HAP were categorical variables. As defined in Equations 1 and 2, we use the values 0 and 1 for packing method I and packing method II, respectively, with the values 1 and 0 for the use of catalyst carrier HAP or not, respectively.

$$X_1 = \begin{cases} 1, \textit{filling method } 1 \\ 0, \textit{filling method } 2 \end{cases} \tag{1}$$

$$X_6 = \begin{cases} 1, \textit{Use of quartz sand as a catalyst carrier} \\ 0, \textit{Use of HAP as a catalyst carrier} \end{cases} \tag{2}$$

The regression model uses the LightGBM (Ke 2017) model, which is a gradient-based tree model that uncovers more hidden non-linear relationships between variables than a simple linear regression model. LightGBM provides the calculation of feature importance and the plotting of the decision tree decision process (Zhanshan Li 2021), which allows us to use the decision process in conjunction with single-factor analysis to the decision process can be combined with single-factor analysis to obtain how catalyst combination and temperature affect ethanol conversion and C4 olefin selectivity.

2.3 *Model solving*

2.3.1 *Single-factor analysis*

Analyzing the control experimental data for each group in Table 1, summarising the data and graphs, we obtained the following relationship between ethanol conversion and C4 olefin selectivity with respect to the catalyst combination (Shaowei Li 2021).

(1) Regarding the packing method: at the same temperature, the ethanol conversion is slightly higher with packing method I, while the C4 olefin selectivity is higher with packing method II.
(2) For Co loading: at lower fixed temperatures (250°C), ethanol conversion decreases first with increasing Co loading and then increases with increasing Co loading; however, at higher fixed temperatures, a Co loading of However, the most significant increase in ethanol conversion was achieved at 2 wt% Co loading when the fixed temperature was increased. In the range of experimental data provided, the selectivity of C4 olefins was best achieved with a Co loading of 1 wt% over the range of data provided.
(3) For the Co/SiO2 loading: When the other variables controlling the catalyst combination were kept constant, the overall ethanol conversion and C4 olefin selectivity increased with and C4 olefin selectivity increased with increasing Co/SiO2 loading. For a particular catalyst combination, the Co/SiO2 charge is for a particular catalyst combination, there is a threshold of Co/SiO2 charge, above which the ethanol conversion and C4 olefin selectivity increase with the charge.
(4) For Co/SiO2 and HAP charge ratios: ethanol conversion decreases with increasing mass ratio of Co/SiO2 to HAP and C4 olefin selectivity decreases with increasing charge volume. The C4 olefin selectivity first increases with the increasing mass ratio of Co/SiO2 to HAP and then decreases with increasing mass ratio after reaching a critical value.
(5) For ethanol concentration: Ethanol conversion decreases with increasing ethanol concentration in the catalyst, while C4 ethylene selectivity increases with increasing ethanol concentration. The ethanol conversion also does not decrease with increasing ethanol concentration and it can be observed that when the ethanol concentration in the catalyst increases to a certain threshold value (It is observed that when the concentration of ethanol in the catalyst increases a certain threshold (around 1.68 ml/min in the experimental data), the ethanol conversion starts to increase.
(6) With or without HAP: HAP is more effective in improving ethanol conversion and C4 olefin selectivity.

2.3.2 *Multi-factor analysis*

The LightGBM model was trained using the mean squared error MSE as the objective function, and we used the root mean squared error RMSE and the maximum absolute error MAE to measure the quality of the regression model. Table 2 gives the errors of the ethanol conversion regression model and the C4 olefin selectivity regression model. Figure 1 shows the significance of the characteristics of the two regression models.

Table 2. Errors of the LightGBM regression model.

Model	RMSE	MAE
Ethanol conversion regression model	0.5244	1.7021
C4 Olefin selectivity regression model	0.7734	2.6900

As seen in Table 2, both regression models have high precision. As seen from the characteristic importance graph 1, among the seven independent variables, the reaction temperature was the most important factor affecting the magnitude of both the ethanol conversion and C4 olefin selectivity. Secondly, the total feed of Co/SiO2 and HAP, the ethanol concentration, and the Co loading also determined to some extent the ethanol conversion and C4 olefin selectivity. The total feed of Co/SiO2 and HAP, ethanol concentration, and Co loading also determined the ethanol conversion and C4 olefin selectivity to some extent. The variables Co/SiO2 and HAP loading ratio and feeding method had less influence on both. The effect of the variables Co/SiO2 and HAP loading ratio and

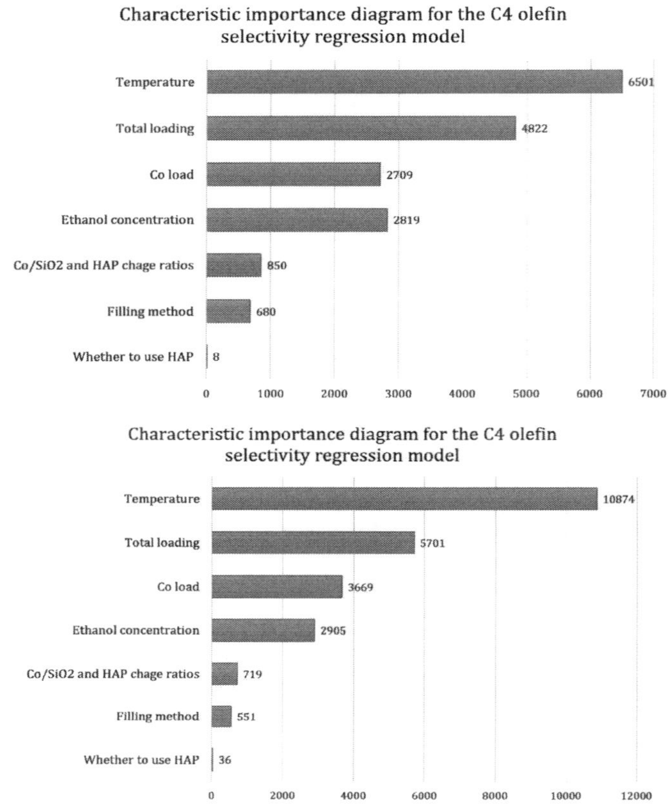

Figure 1. The importance of the LightGBM regression model.

feeding method on both was small, while the use of the catalyst carrier HAP had almost no effect on the target values.

Figures 2 and 3 show the decision process for the first decision tree (Ke 2017) of the ethanol conversion regression and the C4 olefin selectivity regression, respectively. As can be seen from the decision branches of the tree, the majority of the decision process is splitting the leaf nodes based on temperature. The higher the temperature, the higher the ethanol conversion and C4 olefin selectivity. Some of the branch nodes are considering total charge, ethanol concentration, and feeding method. From the branch conditions on the first decision tree and the end of the tree, it can be seen that the higher the charge and the lower the ethanol concentration, the higher the values of the two dependent variables, while ethanol conversion is higher using feeding method I and C4 olefin selectivity is higher using feeding method II. These findings are consistent with the conclusions obtained from the single factor analysis in section 2.3.1.

3 INVESTIGATION OF THE OPTIMUM CONDITIONS FOR THE PREPARATION OF C4 OLEFINS

3.1 *Analysis of the problem*

In Section 2.2, we established a regression model of the ethanol conversion rate and C4 olefin selectivity with respect to the experimental independent variables (catalyst combination and temperature). Aiming at the problem of optimizing the catalyst combination and reaction temperature

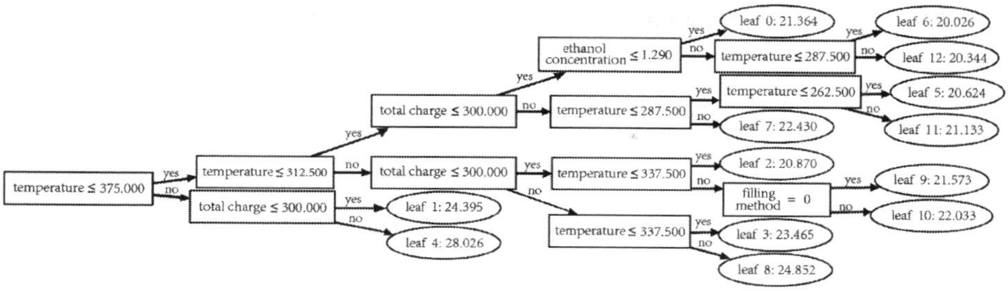

Figure 2. Ethanol conversion rate Decision process for the first decision tree of the LightGBM regression model.

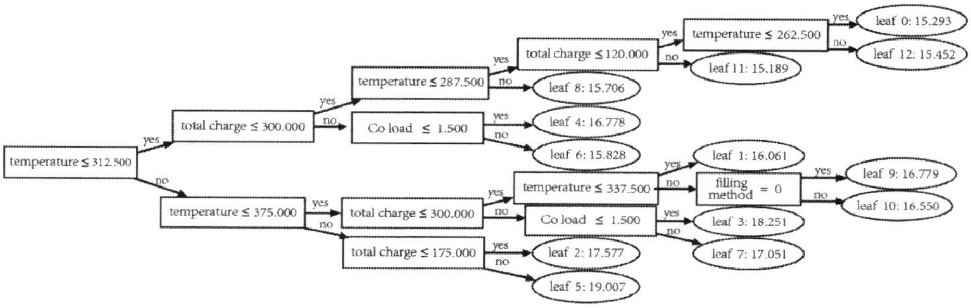

Figure 3. Decision process for the first decision tree of the LightGBM regression model for C4 olefin selection.

for preparing C4 olefins to maximize the yield of C4 olefins, we can use the same idea to establish a regression model of the yield of C4 olefins with respect to the catalyst combination and temperature. Assuming that when considering the catalyst combination and temperature configuration scheme that maximizes the yield of C4 olefins, only the values of the variables in the existing catalyst combination in Annex 1 need to be considered, and only the temperature range given in the data in Annex 1 (250 Degrees to 450 degrees).

The optimization goal for finding the optimal catalyst combination and temperature is to maximize the yield of C4 olefins. After obtaining the regression model of the C4 olefin yield on the independent variables, we can use the grid search method to find the maximum point to confirm the optimal configuration plan. For constrained optimization problems with temperature limitations (the highest temperature is less than 350 degrees), we only need to modify the range of the grid search.

3.2 *Model building*

3.2.1 *Regression model for C4 olefin yields*

As with the method used in Section 2.2, we use the LightGBM method to establish a regression model of the C4 olefin yield with respect to catalyst combination and temperature. Different from the regression model of ethanol conversion rate and C4 olefin selectivity, C4 olefin yield has a more non-linear characteristic relationship with independent variables, so only using variables $X_1, \ldots, X_7$, the regression effect is poor. To this end, we introduced 6 new multiplicative variables $X_8, \ldots, X_{13}$ to enhance the model's ability to fit nonlinear features. The 6 new features are the product of X7 and $X_1, \ldots, X_6$ (ie temperature And the product of 6 catalyst combination variables) (Xianping Li 1997).

$$X_i = X_7 \cdot X_{i-7}, i = 8, 9, \ldots, 13 \tag{3}$$

After adding new features, the LightGBM regression model of C4 olefin yield can achieve higher accuracy, the root means square error RMSE = 7.6291, and the maximum absolute error MAE = 30.1059. Figure 4 shows the importance of features of the C4 olefin yield regression model. It can be seen that the six new multiplicative features play an important role in fitting the C4 olefin yield.

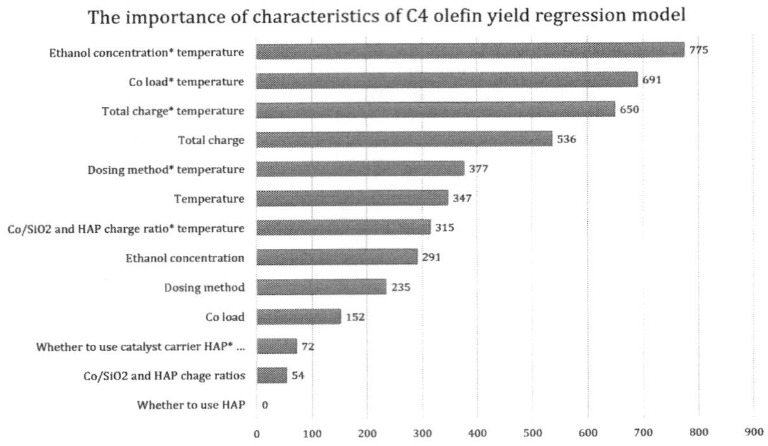

Figure 4. Feature importance of C4 olefin yield regression model.

3.2.2 *Optimization target and grid search method*

Let O_{C4} denote the yield of C4 olefins. It is a regression model of independent variables $X_1, \ldots, X_{13}$ (actually about variables $X_1, \ldots, X_7$). Equation (4) is the optimization problem we need to solve

$$\underset{(X_1,\ldots,X_7)}{\arg\, max} O_{C4} \tag{4}$$

Based on assumptions, we only consider the existing catalyst variable values and temperature values in Annex 1. $X_1, \ldots, X_6$ represent catalyst combination, and X_7 represents temperature. According to Annex 1, X_1 represents the feeding method, only the value $X_1 = \{0,1\}$; X_2 represents the Co load, only the value $X_2 = \{0.5,1,2,5\}$; X_3 represents the total loading of Co/SiO2 and HAP, and only takes the value $X_3 = \{20,50,100,140,150,200,400\}$; X_4 indicates whether to use catalyst carrier HAP, only the value $X_4 = \{0.1\}$; X_5 indicates the ethanol concentration, only the value $X_5 = \{0.3,0.9,1.68,2.1\}$; X_6 indicates the charging ratio of Co/SiO2 and HAP, only The value $X_6 = \{0.5,0.55,1,2\}$.

X_7 represents temperature, and the value range is between 250 and 450. We use sampling every 10 degrees to construct the grid, then X_7 can take the value $X_7 = \{250,260, 450, 440\}$. All possible value combinations of the 7 independent variables are matched to get the search grid Ω, which is easy to get:

$$|\Omega| = 2 \times 4 \times 7 \times 2 \times 4 \times 4 \times 21 = 37632 \tag{5}$$

That is, there are 37632 grid points to be searched, which is allowed in both time complexity and space complexity. Consideration is given to transforming the optimization problem of grid search into (6):

$$\underset{(X_1,\ldots,X_7)\in\Omega}{\arg\, max} O_{C4} \tag{6}$$

It is needed to consider the optimization problem with constraints (the temperature is less than 350 degrees), then X7 = {250, 260, …, 330, 340}, construct a grid with constraintsΩ_C , subsequently, it is easy to get $\Omega_C = 17920$, the optimization problem is deformed as:

$$\underset{(X_1,\ldots,X_7)\in\Omega_C}{\arg\max O_{C4}} \tag{7}$$

Solving (6) and (7) only needs to traverse the grid to find the maximum point.

3.2.3 *Model solving*

Table 3 shows the maximum point of the target O_{C4} obtained by grid search. The results in the analysis table can be obtained:

(1) For the unconstrained optimization problem, the optimal configuration scheme of catalyst combination and temperature is: 200mg 1wt%Co/SiO2- 200mg HAP-ethanol concentration 0.9ml/min, feeding method I, temperature 420 degrees. At this time, the maximum value of C4 olefin yield is 4474.7244.
(2) For constrained optimization problems (temperature less than 350 degrees), the optimal configuration of catalyst combination and temperature is: 200mg 2wt%Co/SiO2- 200mg HAP-ethanol concentration 1.68ml/min, feeding method I, temperature 340 degrees. At this time, the maximum value of C4 olefin yield can be 2541.7747.

Table 3. Solution results of C4 olefin yield optimization problem.

Condition	Grid point set	X_1	X_2	X_3	X_4	X_5	X_6	X_7	Maximum C4 olefin yield
Unconstrained	Ω	0	1.0	400	1	0.90	1.0	420	4474.7244
Constrained	Ω_c	0	2.0	400	1	1.68	1.0	340	2541.7747

In summary, whether it is for unconstrained or constrained C4 olefin yield optimization, the configuration scheme of the catalyst combination is biased toward the use of feeding method I, the loading amount and loading ratio of Co/SiO2 and HAP are both selected 200mg: 200mg, reaction The higher the temperature, the better.

4 TEST RESULTS AND DISCUSSIONS

4.1 *Investigation of the effect of different catalyst combinations and temperatures on the preparation of C4 olefins*

The results showed that temperature was the most important factor affecting the above ethanol conversion and C4 olefin selectivity and that increasing the temperature significantly increased the ethanol conversion and C4 olefin selectivity. The results showed that increasing the temperature significantly improved the ethanol conversion and C4 olefin selectivity. For the catalyst combinations, a high loading (200mg, 100mg) and a low Co loading (0.5wt%, 1wt%) should be chosen as far as possible. (0.5wt%, 1wt%), a balanced Co/SiO2 to HAP charge ratio (1:1) and a moderate ethanol concentration (0.9ml/min, 1.68 ml/min), with little improvement in conversion and selectivity due to the type of feed. Catalyst carrier HAP showed a small increase in conversion and selectivity compared to quartz sand.

4.2 *Investigation of the optimum conditions for the preparation of C4 olefins*

The results show that for unconstrained optimization, the optimum configuration is 200 mg 1 wt% Co/SiO2 - 200 mg HAP - ethanol concentration 0.9 ml/min, feeding method I, temperature 420 degrees. The maximum C4 olefin yield at this point is 4474.7244. If the required temperature is lower than 350 degrees, the optimum configuration is: 200mg2wt% Co/SiO2 - 200mg HAP - ethanol concentration 1.68ml/min. The maximum yield of C4 olefin is 2541.7747.

4.3 *Advantages of the model*

(1) We used a control method for the one-way analysis, and the results of each set of controls are discussed in detail. When considering the effect of catalyst combinations on reactants and products, we used a combination of single-factor analysis based on controlled experiments and multi-factor analysis based on regression models, which complemented each other.
(2) We used LightGBM to build regression models for ethanol conversion, C4 olefin selectivity, and C4 olefin yield. LightGBM learns more non-linear features than linear regression, and we also artificially introduced new variables to increase the ability of the model to fit non-linear features.

4.4 *Disadvantages of the model*

(1) The LightGBM regression model is less interpretative and analytical than the linear regression model, and the optimization problem with the LightGBM regression model as the objective cannot be solved analytically for an optimal solution.
(2) We only consider the catalyst combination variables available in Annex 1 and the combination of equidistant sampling of temperature as the parameter space for the grid search for the optimal configuration solution, which does not allow us to obtain a globally optimal solution based on the LightGBM regression model.

5 CONCLUSION

Based on the results and discussions presented above, the conclusions are obtained as below:

(1) In order to improve the conversion of C4 olefins, we should increase the temperature as much as possible and choose a catalyst combination with a high charge, low Co loading, a balanced Co/SiO2, and HAP charge ratio, and a moderate ethanol concentration.
(2) The maximum value of the C4 olefin yield that could be obtained by grid search was 2541.7747.

REFERENCES

Ke G, Meng Q, Finley T. et al., LightGBM: A Highly Efficient Gradient Boosting Decision Tree. *Advances in Neural Information Processing Systems*, 30 (2017) 3146–3154.

Shaowei Li, an Optimization model for the preparation of C4 olefins by ethanol coupling. *Journal of Taizhou College*, 2021.

Székely GJ, Rizzo M L, Bakirov N K, Measuring and testing dependence by correlation distances. *Annals Statistics*, 2007, 35(6) 2769–2794.

Xianping Li, *Foundations of Probability Theory*. Higher Education Press, 1997.

Zhanshan Li, Feature selection algorithm based on LightGBM. *Journal of Northeastern University*, 2021.

Energy Revolution and Chemical Research – Kok-Keong Chong and Zhongliang Liu (Eds)
© 2023 The Authors, ISBN: 978-1-032-36554-1

Enhancing the production of typical aromatic alcohols with three strategies

Beisong Xu*
Beijing Advanced Innovation Center for Soft Matter Science and Engineering, Beijing University of Chemical Technology, Beijing, China

ABSTRACT: Aromatic compounds of various classes are largely used in industries. Out of all the classes of alcohols, aromatic alcohols serve as an important class of compounds and are used widely in cosmetics, pharmaceuticals, and many more applications. However, the industrial synthesis of aromatic alcohols still relies on natural extracts obtained from plants, which usually have very low production. In the past 5 years, the construction of microbe factories has made it possible to synthesize aromatic alcohols. Renewable carbon sources are converted into targeted products. In this review, we summarize the recent metabolic strategies for synthesizing aromatic alcohols, which make use of decarboxylase, carboxylic acid reductase, and CoA-dependent reductase for synthesizing aromatic alcohol, respectively. Our goal is to enhance the production of aromatic alcohols and provide rational suggestions and strategies to improve the production in the future.

1 INTRODUCTION

Aromatic compounds are a group of chemicals that find different applications in various fields, including pharmaceuticals, food, and cosmetics industries. Among these, aromatic alcohols are one of the most widely used compounds in the industries (Etschmann M 2002). For example, 2-Phenylethanol is used to impart rose-like flavor in wine, cheese, beer, and soy sauce (Jang 2010; Shyamala, B 2007; Yin 2015). Vanillyl alcohol is a natural phenolic compound used for its anti-inflammatory and anti-nociceptive action in mice and other primates. It can also be used as an anti-asthmatic drug in pigs and humans (Hagiwara K 2011; Vázquez-Velasco M 2011). Hydroxy-tyrosol can prevent the formation of low-density lipoproteins, thus slowing the process of aging. The compound is also used as a natural antioxidant to keep foods fresh (Choo 2018).

Most of the aromatic alcohols mentioned above are extracted from natural plants (Etschmann M 2002; Hagiwara K 2011; Jang 2010), 2-phenyethanol from rose, vanillyl alcohol from *Gastrodia elata* Blame, hydroxytyrosol from olive, respectively. However, the production is low and considerable negative effects on climate and environmental destruction are seen. It usually takes several years for plants to flourish and be available for extraction. The process of extracting it from natural plants cannot meet the increasing demand nowadays. For instance, the demand for 2-phenylethonal is around 10,000 tons per year (Etschmann M 2002), and it is increasing at a rate of 15% each year (Etschmann M 2002). As for vanillin and vanillyl alcohol, the demand is 18,653 tons per year with an increasing rate of 6.3% (Shyamala 2007). The demand for hydroxytyrosol is 30,000 tons per year globally (Jang 2010). Alternatively, using microbes to synthesize aromatic alcohols is more promising as it only takes one to four days to achieve the highest production. Recently, there have been some developments in synthesizing aromatic alcohols (*E. coli* & Food Safety 2020; Jin 2020; Lee 2019; Wright 1979). In this review, we summarize three methods of synthesizing

*Corresponding Author: 2021201149@buct.edu.cn

DOI 10.1201/9781003332657-84

aromatic alcohols, namely decarboxylases, carboxylic acid reductase (CARs), and CoA-dependent reductases to synthesize aromatic alcohols, respectively. The emphasis of this review will be put on the enhancement of productions. Suggestions on how to improve the production in the future are given at the end of the paper.

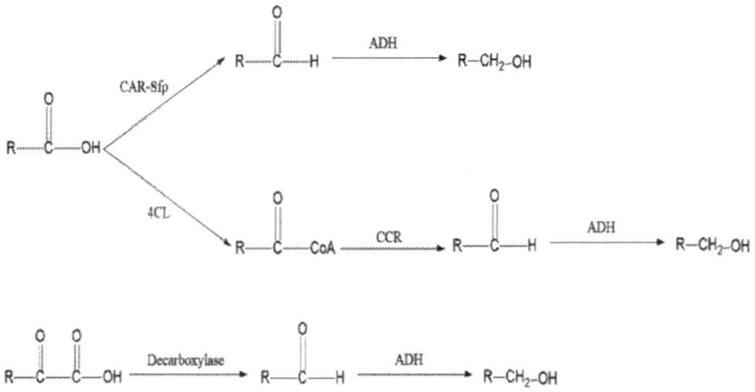

Figure 1. The reaction mechanism of the three strategies for synthesizing aromatic alcohols, namely decarboxylase, carboxylic acid reductase, and CoA-dependent reductase. Adh: aldehyde reductase CCR: cinnamyl CoA reductase, 4CL: 4-cinnamyl ligase. CAR-Sfp: carboxylic acid reductase.

2 SYNTHESIS OF AROMATIC ALCOHOL USING DECARBOXYLASE

2.1 *Benzyl alcohol*

Benzyl alcohol is mostly used in cosmetics in the pharmaceutical and flagrant industries (Scognamiglio 2012; Widhalm 2015). In recent years, benzyl alcohol has gained immense popularity (Mcleish 2003). Mcleish et al. have found a pathway in *Pseudomonas putida*. It can be achieved by using phenylpyruvate as a precursor, which is then converted to S-mandelate acid by hydroxymandelate synthase (hmaS). Sequentially, the two enzymes S-mandelate dehydrogenase (mdlB) and benzoyl formate decarboxylase (mdlC) convert S-mandelate to benzaldehyde. mdlB is responsible for the conversion of S-mandelate to benzoyl formate and mdlC is responsible for the conversion of benzoyl formate to benzaldehyde, respectively. Finally, alcohol dehydrogenase (ADH) is responsible for reducing benzaldehyde to benzyl alcohol. Shawn et al. (Shawn 2015) over-expressed aro10, hmaS, mdlB, and mdlC in *Escherichia coli* and enabled the production of benzyl alcohol of 114 mg/L. Maria et al. (Valera 2020) found that by deleting wild type aro10 and over-expressed hmaS, mdlB, mdlC led to formation of only 5 ug/L benzyl alcohol, suggesting that aro10 is a very important gene within the mandelate pathway. It turns out that aro10 is responsible for controlling the expression of ADH and pyruvate decarboxylase, the two genes that increase the carbon flow in this pathway.

They also found that by deleting extra ald2, sac7, and dld1, the formation of benzyl alcohol is eliminated. ald2, sac7, and dld1 are suggested to be responsible for the expression of ADH and PD, which is the rate-limiting step reaction. Zi et al. (Luo 2020) used *E. coli* as the host to synthesize benzyl alcohol with the same method mentioned above, and then extra NADPH is added from *Nocardia* species. Therefore, the resulting strain produced 336 mg/L benzyl alcohol within 48 h in shake flask culture. It could be speculated that in the future, not only hmaS, mdlB, and mdlC should be over-expressed but also aro10, ald2, sac7, and dld1 must be over-expressed, which could increase the production of benzyl alcohol.

2.2 *2-Phenylethanol*

2-phenyl ethanol imparts rose flavor in food products and is also used in the fragrance industry, especially in perfumes and cosmetics. Heterologous production of 2-phenyl ethanol can be achieved by recruiting L-phenylalanine as a precursor, which is converted to phenylpyruvate by phenylpyruvate synthase. Sequentially, phenylpyruvate decarboxylase is responsible for the conversion of phenylpyruvate to phenylacetaldehyde. Finally, reductases are responsible for converting phenylacetaldehyde into 2-phenyl ethanol (Hua 2011). Kim et al. (2014) reported the production of 2-PE in *Saccharomyces cerevisiae* by over-expressing aro9, aro10, adh1-5, and sfa1. The four genes are responsible for converting phenylpyruvate to 2-PE. The best-engineered strain produced 441.5 mg/L, which is significantly higher than the wild-type strain. In this pathway, ald2 and ald3 are responsible for converting 2-PAA to 2-PAC, which is a competitive pathway for producing 2-PE. Although there are fewer 2-PAC formed, the deletion of ald2 and ald3 did not increase the production as expected. Probably due to the unbalanced co-factor, which is caused by knocking out of the two genes. Yin et al (2015) discovered that the production of 2-PE can be enhanced considerably by additional deletion of tyrB. Gene tyrB is responsible for converting 4-hydroxyphenylpyruvate into tyrosine, which is a competitive pathway upstream. As a result, the production of the best strain reached 2.61 g/L within 48 h in a 2-L fermenter. Kang et al. (2013) found the production of 2-PE in *E. coli* by over-expressing aro9, aro10, and Adh. The three genes are responsible for converting phenylpyruvate to 2-PE, expressing the enzymes to enhance the carbon flow. Moreover, they also knocked out AroF and pheA, both genes are responsible for the conversion of phenylpyruvate to L-phenylalanine, which is a competitive pathway upstream, deletion of the two genes inhibits the competitive pathway. The best strain produced 285 mg/L, which is a bit lower than the production using *S. cerevisiae*. pheA is also responsible for converting prephenate to phenylpyruvate, which reduced the carbon flow upstream. Guo et al. (2018) reported the production of 2-PE in *E. coli* by over-expressing the yjgB gene instead of endogenous Adh, the function of yijB is to convert 2-PAA to 2-PE. The best strain produced 1016 mg/L 2-PE in shake flask conditions, which is the highest production in *E. coli*. To conclude, with the change of enzyme the production enhanced nearly threefold. Adh might be suggested to be one of the bottlenecks in the process. In the future, more studies should be focused on enhancing the activity of yijB.

2.3 *Hydroxytyrosol*

Hydroxytyrosol has a wide range of pharmaceutical applications in treatment of Parkinson's disease, Alzheimer's disease, other forms of dementia, cancer, heart disease, and more (Guo 2018). There are three ways of synthesizing hydroxytyrosol. First, it starts from tyrosine as a precursor, which is converted to tyramine by tyrosine decarboxylase (TDC). Sequentially, tyramine oxidase (TO) is responsible for the transformation of tyramine to 3,4-hydrophenylacetaldehyde (3,4-HPAA). Then endogenous ADH converts 3,4-HPAA into tyrosol. Finally, HpaBC (4-hydroxyphenylacetate 3-monooxygenase) is responsible for the conversion of tyrosol into hydroxytyrosol (Chung 2017). Chuang et al. reported the production of hydroxytyrosol by deleting tyrB and feaB. tyrB is responsible for converting 4-HPP to tyrosine, feaB is responsible for converting 4-HPAA to 4-hydroxyphenylacetic acid.

To enhance the carbon flow Choo et al. (2018) reported the production of hydroxytyrosol in *E. coli* by over-expressing aroGfbr and tyrAfbr (feedback inhibition resistance mutant) and added PCDF to enhance the driving force. aroG is responsible for converting PEP and E4P into 3-deoxy-D-arabino-heptulosonate 7-phosphate, tyrA is responsible for converting chorismate to prephenate, PCDF is a driving force for the two reactions. The two mutant genes are constructed to prevent feedback inhibition. The best strain produced 280 mg/L hydroxytyrosol. The second pathway can be achieved by recruiting tyrosine as a precursor, which is converted to L-DOPA by tyrosine hydroxylase (TH). Second, L-DOPA is converted to dopamine by L-DOPA decarboxylase (DDC). After this, TO is responsible for the conversion of dopamine to 3,4-HPAA. Finally, ADH is responsible for the conversion of 3,4-HPAA into hydroxytyrosol. Yasuharu et al. (Satoh 2012) reported the production of hydroxytyrosol in *E. coli* by over-expressing TH, BH4, DDC, and TYO. The best strain produced 0.08 mM hydroxytyrosol. The study also reported that the bottleneck is the low activity of TH and TYO, which is proved by feeding experiments on tyramine. To help address the issue of bottleneck in the pathway, Yao et al. (Chen 2019) replaced TH with HpaBC mutant 23F9-M4 from

E. coli through structure-guided modeling and directed evolution. They also replaced TYO with VanR from *Corynebacterium glutamicum*. The best strain produced 1.12 mM hydroxytyrosol. The third pathway was recently constructed by Li et al. (2017). It can be synthesized by recruiting chorismate as a precursor, which is converted to 4-hydroxyphenylpyruvate (4-PP) by endogenous enzymes. Sequentially, keto acid decarboxylase (KDC) is responsible for the conversion of 4-PP to 4-hydrophenylacetaldehyde (4-HPAA). Then ADH is responsible for the conversion of 4-HPAA to tyrosol. Finally, tyrosol is converted to hydroxytyrosol by HpaBC. Over-expressing aro10, adh6, TPTA, and deletion of feaB enabled the production of 1456 mg/L, the highest production ever reported. Over-expressing aro10 enhanced the carbon flow upstream, whereas over-expressing adh6 improved the conversion from 4-HPAA to tyrosol. The over-expression of TPTA is to boost the carbon flow through shikimate pathway. The knockout of feaB decreased the conversion from 4-HPAA to 4-hydroxyphenylacetic acid.

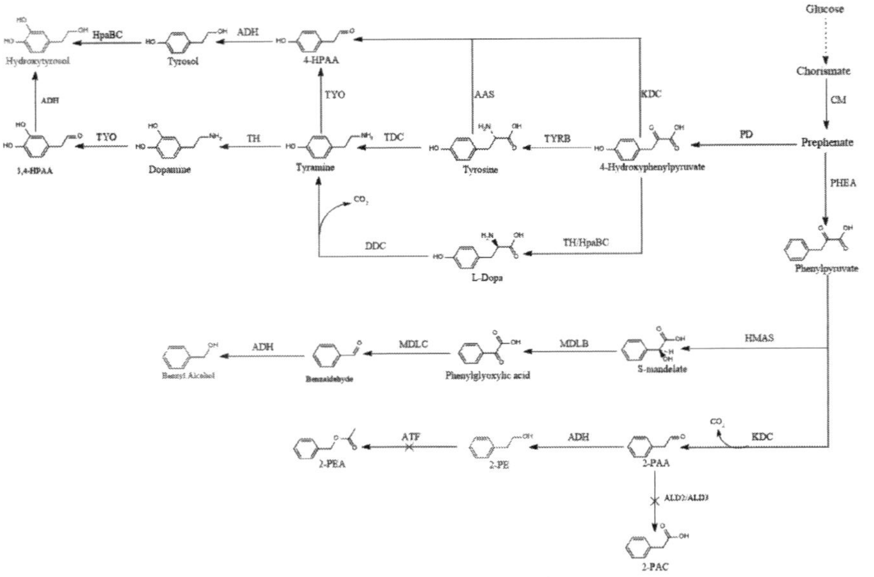

Figure 2. Biosynthesis of aromatic alcohols by decarboxylation of aromatic keto acids. CM, chorismate mutase; PD, prephenate dehydratase; PheA, prephenate dehydrogenase; HmaS, hydroxy-mandelate synthase; MdlB, S-mandelate dehydrogenase; MdlC, phenylglyoxylate decarboxylase; ADH, alcohol dehydrogenase; KDC, 2-keto acid decarboxylase; ATF, alcohol acetyltransferase; ALD2/ALD3, aldehyde dehydrogenases; TyrB, L-tyrosine aminotransferase; TDC, tyrosine decarboxylase; TYO, tyramine oxidase; HpaBC, flavin-dependent monooxygenase; TH, tyrosine hydroxylase; DDC, L-DOPA decarboxylase. 2-PE, 2-phenylethonal; 2-PAA, 2-phenylacetaldehyde; 2-PAC, 2-phenyl acetic acid; 2-PEA, 2-phenylethylacetate. The final products are displayed in red. Arrows with solid line consist of one-step reaction, and arrow with dashed line consists of multi-endogenous enzymes.

3 SYNTHESIS OF AROMATIC ALCOHOLS WITH CARBOXYLIC ACID REDUCTASE

3.1 *Vanillyl alcohol and vanillin*

Vanillyl alcohol and vanillin are mainly used as flavoring agents in food and beverages (B, N, Shyamala 2007; Ni 2015). They can be obtained by recruiting 3-dehydroshikimate acid as the precursor, which is converted to protocatechuic acid by 3-dehydroshikimate decarboxylase (3-DSD). Sequentially, ACAR (aromatic carboxylic acid reductase) is responsible for the conversion of protocatechuic acid to protocatechuic aldehyde. Then, O-methyl transferase (OMT) is responsible for the conversion of protocatechuic aldehyde to vanillin. Hansen et al. (2009) reported the production of vanillin in *E. coli* by over-expressing three enzymes, namely: 3-dehydroshikimate dehydratase (3-DSD) from *Podospora pauciseta*. Aromatic carboxylic acid reductase (CAR) from *E. norcardia* and OMT from Homo sapiens. The three enzymes mentioned above are responsible for converting 3-dehydroshikimate to vanillin. The best strain produced 65 mg/L vanillin. Brochado et al. (Brochado A R 2010) improved the production by using the same pathway, they used a silico design that optimized the cell factory, and two enzymes pyruvate decarboxylase (PDC) and glutamate dehydrogenase (GDH) were inhibited to prevent carbon flow into the TCA cycle. The deletion of PDC1 improved 1.5-fold of production, whereas the deletion of both PDC1 and GDH enhanced 5-fold of vanillin. Both studies mentioned above only used glucose and glycerol as carbon sources while no other sources are tried. After optimizing the conditions, the best strain was obtained. Ni et al. (2015) constructed a pathway that mimics the natural process of plants producing vanillin. The best strains converted from mixed carbon sources produced 90 mg/L vanillin with de novo synthesis. However, the mechanism is not clear so it is hard to improve the production (Chen 2017). Chen et al. reported the synthesis of vanillyl alcohol by trying a different natural pathway in plants and produced mainly vanillyl alcohol. By adding those three heterogeneous enzymes PobA, p-hydroxybenzoate hydroxylase; CAR, carboxylic acid reductase; Sfp, the CAR maturation factor phenylethanoids transferase; COMT, caffeine O-methyltransferase. The ADHs and alcohol dehydrogenases are used in vivo. The three enzymes mentioned above are responsible for converting 4-hydroxybenzylic acid to vanillyl alcohol. The best strain produced 240.69 mg/L vanillyl alcohol, which is the highest production ever reported in the previous studies.

3.2 *4-Hydroxybenzyl alcohol*

4-Hydroxy benzyl alcohol is a very important precursor in the production of gastrodin. Until now, it is the only compound that could be turned into gastrodin. Gastrodin is widely used in pharmaceutical industries, especially for its antioxidant, anticonvulsant, anti-inflammatory, analgesic, and sedative properties, and it can improve learning and memory (Zhang 2015). 4-Hydroxy benzyl alcohol can be achieved by recruiting chorismate as a precursor, which is converted to 4-hydroxybenzoate by UbiC. Sequentially, CAR is responsible for the conversion of 4-hydroxybenzoate into 4-hydroxybenzaldehyde. Finally, aldehyde reductases are responsible for synthesizing 4-hydroxy benzyl alcohol. Fan et al. (2013) first reported the production of 4-hydroxy benzyl alcohol in *Aspergillus foetidus* and *Penicillium cyclopium*. However, both the bacteria do not produce the product in a satisfactory yield. *A. foetidus* produced 13 mg/L and *P. cyclopium* only produced 7 mg/L 4-hydroxy benzyl alcohol, respectively. Moreover, the mechanism is not clear and the price is relatively expensive, so an affordable microbe is used to produce 4-hydroxy benzyl alcohol.

In 2016, Bai et al. (2016) reported the production of 4-hydroxy benzyl alcohol in *E. coli* by overexpressing CAR-Sfp from *Norcardia iowensis* and *Bacillus subtilis* separately. After overexpressing UbiC, CAR-Sfp increases the carbon flow in the shikimate pathway, deleting aroG, adh1 to prevent feedback inhibition of the end product. The best strain produced 127 mg/L 4-hydroxy benzyl alcohol and 545.6 mg/L gastrodin. Despite huge improvements in the artificial approach, both the extra enzymes caused significant harm to the bacteria. Its OD falls from 10 to 6, which is a 40% loss of cells. Yin et al. (Metabolic Engineering of *Saccharomyces cerevisiae* for High-Level Production of Gastrodin from Glucose 2020) reported enhanced production of 4-hydroxy benzyl alcohol and gastrodin in *E. coli* by introducing and overexpressing PPTcg-1syn on CAR-Sfp, Aro4fbr, and K229L to improve the end-product tolerance of *E. coli*. Moreover, the four genes are responsible for the conversion of chorismate to 4-hydroxy benzyl alcohol. The best strain produced 2.1 g/L gastrodin, which is the highest production ever reported.

3.3 *Salicyl alcohol*

Salicyl alcohol and its derivatives are widely used in pharmaceutical industries mostly in the production of aspirin. It can be obtained by recruiting chorismate as the precursor, which is converted to iso-chorismate by iso-chorismate synthase (ICS). Sequentially, iso-chorismate pyruvate lyase (IPL) is responsible for the conversion of iso-chorismate into salicylic acid. Finally, CAR and ADH convert salicylic acid into salicylic alcohol (Catherine 2003; Datsenko, 2000). Lin et al reported the production of salicylic acid in *E. coli* by over-expressing ppsA, aroL, aroF/H/G. The sic genes are responsible for converting chorismate to salicylic acid and increasing the carbon flow upstream. The best strain produced 778.16 mg/L of salicylic acid but no salicylic alcohol was discovered. Hence, it is speculated that there is a barrier for salicylic acid to reduce into salicylic alcohol endogenously. Shen et al. (2018) additionally added entC-pchB, CAR-Sfp, ADH6, which produced 594 mg/L salicylic alcohol, surprisingly when added salABCD it produced 30.1 mg/L gentysl alcohol, which is another important alcohol in the pharmaceutical industry. What's more, it was the first article that ever-produced salicylic alcohol. The production is then improved by Liu et al. (2019). Deletion of pykA/F, tyrA, pheA led to inhibition of the carbon flow of competing pathways of synthesizing tyrosine and phenylalanine. The production of salicylic alcohol was 1560.6 mg/L using glycerol as a simple carbon source, which is the highest of all reports.

3.4 *Cinnamic alcohol*

Cinnamic alcohol is widely used in agricultural industries to prevent damage to plants, mostly produced with the extraction of cinnamic plants. Recently, there has been some progress in microbial synthesis using CAR. It can be obtained by recruiting L-phenylalanine as a precursor, which is converted to cinnamic acid by phenylalanine amino lyase (PAL). Sequentially, CAR is responsible for converting cinnamic acid into cinnamaldehyde. Finally, cinnamaldehyde is converted to cinnamyl alcohol by ADH. Paolo et al. (2018) reported the production of cinnamyl alcohol in *S. cerevisiae*, which is only detectable. Evaldas et al. (2018) over-expressed PAL, CAR, and ADH to enhance the carbon flow of the pathway in *E. coli*. After further optimizing the conditions, it produced 300 mg/L cinnamic alcohol, which is the highest produce using CAR methods.

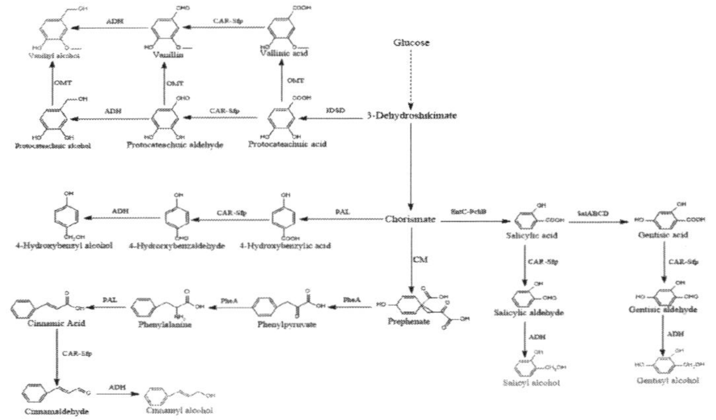

Figure 3. Biosynthesis of aromatic alcohols by carboxylic acid reductase. CM, chorismate mutase; PheA, prephenate dehydrogenase; OMT, O-methyltransferase; ADH, alcohol dehydrogenase; CAR, carboxylic acid reductase; Sfp, encoding CAR maturation factor phosphopantetheinyl transferase; 3-DSD, 3-dehydroshikimate dehydrogenase; CL, chorismate lyase; EntC-PchB, isochorismate synthase, and pyruvate lyase; SalABCD, anthranilate 5-hydroxylase.

4 SYNTHESIS OF AROMATIC ALCOHOLS WITH COA DEPENDENT REDUCTASE

4.1 *Benzyl alcohol*

As mentioned in this paper, benzyl alcohol could also be synthesized with the help of CoA-dependent reductase. However, the reviews of using this method are to be further researched. It can be achieved by recruiting phenylalanine as a precursor, which is converted to cinnamic acid by PAL. Sequentially, acyl-ligase is responsible for the conversion of cinnamic acid into cinnamyl-CoA. Then cinnamyl-CoA is converted to 3-hydroxyphenylpropionic-CoA by CA-CoA hydrolysis. After that, CA-CoA lyase is responsible for turning 3-hydroxyphenyl propionyl-CoA into benzaldehyde. Finally, endogenous ADH is responsible for converting benzaldehyde into benzyl alcohol. Seong et al. (Park 2017) first reported the production of benzyl alcohol with CoA-dependent enzymes in *Populus davidiana*. By HPLC analysis, the production goes to 10 uM/L. However, the price of the microbe is relatively high. Hence, Zi et al. (Ni 2015) used the same pathway in *E. coli*. Moreover, several enzymes are over-expressed, namely PAL, acyl-ligase, CA-CoA hydrolysis, CA-CoA lyase to increase the carbon flow of the pathway, the production of benzyl alcohol goes to 400 mg/L, which is the highest production ever reported.

4.2 *Monolignols*

p-Coumaric alcohol and others are mostly used in pharmaceutical industries to treat AIDS and thrombosis (Morelli 2017). Morelli et al. first reported the synthesis of p-coumaric alcohol in *S. cerevisiae*. The product can be achieved by recruiting phenylalanine as a precursor, which is converted into 4-cinnamic acid by PAL. Sequentially, 4-cinnamic acid hydroxylase (C4H) is responsible for converting 4-cinnamic acid into p-coumaric CoA. Finally, cinnamyl CoA reductase (CCR) and ADH are responsible for the conversion of p-coumaric CoA into p-coumaric aldehyde and p-coumaric alcohol, respectively. Over-expression of PAL, C4H, CCR, and ADH is done to increase carbon flow, but the method only detected 0.97 g/L p-coumaric acid. Production of p-coumaric alcohol is only detectable because the ADH used has low activity. Hence, requiring further improvements.

Frank et al. (Jansen 2014) reported the production of p-coumaric acid in *S. cerevisiae* by over-expression of tyrosine amino lyase (TAL), 4CL, CCR, and ADH. Different from Morelli, Frank over-expressed CAD from *Nocardia* species. The best strain produced 22 mg/L p-coumaric alcohol. It could be speculated that the activity of ADH is key to the production of p-coumaric alcohol. Philana et al. (Summeren-Wesenhagen 2015) improved the production of p-coumaric alcohol by adding co-factors with the same pathway. The best strain produced 52 mg/L p-coumaric alcohol. Although the production increased twofold, it is still low for industrial use, so Chen et al. (2017) found a significant breakthrough with a novel established pathway. It not only produced p-coumaric alcohol but also some other similar derivatives, including caffeol and coniferyl alcohol, by adding the four enzymes it introduced HpaBC, COMT, and CcoAOMT to build a bridge between the compounds. HpaBC helps convert L-tyrosine to L-Dopa, p-coumaric acid to caffeic acid, and p-coumaric alcohol to cafestol. COMT converts caffeic acid to ferulic acid and caffeol to ferulic alcohol. CcoAMOT converts caffeoyl-CoA to ferulic-CoA. The co-existence of the three pathways surprisingly produced 501.8 g/L p-coumaric alcohol is formed. It also produced 232.9 mg/L caffeol and 56.3 mg/L coniferyl alcohol, which is the highest amount of production ever produced. Suggesting the co-expression of the three pathways might help to enhance the production considerably.

4.3 *Cinnamyl alcohol*

Cinnamyl alcohol is used in agriculture to prevent plants from being damaged. It is an important precursor for cinnamaldehyde glycoside. It could also be synthesized by recruiting phenylalanine as a precursor, which is converted to cinnamic acid by PAL. Sequentially, cinnamic acid is converted to cinnamyl-CoA by CoA ligase (4CL). Then CCR is responsible for the conversion of cinnamyl-CoA to cinnamaldehyde. Finally, ADH is responsible for reducing cinnamaldehyde to cinnamyl alcohol. In 2011, Li et al. (Ma 2011) discovered the pathway by using HPLC. However, the production is only detectable using de novo synthesis, and the conversion from feeding cinnamaldehyde is around 80%. Despite the high conversion rate, Mucor is expensive. So, Zhou et al. (Zhou 2017) constructed the artificial pathway in *E. coli* to produce cinnamyl alcohol and cinnamaldehyde glycoside. By overexpressing ATPAL (*A. thaliana*), 4CL, ATCCR (*A. thaliana*), ADH-UGT to increase carbon flow, deleting tyrR, tyrA, trpE inhibiting competing pathways. It produced 197.8 mg/L cinnamyl alcohol, but the deletion of tyrR, tyrA, and trpE reduced the growth of the cells, and the OD value declined from 9 to 6. Future studies of tyrR, tyrA, and trpE should be further investigated. Iman et al. used plant-like cells to reduce the toxicity of cinnamyl alcohol. By feeding 500 mg/L cinnamyl alcohol there is no significant reduction in OD levels. By overexpressing PAL, 4CL, CCR, ADH. The final production reached 263 mg/L within 96 h, which is a bit higher than the previous one. Unfortunately, the bottleneck of the pathway is still not founded; suggesting that there might be another pathway influencing the synthesis of cinnamyl alcohol.

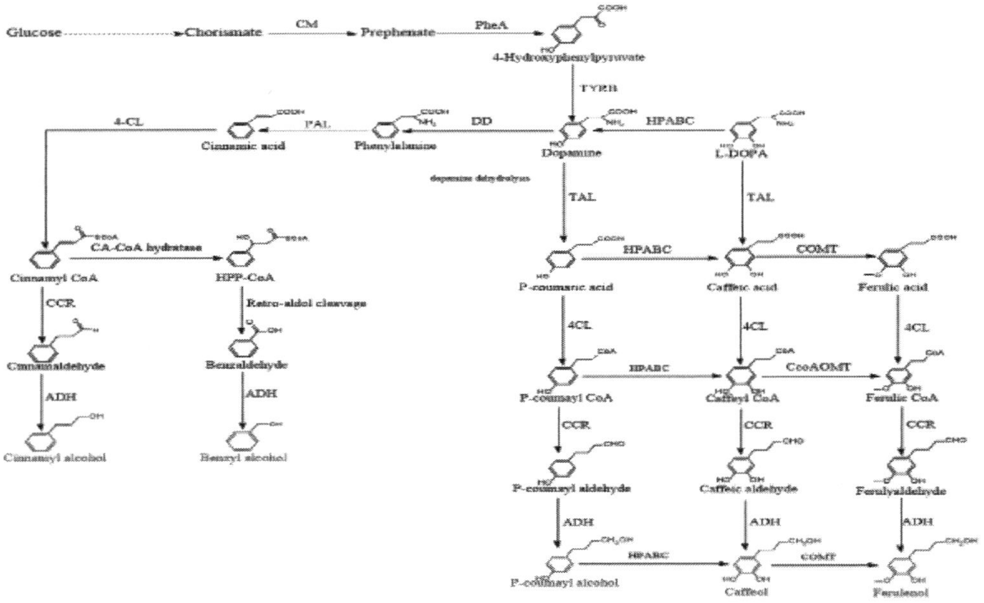

Figure 4. Biosynthesis of aromatic alcohols by carboxyl-CoA reductase. CM, chorismate mutase; PD, PheA, prephenate dehydrogenase; ADH, alcohol dehydrogenase; CCR, cinnamyl CoA reductase; DD, dopamine dehydrolysis, 4CL 4-cinnamyl ligase; PheA, prephenate dehydrogenase; TyrB, L-tyrosine amino transferase; COMT, caffeine O-methyltransferase; TAL, tyrosine amino lyase, CA-CoA, cinnamyl CoA.

5 CONCLUSION

It has been stated previously in this paper that many microbes are used to produce aromatic alcohols. Yet, it is shown that *E. coli* and *S. cerevisiae* are the most commonly used bacteria mentioned in recent research as production hosts due to the relatively simple procedures and fast growth rates. As described above, the production of the two microbes could be enhanced by overexpressing the targeted enzymes. The carbon flow of the two microbes is relatively low without going through any metabolic engineering; the ideal host strain may be constructed by a genetically modified system. Most of the aromatic alcohol is toxic to the host microbes, for instance, phenylacetaldehyde caused 83.3% reduction in OD. Interestingly some of the compounds can be synthesized in various ways, such as benzyl alcohol, cinnamic alcohol, etc. To produce the toxic compounds, a two-phase separation process could be used using organic solvents. Although salicylic acid is formed in huge amounts, as much as 2.1 g/L, the production of salicylic alcohol is still not high probably as the activity of CAR-Sfp is not high. In microbial production of aromatic alcohols, glucose is the most commonly used carbon source, though some other carbon sources, including xylose, glycerol, etc., are used. Future research on other carbon sources should be further investigated, as it is essential for the green process of producing the compounds. In summary, we described the recent progress of de novo synthesis on aromatic products. The microbial production of aromatic alcohol begins from glucose through the shikimate pathway more complex compounds could be synthesized by engineering the shikimate pathway. Three kinds of strategies are summarized in this passage. For further improvements in this area, there are three ways of considerations: (1) Reducing the toxicity of the final product's damage to the cell. (2) Utilization of different pathways of carbon flows to enhance the production. (3) Exploring new pathways with the help of a computer to speculate the optimal pathways. The three key factors are expected to enhance the product of aromatic alcohols with the development of genetic engineering and techniques in the future.

REFERENCES

Bosu, Kim, Bo-Ram, et al. Metabolic engineering of Saccharomyces cerevisiae for the production of 2-phenylethanol via Ehrlich pathway[J]. *Biotechnology and Bioengineering*, 2014, 111(1):115–124.

B, N, Shyamala, et al. Studies on the Antioxidant Activities of Natural Vanilla Extract and Its Constituent Compounds through in Vitro Models[J]. *Journal of Agricultural and Food Chemistry*, 2007, 55(19):7738–7743.

Brochado A R, Matos C, M? Ller B L, et al. Improved vanillin production in baker's yeast through in silico design[J]. *Microbial Cell Factories*, 9, 1(2010-11-08), 2010, 9(1):84–84.

Bai, Yanfen, Yin, Hua, Bi, Huiping. De novo biosynthesis of Gastrodin in Escherichia coli[J]. 2016.

Catherine, G., Cornelia, R., Dieter, H., 2003. Isochorismate synthase (PchA), the first and rate-limiting enzyme in salicylate biosynthesis of Pseudomonas aeruginosa. *J. Biol. Chem.* 278 (19), 16893.

Chen W, Yao J, Meng J, et al. Promiscuous enzymatic activity-aided multiple-pathway network design for metabolic flux rearrangement in hydroxytyrosol biosynthesis[J]. *Nature Communications*, 2019, 10(1):960.

Chen Z, Shen X, Wang J, et al. Establishing an Artificial Pathway for De Novo Biosynthesis of Vanillyl Alcohol in Escherichia coli[J]. *Acs Synthetic Biology*, 2017: acssynbio.7b00129.

Chen, Zhenya, Sun, Xinxiao, Li, Ye,Yan, Yajun, Yuan, Qipeng. Metabolic engineering of Escherichia coli for microbial synthesis of monolignols[J]. *Metabolic engineering*, 2017, 39:102–109.

Choo, HJ, Kim, et al. Microbial synthesis of hydroxytyrosol and hydroxysalidroside[J]. *APPL BIOL CHEM*, 2018, 2018, 61(3) (-):295–301.

Chung, Kim, SY, et al. Production of three phenylethanoids, tyrosol, hydroxytyrosol, and salidroside, using plant genes expressed in Escherichia coli [J]. *SCI REp-UK*, 2017, 2017,7(-):-.

Datsenko, K.A., Wanner, B.L., 2000. One-step inactivation of chromosomal genes in Escherichia coli K-12 using PCR products. *Proc. Natl. Acad. Sci. U. S. A.* 97 (12).

E. coli and Food Safety. [P] CDC.2020

Etschmann M, Bluemke W, Sell D, et al. Biotechnological production of 2-phenylethanol[J]. *Applied Microbiology and Biotechnology*, 2002, 59(1):1–8.

Fan L, Dong Y, Xu T, et al. Gastrodin Production from p-2-Hydroxybenzyl Alcohol Through Biotransformation by Cultured Cells of Aspergillus foetidus and Penicillium cyclopium[J]. *Applied Biochemistry & Biotechnology*, 2013.

Guo, Daoyi, Zhang, et al. Metabolic Engineering of Escherichia coli for Production of 2-Phenylethanol and 2-Phenylethyl Acetate from Glucose[J]. *Journal of Agricultural & Food Chemistry*, 2018.

Hagiwara K, Goto T, Araki M, Miyazaki H, Hagiwara H (2011) Olive polyphenol hydroxytyrosol prevents bone loss. *Eur J Pharmacol* 662:78–84.

Hansen E H, Moller B L, Kock G R, et al. De novo biosynthesis of vanillin in fission yeast (Schizosaccharomyces pombe) and baker's yeast (Saccharomyces cerevisiae).[J]. *Applied & Environmental Microbiology*, 2009, 75(9):2765–74.

Hua D, Xu P. 2011. Recent advances in biotechnological production of 2-phenylethanol. *Biotechnol Adv* 29(6):654–660.

Jang, Y. W., Lee, J. Y., and Kim, C. J. (2010) Anti-asthmatic activity of phenolic compounds from the roots of Gastrodia elata Bl. *Int. Immunopharmacol.* 10, 147–154.

Jin G, Liu J, Wang C, et al. Ir Nanoparticles with multi-enzyme activities and its application in the selective oxidation of aromatic alcohols[J]. *Applied Catalyase B: Environmental*, 2020, 267:118725.

Jansen F, Gillessen B, Mueller F, et al. Metabolic engineering forp-coumaryl alcohol production inEscherichia coliby introducing an artificial phenylpropanoid pathway[J]. *Biotechnology and Applied Biochemistry*, 2014.

Kang Z, Zhang C, Du G, et al. Metabolic Engineering of Escherichia coli for Production of 2-phenylethanol from Renewable Glucose[J]. *Applied Biochemistry and Biotechnology*, 2013.

K Evandelas, Zebec Z, Weise N J, et al. Bio-derived production of cinnamyl alcohol via a three step biocatalytic cascade and metabolic engineering [J]. *Green Chemistry*, 2018: 10.1039. C7GC03325G.

Kang S-Y, Choi O, Lee JK, Hwang BY, Uhm T-B, Hong Y-S (2012) Artificial biosynthesis of phenylpropanoic acids in a tyrosine overproducing Escherichia coli strain. *Microb Cell Factories* 11(1):153.

Luo Z W, Lee S Y. Metabolic engineering of Escherichia coli for the production of benzoic acid from glucose [J]. *Metabolic Engineering*, 2020, 62:298–311.

Li X, Chen Z, Wu Y, et al. Establishing an Artificial Pathway for Efficient Biosynthesis of Hydroxytyrosol [J]. *Acs Synthetic Biology*, 2017: acssynbio.7b00385.

Lin Y, Sun X, Yuan Q, et al. Extending shikimate pathway for the production of muconic acid and its precursor salicylic acid in Escherichia coli[J]. *Metabolic Engineering*, 2014, 23:62–69.

Liu L, Li W, Li X, et al. Constructing an efficient salicylate biosynthesis platform by Escherichia coli chromosome integration [J]. *Journal of Biotechnology*, 2019.

Metabolic Engineering of Saccharomyces cerevisiae for High-Level Production of Gastrodin from Glucose 2020.

Morelli L, Zór K, Bille Jendresen C, Rindzevicius T, Schmidt MS, Nielsen AT, Boisen A (2017) Surface-enhanced Raman scattering for quantification of p-coumaric acid produced by Escherichia coli. *Anal Chem* 89(7):3981–3987.

Ma L, Liu X, Liang J, et al. Biotransformations of cinnamaldehyde, cinnamic acid and acetophenone with Mucor[J]. *World Journal of Microbiology and Biotechnology*, 2011, 27(9):2133–2137.

Mcleish M J, Kneen M M, Gopalakrishna K N, et al. Identification and Characterization of a Mandelamide Hydrolase and an NAD(P)+-Dependent Benzaldehyde Dehydrogenase from pseudomonas putida ATCC 12633[J]. *Journal of Bacteriology*, 2003, 185(8):2451.

Ni J, Tao F, Du H, et al. Mimicking a natural pathway for de novo biosynthesis: natural vanillin production from accessible carbon sources[J]. *Rep*, 2015, 5(1):13670.

Paolo E, Zebec Z, Weise N J, et al. Bio-derived production of cinnamyl alcohol via a three step biocatalytic cascade and metabolic engineering[J]. *Green Chemistry*, 2018: 10.1039.C7GC03325G.

Park S B, Kim J Y, Han J Y, et al. Exploring genes involved in benzoic acid biosynthesis in the Populus davidiana transcriptome and their transcriptional activity upon methyl jasmonate treatment[J]. *Journal of Chemical Ecology*, 2017.

Qian S, Li Y, Cirino P C. Biosensor-guided improvements in salicylate production by recombinant Escherichia coli[J]. *Microbial Cell Factories*, 2019, 18.

Shyamala, B., Naidu, M. M., Sulochanamma, G., and Srinivas, P. (2007) Studies on the antioxidant activities of natural vanilla extract and its constituent compounds throughin vitro models. *J. Agric. Food Chem.* 55, 7738–7743.

S. Lee, Banda C, H. Park. Effect of inoculation strategy of non-Saccharomyces yeasts on fermentation characteristics and volatile higher alcohols and esters in Campbell Early wines[J]. *Australian Journal of Grape and Wine Research*, 2019, 25(4).

Scognamiglio J, Jones L, Vitale D, Letizia CS, Api AM. 2012. Fragrance material review on benzyl alcohol. *Food Chem Toxicol* 50: S140–S160.

Shawn, Pugh, Rebekah, et al. Engineering Escherichia coli for renewable benzyl alcohol production[J]. *Metabolic Engineering Communications*, 2015.

Satoh Y, Tajima K, Munekata M, et al. Engineering of l-tyrosine oxidation in Escherichia coli and microbial production of hydroxytyrosol [J]. *Metabolic Engineering*, 2012, 14(6).

Shen X, Wang J, Gall B, et al. Establishment of Novel Biosynthetic Pathways for the Production of Salicyl Alcohol and Gentisyl Alcohol in Engineered Escherichia coli[J]. *Acs Synthetic Biology*, 2018: acssynbio.8b00051.

Summeren-Wesenhagen P V V, Voges R, Dennig A, et al. Combinatorial optimization of synthetic operons for the microbial production of p-coumaryl alcohol with Escherichia coli[J]. *Microbial Cell Factories*, 2015.

Vázquez-Velasco M, Esperanza Díaz L, Lucas R, Gómez-Martιnez S, Bastida S, Marcos A, Sanchez-Muniz FJ (2011) Effects of hydroxytyrosol-enriched sunflower oil consumption on CVD risk factors. *Br J Nutr* 105:1448–1552.

Valera M J, Zeida A, Boido E, et al. Genetic and Transcriptomic Evidence Suggest ARO10 Genes Are Involved in Benzenoid Biosynthesis by Yeast[J]. *Yeast*, 2020.

Wright D, Atkinson J H. *Process for the production of aromatic alcohol*: CA 1979.
Widhalm J R, Dudareva N. A Familiar Ring to It: Biosynthesis of Plant Benzoic Acids[J]. *molecular biology in plants* 2015, 008(001):83–97.
Yin S, Zhou H, Xiao X, et al. Improving 2-Phenylethanol Production via Ehrlich Pathway Using Genetic Engineered Saccharomyces cerevisiae Strains[J]. *Current Microbiology*, 2015, 70(5):762–767.
Zhang, H. R., Pereira, B., Li, Z. J., Stephanopoulos, G., 2015. Engineering Escherichia coli coculture systems for the production of biochemical products. *P. Natl. Acad. Sci. U.S.A.* 112, 8266–8271.
Zhou W, Bi H, Zhuang Y, et al. Production of Cinnamyl Alcohol Glucoside from Glucose in Escherichia coli[J]. *J. Agric. Food Chem.* 2017.9.

Energy Revolution and Chemical Research – Kok-Keong Chong and Zhongliang Liu (Eds)
© 2023 The Authors, ISBN: 978-1-032-36554-1

A harmlessness and resource treatment of hazardous waste sludge: Wet oxidation of caprolactam sludge

W.H. Ling
Zhejiang Baling Hengyi Caprolactam Co., Ltd., Hangzhou, China

P.F. Guo, X. Zeng, Y.Y. Zhou*, J.F. Zhao & G.D. Yao*
College of Environmental Science and Engineering, Tongji University, Shanghai, China

ABSTRACT: Caprolactam (CPL) is a bulk raw material for polymer and engineering plastics production. The CPL production wastewater contains many toxic substances such as cyclo-hexanone, cyclohexanone oxime, cyclohexane, and benzene. After biochemical treatment, parts of harmful chemicals remain in the sludge, which is regarded as hazardous solid waste. The common treatment for this sludge is incineration after dehydration. However, the moisture content is still high (~70%) after usual dehydration, and incineration requires much fuel, leading to high treatment costs. Herein, the wet oxidation (WO) process was employed as a harmlessness and resource method for the treatment of CPL sludge. The results show the removal rates of total solid (TS) and chemical oxygen demand (COD) can reach 45.3% and 64.6% respectively at 240°C for 1 h with 1 MPa oxygen supply. The organic pollutants were oxidized to small molecular organic acids such as formic acid, acetic acid, and *iso*-butyric acid, which possess good biodegradability. This study provides an efficient approach for harmlessness and resource treatment of hazardous CPL sludge.

1 INTRODUCTION

Caprolactam (CPL) is a bulk raw material for polymer and engineering plastic synthesis which is widely used in textile, electrical, and machinery fields (Gu et al. 2016; Wang et al. 2015). China is the largest producer of CPL in the world. The production capacity in China has reached 4.19 million tons in 2019, accounting for about 51.6% of the world's production capacity (Cui 2021). Due to the raw materials used in the production process, CPL wastewater contains cyclohexanone, cyclohexanone oxime, cyclohexane, benzene, and other harmful substances. The common treatment for this wastewater is a biochemical method combined with physicochemical pretreatment. After biochemical treatment, parts of harmful chemicals remain in the sludge, which is regarded as hazardous solid waste. The common treatment for this sludge is incineration treatment after dehydration. However, the moisture content is still high (~70%) after usual dehydration, and incineration requires much fuel, leading to high treatment costs. Hence, it is necessary to develop a highly efficient and cost-effective approach for CPL sludge treatment.

Wet oxidation (WO) is a technology using oxygen (usually air or oxygen enrichment) to oxidize organic matter in water into low molecular weight organic matter or inorganic matter at high temperatures (120–320°C) and pressure (0.5–20 MPa). The technology applies to the refractory wastewater with a concentration between not easy to be biochemical (too high concentration) and not suitable for incineration (too low concentration). Its research and application can be traced back to the oxidation of papermaking black liquor in the 1950s. In recent years, with the increasing need for treatment of high concentration and refractory wastewater, the technologies of wet air oxidation (WAO) and catalytic wet oxidation (CWO) derived from WO have developed rapidly (Kim & Ihm 2011; M'Arimi et al. 2020). The organics in sludge can be oxidized to low molecular

*Corresponding Authors: zhouyytj@126.com and 2012yao@tongji.edu.cn

 DOI 10.1201/9781003332657-85

weight matters, realizing not only reduction but also harmlessness and resource utilization (Chen et al. 2022). It is an ideal method to treat high moisture content sludge. There are some studies on the treatment of municipal sludge (Jiang et al. 2021) and oil refining sludge by wet oxidation (Jing et al. 2015). Previous studies showed the heat generated by wet oxidation can maintain the reaction temperature of the system at about 200°C when the concentration of organic substances in sludge exceeds 10,000 mg/L. After WO, the chemical oxygen demand (COD) removal rate and total solid (TS) reduction usually reach more than 60% and 45% respectively (Wang et al. 2019). Therefore, WO is suitable for the treatment of high organics concentration and high toxicity industrial sludge. Our group reported WO of pharmaceutical sludge (Guo et al. 2019) and the results showed that the sludge reduction effect was obvious. Herein, the WO of CPL sludge was investigated and TS and COD reduction were analyzed.

2 EXPERIMENTAL SECTION

2.1 *Materials*

The sludge used in this study is taken from the biochemical sludge after pressure filtration in a CPL factory and the moisture content of the sludge is 83%–85%. The COD is 96.7 mg per gram of sludge and volatile solid (VS)/TS = 53.6%. High purity oxygen (99.99%) was used in the lab experiment and industrial oxygen (~90%) was used in the pilot test.

2.2 *Reactor*

The lab experiments were performed in the hydrothermal reactor (YZPR-250; Figure 1) customized by *Shanghai Yanzheng Experimental Instrument Co., Ltd.* The reactor is lined with SUS316 material and the inner volume is 250 mL. The bearing pressure and heating temperatures are 10 MPa and 300°C, respectively. The heating rate of the reactor is 4–5°C/min.

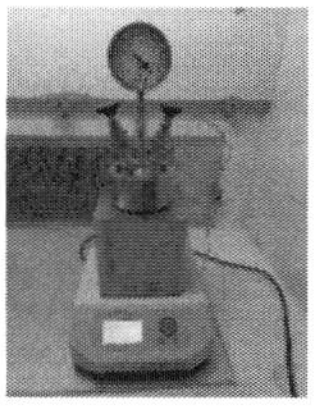

Figure 1. The reactor used in the lab experiment.

The pilot test device consists of four parts: feed tank (with mechanical stirring), reactor (200 L, with magnetic stirring), cooling tower, and three-phase separator (Figure 2). The reactor embeds in heat transfer oil which is heated by six 5-kW heating rods and the set pressure is 6.4 MPa. The heating rate is about 35–45°C/h.

Figure 2. Equipment at the pilot test site.

2.3 *Method*

In the lab experiment, the sludge and water were mixed to a certain moisture content (~90%), and then reacted for 0.5–2 hrs at 170–250°C with a certain oxygen supply (0.5 MPa–2 MPa). After the reaction, the reactor was naturally cooled to room temperature, and the solid sample and clear liquid were collected respectively for the test.

In the pilot operation, a certain amount of sludge and water was added into the feed tank and stirred for 1 min. After mixing, the slurry was pumped into the WO reactor, followed by reacting at the set temperature. After the reaction, the feedstock was discharged through the cooling tower and three-phase separator.

3 RESULTS AND DISCUSSION

3.1 *Compositions of the sludge*

The compositions of the sludge were firstly analyzed by X-ray fluorescence spectrum (XRF). The results showed the sludge mainly contained C (23.8%), N (5.15%), O (26.8%), and Fe (28.8%). The main sources of Fe may come from poly-ferric sulfate addition during the physicochemical treatment process. The content of typical heavy metals such as nickel (0.0446%), copper (0.0676%), zinc (0.0903%), and other heavy metals is low. Therefore, the migration and transformation of heavy metals in sludge before and after the reaction are not involved in this study. The qualitative GC-MS test of the sludge showed the toxic organic substances in the sludge include benzene series (benzene, toluene, biphenyl, etc.), heterocyclic (pyrrole, indole, etc.), alcohols (isopropanol, tertiary-butyl alcohol, etc.) and ketones (acetone, butanone, cyclopentanone, etc.). These organic contaminants should come from the CPL production process and precipitate in the sludge after wastewater treatment.

3.2 *Lab experiment*

The fluidity of slurry is crucial to the mass and heat transfer of reaction and the design of continuous equipment. First, the effect of moisture content on the fluidity of the sludge was investigated. Four moisture content of 88%, 92%, 95%, and 98% were prepared via mixing sludge and water, and sludge pyrolysis was carried out at lower temperatures (<170°C). The results showed that the apparent properties of sludge did not change significantly within 30 minutes when the temperature was lower than 130°C. As the temperature increased to 130°C, the sludge changes obviously (Figure 3) and fluidity became better when the water content is above 90%. However, TS remains stable after reaction at 200°C, indicating VS was not decomposed under present conditions. Thus, the following experiments were conducted at temperatures above 200°C with 90% water content.

Then, the TS reduction effect was studied at 210–250°C in the absence and presence of oxygen supply. As shown in Figure 4, the removal rates of TS with oxygen are higher than that without oxygen at all reaction temperatures. At 250°C for 1 h, the TS removal rate reached 40.8%, whereas only 17.8% was obtained without oxygen. Further determination showed more than 40,000 mg/L COD was in the liquid, indicating that sufficient oxygen supply was key to sludge organics decomposition. In the subsequent lab experiments, the initial oxygen pressure in the reactor was raised to 1 MPa and the reaction temperatures were set to 250°C.

Finally, the influence of reaction time on sludge reduction was studied. As shown in Figure 5, the reaction time was 0.5 h and the TS reduction reached 35.5%, but the COD reduction was only 18.9%. The reason was probably that the mass transfer process of organic matter from the solid phase to the liquid phase was slow and it was not fully oxidized in a short time. With the reaction time increased to 1 h, the reduction of TS raised to 46.4%, while the reduction of COD increased rapidly to 49.9%. The reaction time continued to prolong to 2 hrs, TS removal rate only increased to 51.9%, while COD removal rate increased to 65.7%, indicating that the organic substances in

Figure 3. Apparent properties of CPL sludge at 130°C for 30 min.

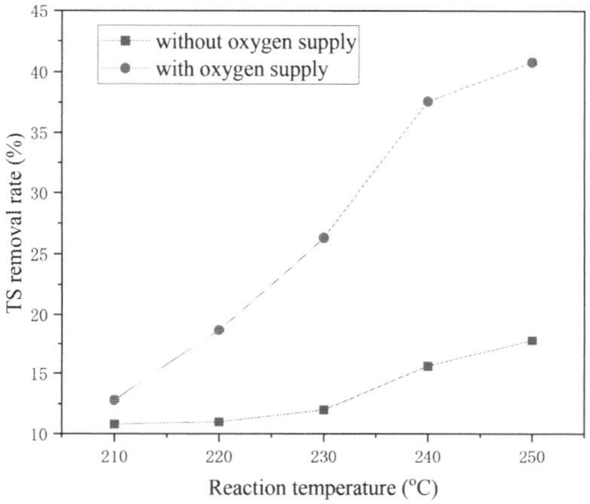

Figure 4. TS removal variation as the reaction temperature increase without and with oxygen supply (0.5 MPa).

the liquid phase were further oxidized and decomposed into inorganic substances. Considering the balance of TS reduction and energy consumption, the reaction time was set as 1 h.

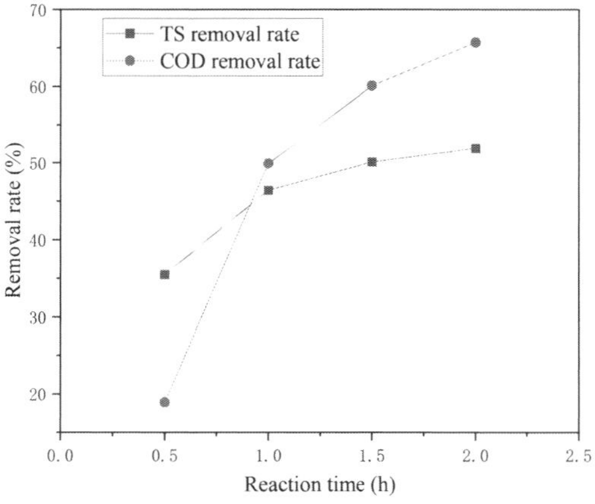

Figure 5. Effect of reaction time on TS and COD reduction (250°C; 1 MPa O_2 supply; 90% of water content).

The liquid products were analyzed by HPLC with a UV-Vis detector. As shown in Figure 6, low molecular weight organic acids such as formic acid, acetic acid, and isobutyric acid were obtained.

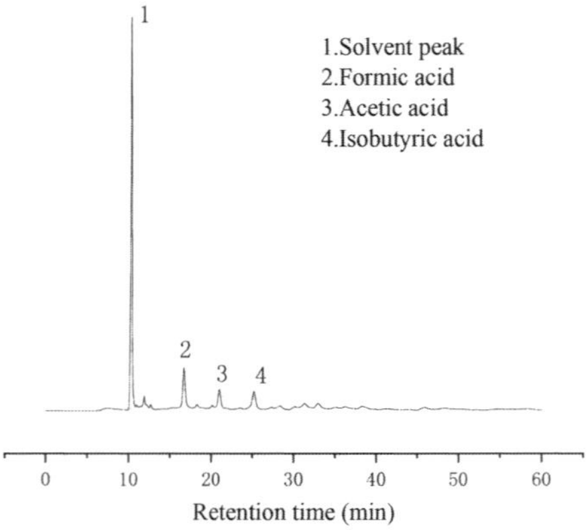

Figure 6. HPLC Chromatograms of the liquid sample (250°C; 1 h; 1 MPa O_2 supply).

These organic acids belonged to WO products of sludge organics. Further quantitative analysis of the product is in progress.

3.3 *Pilot test*

Based on the small-scale study, the main influence factors affecting sludge reduction were determined. A 200 L of WO equipment was designed and built for the pilot test. Effects of reaction temperature and initial oxygen pressure on sludge reduction were investigated. It should be noted that the maximum temperature used in the pilot test was 240°C due to the reactor set pressure limit. In addition, the water content of the input slurry was 95% for facilitating pump delivery. The change in TS reduction and COD removal were similar to the small-scale reaction (Figure 7). Both 45.3% and 64.6% removal rates of TS and COD were achieved respectively at 240°C for 1 h.

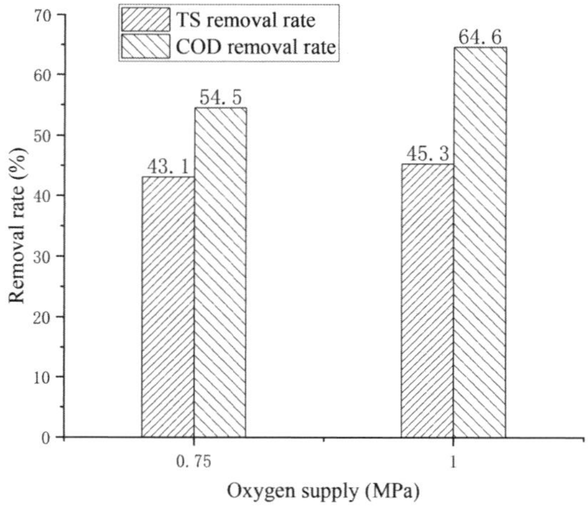

Figure 7. Effects of initial oxygen supply on TS and COD reduction (240°C; 1 h; 95% of water content).

To improve the technical economy, the effects of initial oxygen pressure of 0.75 MPa were studied. As shown in Figure 7, although the oxygen supply decreased by 25%, TS reduction only varied from 45.3% to 43.1%, whereas the COD removal rate changed from 64.6% to 54.5%. The results showed that 0.75 MPa was suitable for TS reduction of CPL sludge. After WO treatment, the water content of sludge approximately decreased to 45% by suction filtration, indicating more than 90% wet sludge reduction was achieved.

4 CONCLUSION

In this study, the WO of CPL sludge was investigated in small-scale and pilot tests. Effects of oxygen supply, reaction temperature, and time on the TS reduction and COD removal were studied. The results showed the removal rates of TS and COD can reach more than 45% and 60% at 240°C for 1 h. The organic pollutants were oxidized to small molecular organic acids such as formic acid, acetic acid, and isobutyric acid. These organic acids are regarded as an excellent carbon source for denitrification in wastewater treatment. Our study findings have provided an efficient approach for harmlessness and resource treatment of hazardous CPL sludge.

ACKNOWLEDGMENT

This paper was supported by the National Key Research and Development Project of China (Grant No. 2018YFC1902103).

REFERENCES

Chen, Z., Zheng, Z., He, C., Liu, J., Zhang, R. & Chen, Q. (2022). Oily sludge treatment in subcritical and supercritical water: A review. *J. hazard. mater.* 433, 128761–128761.

Cui, X. M. (2021). Supply and demand status and development prospect of caprolactam in China and abroad. *Fine Specialty Chem.* 29, 12–15.

Gu, Q., Song, Y., Chen, S., Yang, Y., Chen, Z., Zhou, Z., Chen, X., Wu, K. & Yin, H. (2016). Design and operation of caprolactam wastewater advanced treatment project. *China Water Wastewater* 32, 44–47.

Guo, P., Zeng, X., Yao, G., Zhou, Y. & Zhao, J. (2020). *Wet oxidation treatment of waste activated sludge of a pharmaceutical factory*. 7th Annual International Conference on Material Science and Environmental Engineering (MSEE), Nov 15–16 2019, Wuhan, China.

Jiang, G., Xu, D., Hao, B., Liu, L., Wang, S. & Wu, Z. (2021). Thermochemical methods for the treatment of municipal sludge. *J. Cleaner Product.* 311, 127811.

Jing, G., Luan, M. & Chen, T. (2015). Wet peroxide oxidation of oilfield sludge. *Arabian J. Chem.* 8, 208–211.

Kim, K.H. & Ihm, S.K. (2011). Heterogeneous catalytic wet air oxidation of refractory organic pollutants in industrial wastewaters: A review. *J. Hazard. Mater.* 186, 16–34.

M'arimi, M. M., Mecha, C. A., Kiprop, A. K. & Ramkat, R. (2020). Recent trends in applications of advanced oxidation processes (AOPs) in bioenergy production: Review. *Renew. Sus. Energy Rev.* 121, 109669.

Wang, L., Chang, Y. & Li, A. (2019). Hydrothermal carbonization for energy-efficient processing of sewage sludge: A review. Renew. *Sus. Energy Rev.* 108, 423–440.

Wang, M., Wang, S. Z., Li, Y. H. & Zhang, Z. Q. (2016). *Status and development tendency on treatment technology for caprolactam wastewater*. International Conference on Energy and Mechanical Engineering (EME), Oct 17–18, 2015 Wuhan, China. 651–656.

Energy Revolution and Chemical Research – Kok-Keong Chong and Zhongliang Liu (Eds)
© 2023 The Authors, ISBN: 978-1-032-36554-1

Recovery of 1,1,2,2-tetrafluoroethyl ether from polytetrafluoroethylene waste—A preliminary study

W.K. Lin
Department of Chemical and Petroleum Engineering, Faculty of Engineering Technology & Built Environment, UCSI University, Cheras, Kuala Lumpur, Malaysia

ABSTRACT: Polytetrafluoroethylene, also known as the king of plastics, is an engineering material with excellent properties. With the development of the fluoropolymer industry, much polytetrafluoroethylenes waste is being produced, which resulted in the serious environmental pollution in recent years. Many recycling methods of polytetrafluoroethylene waste have been proposed and achieved remarkable results, but there is a limited report on the economic feasibility of this industrial recycling method. In addition, due to policy and resource constraints, the selling price of 1,1,2,2-tetrafluoroethyl ether has become very high, which has seriously affected the development of downstream products. It is urgent to reduce the manufacturing cost of 1,1,2,2-tetrafluoroethyl ether or find alternatives. This preliminary study aimed to investigate the optimum reaction conditions, i.e., temperature (400–700°C), pressure (0–0.5 MPa), ethanol to polytetrafluoroethylene waste ratio (1:1–5:1), and polytetrafluoroethylene waste mesh size (50–300 mesh), in depolymerizing of polytetrafluoroethylene waste into tetrafluoroethylene. Polytetrafluoroethylene waste is mixed with ethanol then potassium hydroxide is added evenly as a catalyst before pumping into tubular reactors, where the polytetrafluoroethylene waste is depolymerized to form tetrafluoroethylene and simultaneously, tetrafluoroethylene is reacted with ethanol producing 1,1,2,2-tetrafluoroethyl ether. Results revealed that the optimum reaction conditions were: 500°C at 0.2 MPa with a feeding ratio of ethanol to polytetrafluoroethylene waste of 3:1 (w/w) and 200 mesh of polytetrafluoroethylene waste.

1 INTRODUCTION

1.1 *Overview of Polytetrafluoroethylene*

In 1938, Plunkett and his assistant of DuPont first obtained powdered polytetrafluoroethylene from the steel cylinder containing tetrafluoroethylene, then the research on the polymerization conditions, material properties, and application prospects of polytetrafluoroethylene attracted more and more experts together with scholars. Due to the particularity of its molecular structure (only containing carbon and fluorine atoms), polytetrafluoroethylene has the characteristics of good mechanical properties, corrosion resistance, self-lubrication, low friction coefficient, and good biological stability, which is widely used in the fields of precision instruments, automobile transportation, chemical equipment, electronic communication, aerospace, food processing and so on. Some examples are given to show its applications and benefits here. Traumatic dural sinus injuries following penetrating brain injury can be repaired successfully with a polytetrafluoroethylene graft (Abdallah 2022). A circular polytetrafluoroethylene interposition graft can be used to reconstruct inferior vena cava after surgical resection of renal cell carcinoma (Ciancio 2022). Polytetrafluoroethylene can effectively reduce the friction in the coaxial sealing system of the hydraulic cylinder (Deaconescu 2022). The modification of thermoplastic polyurethane with polytetrafluoroethylene can obtain a new polymer material with excellent antifriction performance, which is helpful to reduce friction, vibration, and noise (Li 2022). Through electrospinning technology, polytetrafluoroethylene can be

 DOI 10.1201/9781003332657-86

mixed with other raw materials to make reinforced drug-loaded vascular grafts, implanted in pig's carotid artery, and has no complications, which may be an ideal substitute for vascular reconstruction and bypass surgery in the future (Lee 2022). Polytetrafluoroethylene used for dental caries repair can reduce the extraction number of unhealthy teeth, and the repair effect is good (Klein 2022). Because polytetrafluoroethylene is one of the polymer materials with the lowest dielectric constant, and the integrity of signal transmission in this material is better than in other materials, it has been selected as the mainstream substrate of 5G high-frequency and high-speed copper clad laminate, with the acceleration of the construction of 5G base stations worldwide and Tesla's active layout of lithium battery dry electrode technology (polytetrafluoroethylene can be used to prepare composite isolation membrane for lithium battery), the consumption of polytetrafluoroethylene is expected to usher in explosive growth, and a large number of leftovers participated in waste materials with expired service life will be produced in its processing or use.

1.2 *Treatment Status of Polytetrafluoroethylene Waste*

Polytetrafluoroethylene waste cannot be decomposed by itself under natural conditions. If polytetrafluoroethylene waste cannot be effectively utilized, it will cause serious environmental pollution and this consensus has been widely valued by countries all over the world. Plastic pollution is one of the main challenges facing our times. To meet these challenges, significant changes must be made in science and society to achieve a zero-carbon sustainable bioeconomy (Antranikian 2022). Plastic products such as polytetrafluoroethylene pose a major threat to the marine environment, and active policies are needed to reduce consumption and mismanagement (Delaeter 2022). In recent years, research addressing the issue of plastic pollution is growing (Horton 2022). There has been a lot of research on the recycling of polytetrafluoroethylene waste, but the main recycling method is crushing, melting, and then forming (Yang 1988), which has great industrialization difficulty because the viscosity of polytetrafluoroethylene waste is very high after melting, it hardly flows after reaching the melting point, and it will decompose after heating (Yuan 2013).

Depolymerization and regeneration of polytetrafluoroethylene waste is the best method to treat polytetrafluoroethylene waste. Correlative research began in the 1950s to 1960s have been used by industry and academia (Bawn 1960; Samuel 1952; Williams 1964). The key information of these studies is summarized as follows.

(1) The main product of polytetrafluoroethylene depolymerization is tetrafluoroethylene, accompanied by a small amount of small molecular impurities such as hexafluoropropylene, hexafluorocyclobutane, and octafluorocyclobutane.
(2) The products obtained from the depolymerization of polytetrafluoroethylene have a certain relationship with the pressure in the depolymerization process. In a certain temperature range, the lower the depolymerization pressure, the higher the tetrafluoroethylene yield in the depolymerization products.
(3) Tetrafluoroethylene produced by depolymerization of polytetrafluoroethylene will undergo a secondary free radical reaction. The timely removal of tetrafluoroethylene from the reaction system is conducive to the depolymerization process.
(4) The experimental samples have a side effect, and the depolymerization rate of polytetrafluoroethylene with different thicknesses is inconsistent. The thicker the polytetrafluoroethylene sample is, the lower the depolymerization rate is.

1.3 *Significance of this Study*

The Polytetrafluoroethylene industry is an important part of the world economy (Williams 2022), people can't eliminate the production of polytetrafluoroethylene waste products from the source, but they can solve the problem of the whereabouts of polytetrafluoroethylene waste through scientific and technological power. In view of the serious challenges posed by waste plastic in the environment to sustainable waste management, environment, and public health (Metcalf 2022), this paper aims to provide a solution for polytetrafluoroethylene waste based on previous studies and taking full account of the economic benefits and industrialization feasibility, which is expected to provide a relatively cheap manufacturing process reference for relevant manufacturers.

2 METHODS AND MATERIALS

2.1 *Experimental Principle*

Based on the principle that tetrafluoroethylene can be added with ethanol to form 1,1,2,2-tetrafluoroethyl ether (Kazuya 1998), tetrafluoroethylene produced by depolymerization of polytetrafluoroethylene waste can be reacted in time, the reaction equation is shown in Figure 1.

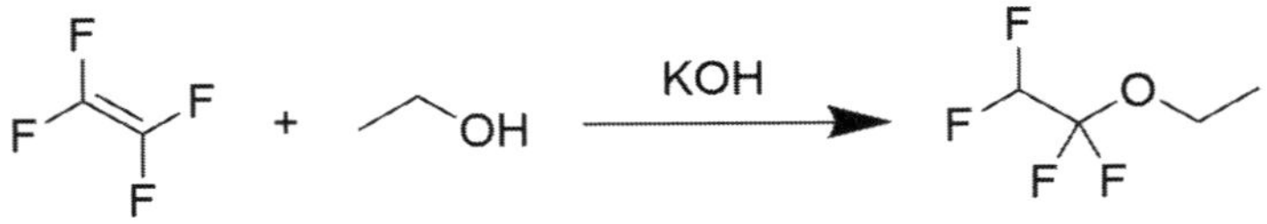

Figure 1. Synthetic route of 1,1,2,2-Tetrafluoroethyl ether.

2.2 *Experimental Conditions*

The experiments are carried out in the device shown in Figure 2. The device consists of a mixing kettle, a rotor pump, tubular reactors (two in series), condensers (three in series), and a product receiving tank. A regulating valve is equipped behind the rotor pump and a back pressure regulating valve is equipped behind the secondary tubular reactor.

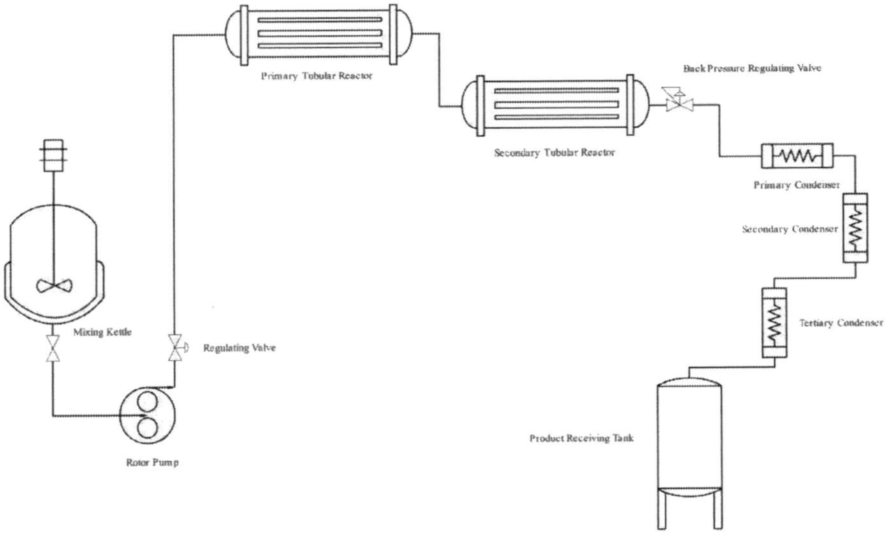

Figure 2. Schematic diagram of experimental device.

Equipment parameters:

Mixing Kettle (volume: 3 m^3; material: 316 L); Rotor Pump (flow: 6.3 m^3/h; lift: 25 m; material: 316 L);

Tubular Reactor (heat exchange area: 50 m^2; material: 316 L); condensers (heat exchange area: 30 m^2; material: 316 L).

The rotor pump is selected because it is suitable for the transportation of slurry or liquid containing solid particles; the motor of the rotor pump is variable frequency to adapt to the changing transmission pressure; the regulating valve and back pressure regulating valve are used to automatically control the pressure required in the reaction process; the tubular reactors are electrically heated to maintain the required reaction temperature; the first and second stage condensers use 35°C circulating water for heat exchange, and the third stage condenser uses –10°C chilled brine for heat exchange; the equipment parameters of the two tubular reactors are the same, and the equipment parameters of the three condensers are also the same.

The particle size of the polytetrafluoroethylene waste sample used for the experiments is 50–300 mesh.

The depolymerization of polytetrafluoroethylene and the addition reaction of tetrafluoroethylene with ethanol were carried out simultaneously at a reaction temperature of 400–700°C and a reaction pressure of 0–0.5 MPa.

The feed ratio of ethanol to polytetrafluoroethylene waste is 1:1–5:1 (w/w).

Potassium hydroxide is the catalyst for the addition reaction, and the dosage is 10% (w/w) of polytetrafluoroethylene waste.

2.3 *Experimental Process*

Polytetrafluoroethylene waste, potassium hydroxide, and ethanol are mixed in the kettle by continuously stirring to obtain a uniform feed solution. The solution is then transferred by a rotor pump to primary and secondary tubular reactors in series. Output stream from the reactors is fed to three condensers in series before entering a receiving tank. The reaction pressure is controlled automatically by the regulating valves and the reaction temperature is controlled by electric heating. The components in the receiving tank consist of 1,1,2,2-tetrafluoroethyl ether, excess ethanol, potassium hydroxide catalyst, un-depolymerized polytetrafluoroethylene waste, and a small amount of impurities.1,1,2,2-tetrafluoroethyl ether is further purified by rectification, and the percent yield is calculated by weighing pure 1,1,2,2-tetrafluoroethyl ether obtained by the formula shown in Figure 3.

$$\text{Yield} = \frac{m_1}{1.46m_2} \times 100\%$$

Figure 3. Formula for calculating the yield of 1,1,2,2-Tetrafluoroethyl ether.

Note: In the above formula, m_1 means the weight of pure 1,1,2,2-tetrafluoroethyl ether; m_2 means the weight of the polytetrafluoroethylene waste (feeding amount).

Polytetrafluoroethylene waste which is not depolymerized is filtered and stored until the next reaction.

2.4 *Experimental Materials*

Polytetrafluoroethylene waste (50–300 mesh, commercially available); potassium hydroxide (AR); ethanol (anhydrous, industrial grade).

3 RESULTS AND DISCUSSIONS

Polytetrafluoroethylene waste can be crushed into different mesh numbers and then tested. If the mesh number is too high, it is easy to carbonize in the reaction process. If the mesh number is too low, it is not conducive to the depolymerization of polytetrafluoroethylene waste. Except for the mesh number of polytetrafluoroethylene waste (A), the ethanol dosage (B), reaction pressure (C), and reaction temperature (D) are also the main factors. To improve the efficiency and accuracy of orthogonal experiments, single-factor experiments are carried out to explore the influence of various factors on the yield of 1,1,2,2-tetrafluoroethyl ether. The results are shown in Table 1.

Table 1. Results of single factor experiments.

Single Factor	Level of Each Factor				The Yield of 1,1,2,2-Tetrafluoroethyl Ether
	A	B	C	D	
A	50	3:1 (w/w)	0.2 MPa	500°C	18.66%
	100	3:1 (w/w)	0.2 MPa	500°C	52.63%
	150	3:1 (w/w)	0.2 MPa	500°C	69.77%
	200	3:1 (w/w)	0.2 MPa	500°C	73.25%
	250	3:1 (w/w)	0.2 MPa	500°C	70.63%
	300	3:1 (w/w)	0.2 MPa	500°C	60.55%
B	200	1:1 (w/w)	0.2 MPa	500°C	45.88%
	200	2:1 (w/w)	0.2 MPa	500°C	60.35%
	200	3:1 (w/w)	0.2 MPa	500°C	73.51%
	200	4:1 (w/w)	0.2 MPa	500°C	66.27%
	200	5:1 (w/w)	0.2 MPa	500°C	53.12%
C	200	3:1 (w/w)	0 MPa	500°C	36.33%
	200	3:1 (w/w)	0.1 MPa	500°C	42.71%
	200	3:1 (w/w)	0.2 MPa	500°C	73.15%
	200	3:1 (w/w)	0.3 MPa	500°C	67.25%
	200	3:1 (w/w)	0.4 MPa	500°C	58.33%
	200	3:1 (w/w)	0.5 MPa	500°C	41.26%
D	200	3:1 (w/w)	0.2 MPa	400°C	49.23%
	200	3:1 (w/w)	0.2 MPa	450°C	64.42%
	200	3:1 (w/w)	0.2 MPa	500°C	72.66%
	200	3:1 (w/w)	0.2 MPa	550°C	70.33%
	200	3:1 (w/w)	0.2 MPa	600°C	63.51%
	200	3:1 (w/w)	0.2 MPa	650°C	61.25%
	200	3:1 (w/w)	0.2 MPa	700°C	50.77%

From the results of single-factor experiments, under the experimental conditions that the mesh number of polytetrafluoroethylene waste is 200, the reaction temperature is 500°C, the reaction pressure is 0.2 MPa and the feed ratio of ethanol to polytetrafluoroethylene waste is 3:1 (w/w), the optimal response value of each factor can be obtained. Based on the single factor experiment, according to the principle of Box-Behnken, the RSM test was carried out by using Design-Expert 12 software to determine the optimal reaction conditions.

Influencing factors and levels of experiments are shown in Table 2.

Table 2. Influencing factors and levels of experiments.

Level	Value of Each Factor			
	A	B	C	D
1	150	2.5	0.15	450
2	200	3.0	0.20	515
3	250	3.5	0.25	580

Taking the yield of 1,1,2,2-tetrafluoroethyl ether as the response value, the orthogonal experiment results are shown in Table 3.

Table 3. Results of orthogonal experiment.

	Value of Each Factor				
Serial No.	A	B	C	D	Yield (%)
1	200	3	0.25	450	69.35
2	150	3.5	0.2	515	66.49
3	250	3	0.2	580	63.26
4	250	3	0.25	515	68.17
5	200	2.5	0.15	515	65.87
6	200	3.5	0.15	515	64.29
7	200	3	0.2	515	72.12
8	200	2.5	0.2	580	67.34
9	150	3	0.2	450	65.23
10	250	3	0.2	450	68.11
11	200	2.5	0.2	450	68.33
12	200	3	0.2	515	71.23
13	200	3	0.25	580	69.21
14	250	3.5	0.2	515	68.19
15	200	2.5	0.25	515	65.29
16	200	3	0.2	515	71.97
17	200	3	0.2	515	71.73
18	150	2.5	0.2	515	69.21
19	200	3	0.2	515	72.01
20	150	3	0.2	580	67.21
21	150	3	0.15	515	65.98
22	250	2.5	0.2	515	66.23
23	200	3.5	0.2	450	68.15
24	200	3.5	0.2	580	67.23
25	250	3	0.15	515	65.44
26	200	3.5	0.25	515	66.13
27	200	3	0.15	450	67.24
28	200	3	0.15	580	64.38
29	150	3	0.25	515	67.15

In the orthogonal experiments, if the level of one factor is properly selected or not, it is directly related to the level of the other factor, then the interaction between the two factors can be judged. The 3D surfaces of the interaction of various factors are shown in Figure 4.

According to the result of variance analysis of the established model, p-value is less than 0.0001, indicating that it is statistically significant and can be used for prediction. In the light of the point prediction results of the software, the best reaction conditions are: the reaction temperature is 504.25°C, the reaction pressure is 0.23 MPa, the mesh number of polytetrafluoroethylene waste is 195.20 and the feed ratio of ethanol to polytetrafluoroethylene waste is 2.94:1 (w/w). The predicted yield of 1,1,2,2-Tetrafluoroethyl Ether is 73.25%. Considering the feasibility of the actual operation, the optimization results are adjusted to: the reaction temperature is 500°C, the reaction pressure is 0.2 MPa and the mesh number of polytetrafluoroethylene waste is 200 and the feed ratio of ethanol to polytetrafluoroethylene waste is 3:1 (w/w).

The adjusted experimental conditions were verified. The actual yield of 1,1,2,2-Tetrafluoroethyl Ether was 73.21%, which showed a smaller error.

4 CONCLUSION

The method described in this paper can depolymerize polytetrafluoroethylene waste, and the tetrafluoroethylene produced by depolymerization can react with ethanol at the same time to

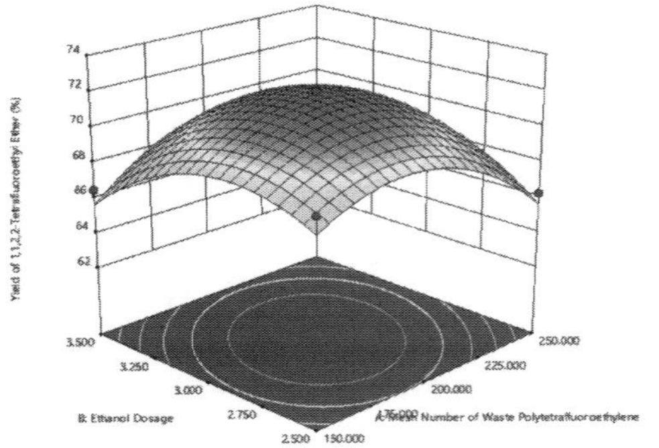

(a) Ethanol Dosage and Mesh Number of Waste Polytetrafluoroethylene

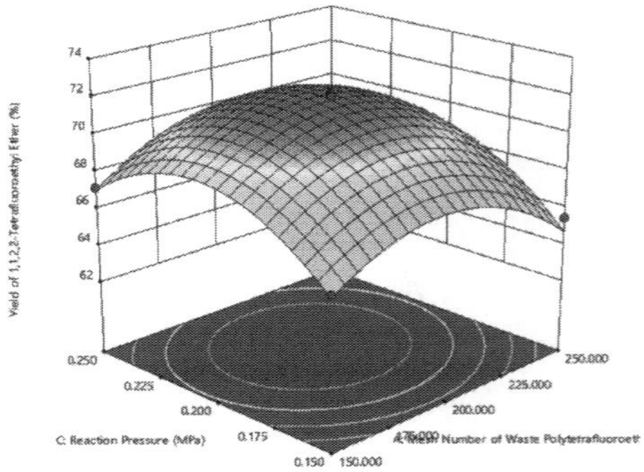

(b) Reaction Pressure and Mesh Number of Waste Polytetrafluoroethylene

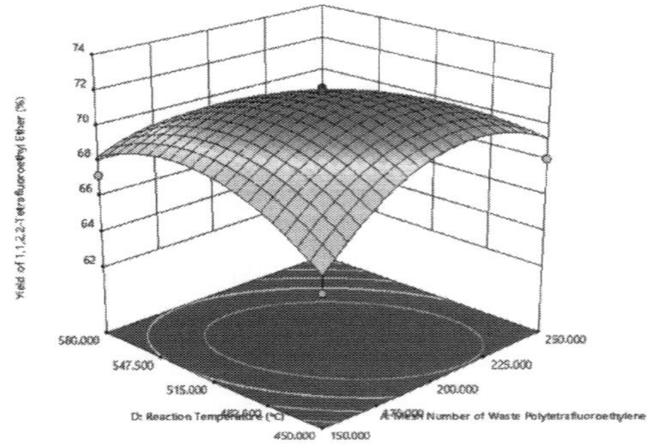

(c) Reaction Temperature and Mesh Number of Waste Polytetrafluoroethylene

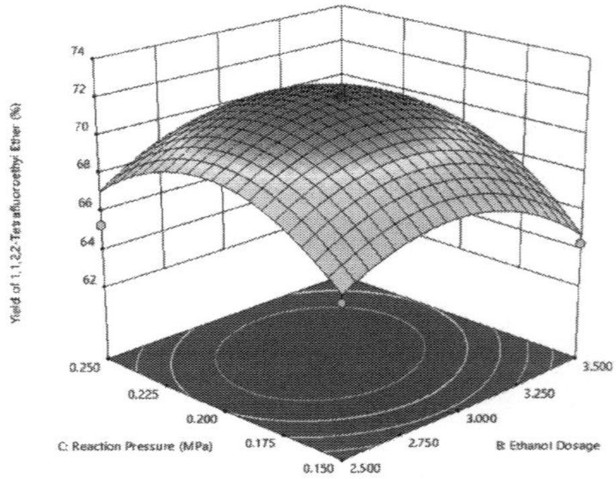

(d) Reaction Pressure and Ethanol Dosage

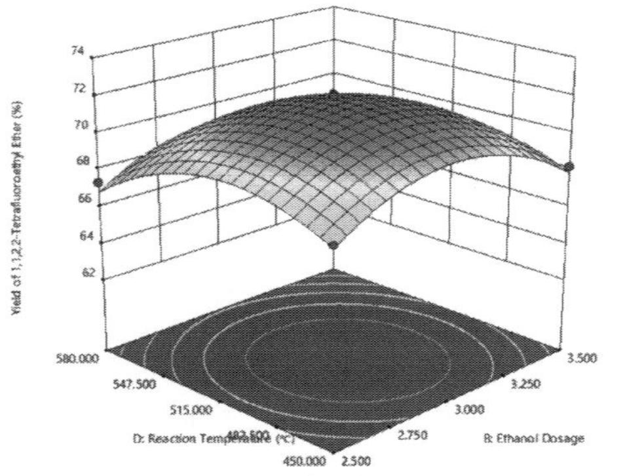

(e) Reaction Temperature and Ethanol Dosage

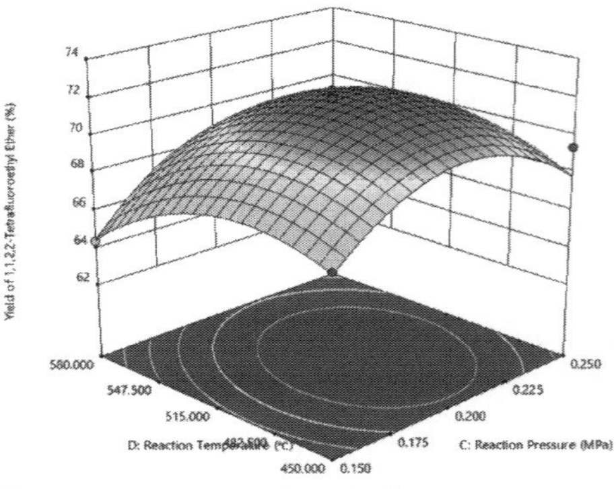

(f) Reaction Temperature and Reaction Pressure

Figure 4. Response surface plots showing the effects of pairwise interactions of the factors on the yield of 1,1,2,2-Tetrafluoroethyl ether.

produce 1,1,2,2-tetrafluoroethyl ether which is a widely used fluorine-containing fine chemical with remarkable economic effect.

Polytetrafluoroethylene waste can be recycled under the following conditions: the reaction temperature is 500°C; the reaction pressure is 0.2 Mpa; the mesh number of polytetrafluoroethylene waste is 200; and the feed ratio of ethanol to polytetrafluoroethylene waste is 3:1 (w/w).

Mainly research contribution of this study is to provide an economically and simple recovery method for polytetrafluoroethylene waste. This method has the prospect of industrialization and is expected to provide a way to reduce the production cost for the manufacturers of 1,1,2,2-tetrafluoroethyl ether.

Improvements will encompass the design of a continuous distillation apparatus for the materials collected in the collection tank, which will further enhance the level of automation and continuity.

REFERENCES

Abdallah, O. I., et al. (2022). "Surgical Reconstruction of a Traumatic Superior Sagittal Sinus Injury Using Synthetic Vascular Graft in a Resource-Limited Civilian Field Hospital During the Syrian Civil War." *World Neurosurg* 159: 126–129.

Antranikian, G. and W. R. Streit. (2022). "Microorganisms harbor keys to a circular bioeconomy making them useful tools in fighting plastic pollution and rising CO2 levels." *Extremophiles* 26(1): 10.

Bawn C. (1960). "Analytical Chemistry of Polymers." *Nature* 186(4718): 52–53.

Ciancio, G. (2022). "Inferior Vena Cava Reconstruction Using a Ringed Polytetrafluoroethylene Interposition Graft and Inferior Vena Cava Filter Placement Following Resection of Renal Cell Carcinoma with a Tumor Thrombus Directly Infiltrating the Inferior Vena Cava." *Vasc Endovascular Surg* 56(1): 5–10.

Deaconescu, T. and A. Deaconescu. (2022). "Experimental Research on Polymer-Based Coaxial Sealing Systems of Hydraulic Cylinders for Small Displacement Velocities." *Polymers* (Basel) 14(2).

Delaeter, C., et al. (2022). "Plastic leachates: Bridging the gap between a conspicuous pollution and its pernicious effects on marine life." *Sci Total Environ*: 154091.

Horton, A. A. (2022). "Plastic pollution: When do we know enough?" *J Hazard Mater* 422: 126885.

KaZuya, O. and Seisaku, K. (1998). "Preparation of Difluoroacetic Acid Fluoride and Difluoroacetic Acid Esters." US 005710317.

Klein, C., et al. (2022). "A quantitative assessment of silicone and polytetrafluoroethylene-based stamp techniques for restoring occlusal anatomy using resin-based composites." *Clin Oral Investig* 26(1): 207–215.

Lee, K. S., et al. (2022). "A Comparative Study of an Anti-Thrombotic Small-Diameter Vascular Graft with Commercially Available e-polytetrafluoroethylene Graft in a Porcine Carotid Model." *Tissue Eng Regen Med.*

Li, N., et al. (2022). "Reinforcement of Frictional Vibration Noise Reduction Properties of a Polymer Material by polytetrafluoroethylene Particles." *Materials* (Basel) 15(4).

Metcalf, R., et al. (2022). "Quantifying the importance of plastic pollution for the dissemination of human pathogens: The challenges of choosing an appropriate 'control' material." *Sci Total Environ* 810: 152292.

Samuel L. Madorsky. (1952). "Rates of Thermal Degradation of Polystyrene and Polyethylene in a Vacuum." *Journal of Polymer Science* (2): 133–156.

Williams, A. T. and N. Rangel-Buitrago (2022). "The past, present, and future of plastic pollution." *Mar Pollut Bull* 176: 113429.

Williams, et al. (1964). "*Thermal Degradation of Organic Polymers*." Interscience.

Yang Zhixin. (1988). "A Study of Regenerating Scrap Material of Polytetrafluoroethylene(polytetrafluoroethylene)." *Materials for Mechanical Engineering* 2(65): 58–60.

Yuan Zongyang, et al. (2013). "Prospect of Recycling of polytetrafluoroethylene and its Composites." *Adhesion* (05): 74–77.

Energy Revolution and Chemical Research – Kok-Keong Chong and Zhongliang Liu (Eds)
© 2023 The Authors, ISBN: 978-1-032-36554-1

Production of ethanol and 1,3-propanediol from raw glycerol by a newly isolated *Klebsiella pneumoniae* from intertidal sludge

Lili Jiang
Liaoning Key Laboratory of Chemical Additive Synthesis and Separation, Yingkou Institute of Technology, Liaoning Yingkou, China

Baowei Zhu & Changqin Li
School of Chemical and Environmental Engineering, Yingkou Institute of Technology, Liaoning Yingkou, China

ABSTRACT: The microorganism that could convert biodiesel-derived raw glycerol to ethanol and 1,3-propanediol (1,3-PDO) was isolated from intertidal sludge. The influence of the initial raw glycerol concentration on the products was analyzed in the batch fermentations with bioreactors. The ability of the isolated strain to convert raw glycerol to ethanol and 1,3-PDO was evaluated in the fed-batch fermentation. The newly isolated strain Y5 from intertidal sludge, identified as *Klebsiella pneumoniae*, could convert raw glycerol to ethanol and 1,3-PDO at high concentration and productivity levels. The highest concentrations of ethanol and 1,3-PDO were 35.04 g L^{-1} and 28.48 g L^{-1} after 36 hrs of cultivation, respectively, in fed-batch fermentation with raw glycerol. The yields and productivities of ethanol and 1,3-PDO were 0.50 mol mol^{-1}, 0.97 g $L^{-1}h^{-1}$, and 0.24 mol mol^{-1}, 0.79 g $L^{-1}h^{-1}$, respectively. Simultaneously, lactate (21.53 g L^{-1}), succinate (4.80 g L^{-1}), acetate (2.48 g L^{-1}), and 2,3-butanediol (1.25 g L^{-1}) were also produced under these conditions. The isolated bacterium effectively converted raw glycerol to ethanol and 1,3-PDO. The findings of this work demonstrated the potential use of this strain for the bioconversion of raw glycerol into value-added products.

1 INTRODUCTION

Biodiesel production has grown rapidly over the last decades since it is an alternative fuel to fossil fuels. At the same time, raw glycerol has exceeded its availability (Anuar & Abdullah 2016). Impurities are unfavorable to the utilization of raw glycerol, and they are also economically unviable for glycerol purification. The disposal or treatment of raw glycerol restricts the development of the biodiesel industry. Therefore, developing processes to directly convert raw glycerol into value-added compounds is sustainable for biodiesel production and the environment (Silva et al. 2020). Owing to the insufficiency of efficient biocatalysts, the biorefinery of raw glycerol has not been feasibly applicable in industry.

Raw glycerol can be utilized and produce biochemicals, such as 1,3-PDO, 2,3-butanediol (2,3-BD), ethanol, and other organic acids, by microbial fermentation, and this process has received increasing attention (Jiang et al. 2021; Monteiro et al. 2018). Among these valuable biochemicals, 1,3-PDO and ethanol have received considerable interest due to their potential applications. 1,3-PDO is widely used in foods, medicines, paints, cosmetics, and lubricants. There is a renewed demand for 1,3-PDO based on the development of polytrimethylene terephthalate (PTT) (Jiang et al. 2017; Zhou et al. 2017). Ethanol is a promising biofuel substitute. Additionally, since the cost of producing ethanol from raw glycerol is approximately 40% lower than that of producing ethanol

 DOI 10.1201/9781003332657-87

from lignocellulosic biomass, the production of ethanol from raw glycerol has gained increased attention (Zhao et al. 2018).

In this study, a novel strain was isolated from intertidal sludge, and its metabolites and their role in the conversion of raw glycerol were analyzed. Based on its metabolite properties, this novel strain was used to produce high-value products from raw glycerol.

2 METHODS AND MATERIALS

2.1 *Media*

The media of seeds and fermentation were determined according to previously described methods (Jiang et al. 2017). Seed medium supplemented with 1.5% agar was used as the solid medium. However, different concentrations of pure or raw glycerol were utilized in the fermentation medium. The composition of raw glycerol was the same according to the research findings by Zhou et al. (2017).

2.2 *Isolation of microorganisms*

Intertidal sludge was collected at the Dalian seashore. Two grams of sample were inoculated into a 250 mL flask sealed with a cotton cap which contained 50 mL seed medium at 37 °C, 200 rpm for 24 hrs. Then, 50 μL of the seed broth diluted 10^5-fold was inoculated on a solid medium at 37°C for 24 hrs. A single colony was selected, and isolated strains were subsequently transferred to a fermentation medium with raw glycerol (40 g L^{-1}).

2.3 *Cultivation conditions*

Seed culture (100 mL) was performed by incubation at 37°C, 200 rpm for 12 hrs in a 250 mL sealed flask. Then, seed cells were inoculated at 10% (v v^{-1}) for batch and fed-batch cultures.

Batch and fed-batch fermentation cultures were carried out according to Jiang et al. (2017). In the fed-batch fermentation, the glycerol level was about 20 g L^{-1} in the fermenter by manually adding raw or pure glycerol during cultivation.

2.4 *Analytics*

Dry cell weight (DCW) was calculated according to Mu et al. (2006). The concentrations of glycerol, ethanol, 1,3-PDO, 2,3-BD, lactate, acetate, succinate, and formate were analyzed by HPLC according to Jiang et al. (2017). The averages of two independent experiments under the same conditions were used as the results.

2.5 *Microbial identification*

The 16S rDNA of the isolated strain was amplified with the primers 1540R: 5′-AGGAGGTGATCCAGCCGCA-3′ and 7F: 5′- CAGAGTTTGATCCTGGCT-3′ and sequenced by Sangon Biotech (Shanghai, China). The 16S rDNA sequences of the isolated strain were submitted to GenBank (accession number KR259313).

3 RESULTS AND DISCUSSION

3.1 *Screening of microorganisms utilizing raw glycerol*

Among the samples tested, 3 microbial strains (named W3, Y5, and Y1) were isolated based on the different types of colonies on solid medium plates. In batch cultures with raw glycerol (40 g

L^{-1}glycerol), strain Y5 produced the highest ethanol amount (10.89 g L^{-1}), with the production of 7.56 g L^{-1} 1,3-PD, 3.22 g L^{-1} lactic acid, 1.23 g L^{-1} acetic acid, 0.99 g L^{-1} succinic acid and 0.34 g L^{-1} formic acid after 9 hrs. Based on its excellent ability to convert raw glycerol into ethanol, strain Y5 was investigated furtherly. Strain Y5 was identified to be *Klebsiella pneumoniae* (99% confidence for strain Y5) based on the 16S rRNA gene sequences. *K. pneumoniae*, as a versatile bacterium, has been employed to convert glycerol to ethanol and 1,3-PDO production (Morcelli et al. 2018; Sun et al. 2018).

3.2 *Batch fermentation with raw glycerol by K. pneumoniae Y5*

The raw glycerol utilization of *K. pneumoniae* Y5 was evaluated with different amounts of initial raw glycerol (20-120 g L^{-1} glycerol). The effects of raw glycerol on the growth of *K. pneumoniae* Y5 are shown in Figure 1. The specific growth rate dramatically declined when the glycerol content was 60 g L^{-1}, and when the concentration of glycerol reached 120 g L^{-1}, the growth of the strain scarcely increased. The inhibitory effect was attributed to the increased osmotic pressure and impurities, such as salts and heavy metals, of the raw glycerol in the fermentation broth (Jiang et al. 2017; Won et al. 2011).

The effects of the raw glycerol content on metabolites in the batch fermentations with *K. pneumoniae* Y5 are shown in Table 1. The main products in all batch fermentations were ethanol and 1,3-PDO. However, the concentration of lactic acid increased significantly when the glycerol concentration was above 40 g L^{-1}. The highest concentrations of ethanol (20.14 g L^{-1}) and 1,3-PDO (21.52 g L^{-1}) were achieved with yields of 0.41 mol mol^{-1} and 0.27 mol mol^{-1}, respectively, at an initial glycerol concentration of 100 g L^{-1}. At the same time, lactate (20.67 g L^{-1}) was the primary byproduct with succinate (4.66 g L^{-1}) and acetate (1.32 g L^{-1}). The results demonstrated that *K. pneumoniae* Y5 had good substrate tolerance in the batch fermentation with raw glycerol, which is consistent with the results of Kumar and Park (2017), who reported that *K. pneumoniae* had better tolerance to high substrate concentrations since its metabolism was not inhibited by impurities in raw glycerol.

The yield of ethanol and 1,3-PDO decreased with increasing raw glycerol concentration. The highest yields of ethanol and 1,3-PDO were 0.56 mol mol^{-1} and 0.43 mol mol^{-1}, respectively, at an initial glycerol concentration of 20 g L^{-1}. This result indicated that *K. pneumoniae* Y5 was competent in converting the majority of raw glycerol to ethanol and 1,3-PDO, and a small amount of raw glycerol was used to produce organic acids. Other authors have also reported that glycerol concentration is the key factor affecting the increased production of ethanol or 1,3-PDO (Jiang et al. 2017; Nunthaphan et al. 2016; Won et al. 2011). The content of glycerol lower than 40 g L^{-1} was considered beneficial to the production of the goal products (Metsoviti et al. 2012; Morcelli et al. 2018; Silva et al. 2020).

Glycerol, which serves as a carbon source, can be utilized by *Klebsiella* under anaerobic conditions through a coupled oxidation-reduction process and has been reviewed in detail by Ahrens et al. (1998). Glycerol is oxidized to dihydroxyacetone (DHA) by glycerol dehydrogenase (GDH) in the oxidative pathway. And then, DHA is conveyed to pyruvate by a series of enzymes. Finally, pyruvate is transformed into various fermentation products, such as ethanol, lactic acid, and 2,3-BD, in addition to energy adenosine triphosphate (ATP) and reducing equivalent (NADH). Glycerol is dehydrated to 3-hydroxypropionaldehyde (3-HPA) in the reductive pathway. Then, 1,3-PDO is produced from 3-HPA with the regeneration of oxidized NAD (NAD^+). The synthesis of ethanol and lactic acid varies according to the microbial species. As shown in Table 1, an increase in lactic acid concentration with an increasing initial glycerol concentration was observed. The results suggested that a high concentration of substrate was conducive to the synthesis of lactic acid, which reduced the yield of ethanol and 1,3-PDO. There have been some strategies reported in the literature that increase ethanol or 1,3-PDO production by deleting the lactate dehydrogenase gene to inhibit lactic acid production (Oh et al. 2012; Rhie et al. 2019; Yang et al. 2007).

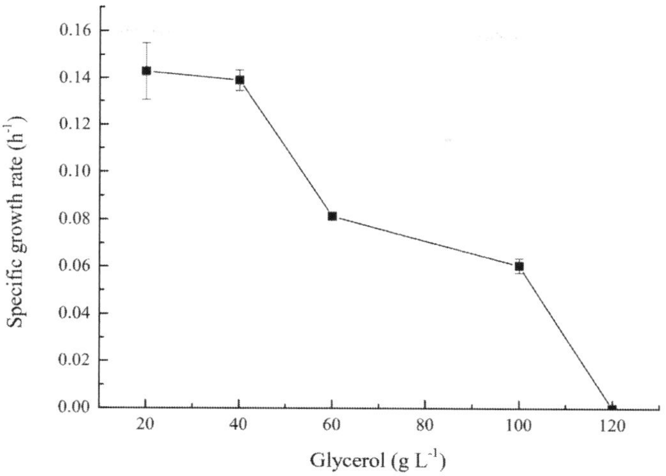

Figure 1. Effect of initial raw glycerol concentration on cell growth of *K. pneumoniae* Y5.

Table 1. Effect of initial raw glycerol concentration on ethanol and 1,3-PDO production in batch fermentations with *K. pneumoniae* Y5.

Glycerol (g L^{-1})	Time (h)	Ethanol (g L^{-1})	1,3PDO (g L^{-1})	By-products (g L^{-1}) Lactate	Succcinate	Formate	Acetate	$Y_{Ethanol}$ (mol mol^{-1})	$Q_{Ethanol}$ (g $L^{-1}h^{-1}$)	$Y_{1,3-PDO}$ (mol mol^{-1})	$Q_{1,3-PDO}$ (g $L^{-1}h^{-1}$)
~20	6.20	5.07	4.89	1.44	0.60	1.64	1.43	0.56	0.82	0.43	0.79
~40	9.00	10.89	7.56	3.22	0.99	0.34	1.23	0.54	1.21	0.24	0.84
~60	13.50	9.88	8.83	8.66	1.70	1.73	4.01	0.49	0.67	0.20	0.65
~100	26.00	20.14	21.52	20.67	4.66		1.32	0.41	0.77	0.27	0.83

3.3 *Fed-batch fermentation with raw glycerol by K. pneumoniae Y5*

To avoid high substrate inhibition, reduce the production of lactic acid, and obtain high ethanol and 1,3-PDO conversion rate, fed-batch fermentation with raw glycerol was performed to evaluate the ability of *K. pneumoniae* Y5 to convert raw glycerol to ethanol and 1,3-PDO. And the results are presented in Figure 2. The main products were ethanol (35.04 g L^{-1}) and 1,3-PDO (28.48 g L^{-1}) with concomitant lactate (21.53 g L^{-1}), succinate (4.80 g L^{-1}), acetate (2.48 g L^{-1}), and 2,3-BD (1.25 g L^{-1}) in the fed-batch fermentation by *K. pneumoniae* Y5 with raw glycerol. 2,3-BD was detected only in the late period of fermentation.

To further investigate the effect of raw glycerol on fermentation by *K. pneumoniae* Y5, fed-batch fermentation with pure glycerol was carried out. A comparison of the products in the fed-batch fermentation by *K. pneumoniae* Y5 with two types of glycerol is summarized in Table 2. As shown in Table 2, lactic acid was the main byproduct in all fed-batch fermentation by *K. pneumoniae* Y5, and in comparison, pure glycerol and raw glycerol were more conducive to ethanol and lactic acid production. The concentration, yield, and productivity of ethanol increased from 33.67 g L^{-1}, 0.49 mol mol^{-1}, and 0.94 g $L^{-1}h^{-1}$ to 35.04 g L^{-1}, 0.50 mol mol^{-1}, and 0.97 g $L^{-1}h^{-1}$, respectively, at the end of the fed-batch fermentation with raw glycerol compared that with pure glycerol. At the same time, the concentration, yield, and productivity of 1,3-PDO decreased from 30.31 g L^{-1}, 0.26 mol mol^{-1}, and 0.84 g $L^{-1}h^{-1}$ to 28.48 g L^{-1}, 0.24 mol mol^{-1}, and 0.79 g $L^{-1}h^{-1}$, respectively. The decline in 1,3-PDO production was most likely influenced by lactic acid, whose concentration increased from 17.42 g L^{-1} to 21.53 g L^{-1}. The effect of raw glycerol on 1,3-PDO production was consistent with the results of Mu et al. (2006), who reported that 1,3-PDO (51.3 g L^{-1}) from

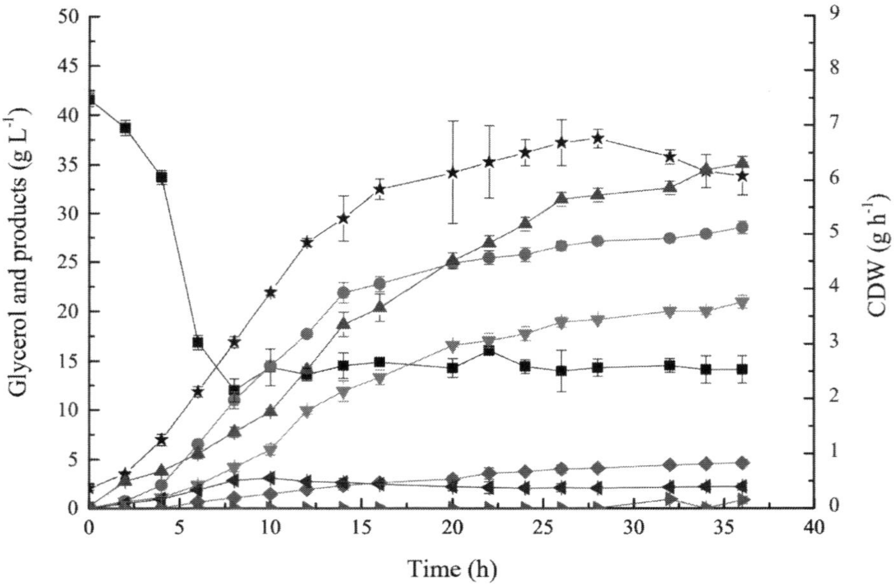

Figure 2. CDW, glycerol, metabolites in fed-batch fermentations with raw glycerol by *K. pneumoniae* Y5. (■) glycerol, (●) 1,3-PDO, (▲) ethanol, (▼) lactate, (◆) succinate, (◀) acetate, (▶) 2,3-BD, (★) CDW.

raw glycerol was lower than that (61.9 g L^{-1}) from pure glycerol. The type of glycerol had little effect on other byproducts, such as succinate, formate, acetate, and 2,3-BD. In addition, similar products were obtained with pure glycerol (Table 2), which showed that *K. pneumoniae* Y5 had good adaptability to raw glycerol. Therefore, raw glycerol as a substrate is preferred according to the experimental results obtained in this work.

Table 2. Products from different types of glycerol in fed-batch fermentation by *K. pneumoniae* Y5.

Glycerol	Ethanol (g L^{-1})	1,3-PDO (g L^{-1})	By-products (g L^{-1}) Lactate	Succinate	Formate	Acetate	2,3-BD	$Y_{Ethanol}$ (mol mol^{-1})	$Q_{Ethanol}$ (g $L^{-1}h^{-1}$)	$Y_{1,3-PDO}$ (mol mol^{-1})	$Q_{1,3-PDO}$ (g $L^{-1}h^{-1}$)
Pure	33.67	30.31	17.42	5.05	0.76	2.23	2.40	0.49	0.94	0.26	0.84
Raw	35.04	28.48	21.53	4.80	0	2.48	1.25	0.50	0.97	0.24	0.79

Lactic acid can be synthesized for sustainable polylactic acid (PLA) and biodegradable and biocompatible plastics, as an important monomer and is extensively applied in the medical, food, and cosmetic industries (Eiteman & Ramalingam 2015). There has been an increase in the literature on the production of lactic acid from raw glycerol. The simultaneous production of lactic acid with 1,3-PDO or ethanol is considered acceptable due to the increased conversion of raw glycerol and the ease of separating compounds from each other (Jiang et al. 2021; Xin et al. 2017).

An interesting result obtained in this study was that the raw glycerol was converted to ethanol with a significant amount of 35.05 g L^{-1} in the fed-batch fermentation; this concentration is comparable to the maximum values reported (Table 3). Studies on improving ethanol production yield or intensity in the literature mainly focus on the following factors: the first factor is screening wild strains and optimizing fermentation conditions. The production of ethanol is affected by the carbon source, substrate concentration, fermentation model, pH, dissolved oxygen, metabolites, temperature, agitation speed, time, and other factors (Adnan et al. 2014; Ju et al. 2020; Morcelli et al. 2018; Nwachukwu et al. 2013; Stepanov & Efremenko 2017;). Nwachukwu et al. (2013) optimized the culture conditions for converting glycerol into ethanol by *Enterobacter aerogenes* S012. Under optimum conditions, the maximum amount of ethanol was 25.4 g L^{-1}, 0.53 g $L^{-1}h^{-1}$, and 1.12 mol mol^{-1}, respectively. The second factor is using genetically engineered bacteria to inhibit the production of byproducts or overexpress the *Adh*E gene (Laxmi et al. 2015; Oh et al. 2012; Silva et al. 2020). Laxmi et al. (2015) improved ethanol production from the metabolic engineering of *E. aerogenes* ATCC 29007. The highest production and yield of ethanol were 38.32 g L^{-1} and 0.96 mol mol^{-1}, respectively, based on the combination of D-*lactate dehydrogenase* (*ldh*A) gene deletion and *alcohol dehydrogenase* (*adh*E) gene overexpression. The third factor is improving the consumption rate of raw glycerol and reducing the fermentation cost by using a microbial community (Nunthaphan et al. 2016).

Table 3. Ethanol production by microorganisms cultivated with glycerol.

Culture	Microorganism	Fermentation type	Glycerol type	Ethanol (g L^{-1})	Yield (mol mol^{-1})	Productivity (g $L^{-1}h^{-1}$)	References
Wild	*Enterobacter aerogenes* S012	Batch	Pure	25.4	1.12	0.53	Nwachukwu et al. (2013)
	Escherichia coli SS1	Batch	Pure	15.72	0.89	0.13	Adnan et al. (2014)
	Pachysolen tannophilus	Batch	Raw	8.3	0.66	0.40	Stepanov and Efremenko (2016)
	K.pneumoniae BLh-1	Fed-batch	Raw	13.2	0.095	0.49	Morcelli et al. (2018)
	K. aerogenes ATCC 29007	Batch	Raw	15.89	0.97	0.83	Ju et al. (2020)
	K. pneumoniae Y5-BH	Fed-batch	Raw	35.04	0.50	0.97	This work
Recombinant	*K. pneumoniae* GEM167	Fed-batch	Raw	31.0	0.89	1.2	Oh et al. (2012)
	E. aerogenes SUMI014	Serum bottles	Pure	38.32	0.51	0.49	Laxmi et al. (2015)
	K.pneumoniae BLh-1	Fed-batch	Raw	17.30	0.08	0.59	Silva et al. (2020)
Community	Anaerobic granule	Serum bottles	Pure	11.1	0.81	0.34	Nunthaphan et al. (2016)

In conclusion, *K. pneumoniae* Y5 was isolated from intertidal sludge, and its ability to produce value-added products, such as ethanol and 1,3-PDO, from raw glycerol was investigated in batch and fed-batch fermentations. The maximum concentrations of 35.05 g L^{-1} ethanol and 28.48 g L^{-1} 1,3-PDO were achieved from raw glycerol in the fed-batch fermentation. The main byproduct lactate was also produced at a high concentration (21.53 g L^{-1}) under the same conditions. Future work will be conducted to determine the optimum culture conditions for the yield of ethanol and 1,3-PDO production by *K. pneumoniae* Y5. In addition, since lactic acid is a major metabolic byproduct, the production of ethanol and 1,3-PDO can be further improved by inhibiting the synthesis of lactic acid through genetic engineering. This study provides a promising biocatalyst for biodiesel-derived glycerol biorefineries.

ACKNOWLEDGMENT

This study was supported by the Liaoning Natural Science Foundation Program, China (Grant No. 2019BS238); the Foundation of Liaoning Key Laboratory of Chemical Additive Synthesis and Separation (Grant No. ZJNK2103); the Program for Excellent Talents of Science and Technology in the Yingkou Institute of Technology (Grant No. RC201906); and Liaoning Province Regional Innovation Joint Fund (Grant No. 2020-YKLH-29).

REFERENCES

Adnan, N.A.A. (2014) Optimization of bioethanol production from glycerol by *Escherichia coli* SS1. *Renew Energy* 66, 625–633.

Ahrens, K. (1998) Kinetic, dynamic, and pathway studies of glycerol metabolism by *Klebsiella pneumoniae* in anaerobic continuous culture: III. Enzymes and fluxes of glycerol dissimilation and 1,3-propanediol formation. *Biotechnol Bioeng* 59, 544–552.

Anuar, M.R. (2016) Challenges in biodiesel industry with regards to feedstock, environmental, social and sustainability issues: A critical review. *Renew Sustain Energy Rev* 58, 208–223.

Eiteman, M.A. (2015) Microbial production of lactic acid. *Biotechnol Lett* 37, 955–972.

Jiang, L.L. (2021) Production of 1,3-propanediol and lactate from crude glycerol by a microbial consortium from intertidal sludge. *Biotechnol Lett* 43, 711–717.

Jiang, L.L. (2017) High tolerance to glycerol and high production of 1,3-propanediol in batch fermentations by microbial consortium from intertidal sludge. *Eng Life Sci* 17, 635–644.

Ju, H.L. (2020) Significant impact of casein hydrolysate to overcome the low consumption of glycerol by *Klebsiella aerogenes* ATCC 29007 and its application to bioethanol production. *Energy Convers Manag* 221, 113181.

Kumar, V. (2017) Potential and limitations of *Klebsiella pneumoniae* as a microbial cell factory utilizing glycerol as the carbon source. *Biotechnol Adv* 36, 150–167.

Laxmi, P.T. (2015) Improved bioethanol production from metabolic engineering of *Enterobacter aerogenes* ATCC 29007. *Process Biochem* 47, 2051–2060.

Metsoviti, M. (2012) Production of 1,3-propanediol, 2,3-butanediol and ethanol by a newly isolated *Klebsiella oxytoca* strain growing on biodiesel-derived glycerol-based media. *Proc Biochem* 47, 1872–1882.

Monteiro, M.R. (2018) Glycerol from biodiesel production: technological paths for sustainability. *Renew Sust Energ Rev* 88, 109–122.

Morcelli, A. (2018) Exponential fed-batch cultures of *Klebsiella pneumoniae* under anaerobiosis using raw glycerol as a substrate to obtain value-added bioproducts. *J Braz Chem Soc* 29, 2278–2286.

Mu, Y. (2006) Microbial production of 1,3-propanediol by *Klebsiella pneumoniae* using crude glycerol from biodiesel preparations. *Biotechnol Lett* 28, 1755–1759.

Nunthaphan, V. (2016) Microbial dynamics in ethanol fermentation from glycerol. *Int J Hydrogen Energy* 41, 15667–15673.

Nwachukwu, R.E.S. (2013) *Optimization of cultural conditions for conversion of glycerol to ethanol by Enterobacter aerogenes* S012. AMB Express 3, 12.

Oh, B.R. (2012) Enhancement of ethanol production from glycerol in a *Klebsiella pneumoniae* mutant strain by the inactivation of lactate dehydrogenase. *Process Biochem* 47, 156–159.

Rhie, M.N. (2019) Recent advances in the metabolic engineering of *Klebsiella pneumoniae*: A potential platform microorganism for biorefineries. *Biotechnol Bioprocess Eng* 24, 48–64.

Silva, V.Z.D. (2020) Construction of recombinant *Klebsiella pneumoniae* to increase ethanol production on residual glycerol fed-batch cultivations. *Appl Biochem Biotechnol* 192, 1147–1162.

Stepanov, N. and Efremenko, E. (2017) Immobilised cells of *Pachysolen tannophilus* yeast for ethanol production from crude glycerol. *New Biotechnol* 34, 54–58.

Sun, Y.Q. (2018) Advances in bioconversion of glycerol to 1,3-propanediol: prospects and challenges. *Process Biochem* 71, 134–146.

Won, J.C. (2011) Ethanol production from biodiesel-derived crude glycerol by newly isolated *Kluyvera cryocrescens*. *Appl Microbiol Biotechnol* 89, 1255–1264.

Xin, B. (2017) Coordination of metabolic pathways: Enhanced carbon conservation in 1,3-propanediol production by coupling with optically pure lactate biosynthesis. *Metab Eng* 41, 102–114.

Yang, G. (2007) Fermentation of 1,3-propanediol by a lactate defificient mutant of *Klebsiella oxytoca* under microaerobic conditions. *Appl Microbiol Biotechnol* 73, 1017–1024.

Zhao, Y. (2018) Bioethanol from corn Stover À a review and technical assessment of alternative biotechnologies. *Prog Energy Combust Sci* 67, 275–291.

Zhou, J.J. (2017) Selection and characterization of an anaerobic microbial consortium with high adaptation to crude glycerol for 1,3-propanediol production. *Appl Microbiol Biotechnol* 101, 5985–5996.

Energy Revolution and Chemical Research – Kok-Keong Chong and Zhongliang Liu (Eds)
© 2023 The Authors, ISBN: 978-1-032-36554-1

Synthesis of 9-Borafluorene derivatives with steric modulation

Li Cong & Xiaodong Yin*
School of Chemistry and Chemical Engineering, Beijing Institute of Technology, Beijing, P.R. China

ABSTRACT: 9-Borafluorene compounds have unique photophysical properties and diverse reactivity, which have attracted much attention in recent years. Herein, a series of 9-borafluorene derivatives of donors modified by different steric hindrances were synthesized and their optoelectronic characters were investigated. The new compounds are light yellow solid powders with relatively high quantum yields in n-hexane solution (up to 69.0% for **BF-MesPhA**), and they have reversible reduction potentials. This work keeps a balance between retaining the intrinsic properties of borafluorene compounds and increasing their stability, and indicates the broad application prospects of borafluorene compounds in the field of optoelectronic materials.

1 INTRODUCTION

Among three-coordinate boranes, boroles have a 4π electronic structure and show anti-aromatic property according to Hückel's rule, resulting in special physical and chemical properties, such as narrow bandgap and strong Lewis acidity (Eisch et al. 1969). The 9-borafluorenes are obtained by fusing two benzene rings based on boroles, which reduce the Lewis acidity and anti-aromaticity of the boroles to some extent, and improve the chemical stability of the system (Su et al 2021). In recent years, to probe the value of 9-borafluorenes in the field of functional materials, various compounds of D-A systems equipped with 9-borafluorene units as acceptors have been researched (Figure 1). In 2002, Yamaguchi synthesized a series of dibenzoborole derivatives with diverse groups such as bithienyl, thienyl and (N, N-diphenylamino) phenyl groups at the 3,7-positions and researched their optical and physical characters (Yamaguchi et al 2002). The new π-electron compounds showed solvatochromic properties in the fluorescence spectra and can react sensitively to fluoride ions. In 2020, Marder and co-workers investigated the photophysical properties of *exo*-aryl functionalized 9-borafluorenes showing thermally activated delayed fluorescence (TADF), which involved 4-CF_3 groups at the biphenyl unit and a -H, -CF_3 or -NMe_2 group at the para positions of *exo*-aryl groups on boron (Rauch et al. 2020). Subsequently, Yin and Marder adopted a method to increase the chemical and thermal stability of 9-borafluorene compounds via introducing an electron donor into the para position of *exo*-aryl groups on boron, separated by a Bis(trifluoromethyl)benzene ring (Chen et al. 2021), and applied these compounds to OLED devices. Recently, 9-borafluorenes have shown potential applications in chemical sensing (Adams & Rupar 2015), transport materials (Chen et al. 2014) and small-molecule activation (Su & Kinjo 2019).

In this work, nitrogen-based electron donors with different steric hindrances were introduced into the para-position of the *exo*-aryl part of 9-borafluorenes. The photophysical and electrochemical properties of these compounds were deeply studied, which can build a foundation for understanding the basic attributes of borafluorene compounds of D–A systems.

*Corresponding Author: yinxd18@bit.edu.cn

DOI 10.1201/9781003332657-88

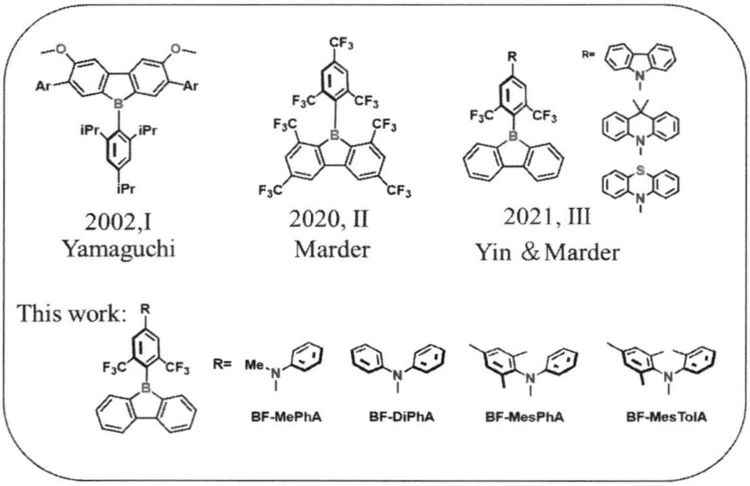

Figure 1. Selected examples of borafluorenes and those designed in this work.

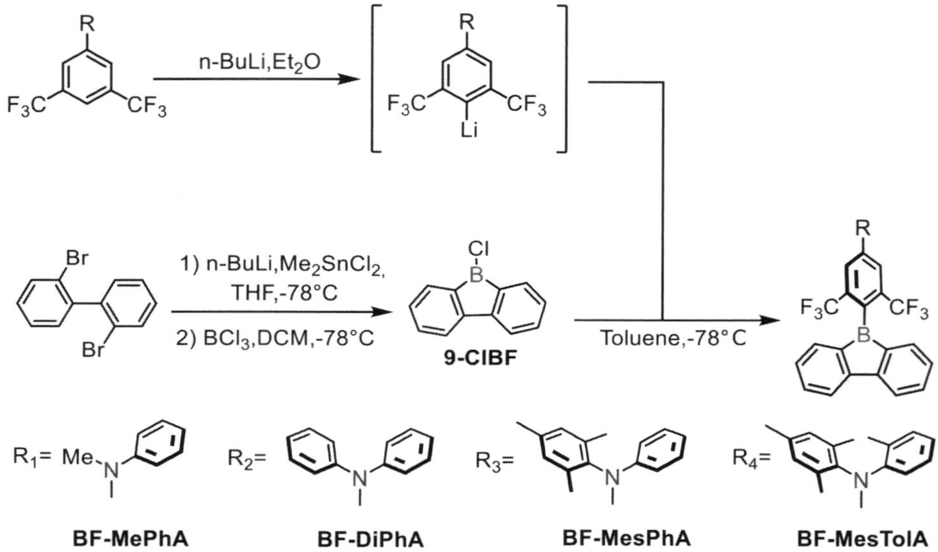

Figure 2. Synthetic route to **BF-MePhA**, **BF-DiPhA**, **BF-MesPhA** and **BF-MesTolA**.

2 METHODS AND MATERIALS

The chemicals and reagents used here are all analytical and chemically pure products on the market and were used directly without further purification unless otherwise noted. The synthetic routes of the new 9-borafluorene derivatives were depicted in Figure 2.

The 2,2'-dibromobiphenyl and n-BuLi were reacted in THF solution at $-78°C$, and Me_2SnCl_2 was added to obtain the corresponding organotin compound. The chloroborane precursor **9-ClBF** was synthesized by tin/boron exchange reactions with BCl_3 (Smith et al. 2016). The Bis(trifluoromethyl)benzene compounds with different steric hindrances were synthesized by Buchwald Hartwig coupling reactions (Cai et al. 2014; Luo et al. 2021; Zhang et al. 2016; Kathewad et al. 2019). The Bis(trifluoromethyl)benzene compounds were then reacted with n-BuLi at −78°C in Et_2O solution for 30 minutes and stirred at room temperature for another three hours to get the corresponding lithiated Bis(trifluoromethyl)benzene compounds. Then the ethyl ether in the reaction flask was removed with a vacuum pump, the corresponding lithium compounds were blended with **9-ClBF** at −78°C in dried toluene solution, and warmed to room temperature to obtain **BF-MePhA**, **BF-DiPhA**, **BF-MesPhA** and **BF-MesTolA**, with total yields of 20, 25, 29 and 27%, respectively. The target compounds are light yellow powders, with better stability towards the atmosphere, which can be purified by column chromatography.

4-(5H-dibenzo[b,d]borol-5-yl)-N-methyl-N-phenyl-3,5-Bis(trifluoromethyl)aniline (BF-MePhA) was characterized by ^{1}H NMR in Figure 3. ^{1}H NMR (400 MHz, CD_2Cl_2)δ 7.50-7.46 (m, 2H), 7.43-7.39 (m, 2H), 7.36-7.32 (m, 2H), 7.31-7.30 (m, 3H), 7.28 (s, 2H), 7.27-7.24 (m, 2H), 7.09-7.06 (m, 2H), 3.46 (s, 3H).

4-(5H-dibenzo[b,d]borol-5-yl)-N,N-diphenyl-3,5-Bis(trifluoromethyl)aniline (BF-DiPhA) was characterized by ^{1}H NMR in Figure 3. ^{1}H NMR (400 MHz, $CDCl_3$)δ 7.47 (s, 2H), 7.40–7.35 (m, 6H), 7.32 (t, J = 7 Hz, 2H), 7.27 (s, 1H), 7.25 (s, 1H), 7.22–7.17 (m, 6H), 7.06 (t, J = 7Hz, 2H).

N-(4-(5H-dibenzo[b,d]borol-5-yl)-3,5-Bis(trifluoromethyl)phenyl)-2,4,6-trimethyl-N-phenylaniline (BF-MesPhA) was characterized by ^{1}H NMR in Figure 3. ^{1}H NMR (400 MHz, $CDCl_3$)δ 7.38-7.36 (m, 4H), 7.33-7.28 (m, 4H), 7.25-7.23 (m, 2H), 7.12-7.09 (m, 2H), 7.07-7.02 (m, 3H), 7.00 (s, 2H), 2.35 (s, 3H), 2.06 (s, 6H).

N-(4-(5H-dibenzo[b,d]borol-5-yl)-3,5-Bis(trifluoromethyl)phenyl)-2,4,6-trimethyl-N-(o-tolyl)aniline (BF-MesTolA) was characterized by ^{1}H NMR in Figure 3. ^{1}H NMR (400 MHz, $CDCl_3$)δ 7.38-7.36 (m, 2H), 7.34-7.27 (m, 3H), 7.25 -7.23 (m, 1H), 7.20-7.11 (m, 3H), 7.09-7.02 (m, 2H), 7.00-6.97 (m, 4H), 6.88-6.85 (m, 1H), 2.34 (s, 3H), 2.16 (s, 3H), 2.03 (s, 6H).

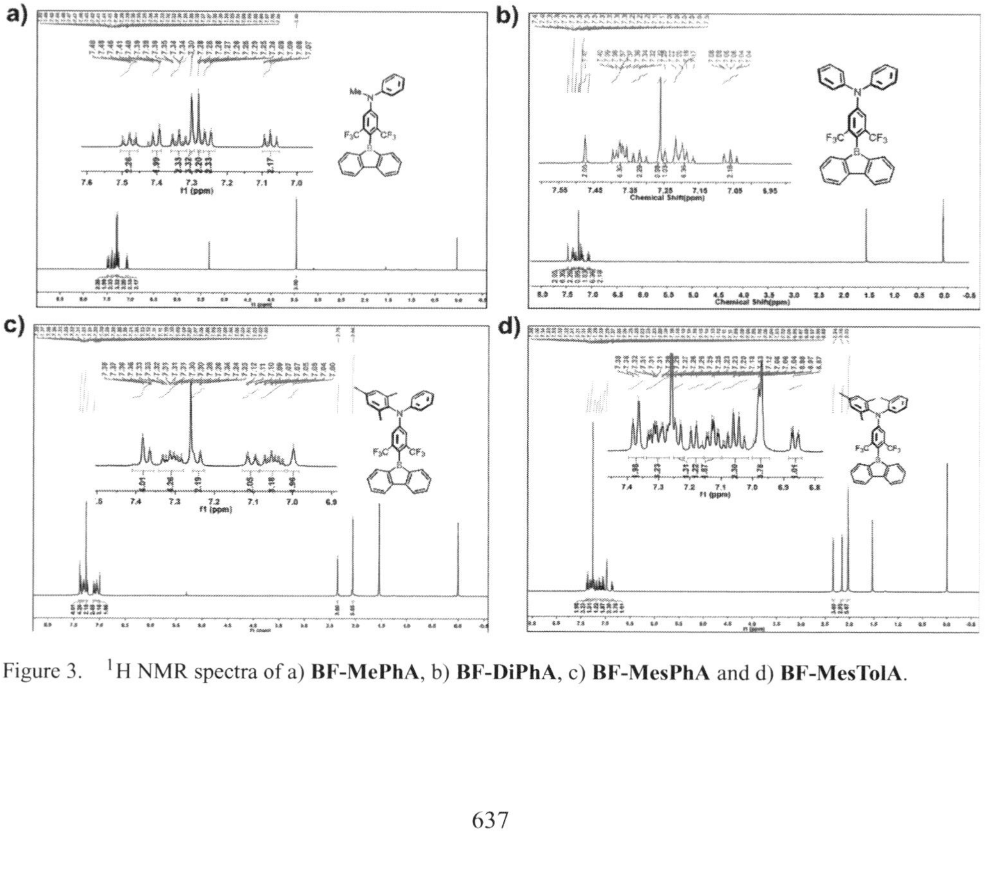

Figure 3. ^{1}H NMR spectra of a) **BF-MePhA**, b) **BF-DiPhA**, c) **BF-MesPhA** and d) **BF-MesTolA**.

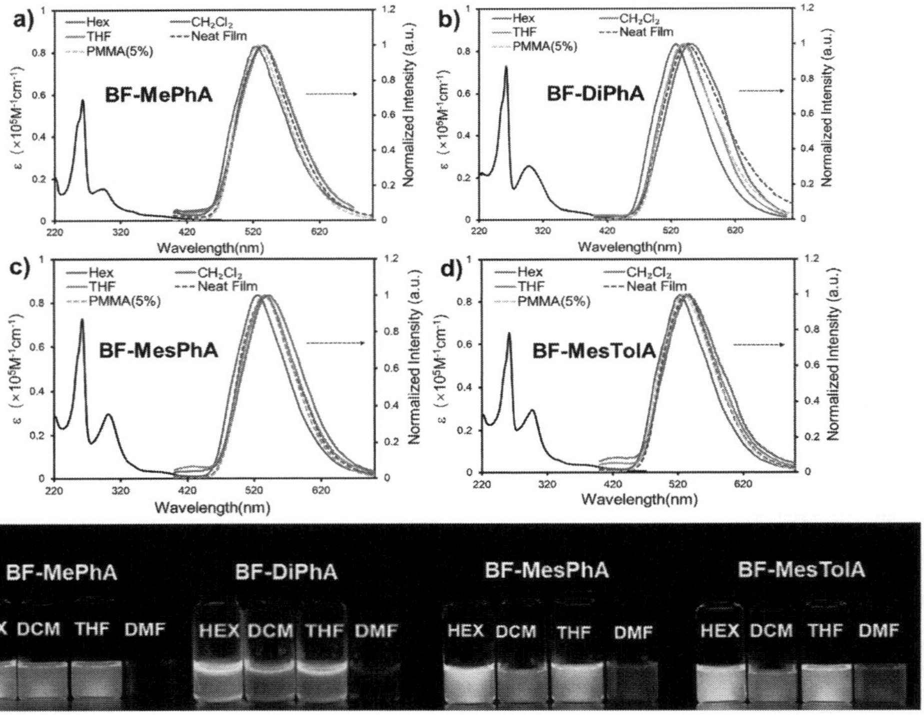

Figure 4. UV-Vis absorption spectra (black line) of functionalized borafluorenes with donors of different steric hindrances a) **BF-MePhA**, b) **BF-DiPhA**, c) **BF-MesPhA** and d) **BF-MesTolA** in CH_2Cl_2 ($c = 1 \times 10^{-5}$ M), and normalized solvent-dependent fluorescence spectra and in the solid state. e) The photo of four compounds in different solvents irradiated by 365-nm light.

In the CH_2Cl_2 solution, the four compounds' UV-Vis absorption spectra were measured. There is a strong absorption band at ca. 260 nm caused by 9-boronfluorene's local transition and a relatively weak absorption band at ca. 280–300 nm (Figure 4).

The fluorescence spectra of **BF-MePhA**, **BF-DiPhA**, **BF-MesPhA** and **BF-MesTolA** were measured in different solvents (Figure 4). In hexane, CH_2Cl_2 and THF solutions, all compounds behave with bright yellow emission. **BF-MePhA** has the maximum emission band at ca. 527 nm (photoluminescence quantum yield (PLQY) = 20.3%) in hexane. **BF-DiPhA** exhibits emission with the maximum band at ca. 527 nm (PLQY = 48.0%) in hexane. **BF-MesPhA** exhibits intense emission with the maximum band at ca. 526 nm (PLQY = 69.0%) in hexane. The fluorescence spectrum of **BF-MesTolA** exhibits emission at ca. 523 nm (PLQY = 42.7%) in hexane. Besides, the fluorescence emission spectra of all compounds in CH_2Cl_2 and THF solutions are slightly red-shifted.

The thin-film quartz plates of the four compounds were made by spin coating, and all the compounds exhibit bright yellow fluorescence. The maximum emission wavelength of **BF-DiPhA** is at ca. 550 nm, and which of the other three compounds are at ca. 535 nm (Figure 4). Among them, **BF-MesTolA** has a relatively high quantum yield of up to 49%. In addition, the quantum yields of 5 wt% of the four compounds doped in poly (methyl methacrylate) (PMMA) are relatively improved, up to 55% of **BF-DiPhA**, possibly due to the weakening of the ACQ effect.

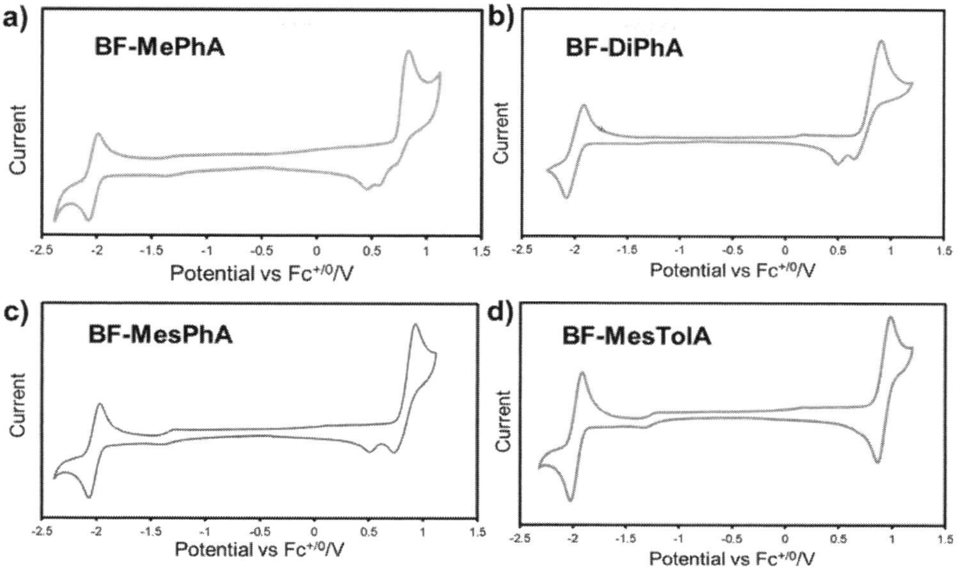

Figure 5. Cyclic voltammograms (CV) of functionalized borafluorenes with donors of different steric hindrances a) **BF-MePhA**, b) **BF-DiPhA**, c) **BF-MesPhA** and d) **BF-MesTolA**.

Utilizing n-Bu_4NPF_6 as the supporting electrolyte and a scan rate of 100 $mV{\cdot}s^{-1}$, the four compounds' cyclic voltammograms (CV) were recorded in CH_2Cl_2 (Figure 5). All compounds have reversible reductive potentials that can be considered as the reduction of boron. The half-wave reductive potentials of the four compounds ranged from -1.95 to -2.03 V versus $Fc^{+/0}$. Only **BF-MesTolA** behaves as a reversible oxidative potential with $E_{1/2}$= +0.93 V versus $Fc^{+/0}$, and the other three behave as irreversible oxidative potentials.

4 CONCLUSION

In brief, a series of 9-borafluorene derivatives featuring donors of different steric hindrances were synthesized. The four compounds exhibit good stability towards the atmosphere and exhibit intense yellowish emission in solutions and the solid state. This work shows the synthetic methods and optoelectronic properties of functionalized borafluorene compounds and offers some proof for the application of this type of compounds in novel optoelectronic materials.

REFERENCES

Adams, I.A. and P.A. Rupar, A Poly(9-Borafluorene) Homopolymer: An Electron-Deficient Polyfluorene with "Turn-On" Fluorescence Sensing of NH3 Vapor. *Macromolecular Rapid Communications*[J], 2015. 36(14): 1336–1340.

Cai, L., X. Qian, W. Song, et al., Effects of solvent and base on the palladium-catalyzed amination: PdCl2(Ph3P)2/Ph3P-catalyzed selective arylation of primary anilines with aryl bromides. *Tetrahedron*[J], 2014. 70(32): 4754–4759.

Chen, D.-M., Q. Qin, Z.-B. Sun, et al., Synthesis and properties of B,N-bridged p-terphenyls. *Chemical Communications*[J], 2014. 50(7): 782–784.

Chen, X., G. Meng, G. Liao, et al., Highly Emissive 9-Borafluorene Derivatives: Synthesis, Photophysical Properties and Device Fabrication. *Chemistry—A European Journal*[J], 2021. 27(20): 6274–6282.

Eisch, J.J., N.K. Hota, and S. Kozima, Synthesis of pentaphenyl borole, a potentially antiaromatic system. *Journal of the American Chemical Society*[J], 1969. 91(16): 4575–4577.

Kathewad, N., A. M. C, N. Parvin, et al., Facile Buchwald–Hartwig coupling of sterically encumbered substrates effected by PNP ligands. *Dalton Transactions*[J], 2019. 48(8): 2730–2734.

Luo, Y.C., F.F. Tong, Y. Zhang, et al., Visible-Light-Induced Palladium-Catalyzed Selective Defluoroarylation of Trifluoromethylarenes with Arylboronic Acids. *Journal of the American Chemical Society*[J], 2021. 143(34): 13971–13979.

Rauch, F., S. Fuchs, A. Friedrich, et al., Highly Stable, Readily Reducible, Fluorescent, Trifluoromethylated 9-Borafluorenes. *Chemistry—A European Journal*[J], 2020. 26(56): 12794–12808.

Smith, M.F., S.J. Cassidy, I.A. Adams, et al., Substituent Effects on the Properties of Borafluorenes. *Organometallics*[J], 2016. 35(18): 3182–3191.

Su, X., T.A. Bartholome, J.R. Tidwell, et al., 9-Borafluorenes: Synthesis, Properties, and Reactivity. *Chemical Reviews*[J], 2021. 121(7): 4147–4192.

Su, Y. and R. Kinjo, Small molecule activation by boron-containing heterocycles. *Chemical Society Reviews*[J], 2019. 48(13): 3613–3659.

Yamaguchi, S., T. Shirasaka, S. Akiyama, et al., Dibenzoborole-Containing π-Electron Systems:? Remarkable Fluorescence Change Based on the "On/Off" Control of the pπ $-$ π* Conjugation. *Journal of the American Chemical Society*[J], 2002. 124(30): 8816–8817.

Zhang, B., G. Chen, J. Xu, et al., Feasible energy level tuning in polymer solar cells based on broad band-gap polytriphenylamine derivatives. *New Journal of Chemistry*[J], 2016. 40(1): 402–412.

Energy Revolution and Chemical Research – Kok-Keong Chong and Zhongliang Liu (Eds)
© 2023 The Authors, ISBN: 978-1-032-36554-1

Preparation and properties of a molybdenum disulfide/graphene lubricating coating for aluminum alloy

Yujing Hu* & Chao Feng
State Grid Hunan Electric Power Company Limited Research Institute, Changsha, China

Jufang Yin, Yan Peng, Xiaolan Tao & Rong Huang
Hunan Xiangdian Experimental Research Institute Co., Ltd, Changsha, China

Yi Xie
State Grid Hunan Electric Power Company Limited Research Institute, Changsha, China

ABSTRACT: There are many problems in practical applications with aluminum alloy material, such as high surface friction coefficient, poor wear resistance, and difficulty in lubrication, which limit its use in some specific environments. Therefore, it is of great practical significance to carry out the modification of wear-resistant and lubrication on the surface of aluminum alloy materials. This work has investigated solid lubricating and antifriction coatings within molybdenum disulfide/graphene for aluminum alloy.

1 INTRODUCTION

Aluminum alloys are used widely in the automotive, aerospace, electrical power systems, and construction sectors due to their light weight, high specific strength, excellent thermal and electrical conductivity, good ductility, and corrosion resistance (Li et al. 2019). However, they have some limitations, such as a high coefficient of surface friction, poor wear resistance, and difficulty in lubrication, which restrict their applications. In special working environments, aluminum alloy surfaces with high hardness, low coefficient of friction, and even self-lubricating properties are required (Ma et al. 2017). Therefore, it is of great significance to modify the surface of aluminum alloy material for wear resistance and lubrication. Furthermore, the development of solid lubricating coatings has attracted broad attention from researchers due to their advantages, such as easy preparation, low cost, and wide application possibilities.

Solid lubrication (Chen et al. 2012) includes the solid lubricating powder, lubricating film, and some integral lubricating material, which can reduce friction and wear between bearing surfaces in relative motion, reduce energy loss in the friction process, and improve the surface quality of components. These characteristics facilitate their widespread use. The main components of lubricating and anti-friction coatings include film-forming substances, solid lubricants, auxiliary additives, solvents, etc. Among them, solid lubricants with low shear resistance play a key role in solid lubricating coatings, which are used to lubricate friction surfaces (Liu et al. 2013; Zhen et al 2017). Molybdenum disulfide (MoS_2) and graphene are two types of solid lubricants commonly used in industry (Gao et al 2016; Suarez et al 2019; Xu et al. 2018). MoS_2 is a black-gray transition metal sulfide with a flaky crystal structure similar to graphite. Due to the weak intermolecular force between layers, it is prone to external force sliding and has a good lubricating effect. However, MoS_2 has the disadvantages of poor thermal conductivity and easy oxidation at high temperatures. As

*Corresponding Author: yjhu1@mail.buct.edu.cn

DOI 10.1201/9781003332657-89

a two-dimensional carbon nano-material, graphene has the advantages of low friction coefficient, good thermal conductivity, good chemical stability, and excellent solvent resistance (Lang et al. 2018). It can be used as a lubricant additive to improve lubricity and anti-wear properties. The high thermal conductivity of graphene can effectively reduce heat accumulation during friction, and its impermeability can slow down the corrosion and oxidation of friction on the surface. Nevertheless, graphene has the problems of easy stacking, poor dispersion, and high cost.

In this paper, by modifying MoS_2/graphene particle and adopting epoxy resin as the matrix, a MoS_2/graphene lubricating coating for aluminum alloy was prepared, and the related properties of the coating film were tested.

2 EXPERIMENTAL SECTIONS

2.1 *Materials*

DGEBA epoxy resin was purchased from Baling Petrochemical. MoS_2 and graphene were provided by Qingdao Suoxin Import & Export Co., Ltd. Aminopropyl triethoxysilane was purchased from Nanjing Youpu Chemical Co., Ltd. Curing agent 593, the adduct of diethylenetriamine and butyl glycidyl ether, and ethylene glycol diglycidyl ether were purchased from Changzhou Lebang Composite Materials Co., Ltd. Tris (dimethylaminomethyl) phenol was obtained from Yu Huntsman. Ethanol, acetic acid, and N,N-dimethylformamide solvent were produced by Sinopharm Reagent Factory. Fumed silica was produced by Shanghai Shanbo Industrial Co., Ltd.

2.2 *Preparation of modified MoS_2/graphene particle (Wang et al. 2022)*

MoS_2 and aminopropyl triethoxysilane were mixed into an ethanol solution (in 1:1 w/w) at a mass ratio of 1:10, and the pH was adjusted to 3.5-5.5 by acetic acid. The solution was stirred in an oil bath at 60°C for 6 hrs. After filtration, the pretreated product of MoS_2 was obtained by washing with ethanol 3–4 times and drying at 60°C for 24 hrs. Then, the above MoS_2 product and the graphene were added into the N,N-dimethylformamide solution at a mass ratio of 3:1, and the graphene was dispersed by ultrasonic for 30 min, and then stirred in an oil bath at 105°C for 6 hrs until the mixture was uniform. Finally, the modified MoS_2/graphene was obtained by filtering, washing 3–4 times with ethanol, and drying at 60°C for 24 hrs.

2.3 *Preparation of MoS_2/graphene lubricating coatings*

The process flow of experimental sample preparation is shown in Figure 1. The coating system shall be mixed according to the following mass ratio: DGEBA epoxy resin in 100, modified MoS_2/graphene particle in 60, ethylene glycol diglycidyl ether in 30, fumed silica in 10, polyetheramine curing agent in 30, tris (dimethylaminomethyl) in 5. And the obtained sample after curing is shown in Figure 2. In addition, the comparison samples were prepared according to the same formula, except that the modified MoS_2/graphene particles were not included.

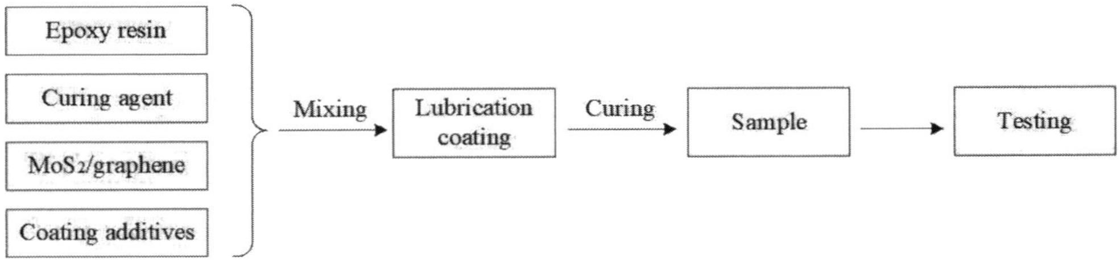

Figure 1. Schematic diagram of MoS_2/graphene lubricating coating preparation.

2.4 *Characterization*

2.4.1 *Coating adhesion*

The coating adhesion of the dry film was measured referring to the GB Standard *Paints and varnishes—Cross-cut test* (Standard code: GB/T 9286-2021). Our first step was to cut a grid pattern with a spacing of 2 mm on the paint film, and then stick it on the adhesive tape parallel to one of the cutting lines. Then, tear the adhesive tape at an angle of 60 degrees within the fixed time. Finally, the adhesion of coatings was measured by peeling tests.

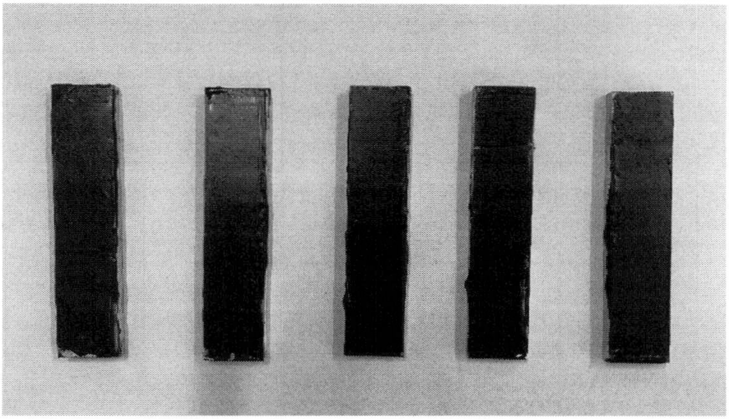

Figure 2. Photo of MoS_2/graphene lubricating coating films.

2.4.2 *Water resistance*

Water resistance of the coating film was tested referring to *Determination of resistance to water of films* (Standard code: GB/T 1733-1993). According to the following test conditions: about two-third of the film's length need to be immersed in deionized water, the test temperature was 23±2°C, and the test time was 8 hours.

2.4.3 *Impact resistance*

The impact resistance was measured by the JB-300B impact testing machine of Jinan Dingshi Testing Equipment Co., Ltd., according to *Determination of impact resistance of coating films* (Standard code: GB/T 1732-2020). A hammer falling on the coating film did not cause it to crack or fall off, as it was qualified.

2.4.4 *Shore hardness*

Referred to *Plastics and ebonite—Determination of indentation hardness by means of a duronmeter (shore hardness)* (Standard code: GB/T 2411-2008), the shore hardness of the coating was obtained by the LX-A shore rubber hardness tester of Shanghai Gaozhi Precision Instrument Co., Ltd.

2.4.5 *Friction and wear performance*

The friction coefficient and wear resistance of the coating were respectively measured according to *Plastics-Test method for friction and wear by sliding* and *JB/T 3578-2007 General technical rules for epoxy coating material on sliding lead-rail* (Standard code: GB/T 3960-2016). The friction and wear tester were produced by Lanzhou Zhongke Kaihua Technology Development Co., Ltd.

2.4.6 *Shear strength*

Referred to *Adhesives—Determination of tensile lap-shear strength of rigid-to-rigid bonded assemblies* (Standard code: GB/T 7124-2008), the shear strength of the coating film was measured at room temperature with the TA Xtc-18 adhesion tester, which is produced by Shanghai Baosheng Industrial Development Co., Ltd. The lap length of the sample is 12.5 mm ± 0.5 mm, and the thickness of aluminum alloy metal sheet is 2.0 ± 0.1 mm.

3 TEST AND RESULTS

3.1 *Effect of curing process conditions on shear strength of the coating*

The curing process of the coating film was confirmed by the shear strength of different curing times. A mixed coating of resin, modified MoS_2/graphene particles, curing agent, and coating additives was applied to the aluminum alloy sample after the resin, particles, and curing agent were mixed thoroughly. Then the sample was cured at room temperature (23°C). The shear strength of the samples with different curing times had been exhibited in Figure 3.

The test showed that with the prolongation of curing time, the shear strength of the test increased, and the shear strength reached a maximum of 23 MPa after curing for 24 hrs. After that, prolonging the curing time had no significant effect on the shear strength of the specimens. Therefore, the curing process of the coating was selected as curing at room temperature for 24 hrs.

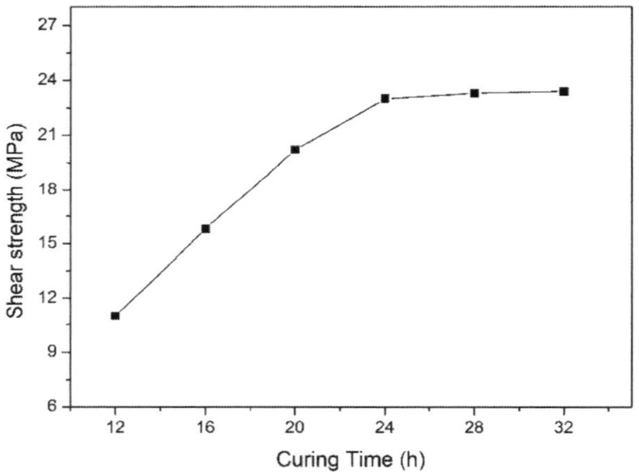

Figure 3. Variation of shear strength of the coating samples with curing time.

3.2 *Friction and wear properties of coatings*

The friction coefficient and the wear loss of the MoS_2/graphene lubricating coating and the comparison example without modified MoS_2/graphene were compared by friction and wear test. The test results are shown in Table 1. It could be found that the friction coefficient of the coating decreased by 80% after adding modified MoS_2/graphene particle, and the wear amount decreased by an order of magnitude. It showed that the use of modified MoS_2/graphene particle had played a versatile synergistic effect and achieved a significant wear reduction effect.

Table 1. The friction coefficient and the wear loss of coatings.

Samples	Friction coefficient	The wear loss (mm^3/N•m)
MoS_2/graphene lubricating coating	0.015	2.36×10^{-8}
Comparison sample	0.068	1.89×10^{-7}

3.3 *Other properties of the coating*

The other performance test results of the MoS_2/graphene lubricating coating were listed in Table 2. It was observed that the coating system could form a good adhesion layer on the aluminum alloy after complete curing, and the coating formed a three-dimensional network structure that could prevent the erosion of water providing good water resistance. The ethylene glycol diglycidyl ether introduced a flexible segment into the coating system, which improved the impact resistance and hardness of the coating to a certain extent. These results had shown the property of the coating system which could achieve the practical application requirements.

Table 2. The results of MoS_2/graphene lubricating coating films.

Test projects	Results
Coating Adhesion of dry films	Grade 0
Water resistance	Unfading
Impact Resistance	No cracking or falling off
Shore hardness (Shore D)	81
Shear strength (MPA)	23

4 CONCLUSION

In this paper, an epoxy resin lubricating coating for aluminum alloy was prepared by using modified MoS_2/graphene as a lubricant. The coating was cured at room temperature for 24 hrs to obtain excellent friction and wear resistance. And through the characterization of adhesion, water resistance, impact resistance, etc., it is demonstrated that the coating had excellent comprehensive performance, which was of great significance and practical application value for the lubrication and wear resistance modification of aluminum alloy surface.

ACKNOWLEDGMENT

This work was supported by State Grid Hunan Electric Power Company Limited (Grant No.5216A521001H).

REFERENCES

Chen F, Feng Y, Shao H, et al. Friction and Wear Behaviors of Ag/MoS_2/G Composite in Different Atmospheres and at Different Temperatures. *Tribology Letters*. 2012, 47(1):139–148.

Gao J, Li B C, Tan J W, et al. Aging of transition metal dichalcogenide monolayers. *ACS Nano*, 2016, 10(2): 2628–2635.

Haojie Lang, Yitian Peng, Xing'an Cao, Kang Yu. Dynamic Sliding Enhancement on the Friction and Adhesion of Graphene, Graphene Oxide, and Fluorinated Graphene. *ACS Applied Materials & Interface*, 2018, 10: 8124–8224.

Jiang yu Wang, Xiaofei Guo, Lei Shi, et al. Effect of molybdenum disulfide-graphene oxide (MOS_2-GO) nanohybrids on anticorrosive waterborne polyurethane acrylate coatings. *Journal of Shanghai University (Natural Science Edition)*, 2022, 28(1):121–131.

Li H , Hu Z , Hu W , et al. Forming Quality Control of an AA5182-O Aluminum Alloy Engine Hood Inner Panel. *JOM: the journal of the Minerals, Metals & Materials Society*, 2019, 71(5): 1687–1695.

Liu E, Gao Y, Jia J, et al. Friction and Wear Behaviors of Ni-based Composites Containing Graphite/Ag_2 MoO_4 Lubricants. *Tribology Letters*. 2013, 50 (3): 313–322.

Mingming MA, Feng Lian, Luping Zang, Qiukuan Xiang, Huichen Zhang. Effect of Dimple Depth on Friction Properties of Aluminum Alloy Under Different Lubrication Conditions. *Acta Metall*, 2017, 53(4): 406–414.

Suarez M P, Marques A, Boing D, et al. MoS_2 solid lubricant application in turning of AISI D6 hardened steel with PCBN tools. *Journal of Manufacturing Processes*, 2019, 47:337–346.

Xu X, Schultz T, Qin Z, et al. Microstructure and elastic constants of transition Metal dichalcogenide monolayers from friction and shear force microscopy. *Advanced Materials*, 2018, 30(39): 1803748.

Zhen J, Zhu S, Cheng J, et al. Influence of graphite content on the dry sliding behavior of nickel alloy matrix solid lubricant composites. *Tribology International*, 2017, 114(1): 322–328.

Energy Revolution and Chemical Research – Kok-Keong Chong and Zhongliang Liu (Eds)
© 2023 The Authors, ISBN: 978-1-032-36554-1

Experimental study on heat transfer of plate pulsating heat pipe with channels of different diameters at the evaporating and condensation ends and channels connected at the evaporating end

Jiale Yuan & Guowei Xiahou
School of Energy and Power Engineering, Changsha University of Science and Technology, Changsha, China

ABSTRACT: To solve the heat dissipation of electronic chips with high heat flux density and cramped space, the parallel channel of this "new heat pipe" underwent a transformation in evaporating and condensation ends with different axial diameters and evaporating ends with the radial connection. Taking the filling rate, inclination angle, and heating power as independent variables, the heat transfer performance of the new heat pipe is further studied and compared with that of the traditional heat pipe. Finally, the following conclusions are drawn: the optimal liquid filling rate of the new heat pipe is 14%, and the optimal inclination angle is 90°C; the new heat pipe's equivalent thermal conductivity is 18,004 W/(m·°C) at a wind speed of 2.0 m/s and heating power of 15 W, which is 73.3% higher than the parallel channel plate pulsating heat pipe. In general, the heat transfer performance of the new heat pipe has been greatly improved compared with the traditional heat pipe.

1 INTRODUCTION

With the advent of the era of the Internet of Things (IoT), to fulfill the demands for high-speed and miniaturization, the integration of electronic chips is increasingly higher, and the heat flux is also increasingly larger. Heat dissipation has become a restrictive factor for the advancement of electronic chips. It is predicted that the average heat flow density of chips will reach 300–500 W/cm^2 in the future, and may exceed 1000 W/cm^2 locally. To support the advent of the new era, it is urgent to invent a device that can solve the heat dissipation of high heat flow chips.

Pulsating heat pipe (PHP) is a two-phase passive heat transfer device based on the phase change phenomenon. Excellent heat transfer performance, compact construction, and no external power requirements make the PHP a promising cooling device. Under the condition of an attempt to employ PHP for chip heat dissipation, it is found that the PHP formed by capillary tubes could not achieve a favorable contact with the chip, while the PHP made of the plate type could achieve a favorable contact with the chip. Therefore, the PHP has evolved from a tube-type (Han 2016; Sun 2019) to a plate-type (Qu 2019; Wang 2011). Wang et al. draw an experimental comparison between the parallel channel type and the loop type pulsating heat pipe (Wang 2011). It is found that when acetone working fluid is adopted, the average thermal resistance of the parallel channel type at 67 W–151.5 W is 0.149°C/W. The average thermal resistance of the loop type at 30 W–103 W is 0.301°C/W; while the thermal resistance jumps to 0.730°C/W at 108 W, with a deterioration in the heat transfer, which indicates that the parallel channel type significantly outperforms the loop type. Sun Qin et al. conducted a comparative study on pulsating heat pipes with equal cross-section and gradual cross-section channels (Sun 2017). Experiments showed that the thermal resistance of the graduated cross-section pulsating heat pipe was 4.2 K/W under the condition of vertical placement and heating power of 6 W, which decreased by 0.6 K/W compared with the equal cross-section pulsating heat pipe. Chien et al. conducted a study on the impact of changes in alternate

DOI 10.1201/9781003332657-90

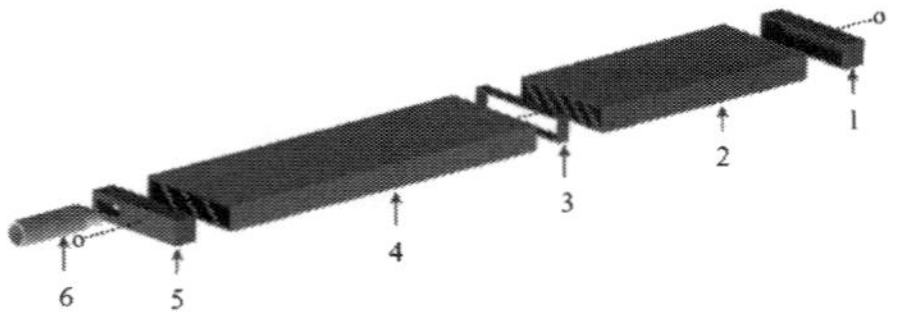

Figure 1. Schematic diagram of the new heat pipe.

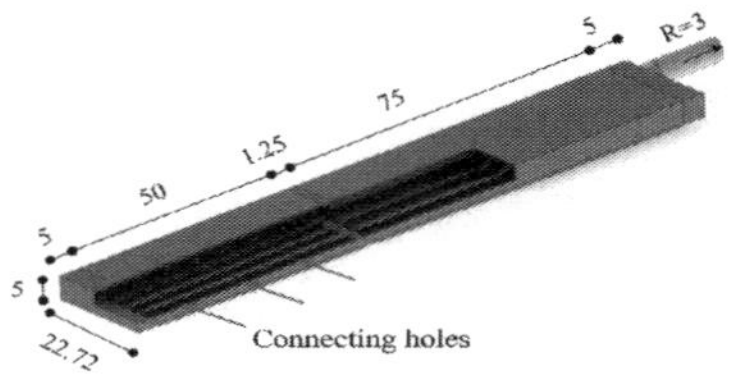

Figure 2. The size of the heat pipe.

pipe diameters on the heat transfer performance of multi-bend loop pulsating heat pipes (Chien 2012). It was found that PHP with alternate pipe diameters can compensate for the loss of gravity by introducing additional unbalanced force, and it can also enhance the start-up performance of horizontal PHP. Ebrahimi et al. set up an oblique coupling between some of the U-shaped tubes in a multi-bend circuit and found that the oblique coupling could reduce the resistance to mass flow and improve the heat transfer performance (Ebrahimi 2015).

In summary, it can be revealed (1) that the heat transfer performance of parallel channel pulsating heat pipes is better than that of loop-type channel pulsating heat pipes; and (2) that the channel change or channel connection of the pulsating heat pipe conduces to the flow of working fluid in the pipe and the reflux of condensate, which can enhance the heat transfer of the heat pipe. Therefore, this paper proposes a plate pulsating heat pipe with a novel channel structure (referred to as the "new heat pipe"). The channels of the new heat pipe are generally parallel, but the channels of the evaporating end and the condensation end are separated, and the equivalent diameters of the evaporating and condensation ends are different, while the evaporating end channels are radially connected. Based on the parallel channel, this change of channel separation, different diameters of evaporating and condensation ends, and interconnection can contribute to the reflux and heat transfer.

2 INTRODUCTION OF THE NEW HEAT PIPE

2.1 *New heat pipe structure*

As shown in Figure 1, this new heat pipe is composed of six parts, namely the connecting cavity (marked as "1"), the evaporating end (marked as "2"), the connecting cavity (marked as "3"), the condensation end (marked as "4"), the connecting cavity (marked as "5") and the liquid-filled short pipe (marked as "6"). The shape of the condensation and evaporating ends of the channel section is an approximate parallelogram. The size is shown in Figure 2. The length of the evaporating end of the optimal heat pipe is 50 mm, the equivalent diameter of the channel is 2.5 mm, and there are six parallel channels. The channel walls have six radial connecting holes at equal pitches, and the equivalent diameter of these holes is 1 mm. The length of the condensation end is 75 mm, the equivalent diameter of the channel is 3.5 mm, and there are four parallel channels.

2.2 *Evaporating and condensation end structure*

The evaporating end is composed of two oblique tooth groove plates with the same structure and length of 50 mm, which are rotated 180°C and embedded into each other. Figure 3(a) presents the oblique tooth groove plate structure of the heat pipe, and Figure 3 (b) presents the cross-sectional view of the evaporating end of the oblique tooth groove plate after mutual embedding. The equivalent diameter of the pulsating channel is 2.5 mm. In addition, to enhance the disturbance and the heat and mass exchange between the channels at the evaporating end, radial connecting holes are arranged on the wall of the channels at the evaporating end at a certain interval and equivalent diameter, which is employed to enhance the heat transfer capacity of the evaporating end.

The structure of the condensation end is similar to that of the evaporating end, consisting of two oblique tooth groove plates with the same structure and a 75 mm length, which are rotated 180°C and embedded in each other. To solve the problem of poor condensate reflux, the equivalent diameter of the cold end channel is appropriately increased on the premise of satisfying the pulsating working behavior, so that the equivalent diameter of the condensation end channel and the evaporating end channel are different. Because the structure of the condensation end is similar to that of the evaporating end, the design drawings are omitted for space-saving.

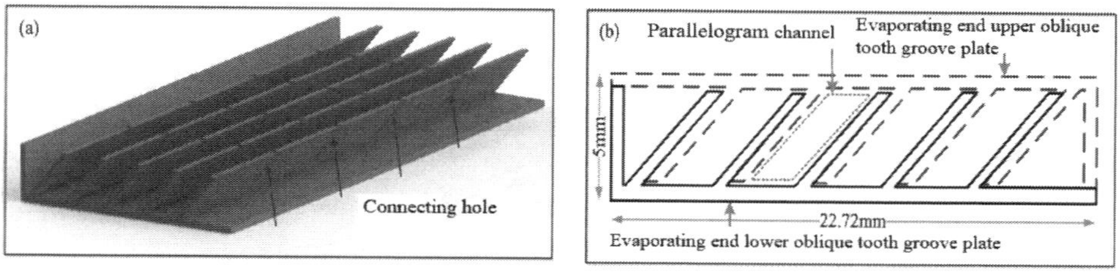

Figure 3. (a) Schematic diagram of the structure of the evaporating end oblique tooth groove plate; (b) nesting diagram of the evaporating end oblique tooth plate.

3 EXPERIMENTAL SETUP AND PROGRAM

3.1 *Experimental setup*

The heat transfer experiment of the new type of heat pipe is carried out, and the device used in the experiment is shown in Figure 4. The experimental device is composed of a power supply system, heating system, air cooling system, measurement, and data acquisition system. Among them, the power supply system is composed of an indoor 220V AC power supply and DJW-10KVA AC stabilized power supply. The heating system is composed of an auto-coupling voltage regulator, electric heating block (analog chip heating), and RK9800N digital power meter. The air-cooling system is composed of Rek-RPS6003C-2 DC stabilized power supply, a T80T12MHA7-52 cooling fan, and a cooling air duct. The measurement and data acquisition system are composed of a K-type thermocouple, Model2700 data collector, KA22 thermal anemometer, and computer. The heat pipe is equipped with 19 temperature measuring points, including six heating blocks, six evaporating ends, six condensation ends, and one room temperature monitoring point (Figure 4).

The temperature in the experiment is measured by a K-type thermocouple, with its uncertainty being ±0.2°C. Based on this, the maximum error of the thermal resistance parameter can be calculated to be 5.21%. The above uncertainties are all within the allowable error range, which shows the validity of the experiment.

3.2 *Experimental program*

The heat transfer performance of the heat pipe is tested by changing the three influencing factors of the liquid filling rate, the inclination angle (the angle between the heat pipe and the horizontal plane when the cold end is on top), and the heating power. The heat pipe heat transfer performance experiment is carried out according to Figure 5. The single variable control method is adopted in the experiment, and the total number of experimental conditions is $6 \times 7 \times 6 = 252$. The experiment is carried out with a room temperature of 18°C and a cooling wind speed of 2.5 m/s. During the experiment, all working conditions are required to reach a stable state. The criterion for the stable state is that the changes in the temperature of all measuring points shall be less than 0.1°C within 10 minutes.

To facilitate the production, the new heat pipe material is made of 304 stainless steel. The working medium is deionized water. The heat pipe needs to be vacuumed before filling the working medium, and the vacuum degree is 4.0×10^{-4} Pa.

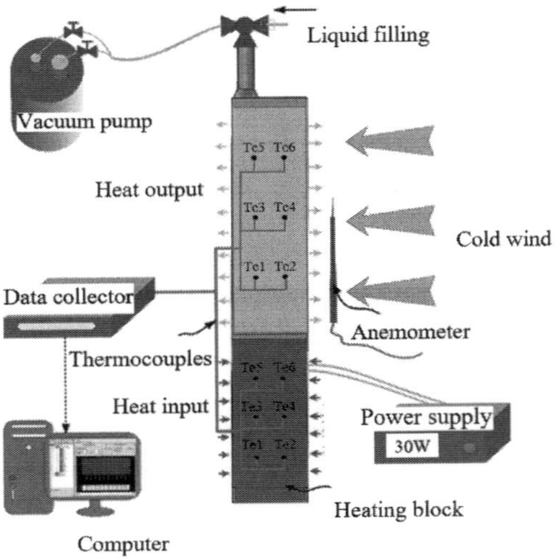

Figure 4. Layout of experimental equipment and measuring point.

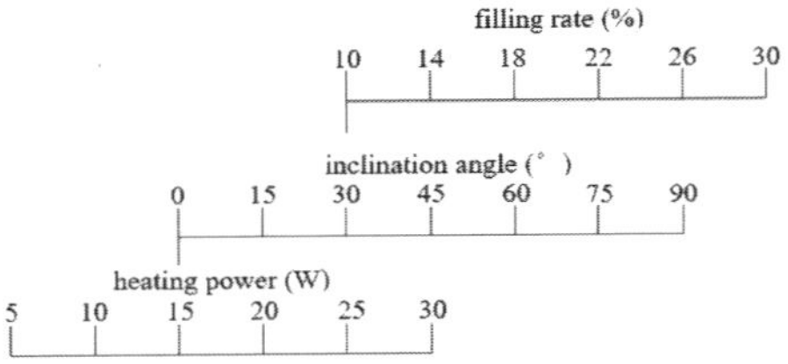

Figure 5. Experimental scheme of the new heat pipe heat transfer.

4 HEAT TRANSFER PERFORMANCE EVALUATION INDEX

Heat transfer performance is evaluated using two indicators: heat source temperature and equivalent thermal conductivity. The definitions of each indicator are as follows:

(1) Heat source temperature refers to the average temperature of each measuring point of the heat source. In general, under the same heat dissipation conditions, the heat source temperature is used to reflect the heat conductivity of the heat pipe. The better the heat conductivity, the lower the heat source temperature.
(2) Equivalent thermal conductivity is employed to evaluate the thermal conductivity of the heat pipe when it is regarded as a composite material. The better the heat transfer performance of the heat pipe, the higher the equivalent thermal conductivity, which is defined as follows:

$$T_e = \frac{1}{n}\sum\nolimits_1^n T_{ei}(i = 1, 2, \cdots, n) \tag{1}$$

$$T_e = \frac{1}{n}\sum\nolimits_1^n T_{ei}(i = 1, 2, \cdots, n) \tag{2}$$

$$Q = UI \tag{3}$$

$$K_{eff} = \frac{QL}{S\,(T_e - T_c)} \tag{4}$$

where T_e represents the average temperature of the evaporating end (°C); T_{ei} represents the temperature of each measurement point of the evaporating end (°C); T_c represents the average temperature of the condensation end (°C); T_{ci} represents the temperature of each measurement point of the condensation end (°C); Q represents the heating power (W); U represents the heating voltage (V); I represents the heating current (A); K_{eff} represents the equivalent thermal conductivity in W/(m·°C); L represents the axial length of the heat pipe (m); S represents the cross-sectional area of the heat pipe (m^2); n represents the number of evaporating end measuring points; m represents the number of condensation end measuring points.

5 HEAT TRANSFER PERFORMANCE OF THE NEW HEAT PIPE

5.1 *Liquid filling rate*

In order to analyze the influence of the filling rate on the heat transfer performance of the optimal heat pipe, the 90°C inclination angle is selected as an example for analysis. Figure 6 presents the curve of the relationship between the average temperature of the heat source and the heating power at different filling rates when the inclination angle is 90°C.

It can be seen from Figure 6 that under the same heating power, the average temperature of the heat source roughly decreases first and subsequently rises with the increase of the filling rate. The lowest temperature can be achieved by most of the heating power when the filling rate is 14%. The second-lowest temperature, slightly below the lowest temperature, can be achieved by only a small part of the heating power. Therefore, 14% can be regarded as the optimal filling rate in this experiment. Subsequent experiments will be carried out based on this filling rate.

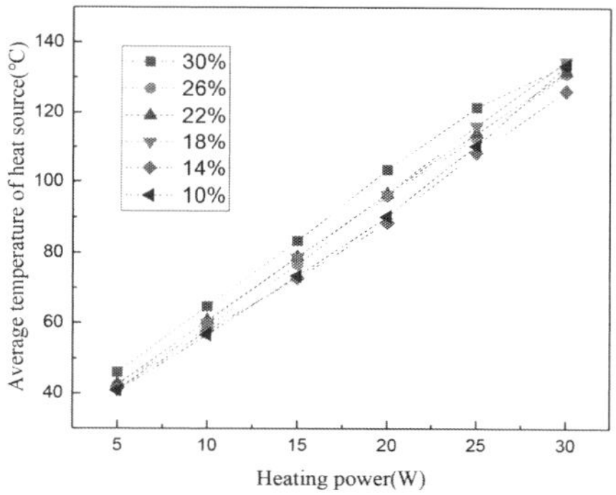

Figure 6. The cooling effect of the new heat pipe with different filling rates at an inclination angle of 90°C.

5.2 *Inclination angle and heating power*

It can be seen from Figure 7 that the thermal conductivity of the new heat pipe increases with the increase of the inclination angle, and the equivalent thermal conductivity of 90°C is the maximum at any heating power. Among them, the thermal conductivity of 90°C and 75°C is significantly higher than that of other angles, and the thermal conductivity of 90°C is significantly higher than that of 75°C. Although the equivalent thermal conductivity of other inclination angles is lower, there is little difference between them. The maximum equivalent thermal conductivity of the best heat pipe in the experimental operating range is 17,679 W/(m·°C), corresponding to a heat source temperature of 121.42°C. This shows that the thermal conductivity of the best heat pipe is extremely good.

It can also be found from Figure 7 that the equivalent thermal conductivity of the optimal heat pipe at each inclination angle increases with the increase of heating power, which is in line with the general law of heat pipe's thermal conductivity.

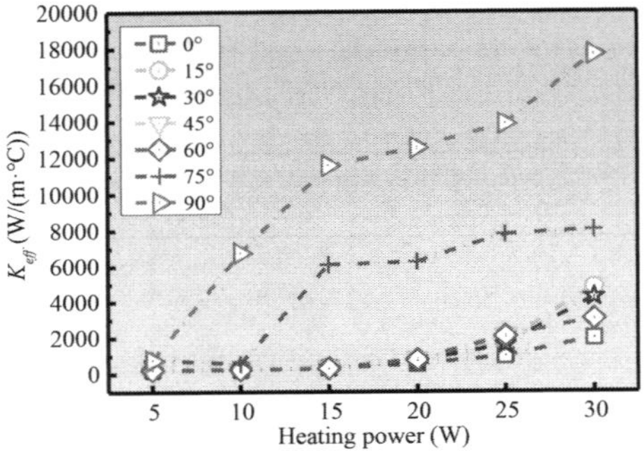

Figure 7. Variation of equivalent thermal conductivity at different inclination angles.

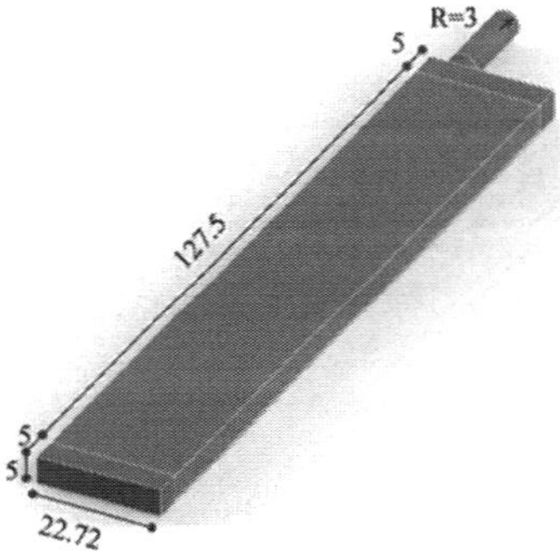

Figure 8. Structure of traditional heat pipe.

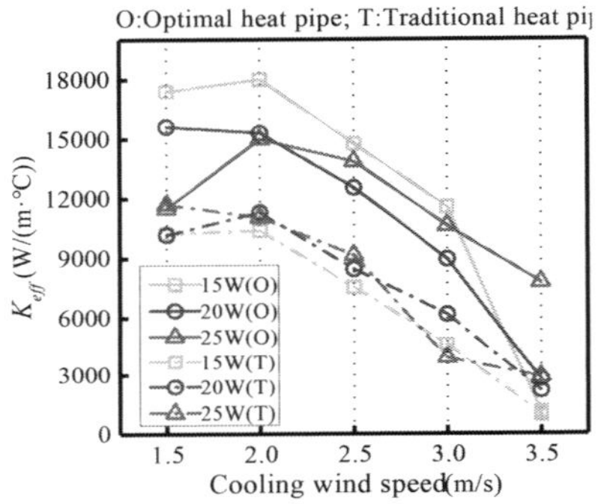

Figure 9. Variation curve of equivalent thermal conductivity of two FP-PHPs under different working conditions.

6 COMPARISON WITH PLATE-TYPE PULSATING HEAT PIPE

Comparison is made with a traditional parallel channel plate pulsating heat pipe of the same dimensions as the optimal heat pipe, which has connecting cavities 1 and 5 but not 3. There are channels with a constant diameter inside the heat pipe, the equivalent diameter of the channel is 2.5 mm, and the evaporating end has no radial connection. The structure and dimensions of the traditional heat pipe are shown in Figure 8.

To test the improvement of the heat transfer performance of the optimal heat pipe, a comparison is drawn between the optimal heat pipe and the parallel channel plate pulsating heat pipe at an inclination angle of 90°C. The equivalent thermal conductivity of both heat pipes is shown in Figure 9. It can be seen from Figure 9 that the equivalent thermal conductivity of the optimal heat pipe under the same conditions is much higher than that of the traditional one. For example, when the wind speed is 2.0 m/s and the heating power is 15 W, the equivalent thermal conductivity of the optimal heat pipe is 18,004 W/(m·°C), while that of the traditional one is only 10,391 W/(m·°C), which increases by 73.3% compared with the traditional one. The results show that the channel improvement of the new heat pipe can greatly enhance the heat transfer performance of the plate pulsating heat pipe. This innovative structural improvement is very valuable.

7 CONCLUSION

Through a summary of the above research on the new heat pipe, the following conclusions can be drawn: (1) the modification of the channel improves the heat transfer of the heat pipe; (2) the optimal filling rate of the new heat pipe is 14%; (3) the inclination angle could exert significant impacts on the heat transfer performance of this new heat pipe, and the optimal inclination angle is 90°C; (4) when the wind speed is 2.0 m/s and the heating power is 15 W, the equivalent thermal conductivity of the optimal heat pipe is 18,004 W/(m·°C), which increases by 73.3% compared with the parallel channel plate pulsating heat pipe.

REFERENCES

Chien K H, Lin Y T, Chen Y R, et al (2012). A novel design of pulsating heat pipe with fewer turns applicable to all orientations. *J. International Journal of Heat and Mass Transfer*, 55(21-22), 5722–5728.

Ebrahimi M, Shafii M B, Bijarchi M A (2015). Experimental Investigation of the Thermal Management of Flat-Plate Closed-Loop Pulsating Heat Pipes with Interconnecting Channels. *J. Applied Thermal Engineering*. 90, 838–847.

Hua Han, Xiaoyu Cui, Yue Zhu, et al (2016). Experimental study on a closed-loop pulsating heat pipe (CLPHP) charged with water-based binary zeotropes and the corresponding pure fluids. *J. Energy*. 109, 724–736.

Jian Qu, Qin Sun, Hai Wang, et al (2019). Performance characteristics of flat-plate oscillating heat pipe with porous metal-foam wicks. *J. International Journal of Heat and Mass Transfer*. 137, 20–30.

Sun Qin, Qu Jian, Yuan Jianping (2017). Comparison of heat transfer characteristics of silicon-based miniature pulsating heat pipes with constant cross-section and variable cross-section channels. *J. Journal of Chemical Industry*. 68(5), 1803–1810.

Wang Yu, Li Weiyi (2011). Experimental study on the operating characteristics of multi-channel parallel loop pulsating heat pipes. *J. Chinese Journal of Power Engineering*. 31(4), 273–278.

Xiao Sun, Sizhuo Li, Bo Jiao, et al (2019). Experimental study on a hydrogen closed-loop pulsating heat pipe with two turns. *J. Cryogenics*. 97, 63–69.

Energy Revolution and Chemical Research – Kok-Keong Chong and Zhongliang Liu (Eds)
© 2023 The Authors, ISBN: 978-1-032-36554-1

Determination of cations in paper-making reconstituted tobacco by ion chromatography

Jieyun Cai, Chunqiong Wang, Haowei Sun, Jie Long, Ke Zhang & Xiaowei Zhang
Yunnan Tobacco Quality Supervision & Test Station, Kunming, China

Chao Li*
China Tobacco Yunnan Industrial Co., Ltd., Kunming, China

ABSTRACT: To monitor the quality of paper-making reconstituted tobacco products, and evaluate the stability of Sodium, Ammonia, Potassium, Magnesium and Calcium content in paper-making reconstituted tobacco, we employ an intra-standard quantitative method for determination of these Contents in paper-making reconstituted tobacco by Ion Chromatography (IC). The LODs and the LOQs of target cations could achieve 0.011–0.028 μg/mL and 0.036–0.093 μg/mL, respectively, while the determination coefficients of all cations are above 99.7%. The Intra-assay and Inter-assay precisions measured by 15 repeat runs are 1.85–3.55% and 2.36%–4.16%, respectively. And with three different added levels, the recovery rates for the target cations lie between 95.7% and 104.6%. The research results showed that our proposed method is environmentally friendly, highly sensitive, and selective. This method is of great significance to the control of paper-making reconstituted tobacco quality.

1 INTRODUCTION

Paper-making reconstituted tobacco, also known as tobacco flake, is a regenerated product made of tobacco dust, stem, broken tobacco leaf, and other raw materials into flake or filamentous, serving as a cigarette filler (Chen 2002; Hu 2010). As an indispensable raw material in cigarette blends, paper-making reconstituted tobacco plays an increasingly prominent role in reducing cigarette cost, reducing tar and harmful components in cigarette smoke, and strengthening cigarette product style (Dai 2013; Han 2007; Miao 2009; Zhao 2014). To adapt to the change in market and brand demand in the tobacco industry, paper-making reconstituted tobacco needs to carry out technological upgrading and innovation to provide support for improving cigarette products.

The content of inorganic cation has an important effect on the quality of paper-making reconstituted tobacco products. The content of K^+ is an important index in the inspection of paper-making reconstituted tobacco products, due to its important effect on the flammability, smoldering retention, and moisture absorption of cigarettes (Yuan 1994; Yang 2011). Moderate Mg^{2+} content can keep the ash intact and not easy to scatter (Zhang 1994). A certain amount of $CaCO_3$ is usually added in the production of paper-making reconstituted tobacco to improve the appearance, improve the capacity of liquid absorption, and reduce the cost and improve the quality of smoking. The content of NH_4^+ had an important effect on the irritability and smoke strength of paper-making reconstituted tobacco (Xie 2011). Therefore, accurate determination of Sodium, Ammonia, Potassium, Magnesium, and Calcium content in paper-making reconstituted tobacco is of great significance for monitoring the quality of paper-making reconstituted tobacco products.

*Corresponding Author: super88man66@126.com

DOI 10.1201/9781003332657-91

Ion chromatography is a liquid chromatography technology developed based on ion-exchange chromatography. At present, Ion chromatography has become the preferred method for the analysis of inorganic anions and cations. In this study, the contents of Sodium, Ammonia, Potassium, Magnesium, and Calcium in paper-making reconstituted tobacco were determined by ion chromatography with plastic containers (Cai 2006).

The cations content of the paper-making reconstituted tobacco is determined by extraction of the sample into a sulfuric acid solution. Ion chromatographic analysis is used to separate ammonium ions from other cationic species. The responses of Sodium, Ammonia, Potassium, Magnesium, and Calcium ions are measured using a conductivity detector and are quantified against an external standard calibration. Results are reported as each cation in micrograms per gram of paper-making reconstituted tobacco (wet weight).

2 MATERIALS AND METHODS

2.1 *Apparatus*

Analytical balance: G204, (METTLER TOLEDO Co., Ltd.), 0.0001 g resolution; syringe filter: 0.45 μm nylon; volumetric flasks of capacities (Brand Co., Ltd.): 100 mL, 250 mL, and 1000 mL; mechanical pipettes with disposable plastic tips (Brand Co., Ltd.): 10 μl–1000 μl; laboratory shaker; polypropylene sample extraction vessels (with caps) of approximately 100 mL volume; weak cation exchange column of mid-capacity (CS12A, 250 mm $\times$ 4 mm, nonmetallic) and cationic protection column (CG12A, 50 mm $\times$ 4 mm); ion Chromatograph (IC), ICS-5000 (ThermoFisher Co., Ltd.), consisting of a conductivity detector, conductivity suppressor and data collection system; polypropylene volumetric flasks, sample flasks, and storage containers should be used to minimize sodium originating from borosilicate glassware.

2.2 *Reagents*

Na^+, NH_4^+, K^+, Mg^{2+} and Ca^{2+} mixed cationic Standard solution (National Non-ferrous Metal and Electronic Materials Analysis and Testing Center, 1000 μg/mL); sulfuric acid (H_2SO_4) > 96 % purity; methanesulfonic acid (MSA) > 99 % purity (ThermoFisher Co., Ltd.); deionized water (resistivity $\geq$ 18.2 MΩ·cm).

2.3 *Preparation of standards and extraction solution*

The steps are as follows: (1) carefully add 1.277 g of H_2SO_4 to approximately 600 mL of deionized water in a 1000 mL polypropylene volumetric flask; (2) then mix and dilute the H_2SO_4 to get a 0.025 N standards and extraction solution with sulfuric acid (0.0125 mol/liter).

2.4 *Preparation of working standards*

The preparation steps are as follows:

(1) Accurately remove 1.0 mL of mixed cationic standard solution (1000 μg/mL) into a 10 mL polypropylene volumetric flask;
(2) Then, add 0.025 N H_2SO_4 and mix thoroughly to the preparation of Stock Solution of 100 μg/mL;
(3) Finally, accurately pipette the specified volumes of Stock Solution into 100 mL polypropylene volumetric flasks for the preparation of Working Standards of 0.100, 0.250, 0.500, 1.00, 5.00, 10.0, and 20.0 μg/mL, respectively.

2.5 *Sample procedure*

The sample of paper-making reconstituted tobacco is milled in powder and filtered through a 40-mesh screen, and the sample is accurate to 0.0001 g.

Weigh approximately 0.100 g ± 0.020 g of the paper-making reconstituted tobacco sample into a 100.0 mL polypropylene extraction vessel and add 50.0 mL of the extraction solution. Place the extraction vessel on a laboratory shaker and shake at a moderate speed (120 r/min) for 40 minutes. Take an aliquot and filter through a 0.45 μm syringe filter and proceed to an analysis by ion chromatography.

Depending on the sodium and other cation contents of the tobacco sample, the extract may require dilution to obtain a chromatographic response covered by the calibration range. If sample dilution is required, the sample should be diluted with 0.025 N Sulfuric Acid. A dilution factor of 10 is sufficient for most samples.

The extracts should be analyzed as soon as possible. Samples are stable for 24 hrs when stored at 4°C ± 2°C. The use of a refrigerated auto-sampler has been shown to extend sample stability during analysis.

2.6 *Sample analysis*

2.6.1 *Example of Ion Chromatography parameters*

Sample analysis should meet the following conditions:

(1) Suppressor current: 88 mA;
(2) Auto-sampler tray temperature: 4°C ± 2°C;
(3) Column temperature: 30°C;
(4) Pressure range: 200 psi (min) to 3000 psi (max);
(5) Flush volume: 250 μl;
(6) Data acquisition time: 0 min to 15 min;
(7) Injection loop: 25 μl;
(8) Injection volume: 25 μl.

Gradient profiles are stated in Table 1 for use with an eluent generator.

Table 1. Preparation of working standards.

Time (min)	Concentration of MSA (mM)	Gradient profile	Flow rate (mL/min)
0.0	10	linear	1.00
9.0	10	linear	1.00
9.5	40	linear	1.00
14.5	40	linear	1.00
15.0	10	linear	1.00

2.6.2 *Calibration of the ion chromatography*

The calibration steps are as follows: (1) inject an aliquot of each Na^+, $NH_4{}^+$, K^+, Mg^{2+}, and Ca^{2+} cationic mixed Standard into the Ion Chromatograph; (2) record the analytic peak area; and (3) plot a calibration curve of the peak area of each cation versus the theoretical concentration in μg/mL. Especially, the calibration curve of the ammonia is fitted by a quadratic function in keeping with weak base chemistry. The response obtained for all test samples should fall within the working range of the calibration curve.

2.6.3 *Calculation expression*

The amount of each cation (μg/g) is calculated using the following expression.

$$C_i = \frac{C_i' \times V}{m} \times DF \tag{1}$$

Where:

C_i = The final concentration of each cation, respectively (μg/g);
C_i' = The concentration of each cation (μg/mL) obtained from the calibration curve;
V = Extraction volume (mL);
m = Mass of the sample (g);
DF = Dilution factor (e.g., 10 would be used if the sample is diluted 10-fold).

3 RESULT AND ANALYSIS

3.1 *Standard operating curves*

Using the method outlined above, we can obtain the chromatogram of the paper-making reconstituted tobacco samples, as shown in Figure 1.

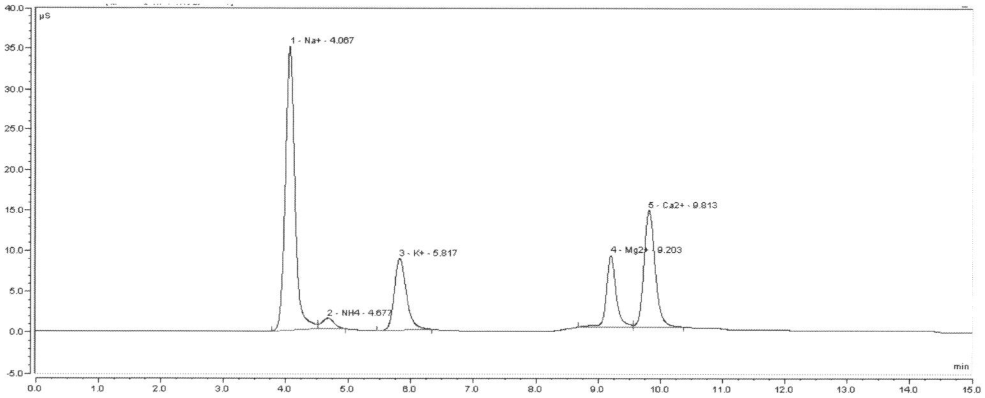

Figure 1. Determination chromatogram of sodium, ammonia, potassium, magnesium, and calcium content in paper-making reconstituted tobacco.

3.2 *Standard operating curves*

Add an aliquot of each cationic mixed Standard (Na^+, NH_4^+, K^+, Mg^{2+}, and Ca^{2+}) to the Ion Chromatograph. Record the analytic peak area. Plot a calibration curve of the peak area of each cation versus the theoretical concentration in μg/mL. Parameters of the cationic standard operating curve are shown in Table 2.

It can be seen from Table 2 that the determination coefficients of all cations are above 99.7% and the correlation is good.

3.3 *Limits of detection (LOD) and limits of quantification (LOQ)*

The lowest concentration of the standard solution was taken for 10 times of parallel determination, and the standard deviation was calculated, with three times of standard deviation as the limit of detection and 10 times of standard deviation as the limit of quantification. The results are shown in Table 3.

Table 2. Parameters of the cationic standard operating curve.

Cation	RSD (%)	Coefficient of determination (%)	Intercept	Slope	Curvature
Na^+	7.5289	99.7888	0.0618	0.5188	0
NH_4^+	5.9779	99.7807	0.0815	0.5135	–0.0268
K^+	8.2963	99.7277	0.0488	0.3238	0
Mg^{2+}	5.3232	99.8633	0.1569	0.9779	0
Ca^{2+}	7.1093	99.7612	0.5860	0.6187	0
Average	6.8471	99.7843	/	/	/

Table 3. LOD and LOQ for cations.

Cation	LOD (μg/mL)	LOQ (μg/mL)
Na^+	0.011	0.036
NH_4^+	0.013	0.043
K^+	0.021	0.070
Mg^{2+}	0.014	0.046
Ca^{2+}	0.028	0.093

Table 3 shows that LOD and LOQ of the cations are 0.011–0.028 μg/mL and 0.036–0.093 μg/mL, respectively.

3.4 *Repetition and reproducibility*

The same sample of paper-making reconstituted tobacco was treated by the above method and the content was determined to evaluate the repeatability and reproducibility. Intra-day repetition is measured by calculating the relative standard deviation (RSD) of the same sample repeated 5 times a day for 3 consecutive days. Inter-day reproducibility was expressed by measuring 5 groups of samples each day for 3 consecutive days and calculating the relative standard deviation. The experimental results were shown in Table 4. The results showed that the RSD of intra-day repetition of ammonia and other cations in paper-making reconstituted tobacco measured by this method were 1.85–3.55% and the RSD of intraday repeatability was 2.36%–4.16%, all less than 5%, indicating that the method had good repeatability and stability and it can meet the quantitative requirements.

Table 4. Inter-day repetition (n = 15) and inter-day reproducibility (n = 15) for cations.

	Intra-assay (n = 15)		Inter-assay (n = 15)	
Cation	Average (μg/g)	RSD (%)	Average (μg/g)	RSD (%)
Na^+	588.32	1.85	583.72	3.46
NH_4^+	152.14	2.52	148.54	3.13
K^+	1634.62	2.81	1626.18	2.52
Mg^{2+}	2786.86	3.55	2778.43	4.16
Ca^{2+}	4789.33	2.63	4898.46	2.36

3.5 *Recovery and precision*

The paper-making reconstituted tobacco with known content was divided into three parts, and the standard substance with six cations was added at low, medium, and high levels respectively. The determination was repeated five times at each level. The spiked samples were treated according to the preceding pre-treatment method and analyzed under the same instrumental analysis conditions.

The average recovery rate and the average relative standard deviation of the spiked values were calculated based on the original contents, added scalar, and measured quantity. The results showed that the recoveries of five cations in the paper-making reconstituted tobacco sample ranged from 95.7% to 104.6%, and the RSD ranged from 1.55% to 4.32%, indicating again that the method could meet the quantitative requirements.

Table 5. Spiked recoveries (n = 5) and precision of the five cations.

		Add 500 μg/g		Add 1000 μg/g		Add 2000 μg/g	
Cation	Content (μg/g)	Recoveries (%)	RSD (%)	Average (μg/g)	RSD (%)	Average (μg/g)	RSD (%)
Na^+	588.32	102.2	2.82	97.4	1.79	97.4	2.52
NH_4^+	152.14	98.8	2.69	102.2	2.13	102.1	1.75
K^+	1634.62	96.7	1.55	97.6	1.64	98.2	1.59
Mg^{2+}	2786.86	96.6	1.70	96.2	2.24	99.1	2.24
Ca^{2+}	4789.33	95.7	2.59	101.7	2.67	104.6	4.32

3.6 *Sample determination*

The paper-making reconstituted tobacco samples are determined using the above method. The determination results are shown in Table 6.

Table 6. Determination of sodium, ammonia, potassium, magnesium, and calcium contents in the paper-making reconstituted tobacco samples.

	Content (μg/g)				
Cation	1#	2#	3#	4#	5#
Na^+	2162.61	462.75	5131.62	2685.26	356.89
NH_4^+	150.56	262.48	1026.51	178.56	408.53
K^+	2234.09	15122.36	2784.27	1607. 34	11856.75
Mg^{2+}	2624.67	3520.67	2475.24	842.63	3687.91
Ca^{2+}	29783.16	41643.21	4264.32	3542.42	45889.63

4 CONCLUSION

To monitor the quality of paper-making reconstituted tobacco products, and evaluate the stability of Na^+, NH_4^+, K^+, Mg^{2+}, and Ca^{2+} content in paper-making reconstituted tobacco, we employ an intra-standard quantitative method for determination of the contents in paper-making reconstituted tobacco by Ion Chromatography.

By collecting a 0.1 g sample with oscillation extraction by 50 mL of the extraction solution, the LODs and the LOQs of target cations could achieve 0.011–0.028 μg/mL and 0.036–0.093 μg/mL, respectively, while the determination coefficients of all cations are above 99.7% and the correlation is good. The Intra-assay and Inter-assay precisions measured by 15 repeats were 1.85–3.55% and 2.36%–4.16%, respectively. And with three different added levels, the recovery rates for the target cations lie between 95.7% and 104.6%. It was proved that the developed method is environmentally friendly, highly sensitive, and selective, and it can better control the quality of paper-making reconstituted tobaccos.

ACKNOWLEDGMENT

This work was funded by the Science and Technology Project of Yunnan Provincial Company of China Tobacco (Grant No. 2021530000241009).

REFERENCES

Cai Junlan. 2006. Advance in Ion Chromatography and its Application in Tobacco Industry. *Tobacco Science & Technology* 39(8): 42–46.

Chen Zugang. 2002. Comparison between Domestic and Foreign Paper-process Tobacco Sheets. *Tobacco Science & Technology* 35(2): 4–7.

Dai Lu. 2013. Research progress of reconstituted tobacco based on papermaking process. *Transactions of China Pulp and Paper* 28(1): 65–69.

Han Wenjia. 2017. The Recent Process of Reconstituted Tobacco by Paper Process. *Heilongjiang Pulp & Paper* 35(4): 47–49.

Hu Huiren. 2010. Comparison of Application of Chitosan and Guar Gum in Tobacco Sheet Manufacture by Papermaking Process. *China Pulp & Paper* 29(7): 32–36.

Miao Yingju. 2009. Present Status of Preparation Technology of Reconstituted Tobacco. *China Pulp & Pape* 28(7): 55–60.

Xie Jianping. 2011. *Chemical components in tobacco and tobacco smoke*. Beijing: Chemical Industry Press.

Yang Haijian. 2011. Rapid Determination of Potassium Content of Paper Making Tobacco Sheet by Sodium Tetraphenylboron-Quaternary Ammonium Salt Volumetric Method. *Chemical Analysis and Meterage* 20(2): 36–38.

Yuan Baosheng. 1994. The Effect of Silicate Bacterial Fertilizer on Raising the Per Unit Yield of Tobacco and Improving the Quality of Tobacco. *Journal of the Hebei Academy of Sciences* (2): 33–43.

Zhang Huailing. 1994. *Tobacco analysis and inspection*. Zhengzhou: Henan Science and Technology Press.

Zhao Qiurong. 2014. Variance analysis on comprehensive quality between domestic and foreign paper-making process reconstituted tobacco. *Journal of Southern Agriculture* 45(7): 1253–1257.

Energy Revolution and Chemical Research – Kok-Keong Chong and Zhongliang Liu (Eds)

Application of ultra-high pressure hydraulic slotting pressure relief and permeability enhancement technology in broken soft coal seam

Lin-Dong Guo
State Key Laboratory of Gas Detecting, Preventing, and Emergency Controlling, Chongqing, China
Chongqing Research Institute Co., Ltd. of China Coal Technology &Engineering Group, Gas Research Branch, Chongqing, China

ABSTRACT: Coal seam pressure relief and permeability enhancement technology is the key technology for coal mine gas disaster prevention and efficient extraction. The 15249N working face of Jiulong belongs to the outburst risk area of 2# coal seam. By analyzing 2# broken soft coal seam and gas occurrence characteristics, the ultra-high pressure hydraulic slit technology is adopted to carry out the coal seam pressure relief and permeability enhancement test. Two groups of cross-layer boreholes in the mining area of the 15249N working face are selected as comparative test boreholes, and efficient matching technology is adopted on the construction site. The application effect is investigated by comparing the gas drainage concentration and purity of test holes. The field test shows that the single knife coal yield of slotted drilling is 0.98 t, the equivalent slotting radius is 2.04–2.57 m, the average single hole gas drainage concentration of slotted drilling is increased by 1.6 times, and the gas drainage purity is increased by 3.57 times compared with ordinary drilling, and the gas drainage effect of the coal seam is significantly improved.

1 INTRODUCTION

China is one of the world's largest coal producers, possessing a vast supply of coal resources. However, coal mining disasters in China happen frequently. High gas leakage accounts for a large proportion of these disasters. With the increase of mining depth, coal seam gas presents the characteristics of "three high and one low", which increases the difficulty of gas control and seriously restricts the safe production of coal mines (Cheng et al. 2009; Guo 2019; Yuan 2021). At present, pre drainage of coal seam gas is the main way of coal seam gas control, and drainage in coal seam construction drilling is the main technical measure (Li 2011; Xu 2014). However, due to the soft coal body and poor permeability of deep coal seams, problems such as less pure gas, low concentration, and fast flow attenuation caused by borehole drainage seriously restrict the gas drainage efficiency, prolong the gas treatment cycle, and cause the tense situation of coal mining and replacement (Tian et al. 2021; Zhai 2018). Therefore, there is an urgent need for an efficient gas control measure to realize the rapid control of coal seam gas.

In recent years, scholars at home and abroad have conducted in-depth research on coal seam pressure relief and permeability enhancement, and efficient drainage technology, forming gas control and efficient drainage technologies such as surface well pre drainage (Cheng 2020), mining protective layer (Wu & Li 2021), adjacent layer pre drainage (Wang 2011), and this coal seam pre drainage (Guo et al. 2018). Due to the long construction period and high engineering cost of surface wells, the scope of popularization and application is less. The influence of pre drainage of gas from the mining protective layer and adjacent layer on gas drainage of this coal seam is limited and the drainage time is long. At present, most coal mines adopt the method of drilling in

*Corresponding Author: guolindong321@outlook.com

DOI 10.1201/9781003332657-92

this coal seam for extraction, but the influence range of single drilling construction measures on coal seam pressure relief and permeability enhancement is limited. By taking hydraulic measures in the borehole to improve the permeability of the coal seam and extraction efficiency. Among them, hydraulic slotting technology is widely used in the field because of its technical advantages of efficient infiltration and uniform pressure relief.

The 15249N working face of Jiulong mine belongs to an outburst dangerous area and does not have protective layer mining. Combined with the coal seam gas storage conditions, the construction through layer drilling in the mining area of the 15249N working face is selected in the lower trench of 15449N working face to carry out the pressure relief and permeability enhancement test of coal seam ultra-high pressure hydraulic slotting technology, to improve the pressure relief and permeability enhancement effect of coal seam and improve the utilization rate of gas drainage.

2 PROJECT OVERVIEW

15249N working face belongs to 2# coal seam, with a ground elevation of + 127.6 to + 136.8 m and underground elevation of - 730 to – 835 m. The design strike length of 15249N working face is 841 m (horizontal distance), the design inclined length is 130 m (horizontal distance), and the recoverable reserves are 707,879 t. The dip angle of the coal seam in the test area is 16° to 22° , and the average thickness of the coal seam is 5.7 m. The highest original gas content is 7.86 m^3/t, the gas pressure is about 0.66 to 0.80 MPa, the permeability coefficient is 0.031 $m^2/(MPa^2 \cdot D)$, and the coal firmness coefficient is 0.4 to 0.5. It is a difficult coal seam. The failure type of coal body is V, and the integrity of the coal body is poor. The working face layout of 15249N is shown in Figure 1.

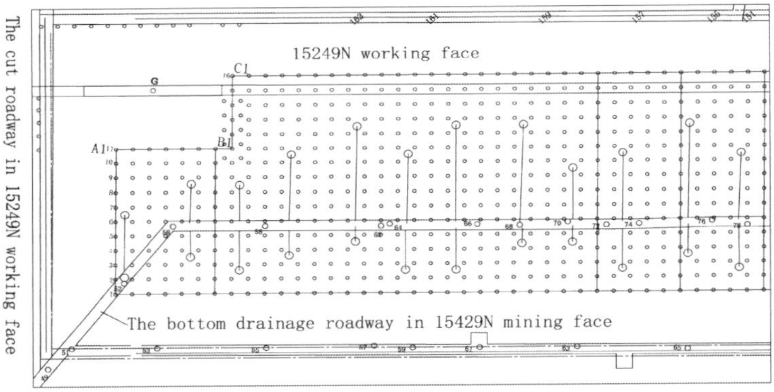

Figure 1. The working face layout of 15249N.

The coal seam working face belongs to the outburst coal seam with high gas and low permeability, and there are abnormal areas of coal seam gas. The coal seam gas content is high, the gas pressure is high, and the coal body is relatively broken and soft. The pressure relief and antireflection effect of the ordinary through layer drilling are very little, the net amount of gas extracted by drilling is less, the concentration is low, and the gas treatment cycle is long.

3 THE ULTRA HIGH-PRESSURE HYDRAULIC SLOTTING TECHNOLOGY AND DEVICE

3.1 *Technical principle*

Ultra-high-pressure hydraulic slotting technology takes high-pressure water as the power source to form a high-pressure water jet to cut the coal around the borehole to form an annular slot, increase the exposed area of the coal and promote the rapid release of gas. Under the action of in-situ stress, the coal deformation around the fracture groove produces spatial movement, increases the gas migration channel, and expands the pressure relief range. Enhancing the gas flow conditions within the coal seam is beneficial since it releases all of the pressure within the coal body, increases the permeability of the coal seam, and enhances gas drainage. The schematic diagram of the ultra-high pressure hydraulic slotting process is shown in Figure 2.

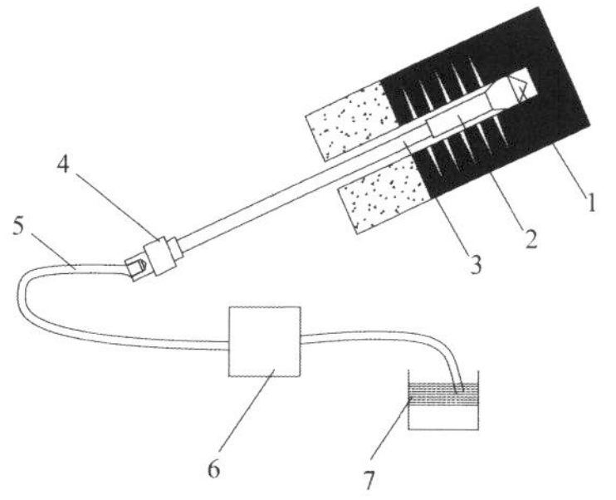

Figure 2. Schematic diagram of ultra-high-pressure hydraulic slotting process.

1—Diamond composite bit; 2—The high and low voltage conversion slotter; 3—The hydraulic slotted shallow spiral integral drill pipe; 4—The ultra-high pressure rotating water tail; 5—The ultra-high-pressure hydraulic hose; 6—The ultra-high-pressure water pump; 7—The water tank

3.2 *Device introduction*

KFSL100-113 water jet crawler seam cutting device is mainly composed of diamond composite bit, hydraulic slotting shallow spiral integral drill pipe, ultra-high pressure rotating water tail, ultra-high pressure clean water pump, high and low-pressure conversion slotter, ultra-high pressure hydraulic hose, etc. The working pressure of the complete set of equipment is up to 100 MPa, integrating drilling and cutting, and has the function of remote control and slotting operation, with a remote-control distance of more than 100 m. The cutting radius of the device can reach 1.5 m to 2.5 m, and the cutting gap width is 2 cm to 6 cm. The key components of the device are shown in Figures 3–8.

Figure 3. The ultra-high-pressure water pump.

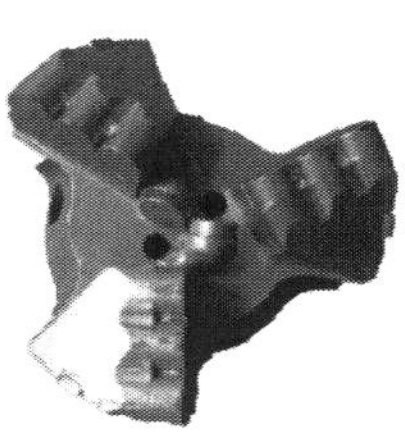

Figure 4. The diamond composite bit.

Figure 5. The hydraulic slit shallow thread integral drill pipe.

Figure 6. The high and low voltage conversion slotter.

Figure 7. The ultra-high pressure rotating water tail.

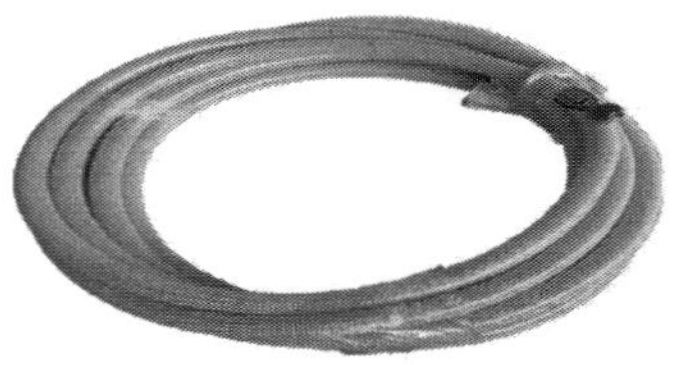

Figure 8. The ultra-high-pressure hydraulic hose.

The working pressure of the ultra-high pressure clean water pump is 100 MPa and the flow is 132 L/min. The hydraulic slotted shallow thread integral drill pipe has a three-stage sealing structure, the pressure bearing is 120 MPa, and has the characteristics of high strength and high torque. The ultra-high pressure rotating water tail adopts the sealing mode of end face seal and rotating dynamic seal, which can ensure the sealed transmission of 150 MPa high-pressure water. The ultra-high pressure hydraulic hose is six layers of steel wire winding and hose sheath, and the working pressure is 120 MPa. At the same time, to improve the safety of the device, a four-fold safety protection system is developed.

3.3 *Construction matching process*

To carry out the standardized management of the underground site and further improve the site construction quality. According to the field construction of ultra-high pressure hydraulic slit, the method of gas separate source extraction is adopted, and two extraction pipelines are laid at the construction site. The completed gas extraction boreholes are connected to the main pipeline for extraction. The construction boreholes are equipped with blowout devices at the orifice, and the gas extraction outlet is connected to the auxiliary extraction pipeline. During the same drilling period, the single drilling pipeline is used to avoid affecting the pumping efficiency of other boreholes. The test drilling adopts the "three plugging and two injections" sealing process. Three sealing bags are under the sealing section of the drilling hole to form two sections of sealing space. The grouting pipe in the sealing space is connected with a blasting valve. When the grouting pressure reaches the threshold value, the blasting valve is opened for hole sealing grouting. When the sealing slurry flows out of the slurry return pipe, the sealing grouting work is completed. The diameter of the casing and slag discharge port in the hole of water slotted drilling BOP is 150 mm, which is convenient for the smooth discharge of a large amount of coal slag during slotting. In addition, the hydraulic slit construction site is also equipped with secondary blowout prevention and slag collection metering devices. The secondary blowout prevention can effectively avoid excessive gas emissions in the hole during slit construction. The slag collection metering device has a volume of 0.7 m^3. In addition, the filter screen structure is adopted on three sides to realize the separation of coal and water, which can realize the rapid collection and measurement of coal slag discharged from the slit, effectively ensure the cleanliness of the construction site, and greatly reduce the labor intensity of on-site construction personnel.

4 THE FIELD TEST AND EFFECT

4.1 *Test scheme*

Based on the coal seam geological conditions of the working face, two groups of cross-layer boreholes in the mining area of 15249N working face are selected as comparative test boreholes, including one group of ordinary cross-layer boreholes and the other group of hydraulic slotting boreholes. There are 7 boreholes in a single group, with the interval between boreholes being 6 m and the interval between final holes being 6 m. The test borehole diameter is 113 mm, the borehole depth is 46 m to 64 m, the borehole azimuth is 310° , and the borehole inclination is 55° to 89° . Among them, the slotting test group has forward slotting, the slot spacing is 1 m, and the number of single hole slotting knives is 6 to 8. The test borehole adopts the "three plugging and two injections" sealing process, and the sealing length is 15 m. The layout section of test boreholes is shown in Figure 9.

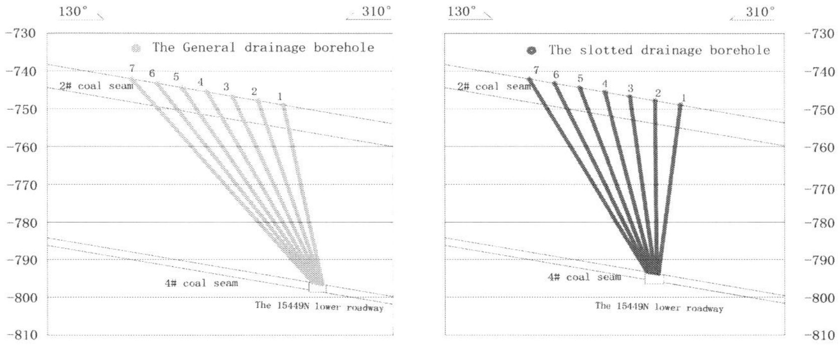

Figure 9. The layout of test boreholes.

4.2 *Analysis of test results*

(1) Analysis of coal output from the borehole

In the experimental construction stage, there are certain changes in the hardness of the coal seam during the seam cutting process. During the seam cutting process, the seam cutting pressure should be adjusted in time according to the hardness of the coal seam. The drilling seam cutting pressure of the seam cutting group is 50 to 90 MPa. Under the condition of ensuring the slotting effect and smooth slag discharge, the coal output of single hole slotting is 6.1 to 8.9 t, and the average coal output of single knife slotting is 0.98 t. The statistical diagram of the coal output of the borehole is shown in Figure 10.

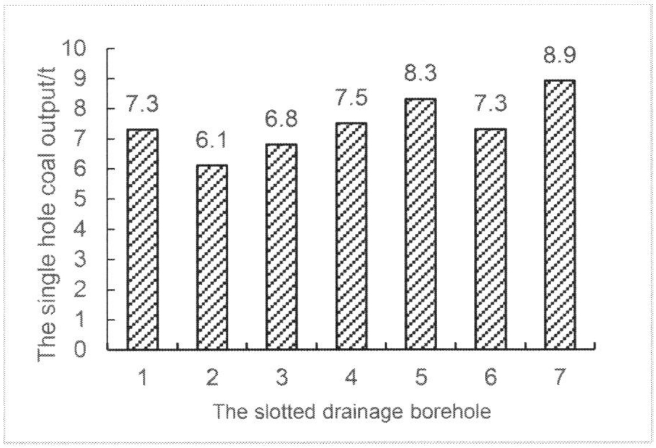

Figure 10. Statistics of coal output from boreholes.

The gap formed by cutting is simplified into cylinder analysis, and the equivalent cutting radius is inversely calculated according to the calculation formula of single knife coal output, which is shown as follows:

$$M = \pi \times r^2 \times h \times K \times \gamma \tag{1}$$

Where:

π —the value is 3.14;

M—the amount of coal dust discharged after slotting (t);

K—unbalanced coefficient of coal loss, ranging from 0.8 to 0.95 ("0.8" is accepted here);

R—equivalent radius of slit after slit cutting (m); the width of the rear slit is from 6 cm to 4 m, and the width of the rear slit is considered as the width of the rear slit;

γ—the unit weight of coal is 1.5 t/m^3.

According to the formula, under the condition that the average amount of coal chips discharged by each knife is m = 0.98 t, the slot is formed after cutting, and the calculated equivalent radius is r = 2.04 to 2.57 M. It can effectively increase the exposed area of coal in the borehole and achieve the goal of rapid pressure relief and permeability enhancement.

(2) Comparative analysis of gas drainage concentration

According to the statistics of gas extraction data for 35 days, the changes in gas extraction concentration in slotted drilling and comparative drilling are shown in Figure 13. Through comparative analysis, it can be seen that the average gas extraction concentration in slotted drilling is 67%, the gas concentration after stable extraction exceeds 70%, the average gas concentration in ordinary extraction is 42.11%, and the gas extraction concentration in drilling after hydraulic slotting is increased by 1.6 times, as shown in Figure 13. At the same time, it also indirectly reflects that the borehole sealing process of "three plugging and two injections" can improve the borehole sealing quality, reduce the attenuation rate of borehole pumping and production, and improve the pumping and production effect. The comparison diagram of average extraction concentration is shown in Figure 11.

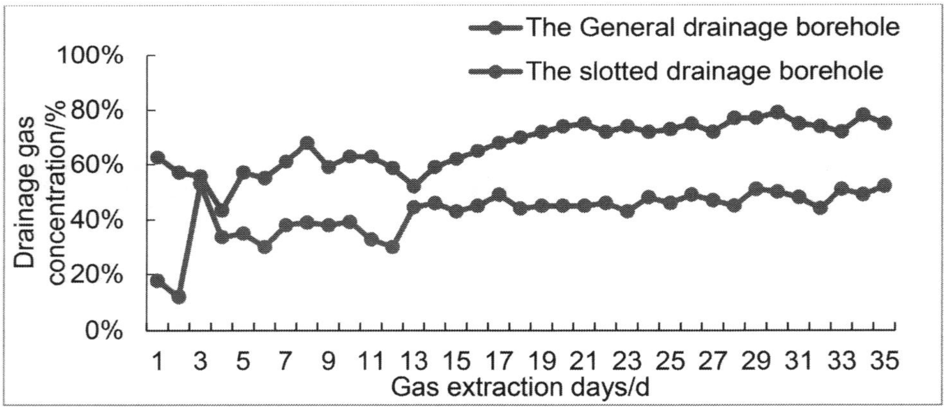

Figure 11. Comparison of average drainage concentration between slotted and normal boreholes.

(3) Comparative analysis of the net amount of gas drainage

As the pumping measurement adopts the unified measurement method of the drilling field hole group, the drilling field hole group data is converted into the average single hole pumping data for investigation and analysis. Within the drainage period of 35 days, the cumulative net amount of gas extracted from 7 slotted boreholes in the slotted drilling yard is 10522.73 m^3, and the average total amount of single-hole drainage is 1503.25 m^3. The cumulative net amount of gas extracted from 7 ordinary boreholes in the ordinary drilling yard is 3052.03 m^3, and the average total amount of single hole extraction is 436.00 m^3. The total amount of slotted drilling is 3.45 times that of ordinary drilling. The average net amount of gas extracted from a single hole in a slotted drilling site is 0.0307 m^3/min, and the average net amount of gas extracted from a single hole in an ordinary drilling site is 0.0086 m^3/min. after hydraulic slotting, the net amount of gas extracted from a borehole is increased by 3.57 times. Through comparative analysis, the effect of gas drainage can be greatly improved after taking slit-cutting measures. The comparison diagram of average single-hole drainage gas purity is shown in Figure 12.

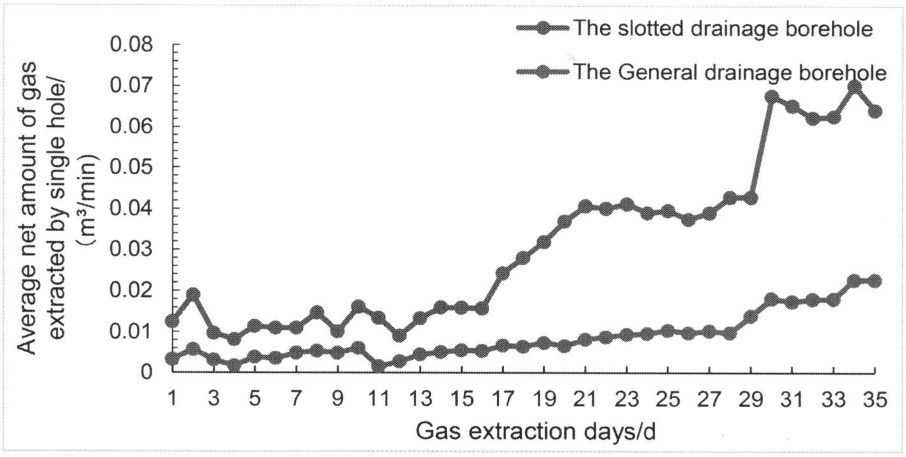

Figure 12. Comparison diagram of average single hole drainage gas purity.

5 CONCLUSIONS

(1) This paper analyzes the 2# fragmentary soft coal seam and gas storage characteristics and the influencing factors of drainage, expounds on the principle and device composition of ultra-high pressure hydraulic slotting technology, and introduces the supporting process methods including gas separate source drainage method, "three plugging and two injections" hole sealing technology and drilling blowout prevention and slag collection device.

(2) Through the analysis of the field hole sealing comparative test results, after the ultra-high pressure hydraulic slotting process measures are adopted, the single knife coal output of slotted drilling is 0.98 t, the equivalent slotting radius is 2.04 to 2.57 M, the average gas extraction concentration of slotted drilling is 67%, and the average gas extraction concentration of ordinary drilling is 42.11%. Compared with ordinary drilling, the average single hole gas extraction concentration of slotted drilling is increased by 1.6 times. The average net amount of gas extracted from a single hole in a slotted drilling site is 0.0307 m^3/min, and the average net amount of gas extracted from a single hole in an ordinary drilling site is 0.0086 m^3/min. after hydraulic slotting, the net amount of gas extracted from a borehole is increased by 3.57 times. The effect of gas drainage has been significantly improved.

(3) The preliminary test preliminarily verified the effect of ultra-high pressure hydraulic slit technology on coal seam pressure relief and increase. In the next step, it is still necessary to optimize the parameters such as slit spacing, slit pressure and single knife slit coal yield, so as to form an ultra-high hydraulic slit pressure relief and permeability enhancement process technology system suitable for gas control in Jiulong coal mine.

ACKNOWLEDGMENTS

This work was supported by the Special key project of science and technology innovation and Entrepreneurship of Tiandi Technology Co., Ltd. (Grant No. 2021-2-TD-ZD008); the key projects of the joint fund for regional innovation and development (Grant No. U21A20110); the General

Project of Chongqing Research Institute of China Coal Industry Group Co., Ltd. (Grant No. 2020YBXM23). The authors also thank the anonymous editors and reviewers very much for their valuable advice.

AUTHOR INFORMATION

Name: Guo Lindong
Tel: 18369909901
Address: No. 55, Shangqiao Third Village, Shapingba District, Chongqing.
About the author: Guo Lindong (1993–), male, was born in Jining, Shandong Province, with a master's degree.
Affiliation: CCTEG Chongqing Research Institute
Research interests: the prevention and control of gas disasters; underground mining.
E-mail: 932062638@qq.com

REFERENCES

Dingqi Li. *Evaluation method and application of gas control status in coal and gas outburst mines* [D] China University of mining and technology, 2011.

Hong Zhai, Jianshe Linghu. Innovative model and practice of gas control in Yangquan mining area [J] *Coal science and technology*, 2018, 46 (02): 168–175.

Hongfeng Wang. Research and practice of gas drainage technology [J] *Shanxi coking coal technology*, 2011, 35 (06): 40–42.

Liang Yuan. Research on coal mine safety development strategy in China [J] *China coal*, 2021, 47 (06): 1–6.

Lindong Guo, Xusheng Zhao, Yongjiang Zhang, Yanbao Liu. Study on integrated sealing technology of borehole sealing for bedding gas drainage [J] *Coal science and technology*, 2018, 46 (05): 114–119.

Lindong Guo. *Study on new technology and reasonable parameters of borehole sealing for soft coal seam gas drainage* [D] General Coal Research Institute, 2019.

Yiquan Wu, Siqian Li. Feasibility study on mining middle protective layer of long-distance coal seam group in deep well [J] *Energy technology and management*, 2021, 46 (03): 10–12.

Yuanping Cheng, Jianhua Fu, Qixiang Yu. Development of coal mine gas drainage technology in China [J] *Journal of mining and safety engineering*, 2009, 26 (02): 127–139.

Zhiheng Cheng, Liang Chen, Shilong Su, Gongda Wang, Yinhui Zou, Yongjiang Zhang, quanle Zou, liming Jiang, Dahe Yan, Zhifeng Du, Xiangdong Wang. Up and down combined outburst prevention mode and its effect dynamic evaluation of short distance coal seam group [J] *Journal of coal*, 2020, 45 (05): 1635–1647.

Zhonglei Tian, Yao Xu, Kaijia Zhang, Xing Ni. Application of hydraulic punching gas control technology in Nuodong coal mine [J] *Energy technology and management*, 2021, 46 (04): 54–57.

Zunyu Xu. Study on gas control technology in the first mining face of coal seam [J] *Shandong coal technology*, 2014 (06): 83-84.

Energy Revolution and Chemical Research – Kok-Keong Chong and Zhongliang Liu (Eds)
© 2023 The Authors, ISBN: 978-1-032-36554-1

Determination of bisphenol in disposable tableware by solid phase extraction and supercritical fluid chromatography

Yue Qiu*, Genrong Li, Jiaxiong Zhao, Mei Long & Chaolan Tan
Chongqing Academy of Metrology and Quality Inspection, Chongqing, China

ABSTRACT: A method for the determination of seven bisphenol compounds in disposable tableware was established based on solid-phase extraction-supercritical fluid chromatography. The effects of extraction conditions and purification conditions on the extraction efficiency of bisphenol from disposable tableware were investigated. Disposable tableware samples were extracted by ultrasonic with acetone as a solvent, purified by HLB solid-phase extraction column, analyzed by supercritical fluid chromatography, and quantified by external standard method. The actual samples of disposable tableware of different materials were analyzed, and it was found that there were residues of BPA, BPF, and BADGE with a content of 0.45-4.10 mg/kg. The method was green and efficient, which was suitable for the determination of bisphenol compounds in disposable tableware.

1 INTRODUCTION

Bisphenols are a class of compounds with similar chemical structures, which are mainly used in the synthesis of epoxy resin and polycarbonate materials. Bisphenol compounds are widely used in the processing and production of food packaging materials. Disposable tableware is widely used in people's life because of its light weight and convenience. Under the conditions of high temperature, microwave heating, cooking, and sun exposure, bisphenol substances in disposable tableware are easy to migrate to food. Bisphenol compounds have a strong estrogen-like effect and interferes with the normal endocrine function of human and animal, which can change some genes of genetic information and lead to cell aberration, tumor, carcinogenic, teratogenic, mutagenic toxicity (Delfosse et al. 2012; Hu et al. 2014; Qiu et al. 2020; Zhang et al. 2014).

GB 9685-2016 stipulates that the specific migration or maximum residue limits of bisphenol A and bisphenol S are 0.6 mg/kg and 0.05 mg/kg, respectively (GB 9685-2016.). At present, the main methods for the determination of bisphenol compounds include gas chromatography-mass spectrometry (García-Córcoles et al. 2018, Talanta 2018), high performance liquid chromatography (Zhou et al. 2018), and liquid chromatography-mass spectrometry (Bruno Alves Rocha et al. 2018). There are many kinds of bisphenol substances with similar properties. However, the research on the detection method of this kind of substance is still insufficient.

In this paper, solid phase extraction and supercritical fluid chromatography method were used for the determination of seven bisphenol compounds in disposable tableware. This method has the advantages of good repeatability, high recovery and less solvent consumption, which can provide data reference for the detection of bisphenol substances in disposable tableware in the future.

*Corresponding Author: qiuyuecqu@foxmail.com

DOI 10.1201/9781003332657-93

2 MATERIALS AND METHODS

2.1 *Materials and reagents*

Bisphenol standard substances (Dr. Ehrenstorfer Germany); chromatographic purity of methanol, acetonitrile and isopropanol (Merck, Germany); 0.22 μm nylon filter membrane.

2.2 *Instruments and equipment*

Ultra-Performance Convergence Chromatography (Waters); SQP electronic analysis balance (Germany Sartorius); N-evap-112 Nitrogen Blower (Organomation, USA); Bilon-2000ct ultrasonic cleaner (Shanghai Bilang Instrument Co., LTD.).

2.3 *Preparation of standard solution*

Each standard substance of 25 mg was accurately weighed and dissolved with methanol, and then prepared into a single standard reserve solution with a mass concentration of 1000 mg/L. Standard working solutions with concentrations of 0.1, 1.0, 10.0, 25.0, 50.0 and 100.0 mg/L were prepared with methanol.

2.4 *Sample pretreatment*

The sample was cut into small fragments with a mass of no more than 0.02 g. The 1.0 g sample was accurately weighed into a 50 mL colorimetric tube, then 25 mL acetone was added and ultrasonic extraction was performed at 30°C for 40 min. The extraction solution was concentrated at 40°C with nitrogen blowing until it was nearly dry. The supernatant was purified by HLB solid phase extraction column and passed through 0.22 μm organic system filtration membrane for UPC^2 analysis.

2.5 *Chromatographic condition.*

The determination was performed on an ACQUITY UPC^2 Torus DEA column (150 mm $\times$ 30 mm, 1.7 μm) with supercritical CO_2 as mobile phase A and methanol as mobile phase B. The column temperature was 35°C, the system back pressure was 1800 psi, the injection volume was 3 μL, and the detection wavelength was 220 nm.

3 RESULTS AND DISCUSSIONS

3.1 *Selection of extraction solvent*

In this paper, bisphenol substances were extracted by ultrasonic extraction. In the experiment, blank samples such as disposable paper cups and disposable lunch boxes were selected for standard addition test, and the extraction effects of acetone, ethyl acetate, methanol and acetonitrile were compared. As shown in Figure 1, when acetonitrile was extracted, the recovery rate of the target substance was relatively the lowest, and when acetone was used as the extraction solvent, the average recovery rate of the seven bisphenol substances was the highest. Therefore, acetone was selected as the best extraction solvent in this experiment.

3.2 *Selection of extraction time*

In order to ensure sufficient extraction of bisphenol substances from disposable tableware, the experiment further investigated the effect of extraction time on the recovery, as shown in Figure 2. The results showed that with the extension of extraction time, the recoveries of 7 kinds of bisphenol

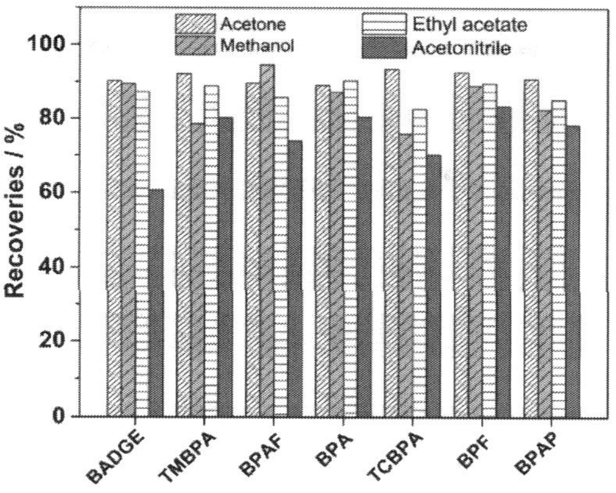

Figure 1. Effect of extraction solvent on recovery of bisphenol substances.

substances were significantly improved and gradually stabilized after 40 min, therefore, 40 min was selected as the best extraction time.

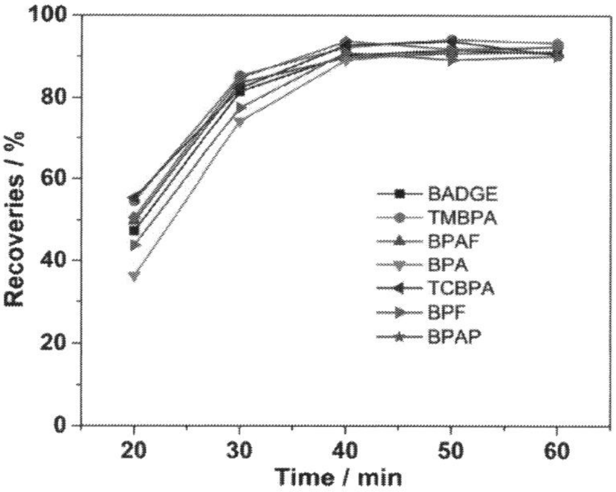

Figure 2. Effect of extraction time on recovery of bisphenol substances.

3.3 *Selection of extraction temperature*

The effect of ultrasonic temperature on recovery was investigated in the range of 10–50°C. As shown in Figure 3, with the increase in ultrasonic temperature, the extraction efficiency of bisphenol substances gradually increased and reached a stable level after 30°C. The recovery rate decreased slightly when the temperature continued to rise, indicating that the excessive temperature was not conducive to the extraction of bisphenol substances, so the optimal ultrasonic temperature was set at 30°C.

3.4 *Optimization of purification conditions*

The experiment compared the effects of HLB, PEP, and Florisil column on the recovery of bisphenol substances in disposable tableware, as shown in Figure 4. The results showed that part of the target substance was lost and the recovery rate was low when the Florisil column was used for

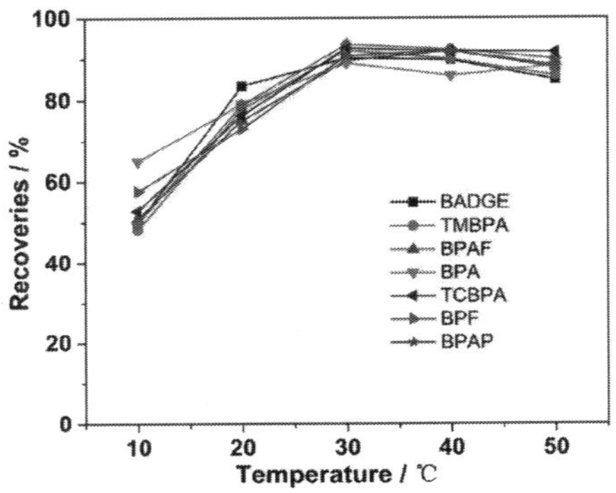

Figure 3. Effect of extraction temperature on recovery of bisphenol substances.

purification. HLB column showed that the best purification effect with recoveries of 7 target substances exceeding 85%. Therefore, the HLB column was selected for purification.

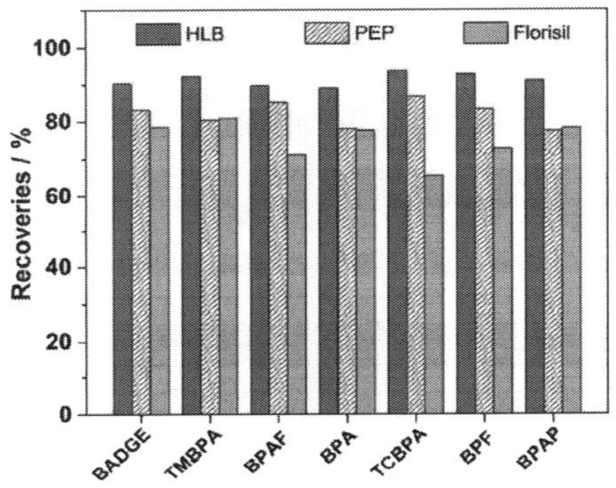

Figure 4. Effect of solid-phase extraction column on recovery of bisphenol substances.

3.5 *Actual sample testing*

Actual samples of ten different disposable tableware (paper cups, meal boxes, straws, etc.) were selected for the detection of bisphenol substances using this method. The results were shown in Table 1. It was found that the residue of bisphenol was found in three of the ten samples, with a content of 0.45–4.10 mg/kg.

4 CONCLUSION

A method for the determination of 7 bisphenol compounds in disposable tableware was established in this paper. In the experiment, bisphenol substances were used acetone as extraction solvent at 30°C for 40 min and purified with an HLB column. Finally, the recovery of 7 target substances could reach over 85%. The method was simple, efficient, accurate, and reliable, which was suitable for daily analysis and detection of bisphenol substances in disposable tableware.

Table 1. Determination of bisphenol substances in actual samples (mg/kg).

Sample	BADGE	TMBPA	BPAF	BPA	TCBPA	BPF	BPAP
Paper cup 1	—	—	—	0.45	—	—	—
Paper cup 2	—	—	—	—	—	—	—
Paper bowl 1	—	—	—	—	—	—	—
Paper bowl 2	—	—	—	2.89	—	—	—
Meal box 1	—	—	—	—	—	—	—
Meal box 2	4.10	—	—	—	—	1.66	—
Straws 1	—	—	—	—	—	—	—
Straws 2	—	—	—	—	—	—	—
Plastic spoon 1	—	—	—	—	—	—	—
Plastic spoon 2	—	—	—	—	—	—	—

ACKNOWLEDGMENT

This research was funded by the Scientific Research Project of Chongqing Market Supervision Administration (Grant No. CQSJKJ2019005).

REFERENCES

Bruno Alves Rocha, Anderson Rodrigo Moraes de Oliveira, Fernando Barbosa Jr. *Talanta*, 2018, 183: 94–101.

Delfosse V, Grimaldi M, Pons J L, Boulahtou A, Maire A, Cavailles V, Labesse G, Bourguet W, Balaguer P. *Proc. Natl. Acad. Sci. USA*, 2012, 109(37): 14930–14935.

García-Córcoles M T, Cipa M, Rodríguez-Gómez R, Rivas A, Olea-Serrano F, Vílchez J L, Zafra-Gómez A. *Talanta*, 2018, 178: 441–448.

GB 9685-2016. Standard for the Use of Additives for Food Contact Materials and Products National Standards of the People's Republic of China.

Hu Xiao-Jian, Zhang Hai-Jing, Wang Xiao-Hong, Ding Chang-Ming, Jin Yin-Long, Lin Shao-Bin. Chinese *J. Anal. Chem.*, 2014, 42(7): 1053–1056.

Qinghua Zhou, Zanhui Jin, Jia Li, Bin Wang, Xiuzhen Wei, Jinyuan Chen. *Talanta*, 2018, 189: 116–121.

Qiu Yue Li Gen-Rong Long Mei Ruan Yan Tan Chao-Lan Zhang Lu Xia Zhi- Ning. Chinese *J. Anal. Chem.*, 2020, 2(48): 255–261.

Zhang Pin, Zhang Jing, Chen Jie-Jun, Duan He-Jun, Shao Bing. *Chinese J. Anal. Chem.*, 2014, 42(12): 1811–1817.

Energy Revolution and Chemical Research – Kok-Keong Chong and Zhongliang Liu (Eds)
© 2023 The Authors, ISBN: 978-1-032-36554-1

Application of modified atmosphere packaging technology in pre-conditioned fish products

Liangzi Zhang, Jiangting Yue, Xin Zhang & Hongbing Dong*
College of Food Science and Technology, Wuhan Business University, Wuhan, China

ABSTRACT: With the continuous improvement of people's living standards, the demand for freshwater fish products has gradually increased, the quality requirements have been continuously improved, and pre-conditioned fish products have emerged as the times require. In order to prolong the storage period of *Erythroculter ilishaeformis* preconditioned fish products, study its storage methods and ensure its quality, the effects of packaging materials (PE, PA, PE+PA composite film) on the quality changes of *Erythroculter ilishaeformis* preconditioned fish products were studied. Different packaging methods were adopted, including room temperature and normal pressure packaging (RTNP), room temperature vacuum packaging (RTVP), low temperature normal pressure packaging (LTNP), and low temperature vacuum packaging (LTVP) (the temperature was at 2 °C). By analyzing the changes of sensory scores, thiobarbituric acid value (TBA value), volatile base nitrogen value (TVB-N value), total bacterial count and pH value during storage, the effects of high-pressure nylon (PA), polyethylene (PE) and composite film (PE + PA) on the preservation of pre-conditioned fish products were determined. The experimental results showed that PE+PA combined with LTVP was the best choice among the four packaging methods. At room temperature and normal pressure, the fish meat samples packaged in composite film (PE+PA) had a lower spoilage degree than the other two packaging materials during the same storage period, and its storage period was also 1-2 days longer than the other two packaging materials. LTVP was better than other packaging methods, the sensory score was higher after 10 days, and the TBA value, TVB-N value, pH value and total number of colonies were lower than the other three packaging methods.

1 INTRODUCTION

Erythroculter ilishaeformis is a high-quality freshwater fish, belonging to the family Cyprinidae, the subfamily Cyprinidae, and the genus Erythroculter, commonly known as big white fish (Chen 2003; Xiao 2016). *Erythroculter ilishaeformis* is widely distributed in Hubei, Anhui, Heilongjiang and other provinces in China and has a large output (Shen 2006). At present, *Erythroculter ilishaeformis* is mostly sold in the fresh form on the market, and the cost is relatively high. If centralized slaughtering, refrigerating and fresh-keeping are adopted, and pre-conditioning fish products, called semi-finished products, are sold in the form of pre-conditioned fish products, it cannot only satisfy the fast lifestyle of people in modern society, but also reduce the waste of fish after slaughter for centralized processing and further processing or utilization.

Common fish packaging technologies include vacuum packaging (Wu 2019), modified atmosphere packaging (Xiao 2020), etc. From the perspective of the principle of bacteriostasis and sterilization, vacuum packaging and modified atmosphere packaging are the best choices for fresh fish packaging (Xie 2011). Vacuum packaging is more conducive to controlling fat oxidation and spoilage (Fan 2016), and the shelf life of salmon vacuum packaging (Gu 2021; Li 2018) is extended

*Corresponding Author: 954797163@qq.com

 DOI 10.1201/9781003332657-94

to 14 days. The PE/PP/PET composite film has a better preservation effect on pork after vacuum packaging according to the literature (Gan 2001).

In this study, the pre-conditioned fish products of *Erythroculter ilishaeformis* were taken as the research object, three packaging materials (PA, PE, PE + PA) and four methods (RTNP, RTVP, LTNP and LTVP) were taken as the control variables. The effectiveness of three packaging materials and methods in the process of packaging and storage were discussed to provide a new reference for freshwater fish preservation technical data.

2 MATERIALS AND METHODS

2.1 *Materials*

Fresh *Erythroculter ilishaeformis* fish, American 3M colony (bacteria) total number test piece rapid detection paper, 2-thiobarbituric acid (TBA), trichloroacetic acid, chloroform (analytical grade), B Diaminetetraacetic acid (EDTA), magnesium oxide, boric acid, hydrochloric acid (HCl), methyl red indicator, bromocresol green indicator, ethanol

2.2 *Experimental methods*

2.2.1 *Storage process of preconditioned fish products*

The process is as follows:

Slaughter → descale, viscera, head and tail → wash → diced → pickled → drained → packaged → stored → measured

After purchasing fresh *Erythroculter ilishaeformis*, transport it to the laboratory, slaughter the fish, dephosphorize, remove the internal organs, remove the head and tail, rinse the fish, cut it into fish pieces (about 100 g ± 5 g), and add salt. Preparation (add 2 g salt per 100g fish meat), drain. A part of the fish pieces was packaged in 3 kinds of packaging bags (PE, PA, PE+PA) and stored at room temperature (16–25°C). The other part of the fish is packaged with composite film (PE+PA), which is packaged at normal pressure and vacuum, and stored in a refrigerator (2°C) and at room temperature (Huang 2022), respectively, forming room temperature normal pressure packaging and room temperature packaging. Vacuum packaging, low temperature and atmospheric pressure packaging, and low-temperature vacuum packaging are four groups of samples. Samples were taken every 2 days to detect indicators.

2.2.2 *Sensory Score*

According to the sensory requirements of SC/T 3018-2011 "Fresh Herring, Grass Carp, Silver Carp, Bighead Carp, Carp" and combined with the sensory evaluation methods of related freshwater fish, the sensory quality of *Erythroculter ilishaeformis* during storage was evaluated. The sensory evaluation team consisted of 10 people. It is composed of trained personnel. Each assessor will score the sample's appearance, smell, tissue state, etc., according to a 10-point system. The scores will be added up to obtain the average personal score of the assessor. The sensory scoring standards are shown in Table 1.

2.2.3 *Determination of pH value*

We take 5 g of meat sample and put it in 25 mL of distilled water, shake well, homogenize at 10,000 r/min for 1min, cool to room temperature and measure its pH value with an acidity meter. This process was repeated three times, and finally, we got the average pH value.

2.2.4 *Determination of TBA value*

The determination of TBA value is based on the thiobarbituric acid method (TBA method).

Table 1. Sensory evaluation criteria.

Score	9–10	6–8	3–5	0–3
Exterior	Normal color, glossy muscle section	Normal color, glossy muscle section	The color is slightly dull, and the muscle section is slightly shiny	Dull color, dull muscle cut surface
Odor	Has the inherent smell of fresh fish, no peculiar smell	Has the inherent smell of fresh fish, basically no peculiar smell	It has the inherent smell of fresh fish and has an obvious peculiar smell	Lose the smell of fresh fish
Organization status	The flesh is tight and intact, elastic	The meat is slightly soft and less elastic	The meat is soft and less elastic	The flesh is soft and inelastic

2.2.5 *Determination of TVB-N value*

The determination of TVB-N value is based on GB 5009.228-2016 Automatic Kjeldahl method. The testing process is as follows: crush each group of fish samples with a food processor; take 10g of the samples into a distillation tube; add 75 mL of distilled water; shake to make the samples fully dispersed in the water; and soak for 30 min.

2.2.6 *Determination of the total number of colonies*

This process is based on the method of combining GB4789.2-2010 with the 3M total bacterial count test piece: weigh 2.5 g of the sample and add 225 mL of normal saline into a sterile homogenizing bag; beat for 1–2 min, and make a 1:100 sample homogenate solution; pipette 1 mL of the sample homogenate solution and drop it vertically on the center of the 3M test piece; drop the upper film, press it gently, and let it stand for more than 1min; with the transparent side of the test piece facing up, put it into a constant temperature incubator, and incubate it at 30° C for 72 h. Finally, the total number of colonies was calculated by the plate count method.

2.2.7 *Data processing and analysis*

Data were analyzed by Excel 2019 and SPSS 26.0 software, and line charts were drawn with OriginPro 9.1.

3 RESULT AND ANALYSIS

3.1 *Effects of packaging materials and packaging methods on sensory quality*

The effects of different packaging materials on the sensory quality of fish samples are shown in Figure 1A, indicating that different packaging treatments have certain effects on the sensory scores of *Erythroculter ilishaeformis*. With the extension of storage time (He 2017), the sensory score of the fish samples became lower and lower, indicating that the fish samples had deteriorated. According to SPSS software analysis, there are significant differences in the effects of the three packaging materials on the sensory score of fish. On the second day of storage, the sensory scores of the three samples in room temperature and atmospheric pressure packaging were all lower than 4, showing obvious changes in appearance, with soft meat, poor elasticity and a peculiar smell. At room temperature and atmospheric pressure, the fish samples with three kinds of packaging materials showed an obvious peculiar smell during the storage period of 2–4 days. However, the sensory score of the PA package was significantly lower than that of the PE package and PE+PA package, while the PE package and PE+PA package had some similar sensory scores. Sensory score records showed that PE and PE+PA packaging materials had a better storage effect than PA packaging materials during 10 days of storage. As can be seen from Figure 1B, with the increase in storage time, certain changes occurred in all samples, and sensory scores gradually decreased.

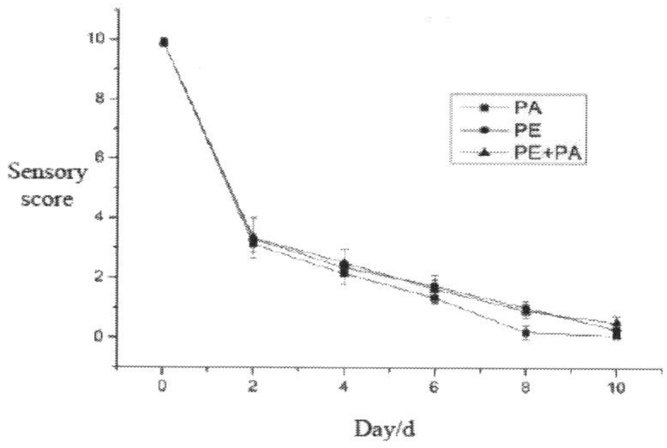

Figure 1A. Effects of packaging materials on sensory quality.

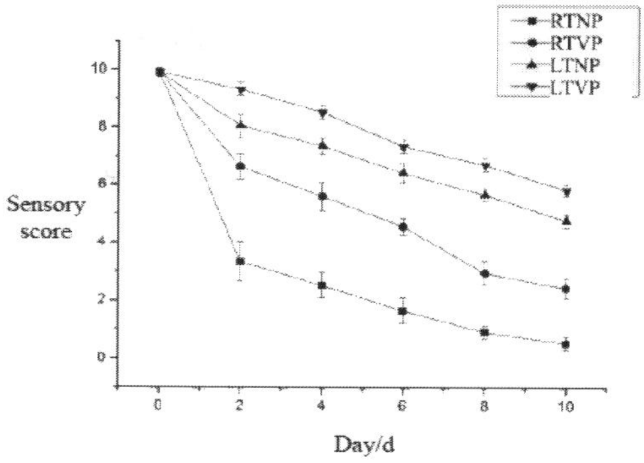

Figure 1B. Effects of packaging methods on the sensory score.

However, the sensory score of the samples stored in a vacuum at low temperature had the smallest range and rate of change, while the sensory score of the samples stored at room temperature and atmospheric pressure had the largest range and rate of change. The four packing methods are in order of the degree of change from small to large, LTVP < LTNP < RTVP < RTNP, low-temperature vacuum packing effect is the most effective among them, and it can prolong the storage period of fish (Yang 2016).

The pH of fish can directly reflect the pH of fish, which is closely related to the freshness of fish (Hu 2014). As can be seen from the variation trend (Figure 2A), the pH value of fish under different packaging treatments decreased first and then increased with the change in storage time. As storage ages, the fish's proteins break down, producing compounds of volatile salts such as trimethylamine, dimethylamine, and ammonia, and increasing alkalinity, which causes the fish's pH to rise. Among the three packaging materials, it was found by comparison that the pH of the sample under PA packaging was higher than that of the other two packaging materials, while the sample difference between PE and the composite film was small. Therefore, the PE bag and composite film (PE+PA) bag packaging storage effect are better than the PA bag. As can be seen, the initial pH value of fish samples is 5.92 (Figure 2B). With the increase of storage time, the pH value of samples decreases first and then increases, which is consistent with the change of muscle pH of aquatic products during storage in a V-shape Consistent. The pH values of room temperature and atmospheric pressure group, room temperature vacuum group, LTNP group and LTVP group reached the lowest values on days 2, 2, 4, and 6, respectively. This indicates that low temperature can delay the occurrence of the minimum pH value. It may be because the reaction of glycolysis and ATP decomposition of fish under low temperature is inhibited (Konno 2011), which slows down the decline of pH value, and microbial growth and reproduction are also inhibited, which affects the rise of pH value. Compared with the other three packaging methods, low-temperature vacuum packaging is better than the other three packaging methods, which can more effectively slow down the increase of pH value of fish, inhibit the growth of microorganisms, and prolong the shelf life of fish storage.

3.2 *Effects of different packaging materials and methods on TBA value*

The degree of fat oxidation is usually an important indicator used to measure the quality of meat products. TBA value can reflect the degree of fat oxidation of aquatic products and become an important indicator to measure the quality of aquatic products (Jeong 2018; Zhao 2012). The trend of TBA values of *Erythroculter ilishaeformis* with storage time under different packaging materials at room temperature and pressure is shown in Figure 3A. During the storage process, the TBA value increases with the increase of storage time, and the TBA value of the sample under PA packaging

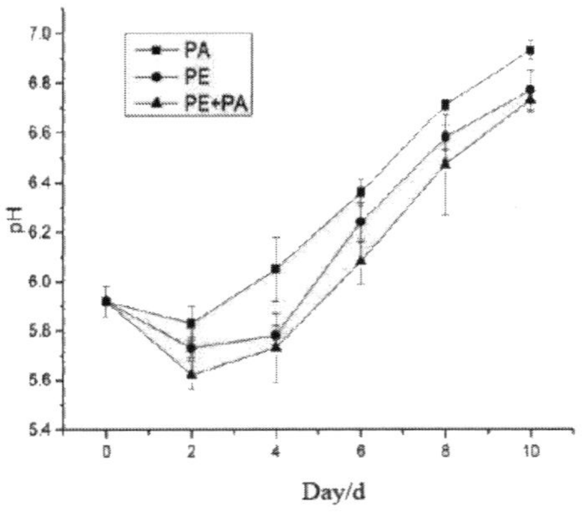

Figure 2A. The influence of packaging materials on pH value.

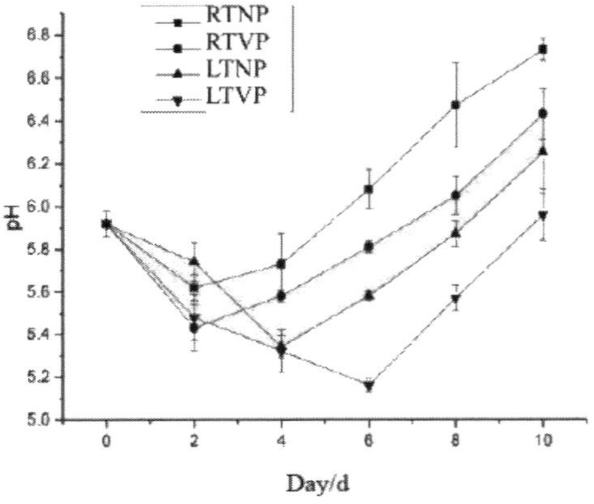

Figure 2B. The influence of packaging methods on pH value.

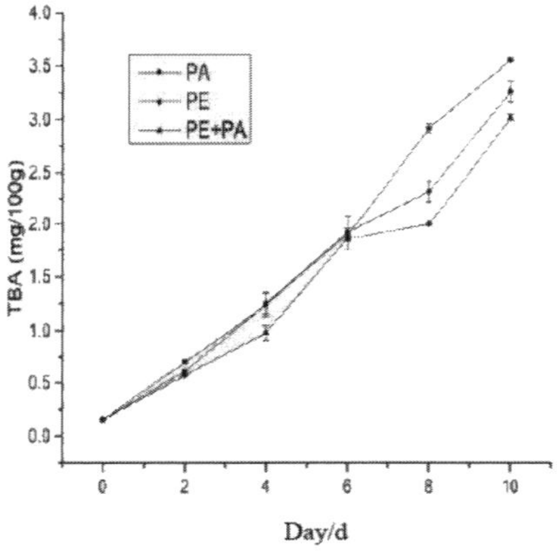

Figure 3A. Changes of TBA value with different packaging materials.

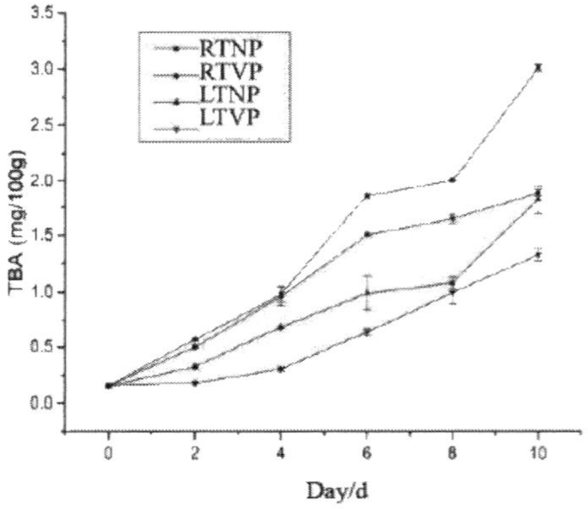

Figure 3B. Effects of packaging methods on TBA value.

changes the most, from 0. 156 mg/100 g to 3.553 mg/100 g. During the storage time of 10 days, the TBA values of the three kinds of packaging materials were significantly different. The TBA values of the fish under the composite film (PE+PA) packaging were significantly lower than those of the other two kinds, and the TBA values of the fish under the PA packaging were the highest. This may be because the composite film (PE+PA) is the most effective isolation of air, avoiding oxygen entry, thereby reducing the oxidation of fat. As shown in Figure 3B, the TBA value of the sample increases gradually with the increase of storage time. Among the four packing methods, the TBA value in room temperature and atmospheric pressure packing changed the most, while the TBA value in low-temperature vacuum packing changed the least. During the storage period, different packaging methods had significant effects on the TBA value of fish, and the TBA value of fish under low-temperature vacuum packaging was significantly lower than that of the other three packaging methods, indicating that low-temperature vacuum packaging could slow down the spoilage of fish.

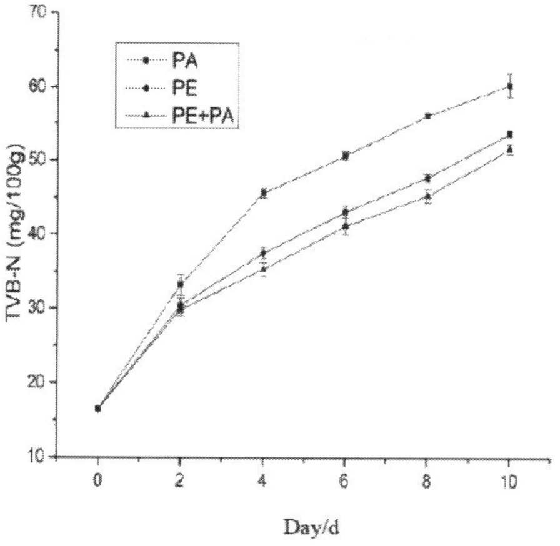

Figure 4A. Changes of the TVB-N value with different packaging materials.

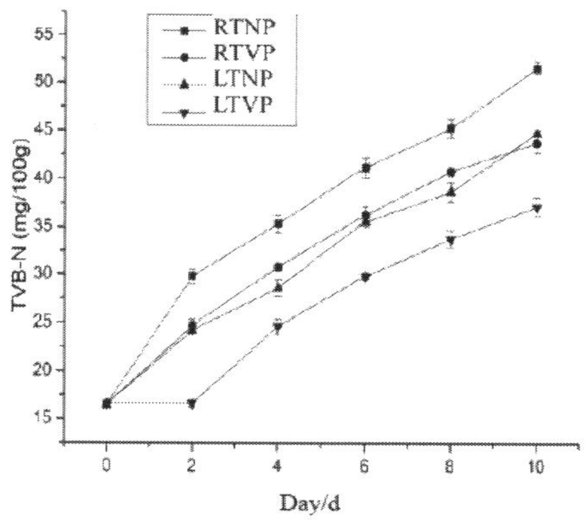

Figure 4B. Effects of different packaging methods on the TVB-N value.

3.3 *Effects of different packaging materials and methods on the TVB-N value*

The variation of the TVB-N value of fish samples stored with different packaging materials is shown in Figure 4A. With the increase in storage time, the TVB-N values of the fish samples under the three packing materials all showed an increasing trend. This is because, during storage, a large number of microorganisms grow and reproduce (Jeong 2018). Under the action of bacterial enzymes, proteins will decompose into volatile base nitrogen. Therefore, the higher the TVB-N value is, the more amino acids in the body are destroyed, and the higher the spoilage degree of meat products will be. However, the TVB-N value of the PA packed sample increased most obviously, and the TVB-N value was the highest after 10 days. The growth rate of the composite membrane (PE+PA) package was the smallest, and the TVB-N value was the smallest after 10 days. This indicates that in the compound membrane packaging environment, protein decomposition is less. The composite membrane packaging bag can better isolate external oxygen, so the composite membrane (PE+PA) can better store *Erythroculter ilishaeformis* fish and prolong its storage time.

As shown in Figure 4B, With the increase of storage time, the TVB-N value of the sample gradually increases. During the 10-day storage period, the TVB-N value content in low-temperature vacuum packaging was lower than that in the other three groups, which was probably because the TVB-N value was related to microbial growth and reproduction. Whether at room temperature or low temperature, the effect of vacuum packaging is better than that of atmospheric pressure, and the TVB-N content is lower, which also reflects that vacuum packaging can inhibit the increase of Sample TVB-N value, inhibit microbial growth in a short-term, and low temperature can effectively inhibit microbial growth. The synergistic effect of low temperature combined with vacuum packaging inhibits the growth and reproduction of microorganisms, reduces the decomposition rate of proteins, and thus reduces the production of basic volatile substances. The experimental data show that low-temperature vacuum packaging can effectively inhibit the spoilage of fish samples.

3.4 *Changes in total bacteria number with different packaging materials and methods*

In the process of fish storage, microorganism growth and reproduction are one of the important factors leading to fish spoilage. The total number of bacteria can be used as an index of microbial contamination of fish and can directly predict the shelf life of food (Zeng 2005). As can be seen from Figure 5A, with the increase in storage time, the total number of bacterial fish samples under the storage of the three packaging materials keeps increasing. The total number of bacteria in the PA packaging group increased rapidly, followed by the PE packaging group, and the composite

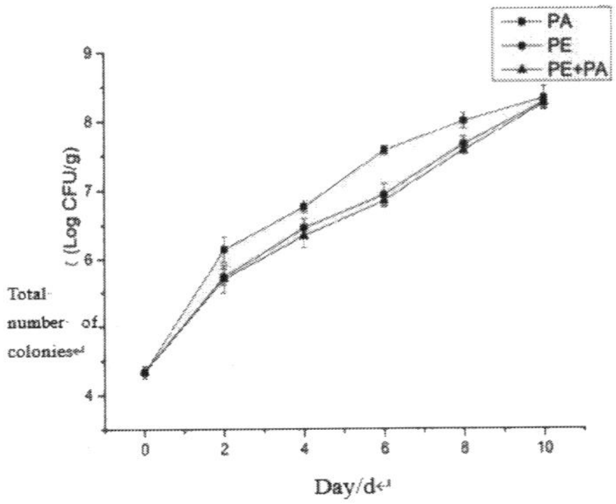

Figure 5A. Effects of packaging materials on the total bacteria number.

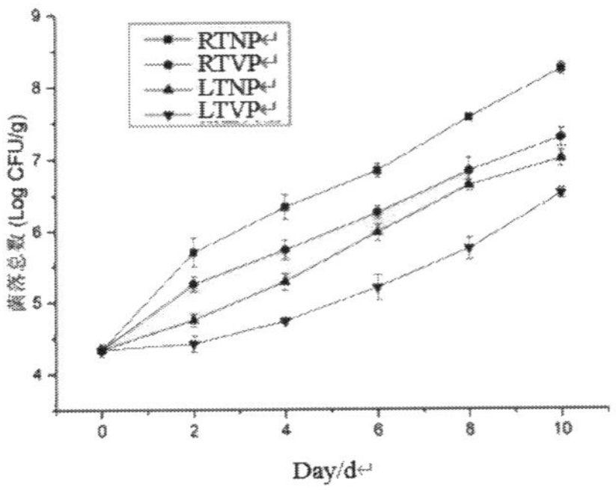

Figure 5B. Effects of packaging methods on the total bacteria number.

membrane group was slower. On the second day of storage, the total number of bacteria in all three groups was close to 6 logCFU/g, indicating that the fish samples had undergone some deterioration. In the storage process, the total number of bacteria in composite membrane packaging is always lower than in PA packaging and PE packaging, indicating that the composite membrane is the best among the three packaging materials in oxygen isolation. Composite membrane packaging material can inhibit bacteria and slow down the growth rate of microorganisms.

As can be seen from Figure 5B with the increase in storage time, the total number of colonies gradually increased. During the 10-day storage period, there was a significant difference in the total number of colonies. The total number of bacteria in low-temperature vacuum packaging was the lowest, which was lower than that in the other three packaging groups. The room-temperature atmospheric pressure packaging and room-temperature vacuum packaging were nearly corrupt on the 2nd and 4th day respectively, while the low-temperature atmospheric pressure packaging and low-temperature vacuum packaging were corrupt on the 8th and 10th day respectively. Under low-temperature vacuum packaging, the total number of bacteria in the sample was less, and the growth and reproduction of the microorganism in the sample were inhibited. Therefore, low-temperature vacuum packaging is better than low-temperature atmospheric pressure packaging, room temperature vacuum packaging, and room temperature atmospheric pressure packaging storage effect.

4 CONCLUSIONS

In conclusion, different packaging materials and packaging methods have significant effects on sensory scores, physical and chemical indexes (pH, TBA, and TVB-N values) and microbial indexes of preprocessed fish products from *Erythroculter ilishaeformis*. PE+PA composite membrane was more effective in inhibiting spoilage of fish samples. Among the four packing methods (room temperature and atmospheric pressure, room temperature vacuum, low temperature and atmospheric pressure, and low-temperature vacuum), the sample under low-temperature vacuum has a low degree of corruption and has a certain inhibitory effect on the growth and reproduction of microorganisms. The spoilage degree of fish samples under low-temperature vacuum packaging is the least, and fish samples can be stored better. Low-temperature vacuum packaging can effectively inhibit spoilage and prolong the shelf life of fish in a certain storage period. Low-temperature vacuum packaging combined with composite film (PE+PA) packaging materials, will get better results.

In addition to vacuum packaging, there are also air-controlled packaging in the market, and so many studies on the effects of freshness preservatives on fish quality (Li 2012), and the synergistic effects of different packaging materials and freshness preservatives. Therefore, there will be other storage methods for pretreated fish products of *Erythroculter ilishaeformis* fish that need to be further studied.

ACKNOWLEDGES

This study was funded by the Natural Science Foundation of Hubei Province, "Research on the structural transformation and mechanism of freshwater fish protein" (Grant No. 2019CFB336), "Undergraduate innovation project" of Hubei Province (Grant No. 202011654085), and Doctoral Fund Project of WBU (Grant No. 2017KB005). The research results were part of the projects.

REFERENCES

Chen, J. M., Ye, J. Y., & Pan, Q. (2003). A nutrition composition analysis of dorsal flesh of erythroculter ilishaeformis. *Journal of Zhejiang Ocean University*.

Fan Kai, Qiao Yu, Liao Li, et al. (2016). Influence of packaging way on the quality of weever during cold storage[J]. *Science and Technology of Food Industry*, 37(7): 316–321.

Gan Bozhong. (2001). The effects of pork fresh-preserving under several plastic film packaging[J]. *Journal of Gansu Agricultural University*, (3): 282–287.

Gu Saiqi; Zou Lin; Zhou Zhenyi, et al. (2021). Effect of different packaging methods on the quality characteristics of dried Engraulis japonicus [J]. *Journal of Fisheries of China*, 45(7): 1054–1065.

He Yanfu, Huang Hui, Li Laihao, et al. (2017). Advances on changes in quality of fish muscle during cold storage and its influence factors: a review[J]. *Journal of Dalian Ocean University*, 32(2): 242–247.

Huang Haiyuan; Shen Siyuan; Shi Wenzheng, et al. (2022). Effect of different packaging materials on the quality of semi-dried black carp during storage[J]. *Food and Fermentation Industries*, 48(2): 110–115.

Hu Jinxin; Li Junsheng; Xu Jing, et al. (2014). Advances on Methods of Freshness Characterization and Evaluation for Aquatic Product[J]. *The Food Industry*, 35(3): 225–228.

Jeong Ah Park, So Young Joo, Mi Sook Cho, et al. (2018). Changes in the physicochemical and microbiological properties of dried anchovy Engraulis japonicus during storage[J]. *Fisheries Science*, 84(6).

Konno, K., Imamura, K., & Yuan, C. H. (2011). Myosin denaturation and cross-linking in alaska pollack salted surimi during its preheating process as affected by temperature. *Food Science and Technology Research*, 17(5), 423–428.

Li Jianrong, Li Tingting, Li Xuepeng, et al. (2012). *Research Progress of Aquatic Products Freshness Evaluation Methods*[C]//The 9th Annual Meeting of CIFST: 264–265.

Li Weili; Wu Xiaoyu; Wang Qinghui. (2018). Effects of Various Packaging on the Shelf-life of Salmon Slice During the Cold Storage[J]. *Journal of Xihua University (Natural Science Edition)*, 37(2): 40–45.

Shen Haiyu, Pan Kaiyu, Lun Feng, et al. (2006). Research Status of Red Culter alburnus in China [J]. *Aquatic Products in China*, (8): 67–69.

Wu Yanyan, Zhao Zhixia, Li Laihao, et al. (2019). Effects of Different Packaging Methods and Storage Conditions on the Quality of Two Low-Salt Cured Tilapia Fillets[J]. *Food Science*, 40(9): 241–247.

Xiao Feng, Li Yanlong, Cheng Weiwei, et al. (2020). Effects of Vacuum and N_2 Packaging on the Freshness of Cyprinus Carpio Haematopterus During Chilled Storage[J]. *Food Science and Technology*, 45(6): 150–155.

Xiao Yingping, Liu Shenggao, Zhang Jun, et al. (2016). Analysis of Nutritional Components in Muscles of Erythroculter ilishaeformis During Different Growth Phases. *Acta Nutrimenta Sinica*, 38(2): 203–205.

Xie Jing, Yang Shengping. (2011). Effects of biopreservative combined with modified atmosphere packaging on shelf-life of trichiutus haumela[J]. *Transactions of the Chinese Society of Agricultural Engineering*, 27(1): 376–382.

Yang Hongxu. (2016). Effects of Various Low-temperature Storage on the Quality Changes of Black Carp Fillet. *Chinese Master's Theses Full-text Database*.

Zeng Qingzhu, Kristin Anna Thorarinsdottir, Guerun Olafsdottir. (2005). Research on quality changes and indicators of Pandalus borealis stored under different cooling conditions. *Journal of Fisheries of China*, (1): 87–96.

Zhao Shue. (2012). TBA model for surimi shelf life predicting[J]. *Jiangxi Food Industry*, (2): 26–27.

Energy Revolution and Chemical Research – Kok-Keong Chong and Zhongliang Liu (Eds)
© 2023 The Authors, ISBN: 978-1-032-36554-1

Optimization and application of parameter identification error algorithm for electro-hydraulic servo and actuator system of the turbine-based on slip window sampling

Longfei Zhu, Paiyou Si, Shuangbai Liu, Changya Xie & Teng Zhang
State Grid Jibei Electric Power Co., Ltd. Research Institute, North China Electric Power Research Institute Co. Ltd., Beijing, China

Yuou Hu
North China Branch of State Grid, Beijing, China

Xiaozhi Qiu
Beijing Jingneng Power Co. Ltd., Beijing, China

ABSTRACT: The parameter identification of steam turbine and its regulating system is of great value to the analysis of power grid stability. Parameter identification of turbine electro-hydraulic servo and actuator systems is an important part, and its identification error is strictly limited by relevant industry standards. The current error algorithm is greatly affected by human judgment and the data processing efficiency is low, which requires improvement. In this paper, the algorithm optimization of the parameter identification error of the turbine electro-hydraulic servo and actuator system is completed based on the slip window sampling method, and the visualization of the program is realized through the Python language. The calculation process of this method is rapid, the calculation results are accurate, and the standard requirements are met.

1 GENERAL INSTRUCTIONS

Carrying out the actual measurement and identification of parameters in the turbine regulation system and establishing the mathematical model of the turbine regulation system required for grid stability research can be used to systematically analyze the frequency response and load response curves of the grid under various disturbance conditions, which is of great practical value for the analysis of grid stability (Li 2011; Wang 2011). The turbine regulation system model is one of the four component models of the power grid, the authenticity of which directly affects the simulation accuracy of the power grid system (Liu 2008, 2012; Sheng 2012; Zhao 2009). Therefore, increasing attention has been paid to the accuracy of model parameter identification.

The parameter identification of the electro-hydraulic servo and actuator system of the steam turbine is an important part of the parameter identification of the steam turbine and its regulating system, and its identification error is strictly limited by the industry standards DL/T 1235-2019 *Guidelines for the Actual Measurement and Modeling of Synchronous Generator Prime Mover and its Regulation System Parameters* (hereinafter referred to as *Guidelines*). At present, manual punctuation is usually used for error calculation of identification results, which is greatly affected by human judgment and has low data processing efficiency. Optimizing the algorithm for identifying errors can effectively improve the efficiency of data processing and the accuracy of processing results.

In this paper, the automatic calculation of the parameter identification error of the electro-hydraulic servo and actuator system of the steam turbine is completed based on the slip window

 DOI 10.1201/9781003332657-95

method (Fu 2006; Yang 2018; 2019) and realized by the visualization program. The calculation process of this method is fast, the calculation results are accurate, and the standard requirements are met.

2 ALGORITHM OPTIMIZATION OF PARAMETER IDENTIFICATION ERROR OF TURBINE ELECTRO-HYDRAULIC SERVO AND ACTUATOR SYSTEM

The test contents of the parameter identification of the turbine and its regulating system include static test and load test. Among them, the purpose of the static test is to carry out the measured modeling of the electro-hydraulic servo and actuator system of the regulating system. The main test contents include large step and small step tests of electro-hydraulic servo and actuator. Usually, the small-step measured data of the turbine high-pressure cylinder regulating valve is selected for simulation.

2.1 *Parameter identification of electro-hydraulic servo and actuator system*

Parameter identification of electro-hydraulic servo and actuator system is to identify PID parameters by using standardized original data and model simulation test and provide accurate parameters for parameter identification of speed control system. The most commonly used model at present is the electro-hydraulic servo and actuator system (GA card) model provided in the *PSD-BPA Transient Stability Program User Manual*, as shown in Figure 1. The relevant parameters of the electro-hydraulic servo and actuator system are obtained through control logic inspection and data collection. The parameter identification process requires human intervention. The specific process is: first, assign the initial value of the parameter to be measured, use the simulation model for simulation calculation, and compare the output waveform continuously to reduce the deviation between the simulated value and the measured value, then simulate the system parameters based on conforming to the identification error standard, and finally select the simulation value with the highest degree of fit as the identification result of the parameter to be measured (Chao 2015; Zhu 2013).

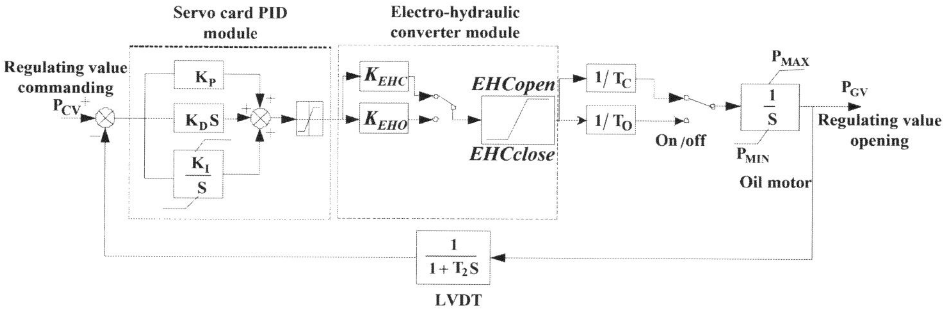

Figure 1. Electro-hydraulic servo and actuator system (GA card) model.

In Figure 1, P_{CV}: valve position command value; K_P, K_D, K_I: PID parameters of high-pressure regulating valve servo card, respectively referring to proportional coefficient, differential coefficient, and integral coefficient; T_C: shutdown time of oil motor; TO: on-time of oil motor; P_{MAX}: maximum output power of prime mover (maximum stroke of oil motor or maximum openness of the valve), per unit; P_{MIN}: minimum output power of prime mover (minimum stroke of oil motor or minimum opening of the valve), per unit; T_2: feedback link of oil motor stroke (LVDT) time; P_{GV}: the openness of high pressure regulating valve.

2.2 *Calculation method of identification result error*

The *Guide* stipulates the allowable deviation of each quality parameter in the comparison between the simulation and the actual measurement of the electro-hydraulic servo and actuator system, as shown in Table 1. In the table, the rise time t_{up} refers to the time required from the addition of the step amount to the change of the controlled amount to 90% of the step amount in the step test; the settling time t_s refers to the minimum time from the start time to the point where the absolute value of the difference between the controlled amount and the final steady-state value always does not exceed 5% of the step amount, as shown in Figure 2.

Table 1. Allowable deviation of each quality parameter.

Quality parameters	Allowed deviation (=Measured value-Simulated value)
Rise time t_{up}	±0.2s
Settling time t_s	±1.0s

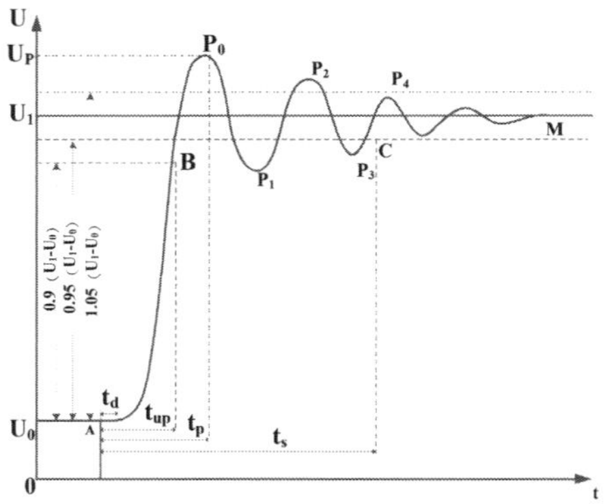

Figure 2. Example curve of the step response characteristic of the actuator.

Since both the measured curve and the simulated curve conform to the basic pattern shown in the example curve in Figure 1, it can be considered that the rise time t_{up} and settling time t_s of the measured curve and the simulated curve satisfy the same calculation method. Taking the measured curve as an example, the calculation method of t_{up} and t_s can be derived. The measured curve is not continuous, but a two-dimensional sequence with a sampling period of 1ms. The abscissa is the time, and the ordinate is the controlled amount (the step amount of the high-profile gate in the static test).

It is assumed that the initial step point of the measured curve is A (t_0, U_0), and the point that first reaches the steady-state M (t_1, U_1). Then the step amount is:

$$\Delta U = U_1 - U_0 \tag{1}$$

The point corresponding to the change of the controlled amount to 90% of the step amount in the curve is B (t_2, $U_0 + 0.9\Delta U$), from which Formula (2) can be obtained:

$$t_{up} = t_2 - t_0 \tag{2}$$

Through programming, the loop function is used to find all the points C_1 (t_{c1}, $U_0 + 0.95\Delta U$), C_2 (t_{c2}, $U_0 + 1.05\Delta U$), C_3 (t_{c3}, $U_0 + 0.95\Delta U$), C_4 (t_{c4}, $U_0 + 1.05\Delta U$) $\cdots$ C_n (t_{cn}, U_n) corresponding to 95% step and 105% step in the curve. The point C_n (t_{cn}, U_n) with the largest abscissa is found, and Formula (3) can be obtained:

$$t_{cn} = \max(t_{c1}, t_{c2}, \cdots) \tag{3}$$

From this, Formula (4) can be obtained:

$$t_s = t_{cn} - t_0 \tag{4}$$

Through the above derivation, the calculation Formulas (5) and (6) of the rise time deviation value Δt_{up} and the settling time deviation value Δt_s can be obtained:

$$\Delta t_{up} = t_{up0} - t'_{up} \tag{5}$$

$$\Delta t_s = t_{s0} - t'_s \tag{6}$$

Where t_{up0} and t_{s0} represent the rise time of the measured curve and the settling time of the measured curve; t'_{up} and t'_s denote the rise time of the simulation curve and the settling time of the simulation curve.

It can be seen that if we want to realize the automatic calculation of the error analysis for the parameter identification of the actuator, it is of great significance to judge the time point A at which the signal step occurs and the time point M at which the system stabilizes after the step. In this paper, the improved slip window method is used to judge the time when the step signal occurs and the system stabilizes again after the step occurs.

2.3 *Slip window sampling*

The slip window sampling method divides the time series samples into continuous-time windows and judges the state change of the system step occurrence and the system stabilization by calculating the normalized standard deviation change of the sample data. The calculation formula is as follows:

$$\mathrm{SSC} = \frac{1}{\bar{\mathrm{x}}}\sqrt{\frac{1}{N-1}\sum_{i=t-N+1}^{t}(\mathrm{x}_i - \bar{\mathrm{x}})^2} \tag{7}$$

where SSC (steady-state criteria) is the sample standard deviation, N is the time window width, and t is the current sample time.

First, the system detection signal samples are selected according to the time window formula, where x is a variable selected for judging quasi-steady-state characteristics, and an appropriate variable needs to be selected for the research object. For the error judgment of the parameter identification results of the turbine actuator, it is recommended to use the opening feedback of the high-pressure cylinder regulating gate. The selection of the time window width N is related to the inertial delay of the process. It can be initially selected according to the actual problem, and then further determined according to the actual calculation results. Usually, it can be taken as 10 s to 20 s (the sample sampling time interval is 0.1 s). After experiments, this paper selects N for 14 s.

2.4 *Improved slip window sampling*

The improved slip sampling window method normalizes the standard deviation of the sample data after obtaining the standard deviation of the sample data in the time window and can obtain the distribution of the standard deviation of the sample data between [0, 1], and then according to the set threshold value, the time point when the step signal occurs and the time point when the system stabilizes again can be determined, as shown in Formula (8).

$$\theta_{low} \le \mathrm{SSC}'_i = \frac{SSC_i}{\max(SSC_i)} \le \theta_{up} \tag{8}$$

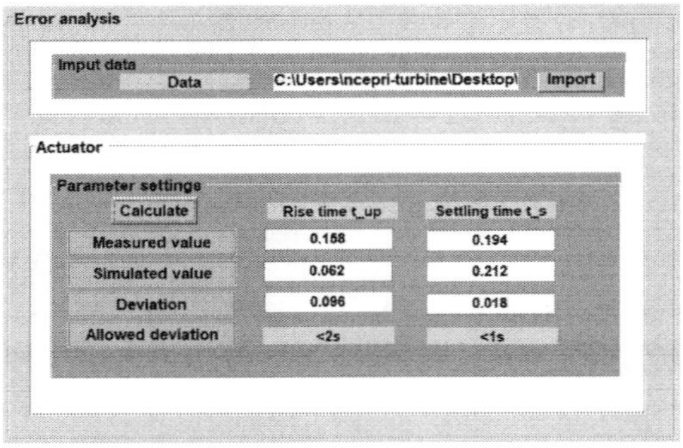

Figure 3. The visual interface of error calculation and testing results.

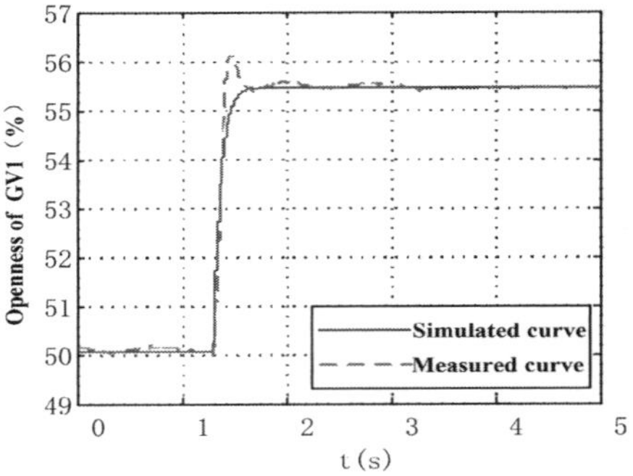

Figure 4. Identification results of GV1 of a 300 MW unit.

Among them, the judgment threshold θ is a dimensionless quantity, which reflects the strictness of the system step judgment. For the judgment of the step occurrence time, the larger θ is, the more obvious the step will be. Therefore, a larger threshold is not better for the initial step time. After the judgment of the initial step time was tested in this paper, θ_{up} was set to 0.95; for the judgment of the time when the system stabilizes again after the step occurs, the larger θ is, the more stable the system will be. However, due to the multiple uncertainties and the non-steady-state in the operation of the system, after the judgment of the post-step stability of the system was tested in this paper, θ_{low} was set to 0.05.

3 IMPLEMENTATION OF THE VISUALIZATION PROGRAM

When designing and developing the visual programming environment of Python language, Python language, and wxPython graphics library are used for design and implementation.

3.1 *Overview of python visual design*

Visual programming, also known as visual programming, is based on the principle of "what you see is what you get" and seeks to visualize the programming work, i.e., the final result can be seen at any time, which can achieve the synchronization between programming and the result.

The visual programming environment of the Python language is designed using the wxPython graphics library. wxPython is a GUI toolbox for the Python programming language, which enables Python programmers to easily create programs with robust and powerful graphical user interfaces.

3.2 *Error analysis module*

wxPython is adopted to visually program the algorithm described in Section 2 of this paper so that it has the following functions:

(1) The original files in TXT format and XLS format can be manually imported;
(2) The calculation of the identification error can be realized with one key, and there is no need to manually input any parameters;
(3) The operation process is fast, and the complete operation process is less than 2 s.

Finally, through programming, the visualization interface of the simulation identification error calculation of the electro-hydraulic servo and actuator system is obtained.

Through testing the program, the operation result is shown in Figure 3, with the operation time taking 0.532 s. It can be seen that the program meets the functional requirements.

4 APPLICATION EXAMPLES

Taking a 300 MW supercritical, one-time intermediate reheat coal-fired generator set as an example, the parameters of its electro-hydraulic servo actuator (high-pressure cylinder regulating valve GV1) were identified, and the measured data and simulated data were obtained as shown in Figure 4. Then, the traditional method and the algorithm in this paper were used to calculate the error. The results are shown in Table 2.

Table 2. Comparison of calculation results of identification error.

	Rise time (s)			Settling time (s)		
	Traditional method	The algorithm in this paper	Deviation margin	Traditional method	The algorithm in this paper	Deviation margin
Measured value	0.161	0.158	1.9%	0.197	0.194	1.5%
Simulated value	0.063	0.062	1.6%	0.214	0.212	0.9%
Deviation	0.098	0.096	2.1%	0.017	0.018	5.5%
Allowed deviation		±0.20		±1.00		

It can be seen from Table 2 that:

(1) The simulated deviations of the two algorithms are less than the allowed deviation, which meets the requirements of the *Guidelines*;
(2) The deviation margins of the two algorithms are small, and the calculation results are accurate.

5 CONCLUSION

Aiming at the problem that the calculation of the parameter identification error of the turbine electro-hydraulic servo and actuator system is greatly affected by human judgment and the data processing efficiency is low, this paper completes the algorithm optimization of the parameter identification error of the steam turbine electro-hydraulic servo and actuator system based on the slip window sampling and implements the program visualization through the Python language. Under the measured data of the static test of the speed control system in a certain unit, the algorithm described in this paper is used to calculate the error of the identification result of the electro-hydraulic servo and actuator system. The results show that the calculation process of the method is rapid, the calculation results are accurate, and the requirements of the standard are met.

REFERENCES

Chao H., Yu H., Xia C (2015). Modeling based on measured values and simulation of the control system of nuclear power unit prime mover. *J. Chinese Journal of Electrical Engineering*, 35 (2), 368–374.

Fu K. C., Dai L. K., Wu T. J (2006). Adaptive steady-state detection based on polynomial filtering algorithm. *J. Chemical Automation and Instrumentation*, 33(5), 5.

Li Y. H., Zhang C. S., Yang T., et al (2011). Experimental study on modeling of turbine speed regulation systems based on grid stability analysis. *J. Turbine Technology*, 53 (4), 291–294.

Liu H., Tian Y. F., Wu T (2008). Turbine governor model and its application considering unit coordination. *J. Power System Automation*, 32 (22), 103–107.

Liu H., Yang Y. P., Tian Y. F., et al (2012). Example of forced power oscillation in power system and analysis of mechanism. *J. Power System Automation*, 36 (10), 113–117.
National Energy Administration (2019). DL/T 1235-2019 *Guidelines for Measurement and Modeling of Synchronous Generator Prime Mover and Its Control System Parameters*. S. Beijing: China Electric Power Press.
Sheng K (2012). Influence of high-pressure cylinder steam volume link model parameters on the speed governing system simulation and calibration for reheat condensing turbine. *J. East China Power*, 40 (11), 2049–2053.
Wang G. H., Huang X (2011). Influence of turbine speed governing system parameters on the damping characteristics of power system. *J. Power Automation Equipment*, 31 (4), 81–90.
Yang Y. P, Li X. E, Yang Z. P, et al (2018). The Application of Cyber-Physical System for Thermal Power Plants: Data-Driven Modeling. *J. Energies*, 11(4), 690.
Yang Z. P, Li K. R., Wang N. L., et al (2019). Economic analysis of peak shaving of coal-fired generating units under the background of big data. *J. Chinese Journal of Electrical Engineering*, 39(16), 11.
Zhao T., Tian Y. F (2009). A mathematical model of electro-hydraulic servo and actuator for grid stability analysis. *J. Power System Automation*, 33 (3), 98–103.
Zhu X. X., Sheng K., Liu L. J (2013). Intelligent identification of turbine and its control system parameters based on integrated algorithm. *J. Power System Protection and Control*, 20, 138–143.

Energy Revolution and Chemical Research – Kok-Keong Chong and Zhongliang Liu (Eds)
© 2023 The Authors, ISBN: 978-1-032-36554-1

Selective synthesis of sulfinic ester from sulfide by photocatalytic oxidation

Qian Li & Xinrui Zhou*
Department of Fine Chemicals, Dalian University of Technology, Dalian, Liaoning, China

ABSTRACT: As the focus on environmental protection and energy crisis worldwide is significant, desulfurization of fuels plays an important role in the petroleum refining process, meanwhile related researches become a trending topic. In this study, using methyl phenyl sulfide on behalf of sulfides, one-step oxidative desulfurization was achieved by the method of adopting mild oxidant, mild and definite wavelength UV light, and performing in a non-polar solvent. The products were methanesulfinic phenyl esters, a *thia*-Baeyer-Villiger type reaction product, signed by the C-S cleavage and oxygen insertion. Methyl phenyl sulfide is transformed to methanesulfinic phenyl esters in 95% conversion and near 100% selectivity. This method is carried out in hydrocarbon solvent, and is more likely to be applied to the desulfurization of light fuels and drive the desulfurization of products to fine chemicals.

1 INTRODUCTION

Sulfides are widely present in gasoline, diesel, and other light petroleum products. SOx generated after combustion leads to acid rain, environmental pollution, and harm to human health. Desulfurization of fuel oil is an important topic in the industry and science, especially whether harmful sulfide can be converted into fine chemical products at the same time as desulfurization at present. Oxidative desulfurization is the most effective method to prepare functionalized sulfur-containing compounds.

Much research on the oxidation reaction of sulfide was published. According to the type of oxidants, there are metal oxide and its salt, halogen and halide, nitric acid and its salt, molecular oxygen oxidation, and peroxide systems (Chen et al. 2004; Dang et al. 2018; Jeon et al. 2014; Xu 2003; Zhou & Ji 2014), etc. Among the above oxidants, hydrogen peroxide is mostly investigated and favored, since it is easily available and environmentally friendly (Dk et al. 2019). Normally, the products are sulfoxide, sulfone, or a series of sulfur-containing oxides. Partial oxidation products pollute the environment and are not conducive to recycling. If thioether compounds are oxidized to sulfinic esters, they can be used as beneficial fine chemicals, and resources can be fully utilized while protecting the environment.

In our previous study, we discovered a *thia*-Baeyer-Villiger (B-V) type reaction based on dibenzothiophene compounds. The reactions were successfully conducted in methanol, at near room temperature and in 3 h. When employed *tert*-butanol peroxide (^{t}BuOOH) as oxidant, metal porphyrin catalyst as photosensitizer, and UV-lamp as light source, dibenzo[1,2]oxathiin-6-oxide compounds, a type of B-V products were selectively obtained (Shi et al. 2020). Unfortunately, this method is not successful on thioether compounds, the examination reactions carried out a series of mixture products but no B-V products. In this study, the method was adjusted according to the results of condition experiments that employed methyl phenyl sulfide (MPS) on behalf

*Corresponding Author: Chinaxinrui@dlut.edu.cn

DOI 10.1201/9781003332657-96

of sulfides. One-step oxidative desulfurization was achieved by this improved method adopting milder oxidant, milder and definite wavelength UV light, and performing in a non-polar solvent. The products were methanesulfinic phenyl esters (MPEs), a type of B-V product, signed by the C-S cleavage and oxygen insertion.

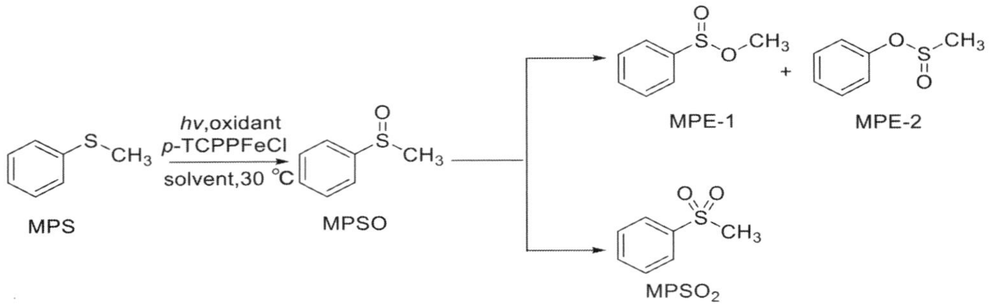

Figure 1. Reaction's pathways.

2 METHODS AND MATERIALS

2.1 *Materials and instruments*

All experiments were performed in the built-in lamp photoreactor with a magnetic stirrer and cooling water condenser. MPS was used as received. Solvents were redistilled. Urea hydrogen peroxide (UHP, 94%, white crystalline powder) was titrated by the iodometry method before being used. Meso-tetrakis(*p*-chlorophenyl)-porphine iron (III) (*p*-TCPPFeCl) were synthesized according to the references reported (Sun et al. 2011; Wang et al. 2006). Hg-lamp (Giguang, 250–450 nm, 125 W) and UV-lamp (Philips TUV-TL, $\lambda = 254$ nm, 6 W) were obtained commercially.

Gas Chromatography (GC) is performed on an Agilent 6890N (G1530N) chromatograph equipped with an FID detector, using an FFAP capillary column, all GC analyses are performed using standard procedures, needle wash solutions A and B are dichloromethanes, the injection volume is set at 3 μL, the shunt ratio is 50:1, the temperature initial value is 60°C, then 20°C isothermal gradients are warmed to 280°C, and the 280°C is run for 5 min.

^{1}H (400 MHz) and ^{13}C (400 MHz) NMR-spectra were recorded on Bruker UltraShield spectrometers. Mass spectra (MS) were recorded at linear ion trap-high resolution liquid-mass spectrometer LTQ Orbitrap XL.

2.2 *Experimental methods*

2.2.1 *General procedure for methyl phenyl sulfoxide (MPSO)*

In a 250-mL three-necked, round-bottomed flask, MPS (0.15 mmol, 19 mg) and *p*-TCPPFeCl (7.2×10^{-3} mmol, 6 mg) in methanol (15.0 mL) are added. UV-lamp was set in a quartz tube and installed in the middle of the round-bottomed flask. The reactor was equipped with a water condenser and immersed in the 30°C water bath with integrated temperature control. Initial 1 Equiv. ^{t}BuOOH (0.15 mmol, 13.5 mg) was added to the solution after switching on magnetic stirring and water recycling. The reactor was covered with tinfoil before turning on the lamp. After irradiating the reaction mixture for 0.5 h, the lamp was turned off, TLC sample and iodometry test were done and ^{t}BuOOH (1 Equiv.) was added into reaction solution only if no ^{t}BuOOH was left. The lamp was restarted to irradiate reaction solution for another 0.5 h. The last step was repeated until the designed time. Starch potassium iodide test paper was used to detect whether there was residual peroxide, quenched with sodium thiosulfate and extracted with dichloromethane. After evaporating the solvent, the crude product was purified by column chromatography with 200-mesh silica gel.

^1H NMR (400 MHz, $CCDCl_3$)δ (ppm): 7.70 - 7.64 (m, 2H), 7.53 (d, J = 7.5 Hz, 3H), 2.73 (s, 3H).

2.2.2 *General procedure for MPEs*

In a 250-mL three-necked, round-bottomed flask, add a solution of MPS (0.75 mmol, 95 mg) and catalyst (7.2×10^{-3} mmol, 6 mg) in solvent (30.0 mL). The light lamp was set in a quartz tube and installed in the middle of the round-bottomed flask. The reactor was equipped with a water condenser and immersed in the 30°C water bath with integrated temperature control. Initial 1 equiv. UHP was added to the solution after being switched on magnetic stirring and water recycling. The reactor was covered with tinfoil before turning on the lamp. After irradiating the reaction mixture for 1 h, the lamp was turned off, the TLC sample was rapidly taken, and iodometry test was performed by adding UHP (1 equiv.) into reaction solution only if no UHP was left. The lamp was restarted to irradiate reaction solution for 1 h more. The last step was repeated until the designed time. The target product is isolated by removing solvent and sublimating urea.

^1H NMR (400 MHz, D_2O) δ (ppm): 7.95 (d, J = 7.6 Hz, 2H), 7.78 (t, J = 7.8 Hz, 3H), 7.66 (t, J = 7.0 Hz, 2H), 7.54 (dt, J = 13.7, 7.0 Hz, 3H), 3.24 (d, J = 0.8 Hz, 3H), 2.79 (d, J = 0.9 Hz, 1H). ^{13}C NMR (400 MHz, D_2O) δ (ppm): 141.56, 137.85, 133.79, 130.82, 128.92, 128.22, 126.09, 124.59, 42.44. The NMR data agrees well with the ones reported (Zhu & Shi 2011).

3 RESULT AND DISCUSSION

3.1 *Condition experiment for complete conversion of MPS into the corresponding sulfoxide*

Previous researches indicate that only sulfoxides can undergo a B-V reaction. Therefore, the conditions for complete conversion of MPS to sulfoxide were first investigated. Normally, the rate constant of conversion from sulfoxide to sulfone (K2) is higher than that from sulfide to sulfoxide (K1, K2 > K1), i.e., the oxidative rate of sulfide is lower than that of sulfoxide, so it is very difficult to keep still on the stage of sulfoxidation. The reaction conditions optimized in the previous work, $^tBuOOH/CH_3OH$/*p*-TCPPFeCl/Hg-lamp, were examined on MPS. The results showed (Table 1, Entry 1) that a mixture of various substances, including MPSO and sulfone (MPSO), was formed. Taking use of natural sunlight, MPS was completely converted into the corresponding sulfoxide at 6 h (Table 1, Entry 2). Further, the light source was changed to UV-lamp with a single wavelength (254 nm) and low watt (6 W), as the results were shown in Table 1 (Entry 3), MPS was completely generated to the corresponding sulfoxide in 4 h, while the products were the mixture of sulfoxide and sulfone when taking use, no light. Therefore, UV-lamp (254 nm, 6 W) was the best light source for sulfoxide formation.

Table 1. Experimental results of preparation of MPSO*.

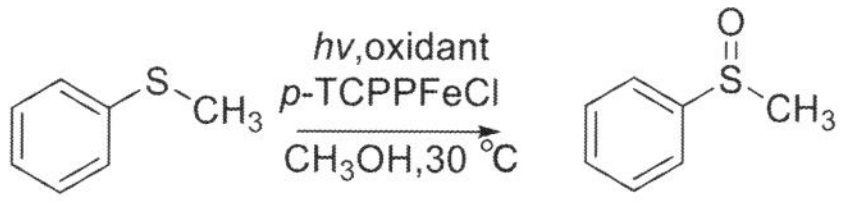

MPS MPSO

Entry	*Tert*-butano-l peroxide [Equiv.]	*Hv* λ [nm]//Watt	Time [h]	Conversion [%]	Selectivity [%]	Product
1	2	Hg-lamp 250-450/125	1	100	–	MPSO, $MPSO_2$ others mixture
2	1	Sunlight	6	100	~100	MPSO
3	2	UVlamp 254/6	4	100	~100	MPSO
4	1	No light in the room	22	100	–	MPSO and $MPSO_2$

*Note: Following synthesis procedure of sulfoxides 2.2.1.

3.2 *Selective generation of sulfinic esters*

As mentioned above, sulfoxide is an intermediate to generate B-V products. To simplify the desulfurization process, the effort was made to combine sulfoxidation and B-V reaction into one step. The above-optimized conditions (^{t}BuOOH/CH_3OH/254 nm) to obtain sulfoxide were introduced, but they just extended the reaction time. However, no desired product was detected by GC spectra until the reaction extended to 20 h (Table 2, Entry 1). Alternatively, water-soluble oxidant, UHP, was used. H_2O_2 was considered first, but researchers indicated it is more suitable for the inorganic catalytic system than for the organic catalytic system because it generally leads to the problem of organic catalyst decomposition due to H_2O_2 easily causing homolytic cleavage to produce strong oxidant, $HO^\bullet$ free radical, which destroys the aromatic system of porphyrins and leads to no selectivity of the products. Urea in UHP partially alleviates the tendency of the homolysis by two-dimension hydrogen bonds (Figure 2). However, the experiments still did not find out the desired products (Table 2, Entry 2). Moreover, a non-polar solvent was introduced to replace methanol, to investigate the effect of solvents, and the possibility of its application in the light fuels' desulfurization. The reaction was conducted in a homogeneous phase, in a mixture of hydrocarbons and alcohol. Ethanol was adopted instead of methanol since it can fully solve in cyclohexane and decalin, which respectively represent gasoline and diesel fuels. Reactions were carried out for 20 h, during the course there were no desired products detected by GC and TLC. Unfortunately, the products were still MPSO and $MPSO_2$ (Table 2, Entry 3, Figures 3 and 4). Finally, pure cyclohexane and decalin were used as solvents, and UHP was used as an oxidant. The results are shown in Table 2 (Table 2, Entry 4, 5). When cyclohexane was used as the solvent, MPS was completely transformed to MPEs. In decalin solvent, MPS is transformed to MPEs in 95% conversion and near 100% selectivity. Because this method is carried out in hydrocarbon solvent, it is more likely to be applied to the desulfurization of light fuels and drives the desulfurization products to the fine chemicals.

Table 2. Selective formation MPEs condition experiment*.

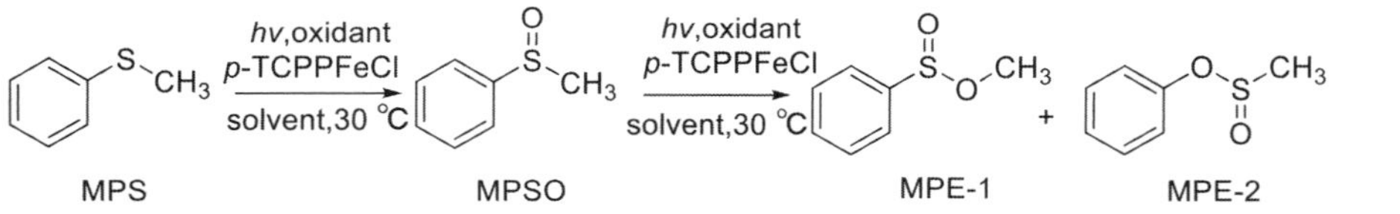

Entry	Oxidant (Equiv.)	Solvent	Time (h)	Conversion (%)	Selectivity (%) MPSO	$MPSO_2$	MPSEs	Unknown product
1	^{t}BuOOH/5	Methanol	20	100	90.37	9.63	–	–
2	UHP/10	Methanol	22	100	74.27	22.37	–	3.36
3	UHP/3	Cyclohexane/EtOH (V/V=1:2)	20	100	64.17	15.97	–	19.86
4	UHP/10	Cyclohexane	44	100	–	–	~100	–
5	UHP/10	Decalin	44	95	–	–	~100	–

*Note: Following the general procedure for MPEs 2.2.2

3.3 *One-step desulfurization and urea recovery*

The study continues to explore the feasibility of one-step desulfurization. As sulfinic ester is generally water-soluble, the use of oil-water two-phase system will separate the product from the oil phase. Moreover, urea of UHP is also transferred to the water phase, which would be recovered and reused. The reaction process depicted in Section 2.2.2 was adopted in the experiments. MPS and cyclohexane solution were used as model fuel, and a small amount of pure water (cyclohexane/water,

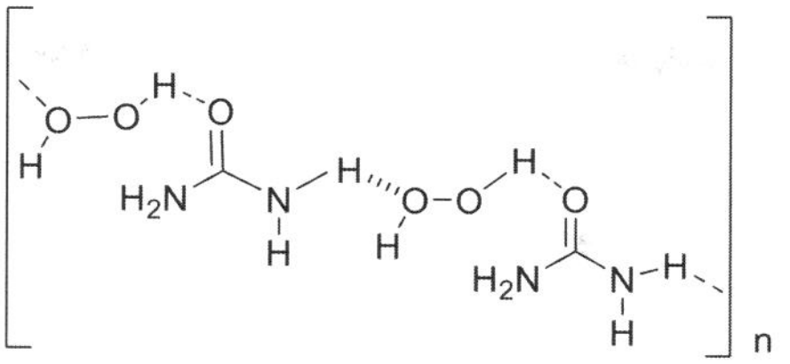

Figure 2. UHP structural formula.

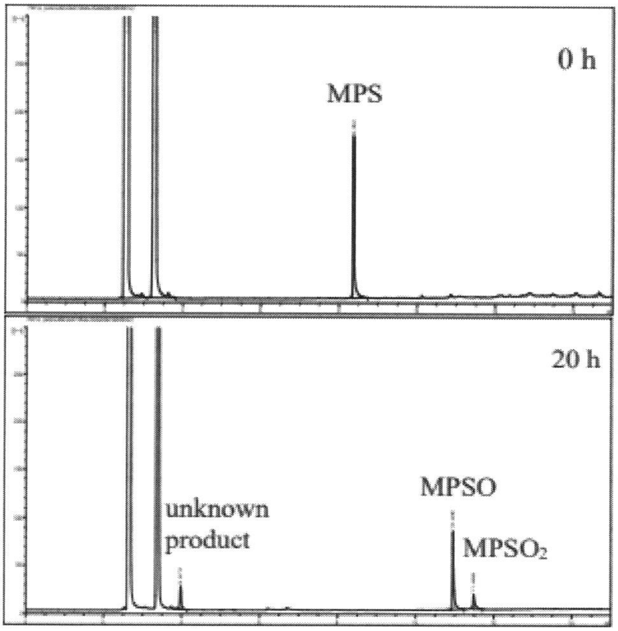

Figure 3-4. MPS in cyclohexane/ethanol photoreaction process GC.

v/v = 60/1) was added. After the reaction, the oil phase sample was tested by GC, and the test results showed that there were no MPS left in the model fuel (Figures 5 to 6). Negotiating the instrument error and raw material impurity peaks, the desulfurization rate was 100%.

Sulfinic ester is organic water-soluble, and might have certain lipophilic properties that normally cause the oil-water interface to be unclear or even show flocculent precipitation. Therefore, the experiments on the oil-water interface were performed according to GB/T19230.2-2003. The reaction liquid was transferred to the measuring cylinder, and the water-oil interface was observed, the results indicated that meet Level 1 of clarity and cleanliness standards.

Urea, a by-product generated by UHP, was obtained by simple sublimation after water removal. This recovery of useful chemicals can reduce the cost of desulfurization and protect the environment. After removing water and urea, the product was detected by ^{1}H-NMR and ^{13}C-NMR, and the results showed that it was MPEs. Since NMR can only detect organic compounds, to verify the formation of SOx in the product, the pH value of the water phase before purification was tested. pH value did not display a visible change, which excludes the formation of small pieces of sulfur oxides.

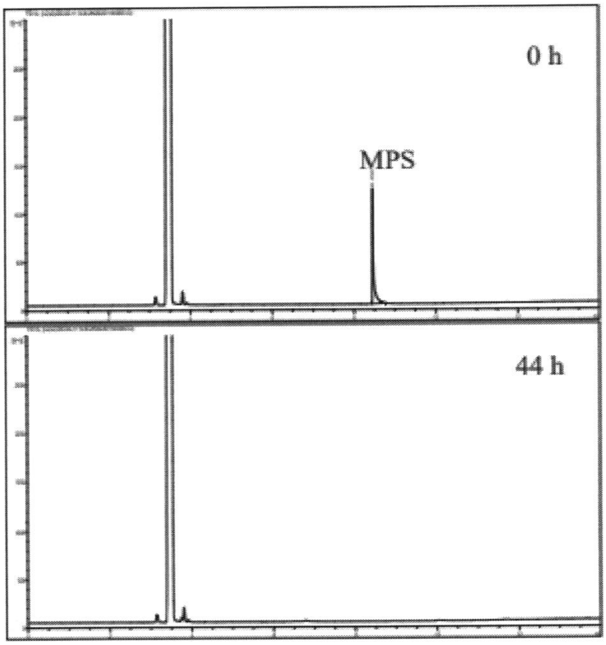

Figure 5-6. MPS in cyclohexane photoreaction process GC.

4 CONCLUSION

We have found a type of B-V product—MPEs in thioether, signed by the C-S cleavage and oxygen insertion. Condition experiment for complete conversion of MPS into the corresponding sulfoxide indicated that UV-lamp (254 nm, 6 W) was the best light source for sulfoxide formation, and the selectivity was nearly 100%. Condition experiment for selective generation of sulfinic esters showed that in hydrocarbon solvent system, using UHP and *p*-TCPPFeCl, MPS was transformed to MPEs under UV-lamp. GC spectra showed that the conversion reached more than 95% and selectivity is closed to 100%. Urea is recycled by sublimation. After removing water and urea, the product was detected by ^{1}H-NMR and ^{13}C-NMR, and the results confirmed it was MPEs. As more attention is paid to environmental protection and the energy crisis, desulfurization of fuel oil has become one of the most urgent topics in industry and science. This encouraging result is more likely to be applied to desulfurization of light fuels and drive the desulfurization of products to fine chemicals. So that sulfur resources can be used meaningfully.

REFERENCES

Chen, M.Y. , Patkar, L.N. & Lin, C.C. 2004. Selective oxidation of glycosyl sulfides to sulfoxides using magnesium monoperoxyphthalate and microwave irradiation. *Journal Organic Chemistry*, 35(8): 2884–2887.

Dang, C., Zhu, L., Guo, H., & Dick, B. 2018. Flavin Dibromide as an Efficient Sensitizer for Photooxidation of Sulfides. *ACS Sustainable Chemistry & Engineering*, 6(11): 15254–15263.

Dk, A., Ar, A. & Sra, B. 2019. Magnetic core-shell nanoparticle-supported Sc(III): A novel and robust Lewis acid nanocatalyst for the selective oxidation of sulfides to sulfoxides by H2O2 under solvent-free conditions. *Catalysis Communications*, 124: 46–50.

Jeon, H.B., Kim, K.T. & Kim, S.H. 2014. Selective oxidation of sulfides to sulfoxides with cyanuric chloride and urea-hydrogen peroxide adduct. *Tetrahedron Letters*, 55(29): 3905–3908.

Shi, W.J. , Zhou, X.R. , Kong, Y. , Li, J. & IE Markó. 2020. Unique Thia-Baeyer-Villiger-Type Oxidation of Dibenzothiophene Sulfoxides Derivatives. *Chemistry-An Asian Journal*, 15(4): 511–517.

Sun, Z.C., She, Y.B., Zhou, Y., Song, X.F. & Li, K. 2011. Synthesis, Characterization and Spectral Properties of Substituted Tetraphenylporphyrin Iron Chloride Complexes. *Molecules*, 16: 2960–2970.

Wang, L., She, Y., Zhong, R., Ji, H., Zhang, Y. & Song, X. 2006. A green process for oxidation of p-nitrotoluene catalyzed by metalloporphyrins under mild conditions. *Organic Process Research & Development*, 10(4): 757–761.

Xu, L. 2003. Chromium(VI) oxide catalyzed oxidation of sulfides to sulfones with periodic acid. *Journal of Organic Chemistry*, 34(42): 5388–5391.

Zhou, X.T. & Ji, H.B. 2014. Highly efficient selective oxidation of sulfides to sulfoxides by montmorillonite-immobilized metalloporphyrins in the presence of molecular oxygen. *Catalysis Communications*, 53: 29–32.

Zhu, R.H. & Shi, X.X. 2011. Practical and highly stereoselective method for the preparation of several chiral arylsulfinamides and arylsulfinates based on the spontaneous crystallization of diastereomerically pure N-benzyl-N-(1-phenylethyl)-arylsulfinamides. *Tetrahedron: Asymmetry*, 22(4): 387–393.

Energy Revolution and Chemical Research – Kok-Keong Chong and Zhongliang Liu (Eds)
© 2023 The Authors, ISBN: 978-1-032-36554-1

Calculation and research on transition process of hydraulic interference in tailrace tunnel hydraulic power generation system shared by two units

Haiyang Liu
Sichuan Dadu River Shuangjiangkou Hydropower Development Co., Ltd., Sichuan, China

ABSTRACT: In the diversion power generation system of the tailrace tunnel shared by two units, a hydraulic disturbance is an inevitable phenomenon in daily operation, and it is of far-reaching significance to study the influence of hydraulic disturbance on the transition process. In this paper, taking the water diversion power generation system with the layout of "single pipe and single unit" and "two units, one tailrace tunnel" as an example, the mathematical model of the water diversion power generation system is constructed by the characteristic line method. The changes in unit output, the relative opening of guide vanes, relative rotational speed, volute end pressure, draft tube inlet pressure, and surge water level in the draft tube surge chamber during the transition of hydraulic disturbance were calculated. The calculation results show that during the hydraulic disturbance transition process of the dual-unit shared tailrace tunnel diversion power generation system in this paper, the maximum surge in the tailrace surge chamber is 2257.9 m, and the extreme value occurrence time is 39.8 s, and the maximum speed increase rate is 44%. The maximum water pressure of the volute is 308.4 m, all of which meet the design requirements.

1 GENERAL INSTRUCTIONS

It is a common development method for hydropower stations to share one hydraulic unit with multiple units. For units developed with multiple units sharing hydraulic units, the problem of hydraulic interference between multiple units will inevitably occur. When some of the units shed load or adjust the load greatly for some reason, the tailwater flow state will fluctuate, which will affect the water head, output, speed, etc. of the associated units, and even impact the power grid. With the increasing single-unit capacity of newly put into production units, the characteristics of hydraulic interference are more worthy of study. Therefore, the research on hydraulic disturbance is also a hot field in the hydropower industry. Fu Liang (2016) used a numerical calculation method to calculate and analyze various factors affecting hydraulic disturbance; Lin Yongbing (2017) studied the mutual influence of hydraulic disturbance when the unit load suddenly changed; Zou Jin (2018)) established a hydraulic interference model for a multi-machine water diversion power generation system based on detailed models of water diversion pipelines, turbines, and generators; Wang Liang (2020) conducted a hydraulic interference test on a pumped-storage power station. The results show that the start-up and shutdown of adjacent units and load shedding have a certain impact on the power of the operating units; Sun Xinzhou (2018) used a numerical calculation method to analyze and study the grid-connected operation mode and hydraulic disturbance transition process of a large hydropower station; Zhang Xianyu (Zhang & Li 2014) used the characteristic line method to analyze the influence of the moment of inertia of the unit on the adjustment quality, which provided a certain basis for choosing the appropriate moment of inertia, etc; This paper takes the hydraulic interference of two units sharing the tailwater hydraulic unit of a power plant as an example to study the influence of hydraulic interference on the units. The power plant is located in the Dadu River

 DOI 10.1201/9781003332657-97

Basin and is a first-class (1)-level project. The main building of the hub is a first-class building, and the secondary buildings are third-class buildings. The plant is equipped with 4 vertical-axis Francis hydro-generator units with a capacity of 500 MW and adopts the layout of "single-unit, single-pipe water supply" and "two-unit, one-room, and one-hole" layout. There are 2 hydraulic units in the whole plant, and 4 pressure pipes are arranged in parallel. The tailwater is led from the tailwater pipe of each unit to the tailwater surge chamber, and every 2 units share a tail water surge chamber and a pressurized tail water tunnel to lead to the downstream river. Due to the prominent features of the power plant layout and unit form, the numerical calculation and analysis of hydraulic interference can provide a reference for the same type of units, which has a certain engineering reference value. The specific operating flow and water head of the power plant are shown in Table 1.

Table 1. Power plant flow and head details.

	parameter	value
Turbine water head	maximum head	251.4 m
	weighted average head	221.2 m
	flood season weighted average water head	228.6 m
	rated water head	215.0 m
	minimum head	161.6 m
	minimum head of initial water storage	155.40 m
Flow	average traffic over the years	502 m^3/s
	the maximum reference flow of the power station	1064 m^3/s

2 THEORETICAL MODELS AND COMPUTATIONAL METHODS

2.1 *Basic assumptions*

It is generally assumed that a one-dimensional unsteady flow with pressure has the following characteristics:

(1) The water flow in the pipe is a one-element flow, and the velocity distribution of the cross-section of the pipe is uniform;
(2) The water flow in the pipe wall and the pipe is linearly elastic, and its stress and strain are proportional;
(3) The resistance loss formula applied to the calculation of constant flow in a pressurized pipeline is still applicable to transient flow, and the water flow in the pipeline system is gradual in each section.

Taking Units 1 and 2 sharing a hydraulic unit as the calculation object, the calculation diagram of the hydraulic unit is shown in Figure 1:

The equation of motion is:

$$\frac{\partial V}{\partial t} + V\frac{\partial V}{\partial x} + g\frac{\partial H}{\partial x} + g\frac{fV|V|}{2D} = 0 \tag{1}$$

The continuity equation is:

$$\frac{\partial H}{\partial t} + V\frac{\partial H}{\partial x} - V\sin\alpha + \frac{a^2}{g}\frac{\partial V}{\partial x} = 0 \tag{2}$$

The positive characteristic line equation is:

$$C^+ : H_P = C_P - BQ_P \tag{3}$$

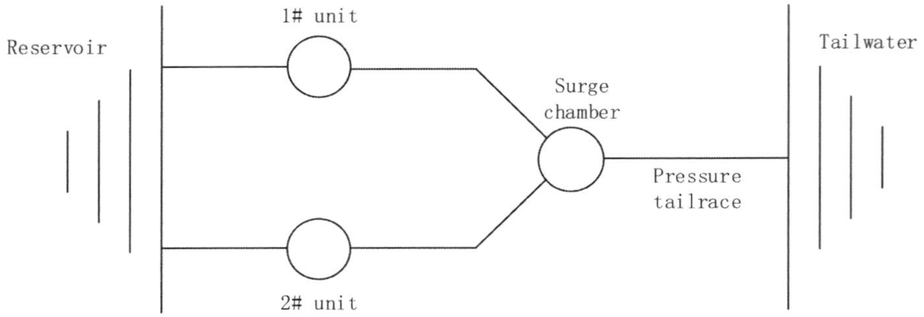

Figure 1. Diagram of the hydraulic unit of a power plant.

The negative characteristic line equation is:

$$C_P = H_A + BQ_A - R|Q_A|Q_A \tag{4}$$

$$C_M = H_A - BQ_B - R|Q_B|Q_B \tag{5}$$

$$B = \frac{a}{gA} \tag{6}$$

$$R = \frac{f\Delta x}{2gDA^2} \tag{7}$$

2.2 *Boundary conditions*

Upstream reservoir boundary conditions are expressed as

$$H_P = H_R - (1 \pm \xi)\frac{Q_P^2}{2gA^2} \tag{8}$$

Turbine boundary conditions are expressed as

$$H_{PU} = C_{P1} - B_1 \cdot Q_P \tag{9}$$

$$H_{PD} = C_{M2} - B_2 \cdot Q_P \tag{10}$$

Boundary conditions of the tail water surge chamber are expressed as

$$C^+ : H_{P4} = C_{P4} - B_4 \cdot Q_{P4} \tag{11}$$

$$C^- : H_{P5} = C_{M5} + B_5 \cdot Q_{P5} \tag{12}$$

The downstream channel boundary conditions are expressed as

$$H_P = H_R - (1 + \alpha)Q_P^2/(2gA^2) \tag{13}$$

$$C^+ : H_P = C_{PS} - B_S \cdot Q_P \tag{14}$$

The governor transfer function is expressed as

$$G(s) = \frac{1 + T_d S}{T_d T_y S^2 + [T_y + T_d b_t]S + b_p} \cdot \frac{1 + T_n S}{1 + K_n T_n S} \tag{15}$$

3 CALCULATION AND ANALYSIS OF TRANSITION PROCESS OF HYDRAULIC INTERFERENCE

3.1 *Calculation conditions*

In the calculation of the hydraulic disturbance transition process, it is assumed that the unit is connected to the infinite power grid and the governor of the running unit participates in the adjustment. The calculation research is carried out according to the working conditions designed in Table 2.

Table 2. Calculation conditions and description of the hydraulic disturbance transition process.

No.	Condition number	Load change	Description of water level combination and working condition change
1	HDT1	2 units → 1 unit	Rated water head, the rated power of 2 units is in normal operation, one of the units accidentally dumps the full load, and the guide vanes are closed in an emergency.
2	HDT2	2 units → 1 unit	The maximum water head, the rated power of 2 units are in normal operation, one unit accidentally dumps the full load, and the guide vanes are closed in an emergency
3	HDT3	1 unit → 2 unit	Rated water head, the rated power of one unit is running normally, and the other unit is increased from no-load to rated power
4	HDT4	1 unit → 2 unit	With the maximum water head, the rated power of one unit is running normally, and the other unit is increased from no-load to the maximum power

The initial conditions for calculating the hydraulic disturbance transition process of Unit 1 are shown in Table 3.

Table 3. Initial conditions for calculation of hydraulic disturbance transition process of Unit 1.

Calculation condition	Upstream water level (m)	Downstream water level (m)	Unit	Initial relay opening (%)	Initial net head (m)	Initial flow (m^3/s)	Initial output (MW)
HDT1	2469.2	2251.12	1#	82.6	215.0	258.4	510.0
			2#	82.6	215.0	258.4	510.2
HDT2	2504.42	2251.12	1#	63.5	250.9	215.8	510.0
			2#	63.5	251.0	215.8	509.8
HDT3	2469.2	2251.12	1#	13.0	217.7	22.8	0.0
			2#	82.6	215.9	259.1	513.2
HDT4	2504.42	2251.12	1#	11.5	253.0	21.7	0.0
			2#	63.2	251.8	215.1	510.1

3.2 *Calculation results*

After the initial conditions are determined, the hydraulic disturbance transition process can be calculated, and the calculation results are shown in Figures 2 to 5. The abscissas in the figure all represent time, and the unit is s; P1 and P2 represent the turbine output of No. 1 and No. 2 units respectively, the unit is MW, the left vertical axis; s1 and s2 represent the relative opening of the guide vane follower of Units 1 and 2 respectively, the unit is %, the right vertical axis (2nd); n1 and n2 represent the relative rotational speed of No. 1 and No. 2 units respectively, in %, on the right vertical axis; HpSC_END1 and HpSC_END2 represent the volute end pressure of Unit 1 and Unit 2 respectively, the unit is m, the left vertical axis; HpDT_IN1 and HpDT_IN2 represent the draft

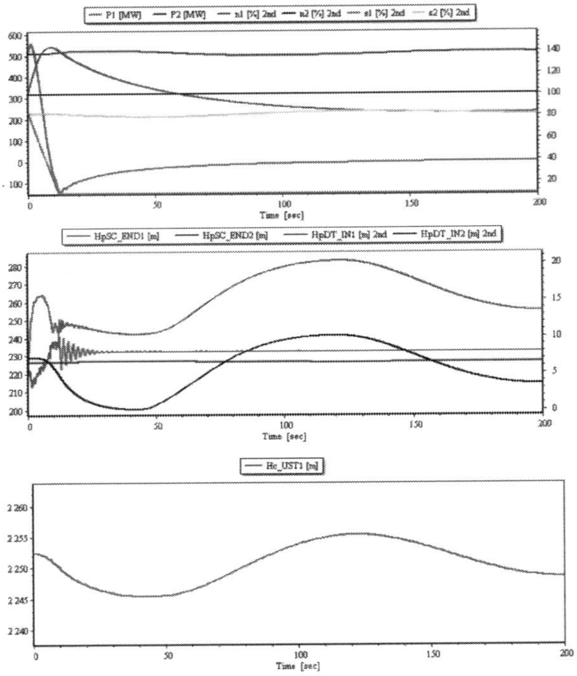

Figure 2. Variation process of each parameter of HDT1 with time.

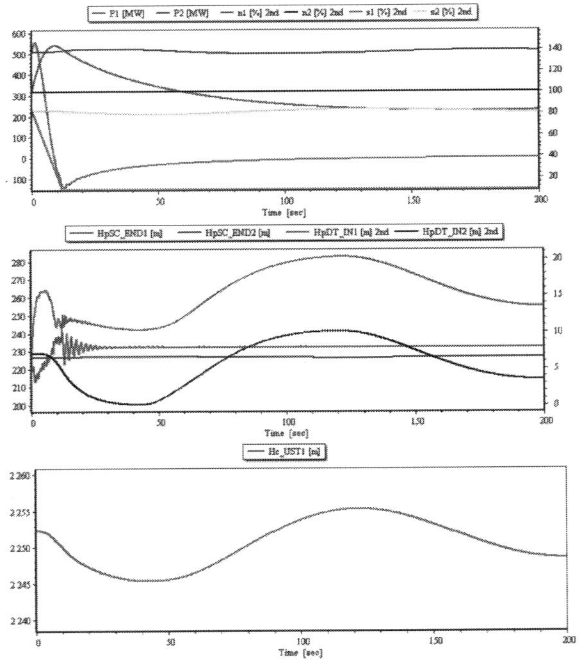

Figure 3. Variation process of each parameter of HDT2 with time.

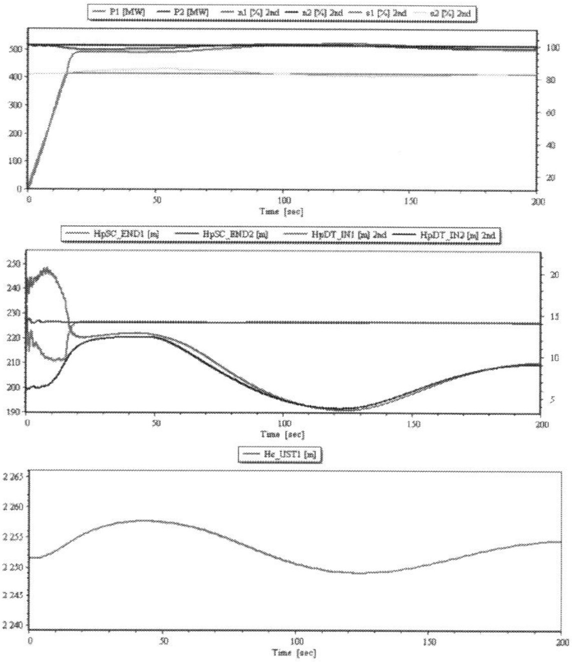

Figure 4. Variation process of each parameter of HDT3 with time.

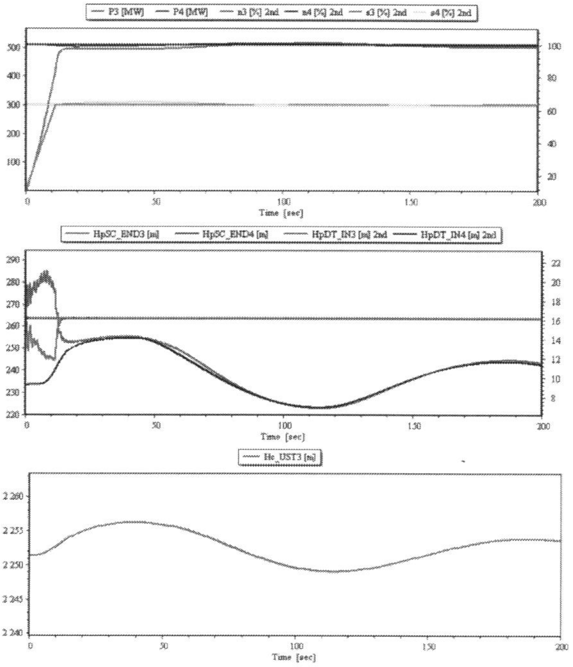

Figure 5. Variation process of each parameter of HDT4 with time.

Table 4. Output change table of Unit 2.

Working condition	Initial output (MW)	Maximum output (MW)	Extreme value occurrence time (s)	Minimum output (MW)	Extreme value occurrence time (s)	Maximum deviation up (%)	Downward maximum deviation (%)
HDT1	510.2	522.2	25.1	499.7	93.2	2.4	2.0
HDT2	509.8	517.2	23.6	503.7	87.4	1.4	1.2
HDT3	513.2	519.9	97.4	499.3	24.8	1.9	2.1
HDT4	510.0	515.5	94.2	503.7	20.5	1.0	1.2

Table 5. Change of water level in the tailwater surge chamber.

Working condition	Initial water levelZ0 (m)	Maximum water levelZ1 (m)	Extreme value occurrence time (s)	Minimum water level (m)	Extreme value occurrence time (s)
HDT1	2252.4	2255.4	117.5	2245.0	37.4
HDT2	2252.0	2255.0	117.4	2245.5	38.3
HDT3	2251.5	2257.9	39.8	2249.0	121.9
HDT4	2251.4	2256.8	39	2248.9	115.8

Table 6. Variation table of unit parameters in the transition process.

Working condition	Unit	Maximum speed increase rate (%)	Extreme value occurrence time (s)	Maximum pressure at the end of the volute (m)	Extreme value occurrence time (s)	Draft pipe inlet minimum pressure (m)	Extreme value occurrence time (s)
HDT1	1#	44.0	8.8	264.5	5.6	3.2	1.8
	2#	/	/	226.9	46.0	0.1	41.3
HDT2	1#	33.0	7.0	308.4	2.5	3.7	1.5
	2#	/	/	263.5	30.0	3.5	42.0
HDT3	1#	/	/	232.4	0.0	3.5	124.3
	2#	/	/	227.7	0.9	3.8	121.4
HDT4	1#	/	/	267.6	0.0	6.6	121.2
	2#	/	/	263.6	106.5	6.8	120.0

pipe inlet pressure of No. 1 and No. 2 units, respectively (unit: m), on the right vertical axis (2nd); Hc_UST1 and Hc_UST3 represent the surge water level in the draft surge chamber of Units 1 and 2, respectively,(unit: m), on the left vertical axis.

Due to the hydraulic interference caused by the load change of the No. 1 unit, the calculation results of the changes in the parameters of the No. 2 unit are shown in Tables 4 to 6.

From the calculation results of the hydraulic disturbance transition process, it can be known that:

(1) When the load of one unit fluctuates, the maximum deviation of the output of the other unit upward is 2.4%, which occurs when one unit is running at full load under the rated water head and the other unit is load-rejected; the maximum deviation of the output downward is 2.1%, which occurs when one unit runs at full load under the rated water head, and the other unit increases from no-load to full-load condition.
(2) The maximum surge water level in the tailwater surge chamber is 2257.9 m, and the occurrence time is 39.8 s. It occurs when one unit is running at full load under the rated water head, and the other unit is increased from no load to full load condition; in the tailwater surge chamber the minimum surge water level is 2245.0 m, and the occurrence time is 37.4 s. It occurs under the rated water head, one unit is running at full load, and the other unit is under load rejection.

(3) The maximum speed increase rate is 44.0%, the occurrence time is 8.8 s, and the occurrence condition is that one unit is running at full load under the rated water head, and the other unit is load shedding.
(4) The maximum pressure of the volute is 308.4 m, and the occurrence time is 2.5 s. The occurrence condition is the maximum water head, one unit is running at full load, and the other unit is load shedding.

4 CONCLUSION

In this paper, combined with the actual engineering data of a dual-unit shared tailrace tunnel diversion power generation system, using the transition process calculation theory of the pressure transient flow characteristic line method, the water diversion of a hydropower station with "single-pipe single-unit" "two units, one tailrace tunnel" is used. A complete mathematical model of the power generation system is established, and the characteristic line method is used to calculate and study the transition process of hydraulic interference. Under the condition of hydraulic interference, the water level in the tailrace surge chamber, the pressure at the end of the volute, and the rotational speed of the unit are calculated and analyzed, and the possible maximum and minimum surge water levels in the tailrace surge chamber are given. The calculation results show that the surge water level in the tailwater surge chamber is convergent, and all parameters of the unit can be within the control value range under the condition of hydraulic interference.

REFERENCES

Fu Liang & Zou Guili & Kou Pangao. (2016) Analysis on the transition process of hydraulic disturbance of two units in a hydropower plant. *J. Hunan Electric Power*, 36(06): 56–9.

Lin Yongbing. (2017) *Research on Hydraulic Interference Problems in One-pipe-Dual-machine System Unit Load Sudden Changes*. D. Xi'an University of Technology.

Sun Xinzhou & Zhang Chao & Diao Xuefen. (2018) Analysis and Research on Grid-connected Operation Mode and Hydraulic Interference of Large Hydropower Stations. *J. Power Grid and Clean Energy*, 34(07): 63–6.

Wang Liang & Deng Lei & Yang Bin. (2020) Experimental study on hydraulic disturbance of Jiangsu Yixing Pumped Storage Power Station. *J. Hydropower and Pumped Storage*, 6(06): 86–92.

Zhang Xianyu & Li Naihui. (2014) Analysis of the influence of the rotational inertia of the unit on the adjustment quality of the unit under hydraulic interference. *J. China Water Transport* (Second Half Month), 14(02): 152–3+6.

Zou Jin & Qin Qiong & Lai Xu. (2018) Simulation Analysis of Hydraulic and Electrical Transition Process in Hydraulic Interference of Multi-machine Water Diversion Power Generation System. *J. China Science and Technology Papers*, 13(07): 813–8.

Energy Revolution and Chemical Research – Kok-Keong Chong and Zhongliang Liu (Eds)
© 2023 The Authors, ISBN: 978-1-032-36554-1

Preparation and characterization of Mo-doped In_2O_3 thin films with magnetron Co-sputtering

Jianping Ma*, Jingjing Chen & Jiayu Qi
Department of Electronic Engineering, Xi'an University of Technology, Xi'an, China

ABSTRACT: Mo was used as a dopant to deposit In2O3 thin films with different doping contents on quartz substrates by magnetron co-sputtering. XRD results show that the structure of the doped In_2O_3 film is still cubic ferromanganese. Scanning electron microscopy observations show that as the power applied to the Mo metal target increases, the film surface is uniform, but the crystallinity becomes poor. The thickness of the prepared sample film is 400 nm. The results of the energy dispersive X-ray show that the Mo element is doped into an In_2O_3 thin film. The electrical performance shows that all deposited films are n-type semiconductors. The resistance of the IMO film is the smallest when the DC power on the Mo target is 15w, its value is $0.97\times10^{-3}\Omega$•cm, and the maximum mobility is 73.6 $cm^2V^{-1}s^{-1}$. The carrier concentration can reach 9.1×10^{19} cm^3 when the DC power is 20 W. The optical band gap of the film increases from 3.746 eV to 3.909 eV with the increase of applied DC power.

1 INTRODUCTION

Indium oxide(In_2O_3), an N-type semiconductor material with a wide band gap (3.7 eV) (Beena et al. 2010), which has high optical transparency in the visible light range (Cui et al. 2018) and high mobility (Zhang et al. 2019). Therefore, it is widely used in transparent electrodes, solar cells, and gas sensors (Krishnan etal. 2018; Liu et al. 2018). In_2O_3 and impurity-doped In_2O_3 (Xu et al. 2018; Yan et al. 2018) have attracted much attention as important functional oxide semiconductors materials. At present, many researchers focus on improving the mobility of doped In_2O_3 films to improve the performance of thin films (Elangovan et al. 2007), so it is very important to choose suitable doping materials.

For high-valent difference doped materials (Pan et al. 2014; Reshmi Krishnan et al. 2019), a small amount of doping can obtain enough carriers. Simultaneous doping reduces the scattering of charged ions and improves carrier mobility and transmittance. High-valence doping further enhances the mobility of the In_2O_3 thin film, as well as their electrical properties without sacrificing optical transmittance, which has caused widespread attention from scholars at home and abroad. Molybdenum has attracted much attention as a high-valence doped material. When molybdenum is selected as the dopant (Meng et al. 2001; Miao et al. 2005), high-valence Mo ions enter the In_2O_3 lattice as doping ions and replace in ions to form molybdenum-doped indium oxide (IMO). There are various preparation methods for high-quality IMO films, including electron beam reactive evaporation (Chen et al. 2011),pulsed-laser deposition technique (Par et al. 2009), activation reaction evaporation technology (Kaleemulla et al. 2008), magnetron sputtering (Elangovan et al. 2008). Among them, the magnetron sputtering method has the advantages of a simple process, good thin film density, and good film uniformity (Kelly & Arnell 2000).

*Corresponding Author: majp@xaut.edu.cn

 DOI 10.1201/9781003332657-98

In this paper, the magnetron co-sputtering method was used to deposit IMO films with different doping concentrations. From the aspects of surface morphology, microstructure, optical and electrical properties, the effects of doping concentration on the properties of the films and the reasons for the influence were studied.

2 EXPERIMENTAL

Deposition of IMO films with different molybdenum contents on quartz substrates was used by a nonmagnetic co-sputtering system (SPC-80C, CSWN Co., Ltd., China), the indium oxide ceramic target (99.9% in purity) and molybdenum metal target (99.99% in purity) were 3 inches and 2 inches in diameter, 4 mm in thickness. In_2O_3 target was sputtered by RF magnetron, and the power was always 100 W, meanwhile, the Mo target was adopted by DC magnetron, and different Mo doped films were prepared by changing the power applied to the Mo target, and the DC power was set in 0 W, 5 W, 10 W, 15 W, 20 W, respectively. Firstly, ultrasonic cleaning was used to remove contaminants on the surfaces of the quartz substrates (10 mm $\times$ 10 mm $\times$ 1 mm). The substrates were ultrasonically cleaned in acetone and ethanol for 10 minutes, and then the surfaces of the substrates were dried using high-pressure nitrogen. Secondly, the quartz substrates were placed in a vacuum chamber, the background vacuum before sputtering was 6.0×10^{-4} Pa. During the sputtering process, a mixed gas of argon and oxygen was used. The ratio of argon to oxygen was 6:1, and the flow rate of the gas was maintained at 70 sccm. The sputtering pressure was 1.0 Pa, and the sputtering time was 4 hours. The thickness value of IMO thin films was about 400 nm, the thickness was monitored by the quartz crystal thickness meter (TM106, SCIENS, China).

X-ray diffraction (UltimaIV, Rigaku, Japan) was used to characterize the crystal structure and preferred growth orientation of IMO thin films. The surface morphology of the films was observed by the scanning electron microscope (Inspect F50, FEI, USA), and in this work, this method was also used to observe the cross-sectional morphology. The composition and atomic percentage of the IMO film under different DC powers were tested by an energy dispersive spectrometer (Quanta 250, FEI, USA). The electrical properties of the IMO samples were measured by the Hall test system (Lakeshore, USA, 7707), which was conducted at room temperature. Optical transmittance and absorption spectrum of the samples in the range of 300 nm ~850 nm were obtained by the ultraviolet spectrophotometer (Lamda950, PerkinElmer, USA), then the optical band gap was calculated, and analyze the optical performance of IMO from the above three aspects.

3 RESULTS AND DISCUSSION

X-ray diffraction (XRD) measurements were carried out to study the effect of molybdenum doping amount on the microstructure of the prepared IMO films.

As shown in figure 1, the XRD patterns of IMO thin films under different Mo DC power are shown. even if the power on the Mo target is different, the spectral lines of the deposited thin films are consistent with the standard characteristic spectral lines of indium oxide. This shows that the incorporation of Mo in the indium oxide film does not change the crystal structure of the film, and the doped IMO film still has a cubic manganite structure. It can be intuitively observed from Figure 1 that the IMO film has 4 distinct diffraction peaks (211), (222), (440), and (622). The films with different amounts of Mo doping have the strongest diffraction peaks at an angle of 2θ of about 30°, under 0 W DC power of the Mo target the deposited In_2O_3 film has the (222) preferred orientation. With the DC power increasing, the diffraction peak of (222) significantly decreased, and there is no obvious crystallization peak in the thin film samples under 15 W and 20 W DC power, it presents an amorphous film state.

Figure 2 shows SEM and cross-section images of IMO films with different DC powers. Figures 2 (a), (b), (c), (d), and (e) show the surface morphology of IMO films deposited at DC power of 0W, 5W, 10W, 15W, and 20W, respectively. It can be seen that the uniformity of the film is good. The

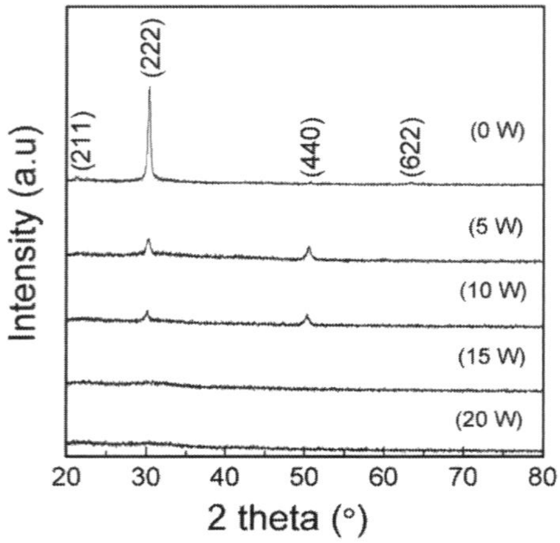

Figure 1. XRD patterns of IMO thin films deposited under different Mo target DC power.

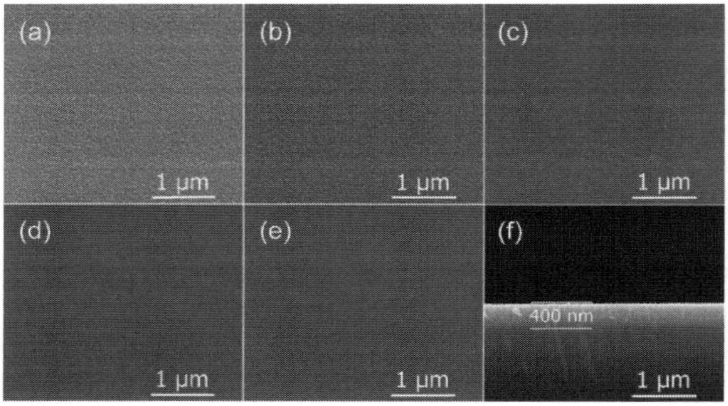

Figure 2. SEM and cross-section images of IMO thin films deposited under different Mo DC power.

crystallization performance of the thin films has deteriorated, which may be due to the reduction of the cell parameters caused by the incorporation. Figure 2(f) is a cross-sectional view of the IMO thin film sample, indicating that the sample thickness is about 400 nm.

EDS is one of the important means to monitor the composition of thin-film samples. Figure 3 is the EDS spectrum of the IMO film prepared when the Mo target has different DC power. obvious peaks of Mo, In, and O elements can be observed from the EDS spectrum, and the peak value of the Mo element also increases obviously with the increase of DC power. This indicates that the deposited Mo-doped In2O3 thin film. The variation trend of Mo and In atomic percentage of the IMO film with different DC power on the Mo target is shown in Figure 4. The DC power increased from 5 W to 20 W, the Mo atomic percentage increased from 1.58% to 4.23%, and the In atomic percentage decreased monotonously.

Figure 5 reveals the change in electrical performance of IMO films with the DC power of Mo target from three aspects of resistivity, carrier concentration, and mobility. It can be intuitively seen from the figure that as the DC power increases, the film carrier concentration also shows an increasing trend. The resistivity of the pure In_2O_3 thin film is $3.8\times10^{-3}\Omega\cdot$cm, and the power increases until 15 W, the resistivity of the films show a downward trend. When the applied power is 15 W, the prepared IMO thin film has the smallest resistivity; its value is $0.97\times10^{-3}\Omega\cdot$cm. The reason for this change is that after the impurity Mo is doped, Mo ions will replace In ions and provide

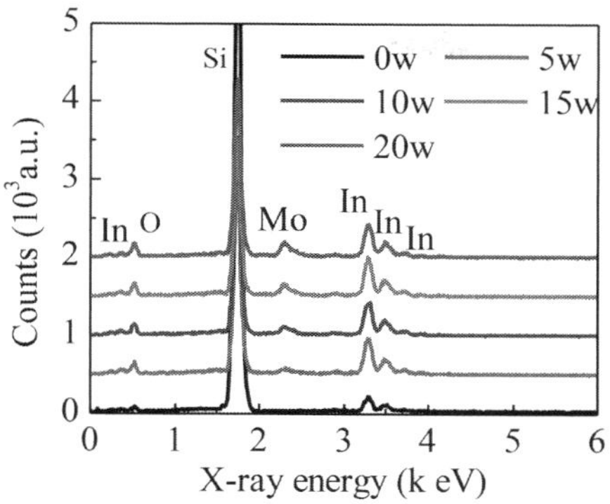

Figure 3. EDS of IMO films deposited under different Mo target DC power.

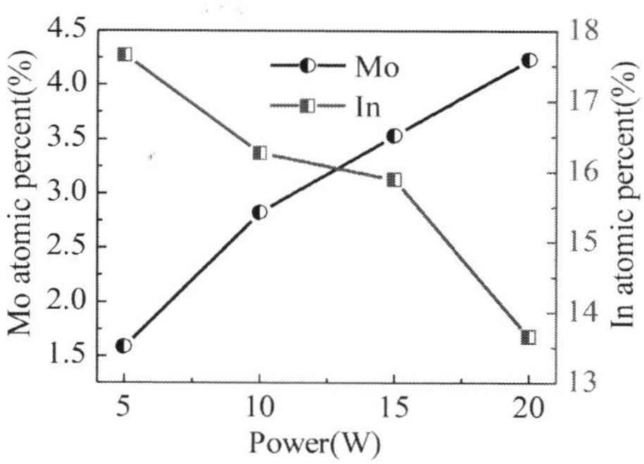

Figure 4. Mo and In atomic percent of IMO films deposited under different Mo target DC power.

three Free electrons, which at the same time cause the film mobility to increase as the applied power increases. But when the applied power continues to increase to 20 W, the resistivity increases, which may be due to the increase in impurity scattering (Kaleemulla et al. 2010; Parthiban et al. 2010). When the applied power is 15 W, the maximum mobility is 72.6 $cm^2V^{-1}s^{-1}$.

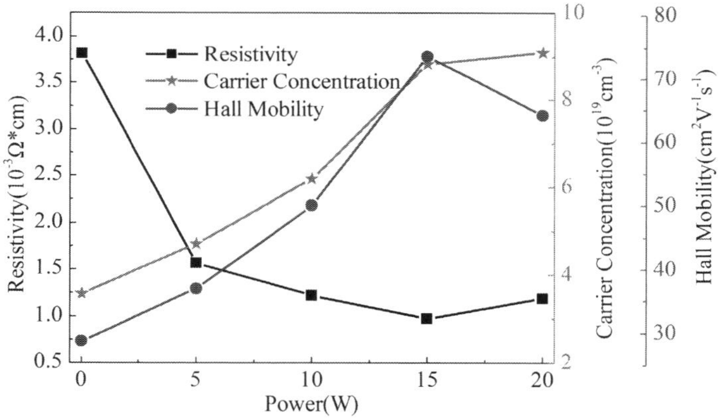

Figure 5. Electrical properties of IMO film deposited under different DC power.

The optical performance of the sample was analyzed by measuring the optical transmittance of the sample and calculating its optical band gap by performing UV spectrum tests on Mo-doped In_2O_3 films with different DC power. Figures 6 to 8 signify the changes in the optical performance of IMO thin films deposited under different DC power of Mo targets in the wavelength range of 300~850 nm from three aspects of optical transmittance, absorbance, and optical band gap. It is observed from Figure 6 that the average transmittance of the films exceeded 80%, showing good optical transparency in the visible range. The absorption coefficient (Shinho 2000) is determined by Equation (1):

$$\alpha = (-\ln T)/d \qquad (1)$$

where d is the IMO film thickness. Figure 7 is an absorption spectrum diagram of Mo-doped In_2O_3 films with different DC powers calculated according to Equation (1).

The results in Figure 7 show that the films have strong absorption performance at 300~400 nm. In the visible light range, the transmittance curve changes gently without obvious absorption.

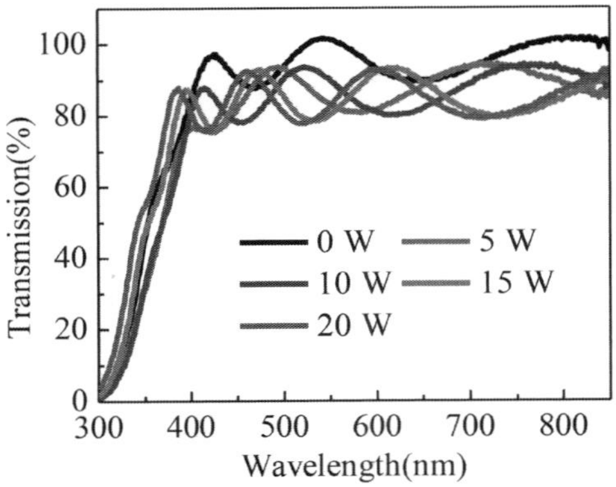

Figure 6. Transmission spectrum of IMO film deposited under different Mo target DC power.

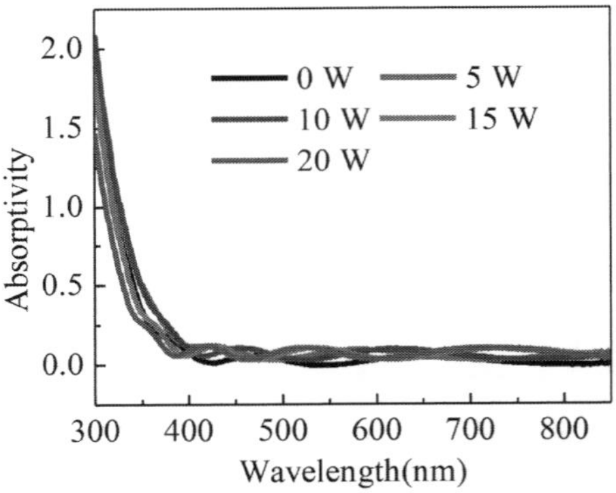

Figure 7. Absorption spectra of IMO film deposited under different Mo target DC power.

These results also show that the incomplete oxide content in the prepared thin film sample is very small because the incomplete oxide has high absorption.

It is of great significance to analyze the optical performance of IMO thin films by analyzing the changes in the optical band gap of thin films under different DC powers. The value of the optical band gap can be calculated by equation (2). The equation (2) is as follows (Tacu et al. 1996):

$$(\alpha h\nu)^2 = \mathrm{A}\left(h\nu - E_g\right) \tag{2}$$

where $h\nu$ is the photon energy and A is a constant (Liu et al. 2019). Use equation (2) to draw the $h\nu$- $(\alpha h\nu)^2$ curves for all samples. As shown in Figure 8, draw the tangent line of each curve when the slope is the largest, which has an intersection point with the $h\nu$-axis. That is the optical band gap Eg of the sample.

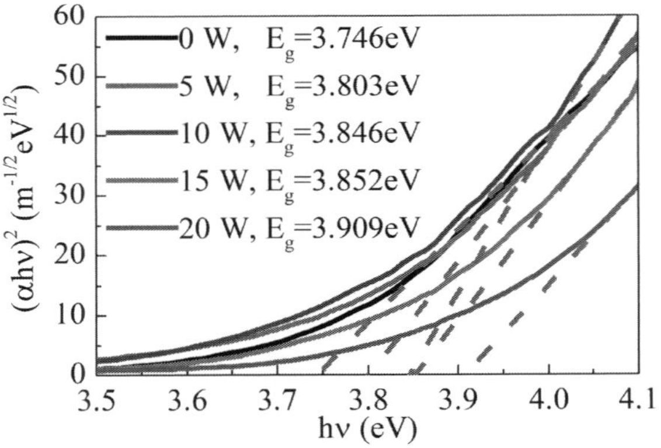

Figure 8. hν- (αhν)2 of IMO film deposited under different Mo target DC power.

It can be seen intuitively from Figure 8 that the DC power on the Mo target increases and the optical band gap of the IMO film increases. When the DC power on the Mo target is 0 W, 5 W, 10 W, 15 W, and 20 W, the optical band gap Eg of the IMO sample is 3.746 eV, 3.803 eV, 3.846 eV, 3.852 eV, 3.909 eV, respectively.

4 CONCLUSION

In this study, a Mo metal target and In_2O_3 ceramic target were co-sputtered to deposit a Mo-doped indium oxide (IMO) film. The Mo content of the deposited films was different with different power on the Mo target, and the properties of the prepared samples were analyzed. The results of the study indicate that the prepared film exhibits good uniformity. The thickness of the IMO film is 400 nm, and the optical transmittance in the visible range exceeded 80%. The minimum resistivity of IMO film is $0.97\times 10^{-3}\Omega\cdot$cm and the maximum Mobility is 73.6 $cm^2V^{-1}s^{-1}$ when the power on the Mo target is 15 W. But when the applied power continues to increase, the resistivity increases and the mobility decreases. The optical band gap of the deposited IMO film increases with the increase of Mo doping content, when the power on the Mo target is 20 W, the maximum band gap of the IMO reaches 3.909 eV.

ACKNOWLEDGMENT

This paper was supported by the Xi'an Key Laboratory of Power Electronic Devices and High Efficiency Power Conversion.

REFERENCES

Beena D, Lethy K J, Vinodkumar R, Detty A P, Mahadevan Pillai V P and Ganesan V 2010 *J. Alloys Compd.* **489** 215–23.

Chen X L, Han D G, Zhang D K, Sun J, Geng X H and Zhao Y 2011 *J.Optoelectronics.Laser* **22(7)** 1022–5.

Cui W, Zhao X L, An Y H, Yao G S, Wu Z P, Li P G, Li L H, Cui C and Tang W H 2018 *J.Nanosci.Nanotechno* **18(2)** 1220–3.

Elangovan E, Marques A, Braz Fernande F M, Martins R and Fortunato E 2007 *Thin Solid Films* **515(13)** 5512–8.

Elangovan E, Marques A, Pimentel A, Martins R and Fortunato E 2008 *Vacuum* **82** 1489–94.

Kaleemulla S, Madhusudhana Rao N, Girish Joshi M, Sivasankar Reddy A, Uthanna S and Sreedhara Reddy P 2010 *J. Alloys Compd.* **504** 351–6.

Kaleemulla S, Sivasankar Reddy A, Uthanna S and Sreedhara Reddy P 2008 *American Institute of Physics Conference Proc.* ed P. Predeep and S. Prasanth et al vol 1004 (American: AIP publishing) pp 316–20.

Kelly P J and Arnell R D 2000 *Vacuum* **56** 159–72.

Krishnan R R, Sanjeev G, Prabhu R and Mahadevan Pillai V P 2018 *JOM* **70(5)** 739–46.

Liu X C, An Y K, Lin Z and Liu J W 2018 *Mod.Phys.Lett.B* **32(24)** 1850284.

Liu Y T, Wang W X, Ma J P, Wang Y, Ye W, Zhang C, Chen J J, Li X Y and Du Y 2019 *J.Adv.Dielectr* **9(6)** 1950048.

Meng Y, Yang X L, Chen H X, Shen J, Jiang Y M, Zhang Z J and Hua Z Y 2001 *Thin Solid Films* **394(1)** 219–23.

Miao W N, Li X F, Zhang Q, Huang L, Zhang Z J, Zhang L and Yan X J 2005 *Thin Solid Films* **500(1)** 70–3.

Pan J J, Wang W W, Wu D Q, Fu Q and Ma D 2014 *J.Mater.Sci.Technol* **30(07)** 644–8.

Park C Y, Yoon SG, Jo Y H and Shin S C 2009 *Appl. Phys. Lett.* **95** 122502.

Parthiban S, Elangovan E, Ramamurthi K, Martins R and Fortunato E 2010 *Sol. Energy Mater. Sol. Cells* **94** 406–12.

Reshmi Krishnan R, Kavitha V S, Chalana S R, Prabhu R and Mahadevan Pillai V P 2019 *JOM* **71(5)** 1885–96.

Shinho C 2000 *Microelectron. Eng* **89(12)** 84–8.

Tacu J, Grigorovici R and Vancu A 1996 *Phys.Sta.Sol* **15(3)** 627–37.

Xu J, Liu J B, Liu B X, Li S N, Wei S H and Huang B 2018 *Adv. Electron. Mater* **4(3)** 1700553.

Yan H Y, Fang F, Chen Z Q, Zhang C M, Niu Q, Xue W and Zhan Z L 2018 *J.Mater.Sci.-Mater.Electron* **29(6)** 1–7.

Zhang X J, Deng N, Chen X J, Yang Y T, Li J, Hong B, Jin D F, Peng X L and Wang X Q 2019 *NANO: Brief Reports and Reviews* **14(3)** 1950040.

Energy Revolution and Chemical Research – Kok-Keong Chong and Zhongliang Liu (Eds)
© 2023 The Authors, ISBN: 978-1-032-36554-1

Determination of fatty acid content in liver and kidney of obese mice induced by a high-fat diet

Huihui Li & Jiayi Zhu
College of Food Technology, Wuhan Business University, Wuhan, Hubei, China

Shenghong Zhou
Hubei Province Intangible Cultural Heritage Research Center, Wuhan, Hubei, China

ABSTRACT: Obesity induced by a high-fat diet was often associated with liver and kidney injury. In this study, GC-FID was adopted to detect the changes in fatty acid content in the liver and kidney of obese mice induced by the high-fat diet. By histopathological examination, obese mice presented a typical characterization of hepatocyte steatosis. Fatty acid detection results showed that the content of total fatty acids in the liver increased in the obesity group, and the metabolic flow of saturated fatty acids was concentrated on C18:0 and its metabolic derivatives. Meanwhile, both n6 and n3 polyunsaturated fatty acids increased significantly in obese mice. In the pathological sections of the kidneys of the obesity group, inflammatory cell infiltration was observed near the glomerulus, which might be an early symptom of obesity-induced liver damage. At the same time, the content of n3 polyunsaturated fatty acids in the kidneys of the obesity group was significantly reduced, and the n6/n3 ratio was also changed. These results were helpful to further understanding the mechanism of liver and kidney injury in obesity and provide a reference for the research and development of weight-loss drugs.

1 INTRODUCTION

Obesity was a chronic metabolic disease caused by the interaction of multiple factors, such as heredity and environment, with excessive accumulation or abnormal distribution of fat in the body as medical symptoms. Obesity could also lead to a series of metabolic syndromes such as cardiovascular disease, type 2 diabetes, and hypertension, which had become one of the most challenging public health problems (Zhou 2019). Therefore, it was of great significance to clarify the mechanism of obesity occurrence and development. The liver played an important role in lipid metabolisms, such as endogenous fat generation, fat transport, and storage in the body, and obese people often suffered from fatty liver. Liver fat deposition and hepatocyte steatosis were closely related to liver oxidative stress injury, inflammation, and lipid metabolism imbalance (Xu 2019). In particular, obesity induced by a high-fat diet could easily lead to non-alcoholic fatty liver disease. It had been reported that about 40% of obese patients had varying degrees of kidney damage (Jiang et al. 2018). The incidence of ORG in China increased from 0.62% to 1% between 2002 and 2006 (Rao et al. 2006). The disorder of lipid metabolism was the initiating link of obesity-related diseases. Therefore, the GC-FID method was used to investigate the effect of obesity induced by a high-fat diet on liver and kidney lipid metabolism from the perspective of fatty acid content changes in this paper.

2 MATERIALS AND METHODS

2.1 *Materials and instruments*

C57BL/6 male mice were purchased from Hunan SJA laboratory animal Co., LTD and the production license number was SCXK (Xiang) 2019-0004. The mice were fed in the Hubei food and

 DOI 10.1201/9781003332657-99

drug safety evaluation center with experimental animal license number SCXK (E) 2015-0018. 45% High-fat feed (TP23000) and normal control feed (TP23302) were purchased from Trophic animal feed High-Tech Co. Ltd., and feed composition was shown in Table 1. The calorie composition of TP23000feed contained 19.4% protein, 35.6% carbohydrate, and 45.0% fat. The feed calorie was 4.50kcal/g, of which sugar calorie accounts for 18.0%, and the cholesterol content was about 141.0mg/kg while the calorie composition of the control TP23302 feed was 19.40% protein, 70.60% carbohydrate, and 10.00% fat, and the calorie was 3.60kcal/g.

Twenty-three carbon fatty acid methyl esters, methyl heptahedron (CAS:1731-92-6), 37 kinds of fatty acid methyl esters, and acetyl chloride were purchased from Sigma-Aldrich, USA. 3,5-di-tert-butyl-4-Hydroxytoluene, methane, chloroform, n-hexane, and potassium carbonate were purchased from Sinopharm Group. Gas Chromatography-Flame Ionization Detector (GC-FID) 8890 purchased in Agilent, USA.

Table 1. Composition of 45% high-fat feed (TP23000).

Composition	Casein	Starch	Maltodextrin	Sucrose	Oil	Lard	Cellulose
Content (g/kg)	240	73	120	203	30	196	60
Composition	Mineral	Vitamin	L-Cystine	Choline Bitartrate		Butylhydroquinone	
Content (g/kg)	59	12	4	3		0.045	

2.2 *Animal experiments*

Animal experimental procedures were performed according to the National Guidelines for Experimental Animal Welfare (MOST of P. R. China 2006). Male C57BL/6 mice aged 4 weeks were purchased and randomly divided into obesity group (HF) and control group (CON), with 8 mice in each group. The mice had free access to food and water in an SPF animal laboratory with a 12 h light/dark cycle at a constant temperature of 20-25 ° C and relative humidity of 40-60%. After 2 weeks of acclimatization, the HF group was given TP23000 high-fat diet, while the control group was given TP23302 basal diet. During the experiment, parameters such as body weight, length, food intake, and water intake of mice were detected every week. After 12 weeks of the experiment, the experimental mice were sacrificed and the weight of liver and kidney was weighed. The liver and kidney tissues of 5 mice in each group were stored in liquid nitrogen for the extraction of fatty acids. The liver and kidney of another 3 mice were placed in a fixative solution for tissue sections.

2.3 *Preparation and observation of mice liver and kidney tissue sections*

The liver and kidney tissues of 3 mice in each group were embedded in wax blocks after being fixed and soaked in formaldehyde for 48 hours, then cut into 3-4 μm thick slices, and stained with hematoxylin-eosin. After dehydrated and sliced, a histopathological examination was performed under a microscope.

2.4 *Fatty acid sample processing and detection*

According to the method of reference (Li et al. 2017), fatty acid was detected. The detection steps are as follows:

(1) Weight about 25 mg of mouse liver and kidney tissue, then added 500 uL of chloroform-methanol solution and homogenized;
(2) Transfer homogenate to Pyrex tube, and add 20 uL of 1 mg/mL internal standard and 2 mg/mL BHT and 1 mL methanol-hexane mixture;

(3) After adding acetyl chloride under a liquid nitrogen ice bath, the mixture was placed in a dark room for 24 h until the methyl esterification reaction was complete. Then add K2CO3 to neutralize the hydrochloric acid produced in the reaction.;
(4) Extract fatty acid three times with 200 μL of n-hexane, and combine the supernatants after centrifuging;
(5) Evaporate the organic solvent to dryness, and add 200 μL of n-hexane to fully dissolve it before GC examination.

Agilent's 8890 GC-FID equipped with DB-225 capillary column was employed in this experiment. The qualitative detection of fatty acids was mainly determined by comparing the retention time of the mix standards. The fatty acid content was mainly quantified by the internal standard method. The experimental conditions for GC-FID were as follows. the temperature of the injection port and the FID detector was 250 ° C. The column temperature was started from 115 ° C for 0.5 min and then increased to 205 ° C at a rate of 30 ° C/min. After being held at 205 ° C for 3 min, increased to 240 ° C (5 ° C/min) and kept for 1.5 min.

2.5 *Statistics of experimental data*

The qualitative detection of fatty acids was mainly determined by comparing the retention time of the mix standards. The fatty acid content was mainly quantified by the internal standard method. The formula for calculating the fatty acid content was as follows: Cx (umol/g) = $(100\times 1000\times Sx)/(S17\times Mx\times mx)$, where Cx was the content of the fatty acid to be detected, Sx was the integral area of fatty acid methyl ester, S17 was the integral area of internal standard methyl heptahedron, Mx was the molar mass of the fatty acid methyl ester, and *mx* is the mass of the sample.

In this experiment, five samples were tested in each group, and the fatty acid content of every group was expressed as mean±SD. T-TEST of irrelevant samples of SPSS19.0 was used to test the differences between groups, and the significance level was set at 5%. "*" refers to $p<0.05$, "**" refers to $p<0.01$, and "***" refers to $p<0.001$ in the following figures and tables.

3 RESULTS AND ANALYSIS

3.1 *Body weight and BMI of obese mice induced by high-fat diets*

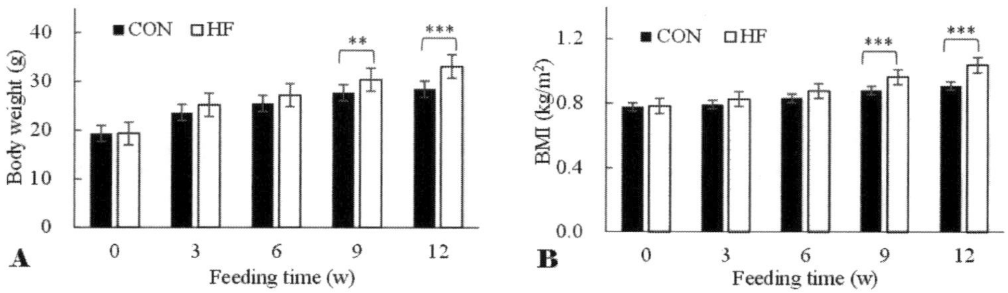

Figure 1. Effect of high-fat diet on body weight (A) and BMI (B) of mice.

Feed and water consumption were measured weekly for 12 weeks of feeding, and no statistical differences were found in the amount of water and food intake between the control and high-fat diet groups at the same time (data not given). As shown in Figure 1, the high-fat diet did not affect the appetite of the mice. From the mice's body weight chart, it was found that the body weight of the mice in the control group increased steadily with the feeding time. After two weeks of acclimation, the feeding period was 0 weeks, and the average body weight of mice aged about 6 weeks was 19.3 g. After 6 weeks of feeding, it was 25.5 g in the normal group and 27.2 g in the high-fat diet group. There was no significant statistical difference in body weight between the two groups. However, from then on, the mice in the high-fat diet group were significantly higher than the control group in terms of body weight and BMI value ($p<0.01$). At 12 weeks of feeding, the high-fat group had an average weight of 33.11 g, compared with 28.47 g in the control group. The body weight in high-fat mice was 16% higher than in the control group. The obesity model induced by high-fat diets was successfully established.

3.2 *Effects of obesity induced by high-fat diet on liver and kidney tissues of mice*

As could be seen from Figure 2, the mean liver weight of the obese mice was 986.6 mg, while the control group was 1060.8 mg. The liver weight of the obese mice was significantly lower than the control group (Figure 2A). Since the obesity group had a higher body weight, the liver organ coefficient of the obesity was significantly lower than the control (Figure 2B). Although there was no statistical difference in kidney weight between the two groups, the organ coefficient of the kidney in the obesity was statistically lower than in the control, which might be due to the higher body weight of the obese mice.

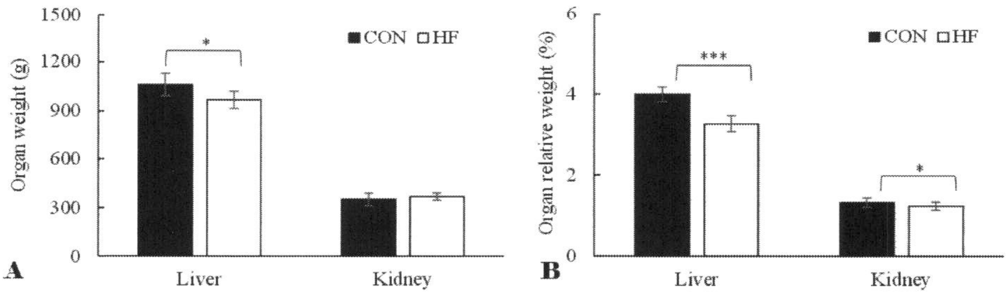

Figure 2. Effect of obesity on liver and kidney tissue weight (A) and organ coefficient (B) of mice.

HE staining of pathological sections showed that the central vein was the core of the liver in the control group which was arranged in a single row into a plate-like structure, and the hepatocyte morphology was normal. The arrangement of hepatocytes in the obesity group induced by high-fat diets was slightly disordered, with numerous round, unequal-sized, and tense vacuoles in the cytoplasm, showing typical characteristics of hepatocyte steatosis (Figure 3A). Figure 3B showed that the glomeruli and surrounding renal tubules in the renal cortex of the control group had clear structures, while the obese mice had obvious inflammatory cell infiltration near the glomeruli, which was consistent with previous reports (Zhang et al. 2009, Lu et al. 2011).

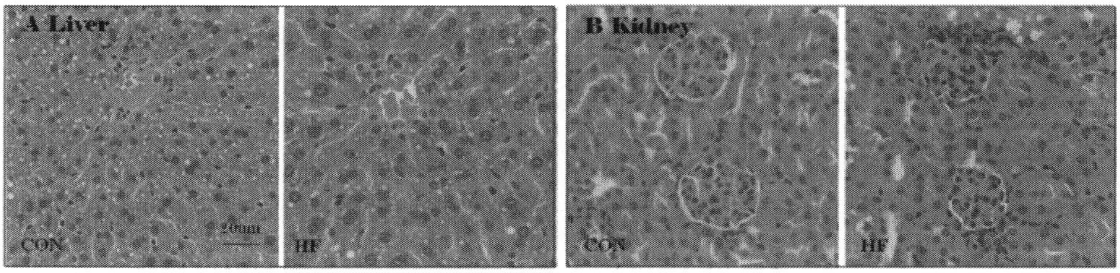

Figure 3. Effect of obesity induced by high-fat diet on liver and kidney tissue structure of mice.

3.3 *Effects of obesity induced by high-fat diet on fatty acid content in liver and kidney of mice*

3.3.1 *Total ion chromatogram of fatty acids in mice liver and kidney*

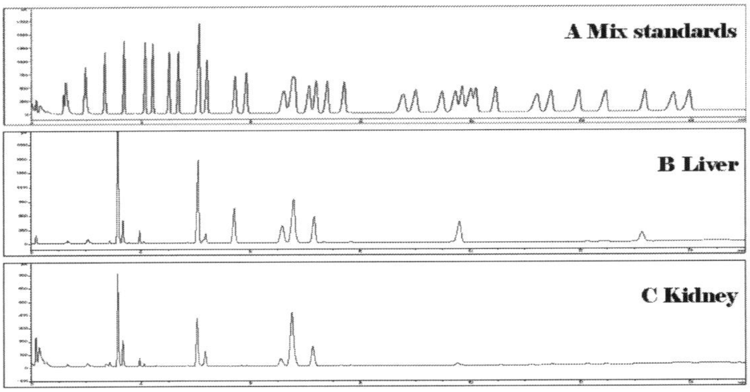

Figure 4. Total ion chromatogram of fatty acid standards (A) and mice liver (B) and kidney (C).

In this paper, the GC-FID method was used to detect the fatty acid content in mouse liver and kidney tissue. Figures 4 and 5 showed that 19 fatty acids were found in mice liver, but only 16 fatty acids in the kidney. Compared with liver fatty acids, long-chain fatty acids including C18:3n6, C20:5n3, and C22:0 were not detected in mice kidneys. According to the peak height and peak area of the total ion chromatogram, the highest content of fatty acids was C16:0 in both the liver and kidney of normal mice, and the proportion of C16:0 in the liver was about 25% while kidney was about 35%. The following higher content is C18:1n9t, C18:2n6. It should be noted that the liver contained approximately 12% C20:4n6, whereas this fatty acid was not detected in the kidney.

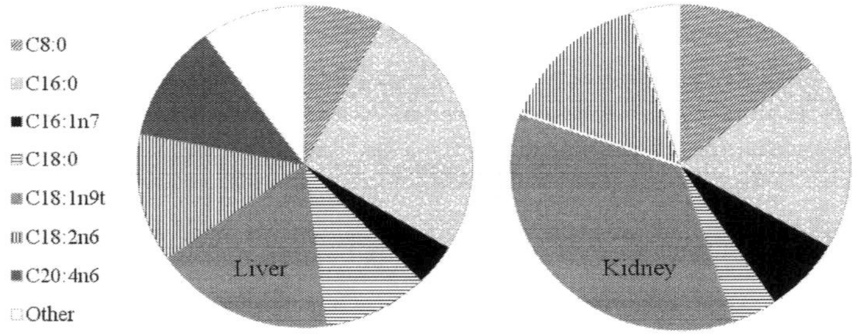

Figure 5. Fatty acid composition of liver and kidney of normal mice.

3.3.2 *Effects of obesity induced by high-fat diet on fatty acids in liver and kidney of mice*

It could be seen from Table 2 that the mean total fatty acid in the liver of the HF group was 88.32 umol/g while the CON group was 61.77 umol/g. The total fatty acid in the liver of obese mice was significantly higher than the control group, which was consistent with the result of lipid deposition in the liver of obese mice in pathological detection. Literature (Chen 2020) showed that obesity would increase the body's fat transport to the liver and cause the accumulation of triglycerides in the liver. At the same time, excessive fat accumulation in the liver might affect fat β-oxidation in the liver, resulting in the production of reactive oxygen species. The resulting oxidative stress might lead to liver damage and contribute to the onset and progression of inflammation. The saturated fatty acids in the obesity group were also significantly higher than in the control group, but the content of C8:0 in the obesity group was lower than that in the control group. There was no statistical

Table 2. Fatty acid contents in liver and kidney tissues of mice.

Retention time (min)	Fatty acid	Liver (umol/g)		Kidney (umol/g)	
		CON	HF	CON	HF
1.98	C8:0	4.62±0.71	3.25±0.29**	13.61±1.43	10.88±1.11*
2.64	C10:0	0.80±0.25	0.67±0.38	0.24±0.12	0.81±0.05***
3.34	C12:0	0.04±0.01	0.04±0.01	0.69±0.16	0.74±0.23
4.06	C14:0	0.30±0.05	0.35±0.10	0.91±0.08	0.84±0.06
4.20	C14:1n5	0.04±0.00	0.04±0.02	0.06±0.00	0.05±0.01
5.05	C16:0	15.43±3.40	20.73±6.93	18.14±1.43	18.37±0.41
5.20	C16:1n7	2.51±0.42	1.41±0.52**	7.19±0.79	5.11±0.86*
6.60	C18:0	6.14±1.34	9.69±2.55*	4.10±0.04	5.08±0.64
6.80	C18:1n9t	10.34±1.85	18.00±5.55	34.01±2.51	34.43±1.80
7.18	C18:2n6	7.71±1.63	11.65±1.56**	14.19±0.86	11.72±0.77*
7.39	C18:3n6	0.17±0.04	0.38±0.10**	ND	ND
7.70	C18:3n3	0.27±0.05	0.23±0.06	0.68±0.09	0.45±0.03**
8.80	C20:1n9	0.09±0.02	0.13±0.05	93.06±10.74	ND
9.83	C20:4n6	7.05±1.68	11.19±1.66**	ND	1.53±0.22
10.08	C20:5n3	ND	0.14±0.05*	ND	ND
11.24	C22:0	0.21±0.05	0.19±0.04	ND	ND
11.50	C22:1n9	0.19±0.02	0.13±0.04*	0.50±0.25	0.43±0.15
13.18	C22:6n3	3.77±1.01	5.93±1.49*	0.57±0.00	0.54±0.13
14.02	C24:1n9	2.21±0.38	2.56±0.00	ND	281.76±0.00
Total fatty acid		61.77±11.22	88.32±15.50*	95.43±4.94	90.95±4.98
Saturated fatty acid		27.47±4.80	38.00±6.64*	37.33±2.53	36.77±1.85
Unsaturated fatty acid		34.29±6.56	50.32±8.88*	58.10±2.44	54.18±3.78
Monounsaturated fatty acid		15.32±2.41	19.71±6.05	42.66±2.29	39.97±2.75
Polyunsaturated fatty acid		18.97±4.34	30.60±2.91**	15.44±0.95	14.21±1.10
n3 polyunsaturated fatty acid		4.05±1.04	6.87±0.83**	1.25±0.10	0.95±0.13*
n6 polyunsaturated fatty acid		14.92±3.32	23.74±3.01**	14.19±0.86	13.25±0.98
n6/n3		3.72±0.21	3.50±0.62	11.36±0.34	13.96±0.86*

Note: ND means not detected in the sample. n6/n3, the ratio of n6 series polyunsaturated fatty acids to n3 series polyunsaturated fatty acids.

difference in C10:0, C12:0, C14:0, C16:0, and C22:0 between the two groups, only the content of C18:0 in the obesity group was significantly higher than in the control group. These indicated that the metabolic flow of saturated fatty acids concentrated on C18:0 and its metabolic derivatives during the occurrence and development of obesity. At the same time, the unsaturated fatty acids in the obesity group were also significantly higher than those in the control group, and there was no significant difference in monounsaturated fatty acids between the two groups. So, the difference mainly came from polyunsaturated fatty acids. Among them, the n6 (C18:2n6, C18:3n6, C20:4n6) and n3 (C20:5n3, C22:6n3) polyunsaturated fatty acids in the obesity group were significantly increased. It was worth noting that the C20:5n3 fatty acid was not found in the control group while it achieved a qualitative change from scratch in the obesity group.

In the kidney of obese mice, the content of n3 series polyunsaturated fatty acid was 0.95 umol/g, which was significantly lower than 1.25 umol/g of the control group, which might be mainly due to the extremely significant reduction of C18:3n3 content. It also led to significant changes in n6/n3. As two subtypes of polyunsaturated fatty acids, n3n6 fatty acids played different roles in fat metabolism and body fat regulation. It was generally believed that n3 fatty acids could reduce the content of intracellular triglycerides, while n6 fatty acids did the opposite. These were related to adipogenesis, lipid homeostasis, brain-gut fat axis, and systemic inflammatory mechanisms (Wang 2015).

4 CONCLUSION

To study the fat metabolism mechanism of obese mice induced by high-fat diets, GC-FID was adopted to detect the changes in fatty acid content in the liver and kidney of obese mice. The main conclusions can be summarized as follows: (1) By histopathological examination, triglyceride droplet deposition was found in hepatocytes of obese mice, presenting a typical characterization of hepatocyte steatosis. Meanwhile, fatty acid detection results showed that the content of total fatty acids in the liver increased in the obesity group, and the metabolic flow of saturated fatty acids was concentrated on C18:0 and its metabolic derivatives. Monounsaturated fatty acids did not change significantly, while both n6 and n3 polyunsaturated fatty acids increased significantly in high-fat diet mice. It was worth noting that the liver weight and organ coefficients of obese mice were significantly lower than the control group, which was worth further investigation. (2) In the pathological sections of the kidneys of the obesity group, inflammatory cell infiltration was observed near the glomerulus, which might be an early symptom of obesity-induced liver damage. Meanwhile, fatty acid detection showed that the content of n3 polyunsaturated fatty acids in the kidneys of the obesity group was significantly reduced, and the n6/n3 ratio was also changed. In terms of future work, according to the "lipid metabolites and pathways strategy", other lipids such as glycerophospholipids, sphingolipids, sterol lipids and prenol lipids should be carried out to enhance our understanding of obesity.

ACKNOWLEDGMENT

We appreciated the financial support from the General Programs of Natural Science Foundation in the Science and Technology Department of Hubei Province (Grant No. 2019CFB764) .

REFERENCES

Chen Q.W. (2020). *Effects of ketogenic diet feeding on body weight and liver lipid metabolism in obese mice*. Diss. Hefei, Anhui Medical University.

Jiang, Y.H. & Jiang, L.Y. & Wu, S. & Jiang, W.J. & Xie, L.F. & Li, W. & Yang, C.H. (2018). *Renal protective effect of Tribulus terrestris extract on obesity-related nephropathy rats*. The 2018 Academic Annual Meeting of the Renal Disease Professional Committee of the Chinese Association of Integrative Medicine.

Li, H.H. &Wang, J. & Zhang, Y. & Sui, H.W. & Xu, M.N. (2017). Detection of fatty acids in pig and beef soup and froth in broth. *Food Sci.Techn.* (10), 5.

Lu, J. & Yuan, W.H. (2011). The pathophysiological mechanism of persistent kidney damage caused by simple obesity. *China Blood Purific.*10(4), 4.

Rao, X.R. & Zhang, G.H. (2006). Obesity and kidney damage. *China J. Integrated.Traditional Chin. West.Med.Nephrol.* 7(10), 616–618.

Wang, L.U. (2015). The role of $\omega 3$ and $\omega 6$ fatty acids in the increase of body fat and the development of obesity. *Chin. J. Hypertens.* 23(1), 9.

Xu, T. (2019). *Dihydromyricetin improves liver fat deposition in high-fat diet-induced obese mice and its mechanism*. Diss. Hengyang, Univerity of South China.

Zhang J & Chu Z.H. & Lu H & Xu W.J. & Ding Y.L. & Yang L & Xie J.X. (2009). Comparison of changes in liver morphology and structure in rats with obesity and type 2 diabetes. *AdvancModern Biomed* (12)4.

Zhou, J.H. (2019). *Effects of EGCG on brown adipose tissue activition and hypothalamic inflammatory pathway in obese mice fed a high-fat diet*. Diss. Hanzhou, Zhejiang University.

Energy Revolution and Chemical Research – Kok-Keong Chong and Zhongliang Liu (Eds)
© 2023 The Authors, ISBN: 978-1-032-36554-1

A conceivable new method of sulfation roasting of spent lithium-ion batteries: Together with industry SO_2 gas

Kaixuan Li, Chuanjin Zhao, Xinyao Zhang, Xuefei Wang, Zichen Tian, Chenguang Ma & Huaqing Ding
School of Metallurgy, Northeastern University, Shenyang, China

ABSTRACT: With the continuous promotion of policies such as "Carbon Peak and Neutrality" and "Circular Economy", the treatment of high-polluting off-gas such as SO_2 and emitted by industry and the utilization of high-value-added "urban minerals" such as spent lithium-ion batteries have become a global research hotspot. Herein, based on the concept of "waste to waste", this paper makes full use of the huge heat carried by the high-temperature SO_2 off-gas emitted by industry to preheat the waste lithium-ion battery and converts lithium into soluble sulfate based on gas-solid sulfation roasting. The Computational Fluid Dynamics (CFD) simulation and calculation results show that by utilizing off-gas of 1000°C to heat cathode powder, after 2 h the whole sample can reach 700°C. Combined with roasting experiments, the recovery rate of Li was more than 99% at 600–800°C in different proportions of SO_2 mixed atmosphere.

Our research provides a probable way for the "sulfation roasting-water leaching" process, to form a higher efficiency, higher recovery rate and lower energy consumption mode. Finally, we considered the promising direction of spent lithium-ion battery recovery processes and comprehensive co-treatment of urban waste in China under the dual carbon target.

1 INTRODUCTION

With the proposal of the strategic goal of "Carbon Peak and Neutrality", the second decade of the 21st century has witnessed the vigorous development of the new energy automobiles and energy-storage systems with lithium-ion batteries (LIBs) as the core components, which will impulse China reducing consumption on fossil fuels, seeking for cleaner greener and more sustainable energy supplies while reducing carbon emissions. In the meantime, the rapid expansion of lithium industries has led to the increasingly serious contradiction between supply and demand in the global lithium resource market. Faced with the gradual exhaustion of primary lithium resources, more and more attention has been paid to the recycling and utilization of polymetallic symbiotic secondary resources such as abandoned lithium-ion batteries. Thus, how to efficiently recycle spent lithium-ion batteries, to form a closed-loop of producing-consumption-regeneration-reuse, is of primary urgency.

SO_2 is one of the main air pollutants, which can cause many environmental problems such as acidic precipitation and heavy pollution weather. At present, sulfur dioxide flue gas is mainly used for waste heat recovery or subsequent acid production process, after desulfurization treatment, after reaching the emission standard, it is discharged into the atmosphere by chimney. For flue gas desulfurization process mainly wet desulfurization, such as limestone-gypsum method, hydrogen peroxide method, organic amine method, and so on, can reach a relatively high desulfurization efficiency. However, there is little research on the innovative application of sulfur dioxide flue gas.

More recently, sulfation-roasting together with water leaching, as a high-selectivity and green process, has experienced ever-growing research in related fields (Li 2019; Lin 2021; Wang 2018), for its lower energy consumption, less waste-acid production, and gas emission. However, this

DOI 10.1201/9781003332657-100

method can't avoid the inherent drawbacks of pyrometallurgy in that the roasting temperature is high and thus lead to large energy consumption. Based on the above-mentioned shortcomings, this paper presents a novel process of recycling Li from cathode powder of spent lithium-ion batteries, using the high-temperature SO2 off-gas co-treatment process. Combined with CFD simulation and the actual roasting experiment, we prove the feasibility of our idea. Finally, the prospect of comprehensive utilization of multi-metal solid waste secondary resources such as the spent lithium-ion battery is prospected, aiming to provide ideas for the development of more green and efficient lithium extraction process in the future.

2 FEASIBILITY-ANALYSIS

The reaction consists of three stages, pre-heating stage(initial stage), reaction stage, end stage, the content ratio of flue gas is set as 10%SO_2-1%O_2-89%Ar, and considering to avoid lithium cobaltate particles being blown away directly, since the particle size of lithium cobaltate after pretreatment is >0.5 mm, the velocity of branch pipe flue gas is designed as V=5 m/s, and the diameter is designed as 300mm, $LiCoO_2$ powder can be approximately unchanged due to the small change of SO_2 content. As the Mach number Ma<0.3, the flue gas is approximately calculated according to the ideal incompressible air, and the waste battery is treated according to the waste 18650 lithium battery. Compared with the direct heating by flame, the energy saved when the waste battery is heated to the optimal reaction temperature of 700 degrees by 1000 degrees of flue gas is calculated.

2.1 *Fluid-structure coupling modeling of branch heating based on finite element method*

Research shows that the reaction product Li_2SO_4 is the most stable at 600°C~800°C (Shi 2019), and the product conversion rate is the highest at the higher partial pressure of SO_2 and O_2 at 700°C when the composition of the atmosphere is certain. Therefore, the modeling selected the preheating stage of the sulfation roasting experiment when all grid cells reached 973 K, assuming that no reaction occurred before this stage. Enlarge the experimental size by 5 times for preliminary simulation. The conclusion of heat saving is obtained, and then the next experiment is conducted in simulation. The flue gas flow should follow the law of conservation of physics, the basic conservation law includes the law of conservation of mass, the law of conservation of momentum, and the law of conservation of energy. Since the flue gas system is approximately treated as incompressible air, the law of conservation of composition can be ignored. For the hypothetical flue gas in this paper, the conservation law is described by the following governing equation.

Mass conservation equation:

$$\frac{\partial \rho_f}{\partial t} + \nabla(\rho_f v) = 0 \tag{1}$$

Momentum conservation equation:

$$\frac{\partial \rho_f v}{\partial t} + \nabla \cdot (\rho_f v^2 - \tau_f) = f_f \tag{2}$$

Where t refers to time; f_f refers to volume force vector; ρ_f refers to the fluid density; v refers to the fluid velocity vector; τ_f refers to shear stress tensor. Then we have:

$$\tau_f = (-p + \mu \nabla \cdot v)I + 2\mu e \tag{3}$$

Where p refers to the pressure of the fluid; μ refers to kinetic viscosity; e refers to velocity stress tensor; $e = \frac{1}{2}(\nabla v + \nabla v^T)$.

The conservation equation for the solid part can be derived from Newton's second law.

$$: \rho_s \ddot{d}_s = \nabla \cdot \sigma_s + f_s \tag{4}$$

Where ρ_s refers to the density of the solid; σ_s refers to Cauchy stress tensor; f_s refers to volume force vector; $\ddot{d}_s$ refers to the acceleration vector where the solid locate.

$$\frac{\partial(\rho h_{hot})}{\partial t} - \frac{\partial p}{\partial t} + \nabla \cdot (\rho_f v h_{hot}) = \nabla \cdot (\lambda \nabla T) + \nabla \cdot (v \cdot \tau) + v \cdot \rho f_f + S_E \tag{5}$$

Where λ refers to thermal conductivity and S_E refers to energy source term.

Because solid particles are not prone to large deformation in contact with flue gas, the influence on fluid can be ignored, that is, the fluid-solid bidirectional coupling can be out of consideration.

At the surface where the fluid and solid are coupled, the following four equations are valid:

$$\begin{cases} \tau_f \cdot n_f = \tau_s \cdot n_s \\ d_f = d_s \\ q_f = q_s \\ T_f = T_s \end{cases} \tag{6}$$

Where f refers to the fluid, while s refers to the solid.

According to the assumptions, the geometric model is established as shown in Figure 1 below:

Due to geometric symmetry, can choose the cylindrical axis symmetry plane and at the same time, cuboids symmetry plane, below 2D calculation domain structure, simplified fluid domain and solid domain as shown in Figure 2, for the finite element analysis, selected the 7,205 nodes in the grid, boundary layer in solid border expansion processing according to the transition of 1.2, Draw the grid as shown in Figure 2:

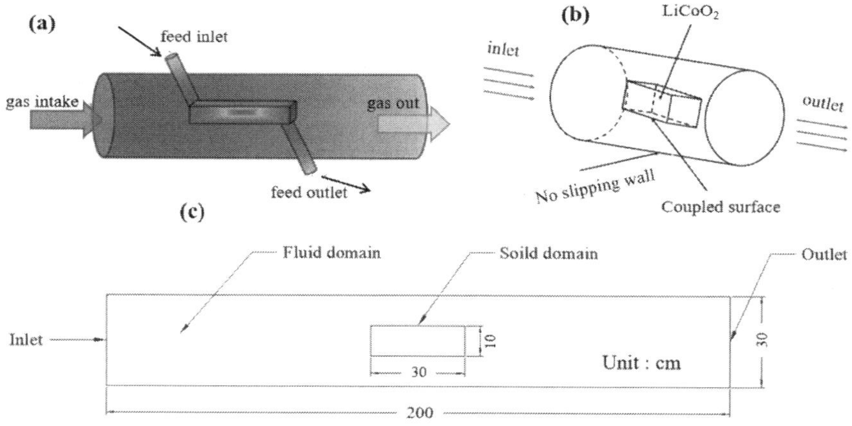

Figure 1. Structure of the geometric model.

2.2 *Simulation calculation of initial parameters*

The flow rate of flue gas was set as 5m/s, and the Coupled algorithm was used to solve the continuity equation, momentum equation, and energy equation simultaneously. Compared with SIMPLE, the convergence and accuracy of solving such problems were better, but the calculation time was longer due to the adoption of multiple iterations, requiring higher computational performance.

To analyze the effect of the overall temperature, the temperature cloud map of 873 K, 973 K, 1,073 K (600°C, 700°C, and 800°C) was drawn, as shown in Figure 3, and the temperature map of the center line as shown in Figure 3(b). Through calculation and simulation, the temperature exceeded 873 K, 973 K, and 1,073 K respectively after 1.53 h, 2.38 h, and 3.71 h, and it was found that the higher the temperature. The longer it takes for the temperature to change. Also, can be seen

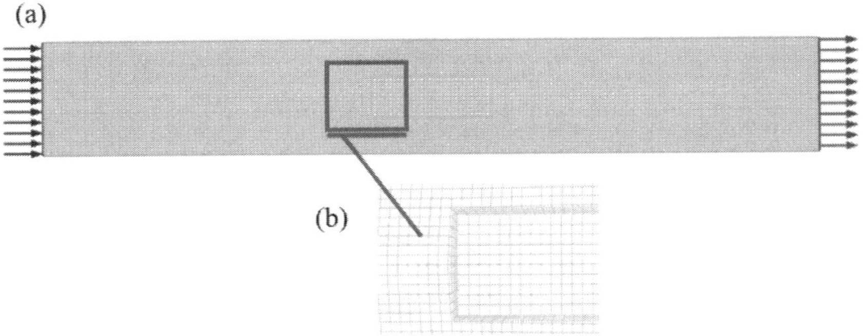

Figure 2. (a) Fluid computational domain and solid lithium cobalt acid computational domain; (b) meshing and boundary layer processing.

from the Figure 3(a), under the influence of temperature to flow, $LiCoO_2$ near the front part of the higher temperature, and the tail as the rear spoiler, temperature rise effect is relatively good, low temperature is located in the center near the location, at the back of the heating effect is relatively good, as shown in Figure 3(b). With the heating process, the temperature curve gradually smooths, making the temperature decrease.

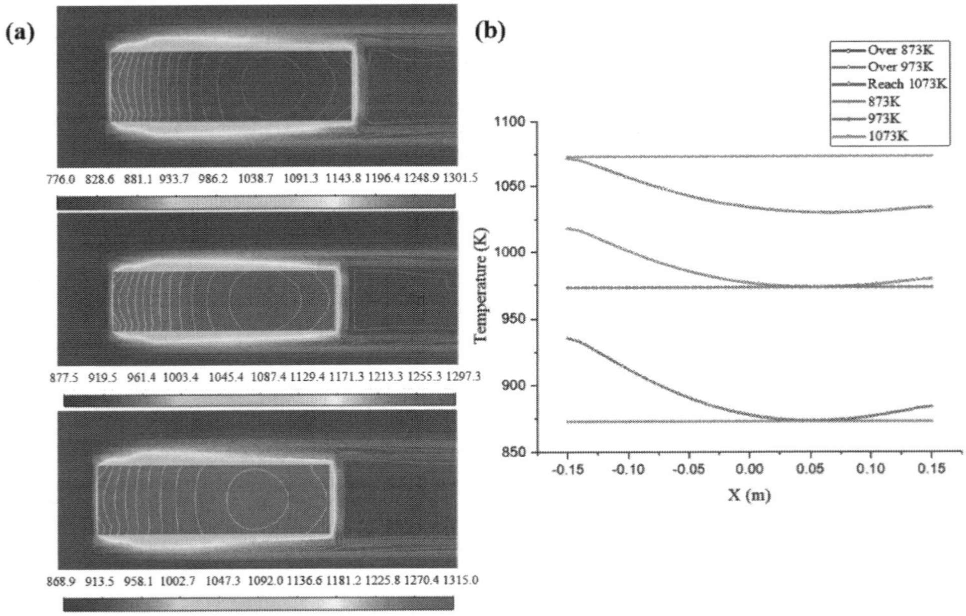

Figure 3. (a) Temperature distribution map at the end of the initial phase; (b) centerline temperature line and reference temperature line.

The above thermal simulation and calculation show that $LiCoO_2$ solid particles are heated by high-temperature SO_2 gas in the pipeline, and the lowest temperature in the solid cubic center can rise to the optimal reaction temperature after about 2 hours. In this process, a series of heat and mass transfer processes are carried out between the flue gas and the cubic $LiCoO_2$, so that the simulated $LiCoO_2$ particles can be acidified and roasted when the temperature rises. The theoretical calculation results fully prove the feasibility of our team's idea.

3 EXPERIMENTS

3.1 *Sulfation Roasting*

The actual experiment is carried out in the horizontal tube furnace. The experimental device is mainly composed of four parts: the Gas intake system, the Reaction system, the Analyse system, and the off-gas Absorption system. To study the reaction state of branch pipes of $LiCoO_2$ in a real factory, the following experimental scheme is set using the similarity principle after the size of the reaction vessel pipe is reduced to 1/5 of the original. As the ratio of SO_2 in the gas emission plant are 8-15% (Yu 2014), and to better simulate industrial off-smoke, we choose 10% SO_2-1% O_2-89%Ar as the atmosphere.

The $LiCoO_2$ powder was placed in a quartz crucible in a horizontal tubular furnace with a diameter of 60mm. The sulfur dioxide mixture gas with a proportion of 10% SO_2-1%O_2-89% Ar is slowly passed through and heated to 1,000°C, and monitored by DFC digital mass flowmeter to control the total gas flow rate of 380mL/min, heating process continues for 180 min. Use a k-type thermocouple to monitor temperature changes, and use a 6 mol/L NaOH solution as an absorption method to capture and detect SO_2 after the exhaust gas of the discharge reaction device is cooled by an ice bath.

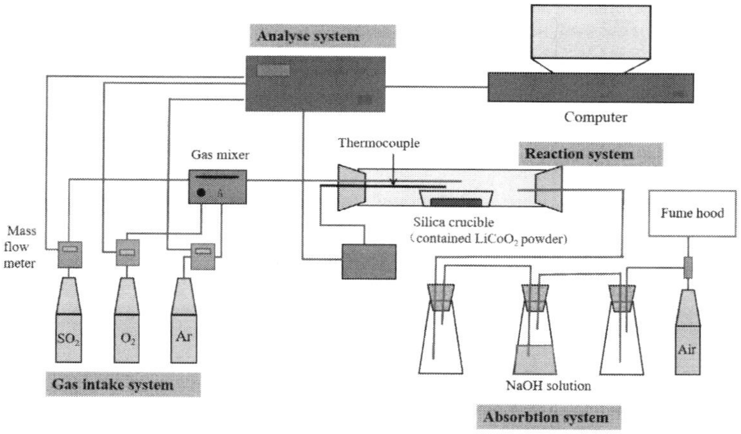

Figure 4. Schematic experiment of reaction in a horizontal tubular furnace.

After roasting, the crucible is quickly removed from the tube furnace and cooled at the ambient temperature. The crucible was weighed on an electronic balance, and the mass change before and after the reaction was recorded.

In the experiment, two atmospheres were selected for comparative experiments to better highlight the synergistic reaction between high-temperature SO_2 gas and samples. The control sample was placed in an air atmosphere and treated at 700°C for 180 min without any change in quality, indicating that the reaction did not take place. After 10 min treatments at 700°C in the mixture of 10%SO_2-1%O_2-89%Ar, the mass of the sample increased by 24.74%, indicating that the reaction had taken place, as shown in Figure 5(a). At the same time, it can be seen that the macroscopic structure of the sample changes significantly along the direction of gas flow, and the color of the part of the Li_2CoO_2sample that first touches the gas changes to nearly yellow, while the color of the sample at the end of the gas flow remains unreacted. Further combined with SEM morphology

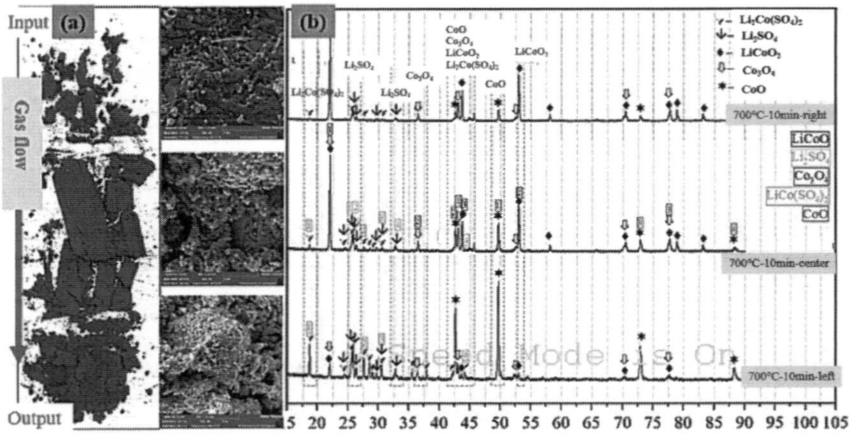

Figure 5. (a) SEM morphology of the $LiCoO_2$; (b) XRD image of the samples during roasting.

in Figure 5(a), it can be seen that the reaction products are mainly granular. To investigate the mechanism of phase change in the sample during the reaction process, the sample was divided into the right, center, and left three sections, and the XRD analysis was carried out on them respectively in Figure 5(b). The results showed that Li_2SO_4 and $Li_2Co(SO_4)_2$ were generated in the reaction products, both of which were water-soluble substances. Therefore, the feasibility of selective lithium extraction by sulfation roasting was proved experimentally.

3.2 *Water-Leaching*

To investigate the leaching effect of lithium metal at different reaction times, water leaching experiments were carried out with distilled water and roasted samples at 25°C, and the results are shown in Figure 6. It can be seen that the leaching rate of lithium can reach more than 99% after the roasting reaction takes place for 30min. With the extension of the reaction time, the leaching rate of lithium remains stable at more than 99%, which proves the correctness of the idea of "Sulfation roasting-water leaching" for lithium extraction in our project.

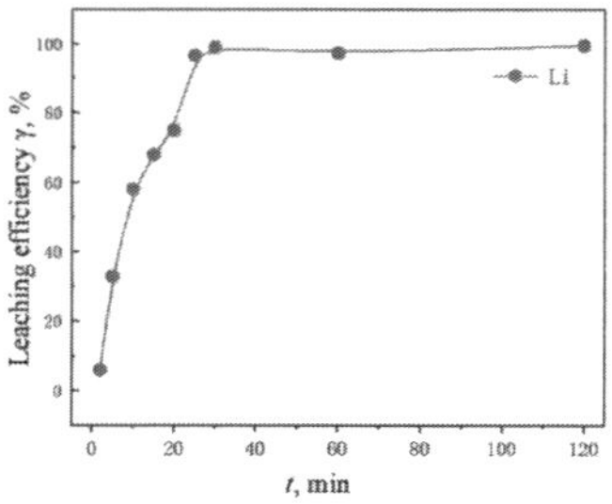

Figure 6. Leaching efficiency of Li.

4 CONCLUSION

1. The CFD simulation of the virtual cubic $LiCoO_2$ shows that by utilizing high-temperature off-gas, the lowest temperature of the solid center line can rise to the optimal reaction temperature

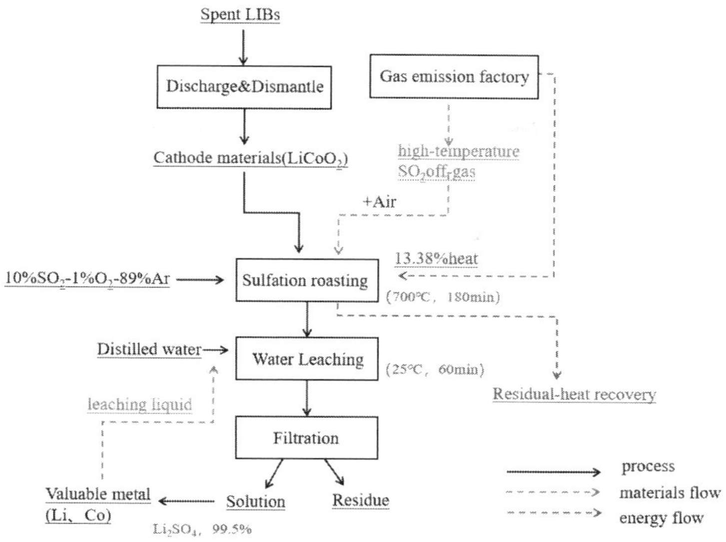

Figure 7. The overall flow sheet of the recycling process.

after about 2 hours, and the overall heat transfer efficiency between 700°C off-gas and $LiCoO_2$is 13.38%.

2. The roasting process with a 10% SO_2 atmosphere at 700°C resulted in a remarkable change in the $LiCoO_2$ sample, after 10 min the mass of the sample increased by 24.74% The SEM image shows that the powder near the direction of the gas takes in changes faster. The major products of our experiment confirmed by the XRD, Li_2SO_4 and $Li_2Co(SO_4)_2$, are watersoluble salts that can be separated in the following water-leaching process.
3. The water leaching experiment was carried out at 25°C in 2 h, and after 30 minutes, the exaction efficiency of Li can reach 99%.
4. This paper presents a novel process of recycling Li from cathode powder of spent lithium-ion batteries, which will utilize the huge heat carried by the high-temperature gas emitted in the industry such as Copper smelting plants, to preheat the waste lithium-ion battery and conduct the sulfation roasting process. A conceivable flow sheet of our process is shown in Figure 7, to form a close-loop of materials and energy recycling and regeneration. This method neatly realizes the two goals of energy conservation and emission reduction, while recycling the valuable metal in the waste battery. Meanwhile, our heat transfer efficiency is only 13.38% by pipe, to achieve the true high-temperature gas heating process, equipment with higher energetic efficiency is essential, such as Heat Exchanger, Fluidized Bed and so on (Wan 2021), and that's what we are struggling. Though this article only provides thought and simulated successfully with CFD software, we consider that it's a promising field that is worth paying attention to. And our team will keep seeking and learning, to learn to make better utilization of multi-metal solid waste secondary resources and waste gas. Even a little difference is made are we satisfied, to provide ideas for the development of more green and efficient lithium extraction process in the future.

REFERENCES

Bhutta, MMA (2012). CFD applications in various heat exchangers design: A review. *J. Applied Thermal Engineering*. 2012, 32: 1–12.

Dang H. (2022). Na2SO4–NaCl binary eutectic salt roasting to enhance extraction of lithium from pyrometallurgical slag of spent lithium-ion batteries. *J. Chinese Journal of Chemical Engineering*, 41(2022), 294–300.

Ji J.Q. (2020). Sodium transformation simulation with a 2-D CFD model during circulating fluidized bed combustion. *J. Fuel*, 2020, 267(C).

Li N. (2019). Aqueous leaching of lithium from simulated pyrometallurgical slag by sodium sulfate roasting. *J. RSC Advances*, 2019, 9(41): 23908–23915.

Meshram P. (2016). Acid baking of spent lithium-ion batteries for selective recovery of major metals: A two-step process[J]. *Journal of Industrial and Engineering Chemistry*, 2016, 43: 117–126.

Shi J.J. (2019). Sulfation roasting mechanism for spent lithium-ion battery metal oxides under SO2-O2-Ar atmosphere. *J. Jom*, 2019, 71(12): 4473–4482.

Tang Y.Q. (2021). Recycling of spent lithium nickel cobalt manganese oxides via a low-temperature ammonium sulfation roasting approach. *Journal of Cleaner Production*, 279(2021)123633.

Wan X. (2021). A potential industrial waste–waste co-treatment process of utilizing waste SO2 gas and residue heat to recover Co, Ni, and Cu from copper smelting slag. *Journal of Hazardous Materials*, 2021, 414: 125541.

Yu D.W. (2014). *Fluidized Bed Selective Oxidation and Sulfation Roasting of Nickel Sulfide Concentrate*. D.University of Toronto, 2014.

Energy Revolution and Chemical Research – Kok-Keong Chong and Zhongliang Liu (Eds)
© 2023 The Authors, ISBN: 978-1-032-36554-1

The economic evaluation of the improvement of the reheating system of the secondary reheat steam turbine

Tongyang Pan
Datang Northeast Electric Power Test & Research Institute Co., Ltd., Changchun, China

ABSTRACT: The definition of ideal cycle thermal efficiency is improved by proposing the concept of ideal heat consumption, the decoupling between the ideal cycle thermal efficiency and the relative internal efficiency of the steam turbine is realized, and the influence of the steam turbine factor in the evaluation of the thermal system is eliminated. This method realizes the correct evaluation of the operating economic status of the thermal system of the double reheat steam turbine. The effects of various factors on the thermal efficiency of the two ideal cycles are calculated through examples, which verify the rationality and accuracy of the method and provide a basis for the next step of the economic diagnosis of the thermal system of the second reheat steam turbine.

1 INTRODUCTION

Research shows that the coal consumption for the power supply of a 1000MW ultra-supercritical unit with secondary reheat is lower than that of a primary reheat unit with the same capacity of 1000 MW (Yin 2013; Zhang 2013). Therefore, secondary reheating technology with clean and efficient characteristics is being more and more widely used (Gao 2014; Zhao 2012; Zhang 2016). At present, the installation and combination of the domestic secondary reheat steam turbine regenerative system are complicated and changeable, and the design is different according to the needs of the unit. The regenerative system and the main equipment of the steam turbine are connected, and the internal parameters of the two influence each other. There is strong coupling, and it is not easy to distinguish the specific factor caused by the change in the thermal economy. Therefore, research on the economic indicators for evaluating the quality of the thermal system of the steam turbine of the secondary reheat unit is of decisive significance for the rationality of its operating economic status, as well as the subsequent fault diagnosis and energy-saving transformation. Research pointed out that for thermal power units with regenerative heat, the ideal cycle thermal efficiency should be selected as the economic evaluation index of the operating state of the thermal system (National Steam Turbine Standardization Technical Committee 2009; ASME 2004).

The article mainly analyzes the shortcomings of the steam turbine ideal cycle thermal efficiency in the evaluation process of the economic state of the secondary reheat thermal system, proposes an improvement plan based on the existing problems, and verifies it through actual calculation examples.

2 THE DEFINITION METHOD OF THE IDEAL CYCLE THERMAL EFFICIENCY OF THE SECONDARY REHEAT UNIT AND ITS EXISTING PROBLEMS

In the power cycle of a coal-fired power plant, the energy generated by fuel combustion is transferred from the boiler to the steam turbine to perform work through the working fluid, and finally converted

DOI 10.1201/9781003332657-101

into the electric power of the generator. The energy transfer equation is:

$$P_{el} = Q_0 \frac{P_t}{Q_Q} \frac{P_i}{P_t} \frac{P_m}{P_i} \frac{P_{el}}{P_m} = Q_0 h_t h_{ri} h_m h_g \tag{1}$$

where Q_0 is the heat consumption of the steam turbine per unit time, kW; P_t is the ideal internal power of the isentropic expansion process of the steam turbine, kW; P_i is the actual power of the steam turbine, kW; P_m is the shaft end power of the steam turbine, kW; P_{el} is the generator output power; η_t,η_{ri},η_m and η_g are the ideal cycle thermal efficiency of the steam turbine thermal cycle, the relative internal efficiency of the steam turbine, the mechanical efficiency and the generator efficiency, respectively.

From Equation (1), it can be concluded that the ideal cycle thermal efficiency of the steam turbine thermal cycle is the ratio of the ideal power of the steam turbine to the heat consumption of the steam turbine:

$$\eta_t = \frac{P_t}{Q_0} \tag{2}$$

For the secondary reheat unit, the ideal power of the steam turbine and the heat consumption of the steam turbine can be expressed as:

$$P_t = D_0\left(1 - \sum_{j=1}^{Z} \alpha_{jt} Y_{jt} - \sum_{k=1}^{z_1} \alpha_{sgkt} Y_{sgkt}\right) \Delta H_t \tag{3}$$

$$Q_0 = D_0(h_0 - h_{fw}) + D_{rh1}(h_{rh1} - h_{vh}) + D_{rh2}(h_{rh2} - h_h) = D_0 Q_h + D_{rh1} Q_{rh1} + D_{rh2} Q_{rh2} \tag{4}$$

Where D_0,D_{rh1} and D_{rh2} are the main steam flow rate, the primary reheated steam flow rate, and the secondary reheated steam flow rate, kg/s; α_{jt} and α_{sgkt} are the fraction of regenerative extraction steam of the j stage heater, the share of the k stage shaft seal and the valve stem leakage steam during the isentropic expansion of the steam turbine; ΔH_t is the ideal enthalpy drop of the steam turbine, kJ/kg; Q_h, Q_{rh1} and Q_{rh2} are the heat absorption of the working fluid in the super heater, primary re-heater and secondary re-heater in the actual process; h_0, h_{fw}, h_{rh1}, h_{rh2}, h_{vh} and h_h are steam turbine main steam enthalpy, feed water enthalpy, primary reheat steam enthalpy, secondary reheat steam enthalpy, ultra-high pressure cylinder exhaust enthalpy, and high-pressure cylinder exhaust enthalpy, kJ/kg.

For a simple Ran-kine cycle thermal power unit, since the ideal cycle thermal efficiency and the relative internal efficiency of the steam turbine body are independent of each other, the ideal cycle thermal efficiency is usually used to evaluate the thermal economy of the thermal cycle. For thermal power units that use regenerative circulation without reheating, when the thermal economy of the steam turbine body changes, the boiler feed water temperature will change, which will cause a change in the ideal cycle thermal efficiency, but this change is relatively small and can be ignored.

For a thermal power unit that adopts a regenerative cycle and has a reheat, once the economy of the steam turbine body changes, it will cause a double change in the temperature of the main feed water and the exhaust enthalpy of the high-pressure cylinder. In this way, when the steam parameter of the reheated steam is maintained at a fixed value, it will cause the change in the heat consumption of the steam in the re-heater and the boiler, and finally, cause the change of the ideal cycle thermal efficiency. Through analysis, it can be found that the change of the steam turbine body factor has a positive effect. The degree of influence of the ideal cycle thermal efficiency of the primary reheat unit is higher than that of the pure regenerative cycle unit.

It can be seen from formulas (2)-(4) that for thermal power units that have regenerative heat and adopt secondary reheat technology, when the thermal economy of the steam turbine body changes, it not only causes the change of the feed water temperature but also causes the change of the exhaust enthalpy of the super high-pressure cylinder and the high-pressure cylinder of the steam turbine.

Under the condition that the reheat steam parameters are kept at fixed values, the total heat consumption of the boiler and the first and second re-heaters will be changed, and finally the ideal cycle thermal efficiency will be changed. Therefore, there is a strong coupling between the ideal cycle thermal efficiency of the thermal system of the secondary reheat thermal power unit and the relative internal efficiency of the steam turbine. The ideal cycle thermal efficiency is affected by the relative internal efficiency of the steam turbine and is not a single-valued function of the operating economic state of the thermal system. It cannot truly evaluate whether the operating state of the thermal system is normal.

Realizing the decoupling between the ideal cycle thermal efficiency of the thermal system and the relative internal efficiency of the steam turbine is a prerequisite for accurately identifying the operating economic status of the thermal system of the secondary reheated thermal power unit during operation.

3 IMPROVEMENT OF IDEAL CYCLE THERMAL EFFICIENCY OF SECONDARY REHEAT STEAM TURBINE

3.1 *Improvement of the definition of ideal cycle thermal efficiency*

To solve the problems found in the above analysis of the method for defining the ideal cycle thermal efficiency of the thermal cycle of the second reheat steam turbine, the definition of the ideal cycle thermal efficiency of the thermal system of the second reheat steam turbine is now improved. The main idea is to minimize the influence of the relative internal efficiency of the super high-pressure cylinder and the high-pressure cylinder of the steam turbine on the ideal cycle thermal efficiency.

Rewrite formula (1) as:

$$P_{el} = Q_0 \frac{Q_{0t}}{Q_0} \frac{P_t}{Q_{0t}} \frac{P_i}{P_t} \frac{P_m}{P_i} \frac{P_{el}}{P_m} = Q_0 \lambda \eta_{tt} \eta_{ri} \eta_m \eta_g \tag{5}$$

Where Q_{0t} is the heat consumption of the steam turbine under the condition of medium entropy expansion of the steam turbine, called the ideal heat consumption, kW; λ is the ratio of the ideal heat consumption of the steam turbine during the isentropic expansion of the steam to the actual heat consumption of the steam turbine during the actual expansion of the steam, which is called the heat absorption coefficient in this article; η_{tt} is the improved ideal cycle thermal efficiency, which is expressed as the ratio between the ideal internal power of the steam turbine and the ideal heat consumption, and its expression is:

$$\eta_{tt} = \frac{P_t}{Q_{0t}} \tag{6}$$

The ideal heat consumption can be expressed as:

$$Q_{0t} = D_0(h_0 - h_{fwt}) + D_{rh1t}(h_{rh1} - h_{vht}) + D_{rh2t}(h_{rh2} - h_{ht}) \tag{7}$$

It also can be expressed as:

$$Q_{0t} = D_0 Q_{ht} + D_{rh1t} Q_{rh1t} + D_{rh2t} Q_{rh2t} \tag{8}$$

Where h_{fwt} is the boiler feed water enthalpy when steam is entropically expanded in the steam turbine, kJ/kg; D_{rh1t} and D_{rh2t} are the primary reheat steam flow rate and the secondary reheat steam flow rate during the isentropic expansion of the steam turbine, kg/s; Q_{ht}, Q_{rh1t} and Q_{rh2t} are the heat absorption of the working fluid in the super-heater, primary re-heater, and secondary re-heater during the isentropic expansion of the steam turbine; h_{vht} and h_{ht} are the exhaust enthalpy of the ultra-high pressure cylinder and the exhaust enthalpy of the high-pressure cylinder during the isentropic expansion of steam in the steam turbine, kJ/kg.

3.2 *Calculation method of relevant parameters in ideal heat consumption*

When steam is entropically expanded in the steam turbine, the outlet water temperature of the heater, that is the ideal feed water temperature of the boiler, is the saturation temperature corresponding to the extraction steam pressure minus the end difference. Ideal feed water enthalpy of boiler can be expressed as:

$$t_{fwt} = t_{set} - \delta t = f(p_{et}, \delta t) \tag{9}$$

According to the feed water temperature and feed water pressure, the ideal feed water enthalpy of the boiler can be obtained as:

$$h_{fwt1} = g(p_{fw}, t_{fwt}) \tag{10}$$

From the heat balance equation of the heater, the ideal regenerative steam extraction amount when the steam is entropically expanded in the steam turbine is obtained as:

$$D_{et} = \frac{D_{fw}(h_{fwt1} - h_{fwt2})}{h_{et} - h_{wt}^{d}} \tag{11}$$

The corresponding parameters under ideal conditions are obtained from the extraction point parameters of the actual expansion process of the steam turbine. It can be expressed as:

$$\frac{D_0 - D_{et}}{D_0 - D_e} = \frac{p_{et}}{p_e}\sqrt{\frac{T_e}{T_{et}}} \tag{12}$$

Then the pressure at the regenerative extraction point during the isentropic expansion of steam in the steam turbine is:

$$p_{et} = p_e \frac{D_0 - D_{et}}{D_0 - D_e}\sqrt{\frac{T_{et}}{T_e}} \tag{13}$$

The ideal regenerative extraction point enthalpy during the isentropic expansion process of steam in a steam turbine is:

$$h_{et} = h_0 - \Delta H_{vt}\eta_{vri} \tag{14}$$

From the properties of water vapor, according to the enthalpy of the ideal regenerative extraction point and the ideal regenerative extraction pressure, the temperature of the ideal extraction point can be obtained as:

$$T_{et} = h(h_{et}, p_{et}) \tag{15}$$

Where h_{fw1},h_{fw2} and h_{w1}^{d} are the first-stage regenerative heater outlet, inlet feed water enthalpy, and its hydrophobic enthalpy, kJ/kg; δt is the upper-end difference of the first stage regenerative heater, °C; p_{e0} andp_e are the exhaust steam pressure of the ultra-high pressure cylinder under design conditions and actual conditions, Pa; T_{e0} and T_e are the exhaust steam temperature of the ultra-high pressure cylinder under design conditions and actual conditions, K; D_{e0} and D_e are the exhaust steam volume of the ultra-high pressure cylinder under design conditions and actual conditions, kg/s; h_e is the exhaust enthalpy of the exhaust steam of the ultra-high pressure cylinder, kJ/kg; $\Delta H_{vt}$is the ideal enthalpy drop of the main steam during the isentropic expansion process in the ultra-high pressure cylinder, kJ/kg; η_{vri} is the relative internal efficiency of the ultra-high pressure cylinder. Because the process is an isentropic expansion process, the relative internal efficiency value is 100%.

By solving the nonlinear equations composed of formulas (9)-(15), the corresponding parameters of the steam in the steam turbine during isentropic expansion can be obtained from the parameters of the actual operating conditions.

The flow rate of reheated steam once is:

$$D_{rh1t} = D_0(1 - \sum_{j=1}^{Z_{vh}} \alpha_{jt} - \sum_{k=1}^{z_{vh1}} \alpha_{sgkt}) \tag{16}$$

The second reheat steam flow rate is:

$$D_{rh1t} = D_0(1 - \sum_{j=1}^{Z_{vh}+Z_h} \alpha_{jt} - \sum_{k=1}^{z_{vh1}+z_{h1}} \alpha_{sgkt}) \tag{17}$$

Where Z_{vh} and Z_{vh1} are the regenerative extraction steam stages, shaft seals, and valve stem leakage stages of the ultra-high-pressure cylinder; Z_h and Z_{h1} are the regenerative extraction steam stages, shaft seals, and valve stem leakage stages of the high-pressure cylinder, respectively.

4 IMPROVED IDEAL CYCLE THERMAL EFFICIENCY ANALYSIS

The steam turbine of a 660 MW ultra-supercritical secondary reheat unit uses 10 stages of regenerative extraction steam as 4 high-pressure heaters. The steam flow system diagram is shown in Figure 1. Taking the 660 MW ultra-supercritical secondary reheat steam turbine as an example, the relative internal efficiencies of the high-pressure and ultra-high-pressure cylinders of the steam turbine, the boiler feed water temperature, and the thermal efficiency of the two ideal cycles are calculated separately.

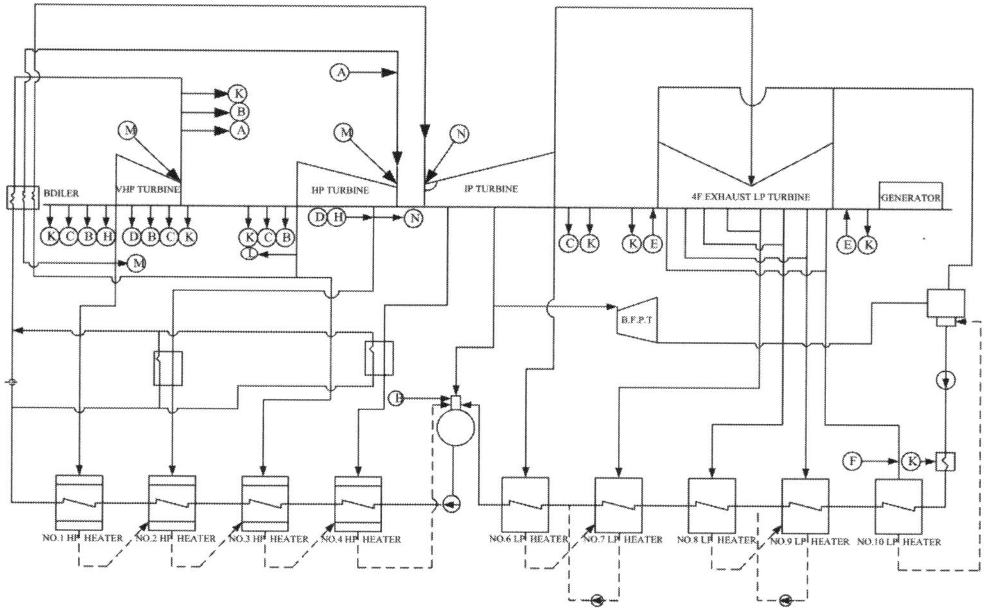

Figure 1. Principle system diagram of a 660MW secondary reheat unit.

For the isentropic expansion of steam in a steam turbine, since the parameters of the regenerative extraction point are not affected by the relative internal efficiency changes of the ultra-high-pressure cylinder, therefore, the ideal feed water enthalpy and ideal feed water temperature are not affected by the relative internal efficiency of the ultra-high-pressure cylinder. By solving the nonlinear equations composed of (9)–(15), we can obtain the regenerative extraction point pressure under the relative internal efficiency of different ultra-high-pressure cylinders, temperature, and actual feed water enthalpy value, actual feed water temperature, and ideal feed water enthalpy value. The change in ideal feed water temperature is shown in Figures 2–5.

The parameters of the first-stage regenerative extraction point all change with the relative internal efficiency of the ultra-high-pressure cylinder, which causes the boiler feed water temperature to

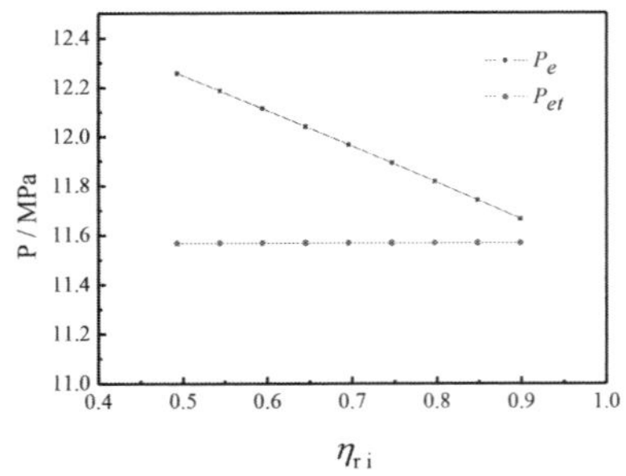

Figure 2. The curve of extraction point pressure with the efficiency of the super high-pressure cylinder.

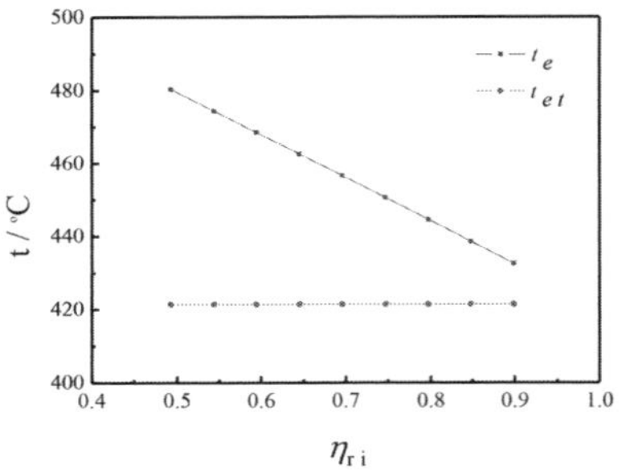

Figure 3. The temperature of the extraction point varies with the efficiency of the ultra-high-pressure cylinder.

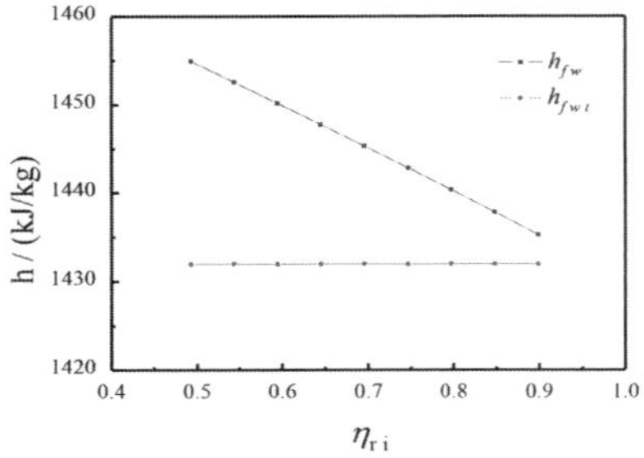

Figure 4. Two kinds of feed water enthalpy change with the efficiency of the ultra-high-pressure cylinder.

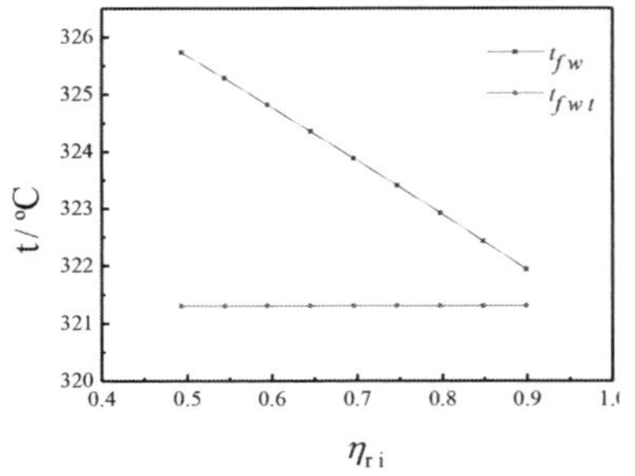

Figure 5. Two kinds of feed water temperature change with the efficiency of the ultra-high-pressure cylinder.

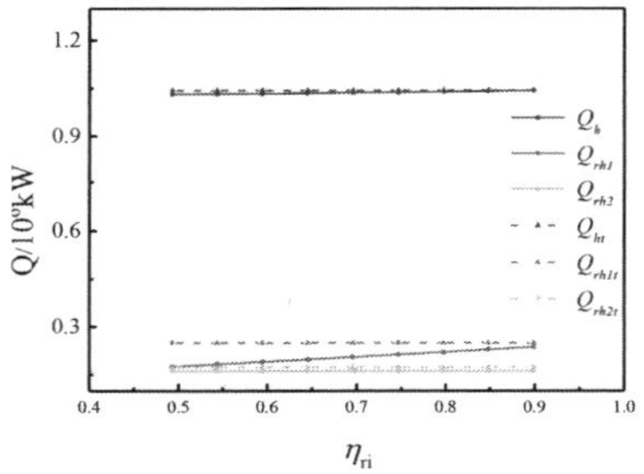

Figure 6. The change curve of the heat absorption of the working fluid in the various equipment of the boiler with the relative internal efficiency of the ultra-high-pressure cylinder.

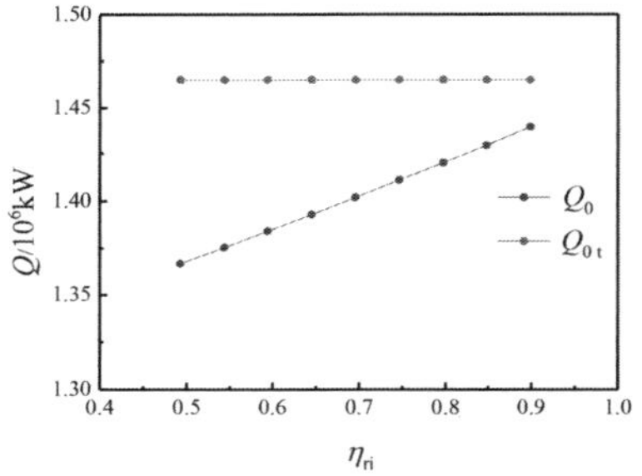

Figure 7. Curve of total heat absorption of the boiler with relative internal efficiency of the ultra-high-pressure cylinder.

change with the relative internal efficiency of the ultra-high-pressure cylinder. In the ideal process, the boiler feed water temperature and feed water enthalpy have nothing to do with the relative internal efficiency of the ultra-high-pressure cylinder. The relative internal efficiency changes of the ultra-high-pressure cylinder and the high-pressure cylinder not only cause changes in the boiler feed water temperature, the exhaust steam enthalpy of the ultra-high-pressure cylinder and the high-pressure cylinder but also cause changes in the flow of reheated steam in the primary and secondary re-heaters, thereby causing the steam to change. This causes a change in the amount of heat absorbed by the steam in the boiler.

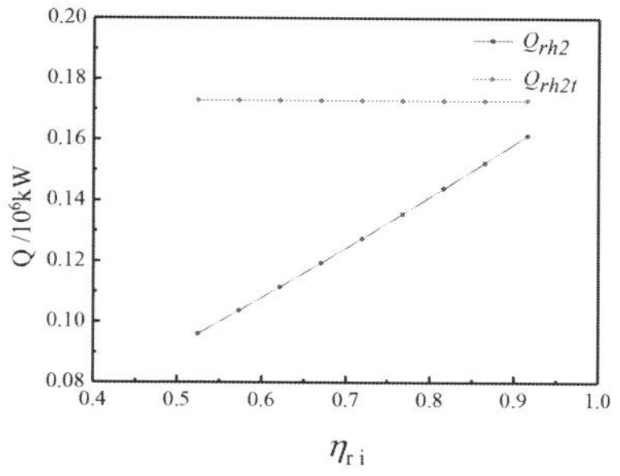

Figure 8. The heat absorption in the secondary re-heater varies with the relative internal efficiency of the high-pressure cylinder.

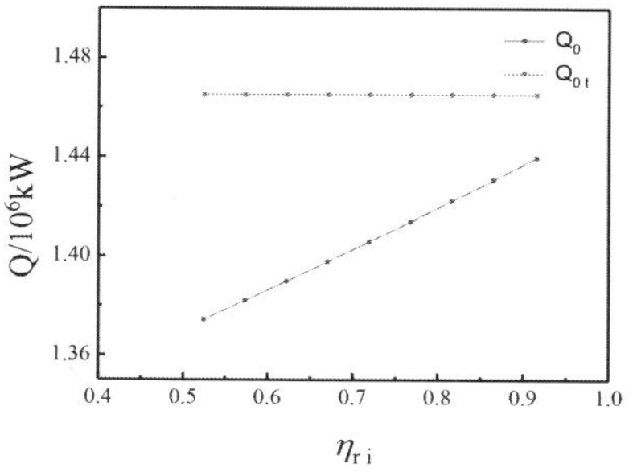

Figure 9. The total heat absorption of the boiler changes with the relative internal efficiency of the high-pressure cylinder.

Through calculations, when the phase efficiency of the super-high-pressure cylinder of the steam turbine changes, the changes in the heat absorption of the steam in the boiler super-heater, primary re-heater, secondary re-heater, and the entire boiler are shown in Figure 6 and Figure 7.

The change in the relative internal efficiency of the high-pressure cylinder of the steam turbine only causes a change in the heat absorption of the steam in the secondary re-heater and the entire boiler. Through calculations, the changes in the heat absorption of the steam in the secondary re-heater and the entire boiler with the relative internal efficiency of the high-pressure cylinder are shown in Figure 8 and Figure 9.

5 CONCLUSION

In the article, the definition of ideal circulatory heat efficiency is improved. By proposing the concept of "ideal thermal consumption", it has completed the decoupling between the ideal circulating heat efficiency and the opposite endurance of the steam turbine, eliminating steam turbine body factors in the thermal system evaluation Influence. The correct evaluation of the operating economy of the secondary reheat power plant turbine thermodynamic system was achieved.

The effects of various factors on the thermal efficiency of the two ideal cycles are calculated through examples to verify the rationality and accuracy of the method in this paper, and provide a basis for the next step of the economic diagnosis of the thermal system of the second reheat steam turbine.

REFERENCES

ASME. ASME PTC 6-2004 Performance Test Codes 6 on Steam Turbine, *New York: ASME Press*, 2004. 58–59.

Gao HT, Fan HJ, Dong JC, et al. The development of supercritical secondary reheat unit. *Boiler technology*, 2014, 45(4): 1–3.

National Steam Turbine Standardization Technical Committee. GB-8117-2008 Steam Turbine Thermal Performance Acceptance Test Regulations, *Beijing: China Standard Press*. 2009.128–134.

Yin YN. Application status and development of secondary reheat ultra-supercritical unit. *Power Plant System Engineering*, 2013, 2:015.

Zhang FW, Liu YY, Tan HZ, et al. Research on Secondary Reheat Technology of Supercritical Thermal Power Generating Unit. *Electric Power Survey Technology*, 2013, (2):34–39.

Zhang W, Yan K, Wang H, et al. Research on Hydrodynamic Characteristics of Water Wall of Supercritical Secondary Reheat Once-through Boiler. *Thermal Power Engineering*. 2016, 31 (8): 75–80.

Zhao ZD, Dang LJ, Liu C, et al. Start-up operation and control of supercritical unit. *Beijing China Electric Power Press*, 2012: 1–310.

Energy Revolution and Chemical Research – Kok-Keong Chong and Zhongliang Liu (Eds)
© 2023 The Authors, ISBN: 978-1-032-36554-1

Analysis of hydrogen liquefaction industry in China

Feng Chen* & Zixuan He*
Faculty III – Department of Process Science, Technical University of Berlin, Berlin, Germany
University of Shanghai for Science and Technology, Shanghai, China

ABSTRACT: Hydrogen energy, as an indispensable part of China's goal of achieving carbon neutrality, is gradually receiving more attention. The hydrogen liquefaction industry in China is relatively backward. This paper mainly introduces the hydrogen liquefaction industry in the world and the conceptual hydrogen liquefaction process in the previous research. The proposals for the hydrogen liquefaction industry were put forward. The LNG (Liquefied Natural Gas) pre-cooled hydrogen liquefaction plant is relevant to the advantages of China's natural gas resources. The proposal of a hybrid liquefaction plant, which makes full use of the abundant renewable energy resources in China, can be verified. The cryogenic part of the plant consumes the largest amount of power, which needs further studies.

1 INTRODUCTION

Compared to the twentieth century, the global average temperature has increased by nearly 1°C, and humanity is facing the risk of global warming (Lindsey & Dahlman 2021). The Paris Agreement points out the general direction of the global low-carbon transition in response to climate change and China will spare no effort to achieve a carbon peak by 2030 and carbon neutrality by 2060 (Xi 2020).

To achieve the zero-carbon transition of the energy structure, renewable energy will be continuously invested in and implemented in China (Wang & Zhang 2020). However, uneven energy supply and demand distribution and the significant seasonal changes in renewable energy would also be challenges (Li et al. 2007) Relying on renewable energy alone would lead to uncertainty in the power supply and cause a significant amount of electricity abandoned every year. Fortunately, the energy storage technology can increase the utilization rate of power equipment and reduce power costs, thereby realizing sustainable, economical, and secure electricity supplies (Wade et al. 2010).

The existing energy storage methods include pumped storage technology, battery storage technology, hydrogen storage technology, etc. (Hadjipaschalis et al. 2009). Hydrogen storage technology has its irreplaceable advantages among these methods: First of all, hydrogen has one of the highest energy density values per mass. In the second place, it is sustainable and nontoxic. Last but not least, the reaction product is water, which is environmentally friendly (Abe et al. 2019). Also, the utilization of hydrogen energy is not limited to power generation. The application of hydrogen energy to fuel cell vehicles can effectively help urban transportation achieve the zero-carbon transition. Also, hydrogen can provide clean energy and raw materials for industrial processes (Liu & Zhong 2019). Thus, hydrogen energy is important in energy and industrial future in China.

Hydrogen production technologies have been studied for years, and the most advanced technology is hydrocarbon reforming (Holladay et al. 2009). However, hydrogen storage technology is still a problem to be overcome. The density of hydrogen is very low at atmospheric pressure, so it

*Corresponding Authors: stefanchen0526@gmail.com and SBC-18-1056@sbc.usst.edu.cn

 DOI 10.1201/9781003332657-102

should be stored under high pressure or in liquid form. Compared with high-pressure gas storage, the advantages of liquid storage are safety, storage with higher energy density, and larger capacity transportation (Abe et al. 2019; Wolf 2003; Züttel 2004). However, the cryogenic temperature liquefaction process would consume approximately 30% of the energy content of hydrogen. Therefore, methods to improve the efficiency of the hydrogen liquefaction process would become the focus of research. This paper will mainly discuss hydrogen liquefaction processes in previous studies and discuss what China could learn from these researches in hydrogen liquefaction technologies.

2 ORTHO-PARA CONVERSION

Before introducing hydrogen liquefaction processes, the quantum effect that hydrogen exhibits should be verified during the liquefaction process. The presence of two distinctive sorts of hydrogen is defined by the distinction in the relative orientation of the nuclear spins of the protons that form the hydrogen particle. As shown in Figure 1, the two proton nuclear spins orient in the same direction for ortho hydrogen and the opposite direction for para hydrogen.

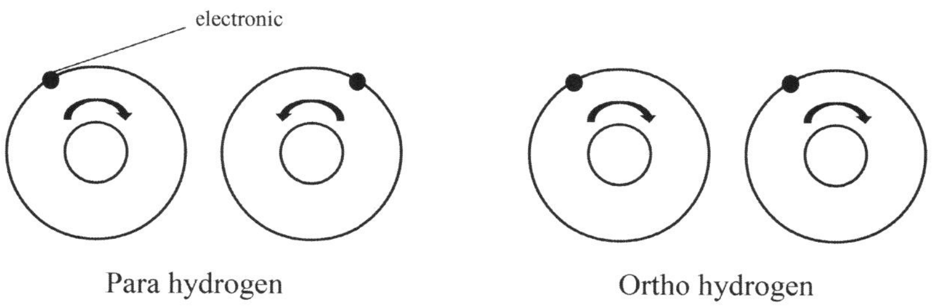

Figure 1. Ortho hydrogen and para hydrogen.

Para hydrogen encompasses a lower energy state than ortho hydrogen, which means the liquid para-hydrogen has a lower enthalpy than liquid ortho hydrogen. During the cooling process of hydrogen, the ortho hydrogen converts to para-hydrogen spontaneously, but the reaction rate is very low without a catalyst. After the hydrogen was liquefied, the liquid hydrogen would still contain a large amount of ortho hydrogen, and the conversion reaction still takes place during the storage period. Also, the heat of the Ortho-para conversion reaction is higher than the condensation heat of hydrogen, which means the boil-off would happen. The vaporization amount of hydrogen would be 18% in one day time and 50% after one week (Baker & Shaner 1978). Therefore, conversion catalysts should be utilized, and additional refrigeration capacity during the storage period would be required (Abe et al. 2019).

3 INITIAL MODEL OF HYDROGEN LIQUEFACTION PROCESSES

Hydrogen is firstly liquefied in 1898 with a flow rate of 4 cc/min. The hydrogen was compressed to 180 bar and precooled to -250°C using the carbonic acid and liquid air as the cooling medium (Dewar 1898). Although the Linde-Hampson process could not be utilized to liquefy hydrogen, the pre-cooled Linde-Hampson process was applied to liquefy hydrogen (Barron 1966). However, the SEC of the pre-cooled Linde-Hampson process is about 70 kWh/kgH2. The pre-cooled Claude cycle was suggested for hydrogen liquefaction, and the SEC was decreased to approximately 35 kWh/kgH2 (Nandi & Sarangi 1993). The flow diagram of the pre-cooled Claude cycle was shown in Figure 2.

Although the liquid hydrogen industry has developed for more than 60 years, high energy consumption in the liquefaction process has not been overcome. Even for the hydrogen liquefaction

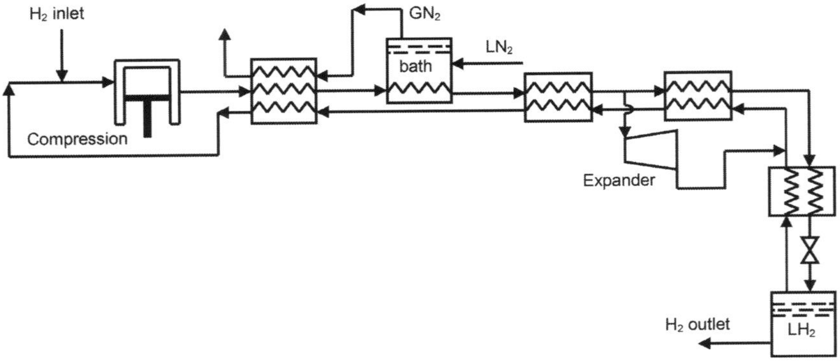

Figure 2. The flow diagram of the pre-cooled Claude cycle.

plant with the best performance in America, the energy consumption of hydrogen liquefaction would be approximately 10 kWh/kgH2, which would be almost 30% of the energy content of hydrogen (Drnevich 2003). Therefore, a lot of research elevating the energy efficiency of hydrogen liquefaction plants has been proposed since 2000 (Krasae-in et al. 2010).

4 CONCEPTUAL PLANTS

After years of exploration and research, the basic model of the hydrogen liquefaction system was gradually determined. The hydrogen liquefaction process should consist of a pre-cooled part and a cryogenic part. In the pre-cooled hydrogen liquefaction systems, the hydrogen would generally be cooled down to -195°C in the pre-cooled part, and liquid nitrogen could work as the refrigerant. After that, hydrogen would be further cooled down and liquified in the cryogenic part. Pure helium or helium-neon mixture was selected as the refrigerant for the cryogenic part (Kuz'menko et al. 2004; Venkatarathnam & Timmerhaus 2008). The pre-cooled part would be necessary for the hydrogen liquefaction process because it cannot only increase the liquid hydrogen yield but also decrease the compression work of the process (Aasadnia & Mehrpooya 2018). Also, the hydrogen gas can be purified in the heat exchanger during the pre-cooling process (Barron 1972).

After that, no effort was spared to improve the exergy efficiency of the liquefaction, no matter in the pre-cooled or cryogenic parts. In this section, different methods to improve the energy efficiency of hydrogen liquefaction plants would be introduced.

4.1 *The pre-cooled hydrogen liquefaction cycle*

In the pre-cooled Claude process, the liquid nitrogen bath was employed to pre-cool the hydrogen (Nandi & Sarangi 1993). Then the process was improved by combining the pre-cooling part and dual pressure stages. Through this method, the SEC (specific energy consumption) of the pre-cooled dual pressure Claude cycle can be reduced to 12.26 kWh/kgH2 (Barron 1972).

The MR (mixed refrigerant) pre-cooled cycle, which utilizes MR instead of pure refrigerant in the pre-cooled part, was suggested in the previous study. The advantage is that the temperature gliding characteristic of mixed refrigerant makes the temperature profiles of hydrogen and mixed refrigerant in the heat exchanger well matched. This implies that the temperature difference in the heat exchanger would be lower, and less exergy destruction would be caused (Walnum et al 2012).

The liquid natural gas pre-cooled system would be a good option for hydrogen liquefaction. Since hydrocarbon reforming reaction is the most convenient method to produce hydrogen, natural gas would be the raw material for producing hydrogen. When the natural gas is transported in liquid form, it should be heated up to gas form to enter the steam reformer process. Therefore, the

LNG could be directly used to pre-cool the hydrogen. For the conceptional hydrogen liquefaction plant with a capacity of 100 TPD (tons per day), the energy expenditure could be reduced from 10 kWh/kgH2 to 4 kWh/kgH2 using LNG for the pre-cooled part (Jan et al. 2006). The CO2 emission from the steam methane reforming process should also be verified. The compromise between the economic benefit and the CO2 emission was comprehensively evaluated and the SEC of the optimized hydrogen liquefaction plant is 11.13 kWh/kgH2 (Bae et al. 2021).

4.2 *The improvement of the cryogenic part*

In the previous studies, helium was selected as the refrigerant for the cryogenic part of the hydrogen liquefaction process. Different versions of the liquefaction processes were evaluated. The SEC of the version with the best performance is 12.7 kWh/kgH2 (Kuz'menko et al. 2004). The optimization was carried out on the hydrogen liquefaction plant process based on liquid nitrogen precooling and helium gas turbine expansion refrigeration. The SEC of the system is 7.13 kWh/kgH2, which is 19.65% lower than the base case (Yinliang & Ju 2019). Instead of using helium as the refrigerant in the cryogenic part, the Joule-Brayton pre-cooled liquefaction plant with helium-neon refrigerant in the cryogenic part was designed and evaluated. The SEC of the plant ranges from 5 to 7 kWh/kgH2 (Quack 2003). The Joule-Brayton cycle could also be applied to the cryogenic part. An MR pre-cooled hydrogen liquefaction plant with four H2 Joule-Brayton cycles in the cryogenic part was designed and optimized. And the SEC of the plant is 5.91 kWh/kgH2 (Krasae-In 2014).

4.3 *Hybrid Plant*

As mentioned in the introduction part, the abandoned electricity from renewable energy power plants could be used to produce hydrogen for energy storage. Besides the hydrocarbon reforming reaction, the electrolysis method, which produces hydrogen from water, has an efficiency of 56-73% (Holladay et al. 2009). The hydrocarbon reforming reaction could be selected as the hydrogen production method in areas rich in natural gas resources. And the electrolysis reaction could be utilized in areas where the supply of natural gas resources is insufficient. No matter which hydrogen production method was chosen, the hydrogen should be liquefied immediately after production to ensure energy storage or transportation. Therefore, it would be attractive if the hydrogen liquefaction system could be combined with renewable energy technology to improve the performance of the liquefaction system (Aasadnia & Mehrpooya 2018).

Geothermal energy can provide heat to an absorption refrigeration system (ARS) to precool the feed hydrogen gas and supply the output work to the hydrogen liquefaction system (Kanoglu et al. 2016). Moreover, the solar-voltaic/thermal design can be combined with geothermal energy to form a triple-effect absorption refrigeration system (Ratlamwala et al. 2012).

The potential for utilizing solar energy for hydrogen liquefaction also was investigated. A low-temperature auto-cascade solar Rankine cycle was designed to supply compression power to an ammonia-water ARS pre-cooled hydrogen liquefaction system (Bao et al. 2011; Kanoglu et al. 2016). A new hybrid hydrogen liquefaction configuration with high efficiency was proposed. The hot water stream from the solar source is used to provide heat to the ammonia-water ARS, and the liquid ammonia is utilized to cool down the nitrogen gas, which would be further employed for interstage cooling of the compressors in the cryogenic part of the hydrogen liquefaction system. The power consumption of the system is 6.47 kWh/kgH2, and the energy efficiency is 45.5% (Aasadnia & Mehrpooya 2018). Also, the liquid ammonia produced from the ARS can be directly used for interstage cooling to decrease the power consumption of the compressors in the cryogenic part of the hydrogen liquefaction system. The energy consumption could be reduced to 4.02 kWh/kgH2, and the exergy efficiency would be 73.57% (Ghorbani et al. 2019).

5 DISCUSSION

Nowadays, the worldwide daily output of liquid hydrogen is more than 350 tons, and North America constitutes more than five-sixths of the total liquid hydrogen production (Fraser 2003). The liquid hydrogen industry in China is relatively lagging. The total domestic production of liquid hydrogen is only 4 TPD, and almost all production is used to supply aerospace engineering. However, the strategic position of hydrogen energy in China's industrial development could not be ignored. In recent years, China has been catching up with the liquid hydrogen industry. In April 2021, the first civil liquid hydrogen plant in China started operation (Hongdaxingye Group Co 2021). In addition, the Air Product company and Jiutai Group Co., Ltd. plan to construct a hydrogen liquefaction plant with a capacity of 30 TPD in Zhejiang Province in China. The project will be completed in 2022 (Tuoketuo People's Government 2021). Furthermore, Hongdaxingye Group Co., Ltd. plans to raise 5.5 billion yuan, mainly for investment in hydrogen energy projects with an annual hydrogen output of 50 thousand tons. The yearly liquid hydrogen output could be 30 thousand tons (Hongdaxingye Group Co 2021). The technologies and the hydrogen liquefaction cycle in the previous studies were discussed. After that, the enlightenment that China can learn from the liquid hydrogen industry abroad would be discussed in this section.

First of all, the LNG pre-cooled system is attractive. In terms of imports, the LNG imports from China have increased every year. However, the proportion of LNG imports has been decreasing from 100% in 2007 to 46.4% in 2014 (Liu 2016). The data shows that China is rapidly developing the domestic LNG industry and heralds the abundance of natural gas resources in China. With sufficient resources, the cooling capacity of LNG can also be fully utilized in the pre-cooled process of hydrogen liquefaction before entering the steam reforming reaction. This kind of creativity truly embodies the concept of sustainable development.

Secondly, the cryogenic part accounts for the largest amount of SEC of the hydrogen liquefaction process. Compared to the existing conventional hydrogen liquefaction plant with the best performance, the SEC of the plant could be nearly 40% if the helium-neon refrigerant or the Joule-Brayton cycle could be employed in the cryogenic part (Drnevich 2003; Krasae-In, 2014; Quack 2003). Therefore, more investigations should be carried out on the cryogenic part to decrease the SEC and the exergy destruction of the plant.

Last but not least, the abundant renewable energy resources in China are opportunities that cannot be ignored. China's continuous investment in renewable energy provides favorable conditions for hybrid hydrogen liquefaction plants. The abandoned electricity could not only be used to produce hydrogen through an electrolysis reaction, but also provide the compression power for the hydrogen liquefaction process. Also, the ammonia-water ARS can be introduced either into the pre-cooling part or the interstage cooling of the compressor to decrease the SEC of the hydrogen liquefaction plant.

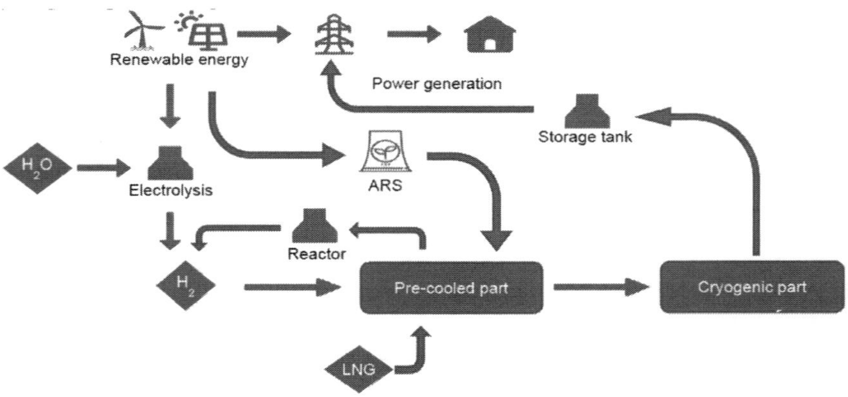

Figure 3. Sustainable energy system based on hydrogen energy storage technology.

The ideal sustainable energy system is shown in Figure 3. Renewable energies are utilized for power generation and abandoned electricity can be used to produce hydrogen through the electrolysis method. The produced hydrogen will then be pre-cooled by LNG or ARS. After that, the heated natural gas can be used to produce hydrogen through a hydrocarbon reforming reaction. On the other hand, the pre-cooled hydrogen will be further cooled and liquefied in the cryogenic part and the product liquid hydrogen will be finally stored in the storage tank for further power generation when the renewable energies are not available. Such a system not only makes full use of renewable energy, but also takes into account different ways of producing hydrogen. It truly satisfies China's view of sustainable development of the energy industry in the context of carbon neutrality.

6 CONCLUSION

Hydrogen energy is an important part of achieving the goal of carbon neutrality in China. And hydrogen liquefaction is a necessary technology to promote hydrogen energy. In this study, different methods to reduce the SEC of the hydrogen liquefaction process were introduced. Recommendations regarding hydrogen liquefaction technology in China also was put forward.

China's current liquid hydrogen industry is relatively backward compared to North America. Therefore, China should combine its resource advantages to find a suitable development path for hydrogen liquefaction technology. As the advantages of natural gas resources in China, liquid natural gas pre-cooled hydrogen liquefaction plant would be a good idea. It not only makes rational use of resources, but also fully conforms to the concept of sustainable development under China's carbon-neutral goals. Based on the accelerating renewable energy in China, it would be optimal to combine the renewable energy technology with the hydrogen liquefaction plant. This solution not only solves the problem of abandoned electricity from the renewable energy power plant but also reduces the SEC of the hydrogen liquefaction plant. And the cryogenic part of the liquefaction plant would still be a direction that requires continuous research.

REFERENCES

Aasadnia M. and Mehrpooya M. (2018) "Conceptual design and analysis of a novel process for hydrogen liquefaction assisted by absorption precooling system," *J. Clean. Prod.*, vol. 205, pp. 565–588, doi: 10.1016/j.jclepro.2018.09.001.

Aasadnia M. and Mehrpooya M. (2018) "Large-scale liquid hydrogen production methods and approaches: A review," *Appl. Energy*, vol. 212, no. September 2017, pp. 57–83, doi: 10.1016/j.apenergy.2017.12.033.

Abe J.O., Popoola A.P.I., Ajenifuja E. and Popoola O.M. (2019) "Hydrogen energy, economy and storage: Review and recommendation," *Int. J. Hydrogen Energy*, vol. 44, no. 29, pp. 15072–15086, doi: 10.1016/j.ijhydene.2019.04.068.

Bae J.E., Wilailak S., Yang, J.H., Yun D.Y., Zahid U. and Lee C.J. (2021) "Multi-objective optimization of hydrogen liquefaction process integrated with liquefied natural gas system," *Energy Convers. Manag.*, vol. 231, p. 113835, doi: 10.1016/j.enconman.2021.113835.

Baker C.R. and Shaner R.L. (1978) "A study of the efficiency of hydrogen liquefaction," *Int. J. Hydrogen Energy*, vol. 3, no. 3, pp. 321–334, doi: 10.1016/0360-3199(78)90037-X.

Bao J.J., Zhao L. and Zhang W. Z. (2011) "A novel auto-cascade low-temperature solar Rankine cycle system for power generation," *Sol. Energy*, vol. 85, no. 11, pp. 2710–2719.

Barron R.F. (1972) "*Liquefaction Cycles for Cryogens," in Advances in Cryogenic Engineering*, Springer, pp. 20–36.

Barron R. (1966) "*Cryogenic Systems.*" McGraw-Hill.

Dewar J. (1898) "Liquid Hydrogen," *Science* (80-.)., vol. 8, no. 183, pp. 3–6, doi: 10.1126/science.8.183.3.

Drnevich R. (2003) "*Hydrogen delivery: liquefaction and compression*," in Strategic initiatives for hydrogen delivery workshop, vol. 7.

Fraser D. (2003) "*Solutions for hydrogen storage and distribution by dynetek industries Ltd.*," The PEI Wind-Hydrogen Symposium. http://www.gov.pe.ca/photos/original/dev_solutions.pdf.

Ghorbani B., Mehrpooya M., Aasadnia M. and Niasar M.S. (2019) "Hydrogen liquefaction process using solar energy and organic Rankine cycle power system," *J. Clean. Prod.*, vol. 235, pp. 1465–1482, doi: 10.1016/j.jclepro.2019.06.227.

Hadjipaschalis I., Poullikkas A. and Efthimiou V. (2009) "Overview of current and future energy storage technologies for electric power applications," *Renew. Sustain. Energy Rev.*, vol. 13, no. 6–7, pp. 1513–1522, doi: 10.1016/j.rser.2008.09.028.

Holladay J.D., Hu J., King D.L. and Wang Y. (2009) "An overview of hydrogen production technologies," *Catal. Today*, vol. 139, no. 4, pp. 244–260, doi: 10.1016/j.cattod.2008.08.039.

Hongdaxingye Group Co. L. (2021) "*Liquid hydrogen industry enters a new stage*" http://www.002002.cn/news/2021-05-17/13043.html.

Hongdaxingye Group Co. L. (2021) "*The first civil liquid hydrogen plant in China*" http://www.002002.cn/news/2020-04-28/8359.html.

Jan G., Huijsmans J. and Austgen D. (2006) "*Clean and Green Hydrogen*," no. June, pp. 1–9.

Kanoglu M., Yilmaz C. and Abusoglu A. (2016) "Geothermal energy use in hydrogen production," *J. Therm. Eng.*, vol. 2, no. 2, pp. 699–708, doi: 10.18186/jte.58324.

Kanoglu M., Yilmaz C. and Abusoglu A. (2016) "Geothermal energy use in absorption precooling for Claude hydrogen liquefaction cycle," *Int. J. Hydrogen Energy*, vol. 41, no. 26, pp. 11185–11200, doi: 10.1016/j.ijhydene.2016.04.068.

Krasae-in S., Stang J. H. and Neksa P. (2010) "Development of large-scale hydrogen liquefaction processes from 1898 to 2009," *Int. J. Hydrogen Energy*, vol. 35, no. 10, pp. 4524–4533, doi: 10.1016/j.ijhydene.2010.02.109.

Krasae-In S. (2014) "Optimal operation of a large-scale liquid hydrogen plant utilizing mixed fluid refrigeration system," *Int. J. Hydrogen Energy*, vol. 39, no. 13, pp. 7015–7029, doi: 10.1016/j.ijhydene.2014.02.046.

Kuz'menko I.F., Morkovkin I.M. and Gurov E.I. (2004) "Concept of Building Medium-Capacity Hydrogen Liquefiers With Helium," *Chem. Pet. Eng.*, vol. 40, no. 1–2, pp. 94–98.

Li Y., Wang Y. and Tang J. (2007) "Temporal and Spatial Variation Characteristics of China's Near-surface Wind Energy Resources," *J. Nanjing Univ.* (Natural Sci., vol. 43, no. 3, pp. 280–291.

Lindsey R. and Dahlman L. "*Climate Change: Global Temperature*," (2021) https://www.climate.gov/news-features/understanding-climate/climate-change-global-temperature.

Liu J.S. (2016) "Evolution of world natural gas trade pattern based on social network analysis," *Econ. Geogr.*, vol. 36, no. 12, pp. 89–95.

Liu J. and Zhong C. (2019) "China's hydrogen energy development status and prospects," *Res. Approach*, vol. 32, no. 05, pp. 32–36.

Nandi T.K. and Sarangi S. (1993) "Performance and optimization of hydrogen liquefaction cycles," *Int. J. Hydrogen Energy*, vol. 18, no. 2, pp. 131–139, doi: 10.1016/0360-3199(93)90199-K.

Quack H. (2003) "*Conceptual design of a high efficiency large capacity hydrogen liquefier*," vol. 255, no. 1, pp. 255–263, doi: 10.1063/1.1472029.

Ratlamwala T.A.H., Dincer I., Gadalla M.A. and Kanoglu M. (2012) "Thermodynamic analysis of a new renewable energy based hybrid system for hydrogen liquefaction," *Int. J. Hydrogen Energy*, vol. 37, no. 23, pp. 18108–18117, doi: 10.1016/j.ijhydene.2012.09.036.

Tuoketuo People's Government. (2021) "*Demonstration project of comprehensive utilization of hydrogen energy*" http://www.huhhot.gov.cn/zwdt/qxqdt/202106/t20210611_967630.html.

Venkatarathnam G. and Timmerhaus K.D. (2008) *Cryogenic mixed refrigerant processes*, vol. 100. Springer.

Wade N.S., Taylor P.C., Lang P.D. and Jones P.R. (2010) "Evaluating the benefits of an electrical energy storage system in a future smart grid," *Energy Policy*, vol. 38, no. 11, pp. 7180–7188, doi: 10.1016/j.enpol.07.045.

Walnum H.T. et al. (2012) "Principles for the liquefaction of hydrogen with emphasis on precooling processes," *Refrig. Sci. Technol.*, vol. 2012, pp. 273–280.

Wang C. and Zhang Y. (2020) "Implementation Pathway and Policy System of Carbon Neutrality Vision," *Chinese J. Environ. Manag.*, vol. 12(6), pp. 58–64, doi: 10.16868/j.cnki.1674-6252.2020.06.058.

Wolf J. 2003 "*Liquid-Hydrogen Technology for Vehicles,*" no. September 2002, pp. 684–687.

Xi Jinping, "*Address by Xi Jinping to the General Debate of the 75th Session of the United Nations General Assembly*," (2020) http://www.xinhuanet.com/politics/leaders/2020-09/22/c_1126527652.htm.

Yinliang L. and Ju Y. (2019) "Process Optimization and Analysis of a Novel Hydrogen Liquefaction Cycle," *Int. J. Refrig.*, doi: 10.1016/j.ijrefrig.2019.11.004.

Züttel A. (2004) "Hydrogen storage methods," *Naturwissenschaften*, vol. 91, no. 4, pp. 157–172, doi: 10.1007/s00114-004-0516-x.

Energy Revolution and Chemical Research – Kok-Keong Chong and Zhongliang Liu (Eds)
© 2023 The Authors, ISBN: 978-1-032-36554-1

Supported PdNPs catalysts prepared by MAO for highly efficient silane oxidation

Li Hua Zheng, Xin Yu Yang, Zi Xuan Wang & Jun Zhou*
Key Laboratory for Anisotropy and Texture of Materials School of Materials Science and Engineering, Northeastern University, Shenyang, Liaoning, China

ABSTRACT: Here, we prepared a catalyst (Pd/MgO) loaded with Pd NPs by micro-arc oxidation technology, and catalyzed silane oxidation with water as an oxidant. In addition to high catalytic activity, Pd/MgO also showed good stability and recovery, even after 10 reaction cycles, it still had high activity. Oxidation of silane by a supported transition metal catalyst was an effective method of preparation of silanols, which had mild reaction characteristics and uncontaminated by-products. Moreover, it also had good catalytic activity for a variety of silanes.

1 INTRODUCTION

Silanol is a compound containing the Si-OH group, which is a very useful molecule in the synthesis of silicon-based polymers and other organic syntheses (Li 2001; Mina et al. 2012; Murugavel, R. et al. 1997). For example, they have been used as nucleophiles for C-C cross-coupling reactions (Chang 2001; Denmark, S. E. et al. 2008; Hirabayashi, K. et al. 2000), organic catalysts (Kondo, S. et al. 2006), or guide groups for C-H activation (Huang 2011). Silanols can be prepared by a variety of methods, such as hydrolysis of chlorosilane, oxidation of organosilanes with stoichiometric oxidants, and substitution of siloxanes by nucleophiles. Among these methods, the Si-H bond activation oxidation method of silane is one of them, and hydrosilane has the advantage of being low in cost and easy to obtain, and the silanol atoms obtained by the oxidation reaction are more economical, so hydrosilane oxidation is usually used to prepare silanol.

Micro arc oxidation (MAO), also known as plasma oxidation (PEO), is developed from traditional anodic oxidation. In this technology, valve metals such as aluminum, titanium, and magnesium are used as anodes, and voltage is applied on the surface through a micro-arc oxidation power supply. When the voltage rises to a certain extent, a plasma discharge effect will occur on the surface of anode metal, discharge sparks will be continuously formed on the metal surface, and then multicomponent oxide coatings with valuable physical and chemical properties will be formed on the surface of the metal substrate. The formation process of the film involves many chemical reactions, including thermochemistry, electrochemistry, and plasma chemistry. This technology is widely used in the field of metal surface anti-corrosion (Saakiyan, L. S. et al. 1994) and wear resistance. With the continuous exploration of micro-arc oxidation technology by researchers, this technical method has also played an important role in the field of biology and catalysis (Cao 2019).

In this paper, we prepared catalysts by micro-arc oxidation technology. The noble metal salt solution could be directly added to the electrolyte system, and magnesium noble metal catalyst could be generated in situ during discharge. The catalyst obtained by this method had a high degree of adhesion between the catalyst film and the matrix, the surface-active particles were not easy to fall off, and the catalyst can be reused.

*Corresponding Author: zhoujun1@mail.neu.edu.cn

DOI 10.1201/9781003332657-103

2 EXPERIMENTAL

2.1 *Materials and methods*

Materials.
Palladium chloride ($PdCl_2$), sodium metasilicate ($Na_2SiO_3 \cdot 9H_2O$), potassium hydroxide (KOH), and potassium fluoride (KF) were purchased from Sinopharm Chemical Reagent Co., Ltd. and used as received.
Characterization.
The surface morphology and pore size of the catalyst were observed by scanning electron microscopy (SU8010). The performance of the catalyst was tested by gas chromatography, and the conversion, yield, and TOF were finally calculated.
Preparation of Pd/MgO Catalyst.
Briefly, magnesium plates (40 mm × 20 mm × 2mm) and stainless-steel plates (100 mm × 80 mm ×2mm) served as the anode and cathode, respectively. They were placed vertically in a beaker filled with electrolyte (4g Na_2SiO_3, 2.5gKF, 3.5gKOH, 2ml$PdCl_2$(0.02g/ml), in 500ml of deionized water) and both electrodes were connected to a micro-arc oxidation power supply (constant voltage model, frequency f = 500 Hz, pulse width pw = 60). Lastly, the sample proceeded to a thermal reduction process in 95% Ar+5% H_2 at 400°C for 2 h.

3 RESULTS AND DISCUSSION

3.1 *Effect of voltage on catalyst performance*

When studying the effect of voltage on catalyst performance, the oxidation time was set to 3s. Through ICP-AES detection, it was found that the Pd amount of the catalyst increased gradually with the increase of micro-arc oxidation voltage, as shown in Table 1. The voltage increased from 180V to 350V, and the Pd amount increased from 9.5μg to 55.4μg.

Table 1. Pd amount of different voltages.

U/V	180	200	250	300	350
ICP/μg	9.5	10.4	25.3	35.1	55.4

The sample was directly used for the silane oxidation reaction. Figure 1 showed the comparison of TOF values of catalyst samples prepared at different voltages for the silane oxidation reaction. With the increase of voltage, the TOF value first increased and then decreased. When the voltage is 250V, the catalyst sample had the highest catalytic activity, and the TOF value is 3330 h^{-1}.

3.2 *Effect of oxidation time on catalyst performance*

According to the study in the previous section, the voltage of micro-arc oxidation was set at 250V, and the influence of oxidation time on catalytic performance would be studied next. By observing Figure 2, it could be found that the size of surface pore size increased with the extension of oxidation time. When the oxidation time was 150 s, the heat could not be released immediately, resulting in cracks on the surface and affecting the performance of the catalyst. The catalytic performance results were shown in Figure 3. With the extension of oxidation time, the TOF value first increased and then decreased. When the oxidation time was 6 seconds, it had the best catalytic activity.

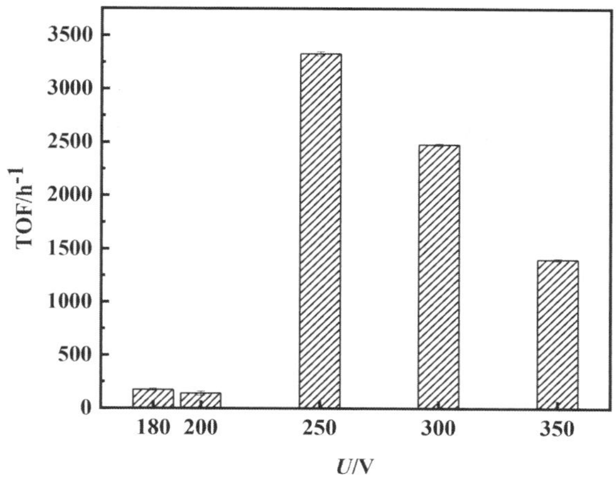

Figure 1. Effects of different voltages on silane reaction.

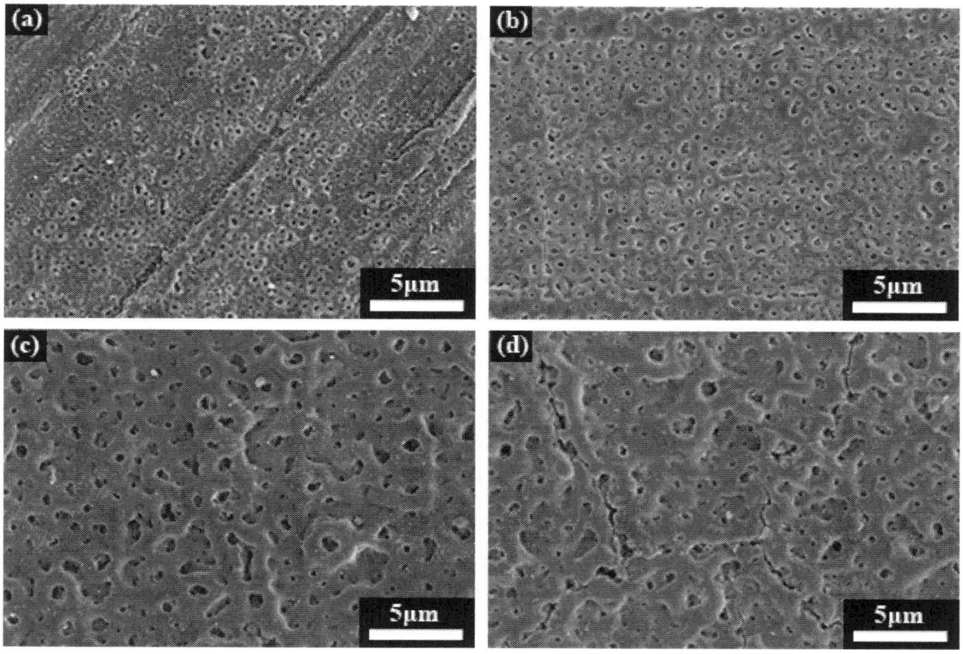

Figure 2. The film morphology of different oxidation times (a) 3s; (b) 6s; (c) 20s; (d) 150s.

3.3 *Catalyzing the oxidation of different silanes*

Table 2 showed the reaction results of catalytic oxidation of different silanes with Pd/MgO catalyst. The results showed that various silanes with different structures could be oxidized to corresponding silanols under the catalytic action of the Pd/MgO catalyst.

Pd/MgO catalyst had an excellent catalytic effect on triethylsilane, tert-butyldimethylsilane, and dimethylphenylsilane. Among them, the reaction rate of triethylsilane was the fastest. Dimethylphenylsilane and tert-butyldimethylsilane reacted at a slower rate than triethylsilane, but

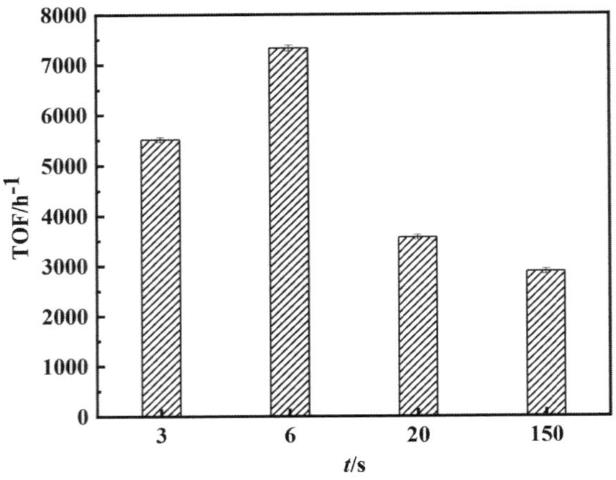

Figure 3. Effects of different oxidation times on silane reaction.

Table 2. Oxidation of various silanes with Pd/MgO catalyst.

Entry	Substrate	Time (min)	Conversion (%)	Yield (%)	TOF (h^{-1})
1	Et_3SiH	120	100	>99	24417
2	t-$BuMe_2SiH$	180	72	58	8431
3	$phMe_2SiH$	180	68	68	11698
4	Ph_2SiH_2	180	16	12	1487
5	Ph_3SiH	180	5	4	1945
6	iPr_3SiH	300	6	5	1598

they also exhibited high activity. Pd/MgO catalyst had a better reaction effect on dimethylphenylsilane in aromaticsilane, while the reaction effect of p-diphenylsilane and triphenylsilane was not as good as that of dimethylphenylsilane. Because the steric resistance of diphenylsilane and triphenylsilane was large, which was not conducive to the reaction.

Analyzing the reasons for the above phenomenon, when the size of the silane molecule was small, it was easier to enter the catalyst channel and react with the active particles in the reaction process. However, when silane molecules with benzene rings or long branched chains have a great steric hindrance effect, the movement rate of silane molecules in the reaction process was slow, and it was relatively difficult to enter the catalyst channel. Therefore, the silane with a short branched-chain had a good reaction effect, while the silane with phenyl or high steric resistance had an average reaction effect.

3.4 *Stability and Reusability*

We investigated the cyclic performance of the Pd/MgO catalyst for the oxidation of dimethylphenylsilane. The cyclic performance test results of the samples were shown in Figure 4, from which it could be seen that the yield of the Pd/MgO catalyst decreased from 100% to 70% after seven consecutive cyclic reactions. Because the sample was continuously used in the reaction, the product could not be completely desorbed from the catalyst surface, resulting in the blockage of the catalyst channel. When such a catalyst was used for silane oxidation reaction, the reactant could not smoothly contact the active particles in the channel, so the activity of the catalyst was greatly reduced. In the experiment, it was found that when the above samples were immersed in acetone solution for one month and then used for silane oxidation reaction, the results showed that their activity would not continue to decline, but increased and basically returned to the initial state. This phenomenon is called the self-cleaning function of the catalyst. Because both dimethylphenylsilane and dimethylphenylsilanol could be dissolved in acetone, soaking the sample in acetone for a long time could release the substances adsorbed on the catalyst into the solution, so that the active sites of the catalyst were re-exposed, and the reactants could be better adsorbed on the catalyst surface and activated. Therefore, the catalyst could be restored to its original state, and the activity after reuse was roughly the same as that of the new catalyst.

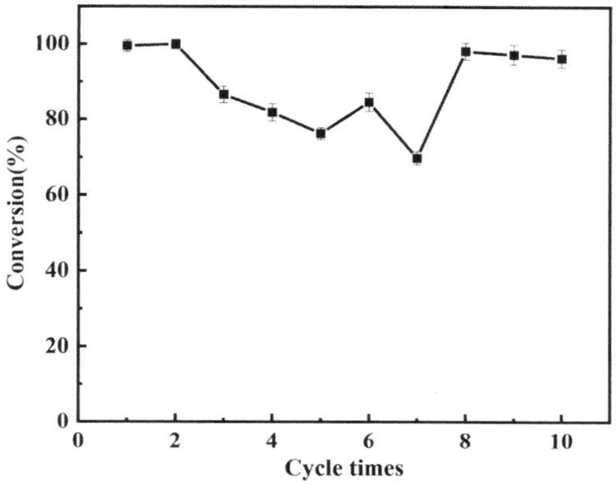

Figure 4. Cycle performance test.

4 CONCLUSIONS

In this paper, the voltage and oxidation time of the Pd/MgO catalyst prepared by the micro-arc oxidation method were studied. By comparing the effect of samples used in silane oxidation reaction and characterization test results, the best preparation process of Pd/MgO catalyst was obtained. The catalyst showed good catalytic performance for the oxidation of various silanes and had good stability and recoverability. A simple, clean and efficient catalyst preparation method was provided.

ACKNOWLEDGMENTS

This work is supported by the Open Fund of State Key Laboratory of Refractories and Metallurgy (Wuhan University of Science and Technology, G202102) and the Natural Science Foundation of Science and Technology Department of Liaoning Province (2020-MS-090).

REFERENCES

Chang, S. Yang, S. H. Lee, P. H. (2001) Pd-catalyzed cross-coupling of alkynylsilanols with iodobenzenes. *Cheminform*, 42: 4833–4835.

Cao, X. Q. Zhou, J. Wang, H. N. et al. (2019) Abnormal thermal stability of sub-10 nm Au nanoparticles and their high catalytic activity. *Journal of Materials Chemistry A*, 7: 10980–10987.

Denmark, S. E. Baird, J. D. Regens, C. S. (2008) Palladium-Catalyzed Cross-Coupling of Five-Membered Heterocyclic Silanolates. *The Journal of Organic Chemistry*, 73:1440–1455.

Hirabayashi, K. Mori, A. Kawashima, J. et al. (2000) Palladium-Catalyzed Cross-Coupling of Silanols, Silanediols, and Silanetriols Promoted by Silver(I) Oxide. *The Journal of Organic Chemistry*, 65: 5342–5349.

Huang, C. Chattopadhyay, B. Gevorgyan, V. (2011) Silanol: a traceless directing group for Pd-catalyzed o-alkenylation of phenols. *ChemInform*, 43: 12406–12409.

Kondo, S. Harada, T. Tanaka, R. et al. (2006) Anion recognition by a silanediol-based receptor. *Organic Letters*, 8: 4621–4624.
Li, G. Wang, L. Ni, H. et al. (2001) Polyhedral Oligomeric Silsesquioxane (POSS) Polymers and Copolymers: A Review. *Inorg. Organomet. Polym.*, 11: 123–154.
Mina. Jeon. Junghoon. et al. (2012) Catalytic Synthesis of Silanols from Hydrosilanes and Applications. *ACS Catal*, 2: 1539–1549.
Murugavel, R. Voigt, A. Walawalkar, M. G. et al. (1997) ChemInform Abstract: Hetero- and Metallasiloxanes Derived from Silanediols, Disilanols, Silanetriols, and Trisilanols. *ChemInform.*, 96: 2205–2236.
Saakiyan, L. S. Efremov, A. P. Soboleva, I. A. (1994) On the micro-arc oxidation and its effect on the corrosion-mechanical behavior of aluminum alloys and coatings. *Prot. Metals.*, 30: 85–88.

Energy Revolution and Chemical Research – Kok-Keong Chong and Zhongliang Liu (Eds)
© 2023 The Authors, ISBN: 978-1-032-36554-1

Carbon-catalyzed etching of porous silicon structures

Zhiyuan Liao
Key Laboratory of Hubei Province for Coal Conversion and New Carbon Materials, School of Chemistry and Chemical Engineering, Wuhan University of Science and Technology, Wuhan, China

Ling Tong
School of Health and Nursing, Wuchang University of Technology, Wuhan, China
Academy of Green Manufacturing Engineering, Wuhan University of science and technology, Wuhan, China

Ying Liu, Baoguo Zhang, Ao Chen & Xiaoyu Yang
Key Laboratory of Hubei Province for Coal Conversion and New Carbon Materials, School of Chemistry and Chemical Engineering, Wuhan University of Science and Technology, Wuhan, China

Ya Hu*
Key Laboratory of Hubei Province for Coal Conversion and New Carbon Materials, School of Chemistry and Chemical Engineering, Wuhan University of Science and Technology, Wuhan, China
Academy of Green Manufacturing Engineering, Wuhan University of science and technology, Wuhan, China

Hailiang Fang*
The State Key Laboratory of Refractories and Metallurgy, Wuhan University of science and technology, Wuhan, China

ABSTRACT: Silicon nanomaterials have remarkable physical and chemical properties, optoelectronic properties, thermal stability, and surface properties, and play an important role in various fields, especially in photovoltaics, thermoelectricity, sensors, and batteries. Therefore, the research on the preparation of silicon nanomaterials has attracted much attention today, and how to quickly and efficiently prepare the required materials is the focus of research. Carbon-catalyzed etching of silicon materials is an emerging approach. Similar to traditional metal etching, it utilizes the higher electronegativity of carbon than silicon to generate electrochemical reactions to obtain the desired silicon nanostructures. Traditional metal etching mostly uses precious metals such as gold and silver, which are expensive, while carbon is more abundant in the earth, relatively easier to obtain, and more economical. In terms of cost, carbon-catalyzed etching has greater advantages. Therefore, carbon catalytic etching has gradually entered the attention of researchers in recent years. In this paper, carbon-catalyzed etching of porous silicon structures was briefly introduced and the current generated by graphite particle etching is higher than that of graphene, and the reaction is more stable.

1 INTRODUCTION

Silicon nanomaterials have remarkable physical and chemical properties, optoelectronic properties, and surface properties, and play an important role in various fields (Hu et al. 2022; Kayes et al. 2005; Peng et al. 2010; Schierning et al. 2011, Yang et al. 2011; Zhang et al. 2011). However, the manufacturing process of silicon nanomaterials is still relatively complex, and there are problems such as complicated control processes, inaccurate preparation of silicon nanostructures, and low preparation efficiency. Now, based on the silicon etching process in oxidized HF aqueous solution, there are also many metal-catalyzed etching processes. A series of research has been carried out

*Corresponding Authors: huya@wust.edu.cn and simbaqiankun@163.com

DOI 10.1201/9781003332657-104

(Hu et al. 2021; Peng et al. 2006; Wang et al. 2018). And in the research, it is found that the corrosive environment is not limited to the aqueous solution, silicon nanostructures can also be prepared in the gas phase environment (Hu et al. 2014).

In the past few years, there has been great interest in the development and application of carbon nanomaterials. We found that carbon materials can be used as a new catalyst to catalyze the preparation of silicon nanomaterials. The electronegativity of carbon is stronger than that of silicon, and the conductive carbon materials are stable cathodes, while silicon is the anode, so in this environment, carbon material can be used as a catalyst to promote silicon etching. These findings provide a new method for the fabrication of silicon micro-nano structures by non-metallic carbon-assisted etching (Hu et al. 2019).

This paper describes a method for carbon etching of silicon in gas-filled HF/H_2O vapor. The promoting effect of carbon materials (including graphite particles and graphene) on silicon corrosion in the HF/H_2O gas phase was investigated. The main research contents are adding carbon with different morphologies in different corrosion environments, and the effects of silicon with different doping degrees on corrosion. In order to find a fast and easy method to prepare silicon nanomaterials, the most suitable corrosion reaction conditions were explored.

2 EXPERIMENTAL

0.1g of uniformly ground graphite particles were dispersed and dissolved in 10mL of deionized water, then a turbid solution of uniformly dispersed graphite particles was prepared by ultrasonic, and consequently, 1mL of the turbid solution of graphite particles were taken and dropped on P-type (100) 1 by drop coating above the 10Ω·cm silicon wafer. The silicon wafer with graphite particles attached to the surface is placed on the PTFE platform, and the platform is placed on a fluoric acid solution with a certain concentration (HF: H_2O_2=1:2, 1:3, 1:4, 1: 5, 1:6 volume ratio) reaction kettle at room temperature for 24 hours (the experiment was carried out in a fume hood). The samples were taken out, cleaned, and prepared, and some samples were left for SEM characterization to observe the surface morphology. All reagents were purchased from Sinopharm Group.

3 RESULTS AND DISCUSSION

3.1 *SEM test results*

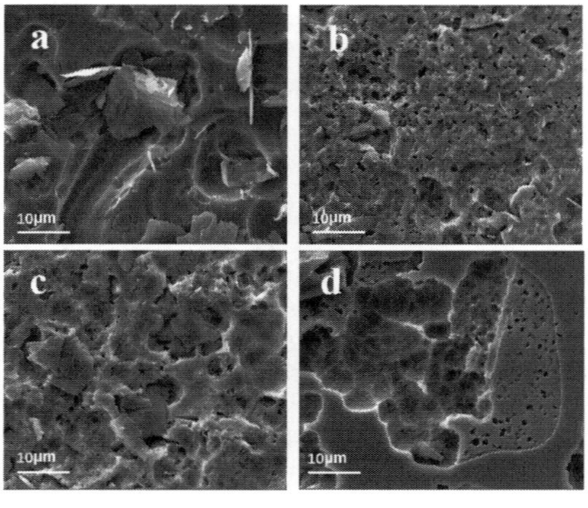

Figure 1. SEM images of carbon-catalyzed etching of silicon materials under different experimental conditions. (a) N-type silicon wafer, HF:H_2O_2=1:4, 24h; (b) P-type silicon wafer, HF:H_2O_2=1:2,24h; (c)P-type silicon wafer, HF:H_2O_2=1:4, 24h; (d) P-type silicon wafer, HF:H_2O_2=1:4,16h.

From the comparison of the four figures, it can be seen that an irregular silicon nanostructure can be obtained by carbon catalyzed etching of the silicon wafer. The carbon suspension is attached to the surface of the silicon wafer by drop coating. The next etching starts from the surface to form the required silicon microstructure. It can be seen that the silicon microstructure obtained by etching is irregular, and the etching conditions are also different due to different experimental conditions.

3.2 *Electrochemical test results and analysis*

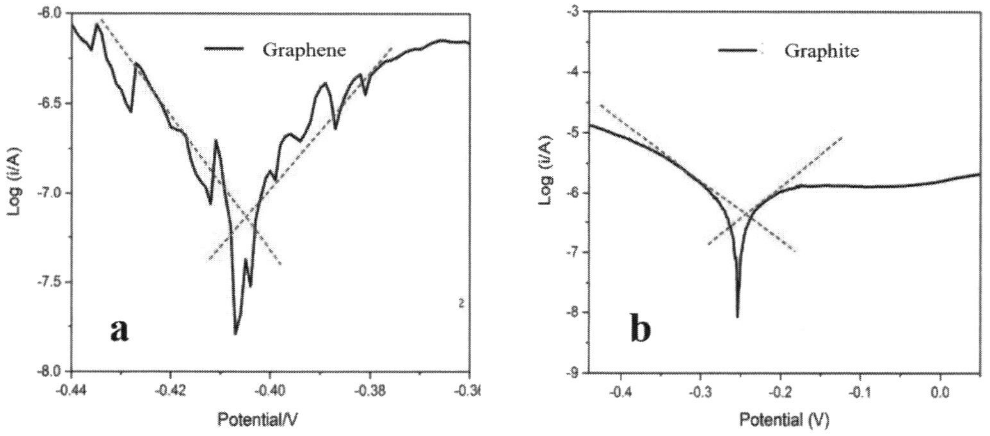

Figure 2. (a) Polarization curve of a silicon wafer after adding graphite. (b) Polarization curve of a silicon wafer after adding graphite particles.

Electrochemical polarization measurements were performed to investigate the anodic and cathodic reactions occurring on the silicon surface, and the effect of graphite particles on graphene etching was investigated. The voltammetry curve measured on a P-type (100) 1–10Ω·cm wafer under the corrosion condition of HF/H_2O_2 is shown in Figure 2. It can be seen from Figure 3.2 that the self-corrosion current of graphite particles is $-0.07\mu A/cm^2$, and the self-corrosion potential is –0.04V; the self-corrosion current of graphene is $-0.04\mu A/cm^2$, and the self-corrosion potential is –0.24V. The left and right branches of these curves correspond to the cathodic hydrogen peroxide reduction reaction and the anodic silicon oxidation reaction in the electrochemical system. In Figure 2a, it can be clearly seen that with graphite particles for etching, the curve is fluctuating, but the general trend is a sharp drop in the left branch and a sharp upward change in the right branch. As can be seen from Figure 2b, when graphene is used for etching, the curve is relatively smooth, and the downward trend of the left branch and the upward trend of the right branch are relatively flat. The comparison of the two figures further shows that the reaction of carbon-catalyzed etching of silicon wafers varies with carbon species, and the silicon etching rate of graphene is slower than that of graphite particles under the same experimental conditions.

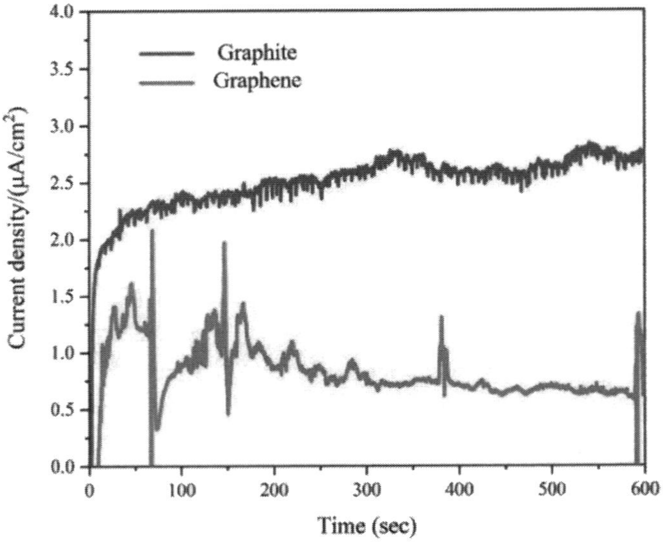

Figure 3. I-T curves of silicon wafers after adding graphene and graphite particles.

Figure 3 is the I-T curve measured by graphite particles and graphene etched silicon wafers. As can be seen from Figure 3, the curves of both are not smooth, which is due to the generation of O_2 during the etching process, which affects the current change during the test process, but the general trend is not affected. Among them, the current generated by the graphene etching process is significantly smaller than that of the graphite particles, and the curve changes between the two are also significantly different. It can be seen that different carbon materials have different rates when etching silicon wafers, the current generated by graphite particle etching is significantly higher than that of graphene, and the reaction is more stable.

The size of the silicon wafer used in the experiment is 1×1cm, and the etching area is 1cm^2. The silicon wafer is etched under different experimental conditions. After the etching is completed,

Table 1. Quality change of P-type silicon wafer before and after etching.

Etching time (h)	HF: H_2O_2 (Volume ratio)	Silicon (g)	Loaded graphite (g)	After etching (g)	After cleaning the graphite (g)
16	1:2	0.1170	0,1178	0.1170	0.1163
	1:3	0.1050	0.1057	0.1049	0.1044
	1:4	0.1180	0.1186	0.1178	0.1174
	1:5	0.1057	0.1063	0.1058	0.1053
20	1:2	0.1141	0.1147	0.1137	0.1133
	1:3	0.1318	0.1325	0.1309	0.1288
	1:4	0.1126	0.1134	0.1125	0.1119
	1:5	0.1062	0.1069	0.1059	0.1053
24	1:2	0.1102	0.1110	0.1091	0.1072
	1:3	0.1189	0.1196	0.1176	0.1153
	1:4	0.1184	0.1192	0.1183	0.1175
	1:5	0.1252	0.1260	0.1244	0.1235

it is taken out for sufficient cleaning and drying, and then analyzed by electronic means. The balance measured the quality of silicon wafers before and after etching under different experimental conditions. It can be seen that the etching speed is the fastest when HF: H2O2=1:2 in volume ratio.

4 CONCLUSION

In this paper, the preparation of carbon-catalyzed etching silicon material is discussed. The effects of different morphologies of carbon and etching solution environment on etching are mainly studied. In the whole experiment process, since there are many experimental conditions that affect the carbon-catalyzed etching of silicon wafers, according to electrochemical and SEM characterization analysis, the best experimental conditions are etching 24 h under HF: H_2O_2=1:2 vapor.

ACKNOWLEDGMENTS

This work was financially supported by the Key Research and Development Program of Hubei Province (Grant No. 2020BAB084 and No. 2021BAA063), the National Natural Science Foundation of China (Grant No.61904130), and the Key Laboratory of Hubei Province for Coal Conversion and New Carbon Materials.

REFERENCES

Hu Y., et al., (2022) Autogenic electrolysis of water powered by solar and mechanical energy[J]. *Nano Energy*, (91):106648.

Hu Y., et al., (2021) Metal Particle Evolution Behavior during Metal Assisted Chemical Etching of Silicon[J]. *ECS Journal of Solid State Science and Technology*, 10(8).

Hu Y., et al., (2014) Metal-Catalyzed Electroless Etching of Silicon in Aerated HF/H_2O Vapor for Facile Fabrication of Silicon Nanostructures[J]. *Nano Letters*, 14(8):4212–4219

Hu Y., et al., (2019) Carbon induced galvanic etching of silicon in aerated HF/H_2O vapor Article) [J]. *Corrosion Science*, 157:268–273

Kayes B.M., et al., (2005) Comparison of the device physics principles of planar and radial p-n junction nanorod solar cells[J]. *Journal of Applied Physics*, 97(11): 114302.

Peng K. Q., et al., (2006) Fabrication of single-crystalline silicon nanowires by scratching a silicon surface with catalytic metal particles[J]. *Advanced materials interfaces*, 16:387–394.

Peng K. Q., et al., (2010) High-performance silicon nanohole solar cells[J]. *Journal of the American Chemical Society*, (132): 6872–6873.

Schierning G., et al., Role of oxygen on microstructure and thermoelectric properties of silicon nanocomposites[J]. *Journal of Applied Physics*, 2011. 110, 11, 113535.

Wang, J., et al., (2018) Oxidant Concentration Modulated Metal/Silicon Interface Electrical Field Mediates Metal-Assisted Chemical Etching of Silicon[J]. *Advanced materials interfaces*, 5(23)

Yang C.C., et al., (2011) Basic Principles for Rational Design of High-Performance Nanostructured Silicon-Based Thermoelectric Materials[J]. *Chemphyschem*, (12): 3614–3618.

Zhang Y., et al., (2011) Recent Progress in Patterned Silicon Nanowire Arrays: Fabrication, Properties, and Applications[J]. *Recent Patents on Nanotechnology*, (5): 62–70.

Energy Revolution and Chemical Research – Kok-Keong Chong and Zhongliang Liu (Eds)
© 2023 The Authors, ISBN: 978-1-032-36554-1

Preparation and property analysis of polyether polyol two-component rubber repair material

Pengfei Nie
Hebei Datang International Wangtan Power Generation Co. Ltd., Tangshan, China

Zhiqiang Liu*
China Electricity Council, Beijing, China

Zhiyuan Liu
Hebei Guohua Dingzhou Power Generation Co. Ltd., Dingzhou, China

ABSTRACT: The invention relates to the preparation of a two-component rubber repair agent and its use method thereof, belonging to the technical field of rubber repair. The modifier is composed of A and B components. Component A can be obtained by blending polyether polyol 330N, chain extender, diethylene glycol, 330NH18 slurry, 330N acetylene black slurry, dioctyl phthalate DOP, catalyst, antioxidant, silane coupling agent, and defoaming agent at room temperature and vacuum. The mixture of polyether polyol GE220, 220H18, and dioctyl phthalate DOP, MDI-3051, and MDI-100LL at room temperature and vacuum was used to obtain component B. When using, components A and B were mixed by 1:0.7-1.3. The repair agent is simple to prepare and easy to use, which can be used for repairing any type of rubber damage. It is not limited by the space and shape of rubber parts. It can be formed in 1 minute, have strength in 8 minutes, reach 80% strength in 2 hours, and put into use.

1 INTRODUCTION

With the rapid development of science and technology, the application of new technology, new process, and new materials is becoming more and more extensive. This paper mainly describes the application of rubber repair agents in equipment maintenance in recent years. Through the improvement of the repair process, it can not only meet the requirements of equipment assembly but also restore the performance of equipment. The composite materials of the two components are developed with high and new technology. It has excellent physical and mechanical properties, wear resistance, corrosion resistance, easiness to use, no heating, no pressure, room temperature operation, no need for special equipment, quick and easy repair, and can work on-site. Rubber repairing adhesive for the same material or material of different metal and nonmetal has high bonding strength, which is widely used in mechanical parts wear resistance, corrosion resistance of the repair, the protective coating and structure edge sealing, fixed, and leakage, insulation, conductive, and is also used in the various defects of repair parts, such as crack, scratch, size out-of-tolerance, casting defects, etc (Nie & Zhang 2012).

At present, a lot of industrial equipment using rubber products does damage to rubber materials such as desulfurization pump shell and pipe internal damage of rubber, rubber conveyor belt injury, damage to cable outer rubber, etc. The existing technology in repair often uses a hot vulcanization process in the repair process, which needs a variety of special materials and special

*Corresponding Author: liuzhiqiang@cec.org.cn

 DOI 10.1201/9781003332657-105

curing equipment, and the process is difficult. The requirements for the professional and technical level of the construction personnel are high, and this process is time-consuming and laborious. The repair work cannot be carried out often because of the lack of equipment or a certain material, or because of the complex structure of the parts to be repaired, there is no corresponding curing equipment, resulting in the repair cannot be carried out. However, the existing repair agent on the market has too long a curing time under light load, which fails to reflect the significance of rapid repair. In order to solve the problem that rubber material damage cannot be repaired quickly and conveniently, it is necessary to develop a kind of convenient and fast repair agent (Ren et al. 2019; Wu et al. 2020).

2 CONSTITUENT MATERIALS

The two-component rubber repair agent based on polyether polyol has the advantages of firm adhesion, fast speed, convenience, and low requirements for technical personnel. The repair agent is simple to prepare, convenient to use, and very short curing time under a light load. The two-component rubber repair agent, composed of components A and B, is suitable for repairing damaged rubber products such as rubber lining, coal conveyance belts, and cables of thermal power plant equipment.

2.1 *Component A material*

A component is represented by weight as follows: polyether polyol 330N, chain extender, diethylene glycol, 330NH18 slurry, 330N acetylene black slurry, dioctyl phthalate DOP, catalyst, antioxidant, silane coupling agent, defoaming agent.

The composition of 330NH18 slurry is polyether polyol 330N and hydrophobic silica: 3.7-2.8 parts.330N acetylene black pulp is composed of polyether polyol 330N and acetylene black.220H18 slurry consists of polyether polyol GE220 and hydrophobic silica. The chain extender was E-300, the defoamer was BMC806, the catalysts were FAE-1, GT-12, and WS8, the antioxidants were 1010, 168, and 1076, and the silane coupling agents were WD52, KH560, and WD60. The hydrophobic silica was HDKH18 (Cao et al. 2019; Xu et al. 2015; Yu et al. 2014).

2.2 *Component B material*

Component B by weight is composed of polyether polyol GE220, 220H18 pulp, dioctyl phthalate DOP, MDI-3051, MDI-100LL.

3 THE PREPARATION METHODS

The preparation method of the above two-component rubber repair agent includes the following steps:

3.1 *Preparation of component A*

Under normal temperature and vacuum conditions, polyether polyol 330N and hydrophobic silica, polyether polyol 330N, and acetylene black were mixed and stirred evenly to prepare 330NH18 slurry and 330N acetylene black slurry. Component A can be prepared by mixing 330NH18 slurry, 330N acetylene black slurry with polyether polyol 330N, chain extender, diethylene glycol, dioctyl phthalate DOP, catalyst, antioxidant, silane coupling agent, and defoaming agent at room temperature and vacuum (Ding et al. 2013; Li P et al. 2017).

The 330NH18 slurry was obtained by mixing 96.3-97.2 parts of polyether polyol 330N and 3.7-2.8 parts of gaseous silica HDK H18 and stirring evenly. The 330N acetylene black paste was obtained by mixing 96.77-97.51 parts of polyether polyol 330N and 3.23-2.49 parts of acetylene black. Polyether polyol 330N: 11.56-12.13 parts, chain extender E-300:18.43-19.06 parts, diethylene glycol: 4.56-4.89 parts, 330NH18 pulp: 48.7-4.95 parts, 330N acetylene black pulp:8.34-8.97 parts, dioctyl phthalate DOP: 5-5.5 parts, catalyst: 0.09-0.12 parts, antioxidant: 0.27-0.33 parts, silane coupling agent: 1.79-1.89 parts, defoamer BMC806:1.26-1.36 parts; they are mixed and stirred evenly to prepare component A (Li et al. 2021).

3.2 *Preparation of component B*

Under normal temperature and vacuum conditions, the 220H18 slurry can be prepared by mixing polyether polyol GE220 and hydrophobic silica evenly. Component B can be prepared by mixing 220H18 slurry with polyether polyol GE220, dioctyl phthalate DOP, MDI-3051, and MDI-100LL at room temperature and vacuum.

To prepare 220H18 slurry, it is needed to mix 96.77-98.16 parts of polyether polyol GE220 and 3.23-1.84 parts of hydrophobic gaseous silica HDK H18 and stir them evenly. Polyether polyol GE220:23.44-23.79 parts, 220H18 pulp: 24.21-25.17 parts, dioctyl phthalate DOP: 4.97-5.56 parts, MDI-3051:23.69-24.33 parts, MDI-100LL:23.69-24.33 parts, they are mixed and stirred evenly to prepare component B (Ameli et al. 2021).

3.3 *Method of use*

The mixing and stirring of each component were carried out under the conditions of room temperature and vacuum, vacuum degree was 0.25-0.3mpa, stirring speed ≥1500 RPM. When using, A and B components were mixed evenly according to the weight ratio of 1:0.7-1.3, which can be used for repairing any type of rubber damage.

4 PERFORMANCE ANALYSIS

For embodiments, the weight fractions of each component in component A and component B are shown in Tables 1-3, and the specific process parameters are shown in Table 4. Components A and B of each embodiment were evenly mixed according to the weight ratio of 1:0.7-1.3 and tested at 25°C and 56% humidity after the performance of the repair agent was stable. The performance test results are shown in Table 5 (Islam et al. 2021; Liang et al. 2015).

Table 1. Composition of 330NH18 pulp, 330N acetylene black pulp, and 220H18 pulp of each embodiment (unit: units).

	330NH18 slurry		330N acetylene black lurry		220H18	
Example	Polyether polyols 330N	HDK H18	Polyether polyols 330N	Acetylene black	Polyether polyols GE220	HDK H18
1	96.3	3.7	96.77	3.23	96.77	3.23
2	97.1	3.0	97.10	2.70	97.76	3.23
3	97.0	3.7	97.44	2.54	97.54	2.04
4	97.2	2.9	96.89	3.15	98.05	2.76
5	96.6	3.5	97.51	2.82	96.99	1.91
6	96.5	2.8	96.82	2.61	97.16	2.38
7	96.9	3.1	97.03	3.19	96.77	2.52
8	96.3	3.6	97.35	2.49	97.92	2.97
9	96.7	3.3	96.77	3.08	97.38	1.84
10	96.4	3.2	97.22	3.23	96.89	3.11
11	96.8	3.4	96.95	2.96	98.16	2.19

Table 2. Component composition of Each embodiment A (Unit: unit).

Example	Polyether polyols 330N	E-300	Diethylene glycol	330NH18 pulp	330N Acetylene black slurry	DOP	catalyst	antioxidants	Silane coupling agent	BMC 806
1	11.56	18.43	4.56	48.7	8.34	5.0	0.09	0.27	1.79	1.26
2	11.71	18.88	4.72	49.2	8.97	5.3	0.11	0.30	1.84	1.31
3	11.99	19.02	4.87	48.9	8.58	5.4	0.09	0.27	1.87	1.36
4	11.64	18.48	4.63	49.5	8.77	5.1	0.10	0.33	1.81	1.29
5	12.13	18.82	4.79	49.1	8.39	5.0	0.12	0.28	1.89	1.28
6	12.08	18.71	4.89	48.8	8.68	5.2	0.11	0.30	1.80	1.35
7	11.79	18.43	4.85	49.0	8.86	5.5	0.12	0.33	1.79	1.33
8	11.87	19.06	4.59	49.3	8.45	5.3	0.10	0.31	1.82	1.26
9	11.56	18.55	4.66	48.7	8.93	5.2	0.12	0.29	1.88	1.30
10	11.58	18.64	4.81	49.4	8.34	5.5	0.10	0.28	1.85	1.27
11	12.11	18.95	4.56	48.8	8.51	5.4	0.09	0.32	1.83	1.34

Table 3. Component composition of Each embodiment B (Unit: unit).

Example	Polyether polyols GE220	220H18pulp	DOP	MDI-3051	MDI-100LL
1	23.44	24.21	4.97	23.69	23.69
2	23.60	25.17	5.16	23.96	24.22
3	23.76	24.36	5.03	24.30	23.69
4	23.63	25.04	4.97	23.71	24.14
5	23.44	24.60	5.25	24.07	23.86
6	23.51	24.25	5.38	24.24	24.33
7	23.70	24.48	5.56	23.85	23.78
8	23.55	25.11	5.10	24.33	24.05
9	23.79	24.90	5.45	23.77	24.97
10	23.67	24.21	4.99	24.18	23.73
11	23.48	24.77	5.52	23.69	24.29

Table 4. Process parameters of each embodiment.

Example	catalyst	antioxidants	Silane coupling agent	vacuum Mpa	Stirring speed /min
1	FAE-1	1010	WD52	0.25	1500
2	GT-12	168	KH560	0.27	1620
3	WS8	1076	WD60	0.26	1540
4	FAE-1	168	WD52	0.29	1560
5	FAE-1	168	KH560	0.28	1500
6	GT-12	1010	WD52	0.30	1550
7	GT-12	1010	KH560	0.30	1600
8	WS8	1010	WD52	0.25	1630
9	WS8	1076	KH560	0.29	1610
10	FAE-1	1010	KH560	0.28	1530
11	FAE-1	1076	WD60	0.27	1500

Table 5. Performance test results of each embodiment.

Example	A: B Weight ratio	appearance	Shore hardness A	Tear strength kg/cm	The tensile strength MPa	elongation %	90° Peel strength N/cm
1	1:0.70	Black rubbery polymer	80	35.2	4.52	289	40.6
2	1:0.8		82	35.8	4.54	282	41.1
3	1:0.9		83	36.2	4.56	273	42.3
4	1:0.95		85	36.7	4.59	269	44.2
5	1:1		86	37.4	4.63	267	46.0
6	1:1.05		86	37.1	4.57	264	45.0
7	1:1.1		88	36.9	4.62	258	45.6
8	1:1.15		89	36.7	4.59	259	46.3
9	1:1.20		84	36.4	4.56	270	43.4
10	1:1.25		83	35.7	4.55	272	42.6
11	1:1.3		81	35.0	4.51	291	41.8

5 CONCLUSIONS

Based on polyether polyol of bicomponent rubber repairing adhesive using acetylene black, gas-phase silica as filler, at the same time, the use of polyether polyols as polymerization material, by adding catalyst, coupling agent, antioxidant, prompting two polyether polyols polymerization reaction, improves the flexibility of the product, the bonding strength, and inhibition of foam. Components A and B can be rapidly polymerized into elastocolloid similar to rubber at room temperature.

Compared with traditional repair technology, this material is more simple and highly effective, the requirements for construction personnel skill levels are not high, and the proportion of components A and B need to be mixed evenly, which can be used for any type of rubber damage repair, without special equipment, without being limited by the space and rubber parts shape, forming the repairing adhesive 1 minutes, for the black rubber polymer, with 8 minutes to achieve strength, 2 hours to achieve 80% strength and being put into use. After the stability of the repair agent, shore hardness A was 82-90 degrees, tear strength ≥35kg/cm, tensile strength ≥ 4.5mpa, elongation ≥250%, and 90° peel strength ≥40N/cm. Through the analysis of the properties of rubber repair material, a reliable and effective surface repair technology was selected to meet the technical requirements of the equipment and restore the normal operation of the equipment, with obvious economic benefits, which provides a reference experience for the popularization and application of rubber repair technology.

The matrix selected in this paper is polyether polyols, and polyether polyols are the basis of all modification processes in the study of properties of different polyether polyols. The next research work should include the influence of different polyether polyols on the effect of rubber repair agents. Due to the limited time in this study, the actual application performance of the repair material was not studied. In the next step, the field test of composite modified rubber repair material should be carried out, and the actual service life of the material should be tracked and investigated.

REFERENCES

Ameli A, Babagoli R, Asadi S, Norouzi N. Investigation of the performance properties of asphalt binders and mixtures modified by Crumb Rubber and Gilsonite[J]. *Construction & building materials*. 2021; 279: 122424.

Cao L, Yang C, Dong Z, Nonde L. Evaluation of crack sealant adhesion properties under complex service ambient conditions based on the weak boundary layer (WBL) theory[J]. *Construction & building materials*. 2019; 200: 293–300.

Ding Z, Lv DL, Zhu YZ, Wu S. Influence Factors and Prediction Model of Viscosity for Crumb Rubber Modified Asphalt[J]. *Advanced materials research*. 2013; 723:376–80.
Islam SS, Singh SK, Ransinchung R.N GD, Ravindranath SS. Effect of property deterioration in SBS modified binders during storage on the performance of asphalt mix[J]. *Construction & building materials*. 2021; 272.
Li P, Ding Z, Zou P, Sun A. Analysis of physicochemical properties for crumb rubber in process of asphalt modification[J]. *Construction & building materials*. 2017; 138:418–26.
Li J, Chen Z, Xiao F, Amirkhanian SN. Surface activation of scrap tire crumb rubber to improve the compatibility of rubberized asphalt[J]. *Resources, conservation, and recycling*. 2021; 169:105518.
Liang M, Liang P, Fan W, Qian C, Xin X, Shi J, et al. Thermo-rheological behavior and compatibility of modified asphalt with various styrene-butadiene structures in SBS copolymers[J]. *Materials & design*. 2015; 88:177–85.
Nie Peng-Fei,Zhang Hong-Yu. Causes Resulting in Stack rainout of Wet Flue Gas Desulphurization for the Thermal Power Plant Without GGH and its Countermeasures, 2012, 38(2): 4–8.
Ren S, Liu X, Fan W, Wang H, Erkens S. Rheological properties, compatibility, and storage stability of SBS latex-modified asphalt[J]. *Materials*. 2019; 12(22): 3683.
Wu S, Liu Q, Yang J, Yang R, Zhu J. Study of adhesion between crack sealant and pavement combining surface free energy measurement with molecular dynamics simulation[J]. *Construction & building materials*. 2020; 240:117900.
Xu M, Liu J, Li W, Duan W. Novel Method to Prepare Activated Crumb Rubber Used for Synthesis of Activated Crumb Rubber Modified Asphalt[J]. *Journal of materials in civil engineering*. 2015; 27(5):4014173.
Yu X, Leng Z, Wei T. Investigation of the Rheological Modification Mechanism of Warm-Mix Additives on Crumb-Rubber-Modified Asphalt[J]. *Journal of materials in civil engineering*. 2014; 26(2):312–9.

Author index